Safety Design for Space Systems

Safety Design for Space Systems

Edited by

Gary Eugene Musgrave, Ph.D.

Axel (Skip) M. Larsen

Tommaso Sgobba

Sponsored by

The International Association for the
Advancement of Space Safety

AMSTERDAM • BOSTON • HEIDELBERG • LONDON
NEW YORK • OXFORD • PARIS • SAN DIEGO
SAN FRANCISCO • SINGAPORE • SYDNEY • TOKYO

Butterworth-Heinemann is an imprint of Elsevier

Butterworth-Heinemann is an imprint of Elsevier
The Boulevard, Langford Lane, Kidlington, Oxford OX5 1GB, UK
30 Corporate Drive, Suite 400, Burlington, MA 01803, USA

First published 2009
Reprinted 2010

Copyright © 2009, Elsevier Ltd. All rights reserved

No part of this publication may be reproduced, stored in a retrieval system or transmitted in any form or by any means electronic, mechanical, photocopying, recording or otherwise without the prior written permission of the publisher

Permissions may be sought directly from Elsevier's Science & Technology Rights Department in Oxford, UK: phone (+44) (0) 1865 843830; fax (+44) (0) 1865 853333; email: permissions@elsevier.com. Alternatively you can submit your request online by visiting the Elsevier web site at http://elsevier.com/locate/permissions, and selecting *Obtaining permission to use Elsevier material*

Notice
No responsibility is assumed by the publisher for any injury and/or damage to persons or property as a matter of products liability, negligence or otherwise, or from any use or operation of any methods, products, instructions or ideas contained in the material herein. Because of rapid advances in the medical sciences, in particular.

British Library Cataloguing in Publication Data
A catalogue record for this book is available from the British Library

Library of Congress Cataloging-in-Publication Data
A catalog record for this book is available from the Library of Congress

ISBN: 978-0-7506-8580-1

For information on all Butterworth-Heinemann publications
visit our website at www.elsevierdirect.com

Printed and bound in the United States of America

10 11 12 13 14 15   10 9 8 7 6 5 4 3 2

Working together to grow
libraries in developing countries

www.elsevier.com | www.bookaid.org | www.sabre.org

ELSEVIER   BOOK AID International   Sabre Foundation

Contents

Preface xxiii
Introduction xxv
About the Editors xxvii
About the Contributors xxxi

CHAPTER 1 Introduction to Space Safety 1
    1.1 NASA and Safety 2
    1.2 Definition of Safety and Risk 3
    1.3 Managing Safety and Risk 3
    1.4 The Book 5
    References 5

CHAPTER 2 The Space Environment: Natural and Induced 7
    2.1 The Atmosphere 8
        2.1.1 Composition 8
        2.1.2 Atomic Oxygen 13
        2.1.3 The Ionosphere 15
    2.2 Orbital Debris and Meteoroids 18
        2.2.1 Orbital Debris 18
        2.2.2 Meteoroids 26
    2.3 Microgravity 31
        2.3.1 Microgravity Defined 31
        2.3.2 Methods of Attainment 34
        2.3.3 Effects on Biological Processes and Astronaut Health ... 40
        2.3.4 Unique Aspects of Travel to the Moon and
               Planetary Bodies 41
    Recommended Reading 41
    2.4 Acoustics 43
        2.4.1 Acoustics Safety Issues 43
        2.4.2 Acoustic Requirements 43
        2.4.3 Compliance and Verification 50
        2.4.4 Conclusions and Recommendations 51
    Recommended Reading 51
    2.5 Radiation 52
        2.5.1 Ionizing Radiation 52
        2.5.2 Radio Frequency Radiation 67
    Recommended Reading 71

2.6 Natural and Induced Thermal Environments 72
    2.6.1 Introduction to the Thermal Environment 72
    2.6.2 Spacecraft Heat Transfer Considerations 72
    2.6.3 The Natural Thermal Environment 73
    2.6.4 The Induced Thermal Environment 80
    2.6.5 Other Lunar and Planetary Environment Considerations 85
2.7 Combined Environmental Effects 86
    2.7.1 Introduction to Environmental Effects 86
    2.7.2 Combined Environments 87
    2.7.3 Combined Effects 88
    2.7.4 Ground Testing for Space Simulation 92
References 94

CHAPTER 3   Overview of Bioastronautics 105
3.1 Space Physiology 106
    3.1.1 Muscular System 106
    3.1.2 Skeletal System 107
    3.1.3 Cardiovascular and Respiratory Systems 108
    3.1.4 Neurovestibular System 110
    3.1.5 Radiation 111
    3.1.6 Nutrition 112
    3.1.7 Immune System 113
    3.1.8 Extravehicular Activity 114
3.2 Short and Long Duration Mission Effects 115
    3.2.1 Muscular System 115
    3.2.2 Skeletal System 116
    3.2.3 Cardiovascular and Respiratory Systems 117
    3.2.4 Neurovestibular System 119
    3.2.5 Radiation 120
    3.2.6 Nutrition 121
    3.2.7 Immune System 121
    3.2.8 Extravehicular Activity 122
3.3 Health Maintenance 123
    3.3.1 Preflight Preparation 123
    3.3.2 In-Flight Measures 126
    3.3.3 In-Flight Medical Monitoring 139
    3.3.4 Post-Flight Recovery 142
3.4 Crew Survival 143
    3.4.1 Overview of Health Threats in Spaceflight 143
    3.4.2 Early Work 144

        3.4.3 Crew Survival on the Launch Pad, at Launch, and During Ascent 145
        3.4.4 On-Orbit Safe Haven and Crew Transfer 150
        3.4.5 Entry, Landing, and Post-Landing 150
  3.5 Conclusion 152
  Acknowledgment 152
  References 153

CHAPTER 4   Basic Principles of Space Safety 163
  4.1 The Cause of Accidents 163
  4.2 Principles and Methods 165
        4.2.1 Hazard Elimination and Limitation 165
        4.2.2 Barriers and Interlocks 166
        4.2.3 Fail-Safe Design 167
        4.2.4 Failure and Risk Minimization 167
        4.2.5 Monitoring, Recovery, and Escape 169
        4.2.6 Crew Survival Systems 169
  4.3 The Safety Review Process 170
        4.3.1 Safety Requirements 170
        4.3.2 The Safety Panels 171
        4.3.3 The Safety Reviews 171
        4.3.4 Nonconformances 173
  References 174

CHAPTER 5   Human Rating Concepts 175
  5.1 Human Rating Defined 175
        5.1.1 Human Rated Systems 175
        5.1.2 The NASA Human Rating and Process 176
        5.1.3 The Human Rating Plan 177
        5.1.4 The NASA Human Rating Certification Process 178
        5.1.5 Human Rating in Commercial Human Spaceflight 178
  5.2 Human Rating Requirements and Approaches 179
        5.2.1 Key Human Rating Technical Requirements 179
        5.2.2 Programmatic Requirements 182
        5.2.3 Test Requirements 183
        5.2.4 Data Requirements 184
  Reference 184

CHAPTER 6   Life Support Systems Safety 185
  6.1 Atmospheric Conditioning and Control 188
        6.1.1 Monitoring Is the Key to Control 188

    6.1.2 Atmospheric Conditioning 190
    6.1.3 Carbon Dioxide Removal 196
  6.2 Trace Contaminant Control 198
    6.2.1 Of Tight Buildings and Spacecraft Cabins................................... 198
    6.2.2 Trace Contaminant Control Methodology..................................... 201
    6.2.3 Trace Contaminant Control Design Considerations 209
  6.3 Assessment of Water Quality in the Spacecraft Environment:
    Mitigating Health and Safety Concerns 211
    6.3.1 Scope of Water Resources Relevant to Spaceflight 211
    6.3.2 Spacecraft Water Quality and the Risk Assessment
       Paradigm 212
    6.3.3 Water Quality Monitoring 217
    6.3.4 Conclusion and Future Directions.. 220
  6.4 Waste Management 220
  6.5 Summary of Life Support Systems.. 221
  References 222

CHAPTER 7 Emergency Systems 225
  7.1 Space Rescue.. 225
    7.1.1 Legal and Diplomatic Basis 226
    7.1.2 The Need for Rescue Capability 226
    7.1.3 Rescue Modes and Probabilities 229
    7.1.4 Hazards in the Different Phases of Flight....................................231
    7.1.5 Historic Distribution of Failures 232
    7.1.6 Historic Rescue Systems..233
    7.1.7 Space Rescue Is Primarily Self Rescue 243
    7.1.8 Limitations of Ground Based Rescue...247
    7.1.9 The *Crew Return Vehicle* as a Study in
       Space Rescue 249
    7.1.10 Safe Haven 255
    7.1.11 Conclusions 256
  7.2 Personal Protective Equipment 256
    7.2.1 Purpose of Personal Protective Equipment 256
    7.2.2 Types of Personal Protective Equipment 257
  References 265

CHAPTER 8 Collision Avoidance Systems................................. 267
  8.1 Docking Systems and Operations... 268
    8.1.1 Docking Systems as a Means for Spacecraft
       Orbital Mating..268

  8.1.2 Design Approaches Ensuring Docking Safety
    and Reliability 270
  8.1.3 Design Features Ensuring the Safety and Reliability of
    Russian Docking Systems 275
  8.1.4 Analyses and Tests Performed for Verification of Safety and
    Reliability of Russian Docking Systems 278
Acknowledgment 280
8.2 Descent and Landing Systems 280
  8.2.1 Parachute Systems 281
  8.2.2 Known Parachute Anomalies and Lessons Learned 296
Acknowledgment 299
References 299

CHAPTER 9   Robotic Systems Safety 301

9.1 Generic Robotic Systems 301
  9.1.1 Controller and Operator Interface 302
  9.1.2 Arms and Joints 302
  9.1.3 Drive System 303
  9.1.4 Sensors 303
  9.1.5 End Effector 303
9.2 Space Robotics Overview 303
9.3 Identification of Hazards and Their Causes 305
  9.3.1 Electrical and Electromechanical Malfunctions 307
  9.3.2 Mechanical and Structural Failures 307
  9.3.3 Failure in the Control Path 307
  9.3.4 Operator Error 307
  9.3.5 Other Hazards 307
9.4 Hazard Mitigation in Design 308
  9.4.1 Electrical and Mechanical Design and Redundancy 308
  9.4.2 Operator Error 308
  9.4.3 System Health Checks 308
  9.4.4 Emergency Motion Arrest 309
  9.4.5 Proximity Operations 309
  9.4.6 Built in Test 310
  9.4.7 Safety Algorithms 310
9.5 Hazard Mitigation Through Training 310
9.6 Hazard Mitigation for Operations 312
9.7 Case Study: Understanding Canadarm2 and Space Safety 313
  9.7.1 The Canadarm2 313
  9.7.2 Cameras 313
  9.7.3 Force Moment Sensor 314

	9.7.4	Training	315
	9.7.5	Hazard Concerns and Associated Hazard Mitigation	316
9.8	Summary		317
References			318

CHAPTER 10 Meteoroid and Debris Protection ... 319

- 10.1 Risk Control Measures ... 319
  - 10.1.1 Maneuvering ... 319
  - 10.1.2 Shielding ... 324
- 10.2 Emergency Repair Considerations for Spacecraft Pressure Wall Damage ... 332
  - 10.2.1 Balanced Mitigation of Program Risks ... 332
  - 10.2.2 Leak Location System and Operational Design Considerations ... 337
  - 10.2.3 Ability to Access the Damaged Area ... 337
  - 10.2.4 Kit Design and Certification Considerations (1 is too many, 100 are not enough) ... 338
  - 10.2.5 Recertification of the Repaired Pressure Compartment for Use by the Crew ... 338
- References ... 339

CHAPTER 11 Noise Control Design ... 341

- 11.1 Introduction ... 341
- 11.2 Noise Control Plan ... 341
  - 11.2.1 Noise Control Strategy ... 342
  - 11.2.2 Acoustic Analysis ... 344
  - 11.2.3 Testing and Verification ... 344
- 11.3 Noise Control Design Applications ... 345
  - 11.3.1 Noise Control at the Source ... 346
  - 11.3.2 Path Noise Control ... 348
  - 11.3.3 Noise Control in the Receiving Space ... 353
  - 11.3.4 Post-Design Noise Mitigation ... 355
- 11.4 Conclusions and Recommendations ... 355
- Recommended Reading ... 356
- References ... 356

CHAPTER 12 Materials Safety ... 359

- 12.1 Toxic Offgassing ... 360
  - 12.1.1 Materials Offgassing Controls ... 361
  - 12.1.2 Materials Testing ... 362

  12.1.3 Spacecraft Module Testing 363
 12.2 Stress-Corrosion Cracking 363
  12.2.1 What Is Stress-Corrosion Cracking? 364
  12.2.2 Prevention of Stress-Corrosion Cracking 364
  12.2.3 Testing Materials for Stress-Corrosion Cracking 366
  12.2.4 Design for Stress-Corrosion Cracking 368
  12.2.5 Requirements for Spacecraft Hardware 369
  12.2.6 Stress-Corrosion Cracking in Propulsion Systems 371
 12.3 Conclusions 373
 References 373

CHAPTER 13 Oxygen Systems Safety 375
 13.1 Oxygen Pressure System Design 375
  13.1.1 Introduction 375
  13.1.2 Design Approach 377
  13.1.3 Oxygen Compatibility Assessment Process 386
 13.2 Oxygen Generators 392
  13.2.1 Electrochemical Systems for Oxygen Production 392
  13.2.2 Solid Fuel Oxygen Generators (Oxygen Candles) 398
 References 401

CHAPTER 14 Avionics Safety 403
 14.1 Introduction to Avionics Safety 403
 14.2 Electrical Grounding and Electrical Bonding 404
  14.2.1 Defining Characteristics of an Electrical
     Ground Connection 405
  14.2.2 Control of Electric Current 406
  14.2.3 Electrical Grounds Can Be Signal Return Paths 406
  14.2.4 Where and How Electrical Grounds Should
     Be Connected 406
  14.2.5 Defining Characteristics of an Electrical Bond 408
  14.2.6 Types of Electrical Bonds 408
  14.2.7 Electrical Bond Considerations for
     Dissimilar Metals 409
  14.2.8 Electrical Ground and Bond Connections
     for Shields 410
 Recommended Reading 410
 14.3 Safety Critical Computer Control 411
  14.3.1 Partial Computer Control 412
  14.3.2 Total Computer Control: Fail Safe 413

- 14.4 Circuit Protection: Fusing 414
  - 14.4.1 Circuit Protection Methods 414
  - 14.4.2 Circuit Protectors 416
  - 14.4.3 Design Guidance 416
- 14.5 Electrostatic Discharge Control 417
  - 14.5.1 Fundamentals 418
  - 14.5.2 Various Levels of Electrostatic Discharge Concern 420
- Recommended Reading 426
- 14.6 Arc Tracking 428
  - 14.6.1 A New Failure Mode 428
  - 14.6.2 Characteristics of Arc Tracking 431
  - 14.6.3 Likelihood of an Arc Tracking Event 432
  - 14.6.4 Prevention of Arc Tracking 432
  - 14.6.5 Verification of Protection and Management of Hazards 433
  - 14.6.6 Summary 433
- 14.7 Corona Control in High Voltage Systems 434
  - 14.7.1 Associated Environments 434
  - 14.7.2 Design Criteria 435
  - 14.7.3 Verification and Testing 436
- Recommended Reading 437
- 14.8 Extravehicular Activity Considerations 437
  - 14.8.1 Displays and Indicators Used in Space 438
  - 14.8.2 Mating and Demating of Powered Connectors 438
  - 14.8.3 Single Strand Melting Points 439
  - 14.8.4 Battery Removal and Installation 441
  - 14.8.5 Computer or Operational Control of Inhibits 442
- 14.9 Spacecraft Electromagnetic Interference and Electromagnetic Compatibility Control 442
  - 14.9.1 Electromagnetic Compatibility Needs for Space Applications 443
  - 14.9.2 Basic Electromagnetic Compatibility Interactions and a Safety Margin 444
  - 14.9.3 Mission Driven Electromagnetic Interference Design: The Case for Grounding 445
  - 14.9.4 Electromagnetic Compatibility Program for Spacecraft 446
- 14.10 Design and Testing of Safety Critical Circuits 450
  - 14.10.1 Safety Critical Circuits: Conducted Mode 450
  - 14.10.2 Safety Critical Circuits: Radiated Mode 456

14.11	Electrical Hazards 461
	14.11.1 Introduction 461
	14.11.2 Electrical Shock 461
	14.11.3 Physiological Considerations 462
	14.11.4 Electrical Hazard Classification 463
	14.11.5 Leakage Current 464
	14.11.6 Bioinstrumentation 464
	14.11.7 Electrical Hazard Controls 465
	14.11.8 Verification of Electrical Hazard Controls 468
	14.11.9 Electrical Safety Design Considerations 468
14.12	Avionics Lessons Learned 469
	14.12.1 Electronic Design 469
	14.12.2 Physical Design 470
	14.12.3 Materials and Sources 471
	14.12.4 Damage Avoidance 472
	14.12.5 System Aspects 472
References 473

CHAPTER 15 Software System Safety 475

15.1	Introduction 475
15.2	The Software Safety Problem 476
	15.2.1 System Accidents 476
	15.2.2 The Power and Limitations of Abstraction from Physical Design 477
	15.2.3 Reliability Versus Safety for Software 479
	15.2.4 Inadequate System Engineering 482
	15.2.5 Characteristics of Embedded Software 484
15.3	Current Practice 486
	15.3.1 System Safety 487
15.4	Best Practice 489
	15.4.1 Management of Software-Intensive, Safety-Critical Projects 490
	15.4.2 Basic System Safety Engineering Practices and Their Implications for Software Intensive Systems 491
	15.4.3 Specifications 493
	15.4.4 Requirements Analysis 494
	15.4.5 Model-Based Software Engineering and Software Reuse 494
	15.4.6 Software Architecture 496
	15.4.7 Software Design 497
	15.4.8 Design of Human-Computer Interaction 500

CHAPTER 16 Battery Safety 507

- 16.1 Introduction 507
- 16.2 General Design and Safety Guidelines 508
- 16.3 Battery Types 508
- 16.4 Battery Models 509
- 16.5 Hazard and Toxicity Categorization 509
- 16.6 Battery Chemistry 509
  - 16.6.1 Alkaline Batteries 509
  - 16.6.2 Lithium Batteries 512
  - 16.6.3 Silver Zinc Batteries 523
  - 16.6.4 Lead Acid Batteries 525
  - 16.6.5 Nickel Cadmium Batteries 527
  - 16.6.6 Nickel Metal Hydride Batteries 528
  - 16.6.7 Nickel Hydrogen Batteries 533
  - 16.6.8 Lithium-Ion Batteries 535
- 16.7 Storage, Transportation, and Handling 544
- References 545

CHAPTER 17 Mechanical Systems Safety 549

- 17.1 Safety Factors 549
  - 17.1.1 Types of Safety Factors 550
  - 17.1.2 Safety Factors Typical of Human Rated Space Programs 551
  - 17.1.3 Things That Influence the Choice of Safety Factors 551
- 17.2 Spacecraft Structures 551
  - 17.2.1 Mechanical Requirements 552
  - 17.2.2 Space Mission Environment and Mechanical Loads 554
  - 17.2.3 Project Overview: Successive Designs and Iterative Verification of Structural Requirements 557
  - 17.2.4 Analytical Evaluations 559
  - 17.2.5 Structural Test Verification 559
  - 17.2.6 Spacecraft Structural Model Philosophy 561
  - 17.2.7 Materials and Processes 562
  - 17.2.8 Manufacturing of Spacecraft Structures 564
- Recommended Reading 566

Also in contents above:
- 15.4.9 Software Reviews 501
- 15.4.10 Verification and Assurance 502
- 15.4.11 Operations 503
- 15.5 Summary 503
- References 503

17.3 Fracture Control 567
    17.3.1 Basic Requirements 567
    17.3.2 Implementation 567
    17.3.3 Summary 568
17.4 Pressure Vessels, Lines, and Fittings 568
    17.4.1 Pressure Vessels 568
    17.4.2 Lines and Fittings 574
    17.4.3 Space Pressure Systems Standards 575
    17.4.4 Summary 575
17.5 Composite Overwrapped Pressure Vessels 576
    17.5.1 The Composite Overwrapped Pressure Vessel System 576
    17.5.2 Monolithic Metallic Pressure Vessel Failure Modes 577
    17.5.3 Composite Overwrapped Pressure Vessel Failure
            Modes 578
    17.5.4 Composite Overwrapped Pressure Vessel
            Impact Sensitivity 579
    17.5.5 Summary 581
17.6 Structural Design of Glass and Ceramic Components for
      Space System Safety 581
    17.6.1 Strength Characteristics of Glass and Ceramics 582
    17.6.2 Defining Loads and Environments 586
    17.6.3 Design Factors 588
    17.6.4 Meeting Life Requirements with Glass and
            Ceramics 589
17.7 Safety Critical Mechanisms 591
    17.7.1 Designing for Failure Tolerance 591
    17.7.2 Design and Verification of Safety Critical
            Mechanisms 594
    17.7.3 Reduced Failure Tolerance 602
    17.7.4 Review of Safety Critical Mechanisms 604
References 605

CHAPTER 18 Containment of Hazardous Materials 607
    18.1 Toxic Materials 610
        18.1.1 Fundamentals of Toxicology 610
        18.1.2 Toxicological Risks to Air Quality in Spacecraft 613
        18.1.3 Risk Management Strategies 618
    18.2 Biohazardous Materials 621
        18.2.1 Microbiological Risks Associated with Spaceflight 621
        18.2.2 Risk Mitigation Approaches 622

        18.2.3 Major Spaceflight Specific Microbiological Risks 623
  18.3 Shatterable Materials 631
        18.3.1 Shatterable Materials in a Habitable Compartment 631
        18.3.2 Program Implementation 631
        18.3.3 Containment Concepts for Internal Equipment 633
        18.3.4 Containment Concepts for Exterior Equipment 636
        18.3.5 General Comments About Working with Shatterable Materials 638
  18.4 Containment Design Approach 639
        18.4.1 Fault Tolerance 639
        18.4.2 Design for Minimum Risk 639
  18.5 Containment Design Methods 640
        18.5.1 Containment Environments 640
        18.5.2 Design of Containment Systems 640
  18.6 Safety Controls 643
        18.6.1 Proper Design 643
        18.6.2 Materials Selection 643
        18.6.3 Materials Compatibility 643
        18.6.4 Proper Workmanship 644
        18.6.5 Proper Loading or Filling 644
        18.6.6 Fracture Control 644
  18.7 Safety Verifications 644
        18.7.1 Strength Analysis 645
        18.7.2 Qualification Tests 645
        18.7.3 Acceptance Tests 646
        18.7.4 Proof Tests 647
        18.7.5 Qualification of Procedures 647
  18.8 Conclusions 648
  References 649

CHAPTER 19 Failure Tolerance Design 653
  19.1 Safe 653
        19.1.1 Order of Precedence 653
  19.2 Hazard 655
        19.2.1 Hazard Controls 655
        19.2.2 Design to Tolerate Failures 656
  19.3 Hazardous Functions 658
        19.3.1 Must Not Work Hazardous Function 658
        19.3.2 Must Work Hazardous Function 659
  19.4 Design for Minimum Risk 659

	19.5 Conclusions	660
	References	660
CHAPTER 20	**Propellant Systems Safety**	**661**
	20.1 Solid Propellant Propulsion Systems Safety	662
	20.1.1 Solid Propellants	662
	20.1.2 Solid Propellant Systems for Space Applications	664
	20.1.3 Safety Hazards	664
	20.1.4 Handling, Transport, and Storage	670
	20.1.5 Inadvertent Ignition	671
	20.1.6 Safe Ignition Systems Design	672
	20.1.7 Conclusions	673
	20.2 Liquid Propellant Propulsion Systems Safety	673
	20.2.1 Planning	675
	20.2.2 Containment Integrity	676
	20.2.3 Thermal Control	677
	20.2.4 Materials Compatibility	678
	20.2.5 Contamination Control	678
	20.2.6 Environmental Considerations	679
	20.2.7 Engine and Thruster Firing Inhibits	679
	20.2.8 Heightened Risk (Risk Creep)	680
	20.2.9 Instrumentation and Telemetry Data	681
	20.2.10 End to End Integrated Instrumentation, Controls, and Redundancy Verification	681
	20.2.11 Qualification	681
	20.2.12 Total Quality Management (ISO 9001 or Equivalent)	682
	20.2.13 Preservicing Integrity Verification	682
	20.2.14 Propellants Servicing	683
	20.2.15 Conclusions	683
	20.3 Hypergolic Propellants	683
	20.3.1 Materials Compatibility	683
	20.3.2 Material Degradation	684
	20.3.3 Hypergolic Propellant Degradation	685
	20.4 Propellant Fire	686
	20.4.1 Hydrazine and Monomethylhydrazine Vapor	687
	20.4.2 Liquid Hydrazine and Monomethylhydrazine	690
	20.4.3 Hydrazine and Monomethylhydrazine Mists, Droplets, and Sprays	691
	References	691

CHAPTER 21 Pyrotechnic Safety 695
- 21.1 Pyrotechnic Devices 695
  - 21.1.1 Explosives 696
  - 21.1.2 Initiators 696
- 21.2 Electroexplosive Devices 696
  - 21.2.1 Safe Handling of Electroexplosive Devices 697
  - 21.2.2 Designing for Safe Electroexplosive Device Operation 700
  - 21.2.3 Pyrotechnic Safety of Mechanically Initiated Explosive Devices 702
- References 704

CHAPTER 22 Extravehicular Activity Safety 705
- 22.1 Extravehicular Activity Environment 705
  - 22.1.1 Definitions 706
  - 22.1.2 Extravehicular Activity Space Suit 708
  - 22.1.3 Sensory Degradation 710
  - 22.1.4 Maneuvering and Weightlessness 710
  - 22.1.5 Glove Restrictions 711
  - 22.1.6 Crew Fatigue 711
  - 22.1.7 Thermal Environment 711
  - 22.1.8 Extravehicular Activity Tools 712
- 22.2 Suit Hazards 712
  - 22.2.1 Inadvertent Contact Hazards 712
  - 22.2.2 Area of Effect Hazards 715
- 22.3 Crew Hazards 716
  - 22.3.1 Contamination of the Habitable Environment 716
  - 22.3.2 Thermal Extremes 716
  - 22.3.3 Lasers 718
  - 22.3.4 Electrical Shock and Molten Metal 718
  - 22.3.5 Entrapment 719
  - 22.3.6 Emergency Ingress 719
  - 22.3.7 Collision 720
  - 22.3.8 Inadvertent Loss of Crew 721
- 22.4 Conclusions 722
- References 722

CHAPTER 23 Emergency, Caution, and Warning System 725
- 23.1 System Overview 725
- 23.2 Historic NASA Emergency, Caution, and Warning Systems 726

  23.3 Emergency, Caution, and Warning System Measures 727
    23.3.1 Event Classification Measures 727
    23.3.2 Sensor Measures 728
    23.3.3 Data System Measures 729
    23.3.4 Annunciation Measures 730
  23.4 Failure Isolation and Recovery... 731
  Reference... 732

CHAPTER 24   Laser Safety 733
  24.1 Background 733
    24.1.1 Optical Spectrum 733
    24.1.2 Biological Effects 734
  24.2 Laser Characteristics... 735
    24.2.1 Laser Principles.......................................735
    24.2.2 Laser Types 737
  24.3 Laser Standards 738
    24.3.1 NASA Johnson Space Center Requirements.....738
    24.3.2 ANSI Standard Z136-1 739
    24.3.3 Russian Standard....................................740
  24.4 Lasers Used in Space 740
    24.4.1 Radars..741
    24.4.2 Illumination...741
    24.4.3 Sensors 741
  24.5 Design Considerations for Laser Safety 742
    24.5.1 Ground Testing.......................................742
    24.5.2 Unique Space Environment 742
  24.6 Conclusions.. 744
  References 744

CHAPTER 25   Crew Training Safety: An Integrated Process 745
  25.1 Training the Crew for Safety.. 746
    25.1.1 Typical Training Flow...............................746
    25.1.2 Principles of Safety Training for the Different
       Training Phases......................................752
    25.1.3 Specific Safety Training for Different Equipment
       Categories 755
    25.1.4 Safety Training for Different Operations Categories...............761
  25.2 Safety During Training... 770
    25.2.1 Overview 770

- 25.2.2 Training, Test, or Baseline Data Collection Model Versus Flight Model: Type, Fidelity, Source, Origin, and Category 771
- 25.2.3 Training Environments and Facilities 775
- 25.2.4 Training Models, Test Models, and Safety Requirements 781
- 25.2.5 Training Model, Test Model, and Baseline Data Collection Equipment Utilization Requirements 795
- 25.2.6 Qualification and Certification of Training Personnel 798
- 25.2.7 Training and Test Model Documentation 799
- 25.3 Training Development and Validation Process 803
  - 25.3.1 The Training Development Process 806
  - 25.3.2 The Training Review Process 807
  - 25.3.3 The Role of Safety in the Training Development and Validation Processes 809
  - 25.3.4 Feedback to the Safety Community from the Training Development and Validation Processes 812
- 25.4 Conclusions 815
- References 815

CHAPTER 26   Safety Considerations for the Ground Environment 817
- 26.1 A Word About Ground Support Equipment 818
- 26.2 Documentation and Reviews 819
- 26.3 Roles and Responsibilities 819
- 26.4 Contingency Planning 819
- 26.5 Failure Tolerance 820
- 26.6 Training 820
- 26.7 Hazardous Operations 821
- 26.8 Tools 822
- 26.9 Human Factors 822
- 26.10 Biological Systems and Materials 823
- 26.11 Electrical 824
- 26.12 Radiation 824
- 26.13 Pressure Systems 825
- 26.14 Ordinance 825
- 26.15 Mechanical and Eelectromechanical Devices 826
- 26.16 Propellants 826
- 26.17 Cryogenics 826
- 26.18 Oxygen 826
- 26.19 Ground Handling 827

26.20	Software Safety	827
26.21	Summary	828

CHAPTER 27 Fire Safety ... 829

- **27.1** Characteristics of Fire in Space ... 830
  - 27.1.1 Overview of Low Gravity Fire ... 830
  - 27.1.2 Fuel and Oxidizer Supply and Flame Behavior ... 831
  - 27.1.3 Fire Appearance and Signatures ... 832
  - 27.1.4 Flame Ignition and Spread ... 836
  - 27.1.5 Summary of Low Gravity Fire Characteristics ... 845
- **27.2** Design for Fire Prevention ... 847
  - 27.2.1 Materials Flammability ... 847
  - 27.2.2 Ignition Sources ... 852
- **27.3** Spacecraft Fire Detection ... 855
  - 27.3.1 Prior Spacecraft Systems ... 855
  - 27.3.2 Review of Low Gravity Smoke ... 858
  - 27.3.3 Spacecraft Atmospheric Dust ... 859
  - 27.3.4 Sensors for Fire Detection ... 860
- **27.4** Spacecraft Fire Suppression ... 864
  - 27.4.1 Spacecraft Fire Suppression Methods ... 864
  - 27.4.2 Considerations for Spacecraft Fire Suppression ... 867
- References ... 877

CHAPTER 28 Safe Without Services Design ... 885

CHAPTER 29 Probabilistic Risk Assessment with Emphasis on Design ... 889

- **29.1** Basic Elements of Probabilistic Risk Assessment ... 889
  - 29.1.1 Identification of Initiating Events ... 890
  - 29.1.2 Application of Event Sequence Diagrams and Event Trees ... 891
  - 29.1.3 Modeling of Pivotal Events ... 893
  - 29.1.4 Linkage and Quantification of Accident Scenarios ... 894
- **29.2** Construction of a Probabilistic Risk Assessment for Design Evaluations ... 894
  - 29.2.1 Uses of Probabilistic Risk Assessment ... 894
  - 29.2.2 Reference Mission ... 896
- **29.3** Relative Risk Evaluations ... 898
  - 29.3.1 Absolute Versus Relative Risk Assessments ... 899

29.3.2 Roles of Relative Risk Assessments in Design Evaluations 900
29.3.3 Quantitative Evaluations 902
29.4 Evaluations of the Relative Risks of Alternative Designs 904
29.4.1 Overview of Probabilistic Risk Assessment Models Developed 904
29.4.2 Relative Risk Comparisons of the Alternative Designs 905
References 911

Index 913

Preface

Bryan O'Connor
Associate Administrator, Safety and Mission Assurance,
Headquarters, National Aeronautics and Space Administration,
Washington, DC

In his book, *To Engineer Is Human*, Henry Petrosky said, "No one wants to learn by mistakes, but we cannot learn enough from successes to go beyond the state of the art." In this elegant statement, he poses both the challenge and the opportunity for the spaceflight system safety engineer. Just how does the engineer facilitate the incorporation of lessons learned from historical failures and close calls into the design of the next spaceflight system?

This book is a compilation of much of the best thinking of the spaceflight safety community. It includes discussion of philosophies, techniques, methods, processes, and standards that over the first 50 years of spaceflight have proven themselves as the basics of the profession. The authors are accomplished practitioners, and acknowledged leaders representing most spacefaring nations of today. They cover a variety of topics relevant to robotic as well as human spaceflight systems. They discuss the environment, both in Earth orbit and deep space, as well as operational hazards both on the ground and in flight. They describe the latest methods and techniques the system safety members of the design team apply to system design, development and testing, as well as integrated hazard and risk methods that the safety integration team applies to the entire system.

If there is a common theme in this comprehensive book, it is very close to the notion captured in Petrosky's quote. Many of the safety engineering tools and techniques of today were spawned as "fixes" to what in retrospect had been inadequate processes leading up to incidents and mission failures. One of the professional challenges of the system safety community is the sure knowledge that the mishap board investigating a failure will almost always have a chapter in their report dealing with the failure of the safety team to prevent the mishap. Clearly, preventing mishaps is the job of everyone, but traditionally, the safety community nearly always finds itself trying to figure out how to do a better job of anticipating, analyzing, predicting, and thus preventing another failure. This book is an attempt to capture the most important aspects of that ongoing improvement process. Use this book to learn your trade and better understand the things your predecessors and peers have learned over the years, often the hard way. And, if you never experience a major failure, you are not off the hook. Take advantage of your close calls, near misses, and high probability risks to continuously improve your trade and your tools. Your learning should never stop, and it will be the basis for future revisions to this book.

Introduction

The development of this textbook, *Safety Design for Space Systems*, has been a labor of dedication by the almost 90 individuals who have labored for almost three years to write, edit, proof, and provide technical support. *Safety Design for Space Systems* is remarkable in that there is no other textbook of its kind anywhere in the world. It provides a detailed look into the discipline of space safety, something that generally is not taught as a curriculum in all but a handful of engineering colleges and universities. Indeed, it is the collective experience of the Editors that safety is something that historically is learned on the job by engineers designing spacecraft and systems. This is a practice that must change. Newly graduated engineers must be able to enter into their career already familiar with the rigors that must be applied in their designs to ensure the safety of spacecraft and systems, the crew, and associated operations.

The Editors and supporters of this project have high hopes that this textbook will be used extensively within academic programs, and by space agencies and commercial space enterprises. This textbook, intended for use at the senior elective or Masters level of an academic engineering program, is a first step in this endeavor in that it provides a sound foundation for the teaching of safety principles and processes to the senior Baccalaureate or Masters student.

The Editors of *Safety Design for Space Systems* express their most sincere appreciation to Professor Joseph Pelton of the George Washington University in Washington, D.C., the chairman of the IAASS Academic Committee, who conceived this project and arranged for its publication; each of the 78 experts in space safety who donated their valuable time to author their respective contributions; to our Editorial Assistant, Wayne Stauffer, for compiling the index, and for cross-checking and obtaining permissions for the reprinting of all copyrighted figures, photos, and tables used within this book; to my wife, Kristine C. Musgrave, who assisted me with the daunting task of proofreading; and to all the other individuals who were involved in the review and approval process for this work.

I was asked if this Preface would be technical or sentimental. It most certainly is the latter. This was my last major project before I retired from NASA, and it has been a worthy, albeit challenging one. I end this section with a bit of an anecdote to illustrate (a bit humorously) the importance of safety in an engineering effort. I like to cook, and television chefs tend to present an attraction to me. One such chef, now deceased, was famous for his down-home Cajun style cooking. On an occasion several years ago, he revealed that for his day job, he was a safety engineer. He said that it was easy to see his dedication to safety because he always wore both belt and suspenders, thus rendering himself two-failure tolerant to a catastrophic event.

—Gary Eugene Musgrave, Ph.D.

About the Editors

Gary Eugene Musgrave, Ph.D.
Chairman (Retired), NASA Payload Safety Review Panel
Johnson Space Center, National Aeronautics and Space Administration
Houston, Texas

Dr. Gary Eugene Musgrave received his undergraduate training at Auburn University, where he received the Baccalaureate in Biological Sciences in 1969, and at the Georgia Institute of Technology, where he studied Electrical Engineering from 1971 until 1973. He received his graduate education at Auburn University, receiving the Master of Science in the field of Pharmacology/Toxicology from the School of Pharmacy in 1976, and the Doctor of Philosophy in the fields of Cardiovascular Physiology and Autonomic Neuropharmacology from the School of Veterinary Medicine in 1979. He was the recipient of a National Institutes of Health postdoctoral fellowship in the field of Clinical Pharmacology and conducted his postdoctoral research on the pharmacological mechanisms involved in the treatment of essential hypertension at the Audie L. Murphy Veterans Administration Hospital, San Antonio, Texas. After completing his postdoctoral research, Dr. Musgrave was appointed Research Assistant Professor in the Department of Medicine at the Medical College of Virginia where he was Co-Investigator and the Engineering Project Director for a NASA sponsored investigation of the baroreflex regulation of blood pressure in astronauts during and after missions in space. This experiment ultimately was flown on the *Spacelab* "Space Life Sciences-1" mission. In 1982, Dr. Musgrave joined the NASA team at the Johnson Space Center in Houston, Texas, as an employee of the Management and Technical Services Company (MATSCO), a subsidiary of the General Electric Corporation, as the contractor manager for NASA's Detailed Science Objective Program, where he was responsible for the development, certification, testing, and flight support for numerous items of medical hardware flown on various Space Shuttle missions. Dr. Musgrave was transferred to the MATSCO office at NASA Headquarters in 1984, where he orchestrated the development of a reference science mission of human, animal, and plant research in support of long duration (years) space flight. The product of his efforts, the *Reference Mission Operational Requirements Document*, provided the initial basis for experimentation and hardware development planning for research on *Space Station Freedom* by the NASA Life Sciences Directorate at NASA Headquarters, Johnson Space Center, Ames Research Center, Marshall Space Flight Center, and Kennedy Space Center. During this time, he was a member of the prestigious Code E (Office of Space Science and Applications) Space Station Planning Group, which was responsible for managing the flow of requirements from the space station user communities into the Level-II space station design structure. During 1990, Dr. Musgrave formally joined NASA as a Level-I

Program Manager for Space Station Freedom utilization at NASA Headquarters, where he was responsible for overseeing the incorporation of user requirements into the Level-II Space Station Freedom design and subsequently as the Branch Chief for Space Station Freedom operations. On returning to the Johnson Space Center in 1994, Dr. Musgrave held a variety of positions, including Program Manager for the ExPRESS rack used to support a variety of experiments on the *International Space Station*, and the International Space Station Program Manager for NASA's *Crew Return Vehicle*. In 1990, he was appointed a seated member of the Payload Safety Review Panel, representing the Safety and Mission Assurance Office of the International Space Station Program and subsequently one of the panel's three chairmen. During 2006, Dr. Musgrave accepted the position of technical assistant to the Manager, Safety and Mission Assurance/Risk Management Office of the International Space Station Program. Dr. Musgrave retired from NASA during 2008 and presently resides in Dayton, Tennessee, where he works as a consultant and educator. He is a member of the International Association for the Advancement of Space Safety, and its Academic Committee.

Axel (Skip) M. Larsen, Jr.
Chairman (Retired), NASA Payload Safety Review Panel
Johnson Space Center, National Aeronautics and Space Administration
Houston, Texas

After graduating from Admiral Farragut Academy on the shores of Toms River, New Jersey, Axel M. (Skip) Larsen entered the U.S. Naval Academy during the summer of 1956. He graduated from the academy in the spring of 1960, being awarded the B.S. degree in Engineering. Mr. Larsen subsequently was commissioned a second lieutenant in the U.S. Air Force. After receiving training as a navigator (Harlingen, Texas), training in electronic countermeasures (Sacramento, California), training in survival (Reno, Nevada), and training on the B-52 (Merced, California), Lt. Larsen was assigned to a USAF B-52H combat ready crew in Sault Ste. Marie, Michigan. By 1967, he had achieved the rank of Navigator Captain and was a Senior Electronic Countermeasures Officer assigned to a select (S) B-52H crew. During the later part of his career in the Air Force, he was an Electronic Countermeasures Standardization and Evaluation Officer with the responsibility of evaluating other B-52 electronic countermeasures officers. On leaving active military service, Mr. Larsen had accumulated 1800 hours of flight time. Mr. Larsen joined NASA at the Johnson Space Center in Houston, Texas, in 1967, where he became a NASA Apollo Flight Controller. During the years from 1967 until 1982, he served as an Operation and Procedures Officer for *Apollo 8*, *Apollo 10*; Skylab TNG, *Apollo 15*; Lead, Skylab Corollary Experiment Officer for *Skylab 1*, *Skylab 2*, and *Skylab 3*; Lead ASTP Experiments Officer, and as a Space Shuttle Program Integration Manager. During 1983, he left NASA to work for Orbital Science Corporation as employee 16 on the Transfer Orbit Stage Program, ultimately becoming Program Manager. Both vehicles manufactured under his tutelage were flown with 100% success. From 1971 to 1985, Mr. Larsen flew in the Texas Air Guard, flying air defense missions from Ellington Field in the F-101 Voodoo (1000 hours) for 10 years and the F-4 for 5 years (400 hours). He retired from the military as a lieutenant colonel. In 1990, he returned to NASA Johnson Space Center as chairman of the Payload Safety Review Panel, a position he held until he retired from the agency in 2006. During his tenure as chairman, his efforts were concentrated on leading safety reviews for experiments from the United States, Europe, Russia, Japan, and Canada conducted on *Space Lab*, *Spacehab*, *Mir*, the Space Shuttles, the *International Space Station*, and the Hubble Space Telescope service missions. He was the longest serving chairman of the panel since its inception. The highlight of Mr. Larsen's career as chairman of the Payload Safety Review Panel is marked by the development of the European Space Agency Franchised Payload Safety Review Panel. Developed in partnership with the European Space Agency, the approval of the European Space Agency Payload Safety Review Panel was signed jointly at NASA Headquarters by representatives of NASA and the European Space Agency. The extreme success of the European Space Agency Payload Safety Review Panel proved that organizations outside NASA can develop and operate safety panels similar to the NASA Payload Safety Review Panel applying the same rigorous discipline, dedication, and responsibility as the NASA panel. Mr. Larsen has been recognized with the Major General K. L. Berry award for lifesaving, the Silver Snoopy Award, the NASA Exceptional Service Medal, and numerous NASA Group Achievement Awards. He is a member of the U.S. Naval Institute, an Associate Fellow with the AIAA, an Associate Fellow with the IAASS, a member of the USNA Alumni Association, an Eagle Scout, and a long time sailor.

Tommaso Sgobba
Head, Independent Flight Safety Office
European Space Research and Technology Center, European Space Agency
Noordwijk, the Netherlands

Tommaso Sgobba is head of the Independent Flight Safety Office, Product Assurance and Safety Department of the European Space Agency. He also chairs the International Space Station Payload Safety Review Panel and the Automated Transport Vehicle Reentry Safety Panel at the European Space Agency. Mr. Sgobba's career spans three decades. Starting as a structural engineer, at the age of 31 he became Chief Inspector of the C27 (then G222) military transport aircraft final assembly line. This position initiated a career in aviation quality management that would include responsibility for the Boeing 767 Advanced Composites Flight Controls Quality Assurance at Aeritalia, and Chief Inspector for the AMX Fighter Spey jet engine at Fiat Avio.

In 1986, Mr. Sgobba moved to the European Space Agency, where he supported the product assurance and safety activities of several European space programs. In 1994, he became Product Assurance Principal Engineer for the European Meteosat

Program. Later as Produce Assurance and Safety Manager of the European Space Agency Microgravity Projects Division, he supported all major European Space Agency crewed missions on the *Space Shuttle* and *Mir*. He was promoted in 1997 to lead the Product Assurance and Safety Office for the European Space Agency International Space Station research facility development projects, a position that later expanded to include the European Space Agency Soyuz missions and International Space Station Operations. In 2007, Mr. Sgobba became head of the newly established Independent Flight Safety Office. Mr. Sgobba is President and co-founder of the International Association for the Advancement of Space Safety. He holds the Master of Science degree in Aeronautical Engineering from the Polytechnic of Turin, Italy, where he was Professor of Space System Safety during the academic years 1999 until 2001.

About the Contributors

John D. Albright
Subsystem Engineer, Space Shuttle Main Propulsion System
Johnson Space Center, National Aeronautics and Space Administration
Houston, Texas

John Albright is the NASA Subsystem Engineer for the Space Shuttle Main Propulsion System in the Energy Systems Division at Johnson Space Center in Houston, Texas. Mr. Albright received the Bachelors degree in Mechanical Engineering from the University of Iowa in 1989 and the Masters degree in Industrial Engineering from the University of Houston in 1995. He has been employed by NASA at the Johnson Space Center since 1991. Prior to becoming the Main Propulsion System NASA Subsystem Engineer, Mr. Albright served as the Pressure Systems Technical Representative to the Space Shuttle Payload Safety Review Panel for the Energy Systems Division. He also worked extensively on the design and testing of Space Shuttle primary thruster propellant valves and X-38 aerosurface electromechanical actuators. Prior to his employment with NASA, Mr. Albright worked as a subsurface engineer for Exxon Company U.S.A. in Corpus Christi, Texas.

Gregg John Baumer
Chairman, International Space Station Safety Review Panel (Retired)
Johnson Space Center, National Aeronautics and Space Administration
Houston, Texas

Gregg John Baumer served as the Chairman of the *International Space Station* Safety Review Panel at the NASA Johnson Space Center from 1996 through 2007. He began his career at NASA as a safety engineer in the payload safety branch. He was responsible for the assuring the safety of numerous Space Shuttle payloads, including *IUS, TDRS, LDEF, LANDSAT, Solar Maximum Satellite, Spacelab 1, Spacelab 2, Spacelab D1,* Solar Maximum repair mission, *Hubble Space Telescope,* and S*pace Station Freedom*. From 1986 through 1993, he represented NASA on the Department of Defense Safety Review Team, which was responsible for the safety review of all military payloads manifested on the *Space Shuttle*. In 1994, he joined the Space Station Program Office as Safety and Mission Assurance Lead for International Partners. In this role, he negotiated the safety and mission assurance requirements for each of the international partner elements of the space station. Throughout his career at NASA, Mr. Baumer has been deeply involved in the development and writing of safety requirements for payloads on the *Space Shuttle* and *International Space Station*. He earned numerous awards for his efforts, including the Silver Snoopy Award, NASA Exceptional Achievement Medal, and the NASA Exceptional Service Medal. He is a 1968 graduate of Tulane University with the Bachelor of Science degree in Physics. He is a military veteran with service in the U.S. Air Force with two tours of duty in Southeast Asia as commander of a KC-135 refueling aircraft. Mr. Baumer is now retired from NASA and resides in Houston, Texas.

Karen S. Bernstein
Structural Engineering Division
Johnson Space Center, National Aeronautics and Space Administration
Houston, Texas

Karen Bernstein (neé Karen Edelstein) is a senior stress analyst at the NASA Johnson Space Center, where she has spent a large part of her career managing spaceflight hardware made of glass. As the Space Shuttle Orbiter Crew Module and Forward Fuselage Subsystem Manager, Ms. Bernstein performed bird strike analysis, micrometeoroid and orbital debris analysis, and damage tolerance studies for the fused silica thermal panes in the *Space Shuttle*. As the International Space Station Windows Subsystem Manager, Ms. Bernstein worked with Boeing during the design phase of the program and the verification of the window hardware. She also worked extensively with her counterparts in Moscow at Rocket Space Corporation-Energia, Khrunichev, and the Institute of Technical Glass to develop damage tolerance data for the Russian windows in the *International Space Station*. For more than 10 years, Ms. Bernstein served as the technical consultant to the Payload Safety Review Panel for issues involving frangible materials, such as glass. In this capacity she reviewed many safety data packages and advised hardware developers on how to manage glass components in NASA's spacecraft safely. Ms. Bernstein received the Bachelor's degree in Aerospace Engineering from the University of Michigan in Ann Arbor in 1984. In 1988, she became a Master of Mechanical Engineering at Rice University in Houston. In 1999, she participated in the NASA Johnson Space Center fellowship program at the University of Houston, where she studied Materials Engineering. In 1997, Ms. Bernstein was honored with a Silver Snoopy Award. She has written several papers concerning damage tolerance in fused silica glass and continues to do research in this area.

Loredana Bessone
Head of Instructional Technologies, Special Skills Training and Exploration Unit
European Astronaut Centre, European Space Agency
Cologne, Germany

Loredana Bessone is head of the Instructional Technologies, Special Skills, and Exploration Unit of the Astronaut Training Division, European Astronaut Center, Cologne, Germany, since 1998. She joined the European Astronaut Center in 1990. As head of Instructional Technologies, Ms. Bessone is responsible for Instructor Training and Certification, for Training Quality Assurance, and for the European Astronaut Center Training Development and Information Systems. In her capacity as head of Special Skills Training, she is responsible for human behavior and performance, robotics, extravehicular activity, and rendezvous and docking training development related to exploration. Ms. Bessone is the European Astronaut Center representative for European Space Agency Exploration activities, including the Aurora Exploration Preparatory Program, where she has been responsible for the Human Mars Mission studies. Ms. Bessone holds the Masters degree in Information Sciences from the University of Turin and the Master of Space System Engineering from the Technical University of Delft, TopTech Studies.

Tony Brown
Materials Evaluation Section
MEI Technologies
Houston, Texas

Tony Brown is a graduate of the University of Arkansas with the Bachelor of Science in Chemical Engineering. He worked for three years in the petrochemical industry, doing research on the extraction of magnesium from seawater and the development of polypropylene film and fiber products. He joined Hamilton Standard in 1984 as a Materials Engineer working on the Space Shuttle space suit. After two years, he worked at ILC on projects relating to the development of space tools for use on the *Hubble Space Telescope* and the Space Shuttle Program. In 1987, he joined the Materials Control Group at Lockheed Martin. During his 18-year tenure at Lockheed Martin, he worked on the STS-26 return to flight, Department of Defense payloads, various IMAX cameras, the Shuttle-Mir program, the International Space Station Program, and materials safety support to the Payloads Safety Review Panel. Mr. Brown was involved in the Interagency Materials Agreement negotiations with NASA's International Partners. For the last two years, he has worked at MEI, Incorporated. Over the years, he has served on configuration change boards and is currently a member of the board that reviews hardware relating to crew life sciences.

Giancarlo Bussu, Ph.D.
Product Assurance and Safety Department
European Space Agency
Noordwijk, the Netherlands

Giancarlo Bussu, after graduating in Aerospace Engineering at Pisa University, Italy, obtained the Doctor of Philosophy at Cranfield University, United Kingdom, in Aerospace Materials and Structural Damage Tolerance. He subsequently worked in the research and development activities for the commercial aircraft industry. He joined the European Space Agency in 2000, where he provides support to all projects in the areas of materials, manufacturing processes, safety, product assurance, and standardization. He is the European Space Agency expert in stress-corrosion cracking and a member of the European Space Agency Payload Safety Review Panel. His work includes a number of publications and a patent in the field.

Victor Chang
Safety and Mission Assurance and Configuration Management
Canadian Space Agency
St. Hubert, Canada

Victor Chang joined the Canadian Space Agency in 1991, supporting the recruitment and training of Canadian astronauts. He also was involved in the design, test, integration, and operation of a variety of payloads and experiments launched and operated by Canadian astronauts. He later became involved in the Space Vision System for the Space Station Robotic Manipulator System and provided support to the design, test, integration, and installation of the system. Afterward, he joined the Safety and Mission Assurance directorate. Mr. Chang is presently the Safety and Mission Assurance and Configuration Management Manager and the North American representative on the International Association for the Advancement of Space Safety Board. Before joining the Canadian Space Agency, Mr. Chang worked at Bombardier Aerospace in the Military Division. He was involved in the design, test, integration, and free flight testing of the CL-227 vertical take off and landing remotely piloted vehicle system. Titles he has held include Test Site Manager, Air Vehicle Operator, Flight Test Conductor, Range Safety Officer, and Chief Crash and Accident Investigator. He participated in flight trials performed at Defense and Research Establishment Suffield, CFB Suffield, Alberta, Fort Huachuca, Arizona; Yuma U.S. Army Missile Test Range, Arizona; test flights aboard the USS *Doyle* frigate off the coast of Florida. As well, Mr. Chang was responsible for developing and conducting a formal training course for the U.S. Navy, at the Naval Air Test Center, and Strike Aircraft at Pawtuxet River, Washington. Mr. Chang attended McGill University and graduated with the Bachelor of Mechanical Engineering degree in 1986.

Antonio Ciccolella, Ph.D.
Directorate of Earth Observation, D/EOP-E
European Space Research Institute, European Space Agency

Antonio Ciccolella received the Doctor of Philosophy degree in Electronic Engineering at the Politecnico di Torino (Italy). He was employed at Alenia Spazio, Space Division, as electromagnetic compatibility engineer from 1986 to 1992. Then, he joined the Electromagnetics Division of the European Space Agency Technical Directorate, where he has been supporting several space projects in electromagnetic compatibility. From 2000 to 2006, he was head of the Electromagnetic Compatibility and Antenna Measurement Section of ESTEC. The section supports European Space Agency project and Industry in the control of electromagnetic noise and interference for all agency spacecraft equipment and systems throughout their life cycle. In 2006, he moved to the Directorate of Earth Observation, where he is coordinator for the space segment of the Global Monitoring for Environment and Security program. His interests mainly are theoretical and numerical electromagnetics, innovative testing techniques, statistical electromagnetics, and transmission line theory.

Jonathan D. Clark, M.D., M.P.H.
Space Medicine Liaison
National Space Biomedical Research Institute, Baylor College of Medicine
Houston, Texas

Jonathan B. Clark is the Space Medicine Liaison, National Space Biomedical Research Institute, at Baylor College of Medicine. He received the Bachelor of Science from Texas A&M, the Doctor of Medicine from Uniformed Services University, and the Master of Public Health from the University of Alabama—Birmingham. He is board certified in Neurology and Aerospace Medicine and is a Fellow of the Aerospace Medical Association. He is President of the Space Medicine Association. He is Clinical Assistant Professor in the Department of Preventive Medicine and Community Health at the University of Texas Medical Branch in Galveston, Texas, and was Co-Director of the University of Texas Medical Branch Aerospace Medicine Residency Program from 2001 to 2003. He was Manager of the Medical Operations Branch and worked as a Space Shuttle Crew Surgeon in the Mission Control Center on shuttle missions STS-92, STS-96, STS-99, STS-102, STS-105, and STS-111. He is on the Spacecraft Survival Integrated Investigation Team at the NASA Johnson Space Center. He flew as a DoD Space Shuttle support flight surgeon for STS-49 and STS-104. He served 26 years in the U.S. Navy, qualifying as a Naval Flight Officer, Naval Flight Surgeon, Navy Diver, U.S. Army parachutist, and Special Forces Military freefall parachutist.

Francis A. Cucinotta, Ph.D.
Chief Scientist, Space Radiation Program
Johnson Space Center, National Aeronautics and Space Administration
Houston, Texas

Francis A. Cucinotta is the Chief Scientist for the NASA Space Radiation Program at the NASA Johnson Space Center. Dr. Cucinotta received the Bachelor of Science in Physics from Rutgers University in New Jersey in 1983, and the Doctor of Philosophy from Old Dominion University in Virginia in 1988. From 1990 and 1997, he served as a Senior Research Scientist at the NASA Langley Research Center in Hampton, Virginia. He has been a Project Scientist and Radiation Health Officer at NASA Johnson Space Center since 1997 and a visiting scientist at the Oncology Center, Johns Hopkins University in Maryland since 1995. He was a visiting scientist at MRC, Radiation and Genome Stability Unit, Harwell, United Kingdom, in 1996, and an instructor in the Department of Mathematics, Christopher Newport University in Virginia during 1993 and 1994. He was awarded NASA Superior Achievement Awards in 2005, 2000, 1995, and 1993; NASA Silver Snoopy Award in 2002; NASA Space Flight Awareness Award in 2001; and the NASA Floyd L. Thompson Fellow Award in 1995. He has served as council member for the National Council on Radiation Protection and Measurements since 2001 and as committee member for the COSPAR Commission F2 for Radiation Effects since 1993. He served as General Councilor for Radiation Research Society from 2003 to 2006 and on the Scientific Committee for National Council on Radiation Protection and Measurements SC 1-7 from 1996 to 2003. He has written more than 200 peer reviewed scientific journal articles, numerous book chapters, over 150 other formal NASA technical reports in the field of space radiation research and contributed numerous scientific presentations at national and international scientific meetings. Dr. Cucinotta's most prominent technical achievements are in the development of the Astronaut Exposure and Risk Database at NASA, discovering the increased risk of cataracts at low cosmic ray fluences in astronauts, the development of state-of-the-art nuclear interaction models, and improving the biophysical assessment of cancer risks and their uncertainties of space radiation.

Volker Damann, M.D.
Head, Crew Medical Support Office
European Astronaut Center, European Space Agency
Cologne, Germany

Dr. Damann graduated from the medical school in Marburg, Germany, in 1985, after which he worked in the University Radiology department and specialized in nuclear medicine. In 1988, together with two colleagues, he established a private radiology practice, offering imaging services in the domains of conventional radiology, CAT, and nuclear medicine. Since 1989, he served for six years as a spaceflight surgeon at the German Aerospace Research Center. During that time, he supported eight Space Shuttle and two Soyuz/Mir missions as crew surgeon. In 1995, he became the Lead Flight Surgeon, and in 1998, Dr. Damann was named head of the Crew Medical Support Office at the European Astronaut Center of the European Space Agency in Cologne, Germany. He became a member of the Aerospace Medical Association (AsMA), the Space Medicine Branch, and the Society of NASA Flight Surgeons in 1996. Building on his acquired operational expertise, Dr. Damann was instrumental in establishing a growing team of physicians and biomedical engineers to staff and operate a European Space Agency medical control center in Cologne. The implementation of new medical technologies, utilization of innovative communication means, and application oriented operational research define major elements of his approach to space medicine. The strengthening of the role and responsibility of a small partner within the International Space Station medical program while maintaining the cultural identity is one of his highest priorities. He is convinced that only a strong international partnership will enable humankind to continue space exploration and expand human knowledge in space medicine.

Michael J. Eiden, Dipl. Ing.
Senior Advisor for Multidisciplinary Mechanical Systems
European Space Research and Technology Center, European Space Agency
Noordwijk, the Netherlands

Michael J. Eiden graduated as Dipl. Ing. in Aeronautic and Aerospace Engineering from the University Stuttgart, Germany. From 1983 through 1990, Mr. Eiden was a Senior Structures Engineer with involvement and responsibilities for structural design assessments of almost all past European Space Agency spacecraft (*Giotto, Ulysses, SOHO, Hubble Space Telescope, Spacelab, ERS, ENVISAT, XMM, Huygens*, and others). In addition, he was responsible for CAE and advanced mission concept assessments of all future science mission studies and their technology development planning, as well as the development of Parachute Systems Design and Analysis Tool. From 1990 to 2003, he headed the European Space Agency Space Mechanisms Section, providing technical leadership and expertise on space tribology and space mechanisms for all European Space Agency space missions and as the European Space Agency technical leader for the Hubble Space Telescope servicing missions 3A and 3B. Since 2003, Mr. Eiden has been Senior Advisor for Multidisciplinary Mechanical Systems in the Technical and Quality Management Directorate at the European Space Research and Technology Center of the European Space Agency. His responsibilities include, among others, the Hubble Space Telescope project manager, advisor in developments of advanced space technology and microtechnology and nanotechnology, development of key space mechanisms hardware telecommunication applications, and technical leadership for parachute recovery systems for sounding rocket payloads recovery.

Lindsay Evans
Senior Robotics Instructor
Canadian Space Agency
St. Hubert, Canada

After graduating from Queen's University in 1997, Lindsay Evans worked for SPAR Aerospace, the leading space robotics company in Canada. Ms. Evans began as a Mechanical Designer working on developing the Special Purpose Dexterous Manipulator (Dextre), a robot that is part of Canada's contribution to the *International Space Station*. Eighteen months later, she went from the mechanical design office into the operations group at MacDonald, Dettwiler, and Associates, and eventually went to work as a contractor at the Canadian Space Agency with responsibilities for performing robotic analysis (Canadarm2) for *International Space Station* assembly flights and supporting the NASA Mission Control community. Six years ago, Ms. Evans accepted a position at the Canadian Space Agency in the Astronaut Training Department, where she currently works. Her responsibilities include instructing astronauts (American, Canadian, Japanese, European) and Russian cosmonauts as well as other ground support personnel, such as mission controllers and flight directors, on how to operate the Canadian robots onboard the *International Space Station*. Over the past four years, she has also been involved in supporting real time robotic Canadarm2 operations from the control center at the Canadian Space Agency. More recently, Ms. Evans has been the Acting Chief Instructor and leads a dynamic team of 10 instructors. She enjoys the international aspect of the job, working with U.S., Japanese, Russian, and European partners.

Simon N. Evetts, B.A. (Hons), M.Sc., Ph.D.
Medical Projects and Technology Lead, Wyle Laboratories GmbH
Crew Medical Support Office
European Astronaut Center, European Space Agency
Cologne, Germany

Dr. Simon N. Evetts started his working life as a British Army officer serving in peacetime and operational circumstances in the United Kingdom and abroad. Since leaving the services and obtaining the Bachelor of Arts in Sport Studies and the Master of Science and the Doctor of Philosophy in applied human physiology related fields, he has been employed at sports medicine institutes, research laboratories, and universities in the United Kingdom, United States, and the Middle East. His teaching and research interests and activities lie in the fields of exercise physiology, space biomedical sciences, and human applied physiology. The research programs within which he actively has been involved are concerned with health and fitness assessment, the provision of basic life support in space, and methods of blood sampling in microgravity and hypogravity environments. Dr. Evetts is the European Space Agency's medical operations Advanced Projects Lead (Wyle GmbH, European Astronaut Center, Cologne, Germany), a member of the KCL/PUCRS MicroG Life Science Research Team, the Research Executive for the U.K. Space Biomedicine Association, and visiting lecturer at King's College London.

Scott C. Forth, Ph.D.
Integration Technical Manager for Fracture Control and Pressure Vessels
Johnson Space Center, National Aeronautics and Space Administration
Houston, Texas

Dr. Scott Forth began his aerospace career at United Technologies Research Center in East Hartford, Connecticut. His responsibilities included evaluating crack growth events for Pratt and Whitney, Sikorsky Aircraft, and Hamilton Sundstrand. Mr. Forth received the United Technologies President's Award for his contributions on the accident investigation of a commercial aircraft. While employed at United Technologies Research Center, he completed the Doctor of Philosophy degree from Clarkson University, Potsdam, New York, in 1999, with a thesis entitled "Mixed Mode Crack Growth: An Experimental and Computational Study." In 2000, Dr. Forth left United Technologies Research Center to join NASA Langley Research Center in Hampton, Virginia, as a materials research engineer. At NASA Langley, his research focused on fracture mechanics models and methods for predicting fatigue crack growth behavior and residual strength of aerospace structures. Dr. Forth led a program with Eclipse Aviation to design the laboratory specimens, understand the relationship between laboratory and aircraft failure modes, and develop the necessary failure data to safely manufacture a jet aircraft using friction stir-welding. Dr. Forth also served as the structures and materials specialist on the NASA Langley Wind Tunnel Model Systems Committee and the Pressure Vessel Systems Committee. In these roles, he developed inspection schedules and crack growth models to resolve continued operational safety issues. In 2006, Dr. Forth transferred to the NASA Johnson Space Center to co-chair the JSC Fracture Control Board and the NASA Fracture Control Methodology Panel. In the Johnson Space Center role, Dr. Forth approves the fracture control certifications for Space Shuttle payloads and International Space Station hardware and provides technical oversight for all fracture critical spaceflight hardware. In the NASA role, he co-chairs the panel responsible for developing and interpreting fracture control requirements for the agency. He led the effort to develop new fracture control requirements for the Constellation Program and championed the importance of fracture control in improving safety within the agency.

Kerry A. George
Section Lead, Radiation Biophysics Laboratory
Wyle Laboratories
Houston, Texas

Kerry George is a Senior Scientist with Wyle Laboratories, and has been providing scientific and technical support to the NASA radiation group at the Johnson Space Center for more than 15 years. She has authored and coauthored more than 50 peer reviewed professional scientific manuscripts on the biological effects of radiation exposure and is a leading expert on radiation biodosimetry. She designed and implemented a successful ongoing radiation biodosimetry program for NASA astronauts involved in long duration space flight missions, and she continues to study and develop new methods and techniques to assess the biological effects of space radiation. Ms. Kerry completed an undergraduate degree in Biology at Napier University in Edinburgh, Scotland. She has been the recipient of many awards, including the Napier University medal for outstanding achievement, several special spaceflight achievement awards from NASA, and Wyle Laboratories' outstanding employee award for excellence in science.

Tateo Goka, Ph.D.
Director of Space Environment Measurement Group
The Institute of Aerospace Technology, Japan Aerospace Exploration Agency
Tsukuba, Japan

Dr. Tateo Goka, who holds the Science Degree in Physics and the Doctor of Philosophy in Systems Management, started in 1970 as a Spacecraft Systems Engineer within the Engineering Test Satellite Design Division at the National Space Development Agency of Japan. He worked within the frame of the ETS-I, ETS-II, MOS-1 programs, ensuring EEE parts reliability and radiation hardness in the H-II rocket and ETS-VI projects. He conducted single event effects research within the Technical Research Division at Tsukuba Space Center in National Space Development Agency of Japan. Since 1994, Dr. Goka has been the Director of Space Environment Measurement Group in National Space Development Agency of Japan and presently in the Institute of Aerospace Technology at the Japan Aerospace Exploration Agency. This section is responsible for the technical support to projects related to the space environment and its effects on crewed and uncrewed space applications, including measurement, evaluation, and assessment of the space environment and development of space environment monitors onboard spacecrafts designed to acquire engineering data useful to the design of future spacecraft to diagnose the anomalies encountered on orbit and collect data to make new radiation belt models. Dr. Goka was principal investigator in neutron energy spectrum measurement using Bonner Ball Neutron Detectors inside the Space Shuttle on STS-89 in 1998, and inside the *International Space Station* through the Human Research Facility project in 2001. Now, he is the principal investigator in the Space Environment Data Acquisition mission of the Japanese Experiment Module Exposed Facility attached payload on *International Space Station*. Dr. Goka is the author of the *Space Environment Risk Lexicon* (book in Japanese) and has been an invited professor in the graduate school of science and engineering at Kagoshima University in Japan since 2001.

Jerry R. Goodman
Chairman (Former), Acoustics Working Group and ISS Acoustics Lead
Johnson Space Center, National Aeronautics and Space Administration
Houston, Texas

Mr. Jerry Goodman has worked on space related programs for 47 years and at NASA for 45 years. He is currently on the staff of the Habitability and Environmental Factors Division, which supports efforts on human factors standards, the Constellation program, and writing. He released his thesis on Apollo crew compartment efforts as a NASA document and is authoring chapters for an aerospace medicine book. He is working on a NASA "lessons learned" manuscript in acoustics, extravehicular activity, and other subjects. Mr. Goodman is an adjunct professor of a graduate level systems engineering course at University of Houston—Clear Lake. Mr. Goodman made significant contributions in four basic areas of human spaceflight: head of Apollo program space suit development; systems engineering and integration efforts on the crew compartment design of the Apollo command module, lunar module, and Space Shuttle; project engineer on extravehicular activity provisions for the Apollo command module, lunar module, and Space Shuttle; and head of the acoustics working groups for Space Shuttle and International Space Station programs, working with United States and international partner hardware developers. He received the B.S.M.E. from Purdue University and the Masters degree in Human Factors from the University of Illinois. He is a recipient of numerous NASA awards and is an Associate Fellow of the American Institute of Aeronautics and Astronautics.

About the Contributors

Russell Graves
Space Exploration Division
Integrated Defense Systems, The Boeing Company
Houston, Texas

Russell Graves is the micrometeoroid and orbital debris protection subsystem lead engineer on the Boeing Company's International Space Station prime contract, a role that he has held since the contract's inception in 1993. Mr. Graves is a nationally recognized expert in the field of impact effects mitigation and assessment (space systems survivability and vulnerability). Additionally, he is Boeing's lead for space station inhabited module pressure wall leak detection and hole repair, an effort that is developing the means for locating atmospheric leaks, applying an intravehicular or extravehicular repair, and ensuring restoration of pressure and structural integrity to allow continued utilization of the station. Beyond his International Space Station roles, Mr. Graves has led and participated in space system survivability efforts for a number of Boeing spacecraft programs, including Boeing Satellite Systems, *Crew Exploration Vehicle*, and the Space Shuttle Program. He also participated in and led proposals for the orbital space plane and *Crew Exploration Vehicle* and acted as principal investigator for a NASA Exploration Systems Mission Directorate broad agency announcement proposal response for an integrated structure for protection and crew exploration. Mr. Graves authored and coauthored several papers on the International Space Station meteoroid and orbital debris mitigation approach. He holds the Master of Science in engineering science and mechanics and the Bachelor of Science in mechanical engineering, both from the University of Tennessee. He was selected as an associate technical fellow for the Boeing Company in 2002.

Nathanael J. Greene
Project Manager
White Sands Test Facility, National Aeronautics and Space Administration
Las Cruces, New Mexico

Nathanael Greene is employed by NASA at the White Sands Test Facility, where he is responsible for composite overwrapped pressure vessel testing and leadership of the White Sands Test Facility composite overwrapped pressure vessel group. This group has been recognized across NASA, industry, and academia for providing test data, analysis, and training for composite overwrapped pressure vessels. Mr. Greene has been involved in the NESC assessments of stress rupture life concerns on Kevlar®-composite overwrapped pressure vessels on the Space Shuttles and for carbon composite overwrapped pressure vessels on the *International Space Station*. He is also an instructor for the Composite Overwrapped Pressure Vessel Visual Inspector Training Course and is currently generating data and providing analysis for the NESC, Crew Exploration Vehicle, International Space Station, and Space Shuttle programs to support definition of design requirements and recertification for flight rationale. In addition, he is involved in industry standards development for safe use of composite overwrapped pressure vessels with the Canadian Space Agency America, Society of Automotive Engineers, and the American Institute of Aeronautics and Astronautics. Mr. Greene received the Bachelor of Science in mechanical engineering from Iowa Sate University in May 2002 and the Master of Science from Iowa State University in December 2004. He is active in the areas of hydrogen combustion hazards testing and assessment and spacecraft fire safety.

Gerald D. Griffith
Chief System Safety Engineer
JAMSS America, Incorporated
Houston, Texas

Gerald D. Griffith has been employed by JAMSS America, a Japanese owned Texas corporation, since his retirement from NASA Johnson Space Center in September 2000. He provides systems safety consultation and support services to the Japanese human spaceflight program, supporting that country's participation as an international partner of the *International Space Station* Program. Mr. Griffith earned the Bachelor of Science in mechanical engineering from Texas A&M University in 1960. In 1961, while serving in the USAF, he earned the Master of Science in mechanical engineering from the University of Illinois—Urbana. Following completion of his USAF service commitment as an Aircraft Systems Test Project Officer at the Air Force Mission Development Center, Holloman AFB, New Mexico, Mr. Griffith began his 36-year NASA career at Johnson Space Center in 1964. Mr. Griffith worked as a flight controller trainer and in mission control for human spaceflight experiments, followed by a 22-year assignment as a technical consultant in the astronaut office. Mr. Griffith received many NASA internal and external awards and recognitions, including the NASA Exceptional Achievement Medal in 1994 for his service as the Astronaut Office Representative on the Space Shuttle Program Payload Safety Review Panel, a position he had occupied for more than 10 years at his retirement.

Ferdinand W. Grosveld, D.E.
Consultant
Hampton, Virginia

Dr. Grosveld has over 30 years' experience in aerospace engineering, including aircraft design and performance, aerodynamics, propulsion, and acoustics. He serves as the manager of Acoustics and Fluid Mechanics for Lockheed Martin Mission Services at the NASA Langley Research Center. Dr. Grosveld has been an acoustics consultant on various aero, structural, architectural, and space station projects. He holds the Doctor of Engineering from the University of Kansas and earned the Master of Science in Aerospace Engineering from the Delft University of Technology. He has taught at the undergraduate and graduate levels at the University of Kansas and Old Dominion University. Dr. Grosveld has co-authored over 130 publications, including book chapters, archival publications, congressional presentations, and technical reports. He co-organized or chaired several symposia, invited lecture series, and technical meeting sessions. He is the co-inventor on two acoustics patents. He was awarded the NASA Special Space Flight Achievement Award for outstanding contributions to International Space Station acoustics, and he received several other NASA, Lockheed Martin, American Institute of Aeronautics and Astronautics and civic, individual, and group honors. He is an Associate Fellow of the American Institute of Aeronautics and Astronautics and the recipient of the Sustained Membership Award. He serves on the national and regional Associate Fellow Selection and Honors and Awards standing committees. He initiated and organized a project for high school students and advisors to design, build, and fly the World's Largest Paper Airplane that was recognized in the Guinness Book of Records 1992-1995. Dr. Grosveld is a licensed private pilot, certified hang glider pilot, accredited tennis instructor, and qualified scuba diver.

Jon Haas
Project Manager, Laboratories Office
White Sands Test Facility
Johnson Space Center, National Aeronautics and Space Administration
Las Cruces, New Mexico

Mr. Haas has been at NASA's White Sands Test Facility since 1996 serving the U.S. space program in several technical and leadership roles. Currently, he manages the Materials and Processes test and evaluation efforts for the Constellation and International Space Station programs in the White Sands Test Facility Laboratories Office. His work centers on crew materials and systems, including the environmental control and life support systems for the *International Space Station*, the *Space Shuttle*, and now the evaluation of designs for NASA's new *Crew Exploration Vehicle*. The *International Space Station* and the former Russian Mir space station (as well as many non-spaceflight systems) use chemical generation as the backup to primary oxygen life support systems. During his time in the White Sands Test Facility Laboratories Office, Mr. Haas lead or otherwise supported several chemical oxygen generation system hardware design efforts, safety analyses, or failure investigations and continues to dedicate time to improving the safety of flight oxygen systems and commercial hardware used inside and external to NASA.

David Hornyak
Lead System Interface Engineer, Payload Engineering Integration
Johnson Space Center, National Aeronautics and Space Administration
Houston, Texas

David Hornyak earned the Bachelors degree in mechanical engineering from Ohio State University. Following a four-year career in the power industry, including two years at a nuclear power station, Mr. Hornyak accepted a job with the Boeing Company on the International Space Station Program in Houston, Texas. After five years supporting the design, development, test and initial launch and operation of the *International Space Station*, Mr. Hornyak moved to Washington, DC, where he spent the next five years supporting launch vehicle development and engineering research and development. During this time, he earned his professional engineering license, the Masters degree in Business Administration, and a certificate in systems engineering. In the winter of 2007, Mr. Hornyak returned to the Johnson Space Center when he accepted a position with NASA on the International Space Station payloads engineering integration team. Mr. Hornyak's work on the space station program includes support to the Payload Safety Review Panel, development of the payload engineering interfaces and the International Space Station caution and warning systems, lead for the payload engineering effort in fire protection for payloads, and participation in the design of the International Space Station vehicle in support of science payload systems. During the course of these activities, Mr. Hornyak won the Space Flight Awareness Award for the 100th Space Shuttle mission, STS-92. In addition to International Space Station vehicle support, Mr. Hornyak supported the design, development, and test of several payload systems and facilities on the space station. His current duties include leading the space station pressurized payload engineering integration effort, the assessment of launch certification for space station systems utilizing multiple launch vehicles, supporting design reviews for the Constellation program and other launch vehicles, and supporting the Space Shuttle program payload engineering integration. Mr. Hornyak is a member of the American Institute of Aeronautics and Astronautics Space Systems Technical Committee.

About the Contributors

John T. James, Ph.D.
Chief Toxicologist, Habitability and Environmental Factors Division
Johnson Space Center, National Aeronautics and Space Administration
Houston, Texas

Dr. John T. James has served as NASA Chief Toxicologist for 18 years and enjoys the diversity of supporting the Space Shuttle, International Space Station, and Constellation Programs. He has been a diplomate of the American Board of Toxicology since 1986. His scientific interests include risk assessment of toxicants in spacecraft air and water, pulmonary toxicity of lunar and martian dusts, development of air quality analyzers for spaceflight, and ethical applications of nanotechnology to disadvantaged people. He is a recipient of the Space Shuttle Program Star Award, of the Astronaut Office Silver Snoopy Award, and of the NASA Exceptional Service Medal.

Judith A. Jeevarajan, Ph.D.
Senior Scientist, Power Systems Branch
Johnson Space Center, National Aeronautics and Space Administration
Houston, Texas

Dr. Judith A. Jeevarajan has been a senior scientist in the energy systems division at NASA Johnson Space Center in Houston, Texas, since 2003. She has worked on several battery projects using various battery chemistries and also represents the battery group at all the NASA safety panels, providing technical design and safety guidance for various projects, including those with the international partners of the *International Space Station*. She also carries out advanced battery technology research as part of the energy storage program for the Constellation program. Dr. Jeevarajan teaches a battery course for engineers involved in battery testing in her organization. From 1998 to 2003, Dr. Jeevarajan worked in the capacity of a senior scientist in the battery group for Lockheed Martin Space Operations, the major contractor at the Johnson Space Center for engineering, testing, and analysis. She was the battery group leader and managed government furnished equipment battery projects. Batteries were designed and safety certified under her guidance, and she was the first to certify and fly a lithium-ion commercial battery in a crewed space environment. Dr. Jeevarajan holds the Master of Science from the University of Notre Dame and the Doctor of Philosophy from the University of Alabama. Her graduate work focused on electrochemistry, and her postgraduate work on battery technology. She has made approximately 50 presentations at conferences, given invited lectures for several organizations, served as session chair at prestigious conferences, and won many NASA awards.

Paul T. Johnson
Safety and Reliability
The Boeing Company
Huntsville, Alabama

Paul T. Johnson has over 43 years of experience in the aerospace, commercial nuclear power plant, and manufacturer's representative industries. With the Bachelor of Science degree in mechanical engineering, he started his career in electromechanical design with the Lockheed Missiles and Space Company in Sunnyvale, California. His career includes both electromechanical and mechanical design, wind tunnel testing, military standard vibration and shock testing, military standard environmental testing, machine shop fabrication liaison, district manufacturers' representative for various companies, commercial nuclear power plant safety certification, spaceflight experiment safety certification, as well as various project engineering and management positions. He currently represents the International Space Station Payloads Project Office on the NASA Johnson Space Center Payload Safety Review Panel. He is employed by the Boeing Company in Huntsville, Alabama.

Joel K. Kearns
Space Operations Mission Directorate
Headquarters, National Aeronautics and Space Administration
Washington, DC

Mr. Joel K. Kearns is currently transition manager in the NASA Headquarters Space Operations Mission Directorate, focusing on the phase-out, transition, and retirement of the Space Shuttle Program. From February 2005 to October 2006, he served as Director of Project Management and Engineering at the Ames Research Center, responsible for all flight aircraft and spaceflight projects and the flight system engineering development assigned to that Ames. He also served as acting manager of the Stratospheric Observatory for Infrared Astronomy Program during its restructure and completion of the 747SP Observatory Aircraft for First Flight. From 1988 to 1995, Mr. Kearns was program manager at NASA headquarters for microgravity material science and biotechnology. He was appointed to the senior executive service at Marshall Space Flight Center in 1995 and served as manager of the national microgravity research program and space product development program from 1995 to 1999. During this time of frequent human spaceflight research missions on the Space Shuttle, *Spacelab*, *Spacehab*, *shuttle-Mir*, and early *International Space Station*, he managed the U.S. programs of spaceflight and Earth based research in low gravity materials science, fluid physics and transport, biotechnology, combustions science, and fundamental physics. Before his first tour at NASA, Mr. Kearns worked at Grumman Aerospace and the Mitre Corporation on a variety of crystal growth process research, space science, Space Shuttle, and International Space Station assignments. Prior to rejoining NASA in 2005, he was Vice President for Engineering and Technology and Director of Crystal Technology at Sumitomo-Mitsubishi Silicon Group USA, part of the second largest silicon wafer manufacturer in the world, SUMCO. He directed research and engineering in dislocation free growth of silicon crystals, the interaction of materials properties with solid state device function, advanced process engineering and manufacturing, material and defect characterization, quality assurance, and new product development. He holds two patents. Mr. Kearns was elected an Associate Fellow of American Institute of Aeronautics and Astronautics in 2005. He earned the Bachelor of Science in mechanical engineering at Worcester Polytechnic Institute in 1983 and conducted graduate studies at that same location. As a graduate student, he led the development of that school's first Get Away Special payload, which successfully flew on STS-40.

Myung-Hee Y. Kim, Ph.D.
Senior Research Scientist
Universities Space Research Association, Division of Space Life Sciences
Houston, Texas

Dr. Myung-Hee Y. Kim is a senior scientist at Wyle Laboratories in Houston, Texas, in the space radiation program at the NASA Johnson Space Center. Dr. Kim received the Doctor of Philosophy in applied sciences from the College of William and Mary in Virginia in 1995. She was awarded a National Research Council Research Fellowship as a postdoctoral fellow at the NASA Langley Research Center from 1995 to 1998. Before she joined Wyle Laboratories, Dr. Kim conducted radiation protection studies at NASA Langley Research Center and in the Department of Chemistry at the College of William and Mary. Her current work at Johnson Space Center focuses on developing and improving radiation transport modeling for the evaluation of radiation protection through the analysis of data on the space environment and vehicle design, developing radiation environmental prediction models, analyzing results from dosimetry and biodosimetry measurements of space missions, and maintaining radiation exposure data and risk assessment for astronauts. She has written more than 50 scientific papers in peer reviewed journals and books in the field of space radiation research and contributed more than 80 scientific presentations at national and international scientific meetings. She received the Excellence in Science Award in 2004 from Wyle Laboratories, the Certificate of Appreciation in 2004 from Langley Research Center, and the Special Space Flight Achievement Award in 2003 from Johnson Space Center.

Paul Kirkpatrick
Chairman, Ground Safety Review Panel
Kennedy Space Center, National Aeronautics and Space Administration
Kennedy Space Center, Florida

Paul Kirkpatrick entered the U.S. Coast Guard Academy in 1972 and graduated in 1976 with the Bachelor of Science in chemistry. After several Coast Guard assignments around the country as a marine safety officer, Mr. Kirkpatrick joined NASA in 1988 as a safety specialist. He was assigned to Launch Pad 39B and solid rocket booster retrieval operations. In 1991, he became a payload safety engineer, eventually becoming the payload safety engineering manager. As a payload safety engineer, he served as the lead ground safety engineer for numerous Space Shuttle payloads and space station elements. In September 2004, he was promoted to Chairman of the NASA Ground Safety Review Panel at Kennedy Space Center. After joining NASA, Mr. Kirkpatrick continued his affiliation with the Coast Guard as a member of the reserve, retiring in 2002 at the rank of captain. Mr. Kirkpatrick was called to active duty after 9/11 to establish the USCG Maritime Homeland Security Organization for the Southeast United States and the Caribbean. He has been awarded the Presidential Meritorious Service Medal, the Coast Guard Commendation and Achievement Medals, the DOT 9-11 Medal, the NASA Exceptional Service Medal, and the NASA Astronaut Corps Silver Snoopy.

Heiner Klinkrad, Ph.D.
Head, Space Debris Office
European Space Operations Center, European Space Agency
Darmstadt, Germany

Dr. Heiner Klinkrad graduated from the Braunschweig University of Technology in aeronautical engineering in 1980, with an intermission for a one-year scholarship at the Florida Institute of Technology in 1975. He also received the Doctor of Philosophy from the Braunschweig University of Technology in 1984. In 1980, he joined the European Space Agency as a visiting scientist at the European Space Operations Center. In 1983, he became a staff member at the European Space Research and Technology Center, where he worked on flight dynamics and mission planning aspects of remote sensing satellite, *ERS-1*. In 1988, he returned to the European Space Operations Center, where he became head of the space debris office in 2006. Apart from astrodynamics in general and the modeling of non-gravitational perturbations, his main interest lies in the area of space debris environment models and the assessment of related collision risks in orbit and impact risks on ground. In his current position, he is the focal point within the European Space Agency for space debris matters, and he represents the European Space Agency in the multinational Interagency Space Debris Coordination Committee. Dr. Klinkrad is a member of the International Academy of Astronautics and an Associate Fellow of the American Institute of Aeronautics and Astronautics. He has served as a member of working groups and panels of the American Institute of Aeronautics and Astronautics, COSPAR, ECSS, IAA, IADC, IAG, ISO, and UNCOPUOS. He became chair of the COSPAR PEDAS panel in 2005 and has since been the main scientific organizer of related COSPAR sessions on space debris. He has been a lecturer on space debris at the Braunschweig University of Technology since 2001, and he wrote a textbook on space debris in 2006.

Professor Nancy G. Leveson
Aeronautics and Astronautics/Engineering Systems
Massachusetts Institute of Technology
Boston, Massachusetts

Nancy G. Leveson is a professor of aeronautics and astronautics and professor of engineering systems at the Massachusetts Institute of Technology, where she teaches and conducts research in the areas of aerospace software engineering, system engineering of software intensive systems, and system and software safety. Results of her research are used in the aerospace and defense communities, and her NASA supported research has been transferred successfully to the commercial sector by a company, Safeware Engineering Corporation, which she co-founded in 1992 with some of her graduate students. Professor Leveson is an elected member of the National Academy of Engineering. She received numerous awards for her research, including the ACM Outstanding Software Research Award, the ACM Allen Newell Award "for establishing the foundations of software safety," the Computing Research Association Nico Habermann Award, and the American Institute of Aeronautics and Astronautics Information Systems Award for technical contributions in space and aeronautics computer technology and science and specifically "for developing the field of software safety and for promoting responsible software and system engineering practices where life and property are at stake." She served as editor-in-chief of *IEEE Transactions on Software Engineering*, has written over 200 research papers in journals and conferences, and authored a book titled *Safeware: System Safety and Computers*. Professor Leveson has served on a variety of NASA advisory committees, including the Aerospace Safety Advisory Panel, and chaired the NASA NRC Study Committee for Review of Oversight Mechanisms for Space Shuttle Flight Software Processes.

About the Contributors

Miguel J. Maes
Flight Systems Test Engineer
Johnson Space Center
White Sands Test Facility, National Aeronautics and Space Administration
Las Cruces, New Mexico

Miguel J. Maes is a NASA Project Manager for NASA White Sands Test Facility near Las Cruces, New Mexico, where he is a propellants and materials project manager in the fields of hypergolics, hydrogen, and oxygen; and he has been the principal investigator and test conductor of both combustion of metals in microgravity and promoted combustion of metals in flowing oxygen. Mr. Maes began his career at White Sands Test Facility in 1996 and has assumed various positions within the laboratories office, including pressure systems engineer, project manager of the IAPU fleet leader test system, technical project manager of the liquid hydrogen recirculation pump program, and acting chemistry laboratory manager. Many other projects that Mr. Maes manages involve determining the compatibility, degradation, and use of materials that are in hypergol service. Mr. Maes earned the Bachelor of Science and the Master of Science in mechanical engineering along with minors in mathematics and environmental engineering from New Mexico State University. He has received numerous NASA performance awards and an Astronaut Office Silver Snoopy Award. Mr. Maes has been published internationally through the ASTM 10th International Symposium on Flammability and Sensitivity of Materials in Oxygen Enriched Atmospheres, the 4th Conference on Aerospace Materials, Processes, and Environmental Technology, and the 33rd and 34th AIAA/ASME/SAE/ASEE Joint Propulsion Conference and Exhibit.

William D. Manha
Senior Specialist, Propulsion and Pressure Systems
Jacobs Engineering
Houston, Texas

Mr. William (Bill) D. Manha has provided engineering support to the Space Shuttle Payload Safety Review Panel to assure compliance with the hazardous materials containment, pressure, and propulsion systems compliance with shuttle and International Space Station payload safety requirements since 1999. From 1988 until 1999, Mr. Manha provided propulsion systems engineering on a Space Shuttle reaction control and orbital maneuvering propulsion systems, Space Station Freedom reaction control propulsion system (monopropellant, hydrazine pressure fed system), slush hydrogen research and development for the national aerospace plane, shuttle extended duration orbiter cryogenics tank module, and global positioning satellite orbit insertion motor (Thiokol *Star-37* solid propellant rocket). From 1962 until 1979, Mr. Manha provided engineering services for the design, development, and verification of Apollo launch escape and Earth landing systems (pyrotechniques, solid rocket motors, and parachutes) and Apollo and Space Shuttle reaction control, service propulsion and orbit maneuvering systems (pressure fed hypergolic propulsion systems).

Jean-Bruno Marciacq
Crew Safety Officer, European Space Agency, Safety Manager
European Astronaut Centre
Cologne, Germany
Centre National d'Etudes Spatiales
Toulouse, France
Institute for Medicine and Physiology in Space
Toulouse, France

Jean-Bruno Marciacq worked for the European Space Agency from 2001 to 2008 as crew safety officer within the Astronaut Division of the Human Spaceflight Department. In this function, he directly contributed to the safety of seven crewed spaceflights, and supported the automated transport vehicle and European Space Agency space station module (*Columbus*) projects. He was responsible for the safety of astronaut training and medical experiments performed on ground, including extravehicular activity training operations in the European Space Agency neutral buoyancy facility. He also acted as European Astronaut Center Safety Manager. Detached since 2001 from MEDES, the French Institute for Space Medicine and Physiology and a subsidiary of the Centre National d'Etudes Spatiales (CNES), to the European Astronaut Center in Cologne, Germany, he represented the interests of his parent agency in crewed spaceflight within the European Astronaut Center Integrated Team, composed of members of the French and German Space Agencies. From 1995 to 2001, he worked respectively as Falcon 50 Deputy Program Manager and Customer Service Manager within the Civil Aircraft Division of DASSAULT Aviation, ensuring the airworthiness of the worldwide fleet of Falcon Business Jets. From 1994 to 1995, he served as a naval officer on the French aircraft carrier *Foch*, for which he was awarded four medals. Mr. Marciacq graduated from E.S.T.A.C.A. University for Aerospace and Automotive Engineering, Paris, and holds a Certificate of Aircraft Design from the Moscow Aviation Institute (M.A.I.). As a complement to his studies, he worked for various aerospace companies such as British Airways, Aérospatiale, Dornier/Deutsche Aerospace, AERO/Sagem Group and the Mudry Aircraft Company.

Yolanda Y. Marshall
Director, Safety and Mission Assurance Directorate
Johnson Space Center, National Aeronautics and Space Administration
Houston, Texas

Ms. Marshall earned the Bachelor of Science degree in biomedical engineering in 1980 from Texas A&M University at College Station. From 2002 until the present, Ms. Marshall has been the Director of Safety and Mission Assurance at the Johnson Space Center. From 1994 through 2002, Ms. Marshall was the Science Applications International Corporation Manager, Shuttle Safety and Mission Assurance Department, and was promoted to Deputy Program Manager for the safety, reliability and quality assurance contract at the Johnson Space Center in 1999. From 1993 through 1994 Ms. Marshall was the Deputy Manager of the Space Station Safety and Mission Assurance Department, and manager of the payload and crew equipment section for Loral Space Information Systems. Ms. Marshall received the Senior Executive Service Presidential Rank Meritorious Executive Award in 2006, the NASA Exceptional Service Medal for management of the Safety and Mission Assurance Directorate during the Columbia recovery efforts and for improving safety on human spaceflight missions in 2004, the Astronaut Office Silver Snoopy Award for her work on STS-27 in 1989, and the NASA Manned Flight Awareness Award for her work on Spacelab 2 in 1985.

Torin McCoy
Habitability and Environmental Factors Division
Johnson Space Center, National Aeronautics and Space Administration
Houston, Texas

Torin McCoy has worked as a space toxicologist with NASA Johnson Space Center since 2003. For much of that time he has been in charge of the Water and Food Analytical Laboratory for analyzing critical water and food samples for the Space Shuttle and *International Space Station* and for evaluating the data from a crew health perspective. This group also is active in providing scientific support for the development of real time water quality monitoring devices for the *International Space Station* with an eye toward remote exploration missions to the Moon and Mars. Mr. McCoy also is actively involved in working with the National Research Council to set air and water quality standards that are appropriate for spaceflight. He previously worked as a regulatory toxicologist with the state of Texas and has a wealth of experience in evaluating environmental data from a toxicological perspective. He received the NASA Spaceflight Awareness Award in 2006 in recognition of his service to NASA. He is a graduate of Texas A&M University and Clemson University.

Ernst Messerschmid, Ph.D.
Institute of Space Systems, Universitaet Stuttgart
Stuttgart, Germany

Ernst Messerschmid, Professor and Deputy Director Space Systems Institute at Universitaet Stuttgart, received the Doctor of Philosophy degree in physics at Freiburg University, Germany. From 1978 to 1983 he worked as a researcher for the German Aerospace Research Establishment, Institute of Communications Technology, Oberpfaffenhofen, in the area of space borne communications and navigation systems. He became a science astronaut at the German Aerospace Research Establishment in Cologne in 1983 and flew on the German Spacelab mission D1, STS-61A with *Challenger*, where he conducted experiments in fluid physics, material science, biology, medicine, and navigation. From 1986 until present, he did research at the Universitaet Stuttgart in the areas of space transportation systems, reentry technology, electric propulsion, microgravity experiments and technology, space stations, and platforms. He teaches space systems engineering, in particular space stations and platforms architectures. From 2000 to 2004, he was head of the European Astronaut Center of the European Space Agency in Cologne (on leave from Universitaet Stuttgart). He holds memberships in the Dutch Royal Academy of Science (since 1985), International Academy of Astronautics (1989), Association of Space Explorers since 1986, and is a Corresponding Member (1986) German Aerospace Society Deutsche Gesellschaft für Luft- und Raumfahrt. Mr. Messerschmid has written more than 150 publications, most of them in journals and conference proceedings, and authored or coauthored 10 books and received 10 patents. He received the NASA Space Flight Medal (1985), NASA Flight Achievement Award (1996), German Cross of Merit First Class (1985), and Golden Hermann Oberth Medal (1986).

About the Contributors

Dean W. Moreland
Executive Officer, Payload Safety Review Panel
Johnson Space Center, National Aeronautics and Space Administration
Houston, Texas

Dean W. Moreland has over 20 years' experience as a member of the NASA Payload Safety Review Panel at the Johnson Space Center in Houston Texas. Currently, he serves a dual assignment as the Executive Officer for both the Payload Safety Review Panel and Space Station Safety Review Panel. He holds the Bachelor's degree in Aerospace Engineering and Mechanics from the University of Minnesota. He is a recipient of the NASA Space Flight Awareness safety award.

John Muratore, P.E.
Research Associate Professor
Aviation Systems and Flight Research, University of Tennessee Space Institute
Tullahoma, Tennessee

John Muratore has 25 years of experience in space and aircraft engineering, testing, and operations. Graduating with the Bachelor of Science in engineering and applied science from Yale, Mr. Muratore entered the U.S. Air Force and completed assignments in launch vehicle processing for USAF and NASA missions at both Vandenberg Air Force Base in California and Cape Canaveral Air Force Station and Kennedy Space Center in Florida. Mr. Muratore then moved to NASA where he has served in a variety of operations, engineering, and program management positions. In operations, Mr. Muratore started as a Space Shuttle flight controller in instrumentation and communications. He then was given increasing responsibility in assignments as division chief responsible for Space Shuttle flight software production, a flight director for four Space Shuttle flights, and division chief for the mission control center. Mr. Muratore earned the Master of Science in computer systems from University of Houston—Clear Lake during this period. At the mission control center, he managed continuing operations of the facility while leading a $250 million overhaul of the facility that transformed it from a Space Shuttle only facility to a combined Space Shuttle—International Space Station—exploration facility. Mr. Muratore made the transition to engineering, where he served as the project manager for the *X-38 Crew Return Vehicle* project. In this role he not only performed programmatic functions but also flew as a flight test engineer on drop flights on the B-52 launch aircraft and served as mission director in the control center. During this period Mr. Muratore attended courses at the civilian National Test Pilot School course for flight test engineers. Following the *Columbia* incident, Mr. Muratore served as manager for Space Shuttle systems engineering and integration and led the rebuilding of the Space Shuttle systems' engineering capability. Following his work as the Space Shuttle lead engineer, Mr. Muratore returned to engineering where he was the NASA lecturer at Rice University teaching instrumentation, data acquisition, flight mechanics, flight test engineering, and systems engineering to undergraduate and graduate students. Mr. Muratore has retired from NASA and is presently a Research Associate Professor at the University of Tennessee Space Institute in Tullahoma.

Kornel Nagy, Ph.D.
Structural Engineering Division
Johnson Space Center, National Aeronautics and Space Administration
Houston, Texas

Dr. Kornel Nagy serves as system manager for structural and mechanical systems for the *International Space Station*, where he is responsible for technical integration of the structural, mechanical, meteoroid and debris protection, and loads and dynamics discipline areas. He served as structures system development manager for Space Station Freedom work package-2. He led design teams for aerobraking orbital transfer vehicle preliminary design, assured crew return vehicle study, the Space Station Freedom pre-integrated truss preliminary design, and the International Space Station redesign option C. He was awarded 10 U.S. patents and 3 international patents. He is a registered professional engineer. Dr. Nagy has authored or co-authored several papers on space vehicle design. He holds the Doctor of Philosophy, the Master of Science, and the Bachelor of Science in mechanical engineering from the University of Houston.

Bryan O'Connor
Associate Administrator, Safety and Mission Assurance
Headquarters, National Aeronautics and Space Administration
Washington, DC

Bryan O'Connor is the Chief, Safety and Mission Assurance, Office of Safety and Mission Assurance, at the National Aeronautics and Space Administration headquarters in Washington, DC. He has functional responsibility for safety, reliability, maintainability, and quality assurance for all NASA programs and institutions. Mr. O'Connor began his career in the U.S. Marine Corps, where he flew as an attack pilot and test pilot in a number of different aircraft. He was selected for the NASA astronaut program in 1980 and served in several positions supporting the first test flights of the Space Shuttle. With more than 20 years of experience with NASA, Mr. O'Connor has guided some of the agency's most important efforts, including the Space Shuttle return to flight after the Challenger disaster, the redesign of the *Space Station Freedom* to the *International Space Station*, and the Space Shuttle Program partnership with Russia on the Shuttle-Mir program. Mr. O'Connor holds a Bachelors degree in engineering from the U.S. Naval Academy and a Masters degree in aeronautical systems from the University of West Florida. He has flown more than 5000 hours in over 40 types of aircraft, and he has almost 400 hours in space on two Space Shuttle flights (pilot on STS-61B and commander for STS-40).

About the Contributors

Thomas Palo
Aerospace Engineer
Kennedy Space Center, National Aeronautics and Space Administration
Kennedy Space Center, Florida

Thomas Palo is employed at NASA's Kennedy Space Center in the Engineering Assurance Branch of the Safety and Mission Assurance Directorate. His work at NASA includes numerous expendable launch vehicle and spacecraft programs, including EO-1, SAC-C, Mars *Odyssey*, Timed, ICESat, CHIPSat, and Gravity Probe-B, as both the Mission Assurance Manager and Safety Officer. Prior to joining NASA, Mr. Palo worked in the Range Safety Office of the U.S. Air Force 45th Space Wing, supporting over 20 Department of Defense, NASA, and commercial launches.

Michael D. Pedley, Ph.D.
Materials and Processes Branch
Johnson Space Center, National Aeronautics and Space Administration
Houston, Texas

Dr. Michael D. Pedley has been a materials engineer at the NASA Johnson Space Center in Houston, Texas, since 1989. He served as section head for Nonmetallic Materials and Branch Chief for Materials and Processes. From 1996 to 2006, he was the Materials and Processes System Manager for the International Space Station Program. He is currently the Materials and Processes lead for the new Constellation program in its work to develop new launch vehicles, a *Crew Exploration Vehicle*, and ultimately a permanent crewed presence on the Moon and crewed missions to Mars. Dr. Pedley has supported the payload safety process for many years and was a key player in the development of materials and processes reciprocal agreements on payloads and other cargo between NASA and the European Space Agency, the Japan Aerospace Exploration Agency, and the Russian Space Program. Specific areas of expertise include materials flammability, oxygen and propellant compatibility, and materials toxic off-gassing. From 1983 through 1989 Dr. Pedley worked for Lockheed Corporation at the NASA White Sands Test Facility and later at Johnson Space Center, where he had numerous responsibilities for materials testing and flight hardware materials certification. Dr. Pedley played a major role in the development of several key NASA requirements and guidelines documents, including NASA-STD-(I)-6016, *Standard Materials and Processes Requirements for Spacecraft*; NASA-STD-6001, *Flammability, Odor, Offgassing, and Compatibility Requirements and Test Procedures for Materials in Environments That Support Combustion*; and JSC 29353, *Flammability Configuration Analysis for Spacecraft Applications*.

Jay L. Perry
Environmental Control and Life Support Systems
Marshall Space Flight Center, National Aeronautics and Space Administration
Huntsville, Alabama

Jay Perry is a 1985 graduate of Vanderbilt University where he earned the Bachelor of Engineering, cum laude, in chemical engineering. Assigned to the NASA Marshall Space Flight Center, Mr. Perry specializes in the design, development, testing, performance analysis, and flight operations and sustaining engineering of cabin air quality control systems for crewed spacecraft. His work has touched on all facets of spacecraft cabin air quality control for the Spacelab and International Space Station Programs, as well as NASA efforts in advanced life support system development for lunar and Mars exploration. Mr. Perry has authored or coauthored more than 60 NASA and technical conference publications in his field and is co-author of a U.S. patent. NASA awarded Mr. Perry the Exceptional Engineering Achievement Medal in 1995 for his contributions to the Spacelab program. Mr. Perry serves as the senior engineer for air quality control and environmental monitoring for the *International Space Station*. In this role, Mr. Perry works with NASA's Russian partner to guide the technical approach to cabin air quality maintenance for the *International Space Station*. This role included serving as the technical lead for a successful joint U.S.-Russian flight experiment flown onboard *Mir* in 1995 to assess cabin atmospheric quality and, since 1994, serves as NASA engineering lead for the multilateral air quality working group for the *International Space Station*. In addition to duties for the *International Space Station* program, Mr. Perry is the senior engineer for the atmosphere revitalization element of the NASA Exploration Life Support Project, which is responsible for identifying and maturing promising atmospheric purification technologies for future spacecraft and surface outposts.

Duane L. Pierson, Ph.D.
Chief Microbiologist, Habitability and Environmental Factors Division
Johnson Space Center, National Aeronautics and Space Administration
Houston, Texas

Dr. Duane Pierson serves as NASA senior microbiologist, and is responsible for formulating, developing, and implementing the NASA microbiology program for current and future human exploration of space. His major responsibilities include both operational and research activities to ensure the health, safety, and optimum performance of the astronauts. He is the Chairman of the Johnson Space Center Biosafety Review Board that reviews all biohazardous materials used in science investigations onboard spacecraft and in the biomedical laboratories at the Johnson Space Center. After earning the Doctor of Philosophy in Biochemistry at Oklahoma State University, he joined Baylor College of Medicine in Houston, and a decade later moved to the NASA Johnson Space Center. He has written numerous articles published in a wide variety of peer reviewed journals, along with book chapters, NASA Technical Briefs, and patents. He was elected in 1998 to Fellow in the American Academy of Microbiology. NASA recognized Dr. Pierson's accomplishments through many awards, including the Medal for Exceptional Scientific Achievement and the Certificate of Commendation. The astronauts recognized his contributions in environmental health with the Silver Snoopy Award. In 2004, he became the NASA Chief Microbiologist. He is a frequent university lecturer and maintains academic appointments with Baylor College of Medicine in Houston, the University of Houston, and the University of Texas Medical Branch in Galveston.

Peter G. Prassinos
Office of Safety and Mission Assurance
Headquarters, National Aeronautics and Space Administration
Washington, DC

Mr. Prassinos has expertise and experience in nuclear safety, probabilistic risk assessment, nuclear engineering, reactor safety research, safety analysis, engineering systems evaluations, transportation safety, and standards development. He has applied probabilistic risk assessment methods and techniques to many systems including nuclear power, hazardous and explosive facilities, and transportation, financial, and security systems. Mr. Peter Prassinos holds the Bachelor of Science in Aerospace Engineering and the Masters of Science in Nuclear Engineering. He is a member of the Safety and Assurance Requirements Division in the Office of Safety and Mission Assurance at the NASA. He has expertise in the application of probabilistic risk assessment and system and facility safety. His present responsibilities include applying probabilistic risk assessment methods and techniques to space systems and is the NASA member on the Interagency Nuclear Safety Review Panel for the Mars Science Laboratory Mission.

Kimberlee S. Prokhorov
NASA Lead, ISS Common Environments Team
Johnson Space Center, National Aeronautics and Space Administration
Houston, Texas

Kimberlee Prokhorov is a 1995 graduate of Iowa State University where she earned the Bachelor of Science in mechanical engineering. Ms. Prokhorov began her career at the NASA Johnson Space Center, working as a contractor teaching the crews of the *International Space Station* about the electrical power, thermal control, and life support systems. In 1998, Ms. Prokhorov took an assignment to assist in technical development of training materials for the Russian segment of the *International Space Station*. She worked closely with Russian instructors in Star City, Russia, and specifically supported Expeditions 2, 4, and 5, and the Space Shuttle crews that were training for early missions to the *International Space Station* prior to Expedition crew flights. From 2000 to 2007, Ms. Prokhorov was a senior subsystem engineer supporting the International Space Station life support equipment on-orbit from a sustaining engineering and failure investigation standpoint. She has been the lead for U.S.-Russian integration of the International Space Station life support systems, as well as the lead for emergency response (fire, depressurization, and toxic atmosphere). In 200, Ms. Prokhorov took a position as a civil servant. NASA awarded the Exceptional Engineering Achievement Medal to Ms. Prokhorov in 2006 for defining the Contingency Shuttle Crew Support assessment and reporting protocol, which defines the number of days the *International Space Station* can support a stranded Space Shuttle crew in addition to the International Space Station crew. Currently, Ms. Prokhorov serves in the International Space Station program vehicle office as the NASA lead of the common environments team, working to assist technical teams resolve issues relating to air quality, water quality, microbiological quality, and other areas which have effects on the whole International Space Station environment.

About the Contributors

Baraquiel Reyna
Systems Architecture and Integration
Johnson Space Center, National Aeronautics and Space Administration
Houston, Texas

Baraquiel Reyna was born in 1978 near the southern tip of Texas. While attending Rice University, he majored in biochemistry and later earned the Masters degree in biomedical engineering at Texas A&M University. He is currently working at the Johnson Space Center leading the Health Maintenance System, part of the International Space Station Crew Health Care System.

Steven L. Rickman
Chief, Thermal Design Branch
Johnson Space Center, National Aeronautics and Space Administration
Houston, Texas

Steven L. Rickman began his career with NASA as a cooperative education student at the Goddard Space Flight Center in 1981, where he gained experience in the areas thermal vacuum and structural dynamic testing. After receiving the Bachelor of Science in aerospace engineering from the University of Cincinnati, he transferred to the Johnson Space Center in 1982, and he subsequently was hired into the Thermal Branch in 1985. He earned the Master of Science in physical science from the University of Houston—Clear Lake in 1993. Mr. Rickman has extensive experience as a thermal analyst, focusing on passive thermal control of orbiting spacecraft. He has served on numerous design study teams, including the Trans-Hab inflatable module project as lead environments engineer and lead thermal analyst. Mr. Rickman was the NASA Technical Manager for the development of the thermal synthesizer system and was the thermal design engineer for the extravehicular charged particle directional spectrometer and the martian radiation environment experiment. He investigated concurrent engineering techniques for the addition of detailed thermal protection system representations in spacecraft thermo-mechanical stress models. Mr. Rickman was named Deputy Chief of the Thermal Branch in 1998 and Chief of the Thermal Design Branch in July 2002. In 2003, he provided thermal expertise to the Columbia accident investigation activities and served as the NASA lead for the flight day-2 object radar team. In 2006, Mr. Rickman assumed leadership of the Tile Overlay Repair development team, focusing on the development of a technique to repair Space Shuttle tile damage on orbit.

Brandan R. Robertson
Chairman, Mechanical Systems Working Group
Johnson Space Center, National Aeronautics and Space Administration
Houston, Texas

Brandan Robertson received the Bachelor of Science in aeronautical and astronautical engineering from Purdue University in 1998 and the Master of Science in physics from the University of Houston—Clear Lake in 2007. While at Purdue, he began working in the cooperative education program of the NASA Johnson Space Center and joined the Mechanical Design and Analysis Branch on receiving his Bachelors degree. He worked on mechanism development for the Space Shuttle, International Space Station, X-38, and Orbital Space Plane programs. In 2004, he became the chair of the Mechanical Systems Working Group, a group of mechanisms specialists that supports the Payload Safety Review Panel, a role he continues to serve in addition to his positions as System Manager for Mechanisms in the Orion program, mechanisms representative to the Constellation integrated loads, structures and mechanisms systems integration group, and member of the NASA engineering and safety center mechanisms super problem resolution team.

Summer L. Rose
System Safety Engineer for the ISS Program
Boeing
Houston, Texas

Summer L. Rose began her career in the space industry as a safety specialist for the Shuttle-*Mir* program in 1997 after graduating from Texas A&M University and working in the medical field. At the end of the *Mir* program, she held the position of Payload Safety Engineer for the NASA Payload Safety Review Panel. She was promoted to Executive Officer of the Payload Safety Review Panel in 2004. In 2006, Ms. Rose became the Executive Officer for the NASA Safety Review Panel. Ms. Rose took a position with the *Crew Exploration Vehicle-Orion* program with Lockheed Martin as a System Safety Engineer at the end of 2006.

Gary A. Ruff, Ph.D.
Advanced Capabilities Project Office
Glenn Research Center, National Aeronautics and Space Administration
Cleveland, Ohio

Dr. Gary A. Ruff is the Project Manager and Technical Leader for the Spacecraft Fire Safety project, conducted at the NASA Glenn Research Center since 2001, with the goal of producing advanced fire safety technology for NASA exploration missions and research in material flammability, fire detection, and fire suppression. Much of the work focuses on how these aspects of fire safety can be made most efficient and reliable in low and partial gravity. While at NASA, Dr. Ruff was the principal investigator or co-investigator on various projects, including droplet combustion in low gravity, spacecraft fire suppression, and fire signatures. He currently serves as Project Scientist on a spaceflight experiment to improve spacecraft fire detection. He participated in the NASA-STD-6001 Topical Working Group that revised the NASA material flammability test methods and played a lead role in the exploration atmospheres working group that recommended atmospheres for the habitable volume of the crew exploration and lunar lander vehicles. From 1990 through 2000, Dr. Ruff was a professor of mechanical engineering at Drexel University in Philadelphia, where he taught courses in basic thermodynamics, fluid mechanics, experimental methods, advanced fluid mechanics, and engineering design. His research areas included sprays and spray combustion, droplet combustion in microgravity, aircraft icing, multiphase flows, turbulence, and convective heat transfer. Prior to 1990, Dr. Ruff held research positions at Arnold Engineering Development Center and the NASA Lewis Research Center. Dr. Ruff received the Bachelor of Science in aerospace engineering from Ohio State University in 1981 and the Master of Science in 1985 from the University of Tennessee. He received the Doctor of Philosophy in aerospace engineering from the University of Michigan in 1990.

Michael K. Saemisch
Lockheed Martin Space Systems Company
Denver, Colorado

Michael K. Saemisch is currently the Safety and Mission Assurance Manager for Project Orion on the Lockheed Martin contract, having responsibility for safety, reliability, and quality assurance. He has over 30 years' experience in System Safety and Product Assurance at Martin Marietta–Lockheed Martin on NASA and Air Force projects. He worked numerous Space Shuttle payloads, including serving as a member of the DoD Space Shuttle Payload Safety Review Team. NASA projects in addition to Project Orion have included the *International Space Station*, Flight Telerobotic Servicer, *Crew Return Vehicle*, and Space Launch Initiative. Air Force projects include DoD payload integration onto the Space Shuttle and Titan program. He established system safety processes for the DoD and Lockheed Martin and participated in key system safety process and requirements documents. Mr. Saemisch earned the Bachelor of Science in chemical engineering from the University of Colorado in 1976.

Juergen Schlutz, Ph.D.
Institute of Space Systems, Universitaet Stuttgart
Stuttgart, Germany

Dr. Juergen Schlutz received a diploma with honors in aerospace engineering at the Universitaet Stuttgart, Germany, with a special focus on space systems and control theory. In 2005, he joined the scientific team around Professor Ernst Messerschmid at the Institute of Space Systems of the Universitaet Stuttgart and is currently working as a research associate in the field of conceptual human space mission design. He is also conducting doctoral work with a concentration on methodology and tools for lunar base design and the characterization and modeling of the lunar surface environment. Since 2007, he has served as a consultant for the European Space Agency in the frame of European exploration architecture studies. Mr. Schlutz holds memberships with the German Aerospace Society-Deutsche Gesellschaft fuer Luft-und Raumfahrt, Space Generation, and the Student Space Exploration and Technology Initiative. He coauthored "Project M3—A Study for a Manned Mars Mission in 2031," *Acta Astronautica* 58 (2006): 88-104, and several conference proceedings and IRS publications. From 1999 through 2005, he studied aerospace engineering with a concentration in space systems and automated control theory at the Universitaet Stuttgart, where he earned a Diploma of Engineering. From 1999 through 2003, he was a student assistant at the Institute of Space Systems, Universitaet Stuttgart, and at the Aerothermic Laboratory of CNRS, France, in plasma research and optical diagnostics. In 2003, he studied at the Alpbach Summer School, Alpbach, Austria, as a participant and team coordinator of Project M3, Working and Living in Space, from ISS to Moon and Mars.

H. F. R. Schöyer
Schöyer Consultancy B.V.
Zoetermeer, the Netherlands

H. F. R. Schöyer studied aircraft engineering at Delft University and got involved in astronautics (sounding rocket design, thesis on the orbital stability of a lunar satellite). After graduating in 1966, he became a conscripted Naval officer. He was a NASA fellow at the Caltech-Jet Propulsion Laboratory from 1969 through 1971 and worked on experimental and theoretical combustion instability in relation to the Mariner Mars G-dot motor. Back in the Netherlands, he taught rocket propulsion at Delft University and performed experimental and theoretical research in combustion, combustion instability, ramjets, rocket design and testing, design and building of large test facilities. He was a North Atlantic Treaty Organization fellow at the Naval Weapons Center in 1981 at China Lake. In 1984, he joined the European Space Agency as head of the Chemical Propulsion Section. In this capacity, he led various industrial studies and developments in the area of rocket propulsion, turbopumps, plug nozzles, engine, propellant, software development, and cryogenic solid propulsion. He also supported projects such as MAGE development and testing, EBM development, and became involved in standardization projects. He taught rocket propulsion at Delft University and supervised thesis work. After his retirement in 2000, he founded Schöyer Consultancy B.V., supporting industry and agencies in space and rocket propulsion, in various studies, standardization, preparation of testing of rocket motors, writing test plans, and designing and supervising the building of test facilities. He is coauthor of *Rocket Propulsion and Spaceflight Dynamics* (Pitman, 1979) and editor of *Combustion Instabilities in Liquid Rocket Engines*, ESA WPP-062, 1993. He authored a large number of technical and scientific publications and holds various patents. H. F. R. Schöyer is cofounder and honorary member of the Dutch Rocket Society, NERO, and an associated fellow of the American Institute of Aeronautics and Astronautics.

About the Contributors

Robert C. Scully, MSEE, PE, NCE
Cochair, Space Shuttle E3 Control Technical Panel and JSC EMC Group Lead
Johnson Space Center, National Aeronautics and Space Administration
Houston, Texas

Since June 2000, Robert C. Scully has been serving as the NASA Johnson Space Center Electromagnetics Compatibility Group Lead Engineer, where he supports multiple programs, including the Space Shuttle and *International Space Station*, and is co-chair of the *Space Shuttle* Electromagnetic Environmental Effects (E3) Control Panel. Mr. Scully holds a Master's degree in electrical engineering and is completing his Ph.D. dissertation in electrical engineering, both from the University of Texas at Arlington. Mr. Scully is a registered Texas Professional Engineer and holds National Association of Radio and Telecommunications Engineers Certification as an Electromagnetic Compatibility Engineer. He also completed an Electromagnetic Compatibility Certification Program with the University of Missouri at Rolla. Within the Electromagnetic Compatibility Society, Mr. Scully currently serves as a member of the Board of Directors and is chair of Technical Committee 4, Electromagnetic Interference Control Technology; vice chair to Technical Committee 1, Electromagnetic Compatibility Management; and secretary to the Technical Advisory Committee. Mr. Scully has over 20 years in military and commercial aviation electrical and electronics engineering and electromagnetic compatibility.

Christopher O. A. Semprimoschnig, Dipl. Ing., Ph.D.
Materials Physics and Chemistry Section, European Space Research and Technology Center
European Space Agency
Noordwijk, the Netherlands

Dr. Christopher O. A. Semprimoschnig has been involved in space activities since the early 1990s in the frame of the European Space Agency HERMES program. He joined the European Space Agency in 1998 as a Materials Science Engineer in the Materials Physics and Chemistry Section within the Materials and Processes Division. His duties focus on space environmental effects (atomic oxygen, ultraviolet, VUV, particle radiation, synergistic effects) on materials and processes and related materials analysis prior to and post space simulation. At the European Space Agency, he is also in charge of the materials thermal analysis laboratory, which supports a wide range of tasks within the section. In addition, he investigates and develops new space environmental simulation methodologies and analysis techniques and performs research into new materials and concepts for space technology. Dr. Semprimoschnig has authored or co-authored about 50 external technical and scientific papers related to space environmental simulation studies for material-environmental interactions, materials analysis prior to and post-space environmental simulation, post-flight material investigations, materials research by thermal analysis, and advanced materials characterization techniques (especially, fracture research).

About the Contributors

Sarah R. Smith
Laboratories Office, White Sands Test Facility
Johnson Space Center, National Aeronautics and Space Administration
Las Cruces, New Mexico

Sarah Smith earned the Bachelor of Science in mechanical engineering from New Mexico State University in 2000. She worked for over seven years as a project leader and engineer in the oxygen hazards analysis group of the NASA White Sands Test Facility. Her work there included the oxygen hazards analysis of many systems and components as well as the development of tests and test systems for evaluating the ignition and combustion of materials in oxygen enriched environments. Ms. Smith is an instructor of courses on oxygen safety, design, maintenance, and operation that are taught throughout NASA as well as in the private sector. In addition, Ms. Smith is an active member in the ASTM Committee G04 on Compatibility and Sensitivity of Materials in Oxygen Enriched Atmospheres. She is co-author of ASTM Manual 36, *Safe Use of Oxygen and Oxygen Systems*, which is the NASA safety standard for oxygen, and wrote numerous papers in the area of oxygen enriched atmospheres. The oxygen hazards group at White Sands Test Facility has been nominated for the Rotary National Award for Space Achievement Stellar Award, and Ms. Smith received a special recognition award from the Food and Drug Administration Center for Devices and Radiological Health. In addition, Ms. Smith received the Astronaut Corps Silver Snoopy award.

Michael G. Stamatelatos, Ph.D.
Office of Safety and Mission Assurance
Headquarters, National Aeronautics and Space Administration
Washington, DC

Dr. Michael Stamatelatos has been Director of the Safety and Assurance Requirements Division in the Office of Safety and Mission Assurance of NASA Headquarters since October 2003. In this capacity, he managed safety, risk, reliability, and quality assurance activities across the agency. Dr. Stamatelatos is a recognized expert in risk and reliability assessment. He came to NASA headquarters in March 2000 as Manager of Risk Assessment in the Office of Safety and Mission Assurance. In that position, he was responsible for the development and application of probabilistic risk assessment policy as well as for coordinating, overseeing, and integrating probabilistic risk assessment programs and activities across NASA. He developed, organized, and taught courses on quantitative risk assessment and risk informed decision making for managers and practitioners. He lectured throughout NASA and outside government to industrial organizations and universities. Over the past 30 years, Dr. Stamatelatos has conducted and managed numerous safety, risk, and reliability studies for many industrial, government, and international applications including aerospace, space nuclear power, chemical munitions demilitarization, petrochemical refineries, commercial electricity generating reactors, and production nuclear reactors. Before joining NASA, Dr. Stamatelatos taught courses on quantitative risk and reliability methods and applications as a university professor and industrial and short probabilistic risk assessment and reliability courses in the United States and abroad. He also taught industrial courses on management decision analysis methods and the statistical design of experiments in the United States and abroad. Dr. Stamatelatos was General Chair of the Second International Conference for Probabilistic Safety Assessment and Management, Technical Co-chair of the Eighth International Conference for Probabilistic Safety Assessment and Management, and General Chair of the Safety Engineering and Risk Analysis Division of the American Society for Mechanical Engineers. Dr. Stamatelatos is author or coauthor of more than 100 technical papers and reports. He is coauthor of a NASA *Procedures Guide for Probabilistic Risk Assessment* and of a NASA *Fault Tree Handbook*. He is also coauthor of two book chapters on probabilistic risk assessment.

Constantinos Stavrinidis, Ph.D.
Head, Mechanical Engineering Department
Directorate of Technical and Quality Management, European Space Agency
Noordwijk, the Netherlands

Dr. Constantinos Stavrinidis earned Bachelor and Master of science degrees in aeronautical engineering from Imperial College, University of London, and obtained Doctoral degrees in Structural Dynamics from University of Stuttgart, Germany, and University of London. He worked in industry with Lloyds Register of Shipping and Kongsberg. He joined the European Space Agency in 1976, where he has been successively in charge of the Structures Section and Structures and Mechanisms Division. In 1997, he became head of the Mechanical Engineering Department, which covers the classical space disciplines structures, mechanisms, optics, optoelectronics, microgravity and life sciences instrumentation, automation and robotics, thermal control and life support, propulsion and aerothermodynamics, and test facilities for verification of space vehicles and subsystems. Dr. Stavrinidis is a fellow of the Royal Aeronautical Society, fellow of the American Institute of Aeronautics and Astronautics, member of the International Academy of Aeronautics and Astronautics, member of the Academie de l'Air et de l'Espace, and Space Branch Chairman of the Council of European Aerospace Societies. Dr. Stavrinidis has written over 80 papers, many in refereed journals.

Christine E. Stewart
Senior Extravehicular Activities Operations Safety Engineer
Science and Applications International Corporation
Houston, Texas

Christine Stewart holds the Bachelor of Science in electrical engineering from Oklahoma State University. She spent two years designing and debugging a missile simulator for the Patriot Advanced Capability III missile at Lockheed Martin in Grand Prairie, Texas. Presently, she works for Science Applications International Corporation in Houston, Texas, the safety contractor for the Space Shuttle and International Space Station Programs. She spent over 7 years supporting the Payload Safety Review Panel as a payload safety engineer, a lead payload safety engineer, and as acting Executive Officer. She led the effort to clarify requirements for on-orbit payload bonding and grounding and worked many international and U.S. payloads in that time. She currently works in extravehicular activity safety operations, supporting the Space Shuttle and *International Space Station*, where she acted as the extravehicular activity safety flight leader for STS-116 and is also the Safety and Mission Assurance Flight Equipment Safety and Reliability Review Panel Executive Officer. She worked real time support of extravehicular activity safety for every Space Shuttle flight since STS-114. Additionally, she was responsible for assuring that all extravehicular activity hazards on the Space Shuttle remote manipulator system and orbiter boom sensor system were appropriately identified and controlled. She has been a member of the International Association for the Advancement of Space Safety since October 2005.

About the Contributors

Joel M. Stoltzfus
Laboratories Office, White Sands Test Facility
Johnson Space Center, National Aeronautics and Space Administration
Las Cruces, New Mexico

Joel M. Stoltzfus is a mechanical engineer at NASA's White Sands Test Facility and manager of the Oxygen Compatibility Assessment and Test Team. Mr. Stoltzfus worked in the oxygen testing and analysis field for 28 years, developing test apparatuses, performing materials and component tests, and reporting test results and writing over 60 technical papers. His group led the development of an oxygen compatibility assessment process that is used within the NASA community to identify and mitigate fire hazards in oxygen enriched environments. The group also developed training classes for designers, operators, and maintainers of oxygen systems that have been presented worldwide.

David E. Tadlock
Manager, Operational Space Systems Support Office (Retired)
Johnson Space Center, National Aeronautics and Space Administration
Houston, Texas

David E. Tadlock is a senior manager of the Avionics Systems Division of NASA. He was the senior avionics safety engineer of the Payload Safety Review Panel since 1984 for all Space Shuttle and *International Space Station* payloads. In this capacity, he conducted seminars on avionics safety design as part of the NASA–European Space Agency joint safety conferences in 2000 and in 2002. Starting in 1965, he applied his Bachelor of Science degree in physics from North Texas State University to guidance and navigation system engineering challenges for *Apollo*, *Skylab*, and the *Space Shuttle*. From 1980 until 2006, he concentrated on avionics safety for the Space Shuttle and International Space Station payloads. He wrote interpretation letters for wiring and circuit protection sizing of electrical power systems and established the basis for allowing computer control of hazardous payload functions. He also established a work instruction document for the analysis of electrical and electronic systems used to control hazardous payloads.

About the Contributors

David L. Urban, Ph.D.
Chief, Combustion and Reacting Systems Branch, Space Processes and Experiments Division
Glenn Research Center, National Aeronautics and Space Administration
Cleveland, Ohio

Dr. David L. Urban earned the Bachelor of Science in Engineering in civil engineering from Princeton University in 1980 and the Doctor of Philosophy in mechanical engineering from the University of California—Berkeley in 1987. Since 1991, Dr. Urban has worked with the Microgravity Combustion Science Branch at the NASA Glenn Research Center, becoming the branch chief in 1997. His responsibilities have included managing and leading the NASA flight and ground based combustion research program, defining the NASA spacecraft fire safety research plan, and supervising researchers in spacecraft fire safety and reacting processes in reduced gravity. Dr. Urban was the project scientist for two combustion spaceflight experiments. His areas of expertise include combusting and reacting systems in low gravity, particulate aerosols, soot and smoke formation, and environmental monitoring and control systems. He has been a principal investigator or co-investigator on numerous microgravity experiments with an emphasis on flame spread, smoldering, gas jet diffusion flames and soot processes, and fire detection. Dr. Urban is the principal investigator of a flight investigation designed to improve our understanding of spacecraft fire detection. He has written over 30 peer reviewed presentations and over 100 conference presentations. In his role as an expert in microgravity combustion and spacecraft fire safety, he was an active member in a team that reviewed the fire safety related design changes to the Russian solid fuel oxygen generator prior to its approval for flight on the *International Space Station*. As a member of the exploration atmospheres working group, he represented the microgravity fire safety research community in the selection of the preferred atmosphere for the crew exploration vehicle and the LSAM.

Marc Van Eesbeek
Materials Physics and Chemistry Section, European Space Research and Technology Center
European Space Agency
Noordwijk, the Netherlands

Marc Van Eesbeek, who holds an engineering degree in industrial chemistry, started in 1977 as a materials engineer within the Materials and Processes Division at the European Space Agency European Space Technology and Research Center in Noordwijk, the Netherlands. Since 1996, he has been the head of the Materials Physics and Chemistry Section within the Materials and Processes Division, responsible for the technical support to projects related to the use of nonmetallic materials in space applications. This includes the evaluation of general materials properties and the assessment of the different environmental effects on materials used in crewed and uncrewed space applications. He has authored or co-authored a large number of papers related to the behavior of materials in natural and induced space environments, on-ground simulation studies for material-environment interactions, post-flight material investigations, contamination and degradation studies on optical and thermo-optical surfaces, and evaluation of contamination effects on spacecraft subsystems, including synergistic effects of contamination with other space environmental parameters.

Keith E. Van Tassel
Pyrotechnics Project Manager and Explosives Safety Officer
Johnson Space Center, National Aeronautics and Space Administration
Houston, Texas

Keith E. Van Tassel has been an engineer at the NASA Johnson Space Center since 1980, where he is currently the Project Manager for Pyrotechnics. He is concurrently the Explosives Safety Officer. Mr. Van Tassel previously has served as the Technical Warrant Holder for Pyrotechnics for the entire agency and is a past chairman of the NASA Pyrotechnics Working Group. As the team leader for pyrotechnics, he managed a team of pyrotechnic engineers that provided pyrotechnic devices and testing for the Space Shuttle, Constellation, and several Jet Propulsion Laboratory programs, including the *Huygens* probe that landed on the Saturn moon Titan, the Mars Surveyor rovers *Spirit* and *Opportunity*, and the deep impact spacecraft that landed on comet Tempel 1. Mr. Van Tassel also served six years as the Space Shuttle Subsystem Manager for pyrotechnics. During that time, he was awarded the Silver Snoopy Award by the Astronaut Office for his work on the STS-95 drag chute door anomaly. During his tenure as the subsystem manager for Space Shuttle pyrotechnics, Mr. Van Tassel also supported the X-38 program by providing numerous pyrotechnic devices for use on the X-38 drop tests. Prior to becoming the subsystem manager for Space Shuttle pyrotechnics, Mr. Van Tassel was the pyrotechnic engineer assigned to the development and qualification of the shuttle drag chute system. Before coming to NASA, Mr. Van Tassel was a fire protection design engineer for Booker Associates in St. Louis, Missouri. While at Booker Associates, he fixed the St. Louis 1904 World's Fair Fountain. Mr. Van Tassel holds the Bachelor of Science in Mechanical Engineering from Washington University in St. Louis, Missouri, and the Bachelor of Arts from Hamline University in St. Paul, Minnesota.

William E. Vesely, Ph.D.
Office of Safety and Mission Assurance
Headquarters, National Aeronautics and Space Administration
Washington, DC

Dr. William E. Vesely has assurance responsibilities for risk assessments and methods and tool developments for risk and reliability assessments for NASA. He lectures in NASA probabilistic risk assessment courses and teaches NASA fault tree courses. He is the principal author of the *Fault Tree Handbook with Aerospace Applications*, published by NASA. He served as a technical coordinator for Space Shuttle probabilistic risk assessment. Dr. Vesely has been in the risk assessment field for over 30 years. He was a principal author of the first major probabilistic risk assessment performed on nuclear plants. He worked at the Nuclear Regulatory Commission as a risk specialist and has been a probabilistic risk assessment consultant for the Department of Defense, Department of Energy, and various national laboratories and companies. Dr. Vesely developed numerous approaches for risk and reliability evaluations, including techniques for data mining, pattern recognition, and risk trending. Dr. Vesely has written over 100 papers and reports on probabilistic risk assessment, statistical analysis, data analysis, and expert systems. He is an adjunct professor for several universities. Dr. Vesely received the Bachelor of Science in physics in 1964 from Case Institute of Technology and the Master of Science and the Doctor of Philosophy in nuclear engineering in 1966 and 1968, respectively, from the University of Illinois.

Joe M. Victor, Ph.D.
Laser Safety Officer
Johnson Space Center, National Aeronautics and Space Administration
Houston, Texas

Dr. Joe Victor graduated from the University of Texas in Austin with the Bachelor of Science in Electrical Engineering in 1962, the Master of Science in Electrical Engineering in 1964, and the Doctor of Philosophy in 1967. His dissertation was on the alternating current residual losses in superconductors. He worked at the Southwest Research Institute in San Antonio from 1966 to 1975, mostly with non-destructive testing devices. He worked from 1975 to 1982 at Petrolite Instruments, designing and building corrosion monitoring equipment for the oil field environment. From 1982 to 1988, he worked for Bio Quantum Technologies, which built a microscope mounted surgical laser tissue welding instrument. He designed a small carbon dioxide laser that was used in the company product. In 1988, he started working for contractors at the NASA Johnson Space Center on a variety of engineering projects, most prominently the trajectory control sensor, which has both a semiconductor modulated continuous wave and a pulse class 3b laser, used as a laser ranger to measure the relative distance and approach speed of the Space Shuttle to both the *Mir* and *International Space Station*. He also supported the Johnson Space Center nanotube project, which uses high power class-4 NdYag lasers to oblate carbon rods in carbon nanotubes. He has 25 years of experience working with class-3 and class-4 lasers and has been an area safety officer for more than 10 years. Dr. Victor also served as a laser technical safety advisor for a number of projects, including the orbiter boom sensor system for surveying Space Shuttle tiles for damage.

Kathryn Anne Weiss, Ph.D.
NASA Jet Propulsion Laboratory, California Institute of Technology
Flight Software and Data Systems Section
Pasadena, California

Dr. Kathryn Anne Weiss is a flight software engineer in the Flight Software Systems Engineering and Architecture Group at the Jet Propulsion Laboratory. Dr. Weiss holds the Bachelor of Science in computer engineering and mathematics from Marquette University (2001). Her advanced degrees include the S.M. (2003) and the Doctor of Philosophy (2006) in aeronautics and astronautics from the Massachusetts Institute of Technology, where her research focused on the intersection between spacecraft, software, systems, and safety engineering with an emphasis on architecture. While at the Massachusetts Institute of Technology, Dr. Weiss won the 2005 American Institute of Aeronautics and Astronautics Foundation Orville and Wilbur Wright Graduate Award for contributions to the evolution of flight.

Johannes Wolf, Dr.-Ing.
Electromagnetics and Space Environment Division (TEC-EEE)
European Space Technology Center, European Space Agency
Noordwijk, the Netherlands

Dr. Johannes Wolf studied at the Dresden University of Technology Department Electrical Engineering-Automatic Control from 1983 through 1988 and was Assistant Professor at the Institute of Automatic Control at Dresden University of Technology from1988 until 1994. In 1994, he earned the Dr.-Ing. with the thesis, *Testing the Immunity of Devices against Impulsive Electromagnetic Disturbances*. From 1994 through 1996, he was a member of the electromagnetic compatibility workgroup within the Institute of Automatic Control at Dresden University of Technology. From 1996 until 1998, Dr. Wolf was a consulting engineer at EMC Baden AG (Group Fribourg), later Montena EMC AG (Montena Group). From 1999 through 2000, he worked as a system engineer in Kayser-Threde GmbH, Munich, where he performed electromagnetic compatibility consulting for several space related projects. Since 2000, he has been an electromagnetic compatibility engineer at the European Space Agency, involved in project reviews, mainly for the *Columbus* module of the *International Space Station*; electromagnetic compatibility test campaigns, troubleshooting on the microgravity science glove box, supporting the European Space Agency Payload Safety Review Panel, and supporting projects and experiments flying on FOTON, International Space Station taxi flight campaigns, and satellite projects. Dr. Wolf's areas of expertise include electromagnetic compatibility design, analysis, simulation, testing, troubleshooting, electrical design, analysis, simulation, power design, digital design, microprocessor technology, and electrical safety.

Stephen S. Woods
Senior Scientist, Jacobs Technology, Incorporated
White Sands Test Facility
Las Cruces, New Mexico

Stephen S. Woods is employed by Jacobs Technology as a Senior Scientist. He holds the Bachelor of Science in physics from George Washington University and the Master of Science in engineering physics from the University of Virginia. During 19 years at the NASA White Sands Test Facility, he has engaged in propellant hazards research with an emphasis on hydrogen hazard assessment, safety, and training. His contributions include guiding development of publications on the hazards of hypergolic propellants, ASTM MNL-36, *Safe Use of Oxygen and Oxygen Systems*; ISO/PDTR 15916, *Basic Considerations for the Safety of Hydrogen Systems*; and ANSI/AIAA G-095-2004, *Guide to Safety of Hydrogen and Hydrogen Systems*. Current efforts include standards development with the American Institute for Aeronautics and Astronautics Hydrogen Committee on Standards, of which he is Chair, and work as a representative for NASA to the ISO Technical Committee 197, Hydrogen Technologies. He is a lead instructor for the NASA hydrogen safety course, and has participated in hazard assessment of numerous aerospace and industry hydrogen systems.

Andrey V. Yaskevich, D.T.S.
Lead, Docking Dynamics Simulation Group
S. P. Korolev Rocket Space Corporation, Energia
Korolev City, Russia

Andrey Yaskevich studied computer and automatic systems at the Civil Aviation Engineers Institute in Riga, Latvia, and attended the postgraduate school of Moscow Technical University, Automatic Control Systems Department, where he earned a Doctor of Technical Science degree in the area of application of computers, math methods, and math simulation in scientific research. He has worked as senior engineer at the Central Research Institute of Computer Aided Control Systems in Aviation, Riga, Latvia, senior engineer for the Krasnoyarsk oil-gas prospecting company, assistant at the Siberian State Aerospace University, as a postgraduate student at Moscow Technical University, and as leader of the Dynamic Simulation Group, Electromechanical System Branch, of Rocket Space Corporation-*Energia*. Dr. Yaskevich's areas of expertise are in mathematical model development and dynamic analysis of docking and robotic systems, ground dynamic test planning, docking flight support in mission control center, and design support of new docking and robotic operations. He received the NASA Space Flight Awareness Team Award for the successful Shuttle-Mir International Space Station docking and undocking during STS-71 flight in 1995, the title of honored specialist of Rocket Space Corporation-*Energia* in 1996, and the medal 100-Year Anniversary of S. P. Korolev's Birth in 2007.

CHAPTER 1

Introduction to Space Safety

Yolanda Y. Marshall
Director, Safety and Mission Assurance Directorate, Johnson Space Center, National Aeronautics and Space Administration, Houston, Texas

CONTENTS

1.1. NASA and Safety 2
1.2. Definition of Safety and Risk 3
1.3. Managing Safety and Risk 3
1.4. The Book 5

"It is immoral to design a product or system for mankind without recognition and evaluation of the hazards associated with that product or station system." This statement is attributed to an anonymous early safety professional.

Human spaceflight has provided mankind the opportunity to experience our universe from a new dimension. From the first orbital flight of Yuri Gagarin, to the early exploration of the Moon, until today as the *International Space Station* (ISS) orbits the Earth, over four decades of experience in space exploration has been amassed. There are now hundreds of individuals who have viewed the Earth from outer space and experienced the effects of microgravity on the human body.

Indeed, the space environment is extreme and harsh, and we are bound by technological limitations. Even so, it is a fact that all accidents that occurred during space missions to date happened because of design and manufacturing errors made well within the knowledge and capabilities of the time for control and prevention. Design errors that occur in space projects essentially can be tracked to the difficult balance between the complexity of space systems and the relative ease by which failures of highly energetic systems can propagate to catastrophic consequences and the sometimes limited systems safety engineering culture of design teams as a whole. Failure to establish this balance can lead to poor communication of risks to management.

The tragic lessons learned pushed the National Aeronautics and Space Administration (NASA) to examine critically its safety culture and renew its organization, technical requirements, and management methods. The NASA effort was not simply to correct specific project problems but to incorporate any lessons learned into agency policy documents, standards, and specifications for use during future projects. This laid the foundation by which space safety engineering is emerging as an established technical discipline worldwide.

1.1 NASA AND SAFETY

On October 1, 1958, NASA was established officially as an organization working toward human spaceflight. The first high profile program of NASA was Project Mercury, which represented an effort to learn if humans could survive in space. This was followed by Project Gemini, which used spacecraft designed for two astronauts to build on the successes of Project Mercury. The human spaceflight efforts of NASA then extended to the Moon with Project Apollo, culminating in 1969 when the Apollo 11 mission first landed humans on the lunar surface. After the Skylab and Apollo-Soyuz Test Projects of the early and mid-1970s, human spaceflight efforts resumed during 1981 when the NASA Space Shuttle Program was initiated. This complex system continues today, helping to build the *International Space Station*.

Additionally, NASA launched a number of important scientific probes, such as the Pioneer and Voyager spacecraft, that explored the Moon, the planets, and other areas of the solar system. As well, NASA sent several spacecraft to investigate Mars. These included the Viking and Mars Pathfinder spacecraft. The *Hubble Space Telescope* and other space science spacecraft enabled scientists to make a number of important astronomical discoveries about our universe.

Much pioneering work was done by NASA in the development and operation of space applications satellites. New generations of communications satellites, such as the Echo, Telstar, and Syncom satellites, were brought about with the help of NASA. As well, the Earth science efforts of NASA literally changed the way our home planet is viewed, with the Landsat and Earth Observing System spacecraft contributing many important scientific findings. Finally, numerous spin-offs from NASA technology found application in widely ranging scientific, technical, and commercial fields. Overall, while the tremendous technical and scientific accomplishments of NASA demonstrate vividly that humans can achieve previously inconceivable feats, the realization that Earth is just a tiny blue marble in the cosmos is humbling.

Along with NASA successes, failures occurred. Some of these resulted in the loss of lives. The corrective measures taken now are seated deeply in the NASA culture. In 1967, an investigation of the Apollo I fire determined that the test conditions at the time of the incident were extremely hazardous. The amount and location of combustibles in the Command Module were not restricted or controlled closely, and there was no way for the crew to egress rapidly from the Command Module during an emergency. There were no emergency procedures or equipment for use by ground support personnel located outside the spacecraft to assist the crew. As well, fire and medical rescue teams were not located in the white room surrounding the Apollo I Command Module.

The committee investigating the incident concluded that a long history of NASA successes in testing and launching space vehicles with pure oxygen environments at 16.7 psi and lower pressures in their crew compartments led to overconfidence and complacency. The recommendations from the investigation resulted in the addition of more adequate safety precautions and the implementation of changes in system design, operations, management, and procedures. The loss of the Space Shuttles *Challenger* and *Columbia* caused, respectively, by a faulty O-ring design and debris hitting the wing was attributed by each of the mishap investigations to faulty management processes for managing risk, ultimately compromising safety.

The historic successes and failures at NASA make it clear that, although space systems and their component parts have become mature and reliable with time and use, any increase in system capability brings with it complexity and uncertainty in performance risk. This risk must be managed. The reality of the complex and uncertain nature of spaceflight manifested throughout its life cycle requires a multifaceted approach to controlling risk and ensuring safety. The risks associated with spaceflight can be managed through the appropriate and effective application of safety methodology.

1.2 DEFINITION OF SAFETY AND RISK

In accordance with *Webster's Unabridged Dictionary*, *safety* is defined as "the condition of being free from undergoing or causing hurt, injury, or loss." Safety in a system can be defined as a quality of a system that allows it to function under predetermined conditions with an acceptable minimum risk of accidental loss (Roland and Moriarty 1990). *Risk*, according to this dictionary, is defined as "the possibility of loss, injury, disadvantage, or destruction." The role of safety is, indeed, to minimize risk.

1.3 MANAGING SAFETY AND RISK

One principle of system safety is that every hazard associated with or presented by a system is known and understood fully before it is allowed to operate. Once the hazards have been explored fully and adequate controls have been implemented, the residual risk of using systems that benefit humankind can be accepted while buffering the disruptive consequences of the occasional accident.

In a study by the NASA Engineering and Safety Center, the following guiding principles were published for use by the teams who are responsible for the safety of a design (NASA 2006):

- *Define a clear and simple set of prioritized program needs, objectives, and constraints, including safety, that form the validation basis for subsequent work;*
- *Manage and lead the program with a safety focus, simple and easy to understand management structures, and clear lines of authority and responsibility among the elements;*

- *Specify safety and reliability requirements through a triad of fault tolerance, bounding failure probability, and adhering to proven practices and standards;*
- *Manage complexity by keeping the primary mission objectives as simple and minimal as possible, and adding complexity to the system only where necessary to achieve these objectives;*
- *Conceive the right system conceptual design early in the life cycle by thoroughly exploring risks from the top down, and using a risk based design loop to iterate the operations concept, the design, and the requirements until the system meets mission objectives at minimum complexity and is achievable within constraints;*
- *Build the system right by applying a multilayered, "defense in-depth" approach of following proven design and manufacturing practices, holding independent reviews, inspecting the end product, and employing a "test like you fly, fly like you test" philosophy; and*
- *Seek and collect warning signs and precursors to safety, mission success, and development risks throughout the life cycle, and integrate those into a total risk picture with appropriate mitigation activities.*

These principles are supported by a foundation of established project management, systems engineering, safety and mission assurance, and operations practices that encourages a safety focus throughout the program life cycle. The safety focus includes the attitude and approach to safety of those conceiving, producing, and operating the integrated system, as well as the individual components and the system elements.

A distinction is made between safety and mission success. A safe system is one in which the survival, health, and well-being of the crew during nominal and off-nominal operational scenarios is ensured during its operation. For such a system, strategies are provided to avoid or deal with unsafe conditions and margin is applied to the system to prevent any limits from being exceeded that could result in harm to the crew. A reliable system is one that assures mission success by functioning properly over its intended life cycle. It has a low and acceptable probability of failure that is achieved through simplicity, proper design, and proper application of reliable parts and materials. In addition to long life, a reliable system is robust and fault tolerant, meaning it can tolerate failures and variations in its operating parameters and environment.

Safety and reliability objectives often work together; however, they also can compete. The two disciplines work together when margins are added to a system to ensure its continued operation from both safety and mission success perspectives. They compete when safety objectives seek to prevent a hazardous condition that also interrupts mission success. For example, a human rated system would establish safety limits, such as engine redlines, prior to the point of failure to allow a crew to react to the situation. In this situation, an early or false abort can occur at the expense of mission success.

The key to managing risk is based on an informed decision making process through cognizance of the risk associated with the technology, and an ability to determine the means to eliminate, mitigate, or accept the risk.

1.4 THE BOOK

The complexity of space systems design, as well as that of the overall organization involved in the realization of the design, requires a deep and widely ranging knowledge of the key principles and techniques of design for safety and a multidisciplinary awareness of the hazards and potential vulnerabilities inherent in the design. The purpose of this book is to provide, from a systems engineering perspective, a compendium of the principles of safety design. By effectively applying these principles to space systems, inherent hazards can be prevented, mitigated, or controlled, thus limiting the risk to crew and to spacecraft to an acceptable level.

This book is organized to present a discussion of the space environment, an overview of the effects of space on the human body, and a presentation of the tools and technology available to ensure an adequate safety design for various spacecraft and associated systems.

REFERENCES

National Aeronautics and Space Administration. (2006) *Design, Development, Test and Evaluation (DDT&E) Considerations for Safe and Reliability Human Rated Spacecraft Systems*. NASA Technical Report RP-06-208. Hampton, VA: National Aeronautics and Space Administration, Langley Research Center.

Roland, H. E., and B. Moriarty. (1990) *System Safety Engineering and Management*. New York: Wiley and Sons.

CHAPTER 2

The Space Environment: Natural and Induced

Gerald Griffith
Chief System Safety Engineer, JAMSS America, Incorporated, Houston, Texas

Tateo Goka, Ph.D.
Director, Space Environment Measurement Group, The Institute of Aerospace Technology, Japan Aerospace Exploration Agency, Tsukuba, Japan

CONTENTS

2.1. The Atmosphere 8
   By Ernst Messerschmid, Prof. Dr. rer. nat., and Juergen Schlutz, Ph.D.

2.2. Orbital Debris and Meteoroids 18
   By D. Heiner Klinkrad, Ph.D.

2.3. Microgravity 31
   By Joel K. Kearns

2.4. Acoustics 43
   By Jerry R. Goodman and Ferdinand W. Grosveld, D.E.

2.5. Radiation 52
   2.5.1. Ionizing Radiation 52
      By Myung-Hee Y. Kim, Ph.D.; Kerry A. George; and Francis A. Cucinotta, Ph.D.
   2.5.2. Radio Frequency Radiation 67
      By Robert C. Scully, MSEE, PE, NCE

2.6. Natural and Induced Thermal Environments 72
   By Steven L. Rickman

2.7. Combined Environmental Effects 86
   By Marc Van Eesbeek and Christopher O. A. Semprimoschnig, Dipl.-Ing., Ph.D.

Human space exploration is an undertaking of high risk. Effective management of the inherent risks of spaceflight to an acceptable level is of paramount importance to the safe and successful outcome of the endeavor. These risks arise predominantly from various elements of the ambient space environment encountered during the operational evolution of a mission. It, therefore, is imperative that a space safety program supporting the design and operation of space systems addresses the entirety of critical space environments.

The intent of this chapter on natural and induced space environments is to provide a discussion of several critical space environments to enable a basic understanding of and appreciation for some of the measures employed to manage these risk factors. Because of space limitations of this book and the breadth of this subject, it is not possible to provide a complete reference text on the entirety of space environments. The subjects selected for inclusion are those with general application to human spaceflight in Earth orbit. Space systems designed for a unique operation, such as an interplanetary space mission, require additional specialized environmental considerations.

The reader specifically should understand that this chapter reflects only a sampling of common critical space environments. For development of an effective operational project, the designers must identify every applicable critical space environment for inclusion in an effective safety program for spacecraft design and operation.

2.1 THE ATMOSPHERE

2.1.1 Composition

An atmosphere is the gaseous envelope surrounding a planetary body such as the Earth. As a working definition for spaceflight, the upper boundary of the atmosphere is that altitude where radiation pressure replaces atmospheric pressure as the dominant perturbing influence. For the Earth, this occurs at an altitude of about 1000 km. The outer regions of the atmosphere consist of the thermosphere, which for the Earth extends from 105 to 750 km, and the exosphere, which is immediately above it (Figure 2.1).

A decisive factor for planning missions in low Earth orbit and for all launches of rockets or other spacecraft is the thorough knowledge and understanding of atmospheric conditions. For example, atmospheric density especially has a major influence on aerodynamic drag and the corresponding torques; therefore, it has considerable influence on the attitude of rockets and spacecraft. Temperature and composition are important characteristics of the atmosphere as well, and an understanding of these factors is useful when judging material degradation caused by atomic oxygen. For that reason, many semiempirical models, each having a different level of accuracy, were developed to derive desired data at any specific location and at any particular point in time. Various systems of differential equations for fluid and thermodynamic processes form the basis of these models. To solve these differential equations, data obtained experimentally are entered as initial values for the model parameters.

Because the atmosphere is subject to the influence of solar radiation and the magnetic field of the Earth, its exact condition at any given location depends on several parameters, including long-term, short-term, and spatial variations. Most of the time, estimations require only a simple model that delivers either mean or minimum and maximum values.

An example of such a model is the CIRA72 atmosphere. Figure 2.2 shows mean values for temperature, T_∞, density, ρ_∞, and composition as a function of altitude. In the upper thermosphere and the exosphere, temperature increases only asymptotically because of extremely effective heat conduction in those regions. The dotted lines in this graph

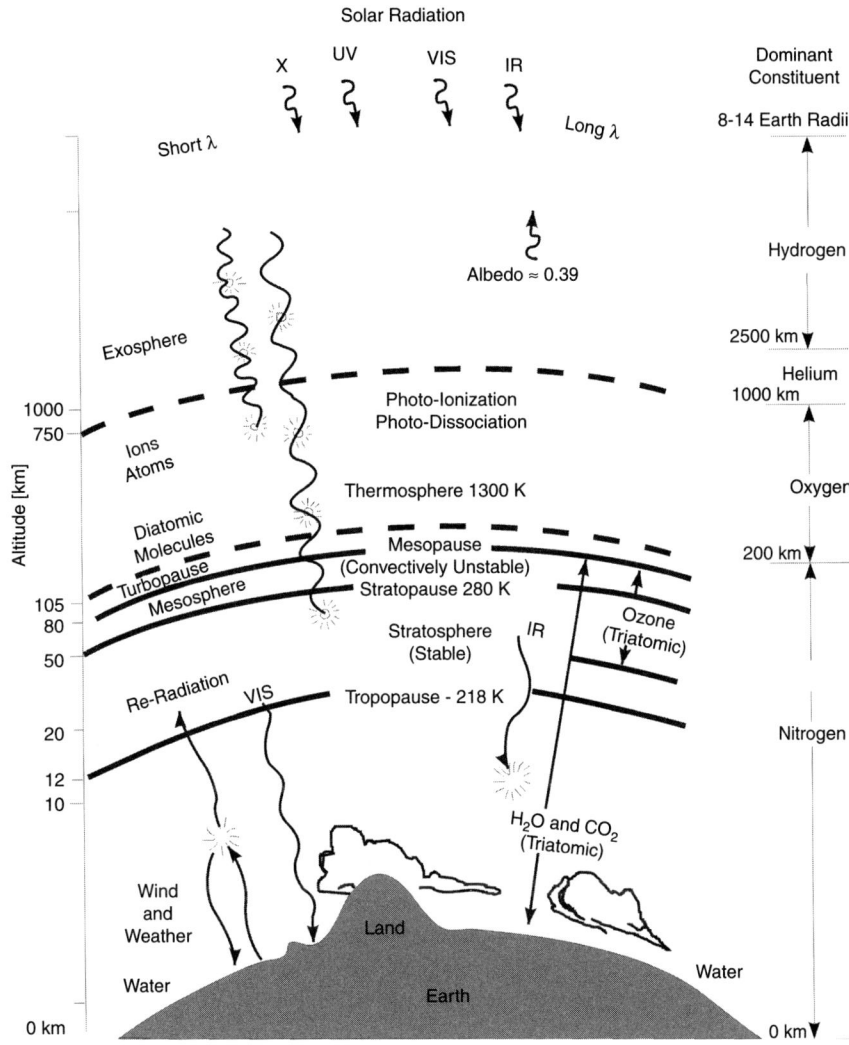

FIGURE 2.1 Atmospheric layers and composition (Messerschmid and Bertrand 1999; graphic by Ernst Messerschmid).

represent the variation range of the data caused primarily by changes in solar activity. On the other hand, local time also influences atmospheric density for both minimum and maximum solar activity (Figure 2.3). Table 2.1 lists density variations, $\Delta \rho$ [%], at altitudes between 150 and 800 km resulting from different effects and different timescales.

Gases within the upper thermosphere are heated by absorption of extreme ultraviolet radiation and at lower altitudes by ultraviolet radiation from the Sun. Heating by radiation,

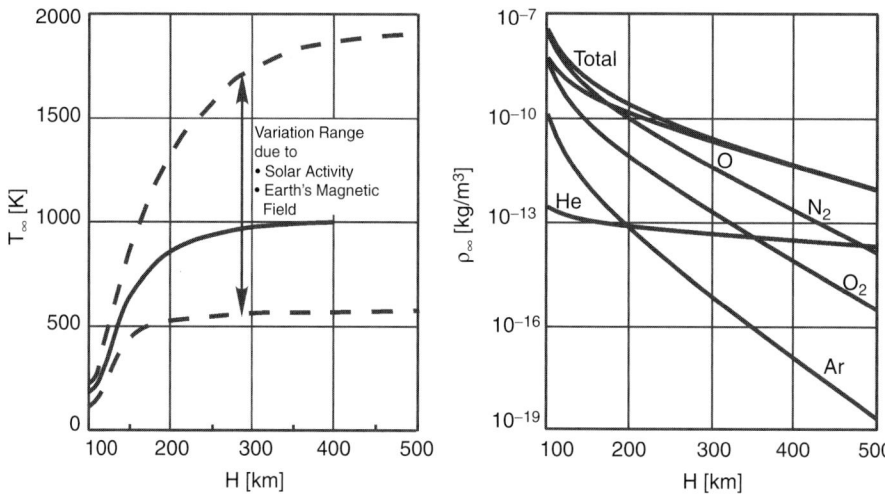

FIGURE 2.2 Average temperature (left), composition with average densities (right) (Messerschmid and Bertrand 1999; graphic by Ernst Messerschmid).

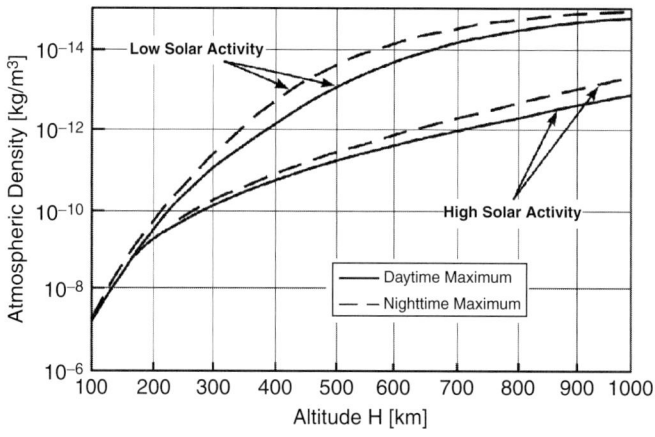

FIGURE 2.3 Diurnal density variations for high and low solar activity (Messerschmid and Bertrand 1999; graphic by Ernst Messerschmid).

of course, occurs only on the dayside of the Earth; however, conductive and convective heat transport distributes the energy within the atmosphere only to a small extent. For that reason, a considerable temperature gradient exists from the transition zone into the eclipse, the gradient increasing in the exosphere to a level of over 200 K (nightside 840 K to dayside 1060 K).

Table 2.1 Density Changes in the Thermosphere for Different Altitudes and Their Time Scales (Messerschmid and Bertrand 1999; Table by Ernst Messerschmid)

Effect	Δρ [%]				Time Scale
	150 km	200 km	400 km	800 km	
Flux (solar cycle)	25	110	1165	3800	Years
Flux (daily)	0	1	5	15	Days
Geomagnetic activity	25	35	60	100	Hours
Local time	10	25	115	230	Hours
Semiannual	15	15	50	80	Months
Latitude	10	15	60	90	Months
Longitude	2	2	5	15	Days

Unlike the relatively uniform composition of the atmosphere from sea level to the altitude of the turbopause, that is, about 100 km, diffusion dominates atmospheric composition at the higher altitudes. Such processes, together with gravitational forces, lead to an effect in which the lighter components, such as hydrogen and helium, accumulate in the higher regions of the atmosphere. At altitudes above 1000 km, these components slowly escape from the gravitational field of the Earth.

Today, highly developed models are used to describe the neutral atmosphere (Skrivanek 1994). Even so, these models are relatively inexact, especially for spaceflight application, because the principal influencing factors listed in Table 2.1 are insufficient. As well, sporadic factors, that is, those influenced by gravitational forces, cannot be reproduced by a model. Some effects of empirical models can be described in more detail by the factors $F_{10.7}$ and K_p, which correlate with electromagnetic and auroral heating. The $F_{10.7}$ radiation, measured at the 10.7-cm wavelength, is caused by other than extreme ultraviolet mechanisms in the solar atmosphere. Because $F_{10.7}$ radiation is not absorbed by the atmosphere of the Earth, it can be used as an approximate indicator for solar activity. The geomagnetic index, K_p, and a further index, a_p, are used to incorporate the global influence of geomagnetism into the models by providing the means to include changes in the magnetic field of the Earth, a parameter measured by a network of magnetometer stations located about the surface of the planet.

The results from most models used at present, J70, MSISE90, GRAM, and the latest, VSH, vary among each other by only 10 to 15%. These models and their updates, for example, NRLMSISE-00, differ from each other mainly in their range of validity at lower altitudes and by the user-friendliness of their interfaces. Figure 2.4 illustrates the standard deviations obtained from the different models as compared to data gauged by electrostatic triaxial acceleration measurement obtained from satellites at an altitude of 250 km and an accuracy of about 1% (Skrivanek 1994). At higher altitudes, the model errors increase. It

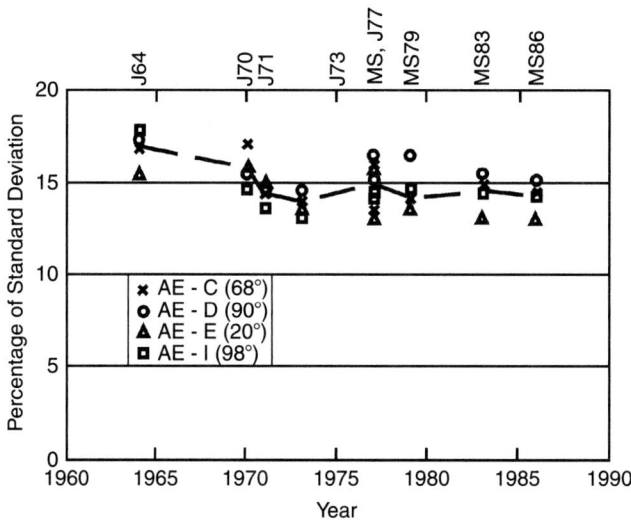

FIGURE 2.4 Statistical comparison of empirical models for the neutral atmosphere (Messerschmid and Bertrand 1999; graphic by Ernst Messerschmid).

is reported, for example, that an error with a 25% standard deviation occurs for the typical altitude of space stations, 400 km. Among other causes, these deviations are due to not having incorporated the velocity of the winds present at high altitudes into the models. Indeed, previous measurements have indicated wind velocities measuring 1.4 km/s at $K_p = 9$. Even though wind velocity is not incorporated, the magnitude of the standard deviation can be reduced to below 5% by measuring the drag coefficient.

The influence of the atmosphere on space vehicles can be seen mainly in the following points:

- At relative velocities of about 8 km/s, considerable impulse and energy exchange takes place between ambient flow and spacecraft. The applied drag is described by means of the drag coefficient, C_D, which as a basis for simple estimations can be set to a range of about 2.2 to 2.5 and where the effective surface is that of the space vehicle normal to the flight direction. Additionally, aerodynamic forces induce torques that can change the attitude of the space vehicle. Both influences must be balanced by active attitude and orbit control systems.
- Gas particles, atomic oxygen in particular, that impact the surface of a space vehicle can have mechanically or chemically erosive effects.
- Sometimes, gas particles in the atmosphere collide with the gases emitted by the vehicle. This can result in contamination of the vehicle surface, for example, deposition of materials on optical glasses.

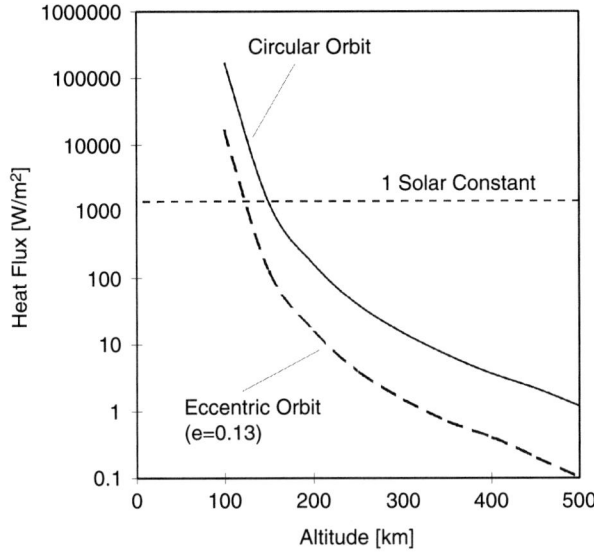

FIGURE 2.5 Aerodynamic heating in low Earth orbits (Messerschmid and Bertrand 1999; graphic by Ernst Messerschmid).

- In very low orbits, aerodynamic heating plays an important role. At an altitude of about 150 km, aerodynamic heating reaches the order of magnitude of the solar constant and quickly decreases with increasing altitudes (Figure 2.5). At 300 km, this drops to about 1% of the solar radiation.

Note that even events induced by a spacecraft itself at least temporarily can have considerable effects on the immediate environment of vehicles such as space stations or orbital platforms.

2.1.2 Atomic Oxygen

At high altitudes, atomic oxygen forms a large part of the atmosphere of the Earth; above 150 km, it is the main constituent, that is, 66% at 200 km and 90% at 500 km. Despite a relatively low particle density at high altitudes, atomic oxygen has considerable erosive effects because of its high chemical reactivity. The importance of the influence of atomic oxygen was discovered during the first Space Shuttle missions when an unexpected phenomenon, the so-called shuttle glow, that is, the shine observed on Space Shuttle surfaces directly exposed to the ambient flow of the residual atmosphere, commonly was observed. The shimmer is the result of chemical reactions between the surfaces of the *Space Shuttle* and atomic oxygen. Certain materials were observed to have experienced severe erosion after these missions; for example, Mylar® foils exhibited a weight loss of 35% after three days of exposure. Further experiments aboard the *Long Duration*

Exposure Facility, the European Retrievable Carrier-1 platform, and during the *D2* mission confirmed the erosive influence of atomic oxygen on all surfaces exposed to the flight direction (Hamacher, Fitton, and Kingdon 1987).

The effects of atomic oxygen depend not only on the altitude but also on solar activity because atomic oxygen is generated mainly by the process of photodissociation. Figure 2.6 shows atomic oxygen flux as a function of altitude for surfaces in the ram direction and for solar viewing surfaces exposed for one year. The rate of erosion caused by atomic oxygen can be very high, so coatings of insufficient thickness easily can be eroded away after only a few days (Tonon et al. 2001). A flow of 10^{22} atoms/cm^2 over a period of 11 a (one solar cycle) forms the basis of assumptions made for the *International Space Station*. Under these circumstances, a supporting truss made from a composite based on graphite would lose more than 30% of its initial wall strength due to atomic oxygen erosion within 15 a, that is, the projected design life of the *International Space Station*. Figure 2.6 also shows the surface recession per year for a typical loss rate (R_e) of 3.0×10^{-24} cm^3/atom of oxygen. Apart from erosion, recombination of oxygen atoms along with a release of bonding energy and a reaction with the wall that produces nonvolatile constituents also occur.

Fortunately, most metals usually are not damaged by atomic oxygen. Indeed, protective oxide layers, as in the case of aluminum, are sometimes formed on them. Exceptions to this, however, are silver, which is used for electrical contacts, and osmium, which is vapor deposited on the surfaces of optical instruments. As well, many polymers used in spaceflight, such as foils, adhesives, paints, and some windows, are subject to degradation. Fibrous composites are susceptible as well. Experiments aboard the European

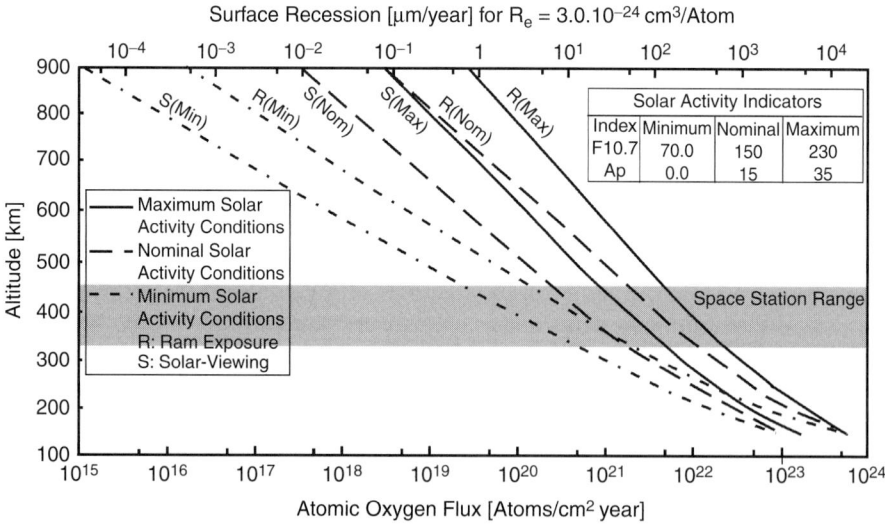

FIGURE 2.6 Flux density of oxygen atoms over a period of one year (Messerschmid and Bertrand 1999; graphic by Ernst Messerschmid).

Table 2.2 Loss Rates for Different Materials (Messerschmid and Bertrand 1999; Table by Ernst Messerschmid)

Material	R_e [10^{-24} cm³/oxygen atom]
Polyethylene	3.7
Kapton®	3.0
Mylar®	2.2
Polyamide	2.2
Teflon®	<0.02
Material on silicone basis	<0.1

Retrievable Carrier-1 and D2 missions demonstrated severe damage to the resin and matrix of some composite materials, and the fibers of the composite occasionally were exposed or even made porous.

The erosive effect of oxygen atoms can be expressed in terms of loss rate, which is indicated by the reaction efficiency, R_e, or the volume of a material that is eroded per impacting atom. Values of loss rates obtained by experiments using different materials are summarized in Table 2.2.

Various coatings can be considered for use as protective measures against atomic oxygen erosion. Examples of such surface treatments are vapor deposited gold used on silver contacts and various atomic oxygen resistant coatings used on fibrous composites. The long-term objective, however, is the development of atomic oxygen resistant polymers (Dueber and McKnight 1993).

Erosion efficiency, η, is defined as the relation of R_e to the atomic volume of typical wall materials with $V_A \approx 3.0 \cdot 10^{-24}$ cm³ as per Equation (2.1):

$$\eta = \frac{R_e}{V_A} \tag{2.1}$$

The erosion rate, r, in m/s is expressed by Equation (2.2):

$$r = \frac{N_O \cdot u \cdot \eta}{N_W} = \frac{\Phi_O \cdot \eta}{N_W} \tag{2.2}$$

where N_O is the particle density of atomic oxygen, N_W is the particle density of the wall material, u is the orbital velocity, and Φ_O is the flux density of the oxygen atoms.

2.1.3 The Ionosphere

Solar radiation contains sufficient energy at short wavelengths to cause considerable photoionization in the upper atmosphere of the Earth. This process is responsible for maintaining the ionosphere, a partially ionized layer mainly located between 50 and 200 km and extending to an upper boundary of about 2000 km from the surface of the Earth.

The ionosphere constitutes only a small part of atmospheric density. Although its upper boundary extends to a maximum on the dayside, it never exceeds 1% of the neutral gas density. The special importance of this conductive layer lies in its capability to reflect impacting radio waves, making possible their distribution around the globe. This phenomenon was demonstrated for the first time by G. Marconi in 1901. Today, the reflective property of the ionosphere is the fundamental basis for research in this area. Below frequencies of 100 MHz, the ionosphere does not permit a radio link to be established between a space vehicle and the Earth. Although this portion of the atmosphere can influence experiments and exposed parts of a spacecraft negatively, it can affect, for example, electromagnetic tethers in a positive way.

Since the 1920s it has been known that the condition of the ionosphere strongly depends on the course taken by the magnetic field lines of the Earth and hence on the geomagnetic latitude. Three regions of high, mean, and low latitude are distinguished. The magnetic field lines of the Earth, as well, are dependent on day, time, and seasons.

Ionospheric Models

The classic model of the ionosphere, which can be attributed partly to S. Chapman, describes the characteristics of this region of the atmosphere at mean geomagnetic latitudes. Four distinct layers are defined according to their electron concentration. These layers are designated as the D (50 to 90 km altitude), the E (90 to 160 km altitude), and the F1 and F2 layers (160 to 900 km altitude). Figure 2.7 shows representatively the distribution of particle concentration and the temperature of electrons, ions, and neutral particles as a function of altitude. Above the F layers and up to an altitude

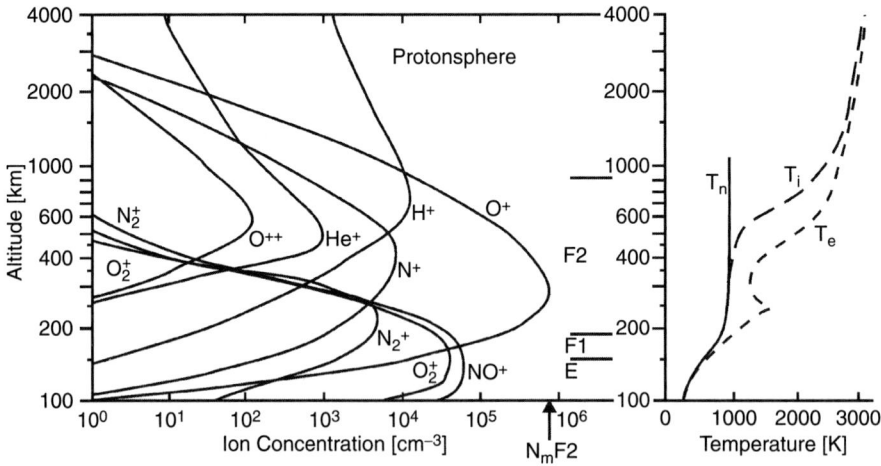

FIGURE 2.7 Particle and temperature profiles in the ionosphere (Messerschmid and Bertrand 1999; graphic by Ernst Messerschmid).

of 1200 km, the upper ionosphere is defined, and above that resides the plasmasphere. This stratification is caused mainly by the following three reasons:

- Depending on their respective absorption characteristics, received solar energy is absorbed in different layers.
- Recombination of electrons and positive ions depends on atmospheric density; this, in turn, depends on altitude.
- The composition of the atmosphere changes with altitude.

In addition to the classic model of the atmosphere just described, a number of other models are in use, such as

- Empirical models, the best known of which is the International Reference Ionosphere, the latest of which is IRI-2001.
- Physical, time dependent models, such as the Utah State University Time Dependent Model of the Global Atmosphere.
- Models coupled with the thermosphere, such as the National Center for Atmospheric Research (NCAR) Global Ionosphere Thermosphere Model.
- Other models, including many that are very complex from a numerical point of view and for which a high-performance computer solves various conservation equations, such as mass, impulse, energy, and species, coupled with equations from the reaction kinetics and transport processes as a function of different influencing factors (Skrivanek 1994).

Variations in the Ionosphere

Only a few minutes after the occurrence of a strong solar flare, the electron density in the D and lower E layers strongly increases. Consequently, radio waves in the high frequency spectrum, which normally would be reflected by the higher layers, are absorbed within the lower layers. The resulting interruptions of communication links, called *sudden ionospheric disturbances*, last for about 1 h. Presumably, sudden ionospheric disturbances are caused by solar X-rays emanating from solar flares and penetrating the lower layers of the ionosphere.

Another kind of communication disturbances are the so-called ionospheric storms. Unlike sudden ionospheric disturbances, ionospheric storms can last for several days. Within the group of ionospheric storms, the polar cap absorption storm and the geomagnetically induced storm can be distinguished. For the case of the polar cap absorption storms, high-energy protons, mainly originating from solar flares, travel along the magnetic field lines of the Earth into the polar regions of the lower ionospheric layers (55 to 90 km), where they cause an increase in electron density due to ionization processes. As a consequence, communication is impaired considerably. The second phenomenon, the geomagnetically induced storm, occurs about 20 h after an actual solar flare event. A geomagnetically induced storm is initiated when low energy electrons and protons penetrate along the magnetic field lines of the Earth into the lower layers of the ionosphere where the result is an increase in electron density, an effect also referred to as an *auroral substorm*. Because of the enormous amount of energy transferred to the magnetic field of the Earth following the occurrence of solar flares, nitrogen and oxygen from the thermosphere (100 to 750 km) rise into the F layer. Consequently, the electron density in the F layers is reduced considerably because of the aforementioned recombination mechanism. The resulting concentration gradient causes considerable turbulence and other perturbations in the ionosphere. These effects can last up to one month at solar minimum.

Behavior of Radio Waves in the Ionosphere

Charged particles within the ionosphere can extract energy from impacting electromagnetic waves, thus weakening or even completely absorbing them. However, an electromagnetic wave moving through a region of constantly varying electron density changes its direction. Under certain circumstances, the direction can change sufficiently for it to be reflected. The reflection of radio waves is based on an effect similar to the refraction of light in transition to an optically thinner medium, that is, the beam angle increases with respect to the local vertical of the interface. Continuous refraction among layers having different electron densities causes diversion of the beam path.

In principle, free electrons absorb the energy of radio waves and re-emit it at the same frequency. If, however, the density of an object with which free electrons collide is high, such as free electrons colliding with neutral particles within the D layer, most of the energy possessed by the electrons is transferred to the high density particles. This energy appears in the form of heat, that is, statistically distributed kinetic energy, and it is lost for signal transmission.

As well, the dispersive properties of the ionosphere decelerate radio waves as a function of their frequency. This effect oftentimes is useful as in the case of satellite and International Space Station supported navigation systems used to compensate relative errors by means of multifrequency transmission techniques. The *International Space Station* supports the navigational and time synchronizing signal techniques in various ways for the purpose of synchronizing clocks, computer networks, and the like. Because of their utility, these ionospheric effects are subject to investigations to be conducted on the *International Space Station* in the future.

2.2 ORBITAL DEBRIS AND METEOROIDS

2.2.1 Orbital Debris

Spaceflight activities since the launch of *Sputnik-1*, in 1957, have generated a human-made particle environment that today must be considered in payload and mission designs to ensure successful operations with an acceptably low risk of losing or degrading a mission or suffering casualties during human spaceflight. Our knowledge of the orbital debris environment is based largely on ground based radar and optical measurements. In low Earth orbit, radar covers size regimes above 5 cm during routine space surveillance and above 2 mm during experiment observations. For geostationary Earth orbit altitudes, telescopes are used to track objects as small as 1 m during routine space surveillance and down to 15 cm during experiment observations. For small debris, in situ impact detectors and retrieved surface material are important information sources. Because of statistical limitations in the achievable impact fluence during constrained exposure times, such data typically become meaningful for object sizes below 1 mm.

Launch and Mission Related Objects

Most of the on-orbit debris mass of about 5000 tons is characterized as launch and mission related objects. A large fraction of these can be observed by the U.S. Space Surveillance Network, which maintains a catalog of objects along with their associated orbits. By May 2005, a total of 4378 launches since 1957 had deployed 17,606 payloads, rocket bodies, expended upper stages, and mission related objects that resulted in 29,916 detectable and trackable objects in various Earth orbits. Of these 29,916 cataloged objects, 19,331 had decayed into the atmosphere, leaving a cataloged on-orbit population of 10,585 objects as of that time. By 1962, the number of new on-orbit objects was observed to be increasing at the near linear rate of about 260 per year. The total number of cataloged objects in the same time frame increased at the rate of about 710 per year. The almost steady state annual launch rate of 110 ± 10 between 1965 and 1990 dropped considerably after the end of the Union of Soviet Socialist Republics to almost 50% of this level. Since the years 2001 and 2002, the rate again has stabilized but at a rate of about 60 launches per year.

When classified by object categories, 31.8% of the cataloged objects in 2005 were payloads, of which, 67% were active satellites. Of the remainder, 17.6% were spent rocket upper stages and boost motors, 10.5% were mission related objects, and about 39.9% were debris mainly resulting from fragmentation events, that is, 28.4% from upper stages and 11.5% from satellites. When classified according to orbit regimes, 69.2% of the cataloged objects were in low Earth orbits at altitudes below 2000 km, and 9.3% were in the vicinity of the geostationary ring. A further 9.7% were in highly eccentric orbits, which include geostationary transfer orbits; 3.9% were in medium Earth orbits located between low Earth orbit and geostationary Earth orbit; and almost 7.8% were outside the geostationary Earth orbit region. Of all this debris, only 160 objects have been injected into Earth escape orbits.

For May 2005, Table 2.3 shows the contribution of launch and mission related objects to the orbital debris population according to the Meteoroid and Space Debris Terrestrial Environment Reference, MASTER-2005 model (Oswald et al. 2006). According to this model, launch and mission related objects prevail only at sizes above 1 m. At 10-cm and 1-cm sizes, the share of launch and mission related objects reduces to 26% and 0.9%, respectively. Occurrences of launch and mission related objects below 1 cm are, for instance, due to the deployment of thin copper wires by the Westford Needles experiments conducted in 1961 and 1963. The altitude distribution of launch and mission related objects is shown in Figures 2.8 and 2.9.

Explosion and Collision Fragments

Orbital debris caused by fragmentation events is the most important source of objects cataloged, with a contribution of 39.9% to the trackable population. By May 2005, a total of 188 on-orbit fragmentation events have been inferred from the detection of new objects and the correlation of their determined orbits with a common source.

Table 2.3 Contributions by Macro-Objects to the MASTER-2005 Reference Population for May 2005, Discriminated by Sources, Size Regimes, and Orbital Regions (Low Earth Orbit: $H < 2286$ km)

Source/Type	Orbit Regime	>1 mm	>1 cm	>10 cm	>1 m
Launch and mission related objects	LEO	4,025	3,214	3,175	2,030
	LEO + MEO + GEO	31,043	5,393	5,354	3,895
Explosions	LEO	6.42e+6	183,179	8,714	379
	LEO + MEO + GEO	1.54e+7	411,226	15,033	764
Collisions	LEO	23,195	453	124	1
	LEO + MEO + GEO	41,391	775	133	1
NaK	LEO	44,935	24,030	0	0
	LEO + MEO + GEO	44,935	24,030	0	0
Solid rocket motor slag	LEO	5.68e+6	16,905	0	0
	LEO + MEO + GEO	1.27e+8	165,493	0	0
Ejecta	LEO	1.23e+6	0	0	0
	LEO + MEO + GEO	3.66e+6	0	0	0
Total count	LEO	1.34e+7	227,782	12,013	2,409
	LEO + MEO + GEO	1.47e+8	606,917	20,520	4,660

GEO = geostationary Earth orbit.
LEO = low Earth orbit.
MEO = medium Earth orbit.

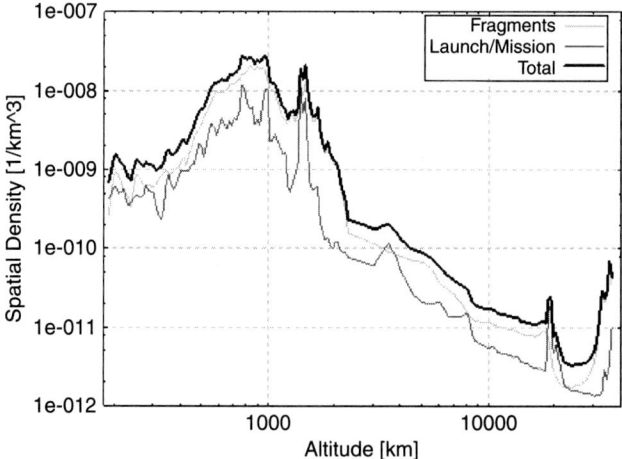

FIGURE 2.8 Spatial density versus altitude for objects of diameters $d > 10$ cm according to the MASTER-2005 model for May 2005. Different sources are indicated in gray scales. The thick black line shows the envelope of the total spatial density.

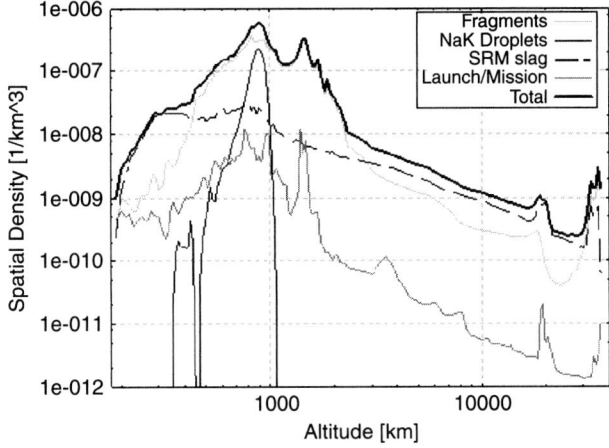

FIGURE 2.9 Spatial density versus altitude for objects of diameters $d > 1$ cm according to the MASTER-2005 model for May 2005. Different sources are indicated in gray scales. The thick black line shows the envelope of the total spatial density.

Approximately 30% of breakups are believed to have been deliberate explosions or collisions that resulted in more than 3000 cataloged fragments. Moreover, another 30% can be attributed to propulsion system explosions, causing more than 4000 cataloged fragments, and an additional 3% can be associated with electrical system failures, mainly battery explosions, causing about 700 cataloged fragments. A further 1% is related to three accidental collisions, the *Cosmos 1934* spacecraft with a *Cosmos 926* mission related object in December 1991, the *Cerise* spacecraft with an *Ariane H-10* fragment in July 1996, and a *Thor* stage with a *CZ-4B* stage fragment in January 2005. The remaining 36% are of unknown origin. With the exception of two known geostationary Earth orbit explosion events, an *Ekran-2* satellite on June 22, 1978, and a *Titan III C Transtage* on February 8, 1994, all known fragmentation events occurred in orbits passing through altitudes below 2000 km. Of these, about 80% were within low Earth orbits, and 17% were within highly eccentric orbits and geostationary transfer orbits. More than 200 and up to 700 trackable fragments were generated for each of 17 of the 188 events. A comparison of experimental observation data with catalog information suggests that several fragmentation events might have occurred unnoticed. Hence, the MASTER-2005 debris environment model added 15 additional on-orbit explosions, with eight of them occurring in geostationary Earth orbit and four in low Earth orbit.

For May 2005, Table 2.3 shows the contribution of fragmentation related objects from explosions and collisions to the orbital debris population according to the MASTER-2005 model. Fragmentation debris prevails at sizes ranging from 1 mm to 10 cm, with shares of 10 to 74%. Their altitude distribution is shown in Figures 2.8, 2.9, 2.10, and 2.11.

22 CHAPTER 2 The Space Environment: Natural and Induced

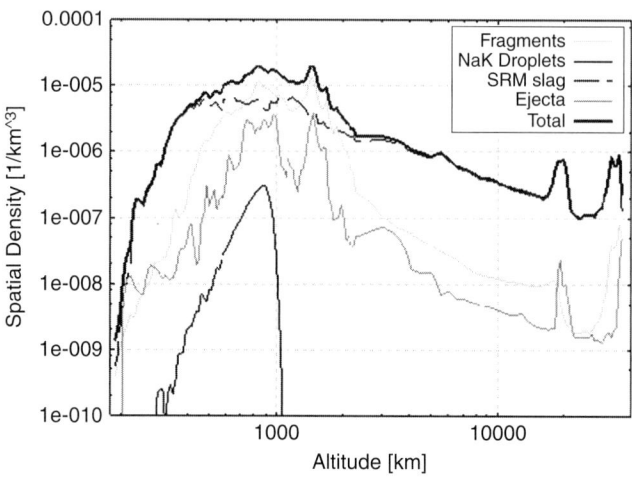

FIGURE 2.10 Spatial density versus altitude for objects of diameters $d > 1$ mm according to the MASTER-2005 model for May 2005. Different sources are indicated in gray scales. The thick black line shows the envelope of the total spatial density.

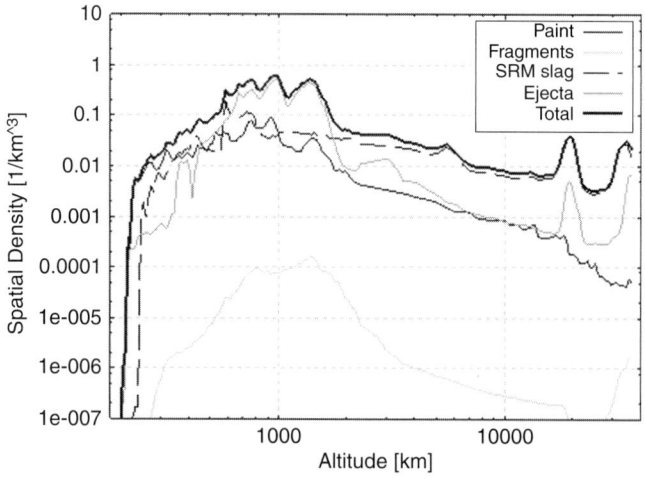

FIGURE 2.11 Spatial density versus altitude for objects of diameters $d > 0.1$ mm according to the MASTER-2005 model for May 2005. Different sources are indicated in gray scales. The thick black line shows the envelope of the total spatial density.

Debris Sources Unrelated to Fragmentation

The most important sources of space debris unrelated to fragmentation are slag and dust particles resulting from solid rocket motor firings. These materials are composed mainly of aluminum oxide and residues of motor liner material. Aluminum powder is added to most solid fuels, typically with a mass fraction of 18%, to stabilize the combustion process. About 99% of this material is assumed to be ejected with the exhaust stream throughout the main thrust phase. The ejecta is in the form of aluminum oxide dust, the particles having diameters largely within the range from 1 to 50 μm. Due to design constraints, several types of solid rocket motors have the nozzle protruding into the burn chamber. This creates a cavity around the nozzle throat. During the burn phase, trapped aluminum oxide, melted aluminum droplets, and parts of released thermal insulation liner material can pool within this area and weld together to form slag particles that can grow to sizes typically ranging from 0.1 to 30 mm in diameter. Indeed, ejected particles as large as 50 mm in diameter have been observed during ground testing of solid rocket motors. The slag particles are released at the end of the main thrust phase as the internal motor pressure decreases. The release depends on the spin rate of the solid rocket motors, a process used to stabilize the motor orientation during the payload injection maneuver.

The number of solid rocket motor firings between 1958 and May 2005 was 1076, with peak rates of up to 47 events per year and a mean annual rate of 23.5. Of the number of injection orbits where solid rocket motors were utilized, 80% are associated with U.S. missions. The size of the solid rocket motors in terms of propellant capacity covers a wide range. The most frequently used solid rocket motors are the Star 37 motors, which have a propellant mass of 1067 kg. These are used, for example, as the final stage of Delta launchers to deploy global positioning system payloads. The payload assist module, containing 2011 kg of propellant mass, is used as the Delta final stage, for instance, for geostationary transfer orbit injections. Another solid rocket motor, the inertial upper stage, deployed from a *Titan IV* or the *Space Shuttle*, is used to inject payloads into geostationary transfer orbit. The inertial upper stage contains a first stage propellant mass of 9709 kg and subsequently delivers the payload into a circular geostationary Earth orbit by using a second stage having a propellant mass of 2722 kg. Still another powerful solid rocket motor, the HS-601 with a propellant mass of 4267 kg, is used by Long March LM-2E launchers for both low Earth orbit and geostationary transfer orbit payload injections.

It is assumed that, for the 1076 solid rocket motor firings that occurred through May 2005, more than 1000 tons of propellant materials were released into space. Based on physical and mathematical model hypotheses, approximately 320 tons thereof were comprised of aluminum oxide dust particles and 4 tons were slag particles comprised of aluminum oxide, metallic aluminum, and motor liner material. Due to orbital perturbations and their different effects on micrometer-size dust and centimeter-size slag particles, merely 1 ton of aluminum oxide dust and 3 tons of solid rocket motor slag particles are believed to remain on orbit. This material ranges in size from 1 μm to 1 cm in diameter and dominates the orbital debris environment in this size regime.

For May 2005, Tables 2.3 and 2.4 show the contribution of solid rocket motor dust and slag particles to the orbital debris population according to the MASTER-2005 model. The altitude distribution of these populations is shown in Figures 2.9, 2.10, and 2.11.

Table 2.4 Contributions by Micro-Objects to the MASTER-2005 Reference Population for May 2005, Discriminated by Sources, Size Regimes, and Orbital Regions (Low Earth Orbit: $H < 2286$ km)

Source/Type	Orbit Regime	>1 μm	>10 μm	>100 μm	>1 mm
Launch and mission related objects	LEO	→	→	5,449	4,025
	LEO + MEO + GEO	→	→	45,931	31,043
Explosion	LEO	→	9.31e+07	9.21e+07	6.42e+06
	LEO + MEO + GEO	→	3.36e+08	3.34e+08	1.54e+07
Collisions	LEO	→	→	378,866	23,195
	LEO + MEO + GEO	→	→	927,728	41,391
NaK	LEO	→	→	→	44,935
	LEO + MEO + GEO	→	→	→	44,935
Solid rocket motor slag	LEO	→	1.13e+11	6.40e+10	5.68e+06
	LEO + MEO + GEO	→	1.02e+13	4.70e+12	1.27e+08
Solid rocket motor dust	LEO	3.23e+13	1.20e+13	0	0
	LEO + MEO + GEO	4.10e+15	4.91e+13	0	0
Paint flakes	LEO	3.23e+11	2.89e+11	4.79e+10	0
	LEO + MEO + GEO	1.71e+12	1.21e+12	1.33e+11	0
Ejecta	LEO	1.67e+13	3.70e+12	2.39e+11	1.23e+06
	LEO + MEO + GEO	2.14e+14	6.24e+13	1.06e+12	3.66e+06
Total count	LEO	4.94e+13	1.61e+13	3.51e+11	1.34e+07
	LEO + MEO + GEO	4.33e+15	1.23e+14	5.88e+12	1.47e+08

GEO = geostationary Earth orbit.
LEO = low Earth orbit.
MEO = medium Earth orbit.
→ indicates that the object count is given by the number to the right, with no contributions from smaller size regimes.

Between 1980 and 1988, out of 31 Russian *Radar Ocean Reconnaissance Satellites* (RORSATs), 16 ejected the fuel cores from their Buk reactors into a deposit orbit having a narrow inclination band close to 65° and an altitude between 900 and 950 km. During each of these ejections, a low melting sodium-potassium alloy escaped from the primary cooling circuit of the reactor, forming sodium-potassium droplets of up to 5.6 cm in size. A total of 208 kg of sodium-potassium coolant is believed to have been released during these 16 events. Thermodynamic considerations suggest that only droplets having diameters greater than 0.1 mm could survive in a stable state for an extended period and smaller objects would evaporate. For sodium-potassium droplets greater than 3 mm in diameter, the time required for evaporation exceeds the orbital lifetime. Hence, since the termination of Buk reactor deployments and the core ejections in 1988, atmospheric drag most likely caused the orbits of all sodium-potassium droplets smaller than 1 mm to decay into the atmosphere by 1993. The population of this debris remaining in orbit has an estimated mass of between 50 and 60 kg.

Ultimately, only droplets greater than 1 cm in diameter maintain a durable population close to the initial release altitudes. Since RORSAT missions were discontinued after 1988, sodium-potassium droplets can be regarded as a historic, nonrenewable source of debris. In May 2005, the sodium-potassium droplets contributed mainly to the size regime of $1 \text{ mm} \leq d \leq 1 \text{ cm}$, and their share of the MASTER-2005 reference debris population was less than 0.03% in total and 0.3% in low Earth orbit. The altitude distribution of the sodium-potassium debris population is shown in Figures 2.9 and 2.10.

In contrast to the sodium-potassium droplets, release products from surface degradation and impact ejecta are a reproductive source. For reasons of thermal control, most spacecraft and rocket upper stages are coated with paint or with thermal blankets. Under the influence of the harsh space environment, the surfaces of space objects tend to erode and do so in particular because of atomic oxygen, aging, and embrittlement caused by extreme ultraviolet radiation and microparticle impacts. This combination of effects was observed on the returned surfaces of the *Long Duration Exposure Facility*, which remained in orbit from April 1984 until January 1990 at altitudes of 470 to 340 km and an inclination of 28.5°. From observations made after its return, atomic oxygen was determined to have caused degradation of the polyurethane binder material in paints and damage to the underlying substrate material by penetrating through cracks in painted surfaces. It is known that extreme ultraviolet radiation can cause defects in polymeric materials, such as Kapton®, Teflon®, and paints. Degradation caused by extreme ultraviolet leads to embrittlement of the material, with subsequent crack formation and possible delamination of layered foils. The effect is enhanced through thermal cycling, mainly caused by the high number of transits made through the shadow of the Earth. Often in combination with microparticle impacts, atomic oxygen corrosion and extreme ultraviolet radiation lead to mass losses from surface coatings and the formation and detachment of paint flakes ranging from micrometer to millimeter in size. Traces of such materials can be found within impact craters on spacecraft. For example, on the average, one Space Shuttle window has to be replaced after each mission because of submillimeter particle impacts.

For May 2005, Table 2.4 shows the contribution of surface degradation products, that is, paint flakes and microparticle impact ejecta, to the orbital debris population according to the MASTER-2005 model. The altitude distribution for these populations is shown in Figures 2.10 and 2.11.

Debris Impact Probability

The statistical behavior of the orbital debris population is represented well by the laws of kinetic gas theory. Hence, the number of collisions, c, encountered by an object of collision cross section, A_c, moving through a stationary debris medium of uniform particle density, D, at a constant velocity, v, during a propagation time interval, Δt, is given by Equation (2.3):

$$c = vDA_c\Delta t \qquad (2.3)$$

where $F = vD$ is the impact flux in units of $m^{-2}s^{-1}$, and $\Phi = F\Delta t$ is the impact fluence in units of m^{-2}. The collision probability follows a binomial law that can be approximated well by a Poisson distribution, generating the probability $P_{i=n}$ of n impacts, and $P_{i=0}$ of no impact as stated by Equation (2.4):

$$P_{i=n} = \frac{c^n}{n!}\exp(-c) \mapsto P_{i=0} = \exp(-c) \qquad (2.4)$$

The probability of one or more impacts is, hence, the complement of no impact as expressed by Equation (2.5):

$$P_{i\geq n} = P = 1 - \exp(-c) \approx c \mapsto P \approx vDA_c\Delta t \qquad (2.5)$$

The demanding part in this equation is the determination of the particle flux, $F = vD$. In the MASTER-2005 model, three-dimensional, time dependent spatial object density distributions are established for a grid of volume elements covering the entire Earth environment from low Earth orbit to geostationary Earth orbit altitudes (Figures 2.12 and 2.13). Contributions from each member of the orbital debris population go into this distribution. For each of these objects, the velocity, magnitude, and direction is maintained for each volume element. This information is later retrieved to determine relative impact velocities with respect to a target object passing through the volume grid (Klinkrad 2006). For typical target orbits given in Table 2.5, the mean times between impacts by orbital debris of different sizes are listed in Table 2.6.

2.2.2 Meteoroids

Meteoroid Population

During almost 50 years of space activities, approximately 27,000 tons of human-made material reentered the atmosphere of the Earth. This compares with an estimated 40,000 tons of natural meteoroid material that also reaches the atmosphere each year. In contrast with orbital debris, however, most of the meteoroid mass is composed of small particles with diameters of about 200 μm and with corresponding masses of about 1.5×10^{-5} g. The resulting risk to operational spacecraft is generally low as compared

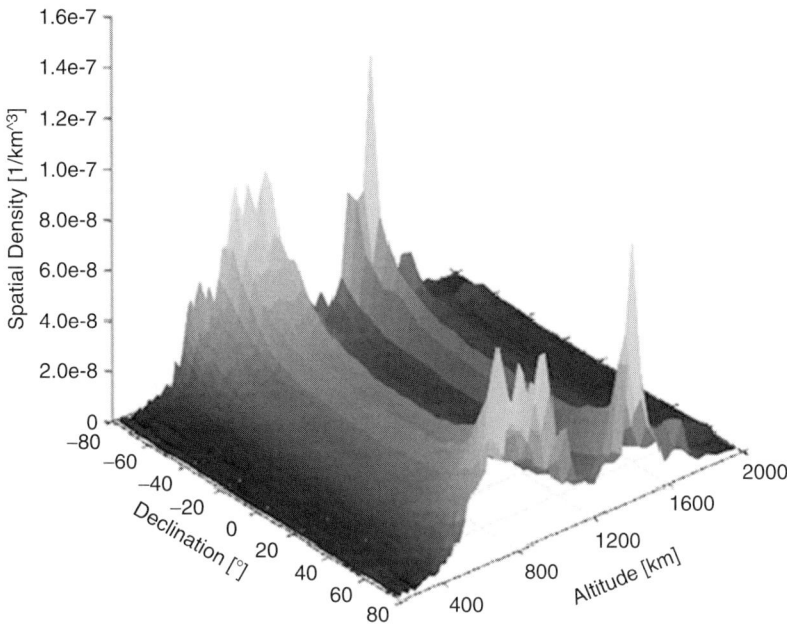

FIGURE 2.12 Spatial density in the low Earth orbit region versus altitude and declination for orbital debris of diameters $d > 10$ cm according to the MASTER-2005 model for May 2005.

with space debris, despite the much higher meteoroid impact velocities of up to 72 km/s (corresponding to a heliocentric escape orbit of 42 km/s intercepted by the orbit of the Earth at 30 km/s). Comparatively, the meteoroid environment of the Earth is well known from ground based photographic and radar observations, space borne detector experiments, retrieved space surfaces, meteoritic material captured by high flying airplanes, and from microcraters found on returned lunar rocks. In contrast with the highly dynamic space debris environment, the dominant part of the meteoroid environment can be assumed to be sporadic and invariant over time. However, seasonally recurring meteoroid stream events are associated with material in cometary orbits that give noticeable contributions, particularly in the case of meteoroid storms.

The meteoroid environment can be modeled by two main avenues. These involve the use of engineering equations for determining the omnidirectional impact flux (Grün et al. 1985) or more detailed models that resolve flux in terms of spatial densities and velocities with contributions from different families of meteoroids (Divine 1993; Grün et al. 2001). The latter approach is implemented in the MASTER-2005 model (Oswald et al. 2006). Both approaches require corrections of the baseline flux at 1 AU, that is, the mean radius of the orbit of the Earth, to account for geometric shielding by the Earth for some approach directions, gravitational focusing due to the geopotential field, and the relative motion of a target object with respect to the heliocentric system.

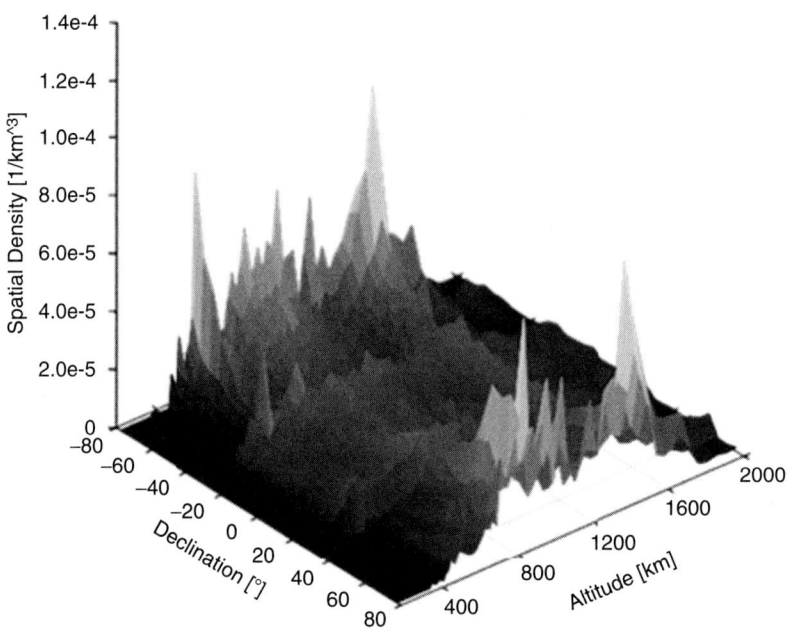

FIGURE 2.13 Spatial density in the low Earth orbit region versus altitude and declination for orbital debris of diameters $d > 1$ mm according to the MASTER-2005 model for May 2005.

Table 2.5 Representative Target Orbits Used in a Debris Flux Analysis with the MASTER-2005 Space Debris and Meteoroid Environment Model

Orbit	H_{pe} (km)	H_{ap} (km)	i (°)
International Space Station	356.0	364.1	51.6
Envisat	773.5	789.2	98.6
Globalstar	1,400.0	1,400.0	52.0
Global positioning system	20,000.0	20,000.0	55.0
Geostationary transfer orbit	560.0	35,786.0	7.0
Geostationary earth orbit	35,786.0	35,786.0	0.1

Meteoroid models are averaged in time. Hence, they contain information about seasonally recurring meteoroid streams, which generally contribute less than 5% to the overall meteoroid flux. If so required, seasonal meteoroid stream events can be modeled for certain user defined time spans (Jenniskens 1994; McBride 1997). In contrast to the background meteoroid flux, these streams, which are associated with trails of cometary dust, have strong

Table 2.6 Mean Time Between Collisions of a 1-m² Spherical Target with the MASTER-2005 Orbital Debris Population of Diameters Larger Than a Given Size Threshold and in Orbits Specified in Table 2.5

Orbit	$d > 0.1$ mm (days)	$d > 1$ mm (years)	$d > 1$ cm (years)	$d > 10$ cm (years)
International Space Station	29.5	1,480	191,791	3.03e+6
Envisat	2.3	170	7,267	82,440
Globalstar	3.5	287	18,942	309,597
Global positioning system	98.2	10,787	8.27e+6	1.41e+9
Geostationary transfer orbit	50.3	4,559	444,049	7.40e+6

directional properties and fixed velocities. Whereas the effect of such streams is generally minor, the related impact risk can increase by orders of magnitude during meteoroid storms because they occur at times when the source comet is passing through its perihelion. The time at which a meteoroid storms can occur and, to a lesser extent, the associated particle flux level (in terms of zenithal hourly rates of meteor counts) can nowadays be predicted sufficiently well to issue warnings to spacecraft operators. The flux peaks mostly are limited to a few hours' duration. Even in worst-case scenarios, such as the Leonids in 1966, the integrated flux from such a storm is compensated by just a few extra days of exposure to the random background meteoroid flux.

Meteoroid Impact Probability

To determine the probability of impacts by meteoroids, the same principles apply as for orbital debris. The impact probability, P, for an object of collision cross section, A_c, moving through a stationary meteoroid medium of uniform particle density, D, at a constant velocity, v, during a propagation time interval, Δt, is given by Equation (2.6):

$$P \approx vDA_c\Delta t \qquad (2.6)$$

where $F = vD$ is the impact flux in units of m^{-2}s^{-1} and $\Phi = F\Delta t$ is the impact fluency in units of m^{-2}. Engineering models directly provide the flux $F = vD$ at 1 AU from the Sun (Grün et al. 1985). Models that are more refined resolve the spatial density and velocity contributions for different meteoroid families (Divine 1993; Grün et al. 2001). Figure 2.14 depicts the meteoroid flux level at 1 AU as a function of the particle mass for the Divine model. For typical target orbits as specified in Table 2.5, Table 2.7 indicates mean times between meteoroid impacts as a function of the meteoroid sizes. These results take into account geometric shielding by the Earth (scaling factor: $0.5 < \eta_s < 1$), gravitational focusing by the Earth (scaling factor: $1 < \eta_f < 2$), and the relative velocities of the target objects (Klinkrad 2006).

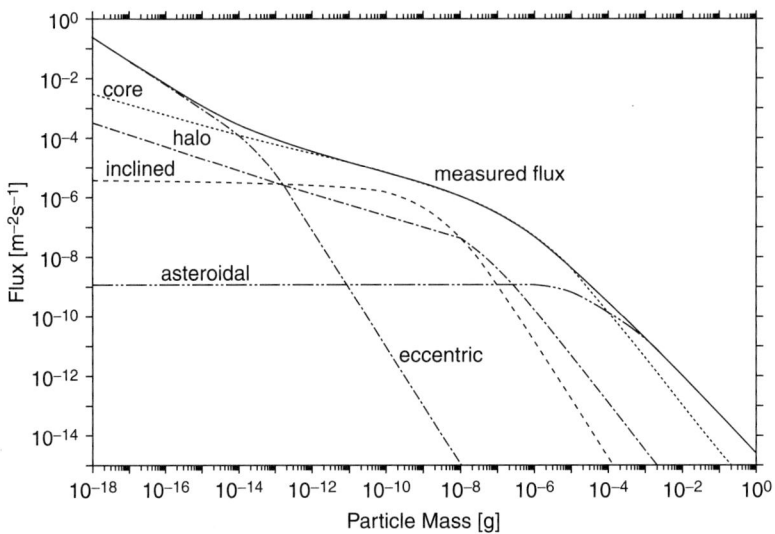

FIGURE 2.14 Meteoroid flux at a heliocentric distance of 1 AU according to the Divine model (Divine 1993; Grün et al. 2001). The curves show the cumulated flux, $F(m > m_{ref})$, from different meteoroid families and the envelope of the total cumulated flux. (Reprinted from Klinkrad 2006 with permission of Springer Science+Business Media).

Table 2.7 Mean Time Between Collision of a 1 m² Spherical Target with the MASTER-2005 Meteoroid Population (Divine 1993; Grün et al. 2001) of Diameters Larger Than a Given Size Threshold, in Orbits as Specified in Table 2.5

Orbit	$d > 1$ μm (days)	$d > 10$ μm (days)	$d > 100$ μm (days)	$d > 1$ mm (years)	$d > 1$ cm (years)
International Space Station	0.10	0.49	46.34	187.0	1.59e+6
Envisat	0.09	0.45	43.69	180.8	1.57e+6
Globalstar	0.09	0.46	43.76	183.9	1.54e+6
Global positioning system	0.12	0.60	56.50	263.9	2.38e+6
Geostationary transfer orbit	0.12	0.59	57.41	254.7	2.42e+6
Geostationary earth orbit	0.12	0.63	61.41	282.0	2.59e+6

2.3 MICROGRAVITY

Microgravity is the condition of apparent weightlessness provided in a free fall state. The body force fields experienced during free fall are complex in nature, and the instrumentation to measure the environment has required decades of development to mature. The free fall state has been used to study basic physiochemical and biological processes. Human health and performance have been studied since the first piloted missions into space, and some deleterious effects to human health caused by prolonged exposure have been identified. Because humans are expected to operate within an unusual environment, events can proceed in a nonintuitive nature, thus creating hazards where none would be anticipated when operating in an Earthlike acceleration environment.

2.3.1 Microgravity Defined

The word *microgravity* is a colloquialism. It is often used as a synonym with *weightlessness* or *zero gravity* (*zero g*). The term commonly is applied to the condition of free fall within a gravitational field, wherein the apparent weight of an object is less than that of its weight at rest on Earth. The term also is applied to a state where gravity is reduced to seemingly negligible magnitudes, literally, one millionth of the gravity magnitude at the surface of the Earth. Neither weightlessness nor zero *g* can be encountered in practice.

Gravitation

The English word *gravity* and the related word *gravitation* come from the Latin word *gravitas*, from gravis or heavy. The modern meaning of the word has been in use since the time of Sir Isaac Newton.

Newton defined the law of universal gravitation, which stipulated that each object in the universe attracts every other object with a force that is inversely proportional to the square of the distance between the two objects and that is proportional to the product of their masses as per Equation (2.7). Here, the property, mass, in question is gravitational mass by definition. The direction of this force is along the line joining the mass centers of the two objects as per Equation (2.7):

$$F_g = G \frac{m_1 m_2}{r^2} \tag{2.7}$$

The force so defined is a body force; as long as the masses and the distances between the masses remain constant, this force does not vary.

The constant of proportionality, G, is the universal gravitational constant. The value of G has been calculated to be approximately 6.7×10^{-11} Nm^2/kg^2. This formulation of universal gravitation is similar in form to Newton's laws of motion. In both, the quantity not related to the motion of the physical objects in question is mass. The apparent weight can be shown to be the gravitational force exerted on an object of defined mass. Mass is a measure of the amount of material in an object, whereas weight is the force due to gravitation affecting that object. Simplifying this formulation for an object on the surface of the planet Earth yields Equation (2.8):

$$F_g = W = G\frac{m_1 m_2}{r^2} = G\frac{M_e m_2}{r_e^2} = m_2 g_e \qquad (2.8)$$

where

$$g_e = G\frac{M_e}{r_e^2} \qquad (2.9)$$

For an object on the surface of the Earth, the value of g_e is approximately 9.8 m/s^2. The quantity g_e is defined as the acceleration due to gravity at the surface of the Earth. Although formulated here as a scalar, both acceleration and force are vector quantities, possessing both magnitude and direction. Examination of Equation (2.7) illustrates how apparent weight varies for an object located on the surface of different planets. For each planet, m_1 would be the mass of that particular planet, and different local gravitational constants can be calculated for each planet with known planetary mass. As well, gravitational force and acceleration can be calculated for interaction with extremely low mass bodies, such as asteroids.

Gravitational mass (m_2) is the parameter of the object that defines the gravitational force between that object and another (in the preceding discussion, a planet). The inertial mass of an object defines the magnitude of force required to produce an acceleration of that object. These two quantities are not related mathematically in Newtonian mechanics; however, it can be observed experimentally that the magnitudes of the two masses are identical for a given object. The relationship is explained by Einstein's theory of general relativity. Although the two masses are equivalent in magnitude, note that the forces associated with the different type masses of each object, that is, the force to accelerate an inertial mass, and the force of gravity related to proximity of objects with gravitation mass are separate phenomena. It is possible to realize one force while the value of the other force is 0. That is, in an environment with low or no gravitational force, substantial force still can be required to accelerate or decelerate a floating object possessing large inertial mass.

Free Fall in Planetary Orbit

Humans have traveled to the surface of the Moon, but the majority of human activity in space has taken place in orbit around the Earth. One obvious difference between space and Earth can be seen in video and photographs taken of astronauts and recognizable objects floating while in space. The astronauts and objects seen in these images have no or low apparent weight. However, piloted spacecraft orbit the Earth at relatively low altitude. Naumann and Herring (1980) point out that a spacecraft orbiting at 400-km altitude is only approximately 6% farther from the mass center of the Earth than an object located on the surface of the Earth. The gravitation force on an orbiting vehicle at that altitude calculated from Equation (2.8) would be only approximately 12% smaller than at the surface of the Earth. The state of weightlessness occurs instead because the spacecraft and all its contents are falling in quasicircular orbit around the Earth, that is, in a state of free fall. Orbiting the Earth at the high velocity mandated by orbital mechanics, the spacecraft and all its contents are falling free relative to the motion of the inertial mass center of the system. The objects very near the mass center of the spacecraft are orbiting the Earth in nearly the same orbit and appear to an observer on the spacecraft to float suspended with no relative motion. Einstein's principle of equivalence demonstrated that the physical behavior inside a system in free fall is identical to that inside a system extremely remote from gravitational matter.

The magnitude of acceleration imparted on an object in free fall is not zero. Accelerations can be induced from noncircular orbits or interaction with external forces from radiation pressure or residual atmospheric drag. The magnitude of these body forces can be on the order of $1 \times 10^{-6} g_e$. Body rotation induced acceleration often is calculated in two components, centripetal and tangential. For spacecraft that adopt a local vertical local horizontal attitude, that is, one in which the attitude with respect to the horizon of the Earth is fixed, the nominal rotational velocity is equal to the orbit rate in a direction perpendicular to the orbit plane.

When in orbit around a planetary body, a gradient in acceleration is caused by the difference in gravitational force experienced at a point on an object because of its spatial difference from the inertial mass center of the object and the mass center of the planetary body (Waters, Heck, and DeRyder 1988). This gravity gradient produced acceleration also is dependent on the relative rotational motion of the object in orbit around the Earth, whether the attitude of the object is constrained to follow the local vertical of the horizon of the Earth, fixed in the inertial frame of reference with respect to the center of mass of the Earth, or rotates in a different manner than either of these as per Equation (2.10):

$$\vec{a}_{gg} = \mu \left| \frac{\vec{r}_p}{r_p^3} - \frac{\vec{r}_g}{r_g^3} \right| \tag{2.10}$$

where a_{gg} is the acceleration due to the gravity gradient, μ is the gravitational constant for the Earth that is approximately equal to 3.9×10^{14} m^3/s^2, r_p is the distance from the mass center of the Earth to the point of interest on the rigid body in orbit around the Earth, and r_g is the distance from the mass center of the Earth to the mass center of the rigid body in orbit around the Earth. These positions and the resulting acceleration gradient are vector quantities. For a given orbit altitude and an assumed distance from a point of interest such as a crewmember or item of equipment, to the inertial mass center of the spacecraft, the body force generated from this gravity gradient acceleration can be calculated. Isoacceleration contour profiles can be drawn based on this calculation to define volumes above or below a specified acceleration magnitude of interest. Bodies of finite length experience a gradient of acceleration along their length. This acceleration due to gravity gradient is the major quasisteady acceleration acting at points of interest remote from the spacecraft mass center. The magnitude of gravity gradient acceleration along the line between the mass center of the planetary body and the mass center of the spacecraft can be seen to be twice the magnitude as other orthogonal directions, that is, along velocity vector or perpendicular to orbit plane. For a spacecraft in local vertical local horizontal attitude, the centripetal acceleration components produced from the spacecraft attitude inertial rotation are added to those from gravity gradient acceleration.

Time Dependence of Acceleration Body Fields

The preceding discussion is focused on accelerations idealized as time invariant. The specific accelerations are more accurately quasisteady with their magnitudes, changing only within an order of magnitude and then slowly over a period of tens of minutes. The actual acceleration environment in a spacecraft in free fall or on an interplanetary trajectory is complex and changes with time. The acceleration is three-dimensional and can have quasisteady, periodic, and random time variant components. The phenomena of periodic or random accelerations on orbit due to crew or equipment disturbances were designated as *g*-jitter when first observed on *Skylab*. These are the subject of much theoretical and experimental study because of their potential to disturb acceleration sensitive physiochemical experiments. Events can occur because of operating equipment, crew disturbances, venting of gases, reaction control jet firings to maintain attitude, delta-v maneuvers, and other sources. Disturbances from these sources are both vehicle and mission specific. There is no generic microgravity environment.

Both the measurement and data display of this complex acceleration environment have been the foci of considerable study. Depending on the sensitivity of a particular process or subject, the acceleration field components of interest can be those that are quasisteady below a specific cutoff frequency, a periodic impulse, or some random magnitude or direction. For structural environmental testing, induced environments often are specified by using power spectral density descriptions that show the relationship of power magnitude to frequency for either input or output acceleration. These usually are specified for each of three orthogonal axes or as a resultant. The microgravity acceleration environment can be specified by similar means.

Displays incorporating various combinations of magnitude, direction, frequency, and time have been created for specific systems or applications in attempts to portray the characteristics of the complex environment of interest. DeLombard et al. (1999) proposed that principal component spectral analysis and quasisteady three-dimensional histogram techniques, together, provide a satisfactory means to describe on a single plot the microgravity acceleration environment over a long period of time, such as a day, week, or month. This provides for a straightforward comparison of the microgravity environment among operating missions on various space vehicles, locations within the vehicles, or when using different operating conditions.

2.3.2 Methods of Attainment

Although obtainable during orbital or interplanetary spaceflight, free fall can be generated for short periods in certain Earth based facilities. A variety of aircraft has been modified by NASA, including KC-135s, Lear jets, the DC-9, and the C-9 to produce free fall flight profile conditions. One embodiment is to fly these aircraft repetitively in parabolic arcs that are 6 miles in length, first climbing in altitude, then falling in such a way that the flight path and speed correspond to that of an object without propulsion and not experiencing air friction. Crew and equipment float in free fall for a period of 12 to 25 s. Typically, one of these flights lasts about 3 h, during which up to 50 parabolas are flown. The European Space Agency (ESA), the Russian Space Agency (RSA), and more recently American private companies utilize similar aircraft for this purpose. Suborbital rockets also can be used to carry experiment payloads and subjects in a longer duration, higher altitude version of the aircraft parabolic arc.

Long vertical shafts into which equipment can be dropped to create a free fall condition for short periods have been utilized by NASA and other space agencies. These facilities typically are referred to as *drop towers*. A 145-m vertical vacuum environment tube into which an experiment package can fall for up to 5.2 s and for a distance of 132 m is operated by the NASA Glenn Research Center. At the bottom of the tube, the experiment package is decelerated to a stop by a bed of polystyrene pellets, where they experience a peak deceleration rate of 65 g_e. Also at the Glenn Research Center is a second drop tower that is 24 m tall and provides 2.2 s of free fall without vacuum. Additionally, the NASA Marshall Space Flight Center utilizes a drop tube facility that is 105 m tall to provide 4.6 s of free fall under near vacuum conditions for a molten material sample. Humans cannot experience free fall in these ground based drop towers because of the high deceleration experienced at the conclusion of the drop.

Another type of apparent weightlessness can be simulated by the use of neutral buoyancy. Unlike free fall conditions, human subjects and experiment apparatus are submerged in water and weighted until they hover, neither rising nor sinking, at a predefined location. Neutral buoyancy produces suspension of the object in the water medium by a balance of forces, that is, weight balanced against buoyancy, acting on the surface of the floated object. As such, the same body acceleration field (or body force field) as free fall is not produced; however, an object or person is suspended above the nominal floor. Neutral buoyancy is used to provide simulated conditions for astronaut training for extra-vehicular activity, and the process can be used to suspend large objects for manipulation by astronauts wearing their spacesuits. Although neutral buoyancy effectively simulates weightless conditions, movement of large objects is affected by viscous forces generated in the suspension medium.

Instrumentation Used to Measure the Acceleration Environment

Accelerometers were flown originally on space launch vehicles to quantify launch loads and vehicle staging accelerations. As biology, physics, and chemistry experiments began to be conducted in space, accelerometer packages were customized with filtering and signal conditioning to detect low magnitude and low frequency accelerations. Each experimenter constructed a unique transducer and data acquisition system for use on a particular mission. Hamacher (1996) reviewed the microgravity environment of *Spacelab* as measured by acceleration measurement apparatus developed by the ESA.

In 1986, NASA initiated a project to acquire a standardized accelerometer system, the space acceleration measurement system, to support research experiments on microgravity platforms. The space acceleration measurement system was developed as a multiple mission, multiple sensor apparatus to serve the needs of microgravity research community.

The first flight of a space acceleration measurement system was on STS-40 during 1991. Since then, seven space acceleration measurement systems have flown on over 20 Space Shuttle missions in the *Spacelab* modules, in the middeck, on the multipurpose experiment support structure cargo bay carriers, and in the *Spacehab* modules. One system was modified for operation on *Mir*, and spent 3½ years there as part of the NASA-*Mir* science program. The longest residing U.S. apparatus onboard, it was returned to Earth before the *Mir* was allowed to deorbit. Data from the Space Shuttle atmospheric research accelerometer systems, the orbital acceleration research experiment from the NASA Langley Research Center, and the high resolution accelerometer package from the NASA Johnson Space Center were integrated analytically with the space acceleration measurement system data to provide a full spectrum of acceleration data. The orbital acceleration research experiment sensor subsystem is used to measure low range frequencies up to 1 Hz. The high resolution accelerometer package is used to characterize the International Space Station vibratory environment from 0.01 to 100 Hz. Both of these packages subsequently were integrated into a complementary system for the *International Space Station*.

The NASA Marshall Space Flight Center acceleration characterization and analysis project was transferred to the NASA Glenn Research Center in early 1990 and restructured as the principal investigator microgravity services project. Data analysis, distribution, and archival functions were consolidated with the instrumentation development organization at Glenn Research Center.

After development of the space acceleration measurement system, individual investigators continued to develop customized acceleration data acquisition systems. These developments were performed to provide an accelerometer having specific frequency response characteristics to fit within an unusual enclosure, such as the getaway special canister, or to acquire the expertise to develop these systems and interpret the resulting signals and data. For example, the German Aerospace Research Agency (DLR) in 1977 developed a quasisteady acceleration measurement system for *Spacelab* missions to detect steady, very low frequency, residual accelerations between 0 and 0.02 Hz. To achieve this, the measurement signal was modulated by rotating a sensor about its sensitive axis. The system employs four rotating sensors to allow three-dimensional acceleration detection.

An updated space acceleration measurement system was developed for the *International Space Station*. The space acceleration measurement system-II (SAMS-II) measures vibrations from vehicle acceleration, systems operations, crew movements, and thermal expansion and contraction. Together, the SAMS-II and the quasisteady microgravity acceleration measurement system collect acceleration data from the *International Space Station*. Those data are then published in near real time on the Internet and in reports. Much of this data has been analyzed by investigators, International Space Station systems organizations, Russian investigators, and other international investigators. Over 50 GB of data were collected from over 20 Space Shuttle flights and from *Mir*, and several terabytes of data have been collected from the *International Space Station* through early 2007 as well.

Several SAMS-II remote triaxial sensor systems are used to monitor individual research apparatus requiring direct monitoring. Each remote triaxial sensor is designed to measure g-jitter at frequencies between 0.01 Hz and beyond 300 Hz. The remote triaxial sensor enclosure is placed as close to the experiment hardware as possible for the purpose of converting vibratory acceleration information into a digital signal. The companion remote triaxial sensor electronic enclosure provides power and command signals for up to two remote triaxial sensor enclosures and receives vibration data from the sensors.

The *International Space Station* based microgravity acceleration measurement system consists of a low frequency triaxial accelerometer, the miniature electrostatic accelerometer, a high frequency accelerometer, the high resolution accelerometer package, and associated computer, power, and signal processing subsystems in the *U.S. Laboratory Module, Destiny*. The microgravity acceleration measurement system is operated only during special events, such as an International Space Station orbit altitude reboost, or a *Progress, Soyuz*, or Space Shuttle docking and undocking. The microgravity acceleration measurement system records low frequency accelerations. These forces are related to aerodynamic drag, gravity gradient and rotational effects, venting of gas or water, and mechanical movement, such as that of the solar photovoltaic arrays and the communications antennas. During measurements taken on the *International Space Station*, signatures were uncovered in the data whose origins were not obvious. Movement of the International Space Station Ku-band antenna eventually was identified as the source of the unusual characteristics in the quasisteady data collected by the microgravity acceleration measurement system (Del Basso et al. 2002).

NASA Research Related to Microgravity

Early in the design of liquid fuel rockets, NASA sponsored both fundamental and engineering development investigations to understand fluid physics and dynamics and fluid management in free fall conditions. Fluid fuel and oxidizer must be moved to rocket engines while an upper stage is falling free after staging or while it is in a parking orbit. From 1969 to 2004, NASA sponsored progressively sophisticated scientific investigations related to physical and chemical phenomena in microgravity. Initial experiments to test concepts for the materials processing in space program on *Apollo 14, Skylab*, and the *Apollo-Soyuz Test Project* were carried out in parallel with theory driven experiments by the physics and chemistry in space effort. These investigations, which were combined organizationally in 1986, conducted experiments that tested theories of the behavior of solids, liquids, and gases while in free fall and under a variety of temperature and pressures.

One focus of this research was the study of single phase, single crystal growth processes, both as a study of the kinetics and transport processes and for the potential to produce unique benchmark materials in free fall. One example of crystal growth studies was the directional solidification of bulk mercury-cadmium-telluride solid solution. This process requires that the temperature gradient to crystal growth rate ratio be high to avoid constitutional supercooling. Any fluid flow of the melt with a velocity comparable to or higher than the very low crystal growth rates employed (0.2 μm/s) would dominate the solidification characteristics, particularly the compositional distribution in an alloy such as mercury-cadmium-telluride. Gilles et al. (1997) found that different Space Shuttle attitudes during the mission produced differences in the resultant residual acceleration vector in both magnitude and direction, and these differences caused large compositional variations across the radii of the crystal and along its surface. The differences in the resultant residual acceleration vector were due to Space Shuttle attitude changes in the quasisteady acceleration field produced from drag, centripetal acceleration, and gravity gradient accelerations.

In many early physical sciences investigations, almost as enlightening as the experiments themselves were data gathered about the performance of the experiment apparatus during extended operation in free fall. Compared to the conditions of Earth based preflight tests, a great reduction in buoyancy driven convection occurred during spaceflight that affected conditions for heat transfer. This resulted in the failure of several pieces of the early apparatus because of local heating and expansion of precision fit mechanisms. As well, it was found that metal particles, which settle in avionics assemblies during manufacture, float in free fall and migrate in the direction of the air cooling return vent. Alternately, these particles would migrate into electronic assemblies and cause failures in avionics operation. Scientific investigations on *Spacelab* demonstrated that free falling dust of various particulate sizes agglomerate, based on their charge state. Detailed reports of these experiments and the phenomena just described are preserved in the peer reviewed archival literature, the NASA post-mission reports, the microgravity research experiment (MICREX) database maintained by NASA, and the annual program tasks and bibliography book. The last document contains reports of all peer reviewed projects funded by NASA biological and physical sciences programs until 2004. Internal research conducted at NASA field centers and research supported by the National Space Biomedical Research Institute are included in this book as well (NASA 2007).

Although modifications to heat transfer behavior in free fall should invite caution in designing systems, it was found that combustion process behavior was of immediate concern. Combustion scientists at NASA summarized the key findings from experiments on fire and combustion conducted in microgravity from 1981 to 2000 (Friedman and Urban 2000). From these data, it was concluded that ignition is promoted in free fall microgravity conditions as the result of thermally stressed components overheating rapidly, particulate spills forming flammable aerosols that persist for long periods, and burning plastics ejecting hot material randomly and violently. Flame appearance is altered. In quiescent environments, flames often assume a symmetrical shape and become nearly invisible in ambient light. Under low rates of imposed airflow, such as from the environmental control fans, the flames can intensify, becoming bright and sooty. Flammability and flame spread rate were studied. Quiescent conditions increased in some cases to match or exceed those when in normal gravity. It was found that low rate ventilating flows stimulate low gravity fires and greatly extend their flammability range and flame spread rates. Freely propagating flames tend to spread toward the wind or into the oxygen source. Detection signatures also are altered. In free fall, flames often are cooler and less radiant, and the average size and range of soot particles is greater. The nature and quantities of combustion products are altered as well.

During February 1997, a problem with an oxygen generating device located in the Kvant-1 module on the Mir space station generated dense smoke and flame and caused damage to some hardware. A chemical oxygen generator is a device that releases oxygen by means of a chemical reaction. The oxygen generating lithium perchlorate candles usually burn from 5 to 20 min. When more than three people were onboard the Mir space station, these candles were burned to generate supplemental oxygen. During the exchange of an air filter, a failed chemical oxygen generator ejected sparks and a jet of molten metal across the *Kvant-1*. The defective oxygen candle burned for 14 min and blocked the escape route to one of the Soyuz spacecraft. The accident was caused by leaking lithium perchlorate. Damage to some of the Mir hardware was caused by exposure to excessive heat rather than from open flame. The heat destroyed the hardware in which the device was burning, as well as the panel covering the device. The crew also reported that the outer insulation layers of various cables had melted. Because the vehicle was in free fall, smoke fully filled the volume of the Kvant-1 module rather than rising to the ceiling as buoyancy would have caused it to do in Earth gravity. It should be noted that the chemical oxygen generator was attached to a structural support. Had the generator been floating freely, it could have contacted the wall of the spacecraft where the flame and high temperature would have caused the pressure vessel wall to rupture.

Flames consume solid or liquid fuels, create gaseous combustion products, and produce solid materials. The solid byproducts of combustion are called *soot*. The mechanisms by which flames create soot are among the most important unresolved problems of combustion science. The heat that one feels from a fire is due principally to soot. In addition to contributing to pollution, soot enhances the emission of other pollutants from flames, such as carbon monoxide; causes undesirable radiative heat loads to combustion chambers; enhances the spread of unwanted fires due to radiant emission; and hampers firefighting efforts by obscuring flames. Gerard Faeth of the University of Michigan was the principal investigator for the laminar soot processes microgravity flight experiment conducted to help improve the understanding of soot processes in flames. The experiment was conducted in the combustion module that flew on the STS-83 and STS-94 missions of *Space Shuttle Columbia* in 1997 and mission STS-107 in 2003.

The comparative soot diagnostics experiment flew aboard STS-75 in February 1996. This experiment was designed to analyze smoke produced from several sources and evaluate the responsiveness of the smoke detection systems used on the *International Space Station* and on the Space Shuttle fleet. Among the combustible sources tested were wire coated with Teflon® and Kapton®, paper, and silicone rubber.

The comparative soot diagnostics experiment showed that smoke produced in low gravity is different from that produced in normal gravity. In microgravity, smoke particles are larger. Because the smoke detectors used by NASA are designed to detect smoke particles in particular size ranges, they respond differently when used in the microgravity environment than they do on Earth. This suggests that the level of fire protection on a spacecraft is also different from what was believed previously.

The smoke detectors used in the *International Space Station* apply photoelectric technology, whereas those used in the *Space Shuttle* are ionization detectors. In general, the light scattering laser detector used on the *International Space Station* detected smoke during the majority of tests. In other tests, the ionization detector was observed to be less sensitive to smoke. Specifically, the International Space Station detector performed much better in its ability to recognize Kapton® smoke in microgravity, and the Space Shuttle detector performed less well in recognizing paper and silicone rubber smoke in low gravity. Because both detectors responded equally well to the smoke from these sources in preflight normal gravity tests, it appears that larger smoke particle sizes are produced in microgravity. This is because smoke particles spend more time in smoke production regions of the combustion event when in microgravity.

Once the comparative soot diagnostics experiment demonstrated that microgravity affects smoke particle size, the phenomenon had to be quantified precisely so that better detectors could be designed and built for spacecraft. The International Space Station smoke aerosol measurement experiment was developed for this purpose, and it tests how well the smoke detectors perform when exposed to the smoke particles produced by material commonly found on a spacecraft, such as Teflon®, silicon, and cellulose. As well, the smoke aerosol measurement experiment provides an evaluation of the smoke detectors used on the *Space Shuttle* and the *International Space Station* and identifies any design options for their improvement.

2.3.3 Effects on Biological Processes and Astronaut Health

Exposure to weightlessness is demonstrated to have some deleterious effects to human health and performance. In addition, cells and nonhuman organisms have exhibited similar or additional effects. Humans are adapted well to the physical conditions prevailing at the surface of the Earth, including the body acceleration due to gravity, g_e. When weightless, certain physiological systems begin to alter, and both temporary and long-term health issues occur.

The most common initial condition experienced by humans after several hours of exposure to free fall acceleration conditions is the space adaptation syndrome, commonly known as *space motion sickness*. Its symptoms include general nausea, headaches, vertigo, vomiting, and an overall lethargy. The duration of space adaptation syndrome varies, but usually it is not more than 72 h. The first case was reported by Russian cosmonaut Gherman Titov in 1961, who piloted the Vostok spacecraft. Later, it was reported by American astronauts flying in the Apollo Command Module. Since then, roughly 45% of all people to experience long-term free fall have suffered from this condition. Early in the Space Shuttle Program, missions to deploy satellites were envisioned as being of relatively short duration, that is, less than 7 d. For these missions, space adaptation syndrome became a concern in that, for much of the mission, the crew would operate at less than peak performance. Later, as the *Space Shuttle* was refocused on longer duration missions and space adaptation syndrome became better understood and partially effective countermeasures were developed, this concern receded.

Skeletal decalcification and muscle atrophy are the most important adverse effects of long-term free fall exposure. Countermeasures investigated on long duration Skylab, Salyut, and Mir missions indicate that the rate of their cumulative effects can be minimized through a regimen of exercise. Other major physiological effects include fluid redistribution, weakening of the cardiovascular system, decreased production of red blood cells, balance disorders, and diminished effectiveness of the immune system. Lesser symptoms include loss of body mass, nasal congestion, sleep disturbance, excess flatulence, and puffiness of the face. These effects are reversible on return to Earth. Many of the conditions caused by exposure to weightlessness are similar to those resulting from aging.

Because a spacecraft is a closed environment, spilled fluid that makes contact with a surface spreads across that surface, given appropriate wetting behavior and surface energy. Fluids can migrate behind equipment racks and containers, making cleaning difficult. Exercising astronauts routinely build up a coating of sweat because the fluid does not run off or drip from the body, and the standard pressure and temperature conditions of the cabin do not encourage quick evaporation. For the fluid that does evaporate, it slowly diffuses away unless the astronaut is located in an airflow pathway.

As stated in the first subsection, substantial force to accelerate or decelerate a floating object with large inertial mass is required in the free fall environment, just as it is on Earth. Astronauts in the shirtsleeve environment of a spacecraft often must move massive objects, such as equipment racks or lockers. During an extravehicular activity, suited crewmembers must position and restrain themselves adequately before moving large items of equipment; otherwise, the astronaut rather than the desired item will translate or rotate, or the motion of the item will be affected in some unexpected way. Translation and rotation of massive objects are hazardous operations; therefore, these activities require diligent planning and careful execution.

2.3.4 Unique Aspects of Travel to the Moon and Planetary Bodies

The in-space microgravity environment experienced since the end of the Apollo program in 1972 is the free fall condition of low Earth orbit. As humans travel back to the Moon and on to other planetary bodies, they will experience transfer orbit trajectories similar to that of the Apollo expeditions and for greatly increased durations. While still a free fall environment, the increasing separation distance from the Earth changes the acceleration environment. This directly changes the acceleration gradients as the spacecraft is increasingly distant from Earth, and the acceleration gradients have different magnitudes based on the local gravitational field of the target body. The research conducted in Earth orbit since 1972 has been conducted in the acceleration gradients produced by an Earth orbiting spacecraft at relatively low altitude. The general phenomena observed for a half century holds true, but new environments can elucidate new, subtle effects. The effects of small, complex accelerations should be examined in advanced materials systems as they are used in future aerospace vehicles.

RECOMMENDED READING

Alexander, J. I. D. (1995) *Passive Accelerometer System: Measurements on STS-50 (USML-1)*. Huntsville, AL: The University of Alabama at Huntsville, Center of Microgravity and Materials Research.

Clement, G. (2003) *Fundamentals of Space Medicine*. Norwell, MA: Kluwer Academic Publishers.

DeLombard, R. (2000) Disturbance of the microgravity environment by experiments. *Proceedings of the AIP Space Technology and Applications Conference* 504, no. 1: 614-618.

DeLombard, R., K. Hrovat, E. M. Kelly, and B. Humphreys. (2005) Interpreting the International Space Station microgravity environment. AIAA 2005-0727. *Proceedings of the 43rd AIAA Aerospace Sciences Meeting and Exhibit*. Reno: American Institute of Aeronautics and Astronautics.

DeLombard, R., K. Hrovat, M. E. Moskowitz, and K. M. McPherson. (1998) Comparison tools for assessing the microgravity environment of space missions, carriers, and conditions. *Proceedings of the Visual Information Processing Conference 7*, International Society for Optical Engineering, 3387: 394-402.

DeLombard, R., E. M. Kelly, K. Hrovat, E. S. Nelson, and D. R. Pettit. (2005) Motion of air bubbles in water subjected to microgravity accelerations. AIAA 2005-722. *Proceedings of the 43rd AIAA Aerospace Sciences Meeting and Exhibit*. Reno: American Institute of Aeronautics and Astronautics.

Fox, J. C., J. E. Rice, J. P. Ebling, and W. O. Wagar. (1998) The International Space Station microgravity acceleration measurement system. AIAA 98-0459. 37th AIAA Aerospace Sciences Meeting and Exhibit. Reston, VA: American Institute of Aeronautics and Astronautics, pp. 1-10.

Jules, K., K. Hrovat, and E. Kelly. (2002) *International Space Station Increment-2: Quick Look Report*. NASA Technical Memorandum TM-2002-211200. Cleveland: National Aeronautics and Space Administration, Glenn Research Center.

Jules, K., K. Hrovat, E. Kelly, K. McPherson, and T. Reckart. (2002) *International Space Station Increment-2: Microgravity Environment Summary Report*. NASA Technical Memorandum TM-2002-211335. Cleveland: National Aeronautics and Space Administration, Glenn Research Center.

Jules, K., K. Hrovat, E. Kelly, K. McPherson, T. Reckart, and C. Grodsinksy. (2002) *International Space Station Increment-3: Microgravity Environment Summary Report*. NASA Technical Memorandum TM-2002-211693. Cleveland, OH: National Aeronautics and Space Administration, Glenn Research Center.

Jules, K., K. Hrovat, E. Kelly, and T. Reckart. (2006) *International Space Station Increment 6/8: Microgravity Environment Summary Report*. NASA Technical Memorandum TM-2006-213896. Cleveland: National Aeronautics and Space Administration, Glenn Research Center.

Kaldis, E. (1989) *Microgravity*. London: Elsevier.

Monti, R., and R. Savino. (1996) Microgravity experiment acceleration tolerability on space orbiting laboratories. *Journal of Spacecraft and Rockets* 33, no. 5: 707-716.

Naumann, R. (1999) An analytical model for transport from quasi-steady and periodic accelerations on spacecraft. AIAA 99-1028. 37th AIAA Aerospace Sciences Meeting and Exhibit. Reston, VA: American Institute of Aeronautics and Astronautics, pp. 1-8.

Prisk, G. K., and M. Paiva (eds.). (2001) *Gravity and the Lung: Lessons from Microgravity*. New York: Marcel Dekker.

Rogers, M. J. B., J. I. D. Alexander, and J. Schoess. (1993) Detailed analysis of Honeywell in-space accelerometer data—STS-32. *Journal of Microgravity Science and Technology* 6, no. 1: 28-33.

Ross, H. D. (ed.). (2001) *Microgravity Combustion: Fire in Free Fall*. Princeton, NJ: Academic Press.

Ruff, G. A., D. L. Urban, and M. K. King. (2005) A research plan for fire prevention, detection, and suppression in crewed exploration systems. AIAA 2005-341. *Proceedings of the 43rd AIAA Aerospace Sciences Meeting and Exhibit*. Reno: American Institute of Aeronautics and Astronautics.

Urban, D., D. Griffin, G. Ruff, T. Cleary, J. Yang, G. Mulholland, and Z. Yuan. (2005) Detection of smoke from microgravity fires. SAE 2005-01-2930. *Proceedings of the International Conference on Environmental Systems*. Warrendale, PA: Society of Automotive Engineers.

2.4 ACOUSTICS

2.4.1 Acoustics Safety Issues

The acoustics environment in space operations is important to maintain at manageable levels so that the crew can remain safe, functional, effective, and reasonably comfortable. High acoustic levels can produce temporary or permanent hearing loss or cause other physiological symptoms, such as auditory pain, headaches, discomfort, strain in the vocal cords, or fatigue.

Noise is defined as undesirable sound. Excessive noise can result in psychological effects, such as irritability, inability to concentrate, decrease in productivity, annoyance, errors in judgment, and distraction. A noisy environment also can result in the inability to sleep or sleep well. Elevated noise levels can affect the ability to communicate, understand what is being said, and hear what is going on in the environment; it can degrade crew performance and operations and create habitability concerns. Superfluous noise emissions also can create the inability to hear alarms or other important auditory cues, such as the sound of an equipment malfunction. Recent spaceflight experience, evaluation of the requirements in crew habitable areas, and lessons learned (Allen and Goodman 2003; Goodman 2003; Grosveld, Goodman, and Pilkinton 2003; Pilkinton 2003) show the importance of maintaining an acceptable acoustics environment. This is best accomplished by having a high quality set of limits and requirements early in the program, that is, the designing-in of acoustics in the development of hardware and systems, and the monitoring, testing, and verifying sound levels to ensure that they are acceptable.

2.4.2 Acoustic Requirements

Requirements are a key pillar to successful design and need to be as well defined and as clear as possible at the beginning of a program. To be successful in meeting the requirements, acoustics needs to be treated as a technical specialization on par with other design disciplines, and experienced and knowledgeable personnel need to be assigned to implement the defined requirements. The following factors should be considered when tailoring requirements to meet a specific application:

- Type of mission.
- Mission duration.
- Number and characteristics of crew occupants.
- Size, function, number, and type of hardware systems that make up the crewed vehicle, module, or enclosure and the supplementary hardware, such as payloads and supplementary government furnished equipment.
- Whether single or dual shift operations are to be used.
- Distance between crewmembers required for good communications.
- Quality of the communications needed.

All requirements presented in this section apply throughout the crew habitable volume. Separate acoustic restrictions need to be applied to areas that are outside of this volume but which can be accessed for short-term use during equipment change out or maintenance. Special consideration should be given to the acoustic levels allowed in the habitable volume should such access require leaving open access doors, panels, or other means for sound to enter. The terms applied to the habitable volume in a crewed spacecraft, module, or other type of crewed enclosure used in space are the crew compartment or the habitat. Use of design goals instead of firm requirements is not recommended, because they set the stage for efforts that are essentially "do what you can do" and imply that efforts should be limited to those objectives that readily can be met or can be interpreted thusly. Some important acoustic safety requirements currently employed by NASA and its international partners in crewed spacecraft applications are discussed in the following subsections.

Continuous Noise

Spaceflight missions typically range in duration from several days to many months. Special requirements are needed to administer safely the 24 h per d, 7 d per week exposure to noise in space vehicle environments. Noise sources operating for more than 8 h in any 24-h period are classified as those producing continuous noise. In 1972, NASA adopted noise criteria (NC) curves as the acoustic noise criteria standard used to manage continuous noise in crewed spacecraft (NASA 1972). The noise criteria curves specify the octave band limits of the acceptable noise levels in habitable environments while all systems are operating. As well, they commonly are used in industry for defining the ratings used for control of ambient noise in buildings. The acoustic environment, with the integrated government furnished equipment as part of the habitable space, is limited by the NC-50 curve shown in Figure 2.15. These curves are extrapolated to include the 16-kHz octave band to cover the audible range better at the higher frequencies.

An appropriate limit or suballocation must be applied to the composite of other noteworthy hardware located within the crew compartment or habitat that is not required for the basic functioning of the spacecraft, module, or enclosure systems. In the past, this category included such items as payloads, nonintegrated government furnished equipment, experiments, cargo, or other classifications of hardware. If these payloads and other types of hardware, together, amount to a considerable acoustics contribution, as have the payload racks in the *International Space Station*, then a suballocation for them as a complement should be made and it should be restricted to NC-48 in consonance with the limits applied to the integrated system and crew compartment or habitat (NASA 2003). Each individual rack equivalent item should meet the NC-40 per Figure 2.15, or lower, as applicable (NASA 2000). Appropriate suballocations need to be given to hardware components that make up payload rack type hardware to ensure that the rack limit is controlled. This is especially true if rack makeup or components are changed out during the operational life of the rack or hardware. An individual hardware item that is of lower complexity than a payload rack and similar to an item of hardware mounted in the aisle should either fit into the complement total limit or meet a lower limit itself. The continuous acoustic levels for the integrated systems affecting the crew compartment or habitat, including the noise from supplementary hardware, such as payloads, nonintegrated government furnished equipment,

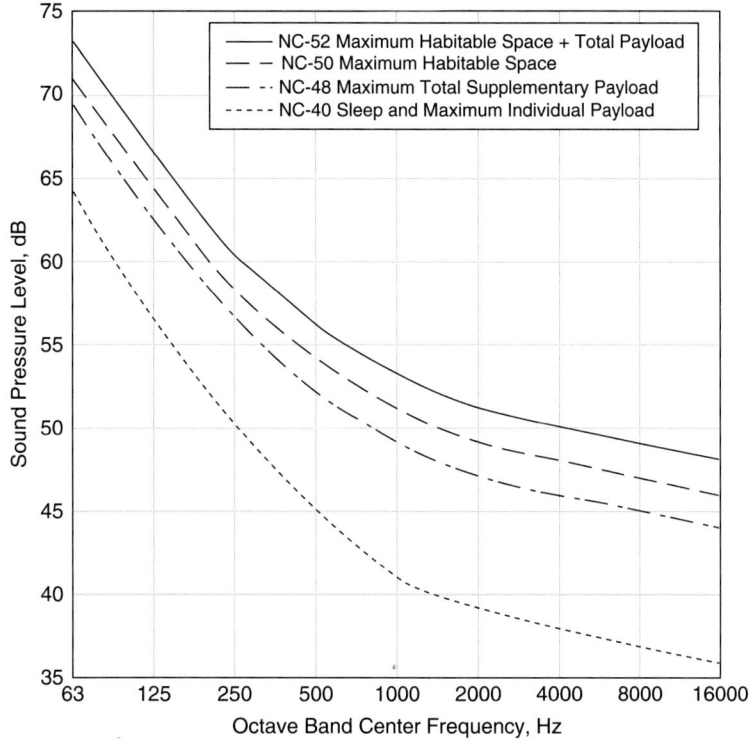

FIGURE 2.15 Extended continuous noise criteria specifications (Courtesy of NASA).

or other classifications, is then limited to the NC-50 + NC-48, or the approximate NC-52 level shown in Figure 2.15. If the supplementary hardware is considerably less complex in nature than the International Space Station payload rack hardware and it does not merit the NC-48 level allotment, the total noise in the crew compartment or habitat should be at the NC-50 rather than the NC-52 level. The NC-50 specification, which is preferred over the NC-52 level because it provides for improved quality of communications and word intelligibility, is recommended for crewed spacecraft in general, such as the *Space Shuttle* and the *International Space Station* (NASA 1972; Pearsons 1975; Piland 1980; CHABA 1987). Figure 2.16 shows the quality of face-to-face communications expected for vocal effort and separation distance in terms of preferred speech interference level and dBA levels (NASA 1995).

The preferred speech interference level was established to determine the effect of continuous background noise on speech communications in a work environment. It is defined as the arithmetic average of the sound pressure levels in the 500-Hz, 1000-Hz, and 2000-Hz octave bands. Figure 2.16 also shows the same evaluation against an A-weighted background noise level, which uses one third octave band dependent A-weighting to better evaluate subjectively the response of the human ear. Figure 2.17

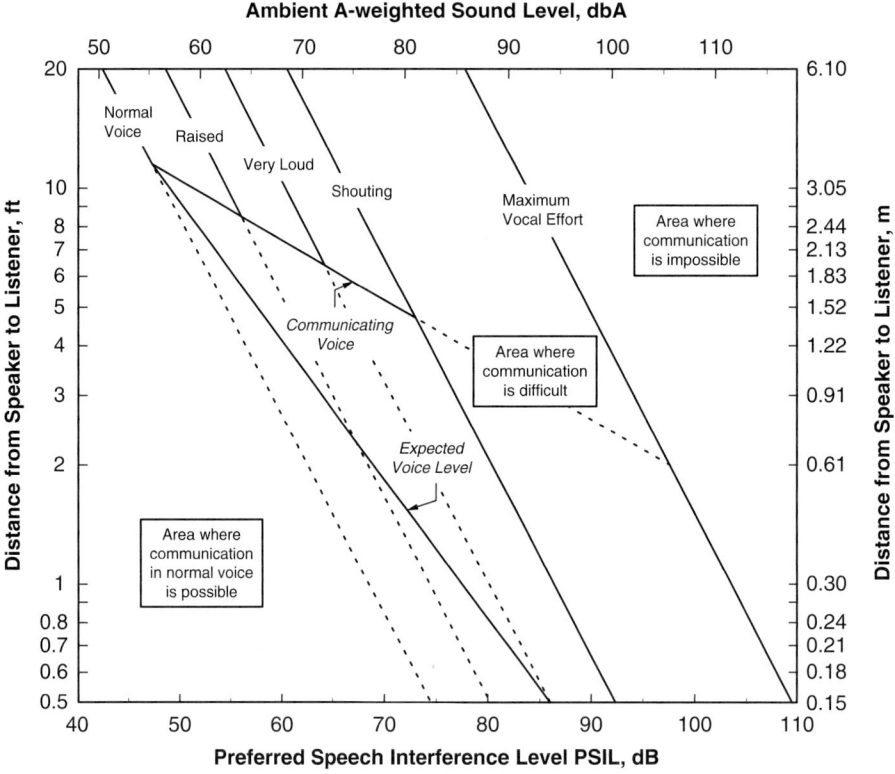

FIGURE 2.16 Effectiveness of voice communications as a function of the preferred speech interference level or the dBA noise level and the distance from speaker to listener (Courtesy of NASA).

shows the percent intelligibility levels plotted versus the noise criteria ratings (dBA levels) for crew to crew communication distances from 5 to 8 ft. Improvement in intelligibility is shown by the use of NC-50 versus the NC-52 rating (Pearsons 1975). The minimum percentage of intelligibility is recommended by NASA to be 75% for the satisfactory communication of most messages (Figure 2.17). An intelligibility of 95% is recommended for sentences spoken under normal vocal effort with the talker and listener being visible to each other (CHABA 1987). Note that the data used for this curve are based on communications between males conversing in the English language and does not take into account female voices or foreign dialects.

The crew needs a reasonable limit for the acoustic levels present during their sleep periods so that they can obtain necessary rest and recover from any high noise exposure during their activity periods. Where the crew compartment or habitat design permits, the sleeping area should be an accommodation separate from areas of work and higher noise. The crew sleeping area should not exceed NC-40, as shown in Figure 2.15 (NASA 1972,

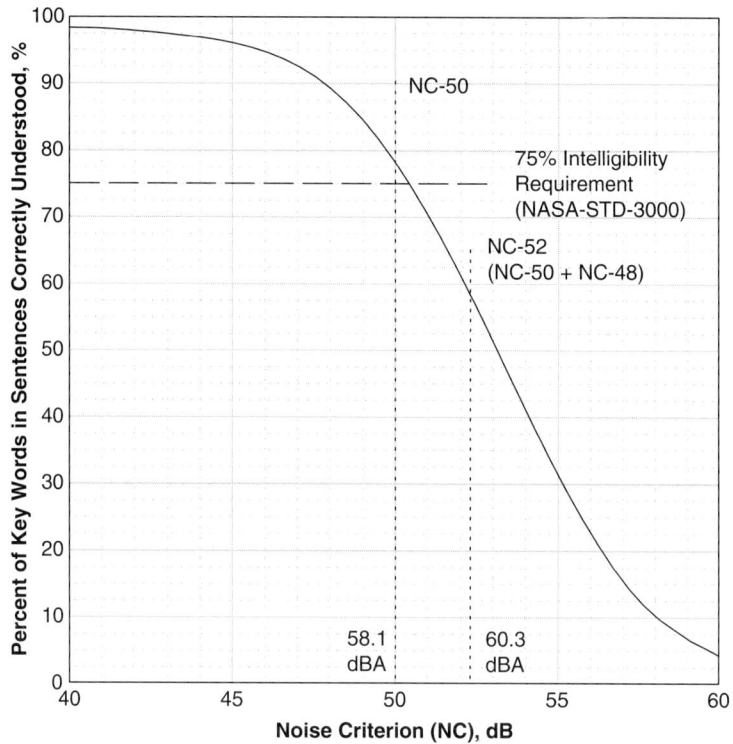

FIGURE 2.17 Percent intelligibility level versus the noise criteria rating for crew to crew communication at distances from 5 to 8 ft (Courtesy of NASA).

1995). To preclude awakening of sleeping crewmembers, impulse or transient noises in the sleeping area should be limited to less than 10 dB above the background noise (NASA 1972, 1995).

Intermittent Noise

Intermittently, that is, less than 8 h in any 24-h period, operating hardware can be very disturbing, wake crewmembers, and interfere with sleep or nominal operations. Supplementary hardware, such as that found in the payload rack classification, should limit intermittent A-weighted acoustic emissions to the levels and durations defined in Table 2.8 with measurements taken 0.6 m from the loudest point on the hardware (NASA 2000).

Some crew compartment hardware, such as toilets, pressurized gas systems, or other stand-alone hardware of acoustic importance, should be controlled similarly. Most exercise equipment, for example, treadmills and ergometers, can be difficult to control to these limits, and depending on crew size and the like, they can produce loud acoustic levels over time. It is suggested that, if possible, the exercise area be allotted separate quarters from other habitable areas in the crew compartment or habitat.

Table 2.8 Intermittent A-weighted Overall Sound Pressure Levels and Corresponding Operational Limits for Supplementary Hardware, Such as Rack Mounted Payload Hardware and Nonintegrated Government Furnished Equipment (Courtesy of NASA)

Maximum Noise Duration (per 24 h)	A-weighted Overall Sound Pressure Level (dBA)
8 h	49
7 h	50
6 h	51
5 h	52
4.5 h	53
4 h	54
3.5 h	55
3 h	57
2.5 h	58
2 h	60
1.5 h	62
1 h	65
30 min	69
15 min	72
5 min	76
2 min	78
1 min	79
Not allowed	80

Narrowband Components

A narrowband component is a simple or complex tone or a line spectrum having intense and steady state frequency components in a very narrow band, that is, 1% of an octave band or 5 Hz, whichever is less, and is heard as a musical sound, either harmonic or discordant. The maximum sound pressure level of any narrowband component should be at least 10 dB less than the sound pressure level of the octave band that contains the component (NASA 1972, 1995).

Ultrasound and Infrasound

Ultrasound is high frequency sound, above 15 to 20 kHz, that is inaudible to the human ear. Ultrasonic sound can have physiological effects on humans, and it should be addressed as part of the acoustic environment. It is thought, however, that pertinent concerns regarding ultrasound should be focused on direct body contact and any audible noise associated with the subharmonics of the hardware that produces it. Ultrasonic noise can be generated by electrical converters, battery chargers, and other types of equipment. Two concerns are of importance when dealing with this type of noise:

- It is difficult and costly to predict whether the hardware produces levels in the crew compartment or habitat sufficient to be of concern or that exceed defined limits.
- The hardware required to measure ultrasonic emissions commonly is not available or used.

Use of the extended noise criteria curves to 16 kHz (Figure 2.15) helps to understand most subharmonic effects in the audible range, but it is recommended that some screening be used to determine if the resultant ultrasonic levels in the crew compartment or habitat are of concern or exceed any recommended threshold limit value, as shown in Table 2.9 (ACGIH 2004).

Infrasound constitutes acoustic emission below the audible range of human hearing. Infrasound in the crew compartment or habitat should be limited to 120 dB within the frequency range of 1 to 16 Hz for a 24-h exposure (NASA 1995).

Hazardous Overall Noise Limits

Excessively loud overall noise levels can cause harm to the hearing abilities of crewmembers and should be limited. The continuous noise level during flight in the integrated

Table 2.9 Threshold Limit Value for Ultrasonic Sound in Air (Courtesy of NASA)

One-Third Octave Band Center Frequency (kHz)	Ceiling Values (dB)	8-h Time Weighted Average (dB)
10	105	89
12.5	105	89
16	105	92
20	105	94
25	110	—
31.5	115	—
40	115	—
50	115	—
63	115	—
80	115	—
100	115	—

crew compartment or habitat is limited to a maximum of 85 dBA at the crewmember's ear (NASA 1995). Noise from hardware associated with cabin depressurization, repressurization, or similar activities should be limited to 105 dBA at the crewmember's ear during these types of operations (NASA 2006). If such activities recur very often, then limits on the noise dose should be considered.

Reverberation Time

Reverberation time is the time required for the energy density in an acoustic field to reduce to a level 60 dB below its steady state value. Reverberation time has a pronounced effect on speech intelligibility. Because it is an important criterion for conversational speech, the reverberation time should be adjusted to the volume of the crew compartment or habitat.

Alarms

Alarm signals used within the crew compartment or habitat should be heard readily and easily be discernible by crewmembers when working or sleeping. Signals from local loudspeakers or that emanate from other locations within a spacecraft, such as adjacent crew compartments or modules, should possess a sufficient signal-to-noise ratio to be heard over the local background noise.

2.4.3 Compliance and Verification

It is intended that acoustics requirements and limits be met without the attenuation afforded by hearing protection, communication headsets, or other coverings, except during launch, entry, burn, or other short-term limited phases of a mission. An example of a limited phase would be that which occurs during cabin depressurization. Meeting the acoustics limits ensures a safe and habitable environment and precludes the use of the hearing protection and other means noted from being imposed upon the crew and their subsequent reliance on it rather than using the actual design implementation for protection.

Frequently, acoustic requirements at the beginning of a program are challenged. They typically are regarded as too strict and are considered to lead to unacceptable impacts. However, the previously discussed requirements and limits can be met if the appropriate resources and efforts, experience, and expertise are applied, especially if addressed early in the program at hand. Verification, another key pillar to a good design, is a process that defines what needs to be completed and how this is to be done to prove that requirements have been met. It is usual practice to have companion verification procedures written by the originator of the requirements. These procedures ensure that every verification includes how to test, demonstrate, inspect, or analyze the system to show that the requirements have been satisfied. To be effective, the verification procedures need to be stated as precisely as possible and as well should define the system test success criteria and the use of necessary equipment.

An acoustic noise control plan is required to define the basic efforts necessary to ensure compliance to the requirements. The noise control plan should include the selection or development of quiet noise sources and the procedures employed to determine and control their levels. As well, it should include plans for development and verification testing and an acoustic analysis approach. The noise control plan should be updated throughout the life of the program to reflect completed and current status of efforts to implement it. By monitoring the progress of the noise control plan and through oversight of the associated design and development efforts, an understanding and agreement with the efforts contribute to full compliance with the requirements. When requirements are not met, one aspect in a possible waiver or deviation assessment should be to address whether early and reasonable efforts toward compliance have been applied. If proper monitoring of the design and development process is performed, then reasonable efforts are addressed and attended to as early as possible in the program. Requirements might be written perfectly, but if they are not implemented and verified correctly, and with the right equipment, methods, and experience, then the purpose of the requirements cannot be achieved.

2.4.4 Conclusions and Recommendations

Stringent acoustic requirements are considered necessary for current and future spaceflights for the protection of the safety and well-being of individual crewmembers and the successful completion of their intended missions. The acoustic requirements applicable to the habitable volume and other areas accessible to the crew, the integrated hardware, the supplementary government furnished equipment, and other payloads need to be defined early in the program cycle, implemented correctly, and verified. The requirements are uniquely dependent on the character, duration, frequency content, and level of the noise source emission. A noise control plan strongly is recommended, and it should be updated throughout the design, the manufacturing stages, and all flight phases of the space vehicle. The noise control plan, in combination with monitoring and oversight of the design, development, and verification efforts, is essential to achieve full compliance with the determined acoustic requirements.

RECOMMENDED READING

Berglund, B., T. Lindvall, and D. H. Schwela (eds.). (1999) *Guidelines for Community Noise* [e-Book]. World Health Organization, available at http://whqlibdoc.who.int/hq/1999/a68672.pdf (cited February 19, 2007).

Bruel & Kjaer. (2007) Acoustics literature, available at www.bksv.com/default.asp?ID=17 (cited February 19, 2007).

Crocker, M. J. (1998) *Handbook of Acoustics*. New York: Wiley and Sons.

Kelso, D., and A. Perez. (1983) *Noise Control Terms Made Somewhat Easier*. Available at www.nonoise.org/library/diction/soundict.htm (cited February 19, 2007).

Kinsler, L. E., A. R. Frey, A. B. Coppens, and J. V. Sanders. (1982) *Fundamentals of Acoustics*. New York: Wiley and Sons.

NASA Johnson Space Center, Habitability and Environmental Factors Division, Acoustics Office. Available at http://hefd.jsc.nasa.gov/acoustics.htm (cited February 19, 2007).

2.5 RADIATION

2.5.1 Ionizing Radiation

Sources of Ionizing Radiation

Cosmic rays have been studied for a long time, and many references describe their characteristics (Ginzburg 1958; Colgate, Grasberger, and White 1963; Haffner 1967; Meyer 1969). During the last five decades, the study of cosmic rays has been included in an emerging field of nuclear science that perhaps most descriptively is called *radiation bioastronautics*. Cosmic rays, which have extraordinary penetrating power and fall continuously upon the Earth from somewhere beyond, can be the major radiation hazard in crewed space missions. Yet, as scientists and engineers strive to implement the NASA Vision for Space Exploration, cosmic rays are only one of many challenges that must be overcome to ensure that crewmembers are given proper protection against space radiation hazards.

Most cosmic rays probably originate from within our galaxy, especially from supernova explosions (Ginzburg 1958; Colgate et al. 1963), although the highest energy components, $\geq 10^{17}$ eV amu^{-1}, likely are of extragalactic origin (Meyer 1969). The Sun contributes considerably to the flux of low energy cosmic rays, below 0.5 GeV amu^{-1}, arriving at the Earth. Disturbed regions on the Sun sporadically emit bursts of energetic charged particles into interplanetary space.

The types and energies of particle radiation in space are summarized in Figure 2.18. The predominant types of particle radiation in the Earth environment are solar wind protons, auroral electrons, solar storm protons, trapped protons, trapped electrons, solar protons, and galactic cosmic radiation. There are temporal variations as well as spatial

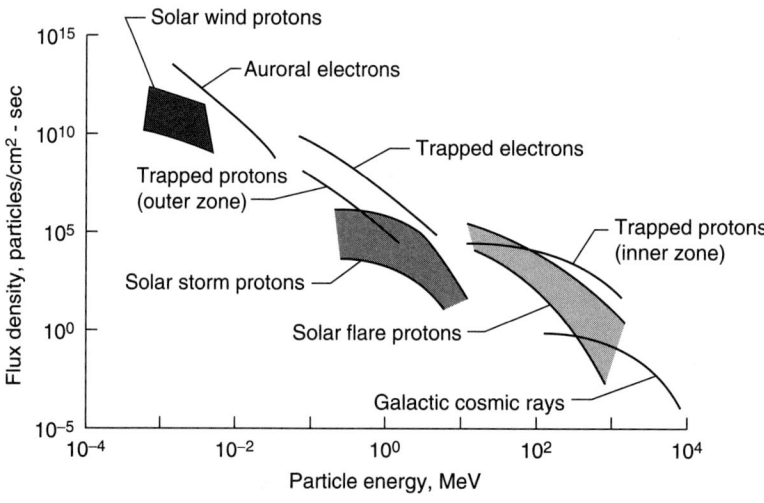

FIGURE 2.18 Space radiation environment (Wilson et al. 1991, courtesy of NASA).

distributions for each type of radiation. It is convenient to consider the particulate radiation in space as arising from these distinct sources defined by their location, that is, solar particle radiation, galactic cosmic radiation, and trapped particle radiation.

Solar Particle Radiation   The solar wind is plasma of both positive and negative particles trapped in a magnetic field emanating from the Sun. The solar wind is really an extension of the solar corona to at least several astronomical units from the Sun (1 AU $\approx 1.5 \times 10^8$ km). It is composed mostly of protons and persists through variable parts of the output of the quiet Sun. The solar wind protons have thermal energies of from approximately 1 keV to 10 keV. Except when the Sun is active, the solar wind constitutes the most important particulate solar radiation.

A solar flare is an intense local brightening on the face of the Sun close to a sunspot. The solar abnormality results in an alteration of the general outflow of solar plasma at moderate energies and the local solar magnetic fields carried by that plasma. As the solar plasma envelops the Earth, the magnetic screening effects inherent in plasmas act to shield the Earth from galactic cosmic radiation, a phenomenon known as a *Forbush decrease* (Forbush 1937), while contributing far more radiation of their own. When the solar plasma interacts with the geomagnetic field, a disturbance or storm occurs. During an intense magnetic disturbance, the magnetic fields of the Earth, that is, the Van Allen radiation belts, are compressed into the atmosphere of the Earth at the polar regions and trapped electrons in the belt are lost. These auroral electrons are seen only in the polar regions and are associated with coronal mass ejection after solar flares.

The solar protons tend to be eliminated from equatorial regions of the magnetosphere because they are deflected by the horizontal geomagnetic field lines into space. However, solar primary particles arrive at the poles by moving along the near vertical geomagnetic field lines and are therefore not deflected. When the low energy solar storm protons are channeled into the polar regions by the magnetic field of the Earth, radio blackouts are produced in the lowest ionospheric region after certain solar flares occur. This is called a *polar cap absorption event* (Kundu and Haddock 1960).

Radiations with energies below 100 keV, such as solar wind protons and auroral electrons, and the solar storm protons with energies below 10 MeV are considered biologically unimportant because they are shielded against by even gaseous barriers.

In association with many of the optical flares occurring from time to time on the solar surface, large fluxes of solar energetic particles are sometimes accelerated and emitted, and these emissions of solar cosmic radiation are designated solar particle events. These events, having periods of hours to days, represent one of several short-lived manifestations of the active Sun. The solar wind and the solar particle events are composed of the same types of particles, mostly protons, with the next major component being alpha particles (Biswas, Fichtel, and Guss 1962; Biswas et al. 1963; Freier 1963). These two groups of particles are distinguished by their numbers and their speed or energy. Heavier nuclei, mostly in the carbon, nitrogen, and oxygen groups (Biswas, Fichtel, and Guss 1966; Durgaprasad et al. 1968), and even heavier particles, that is, those with an atomic charge number (Z), between 22 and 30 (Bertsch, Fichtel, and Reames 1969), have been observed from major solar particle events as well. Rare clusters of high intensity (several orders of magnitude) and high-energy events are critical to spaceflight and extravehicular activity because the large events alone determine the yearly fluences of solar particles, and the dose rate effect is much higher during the short period of peak (Kim, Cucinotta, and Wilson 2006a).

By placing all the available flux and fluence data from past Solar Cycle 19 through Solar Cycle 21, that is, 1955 through 1986, Shea and Smart (1990) assembled a list of major solar particle events and proton fluences into a useful continuous database. From 1986 to the present, that is, Solar Cycle 22 and Solar Cycle 23, a solar particle event list and the geostationary operational environmental satellite spacecraft measurements of the 5-min average integral proton flux can be obtained through direct access to the National Geophysical Data Center of the National Oceanographic and Atmospheric Agency (NOAA). Table 2.10 lists the large solar particle events in the past five solar cycles for which the omnidirectional proton fluence with energy above 30 MeV, Φ_{30}, exceeded 10^9 protons/cm^2.

As seen in Figure 2.19, the frequency of solar particle event occurrence recorded by the NOAA geostationary operational environmental satellites for Solar Cycle 23 is shown for 3-mo periods. The monthly mean number of sunspots is included in the figure to show the association between solar particle event occurrence and solar activity, and the times at which five large, that is, $\Phi_{30} > 10^9$ protons/cm^2, solar particle events occurred are marked with arrows. An increase in solar particle event occurrence has been seen with increasing solar activity. Although, no recognizable pattern has been identified, large events definitely

Table 2.10 Large Solar Particle Events During Solar Cycle 19 Through Solar Cycle 23, with $\Phi_{30} > 10^9$ Protons/cm^2 (Courtesy of NASA)

Solar Cycle	Solar Particle Event	Φ_{30}, Protons/cm^2
19	11/12/1960	9.00×10^9
20	8/2/1972	5.00×10^9
22	10/19/1989	4.23×10^9
23	7/14/2000	3.74×10^9
23	10/26/2003	3.25×10^9
23	11/4/2001	2.92×10^9
19	7/10/1959	2.30×10^9
23	11/8/2000	2.27×10^9
22	3/23/1991	1.74×10^9
22	8/12/1989	1.51×10^9
22	9/29/1989	1.35×10^9
23	1/16/2005	1.04×10^9
19	2/23/1956	1.00×10^9

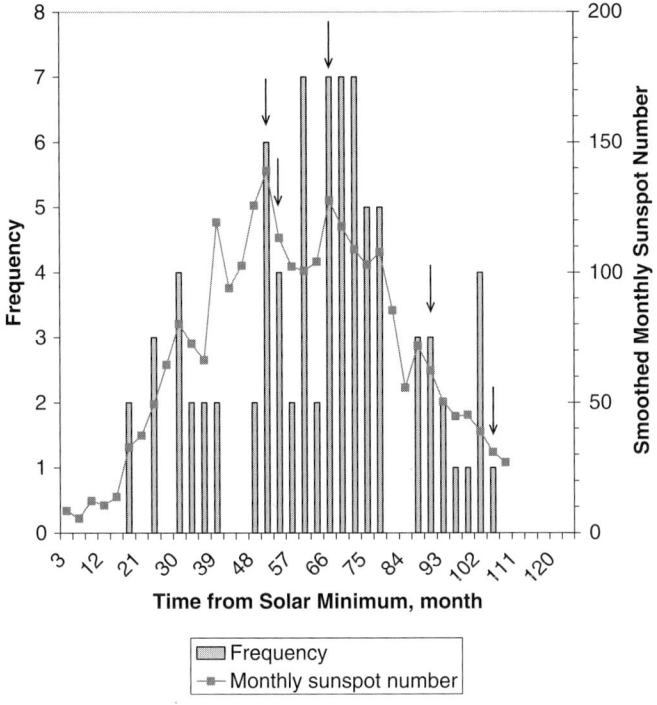

FIGURE 2.19 Frequency of solar particle event occurrence in 3-mo periods of Solar Cycle 23. The arrows indicate the times when large solar particle events with $\Phi_{30} > 10^9$ protons/cm² occurred. From left: July 2000, November 2000, November 2001, October 2003, and January 2005 (Courtesy of NASA).

occurred during solar active years; however, they did not occur exactly during the months of solar maximal activity. Moreover, they are more likely to occur in the ascending or declining phases of the solar cycle (Goswami et al. 1988). This sporadic behavior of solar particle event occurrence is a major operational problem encountered in planning for missions to the Moon and Mars.

The shapes of the energy spectra, as well as the total fluences vary considerably from event to event (Biswas et al. 1962; Freier and Webber 1963; King 1972; Shea and Smart 1990; Kim, Cucinotta, and Wilson 2006b; NOAA 2006). Figure 2.20 shows the energy spectra of the solar particle event of January 16, 2005. There was a sudden increase in proton flux, especially for particles with energies greater than 50 MeV. Protons having energies greater than 100 MeV increased by as much as four orders of magnitude after declining from the major pulse. Although, during this sharp commencement, the fluence did not reach the value obtained at the major peak intensities, this type of sudden increase in high-energy particles can pose a greater threat than the major particle intensities. Total fluence for a solar particle event is the representative indicator of a large solar particle event, and the detailed energy spectra for a large solar particle event, especially at high

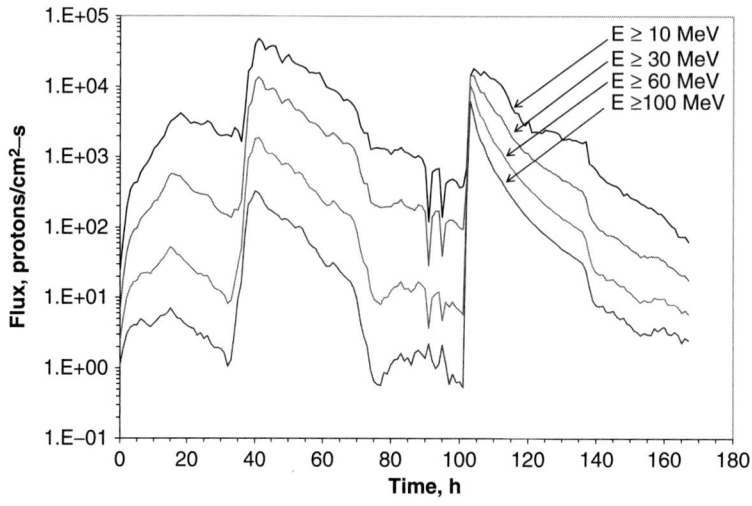

FIGURE 2.20 Hourly averaged proton flux of geostationary operational environmental satellite measurements during the solar particle event of January 16 through January 22, 2005 (Courtesy of NASA).

energies, is the most important parameter for assessment of the risk of radiation exposure (Kim et al. 2006a).

An example of detailed temporal analysis of the dose rate at the blood forming organs is given in Figure 2.21 for the August 1972 solar particle event. This was one of the largest solar particle events in the modern era and had the highest dose rate at its peak. During peak times, the rather heavy shielding, that is, up to 30 g/cm^2 of aluminum, provided by a spacecraft is not enough to reduce the dose rate to the blood forming organs to 1 cGy-Eq/h, where a pivotal transition from low to high dose rates would start. The temporal behavior shown in this figure suggests that considerable biological damage would be incurred during the first major peak times. Biological effects are expected to increase considerably for dose rates above 5 cGy/h. As shown in Figure 2.22, the current recommended 30-d exposure limit at the blood forming organs, 25 cGy-Eq (NCRP 2000), is exceeded easily, and early effects from acute exposure cannot be avoided when only a conventional amount of spacecraft material is provided to protect the blood forming organs from this class of solar particle event. To avoid placing unrealistic mass on a space vehicle while increasing the safety factors for astronauts, one solution for shielding against solar particle events would be to select optimal materials for vehicle structure and shielding. Indeed, it has been shown that materials having lower atomic mass constituents have better shielding effectiveness (Wilson et al. 1995a; Cucinotta et al. 2000).

An event of special interest occurred on February 23, 1956 (Wilson et al. 1999a), where a striking feature was a large number of high-energy particles early in the event. However, the accuracy of spectral determinations for this event still is debated because

2.5 Radiation

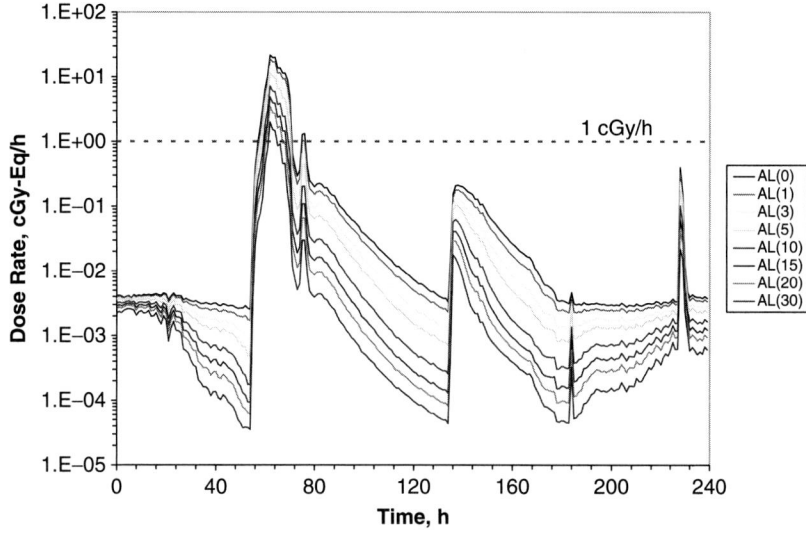

FIGURE 2.21 Dose rate to blood forming organs behind various aluminum thicknesses during the solar particle event of August 2 through August 11, 1972 (Courtesy of NASA).

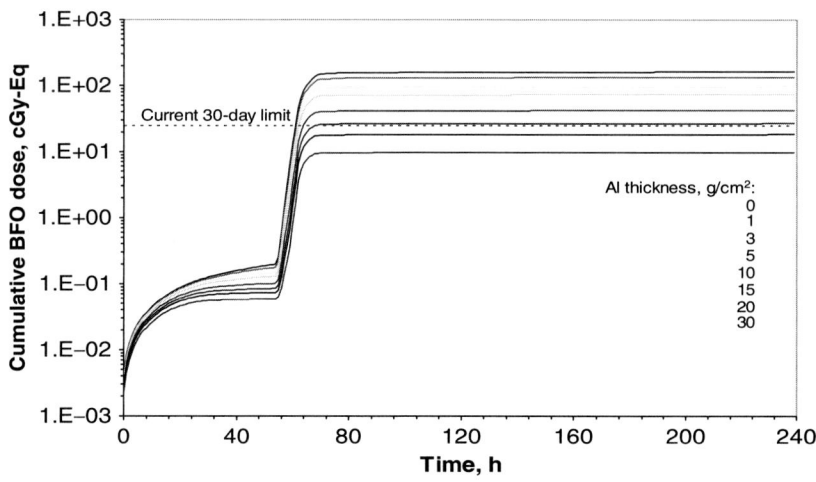

FIGURE 2.22 Cumulative dose to blood forming organs behind aluminum of various thicknesses during the solar particle event of August 2 through August 11, 1972 (Courtesy of NASA).

only ground based neutron monitors were available. Large uncertainties exist in the determination of spectra because of the atmospheric propagation calculations required to unfold the spectra. Overall exposure levels from this specific event have been estimated to be greater than 10 cSv (10 rem) at sensitive sites, whereas those from other large solar particle events recorded in the modern era can be reduced to below 10 cSv when heavily shielded storm shelters are added to a typical spacecraft (Kim, Hu, and Cucinotta 2005). However, this result should come with the caveat that considerable uncertainties are inherent in the determination of the source spectra of protons.

Galactic Cosmic Radiation In addition to radiation from the Sun, the Earth is bombarded with charged particles from outside the solar system, that is, galactic cosmic radiation. These particles appear to pervade isotropically, at least in the near Earth environment, and have a range of energies that exceeds 10 GeV per nucleon. Galactic cosmic radiation is comprised of fully ionized nuclei, the electrons having been stripped from the atoms during their acceleration to galactic cosmic radiation energies. The region outside the solar system in the outer part of the galaxy is believed to be filled uniformly with galactic cosmic radiation, whose nuclei constitute approximately one third of the energy density of the interstellar medium; and on a galactic scale, they form a relativistic gas whose pressure is important to take into account in the dynamics of galactic magnetic fields. The galactic cosmic radiation nuclei are the only direct and measurable sample of matter from outside the solar system. It is a unique sample because it includes all of the elements from hydrogen to the actinides. Galactic cosmic radiation arriving from beyond the magnetic field of the Earth at the distance of the Earth from the Sun are composed of ~98% nuclei and ~2% electrons and positrons (Simpson 1983a). Within the energy range of 10^8 to 10^{10} eV amu^{-1}, where it has its highest intensity, the nuclear component consists roughly of 87% protons, ~12% helium nuclei, and a total of ~1% for all of the heavier nuclei from carbon to the actinides (Simpson 1983a).

At 1 AU, the galactic cosmic radiation flux is affected by solar activity because it interacts with the solar plasma emitted into interplanetary space. However, it is out of phase with the activity of the Sun, in that the more active the Sun, the lower the galactic cosmic radiation flux at the Earth. The intensity of the galactic cosmic radiation flux varies over the approximate 11-a solar cycle due to changes in the interplanetary plasma that originate in the expanding solar corona (Bobcock 1961; Badhwar and O'Neill 1992). The galactic cosmic radiation flux that reaches Earth is smaller during intense sunspot activity because the low energy galactic cosmic radiation particles are deflected by the enhanced magnetic field of the Sun carried by the expanding solar plasma. The maximum dose received occurs at solar minima because of the lower solar plasma output. Measurements at solar minimum modulation, in which major solar particle events usually are absent, show the greatest extent of galactic cosmic radiation exposure (Badhwar 1999).

Galactic cosmic radiation turned out to be a vital contributor to our understanding of high-energy phenomena in our galaxy. Whereas protons carry most of the galactic cosmic radiation energy, heavy particles give information on composition and propagation. Although galactic cosmic radiation probably includes every natural element, not all are important for the purposes of space radiation protection. The elemental abundances for species heavier than iron ($Z > 26$) typically are two to four orders of magnitude smaller than that for iron (Adams, Silberberg, and Tsao 1981). In the solar system, some elements such as the L nuclei, that is, lithium, beryllium, and boron, as well as fluorine and several nuclei between silicon and iron are quite rare (Simpson 1983b; Cucinotta et al. 2006b). However, within the galactic cosmic radiation flux, nuclei of these elements are almost as commonly present as those of their neighbors (Simpson 1983b). This shows that they originate in the breakup of heavy particles during galactic cosmic radiation propagation and would not be present in galactic cosmic radiation from stellar sources (Parker 1965; Webber et al. 1990a; Fields, Olive, and Schramm 1994).

Experimental studies of high-energy, high-charge particles were made on the Pioneer, Voyager, and Ulysses spacecraft to measure the isotopic composition of galactic cosmic radiation elements near Earth and in deep space (Wiedenback and Greiner 1981; Wiedenback 1985; Webber, Kish, and Schrier 1985; Webber et al. 1990a; Hesse et al. 1991; Lukasiak et al. 1993, 1995). These data have been used to develop predictions of galactic cosmic radiation spectra behind shielding as an important goal for the NASA space radiation research program (Cucinotta et al. 2006b). Examples of the galactic cosmic radiation energy spectra for hydrogen and helium isotopes are shown in Figure 2.23 at solar minimum and solar maximum and for neon, silicon, and iron isotopes in Figure 2.24 at solar minimum. These figures show the contribution of different isotopes to primary galactic cosmic radiation composition (Cucinotta et al. 2006b). In recent years, new data obtained from the galactic cosmic radiation near Mars were collected by the Martian radiation environment experiment (MARIE) on the Mars Odyssey spacecraft (Zeitlin et al. 2004) to be used in planning the design of future crewed spacecraft and missions to the Moon and Mars.

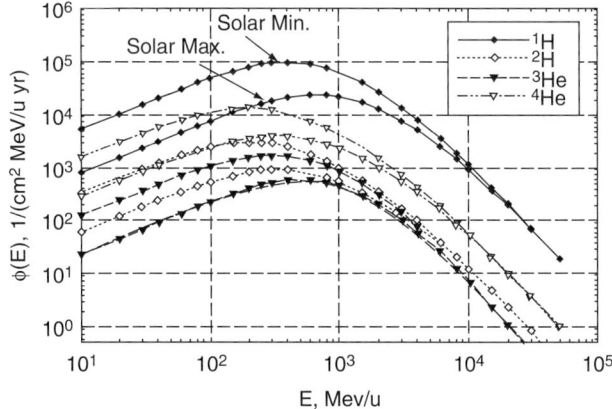

FIGURE 2.23 Energy spectra for hydrogen and helium isotopes at solar minimum and solar maximum (Courtesy of NASA).

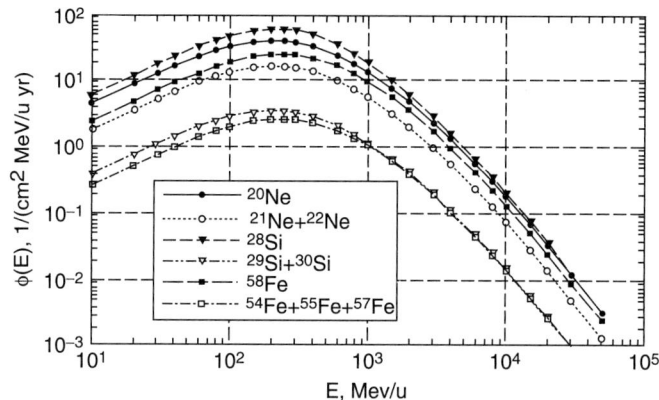

FIGURE 2.24 Energy spectra for neon, silicon, and iron isotopes at solar minimum, showing contributions from different isotopes to primary galactic cosmic radiation composition (Courtesy of NASA).

The propagation of galactic cosmic ions through matter has been studied by many researchers as a way to determine the origin of these ions, as well as to evaluate required shielding. As the galactic cosmic components are transported through target media, their energies are attenuated by two distinct mechanisms:

- Electromagnetic interactions, resulting in ionization and excitation.
- Nuclear interactions resulting in the generation of a multitude of cascading secondary particles from all subsequent generation collisions (fragmentation).

In the many collisions and electromagnetic interactions along the path of the incident particle, energy is lost in extremely small increments, and the average rate of energy loss per unit path length, $MeV/(g/cm^2)$, is expressed by stopping power. Comprehensive tables of stopping power versus particle energy are available in the literature (Barkas and Berger 1964; Janni 1966; Williamson, Boujot, and Picard 1966; Steward 1968; Bichsel 1969).

For the strong nuclear interactions of an incident particle with a nucleus of the target medium, the quantum multiple scattering of heavy ion fragmentation describes the physics of the abrasion ablation model of fragmentation (Cucinotta, Townsend, and Wilson 1992; Cucinotta et al. 1997a; Cucinotta, Wilson, and Townsend 1997b; Cucinotta et al. 1998; Cucinotta and Dubey 1994) and agrees well with data from experiments (Brechtmann and Heinrich 1988; Webber, Kish, and Schrier 1990b; Knott et al. 1996, 1997; Zeitlin et al. 1997, 2001). The theoretical calculation of the fragmentation cross sections involves the following areas:

- Description of the probability of removing a given amount of mass and charge.
- Description of the distribution of prefragment excitation energies formed in the abrasion step.
- Description of the statistical decay of the prefragments to form the final fragment distribution.

Any particular final nuclide that occurs as a result of the de-excitation of a primary residue is a nuclear fragment, sometimes referred to as a *secondary product*. Customarily in cosmic ion transport studies, the fragment velocities are assumed to be equal to the fragmenting ion velocity before collision at the interaction site (Townsend et al. 1993; Wilson et al. 1993a); however, the momentum spread of light nuclear fragments becomes increasingly important in thick shields (Shavers, Cucinotta, and Wilson 2001).

During the last 25 a, the description of galactic cosmic radiation transport in shielding has improved dramatically for the nuclear interactions and propagation of protons, heavy ions, and their secondaries. Major milestones include the development of an accurate free space galactic cosmic radiation model (Badhwar and O'Neill 1992), the HZETRN code (Wilson 1977; Wilson et al. 1991), the measurement of a considerable number of fragmentation cross sections (Brechtmann and Heinrich 1988; Webber et al. 1990b, 1998; Knott et al. 1996, 1997; Zeitlin et al. 1997, 2001), and the development of an accurate nuclear fragmentation model (Cucinotta et al. 1997b, 1998). Laboratory (Schimmerling et al. 1989) and spaceflight (Badhwar and Cucinotta 2000) validation data also have become available. The combination of the galactic cosmic radiation model of Badhwar and O'Neill, the quantum multiple scattering of heavy ion fragmentation cross section database, and the HZETRN transport code has been shown to agree with flight measurements of the galactic cosmic radiation dose and dose equivalent within $\pm 15\%$ on several space vehicles (Cucinotta et al. 2006b). However, further spectral data sets, both in space and at heavy ion accelerators, are needed to validate these codes fully.

The implementation of heavy ion transport models has progressed from models that did not satisfy unitarity (Letaw, Tsao, and Silberberg 1983) to the current fully energy dependent models with accurate absorption cross sections (Cucinotta 1993; Wilson et al. 1993b; Shinn, Wilson, and Badavi 1994). Future work still is required for light particle (n, p, d, t, h, α, and mesons and their decays) transport, including the establishment of production cross section models and data, and for understanding the role of angular deflections that are more important for transport of protons and neutrons than transport of heavy ions. However, the heavy ion problem is in much better shape with many of the remaining tasks of implementation. One exception can be improvements in fragmentation cross sections and laboratory validation for $Z = 1$ to 5 nuclei produced from the heavier projectile nuclei ($Z > 10$).

Trapped Radiation Belts All the magnetized planets have populations of highly energetic particles trapped in their planetary magnetic fields. The most extensively studied of these trapped populations are the radiation belts of Earth, that is, the Van Allen belts, and the radiation belts of Jupiter. The magnetic field surrounding the Earth is roughly in a dipole configuration. Charged particles are trapped in the geomagnetosphere (Van Allen et al. 1958), where two radiation belts with high radiation intensities are trapped geomagnetically. The stable trapped radiation in the inner zone, consisting mostly of protons with a small percentage of electrons, primarily is centered at an altitude of 2000 km at the equator. The radiation in the transient outer zone, consisting mostly of low energy electrons with a small percentage of protons, is centered at 20,000 km (Parker and West 1973). Energies range from ~100 keV to more than 400 MeV for protons, and from tens of kiloelectron volts to in excess of 10 MeV for electrons. The particles undergo three distinct motions:

- A spiraling around the magnetic field lines in a helical motion having a typical spiraling period of 10^{-6} s for electrons and 10^{-3} s for protons.
- A bouncing back and forth along the field lines between mirror points, having a typical bounce period of from 0.1 to 2 s, depending on energy and particle.
- A drifting around the Earth with typical drift period of from 1 to 10 h for electrons and from 5 s to 30 min for protons, depending on energy.

These naturally trapped Van Allen belts, which are in a plane normal to the solar wind direction, are distorted spatially by the presence of the solar wind pressing on the geomagnetosphere.

The magnetic field of the Earth is not centered at the geographic center of the planet. The main dipole moment along the principal axis of the magnetic field is tilted with respect to the rotational axis of the Earth. Thus, the geomagnetic field is not symmetrical with respect to geographic coordinates. An interesting combination of two geomagnetic features, the effective dipole displaced away from Brazil and the local distortion of geomagnetic field in South Africa, called the *Capetown anomaly*, cause trapped particles of the inner belt to dip close to the surface of the Earth in the region of the South Atlantic Ocean between Brazil and South Africa. This increase in particle flux at low altitude has been called the *South Atlantic anomaly*. Most of the radiation encountered by satellites in low inclination orbits comes from the South Atlantic anomaly.

Crewed missions in low Earth orbit are flown at altitudes below the inner belt. However, the 51.6° inclination orbit of the *International Space Station* takes high geomagnetic latitudes, where exposure to the increased number of relativistic electrons in the radiation belt and solar particle event fluxes during major solar disturbances are unavoidable. This is as well the case for the higher galactic cosmic radiation fluxes. Reduction of radiation risk in International Space Station orbit can be relatively easy, and recommendations can be found in a National Research Council (NRC) report (NRC 2000). However, astronauts embarking on or returning from journeys to the Moon or Mars must pass through the Van Allen belts and be exposed for brief periods to high levels of radiation.

Space Radiation Protection Issues

Radiation exposure limits for humans in space (Cucinotta and Durante 2006) are adhered to by NASA, and appropriate risk mitigation measures to ensure that humans can safely live and work in the space radiation environment anywhere and anytime are implemented. In the context of the radiation protection principle of "as low as reasonably achievable" (ALARA), the term *safety* means that acceptable risks are not exceeded during the lifetimes of crewmembers, where acceptable risks include limits on post-mission and multimission consequences. The most important types of radiation for biological consideration are the trapped protons in the inner zone, the trapped electrons in both the inner and the outer zones, solar particle events, and especially the galactic cosmic radiation (Wilson et al. 1991).

For past short duration exploratory missions, the main radiation hazards were considered to be the more intense components of space radiation, such as solar particle events and trapped radiation, because the continuous galactic cosmic radiation background exposures are of low intensity. Career radiation limits were based on fatal cancer risks, and increased lifetime cancer risk above the natural incidence is limited to 3% for NASA missions at low Earth orbit as recommended by the National Council on Radiation Protection and Measurements (NCRP) (NCRP 2000).

Safety concerns for long-term space explorations include carcinogenesis, degenerative tissue effects such as cataracts (Cucinotta et al. 2001) or heart disease (Yang and Ainsworth 1982; Preston et al. 2003; Howe et al. 2004), and acute radiation syndrome (NCRP 2000). Other risks, such as damage to the central nervous system, are a concern for high-energy nuclei (NAS 1996) because of their unique pattern of energy deposition on the microscopic scale of living cells. Standards for lunar missions are under review at this time and it is expected that cancer risks are to be the major component of radiation limits. Even so, new knowledge about chronic noncancer risks from radiation is needed.

Because the abundance of some heavy ions from major solar particle events can increase rapidly by three or four orders of magnitude above galactic cosmic radiation background for periods of several hours to days, solar particle events present the most important risk for short stay lunar missions, that is, <90 d. The primary radiation protection from solar particle events is intended to control early somatic radiation effects that can have an impact on mission safety. However, effective mitigation against solar particle events becomes viable by implementing several options:

- A spacecraft devised using high-performance structural material, such as carbon composite with high hydrogen content having effective radiation shielding properties.
- Adequate mission planning for timing and location.
- Seeking a shelter and using personal localized shielding in a timely manner with the warning system developed (Cucinotta, Kim, and Ren 2006a).

The risk of acute death from any known large solar particle event has been assessed as being extremely unlikely and acute radiation sickness extremely improbable inside exploratory spacecraft or lunar habitation modules except when extravehicular activity is performed during a major solar particle event for more than 2 h (Cucinotta et al. 2006a). Skin damage (Kim et al. 2006c) and cataracts (Cucinotta et al. 2001) give cause for special concern because of the dose rate effect.

In contrast, for long-term missions, such as long duration lunar (>90 d) or Mars missions, risk from galactic cosmic radiation can exceed the acceptable radiation risk limits. The unusually high specific ionization of high-energy nuclei in galactic cosmic radiation is the ultimate limiting factor for long-term space operations, because although their relative dose contributions are comparable to those of light particles, their biological effects, which as of yet are poorly understood, are far more serious (Cucinotta and Durante 2006).

For use in the effort to make accurate projections of radiation doses to astronauts, which are required for planning of future exploration class and long duration space missions (Cucinotta and Durante 2006), a solar cycle statistical model has been developed (Wilson et al. 1999b; Kim and Wilson 2000; Kim, Wilson, and Cucinotta 2004, 2006d). A systematic method of making short-range projections of future levels of solar cycle activity was established by quantifying the progression level of sunspot numbers within the solar cycle. The resultant solar activity levels were coupled to galactic cosmic radiation deceleration potential (Φ) and the mean occurrence frequency of solar particle events (ν) for projection of the future space radiation environment.

The galactic cosmic radiation deceleration parameter, $\Phi(t)$, represents the temporal galactic cosmic radiation environment of interplanetary space, and the calculated values are shown as a function of time in the upper graph of Figure 2.25. The point dose equivalents inside a typical equipment room of a spacecraft (5 g/cm^2 of aluminum shielding) are calculated from the galactic cosmic radiation environment in interplanetary space, which is determined as a function of $\Phi(t)$ and at low Earth orbit by using the HZETRN computer model (Wilson et al. 1995b) as shown in the lower graph of Figure 2.25. This calculation shows that, within a simple and representative spacecraft configuration, the galactic cosmic radiation exposure levels simply are affected by solar modulation in interplanetary space by a factor of three and at low Earth orbit by about a factor of two due to the further modification of the galactic cosmic radiation environment by geomagnetic fields and atmospheric shielding.

Although no definite pattern of solar particle event occurrence has been observed in the past solar cycles, large solar particle events have been recorded during the solar active years, as shown in Figures 2.19 and 2.26. This strong possibility of large solar particle event occurrences at high $\Phi(t)$ and the multiple annual occurrences of medium to large solar particle events per year during the solar active years in future cycles, as shown in Figure 2.27, are definitive major operational problems for planning and protecting astronauts on future missions to the Moon and Mars. Because the sporadic nature of solar

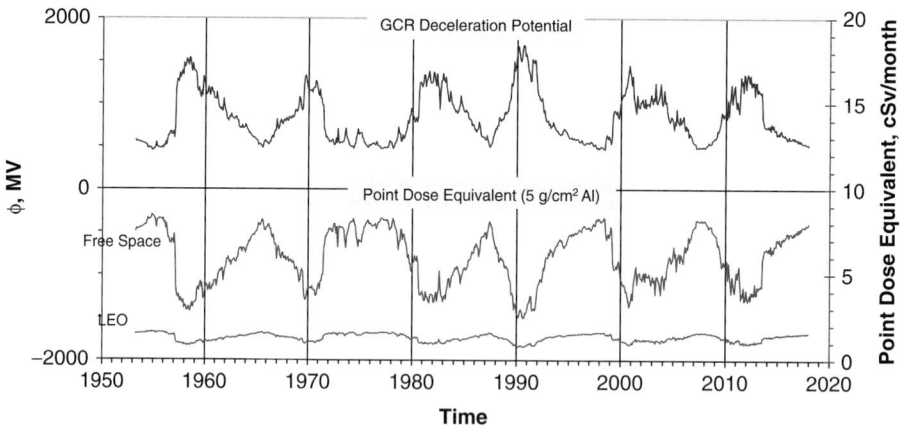

FIGURE 2.25 Galactic cosmic radiation deceleration potential as a function of time (upper graph), calculated from the neutron monitor rate measurements and from the projected neutron monitor rates in the future with the statistical model (Kim et al. 2006d) and point dose equivalents from the galactic cosmic radiation inside spacecraft (lower graph) shielded with 5 g/cm^2 of aluminum, calculated with HZETRN (Wilson et al. 1995b) (Courtesy of NASA).

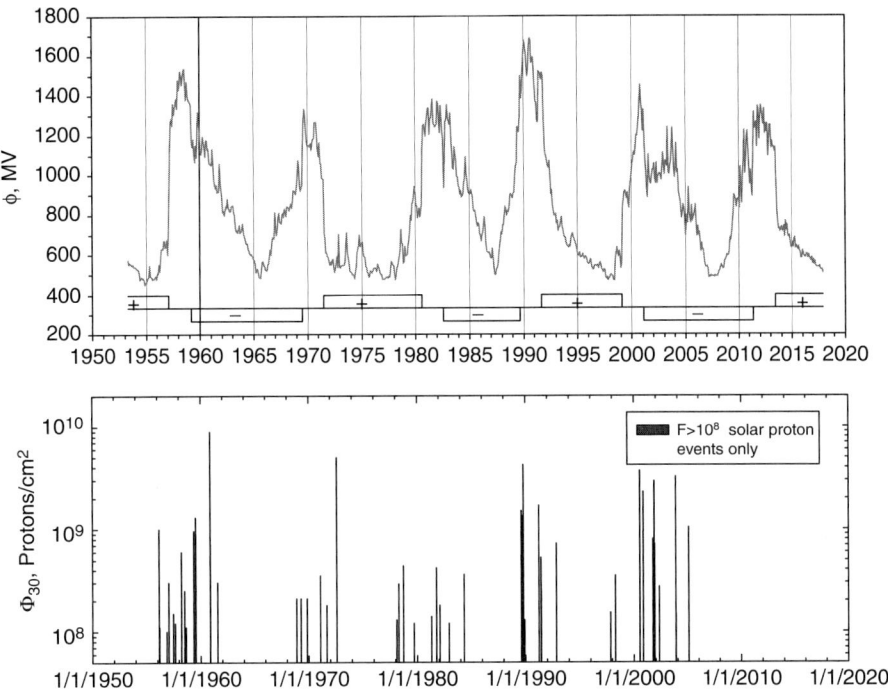

FIGURE 2.26 Galactic cosmic radiation deceleration parameter (upper graph) and proton fluence of large solar particle event occurrences (lower graph) (Shea and Smart 1990; NOAA 2006) (plotted for events of size $(\Phi_{30}) > 1 \times 10^8$ protons/cm² only) as a function of time (Courtesy of NASA).

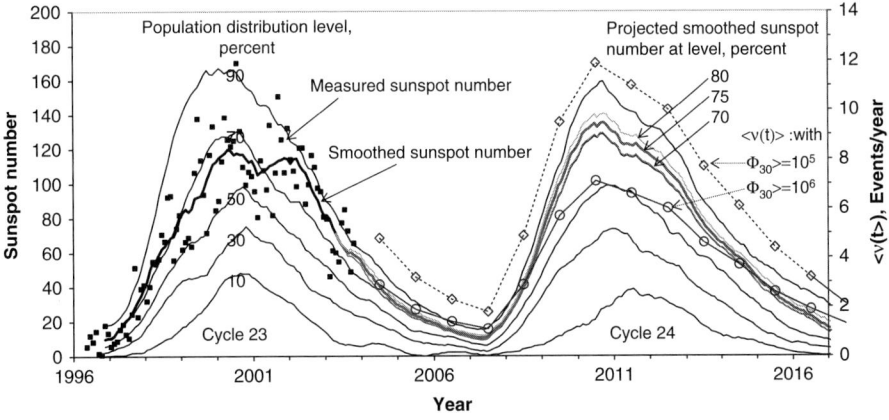

FIGURE 2.27 Sunspot sampling distribution and projections of solar cycles and mean occurrence frequencies of solar particle events ($< v(t) >$) plotted for two event sizes (Φ_{30}) (Kim et al. 2006a) (Courtesy of NASA).

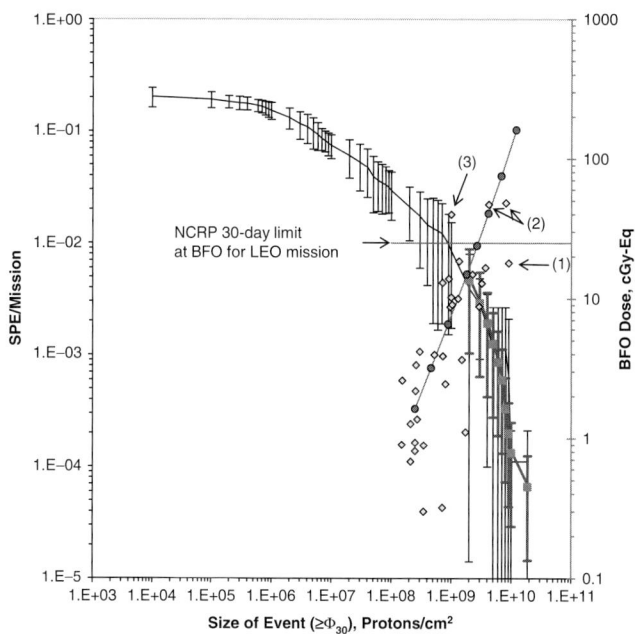

FIGURE 2.28 Probability that a solar particle event will occur during a 1-wk mission (thin line: average probability of a solar particle event and statistical fluctuation during the space era; thick line: extended average probability including impulsive nitrate events [McCracken et al. 2001]; filled squares: probability of impulsive nitrate events), and dose to blood forming organs inside a spacecraft of 5 g/cm² aluminum (line with filled circles: dose to blood forming organs for the worst-case solar particle event model [Xapsos et al. 2000]; filled diamonds: dose to blood forming organs from 34 large solar particle events during the space era). Note 1: The largest event recorded in the space era (Φ_{30}) resulted in an estimated dose to blood forming organs inside a spacecraft (5 g/cm² aluminum) that was lower than the current National Council on Radiation Protection 30-d limit for low Earth orbit missions (NCRP 2000). Note 2: $\Phi_{30} > 2 \times 10^9$ protons/cm²; the dose to blood forming organs was over the current limit. Note 3: Φ_{30} at 70% confidence level of the worst-case solar particle event; the dose to blood forming organs is over the current limit (Courtesy of NASA).

particle events makes it impossible to pinpoint the exact time of future large solar particle event occurrences, the calculated probabilities of solar particle events occurring in a short mission period and the exposure levels from various solar particle events are shown in Figure 2.28 for guidance in the design of protection systems.

Summary

Considerable efforts and improvements have been made in the study of ionizing radiation exposure occurring in various regions of space. Satellites and spacecraft equipped with innovative instruments continually are refining particle data, and they are providing more accurate information about the ionizing radiation environment. The major problem in accurate spectral definition of ionizing radiation seems to be that of obtaining the detailed energy spectrum, especially at high energies, which is the important parameter for accurate radiation risk assessment. Prediction of the magnitude of risks posed by exposure to radiation in future space missions is subject to the accuracies of predictive forecast of event size of solar particle events, the galactic cosmic radiation environment, geomagnetic fields, and the atmospheric radiation environment. Although heavy ion fragmentations and interactions are resolved adequately through laboratory study and model development, improvements in fragmentation cross sections for the light nuclei produced from high-energy nuclei and their laboratory validation remain required to achieve the principal goal of planetary galactic cosmic radiation simulation at a critical exposure site. A more accurate procedure for predicting the ionizing radiation environment can be developed with a better understanding of solar and space physics, fulfillment of required measurements for nuclear and atomic processes, and validation and verification of models with data collected from spaceflights and heavy ion accelerator experiments. It is certainly true that continued advancements in solar and space physics, combined with physical measurements, strengthen confidence in the ability of humans to conduct safely future exploration of the solar system. Advancements in radiobiology surely give meaningful radiation hazard assessments for short- and long-term effects so that appropriate and effective mitigation measures can be put into place to ensure that humans can live and work safely in space anywhere and at anytime.

2.5.2 Radio Frequency Radiation

The total radio frequency environment for a space vehicle is comprised of various individual sources, including

- Local facility emitters, such as handhelds and frequency agile trunking radios.
- Terrestrial low and high power emitters, such as commercial broadcast, government, and military communications, tactical systems, and range safety tracking radars.
- Natural and triggered lightning.
- Vehicle borne emitters.
- Orbiting communications and geolocation satellite traffic.
- Emissions associated with solar flare and sunspot activity.

Hardware exposure to radio frequency radiation can occur at any time during the vehicle life cycle, and therefore protection from such exposure must be endemic to the overall design and operational considerations for the vehicle and its supporting elements. For example, ground processing operations can extend well beyond the final assembly area, and potential exposures during fabrication and shipping from distant locations, as well as during recovery operations following landing at remote sites, must be considered. This subsection attempts to provide some description of the various elements of the radio frequency environment in which a modern space vehicle is expected to operate. The techniques of design for protection related to hardware are contained in other

68 CHAPTER 2 The Space Environment: Natural and Induced

sections. Operational constraints and rules for use are considered to be self-explanatory methods of control and or protection and are not elaborated here.

Sources

The radio frequency environments comprise both intentional and unintentional radiation sources and are often characterized by what some engineers call a *skyscraper profile*. An intentional source is one that has a designed transmit capability from either an integral or externally connected antenna. A cell phone is an example of an intentional source having an integral antenna. An unintentional source has no designed transmit capability and no obvious antenna, either integral or externally connected. However, most electrical and electronic systems are interconnected by a wiring harness of one or more types, from simple to complex, unshielded and shielded; this harness can act as an antenna element under the right conditions. Some equipment unintentionally radiates directly from the case, depending on its function and design characteristics. Laptop computer screens have been known to radiate across a wide range of frequencies, from hundreds to thousands of kilohertz. These frequencies are within the AM broadcast band and can interfere easily with any operational AM communications systems if the laptop is located close enough to the victimized receiving equipment. Typical radio frequency environments taken from Mil-STD-464 (DoD 2002) are shown in Figure 2.29. This figure shows three radio frequency environments. The most stringent of these is the baseline environment. This environment covers all known emitters not otherwise accounted for in the others and

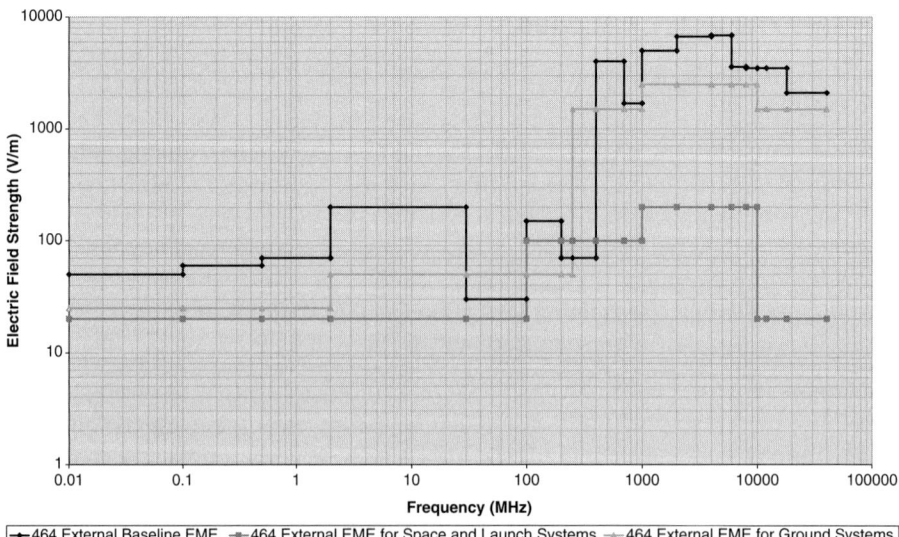

FIGURE 2.29 Typical Mil-STD-464 radio frequency environments (Courtesy of the U.S. Department of Defense).

is considered to be generally applicable if no other environment can be identified as appropriate.

The other two environments are probably closer in a baseline sense for space vehicle applications. One environment is for ground operations, and the second, albeit less stringent, is for the vehicle itself while in flight. These radio frequency environments should serve as a starting point for tailoring to a particular space vehicle, with known emitters replacing or adding to the environments until all are accounted for. It is left to engineering judgment to determine the appropriate setting for the radio frequency environments in regions of frequency where no known emitters exist. Generally, for space vehicles, separate radio frequency environments are published for ground processing, launch and landing, and orbital operations. The final radio frequency environments should be developed carefully, for they typically are used to establish operational maximum allowable limits for surrounding infrastructural components like weather monitoring and vehicle tracking radars. Precisely this type of formally documented arrangement is held by NASA with both the Western and Eastern Test Ranges, located at Vandenburg Air Force Base, California, and Cape Canaveral, Florida, respectively. Formally documented control agreements with commercial, government, and military terrestrial and space borne emitters can be, and typically are, based on similar radio frequency environments as well.

Local Facility Sources Local facility sources comprise many intentional radiators, such as handheld, fixed, and mobile general communications radio equipment used by security and emergency services; cell phone transceiver towers and microwave telephone relay stations; local area networks such as IEEE 802.11G and Bluetooth® capable wireless computer links and accessory devices; portable personal electronic devices like Blackberries and related cell phone accessories; and navigational radio equipment in locations near aircraft operations. Unintentional sources include welding operations, heavy machinery operations, and electric power switching and distribution equipment installations. Protection from local facility sources is a function of hardware design, and the judicious application of appropriate operational constraints dictate when, where, and at what times different types of emitters can be utilized. An important aspect of this control is that, when new emitters are introduced into the environment, they must undergo a threat assessment to determine what impact they can present to existing infrastructure, vehicle components, and operations. Frequencies range from hundreds of hertz to gigahertz, and field strengths can range from microvolt per square meter to hundreds of volts per square meter for the line of sight of communications transmitting equipment.

Terrestrial Emitters Terrestrial emitters of concern include both low and high power emitters, such as commercial broadcast, government, or military communications and tactical systems and range safety tracking radars. Most of these are known transmitter threats, and they are fixed services with known areas of influence that do not vary over time. Commercial broadcast stations have strong signals but generally do not pose a threat to orbiting vehicles because they are neither highly directional nor are they designed to radiate efficiently upward into the atmosphere. Most government and military communications and tactical systems operate similarly, with notable exceptions including over-the-horizon high-frequency ionospheric reflectors, tropospheric scatter systems, and tracking radar installations used for aircraft and missile surveillance and observation. Range safety radars typically operate on or very close to launch sites, and they are designed primarily to track a vehicle as it ascends so that, in the event something goes awry, a destruct signal can be sent to it to prevent a downrange incident. As before, protection from terrestrial emitters is a function of hardware design, and the judicious application of appropriate operational constraints dictate when, where, and at what times different types of emitters can be utilized. Exposure to terrestrial emitters typically is controlled through operational agreements that constrain emission frequencies, levels, and in appropriate cases, direction of radiation as a function of time or vehicle orbital location. Frequencies range from thousands of kilohertz to tens of gigahertz, and field strengths can range from tens to thousands of volts per square meter in the line of sight of the transmitting equipment.

Natural and Triggered Lightning Natural and triggered lightning can be a very strong source of radio frequency interference, particularly if it occurs within a few kilometers of the vehicle. The radio frequency bandwidth of lightning generally ranges from a few hertz up to a few thousand megahertz, with strong content in the 30 to 300 MHz spectrum. Electrostatic and very low frequency electric field strengths near lightning channels can be extremely high in amplitude, approaching several thousand volts per square meter associated with moderately strong events. Measured radio frequency amplitudes range considerably over frequency and generally show a tendency to exhibit an inverse frequency relationship with the highest amplitudes occurring below 100 kHz. Control of the radio frequency effects of lightning is accomplished through protective measures designed into the hardware and operational constraints that dictate when and where equipment can be moved or launched. Weather forecasting is a major part of the protection of space vehicles from the effects of lightning.

Vehicle Borne Emitters Vehicle borne emitters generally are low to medium strength communication and navigation systems. Protection from the effects of vehicle borne emitters is again a function of hardware design, and the judicious application of appropriate operational constraints dictating when, where, and at what times different types of emitters can be utilized. It must be kept in mind that emitters onboard a vehicle are less likely to pose a threat to other platform located hardware, but emitters on a visiting vehicle or a vehicle being visited are likely to pose a considerable threat if proper precautions are not observed in operations. Frequencies range from a few hundred megahertz to a few tens of gigahertz, and field strengths generally range from volts per square meter to hundreds of volts per square meter in the line of sight of the transmitting equipment.

Satellite Sources Satellite sources are similar to vehicle borne emitters, with the exception that they are directed toward the Earth or in some cases toward the vehicle itself, such as the *Tracking and Data Relay Satellite*. Satellite sources usually are not an important radio frequency radiation threat. Protection from the effects of satellite sources is primarily a function of hardware design because it generally is not possible to control the satellite emissions themselves. Frequencies range from a few hundred megahertz to a few tens of gigahertz, and field strengths generally range from volts per square meter to tens of volts per square meter in the line of sight of transmitting equipment.

Solar Flares and Sunspots Radio emissions from the Sun typically occur in radio wavelengths from centimeters to decameters, under both quiet and disturbed conditions. The four major types or categories of solar radio emissions are as follows:

- Type I represents a noise storm composed of many short, narrowband bursts in the frequency range of 300 MHz to 50 MHz.
- Type II represents narrowband emissions that begin in the meter range around 300 MHz and sweep slowly over tens of minutes toward decameter wavelengths around 10 MHz. Emissions of this category often occur in loose association with major solar flares and are indicative of a shock wave moving through the solar atmosphere.
- Type III represents narrowband bursts that sweep rapidly over seconds from decimeter to decameter wavelengths in frequencies from 500 MHz to 0.5 MHz. They often occur in groups and are an occasional feature of complex solar active regions.
- Type IV represents a smooth continuum of broadband bursts primarily in the meter range of frequencies from 300 MHz to 30 MHz. These bursts are associated with some major flare events beginning 10 to 20 min after the flare maximum and can last for hours.

Radio interference from solar events can be a major problem and has been known to damage sensitive global positioning systems receiving equipment. In general, protection from the effects of solar activity is primarily a function of hardware design and avoidance, because it is not possible to control solar emissions.

RECOMMENDED READING

References on the topic of radio frequency environments are not voluminous, but they do exist in the literature, primarily in military standards, texts specifically on the topic, texts of a more general nature dealing with electromagnetic compatibility engineering, and technical papers written for and published by organizations such as the International Electrical and Electronic Engineers (IEEE). Rather than attempt to provide a comprehensive listing, the following is a suggested subset of recommendations for both the interested reader and the serious student that can serve as foundational material leading to additional advanced reading and study.

AFSC. (1984) *Electromagnetic Compatibility*. Air Force Systems Command.

Kaiser, K. L. (2004) *Electromagnetic Compatibility Handbook*. Boca Raton, FL: CRC Press.

Keiser, B. (1987) *Principles of Electromagnetic Compatibility*. Norwood, MA: Artech House.

Mardiguian, M. (1999) *EMI Troubleshooting Techniques*. New York: McGraw-Hill.

Mills, J. P. (1993) *Electromagnetic Interference Reduction in Electronic Systems*. Englewood Cliffs, NJ: Prentice-Hall.

Violette, J. L. N., D. R. J. White, and M. F. Violette. (1987) *Electromagnetic Compatibility Handbook*. New York: Van Nostrand Reinhold.

2.6 NATURAL AND INDUCED THERMAL ENVIRONMENTS

2.6.1 Introduction to the Thermal Environment

A spacecraft in proximity to a planet or a moon experiences natural environmental heating from three sources: the Sun; solar energy reflected from the planet, that is, albedo; and infrared energy emitted from the planet, that is, planetary infrared or outgoing long wave radiation. These environmental heating sources, in concert with orbital parameters, spacecraft attitude, and the design of the vehicle, determine the induced thermal environment and, hence, the thermal response of the spacecraft.

This section provides the reader with an introduction to the natural and induced thermal environments for orbiting spacecraft. Whereas the focus is primarily on Earth orbiting spacecraft, extensions to the theory provided in this section permit an understanding of the thermal environments experienced while in proximity to other bodies. Furthermore, planetary surface environments are discussed.

2.6.2 Spacecraft Heat Transfer Considerations

Heat transfer within a spacecraft can be accomplished with conduction, convection, and radiation. Spacecraft external heat rejection usually is governed by radiation heat transfer. Both conductive and convective heat transfer are proportional to the temperature difference between the two objects exchanging heat as per Equation (2.11), whereas radiative heat transfer is proportional to difference of the fourth power of absolute temperatures as per Equation (2.12):

$$\dot{Q} = G(T_1 - T_2) \tag{2.11}$$

$$\dot{Q} = G_{\text{RAD}}\left(T_1^4 - T_2^4\right) \tag{2.12}$$

In these equations, $\dot{Q}$ is heat transferred per unit time, G is the conductance (conductive or convective), T_1 and T_2 are the temperatures of the two objects exchanging heat, and G_{RAD} is the radiation conductance.

Thermo-Optical Properties

Radiant energy incident on a spacecraft surface can be absorbed or reflected from its surface, or if the surface is at all transparent, it can be transmitted. These effects can occur in any combination, and due to the law of conservation of energy, the sum of the absorbed, reflected, and transmitted energy must equal the incident energy.

All materials absorb, reflect, and transmit energy differently. Indeed, the properties of thermal control surface materials are exploited during spacecraft design. A good thermal design utilizes environmental heating sources where heat is needed and desensitizes other areas to heat absorption while maximizing heat rejection when the spacecraft possesses an excess amount of heat.

For a given wavelength, the ability of a surface to absorb is equivalent to its ability to emit as per Equation (2.13):

$$\alpha_\lambda = \varepsilon_\lambda \quad (2.13)$$

Here, α_λ and ε_λ are the absorptance and emittance at wavelength λ, respectively. For spacecraft applications, incoming thermal radiation is a mixture of solar and infrared spectrum energy, whereas the energy emitted from the spacecraft, in the form of waste heat, is in the infrared spectrum. It is common practice within the spacecraft thermal community to utilize the symbol α for the absorptance of solar spectrum energy, and ε for the absorptance and emittance of infrared energy. In these instances, the λ subscript is dropped. These terms, expressed as numbers between zero and unity, are a measure of the ability of an object to absorb or emit energy compared to a blackbody at the same temperature. For example, a surface with $\varepsilon = 0.7$ can absorb or emit infrared energy only 70% as well as a blackbody at the same temperature.

Overall Spacecraft Heat Balance

The spacecraft reaches steady state temperature conditions when the heat radiating from the spacecraft is equal to the heat absorbed from the environment plus the heat generated within the spacecraft as per Equation (2.14):

$$\dot{Q}_{OUT} = \dot{Q}_{GEN} + \dot{Q}_{IN} \quad (2.14)$$

Here, $\dot{Q}_{OUT}$ is the heat per unit time radiated; $\dot{Q}_{GEN}$ is the power generated within the spacecraft, that is, waste heat from electronics and the like; and $\dot{Q}_{IN}$ is the heat per unit time absorbed from the environment. $\dot{Q}_{IN}$ is comprised of heating from the natural and induced thermal environments. Often, the steady state condition is one of cyclic equilibrium in which the constantly varying on-orbit thermal environment induces a periodic cycling of temperatures between one extreme and another. For the transient case of a flat, unshadowed plate, $\dot{Q}_{IN}$ from Equation (2.14) becomes Equation (2.15):

$$\dot{Q}_{IN}(t) = A[\alpha\cos\varphi(t)\dot{q}_{SOLAR}(t) + \alpha FF \dot{q}_{ALBEDO}(t) + \varepsilon FF \dot{q}_{PLANET}(t)] \quad (2.15)$$

where A is the surface area, $\varphi(t)$ is the angle between the incoming sunlight and the flat surface normal vector, FF is the form factor between the plate and the planet, and $\dot{q}_{SOLAR}$, $\dot{q}_{ALBEDO}$, and $\dot{q}_{PLANET}$ are the solar, reflected solar, and planetary heating fluxes (measured in units of energy per unit time per unit area, or simply power per unit area) as a function of time, t, respectively. For an Earth orbiting spacecraft, $\dot{q}_{SOLAR}$ and $\dot{q}_{PLANET}$ are often considered constant over an orbit and the time dependence can be dropped.

2.6.3 The Natural Thermal Environment

Space is very cold and serves as a sink for spacecraft rejecting heat. Effectively, space acts as a blackbody with a temperature of nearly 3 K. However, for virtually all spacecraft thermal analyses, space can be considered to be at a temperature of absolute zero when spacecraft temperatures are $\gg$ 3 K. In this case, Equation (2.15) reduces to the simpler expression, Equation (2.16):

$$\dot{Q} = G_{\text{RAD}} T^4 \tag{2.16}$$

Even though space is an excellent heat sink, specific attention must be paid to those natural environmental heating sources discussed in the following subsections.

Solar Flux

The Sun, a nuclear furnace that converts hydrogen to helium through the process of thermonuclear fusion, lays at the heart of our solar system. The fusion reaction yields vast amounts of energy and results in streams of electromagnetic radiation and particles emanating from the Sun. Of great interest to thermal engineers is the amount solar heating incident on a spacecraft as it travels throughout its orbit.

The solar irradiance, $E_{b\lambda}$ profile closely approximates that of a Planck blackbody at 5777 K (Anderson, Justus, and Batts 2001) and is calculated using Equation (2.17):

$$E_{b\lambda} = \frac{2\pi h c^2 \lambda^{-5}}{e^{hc/\lambda kT} - 1} \tag{2.17}$$

where h is Planck's constant (6.626×10^{-34} J/s), c is the speed of light (2.998×10^8 m/s), and k is Boltzmann's constant (1.38×10^{-23} J/K) (Holman 1981). Once corrected for the solar distance, the flux contribution as a function of wavelength is in good agreement with the measured solar flux (RReDC n.d.) as shown in Figure 2.30.

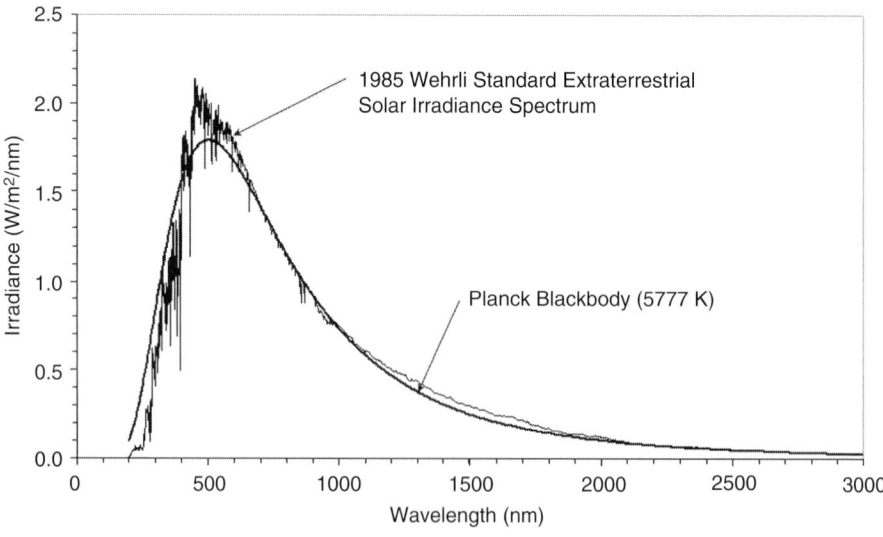

FIGURE 2.30 Solar spectral irradiance as a function of wavelength compared to a Planck blackbody at 5777 K (RReDC n.d.) (Courtesy of NASA).

For the Planck blackbody temperature approximation, the wavelength corresponding to the peak power output, λ_{Max}, is 501 Nm as determined by Wien's displacement law, Equation (2.18):

$$\lambda_{Max} T = 2.897 \times 10^6 \text{Nm} \cdot \text{K} \qquad (2.18)$$

To determine the amount of solar flux incident on a spacecraft at a given solar distance, the Sun is considered to be a point source with a power output, $\dot{S}$, broadcasting uniformly in all directions. Next, a sphere of radius, r_1, is constructed with the Sun at the center. At this distance, the amount of flux, $\dot{q}_{SOLAR}(r_1)$ crossing the spherical surface area is given by Equation (2.19):

$$\dot{q}_{SOLAR}(r_1) = \frac{\dot{S}}{4\pi r_1^2} \qquad (2.19)$$

Similarly, if another sphere of a different radius, r_2, is constructed, the amount of flux crossing the sphere is given by Equation (2.20):

$$\dot{q}_{SOLAR}(r_2) = \frac{\dot{S}}{4\pi r_2^2} \qquad (2.20)$$

From the law of conservation of energy, the amount of energy per unit time crossing the inner sphere must be equal to the amount crossing the outer sphere. Combining Equations (2.19) and (2.20) yields Equation (2.21):

$$\dot{S} = \dot{q}_{SOLAR}(r_1) 4\pi r_1^2 = \dot{q}_{SOLAR}(r_2) 4\pi r_2^2 \qquad (2.21)$$

Equating the last two terms provides an expression for determining the solar flux at any distance from the Sun, r_2, if the solar flux is known at a given location, r_1. The expression becomes Equation (2.22):

$$\dot{q}_{SOLAR}(r_2) = \frac{\dot{q}_{SOLAR}(r_1) r_1^2}{r_2^2} \qquad (2.22)$$

At a distance of $r_1 = 1$ AU, the accepted solar flux value is 1367 ± 5 W/m² (Anderson et al. 2001). We now use Equation (2.22) to determine how the solar flux varies. Consider, for example, that the slightly elliptical orbit of Earth places its closest point to the Sun, the perihelion, at $r_2 = 0.9833$ AU. At this distance, the solar flux is 1414 W/m². At aphelion, $r_2 = 1.0167$ AU and results in a minimum solar flux value of 1322 W/m². It is important to note that the solar flux neither determines the seasons nor does the maximum or minimum flux correspond with the equinoxes or solstices. Earth reaches aphelion, and hence minimum solar flux, on or about July 4 of each year during summer in the northern hemisphere. Perihelion, corresponding to the maximum solar flux, occurs on or about January 4, weeks after the beginning of northern winter.

Planetary Infrared Flux or Outgoing Long Wave Radiation

In the previous subsection, we saw that the spectrum of energy radiated from the Sun is related to its absolute temperature. This is true for all objects and serves as the foundation for the calculation of planetary infrared radiation.

When the solar spectrum energy radiated from the Sun strikes Earth, some flux is absorbed and some is reflected. The absorbed energy heats the atmosphere, water, and ground, resulting in an overall temperature rise. This energy is reradiated in the infrared spectrum, and it then impinges on the orbiting spacecraft affecting its thermal response.

If we assume that the absorbed solar energy is evenly distributed about the planet due to heat transfer in the atmosphere and a rapid rotation rate, we can establish a heat balance that is useful in calculating the planetary infrared heating component. At steady state, the heat radiated by Earth is equal to the heat absorbed from the Sun as per Equation (2.23):

$$\dot{Q}_{RADIATED} = \dot{Q}_{ABSORBED} \tag{2.23}$$

The absorbed solar heating is defined by Equation (2.24):

$$\dot{Q}_{ABSORBED} = \dot{q}_{SOLAR}(1-a)\pi r_e^2 \tag{2.24}$$

where $\dot{q}_{SOLAR}$ is the solar flux at a specified solar distance calculated using Equation (2.22), a is albedo of the Earth, and r_e is radius of the Earth. Because the sunlight intensity varies at different locations on the illuminated half of the Earth, the projected area (πr_e^2) is used instead of half the surface area of the sphere.

The outgoing energy is given by Equation (2.25):

$$\dot{Q}_{RADIATED} = 4\pi r_e^2 \varepsilon_{EARTH} \sigma T^4 \tag{2.25}$$

where ε_{EARTH} is average infrared emittance of the Earth, T is the effective absolute temperature of the Earth, and σ is the Stefan-Boltzmann constant (5.669×10^{-8} W/m² K⁴). Note that Earth is assumed to radiate the distributed heat uniformly in all directions, that is, from the entire surface area, because of its fast spin and heat transfer through the atmosphere. Therefore, the expression for surface area, $4\pi r_e^2$, is used in Equation (2.25).

Combining Equations (2.24) and (2.25) and solving for T yields Equation (2.26):

$$T = \sqrt[4]{\frac{\dot{q}_{SOLAR}(1-a)}{4\varepsilon_{EARTH}\sigma}} \tag{2.26}$$

This intermediate result gives us an idea of the effective radiation temperature of the Earth. If we substitute typical average values for $\dot{q}_{SOLAR} = 1367$ W/m², $a = 0.3$ (a spatial and temporal average), and $\varepsilon_{EARTH} = 1.0$, we see that the effective radiation temperature of Earth is 255 K, which is a temperature below the freezing point of water. Students are often surprised to see this result because they expect Earth to be much warmer. It is also interesting to note that r_e does not appear in the final equation.

The planetary infrared flux, then, is simply Equation (2.27):

$$\dot{q}_{PLANET} = \varepsilon_{EARTH}\sigma T^4 \tag{2.27}$$

For an effective radiation temperature of 255 K, the value for $\dot{q}_{PLANET}$ is 239 W/m². The measured mean value is approximately 234 W/m² (Stephens, Campbell, and Vonder Haar 1981), which is in good agreement with the approximate calculation. However, what about the heat generated in the core of the Earth caused by the decay of radioactive core elements?

As it turns out, the contribution of the energy flux of the core is much less than 1% of that received from the Sun. This suggests that the steady state heat balance approximation for Earth is reasonable. Even so, this simple heat balance is not applicable for every planetary body in the solar system. Jupiter, for example, actually radiates more energy than it takes in from the Sun, a phenomenon caused by radiated gravitational energy from its collapse (Beatty and Chaikin 1990).

Albedo Flux

When calculating planetary infrared flux, we saw that only a fraction of the sunlight incident on the Earth is absorbed and reradiated subsequently to space as infrared flux. The remaining amount is reflected from the land surface, oceans, atmosphere, and cloud tops; and it changes with seasons, local geography, and weather. Whereas some incoming wavelengths are absorbed, the reflected energy is still largely representative of the solar spectrum.

Unlike the assumption made for planetary infrared flux, the albedo flux varies with orbital position. This results from variation in solar zenith angle, ζ, defined as the angle between the line connecting the center of the Earth and the spacecraft and another line connecting the centers of the Earth and the Sun, as well as the albedo factor as a function of incidence angle. This causes nonuniform illumination of the planet. At the subsolar point where $\xi = 0°$, that is, the point on the ground directly beneath the incoming solar flux, direct illumination is at a maximum, and the local albedo flux can be approximated well by Equation (2.28):

$$\dot{q}_{ALBEDO} = a\dot{q}_{SOLAR} \qquad (2.28)$$

Moving away from the subsolar point, the value of ζ increases, and reaches a value of 90° at the terminator, the line dividing the illuminated and darkened halves of the planet. If ζ is the only variation considered, Equation (2.28) can be expanded to the more general expression Equation (2.29):

$$\dot{q}_{ALBEDO}(\zeta) = a\dot{q}_{SOLAR}\cos(\zeta) \qquad (2.29)$$

Additionally, to account for the albedo factor variation with ζ (in degrees), Equation (2.30) can be used for sunlit portions of Earth orbits over the latitude band of ±30° (Anderson et al. 2001):

$$a(\zeta) = a(0) + 4.9115 \times 10^{-9}\zeta^4 + 6.0372 \times 10^{-8}\zeta^3 \\ - 2.1793 \times 10^{-5}\zeta^2 + 1.3798 \times 10^{-3}\zeta \qquad (2.30)$$

where $a(0)$ is the albedo at $\zeta = 0°$.

The Planetary Form Factor

The planetary and albedo fluxes previously calculated established heating fluxes radiating from the planet. Spacecraft surfaces directly facing the planet receive more of these flux components than those facing away from it. In addition, spacecraft in close proximity to the planet receive more flux than those more distant. To determine the amount of variation from these effects, it is necessary to consider the form factor to the planet.

CHAPTER 2 The Space Environment: Natural and Induced

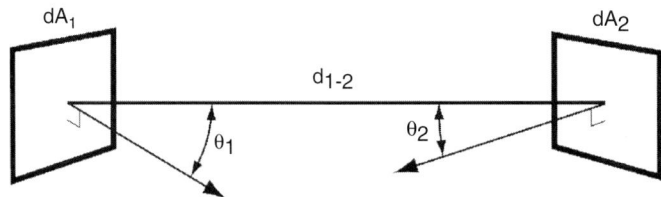

FIGURE 2.31 Differential areas and parameters for the form factor calculation (Graphic by F. A. Brown, courtesy of NASA).

The form factor between two differential areas (Figure 2.31) is defined by Equation (2.31):

$$FF_{1-2} = \frac{1}{A_1} \iint \frac{\cos\theta_1 \cos\theta_2}{\pi d_{1-2}^2} dA_1 dA_2 \tag{2.31}$$

Solving Equation (2.31) for all but a few simplified geometries requires numerical solution techniques. For the case of a nadir facing plate at an altitude, h, above the planet surface and where the surface normal is parallel to a line connecting the plate to the planet center, the form factor between the plate and the planet is given by Equation (2.32):

$$FF_\parallel = \left(\frac{r_e}{r_e + h}\right)^2 \tag{2.32}$$

For the case where the plate surface normal faces perpendicular to nadir, the expression becomes more complex, and Equation (2.33) can be derived from Ballinger et al. (1960):

$$FF_\perp = \left(\frac{1}{2\pi}\right)\left\{\pi - 2\sin^{-1}\left(\sqrt{1 - \left(\frac{r_e}{r_e + h}\right)^2}\right) - \sin\left[2\sin^{-1}\left(\sqrt{1 - \left(\frac{r_e}{r_e + h}\right)^2}\right)\right]\right\} \tag{2.33}$$

For other shapes and orientations, numerical solutions are recommended.

Finally, the incident flux values for $\dot{q}_{PLANET}$ and $\dot{q}_{ALBEDO}$ presented by Equations (2.26) and (2.29), respectively, are multiplied by the appropriate form factor to obtain the incident flux from these heating components.

Combined Albedo and Planetary Infrared Effects

When deriving the planetary infrared flux, we used an idealized heat balance to facilitate the calculation. In reality, Earth, or any body for that matter, experiences local variations in the overall radiation heat balance caused by variations in the local ground albedo and cloud cover, as well as seasonal effects such as snow cover and weather. These variations give rise to deviations from the average heat balance and result in locally varying albedo and planetary infrared fluxes.

2.6 Natural and Induced Thermal Environments

The Earth radiation budget experiment measured these local variations utilizing a series of radiometers (Barkstrom 1984). By processing the local albedo and planetary infrared flux combinations calculated from orbital data collection, a database of combinations of albedo and infrared flux was compiled. The collected data were time averaged over a variety of intervals so that natural environmental flux values pertinent to spacecraft component time response could be generated (Anderson et al. 2001). A representative plot is shown in Figure 2.32. The shaded region represents observed combinations of albedo and planetary flux measured over intervals of 128 s and allows the environmental parameters to be analyzed statistically. Thermal engineers utilize these data when formulating the spacecraft design thermal analysis case matrix.

Selecting proper albedo and outgoing long wave radiation combinations for thermal analysis can be a complicated process, for consideration must be given to the characteristic reaction time of the spacecraft components to the changing environment. If the reaction time is short, a set of values must be selected that includes the variation over the same time frame as the spacecraft response. Short period albedo and outgoing long wave radiation variations often lead to excursions far from the mean albedo and outgoing long wave radiation values. For example, instantaneous Earth radiation budget experiment data results in albedo measurements as low as 0.06 and as high as 0.50. Similar swings in outgoing long wave radiation are observed over short time periods with values ranging from 108 W/m^2 to 332 W/m^2 (Anderson et al. 2001).

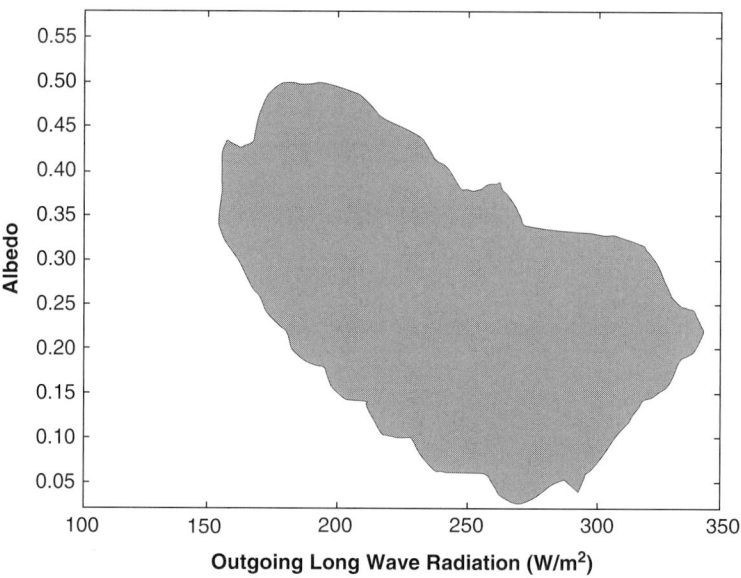

FIGURE 2.32 Albedo and outgoing long wave radiation correlation for high inclination orbits using 128-s averages (Adapted, with permission, from Anderson et al. 2001, courtesy of NASA).

2.6.4 The Induced Thermal Environment

The preceding sections focused on deriving the magnitudes of the natural environmental heating sources. The fluxes that arise from these sources, however, do not result in a uniform spacecraft environment. Rather, they act only on the spacecraft surfaces on which they impinge. A spacecraft surface facing away from the Sun or located within the shadow of a planet experiences no direct solar flux.

This section focuses on the effects of spacecraft attitude and the constantly changing orbit orientation with respect to the Sun as characterized by the β-angle parameter and spacecraft geometric effects.

Spacecraft Attitude Considerations

Every spacecraft is designed to provide a desired function, whether it is to map the planet beneath or transport humans to the Moon. An ideal spacecraft thermal design is one in which the spacecraft maintains component temperatures within acceptable limits in any orientation. This rarely is achieved because of heat rejection requirements, power generation needs, and instrumentation viewing requirements. Although engineers focus on providing as much attitude flexibility as possible, it is important to recognize that such aspirations have practical limits.

These limits often manifest themselves as attitude constraints driven by spacecraft operational needs, such as the requirement to point a sensor at a given target or maintain an orientation that permits communication. Excursions to other attitudes can be required to align a star tracker or provide the desired thrust vector for a propulsive burn. Even so, attitudes can be constrained as well for thermal reasons. Thermal control system radiators must have a good view to a cold sink temperature for them to reject heat effectively. Solar panels require a favorable orientation with respect to the incoming solar flux. It may not be practical for equipment to be designed to operate in full sunlight or when facing the extreme cold of deep space.

The Orbit β-Angle

A spacecraft experiences perturbations that change the orientation of its orbital path over time. At the same time, the orbited body makes it own path around the Sun. The combination of these effects gives rise to changes in the thermal environment, captured in the β-angle parameter, which is defined as the angle between the solar vector and its projection onto the orbit plane.

In a celestial inertial coordinate system, the path of the Sun is tilted at an angle, E, the obliquity of the ecliptic. In this plane, the Sun moves through an angle, Γ, the ecliptic true solar longitude. These parameters are depicted in Figure 2.33. For Earth, E is presently 23.45°, and Γ changes on average at the rate of 0.986°/d, such that a full ecliptic circuit is completed in approximately 365.25 d. The location where the solar path on the ecliptic plane crosses the celestial equator, moving from the southern to the northern hemisphere, is called the *vernal equinox*. This also is known as the first day of northern spring, and for Earth, this occurs on or about March 20 of each year. To calculate β, we must first define a vector pointing to the Sun and another defining the perpendicular to the orbit plane.

2.6 Natural and Induced Thermal Environments

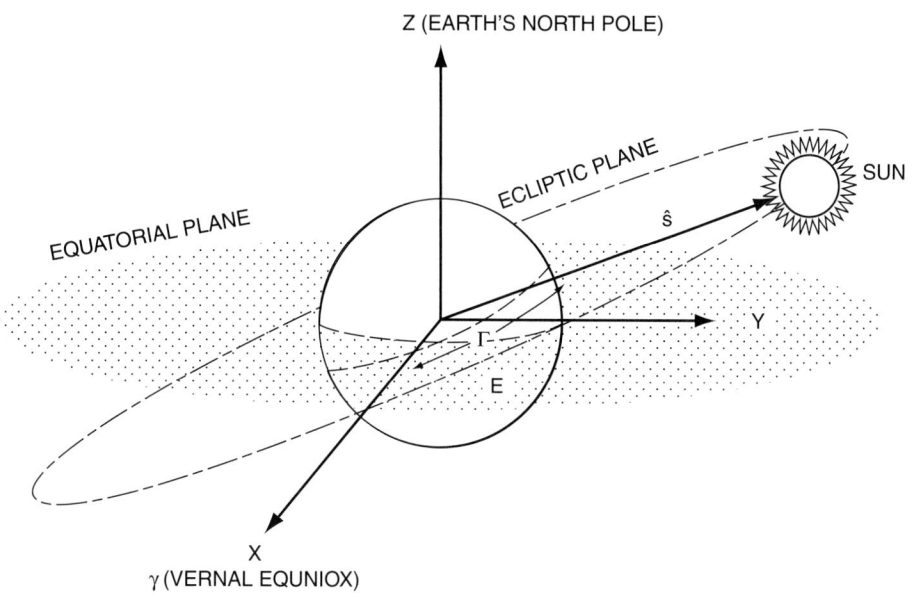

FIGURE 2.33 Parameters used for calculation of the solar vector (Graphic by F. A. Brown; courtesy of NASA).

The solar vector, ŝ, is formed from Euler angle transformations, as shown in Equation (2.34):

$$\{\hat{s}\} = \begin{bmatrix} \cos \Gamma & -\sin \Gamma & 0 \\ \sin \Gamma & \cos \Gamma & 0 \\ 0 & 0 & 1 \end{bmatrix} \begin{bmatrix} 1 & 0 & 0 \\ 0 & \cos E & -\sin E \\ 0 & \sin E & \cos E \end{bmatrix} \begin{Bmatrix} 1 \\ 0 \\ 0 \end{Bmatrix} = \begin{Bmatrix} \cos \Gamma \\ \sin \Gamma \cos E \\ \sin \Gamma \sin E \end{Bmatrix} \quad (2.34)$$

Next, to determine the angle between the solar vector and its projection onto the orbit plane, it is necessary to describe the orbit plane in the same celestial inertial coordinate system (Figure 2.34).

Planes are defined by their normals. Here, we form a vector of unit length and perpendicular to the orbit plane, ô, where the orbit normal vector is tilted an inclination angle, i, with respect to the planet equatorial plane. The orbit plane is shifted at an angle, Ω, which is the right ascension of the ascending node as per Equation (2.35):

$$\{\hat{o}\} = \begin{bmatrix} \cos \Omega & -\sin \Omega & 0 \\ \sin \Omega & \cos \Omega & 0 \\ 0 & 0 & 1 \end{bmatrix} \begin{bmatrix} 1 & 0 & 0 \\ 0 & \cos i & -\sin i \\ 0 & \sin i & \cos i \end{bmatrix} \begin{Bmatrix} 0 \\ 0 \\ 1 \end{Bmatrix} = \begin{Bmatrix} \sin \Omega \sin i \\ -\cos \Omega \sin i \\ \cos i \end{Bmatrix} \quad (2.35)$$

With both unit vectors expressed in the same coordinate system, the cosine of the angle between them is determined by taking their dot product. But this angle is measured between ŝ and the orbit normal vector, ô. Because the orbit plane and the normal vector, by definition, are orthogonal to one another, the angle between ŝ and the orbit plane,

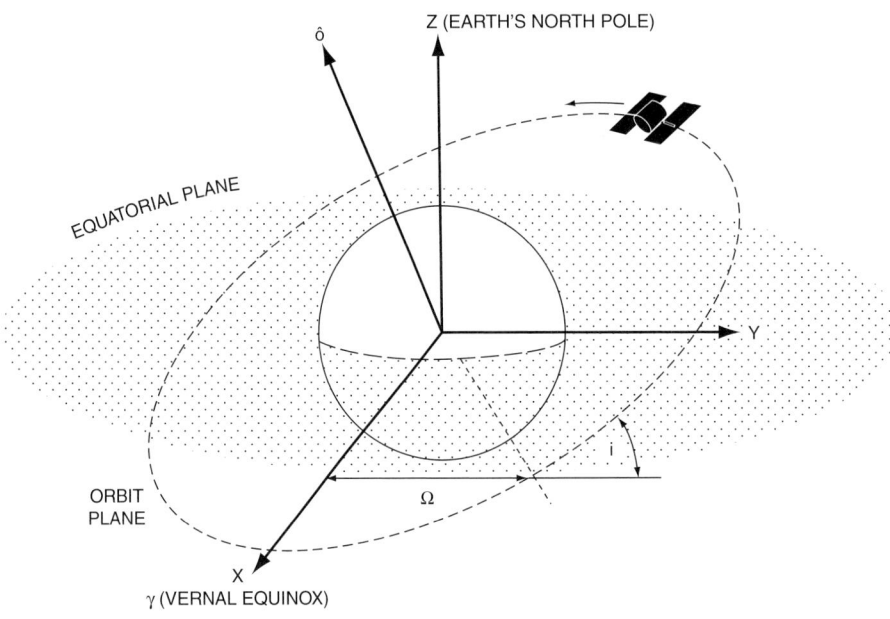

FIGURE 2.34 Parameters used for the calculation of the orbit normal vector (Graphic by F. A. Brown; courtesy of NASA).

therefore, can be determined by taking the arcsine of the dot product of the two vectors as per Equation (2.36):

$$\beta = \sin^{-1}(\hat{o} \cdot \hat{s}) \\ = \sin^{-1}(\cos \Gamma \sin \Omega \sin i - \sin \Gamma \cos E \cos \Omega \sin i + \sin \Gamma \sin E \cos i) \qquad (2.36)$$

As the Sun progresses along the ecliptic plane, Γ undergoes change. As the orbit plane precesses about the central body, Ω changes. As well, β varies with time, and this variation leads to a number of effects on the spacecraft thermal environment.

The first effect is seen in the amount of direct solar radiation impinging on a spacecraft surface. For a given orientation, the amount of sunlight incident on that surface changes as β changes. Consider the simplified orbit and spacecraft geometry presented in Figure 2.35. Surface A experiences a reduction in peak incident solar flux as β increases. At the same time, surface B experiences an increase in peak incident solar flux.

The change in β is dependent upon orbit parameters affecting the rate of change of $\dot{\Omega}$, primarily altitude and orbit inclination. If $\dot{\Gamma}$ is assumed to change at a constant rate (a good approximation for Earth but not necessarily so for other planetary bodies) and a typical International Space Station circular orbit is assumed, the change in β over time is represented by a plot similar to that shown in Figure 2.36.

2.6 Natural and Induced Thermal Environments

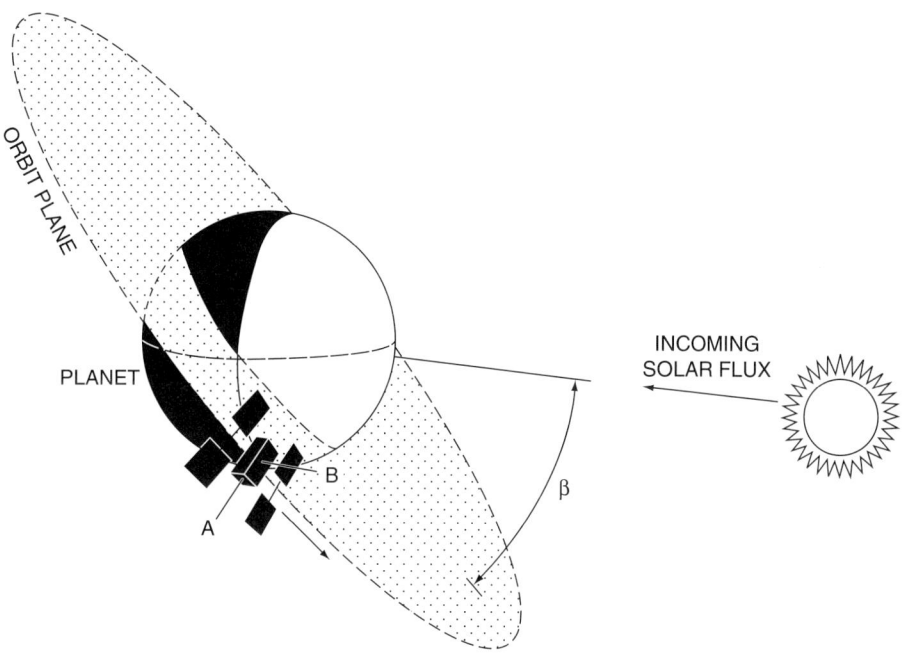

FIGURE 2.35 Variation of the incident solar flux on an orbiting spacecraft with changing β-angle (Graphic by F. A. Brown; courtesy of NASA).

The second effect of a changing β is to affect the amount of time spent on orbit eclipse directly. Consider the geometry presented in Figure 2.37.

By drawing lines of sight tangent to both the planet and the Sun, the boundaries of umbral and penumbral shadows emerge. A spacecraft passing through the penumbra experiences a reduction in direct solar flux due to partial obscuration by the planet. When in the umbral shadow, the spacecraft is blocked completely from solar flux. For orbits close to the planet, such as those of the Space Shuttle and International Space Station missions, the amount of time spent in the penumbral shadow is negligible. Furthermore, at the distance of the Earth from the Sun, the high aspect ratio of the umbral cone can be approximated by a cylinder. For circular orbits in which the orbit altitude is much less than the planet radius, that is, $h \ll r_e$, the fraction of time spent in the planetary umbral shadow, F, is given by Equation (2.37):

$$F = 1 - \frac{\sin^{-1}\left\{\sqrt{\frac{1}{\cos^2\beta}\left[\left(\frac{r_e}{r_e+h}\right)^2 - \sin^2\beta\right]}\right\}}{\pi} \qquad (2.37)$$

The onset of a fully illuminated orbit for a variety of circular orbit altitudes is depicted in Figure 2.38. A discussion of shadowing for elliptical orbits can be found in Mullins (1991).

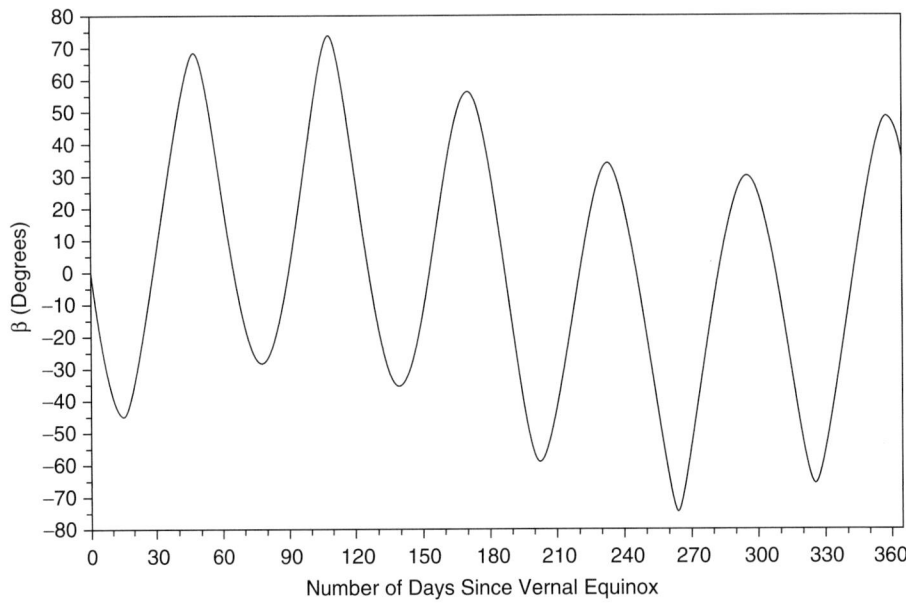

FIGURE 2.36 Representative β-angle profile for the *International Space Station*, assuming a circular orbit with $h = 408$ km, $i = 51.6°$, $\dot{\Gamma} = 0.986°/d$, and $\dot{\Omega} = -4.806°/d$ (Courtesy of NASA).

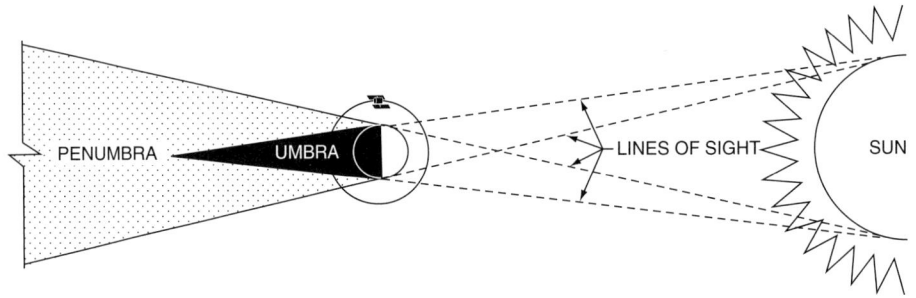

FIGURE 2.37 Sun-planet geometry and the consequence of umbral and penumbral shadows (Graphic by F. A. Brown; courtesy of NASA).

Spacecraft Geometric Effects

The spacecraft configuration, itself, can have a profound effect on its own thermal response. Heating flux from the environment incident on a spacecraft surface can be reflected so that other components receive additional heating flux.

Consider, for example, a solar panel on the side of a spacecraft. If the solar absorptance, α, is less than unity, some energy is reflected and can impinge on other spacecraft

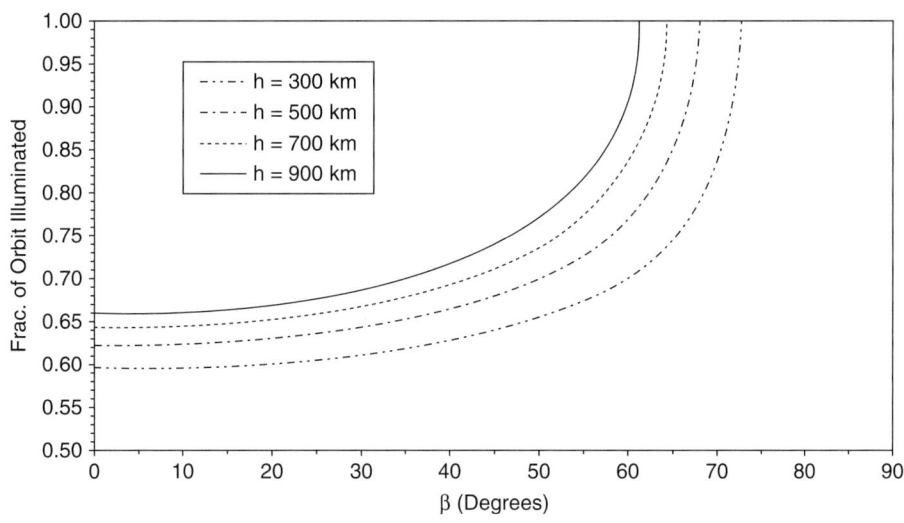

FIGURE 2.38 Variation of the illuminated portion of the orbit as a function of the β-angle. Circular orbit at altitude, h, about Earth (r_e = 6378.14 km) using a cylindrical shadow assumption (Courtesy of NASA).

surfaces. Because thermal response is driven by the incident environment, the additional solar flux incident on the component raises its temperature. Solar panels typically are hot components with highly emissive backside surfaces. These components can contribute a considerable amount of infrared heating.

Another example of the induced thermal environment is seen in the case of radiators. To reject waste heat, radiators typically have a high infrared emittance, ε. Waste heat rejected by the radiator can impinge on other spacecraft surfaces, resulting in increased infrared heat input into other components. Good thermal designs must account for not only the natural environment but the induced environment as well. This is why spacecraft radiators are located with a good form factor to space.

Care must be exercised when considering the spacecraft layout to avoid areas of solar entrapment. Solar energy incident on a cavity can provide a means for solar energy to be trapped within and result in only a limited ability to reradiate the heat generated as a result of its absorption.

2.6.5 Other Lunar and Planetary Environment Considerations

The preceding discussion focused on Earth orbiting spacecraft. The nearly circular orbit of the Earth about the Sun results in relatively minor variations in the solar flux throughout the year. Other planetary bodies, take Mars for example, follow more elliptical paths and experience wider variations in solar flux throughout their orbits.

Spacecraft in orbit about the Moon experience the same variation in solar flux as an Earth orbiting spacecraft. However, because the Moon rotates very slowly with respect to the Sun (29.53 d compared to 1 d for Earth) and is devoid of any atmosphere to spread heat, the previously discussed techniques for calculating planetary infrared flux are not applicable.

For spacecraft on the lunar surface, dayside temperature extremes reach as high as 396 K, hot enough for water to boil in 1 atm. Conversely, during the 2-wk lunar night, surface temperatures dip as low as 116 K. In permanently shadowed craters near the poles, temperatures can be as low as 40 K (Roncoli 2005). Spacecraft designed to survive these challenging environments must be able to remain cool in the daytime heat while staying warm throughout the frigid night.

Additional environmental heating considerations are required when the planetary body has an atmosphere. In such cases, natural and forced convective heat transfer with the atmosphere occurs, and there must be an accounting of atmospheric extinction and scattering of the incoming solar flux.

It is also important to consider the local surface geography. Mountainous terrain in close proximity to the landing site can obscure the Sun when it is low in the sky. If a spacecraft resides in a crater, this also can affect the amount of direct solar flux, the local albedo flux, and the planetary infrared heating, as well as its view to space.

2.7 COMBINED ENVIRONMENTAL EFFECTS

2.7.1 Introduction to Environmental Effects

During its lifetime, space hardware can be subjected to a whole range of environmental parameters, sometimes in a simultaneous and sometimes in a sequential fashion. Each of these environmental parameters, on its own or as a result of the interaction with other parameters, can have an effect on possible changes occurring in that hardware. Therefore, in view of the suitability of an item for its application, it is necessary to understand and evaluate not only the effect of each single parameter but also look at the impact of the combination these parameters.

This is not an easy task and requires a thorough understanding of the various mechanisms and synergisms involved in the interaction between the different environments and the hardware. In a general way, the following three environments to which an item is subjected can be differentiated, each having its own specificity:

- Ground.
- Launch and reentry.
- Space.

During the ground phase, hardware is exposed to all kinds of manufacture, assembly, integration, and test activities, as well as storage and transport. As a rule, space hardware is kept as much as possible in a mild environment, such as a clean room with controlled particulate and molecular contamination levels, relative humidity, temperature, and possibly bioburden. Nevertheless, a device experiences periods of high stress when it is tested. It is subjected to vibrations up to levels similar to that expected during launch, and it also experiences the thermal cycling environment it is to encounter in space. During transport, and to a lesser extent during storage, the environment within which the hardware is kept could be less guarded, resulting in mechanical shocks, temperature variations, or humid conditions.

Launch and reentry environments are extremely harsh, and their effects can be pronounced even though the space hardware is exposed to them for only a comparatively short time. Many problems encountered during operation in space originate from the environment to which the hardware was subjected during launch. Structures, fasteners, and adhesive bonds suffer from the vibration and acoustic stresses, and the resulting defects can worsen due to space thermal cycling, atomic oxygen penetration into microcracks, or other causes. Unfortunately, catastrophic launch and reentry failures are always spectacular and especially tragic when crewmembers are involved. Such failures are exemplified by the launch failure of the *Space Shuttle Challenger* in 1986, and the reentry failure of the *Space Shuttle Columbia* in 2003. In the first case, the low atmospheric temperature at the launch pad caused the sealing property of the O-ring in one of the joints of a solid rocket booster to fail. The cause was embrittlement of its elastomeric material at temperatures below the glass transition region (Presidential Commission 1986). In the case of *Columbia*, damage was caused on the ascent by foam insulation shedding from the external tank, which resulted in tile damage on the leading edge of a wing. The subsequent catastrophic failure of the vehicle on reentry was due to high thermomechanical stresses at the point of damage (CAIB 2003).

Finally, the space environment itself widely is varied, as discussed earlier in this chapter. It is important to take into account a specific mission profile to assess the individual contribution of each distinct environment that can affect the performance or behavior of an equipment item or the spacecraft. Well-known examples are the low Earth orbit missions, where spacecraft perform some 5800 orbits each year. This number of orbits necessarily entails a high number of thermal cycles, each with typical temperature extremes varying over a range of more than 200°C.

In the following subsections, emphasis is placed on the combined effect of the environments observed during spaceflight. The final subsection of this chapter addresses some aspects of ground testing in which the combined space environmental parameters are simulated.

2.7.2 Combined Environments

Certain external systems and components such as thermal control surfaces, multilayer insulation, and radiators, or power generator surfaces, such as solar arrays, are exposed fully to the space environment. Most other subsystems are shielded partially or fully against environmental factors such as ultraviolet light, low- to medium-energy electrons and protons, atomic oxygen, and even extreme thermal excursions.

Various combinations of environmental factors most often affect adversely the external parts of a spacecraft. A few examples of the effects are discussed later in this chapter. Although the material presented elaborates on the primary consequences of environmental

88 CHAPTER 2 The Space Environment: Natural and Induced

exposure, such as direct material degradation, the student should understand that ultimately the implications of these effects is manifested in the performance of higher level subsystems and systems.

2.7.3 Combined Effects

Electrical charging of a satellite is one of the most common anomalies caused by radiation (Leach and Alexander 1995; Lévy 2002). This phenomenon can occur when a satellite moves through a region of charged particles, such as the magnetically trapped particles in the Van Allen belts. It also can occur when a satellite is bombarded by charged particles from the solar wind, higher energetic particles from solar events, or even from a photoelectric effect such as photoemission of electrons.

Large variations in potential can be induced on surfaces with different electrical or dielectric characteristics. An accumulated electrostatic charge ultimately results in a discharge or arc when the breakdown voltage of a material or between two adjacent surfaces is exceeded. Discharges such as these can damage materials severely, cause spurious circuit switching, or result in false sensor readings. In some cases, a discharge is sustained for longer periods and results in major damage. A spectacular example of such an effect was found on the European Retrievable Carrier-1 solar panel (Figure 2.39), where an arc that discharged the batteries was sustained for a sufficient amount of time and with such

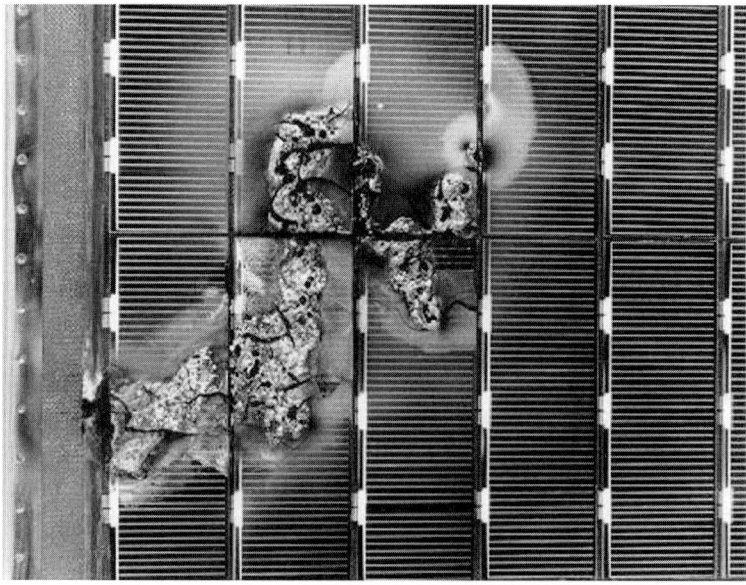

FIGURE 2.39 Solar cells on the European Retrievable Carrier-1 (after retrieval). Sustained arcing in space led to severe damage to the solar panel (Photo: ESA).

intensity that the silicon cell and its cover glass material melted, Kapton films were charred, and the underlying honeycomb of the array was damaged (Bogus 1996).

Combined radiation is probably the most damaging environmental factor for paints used on external surfaces of spacecraft, because high-energy particles and ultraviolet light tend to cause embrittlement of the paint layers. The main effect, however, is degradation of optical properties (Marco and Remaury 2004). The effects of combined radiation can best be studied by measurements taken under vacuum because atmospheric gases can bleach the defects created in the paint. The increase in absorptance observed is due to changes in both pigment and binder. As to the effect on pigment, color centers are created that absorb at specific wavelengths. The effect on the binder is seen by the absorptance edge of the ultraviolet side moving toward the longer wavelengths and the occasional appearance of new bands.

A well-known combined effect of the environments is charge or photo-enhanced deposition of contaminants. The rate at which a material is outgassing increases with temperature, and at first instance, one would expect surface condensation to be dependent on the inverse of the temperature. However, because of the combination of various environmental factors, this is not always the case. In-flight results from the spacecraft charging at a high altitudes experiment (Clark and Hall 1981) demonstrated that an electrostatic reattraction process was also responsible for contaminant deposition. Furthermore, ground experiments showed that condensation rates increase due to ultraviolet radiation (Hall and Stewart 1985; Pereira et al. 2003). For radiators built with optical solar reflectors, the main cause of performance loss is not degradation of the glass substrate or metallic coating caused by proton and electron bombardment but darkening from ultraviolet radiation induced polymerization of the contaminants deposited on surfaces through these processes (Paillous 1985; Faye and Marco 2003).

Rear side silverized or aluminized perfluorinated ethylenepropylene film is a material very regularly used as the external layer of multilayer insulation because of its attractive thermo-optical properties, low solar absorptance, and high thermal emittance. Due to the steric protection of the carbon backbone by the fluorine atoms, it also has a low sensitivity to atomic oxygen erosion (Levadou et al. 1992). Thanks to the post-flight investigations conducted after various Space Shuttle flights, the retrieval of the *Long Duration Exposure Facility* (LDEF), the *Hubble Space Telescope* and space telescope solar array service missions, and the *European Retrievable Carrier-1*, this material has been the subject of a large number of degradation studies (LDEF 1992, 1993, 1995; EURECA 1994; HST 1995).

A few quite interesting results have been reported. First of all, it appeared that the erosion rate depended greatly on the atomic oxygen–ultraviolet ratio. During Space Shuttle flights conducted at relatively low altitude and where the atomic oxygen–ultraviolet ratio is high (1×10^{20} oxygen atoms/cm^2, 44 e.s.h.), erosion rates of $<5 \times 10^{-26}$ cm^3/oxygen atom were observed. These rates are approximately 10 times lower than calculated from the *Long Duration Exposure Facility* (3.8×10^{-25} cm^3/oxygen atom or 9×10^{21} oxygen atoms/cm^2, 11,200 e.s.h.) (Silverman 1995) and more than 100 times lower than found during ground simulation tests conducted in facilities equipped with atomic oxygen sources producing ultraviolet radiation as a side product. Major contributors to the degradation mechanisms are vacuum ultraviolet radiation because of the high extinction coefficient of perfluorinated ethylenepropylene at these wavelengths, and soft X-rays emitted by the Sun during solar flare events (Weihs and Van Eesbeek 1994; Rohr and Van Eesbeek 2005). Recent work by Fischer and Semprimoschnig (2006) based on microthermal analysis concluded that the surface regions of perfluorinated ethylenepropylene are affected severely when exposed to vacuum ultraviolet but less affected when exposed to ultraviolet radiation.

The next quite intriguing point is the severe loss in mechanical integrity of perfluorinated ethylenepropylene retrieved from the *Hubble Space Telescope*, as well as from some parts of the Long Duration Exposure Facility blanket (Figure 2.40). X-rays, electrons, and protons have been reported to cause the larger part of the bulk effects, that is, considerable degradation in elongation and strength of the materials (Van Eesbeek, Lavandou, and Milintchouk 1995; Zuby, Groh, and Smith 1995; Groh et al. 2006).

A typical combined atomic oxygen and thermal cycling effect is the degradation of silver interconnects. Under the influence of atomic oxygen, the silver surface oxidizes. The resulting silver oxide layer has an apparent density three to four times lower than the metal. Because of the porosity of the silver oxide layer, atomic oxygen can further penetrate and continue to react with fresh silver. If this process occurs in an isothermal environment, it will slow down and eventually stop. However, thermal cycling induces stresses in the silver oxide layer because of the difference in the relative rates of

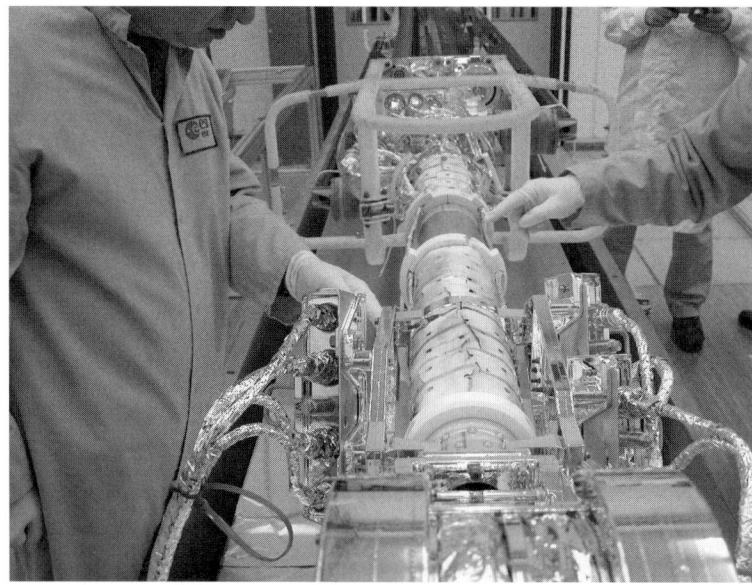

FIGURE 2.40 Postflight investigation at ESA/ESTEC of the *Hubble Space Telescope* in 2002. Severe degradation of perfluorinated ethylenepropylene multilayer insulation is noticeable, including changes in thickness and cracks (Photo: ESA).

contraction and expansion between it and the silver substrate. The oxide layer subsequently flakes off, thereby exposing fresh silver. Eventually, all the silver is converted to silver oxide (Rooij 1989).

Under atomic oxygen exposure, silicon based materials can form surface oxides, such as silicon dioxide, that affect the optical transparency of solar cells (Dever et al. 1994). However, as reported by Moser et al. (2006), a layer of silicon based contamination, once formed, was observed to have protected certain areas of materials retrieved from the *Hubble Space Telescope* from further erosion by atomic oxygen. Even so, diffusion of atomic oxygen through such silicon dioxide layers also has been described (Tagawa et al. 2000). Similar effects have been observed for materials exposed to either ultraviolet illumination or simultaneous exposure to atomic oxygen and ultraviolet rays (Coleman and Luey 1998; Shiloh et al. 1998).

Under certain conditions, there can be some beneficial or restorative effects of atomic oxygen erosion. A good example of this was reported (Roussel et al. 2004) from the French Earth observation satellite, *Satellite Pour l'Observation de la Terre* (SPOT). From a study of the temperature variation among different surfaces, a cyclic change in solar absorptivity was deduced. This finding correlated with the solar cycle, and the concomitant change in atomic oxygen density; that is, at 800 km, there is an increase in the atomic oxygen flux of more than three orders of magnitude between solar minimum and solar maximum. It was concluded that contaminants deposited during solar minimum were subsequently eroded by atomic oxygen during periods of high solar activity.

Highly energetic protons and electrons can create color centers in radiation susceptible optical glasses. Temperature and even solar radiation can have a bleaching effect. Similar effects can occur in paint pigments. Discoloration or bleaching due to proton or electron bombardment is caused by ultraviolet radiation.

Dielectric surfaces are susceptible to electrostatic charging in the space environment. Kapton and perfluorinated ethylenepropylene, both of which are used on external surfaces, have dielectric properties, and an electrical charge of up to several thousands volts can develop across their opposing surfaces. Kapton, however, exhibits a property known as *photoinduced conductivity*, which results in a much lower equilibrium voltage when sunlit. Even more pronounced, irradiation of Kapton by energetic particles increases the dark conductivity of the material, thus ensuring an increased charge transport when the material is shadowed (Lévy et al. 2006).

Impacts from micrometeoroid and orbital debris not only create instantaneous catastrophic incidents but also long-term degradation of equipment performance caused by damage to protective coatings. Once the integrity of these coatings has been compromised, further degradation of the underlying substrate by atomic oxygen (Figure 2.41) is inevitable (Banks, Steuber, and Norris 1998; Banks, Miller, and Groh 2004; LDEF 1992, 1993, 1995). Evolution of the physical and mechanical properties of materials used in a space environment has been found to occur mostly as a consequence of the combined and competitive effects of erosion etching and thermal cycling (Issoupov et al. 2000).

Stress-corrosion cracking is another failure phenomenon that can occur in engineering materials. Typically a process seen in metals, it is seen in ceramic and polymeric materials

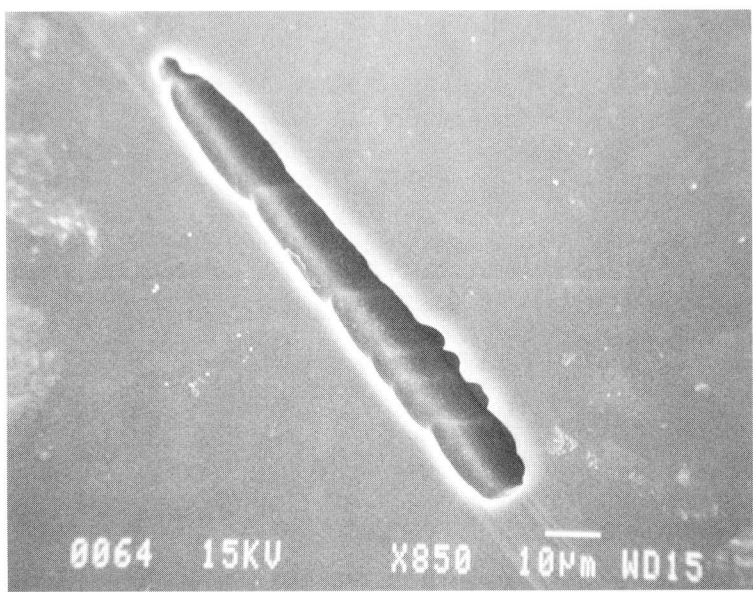

FIGURE 2.41 Undercut cavity in Kapton after removal of aluminized coating (Banks et al. 1998; courtesy of NASA).

as well. It is characterized by slow environmentally induced crack propagation that results from the combined synergetic interaction of mechanical stress and corrosion (Bussu and Dunn 2002).

2.7.4 Ground Testing for Space Simulation

The *Long Duration Exposure Facility* contained several experiments to evaluate specific environmental effects. In a more recent past, other exposure experiments, such as the materials International Space Station experiment (MISSE) (Groh et al. 2001; Soares et al. 2004) and the Japanese space exposed experiment developed for students (SEEDS) (Imai and Imagawa 2003), flew on the *International Space Station*. In addition, active experiments such as the material exposure and degradation experiment (MEDET), are enlarging our knowledge of combined environmental interaction (Tighe et al. 2002).

However, because of cost and limited flight opportunities, the various space environmental parameters needed to understand the behavior of a material, in most cases, are conducted through ground simulation. A difficulty is that the validity and reliability of results obtained through simulation seldom are verified while in flight. Generally, there is a need to reduce exposure time by performing accelerated testing or extrapolation of results. In such cases, real time testing should provide the baseline data against which all accelerated work should be validated (Gates 2003). When large numbers of parameters are involved, the complexity of a ground simulation is such that carefully considered simplification of the space environment is needed. This is asking for a differentiation between primary and secondary actors, and the process depends on the property being evaluated. Identification of the environment parameters expected to have an impact on the properties studied is imperative. For example, acceleration should be limited to such a level that an observed property change is related only to fluence and not flux. As well, conditions should not result in overstressing another environment factor, such as excessive electron or proton flux that results in overheating the sample (this also applies to ultraviolet radiation). Moreover, the cost and complexity of the setup restrict the number of parameters that can be studied in a single facility. The test environment itself can have an adverse effect on the result obtained. A typical example showing these complexities is the evaluation of contamination deposition during an ultraviolet test in which much greater degradation of thermo-optical properties could take place than would be expected to occur in space. Some property under evaluation also could be influenced by the environment in which the measurement takes place; for example, in the case of ex situ measurement, bleaching can occur, resulting in lower solar absorption than otherwise would be measured in situ.

A few dedicated research centers around the world operate ground facilities for conducting space simulations specifically for the evaluation of environmental effects on materials. These facilities are as diverse as the space environment. They simulate various aspects of the space environment, like high vacuum, atomic oxygen, ultraviolet, vacuum ultraviolet, X-ray, electron and protons of various energies, and even micrometeoroid and orbital debris impact, for testing. As in the case of atomic oxygen simulation, different techniques to simulate the environment in low Earth orbit are used. Some of the facilities are easy to use and require no expensive investment. Such facilities are often used for screening evaluations. Others are rather complex to operate and require large capital investment. These facilities are used for qualification testing because the results obtained from their testing programs appear to be very consistent with the data obtained from material retrieved from space. The reader is referred to the references for an overview of different atomic oxygen sources (Kleiman et al. 2003) and some test results from people operating atomic oxygen sources (Hoflund and Weaver 1994; Grossman et al. 2000; Finckenor et al. 2002; Minton and Garton 2003). Another type of facility specializes in the simulation of the response of materials to electromagnetic energy (Marco, Paillous, and Levadou 1988). The most common type covers ultraviolet facilities, but other wavelength ranges are used as well. Sometimes these facilities need to accelerate particles to rather high solar constants to obtain the required e.s.h. dose. Recent work driven by inner solar system mission requirements are compelled to use such acceleration levels even at temperatures of up to 350°C and with intensities up to 20 solar constants (Heltzel, Semprimoschnig, and Van Eesbeek 2006; Semprimoschnig et al. 2004, 2006). These studies also can elucidate the combined effect of ultraviolet and temperature versus temperature alone. Other combined environmental simulation facilities can embrace various particle sources. So that primary and secondary radiation (up to the megaelectronvolt energy range) do not escape the facility, radiation shielding becomes an important safety consideration. This necessarily drives up size and cost requirements for the facility and the simulations.

REFERENCES

Adams, J. H., R. Silberberg, and C. H. Tsao. (1981) *Cosmic Ray Effects on Microelectronics. Part I. The Near Earth Particle Environment.* Memorandum Report MR-4506-pt. I. Washington, DC: U.S. Navy, Naval Research Laboratory.

Allen, C. S., and J. R. Goodman. (2003) Preparing for flight—The process of assessing the International Space Station acoustic environment. In *Proceedings of NOISE-CON 2003*. Washington, DC: Institute of Noise Control Engineering.

American Conference of Government Industrial Hygienists. (2004) *Threshold Limit Values and Biological Exposure Indices (BIEs)*. Cincinnati, OH: ACGIH.

Anderson, B., C. Justus, and G. Batts. (2001) *Guidelines for the Selection of Near Earth Thermal Environment Parameters for Spacecraft Design*. Technical Memorandum TM-2001-211221. Huntsville, AL: National Aeronautics and Space Administration, Marshall Space Flight Center, pp. 2-17.

Badhwar, G. D. (1999) Radiation dose rates in Space Shuttle as a function of atmospheric density. *Radiation Measurement* 30: 401-414.

Badhwar, G. D., and F. A. Cucinotta. (2000) A comparison on depth dependence of dose and linear energy transfer spectra in aluminum and polyethylene. *Radiation Research* 153: 1-8.

Badhwar, G. D., and P. M. O'Neill. (1992) An improved model of GCR for space exploration missions. *Nuclear Tracks and Radiation Measurements* 20: 403-410.

Ballinger, J., J. Elizalde, R. Garcia-Varela, and E. Christiansen. (1960) *Environmental Control of Space Vehicles. Part II. Thermal Environment of Space*. Internal Report ERR-AN-016. General Dynamics Corporation, Convair.

Banks, B. A., S. K. Miller, and K. de Groh. (2004) *Low Earth Orbital Atomic Oxygen Interactions with Materials*. Technical Memorandum TM-2004-213223. 2nd AIAA IECEC. Cleveland, OH: National Aeronautics and Space Administration, Glenn Research Center.

Banks, B., T. Stueber, and M. Norris. (1998) *Monte Carlo Computational Modeling of the Energy Dependence of Atomic Oxygen Undercutting of Protected Polymers*. Technical Memorandum TM 1998-207423. ICPMSSE-4. Cleveland, OH: National Aeronautics and Space Administration, Glenn Research Center.

Barkas, W. H., and M. J. Berger. (1964) Tables of energy losses and ranges of heavy charged particles. In *Studies in Penetration of Charged Particles in Matter*. NAS-NRC Publication 1133. Washington, DC: National Academy of Science, National Research Council, pp. 103-172.

Barkstrom, B. (1984) The Earth radiation budget experiment (ERBE). *Bulletin of the American Meteorological Society* 65, no. 11: 1170-1185.

Beatty, J., and A. Chaikin. (1990) *The New Solar System*. Cambridge, England: Cambridge University Press and Sky Publishing, p. 135.

Bertsch, D. L., C. E. Fichtel, and D. V. Reames. (1969) Relative abundance of iron group nuclei in solar cosmic rays. *Astrophysical Journal* 157: L53-L56.

Bichsel, H. (1969) *Passage of Charged Particles through Matter*. Publication USC-136-150. Los Angeles: University of Southern California, Department of Physics.

Biswas, S., C. E. Fichtel, and D. E. Guss. (1962) Study of the hydrogen, helium, and heavy nuclei in the November 12, 1960 solar cosmic ray event. *Physical Review* 128, no. 6: 2756-2771.

Biswas, S., C. E. Fichtel, and D. E. Guss. (1966) Solar cosmic ray multiply charged nuclei and the July 18, 1961, solar event. *Journal of Geophysical Research* 71: 4071-4077.

Biswas, S., C. E. Fichtel, D. E. Guss, and C. J. Waddington. (1963) Hydrogen, helium and heavy nuclei from the solar event on November 15, 1960. *Journal of Geophysical Research* 68: 3109-3122.

Bobcock, H. W. (1961) The topology of the Sun's magnetic field and the 22 year cycle. *Astrophysical Journal* 133, no. 2: 572-587.

Bogus, K. (1996) Solar array environmental interactions. In *Space Environment: Prevention of Risks Related to Spacecraft Charging*. Toulouse, France: Cépaduès-Éditions.

Brechtmann, C., and W. Heinrich. (1988) Fragmentation cross sections of 32S at 0.7, 1.2, and 200 GeV/nucleon. *Zeitschrift für Physik* A331: 463-472.

Bussu, G., and B. D. Dunn. (2002) ESA approach to the prevention of stress-corrosion cracking in spacecraft hardware. *Proceedings of the 1st Joint ESA-NASA Space Flight Safety Conference*. Noordwijk, the Netherlands: European Space Agency, European Space Research and Technology Center.

CHABA. (1987) *Guidelines for Noise and Vibration Levels for the Space Station*. NASA Contractor Report CR-178310. Washington, DC: Committee on Hearing, Bioacoustics, and Biomechanics, Commission on Behavioral and Social Sciences and Education, National Research Council.

Clark, D. M., and D. F. Hall. (1981) Flight evidence of spacecraft surface contamination rate enhancement by spacecraft charging obtained with a quartz crystal microbalance. Conference Publication CP 2182. *Proceedings of the 3rd Air Force/NASA Spacecraft Charging Technology Conference*. Cleveland, OH: National Aeronautics and Space Administration, Glenn Research Center.

Coleman, D. J., and K. T. Luey. (1998) Photochemical deposition of spacecraft material outgassing products. *Proceedings of the SPIE International Symposium on Optical Science, Engineering and Instrumentation*. San Diego, CA: International Society for Optical Engineering, p. 336.

Colgate, S. A., W. H. Grasberger, and R. H. White. (1963) The dynamics of a supernova explosion. *Journal of the Physical Society of Japan*, 17, no. A-III: S157.

Columbia Accident Investigation Board. (2003) *Report of the Columbia Accident Investigation Board*. Washington, DC: National Aeronautics and Space Administration, Headquarters.

Cucinotta, F. A. (1993) *Calculations of Cosmic Ray Helium Transport in Shielding Materials*. Technical Publication TP-3354. Houston, TX: National Aeronautics and Space Administration, Johnson Space Center.

Cucinotta, F. A., and R. R. Dubey. (1994) Alpha cluster description of excitation energies in 12C(12C, 3a)X at 2.1 GeV. *Physical Review* C50: 979-984.

Cucinotta, F. A., and M. Durante. (2006) Cancer risk from exposure to galactic cosmic rays: Implications for space exploration by human beings. *Lancet Oncology* 7: 431-435.

Cucinotta, F. A., M. Y. Kim, and L. Ren. (2006a) Evaluating shielding effectiveness for reducing space radiation cancer risks. *Radiation Measurement* 41: 1173-1185.

Cucinotta, F. A., J. W. Wilson, P. Saganti, X. Hu, M. Y. Kim, T. Cleghorn, C. Zeitlin, and R. K. Tripathi. (2006b) Isotopic dependence of GCR fluence behind shielding. *Radiation Measurement* 41: 1235-1249.

Cucinotta, F. A., F. K. Manuel, J. Jones, G. Iszard, J. Murrey, B. Djojonegro, and M. Wear. (2001) Space radiation and cataracts in astronauts. *Radiation Measurement* 156: 460-466.

Cucinotta, F. A., L. W. Townsend, and J. W. Wilson. (1992) Multiple scattering effects in quasi-elastic a-4He scattering. *Physical Review* C46: 1451-1456.

Cucinotta, F. A., J. W. Wilson, J. L. Shinn, R. K. Tripathi, K. M. Maung, F. F. Badavi, R. Katz, and R. R. Dubey. (1997a) Computational procedures and database development. In J. W. Wilson, J. Miller, A. Konradi, and F. A. Cucinotta (eds.), *NASA Workshop on Shielding Strategies for Human Space Exploration*.

Conference Publication CP-3360. Houston, TX: National Aeronautics and Space Administration, Johnson Space Center.

Cucinotta, F. A., J. W. Wilson, and L. W. Townsend. (1997b) Abrasion-ablation model for neutron production in heavy ion collisions. *Nuclear Physics* A619: 202-212.

Cucinotta, F. A., J. W. Wilson, R. K. Tripathi, and L. W. Townsend. (1998) Microscopic fragmentation model for galactic cosmic ray studies. *Advances in Space Research* 22: 533-537.

Cucinotta, F. A., J. W. Wilson, J. R. Williams, and J. F. Dicello. (2000) Analysis of Mir-18 results for physical and biological dosimetry: Radiation shielding effectiveness in LEO. *Radiation Measurement* 32: 181-191.

Del Basso, S., M. Laible, E. O'Keefe, A. Steelman, S. Scheer, and S. Thampi. (2002) Capitalization of early ISS data for assembly complete microgravity performance. AIAA 2002-606. *Proceedings of the 40th AIAA Aerospace Sciences Meeting and Exhibit*. Reno, NV: American Institute of Aeronautics and Astronautics.

DeLombard, R., K. Hrovat, and K. McPerson. (1999) Comparison tools for assessing the microgravity environment of orbital missions, carriers and conditions. *Proceedings of the 16th IEEE Instrumentation and Measurement Technology Conference* 2, pp. 929-934.

Dever, J. A., E. J. Bruckner, D. A. Scheiman, and C. R. Stidham. (1994) *Combined Contamination and Space Environment Effects on Solar Cells and Thermal Control Surfaces*. AIAA 94-2627. *Technical Memorandum TM 106592*. Cleveland, OH: National Aeronautics and Space Administration, Glenn Research Center.

Divine, N. (1993) Five populations of interplanetary meteoroids. *Journal of Geophysical Research* 98: 17029-17048.

Department of Defense. (1999) *Department of Defense Design Criteria Standard: Human Engineering*. MIL-STD-1472F. Washington, DC: U.S. Department of Defense.

Department of Defense. (2002) *Interface Standard, Electromagnetic Environmental Effects Requirements for Systems*. MIL-STD-464A. Washington, DC: U.S. Department of Defense.

Dueber, R. E., and D. S. McKnight. 1993. *Chemical Principles Applied to Spacecraft Operations*. Malabar, FL: Krieger.

Durgaprasad, N., C. E. Fichtel, D. E. Guss, and D. V. Reames. (1968) Nuclear charge spectra and energy spectra in the September 2, 1966, solar particle event. *Astrophysical Journal* 154: 307-315.

EURECA. (1994) *EURECA, the European Retrievable Carrier*. Technical Report WPP069. Paris: European Space Agency.

Faye, D., and J. Marco. (2003) Effects of ultraviolet and protons radiation on thermal control coatings after contamination. SP-540. *Proceedings of the 9th International Symposium on Multimedia Software Engineering*. Noordwijk, the Netherlands: European Space Agency.

Fields, B. D., K. A. Olive, and D. N. Schramm. (1994) *Cosmic Ray Models for Early Galactic Lithium, Beryllium and Boron Production*. Fermilab-Pub-94/010-A. Batavia, IL: U.S. Department of Energy, Fermi National Accelerator Laboratory.

Finckenor, M. M., D. L. Edwards, J. A. Vaughn, T. A. Schneider, M. A. Hovater, and D. Hoppe. (2002) *Test and Analysis Capabilities of the Space Environment Effects Team at Marshall Space Flight Center*. Technical Publication TP-2002-212076. Huntsville, AL: National Aeronautics and Space Administration, Marshall Space Flight Center.

Fischer, H., and C. O. A. Semprimoschnig. (2006) Analysis of the durability of polymeric materials in space—Application of scanning thermal microscopy. *Proceedings of the 10th International Symposium on Multimedia Software Engineers, and the 8th International Conference on Protection of*

Materials and Structures in a Space Environment, Collioure, France. SP-616. Noordwijk, the Netherlands: European Space Agency.

Forbush, S. E. (1937) On the effects in cosmic ray intensity observed during the recent magnetic storm. *Physical Review* 51: 1108-1109.

Freier, P. S. (1963) Emulsion measurements of solar alpha particles and protons. *Journal of Geophysical Research* 68: 1805-1810.

Freier, P. S., and W. R. Webber. (1963) Exponential rigidity spectrums for solar flare cosmic rays. *Journal of Geophysical Research* 68: 1605-1629.

Friedman, R., and G. Urban. (2000) *Spacecraft fire safety: Key features of fires in low gravity and microgravity.* Available at http://exploration.grc.nasa.gov/life/spacecraft_fire_safety/features.htm (cited June 25, 2007).

Gates, T. S. (2003) *On the Use of Accelerated Test Methods for Characterization of Advanced Composite Materials.* Technical Publication TP-2003-212407. Houston, TX: National Aeronautics and Space Administration, Johnson Space Center.

Gilles, D., S. Lehoczky, F. Szofran, D. Watring, H. Alexander, and G. Jerman. (1997) Effect of residual accelerations during microgravity directional solidification of mercury-cadmium-telluride on the USMP-2 mission. *Journal of Crystal Growth* 174, nos. 1-4.

Ginzburg, V. L. (1958) The origin of cosmic radiation. In J. G. Wilson and S. A. Wouthuysen (eds.), *Progress in Elementary Particle and Cosmic Ray Physics*, vol. IV, p. 339. New York: Wiley and Sons.

Goodman, J. R. (2003) International Space Station acoustics. Inc03_125. *Proceedings of NOISE-CON 2003.* Cleveland, OH: Institute of Noise Control Engineering.

Goswami, J. N., R. E. McGuire, R. C. Reedy, D. Lal, and R. Jha. (1988) Solar flare protons and alpha particles during the last three solar cycles. *Journal of Geophysical Research* 93, no. A7: 7195-7205.

Groh, K. K. de, B. A. Banks, A. Hammerstrom, E. Youngstrom, C. Kaminski, L. Marx, E. Fine, J. D. Gummow, and D. Wright. (2001) *MISSE PEACE Polymers: An International Space Station Environmental Exposure Experiment.* AIAA 2001-4923 and Technical Memorandum TM-2001-211311. Cleveland, OH: National Aeronautics and Space Administration, Glenn Research Center.

Groh, K. K. de, J. A. Dever, A. Snyder, S. Kaminski, C. E. McCarthy, A. L. Rapoport, and R. N. Rucker. (2006) Solar effects on tensile and optical properties of the Hubble Space Telescope silver-teflon insulation. *Proceedings of the Materials Research Society Symposium.*

Grossman, I., G. Gouzman, M. Lempert, Y. Shiloh, Y. Noter, and E. Lifschitz. (2000) Advanced simulation facility for in situ characterization of space environmental effects on materials. *Proceedings of the Eighth International Symposium on Multimedia Software Engineers, and the Fifth International Conference on Protection of Materials and Structures in a Space Environment.* Noordwijk, the Netherlands: European Space Agency.

Grosveld, F. W., J. R. Goodman, and G. D. Pilkinton. (2003) International Space Station acoustic noise control—Case studies. Inc03_117. *Proceedings of NOISE-CON 2003.* Cleveland, OH: Institute of Noise Control Engineering.

Grün, E., H. Zook, H. Fechtig, and R. Giese. (1985) Collisional balance in the meteoritic complex. *Icarus* 62: 244-272.

Grün, E., B. Gustofson, S. Dermott, and H. Fechtig. (2001) *Interplanetary Dust.* Berlin/Heidelberg/New York: Springer Verlag.

Haffner, J. W. (1967) *Radiation and Shielding in Space.* Nuclear Science and Technology, Series 4. New York: Academic Press.

Hall, D. F., and T. B. Stewart. (1985) Photo-enhanced spacecraft contamination deposition. AIAA-85-0953. *Proceedings of the 20th Thermophysics Conference*. Long Beach, CA: American Institute of Aeronautics and Astronautics.

Hamacher, H. (1996) Spacelab's microgravity environment—A characterization based on data of previous missions. Meeting Papers on Disc 96-2229. *The 19th Advanced Measurement and Ground Testing Technology Conference*. New Orleans: American Institute of Aeronautics and Astronautics, pp. 17-20.

Hamacher, H., B. Fitton, and J. Kingdon. (1987) The environment of Earth orbiting systems. In H. U. Walter (ed.), *Fluid Sciences and Material Sciences in Space: A European Perspective*. Berlin: Springer, pp. 1-50.

Heltzel, S., C. O. A. Semprimoschnig, and M. R. J. Van Eesbeek. 2006. Environmental testing of thermal control material at elevated temperature and high intensity radiation. SP-616. *Proceedings of the 10th International Symposium on Multimedia Software Engineers, and the 8th International Conference on Protection of Materials and Structures in a Space Environment*. Noordwijk, the Netherlands: European Space Agency.

Hesse, A., B. S. Acarya, U. Heinbach, W. Heinrich, M. Henkel, C. Koch, B. Luzzietti, C. Pfeiffer, M. Simon, J. A. Esposito, V. K. Balasubrahmanyan, L. M. Barbier, E. R. Christian, J. F. Ormes, and R. E. Streitmatter. (1991) The isotopic composition of silicon and iron in the cosmic radiation as measured by the ALICE experiment. *Proceedings of the 22nd International Conference*. Dublin: The Dublin Institute for Advanced Studies, pp. 596-599.

Hoflund, G. B., and J. F. Weaver. (1994) Performance characteristics of a hyperthermal oxygen atom generator. *Measurement Science and Technology* 5, no. 3: 201-204.

Holman, J. (1981) *Heat Transfer*. New York: McGraw-Hill, p. 310.

Howe, G. R., L. B. Zablotska, J. J. Fix, J. Egel, and J. Buchanan. (2004) Analysis of the mortality experience amongst U.S. nuclear power industry workers after chronic low dose exposure to ionizing radiation. *Radiation Research* 162: 517-526.

Hubble Space Telescope. (1995) Solar array workshop. ESA-WPP-77. *Proceedings of the Hubble Space Telescope Solar Array Workshop*. Noordwijk, the Netherlands: European Space Agency, European Space Research and Technology Center.

Imai, F., and K. Imagawa. (2003) NASDA's space environment exposure experiment on ISS—First retrieval of SM/MPAC and SEED. SP-540. *Proceedings of the Ninth International Symposium on Multimedia Software Engineering*. Noordwijk, the Netherlands: European Space Agency.

International Organization for Standardization. (2003) *Ergonomics—Danger signals for public and work areas—Auditory danger signals*. ISO Standard 7731:2003. Geneva: International Organization for Standardization.

Issoupov, V., O. V. Startsev, A. Paillous, V. Viel, J. Siffre, and E. F. Nikishin. (2000) Generalized conclusions made on the basis of in-flight and ground based simulation experiments. *Proceedings of the 10th International Symposium on Multimedia Software Engineers, and the 8th International Conference on Protection of Materials and Structures in a Space Environment*. Arcachon, France; Noordwijk, the Netherlands: Centre National d'Etudes Spatiales, European Space Agency.

Janni, J. F. (1966) *Calculations of Energy Loss, Range Path Length, Straggling, Multiple Scattering, and the Probability of Inelastic Nuclear Collisions for 0.1 to 1000 MeV Protons*. Report NWL-TR-65-150. Wright-Patterson Air Force Base, OH: Air Force Weapons Laboratory.

Jenniskens, P. (1994) Meteor stream activity—I. The annual meteor streams. *Journal of Astronomy and Astrophysics* 287: 990-1013.

Kim, M. Y., F. A. Cucinotta, and J. W. Wilson. (2006a) Mean occurrence frequency and temporal risk analysis of solar particle events. *Radiation Measurement* 41: 1115-1122.

Kim, M. Y., F. A. Cucinotta, and J. W. Wilson. (2006b) A temporal forecast of radiation environments for future space exploration missions. *Radiation and Environmental Biophysics* 46, no. 2: 95-100.

Kim, M. Y., K. A. George, and F. A. Cucinotta. (2006c) Evaluation of skin cancer risk for lunar and Mars missions. *Advances in Space Research* 37: 1798-1803.

Kim, M. Y., X. Hu, and F. A. Cucinotta. 2005. Effect of shielding materials from SPEs on the Lunar and Mars surface. AIAA 2005-6653. *Space 2005*. Long Beach, CA: American Institute of Aeronautics and Astronautics.

Kim, M. Y., and J. W. Wilson. (2000) *Examination of Solar Cycle Statistical Model and New Prediction of Solar Cycle 23*. Technical Publication TP-2000-210536. Houston, TX: National Aeronautics and Space Administration, Johnson Space Center.

Kim, M. Y., J. W. Wilson, and F. A. Cucinotta. (2004) *An Improved Solar Cycle Statistical Model for the Projection of Near Future Sunspot Cycles*. Technical Publication TP-2004-212070. Houston, TX: National Aeronautics and Space Administration, Johnson Space Center.

Kim, M. Y., J. W. Wilson, and F. A. Cucinotta. (2006d) A solar cycle statistical model for the projection of space radiation environment. *Advances in Space Research* 37: 1741-1748.

King, J. H. (19720) Unpublished records of a NASA workshop held several weeks after the August 1972 event. Correspondence dated October 24, from J. H. King at Goddard Space Flight Center to A. C. Hardy at Johnson Space Center. Houston, TX, National Aeronautics and Space Administration, Johnson Space Center.

Kleiman, J., Z. Iskanderova, Y. Gudimenko, and S. Horodetsky. (2003) Atomic oxygen beam sources: A critical overview. SP-540. *Proceedings of the Ninth International Symposium on Multimedia Software Engineering*. Noordwijk, the Netherlands: European Space Agency.

Klinkrad, H. (2006) *Space Debris: Models and Risk Analysis*. Springer-Praxis Books in Aeronautical Engineering. Berlin-Heidelberg-New York: Springer Verlag.

Knott, C. N., S. Albergo, Z. Caccia, C.-X. Chen, S. Costa, H. J. Crawford, M. Cronqvist, J. Engelage, P., Ferrando, R. Fonte, L. Greiner, T. G. Guzik, A. Insolia, F. C. Jones, P. J. Lindstrom, J. W. Mitchell, R. Potenza, J. Romanski, G. V. Russo, A. Soutoul, O. Testard, C. E. Tull, C. Tuve, C. J. Waddington, W. R. Webber, and J. P. Wefel. (1996) Interactions of relativistic neon to nickel projectiles in hydrogen, elemental production cross sections. *Physical Review* C53: 347-357.

Knott, C. N., S. Albergo, Z. Caccia, C.-X. Chen, S. Costa, H. J. Crawford, M. Cronqvist, J. Engelage, L. Greiner, T. G. Guzik, A. Insolia, P. J. Lindstrom, J. W. Mitchell, R. Potenza, G. V. Russo, A. Soutoul, O. Testard, C. E. Tull, C. Tuve, C. J. Waddington, W. R. Webber, and J. P. Wefel. (1997) Interactions of relativistic 36Ar and 40Ar nuclei in hydrogen: Isotopic production cross sections. *Physical Review* C56: 398-406.

Kundu, M. R., and F. T. Haddock. (1960) A relation between solar radio emission and polar cap absorption of cosmic noise. *Nature* 186: 610-613.

LDEF. (1992) LDEF—69 months in Space. *Proceedings of the First Post Retrieval Symposium. Conference.* Publication CP3134. Langley, VA: National Aeronautics and Space Administration, Langley Research Center.

LDEF. (1993) LDEF—69 months in Space. *Proceedings of the Second Post Retrieval Symposium. Conference.* Publication CP3194. Langley, VA: National Aeronautics and Space Administration, Langley Research Center.

LDEF. (1995) LDEF—69 months in Space. *Proceedings of the Third Post Retrieval Symposium. Conference.* Publication CP3275. Langley, VA: National Aeronautics and Space Administration, Langley Research Center.

Leach, R. D., and M. B. Alexander. (1995) *Failures and Anomalies Attributed to Spacecraft Charging*. Research Paper RP-1375. Huntsville, AL: National Aeronautics and Space Administration, Marshall Space Flight Center.

Letaw, J., C. H. Tsao, and R. Silberberg. (1983) Matrix methods of cosmic ray propagation. In M. M. Shapiro (ed.), *Composition and Origin of Cosmic Rays*. Dordrecht, the Netherlands: D. Reidel Publishing Company, pp. 337-342.

Levadou, F., M. Frogatt, M. Rott, and E. Schneider. (1992) Preliminary investigations into UHCRE thermal control materials. *Proceedings of the First Post Retrieval Symposium*. Conference Publication CP3134. Langley, VA: National Aeronautics and Space Administration, Langley Research Center, pp. 875-898.

Lévy, L. (2002) Material charging. In *Space Environment: Prevention of Risks Related to Spacecraft Charging*. Toulouse, France: Cépaduès-Éditions, pp. 241-265.

Lévy, L., B. Dirassen, R. Reulet, M. Van Eesbeek, and P. Molinié. (2006) Dark and radiation induced conductivity on space used external coatings. SP-616. *Proceedings of the 10th International Symposium on Multimedia Software Engineers, and the 8th International Conference on Protection of Materials and Structures in a Space Environment, Collioure, France*. Noordwijk, the Netherlands: European Space Agency.

Lukasiak, A., P. Ferrando, F. B., McDonald, and W. R. Webber. (1993) Cosmic ray composition of 6<Z<8 nuclei in the energy range 50-150 MeV/n by the *Voyager* spacecraft during the solar minimum and maximum periods. *Proceedings of the 23rd International Cosmic Ray Conference*, Calgary, Canada, pp. 539-542.

Lukasiak, A., F. B. McDonald, W. R. Webber, and P. Ferrando. (1995) Voyager measurements of the isotopic composition of Sc, Ti, V, Cr, Mn, and Fe nuclei. *Proceedings of the 24th International Cosmic Ray Conference,* Rome, Italy, pp. 576-579.

Marco, J., A. Paillous, and F. Levadou. (1988) Combined radiation effect on optical reflectance of thermal control coatings. *Proceedings of the Fourth European Symposium on Spacecrafts Materials in Space Environment,* Toulouse, France, p. 121.

Marco, J., and S. Remaury. (2004) Evaluation of thermal control coatings degradation in simulated geo-space environment. *High Performance Polymers* 16, no. 2: 177-196.

McBride, N. (1997) The importance of the annual meteoroid stream to spacecraft and their detectors. *Advances in Space Research* 20, no. 8: 1513-1516.

McCracken, K. G., G. A. M. Dreschhoff, E. J. Zeller, D. F. Smart, and M. A. Shea. (2001) Solar cosmic ray events for the period 1561-1994 1. Identification in polar ice, 1561-1950. *Journal of Geophysical Research* 106, no. A10: 21585-21598.

Messerschmid, E., and R. Bertrand. (1999) *Space Stations: Systems and Utilization*. Berlin-Heidelberg-New York: Springer.

Meyer, P. (1969) Cosmic rays in the galaxy. *Annual Review of Astronomy and Astrophysics* 7: 1.

Minton, T. K., and D. J. Garton. (2003) Dynamics of atomic oxygen induced polymer degradation in low Earth orbit. In R. A. Dressler (ed.), *Advanced Series in Physical Chemistry: Chemical Dynamics in Extreme Environments*. Singapore: World Scientific, pp. 420-489.

Moser, M., C. O. A. Semprimoschnig, M. R. J. Van Eesbeek, and R. Pippan. (2006) Post flight investigations on thermal control material from HST. SP-616. *Proceedings of the 10th International Symposium on Multimedia Software Engineers, and the 8th International Conference on Protection of Materials and Structures in a Space Environment*. Collioure, France. Noordwijk, The Netherlands: European Space Agency.

Mullins, L. (1991) Calculating satellite umbra/penumbra entry and exit positions and times. *Journal of Astronautics Sciences* 39, no. 4: 411-422.

National Academy of Sciences. (1996) *Report of the Task Group on the Biological Effects of Space Radiation: Radiation Hazards to Crews on Interplanetary Mission*. Washington, DC: National Academy of Sciences, Space Science Board.

National Aeronautics and Space Administration. (1972) *Acoustic Noise Criteria*. Design and Procedural Standard 145. Huntsville, AL: National Aeronautics and Space Administration, Marshall Space Flight Center.

National Aeronautics and Space Administration. (1995) *Man-System Integration Standard*. STD-3000, vol. I, Part A. Houston, TX: National Aeronautics and Space Administration, Johnson Space Center.

National Aeronautics and Space Administration. (2000) *Pressurized Payload Interface Control Document*. Space Station Program Document, SSP 57000, Rev E. Houston, TX: National Aeronautics and Space Administration, Johnson Space Center.

National Aeronautics and Space Administration. (2003) *Payload Verification Program*. Space Station Program Document SSP 57011-B. Houston, TX: National Aeronautics and Space Administration, Johnson Space Center.

National Aeronautics and Space Administration. (2004) *International Space Station Flight Integration Standard STD-3000/T*. Space Station Program Document SSP 50005, Rev. D. Houston, TX: National Aeronautics and Space Administration, Johnson Space Center.

National Aeronautics and Space Administration. (2007) *Program tasks and bibliography*. Available at http://peer1.nasaprs.com/peer_review/taskbook/taskbook.html (cited: June 25, 2007).

National Council on Radiation Protection. (2000) *Radiation Protection Guidance for Activities in Low Earth Orbit*. Report No. 132. Bethesda, MD: National Council on Radiation Protection and Measurements.

National Oceanographic and Atmospheric Administration. (2006) *GOES Space Environment Monitor (SEM) Data*. Boulder, CO: National Oceanographic and Atmospheric Administration, National Geophysical Data Center.

National Research Council. (2000) *Radiation and the International Space Station: Recommendations to Reduce Risk*. National Research Council. Washington, DC: National Academy Press.

Naumann, R. J., and H. W. Herring. (1980) *Materials Processing in Space: Early Experiments*. Special Publication SP-443. Washington, DC: National Aeronautics and Space Administration, Headquarters.

Oswald, M., S. Stabroth, C. Wiedemann, P. Wegener, and C. Martin. (2006) Upgrade of the MASTER Model. In *Final Report. ESA contract 18014/03/D/HK*. Braunschweig, Germany: European Space Agency.

Paillous, A. (1985) Long term tests of contaminated OSRs under combined environment. *Proceedings of the Third European Symposium on Materials in the Space Environment*. Noordwijk, the Netherlands: European Space Agency, European Space Research and Technology Center, p. 245.

Parker, E. N. (1965) The passage of energetic charged particles through interplanetary space. *Planetary and Space Science* 13: 9–49.

Parker, E. N., and V. R. West. (1973) *Bioastronautics Data Book*. Special Publication SP-3006, Springfield, VA: National Technical Information Service.

Pearsons, K. S. (1975) *Recommendations for Noise Levels in the Space Shuttle*. Job No. 157160. Houston, TX: National Aeronautics and Space Administration, Johnson Space Center.

Pereira, A., J-F. Roussel, M. Van Eesbeek, J. M. Guyt, O. Schmeitzky, and D. Faye. (2003) Study of the UV enhancement of contamination. SP-540. *Proceedings of the Ninth International Symposium on Multimedia Software Engineering*. Noordwijk, the Netherlands: European Space Agency.

Piland, R. O. (1980) *Evaluation of OV-102 Acoustical Noise Test at KSC*. Memorandum SD3-80-278. Houston, TX: National Aeronautics and Space Administration, Johnson Space Center.

Pilkinton, G. D. (2003) ISS acoustics mission support. *Proceedings of NOISE-CON 2003*. Washington, DC: Institute of Noise Control Engineering.

Presidential Commission. (1986) *Report of the Presidential Commission on the Space Shuttle Challenger Accident*. Washington, DC: U.S. Government Printing Office.

Preston, D. L., Y. Shimizu, D. A. Pierce, A. Suyumac, and K. Mabuchi. (2003) Studies of mortality of atomic bomb survivors. Report 13. Solid cancer and non-cancer disease mortality: 1950–1977. *Radiation Research* 160: 381–407.

Rohr, T., and M. Van Eesbeek. (2005) Polymer materials in the space environment. *Proceedings of the Eighth International Symposium on Polymers for Advanced Technologies*. Budapest, Hungary.

Roncoli, R. (2005) *Lunar Constants and Models Document*. Document D-32296. Pasadena, CA: California Institute of Technology, Jet Propulsion Laboratory, p. 22.

Rooij, A. de. (1989) The oxidation of silver by atomic oxygen. *ESA Journal* 13: pp. 363.

Roussel, J. F., I. Alet, D. Faye, and A. Pereira. (2004) The effect of space environment on spacecraft thermal control coatings on Sun synchronous orbits. *Journal of Spacecraft and Rockets* 41, no. 5: 812–820.

RReDC. (n.d.) *Wehrli 1985 AM0 spectrum*. Available at http://rredc.nrel.gov/solar/spectra/am0/wehrli1985.new.html (cited November 17, 2006).

Schimmerling, W., J. Miller, M. Wong, M. Rapkin, J. Howard, H. G. Spieler, and J. V. Blair. (1989) The fragmentation of 670AMeV neon-20 as a function of depth in water. *Radiation Research* 120: 36–51.

Semprimoschnig, C. O. A., S. Heltzel, A. Polsak, and M. Van Eesbeek. (2004) Space environmental testing of thermal control foils at extreme temperatures. *High Performance Polymers* 16, no. 2: 207–220.

Semprimoschnig, C. O. A., S. Heltzel, M. Van Eesbeek, J. R. Williamson, A. P. Tighe, and A. Polsak. (2006) The ESA Venus Express mission—from a materials engineering perspective. ESA SP-616. *Proceedings of the 10th International Symposium on Multimedia Software Engineers, and the 8th International Conference on Protection of Materials and Structures in a Space Environment*. Noordwijk, the Netherlands: European Space Agency.

Shavers, M. R., F. A. Cucinotta, and J. W. Wilson. (2001) HZETRN: Neutron and proton production in quasi-elastic scattering of GCR heavy ions. *Radiation Measurement* 33: 347–353.

Shea, M. A., and D. F. Smart. (1990) A summary of major proton events. *Solar Physics* 127: 297–320.

Shiloh, M., E. Grossman, Y. Noter, M. Murat, and Y. Lifshitz. (1998) Evaluation of silicone paint contamination effects on optical surfaces exposed to the space environment. *Proceedings of the International Symposium on Optical Science, Engineering and Instrumentation*. San Diego, CA: International Society for Optical Engineering, p. 348.

Shinn, J. L., J. W. Wilson, and F. F. Badavi. (1994) *Fully Energy Dependent HZETRN*. Technical Publication TP-3243. Houston, TX: National Aeronautics and Space Administration, Johnson Space Center.

Silverman, E. (1995) *Space Environmental Effects on Spacecraft: LEO Materials Selection Guide*. Contractor Report CR-4661. Cleveland, OH: National Aeronautics and Space Administration, Glenn Research Center.

Simpson, J. A. (1983a) Introduction to the galactic cosmic radiation. In M. Shapiro (ed.), *Composition and Origin of Cosmic Rays*. Dordrecht, the Netherlands: D. Reidel, pp. 1–24.

Simpson, J. A. (1983b) Elemental and isotopic composition of the galactic cosmic rays. *Annual Review of Nuclear and Particle Science* 33: 323–381.

Skrivanek, R. A. (1994) *Contemporary Models of the Orbital Environment*. Special Project Report SP-069-1994. Reston, VA: American Institute of Aeronautics and Astronautics, p. 118.

Soares, C., R. Mikatarian, D. Schmidl, M. Finckenor, M. Neish, K. Imagawa, M. Dinguirard, M. Van Eesbeek, S. F. Naumov, A. N. Krylov, L. V. Mishina, Y. I. Gerasimov, S. P. Sokolova, A. O. Kurilyonok, and

N. G. Alexandrov. (2004) Overview of International Space Station orbital environments exposure flight experiments. *Proceedings of the SPIE International Symposium on Optical Science, Engineering and Instrumentation*. International Society for Optical Engineering.

Stephens, G., G. Campbell, and T. Vonder Haar. (1981) Earth radiation budgets. *Journal of Geophysical Research* 86 no. C10: 9739–9760.

Steward, P. G. (1968) *Stopping Power and Range for Any Nucleus in the Specific Energy Interval 0.01- to 500 MeV/amu in Any Non-Gaseous Material*. UCRL 18127. Livermore, CA: University of California, Lawrence-Livermore National Laboratory.

Tagawa, M., K. Yokota, N. Ohmae, and H. Kinoshita. (2000) Diffusion of atomic oxygen in SiO_2 protective coatings. *High Performance Polymers* 12, no. 1: 53–63.

Tighe, A. P., M. Dinguirard, J. Mandeville, M. Van Eesbeek, C. Durin, S. Gabriel, and D. Goulty. (2002) Materials exposure and degradation experiment (MEDET). AIAA-2001-5070. *Proceedings of the International Space Station Utilization Conference*. Cocoa Beach, FL: National Aeronautics and Space Administration, Kenney Space Center.

Tonon, C., C. Duvignacq, G. Teyssedre, and M. Dinguirard. (2001) Degradation of the optical properties of ZnO based thermal control coatings in simulated space environment. *Journal of Physics D: Applied Physics* 34: 124–130.

Townsend, L. W., J. W. Wilson, R. K. Tripathi, J. W. Norbury, F. F. Badavi, and F. Khan. (1993) *An Energy Dependent Semiempirical Nuclear Fragmentation Model*. Technical Publication TP-3310. Houston, TX: National Aeronautics and Space Administration, Johnson Space Center.

Van Allen, J. A., G. H. Ludwig, E. C. Ray, and C. E. McIllwain. (1958) Observation of high intensity radiation by satellites. *Alpha and Gamma. Jet Propulsion* 28: 588–592.

Van Eesbeek, M., F. Lavandou, and A. Milintchouk. (1995) A study of FEP behavior in space environment. *Proceedings of the 25th International Conference on Environmental Systems,* San Diego, CA.

Waters, L., M. Heck, and L. DeRyder. (1988) *Steady state micro-g environment on space station*. AIAA 88-2462. Williamsburg, VA: American Institute of Aeronautics and Astronautics, SDM Conference on Issues of the International Space Station.

Webber, W. R., J. C. Kish, and D. A. Schrier. (1985) Cosmic ray isotope measurements with a new Cerenkov total energy telescope. *Proceedings of the 19th International Cosmic Ray Conference, La Jolla, CA*, pp. 88–95.

Webber, W. R., A. Southoul, P. Ferrando, and M. Gupta. (1990a) The source charge and isotopic abundances of cosmic rays with Z = 9–16: A study using new fragmentation cross sections. *Astrophysical Journal* 348: 611–620.

Webber, W. R., J. C. Kish, and D. A. Schrier. (1990b) Individual isotopic fragmentation cross sections of relativistic nuclei in hydrogen, helium and carbon targets. *Physical Review* C41: 547.

Webber, W. R., L. A. Soutoul, J. C. Kish, J. M. Rockstroh, Y. Cassagnou, R. Legrain, and O. Testard. (1998) Measurements of charge changing and isotopic cross sections at ~600 MeV/nucleon from the interactions of ~30 separate beams at relativistic nuclei from 10 B to 55 Mn in a liquid hydrogen target. *Physical Review* C58: 3539–3552.

Weihs, B., and M. Van Eesbeek. (1994) Secondary VUV erosion effects on polymers in the ATOX atomic oxygen exposure facility. *Proceedings of the Sixth International Symposium on Multimedia Software Engineers*. Noordwijk, the Netherlands: European Space Agency, pp. 277–2784.

Wiedenback, M. E. (1985) The isotopic composition on cosmic ray chlorine. *Proceedings of the 19th International Cosmic Ray Conference,* La Jolla, CA, pp. 84–87.

Wiedenback, M. E., and D. E. Greiner. (1981) High resolution observations of the isotopic composition of carbon and silicon in the galactic cosmic rays. *Astrophysical Journal* 247: L119–L122.

Williamson, C. F., J. P. Boujot, and J. Picard. (1966) *Tables of Ranges and Stopping Power of Chemical Elements for Charged Particles of Energy 0.05 to 500 MeV.* Report CEA-R3042. Saclay, France: Centre d'Etudes Nucleaires de Saclay.

Wilson, J. W. (1977) *Analysis of Theory of High Energy Transport.* Technical Note D-8381. Houston, TX: National Aeronautics and Space Administration, Johnson Space Center.

Wilson, J. W., L. W. Townsend, W. Schimmerling, G. S. Khandelwal, F. Khan, J. E. Nealy, F. A. Cucinotta, L. C. Simonsen, J. L. Shinn, and J. W. Norbury. (1991) *Transport Methods and Interactions for Space Radiations.* NASA Research Paper RP-1257. Springfield, VA: National Technical Information Service.

Wilson, J. W., S. Y. Chun, F. F. Badavi, and S. John. (1993a) *Coulomb Effects in Low Energy Nuclear Fragmentation.* Technical Publication TP-3352. Houston, TX: National Aeronautics and Space Administration, Johnson Space Center.

Wilson, J. W., S. A. Thibeault, J. E. Nealy, M. Y. Kim, and R. L. Kiefer. (1993b) Studies in space radiation shield performance. *Proceedings of the Engineering and Architecture Symposium.* Prairie View, TX: Prairie View A&M University, pp. 169-176.

Wilson, J. W., M. Kim, W. Schimmerling, F. F. Badavi, S. A. Thibeault, F. A. Cucinotta, J. L. Shinn, and R. Kiefer. (1995a) Issues in space radiation protection. *Health Physics* 68: 50-58.

Wilson, J. W., F. F. Badavi, F. A. Cucinotta, J. L. Shinn, G. D. Badhwar, R. Silberberg, C. H. Tsao, L. W. Townsend, and R. K. Tripathi. (1995b) *HZETRN: Description of a Free Space Ion and Nucleon Transport and Shielding Computer Program.* Technical Publication TP-3495. Hampton, VA: National Aeronautics and Space Administration, Langley Research Center.

Wilson, J. W., F. A. Cucinotta, J. L. Shinn, L. C. Simonsen, R. R. Dubey, W. R. Jordan, T. D. Jones, C. K. Chang, and M. Y. Kim. (1999a) Shielding from solar particle event exposures in deep space. *Radiation Measurement* 30: 361-382.

Wilson, J. W., M. Y. Kim, J. L. Shinn, H. Tai, F. A. Cucinotta, G. D. Badhwar, F. F. Badavi, and W. Atwell. (1999b) *Solar Cycle Variation and Application to the Space Radiation Environment.* Technical Publication TP-1999-209369. Houston, TX: National Aeronautics and Space Administration, Johnson Space Center.

Xapsos, M. A., et al. (2000) Characterizing solar proton energy spectra for radiation effects applications. *IEEE Transactions in Nuclear Science,* 47, no. 6: 2218-2223.

Yang, V. V., and E. J. Ainsworth. (1982) Late effects of heavy charged particles on the fine structure of the mouse coronary artery. *Radiation Research* 91: 135-144.

Zeitlin, C., L. Heilbronn, J. Miller, S. E. Rademacher, T. Borak, T. R. Carter, K. A. Frankel, W. Schimmerling, and C. E. Stronach. (1997) Heavy fragment production cross sections from 1.05 GeV/nucleon ^{56}Fe in C, Al, Cu, Pb, and CH_2 target. *Physical Review* C56: 388-397.

Zeitlin, C., A. Fukumura, L. Heilbronn, Y. Iwata, J. Miller, and T. Murakami. (2001) Fragmentation cross sections of 600 MeV/nucleon ^{20}Ne on elemental targets. *Physical Review* C64: 24902/1-16.

Zeitlin, C., T. Cleghorn, F. Cucinotta, P. Saganti, V. Andersen, K. Lee, L. Pinsky, W. Atwell, R. Turner, and G. Badhwar. (2004) Overview of the Martian radiation environment experiment. *Advances in Space Research* 33: 2204-2210.

Zuby, T. M., K. K. de Groh, and D. C. Smith. (1995) *Degradation of FEP Thermal Control Materials Returned from the Hubble Space Telescope.* Technical Memorandum TM-104627. Cleveland, OH: National Aeronautics and Space Administration, Glenn Research Center.

CHAPTER 3

Overview of Bioastronautics

Simon N. Evetts, B.A. (Hons), M.Sc., Ph.D.
Medical Projects and Technology Lead, Wyle Laboratories GmbH, Crew Medical Support Office, European Astronaut Center, European Space Agency, Cologne, Germany

CONTENTS

- 3.1. Space Physiology 106
  By Simon N. Evetts, Ph.D., and Volker-Damann, M.D.
- 3.2. Short and Long Duration Mission Effects 115
  By Simon N. Evetts, Ph.D., and Volker-Damann, M.D.
- 3.3. Health Maintenance 123
  By Simon N. Evetts, Ph.D., and Volker-Damann, M.D.
- 3.4. Crew Survival 143
  By Jonathan D. Clark, M.D., M.P.H.
- 3.5. Conclusion 152

The human body is capable of rapid adaptation to environmental circumstances, an ability that has been crucial to the success of the human species. This capability to adapt is such that the physical structure and function of many tissues, organs, and systems of the body rapidly alter to enable life to proceed in space in an efficient and economical manner. Thus, the changes seen over hours, days, weeks, and even months are a positive response to the space environment and in particular to the absence of gravity. These responses, however, prove problematic if the gravity vector is reimposed or when Earth related achievement standards are required while in flight.

In addition to space adaptation, the human body is exposed to other stimuli and forces in space for which adaptation does not necessarily occur but from which harm can. These circumstances, in particular radiation, confinement, stress, and extravehicular activity, pose challenges to astronauts who undertake long duration missions that are on the order of months to years, but also must be taken in to account for many of the more arduous short duration missions as well. As a result of these hazardous activities and circumstances, and in an attempt to retain earthbound levels of physical capability, health and fitness related countermeasures, procedures, and programs are required. These, which can be collectively termed *health maintenance*, are necessary for astronauts to be able to conduct planetary exploration related activities, react successfully to emergency situations, and remain sufficiently functional in space to ensure minimal post-flight rehabilitation is required on their return to Earth.

Naturally, the ability to contend with the effects and hazards of the space environment becomes insignificant if a crew is unable to survive any extreme events associated with a mission. Crew survival must be considered a primary mission success criterion and should be a principal consideration in vehicle design and mission architecture. Equipment, facilities, operational circumstances, and procedures all should meet the requirements declared in advance as being necessary for crew survival under circumstances identified as being potentially life threatening.

3.1 SPACE PHYSIOLOGY

3.1.1 Muscular System

Human muscularity to a large extent exists as a result of the action of regular and repetitive stimuli. Although genetics governs the baseline level of musculature a body possesses, the environment in which the body lives dictates to what extent the physical state departs from this baseline. The need to move body mass and the mass of other objects against the action of gravity hundreds of times each day is a stimulus for the maintenance of appropriate muscle structure and function for the environment in question. During any movement undertaken, several forms of muscle contraction take place. For one form of contraction, that is, eccentric, the elongation of muscle during contraction acts as one of the primary stimuli for muscle growth (Yi-Wen et al. 2003). In microgravity, the loss of the requirement to work against gravity means that a major stimulus for maintaining muscle composition is lost, the result being that muscle structure alters and muscle mass decreases. Further, the muscular activity that does take place in space invariably does not require eccentric contraction, a component of work normally brought about by the presence of gravity. The changes in muscle biochemical structure and the reduction in mass, that is, atrophy, that ensue constitute two of the major elements of what can be termed *space deconditioning syndrome*. If left unchecked, this condition severely weakens the astronaut, thus potentially threatening the mission under circumstances for which strength or fitness is required.

The biochemical changes to muscle resulting from disuse deconditioning such as that occurring in space cause the aerobic (endurance) and anaerobic (power) capabilities of the tissue to deteriorate (Tesch and Berg 1998). Decreases in enzyme levels, the quantity of blood capillaries, and the size and number of the energy producing mitochondria are all noted (Neufer 1989). The result is a decreased muscle cross-sectional area and a loss of force production and endurance capabilities. That is to say, the muscles affected become smaller, weaker, and tire more easily.

There is some debate as to whether the different forms of muscle fiber within the tissue are affected equally or not. A number of studies have produced results suggesting that the highly aerobic slow twitch Type I fibers decondition faster than the more anaerobic fast twitch Type II fibers. These same studies indicate that Type I fibers can in some cases change their morphological characteristics to those of Type II (Ohira et al. 1992; Edgerton et al. 1995; Ohira 2000). Such an adaptation alters the ratio of these fibers and consequently the characteristics of the muscle itself. This is to say that the muscle becomes less aerobic and more anaerobic in nature. Even if this difference is found to be significant, such an effect, however, is overshadowed greatly by the detriment in function that occurs with deconditioning atrophy.

Exposure to microgravity certainly causes human muscle groups to atrophy. The rate at which this occurs differs according to whether or not they normally are active at $+1$ Gz (Fitts, Riley, and Widrick 2000), that is, Earth gravity acting in the direction from head to foot. This is the case for the so-called antigravity muscles, the primary ones of which are located in the thigh (hamstring), calf (soleus and gastrocnemius), and back (erector spinae and iliopsoas) and used principally for posture and locomotion. These muscles tend to decondition faster than the remainder (Convertino 1996a). Also, the antigravity muscle groups tend to contain the greatest quantity of aerobic Type I fibers, a fact supporting the contention that Type I fibers appear to be affected to a greater degree by microgravity than Type II. Furthermore, when studies indicated a decrease in Type I and an increase in Type II fibers, which suggests a transition in characteristics, this phenomenon has been most evident from the antigravity muscles (Jiang et al. 1992; Ohira et al. 1992; Widrick et al. 1999). The decrements in strength and local muscular endurance that occur during deconditioning are derived not only from muscle loss but also from detriment of neural stimulation. These changes do not continue indefinitely but lead to a new baseline, after which muscle mass and muscular capabilities level off in line with the new normal workload (Baldwin and Haddad 2002). In microgravity, however, this new baseline is considerably less than that which is normal in the $+1$-Gz state.

3.1.2 Skeletal System

It is accepted widely that the skeletal system acts as our internal support structure and to some extent as protection for our delicate internal organs. Human bones also offer attachment points for muscles and tendons to facilitate movement. A third, less commonly regarded role of the skeletal system is that of a mineral reservoir whereby the skeleton helps maintain blood levels of the physiologically important mineral calcium. The integrity of bone is sustained by a cycle of constant formation of new bone and resorption of the old. Viewing bone as dead is a misconception. It is a living, metabolically active tissue.

As with muscle, repeated and regular mechanical stimuli, possibly detected by certain bone cells (Backup et al. 1994), resulting from activity in a $+1$-Gz environment, affects the structure and function of bone. A balance between formation and resorption of bone tissue is controlled by daily stimuli such that a commensurate density and mass of bone exists for those individuals living in some particular environment within which these stimuli occur. In addition, dietary intake, lighting, ambient levels of carbon dioxide, and the degree and nature of forces acting on the skeletal system, such as muscular contraction, also affect the nature of bone.

It appears that the primary effect of living in the space environment on bone is due to the loss of gravity and associated mechanical loading of the skeletal system. The absence of gravity is detected by the bone, causing a decrease in bone formation and an increase in bone resorption (Caillot-Augusseau et al. 1998; LeBlanc, Shackelford, and Schneider 1998).

Subsequently the affected bones lose both physical mass and mineralization, that is, mineral density is reduced. The degree of demineralization varies from person to person, potentially being more of a concern for women, and similarly varies from bone to bone. Indeed, the weight bearing antigravity bones, such as those of the lumbar vertebrae, pelvis, and femur, are more affected than those that do not bear weight, such as arms, ribs, and upper vertebrae (Vico et al. 2000).

The principal mineral lost from bone during the deconditioning process is calcium (Greenleaf 2004). Increased in-flight urinary calcium levels have been detected in astronauts from as early as the Gemini missions conducted during the 1960s (Lutwak et al. 1969), in which space sojourns lasted only days. Calcium is integral to bone formation, in that the organic matrix or osteoid laid down during the process is calcified by the deposition of apatite mineral deposits. During the ongoing process of bone resorption, calcium is released, and subsequently the mineral becomes part of the process for stimulating bone remodeling (Rodan 1996). This balanced cycle of calcium as it is involved in bone formation, when disrupted by a decrease in mechanical loading as seen in space, leads to increased calcium loss from the body. This loss is detectable in the urine.

The consequences of the microgravity induced changes in bone structure and function are that bone strength first is reduced, thus increasing the risk of fracture in physically strenuous circumstances. Second, alterations in renal function, fluid redistribution, bone loss, and the accompanying muscle atrophy heighten urinary calcium levels, thus increasing the risk of renal (kidney) stone development during and immediately after a mission. This particular condition is usually very painful, can be debilitating, and often requires medical treatment (Buckey 2006).

3.1.3 Cardiovascular and Respiratory Systems

A number of physiological changes occur as a result of exposure to the space environment that can be categorized primarily as cardiovascular in nature. These changes can be seen as alterations in cardiorespiratory (aerobic) fitness and function, altered blood pressure control, and structural and functional changes in the heart.

Presently, the most widely accepted method for assessing the cardiorespiratory endurance capability of an individual is the measurement of maximum oxygen uptake or VO_2 max (Day et al. 2003). Knowledge of this value enables the comparison of fitness between subjects and allows fitness across time to be monitored for a single subject. The previously mentioned lack of normal $+1\text{-}Gz$ stimuli in the absence of gravity results in the deterioration of many cardiorespiratory and vascular adaptations to life on Earth. This in effect is a state of deconditioning. Under these circumstances, it has been observed that the size of the myocardium, or heart muscle, decreases, as does its force output. Additionally, in the deconditioned state, blood plasma volume and red blood cell mass decrease right along with the size of blood capillary beds and the quantity of oxidative enzymes available for aerobic metabolism (Neufer 1989). Along with these changes and the myocardial deconditioning that occurs comes diminished maximum cardiac output. In line with these and certain other adaptations, decreased aerobic fitness occurs as well (Neufer 1989).

Also aligned with these effects of microgravity is a shift of blood from the lower body into the lung circulation. This pulmonary engorgement and the upward movement of the abdominal contents, which cause the diaphragm to bulge into the thoracic cavity, reduce lung volume. Additionally there appears to be some alteration in the control of ventilation that, under certain circumstances, such as in reduced oxygen conditions, can diminish physical performance (Prisk et al. 2003).

Space anemia is an adaptation to spaceflight related to cardiovascular deconditioning. Even though plasma volume decreases in space, the concentration of red blood cells in the bloodstream remains about the same as that seen on Earth (Alfrey et al. 1996). For this to be the case, red blood cells must be destroyed, the production of new red blood cells must diminish, or both eventualities must exist in microgravity. The exact cause and nature of this condition is not certain, but the downregulation of the red blood cell production system, that is, erythropoiesis, is apparent. The result of this condition is that less oxygen can be carried through the body during intensive exercise, a condition that contributes to fatigue when work levels are high.

A major concern is the effect that cardiovascular deconditioning has on blood pressure control mechanisms upon completion of a mission. The ability to withstand the physiological strain of standing in +1 Gz is known as *orthostatic tolerance*, which is a manifestation of the effect of confinement and deconditioning found in the majority of astronauts after long duration missions. Indeed, indications of this condition have been seen in some after as little as 34 h (Hoffler, Wolthuis, and Johnson 1974). In space, the lack of a gravity induced hydrostatic pressure gradient means that the body has little difficulty in returning blood through the venous system to the heart and then providing it to essential organs, such as the brain. Adaptations that enable this process to happen on Earth, where such a gradient exists, diminish during exposure to microgravity. This, coupled with other space induced changes, results in blood pressure control difficulties when post-mission hydrostatic gradients are reimposed (Greenleaf 2004). The orthostatic intolerance is evidenced by elevated heart rates, greater decreases in blood pressure, inadequate peripheral vascular resistance, and increased pooling of blood in the lower body during orthostatic stress, such as that caused by standing during the first few hours after flight (Fritsch-Yelle et al. 1996). The result of this condition for some of those who attempt standing during the post-fight period is loss of consciousness and collapse. Fortunately, preflight levels of orthostatic tolerance usually are regained within a few days, irrespective of the duration of the mission.

A number of postulated mechanisms lie behind this condition. Changes in baroreceptor reflexes, which constitute a major element of the blood pressure control system, are brought about by the headward fluid shifts that occur during spaceflight (Fritsch et al. 1992). Similarly and concurrently, reduced blood volume, decreased sensitivity to stress hormones (Fritsch-Yelle et al. 1996), and altered blood vessel contractile capabilities (Arbeille et al. 1996) contribute to the orthostatic intolerance. Another related and possible contributing factor is the altered structure and function of the heart (Philpot, Kato, and Miquel 1992). The initial few hours of exposure to microgravity leads to expansion of the heart chambers. In the longer term, however, the heart muscle atrophies, and its structure can become altered such that the power and ability of the heart to empty in accord with the pressure of chamber filling is affected adversely (Philpot et al. 1990; Perhonen et al. 2001). These changes, in a number of ways, contribute to reduced cardiorespiratory fitness during and after spaceflight and the decreased orthostatic tolerance observed post-flight.

3.1.4 Neurovestibular System

On Earth, the human body has a refined and sophisticated system for sensing position, limb movement, and balance. The system is dynamic, fast acting, and relies on inputs to the brain from various subsystems, these being primarily

- Vestibular, that is, balance and movement organs in the inner ear.
- Proprioceptor, that is, position sensing receptors in muscles, tendons, and joints.
- Three of the primary five senses: touch, hearing, and sight.
- Associated neural tracts.

Continuous and varying inputs originate from each of the biologic sensors in accord with the environment and physical condition within which we find ourselves. This complementary plethora of inputs forms a reference within the brain that continually is used as a basis of comparison as the physical or environmental state changes and new integrated sensory inputs are produced.

During spaceflight, the stimulus to many of these sensors changes. The vestibular organs are not being stimulated in the normal +1-Gz manner, proprioceptors suddenly are receiving less stimulation, and all receive stimuli that normally do not occur in the circumstances that the eyes of the astronaut are reporting. That is to say, if the eyes report that the astronaut is upright, the neural inputs coming from the various sensors are not those expected for the upright position. Of particular note is that, on Earth, the +1-Gz vector is the primary means by which aspects of the vestibular system are calibrated. Without this factor, the resultant disparity between expected neural input and actual neural input, according to the current leading theory, is likely to be the primary factor leading to the condition of space motion sickness for many sufferers (Crampton 2000). It is possible that this neural conflict leads to a coordinated sensory input similar to that experienced during the early stages of neurotoxin poisoning (Treisman 1977; Crampton 2000). Consequently, one suffering from space motion sickness experiences the common signs and symptoms of poisoning (Gorgiladze and Brianov 1989). The symptoms include pallor, lethargy, malaise, nausea, and in many cases vomiting.

Approximately two thirds of all astronauts entering microgravity suffer from space motion sickness to some degree (Davis et al. 1988). It is possible that microgravity associated effects, such as the alterations in the hydrostatic gradient, and in particular central spinal fluid and inner ear fluid pressures, can combine with the postulated neural conflict to produce space motion sickness (Parker 1977; Crampton 2000). The severity is such that 1 in 10 can become incapacitated if no prophylactics are used and no treatment is available (Jennings 1988). The condition tends to last from 1 to 3 d, subsiding after this time to a point in which signs and symptoms no longer occur (Shupack and Gordon 2006). As a result of the possible 3-d period of affliction, extravehicular activity is not scheduled during the first 72 h of a mission. Furthermore, mission lengths, by design, are greater than 72 h to ensure that nominal reentry does not occur while the astronauts are suffering from space motion sickness, a condition to be avoided if, for example, Space Shuttle piloting skills are to be optimal. Although the prediction of susceptibility to this infliction is difficult, some understanding of the conditions required to elicit space motion sickness gradually is emerging. It is now known that a certain degree of preadaptation through prior exposure of astronauts to simulated weightless conditions, such as during parabolic flight and underwater training, can be beneficial.

Adaptations of the neurovestibular and associated systems that occur during a mission, in particular the unloading of the otolith organs in the inner ear caused by loss of the gravity vector, lead to a condition for some in which vision, balance, and locomotion are affected adversely (Black et al. 1999). Aligned conditions are in-flight and post-flight postural instability and visual disturbances (Nicogossian, Huntoon, and Pool 1994). Many crewmembers experience illusions of object or self motion. Visual oddities, such as vertical shortening and incidences of illusory translation, appear to increase with flight duration. These conditions, although not clinically serious, offer the potential for decreased performance capabilities both while in-flight and after flight and thus increase the risk of accidents.

3.1.5 Radiation

The radiation to which astronauts are exposed during spaceflight consists of solar particles ejected into space during solar flares, energetic particles trapped within the magnetic field of the Earth (primarily within the Van Allen belts), and the high-energy protons and heavy ions emanating from outside of the solar system, that is, the galactic cosmic radiation. When in low Earth orbit, the magnetic field of the Earth provides a good degree of protection against solar particles and, to some extent, galactic cosmic radiation.

Each of these three forms of space radiation can be termed *ionizing*, in that they can deliver enough energy to ionize an atom, that is, strip an electron from its orbit. If the body of an astronaut is exposed to ionizing radiation, the energy imparted is absorbed by the molecules of the various tissues affected. The major interaction is between the high-energy radiation particles and the electrons and nuclei of the cells of the tissue involved. The ionization of certain cell components, in particular water, can lead to deoxyribonucleic acid damage at sites close to the location of the radiation hit. A primary effect such as this can lead to the death or mutation of the cell. If sufficient radiation is absorbed over a relatively short period of time, the ionization damage can lead subsequently to tissue and organ damage, cancer, and even death. In addition to this direct radiation effect, the passing of radiation through substances, such as some forms of shielding, can cause a secondary effect, in that energetic particles can be emitted from the shielding material as a result of the bombardment and cause damage similar to that of the original ionizing radiation itself. Indeed, it is possible for secondary particles to prove even more destructive than primary particles on occasion (George et al. 2002).

The magnitude or quantity of the radiation present is not of primary importance to the astronaut but rather the amount of radiation actually absorbed by the tissues.

Furthermore, not all radiation particles show the same ability to do damage. Some larger particles, having a commensurate cross-sectional area, can damage more tissue as they pass than smaller, equally energetic particles. To understand the radiation burden to which a crew might be exposed, it is useful to consider that the effects of radiation can vary according to the type (high or low linear energy), the nature of the radiation field (the mixture of different types of radiation), the dose or magnitude, the dose rate or dose over time, and the duration of exposure.

The standard unit of absolute quantity of energy imparted is the Gray (Gy). The allocation of a qualitative numerical value to a particle according to its ability to damage, however, offers a means to express radiation doses in a more meaningful manner. To accomplish this, the value for the absolute quantity of energy is multiplied by a value representing the capability of a particle to damage biological tissue. The measure produced, termed the *dose equivalent*, is expressed in Sieverts (Sv), where 1 Sv = 100 rem, which under certain circumstances also can equal 1 Gy.

The health effects resulting from radiation exposure can be categorized as acute and delayed. High dose, whole body exposure, such as that experienced during a solar particle event, can lead to acute effects, such as nausea, vomiting, and diarrhea, that are noticeable within hours, days, or weeks. A number of months or even years later or as a result of low dose long duration exposure such as that delivered by galactic cosmic radiation over time, the delayed effects of radiation can be evidenced by tissue damage, eye cataracts, cancers, neurological disorders, and fertility impairment (Pecaut, Gridley, and Nelson 2003; Blakely and Chang 2004; Durante 2004). The primary concerns with regard to the hazards of radiation exposure are absolute dose, tissue susceptibility, and rate of application. Both acute and chronic exposure to ionizing radiation can be damaging, depending on the balance of these factors.

3.1.6 Nutrition

The variety and quality of food available during spaceflight affects not only the physical health but also the mental health and overall performance of the crew. To ensure crew health, nutritional concerns, such as sufficient calorific intake, nutritional density, food palatability, variety, and cultural choices of preferred food must be considered adequately (Ball and Evans 2001). A number of the physiological adaptations to spaceflight, such as bone loss, muscle atrophy, and alterations in hematological variables, have nutritional implications (Lane and Schoeller 2000). Routinely astronauts lose body mass during space missions (Yegorov et al. 1981; Smith et al. 1999a; Lane and Schoeller 2000). Although and to a large extent the initial drop in mass can be attributed to decreased body fluid volumes, as the duration of the mission lengthens, the contribution of mass loss from muscular and skeletal atrophy increases (LeBlanc et al. 1996). The decrease in mass resulting from these aspects of deconditioning is not purely the result of exposure to microgravity per se but also to reduced calorific intake (Smith et al. 2001).

The stress of operational space missions, anorexia associated with space motion sickness, and at times simply the time demands of the mission, often have a negative impact on food and fluid intake. In many cases, this results in a considerable calorie deficit when the operational activities and exercise countermeasures of a mission are taken into account. Ground based studies have shown that physiological stress leads to an increased rate of whole body protein turnover (Wolf, Jahoor, and Hartl 1989). An effect such as this requires both energy and nutrients if a detriment in muscle condition is to be avoided. Therefore, although the requirement for energy to fuel muscular movement is less in microgravity, other physiological processes that require more fuel are accelerated. The lack of sufficient nutrients results in a negative energy balance, a state within which many astronauts find themselves whereby they burn off more calories than they ingest (Stein 2000).

Early in a space mission, fresh fruit and vegetables are provided as part of the stores. However, these and their associated nutritional content quickly become unavailable as limited supplies of the perishable items become exhausted. As supplies dwindle, the ability to meet the general health requirement for the astronauts by having them eat a balanced diet is compromised. The relatively high bulk and low density of foods appropriate for a high fiber, low fat diet can conflict with the need for high density, low bulk foods that can be stored for longer duration space missions. The resolution of such a conflict often requires compromise, which results in a somewhat less than optimal diet. Because of the need to minimize the possibility of cardiovascular disease over the years that a long duration mission, such as Mars exploration, might entail, this aspect of nutrition is of high importance.

3.1.7 Immune System

The activities involved with traveling to and spending time in space present challenges to the human body. As well, the operational requirements of conducting a space mission place further stresses on the cognitive and physical capabilities of the astronaut. Stress, both physical and emotional, adversely can affect the effectiveness of the human immune response through its effect on endocrinological substances, such as the stress hormones (catecholamines) (Shimamiya et al. 2005). These factors coupled with the deconditioning effect of and exposure to the space environment lead to a condition of reduced immunity to disease (Stowe, Sams, and Pierson 2003).

Humans carry viruses in their bodies. These microorganisms are kept under control by the immune system. Spaceflight and its associated stresses appear to reactivate some of these latent viruses. Reactivation of latent viruses, a condition commonly termed *viral shedding*, causes astronauts to become susceptible to viral infections that they otherwise would not have developed back on Earth (Payne et al. 1999). An example of such a viral infection is caused by the Epstein-Barr virus, a deoxyribonucleic acid bound virus that infects 60 to 90% of the human population (Braude, Davis, and Fierer 1986; Svahn et al. 2006). The Epstein-Barr virus often becomes latent once its acute infectious stage is over. On Earth, any reactivation of Epstein-Barr virus normally is controlled by the immune system, in particular by the cell mediated immune element of the system (Simon 1997). With diminished immune function, which is a consequence of spaceflight, the risk of reactivation of this and other latent viruses is increased. The potential detrimental effect of this personal condition is compounded by the confined environment of a space vehicle and the use of recycled air, which together increase the risk of spreading illness among the crew. Consequently, a potential for multiple crewmember illness arises, which increases the possibility of a deleterious impact on mission operations. This is an eventuality that cannot be ignored, especially when exploration class missions are being considered.

In addition to the risks imposed by a generally reduced immune response and viral shedding, it also is apparent that some species of bacteria proliferate during spaceflight (Manko et al. 1986; Takahashi et al. 2001). Although few species pose a threat to humans, a number of bacteria are capable of adversely affecting health, such as *E. coli*. This factor also is exacerbated by the confined nature of the spaceflight habitat and difficulties in maintaining a high degree of hygiene in a permanently warm, closeted environment (Moatti et al. 1986).

3.1.8 Extravehicular Activity

Since the first space walk by the Russian cosmonaut Aleksei Leonov from the *Voshkod-2* spacecraft in 1965, extravehicular activity has been and remains a fundamental element of human spaceflight. Progressing from this event, the number, duration, and complexity of extravehicular activities have increased exponentially. When we consider the amount of extravehicular activity presently involved in building the *International Space Station* and those extravehicular activities that will be undertaken in the process of establishing a lunar base and for Mars exploration, it is quite clear that extravehicular activity is to remain a major component of space activities well into the future.

When an extravehicular activity is undertaken, the principal tasks invariably require upper body physical activity. To enable adequate movement within an individual space suit assembly and extravehicular mobility unit, suit pressure is reduced to a level considerably lower than that normally experienced on Earth; for example, in the case of the U.S. suits, pressure is reduced from 760 mmHg (101 kPa) to 222 mmHg (29.6 kPa). Rather than the usual process for removal of gases dissolved in body fluids via the lungs, reductions in atmospheric or suit pressure, if too precipitous, can cause bubbles or emboli to form within the tissues. The gas of importance in this situation is nitrogen, the primary cause of decompression sickness. Under these conditions, nitrogen readily comes out of solution to form emboli and circulates around the body in the blood. A sufficient number of emboli, as well as their relative size, can lead to decompression sickness symptoms of joint stiffness and pain, neurological damage, and potentially cardiovascular collapse (Heimbach and Sheffield 1985; Cowell et al. 2002). Typically, decompression sickness symptoms are divided into Type I, which is presented by joint pain, possibly with cutaneous or lymphatic involvement, and the more serious Type II, which involves the nervous and respiratory systems, that is, loss of sensation, respiratory distress, and paralysis (Horrigan, Waligora, and Bredt 1989). The potential for suffering decompression sickness exists not only as a direct function of the ratio between the absolute nitrogen tissue pressure and the ambient pressure but also in relation to the duration of an extravehicular activity, in that long duration extravehicular activity can lead to fatigue, that is, a greater relative workload, which heightens the risk of decompression sickness

(Newman and Barratt 1997). Currently extravehicular activities of 5 to 8 h in duration are viewed as acceptable; however, as duration exceeds the upper limit of this range and fatigue builds, the risk of decompression sickness increases (Newman and Barratt 1997).

With increased cardiovascular fitness comes the added benefit of an improved thermoregulatory capability, that is, an improvement in the ability to control the body core temperature. The opposite is also true, that is, deconditioning reduces thermoregulatory ability. Aligned with these responses are the findings that thermoregulation appears to be impaired as a consequence of exposure to spaceflight or microgravity simulation scenarios (Fortney 1991; Crandall et al. 1994; Cowell et al. 2002). Due to the thermoregulatory limitations of present day space suits (Cowell et al. 2002), the potential for an astronaut to overheat and consequently suffer from heat stress during extravehicular activity increases with time in space if cardiovascular deconditioning occurs. The potential risk for heat stress is exacerbated by the lack of convection in microgravity. If physical strength diminishes during the course of a mission and if a crewmember finds any extravehicular activity that is resistive in nature to be physically more demanding than would otherwise be expected, greater body heat production would be the consequence.

3.2 SHORT AND LONG DURATION MISSION EFFECTS

3.2.1 Muscular System

During routine activities in space, the human musculature is not required to produce as much force as on Earth. As a result, the condition is rather like partial immobilization, in that the muscles are to a large degree in a state of rest for considerably longer periods of time than is seen at $+1$ Gz. Immobilization of a limb or the cessation of training causes any affected muscles to lose strength and mass (Greenleaf 2004). Muscle peak torque has been noted to decrease by up to 22% after as few as 9 d of immobilization (Miles et al. 1994). After as little as 5 or 6 d on orbit, deficits in muscle structure and function have been noted in both animal and human subjects (Baldwin and Haddad 2002). The general consensus is that short periods of spaceflight, from a few days to 1 or 2 wk, although causing the neuromuscular system to decondition, do not lead to sufficient reductions in muscular strength and endurance to be considered hazardous to an individual or mission. Thus, they are not normally of excessive concern post flight. As the duration of exposure to the microgravity environment increases, however, so do the detrimental effects.

Bed rest simulates some of the effects of the space environment and can be used to provide a means of deconditioning that closely resembles that observed in space. Bed rest studies have shown reductions in knee and ankle extensor maximum force output of about 20% over a month and ankle flexor loses close to 40% after 3 mo (Convertino 1996a). Goubel (1997) observed maximum voluntary calf plantar flexion to decrease by up to 37.6% after 1 to 6 mo of spaceflight. It appears that, for a limited period of several months, strength reductions, if not countered, can be in the region of from 2 to 5% per week depending on the site and function of the muscle (Narici et al. 1989; Tesch and Berg 1998). The upper body musculature (generally not locomotive muscles) tends to exhibit far less deconditioning than that of the lower body (Johnson and Dietlein 1977). During a study of *Skylab* astronaut muscle force production, it was noted that, when considering six astronauts from across all three missions, four subjects showed a mean loss of arm strength of between 2.5 and 10%, whereas five subjects showed losses of leg strength in the region of 10 to 20% of their preflight maximal capability (Johnson and Dietlein 1977). Also of note, however, is that interindividual differences and the effects of countermeasures are such that for some astronauts it is possible for little or no detriment to occur (Goubel 1997).

Even though, in many ways, complete bed rest forces muscle groups to work less than they would in space, the indications are that a given duration in space can lead to a greater volume loss for some muscles than that seen during an equal duration of bed rest immobilization (Buckey 2006). Although the sources of this effect can be centered more operationally and psychologically on stress and a negative energy balance, the fact remains that living and working in space causes muscles to decondition in a manner that many liken to an accelerated aging process. Ultimately, the result is a physical state that is less capable than that experienced on Earth along with a concomitant increase in risk during emergencies, such as self rescue from a damaged vehicle.

3.2.2 Skeletal System

Urinary calcium levels were seen to increase rapidly following orbital insertion during the *Skylab* missions of the 1970s (Johnson and Dietlein 1977). Bone calcium loss can be seen within days after a space mission (acute effect) and has been recorded to be still occurring 6 mo later (chronic effect) (Vico et al. 2000). During the *Skylab* missions, the calcium lost in the urine appeared significantly greater than ground based bed rest studies predicted. This indicated a potential for a 25% bone mineral loss from some bones over a year should these rates continue unabated (Rambaut and Johnson 1979). During missions of even greater length, Soviet scientists noted that durations of close to a year also have been associated with negative calcium balance (Vorobyov et al. 1983). More recently, spaceflight induced increases in urinary calcium levels have been measured to reach values in the region of 50% during a 115-d mission (Smith et al. 1999b).

Increased urine calcium output and the concomitant increases in fecal and intestinal calcium levels (Lutwak et al. 1969; Rambaut and Johnson 1979) are indicative of a negative calcium balance. Bone mineral density measurement supports the implications of the calcium studies. Vico and coworkers (2000) were able to show mean bone mineral density losses of between 0.9 and 1.8% per month over 1- to 6-mo sojourns on *Mir*. Currently it is accepted that the average reductions in bone mass and mineralization appear to be on the order of from 1 to 2% per month; however, this value varies markedly between measurement sites (McCarthy 2005). For example, astronauts have exhibited little or no loss of mineral content from their arms (radius) after several months (Smith et al. 1977), whereas weight bearing sites such as the heel (calcaneus) and pelvis have been seen to lose 1.5% per month over extended durations (Le Blanc et al. 1996).

A recent examination of International Space Station astronauts shows that, even with an active exercise and pharmacological countermeasure program in place, bone mineral

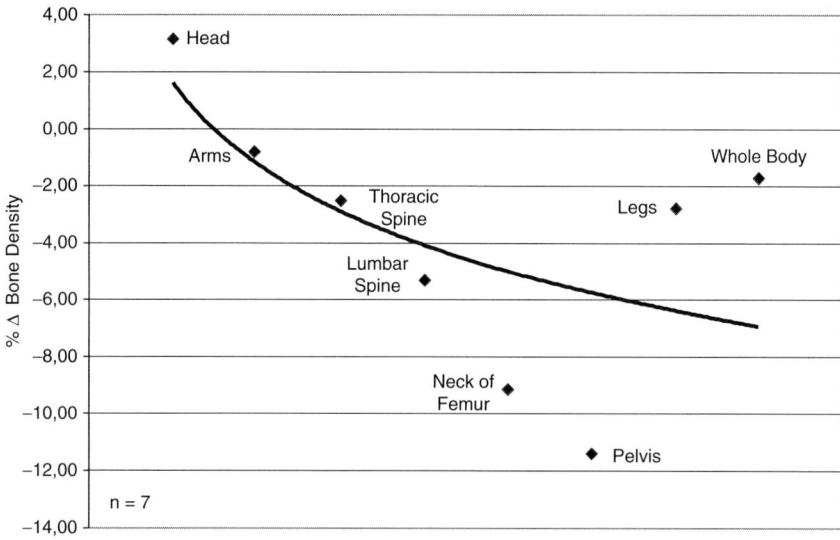

FIGURE 3.1 Regional bone loss for seven cosmonauts after International Space Station long duration missions (Graphic reproduced by permission of V. Oganov, Institute of Biomedical Problems, Moscow).

density was lost at a rate of around 0.9% per month in the lower spine and at 1.4% per month in the upper thigh (Lang et al. 2004, 2006). Some astronauts who undertook missions of several months' duration have been seen to lose in the region of 20% of their preflight lower extremity bone density on a return to Earth (Figure 3.1) (Oganov et al. 1992). Additionally, they exhibited indications of continued bone loss for several months thereafter (Vico et al. 2000). Also of note is that some individuals had not regained their preflight bone density levels many months after the completion of their mission (Tilton, Degioanni, and Schneider 1980).

As with muscle, the degree and rate of bone atrophy and demineralization vary considerably from person to person and, as has been noted, from site to site. There are strong indications that some women can be more prone to this effect than many men. Those astronauts who are susceptible and who cannot counter the effect not only run the risk of being inflicted by the condition of kidney stones but also are exposed to a greater risk of bone fracture should they be subjected to any appreciably arduous conditions, such as during planetary exploration or emergencies. Furthermore, if bone atrophy and demineralization are pronounced, there exists a potential for not achieving full recovery.

3.2.3 Cardiovascular and Respiratory Systems

The equilibration of the hydrostatic pressure gradient of the body and thus the headward movement of body fluid is an immediate effect of entry into microgravity. This fluid shift reduces lung volume and causes sensors in the upper body to detect what is interpreted mistakenly as hypervolemia, that is, high body fluid volume. The reflex responses normally resulting from such a stimulus lead to increased sodium and fluid loss in the urine, thereby causing body fluid levels to decrease to a state of relative hypovolemia or low body fluid volume. During spaceflight, however, this response occurs against a background level of dehydration brought about by the launch conditions, such as lengthy periods in which crewmembers are in a head down position and the desire of an astronaut to limit fluid intake so as to avoid the need to urinate once strapped in for launch. As a consequence, directly observed increases in body fluid loss, that is, increased urine output, during spaceflight usually are not observed even though body fluid levels are seen to be lower than observed preflight.

Measurements taken within the first few flight days have shown astronauts to be fluid depleted by between 10 and 20% (Convertino 1996b). Initial adaptations such as this affect the aerobic capability of astronauts in space to some degree; however, if countermeasures are not sufficient to enable aerobic fitness to be retained, physiological adaptations gained during exercise training prior to the mission are lost. One of the space induced adaptations that prove detrimental in this respect is space anemia. During the first few days in space, a large proportion of young red blood cells disappear from the blood. This loss of blood cells results in a decreased red blood cell count in the range of 10 to 15% (Alfrey et al. 1996), with an associated reduction in oxygen carrying capacity. An examination of Apollo astronaut oxygen uptake levels at a heart rate of 160 beats per minute showed that space missions in the region of a week have lead to an average reduction in aerobic capability of 17% (Berry 1970). More recently, 9 to 14 d of spaceflight have shown VO_2 max reductions in the region of 22% (Levine et al. 1996). These spaceflight induced losses can be regained quite quickly after short missions (Levine et al. 1996); however, for longer missions in which countermeasures have proven to be inadequate, some weeks can be required to regain preflight levels of aerobic capability.

Astronauts who exhibit appreciably reduced cardiovascular fitness are unable to maintain high intensity activity for short periods of time or moderate activity levels for longer periods. They are unable to react quickly and effectively during crisis due to their weak and fatigued state and subsequently are susceptible to an increased risk of injury in stressful circumstances. These effects are not overly noticeable during routine microgravity operations; however, they become increasingly apparent as the intensity of physical activity increases or as gravity is reimposed.

Although blood pressure is affected little by adaptations to microgravity, blood volume is decreased, and the normal reference set points used by blood pressure control systems, such as the baroreflexes, appear to be reset to enable appropriate blood pressure control in microgravity (Fritsch-Yelle et al. 1994, 1996). Further, as days turn to weeks and weeks to months, the structure and function of the blood vessels of the lower body and the heart alter, that is, decondition, resulting in a lesser ability to react to stress hormones and thus aid blood perfusion. Such deconditioning contributes to excessive blood pooling in the lower body and orthostatic intolerance on a return to Earth (Philpot et al. 1992; Fritsch-Yelle et al. 1996; Perhonen et al. 2001). In a classic study of this condition, 9 of 14 (64%) crewmembers after spaceflight could not complete 10 min of quiet standing after a 1- to 2-wk stay on orbit (Buckey et al. 1996). The incidence of orthostatic intolerance varies to some degree according to mission duration and the effectiveness of any inherent countermeasures. It has been seen to be as high as 64% for short missions (Buckey et al. 1996) and greater than 90% for some long stays on orbit (Vorobyov et al. 1983). Naturally, the concern regarding this condition is that during a postmission or planetary landing emergency, orthostatic stress brought about by the imposition of acceleration forces can lead to fainting during a mission or other life threatening circumstances.

3.2.4 Neurovestibular System

An oftentimes debilitating condition, space motion sickness characteristically occurs in situations of conflicting sensory input, such as that found in space. Although it is becoming increasingly clear which stimuli can lead to space motion sickness, the underlying determinants of individual susceptibility are still under investigation. With regard to motion sickness as opposed to space motion sickness, a genetic factor appears to be involved, whereby it is likely that there is a link with age and other parameters. For example, children are more susceptible than adults. Women and those who are aerobically fit can suffer more readily as well (Reavley et al. 2006; Shupak and Gordon 2006).

The incidence of space motion sickness appears to be something on the order of 70% of all astronauts entering microgravity, with about half of these classified as having moderate to severe symptoms (Jennings 1998). In a study of some of the earlier Space Shuttle missions, 13% of astronauts exhibited severe symptoms (Davis et al. 1988). The most disabling form of this condition is chronic motion sickness in which those afflicted do not adapt. These subjects remain sick for long periods or even for the entire time spent in the environment that elicited the response. It is likely that less than 5% of those susceptible to motion sickness will suffer from the chronic condition. For those who do, it is debilitating, and the condition can affect mission effectiveness (Reason and Brand 1975; Buckey 2006). The balance between the severity of space motion sickness and the nature of the activity required of the sufferer is the key determinant of the degree to which mission operations are affected. Performance appears to be poorest for complex tasks requiring sustained attention, whereas simple tasks for which the sufferer need not be overly attentive are accomplished more easily (Hettinger, Kennedy, and McCauley 1990).

As discussed in the previous section, the vast majority of space motion sickness sufferers adapt to the new environment within a few days, and the signs and symptoms abate. Although adaptation to the condition allows for uninterrupted functioning of the affected crewmembers during a mission, similar symptoms often are detected on the return to Earth because the now seemingly alien environment of Earth is one to which the crew must re-adapt. This recurrence of space motion sickness type symptoms constitutes the so-called readaptation syndrome noted in a majority of astronauts (Bacal, Billaca, and Bishop 2003).

In addition to the post-flight readaptation condition, the visual disturbance and postural instability that some crewmembers experience when returning to the $+1$-Gz environment are of concern (Nicogossian et al. 1994; Black et al. 1999). An examination of postural instability relative to preflight in four Space Shuttle astronauts after an 8-d mission indicate that inadequate vestibular feedback could have been the most important sensory deficit contributing to the condition (Black et al. 1999). Astronauts suffering from a post-flight postural instability walk with a gait and balance that can be hazardous to themselves under certain circumstances. Still more important, if they are required to drive or fly a vehicle, their affected sense of balance and attitude is hazardous to the mission and crew.

3.2.5 Radiation

In low Earth orbit, measurements have indicated daily radiation rates in the region of 300 Gy (Badhwar et al. 1996), which translates to something in the region of 0.002 to 0.003 Sv per day. Medium duration mission levels in the region of 0.15 Sv (Buckey 2006) and 0.4 Gy for astronauts staying for five or so months on the old Soviet space station *Mir* also have been noted (Testard et al. 1996). These levels can be compared to a single solar particle event that potentially can emit from 0.15 to 0.3 Sv of radiation per hour (Buckey 2006). Whole body exposure to the sorts of radiation doses expected from large solar particle events or clinical treatments is likely to have adverse long-term effects on immune system function (Pecaut et al. 2003).

Levels of whole body radiation exposure of up to 2 or 3 Gy that adversely effect the immune system are considerably larger than accepted occupational limits and quantifiably similar to those observed as a result of one or more clinical treatments (Pecaut et al. 2003). These doses, however, if experienced over sufficient duration, prove not to be appreciably damaging. It has been estimated that a round trip to Mars, taking over 3 a, will expose astronauts to approximately 1.0 Sv (Ponomarev, Nikjoo, and Cucinotta 2005). Presently, these sorts of exposures are accepted to increase the risk of lifetime cancer mortality for an astronaut by around 2%, that is, from the typical 3% to 5% (Cucinotta and Durante 2006). It can be that the primary concern for astronauts undertaking missions of this nature, therefore, is not the risk of acute radiation effects but delayed effects. Examples of the late effects of concern include secondary cancers, cataract, fibrosis, neurodegeneration, blood vessel damage, and immunological, endocrine, and genetic effects (Blakely and Chang 2004).

During a 3-a Mars mission, every cell nucleus in the body of an astronaut will be hit by a proton or secondary electron (Cucinotta, Nikjoo, and Goodhead 1998). Total body exposures are estimated to be approximately 0.5 to 1.0 Sv due to galactic cosmic radiation and background solar activity (Hoffman and Kaplan 1997). Depending on the shielding employed, large solar particle events can add another 2 or 3 Gy of proton exposure (Setlow et al. 1996). Examples of recommended maximal limits of exposure for terrestrial radiation exposed workers are listed in Table 3.1.

Although the international space community is not harmonized fully in its recommendations for radiation limits, it is agreed generally that maximum career dose equivalents of between 0.5 and 1.0 Sv can be assumed. Data such as this suggest that, according to some current Mars mission models, astronauts reasonably could not be expected to undertake more than one mission in a career. The physiological effects of radiation are not a problem in the short term if sufficient shielding is present or if the mission of an astronaut can be timed to avoid solar particle events in the longer term. The question is, just how much is the risk of developing delayed, long-term effects increased by chronic exposure to ionizing radiation?

Table 3.1 Exposure Limits for Ionizing Radiation (Sv)* (Table reprinted with permission by Dr. Townsend, University of Tennessee, Knoxville)

Exposure	BFO	Eye	Skin
30 days	0.25	1	1.5
Annual	0.5	2	3
Career	1–4	4	6

*The career dose equivalent, Sv, is based upon a maximum 3% lifetime risk of cancer mortality.
BFO = Blood forming organs.

3.2.6 Nutrition

In a study of two *Mir* cosmonauts, it was noted that only 40 to 50% of the World Health Organization predicted energy requirement was consumed and that, after 4 mo, the crewmembers had lost over 10% of their preflight body mass (Smith et al. 2001). Indeed, some space mission crews have been found to show a negative energy balance of approximately 20% (Heer, Elia, and Ritz 2001).

One primary source of body mass loss is skeletal muscle protein. The major site of protein loss appears to be the antigravity muscles, which exhibit the greatest rate of atrophy during long duration flights (Stein, Leskiw, and Schluter 1996). Astronauts tend to eat less than is optimal for nutritional requirements. However, studies of the *Skylab* crews in the 1970s show that, when sufficient emphasis is placed on nutrition monitoring and nutrition related flight elements, dietary recommendations can be met and strength, fitness, and body mass levels can be maintained close to those in a 1-g environment. For some, these values can even be exceeded (Rambaut, Leach, and Leonard 1977).

Clearly, there is a close relationship between nutrition and countermeasures with regard to the maintenance of astronaut health. Naturally, the facilities available for the appropriate storage and preparation of food are a vital component of the spacecraft. These facilities, coupled with the quantity and nature of the diet provided, that is, vitamins, minerals, and macronutrients, must complement the operational tasks and the countermeasures program to which the crew are subjected. The effectiveness of the countermeasures program is optimized if appropriate dietary content is available, which in turn optimizes the operational capability of the crew.

3.2.7 Immune System

Although the health optimization procedures used in the preparation of a mission, that is, minimal exposure to nonessential personnel, minimize the possibility of acquiring an infection, the risk of contamination is not absent. It is apparent that the *Apollo 7* mission upper respiratory tract illness widely reported at the time was acquired by one or more of the crew in the lead-up to the flight (Berry 1969). Furthermore, intrinsic latent viruses pose a risk to astronauts, in that they can be reactivated if the immune system is affected adversely by the environment and stresses imposed by the mission. Ground based simulations have shown alterations in animal immune system function in as little as 48 h (Aviles et al. 2005) and changes in human subject immune status after just 10 d (Shimamiya et al. 2005). Therefore, the incidence of suppressed immune system function clearly is not a direct product of the microgravity environment. Although it appears in some respects that the duration of a space mission can be related positively to the effect on the immune system, indications are that the cause of the condition is related more to the many stresses involved than can be attributed to microgravity per se (Payne et al. 1999).

To date it has not been possible to attribute positively the few clinical cases of infection to diminished immune system capabilities. Insufficient data exist to quantify properly the risks to either individual astronaut health or mission safety. In the future, it might be possible to minimize or even prevent this condition by ensuring appropriate mission conditions, such as adequate sleep, minimal stress, and a suitable psychosocial environment. However, a space mission of this nature is unlikely to be seen in the short or medium term. To complicate the issue, the nature of space missions in the future, that is, long duration exploration class missions, is such that other factors, such as increased exposure to ionizing radiation, greater physical and psychological separation from family, and heightened expectations of crew performance, will impact the immune system in a manner that can lead to greater reductions in immune system functionality than currently are noted (Sonnenfeld 2006). With this in mind, the present perception of altered immune system function as being low risk and low hazard is not appropriate for future space missions, and increased measures designed to limit or prevent this condition are required.

3.2.8 Extravehicular Activity

Although procedures have been adopted to reduce body nitrogen levels in preparation for extravehicular activity decompression, there remains a risk of decompression sickness. Tests involving nitrogen tissue to ambient pressure ratios that are reflective of those from Space Shuttle extravehicular activities suggest that this risk can be as high as 25% for Type I symptoms and 5% for Type II (Waligora, Horrigan, and Nicogossian 1991). These magnitudes, if correct, will result in some astronauts suffering from a degree of mild decompression sickness during the lifetime of the *International Space Station* and quite possibly moderate to severe symptoms at some time in the future. To date, however, no clear incidence of decompression sickness has been reported after any NASA extravehicular activity. It is possible, however, that there is a tendency for astronauts not to report mild Type I symptoms, particularly if such symptoms resolve quickly (Buckey 2006).

A simulation of the pressure profile associated with extravehicular activity was performed during the late 1980s (Waligora, Horrigan, and Conkin 1987). The subjects of the study prebreathed oxygen for 6 h before an exposure to 29.6 kPa cabin pressure. Upper body exercise was performed during this exposure to simulate the physicality of extravehicular activity. Detectable venous gas bubbles were noted in 47% of the subjects, and 11% developed symptomatic decompression sickness. From this data, it seems that the risk of decompression sickness is increased with exercise during and immediately after extravehicular activity (Pilmanis et al. 1999). Naturally, the many activities to be undertaken during extravehicular activity, in particular during future planetary operations, are clearly forms of physical exercise, and the potential risk these activities pose must be understood. A 65% incidence of venous gas emboli without symptoms and a 30% incidence of Type I symptoms were recorded in subjects performing upper body exercise designed to simulate extravehicular activity activities at 29.6 kPa (Conkin et al. 1987).

What is of note is that the general physical and physiological effects of undertaking long duration missions affect the astronaut in a manner that increases the risk of decompression sickness during and after extravehicular activity. Decreased fluid volume levels, decreased muscle strength and endurance, diminished thermoregulatory capability, and other common adaptations to the space environment, all heighten the potential for decompression sickness as a result of extravehicular activity (Newman and Barratt 1997; Cowell et al. 2002). After a year or more into a mission, an astronaut, for whom the prescribed countermeasure program has not been optimal, finding himself or herself in a situation where an emergency extravehicular activity must be performed that requires considerable physical effort, is exposed to a high risk of developing some degree of decompression sickness.

3.3 HEALTH MAINTENANCE

Crew health is crucial for human space mission success. Maintaining the physical and mental well-being of astronauts during all mission phases and protecting them from the detrimental effects of the space environment are the primary responsibilities of the medical operations team. They include a comprehensive health conditioning, countermeasures, and a rehabilitation program, as well as provision for adequate in-flight medical capabilities for diagnostics and treatment.

3.3.1 Preflight Preparation

Preflight operational requirements, if too onerous, can lead to the crew entering orbit in a physically and mentally suboptimal state, that is, not well prepared. The mission buildup phases should allow sufficient time for the crew to maintain or even gain the appropriate elements of fitness: endurance, strength, and flexibility. Time for exercise training should be an integral part of the preparatory phase, running in series with baseline data collection and other training. A number of physical, behavioral, clinical, and cognitive assessments also are conducted (Table 3.2) to provide a preflight baseline against which the effects of the mission and the post-flight rehabilitation program can be assessed. Finally, preparatory activities can be conducted to stabilize health and ameliorate or prepare for some of the effects of spaceflight, such as vestibular system preconditioning by means of parabolic flight (Figure 3.2) or rotation/tilt chairs such as the U.S. tilt translation device. Crew health stabilization protocols vary somewhat from mission to mission. However, such protocols are centered on mitigating the risk of crewmembers contracting infections by minimizing the number of people to whom they are exposed in the lead-up to launch and undertaking health checks for all those with whom they need to come into contact

Table 3.2 Crew Health Assessment and Monitoring (Courtesy of ESA)

	Preflight	In Flight	Post-Flight
Clinical assessment and monitoring	Crew surgeon clinical assessment	Health status evaluations	Crew surgeon clinical assessment
	Hearing assessment	Hearing assessment	Hearing assessment
	Neurological assessment	Body mass measurement	Neurological assessment
	Neurovestibular platform test		Neurovestibular platform test
	Resting electrocardiogram		Resting electrocardiogram
	24-h ambulatory electrocardiogram		24-h ambulatory electrocardiogram
	Dental and ophthalmology examination		Dental and ophthalmology examination
	Bone densitometry		Bone densitometry
	Ultrasound imaging (sonography)		Ultrasound imaging (sonography)
		Private medical conference	Photo documentation of skin
Laboratory	Laboratory testing	Laboratory testing	Laboratory testing
	Helicobacter pylori test	Clinical laboratory testing (ESA)	
Radiation	Biodosimetry	Personal dosimetry	Biodosimetry
Cardiovascular	Aerobic fitness/cycle ergometer	Aerobic fitness/cycle ergometer	Aerobic fitness/cycle ergometer
	Treadmill testing	Treadmill testing	Active postural stand test
Exercise and fitness	Functional fitness assessment	Calf volume measurement	*Functional fitness* assessment
	Isokinetic assessment	Periodic fitness evaluation	Isokinetic assessment
Extravehicular activity	Arm ergometry (Orlan)	Arm ergometry (Orlan)	
		Pre-/post-extravehicular activity crew medical officer exam	
		Monitoring during extravehicular activity	

3.3 Health Maintenance

Table 3.2 Crew Health Assessment and Monitoring (Courtesy of ESA)—cont'd

	Preflight	In Flight	Post-Flight
Psychological/behavioral	Preflight behavioral health status checks		
	Preflight psychological evaluation	Private psychological conferences	Post-flight psychological evaluation
	Mood assessment	Mood assessment	Mood assessment
	Neurocognitive assessment	Neurocognitive assessment	Neurocognitive assessment
Nutrition	Nutritional assessments	Nutritional assessments	Nutritional assessments

FIGURE 3.2 Microgravity by means of parabolic flight (preadaptation) (Photo: ESA).

with (Nicogossian et al. 1994). Furthermore, cardiovascular and muscular fitness evaluation, bone densitometry, and exercise prescription are employed to ensure optimization of crew health prior to launch.

Preparatory training is one of the crucial elements of preparing for a mission that has an impact not only on mission success but also on crew mental well-being. Operational tasks, procedures, and scenarios are practiced and thoroughly exercised prior to launch. The training program must be thorough, progressive, and appropriate to ensure that mission requirements can be met with a high degree of confidence. Simulations should be used as both training aids and assessments to monitor progress and identify weaknesses that require further work.

For a crew to undertake a space mission, and in particular a long duration orbital stay or a planetary exploration mission, they must be free from worries concerning social and family matters. For this not to be the case is to embark on a complex, expensive, and often globally important endeavor with a vital element of the mission, that is, the crew, being possessed with an increased risk of breakdown. The psychosocial support structure employed should be in place and active several years in advance of the mission. Appropriate counseling, as well as child and marriage support, should be provided to help the crew reach the launch day with family and outside of work commitments and responsibilities balanced and in order. Services such as these should be maintained throughout the mission and well into the postmission period. These services currently are provided by the medical support offices of the relevant space agency.

Long duration missions can be stressful. These missions receive minimal external support, thus requiring the crew to function effectively and smoothly by means of their own devices. For this to be possible, long duration mission crews should be trained to understand team dynamics. They should be able to identify and deal with individual and team stressors effectively, especially through the use of conflict resolution tools. The crew also must be able to minimize the possibility of depression and be able to treat any occurrence (Nicholas and Foushee 1990).

3.3.2 In-Flight Measures

Exercise Countermeasures

In his review of physical conditioning and deconditioning, Greenleaf (2004) suggests that the physical and physiologic states of athletes and the elderly can be seen to occupy polar opposite points on the same continuum, with the normal or fit state being located at the center (Figure 3.3). Acceptance of this concept shows us that time spent in space is in many ways like accelerated aging or chronic bed rest. Many of the fundamental aspects of aging and most of the responses to bed rest can be arrested or at least slowed with regular, intermittent, and appropriate exercise (Greenleaf 2004). In a similar manner, many of the effects of chronic exposure to the space environment can be prevented or reduced with exercise. As a consequence, the most prevalent form of countermeasure used in space is exercise, which at the time of writing, has a mandatory requirement to be performed for 2 h every day.

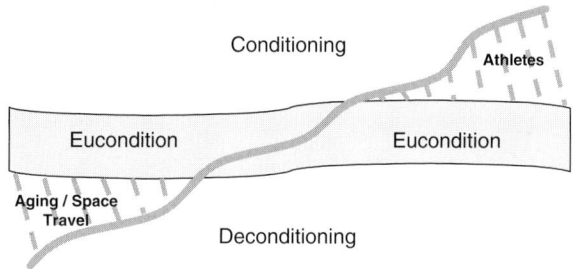

FIGURE 3.3 Greenleaf conditioning-deconditioning continuum model (Copyright 2004 from *Deconditioning and Reconditioning* by J. L. Greenleaf. Reproduced by permission of Taylor & Francis, a division of Informa plc).

Vibration Isolation and Stability Systems

Vibration isolation and stability systems exist in various forms but commonly act as an attachment between a piece of equipment and the space vehicle. Equipment that involves the movement of a moderate to large mass must be isolated from the space vehicle in a manner that prevents the transmission of vibrations through the hull, which if allowed can magnify causing disruption to sensitive scientific research and potential damage to operational systems and vehicle structure.

Cardiovascular Exercise

On Earth, repetitive, moderate to high intensity exercise utilizing a large volume of skeletal muscle typically leads to physiological adaptations that increase cardiovascular fitness, that is, an increased VO_2 max (Neufer 1989; Abernethy, Thaer, and Taylor 1990). The adoption of similar exercise routines during spaceflight tends to attenuate the loss of these adaptations, minimizing the cardiovascular deconditioning normally seen and potentially helping to minimize spaceflight induced orthostatic intolerance by means of reduced loss or maintenance of blood volume (Convertino 1991; Van Lieshout 2003). With an appropriate exercise prescription, an increase in VO_2 max for some can be observed. Furthermore, the stimulus presented to the muscles by repeated muscular contraction acts positively on protein turnover to reduce the rate of spaceflight induced muscle atrophy. Another crucial benefit of this form of exercise is the repeated application of force loads along the long axis of the antigravity bones. Here, exercise acts to stimulate the bone formation and resorption balance in a manner that is likely to reduce microgravity induced bone demineralization and atrophy (Layne and Nelson 1999). Although these primary positive effects and other lesser effects are derived from cardiovascular exercise, at present exercise countermeasure programs of this nature do not arrest fully the loss of fitness, muscle, and bone.

Treadmills Several treadmill designs exist, and these incorporate either passive or active modes or both. The treadmill allows the user to undertake aerobic exercise involving the major +1-Gz locomotive muscles. For the exercise to be effective, the user must be

128 **CHAPTER 3** Overview of Bioastronautics

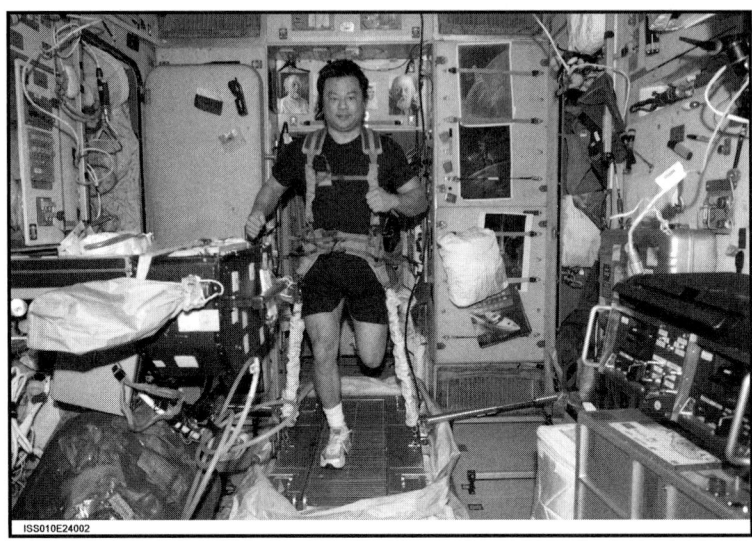

FIGURE 3.4 International Space Station treadmill with vibration isolation and stability system (Courtesy of NASA).

pulled down onto the track surface, a feat accomplished by an elasticized bungee system or the more elaborate subject loading device that involves a harness and force production system.

Treadmill with vibration isolation and stability system. The U.S. powered treadmill currently in use on *International Space Station* (Figure 3.4), that is, the treadmill with vibration isolation and stability, can be used with either bungees or subject loading device systems and in either passive, that is, the user moves the track with his or her own running action, or active, that is, powered and capable of a maximum of 16 km/hr, modes. Additionally when in passive mode, and in particular during the initial weeks of a mission, a subject positioning device attached to the treadmill and to the waist of a subject is used to ensure the subject maintains a central position on the track. A second generation treadmill with vibration isolation and stability, currently being built by NASA and ESA, incorporates similar attributes as the treadmill with vibration isolation and stability but with a more robust subject loading device system. The second treadmill is required for the expected increase from three to six crewmembers for the *International Space Station*.

Russian treadmill. The Russian treadmill functions as a contingency running belt on the *International Space Station* should the treadmill with vibration isolation and stability breakdown. It is a passive system built on the original *Mir* space station design. It is placed on the track of the primary treadmill in the event of a belt failure.

Contingency exercise surface. In the event of multiple treadmill failures, the contingency exercise surface offers a low friction surface on which the user can run if wearing appropriate running boots. The user still must be pulled down onto the surface by a bungee or subject loading device system, but otherwise there are no other moving parts.

Cycle Ergometers

Cycle ergometer with vibration isolation and stability system. The U.S. manufactured cycle ergometer with a vibration isolation system is the primary cycle exercise equipment on the *International Space Station* (Figure 3.5). It allows for measured and controlled workloads to be used within an exercise protocol. The system can be altered to offer upper body exercise by means of arm cranking. The user does not use a seat but instead is supported by a backrest and belt. The cycle ergometer with vibration isolation system is capable of offering up to 250 W of resistance and can run at a maximum speed of 120 rpm. Currently the cycle ergometer with vibration isolation system is being used in conjunction with a joint NASA/ESA system designed to measure oxygen uptake to

FIGURE 3.5 International Space Station cycle ergometer with vibration isolation and stability system (Courtesy of NASA).

FIGURE 3.6 International Space Station Russian Velo ergometer (Courtesy of NASA).

enable VO_2 max during flight to be ascertained, a vital capability that until now has not been possible.

Velo ergometer. The Russian equivalent to the cycle ergometer with a vibration isolation system, the Velo ergometer (Figure 3.6), allows the user to exercise the upper body while cycling (note that in the case of the cycle ergometer with a vibration isolation system, it is either/or) owing to the use of pulleys or force loaders routed through castors. This system provides resistance of up to 30 kgf against which the user works. Velo is capable of 250 W and 80 rpm. The opportunity to perform upper extremity exercise is particularly useful in preparation for extravehicular activity.

Resistance Exercise for Muscle Growth

The optimal stimulus for the maintenance or growth of muscle is high intensity muscular contraction, in particular eccentric contraction, that is, the muscle lengthens while contracting (Friden et al. 1983; Boppart et al. 2001). Several sets of a small number of high intensity contractions (in the region of 6 to 12) stimulates muscle growth more effectively than many low intensity contractions such as those that occur during running. To achieve this form of exercise requires the subject to act against some resistance; hence, the term *resistance exercise* is applied. An appropriate program of resistance exercise can minimize, or in some cases even prevent, skeletal muscle atrophy during spaceflight. The ideal form of resistance exercise should reproduce the muscle contraction profile seen at +1 Gz when lifting and lowering heavy weights. Of particular note is the degree to which an exercise device can offer eccentric loads in addition to concentric loads (Layne, Forth, and Abercromby 2005). The following four systems have been shown to provide beneficial concentric to eccentric resistance ratios.

Interim resistance exercise device. The interim resistance exercise device is the primary resistance exercise device (Figure 3.7) on the *International Space Station* at present (Schneider et al. 2003). The system works by means of two cylinders of rubber disks arranged in a circular fashion around a central axle. When the subject pulls on the pulleys, the rubber disks are deformed. The system can be set to offer different levels of resistance, thus enabling exercise against simulated and differing weights. The interim resistance exercise device allows most major resistance exercises to be undertaken, including in particular those requiring work from the antigravity muscles, such as squats, back extensions, and heel raisers. Around 26 exercises are possible; however, because it is hard mounted without a vibration isolation system, there are limitations on the speed and rate at which repetitions can be conducted, for example, 3 s per repetition.

Advanced resistance exercise device. The advanced resistance exercise device is a more complete and versatile version of the interim resistance exercise device, designed to be used onboard the *International Space Station* from 2009. The advanced resistance exercise device can do all that the interim resistance exercise device can; however, it is built to be more durable, require less maintenance, and offer greater resistance loads to the user.

Flywheel exercise device. The ESA flywheel exercise device is designed to enable the performance of resistance exercises similar to those undertaken with the interim resistance exercise device. This device relies on the ability of the subject to act against the inertia of a massive flywheel when moving a pulley in one direction and against the kinetic energy of the rotating wheel when allowing the pulley to return to its starting position. The device is smaller and less massive than the other resistance exercise device systems; however, it has yet to be determined whether the muscle contraction resistance profile from using the flywheel exercise device is as effective as that induced by the other resistance exercise device systems.

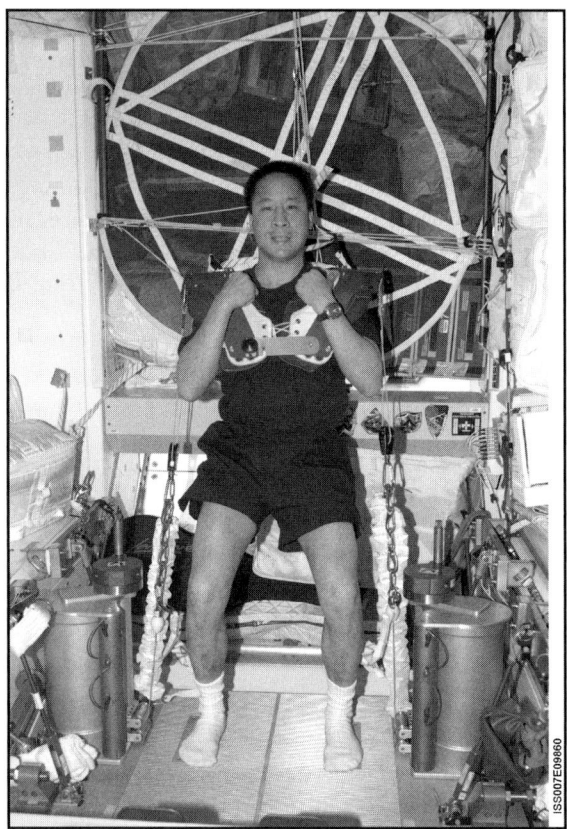

FIGURE 3.7 Interim resistance exercise device in use aboard the *International Space Station* (Courtesy of NASA).

The flywheel exercise device is situated onboard the *International Space Station* in the European *Columbus* laboratory.

Elasticized bands and expander spring systems. Various types of elasticized bungee and spring systems that offer a resistance when stretched have been evaluated, and some are available for use while in flight. The International Space Station expander bands are elastic lengths with handles at each end. For the longer models, shoulder pads are positioned centrally. Short lengths are used for shoulder exercises and the longer lengths for trunk and leg training. The muscular contraction profile elicited, however, is not the same as that seen during earthbound weight lifting, primarily because the resistance increases with stretch. Consequently, the force-velocity relationship is not optimal. For this reason, most of these systems simply are used in support of the primary systems, such as the interim resistance exercise device.

Training for Bone Growth

Cardiovascular and resistance training programs, such as those just outlined, each to some degree offers a stimulus for bone growth or bone mass maintenance (Layne and Nelson 1999; Langberg et al. 2000). At present, however, these forms of exercise are not optimal in this regard (Schneider et al. 2003). Research indicates that activities involving +1-Gz oriented force or impact along the long axis of the antigravity bones can be of most benefit in counteracting the bone wasting seen predominantly in the lower body during spaceflight (Vernikos et al. 1996). The exact manner, that is, magnitude of force, frequency, and duration, in which forces should be applied for optimal stimulation of bone is still under investigation.

Pharmacological and Nutritional Countermeasures

A number of pharmacological agents have been examined for use as prophylactic agents for the prevention or minimization of the effects of space travel. A standard for the use of drugs within a countermeasures program does not exist at present. Still, it is likely that certain agents will be used at some point in the future as an integral part of a countermeasures program for long duration exploration class missions. Some of the more prevalent examples are outlined.

Space motion sickness. A number of commonly used motion sickness drugs are effective prophylactics against the symptoms of motion sickness. Probably the most common are scopolamine and promethazine, where promethazine is the preferred drug for use during spaceflight. One side effect associated with many of these drugs is tiredness, which has in some circumstances led to the concurrent use of an amphetamine. This side effect, however, is considered beneficial, in that it can provide for a good quality of sleep during the first days on orbit. A further prophylactic measure is to minimize provocative stimuli such as head movement and unusual orientation. Consequently, it has proven beneficial during the initial hours and days of spaceflight for the crew to adopt the same orientation with respect to the spacecraft and each other and to minimize any extravagant motions, such as free spinning.

Body fluid levels. A common physiological adage is that "where salt goes, water follows." It is a nominal procedure for a crew to ingest sodium containing fluids and foods about 1 h before reentry. The fluid loading is an attempt to increase blood volume levels as an aid during orthostatic stress.

Orthostatic intolerance. Drugs that augment blood vessel constriction potentially can reduce the incidence of orthostatic intolerance after spaceflight. Midodrine has been used as a treatment for patients on Earth who suffer from poor blood pressure control. It appears to be effective against orthostatic intolerance for astronauts on their return to Earth (Kaufmann, Saadia, and Voustianiouk 2002).

Bone. Bisphosphonates, such as alendronate, are inhibitors of the process of bone resorption and appear to affect bone formation positively (LeBlanc et al. 2002). The manner in which these substances can be taken and their effectiveness of use are still under investigation. It has been noted that ingestion of bisphosphonates not always is tolerated well by the body and can result in some side effects. Even so, it is quite possible that these substances will prove to be a valuable countermeasure for future long duration missions. Other substances that can benefit bone formation are calcium, phosphorus, and vitamin D supplements. It is likely that the use of these substances will be of benefit to the crew, who through the rigors of the mission operational profile is likely to be ingesting too few micronutrients to maintain their calcium balance.

Radiation. Antioxidants are to some extent able to react with the free radicals produced when body tissues are bombarded by radiation, thus positively modulating the cellular response to radiation and potentially reducing damage (Saliou et al. 1999; Fang, Yang, and Wu 2002). Antioxidants are both naturally occurring, such as vitamins A, C, and E, which are found specifically in fresh fruit and vegetables, and manufactured, such as amifostine.

Muscle. Antioxidants, as well, can be beneficial as a countermeasure to muscle atrophy; however, sufficient work has yet to be undertaken to prove positively their potential. Any beneficial effects will be accrued automatically if they are used as an element of the radiation countermeasures program.

Artificial Gravity

The provision of an acceleration vector equivalent to that of the +1-Gz gravity experienced on Earth prevents most adverse effects related to microgravity (Vernikos et al. 1996; Fong 2004). The difficulty of doing so, of course, lies in the resources and technical capabilities required to build a spacecraft that can rotate and thus produce centrifugal force equivalent to Earth gravity. At present, it is likely to be exorbitantly expensive to build a whole body rotation spacecraft, although a number of designs are being examined. One major consideration affecting the choice of design is the diameter or lever arm length to revolution speed ratio. A long arm centrifuge can produce appropriate centrifugal forces at slow revolutions, whereas short arm devices require faster revolutions to do the same. Long arm designs can require the whole vehicle to rotate, whereas short arm devices can reside within the vehicle. The downside of short arm centrifuge use is that considerable Coriolis effects can be introduced. These negatively affect the vestibular system and most likely introduce nausea and other motion sickness related effects (Graaf and Roo 1996).

Balancing Velocity and Radius Smaller, less expensive rotation devices and spacecraft must be spun faster to produce sufficient gravitational force for use as an effective countermeasure. In so doing, however, the incidence of motion sickness is increased. Generally speaking, acceleration to a maximum rotation of about 3 rpm is tolerable. If, however, crewmembers are accelerated to rotation speeds beyond this level, the resulting Coriolis effect leads to motion sickness for most. A gradual ramped acceleration profile, that is, accelerate, hold, accelerate, hold, and so forth, can allow users to tolerate between 7 and 10 rpm. Even so, the fact remains that a balance must be struck between the cost and size of the device and the rotation speed required to offer appropriate centrifugal force without undue risk of motion sickness. In simple terms, this demands the construction of either large, slowly rotating vehicles or small, rapidly spinning human centrifuges that can be contained within more conventional spacecraft.

Large rotational spacecraft would offer the potential for the crew to retain a +1-Gz equivalent stimulus for the duration of a mission and, thus, negate the need for most other countermeasures. If this is not possible, the smaller human rated centrifuge might be used by the crew on an intermittent basis. In this scenario, a crewmember might, for example, undertake a 1-h session at +2.5 Gz each day or perhaps a 2-h session at +1 Gz might be sufficient to prevent serious deconditioning of the primary systems of the body. The exact magnitude to quantity ratio, and thus the nature of the centrifugation prescription required, is still under review.

Short Arm Centrifuge

Passive. Short radius centrifuges requiring the user to lie passively while being spun have been examined, for example, by the NASA University of Texas Medical Branch/Wyle Artificial Gravity Project, and appear to offer the potential for considerable beneficial effects. The use of simultaneous exercise during centrifugation, however, is to offer the greatest benefits when the adverse effects of long duration spaceflight are considered.

Exercise + centrifugation. A number of modes of exercise during short arm centrifugation have been examined (Greenleaf et al. 2001; Edmonds, Jarchow, and Young 2007). Impact exercise can prove to be highly beneficial with regard to bone protection; however, the harnessing of human energy by means of a cycling or running action to power the centrifuge is also of great interest. Human powered centrifuges have the advantage of using the work done by the astronaut user to rotate the device (Figure 3.8), thus conserving spacecraft power (Greenleaf et al. 2001). Further, the physical activity in question (normally a cycling motion) enables two countermeasures to be combined into one, in that exercise is undertaken that positively affects cardiovascular and muscular deconditioning and acceleration forces are imposed that can offset orthostatic intolerance and, to some extent, bone atrophy.

Long Arm Centrifuge These forms of centrifuge design are clearly the most beneficial physiologically, because if crewmembers are not exposed to the rigors of microgravity, they will not decondition in the manner outlined in the previous sections. The potential for motion sickness, however, still exists, and little work has been done to evaluate fully the long-term effects of exposing the human body to centrifugation. One other advantage

FIGURE 3.8 ESA short arm human centrifuge at MEDES, Toulouse, France. The four arm short arm human centrifuge is able to accept countermeasure (vibration) and exercise attachments (Photo: ESA).

for this form of centrifuge over that of the short arm device is that the gravity gradient along the body is not as extreme as that seen in, for example, a 6-m diameter device that can produce +5 Gz at the feet, +2 Gz at the hips, and +1 Gz at the head.

Other

Lower Body Negative Pressure Lower body negative pressure entails placing the lower body in an enclosed compartment and depressurizing the closed system. The resulting negative pressure inside the compartment causes blood in the upper body to be drawn in to the lower body in a fashion similar to that seen during standing on Earth. Although the pressure gradients are not the same as that produced by standing at +1 Gz, much of the resulting stimulus to the body is. Regular use of lower body negative pressure while in flight appears to reduce the severity of spaceflight induced orthostatic intolerance for some users. There remains some debate as to the effectiveness of this system; however, the Russian program incorporates its use onboard the *International Space Station* by means of the Chibis suit (Convertino 2001).

Chibis Chibis is a Russian lower body negative pressure device that encompasses the waist, hips, and legs of an astronaut individually (Figure 3.9). Negative pressures of between −1 and −60 mmHg can be applied during lower body negative pressure sessions. These normally are undertaken as an element of the countermeasures program prescribed in preparation for landing and are designed to reduce the incidence of orthostatic intolerance when at +1 Gz.

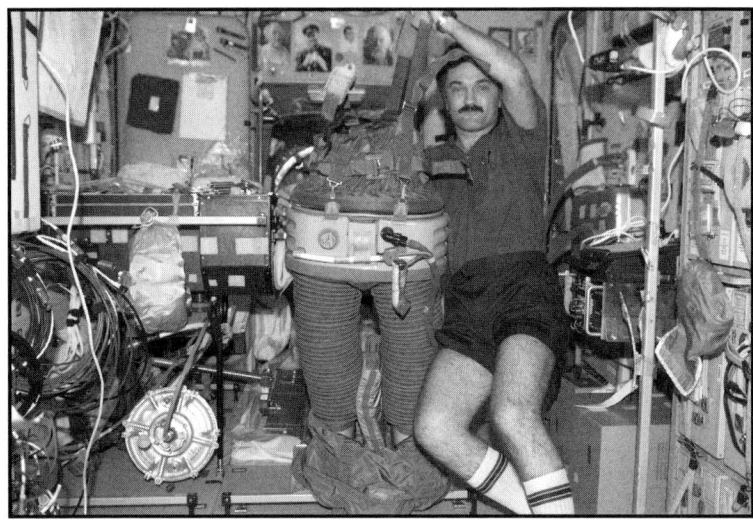

FIGURE 3.9 Russian lower body negative pressure device (Chibis) used on *International Space Station*. (Courtesy of NASA).

Fluid Loading One of the means for aiding post-flight blood pressure control is to increase blood volume levels. Currently, it is common for a regimen of fluid loading to be adopted in which 1 L of isotonic saline (salt) solution is imbibed approximately an hour before reentry (Clément 2005). This fluid loading protocol, however, does not protect from the orthostatic intolerance experienced once back on Earth. Instead it has become common practice to provide an intravenous line for every long duration crewmember that is used for infusing large amounts of fluid (target is about 1 to 2 L of sodium chloride solution) immediately post-landing.

Electrostimulation The Russian Space Agency uses a system called the *myoelectrostimulator* as a countermeasure to muscle atrophy for their cosmonauts. Electrodes contained within a compliant suit are placed on the thighs, calves, abdomen, back, and shoulder (trapezius) muscles. Electrical stimulation causes the muscles to contract, thus potentially providing a stimulus for muscle growth. This system is not used by all cosmonauts, but its employment is beneficial, in particular during periods in which the primary exercise modes, for whatever reason, cannot be employed.

Compression/Load Clothing
- **Penguin suit.** The penguin suit is a Russian manufactured garment containing elastic elements designed to compress along the long axis of the body. It is designed to be used for limited periods of time during normal daily activities and offers a force against which the antigravity muscles and bones must work. The constraints imposed by the suit are such that some cosmonauts prefer not to use it.

Occlusion cuffs. The placement of adjustable cuffs (bracelets) around the upper sections of the legs, if applied at a pressure slightly greater than venous blood pressure, reduces venous outflow from the lower body, thus causing blood pooling in the lower body. The pooling increases venous pressure and can lead to an increase in stimulation of the receptors associated with lower limb blood vessels. Through the use of these devices, the potential exists to retrain the lower limb vascular constriction system, therefore aiding blood pressure control when standing post-flight. Furthermore, the use of thigh bracelets during the early period of adaptation to microgravity can ameliorate some of the adverse effects caused by the headward movement of fluids that occurs at this time.

Compression garments. Lower body compression garments such as the Russian *kentavr* suit and the U.S. antigravity suit can be worn on reentry and landing to decrease lower limb compliance. These devices compress tissues by means of inflatable cuffs or adjustable, elasticized straps to minimize lower body blood pooling; thus, they aid blood pressure control when standing immediately post-flight.

Extravehicular Activity Preparation The primary means for decreasing the risk of decompression sickness during or after extravehicular activity is to reduce the pressure of dissolved nitrogen in the tissues beforehand. Tissues can be denitrogenated by pre-breathing 100% oxygen (preoxygenation) or by being exposed to low ambient pressure (nitrogen equilibration) in the lead-up to extravehicular activity. No single, 100% effective protocol for the preparation of crew for extravehicular activity exists; however, various combinations of preoxygenation and nitrogen equilibration durations, gas partial pressures, and exercise either have been examined or are used (Waligora et al. 1991; Pilmanis et al. 2004). At the time of this writing, for all extravehicular mobility unit based extravehicular activities, an exercise prebreathing protocol, a 4-h in-suit prebreathing protocol, and a 9-h campout style prebreathing protocol have been accepted for *International Space Station* use with varying degrees of decompression sickness risk.

Shielding The only effective countermeasure against high dose acute radiation exposure is shielding. Traditionally bulk mass shielding composed of materials, such as aluminum, has been used for passive shielding in space. However, for spaceflight, these materials pose weight problems. In recent years, it has become evident that light, highly hydrogenated materials, such as polyethylene, are ideal for space use (Shavers et al. 2004; Zeitlin et al. 2005), particularly if used as an element of a storm shelter designed to act as a last redoubt during high radiation events. An attractive alternative to passive shielding is active shielding, whereby electrostatic fields, plasma, or magnetic fields can be used to deflect particles (Spillantini et al. 2007); however, the realistic use of such technology is still someway off.

Combination Promising combinations of countermeasures are lower body negative pressure and exercise (Smith et al. 2003) and exercise and centrifugation, as already discussed (Greenleaf et al. 2001). In the former case, the use of lower body negative pressure while pulling the user down onto a treadmill belt as he or she runs has indicated that some benefit can be accrued by the long bones of the lower body (Schneider et al. 2002). This combination is likely to prove to be a useful countermeasure against bone and muscle atrophy and orthostatic intolerance when an appropriate intensity, duration, and frequency profile has been prescribed.

3.3.3 In-Flight Medical Monitoring

In-Flight Medical, Psychological, and Biomedical Monitoring

A number of crew physiological variables are monitored during certain scientifically and operationally sensitive periods of a mission. An example of a procedure likely to be adopted for the *International Space Station* is that of fitness evaluation using oxygen uptake measurement (Figure 3.10). Another area of importance is that of radiation monitoring. Passive dosimeters, such as the European crew personal dosimeter, are for use to provide crewmembers with measures of personal radiation exposure. Active devices developed by NASA and JAXA are also in use in and around the *International Space Station* to offer real time viewing and data collection of radiation fields at different locations.

Flight environmental conditions and crew psychosocial interactions also are monitored for any off nominal circumstances. Routine monitoring activities and certain emergency procedures are conducted by biomedical engineers at a console using telemetry

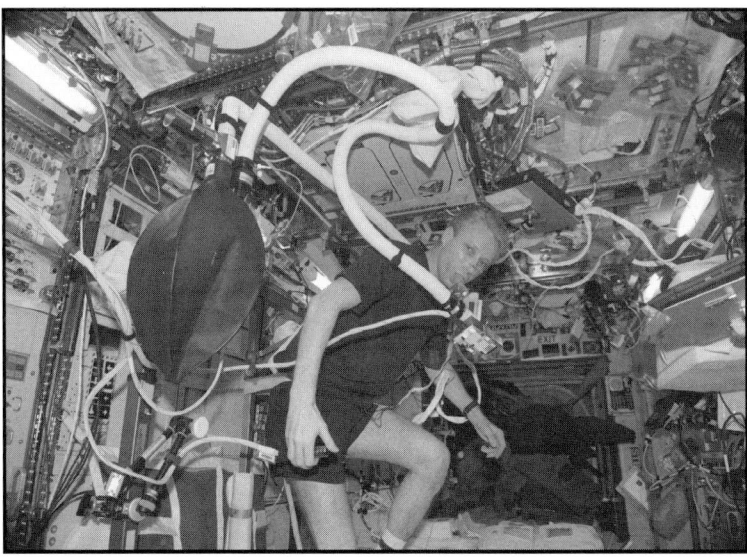

FIGURE 3.10 International Space Station periodic fitness evaluation with oxygen uptake measurement (Photo: ESA).

FIGURE 3.11 European astronaut center, medical operations console, Cologne, Germany (Photo: ESA).

(Figure 3.11), whereby cardiovascular fitness is evaluated by means of a metabolic gas analysis system. For psychological and behavioral assessment, agency behavioral scientists assess crew and their personal activities. For critical and potentially dangerous periods, flight surgeons also staff the console to provide medical support and advice when it is needed. Additionally, on arrival at the *International Space Station*, crewmembers begin daily private medical conferences with their flight surgeon. These conferences reduce to weekly after day 5. Furthermore, private medical conferences also are held the day before and the day after an extravehicular activity and whenever requested by a crewmember.

In-flight monitoring will be extremely important during long duration exploration class missions. During these sojourns, it is quite likely that stress, depression, and potential crew conflict could be encountered. While in flight, aids to emotional health, such as regular communication with family and friends and personally tailored recreational facilities can be invaluable. It is expected that the crew psychosocial support and countermeasures program will become an increasingly important element of the health maintenance program in the future.

During the planning process for the International Space Station onboard medical services and resources, the international space medicine community followed the paradigm "stabilize and transport." This is a process followed routinely in terrestrial medical contingencies to provide the infrastructure, resources, and trained personnel to evaluate the criticality of a medical incidence and then to provide aid to stabilize the vital functions of the patient. Finally, some transportation method, such as an ambulance, is used to deliver the stabilized patient to a definite health care unit, where further diagnosis and treatment can be applied. Such processes do not require a full-fledged emergency medical care unit on orbit but only the resources required to bridge the gap between incident and arrival at the hospital. With the deletion of a space ambulance, that is, the *Crew Return Vehicle*, from the International Space Station Program, the medical support paradigm was changed to a stand and fight philosophy, because the onboard medical infrastructure was not enhanced to provide medical support beyond stabilization of major vital body functions. Within the current circumstances, medical resources are limited on the *International Space Station*, and major medical incidences can be presented in which treatment options are reduced considerably.

Nevertheless, the *International Space Station* provides a crew health care system comprising not only a basic medical infrastructure but also some advanced health care systems. It consists of several modular medical supply and diagnostic kits, emergency medical kits, and includes comprehensive medical checklists, procedures, and decision making algorithms in a suitable format to enable non-physicians to apply appropriate measures. Further, environmental monitoring devices to take surface and air samples for microbiological contamination analysis are included.

Some astronauts receive specific training to act as a crew medical officer who can function as a paramedic. On orbit, the crew medical officer can be called to provide medical services and to interact with the flight surgeon on the ground. The crew medical officer has some freedom to prescribe drugs and initiate medical protocols during a mission, as well.

Some human life science experiments include devices, such as ultrasound equipment, that potentially can be used for medical care purposes. The international medical community is evaluating these resources and seeking agreements with relevant scientists with a view to implementing appropriate processes and crew training for the nonscientific use of this equipment.

Communication and access to medical knowledge are other important aspects of health maintenance. A backbone for the distribution of secure medical data and implementation of state-of-the-art database and collaboration tools has been established. Work is progressing to both augment these capabilities and extend accessibility not only globally but possibly to potential clients on orbit. Specifically, ESA has designed and implemented the core of the space medicine information system that globally connects all medical international partners of the *International Space Station*.

Countermeasure Prescription

The cornerstones of the countermeasures program currently employed onboard the *International Space Station* are the treadmill with vibration isolation and stability, the interim resistance exercise device, the cycle ergometer with a vibration isolation system, and the Velo ergometer. Standard is 2½ h for the preparation for and conduct of exercise; however, the crew usually chooses to split this duration into two sessions of 1 h and 1½ h, respectively. Daily exercise report summaries show the quantity of exercise and the mode of equipment used by each crewmember (Table 3.3).

The intensity of effort as evidenced by the workload or heart rate to be achieved for cardiovascular exercise is predetermined from preflight baseline measures. The program developed and implemented is progressive in an attempt not only to slow any effects of

Table 3.3 Example of International Space Station Crew Exercise Routine for One Day (Courtesy of NASA)

	CDR (h)	FE-1 (h)	FE-2 (h)
Treadmill with vibration isolation system	1.3		1.0
Resistance exercise device		1.0	
Cycle ergometer with vibration isolation system		1.5	
Velo ergometer	1.0		1.5
Treadmill + cycle ergometer			
Velo + resistance exercise			

CDR = commander.
FE = flight engineer.

spaceflight but also to prevent the effects if possible. With regard to resistance exercise, all crewmembers undertake squats, heel raisers, and dead lifts. The choice, quantity, and frequency of other exercises is determined between the crewmember and the strength and conditioning coach prior to launch and are adapted throughout the mission in accord with the physical effects noted.

Some differences do exist between the U.S., European, and Russian crew countermeasure programs. For instance, Russian crews undergo in-flight orthostatic tolerance assessments using Chibis and allocate and adjust the intensity and duration of cardiovascular exercise in a different manner than the U.S. and European programs.

3.3.4 Post-Flight Recovery

Routinely, a 45-d recovery and rehabilitation period is implemented for all long duration flyers. The emphasis of post-flight recovery procedures focuses on the rehabilitation of the crew from the rigors of the mission. Post-flight physical examinations and baseline assessments are conducted to provide accurate measures of the magnitude of any mission related effects in comparison with baseline preflight assessments. In addition to any psychosocial support deemed necessary, the post-flight program should offer treatment for cardiovascular, muscle, bone, and neurovestibular changes. The post-flight recovery programs currently used offer individually prescribed elements according to the specific effects the mission has had on a crewmember. Fitness, flexibility and strength training, neuromuscular coordination training, and in the event of very long duration missions, social reintegration tutoring can all be required. Striking a balance between the requirement to address these items as soon as possible on a return from a mission and allowing crewmembers time to relax with friends and family is a critical element.

3.4 CREW SURVIVAL

This section addresses the role that crew survivability plays in human space exploration. Specific design trade-offs for crew survivability; that is, the capabilities and limitations and human health threats associated with the various phases of spaceflight, are discussed. Characteristics of human crewed spacecraft include capability, sustainability, affordability, and survivability. Crew survival is the collective implementation of abort, escape, emergency egress, safe haven, emergency medical, and rescue capabilities throughout a mission to strengthen the capability to keep the crew alive and return them to Earth safely in response to some imminent catastrophic condition. Crew survivability, which is about options, has been a fundamental requirement for human space systems since the dawn of human spaceflight.

3.4.1 Overview of Health Threats in Spaceflight

Human spaceflight is known to be extremely risky. Threats to crew health have occurred during every phase of a mission, including prelaunch, launch and ascent, on orbit, reentry and landing, and post-landing. Spaceflight related human health threats include the space environment (microgravity, vacuum, and radiation), spacecraft environment (noise, closed life support), and mission (circadian disruption, sleep deprivation) effects. Humans exhibit considerable adaptive responses to microgravity, such as neurovestibular, musculoskeletal, and cardiovascular changes. The spacecraft environment as well can expose crew to toxins and other hazardous materials. Confinement, isolation, and intense workload have created psychosocial adaptation issues.

Training and ground checkout is risky, in that 3 Russian and 7 U.S. fatalities occurred while preparing for space operations. An additional 3 Russian fatalities related to space operations occurred on reentry and landing, and 15 U.S. fatalities, of which 7 occurred on ascent (*Space Shuttle*) and 8, including one X-15 pilot who qualified for astronaut wings, on reentry (Space Shuttle). Catastrophic loss of crewed launch vehicles have occurred on the pad (*Soyuz 18A*) and on ascent (*Soyuz T10A*, STS-51L), with both *Soyuz* crews surviving (Hall and Shayler 2003).

Reentry anomalies have occurred frequently, often due to vehicle configuration or faulty separation from modules. Anomalies in the crew cabin environment during descent have resulted in death and serious injury. Landing and post-impact issues also have occurred. These include hard impact injuries and injuries that have been the result of an inability of rescue forces to reach a crew in a timely fashion.

The conduct of an extravehicular activity represents one of the most dangerous endeavors for a flight crew (Shayler 2000). During an extravehicular activity, astronauts and cosmonauts have been exposed to thermal injury (*Gemini 9*), separation from the spacecraft (*Salyut 6*), a suit leak (STS 37), contact with toxic substances (STS 98), an ocular foreign body (STS 100), and severe pain due to an improper boot fit. Additionally, radiation, retinal injury from sunlight, life support system failures, work site injury (crush or electrical), hypobaric decompression sickness, and spacesuit pressure loss due to micrometeoroid and orbital debris are potential concerns for crewmembers working outside the spacecraft.

Spaceflight emergencies occurring on orbit (Shayler 2000) include cabin pressure loss (1997), fire (1971, 1977, 1988, and 1997), and a toxic environment (1997). Human factors errors in both space flyers and ground controllers have affected mission milestones, and these have come close to catastrophe. Loss of vehicle control has occurred on ascent, on orbit, and during reentry on X-15 flight 191, *Gemini 8*, the *Apollo 10* lunar module, *Apollo 13* command and service modules, STS 25 (51-L), STS 32, *Mir* following the *Progress M-34* collision, STS 107, and Space Ship One (the X Prize qualifying flight).

Medical events have occurred in space, and these affected mission objectives. Indeed, medical evacuation from space occurred three times, the first in 1976 for intractable headaches, again in 1985 for urinary infection, and most recently in 1987 for a cardiac irregularity (Newkirk 1990). Medical evacuation was in process on three other occasions when the medical condition stabilized or resolved.

Crew survival must take into consideration the phase of mission coverage and the survival concept of associated operations. Mission operations coverage spans launch pad and ground egress, crew escape during launch and reentry, on-orbit operations, in-flight safe haven, crew transfer or rescue, mission abort or emergency crew return, and crew survival after landing (OTA 1989). To ensure crew survival in the event of some catastrophic failure, the concept of operations should account for vehicle autonomy and life span, crew size and length of time on orbit, life support capability, and individual crew protective systems.

Design and inclusion of an adequate crew survival system within a spacecraft presents a number of challenges and limitations to vehicle design and performance, such as weight, center of gravity, and aerodynamics. A crew survival system should neither create additional risks to a crewmember nor unduly limit spacecraft capability, affordability, and sustainability. Survival and escape systems that jeopardize nominal operations can defeat overall mission success.

3.4.2 Early Work

As part of the necessary preparation for sending humans to space, life support and survival systems were developed and tested using animals and human subjects. Both the Russian and U.S. military used human subjects in high altitude balloon parachute jumps. The U.S. Air Force conducted high altitude balloon studies on humans. On August 6, 1960, during the third flight in Project Excelsior, Captain Joe Kittinger parachuted in free fall from an open gondola from 102,800 ft for 4 min, 36 s, attaining a maximum speed of 714 mph until the main parachute opened. His right glove lost partial pressure, and his hand became swollen, causing extreme pain; however, it had returned completely to normal by 3 h after landing. On a previous jump during Project Excelsior, a drogue chute entangled him. He went into a high speed spin and became unconscious.

The Russians also conducted a high altitude parachute program to test their full pressure suit for the Vostok space program. Two crewmembers ascended in a pressurized gondola on the Volga balloon on November 1, 1962. Colonel Pyotr Dolgov exited at

86,156 ft, and his parachute opened immediately. As he exited the balloon, his helmet visor hit an attachment and cracked. During descent, his suit depressurized, and he was found dead after landing.

An American civilian high altitude parachute test series, Project Strato Jump, was conducted in 1966. Nick Piantanida reached 123,500 ft but was unable to disconnect from onboard oxygen. Ground control cut the gondola from the balloon. Later in 1966, he again attempted a record leap. While ascending through 57,600 ft, ground control heard a sudden rush of air and heard him scream, "Emergency!" The gondola was cut from the balloon and landed 25 min later. He was unconscious on landing, and died 4 mo later in a hospital.

3.4.3 Crew Survival on the Launch Pad, at Launch, and During Ascent

Explosion, blast overpressure, flying debris, toxic fumes, cryogenic fluids, fire, falls, and noise are among the more important dangers that threaten crew and personnel at the launch pad (Table 3.4). To overcome these threats, a launch escape system is required to separate a crew from the hazard rapidly. The launch escape system must be able to pull the crew capsule from an exploding launch vehicle during prelaunch and launch ascent; separate the launch escape system from crew capsule once free of the catastrophic environment; provide survivable parameters for the crew while considering dynamic pressure, Mach number, and flight path angle; and provide an appropriate crew capsule attitude for entry and parachute deployment.

During Project Mercury, a mobile service structure and a remote controlled maneuverable arm cherry picker were used for pad aborts. The Mercury capsule launch escape system underwent high-Q abort tests between August 1959 and April 1961, when seven uncrewed Little Joe rocket boosters were launched from Wallops Island, Virginia.

During Project Gemini, ejection seats were used to reduce the launch weight associated with the capsule launch escape system used on Mercury. The Gemini ejection seat system underwent extensive qualification testing. The high altitude ejection test program was used to demonstrate the functional reliability of the Gemini personnel recovery system up to 40,000 ft and Mach 1.7. This system came close to being used during the *Gemini 6* launch abort on December 12, 1965, when the main engines shut down 1 s after ignition. The booster settled onto the launch pad unsecured, and although the criteria for ejection had been met, the crew rapidly assessed the situation and decided not to eject.

Pad egress from the launch umbilical tower during Project Apollo included a slide wire egress system, a high speed elevator, an emergency egress escape tube, and a rubber room capable of sustaining 20 people for 24 h. The Apollo capsule launch escape system was tested extensively (Young 2005). Four test flights of the Apollo capsule launch escape system were conducted from January 1964 to January 1966 using the Little Joe II rocket booster at White Sands, New Mexico. On August 28, 1963, the booster failed 25+ s into flight, but the launch escape system functioned normally.

Table 3.4 Atmospheric Entry Human Health Threats

Entry Phenomenon	Physical Characteristics	Altitude Range	Biologic Effect	Countermeasure
Plasma	Ionized molecular oxygen	>130,000 ft	Chemical or thermal burn	Thermal protective system
High gravity forces	g force in X, Y, and Z linear or angular planes	Entry interface (400 K to ~10 K)	Organ damage, body fragmentation	Crew compartment stability system and axial restraint system
Hot gas	°C	<150,000 ft	Thermal burn	Thermal protective system
Dynamic heating	°C		Thermal burn	Thermal protective system
Atmospheric pressure	Atmospheric absolute (ATA)	<63,000 ft (0.06 ATA)	Tissue water to gas resulting in embolism	Pressure vessel, pressure suit, and pressure breathing mask
		<18,000 ft (0.5 ATA)	Evolved tissue nitrogen resulting in decompression sickness	Pressure vessel, pressure suit, and pressure breathing mask
Pressure differential	Delta pressure (dp/dt)	<18,000 ft (0.5 ATA)	Barotrauma in gas filled spaces (pulmonary, otic, dental, gastrointestinal)	Pressure vessel and pressure suit
Oxygen partial pressure	ppO_2	>10,000 ft	Hypoxia and asphyxiation	Pressure vessel, pressure suit, and supplemental oxygen
Intrusion of habitable space			Fatal or severe organ damage due to penetration	Crew compartment protective system
Terrain impact	g force in X, Y, and Z linear planes	Surface	Injury due to deceleration	Parachute, lifting body, airbag, and braking rocket

A pad explosion and the associated fireball requires that an escape system be able to protect the crew from heat, overpressure, and an unstable launch platform. A launch pad abort with activation of a launch escape system occurred on September 26, 1983, at the Baikonur Cosmodrome. A minute and a half before the liftoff of *Soyuz T-10A* carrying two cosmonauts, a fuel valve failed and leaked fuel onto the launch pad. A fire ensued at the base of the rocket. The launch control team fired the launch escape system, carrying the capsule several miles away for a parachute landing. The crew experienced over 15 g when the escape rocket fired. Another ascent abort occurred on April 5, 1975, on *Soyuz 18-1* with two crewmembers aboard. The third stage ignited but failed to separate from the second stage. With the launch vehicle gyrating wildly, mission control commanded separation. The crew sustained over 20 g on reentry. One crewmember sustained internal injuries, and neither flew in space again.

Following the *Challenger* incident, the Rogers Commission recommended that a system be developed to ensure crew egress and escape (NASA 1986). According to this recommendation, crew escape and egress procedures, together with other elements of the egress and escape system, should enhance the capability for all crewmembers to escape safely from a disabled *Space Shuttle* on the pad, during subsonic flight, and on the ground (NASA 1986). Furthermore, controlled subsonic gliding flight conditions must be required for use of the crew escape and egress system during flight. As well, the crew escape and egress system should provide the necessary protective and survival equipment to sustain the crew at or below 100,000 ft altitude and ensure the safety of the crew after vehicle egress until the danger is past or they are removed to a safe area (Miller 1990).

Presently, the ground abort modes for Space Shuttle operations include prelaunch egress and escape from the launch pad (Modes 1 through 4) and post-landing egress and escape on or off the runway (Modes 5 through 7). Space Shuttle launch pad egress from the 196-ft level is a 30-s descent down a 1200-ft slide wire. Seven slide wire baskets can each carry three people down to a concrete bunker. An armored personnel carrier is used to escape to a prearranged rescue site.

A launch abort and pad fire occurred on June 26, 1984, with the *Space Shuttle Discovery* STS-41D (STS-12). The Space Shuttle's main engine experienced an out of sequence start that resulted in an abort shutdown about 4 s before solid rocket booster ignition. After the abort, a hydrogen fire occurred on the starboard body flap and burned for about 12 min. The aft base heat shield water deluge system was activated repeatedly. Following a contingency landing, the crew can egress via the pyrotechnically jettisoned side hatch and inflatable descent slide or an overhead hatch and sky genie rope descender (SFOC 2000). However, during this event, the launch control center elected to keep the crew inside the vehicle.

The abort modes during launch and ascent include return to launch site, transoceanic abort landing, abort to orbit, and abort once around. An abort to orbit occurred on July 29, 1985 on the *Space Shuttle Challenger* (STS-51F) when one Space Shuttle main engine shut down prematurely at $T + 5{:}45$ into the flight. The Space Shuttle in-flight abort mode (Mode 8) calls for an in-flight bailout from controlled gliding flight. The Space Shuttle crew escape system was designed and certified for use during controlled gliding flight (Ondler 1987).

Following the Challenger accident (STS-51L), astronauts Bill Shepherd and Mike Foale reviewed data regarding the feasibility of using the Space Shuttle crew escape system equipment for egress during loss of control or breakup (Foale and Shepherd 1989). Their study evaluated the forces leading to vehicle breakup, crew module dynamics following breakup, crew module survival, crew module stable attitudes, and time available to egress.

This study determined that the primary egress concern was related to exactly when to leave the *Space Shuttle*. Several crew egress procedures and techniques were recommended, among which is that vehicle egress should begin once the crew module is below 40,000 ft. It was indicated that the cues to egress are diminishing g forces and suit depressurization. These would afford the crew approximately 80 to 90 s to egress. Although the contingency egress during a loss of control following vehicle breakup is not a defined egress mode, it tentatively has been called a *Mode 9 bailout*. Their conclusion was that the crew module could withstand reentry heating for an ascent breakup occurring at 280,000 ft and below. Forces causing vehicle breakup in the Challenger accident, which were estimated to be of the range from 12 to 20 g, were short duration and nonlethal (Kerwin 1986).

Various crew escape systems, including bailout, ejection seats, extraction rocket system, encapsulated seats, modular separation, or a hybrid combination of these systems, have been considered to provide coverage during ascent and reentry (Whitehurst 1987, Miller 1990). A recovery system, such as a parachute, must provide the required deceleration, meet mass and volume requirements, and should be protected from vacuum, the chemical effects of rocket motor fuels, and heating from atmospheric entry. It also must account for the wake effect of large spacecraft on drag area, avoid collision between jettisoned hardware and the parachute, minimize shock loads during parachute deployment, disengage from crewmembers immediately after impact to avoid unwanted dragging of payload, and provide redundancy for subsystems failure.

Following the reentry breakup of *Columbia* (STS-107), a crew survival working group was appointed to assist the Columbia Accident Investigation Board. The *Columbia* vehicle breakup occurred at approximately 180,000 ft, with the crew module breakup occurring between 148,000 and 138,000 ft. The maximum survivable altitude for a breakup occurring during entry has not been determined. As little as 4 g could separate the crew module from the rest of the vehicle. After the Columbia accident, NASA associate administrator Bill Ready said "I do not anticipate that the next system will talk about 'crew escape'—but rather 'crew survival'." The Columbia Accident Investigation Board, in observation O-10.2-1, stated, "Future crewed vehicle requirements should incorporate the knowledge gained from the Challenger and Columbia accidents in assessing the feasibility of vehicles that could ensure crew survival even if the vehicle is destroyed" (CAIB 2003).

Several Space Shuttle escape studies have been performed through the years. Initial design considerations in 1971 determined that the only system that could provide protection for more than the two-person experimental flight crew was a separable crew compartment, which would add substantial weight and development cost. The National Space Transportation System crew egress and escape study was conducted after the Challenger accident in 1986 to review past studies and identify new and innovative concepts. Low cost options provided less coverage, whereas more costly concepts severely impacted the Space Transportation System performance capability. In 1989, a study was conducted to assess the impacts of retrofitting a crew escape module into the *Space Shuttle*. It was concluded that this would entail a design that would be equivalent to initiating a new Space Shuttle Program. The Space Shuttle Evolution-II crew escape study undertaken in 1991 assessed the impacts of incorporating ejection seats and extraction seat concepts into the existing *Space Shuttle* (Whitehurst

1987). The ejection seat concept was the lowest risk option. It provided for the escape of up to five crewmembers and considerably reduced Space Shuttle impacts.

NASA headquarters commissioned the access to space study summary report in 1994. According to this report, the high cost of incorporating additional escape capabilities combined with the considerable impact on vehicle capabilities did not warrant the addition of escape seats or an escape pod. The space transportation architecture study blue team review, conducted in 1999, assessed crew extraction and crew escape modules. It brought forward the concept of a passive crew survival system that relied on the modular nature of the crew cabin and its ability to survive an accident relatively intact. From 1999 through 2001, both NASA and contractor crew escape studies developed multiple extraction concepts utilizing ejection seats, modular separation, and hybrid concepts that focused on ascent. As part of these studies, cost, weight, and schedule estimates were developed for each. None of the candidate concepts met all of the technical guidelines of the Space Shuttle Program.

Following the *Columbia* accident in 2003, the Space Shuttle crew escape system report (NASA 2003) considered escape concepts for seven crewmembers from 0 to 150,000 ft during ascent and with a 90% probability of egress without major injury, while at the same time having no impact on the Space Shuttle outer mold line. No guidelines were provided for reentry. After review of the report, the NASA Astronaut Office issued a position on crew survival during ascent and entry. It stated that "a significant advance in crew survival can only be achieved by providing an effective escape or abort capability throughout ascent and entry (full envelope capability)." It also declared that "retrofitting this full envelope capability into the *Space Shuttle* does not appear feasible with current technology" and that "the Astronaut Office concurs with the current plan to return the *[Space] Shuttle* to flight without a new crew escape system" (Cabana 2003).

Ejection seats were used in the U.S. Gemini capsule, the first four Space Shuttle flights, and the Russian *Vostok* capsule. The six cosmonauts on the Vostok series all ejected prior to the capsule landing. Ejection seats have been considered in the Space Shuttle escape study (Siddiqi 2000). The injury potentials while using ejection seats are related to ejection seat g forces, fouling with seat or cockpit structure, windblast, flail, wind drag deceleration, and parachute opening shock. The incidence of major injuries and fatalities is projected to increase sharply at ejection speeds over 600 kn airspeed. These injuries would be primarily from flailing.

Parachute failures occurred on *Soyuz 1* in 1967 (fatal) and *Apollo 15* in 1971. The parachute riser that tangled on *Soyuz 5* might have been similar to that used on the fatal Soyuz-1 parachute failure. The failure of one of the three *Apollo 15* chutes resulted in a slightly harder landing (32 fps versus 28 fps), but the seat stroking mechanisms did not reach activation threshold and no injuries occurred.

A crew pod escape system has been considered for spacecraft (Miller 1990). Single person escape pods were used on the B-58 and the XB-70 and were designed for up to Mach 2 and 70,000 ft. When used, the crewmember remained in the capsule until touchdown. Shock absorbers eased the impact, and flotation was provided for water landing. Modular systems, where the crew remains in a vehicle substructure, have been used successfully in the F-111 fighter bomber. A modular system was considered for the *Space Shuttle*, but it required sophisticated pyrotechnic cutters, parachutes, and air bags.

3.4.4 On-Orbit Safe Haven and Crew Transfer

An on-orbit safe haven must provide a life sustaining environment for a crew until rescue or return is possible in the event the primary spacecraft is rendered uninhabitable. This was put to the test when the *Lunar Excursion Module Aquarius* became the lifeboat for the three Apollo 13 crew for 100 h, that is, the time from the initial problem until reentry. Normally, the lunar excursion module carries 45 h of consumables for two crewmembers. If the *Space Shuttle* encounters a major anomaly or is not capable of an on-orbit repair or successful reentry, it is to be docked to the *International Space Station*, and a contingency Space Shuttle crew support event is declared. Once the Space Shuttle consumables are depleted, the crew is to evacuate to the *International Space Station*. The *Space Shuttle* is then to be undocked and undergo an uninhabited, controlled reentry. The International Space Station safe haven uses available onboard capabilities to assure a safe, survivable environment for the Space Shuttle crew while awaiting rescue in emergency cases that otherwise would require their evacuation.

Another crew survival requirement is for a vehicle to vehicle crew transfer and rescue process that provides an ability to transfer crew from a safe haven or lifeboat to the rescue or return vehicle if a docked personnel transfer is not possible. Before the *Challenger* accident, Space Shuttle crews did not wear launch and reentry pressure suits and hence had no ability to move from one vehicle to another should a rescue mission by another *Space Shuttle* be possible. An 86-cm diameter rescue sphere, the personal rescue enclosure, was devised. A single person is to climb inside, assume a fetal position, and be zipped up by another crewmember. While inside, wearing a mask connected to a carbon dioxide scrubber and 1 h of oxygen, the crewmember is to be transferred to the rescue *Space Shuttle* by a suited astronaut. In the Columbia Accident Investigation Board report (2003), consideration was made for a vehicle to vehicle transfer using the extravehicular mobility unit.

During the Skylab program, a crew rescue capability was developed using an Apollo command module that would hold five crew couches in the event the Skylab crew could not return in their Apollo capsule. The rescue capsule would dock with the second port on the Skylab docking module. During *Skylab 3*, a reaction control system thruster on the Apollo capsule developed a leak. If a second thruster had failed, the spacecraft would not have been maneuverable for reentry. A Skylab rescue mission was in preparation when the thruster issue was resolved and the rescue mission became unnecessary.

3.4.5 Entry, Landing, and Post-Landing

The design of crew survival systems must consider the ability to return a crew safely to Earth in the event the primary spacecraft is rendered incapable of safe return. Following the Columbia accident, a launch on a crew rescue plan was initiated as part of the return to flight effort. Called STS-300, this crew rescue mission would return a stranded crew of 7 from the *International Space Station* safely within 90 d using the next available *Space Shuttle*.

3.4 Crew Survival

Reentry vehicle design factors include crew number and configuration, reentry vehicle type (blunt cone, biconic body, or winged vehicle), acceleration tolerance, crew health status, environmental systems, suit requirements, crew interface and controls, vehicle systems autonomy, landing accuracy, landing impact forces, and post-landing activities and requirements. The size of the entry corridor depends on deceleration, heating, and accuracy, with a specific drag profile flown to stay within the corridor.

A history of crewed capsule failures is depicted in Figure 3.12. Unstable reentry attitude due to incomplete separation of the crew module from the service module occurred on *Vostok 1*, *Vostok 2*, *Vostok 5*, *Voskhod 2*, and *Soyuz 5* (Siddiqi 2000). During the *Soyuz 5* reentry on January 18, 1969, the service module failed to separate. The spacecraft sought the most aerodynamically stable position in which its hatch was forward facing into the thermal and aerodynamic load. As the entry g forces were reversed, the cosmonaut hung in the straps. The service module finally separated, and the descent module turned around in the proper direction. At touchdown, the braking rockets failed, resulting in a hard landing in which the cosmonaut broke his front teeth. On October 16, 1976, *Soyuz 23* underwent early reentry when its rendezvous system failed. The capsule landed at night in a snowstorm, in subzero weather on a frozen lake. Recovery crews did not find the capsule until the next morning and were surprised to find the crew alive. Obviously,

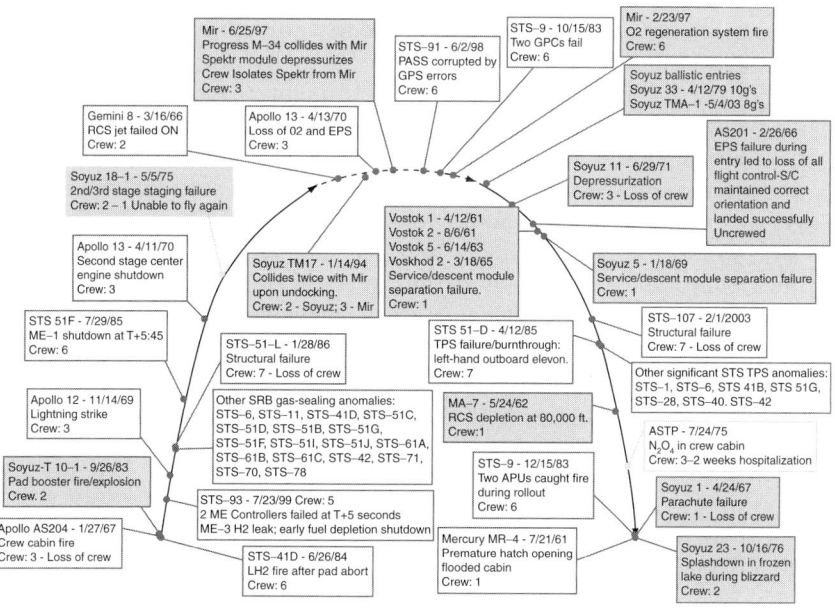

FIGURE 3.12 History of crewed capsule failures (Courtesy of NASA).

crew survival must also include consideration of the post-landing environment, and protective measures must remain in effect until recovery forces have arrived and are operational.

3.5 CONCLUSION

Space is inherently dangerous to the human being. Human bodies have evolved over millennia to work efficiently in the presence of gravity, minimal radiation, and in relatively benign ambient conditions. Humans are capable, however, of rapid adaptation when the environment changes, a fact borne out by the space adaptation syndrome seen in astronauts. The use of appropriate countermeasures, in particular those associated with exercise, minimizes these adaptations and counters some of the hazardous aspects of spaceflight, thus enabling the astronaut to retain a considerable degree of Earth based capability. The retention of such ability (physical, cognitive, and emotional) not only offers a distinct advantage in space when highly intensive physical and psychological stresses are imposed but also proves of great benefit when acceleration forces are reimposed on the body, such as during reentry or on a planetary surface. At these times, it is best to be strong, fit, and emotionally capable, to enable the arduous tasks to be accomplished and to react successfully to emergency circumstances.

Six decades have passed since the first foray of humanity into space, and considerable energy and sacrifice have been expended in understanding the nature of space and the place of humanity in it. The human capability to offset most of the negative effects and compensate for many others has improved to the point at which interplanetary travel and exploration are within reach. As humans progress toward this goal, the events of previous space mishaps, close calls, and crises should be reevaluated to improve crew survival in the future. Crew survival should incorporate advanced technologies where feasible, and the means should be simple, reliable, and attainable to address catastrophic failure modes. Worst case failures should be considered when establishing crew survival design requirements, and preparatory design for crew survival must be deemed more effective than postproduction modification of a spacecraft. The dangers are great, and there is some way to go before humans can be completely confident of their abilities. The rewards of success and the achievements derived and continue to be derived from human spaceflight are such that the efforts of the past and those of today and the future are justified.

ACKNOWLEDGMENT

The author wishes to acknowledge the assistance of Dr. Bret Palmer, Dr. Ulrich Straub, Dr. Patrik Sundblad, Dr. Oliver Ongerer, and Daniela Petrova in the preparation of this manuscript.

REFERENCES

Abernethy, P. J., R. Thaer, and A. W. Taylor. (1990) Acute and chronic responses of skeletal muscle to endurance and sprint exercise. *Sports Medicine* 10, no. 6: 365-389.

Alfrey, C. P., M. M. Udden, C. L. Huntoon, and T. Driscoll. (1996) Destruction of newly released red blood cells in space flight. *Medical Science and Sports Exercise* 28, no. 10: S42-S44.

Arbeille, P., F. Achaïbou, G. Fomina, J. M. Pottier, and M. Porcher. (1996) Regional blood flow in microgravity: Adaptation and deconditioning. *Medical Science and Sports Exercise* 28, no. 10: S70-S79.

Aviles, H., T. Belay, M. Vance, and G. Sonnenfeld. (2005) Effects of space flight conditions on the function of the immune system and catecholamine production simulated in a rodent model of hind limb unloading. *Neuroimmunomodulation* 12, no. 3: 173-181.

Bacal, K., R. Billica, and S. Bishop. (2003) Neurovestibular symptoms following space flight. *Journal of Vestibular Research* 13, no. 2: 93-102.

Backup, P., K. Westerlind, S. Harris, T. Spelsberg, B. Kline, and R. Turner. (1994) Spaceflight results in reduced mRNA levels for tissue specific proteins in the musculoskeletal system. *American Journal of Physiology* 266, no. 4, pt. 1: E567-E5673.

Badhwar, G. D., M. J. Golightly, A. Konradi, W. Atwell, J. W. Kern, B. Cash, E. V. Benton, A. L. Frank, D. Sanner, R. P. Keegan, L. A. Frigo, V. M. Petrov, I. V. Tchernykh, Y. A. Akatov, V. A. Shurshakov, V. V. Arkhangelsky, V. V. Kushin, N. A. Klyachin, N. Vana, and K. M. Baldwin. (1996) Effect of spaceflight on the functional, biochemical and metabolic properties of skeletal muscle. *Medical Science and Sports Exercise* 28, no. 8: 983-987.

Baldwin, K. M., and F. Haddad. (2002) Skeletal muscle plasticity: Cellular and molecular responses to altered physical activity paradigms. *American Journal of Physical Medicine and Rehabilitation* 81, no. 11: S40-S51.

Ball, J. R., and C. H. Evans. (2001) *Safe Passage: Astronaut Care for Exploration Class Missions*. Washington, DC: Institute of Medicine, National Academy Press.

Berry, C. A. (1969) Preliminary clinical report of the medical aspects of Apollos VII and VIII. *Aerospace Medicine* 40, no. 3: 245-254.

Berry, C. A. (1970) Summary of medical experience in the Apollo 7 through 11 manned spaceflights. *Aerospace Medicine* 41, no. 5: 500-519.

Black, F. O., W. H. Paloski, M. F. Reschke, M. Igarashi, F. Guedry, and D. J. Anderson. (1999) Disruption of postural readaptation by inertial stimuli following space flight. *Journal of Vestibular Research* 9, no. 5: 369-378.

Blakely, E. A., and P. Y. Chang. (2004) Late effects from hadron therapy. *Radiotherapy and Oncology* 73: S134-S138.

Boppart, M. D., M. F. Hirshman, K. Sakamoto, R. A. Fielding, and L. J. Goodyear. (2001) Static stretch increases c-Jun NH2-terminal kinase activity and p38 phosphorylation in rat skeletal muscle. *American Journal of Physiology: Cell Physiology* 280: C352-C358.

Braude, A., C. Davis, and J. Fierer (eds.). (1986) *Infectious Diseases and Medical Microbiology*. Philadelphia: W. B. Saunders.

Buckey, J. C. (2006) *Space Physiology*. Oxford, England: Oxford University Press.

Buckey, J. C., L. Lane, B. Levine, D. Watenpaugh, S. Wright, W. Moore, F. Andrew-Gaffney, and G. C. Blomqvist. (1996) Orthostatic intolerance after space flight. *Journal of Applied Physiology* 81: 7-18.

Cabana, R. D. (2003) *Astronaut Office Position on Crew Survivability during Ascent and Entry*. Letter CA-03-31. Houston, TX: National Aeronautics and Space Administration, Johnson Space Center.

Columbia Accident Investigation Board. (2003) *The CAIB Report*, vol. 1. Washington, DC: National Aeronautics and Space Administration, Headquarters.

Caillot-Augusseau, A., M. H. Lafage-Proust, C. Soler, J. Pernod, F. Dubois, and C. Alexandre. (1998) Bone formation and resorption biological markers in cosmonauts during and after a 180-day space flight (Euromir 95). *Clinical Chemistry* 44: 578-585.

Clément, G. (2005) The maintenance of physiological function in humans during spaceflight. *International Journal of Sports Medicine*, 6, no. 4: 185-198.

Conkin, J., D. Waligora, D. Horrigan, and A. Hadley. (1987) Effects of exercise on venous gas emboli and decompression sickness at 4.3 psi. Personal communication AO2/MF AO1. Houston, TX: National Aeronautics and Space Administration, Johnson Space Center.

Convertino, V. A. (1991) Blood volume; it's adaptation to endurance training. *Medical Science and Sports Exercise* 23, no. 12: 1338-1348.

Convertino, V. A. (1996a) Exercise and adaptation to microgravity environments. In M. J. Fregly and C. M. Blatteis (eds.), *Handbook of Physiology*, Section 4, Environmental Physiology. New York: Oxford University Press, pp. 815-844.

Convertino, V. A. (1996b) Clinical aspects of the control of plasma volume at microgravity and during return to one gravity. *Medical Science and Sports Exercise* 28, no. 10: 45-52.

Convertino, V. A. (2001) Lower body negative pressure as a tool for research in aerospace physiology and military medicine. *Journal of Gravitational Physiology* 8, no. 2: 1-14.

Cowell, S. A., J. M. Stocks, D. F. G. Evans, S. R. Simonson, and J. E. Greenleaf. (2002) The exercise and environmental physiology of extravehicular activity. *Aviation, Space and Environmental Medicine* 73, no. 1: 54-67.

Crampton, G. H. (2000) *Motion and Space Sickness*. Boca Raton, FL: CRC Press.

Crandall, C. G., J. M. Johnson, V. A. Convertino, P. B. Raven, and K. A. Engelke. (1994) Altered thermoregulatory responses after 15 days of head down tilt. *Journal of Applied Physiology* 77: 1863-1867.

Cucinotta, F. A., and M. Durante. (2006) Cancer risk from exposure to galactic cosmic rays: Implications for space exploration by human beings. *Lancet Oncology* 7: 431-435.

Cucinotta, F. A., H. Nikjoo, and D. T. Goodhead. (1998) The effects of delta rays on the number of particle track transversals per cell in laboratory and space exposures. *Radiation Research* 150: 115-119.

Davis, J. R., J. M. Vanderploeg, P. A. Santy, R. T. Jennings, and D. F. Stewart. (1988) Space motion sickness during 24 flights of the Space Shuttle. *Aviation, Space and Environmental Medicine* 59: 1185-1189.

Day, J. R., H. B. Rossiter, E. M. Coats, A. Skasick, and B. J. Whipp. (2003) The maximally attainable VO_2 during exercise in humans: The peak vs. maximum issue. *Journal of Applied Physiology* 95: 1901-1907.

Durante, M. (2004) Heavy ion radiobiology for hadron therapy and space radiation protection. *Radiotherapy and Oncology* 73: S158-S160.

Edgerton, V. R., M. Y. Zhou, Y. Ohira, H. Klitgaard, B. Jiang, G. Bell, B. Harris, B. Saltin, P. D. Gollnick, R. R. Roy, M. K. Day, and M. Greenisen. (1995) Human fiber size and enzymatic properties after 5 and 11 days of spaceflight. *Journal of Applied Physiology* 78: 1733-1739.

Edmonds, J. L., T. Jarchow, and L. Young. (2007) A stair stepper for exercising on a short radius centrifuge. *Aviation, Space and Environmental Medicine* 78, no. 2: 129-134.

Fang, Y., S. Yang, and G. Wu. (2002) Free radicals, antioxidants, and nutrition. *Nutrition* 18, no. 10: 872-879.

Fitts, R. H., D. R. Riley, and J. J. Widrick. (2000) Physiology of a microgravity environment: microgravity and skeletal muscle. *Journal of Applied Physiology* 89: 823-839.

Foale, M., and W. Shepard. (1989) *Crew Bailout Cue Card for LOC/Breakup*. Study Report, Can. Houston, TX: National Aeronautics and Space Administration, Johnson Space Center.

Fong, K. (2004) The next small step. *British Medical Journal* 329: 1441-1444.

Fortney, S. M. (1991) Exercise thermoregulation: Possible effects of spaceflight. *Proceedings of the 21st International Conference on Environmental Systems*. Special SAE Report 911460. San Francisco: International Society of Automotive Engineers.

Friden, J., J. Seger, M. Sjostrom, and B. Ekblom. (1983) Adaptive response in human skeletal muscle subjected to prolonged eccentric training. *International Journal of Sports Medicine* 4, no. 3: 177-183.

Fritsch, J. M., J. B. Charles, B. S. Bennett, M. M. Jones, and D. L. Eckberg. (1992) Short duration spaceflight impairs human carotid baroreceptor cardiac reflex responses. *Journal of Applied Physiology* 73: 664-671.

Fritsch-Yelle, J. M., J. B. Charles, M. M. Jones, L. A. Beightol, and D. L. Eckberg. (1994) Spaceflight alters autonomic regulation of arterial pressure in humans. *Journal of Applied Physiology* 77: 1776-1783.

Fritsch-Yelle, J. M., P. A. Whitson, R. L. Bondar, and T. E. Brown. (1996) Subnormal norepinephrine release relates to presyncope in astronauts after spaceflight. *Journal of Applied Physiology* 81, no. 5: 2134-2141.

George, K., V. Willingham, H. Wu, D. Gridley, G. Nelson, and F. A. Cucinotta. (2002) Chromosome aberrations in human lymphocytes induced by 250 MeV protons: effects of dose, dose rate and shielding. *Advances in Space Research* 30, no. 4: 891-899.

Gorgiladze, G. I., and I. I. Brianov. (1989) Space motion sickness. *Kosmicheskaia Biologiia Aviakosmicheskaia Meditsina* 23, no. 3: 4-14.

Goubel, F. (1997) Changes in mechanical properties of human muscle as a result of spaceflight. *International Journal of Sports Medicine* 18: S285-S287.

Graaf, B. de, and A. J. de Roo. (1996) Effects of long duration centrifugation on head movements and a psychomotor task. *Journal of Vestibular Research* 6, no. 1: 23-29.

Greenleaf, J. E. (2004) *Deconditioning and Reconditioning*. New York: CRC Press.

Greenleaf, J. E., S. R. Simonson, J. M. Stocks, J. Evans, C. F. Knapp, S. A. Cowell, K. N. Pemberton, H. W. Wilson, J. M. Vener, S. N. Evetts, P. A. Hardy, R. E. Grindeland, H. Hinghofer-Szalkay, S. M. Smith, M. G. Ziegler, D. R. Brown, D. G. Evans, F. B. Moore, and D. T. Quach. (2001) *Effect of Exercise Training and +Gz Acceleration Training on Men*. TM-2001-210926. Moffett Field, CA: National Aeronautics and Space Administration, Ames Research Center.

Hall, R., and D. Shayler. (2003) *Soyuz: A Universal Spacecraft*. Berlin: Springer Praxis.

Heer, M., M. Elia, and P. Ritz. (2001) Energy and fluid metabolism in microgravity. *Current Opinion in Nutrition and Metabolic Car*, 4, no. 4: 307-311.

Heimbach, R. D., and P. J. Sheffield. (1985) Decompression sickness and pulmonary overpressure accidents. In R. L. DeHart (ed.), *Fundamentals of Aerospace Medicine*. Philadelphia: Lea and Febiger, pp. 132-161.

Hettinger, L. J., R. S. Kennedy, and M. E. McCauley. (1990) Motion and human performance. In G. H. Crompron (ed.), *Motion and Space Sickness*. Boca Raton, FL: CRC Press, pp. 411-442.

Hoffler, G. W., R. A. Wolthuis, and R. L. Johnson. (1974). Apollo space crew cardiovascular evaluations. *Aerospace Medicine*, 45, no. 8: 807-823.

Hoffman, S. J., and D. I. Kaplan. (1997) *Human Exploration of Mars: The Reference Mission of the NASA Mars Exploration Study Team*. Special Publication SP-6107. Houston, TX: National Aeronautics and Space Administration, Johnson Space Center.

Horrigan, D. J., J. M. Waligora, and J. H. Bredt. (1989) Extravehicular activities. In A. E. Nicogossian, C. Leach-Huntoon, and S. L. Pool (eds.), *Space Physiology and Medicine*. Philadelphia: Lea and Febiger, pp. 121-135.

Jennings, J. T. (1998) Managing space motion sickness. *Journal of Vestibular Research* 8, no. 1: 67-70.

Jiang, B., Y. Ohira, R. R. Roy, Q. Nguyen, E. I. Ilyina-Kakueva, V. Oganov, J. F. Marini, and V. R. Edgerton. (1992) Adaptations of fibers in fast twitch muscles of rats to spaceflight and hind limb suspension. *Journal of Applied Physiology* 73: 58S-65S.

Johnson, R. S., and L. F. Dietlein. (1977) *Biomedical Results from Skylab*. Special Publication SP-377. Houston, TX: National Aeronautics and Space Administration, Johnson Space Center.

Kaufmann, H., D. Saadia, and A. Voustianiouk. (2002) Midodrine in neurally mediated syncope: A double blind, randomized, crossover study. *Annals of Neurology* 52, no. 3: 342-345.

Kerwin, J. P. (1986) Concerning the cause of death of the Challenger crew. Letter to Rear ADM Richard Truly, 28 July. Washington, DC: National Aeronautics and Space Administration, Headquarters.

Lane, H. W., and D. A. Schoeller. (2000) *Nutrition in Spaceflight and Weightlessness Models*. Washington, DC: CRC Press.

Lang, T. F., A. LeBlanc, H. Evans, Y. Lu, H. Gennant, and A. Yu. (2004) Cortical and trabecular bone mineral loss from the spine and hip in long duration spaceflight. *Journal of Bone and Mineral Research* 19, no. 6: 1006-1012.

Lang, T. F., A. D. LeBlanc, H. J. Evans, and Y. Lu. (2006) Adaptation of the proximal femur to skeletal reloading after long duration spaceflight. *Journal of Bone and Mineral Research* 21, no. 8: 1224-1230.

Langberg, H., D. Skovgaard, S. Asp, and M. Kjaer. (2000) Time pattern of exercise induced changes in type I collagen turnover after prolonged endurance exercise in humans. *Calcified Tissue International* 67: 41-44.

Layne, C. S., K. E. Forth, and A. F. Abercromby. (2005). Spatial factors and muscle spindle input influence the generation of neuromuscular responses to stimulation of the human foot. *Acta Astronautica* 56: 809-819.

Layne, J. E., and M. E. Nelson. (1999) The effects of progressive resistance training on bone density: A review. *Medical Science and Sports Exercise* 31, no. 1: 25-30.

LeBlanc, A. D., V. Schneider, L. Shackelford, S. West, V. Oganov, A. Bakulin, and L. Veronin. (1996) Bone mineral and lean tissue loss after long duration space flight. *Journal of Bone and Mineral Research* 11: S323.

LeBlanc, A. D., L. Shackelford, and V. Schneider. (1998) Future human bone research in space. *Bone* 22, no. 5: 113S-116S.

LeBlanc, A. D., T. B. Driscoll, L. C. Shackelford, H. J. Evans, N. J. Rianon, S. M. Smith, D. L. Feeback, and D. Lai. (2002) Alendronate as an effective countermeasure to disuse induced bone loss. *Journal of Musculoskeletal Neuronal Interaction* 2, no. 4: 335-343.

Levine, B. D., L. D. Lane, D. E. Watenpaugh, F. A. Gaffney, J. C. Buckey, and C. G. Blomqvist. (1996) Maximal exercise performance after adaptation to microgravity. *Journal of Applied Physiology* 81: 686-689.

Lutwak, L., G. D. Whedon, P. A. LaChance, J. M. Reid, and H. Lipscomb. (1969) Mineral, electrolyte and nitrogen balance studies of the Gemini VII 14-day orbital space flight. *Journal of Clinical Endocrinology and Metabolism* 29: 1140.

Manko, V., V. Kordyum, L. Vorobyev, N. Konshin, and G. Nechitaylo. (1986) Changes over time in *Proteus vulgaris* cultures grown in the ROST-4M2 device on the Salyut-7 space station. In K. Sytnik (ed.), *Space Biology and Biotechnology: A Collection of Papers*. Kiev, Russia: Naukov Dumka, pp. 3-10.

McCarthy, I. D. (2005) Fluid shifts due to microgravity and their effects on bone. *Annals Biomedical Engineering* 33, no. 1: 95-103.

Miles, M. P., P. M. Clarkson, M. Bean, K. Ambach, J. Mulroy, and K. Vincent. (1994) Muscle function at the wrist following 9 d of immobilization and suspension. *Medical Science and Sports Exercise* 26: 615-623.

Miller, B. A. (1990) Spacecraft crew escape. *SAFE Journal* 20, no. 2: 10-14.

Moatti, N., L. Lapchine, G. Gasset, G. Richoilley, J. Templier, and R. Tixador. (1986) Preliminary results of the "antibio" experiment. *Naturwissenschaften* 73: 413-414.

Narici, M. V., G. S. Roi, A. E. Minetti, and P. Cerretelli. (1989) Changes in force, cross sectional area and neural activation during strength training and detraining of the human quadriceps. *European Journal of Applied Physiology* 59: 310-319.

NASA. (1986) *STS Crew Egress and Escape Study*. JSC 22275. Houston, TX: National Aeronautics and Space Administration, Johnson Space Center.

NASA. (2003) *Space Shuttle Crew Escape System Report*. Space Transportation System Document, NSTS 60518. Houston, TX: National Aeronautics and Space Administration, Johnson Space Center.

NASA History Office. (1986) *Report of the Presidential Commission on the Space Shuttle Challenger Accident*. Washington, DC: National Aeronautics and Space Administration, Headquarters.

Neufer, P. D. (1989) The effect of detraining and reduced training on the physiological adaptations to aerobic exercise training. *Sports Medicine* 8, no. 5: 302-321.

Newkirk, D. (1990) *Almanac of Soviet Manned Space Flight*. Houston, TX: Gulf Publishing Company.

Newman, D., and M. Barratt. (1997) Life support and performance issues for extravehicular activity. In S. E. Churchill (ed.), *Fundamentals of Space Life Sciences*. Malibu, FL: Krieger, pp. 337-364.

Nicholas, J. M., and H. C. Foushee. (1990) Organization, selection and training of crews for extended spaceflight: findings from analogs and implications. *Journal of Space Rockets* 27: 451-456.

Nicogossian, A. E., C. Huntoon, and S. L. Pool. (1994) *Space Physiology and Medicine*. Philadelphia: Lea and Febiger.

Oganov, V. S., A. I. Grigorev, L. I. Voronin, A. S. Rakhmanov, A. V. Bakulin, V. S. Schneider, and A. D. LeBlanc. (1992) Bone mineral density in cosmonauts after flights lasting 4.5-6 months on the Mir orbital station. *Aviakosm Ekolog Meditsina* 26, nos. 5-6: 20-24 [In Russian].

Ohira, Y. (2000) Neuromuscular adaptation to microgravity environment. *Japanese Journal of Physiology* 50, no. 3: 303-314.

Ohira, Y., B. Jiang, R. R. Roy, V. Oganov, E. I. Ilyina-Kakueva, J. F. Marini, and V. R. Edgerton. (1992) Rat soleus muscle fiber responses to 14 days of spaceflight and hind limb suspension. *Journal of Applied Physiology* 73: 51S-57S.

Ondler, R. M. (1987) *Phase II Escape—Trajectory Studies for Out of Control Orbiter Configurations*. Memo ED3/8707-108. Houston, TX: National Aeronautics and Space Administration, Johnson Space Center.

OTA. (1989) *Round Trip to Orbit: Human Spaceflight Alternatives*. Special report. OTA-ISC-419. Washington, DC: Office of Technology Assessment, United States Government Printing Office.

Parker, D. E. (1977) Labyrinth and cerebral spinal fluid pressure changes in guinea pigs and monkeys during simulated zero G. *Aviation, Space and Environmental Medicine* 48: 356-361.

Payne, D. A., S. K. Mehta, S. K. Tyring, R. P. Stowe, and D. L. Pierson. (1999) Incidence of Epstein-Barr virus in astronaut saliva during spaceflight. *Aviation, Space and Environmental Medicine* 70, no. 12: 1211-1213.

Pecaut, M. J., D. S. Gridley, and G. A. Nelson. (2003) Long term effects of low dose proton radiation on immunity in mice: Shielded vs. unshielded. *Aviation, Space and Environmental Medicine* 74: 115-124.

Perhonen, M. A., F. Franco, L. D. Lane, J. C. Buckey, C. G. Blomqvist, J. E. Zerwekh, R. M. Peshock, P. T. Weatherall, and B. D. Levine. (2001) Cardiac atrophy after bed rest and spaceflight. *Journal of Applied Physiology* 91: 645-653.

Philpot, D. E., K. Kato, and J. Miquel. (1992) Ultrastructure and cellular mechanisms in myocardial deconditioning in weightlessness. *Advances in Space Biology and Medicine* 2: 83-112.

Philpot, D. E., I. A. Popova, T. K. Kato, J. Stevenson, J. Miquel, and W. Sapp. (1990) Morphological and biochemical examination of Cosmos 1887 rat heart tissue: Part I—Ultrastructure. *Journal of the Federation of American Societies for Experimental Biology* 4: 73-78.

Pilmanis, A. A., R. M. Olson, M. D. Fischer, J. F. Wiegman, and J. T. Webb. (2004) Exercise induced altitude decompression sickness. *Aviation, Space and Environmental Medicine* 70: 22-29.

Ponomarev, A. L., H. Nikjoo, and F. A. Cucinotta. (2005) NASA radiation track image GUI for assessing space radiation biological effects. Space Conference—2005. Long Beach, CA: American Institute of Aeronautics and Astronautics.

Prisk, G. K., A. R. Elliott, M. Paiva, and J. West. (2003) Sleep and respiration in microgravity. In J. Buckey and J. Homick (eds.), *The Neurolab Spacelab Mission: Neuroscience Research in Space*. Special Publication SP-2003-535. Houston, TX: National Aeronautics and Space Administration, Johnson Space Center, pp. 223-232.

Rambaut, P. C., and R. S. Johnson. (1979) Prolonged weightlessness and calcium loss in man. *Acta Astronautica* 6: 113-123.

Rambaut, P. C., C. S. Leach, and J. I. Leonard. (1977) Observations in energy balance in man during spaceflight. *American Journal of Physiology* 277:R1-R10.

Reason, J. T., and J. J. Brand. (1975) *Motion Sickness*. London: Academic Press.

Reavley, C. M., J. F. Golding, L. F. Cherkas, T. D. Spector, and A. J. MacGregor. (2006) Genetic influences on motion sickness susceptibility in adult women: A classical twin study. *Aviation, Space and Environmental Medicine* 77, no. 11: 1148-1152.

Rodan, A. G. (1996) Coupling of bone resorption and formation during bone remodeling. In R. Marcus, D. Feldman, and J. Kelsey (eds.), *Osteoporosis*. Colorado Springs, CO: International Academic Publishers.

Saliou, C., M. Kitazawa, L. McLaughlin, J. P. Yang, J. K. Lodge, T. Tetsuka, K. Iwasaki, J. Cillard, T. Okamoto, and L. Packer. (1999) Antioxidants modulate acute solar ultraviolet radiation induced NF-kappa-B activation in a human keratinocyte cell line. *Free Radical Biology and Medicine* 26, nos. 1-2: 174-183.

Schneider, S., D. E. Watenpaugh, S. M. Lee, A. C. Ertl, W. J. Williams, R. E. Ballard, and A. R. Hargens. (2002) Lower body negative pressure exercise and bed rest mediated orthostatic intolerance. *Medical Science and Sports Exercise* 34, no. 9: 1446-1453.

Schneider, S. M., W. E. Amonette, K. Blazine, J. Bentley, S. M. Lee, J. A. Loehr, A. D. Moore Jr., M. Rapley, E. R. Mulder, and S. M. Smith. (2003) Training with the International Space Station interim resistive exercise device. *Medical Science and Sports Exercise* 35, no. 11: 1935-1945.

Schoner, W. (1996) In-flight radiation measurements on STS-60. *Radiation Measurements* 26, no. 1: 17-34.

Setlow, R., J. F. Dicello, R. J. M. Fry, J. B. Little, R. J. Preston, J. B. Smathers, and R. L. Ullrich. (1996) *Radiation Hazards to Crews of Interplanetary Missions: Biological Issues and Research Strategies*. Washington, DC: Space Studies Board, National Research Council and National Academy Press, pp. 13-34.

SFOC. (2000) *Crew Escape Systems 2002*. SFOC FL 0236, Rev. A. Houston, TX: National Aeronautics and Space Administration, Johnson Space Center.

Shavers, M. R., N. Zapp, R. E. Barber, J. W. Wilson, G. Qualls, L. Toupes, S. Ramsey, V. Vinci, G. Smith, and F. A. Cucinotta. (2004) Implementation of ALARA radiation protection on the International Space Station through polyethylene shielding augmentation of the service module crew quarters. *Advances in Space Research*, 34, no. 6: 1333-1337.

Shayler, D. (2000) *Disasters and Accidents in Manned Spaceflight*. Berlin: Springer-Praxis.

Shimamiya, T., N. Terada, S. Wakabayashi, and M. Mohri. (2005) Mood change and immune status of human subjects in a 10-day confinement study. *Aviation, Space and Environmental Medicine* 76, no. 5: 481-485.

Shupak, A., and C. R. Gordon. (2006) Motion sickness: Advances in pathogenesis, prediction, prevention and treatment. *Aviation, Space and Environmental Medicine* 77, no. 12: 1213-23.

Siddiqi, A. (2000) *Challenge to Apollo: The Soviet Union and the Space Race, 1945-1974*. Special Publication SP-2000-4408. Houston, TX: National Aeronautics and Space Administration, Johnson Space Center.

Simon, M. W. (1997) Manifestations of relapsing Epstein-Barr virus illness. *Journal of the Kentucky Medical Association* 95: 240-243.

Smith, M. C., P. C. Rambaut, J. M. Vogel, and M. W. Whittle. (1977) Bone mineral measurement (experiment M078). In R. S. Johnston and L. F. Dietlein (eds.), *Biomedical Results of Skylab. SP-377*. Washington, DC: National Aeronautics and Space Administration, Headquarters, pp. 183-190.

Smith, S. M., J. E. Davis-Street, B. L. Rice, J. L. Nillen, P. L. Gillman, and G. Block. (1999) Nutritional status assessment in semiclosed environments. *Journal of the Federation of American Societies for Experimental Biology* 13: A265-A266.

Smith, S. M., M. E. Wastney, B. Morukov, I. M. Larina, L. E. Nyquist, S. A. Abrams, E. N. Taran, C. Shih, J. L. Nillen, J. E. Davis-Street, B. L. Rice, and H. W. Lane. (1999) Calcium metabolism before, during, and after a 3 month spaceflight: Kinetic and biochemical changes. *American Journal of Physiology: Regulatory, Integrative and Comparative* 277, no. 1: R1-R10.

Smith, S. M., J. E. Davis-Street, B. L. Rice, J. L. Nillen, P. L. Gillman, and G. Block. (2001) Nutritional status assessment in semiclosed environments: Ground based and space flight studies in humans. *Journal of Nutrition* 131: 2053-2061.

Smith, S. M., J. E. Davis-Street, J. Vernell-Fesperman, D. Calkins, M. Bawa, B. R. Macias, R. S. Meyer, and A. R. Hargens. (2003) Evaluation of treadmill exercise in a lower body negative pressure chamber as a countermeasure for weightlessness induced bone loss: A bed rest study with identical twins. *Journal of Bone and Mineral Research* 18, no. 12: 2223-2230.

Sonnenfeld, G. (2006) Exploration class missions and return: Effects on the immune system. *Gravity, Space and Biology Bulletin* 19, no. 2: 45-48.

Spillantini, P., M. Casolino, M. Durante, R. Mueller-Mellin, G. Reitz, L. Rossi, V. Shurshakov, and M. Sorbi. (2007) Shielding from cosmic radiation for interplanetary missions: active and passive methods. *Radiation Measurements* 42: 14-23.

Stein, T. P. (2000) The relationship between dietary intake, exercise, energy balance and the space craft environment. *European Journal of Physiology* 441, nos. 2-3: S21-S31.

Stein, T. P., M. J. Leskiw, and M. D. Schluter. (1996) Diet and nitrogen metabolism during spaceflight on the shuttle. *Journal of Applied Physiology* 81, no. 1: 82-97.

Stowe, R. P., C. F. Sams, and D. L. Pierson. (2003) Effects of mission duration on neuroimmune responses in astronauts. *Aviation, Space and Environmental Medicine* 74: 1281-1284.

Svahn, A., J. Berggren, A. Parke, J. Storsaeter, R. Thorstensson, and A. Linde. (2006) Changes in seroprevalence to four herpes viruses over 30 years in Swedish children aged 9-12 years. *Journal of Clinical Virology* 36, no. 2: 118-123.

Takahashi, A., K. Ohnishi, S. Takahashi, M. Masukawa, K. Sekikawa, T. Amano, T. Nakano, S. Nagaoka, and T. Ohnishi. (2001) Differentiation of dictostelium discoideum vegetative cells into spores during earth orbit in space. *Advances in Space Research* 28: 549-553.

Tesch, P. A., and H. E. Berg. (1998) Effects of spaceflight on muscle. *Journal of Gravitational Physiology* 5, no. 1: 19-22.

Testard, I., M. Ricoul, F. Hoffschir, A. Flury-Herard, B. Dutrillaux, B. Fedorenko, V. Gerasimenko, and L. Sabatier. (1996) Radiation induced chromosome damage in astronauts' lymphocytes. *International Journal of Radiation Biology* 70, no. 4: 403-411.

Tilton, F. E., J. J. C. Degioanni, and F. S. Schneider. (1980) Long term follow up of Skylab bone demineralization. *Aviation, Space and Environmental Medicine* 51: 1209-1213.

Treisman, M. (1977) Motion sickness, an evolutionary hypothesis. *Science* 197: 493-495.

Van Lieshout, J. J. (2003) Exercise training and orthostatic intolerance: A paradox? *Journal of Physiology* 551, Pt. 2: 401.

Vernikos, J., D. A. Ludwig, A. C. Ertl, C. E. Wade, L. Keil, and D. O'Hara. (1996) Effect of standing and walking on physiological changes induced by head down bed rest: Implications for spaceflight. *Aviation, Space and Environmental Medicine* 67, no. 11: 1069-1079.

Vico, L., P. Collet, A. Guignandon, M. A. Lafage-Proust, T. Thomas, M. Rehailia, and C. Alexandre. (2000) Effects of long term microgravity exposure on cancellous and cortical weight bearing bones of cosmonauts. *The Lancet* 355, no. 9215: 1607-1611.

Vorobyov, E. I., O. G. Gazenko, A. M. Genin, and A. D. Egorov. (1983) Medical results of Salyut-6 manned space flights. *Aviation, Space and Environmental Medicine* 54, no. 12: S31-S40.

Waligora, D., D. Horrigan, and J. Conkin. (1987) The effect of extended O_2 prebreathing on altitude decompression sickness and venous gas bubbles. *Aviation, Space and Environmental Medicine* 59, no. 9: A110-A112.

Waligora, J., D. Horrigan, and A. Nicogossian. (1991) The physiology of spacecraft and spacesuit atmosphere selection. *Acta Astronautica* 23: 171-177.

Whitehurst, T. N. (1987) Space Shuttle orbiter ejection seat survey. *SAFE Journal* 17, no. 1: 34-39.

Widrick, J. J., S. T. Knuth, K. M. Norenberg, J. G. Romatowski, J. L. Bain, D. A. Riley, M. Karhanek, S. W. Trappe, D. L. Costill, and R. H. Fitts. (1999) Effect of 17-day spaceflight on contractile properties of human soleus muscle fibers. *Journal of Physiology* (London) 516: 915-930.

Wolf, R. R., F. Jahoor, and W. H. Hartl. (1989) Protein and amino acid metabolism after injury. *Diabetes and Metabolism Review* 5: 149-164.

Yegorov, A. D., I. I. Kasyan, A. A. Zlatorunskiya, S. F. Khlopina, V. A. Talavrinov, I. A. Yevdokimova, Y. M. Romanov, and V. I. Somov. (1981) Changes in body mass of cosmonauts in the course of a 140-day space flight. *Space Biology and Aerospace Medicine* 15, no. 1: 49-51.

Yi-Wen, C., J. Monica, J. Hubal, E. P. Hoffman, P. D. Thompson, and P. M. Clarkson. (2003) Molecular responses of human muscle to eccentric exercise. *Journal of Applied Physiology* 95: 2485-2494.

Young, A. (2005) Boost testing the crew exploration vehicle. *Space Times* (September-October): 21-22.

Zeitlin, C., S. Guetersloh, L. Heilbronn, and J. Miller, J. (2005) Shielding and fragmentation studies. *Radiation and Proton Dosimetry* 116, nos. 1-4, Pt. 2: 123-124.

CHAPTER 4

Basic Principles of Space Safety

Tommaso Sgobba
Head, Independent Flight Safety Office, European Space Research and Technology Center,
European Space Agency, Noordwijk, the Netherlands

Axel M. (Skip) Larsen, Jr.
Chairman (Retired), NASA Payload Safety Review Panel, Johnson Space Center,
National Aeronautics and Space Administration, Houston, Texas

Gary E. Musgrave, Ph.D.
Chairman (Retired), NASA Payload Safety Review Panel, Johnson Space Center,
National Aeronautics and Space Administration, Houston, Texas

CONTENTS

4.1. The Cause of Accidents	163
4.2. Principles and Methods	165
4.3. The Safety Review Process	170

4.1 THE CAUSE OF ACCIDENTS

Notoriously, space missions are risky. The risks of spaceflight, in a number of cases, are related to the extreme nature of the space environment and existing technological limitations. However, it is a fact that all accidents that occurred during space missions to date have happened because of design and manufacturing decisions that were made and were within the technological knowledge and capabilities that existed at the time.

Accidents happen because of failures, malfunctions, operational errors, or combinations of these. In addition, lack of or inadequate crew survival provisioning, such as a crew escape system, can contribute greatly to the fatal outcome of an incident (Hammar 1993). Although the realization of adequate systems reliability is paramount for safety, reliability alone is not sufficient to prevent an accident.

The main causes for failures and malfunctions are design and manufacturing errors. Design errors are generally of three types:

- Wrongly assumed or underestimated environmental conditions, such as limit loads or worst cases.

- Deficient control of intrinsic hazardous characteristics, such as flammable materials or stored energy.
- Incorrect or inaccurate detailed design.

Historically, a number of catastrophic incidents occurred within the space programs of the United States and Russia. Although not all these incidents resulted in death, each placed the crew in a position of immanent danger. In every case, strict adherence to the principles and practices of safety in the design of the spacecraft involved in the incidents would have minimized the danger to crew or even prevented the incident altogether.

In 1967, the Apollo 1 incident that occurred during ground testing was caused by the use of a 100% oxygen atmosphere in the presence of combustible materials, vulnerable power wiring, plumbing carrying a combustible and corrosive coolant, and finally, but not unimportant, inadequate provisioning for crew escape. To open the hatch of the Apollo 1 capsule, the interior cabin pressure, which was by design higher than the ambient atmospheric pressure, first had to be equalized across the hatch by venting, and a number of latches then operated. Under ideal conditions, a minimum of 90 s was required for opening the Apollo 1 hatch. When fire erupted, it took approximately 5 min for test personnel to open the hatch. This amount of time was much too long for the crew to have had any chance for survival.

During 1971, loss of the Soyuz 11 crew resulted from spacecraft decompression during reentry. This incident was caused by a malfunction of a pressure equalization valve. Although the valve was designed to be closed manually during an emergency (as one of the cosmonauts had attempted), it was later determined that it could never have been closed fast enough to prevent the crew from becoming incapacitated by hypoxia. In this case, a design lacking in an adequate consideration of biomedical factors defeated the purpose of an emergency operation.

The Space Shuttle Challenger disaster in 1986 was caused by the failure of a joint in one of the solid rocket motors. The failure was determined to be due to several concurrent factors. Among these major factors contributing to the failure were the poorly designed joint that permitted exposure of the redundant O-ring seals to hot gases and a further loss of O-ring sealing capability caused by temperatures that exceeded the low temperature qualification limit on the day of launch. As with the Apollo 1 fire, lack of escape provisions eventually doomed those crewmembers still alive after the explosion. Ejection seats that could have been used had been removed from all *Space Shuttles* after the first four qualification flights, when the number of crew was increased from two to seven.

Prevention of human error is an integral part of a good design. Indeed, operational errors can result in inadvertent commanding or the execution of operations that exceed the qualification envelope of the hardware. In the past, human operational errors were controlled mainly through instruction and training. Nowadays, any foreseeable mistake that is not prevented adequately by design is considered in every respect to be a design error, such as the wrong mating of connectors or the accidental activation of a switch.

A study in which four decades of major space system failures were analyzed (Newman 2001) concluded that manufacturing, production, and assembly errors have been important causes of these incidents. Such errors, which can go undetected through the usual verification processes, can manifest a myriad of consequences. For example, the root cause of the Apollo 13 liquid oxygen tank explosion was an overlooked design change that occurred during manufacturing. In this case, the existing wiring in the oxygen tank, which was designed originally to carry 28 V, was replaced so that it would be capable of carrying higher voltage for use during ground processing. However, the specification change failed to address the thermostat, which was rated to operate using 28 V. An undetected failure of the thermostat occurred during ground processing, and the stuck relay contacts resulted in high temperature damage to the internal tank wiring. Eventually, during the Apollo 13 mission, the circuit with the damaged wiring was energized, and a spark resulted in ignition and tank explosion.

4.2 PRINCIPLES AND METHODS

Design errors in space projects are a consequence of the complexity of space systems and the relative ease by which failures of highly energetic systems can propagate to catastrophic consequences. The difficulty of minimizing the occurrence of design errors is exacerbated by the limited systems safety engineering culture of design teams as a whole. The complexity of space systems design, as well as that of the organizations involved in its realization, demands broad knowledge of the key principles and techniques of safety engineering and a multidisciplinary awareness of the associated hazards and potential vulnerabilities inherent in the system. Unfortunately, safety design methods and hazard analysis techniques generally are not taught in engineering schools. In the aerospace industry, on-the-job training in systems safety usually is relegated to small, specialized groups of safety engineers, who often lack the in-depth knowledge of all systems necessary for them to become an integral part of the design teams.

The following are basic safety principles and methods common to several industries. These are used to minimize the possibility of accidents and reduce the consequences of an accident should one occur:

- Hazard elimination and limitation.
- Barriers and interlocks.
- Fail-safe design.
- Failure risk minimization.
- Monitoring, recovery and escape.

Often in combination, these principles and methods, which are the basic elements of a comprehensive safety design program, are used as deemed appropriate.

4.2.1 Hazard Elimination and Limitation

Hazards oftentimes can be eliminated by the proper selection of a design solution. For example, the choice of a nominal air atmosphere for a spacecraft instead of one of enhanced or pure oxygen greatly diminishes the risk of fire. As another example, designing a pressure vessel to leak-before-burst requirements avoids the potentially violent rupture and fragmentation that can cause additional damage, disable redundant systems, or injure a crewmember. Often, however, a hazard cannot be eliminated without concomitant loss of some major system functionality. Still, the level of a hazard can be limited by proper selection of design parameters. For electric power distribution efficiency, a high voltage system can be selected. However, if the power distribution system is designed so that the power is converted locally to provide low voltage at most electric outlets utilized by the crew, the risk of electric shock is limited.

4.2.2 Barriers and Interlocks

Barriers, sometimes called *inhibits* for certain applications, are a means for physically isolating a hazard. A barrier can be a physical interruption between an energy source and some function, a means of separating incompatible materials, or a means of isolating materials that when mixed would constitute a hazard.

Shields used to protect a spacecraft from meteoroids and orbital debris are safety barriers. They perform an energy absorption function, thus preventing impact damage to the vehicle and its systems. This is similar to the concept of structural containment, which is used as a means of protection for rotating items. Other examples of barriers are the use of relays between a battery and a pyrotechnic initiator and an isolation valve installed between a propellant tank and thrusters. It is important to realize, however, that commands, personnel, computers and software, panel switches, and procedures are not barriers and should not be considered as such in any design.

In the case of fire, combustion requires the presence of a fuel, an oxidizer, and an ignition source. Isolating any one of the three from the others would prevent effectively a fire. The use of a slightly pressurized inert gas, such as nitrogen, in a leak tight fuel container represents an effective barrier to prevent air from coming into contact with the combustible fuel.

Some materials and compounds possess harmful characteristics, such as toxicity or radioactivity. In such cases, barriers selected during design can be in the form of containers (isolating the material) or masks and other protective gear and equipment (isolating the crew). In a microgravity environment, metallic chips and glass fragments are potentially harmful if ingested by inhalation. In such cases, barriers can take the shape of filters with adequately fine meshes. Additionally, metallic debris can cause shorts in avionics equipment or become lodged in critical mechanisms. The barriers applied in this situation are comprised of layers of conformal coatings over electronic circuitry or guards and enclosures placed over mechanisms.

Sometimes barriers are intended to be temporary, such as the use of interlocks to prevent inadvertent access or exposure when a hazard source is present. An interlock can be used to prevent access to an energized laser, rotating equipment, or high temperature surfaces by locking covers or access doors while power is applied. As well, interlocks can function by automatically removing the hazard source, such as power to a laser when a protective cover (a barrier) is removed.

4.2.3 Fail-Safe Design

The purpose of fail-safe design, basically, is to place a system in a safe mode following some failure. In this safe mode of operation, some system functionality can be lost. Nevertheless, the primary intent is to prevent harm to crew and secondarily to prevent further system damage.

The fail-safe design approach applies mainly to system functions that are not essential for crew life support. Examples of these so-called must-work functions are the systems that regulate pressure and atmospheric composition in a spacecraft.

There are three fail-safe design approaches: fail passive, fail active, and fail operational. The option selected is determined by the specific purpose and functionality of the system.

- **Fail passive.** The equipment automatically is de-energized and ceases operation until corrective actions are taken. Fuses and circuit breakers are typical fail passive devices.
- **Fail active.** The equipment remains energized. The design includes standby redundancy that maintains the system in a safe mode until corrective action is taken. Redundant fastener arrangements in structures are examples of fail active design.
- **Fail operational.** The failure causes the equipment to revert to a mode that allows continued operation in a safe manner; however, functionalities that would otherwise present an unsafe situation are lost.

Sometimes the term *fail-safe* is used as a synonym for *redundancy*, although in general these are different concepts. A fail-safe design does not maintain or ensure safety by enhancing system reliability. A fail-safe design, indeed, can be unreliable and yet ensure safety.

4.2.4 Failure and Risk Minimization

Fault Tolerant Design

Fault tolerance is defined as the designed-in characteristics that maintain prescribed functions or services to users despite the existence of one or more faults. Fault tolerance is implemented by redundancy, fault detection, and response capability.

Redundancies can be established at the level of detailed design within a subsystem or system. Such is the case when multiple systems or subsystems are designed into the hardware to accomplish a particular function. Redundancy also can be established at the functional level. In this case, the system is designed so that more than one means is available for accomplishing a function and more than one failure must occur before the function is lost.

A redundancy is called *hot* (active) when redundant elements nominally are energized fully and it is not necessary to switch in the redundant element or switch out the failed one. Conversely, a redundancy is referred to as *cold* (standby) when secondary or tertiary redundant elements are nonoperative until they are switched intentionally into operation upon failure of a primary element.

An important general principle is that all redundancies and inhibits included within a design must be verifiable. As well, it is essential that redundancies and inhibits be independent, in the sense that no single credible failure, event, or environment can eliminate more then one.

Redundancies are used to enhance mission success as well as safety. It generally is accepted that, by implementing a number of independent safety redundancies as a protection against the possible consequences of a failure, that is, whether they are catastrophic or critical, the overall risk associated with operating the system becomes acceptable because the event probability becomes more remote. Four fault tolerant design criteria generally are in use:

- Safety critical system functions must be two-failure tolerant. This criterion applies to those systems that absolutely must work, for example, those ensuring crew survival such as life support systems or public and ground personnel safety, such as flight termination systems.
- Inadvertent operation of safety critical system functions must be controlled by a minimum of three inhibits. One of these inhibits must preclude any operation by a radio frequency command. The ground return for the circuit of the safety critical function must possess the capability to be interrupted by one of the three inhibits. As well, at least two of the three inhibits must be monitored.
- No combination of two failures or operator error shall have catastrophic consequences. *Catastrophic consequences* are defined as those causing loss of life, a life threatening or permanently disabling injury, or an occupational illness. Loss of major flight elements or ground facilities also can be classified as catastrophic.
- No single failure or operator error shall have critical consequences. *Critical consequences* are defined as those causing temporary disabling injury or occupational illness, use of contingency or emergency procedures, or damage to equipment.

Design for Minimum Risk

Hazards controls based on compliance with specific safety requirements other than those utilizing fault or failure tolerance fall collectively within a category called *design for minimum risk*. These requirements rely on safety factors and safety margins established by analysis and test, past experience, and international safety standards to ensure an established level of acceptable risk. Cases in which design for minimum risk is appropriate for use include structures, pressure vessels, pyrotechnic devices, and design for electromagnetic compatibility.

The concept of design for minimum risk is probably the earliest design method used to minimize failures. It typically results in overdesign to account for various uncertainties encountered, such as environmental conditions, analyses and test methods, materials variability, and manufacturing processes variances. For the engineer, overdesign essentially is a way for distancing the statistical distributions for cumulative stress and strength curves that are not known with requisite precision, thus preventing them from overlapping. Safety factors and safety margins change and are refined as our knowledge in pertinent areas advances.

The equivalent concept in electronic design is sometimes called *fault avoidance*. It consists of reducing the possibility of a system failure by increasing the reliability of individual items through the use of electrical design margins, derating criteria, high reliability components, application of work standards, and the like.

4.2.5 Monitoring, Recovery, and Escape

Any malfunction, failure, or error can lead to a contingency that represents an off nominal and hazardous situation that has yet to cause extensive system damage or injury or death to a crewmember. As an integral part of system design, provision must be made for the timely identification of contingencies, use of recovery means, or escape. For example, in the close environment of a spacecraft, the release of even small amounts of a toxic substance can become extremely hazardous. Therefore, provision must be made for removal of toxic materials either manually or automatically. As an example, the occurrence of a fire requires not only the removal of combustion by-products but also the suppressant, which must be selected to be effective and easy to clean up.

Considerations should be made for monitoring critical parameters and functions that have a potential to endanger life. Through such monitoring, a crew can be alerted to a hazardous situation and be able to conduct recovery procedures. The emergency, caution, and warning system consists of sensors used to detect malfunctions or out-of-limits conditions, monitor connections to redundant power sources and dedicated data storage and transmissions systems, and finally to provide for audible and visual signaling capability.

4.2.6 Crew Survival Systems

Deterioration of a contingency can continue until the point is reached when it becomes necessary to escape or abandon the spacecraft to ensure the survival of the crew. Escape equipment and systems are of fundamental importance in the daily life of the space crew and paramount in any hazardous environment. The importance of these systems has been demonstrated by some tragic accidents. Any functional spacecraft design must include crew escape and rescue capabilities for all mission phases, including those of ground testing and on-pad operations. In certain cases, safe haven capabilities can be considered as a means to sustain crew life until escape or rescue can be accomplished.

The design of escape systems for spacecraft must consider redundancy and design safety. The importance of a robust design was demonstrated during the launch of a Soyuz spacecraft in 1983. A valve in the propellant line failed to close 90 s from launch, causing a large fire to start at the base of the launch vehicle. The fire quickly engulfed the rocket, and the automatic abort sequence failed as the wires involved were burned through. The launch controller remotely aborted the mission by sending radio commands from the launch blockhouse, and the Soyuz descent module was pulled clear by the launch escape system. The crew landed safely some 4 km from the launch vehicle, which exploded seconds after separation.

4.3 THE SAFETY REVIEW PROCESS

In addition to the basic principles of safety described within this book other tools are available to assist an engineer in the design of safe spacecraft, payloads, and all related hardware. Among these, one stands out that is worthy of overview. The tool is often referred to as the *safety review process*. It is an essential element of the overall design effort to assure development and operation of safe space vehicles and hardware. Although not their primary role in the design cycle, programs such as quality, reliability, risk management, probabilistic risk assessment, and others are used to assess the risks associated with operating and living onboard a spacecraft. These are not safety programs in the most strict sense; however, they, indeed, are important elements of the safety process and, as such, routinely contribute to the overall safety of the vehicle, its payloads, and the crew.

4.3.1 Safety Requirements

The safety review process is based on a cadre of comprehensive and well accepted safety requirements. Over the years, NASA, ESA, RSA, JAXA, CSA, and other space agencies developed and refined safety requirements that are supportive of the design and operational philosophies of the respective agencies. Cooperation among the various space agencies developed through participation in jointly crewed programs, such as *Spacelab*, *Shuttle-Mir*, and the *International Space Station*. Because requirements essentially are driven by the particular launch vehicle employed, hardware launched on U.S. or Russian vehicles is designed to comply with the requirements of the agency launching the mission. Fortunately, for hardware designers, safety is safety, and the basic principles are the same everywhere. The safety requirements used by RSA and by NASA are quite similar, although differences do exist.

With partnerships established among the various space agencies for building and operating the *International Space Station*, NASA was designated as the agency responsible for the overall safety in this endeavor. Various sets of requirements applicable to payloads and the International Space Station vehicle were adapted from those of the Space Shuttle Program (NASA 1995a, 1995b, 1996). These requirements are used by not only NASA designers and developers but also those of ESA, JAXA, CSA, and others participating in the International Space Station Program. Although RSA continues to utilize its own safety requirements, their similarity with those of NASA presents few problems. Where differences exist, they are discussed and resolved by various joint working groups.

Whatever the source for safety requirements, it is essential that they be comprehensive, well established, and widely accepted. As such, these requirements are not only a basis for safe design but also the foundation on which the safety review process functions. The safety processes and requirements described in this chapter have been improved and upgraded over four decades of continuous successful use.

4.3.2 The Safety Panels

Two entities are involved in conducting a safety review, each having the common goal of designing and delivering a spacecraft and associated hardware that are certified as safe to crew and vehicle. The hardware developer employs a safety team to produce a safety assessment demonstrating design compliance to applicable safety requirements and to present and then defend this assessment to the Safety Review Panel, which is only one such panel used by NASA to review an assess safety compliance for International Space Station hardware. The Shuttle Safety Panel performs a similar assessment and approval for hardware operating on or transported in the *Space Shuttle*. Payloads flying aboard the *Space Shuttle* and the *International Space Station* are reviewed by the Payload Safety Review Panel. As well, the Ground Safety Review Panel reviews hardware and associated ground operations to ensure safety at the launch site. In addition, NASA has established with ESA a franchise Payload Safety Review Panel and, at the time of this writing, is in the process of concluding an agreement establishing a similar franchise agreement with JAXA. The RSA has its own process for conducting safety assessments; however, NASA exercises final approval authority for the safety of Russian hardware and payloads, as it does for those reviewed by the ESA and JAXA franchises.

Each of the NASA safety panels is charged with the responsibility to evaluate the safety assessment conducted by the safety team of the hardware design organization. It is important to note that the safety panels conduct safety reviews and not audits. The safety review is not a process by which strict compliance to each and every safety requirement is determined. Instead, the purpose of a safety panel is to evaluate and approve the overall design approach selected by the developer to meet the applicable safety requirements, thus assuring safety of the crew and vehicle.

As is typical of the other panels, the Payload Safety Review Panel is comprised of official representatives from key NASA disciplines, such as the Engineering Directorate, Mission Operations Directorate, the Extravehicular Activity Office, the Astronaut Office, the Space Shuttle Program Office, the International Space Station Program Office, the Life Sciences Directorate, and the Panel Chairman and Executive Officer. It is supported by a cadre of technical experts from a variety of disciplines such as electrical systems, toxicology, structures, materials, radiation, pressurized systems, and biohazards. In addition, the Chairman administratively and technically is supported and staffed by a technical writer for documenting meeting minutes and a cadre of safety engineers from the Safety and Mission Assurance Directorate, who are responsible for receiving, distributing, reviewing, assuring U.S. export control compliance, and providing preliminary assessment of safety data packages and hazard reports. The makeup of each of the NASA safety panels is somewhat different; however, the overall procedures and processes followed are the same.

4.3.3 The Safety Reviews

During the process for design and development of spaceflight hardware, whether an experiment, payload, or spacecraft systems or subsystems, the project must be prepared to conduct up to three major safety reviews. The first review for the hardware is expected to occur around the time of the preliminary design review. This is the Phase I safety review. The second, the Phase II safety review, normally occurs around the time of the critical design review. The third and final Phase III safety review must be completed prior to the hardware being shipped to the launch facility. The phased ground safety reviews are conducted similarly.

For a major experiment, such as the *Hubble Space Telescope* or an International Space Station module, each review can require several weeks to conduct fully. Seldom does a review for this class of hardware take less than a week.

In preparation for a safety review, the hardware developer or the project is required to prepare a safety data package. This document, which easily can exceed 1000 pages, describes in detail the purpose and operation of the hardware and provides a detailed technical description of the design, as well as various analyses that have been conducted. The safety data package must contain an identification, description, and assessment of all applicable hazards and their causes and any design or operational solutions used to mitigate the hazards for the hardware. The hazard information, as well as the identification of any hazard controls and the associated verification procedures, typically is contained within a number of formal hazard reports, each hazard occupying a single, unique report. The content and structure for a safety data package and associated hazard reports are specified typically within the safety requirements documents (NASA 1995a, 1995b, 1996).

The safety data package and associated hazard reports are required to be delivered by the developer approximately 45 d prior to each of the scheduled Phase I, II, and III safety reviews. It is delivered to the Executive Officer, who logs it in and subsequently assigns it to a safety engineer. Once assigned, the safety engineer distributes the safety data package and hazard reports to each of the safety panel members and the technical support staff for their review. The safety engineer receives any comments from the various individuals reviewing the material, compiles them, and works with the developer to resolve any issues identified prior to the scheduled formal review.

The formal safety review is opened by the Chairman, after which the developer takes the floor and presents to the full panel a detailed presentation of the hardware, its function, its purpose, and any particular nuances of the design that have potential for being of interest to the panel. During the presentation, the panel members, as well as the technical experts supporting the panel, question the design team regarding any issues not fully illuminated in the safety data package. Once the presentation is competed, the developer presents each of the hazard reports to the panel. During the Phase I safety review, the hazard reports are reviewed in minute detail to ensure that all pertinent hazards and their associated causes are identified and clearly stated. At the Phase II safety review, the hazard reports are scrutinized to ensure that appropriate controls and their verification methods are identified and stated clearly. The final status of verification for the hazard controls is confirmed during the Phase III safety review. At any phase of the safety review process, hazards, causes, controls, and verifications can be altered by either the safety panel or the hardware developer. Any such alteration must be approved by the safety panel, however. The Chairman of the Payload Safety Review Panel is authorized by the International Space Station and the Space Shuttle program managers to indicate approval of the hazard reports by signature at the conclusion of each phase of the safety review process. Once the Chairman has signed the Phase III hazard reports, the hardware is deemed to have completed successfully the safety review process and is considered safe to launch and operate.

4.3.4 Nonconformances

During the course of a safety review, an occasion can arise when, despite the best effort of the design engineers, adequate control of a hazard cannot be realized. Typically referred to as a *noncompliance*, such a case requires the developer to initiate a special nonconformance report describing the hazard, its cause, and most importantly, all proposed acceptance rationale that can be used by NASA management in the decision process for approval.

The need for a nonconformance report can be recognized at any phase of the safety review process. In fact, the earlier the necessity for a nonconformance report is recognized, the better it is for the developer because more time remains available to eliminate the hazard, obtain approval, or seek more exotic and oftentimes expensive means for controlling the hazard. Once the developer and the Chairman have agreed that a nonconformance report is necessary and the developer has initiated the nonconformance report, it is presented formally to the safety panel at a special meeting scheduled specifically for this purpose. A fully detailed description of the system or subsystem in question, the hazard, the failed attempts to achieve a design solution, and the acceptance rationale for the nonconformance report are provided by the developer. With consensus of the safety panel, the nonconformance report can be signed by the Chairman. Should the safety panel not be satisfied by the rationale for the nonconformance report or the magnitude of the risks involved, the developer is then directed by the Chairman to find an acceptable design solution to control the hazard. If none can be found and the nonconformance report remains unapproved, then the hardware cannot be deemed sufficiently safe for flight.

For some rather benign cases in which the risk associated with acceptance is low, the approval of a deviation-nonconformance report by the Chairman is binding on the program. Those nonconformance reports that entail acceptance of higher degrees of risk necessarily have to be forwarded to upper program management for final disposition. For these cases, once the Chairman has indicated that a waiver-nonconformance report can go forward, the developer proceeds to a board chaired by the program manager. Here, the waiver-nonconformance report is again presented and scrutinized, and upper level management must decide whether the risk presented by the request is of an acceptable or manageable level. If the decision is negative, then the developer has the option of implementing a new, more difficult, and routinely more expensive design solution. Alternatively, the hardware cannot be approved for flight.

It is always best for the developer to avoid a nonconformance report situation by making every reasonable effort to find a way to eliminate a hazard or provide adequate controls for it. Despite the preparation and submittal of a nonconformance report, there is no guarantee of its approval by either the safety panel Chairman or upper management, either of which can believe that the risk incurred by its acceptance is simply too great to accept.

REFERENCES

Hammar, W. (1993) *Product Safety Management and Engineering*. Des Plaines, IL: American Society of Safety Engineers Press.

NASA. (1995a) *Safety Policy Requirements for Payloads Using the International Space Station*. Space Shuttle Document NSTS-1700B (ISS Addendum). Houston, TX: National Aeronautics and Space Administration, Johnson Space Center.

NASA. (1995b) *Safety Requirements Document for the International Space Station*. Space Station Document SSP-50021. Houston, TX: National Aeronautics and Space Administration, Johnson Space Center.

NASA. (1996) *Safety Policy and Requirements for Payloads Using the STS*. NASA Space Shuttle Document NSTS-1700B. Houston, TX: National Aeronautics and Space Administration, Johnson Space Center.

Newman, J. S. (2001) Failure in space: A systems engineering look at 50 space system failures. *Acta Astronautica* 48, pp. 5–12.

CHAPTER 5

Human Rating Concepts

Michael K. Saemisch
Lockheed Martin Space Systems Company, Denver, Colorado

CONTENTS

5.1. Human Rating Defined 175
5.2. Human Rating Requirements and Approaches 179

This chapter addresses the concept of human rating as applied to space systems. As defined here, this process recently has become a formal process for NASA programs and, within the U.S. law, for non-NASA American operators. This now formal NASA process captures and consolidates the past successful practices of NASA for application in all future NASA programs.

Human rating is an encompassing effort that affects many aspects of a program, from its earliest phases and throughout its life. Technical human rating requirements drive the design of flight and ground systems. In contrast, human rating programmatic requirements drive processes such as management involvement and detailed analysis, and they generate the data to support formal human rating certification.

5.1 HUMAN RATING DEFINED

5.1.1 Human Rated Systems

NASA defines a *human rated space system* as one that "incorporates those design features, operational procedures, and requirements necessary to accommodate human participants." Accordingly, a human rated system (NASA 2005) is one in which

- Risks have been evaluated and either eliminated or reduced to acceptable levels.
- Human performance and health management and care have been addressed appropriately such that the system has been certified to support human activities safely.
- The capability to conduct crew tended operations safely has been provided, including safe recovery from any credible emergency situation.

A fundamental concept conveyed in this definition, is that the human rated designation is applied to a space system. In this context, a *space system* is defined to be "any system developed or operated that supports activity in space, including but not limited to, subsystems supporting launch, mission control, and on-orbit operations" (NASA 2005). For the NASA Orion project, the space system includes the entire launch stack, that is, the ARES-1 launch vehicle and the Orion crew vehicle. These elements that comprise the space system are to be human rated. Should *Orion* be launched on a different launch vehicle or if the *ARES-1* were to launch a different crewed vehicle, the entire new stack would need to be human rated as a new space system, independent of the former.

It follows that the various elements of a crewed space system individually are not human rated; however, they must meet the applicable portion of the human rating requirements for the fully integrated system. It also follows that subsystems or components from one human rated system might not be adequate for use in another because they are not given a human rating on an individual basis. As an example, if an engine from the human rated *ARES-1/Orion* stack is to be used in a new or a different crewed launch vehicle, human rating would have to be established for the new launch stack as a system. Human rating requirements would need to be determined for the engine, as well as all other elements, for the new stack to be deemed acceptable for crewed operations. It is likely that much of the data from the previous human rating process would be applied. The entire process must be completed, however, because the elements might be used in a different capacity and could have different criticality when used in the new space system. As an example, the Orion space system is able use a procedural workaround to meet certain human rating requirements, whereas for the new stack such a workaround is not acceptable. This increases the criticality, which influences the applicability of the requirements used to certify components from the original space system.

5.1.2 The NASA Human Rating and Process

Human rating is the term applied to the process of imposing certain requirements on the design of a space system and verifying the application of these requirements and the overall effectiveness of the design. The objective of this process is to complete the NASA human rating process successfully. Today, the term *human rated* is a designation that the human rating process for a system has been completed and human rating certification has been obtained through a formal process. In the past, crewed programs were not subjected to any formal process specifically designated to grant a human rating certification. The term *human rating* simply was applied to all systems designed to carry humans into space or support humans when in space, such as the *Space Shuttle* or *Apollo*.

The human rating process and all associated requirements are defined by NASA in NPR 8705.2A, *Human Rating Requirements for Space Systems* (NASA 2005). The stated purpose of NPR 8705.2A is that it is "NASA's policy is to protect the health and safety of humans involved in or exposed to space activities, specifically the public, crew, passengers, and ground personnel. NASA will fulfill all requirements of NPR 8705.2A for all space systems involving humans or interfacing with human space systems prior to becoming operational and throughout their use. A program is eligible for human rating certification only if it meets engineering requirements, health requirements, and safety requirements contained in this NASA programmatic requirements document. Human rating certification provides the maximum reasonable assurance that a failure will not result in a crew or passenger fatality or permanent disability."

5.1.3 The Human Rating Plan

A formal human rating plan is the initial step of the human rating process as required by NPR 8705.2A for all NASA programs (NASA 2005). New NASA programs, such as Constellation, are now required to establish a human rating plan applicable to the entire program and includes all projects within the program. The human rating plan consists of three volumes:

- Volume I provides traceability for each requirement stated in NPR 8705.2. It includes a tracking matrix describing the process the program is to use to comply with each requirement assigned. It identifies where each requirement is to be incorporated into program documentation or levied onto a contractor. Any tailoring and exceptions to these requirements and explicitly stated justification for approval are to be included. For programs consisting of multiple projects or elements, traceability is provided in Volume I to ensure implementation of the requirements for each of these facets. Volume I also must include a set of applicable standards approved by the NASA Independent Technical Authority; the duration of the human rating certification for the program is documented as well. Volume I of the human rating plan is generated and approved prior to the system requirements review of a program. Note that any human rating requirements specified by the human rating plan can be tailored or exempted for the program, or an element may be tailored or otherwise altered, by approving management depending on the level of acceptable risk involved.
- Volume II provides a description of objective evidence used to demonstrate compliance with each human rating requirement and identifies and describes all critical functions for the space system. Volume II of the human rating plan must be generated and approved before the preliminary design review of a program is conducted. The human rating plan is reviewed during the preliminary design review; and the adequacy of all planned verification tasks, such as qualification and flight testing, is assessed.
- Volume III of the human rating plan describes the performance criteria for each critical function of the space system and the means by which these criteria are to be ensured through analysis, test, inspection, and demonstration. Finally, Volume III establishes the processes used by the program to ensure that the space system is maintained in the "as certified" condition. Volume III must be approved before the critical design review of the program is conducted.

Every human rating plan is reviewed and approved internally by NASA. Therefore, all human rating requirements and processes become transparent to the element and project element contractors. At these levels, every requirements document provided by NASA contains the human rating requirements integrated with other requirements of similar type, technical or programmatic. Data delivery requirements levied onto the contractor ensure that all data needed by NASA to satisfy the internal human rating process are obtained.

5.1.4 The NASA Human Rating Certification Process

The objective of the internal NASA human rating certification process is "to document that all critical engineering requirements, health requirements, and safety requirements have been met for a space system that provides maximum reasonable assurance that the system's failure will not result in a crew or passenger fatality or permanent disability" (NASA 2005). A NASA program manager is responsible for ensuring that a program complies with the requirements set forth in NPR 8705.2 and the space system is qualified for human rating certification. The NASA Chief Safety and Mission Assurance Officer serves as the official of primary responsibility for the human rating requirements. The NASA human rating board performs executive level activities for the human rating requirements. The NASA Associate Administrator for Space Operations and the Associate Administrator for Exploration Systems are Co-Chairmen of the human rating independent review team. The human rating independent review team supports the human rating board by providing insight into the development and implementation of the human rating plan and thus the human rating certification of the system. The Chief Engineer and the Chief Health and Medical Officer (both are independent technical authorities) ensure that tailoring, exceptions, deviations, and waivers technically are correct and have adequate justification. The Associate Administrator for Space Operations, Associate Administrator for Exploration Systems, the Chief Engineer, and the Chief Health and Medical Officer all must approve the human rating plan with concurrence from the Chief Safety and Mission Assurance Officer, who provides verification that objective quality evidence has been assessed and has demonstrated compliance with human rating requirements. The Associate Administrator for Space Operations provides the human rating certification for each space system.

5.1.5 Human Rating in Commercial Human Spaceflight

Human rating for commercial systems operated by citizens of the United States anywhere in the world must utilize an equivalent of the NASA human rating requirements and processes defined and administered by the Federal Aviation Administration, whose process, however, does not apply to launches conducted by the government or on the behalf of the government. Instead, such launches are encompassed by the NASA process.

The Federal Aviation Administration human spaceflight requirements for crew and spaceflight are contained within the Code of Federal Regulations, Parts 401, 415, 431, 435, 440, and 460, which are available at www.faa.gov. The rule can be summarized as follows: "The FAA is establishing requirements for human spaceflight as required by the

Commercial Space Launch Amendments Act of 2004, including rules on crew qualifications and training, and informed consent for crew and spaceflight participants. These requirements should provide an acceptable level of safety to the general public, and ensure individuals onboard are aware of the risks associated with a launch or reentry." The rule also applies existing financial responsibility and waiver of liability requirements to human spaceflight and experiment permits. Experiment permits, as well, are the subject of separate rule making.

5.2 HUMAN RATING REQUIREMENTS AND APPROACHES

The intent of the NASA technical human rating requirements is to define a minimum acceptable level of risk to which crews, flight and ground, as well as the public are exposed. These requirements are not intended to assure mission success. That process is left to the programs through application and enforcement of other requirements and processes. The approach to obtaining human rating certification is to achieve an acceptable level of risk through the imposition of technical requirements that allow the widest possible latitude for NASA programs to develop both design and operational methods.

Acceptable levels of risk are not defined numerically, such as by specifying a minimum likelihood for crew loss, but rather are defined in terms such as the minimum level of failure tolerance (two or three) or in terms of equivalent requirements for different types of risks. The level of risk is further mitigated through the imposition of requirements to provide alternate means for ensuring crew safety should the specified risk mitigation approaches fall short.

5.2.1 Key Human Rating Technical Requirements

This section addresses many key technical requirements involved in human rating. Reference should be made to NPR 8705.2A (NASA 2005) for the complete set of human rating requirements. Several of these key requirements are used to define minimum acceptable levels of risk. The many corollaries to these requirements are used to achieve similar levels of risk, albeit through alternate approaches.

Failure Tolerance

The first and foremost human rating technical requirement is stated in paragraph 3.1.1 of NPR 8705.2A: "Space systems shall be designed so that no two failures result in crew or passenger fatality or permanent disability (Requirement 34419)" (NASA 2005). This requirement perspicuously signifies that the minimum acceptable level of risk for systems is the toleration of two credible failure modes without incurring the stated undesired effects. To meet this requirement, systems designers must identify various scenarios that can result in hazards or other undesired end effects and implement appropriate design solutions, such as triple redundancy or the use of multiple series inhibits, to prevent occurrence of these hazards.

Design for Minimum Risk

Failure or fault tolerance is neither technically feasible to control every hazard cause nor is it always the most appropriate approach to control a hazard. For example, hazards associated with the use of fully charged pressure vessels containing hazardous materials, which if released could cause crew injury or death, more appropriately are controlled through assuring the strength of the containment vessel rather than implementing an approach to tolerate rupture by utilizing secondary containment. A similar approach applies to various structures, materials, passive shielding, and the like.

For design cases where the application of full failure tolerance is not feasible, the approach taken is known as *design for minimum risk*. The essential criteria for the application of design for minimum risk rather than a failure tolerance approach are controlled formally by the human rating process. Areas of proposed minimum risk must be identified, and formal approval to classify them as such must be obtained by the design team. Indeed, NASA has encouraged the use of past successful design approaches, so such heritage can be cited as a basis for approval for many cases where design for minimum risk is proposed. The design for minimum risk approach can involve the use of some level of failure tolerance; however, the application of full two-failure tolerance in these situations is a less than suitable design solution. For example, the use of redundant pyrotechnic chains is inappropriate considering the historically high reliability of pyrotechnics.

Human-System Interactions

A human rated system must provide the ability for a crew to interact with and control critical elements of that system. To that end, NASA established the requirement in section 3.2 of NPR 8705.2A (NASA 2005), which states that "a space system shall be designed to allow the crew to monitor and operate all critical functions, including automatic overrides." In event of a catastrophic event or situation, the final decision for crew safety, therefore, is placed in the hands of those who are most affected by the decisions, that is, the crew.

NPR 8705.2A, however, specifies neither specific design solutions nor specific requirements to influence design in this area. In general, NASA seeks the maximum possible amount of crew interaction in the process to maximize the number of available options that affect crew safety. In this way, the greatest amount of insight into each hazardous subsystem is provided, the objective being to provide all viable options to the crew when hazardous conditions arise. For example, inhibits in hazardous subsystems should be monitored and the status made available to the crew. Through knowledge of the status of the various inhibits, a basis for possible control actions or mission decisions is established. Note, however, that monitoring is not part of a failure tolerance scheme, but it is indeed a critical element for assessing failure tolerance status.

Safety of Flight Control Systems and Software

Computer based control, including the use of resident software, requires a unique requirements approach, because failure tolerance cannot be applied to software, such as three independent software programs controlling the same function. In addition to applying rigor to the development and verification of software and software requirements, NASA engaged the use of backup flight systems to achieve a degree of independence from the primary software through independent development and dissimilarity. An example of this is the backup flight control system on the *Space Shuttle*. Similarly, the *International Space Station* possesses some ability to control the station should the primary computers fail (a situation that, indeed, has occurred).

The NASA human rating requirements in NPR 8705.2A, section 3.10.1 (NASA 2005) mandate that designers apply and achieve a degree of backup and dissimilar flight control in crewed spacecraft. The text of the requirement addresses the concern that has driven the design and implementation of past space systems but does not specify any specific design solution: "The system design shall prevent or mitigate the effects of common cause failures in time critical software, e.g., flight control software during dynamic phases of flight such as ascent."

As stated, a wide possible set of design solutions, ranging from rigorous design and testing of a single software solution, to the use of a full independently specified, developed, and implemented backup system, is permitted. From all available options, designers must determine a suitable approach and justify the degree of dissimilarity (or lack thereof) to NASA. It should be assumed that, as a minimum, some degree of independent backup to assure crew survival is to be required. Indeed, NASA does offer that "implementation of this requirement can take different forms." The following methods have been used in human rated systems, and they meet the intent of this requirement (NASA 2005):

- Redundant independent software running on a redundant identical flight computer.
- Use of an alternate guidance platform, computer, and software, such as using the space craft guidance to control a booster.
- Use of nearly identical source code uniquely compiled for different dissimilar processors.

Applicable Design Standards

Successful human rating approaches applied to NASA programs also have been captured in detailed specifications that address topics such as structural design and verification. NASA requires implementation of certain standards as specified in Chapter 2 of NPR 8705.2A. In addition, programs must determine whether any additional standards, such as any established with the NASA Independent Technical Authority, apply.

Crew Survival

The NASA human rating requirements state that the means for crew and passenger survival in the event of a mishap should be designed into spacecraft. Implementation of crew survival capability has been determined by NASA to be necessary through evaluation of the demonstrated reliability that has been experienced with launch vehicles. To that end, section 3.9 of NPR 8705.2A (NASA 2005) institutionalizes the requirements that provide for crew removal during emergencies and for emergency modes that can be used to assure crew safety should major system failures occur. These requirements are separate and in addition to existing design failure tolerance and design for minimum risk requirements. These crew survival requirements cannot be counted toward meeting the failure tolerance requirements.

Crew survival requirements address the most hazardous phases of the flight, when crew and passengers are aboard a human rated system, and apply during on-pad operations through launch, ascent, and descent. These requirements impose provisions to ensure crew survival, such as a launch abort system similar to that used on *Apollo* to remove the crew vehicle from the launch vehicle or by providing for an escape method from the vehicle during ascent. For descent, these requirements influence the design to include further crew survival provisioning, such as a simple and independent backup flight system should the primary system fail, or influence the shape of the vehicle to permit safe reentry in an unguided mode should guidance systems fail during ascent (a ballistic reentry). Additionally, crew survival requirements further provide for crew interaction to initiate or inhibit automatic abort or escape systems.

5.2.2 Programmatic Requirements

In addition to the programmatic requirements associated with building and implementing a human rating plan, specific programmatic requirements must be implemented by a program seeking NASA human rating certification.

Risk Quantification

As stated earlier, NASA human rating requirements neither quantify what is to be considered minimum acceptable safety nor state specific quantitative safety requirements. Instead, there is reliance on the minimum levels of safety specified through other requirements, such as failure tolerance. Programs must analyze the probability for an occurrence of catastrophic events and use the analysis for related design and operational trade studies (NASA 2005). Furthermore, if NASA does establish a relative risk goal or requirement, probabilistic risk assessment, as defined in NPR 8705.5, *Probabilistic Risk Assessment Procedures for NASA Programs and Projects*, must be used to show compliance.

Critical Functions

The NASA human rating requirements state that a program must define and verify critical functions throughout its life cycle. These requirements are not detailed particularly as to the rationale and approach for implementing this programmatic instruction. Reference should be made to NPR 8705.2A for the documentation requirements associated with critical functions, defined as "capabilities or functions that are essential to the safety of the crew and passengers, that if lost would cause loss of life or permanent disability."

System Safety and Mission Assurance

NASA strongly believes in the necessity for applying the disciplines of safety, reliability, and quality engineering to achieve acceptably safe and human ratable systems. Therefore, section 1.6.3.1 of NPR 8705.2A requires implementation of NASA procedural requirements relative to human health and safety to identify, analyze, track, and eliminate or mitigate hazards and risks throughout the life of the program (NASA 2005). Hazard identification is also essential for the application of failure tolerance and other requirements of NPR 8705.2A.

Human Factors Engineering

In addition to imposing NASA process requirements for safety and mission assurance functions, NASA also imposes human factors engineering to be implemented, beginning with concept development and continuing throughout the life cycle of the system. The human rating requirements specify that the program manager must establish human performance criteria and system usability requirements. The NASA human rating requirements do not specify these requirements but contain an order of precedence for human error management in paragraph 3.1.5, along with a listing of possible methods to prevent human error (NASA 2005).

5.2.3 Test Requirements

Verification of compliance to and effectiveness of human rating features must be accomplished by the typical requirements verification methods of test, analysis, and inspection. However, a stronger emphasis is placed on verification through testing. Although a specific definition for human rating verification testing is not contained in the human rating requirements, one should be derived by the program that would be considered extensive and perhaps even include one or more test flights conducted throughout the mission profile.

Human-in-the-Loop Testing

Because the system is designed to be rated for carrying humans, testing must involve humans to ensure that the systems can be operated as designed from both an operational standpoint and a safety standpoint (and for contingency operations training). NASA imposes general requirements that establish the definition of detailed human-in-the-loop testing for a program. As stated within section 1.6.6 of NPR 8705.2A, "The Program Manager shall perform usability testing of human system interfaces for the critical functions using support from the user community including the Astronaut Office, ground processing crew, and mission control crew to verify that the system design meets the human performance requirements during system operation and in-flight maintenance consistent with the anticipated mission operations concept and anticipated mission duration. The Program Manager shall use an iterative design process, where the results of the usability testing are incorporated into the design" (NASA 2005).

Software Testing

Considering the criticality of software and the potential consequences of failure with regard to the hazardous operations involved in space travel, software presents the greatest challenge to safety design and verification in all possible situations. NASA imposed stringent general requirements for software testing that include a need to test the software through the entire performance of the flight envelope. As stated in paragraphs 1.6.7.1–.3 of NPR 8705.2A (NASA 2005), this testing program could involve specific non-crewed test flights, flight equivalent test beds, and ground system computer platforms that can be used to support space missions. Specific testing requirements, however, must be derived by the program.

Flight Testing

In Volume III of the human rating plan, the program manager is required to propose and document the type and number of flight tests that are to be performed across the mission profile under actual and simulated conditions to achieve human rating certification. The type and number of flight tests are approved as part of the human rating plan approval process.

Per NPR 8705.2A (NASA 2005), the flight test program provides two objectives. First, the flight test program utilizes testing to validate the integrated performance of the space system hardware and software in the operational flight environment. Second, the flight test program uses testing to validate the analytical math models, which are the foundation of all other analyses, including those used to define operating boundaries that are not expected to be approached during normal flight:

- "Flight and ground tests are needed to ensure that the data for the analytical models can be interpolated to predict confidently the performance of the space systems at the edges of the operational envelopes and to predict the margins of the critical design parameters; and
- In order to minimize risk to the flight test crew, it is preferred that an unmanned flight test be conducted prior to a manned flight test. It is acknowledged that this may not be feasible for all phases of flight and may not be necessary at all for some systems (e.g., rovers)."

5.2.4 Data Requirements

The human rating requirements defined in NPR 8705.2A (NASA 2005) typically are integrated and implemented into the program requirements. Any associated technical requirements are implemented in program technical requirements documents. For example, the technical human rating requirements from the human rating plan can be contained in an architecture requirements document, which is the top technical specifications for a program. Each project in the program then generates its own system requirements document, which contains the technical requirements applicable to the specific project. The system requirements document is imposed on contractors via a contract as an integrated set of technical requirements. Programmatic and process requirements are implemented in NASA program and project plans along with other process requirements. These requirements are imposed on contractors by defining specific tasks in the statement of work imposed by contract. An example of this type of task is the performance of safety analysis.

REFERENCE

NASA. (2005) *Human Rating Requirements for Space Systems.* NASA Programmatic Requirements NPR-8705.2A. Washington, DC: National Aeronautics and Space Administration, Office of Safety and Mission Assurance, Headquarters.

CHAPTER

Life Support Systems Safety

Kimberlee S. Prokhorov
NASA Lead, ISS Common Environments Team, Johnson Space Center, National Aeronautics and Space Administration, Houston, Texas

CONTENTS

6.1. Atmospheric Conditioning and Control 188
   By Kimberlee S. Prokhorov
6.2. Trace Contaminant Control 198
   By Jay L. Perry
6.3. Assessment of Water Quality in the Spacecraft Environment: Mitigating Health and Safety Concerns 211
   By Torin McCoy
6.4. Waste Management 220
   By Kimberlee S. Prokhorov
6.5. Summary of Life Support Systems 221
   By Kimberlee S. Prokhorov

The ability to sustain human life in a hostile space environment for prolonged periods is perhaps the most complex challenge to be encountered in spaceflight (Johnston and Michel 1962). The life support systems of a spacecraft make it possible for humans to live onboard. These systems, however, present many safety challenges. Their design and operation must ensure that, while providing essential services, their operation does not place the health and safety of the crew, vehicle, or mission in a position of unacceptable risk.

The main functions of the life support systems can be defined as follows:

- Atmospheric conditioning and control:
  Controlling atmospheric pressure.
  Conditioning the atmosphere (maintaining total pressure, providing oxygen, removing carbon dioxide and trace contaminants, and controlling temperature and humidity).
- Respond to emergency conditions:
  Fire.
  Rapid decompression.
  Hazardous atmosphere.
- Provide water.

- Manage crew waste.
- Provide support for extravehicular activity operations.

The interaction and interdependency of the life support subsystems and their functions are depicted in Figure 6.1.

Life support subsystems are interrelated tightly, and a malfunction of any one subsystem can have detrimental effects on others. Similar to the ecosystems on Earth, the relationships among life support subsystems are symbiotic. Although essentially unlimited buffering capacities exist within ecosystems, this is not the case for the habitable environment of a spacecraft. Indeed, for even a single small change to this unique environment, the magnitude of the resulting effects and the speed at which they occur are pronounced (Eckart 1996).

Certain design trades can be made to improve one system over another, and great care and detailed analysis should be given to such studies. Unanticipated effects of a design trade unknowingly can place crew safety in a position of unacceptable risk. Throughout the history of human spaceflight, trades have been made for the benefit of one system while putting others at risk of affecting crew safety and mission assurance. The most difficult aspect of dealing with life support systems is the ability to strike a balance between system design trades and acceptable risk to crew safety.

During the early ground tests of the life support and suit systems of the Mercury vehicle, nitrogen gas was found to concentrate within the circuit of the pressure suits. This occurred because the flow of oxygen into the suit was initiated by a slight negative

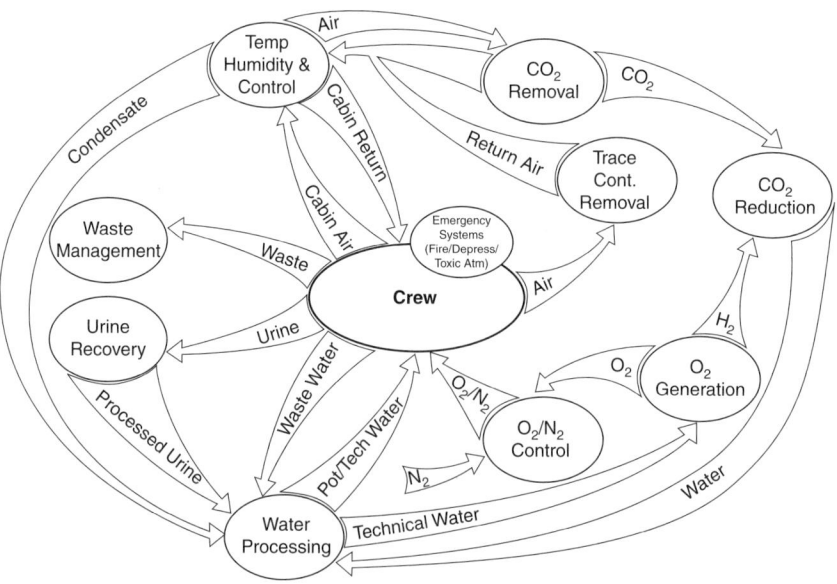

FIGURE 6.1 Life support subsystem relationships.

pressure on a demand regulator. Consequently, the cabin atmosphere was changed to 100% oxygen to alleviate this problem. Special emphasis was placed on material selection and quality control to eliminate the potential fire hazard resulting from the pure oxygen atmosphere (Link 1965).

Bromotrifluoromethane is considered to be a good fire suppressant on Earth, and for this reason, it was chosen for use on the Space Shuttle. However, this suppressant is not approved for use onboard the *International Space Station*. Once released, it is extremely difficult to scrub from the atmosphere of a spacecraft cabin, and as well, there is a risk of pyrolysis should the suppressant come into contact with any high temperature components of systems or payloads. Because of these factors, bromotrifluoromethane is acceptable for use only on a vehicle that possesses an efficient and effective capability to purge its atmosphere.

Although used on the Space Shuttle, bromotrifluoromethane should not be used on a vehicle that docks to a long-term permanently orbiting vehicle. Even though it has never been expelled to suppress an on-orbit fire on the Space Shuttle, trace amounts of the suppressant have been found in random air samples taken from the *International Space Station*. Furthermore, spikes in the level of bromotrifluoromethane have been observed from the ground analysis of air samples collected from the *International Space Station* during operations in which the Space Shuttle is docked.

The U.S. segment of the International Space Station specifies the use of a carbon dioxide fire suppressant, whereas water based fire extinguishers are required by the Russian segment. Technical teams representing the U.S. and Russian segments have discussed the appropriate use of these extinguishers in the event of a fire within the *International Space Station*. Both sides agreed that the water based suppressant is better suited for use in the open cabin area, where there are free flames. However, the carbon dioxide extinguisher was determined to be better suited for use within an enclosed volume, such as behind the rack panels, where it might be difficult for the water based suppressant to access the exact location of smoldering wires or active combustion. As well, the voltage used on the U.S. segment is 120 VDC as compared to the 28 VDC used on the Russian segment. For this reason, the use of the Russian water based suppressant on the U.S. segment was believed to pose too great of a shock hazard to the crew because the area surrounding the event cannot be guaranteed to be unpowered during a fire (Curry et al. 2003). Had both segments of the *International Space Station* been designed to 28 VDC, the Russian fire extinguisher could have been used within the U.S. segment. This would have eased the use of equipment in each segment, given that a common fire extinguisher interface port and nozzle is in use. Figure 6.2 shows the fire extinguishers utilized on the *International Space Station*.

Because the combined effects of design trades can be synergistic or antagonistic, great attention must be paid to any potential subsystem interactions and the operational implementation of trade results before implementing design solutions. This is especially true for spacecraft life support systems. Conscientious systems engineering and a broad view of the design are essential to ensuring mission safety and success.

Systems typically associated with life support systems, such as emergency, caution and warning, fire safety, and oxygen systems, are covered in more detail elsewhere within this book. This chapter focuses on those portions of life support systems relating

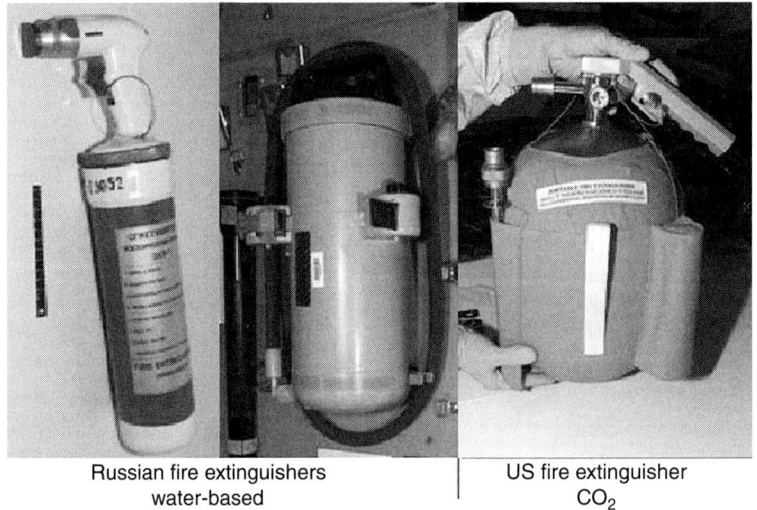

| Russian fire extinguishers water-based | US fire extinguisher CO_2 |

FIGURE 6.2 International Space Station fire extinguishers (Courtesy of NASA).

to atmospheric conditioning and control, trace contaminant control, toxic material removal, water quality, and waste management systems.

6.1 ATMOSPHERIC CONDITIONING AND CONTROL

The systems that provide atmospheric conditioning and control influence a number of critical functions on a spacecraft. Atmospheric conditioning and control systems maintain total and partial pressures for atmospheric gases to levels safe for crew inhabitance and hardware operation. Additionally, they maintain cabin temperature and humidity within a reasonable range so that the crew can function with no adverse effects to health, and they ensure the proper operation of systems, subsystems, and payloads.

6.1.1 Monitoring Is the Key to Control

Reliable control of the habitable environment of a spacecraft demands the use of dependable instruments to measure the various parameters of the cabin atmosphere. Instrumentation employed to monitor atmospheric composition must be inexpensive, simple, and lightweight. More important, it must provide robust performance to ensure an environment that promotes human safety and health and can be maintained with a high degree of confidence.

Monitoring instruments are used as well to provide for a high degree of autonomous control of the life support system to minimize the need for the crew or mission control personnel to intervene in its operation. This capability is necessary for long duration missions and those conducted distant from the Earth (Tatara and Perry 2004).

Atmospheric monitoring on a spacecraft entails the measurement of not only the major atmospheric constituents but also any trace contaminants in the atmosphere. Throughout the years, vehicle architecture and mission design have influenced the choice of atmospheric constituents and trace contaminants monitored during spacecraft operations.

Mercury and Gemini atmospheres consisted of 100% oxygen maintained at a pressure of 34.5 kPa (5 psia). The Apollo Program used a two-gas atmosphere of 60% oxygen and 40% nitrogen at 103 kPa (15 psia) during launch operations. While in flight, the total cabin pressure was allowed to decay to 34.5 kPa. Oxygen was used as the primary gas for breathing. It was also used to compensate for overboard leakage. In this regard, as the cabin pressure dropped, the concentration of oxygen in the total volume increased throughout the mission, eventually increasing to 100%. All Apollo missions carried a gas chromatograph that was used to measure the concentration of major atmospheric constituents and certain trace contaminants. Typically, nitrogen, oxygen, carbon dioxide, carbon monoxide, hydrogen, methane, ammonia, and water vapor were monitored (Diamant and Humphries 1990). For the Apollo missions, as well as for those of *Mercury* and *Gemini*, the total pressure of the cabin atmosphere and the carbon dioxide concentration were considered the most critical parameters to monitor.

Skylab also utilized a two-gas atmosphere at 34.5 kPa (5 psia). A mass spectrometer, which was onboard to support experiments conducted to evaluate human performance in space (Martin 1992), was used to monitor the major atmospheric constituents. Oxygen was maintained near a concentration of 72% by volume. Carbon dioxide and total pressure were monitored as well. Nitrogen was used to maintain the total pressure of the cabin.

The Space Shuttle was built to simulate earthlike environmental conditions in the crew cabin. The life support system is designed so that oxygen and nitrogen concentrations are maintained respectively at 21 and 79% and total cabin pressure is maintained near 101 kPa (14.7 psia). The partial pressures of oxygen, carbon dioxide, and water vapor are monitored, as is total atmospheric pressure.

Throughout the history of the Russian space program, monitors for oxygen and carbon dioxide have been employed on every vehicle from *Vostok* through to the *International Space Station*. *Soyuz* included the addition of a water vapor sensor. Sensors for hydrogen and carbon monoxide were included in the *Mir*, and they have been incorporated into the Russian segment of the *International Space Station* as well.

Similar to the *Space Shuttle*, the *International Space Station* maintains the total pressure of the cabin near 101 kPa, the nominal range being from 97.9 to 102.7 kPa (14.2 to 14.9 psia). Oxygen and nitrogen partial pressures are maintained near 21 and 79%, respectively. Higher percentages of oxygen are permitted in the small enclosed area of the U.S. Joint Airlock but only during extravehicular activity preparation.

A mass spectrometer called the *major constituent analyzer* is used in the U.S. segment to measure atmospheric concentrations of nitrogen, oxygen, carbon dioxide, hydrogen, methane, and water vapor. Russian provided gas monitoring equipment is used to monitor oxygen, carbon dioxide, hydrogen, carbon monoxide, and water vapor concentrations as well. The sensor used for carbon monoxide has yet to be proven accurate, as is the case when it was used on the *Mir*. These primary monitors of atmospheric constituents are supplemented by a variety of portable monitors designed to measure oxygen, carbon dioxide, carbon monoxide, combustion products, and the dew point. Many of these portable monitors are commercial off-the-shelf devices that have been modified for spacecraft application.

Many challenges are inherent to the utilization of measurement technologies during spaceflight. The instruments must survive the launch environment, be reliable, and function for long periods without being recalibrated. Although the size of monitoring equipment is an important consideration for all vehicles, in some cases, a portable monitor exhibits great advantages over one that is stationary. Even so, portable equipment requires the use of batteries, which can be a daunting resupply problem over the life of the *International Space Station*.

Today, sensors remain the Achilles' heel of atmospheric monitoring and control. There is no perfect sensor that can be applied to the wide range of requirements associated with monitoring the space environment. Depending on the application, various technologies from which a sensor can be selected are available. From these, an appropriate sensor for a particular application can be identified by keeping in mind the parameter it is intended to monitor and the capabilities and constraints associated with its use.

Many sensing devices used throughout the history of spaceflight are electrochemical in nature. Additionally, colorimetric tubes are employed for certain trending measurements and are beneficial for applications in off nominal situations. Through constant use during an off nominal event, an electrochemical sensor can be consumed, and as the chemical constituents are depleted, it will begin to underreport levels of the constituent being monitored. Because the colorimetric tube is exposed only at the time of monitoring, it provides an accurate measurement of the constituent. Even so, these devices provide only a single value for a point in time.

The major constituent analyzer used on the *International Space Station* suffers the effects of microgravity near the end of life of its ion pump. Although the major constituent analyzer has proven generally reliable, it can have upsets (Reysa et al. 2004). Except for the Russian segment, the major constituent analyzer is plumbed to monitor the various modules of the *International Space Station*. When disparities in data obtained from sampling different areas of the *International Space Station* are observed, a portable device must be used to investigate the conflict.

Atmospheric monitoring must be based on well-defined specifications and should be included as a major part of the system design trade studies. New technologies, such as solid state sensors and wireless systems, should be considered for application. Indeed, these can be a welcome benefit to resolving a difficult problem. By evaluating the historical aspects of spacecraft cabin atmospheric monitoring and control, it is apparent that data indicative of spacecraft health can be provided adequately by monitoring only a limited number of major atmosphere constituents (Tatara and Perry 2004).

6.1.2 Atmospheric Conditioning

Atmospheric temperature and humidity must be maintained within a specified range for the spacecraft environment to be habitable. Heat from the operation of equipment, along

6.1 Atmospheric Conditioning and Control

with heat and water vapor produced by the crew, must be managed to provide a comfortable and healthy environment within which to live and work. An environment that is too hot or too cold soon becomes uncomfortable to the crew and, over time, can cause detrimental effects to their health. The crew can become dehydrated if the environment is too dry. If the humidity is too high in the spacecraft cabin, condensation can occur in undesirable areas, causing hardware problems or microbiological safety and health issues.

Because natural air convection does not occur in the microgravity environment, forced ventilation is required to circulate the air to ensure atmospheric conditioning and control of contaminants. For these life support processes to be effective, it is essential to maintain a minimum air velocity within the habitation area of the vehicle. The maximum atmospheric velocity requirements for the face have been established by NASA medical doctors at 0.2 m/s to prevent drafts. In general, the temperature should be maintained between 18 and 27°C, with an associated relative humidity between 25 and 70% (Eckart 1996). All spacecraft flown to date have used either condensing heat exchangers or heat exchangers along with absorbents and cold plates to control cabin temperature and to remove humidity.

The *International Space Station* is required to maintain temperature within a range from 18 to 28°C and relative humidity between 30 and 70%. Additionally, the dew point within the *International Space Station* is required to be within a range of 4.4 to 15.6°C. Figure 6.3 shows the average temperature and the dew point on the *International Space Station* from March 2001 to May 2007. In this graph, cabin temperature is represented by the upper line and dew point by the lower one.

Ventilation

Successful maintenance of spacecraft cabin atmosphere is dependent on forced ventilation, because passive air convection is not a natural process on orbit. Adequate mixing

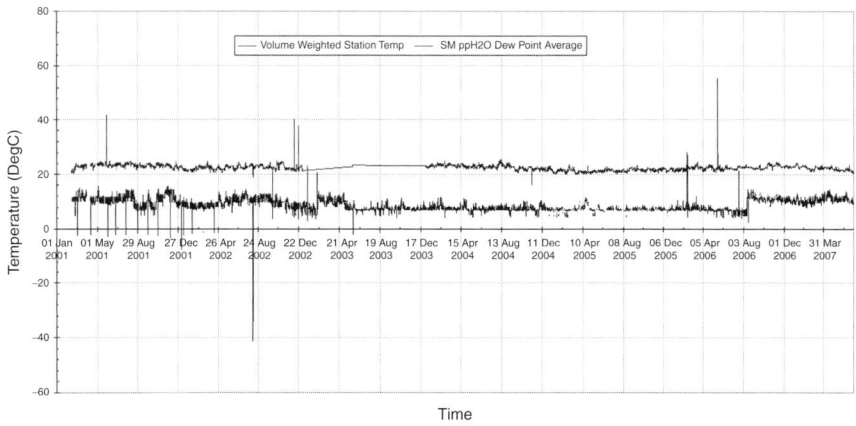

FIGURE 6.3 Average temperatures and dew points of the *International Space Station* from 2001 to 2007 (Courtesy of NASA).

of the atmosphere is essential to the ability of the life support system to prevent thermal gradients, distribute gases being added to the environment, and remove contaminants. For these reasons and for crew comfort, there must be a detectable atmospheric flow; however, the velocity should not be so great that the flow would be considered a draft. The *International Space Station* is required to maintain the circulated atmospheric velocity within the range from 0.05 to 0.20 m/s to meet this need (NASA 2001).

When designing ventilation systems, all maintenance configurations for the spacecraft should be analyzed to ensure that the environment of the work area complies with the required atmospheric quality. During the early assembly of the *International Space Station*, the crew of Flight 2A.1 experienced headaches and light headedness. These symptoms, which were attributed to poor air quality, were most severe after the crew had been conducting maintenance operations for many hours in the functional cargo block, a Russian module, when certain panels were open (Reuter 2000). The results of air samples taken on orbit were nominal. It finally was concluded that the open panels within the functional cargo block short circuited the ventilation. Based on this conclusion, corrective measures were put in place. Today, the crew must use personal fans when working within the functional cargo block when these panels are open. Limits were placed on the amount of time a crew can use to conduct maintenance operations on the Russian segment. Finally, portable equipment was built to provide an easy means to assess the environment by monitoring carbon dioxide concentration and air velocity in an area where the crew is working and is now onboard the *International Space Station*.

On the *International Space Station*, a number of flexible ducts are in use. On a few occasions, these ducts have collapsed, presumably because of age and the constant and repetitive manipulation and stretching to which they had been subjected (Williams and Gentry 2006). When a collapse occurs, ventilation is affected adversely by the obstruction. If the blockage is severe, contaminants can build up in the atmosphere of the affected module or section. Flexible ducting is simple to use in a spacecraft of modular design and can be used easily when vehicles are docked to a module. Even so, the design of these ducts and the materials of which they are constructed should be sufficiently robust to prevent collapse during frequent use and manipulation on orbit. To ensure safety during their use, they should be compact enough to be stowed until needed, sufficiently flexible that they can be routed easily, and adequately rigid that they do not collapse when in place.

The *International Space Station* has no closets. This design deficit has created a problem for ground engineers needing to place large amounts of stowage on orbit. Although there are a limited number of areas located behind panels where hardware can be stowed, other locations on the *International Space Station*, which were never intended to be used for this purpose, have become permanent stowage areas. On the *International Space Station*, there are modules having one or more walls completely covered by stowage. The pressurized mating adapter, a tunnel connecting the U.S. and Russian segments, is lined with stowage bags (Figure 6.4). The water wall, that is, a wall of bags filled with stores of water, is located in the hatch area of *Node-1* (Figure 6.5). Figure 6.6 is a photograph taken from another vantage point of the equipment stowed in Node-1. The entire floor of the *Functional Cargo Block* is covered with stowage (Figure 6.7), leaving only the upper half of the aisle available for traversing the module.

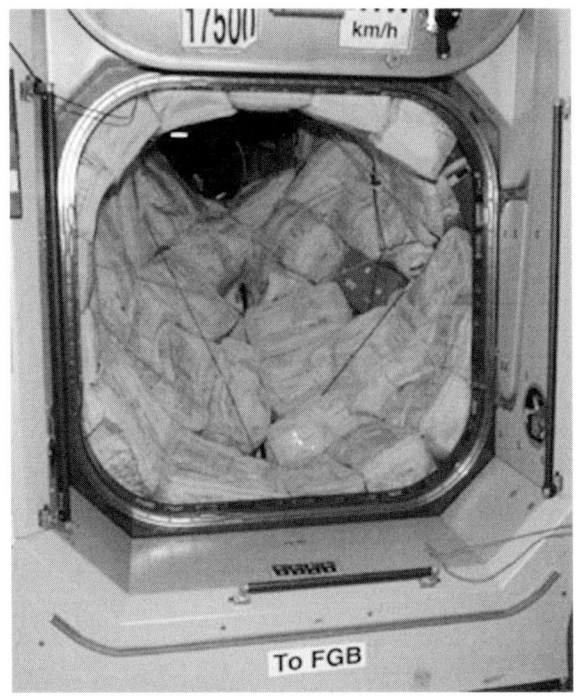

FIGURE 6.4 Stowage bags in the International Space Station pressurized mating adapter (Courtesy of NASA).

FIGURE 6.5 Stowage of water bags in the hatch area of *Node-1* of the *International Space Station* (Courtesy of NASA).

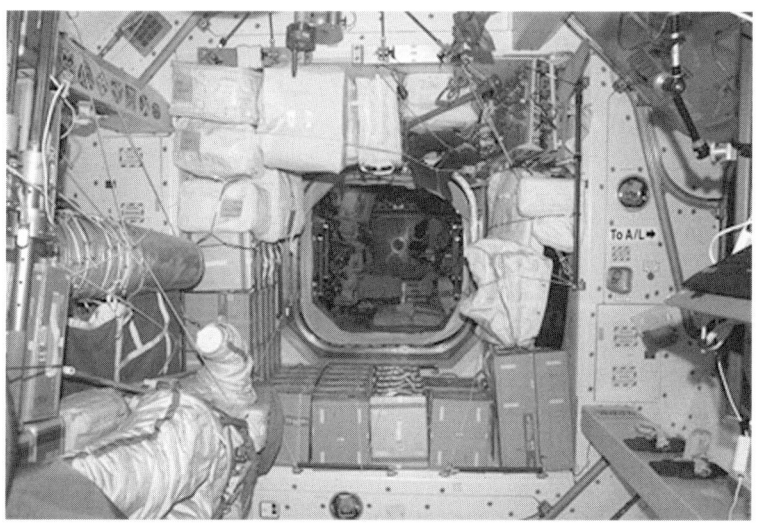

FIGURE 6.6 International Space Station stowage in *Node-1* looking forward into the *U.S. Laboratory Module* (Courtesy of NASA).

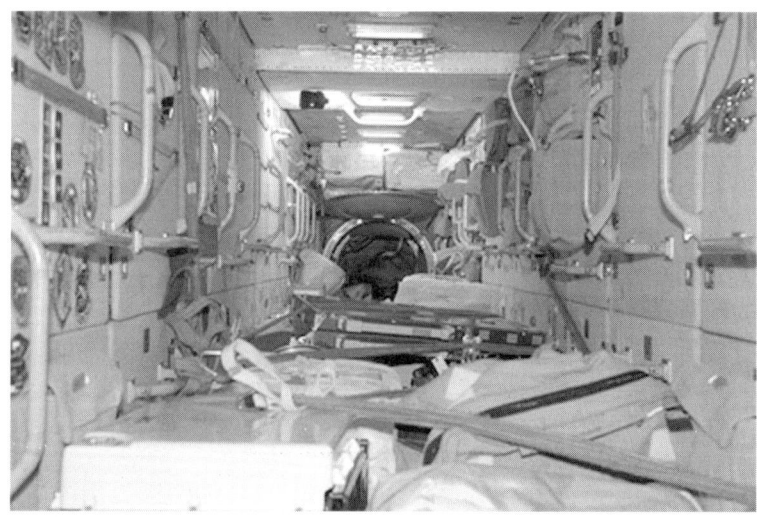

FIGURE 6.7 International Space Station stowage on the *Functional Cargo Block* floor (Courtesy of NASA).

The crew has been advised not to stow items in certain locations because to do so would affect the function of life support systems. These areas generally are located around air intakes and diffusers, smoke detectors, emergency equipment, and the like. Ground engineers constantly remind the crew that these areas should be avoided for use as stowage and these areas should be unblocked when stowage has been placed in violation of the restrictions. The amount of stowage on the *International Space Station* is nearing a critical level at the time of this writing. It is only a matter of time before the number of additional items stowed within the habitable volume of *International Space Station* has an adverse impact on life support systems, such as short circuiting the ventilation. It is important that ground engineers have sufficient functional knowledge of the operational scenarios of a mission to be able to avoid everyday activities from affecting designed performance.

The initial design of spacecraft for future long duration missions should include adequate areas for stowage. The problems with which ground engineers presently are coping, at least in part, would have been alleviated if adequate stowage volume had been included in the *International Space Station* design. By providing for adequate stowage accommodation in the spacecraft design and preparation for a long duration mission, design engineers can avoid ventilation and other problems caused by interference with the operation of the life support systems, thus minimizing risk to crew health and safety.

Dust in the Wind

Other environmental parameters can provide insight into the health of the ventilation system. A difference in the partial pressures of carbon dioxide measured in the U.S. and Russian segments was observed to increase slowly over time from late 2002 to early 2003. An on-orbit test of intermodule ventilation from the Russian segment to the U.S. segment was performed using a device flown after the Flight 2A.1 ventilation issue. The test indicated that a flow obstruction existed. After a considerable amount of lint was removed from the intermodule ventilation fan flow straighteners (Figure 6.8), flow and mixing of the atmosphere returned to nominal (Williams, Lewis, and Gentry 2003). The air flowing to this fan originates in the Russian segment. It is not filtered to the fine particulate level that is standard for the U.S. segment. Through an oversight in the integrated design of the system, there is no requirement for the air supplied to this fan from the Russian segment to be filtered to the same gradient as specified for the U.S. segment. An operational solution was implemented in which maintenance on this fan is provided when needed.

Smoke detectors on the *International Space Station* have elicited false alarms when cargo is moved around or during maintenance performed on a rack (Williams and Gentry 2004). The Node-1 smoke detector, which is most readily accessible to the cabin, accumulates dust rapidly. The accumulation of dust usually increases during missions in which cargo is transferred through Node-1 from the mini-pressurized logistics module. Figure 6.9 shows the accumulated dust on the Node-1 smoke detector, which was returned to Earth for analysis and testing.

As shown by these examples, dust is a continuing issue onboard the *International Space Station*. There is little doubt that this is a problem with which to contend during future lunar and other planetary missions.

FIGURE 6.8 Lint removed from the U.S. intermodule ventilation fan (Courtesy of NASA).

FIGURE 6.9 Dust found on International Space Station Node-1 smoke detector during postflight ground inspection (Courtesy of NASA).

6.1.3 Carbon Dioxide Removal

Each human produces approximately 1 kg of carbon dioxide per day (Eckart 1996). The carbon dioxide produced by the crew and any animals present must be removed from the habitable environment of a spacecraft to prevent the concentration of the gas in the atmosphere from increasing to a toxic level. Spacecraft rely on the use of a scrubber system to remove excess carbon dioxide from the atmosphere. During the history of spaceflight, only a few types of scrubber systems have been used. These systems rely on various removal techniques that employ different architectures and media: permeable membranes, liquid amine, adsorbents, and absorbents.

Sorbent systems have been used for carbon dioxide removal since the first crewed missions. Early sorbent systems were non-regenerative and discarded once spent. Although bulky and heavy, they were an acceptable solution for short duration missions. However, as mission duration increased over time, it became cost prohibitive to employ these carbon dioxide removal systems uniquely. As testament to their effectiveness, canisters of lithium hydroxide were used during the early U.S. space programs, and they continue to be used today on the *Space Shuttle*. Russian vehicles also used lithium hydroxide canisters beginning with the Soyuz missions (Eckart 1996). Even though different primary scrubber systems are used in the U.S. and Russian segments of the *International Space Station*, the sorbent system is used as a backup when they fail or require maintenance. Regardless of the vehicle employing the sorbent system, a proper rate of airflow is essential for the lithium hydroxide to be utilized in the most effective manner.

A molecular sieve was used to remove carbon dioxide from the atmosphere of *Skylab*, and a similar, albeit improved design of this system is for use on the *International Space Station* (El Sherif and Knox 2005). Early Russian scrubber systems functioned by reacting carbon dioxide with potassium hydroxide in the oxygen regenerator to form potassium carbonate and water. Used on the *International Space Station*, and previously on *Mir*, the Vozdukh is an adsorption type system that is regenerated on orbit (Diamant 1990).

More advanced systems have been developed that reuse the sorbent by regenerating the material by means of heat, pressure, or a combination of the two. The International Space Station carbon dioxide removal assembly, a system with a four-bed molecular sieve, uses heat and vacuum to regenerate the zeolite sorbent material. This system selectively removes carbon dioxide from the cabin atmosphere and is designed to discharge the gas overboard, that is, an open loop operation, or route it for downstream reduction, that is, a *closed loop* operation (El Sherif and Knox 2005). Although the carbon dioxide reduction unit is no longer in the International Space Station baseline, it might be launched as a payload to test and further the technology. Methods that can be employed for carbon dioxide reduction can be found in Eckart (1996).

During its use on *Skylab*, the molecular sieve system was found to be sensitive to water, in that water adsorbed onto the molecular sieve reduces the rate of carbon dioxide adsorption. The carbon dioxide removal assembly design implements desiccants on the inlet and outlet, and the water vapor adsorbed during the previous half cycle is returned to the cabin. Even so, the zeolite remains sensitive to water breakthrough.

Other methods, such as electrochemical depolarization concentration, air polarized concentration, membrane removal, and other regenerative technologies for removing carbon dioxide from the atmosphere can be applied to spaceflight. Although listed in Eckart (1996), these so far have not been the best choices to use on spacecraft. They could be considered more favorably for this application if the technologies advance sufficiently.

Safety issues associated with regenerable carbon dioxide removal systems are the high temperatures required for these units to bake out the carbon dioxide to vacuum and the vacuum interface. For the latter, two valves to vacuum are utilized, and each must fail closed to be able to isolate the cabin from space in case of a failure.

The medical community currently is reviewing the established level of carbon dioxide in the spacecraft environment that should be maintained when on orbit. Current International Space Station rules state that the partial pressure of carbon dioxide should be maintained below 5.3 mmHg. Data being collected from the International Space Station crews might lead to a new acceptable level of carbon dioxide in the cabin atmosphere that future systems might be required to support.

6.2 TRACE CONTAMINANT CONTROL

It is understood that spacecraft cabin atmospheric composition, temperature, pressure, and humidity are important parameters that must be addressed to ensure human health and safety, as well as mission success. Therefore, it is natural to expect that considerable attention be applied toward controlling these parameters during all phases of cabin design for a crewed spacecraft. Subtle design and in-flight operational choices directly influence the trace constituent content of the cabin atmosphere. Commonly referred to as *trace contaminants*, these atmospheric components have profound effects on habitability, and they must be considered at all vehicle design and operational phases.

6.2.1 Of Tight Buildings and Spacecraft Cabins

Indoor air quality is a pervasive issue having a variety of consequences if not properly maintained (Hines et al. 1993). A spacecraft cabin possesses similarities to a terrestrial tight building in an extreme sense (Limero et al. 1990). Very low atmospheric gas leakage and replacement rates, a small volume to occupant ratio, and widespread use of nonmetallic materials throughout the cabin accelerate the rate at which airborne chemical and particulate matter contamination accumulates. Unless a means for purifying the atmosphere is provided, crewmembers can experience a number of acute or chronic health effects, commonly and collectively known as *sick building syndrome*. The symptoms of sick building syndrome, as defined by the World Health Organization, include nasal and eye irritation, dry mucous membranes, fever, joint and muscle pain, lethargy, nosebleed, dry skin or skin rash, and headache (Molhave 1992). Symptoms can be present alone or in combination. Although people suffering from sick building syndrome symptoms in a terrestrial setting usually recover once exposure to the contamination ends, it can recur or even grow worse with subsequent exposure (Block 1991). The situation aboard a spacecraft is no different, except that there is no way for crewmembers to leave their environment during a mission to seek relief. Designing for good cabin atmospheric quality is imperative because the consequences place at risk crew health and mission success.

Trace contaminants in the cabin atmosphere of a spacecraft can be found in the form of chemical vapors, solid aerosols, and liquid aerosols. Chemical vapors from equipment off gassing and human metabolism are the most prevalent trace contaminant form. Solid aerosols originating from crewmembers and their in-flight activities, as well as dislodged foreign object debris from vehicle manufacturing and ground processing are the typical forms of particulate matter encountered. Sources can be localized and easy to characterize or distributed and quite difficult to portray. Human generated chemical and particulate matter on top of equipment generated contaminant loads merely adds complexity and unpredictable variability to the design and operational challenges of the vehicle as they relate to atmospheric quality control. For this reason, studying cabin atmospheric quality data acquired from operational crewed spacecraft and space stations is essential to specifying effective design and operational solutions.

The atmosphere of a spacecraft cabin, such as that found in the *International Space Station*, typically contains approximately 34 mg/m^3 of trace chemical contaminants. Figures 6.10 and 6.11 show relative percentages of trace chemical contaminants typically observed onboard the *International Space Station*, with and without methane included. Examining Figure 6.10, one finds that methane, accounts for nearly two thirds of the total trace chemical contaminant concentration onboard the *International Space Station*, whereas non-methane volatile organic compounds, representing aldehyde, alcohol, aromatic, halocarbon, ester, and ketone functional classes, account for approximately one quarter of the observed contamination load. Hydrogen, carbon monoxide, and polymethylcyclosiloxanes account for the remaining trace gaseous contaminants. When non-methane volatile organic compounds are considered alone, Figure 6.11 shows that the alcohol class dominates. This differs from terrestrial indoor air quality, in which the aromatic class dominates, formaldehyde being the single most prevalent pollutant (Hines et al. 1993).

Between the mid-1980s and early 1990s, indoor air quality researchers studied perceived air quality associated with non-methane volatile organic compound mixtures up to 25 mg/m^3 (Molhave, Bach, and Pedersen 1986; Otto et al. 1992). The non-methane

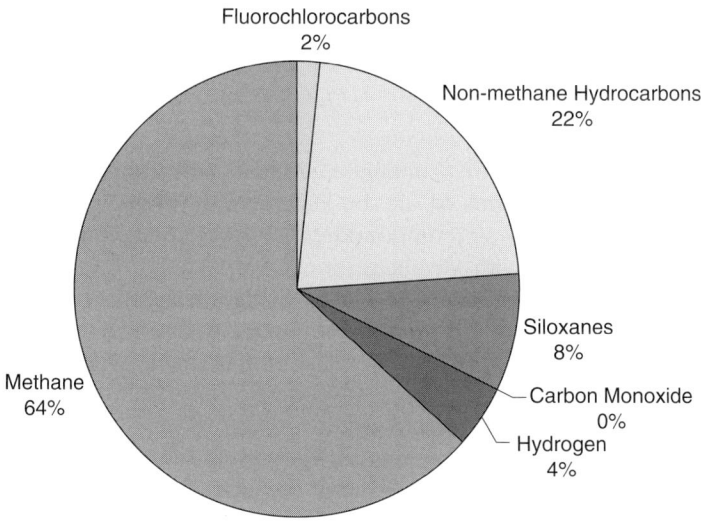

FIGURE 6.10 Total trace atmospheric component incidence aboard *International Space Station* (Graphic adapted from Perry and Peterson 2003; Courtesy of NASA).

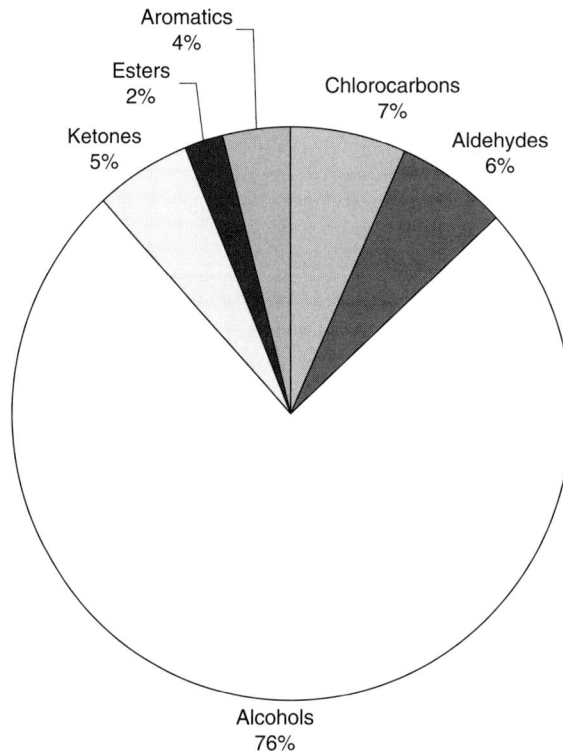

FIGURES 6.11 Non-methane volatile organic compounds distribution by functional class aboard the *International Space Station*.

volatile organic compound mixtures contained 22 chemical compounds representative of nonindustrial indoor air. Incidence of sick building syndrome symptoms correlated with total non-methane volatile organic compound concentration. Interestingly, this research found a range of sensitivity with strong responses reported at 5 mg/m^3 as well as at 25 mg/m^3. Spaceflight toxicologists use 25 mg/m^3 as a guideline for the total non-methane volatile organic compound limit in combination with the spacecraft maximum allowable concentration limits for individual chemical contaminants. Figure 6.10 shows that the non-methane volatile organic compound concentration onboard the *International Space Station* averages approximately 8.5 mg/m^3.

Particulate matter concentration in crewed spacecraft cabin atmospheres very rarely has been characterized. An in-flight evaluation of suspended particulate matter conducted in 1990 during the STS-32 mission found 56.4 μg/m^3 in the cabin atmosphere, with a bimodal distribution across particle aerodynamic diameter ranges of interest (Liu et al. 1991). The STS-32 study found 4% of the particulate matter was <2.5 μm, 33% was between 2.5 and 10 μm, 10% was between 10 and 100 μm, and 53% was >100 μm.

For human health, the fraction <10 µm in aerodynamic diameter, which comprises respirable particulates, is of most interest (Hines et al. 1993). The particle mass concentration measured during STS-32 was observed to be lower than that of average terrestrial indoor air for the fraction <2.5 µm but between approximately 3 and 16 times higher than indoor air for the larger size fractions. The lack of sedimentation due to the free fall conditions during flight permits larger particulate matter to remain suspended in the cabin atmosphere. A study conducted in 2005 to characterize the particulate matter background onboard the *International Space Station* to facilitate improved smoke detector design found a much different situation. In this case, measurements from a light scattering instrument reported the particulate matter concentration to be <5 µg/m^3 (Urban et al. 2005). The primary difference between the Space Shuttle and the *International Space Station* are crew size and cabin filtration level. The Space Shuttle normally has a crew of seven and employs filters rated at 280 µm, whereas the *International Space Station* typically has a crew of three and employs high efficiency particulate air filters rated to 99.97% efficient for 0.3-µm particulate matter. This result indicates that the combination of a smaller crew size combined with more effective filtration yields a lower concentration of suspended particulate matter.

Beyond human generated particulate matter lies a myriad of questions about surface dust intrusion during lunar and Mars exploration missions (Agui 2007). Handling the surface dust intrusion load, which overlays the human generated load, presents challenges to atmospheric quality control system designers. Characterizing dust, particularly in the <10 µm range (which presents the greatest risk to human health), is most important for ensuring acceptable cabin atmospheric quality in surface habitats. Whereas the particulate chemical composition plays a role in human health effects, spacecraft designers and operations personnel need to consider only chemical composition when evaluating the environmental effects that liquid aerosols present. This is expected to change as lunar and Mars surface exploration proceeds and knowledge about surface dust properties expands.

6.2.2 Trace Contaminant Control Methodology

Good cabin atmospheric quality is not an accident but the direct result of sound design practice during all phases of cabin design for a spacecraft. For a stable, safe cabin environment to exist, contamination generation sources and removal processes (Figure 6.12) must be in balance (Perry 1998). This balance can be upset at any time by an increase in any of the generation sources. Sources are numerous and vary in frequency, duration, and magnitude.

A disciplined approach to trace contaminant control reduces the likelihood of an imbalance between sources and removal processes. Such an approach considers the factors that contribute to spacecraft cabin atmospheric quality, defines passive methods for minimizing contamination source magnitude, selects active control means for in-flight operations, identifies potential hazards that can have environmental and equipment impact, and determines the design margin magnitude. The following discussion defines these elements of contamination control methodology.

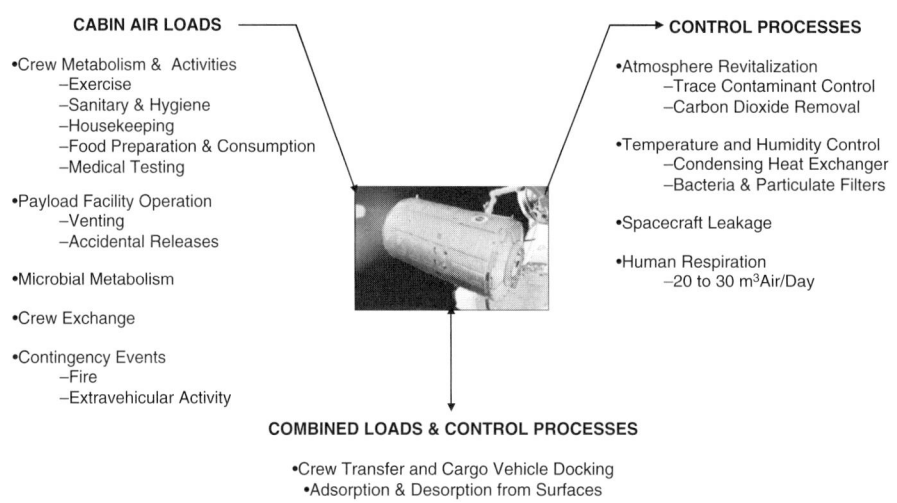

CABIN AIR LOADS

- Crew Metabolism & Activities
  - Exercise
  - Sanitary & Hygiene
  - Housekeeping
  - Food Preparation & Consumption
  - Medical Testing

- Payload Facility Operation
  - Venting
  - Accidental Releases

- Microbial Metabolism

- Crew Exchange

- Contingency Events
  - Fire
  - Extravehicular Activity

CONTROL PROCESSES

- Atmosphere Revitalization
  - Trace Contaminant Control
  - Carbon Dioxide Removal

- Temperature and Humidity Control
  - Condensing Heat Exchanger
  - Bacteria & Particulate Filters

- Spacecraft Leakage

- Human Respiration
  - 20 to 30 m^3Air/Day

COMBINED LOADS & CONTROL PROCESSES

- Crew Transfer and Cargo Vehicle Docking
- Adsorption & Desorption from Surfaces

FIGURE 6.12 Factors affecting spacecraft cabin atmospheric quality (Graphic adapted from Perry 1998; courtesy of NASA).

Factors Contributing to Atmospheric Quality

Attention to contamination control when selecting materials of construction, processing fluids, and when performing manufacturing, ground processing, and in-flight operations (among many others) is a key component to overall trace contaminant control. No single factor can be safely assumed to be exclusive of others. This is illustrated by Figure 6.13, which emphasizes that good cabin atmospheric quality is a target reached while working within a boundary defined by the atmospheric quality standards (Perry 1998). Atmospheric quality standards are set first and foremost to ensure the health and safety of the crew. Crewmembers suffering from sick building syndrome symptoms cannot execute mission tasks safely, so atmospheric quality standards, known as *spacecraft maximum allowable concentrations*, are set by industrial hygiene and toxicology experts such that the most sensitive individual can experience only minimal effects during continuous exposure. Note that aspects of the space mission operational concept, vehicle design, and manufacturing processes contribute to the end product, that is, a safe, high quality atmosphere.

Passive Control Methods

Passive control methods address what can be done during spacecraft design specification, mission operations concept development, manufacturing, and in-flight operations to limit contamination sources and their magnitude. Spacecraft and mission designers implement passive controls on the ground before the vehicle is launched to lessen the contamination load in the cabin. An added benefit is that a smaller contamination load leads to an increased operational margin for cabin atmospheric purification equipment or even to smaller and lighter equipment, depending on the outcome of trade study assessments as design of the spacecraft progresses.

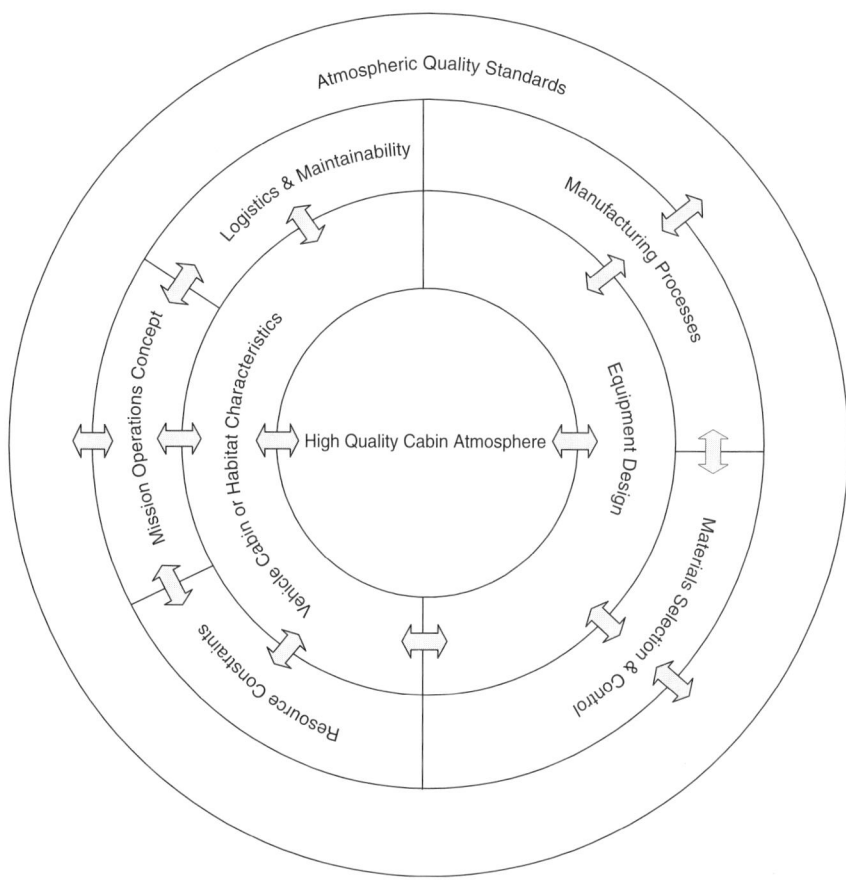

FIGURE 6.13 Interaction of factors contributing to atmospheric quality (Graphic adapted from Perry 1998; courtesy of NASA).

Selecting materials of construction that produce very low amounts of offgassed products is the cornerstone of passive control. By selecting materials with low offgassing characteristics, the total load of non-methane volatile organic compounds is reduced to low levels. Offgassing data collected from bulk materials and assembled components allow the vehicle contamination load to be defined more rigorously and ultimately serve as the design basis for active contamination control equipment deployed onboard the vehicle. In addition, the physical containment of chemicals used in onboard systems and payload facilities, combined with the application of appropriate operational protocols that address in-flight maintenance and handling, provide the primary means of control for accidental releases. Designing systems and payload equipment to limit fugitive emissions from fluid couplings is also an important consideration.

Material selection and control plays an important role in limiting particulate matter generation within a spacecraft cabin. Identifying frangible materials and avoiding or limiting their use as materials of construction considerably reduces particulate matter generation sources. As well, identifying dust generating in-flight operations and providing a means to contain the work area benefits the cabin environment.

Conducting all manufacturing activities for spacecraft equipment according to a rigorous contamination control plan to limit and remove any foreign object debris from the spacecraft during cabin manufacturing, outfitting, and preflight processing is another important passive contamination control method. Without a contamination control plan in place during manufacturing, foreign object debris can become trapped in crevices and inaccessible areas within the cabin. Trapped foreign object debris becomes dislodged by vibration during launch and floats freely in the cabin atmosphere. In this state, it becomes a safety hazard to the crew and vehicle systems.

Active Control Methods

Because people live and work in the spacecraft cabin and no material is truly zero offgassing, some means for preventing contaminant buildup to undesirable concentrations must be a part of the overall contamination control approach. Equipment offgassing and human metabolic processes are continuous and thus form a basis for design. All other sources overlay this basis and must be subjected to risk assessments to determine if they warrant inclusion into the design specification for active contamination control equipment. Typically, transient loads serve to define design margins. Table 6.1 provides a recommended design load model based on equipment offgassing and human metabolic sources (Perry 1995, 1998). Typical design margins for contamination control equipment established for upset conditions and that contribute to design robustness can range from 25 to 50%, minimum.

Active contamination control equipment is simple in design, with most approaches employing a packed bed of granular activated carbon as the primary active medium. The granular activated carbon, which removes non-methane volatile organic compounds by physical adsorption, can be supplemented with specialized treatments to target ammonia or formaldehyde. Such treatments irreversibly react with the target compound. Stages downstream of the granular activated carbon bed can include an ambient temperature catalyst, for converting carbon monoxide to carbon dioxide, or a thermal catalytic oxidation reactor that provides broader control for longer duration missions. The catalyst is usually a platinum group metal supported on an inert substrate that can be a monolith, pellet, or granule. When thermal catalytic oxidation is used, attention must be given to post-processing, because the potential exists to convert some chemical contaminants, particularly halocarbons, to acidic oxidation products, such as hydrogen chloride and hydrogen fluoride. Placing the processing stages in an order that prevents or minimizes the incidence of reactive hazards within the contamination control and other cabin atmospheric purification equipment requires close attention during the design and verification stages of a spacecraft.

Two active trace contaminant control units purify the cabin atmosphere onboard the *International Space Station*. One, located in the *U.S. Laboratory Module, Destiny*, and

Table 6.1 Trace Chemical Contaminant Load Model for Cabin Air Quality Maintenance (Courtesy of NASA)

Contaminant	Maximum Concentration (mg/m^3)[a]	Generation Rate		
		Off Gassing (mg/d/kg)	Metabolic (mg/d/person)	System (mg/d)
Methanol	9	1.3×10^{-3}	0.9	0
Ethanol	2000	7.8×10^{-3}	4.3	1000
n-Butanol	80	4.7×10^{-3}	0.5	0
Methanol (formaldehyde)	0.1[c]	4.4×10^{-6}	0.4	0
Ethanol (acetaldehyde)	4	1.1×10^{-4}	0.6	0
Benzene	1.5	2.5×10^{-5}	2.2	0
Methylbenzene (toluene)	30[c]	2×10^{-3}	0.6	0
Dimethylbenzenes (xylenes)	220[c]	3.7×10^{-3}	0.2	0
Furan	0.07	1.8×10^{-6}	0.3	0
Dichloromethane	50	2.2×10^{-3}	0.09	0
2-Propanone (acetone)	52	3.6×10^{-3}	19	0
Trimethylsilanol	3.7[c]	1.7×10^{-4}	0	0
Hexamethylcyclotrisiloxane	90	1.7×10^{-4}	0	0
Ammonia	5[c]	8.5×10^{-5}	90[c]	175[c]
Carbon monoxide	11[c]	2×10^{-3}	18	0
Hydrogen[b]	340	5.9×10^{-6}	42	0
Methane[b]	3800	6.4×10^{-4}	329	0

[a] 7-d SMAC from JSC 20584, dated June 1999.
[b] For mission operations concepts >1-mo duration.
[c] Under review by NASA.

known as the *trace contaminant control subassembly*, employs physical adsorption, thermal catalytic oxidation, and posttreatment of the catalytic oxidation effluent (Weiland 1998a). Figure 6.14 shows this unit mounted in the atmosphere revitalization system rack and a process flow diagram. The second unit, located in the Russian module, *Zvezda*, and known as the *block for micropurification*, employs physical adsorption and ambient temperature carbon monoxide oxidation. It differs from the U.S. trace contaminant control subassembly by using a vacuum swing process to regenerate its activated carbon beds. Figure 6.15 shows a block for micropurification process flow diagram (Weiland 1998b;

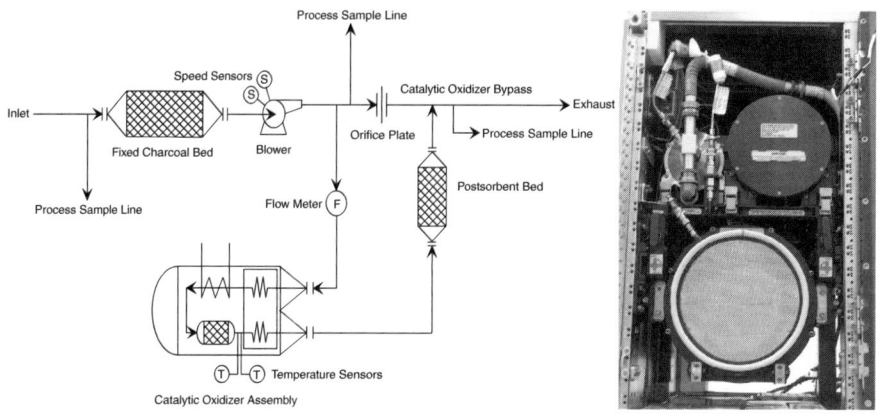

FIGURE 6.14 Trace contaminant control subassembly (Schematic: Perry et al. 1998; photo: Perry, Carrasquillo, and Harris 2006) aboard the International Space Station (Courtesy of NASA).

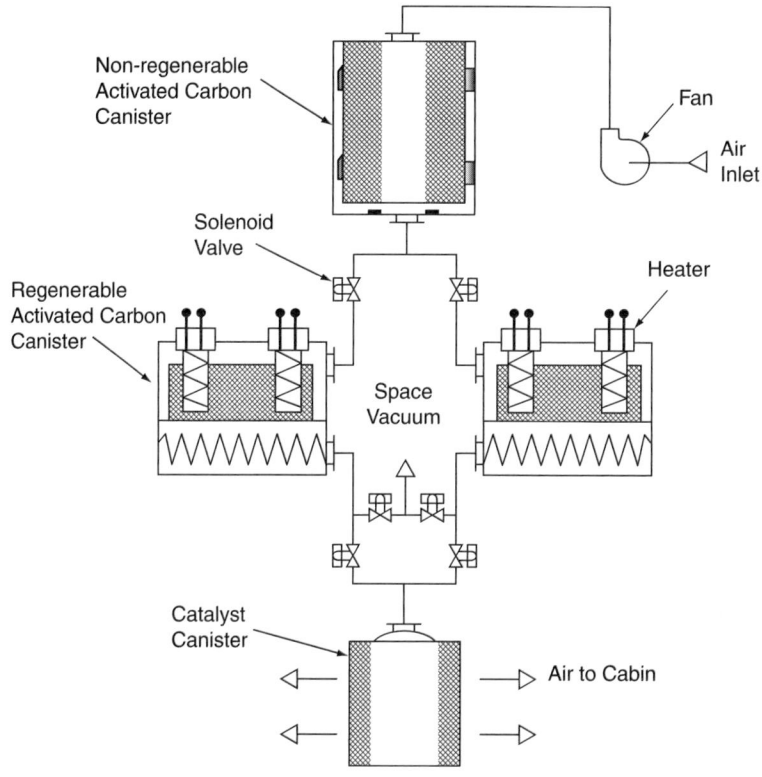

FIGURE 6.15 Russian trace contaminant control equipment process diagram (Curtis, Perry, and Abramov 1997) (Courtesy of NASA).

Mitchell et al. 1994). Russian designers added a supplemental thermal catalytic oxidation reactor to the block for micropurification after several years in service onboard the *International Space Station*. This supplemental reactor targets methane.

Filters of varying efficiency remove particulate matter from the cabin atmosphere. Early spacecraft employed debris filters rated to 280 μm nominal and 300 μm absolute. The Space Shuttle provides for supplemental filtration to 40 μm. Whereas these filtration approaches remove most of the particulate matter that can foul heat exchangers or damage fans, they do not address the particulate matter fraction <10 μm. The *International Space Station* employs HEPA filtration, and future spacecraft are likely to employ HEPA filtration at a minimum to meet the increasingly stringent and challenging particulate matter control needs for lunar and Mars surface exploration. Figure 6.16 shows a filter element used onboard the *International Space Station* and some typical debris found in the cabin atmosphere. The crew generates most of the debris.

Guidelines for particulate matter control are <0.2 mg/m^3 (24-h average) for particulate matter between 0.5 and 100 μm in aerodynamic diameter, with a ceiling of 2 mg/m^3 (Perry 1998). Lunar dust concentration guidelines are to be lower and present a design and environmental quality verification challenge.

FIGURE 6.16 International Space Station cabin particulate filter (top) and typical debris (bottom) (Courtesy of NASA).

Equipment and Environmental Impact Assessment

System, payload, and housekeeping fluids and chemicals present potential hazards to the cabin environment of a spacecraft. Although materials selection and control limits chemical emissions into the cabin atmosphere, the potential for hazards from fluid and chemical leaks, spills, and residues remain. System and payload safety reviews play an important role by assuring that proper use and containment protocols are in place. Operational guidelines and rules for crew and mission controllers to follow are established from these containment protocols. These protocols address various aspects of the impact of a fluid or chemical on the cabin environment and their compatibility with the life support system of the spacecraft. Impact assessments evaluate the potential to degrade life support system function, how easily and rapidly the cabin atmosphere can be restored to its original state, and the magnitude of in-flight resources necessary to restore a safe cabin environment. Figure 6.17 summarizes the basic information flow and the areas for investigation by life support system discipline experts. Toxicological, flammability, and biohazard characteristics also are evaluated by discipline experts.

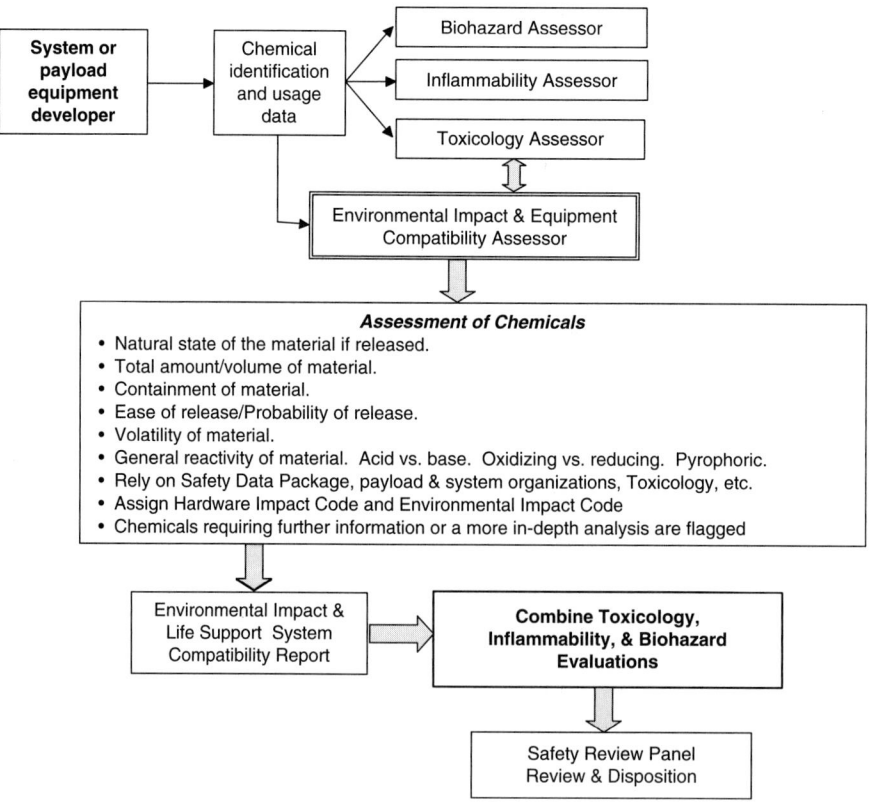

FIGURE 6.17 Simplified environmental impact assessment process (Courtesy of NASA).

Life support system discipline evaluation consists of two parts and uses information collected by the process depicted by Figure 6.17. The first part addresses the potential for a leaked fluid or chemical to degrade the functional performance of the life support system. Emphasis is on effects to the atmospheric and water purification equipment. The latter impact is more important aboard spacecraft or in space habitats that recycle humidity condensate for human consumption.

The evaluation also investigates the potential for the fluid or chemical to react or decompose under process conditions found in the life support system equipment. Reaction or decomposition to more hazardous products increases the safety hazard of the fluid or chemical. It is very possible for a fluid or chemical that might be considered benign in its own right to react to form a more hazardous product. Such reactive and decomposition hazards must be identified and avoided.

The second part assesses the capability and suitability of life support system processes to remove contaminants, including hazardous decomposition and reaction products, from the cabin atmosphere. Areas considered in this part of the assessment include impacts on the operational margin and use of expendable resources. Impacts to both operational margin and expendable resources adversely affect environmental suitability and can lead to an unsafe situation onboard the spacecraft. Life support system experts use both engineering analysis and testing to address this part of the assessment. When combined with toxicological, flammability, and biohazard assessments, the composite evaluation serves as the basis for establishing containment, handling, and spill remediation protocols.

6.2.3 Trace Contaminant Control Design Considerations

Life support system designers must consider a number of factors that can affect cabin atmospheric quality. Ventilation rate and quality, accumulation of low level sources that material screening might not adequately identify, quiescent periods with no ventilation, environmental upsets from fire or chemical leaks, and cabin temperature and humidity extremes are among the factors that influence cabin atmospheric quality. For example, the accumulation of low level formaldehyde generation sources, combined with degraded ventilation flow between modules onboard the *International Space Station* caused increased cabin concentrations that briefly peaked above the allowable concentration in the *U.S. Laboratory, Destiny* (Perry 2005). Figure 6.18 shows the increasing gradient between *Destiny* and the Russian service module, *Zvezda*. Investigation found that lint accumulation over time on a ventilation fan flow straightener had reduced severely the exchange of atmosphere between the modules (Williams et al. 2003). Cargo delivery to the station was very active at that time. The gradient moderated after removing the lint from the ventilation fan. Later, this situation repeated but to a lesser degree, indicating that the formaldehyde source had moderated over time.

Ventilation upsets during maintenance operations can cause locally degraded atmospheric quality. For example, during early *International Space Station* assembly, a crewmember opened an access panel to work on equipment. Opening the panel disrupted the ventilation flow. During the maintenance operation, the local atmospheric quality degraded and caused the crewmember to feel ill (Dunaway et al. 2000; Reuter 2000).

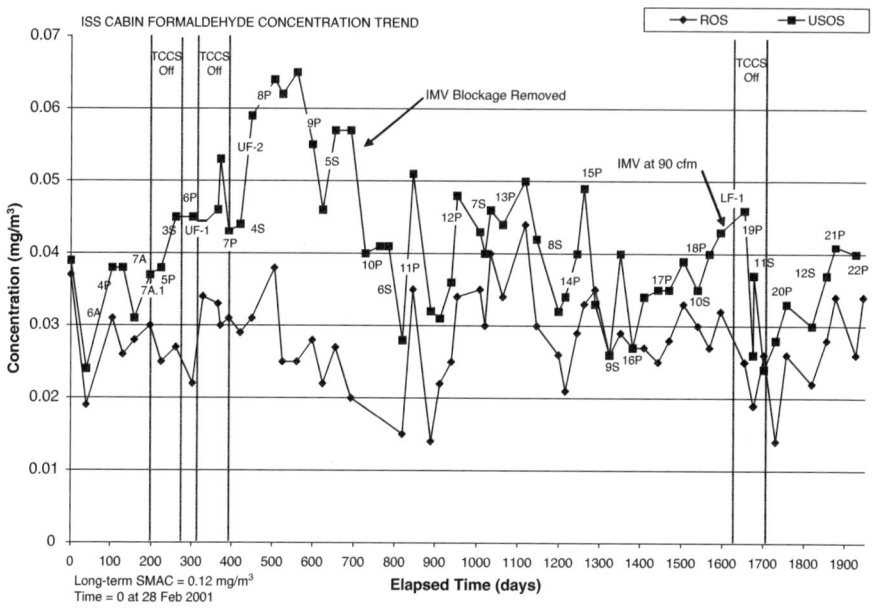

FIGURE 6.18 Formaldehyde concentration gradient observed aboard the *International Space Station* (Perry 2005) (Courtesy of NASA).

Setting up local ventilation fans during maintenance operations helps prevent locally degraded atmospheric quality.

Trace contaminant buildup during quiescent periods must be considered for safe entry into isolated volumes (Perry 2001). Two or more months routinely elapse between closing the hatch on the ground and the first crew entry during flight for new elements and logistics modules and elements delivered to the *International Space Station*. Contaminants accumulate during this time. Although passive contamination control limits the buildup rate, contaminant concentrations still can rise to unacceptable levels. Active contamination control approaches, including dilution and operating scrubbing systems before crew entry, help reduce the health and safety hazard. Any isolated module or volume should be treated as though it is contaminated, and protocols must be implemented to ensure human health and safety on entry.

Monitoring, by using either archival sampling or near real time techniques, verifies acceptable cabin atmospheric quality. Instruments for monitoring cabin pressure, oxygen partial pressure, carbon dioxide concentration, and combustion products are well suited for near real time environmental monitoring. Monitoring trace chemical contaminants and particulate matter in near real time is difficult and can contribute to information overload. Monitoring non-methane volatile organic compounds as a class to provide a simple result to crewmembers and flight controllers, as well as to serve as a life support system process control signal, is a continuing challenge. Traditional trace chemical contaminant monitoring techniques rely on archival sampling techniques and complex analytical methods (Perry et al. 1997). Data reported from these methods are better suited to interpretation by discipline experts and are not amenable for process control. Much improvement in trace chemical contaminant monitoring is required before space exploration becomes truly autonomous.

Monitoring limitations, accumulation of low level generation sources, and environmental upsets drive the design margin for atmospheric purification equipment. The design margin magnitude must be reasonable. Otherwise, the atmospheric purification equipment can grow to an enormous size. Balancing the needs for basic trace contaminant control and the magnitude of potential upset events relies on hazard and risk assessment inputs into well-defined trade studies. Understanding the hazards and risks associated with environmental upsets and their likelihood and impact leads directly to a robust atmospheric purification system design that ensures crew health and safety as well as mission success.

6.3 ASSESSMENT OF WATER QUALITY IN THE SPACECRAFT ENVIRONMENT: MITIGATING HEALTH AND SAFETY CONCERNS

6.3.1 Scope of Water Resources Relevant to Spaceflight

Water, as a critical spaceflight resource, must be managed carefully from a crew health and safety perspective. The intent of this section is to present the complexity of this challenge and describe the measures that have been taken by NASA to ensure spaceflight water quality. Just as on Earth, there is a variety of uses for water in the context of human spaceflight. Obviously, sufficient water intake is necessary to support the crew and maintain health. Water also is used by crewmembers for food rehydration, beverage preparation, and personal hygiene. However, water further can be used for technical purposes, for example, as a coolant for various systems, as flush water for sanitary purposes, and even as a source of oxygen (from the electrolytic splitting of water).

Whereas water quality is a consideration even for technical water, for example, high levels of solids or free gas can affect adversely certain system functions, this chapter is focused on the toxicological and crew health aspects of maintaining the quality of water utilized for potable purposes by the crew. Even when limiting the discussion to potable water, one finds a complex picture. The current situation on the *International Space Station* provides a telling example. Potable water can be derived from the following sources:

- Ground supplied water launched in a Russian *Progress* vehicle and delivered to the *International Space Station* in twin 210-L Rodnik tanks.
- Recycled humidity condensate from the *International Space Station* that is reclaimed as potable through elaborate processing by on-orbit reclamation systems.
- Potable water transferred to the *International Space Station* from the *Space Shuttle* via contingency water containers, that is, the 44-L bladders that hold water generated as a byproduct of the Space Shuttle fuel cell power system.

On top of this, NASA currently is engaged in efforts with the International Space Station international partners to deliver ground supplied water on their planned launch vehicles, that is, the ESA *Automated Transfer Vehicle* and the JAXA *H-II Transfer Vehicle*. A urine processor developed by NASA that recently has been deployed to the *International Space Station* will provide for the recovery of water. Although the figure varies by specific International Space Station increment, recycled humidity condensate accounts for roughly half the potable water used on the *International Space Station*, with the remainder being supplied by the other sources. Each of these water sources carries its own unique considerations, advantages, and challenges. These are described in more detail through the discussions in the following sections. Although new challenges inevitably arise, many of these considerations can be considered timeless, that is, relevant in the days of *Apollo* and still visible through the telescope of exploration.

6.3.2 Spacecraft Water Quality and the Risk Assessment Paradigm

This section describes the safety considerations associated with assessing the physical chemical properties of the potable water. Whereas water microbiological issues are reserved for a separate discussion, the identification of organic and inorganic constituents of potential concern is a focus, and concepts in toxicology and environmental chemistry play major roles in crew health decision making. A tenet of environmental risk assessment is that the inherent toxicity of a given compound and the level of exposure, that is, the concentration and amount ingested, are together critically important to establishing overall health risk. In describing the assessment of water quality risks, the discussion focuses on the two general categories of identification of constituents of concern and establishment of levels of exposure that are health protective for the spacecraft environment.

Identification of Chemical Constituents of Concern

If a scientist has experience in assessing risks to groundwater or surface water resources on Earth, it might be reasonable for him or her to expect that the constituents of concern in a spacecraft environment would be similar. However, this is not the case. For example, the use of recycled spacecraft humidity condensate as potable water opens the door to a world of constituents that might never cross the desk of a laboratory manager who deals with terrestrial water samples. This can be a challenge, for it requires a change in technical perspective. However, this is a part of what makes space environmental science challenging and unique (Figure 6.19).

Pollutants that pose water quality challenges can be derived from several different sources. For ground supplied water, pollutants can be present in the source water, introduced unintentionally during processing, derived from the material that composes the bladder and on-orbit delivery tank, or transferred by a dispensing system. Recycled humidity condensate is somewhat unique, in that it also can contain water soluble chemicals transferred from the atmosphere of the spacecraft vehicle. Because the spacecraft is basically a closed loop system, there is a variety of potential challenges to maintaining water quality (Schultz, Plumlee, and Mudgett 2006). For instance, although carefully controlled, the list of compounds that potentially could be released into the atmosphere is quite

6.3 Assessment of Water Quality in the Spacecraft Environment

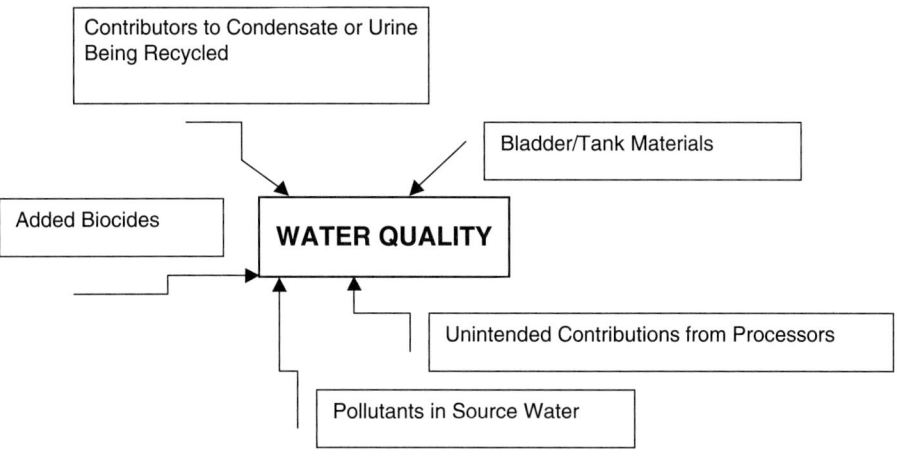

FIGURE 6.19 Diverse challenges to water quality in spaceflight.

extensive, especially given the diversity of chemical containing spacecraft payloads and materials. Human metabolic products, such as organic acids and esters, are also important contributors, and impacts from pharmaceutical agents and drug metabolism must be considered as well.

Through application of a tremendous amount of on-the-ground and in-flight experience, NASA has developed a target list of compounds to evaluate when assessing spaceflight water quality. As discussed already, the condensate of the *International Space Station* probably represents the biggest challenge in terms of identifying constituents of concern. Our current International Space Station target list evolved over time as new information became available. The sources from which target compounds were identified included our evaluation of Space Shuttle condensate and our experiences in working with our Russian colleagues during Space Shuttle–Mir collaborations. Of course, identification of target compounds is not static, and NASA continues to develop methods and take advantage of sensitive analytical instruments that allow us to better characterize chemical unknowns in our water samples.

It also must be considered that some compounds are added purposefully to spacecraft water. To ensure that microbes are controlled, either iodine or silver is added because of their biocidal properties. As is the case with most biocides, however, a human health concern can be associated with excessive exposure to both iodine, such as thyroid function, and silver, such as argyria. Minerals, such as calcium and magnesium, also are added in some spacecraft water sources because some feel that their presence improves the palatability of the drinking water or offers dietary enhancement.

Table 6.2 includes a list of primary contaminants that are particularly relevant to the various water sources. This unranked list was generated by qualitatively considering the frequency of detection, magnitude of detection, toxicological characteristics, and other factors. Regarding International Space Station humidity condensate, it should be noted

Table 6.2 Comparison of Contaminants of Concern for Different Potable Waters

International Space Station U.S. Laboratory Condensate	Space Shuttle Generated Fuel Cell Water	Ground Supplied Russian Rodnik Water (Progress Launched)
Benzyl alcohol	Nickel[b]	Chloroform
Ethanol	Ethanol[a]	Manganese
Methanol	Iodine[a]	Silver[a]
Acetate	Free gas	
Formate	Cadmium[b]	
Propionate	Lead[b]	
Zinc[b]	Caprolactam[c]	
Nickel[b]		
Formaldehyde		
Ethylene glycol		
Propylene glycol		

[a]Related to biocide addition.
[b]Generally resulting from releases from metallic heat exchanger coatings or dispenser parts.
[c]Resulting from leaching of bladder material.

that this list is specific to raw condensate, which is not consumed by the crew until it has been subjected to an extensive and extremely effective process to remove any pollutants of concern.

Establishing Exposure Limits

Once pollutants and compounds of concern contained in spacecraft potable water are identified, the setting of appropriate health protective water quality limits generally is necessary. Setting these limits involves a synthesis of knowledge about crew exposures to a pollutant in the water and a quantitation of the toxicity of that pollutant. Limits that are too high can lead to an immediate (acute) or long-term (chronic) risk to crew health. Conversely, inappropriately low limits result in a different set of problems. These include unnecessary system designs, such as incorporation of specially designed filters, excluding certain materials from use in construction, reprocessing or discarding vital water resources, and incorrectly focusing limited crew and ground support time and resources to address benign issues. Whereas both types of errors must be avoided, there is an intentional bias toward crew health protection in setting exposure limits. This is accomplished through the incorporation of uncertainty or modifying factors that decrease the allowable level of a pollutant in water when data are sparse and assumptions must be made, such as extrapolating from animal data because limited amounts of human toxicity data exist.

Risk Assessment Assumptions

It is beyond the scope of this text to present fully the variety of risk assessment considerations and toxicity evaluation strategies employed in setting water quality limits. A substantial number of other toxicological considerations must be taken, and *Methods for Developing Spacecraft Water Exposure Guidelines* (NRC 2000) is an excellent resource on this subject. However, a general description of the process and a discussion of certain key assumptions is warranted because it is pertinent to a working understanding of how water safety is addressed in spaceflight.

In assessing water quality on Earth, health based limits for many compounds and pollutants have been set by the U.S. Environmental Protection Agency in the form of maximum contaminant levels or health advisory levels (EPA 2006). These limits are established, along with incorporation of particular uncertainty or modifying factors and other measures, to protect the general population, including the elderly, infants, and people with preexisting health conditions. In addition, these limits are intended generally to address long-term, if not lifetime, exposures. In contrast, NASA is charged with protecting a healthy prescreened astronaut corps that by and large is exposed for much shorter durations, such as 10 to 14 d for a Space Shuttle crewmember as compared to a maximum of 180 d on the *International Space Station*. Although the astronaut population is healthier than the general population, unique physiological changes and challenges are associated with spaceflight that need to be considered. As an example, spaceflight can result in the crew experiencing reduced red blood cell mass, a factor that should not be ignored when determining a water quality limit for a pollutant that can cause anemia. Hence, whereas there are instances where maximum contaminant levels are utilized as a guideline, it is much preferred to have spaceflight limits that fully consider the range of factors that can make crews more or less susceptible to health impairment.

Accordingly, NASA has established a relationship with the National Research Council. This association resulted in the formation of a Subcommittee on Spacecraft Exposure Guidelines. This subcommittee is composed of experts in the fields of toxicology, pharmacology, statistics, epidemiology, physiology, and other relevant disciplines. With respect to water quality, the National Research Council subcommittee is working with NASA to develop water quality limits referred to as *spacecraft water exposure guidelines*. To this end, NASA toxicologists research the scientific literature to bring all pertinent data to light in regard to a compound of interest. Combined with knowledge of crew exposure conditions, NASA proposes spacecraft water exposure guidelines for specific time frames, that is, 1, 10, 100, and 1000 d. The proposed spacecraft water exposure guidelines are presented and approved by the National Research Council subcommittee before being made final and published (NRC 2004). In many cases, spacecraft water exposure guidelines decrease as the assumed duration of exposure increases, until a threshold concentration is reached, below which no adverse effects are expected regardless of the duration of exposure. Table 6.3 lists the spacecraft water exposure guidelines approved to date, along with the U.S. Environmental Protection Agency maximum contaminant levels or health advisory levels where available. Through this table, one can note the compounds of most interest from a spacecraft water quality perspective, observe general trends with respect

Table 6.3 Listing of Spacecraft Water Exposure Guidelines (mg/L) (Courtesy of NASA)

Compound	SWEG 1 d	SWEG 10 d	SWEG 100 d	SWEG 1000 d	U.S. EPA Maximum Contaminant or Health Advisory Level
Acetone	3500	3500	150	15	N/A
Alkylamines (di)	0.3	0.3	0.3	0.3	N/A
Alkylamines (mono)	2	2	2	2	N/A
Alkylamines (tri)	0.4	0.4	0.4	0.4	N/A
Ammonia	5	1	1	1	N/A
Barium	21	21	10	10	2
Cadmium	1.6	0.7	0.6	0.02	0.005
Caprolactam	200	100	100	100	N/A
Chloroform	60	60	18	6.5	0.08
Di (2EH) Phthalate	1800	1300	30	20	0.006
Di-n-butyl Phthalate	1200	175	80	40	N/A
Dichloromethane	40	40	40	15	0.005
Formaldehyde	20	20	12	12	7
Formate	10,000	2500	2500	2500	N/A
Manganese	14	5.4	1.8	0.3	0.05 (aesthetics)
2-Mercapto-benzothiazole	200	30	30	30	N/A
Nickel	1.7	1.7	1.7	0.3	0.1
Phenol	80	8	4	4	2
n-Phenylbeta-naphthylamine	1600	1600	500	260	N/A
Silver	5	5	0.6	0.4	0.1
Total organic carbon	N/A	N/A	3	3	N/A
Zinc	11	11	2	2	10

SWEG = Spacecraft water exposure guidelines.
N/A signifies that a limit has not been set for this compound.

to spacecraft water exposure guidelines across exposure time frames, and note inherent differences between them and municipal water supply limits.

In calculating spacecraft water exposure guidelines, several exposure parameters are standardized and applied for every compound. For example, a standard adult bodyweight of 70 kg (154 lb) is utilized. This is because the lower the body weight, the higher the dose, when given an equivalent volume of intake. In regard to potable water ingestion, it is assumed that a crewmember consumes 2.8 L of water per day on average, including its use in food rehydration and so forth. The spacecraft water exposure guidelines are focused on addressing the ingestion of water by the spacecraft crew, because that is clearly the most relevant and important exposure pathway in that environment. Although other pathways, such as dermal absorption of compounds in water, can have some theoretical applicability, they are inconsequential from a spacecraft exposure standpoint.

6.3.3 Water Quality Monitoring

Materials Evaluation and Testing

Preflight efforts to ensure water quality are critical and cannot be overstated. Materials and payloads employed to contain water in spaceflight carefully are evaluated by water quality engineers, NASA toxicologists, and the rest of the safety community to ensure that any adverse water quality effects are anticipated and mitigated. Water recovery systems are required to demonstrate their performance during extensive ground based testing. Containers used to transfer potable water are subjected to longevity studies and tests designed to reveal unacceptable levels of component leaching. These measures are examples of proactive safety considerations. Whereas testing often is conducted specifically to ensure that no water quality concerns exist, evaluations by analysis also are conducted to ensure, for example, that materials are selected from a list that incorporates compatibility considerations.

Preflight Monitoring

For potable water that is prepared on the ground and loaded for flight, comprehensive sampling is conducted prior to launch, with time typically allowed to take mitigation steps if any concerns are identified. A certain amount of water is loaded for use before fuel cell generated water becomes available during flight. Extensive Space Shuttle preflight water sampling is conducted by NASA, with samples initially collected 15 d and again 3 d before launch to ensure that water quality requirements are met (Hwang, Schultz, and Sumner 2006). Russian ground loaded water also is tested and evaluated prior to flight. In both instances, it is critical that problems be identified as early in the process as possible to maximize flexibility for any mitigation efforts.

In-Flight Monitoring

Despite the importance placed on preflight evaluations, these measures are not meant to eliminate the need for post-launch water quality monitoring as a foundation of spaceflight safety. Real time monitoring involves the utilization of technologies focused on generating data for a specific pollutant that can be evaluated quickly during a flight. Ideally, all monitoring information would be available in real time so that health and safety decisions can be made in an expeditious manner and with minimal uncertainty. However, for a variety of reasons, real time monitoring capabilities are limited. To someone uninformed of the complexity of spaceflight, it can be difficult to comprehend these limitations. After all, a quick Internet search can reveal a number of technologies that can provide some form of real time data for a specific pollutant of interest. In practice, however, the problems encountered by adapting these technologies to spaceflight quickly become evident. These factors largely account for our current inability to generate real time water quality data. These considerations include questions such as:

- Can the technology meet required analytical detection limits?
- Does the technology require resupply of consumables?
- Does the technology utilize chemicals or reagents that can pose a crew health concern in a closed loop spacecraft environment?
- Does the technology have the specificity to handle mixtures of pollutants without affecting the reliability of the individual results?
- Can the technology be adapted to the uniqueness of a zero-gravity environment?
- Is it rugged enough to perform in that environment?
- Does the technology minimize critical crew time?
- Are the weight and power needs within practical limits?

In the absence of specific technologies for all pollutants, total organic carbon is a general chemical parameter that has been valuable to NASA for evaluating water quality in the spacecraft environment, at least with respect to organic compounds. A total organic carbon analyzer has been employed on the *International Space Station* in the past, and a next generation unit is in development. Although not compound specific, that is, total organic carbon data reflect any organic carbon present in a sample, it can serve as a useful screening tool. By assuming that a reasonable worst-case compound is contributing to and responsible for an elevated total organic carbon level, an appropriate screening limit can be established. Real time decisions then can be made as to whether the organic pollutant load is low enough to be health protective. Of course, by following this screening approach, false positives can occur; that is, the total organic carbon level can be elevated by a low toxicity compound. This is the trade-off for not having specific data for each pollutant.

Other technologies that can provide real time data on specific organic and inorganic compounds in water are also in development, such as colorimetric techniques that can provide data on iodine, silver, and other compounds. Indeed, NASA participated with developers of promising technologies in efforts to prepare and adapt them to the spacecraft environment. A hallmark of this support has been the arrangement of parabolic flights for these developers, where the monitoring technology is tested in microgravity conditions using the NASA KC-135 (Figure 6.20). Critical knowledge has been gained as the result of these flights. For example, it clearly has been observed that air bubbles do not separate within a water sample in microgravity in the same manner that they do on

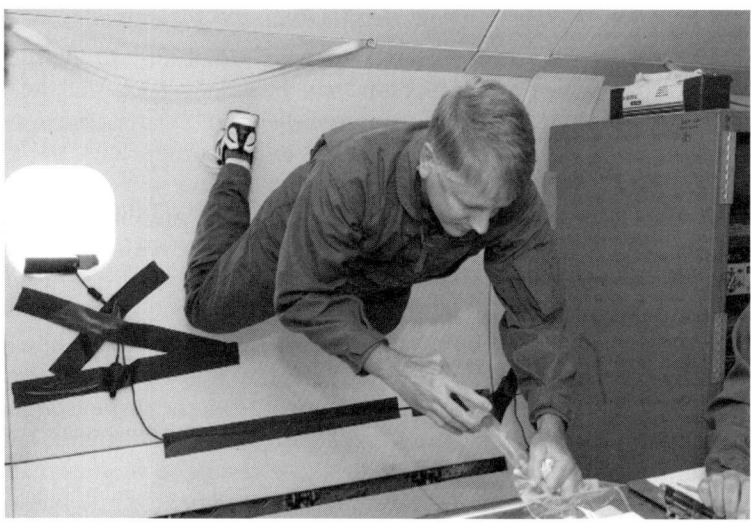

FIGURE 6.20 "Vomit comet" participant participates in microgravity testing of real time water quality monitoring technology (Courtesy of NASA).

Earth. Accordingly, the sensitivity of a monitoring technology for air-water separation is an important factor to consider, and a focus on developing effective bubble mitigation techniques that can preserve analytical performance has resulted.

Archive or Post-Flight Monitoring

A second general class of monitoring, the collection of archive samples, is currently the most utilized method and presents a number of advantages and disadvantages. With archive sampling, water samples are collected in-flight and returned to Earth in specially designed sampling bags to preserve their integrity. These samples are then analyzed in ground based laboratories. The main disadvantage is, of course, the inability to make real time decisions based on the data. However, a much more complete list of analytes can be tested than is ever possible using real time monitors. Heavy reliance on archive samples does place greater importance on the regular return of those samples for analysis. Because there is tremendous competition for crew time resources and down mass, this is an area that requires vigilant advocacy to ensure that adequate data are available to make crew health decisions reliably.

Post-flight sampling also is utilized in some instances. A good example is from the *Space Shuttle*, where post-flight samples are collected from the galley (Hwang et al. 2006). Such data can be useful in identifying potential concerns and verifying nominal system performance.

Application of Monitoring Data in Crew Health Decision Making

It can be beneficial to provide a recent practical example of how preflight and archive water quality data can be used together in crew health decision making. Consistent with normal practice, archive samples taken from the stored water system on the *International Space Station* and dispensed through a Russian system called the *Russian segment water dispenser* were returned and analyzed for a suite of contaminants (Straub, Plumlee, and Schultz 2006). Cadmium was observed to be present at levels that were not a health threat but definitely higher than normally observed on the *International Space Station*. The logical first step was to evaluate existing data on the source water for that system, that is, ground loaded Rodnik water. This evaluation showed cadmium to be below analytical detection limits and suggested that the dispensing system was introducing the low levels of cadmium being detected. Subsequent observations confirmed that cadmium levels in an initial flush of water from the dispenser, that is, water that had a longer residence time in the system contained much higher levels than in samples taken after the flush. After presenting these results to the International Space Station international partners, it was decided that replacement of the Russian segment water dispenser was warranted.

6.3.4 Conclusion and Future Directions

Managing spacecraft water quality is diverse and challenging and requires a scientific perspective quite different from that taken with water resources on Earth. As NASA plans for longer-term and more remote missions to the Moon and Mars, there is likely to be an increased focus on the recycling of water, real time monitoring capabilities, and possibly even water quality issues that pertain to our use of in-situ extraterrestrial resources. It is safe to say that crewed spaceflight will always require a reliable source of potable drinking water, along with creative scientists and engineers who are up to the challenge of ensuring its safety.

6.4 WASTE MANAGEMENT

The waste management subsystem covers the collection, treatment, storage, and disposal of crew wastes. The reprocessing of consumables also can be included in this list for long duration missions, on which frequent resupply is impractical or impossible. The processes of waste management are (in sequence) collection, segregation, fractionation, stabilization, storage, and when appropriate, recycling (Eckart 1996).

Throughout the history of spaceflight, all waste management systems have stored waste and subsequently delivered it to the ground or placed it into spent cargo vehicles that are burned up on reentry. These processes remain in use today. Newer systems being developed have yet to be tested thoroughly, and it will be some time before they can be utilized in future programs. The general challenges of these new systems are covered here.

First, wastes must be collected, sorted, and stabilized. Collection methods have been similar among all vehicles used since humans first orbited the Earth. Human waste is collected either manually in bags or containers or with the assistance of a blower or fan to direct it into a collection device.

Once collected, waste must undergo a process of stabilization. The methods in use for this purpose are

- Condensation.
- Drying, using dry or wet heat.
- Osmotic pressure.
- Freezing.
- Extreme pH levels.
- Metallic or organic toxins.
- Silver-(II) oxidation.

After stabilization, wastes are stored or further decomposed. For solid waste treatments, three principal methods are used: wet oxidation, combustion or incineration, and biological treatment. Wet oxidation involves the oxidation of either a dilute or concentrated slurry (1 to 10% solids) of organic materials at an elevated pressure and temperature and uses water in its supercritical state as part of the process. The high operating temperatures and pressures inherent in this process are safety concerns, as are material corrosion problems and the removal of any toxic gases produced. Despite these safety risks, the wet oxidation process affords the recovery of useful water, and the volume and mass of unwanted solid wastes is reduced.

Starved air combustion and incineration are believed to be the best methods for decomposition of solid waste. Because of the lower operating temperatures employed, these processes are considered to be the only methods practical enough to use (Eckart 1996). Bioregenerative systems are also effective; however, much more development is required before they can be used in spacecraft.

6.5 SUMMARY OF LIFE SUPPORT SYSTEMS

Konstantin Tsiolkovsky (1857–1935) once stated, "The Earth is the cradle of mankind, but mankind cannot stay in the cradle forever." Humankind has always been determined to climb out of the cradle, but the challenge remains in learning to survive. Throughout the history of spaceflight, life support systems evolved. However, space travelers have not yet achieved the goal of continued long-term survival without replenishment of critical resources from Earth. The *International Space Station* represents an excellent test bed for the advancement of the regenerable life support systems that evolved from the non-regenerable systems used on previous spacecraft. These regenerable systems are crucial to our ability to go farther and stay longer in space. If humankind is able to achieve sustainable survival outside the cradle of Earth, then Tsiolkovsky's vision will have been realized. This vision does not come without risk, and safety is of the utmost importance to ensure that risk is mitigated to the extent reasonably possible for such an endeavor.

REFERENCES

Agui, J. (2007) Lunar dust characterization for exploration life support systems. Paper presented at the 45th AIAA Aerospace Sciences Meeting and Exhibit. Reno, NV.

Block, S. S. (1991) Microorganisms, sick buildings, and building related illnesses. In S. S. Block (ed.), *Disinfection, Sterilization, and Preservation*. Philadelphia: Lea and Febiger, pp. 1107-1119.

Curry, K., K. Prokorov, E. Turner, and S. Blaha. (2003) International Space Station automated safing responses to fire emergencies. Paper 2003-01-2595. The 33rd International Conference on Environmental Systems. Vancouver, BC: Society of Automotive Engineers.

Curtis, R. E., J. L. Perry, and L. H. Abramov. (1997) Performance testing of a Russian *Mir* space station trace contaminant control assembly. SAE 972267. 27th International Conference on Environmental Systems. Lake Tahoe, NV: Society of Automotive Engineers.

Diamant, B. L., and W. R. Humphries. (1990) Past and present environmental control and life support systems on manned spacecraft. Paper 901210. Warrendale, PA: Society of Automotive Engineers.

Dunaway, B., M. Edeen, C-F. Tsai, and E. Turner. (2000) Integrated orbiter/International Space Station air quality analysis for post-mission 2A.1 risk mitigation. The 30th International Conference on Environmental Systems. Toulouse, France: Society of Automotive Engineers.

Eckart, P. (1996) *Spaceflight Life Support and Biospherics*. Boston: Microcosm Press, Kluwer Academic Publishers.

El Sherif, D., and J. C. Knox. (2005) International Space Station carbon dioxide removal assembly concepts and advancements. Paper 2005-01-2892. The 35th International Conference on Environmental Systems. Warrendale, PA: Society of Automotive Engineers.

Environmental Protection Agency. (2006) *Drinking Water Standards and Health Advisories*. EPA 822-R-06-013. Washington, DC: U.S. Environmental Protection Agency, Office of Water.

Hines, A.L., T. K. Ghosh, S. K. Loyalka, and R. C. Warder. (1993) *Indoor Air Quality and Control*. Englewood Cliffs, NJ: Prentice-Hall, pp. 1-10.

Hwang, M., J. Schultz, and R. Sumner. (2006) Shuttle potable water quality from STS-26 to STS-114. Proceedings of the 2006 International Conference on Environmental Systems. Norfolk, VA: Society of Automotive Engineers.

Johnston, R. S., and E. L. Michel. (1962) *Spacecraft Life Support Environment*. Fact Sheet 115. National Aeronautics and Space Administration. Manned Spacecraft Center.

Limero, T. F., R. D. Taylor, D. L. Pierson, and J. T. James. (1990) Space Station *Freedom* viewed as a "tight building." The 20th International Conference on Environmental Systems. Williamsburg, VA: Society of Automotive Engineers.

Link, M. M. (1965) *Space Medicine in Project Mercury*. NASA Special Publication SP-4003. Washington, DC: U.S. Government Printing Office.

Liu, B. Y. H., K. L. Rubow, T. J. McMurry, T. J. Kotz, and D. Russo. (1991) Airborne particulate matter and spacecraft internal environments. The 21st International Conference on Environmental Systems. San Francisco, CA: Society of Automotive Engineers.

Martin, C. E. (1992) Environmental control and life support system historical data summary—Apollo through Space Station *Freedom*. MDC 92W5038. Huntsville, AL: McDonnell Douglas Space Systems Co., pp. A-13, A-24, B-52, B-72-B-75, C-28-C-29.

Mitchell, K. L., R. M. Bagdigian, R. L. Carrasquillo, D. L. Carter, G. D. Franks, D. W. Holder, C. F. Hutchens, K. Y. Ogle, J. L. Perry, and C. D. Ray. (1994) *Technical Assessment of Mir-1 Life Support Hardware for the International Space Station*. Technical Manual TM-108441, Huntsville, AL: National Aeronautics and Space Administration, Marshall Space Flight Center, pp. 39-52.

Molhave, L. (1992) Volatile organic compounds and the sick building syndrome. In M. Lippmann (ed.), *Environmental Toxicants—Human Exposures and Their Health Effects*. New York: Van Nostrand Reinhold, pp. 633-646.

Molhave, L., B. Bach, and O. F. Pedersen. (1986) Human reactions to low concentrations of volatile organic compounds. *Environmental International* 12: 167-175.

National Aeronautics and Space Administration. (2001) *International Space Station Specifications*. Space Station Program Document SSP 41000AF and Boeing 2001(CI #2B945). Houston, TX: National Aeronautics and Space Administration, Johnson Space Center and The Boeing Company.

National Research Council. (2000) *Methods for Developing Spacecraft Water Exposure Guidelines*. Washington, DC: National Research Council, National Academy Press.

National Research Council. (2004) *Spacecraft Water Exposure Guidelines for Selected Contaminants*, vol. 1. Washington, DC: National Research Council, National Academy Press.

Otto, D. A., H. K. Hudnell, D. E. House, L. Molhave, and W. Counts. (1992) Exposure of humans to a volatile organic mixture. *Archives of Environmental Health* 2 (Sensory), no. 47: 31-38.

Perry, J. L. (1995) *Trace Chemical Contaminant Generation Rates for Spacecraft Contamination Control System Design*. Technical Manual TM-108497. Huntsville, AL: National Aeronautics and Space Administration, Marshall Space Flight Center, pp. 1-13.

Perry, J. L. (1998) *Elements of Spacecraft Cabin Air Quality Control Design*. Technical Publication TP-1998-207978. Huntsville, AL: National Aeronautics and Space Administration, George C. Marshall Space Flight Center, pp. 3-33.

Perry, J. L. (2001) Predictive techniques for spacecraft cabin air quality control. Paper presented at the 31st International Conference on Environmental Systems. Orlando, FL: Society of Automotive Engineers.

Perry, J. L. (2005) Formaldehyde concentration dynamics of the International Space Station cabin atmosphere. The 35th International Conference on Environmental Systems and 8th European Symposium on Space Environmental Control Systems. Rome: Society of Automotive Engineers.

Perry, J. L., and B. V. Peterson. (2003) Cabin air quality dynamics onboard the International Space Station. Conference Publication 2003-01-2650. Proceedings of the 33rd International Conference on Environmental Systems. Vancouver, BC: Society of Automotive Engineers.

Perry, J. L., R. L. Carrasquillo, and D. W. Harris. (2006) Atmosphere revitalization technology development for crew space exploration. AIAA 2006-140. 44th AIAA Aerospace Sciences Meeting and Exhibit. Reno, NV: American Institute of Aeronautics and Astronautics.

Perry, J. L., J. T. James, H. E. Cole, T. F. Limero, and S. W. Beck. (1997) *Rationale and Methods for Archival Sampling and Analysis of Atmospheric Trace Chemical Contaminants on Board Mir and Recommendations for the International Space Station*. Technical Manual TM-108534. Huntsville, AL: National Aeronautics and Space Administration, Marshall Space Flight Center.

Perry, J. L., R. E. Curtis, K. L. Alexandre, L. L. Ruggiero, and N. Shtessel. (1998) Performance testing of a trace contaminant control subassembly for the International Space Station. SAE 981621. 28th International Conference on Environmental Systems. Danvers, MA: Society of Automotive Engineers.

Reuter, J. L. (2000) International Space Station environmental control and life support system status: 1999-2000. The 30th International Conference on Environmental Systems. Toulouse, France: Society of Automotive Engineers.

Reysa, R., J. Granahan, G. Steiner, E. Ransom, and D. E. Williams. (2004) International Space Station major constituent analyzer on-orbit performance. Paper 2004-01-2546. The 34th International Conference on Environmental Systems. Colorado Springs, CO: Society of Automotive Engineers.

Schultz, J., D. Plumlee, and P. Mudgett. (2006) Chemical characterization of U.S. lab condensate. Proceedings of the 2006 International Conference on Environmental Systems. Norfolk, VA: Society of Automotive Engineers.

Straub, J., D. Plumlee, and J. Schultz. (2006) ISS expeditions 10 and 11: Potable water sampling and chemical analysis results. Proceedings of the 2006 International Conference on Environmental Systems. Norfolk, VA: Society of Automotive Engineers.

Tatara, J. D., and J. L. Perry. (2004) *Performance of Off-the-Shelf Technologies for Spacecraft Cabin Atmospheric Major Constituent Monitoring*. Technical Manual TM-2004-213392. Huntsville, AL: National Aeronautics and Space Administration, Marshall Space Flight Center.

Urban, D., G. Griffin, G. Ruff, T. Cleary, J. Yang, G. Mulholland, and Z. Yuan. (2005) Detection of smoke from microgravity fires. The 35th International Conference on Environmental Systems and 8th European Symposium on Space Environmental Control Systems. Rome: Society of Automotive Engineers.

Weiland, P. O. (1998a) *Living together in Space: The Design and Operation of Life Support Systems on the International Space Station*, vol. 1. NASA Technical Manual TM-1998-206956. Huntsville, AL: National Aeronautics and Space Administration, Marshall Space Flight Center, pp. 147-154.

Weiland, P. O. (1998b) *Living together in Space: The Design and Operation of Life Support Systems on the International Space Station*, vol. 2. Technical Manual TM-1998-206956. Huntsville, AL: National Aeronautics and Space Administration, Marshall Space Flight Center, pp. 66-68.

Williams, D. E., and G. Gentry. (2004) International Space Station environmental control and life support system status: 2003-2004. Paper 2004-01-2382. The 34th International Conference on Environmental Systems. Rome: Society of Automotive Engineers.

Williams, D. E., and G. Gentry. (2006) International Space Station environmental control and life support system status: 2005-2006. Paper 2006-01-2055. The 36th International Conference on Environmental Systems. Norfolk, VA: Society of Automotive Engineers.

Williams, D. E., J. F. Lewis, and G. Gentry. (2003) International Space Station environmental control and life support system status: 2002-2003. The 33rd International Conference on Environmental Systems. Vancouver, BC: Society of Automotive Engineers.

CHAPTER 7

Emergency Systems

John Muratore, P.E.
Research Associate Professor, Aviation Systems and Flight Research,
University of Tennessee Space Institute, Tullahoma, Tennessee

CONTENTS

7.1. Space Rescue 225
  By John Muratore, P.E.
7.2. Personal Protective Equipment 256
  By Baraquiel Reyna

7.1 SPACE RESCUE

Space rescue has been a topic of speculation for a wide community of people for decades. Astronauts, aerospace engineers, diplomats, medical and rescue professionals, inventors, and science fiction writers all speculated on this problem. Martin Caidin's (1964) novel, *Marooned*, dealt with the problems of rescuing a crew stranded in low Earth orbit. Legend at the Johnson Space Center says that Caidin's portrayal of a Russian attempt to save the American crew played a pivotal role in convincing the Russians to join the joint Apollo-Soyuz mission. Space rescue has been a staple in science fiction television and movies, portrayed in programs such as *Star Trek*, *Stargate-SG1*, and *Space 1999*, as well as movies such as *Mission to Mars* and *Red Planet*. As dramatic and difficult as rescue appears in fictional accounts, in the real world, it has even greater drama and greater difficulty.

Space rescue is still in its infancy, and much remains to be done as it matures as a discipline. Issues associated with space rescue and the work done so far in this field are presented in this chapter. Here, the term *space rescue* refers to any system that allows for rescue or escape of personnel from situations where human life is endangered during a spaceflight operation, as defined by the period from crew ingress prior to flight through crew egress post-landing. Additionally, the term *primary system* refers to a spacecraft system from which a crew is either attempting escape or to which an attempt is being made to rescue the crew.

7.1.1 Legal and Diplomatic Basis

Article V of the United Nations *Treaty on the Peaceful Uses of Outer Space* asserts that

- "States Parties to the Treaty shall regard astronauts as envoys of mankind in outer space and shall render to them all possible assistance in the event of accident, distress, or emergency landing on the territory of another State Party or on the high seas. When astronauts make such a landing, they shall be safely and promptly returned to the State of registry of their space vehicle."
- "In carrying on activities in outer space and on celestial bodies, the astronauts of one State Party shall render all possible assistance to the astronauts of other States Parties."
- "States Parties to the Treaty shall immediately inform the other States Parties to the Treaty or the Secretary General of the United Nations of any phenomena they discover in outer space, including the moon or other celestial bodies, which could constitute a danger to the life or health of astronauts."

7.1.2 The Need for Rescue Capability

Space rescue requires attention because spaceflight currently is considerably more dangerous than other types of flight. To ameliorate the dangers of spaceflight to the crew, the catastrophic failure modes of spacecraft systems must be understood well and rescue systems implemented accordingly. Although the need for this capability easily can be demonstrated, the actual implementation of rescue systems in a spacecraft often is affected by numerous emotional, budgetary, and technical issues.

Development of all space rescue systems eventually turns into a risk versus cost discussion. Due to the nature of the technology, space rescue systems are usually complex, expensive, and difficult to test. When primary space vehicle development programs get into cost, schedule, or technical difficulty, the requirements for space rescue often are challenged because they involve substantial investment in a system that is intended for use only when all other aspects of primary space vehicle design have failed. It often is argued that resources should be used to improve the reliability of the primary system being developed to eliminate the need for a rescue system rather than in implementing a difficult and costly rescue option. In discussing the necessity for space rescue systems, the lines of reason revolve around an argument based on commercial airliner flight; that is, at one time, commercial airliners carried parachutes but this is no longer the practice. Why then should astronaut crews have an escape option that airline passengers do not? Similar arguments have been made regarding the risks associated with flight in military transport aircraft and helicopters in a combat zone.

The Safety and Mission Assurance Directorate of the NASA Johnson Space Center recently compared the dangers associated with different types of flight. As a relative order of magnitude metric, the risks for various classes of aircraft assessed by this study are presented in Table 7.1. At this time, the risk associated with human spaceflight is radically higher than any other type of flight, and these statistics represent the best counter to the "rescue systems aren't required" argument.

Table 7.1 Risk of Loss of Life During Flight

Commercial airplane	1 in 1,000,000 flights
Military aircraft	1 in 100,000 flights
Combat in a military jet aircraft	1 in 10,000 flights
Human spaceflight	1 in 100 flights

The very high risk of human spaceflight is driven by many factors. The state of maturity for the technology is a major contributor. Compared to commercial or military aviation, human spaceflight has accumulated relatively few hours of operation and has had very few generations to accomplish the design evolution necessary for a highly reliable system. Space systems also operate in a very demanding environment due to the speeds required to reach orbital velocity and the amount of fuel and oxidizer mass required at launch. This forces tremendous efficiency in structural design and propulsion. For example, the wall of the Space Shuttle external tank, if scaled down to handheld size, would be considerably thinner than that of a soda can. The Space Shuttle main engine is one of the most efficient power plants ever produced by humans. Robert Ryan, James Blair, and Luke Schutzenhofer of the NASA Marshall Space Flight Center developed several interesting metrics to help people understand the design challenges associated with space launch. They compared the horsepower to weight ratio for several different types of engines, and the comparisons are shown in Table 7.2.

It is amazing to realize that, based on the horsepower to weight ratio, there are three orders of magnitude between a car engine and the Space Shuttle main engine. There is even a factor of 6 between this metric for the Space Shuttle main engine and the only comparable mass produced system, a large jet engine. Ryan, Blair, and Schutzenhofer further went on to compare the propulsion power to structural system weight ratio and the propellant mass fraction of representative vehicles; the results of this comparison are presented in Table 7.3.

Accordingly, an automobile designed to the same propulsion power to structural weight ratio as the Space Shuttle main engine would weigh only 2.5 lb. Spaceflight

Table 7.2 Comparison of Engine Weight Versus Horsepower

Engine	Weight (lb)	Horsepower	Horsepower : Weight
Automobile	370	200	0.54
Indy 5600 racing engine	275	800	2.91
Small jet engine	2,890	52,900	18.3
Large jet engine	13,065	1,950,300	149.3
Space Shuttle main engine	7,480	6,786,981	907.4

Table 7.3 System Power-to-Weight Ratio and Propellant Mass Fraction

System	Propulsion Power : System Dry Weight	Propellant Mass Fraction
Automobile (1995 Mustang)	40.5	<0.1
Commercial airliner (747 and 737)	326.8	0.4
Launch vehicle (*Saturn*)	76,700	0.9

systems clearly operate at the extreme end of human design capability for both propulsion and structures in a highly demanding environment.

The spaceflight environment presents many hazards not present in the terrestrial environment. For example, the crew must be protected from the vacuum of space, and many materials used in construction of space systems perform differently in vacuum than at sea level pressure. The crew must be protected from extremes of temperature present on orbit. It is not unusual to have a 400°F temperature gradient within a few inches when traveling from full sunlight to full shadow during on-orbit flight. The aerodynamic loads on a vehicle during ascent can be large, ranging between 700 and 800 lb/ft^2 for the *Space Shuttle*. The heating and aerodynamic loads during Space Shuttle reentry are equally severe, with aerodynamic loads nearing 500 lb/ft^2 while temperatures increase to near 2800°F.

On orbit, there are unique environmental hazards, such as solar radiation and micrometeoroid and orbital debris, due to the lack of atmospheric protection. Indeed, several Space Shuttle outer windows and thermal radiators have suffered impact damage from micrometeoroid and orbital debris. These combined environment stresses are unique to the aerospace environment.

Perhaps one of the most difficult things about dealing with extreme performance requirements and hazardous environments is that there are relatively few opportunities for engineers to take advantage of lessons learned through new spacecraft designs. Most of the hardware used in the U.S. and Russian programs are representative of a slow evolution of design. In the United States, opportunities to design and test new human space vehicles are separated by decades.

Aerospace engineers also struggle with the fact that, even on the rare occasion for development of a new system, it is very hard to ground test such systems in the laboratory because of the combined environments problem. It is almost impossible to test new designs with all the loading factors applied simultaneously, as would be the condition during actual flight. Aerospace engineers most often are constrained to observe the performance of their design while being tested using one loading environment at a time. The effect of temperature, inertial loads, aerodynamic loads, internal pressure loads, vibration, shock, and the like are assessed independently and the results combined through computer modeling to ascertain the adequacy of their designs.

Unlike aviation, spaceflight provides very few opportunities for buildup testing. In an aircraft development program, the vehicle is first flown at low dynamic pressures and Mach numbers while the aircraft is monitored for performance. As the aircraft safe performance limits are determined, the allowable flight envelope is expanded slowly as test data are compared to preflight predictions and adjustments are made in the test program. This can occur over a sequence of perhaps 100 flights. Testing of this type is very hard to do for space systems, as most operate in an all or nothing environment. Once a new space system is launched, it usually has to go through its entire flight envelope during its first flight. This, along with the problems and difficulties of conducting a combined environment test program, explains why so many new rocket systems fail during the first attempt at flight. In contrast, today, a first flight failure almost is unknown in the aircraft industry. It is worth pointing out that the first *Space Shuttle* went supersonic 60 s into its first flight and hypersonic 2 min after that. There was no opportunity to look at the data, decide that things were not as expected, and return to base for a quiet examination of the flight performance.

The high cost for current spaceflight systems provides for very few opportunities to perform dedicated test flights. Unlike an aircraft certification program, which for all practical purposes cannot be conducted for a new aircraft in less than 100 flights, space systems routinely are declared operational after only 1 or 2 flights. As such, the majority of the certification program for space systems is performed through model predictions of performance. Subsequently, data from the models are validated by monitoring the performance of the spacecraft during a limited number of flights.

Due to the hostile environment within which human spaceflight must operate, the high performance required of space vehicles, and the limited opportunities for new development and testing of these spacecraft, human spaceflight will remain the most risky mode of human flight for many years to come. This being the case, it is important to understand fully the limitations and capabilities of the spacecraft and consider the need for and ability to rescue the crew during all phases of human spaceflight.

7.1.3 Rescue Modes and Probabilities

The probability of crew survival in a space system with a rescue capability can be computed from Equation (7.1):

$$P_{\text{crew survival}} = 1.0 - (P_{\text{primary failure}})(P_{\text{rescue failure}}) \tag{7.1}$$

which is equivalent to Equation (7.2):

$$P_{\text{crew survival}} = 1.0 - (1.0 - P_{\text{primary success}})(1.0 - P_{\text{rescue success}}) \tag{7.2}$$

From these equations, it can be seen that the primary value of a rescue system from the perspective of crew survival is that it enables a higher probability of crew survival and does so without having to design higher levels of reliability into the primary system. Reliability is usually a major cost driver in systems development. Usually, a system with .99 reliability is considerably more expensive than a system with .9 reliability. The cost associated with increasing system reliability from .99 to .999 is even higher in terms of the percentage of the cost associated with the increase. It can be seen from these equations that the probability of crew survival using a system, for example, with .9 reliability can be increased to .99 by employing a rescue system of .9 reliability.

This type of increase is particularly critical when trying to achieve high reliability rates for vehicles with long mission lifetime. For these vehicles, it is very hard to design in high levels of reliability because of the effect of an extended mission duration on the probability of failure. If the failure rate of components is expressed in failures per unit time, then the failure probability is defined by Equation (7.3):

$$1.0 - e^{-\lambda t} \tag{7.3}$$

where λ is the mean time between failures. Using this in an example, consider a system having a component with a mean time between failures of 500 or 1000 h. Table 7.4 shows the expectation for failure after increasing the numbers of hours of operation.

We can see from Equation (7.3) that it can be difficult for components to meet high reliability requirements when the mission duration is sufficiently long. In these cases, the normal design strategy is for the application of redundancy and in-flight maintenance. This, however, can be problematic when the logistics depot is located on Earth and the operational location is in Earth orbit or even deeper in space; it is also true when the dynamic nature of a flight, such as descent to a planetary surface after a long sojourn, does not provide the opportunity for maintenance. For these cases, a rescue system can provide an alternative to the engineering of increasingly higher reliability into the primary system. The addition of a rescue or escape system into these types of spacecraft can have a large effect on the probability of crew survival.

It is interesting to note that the calculation of expectation for crew survival also can be used to determine the effectiveness of a crew escape system used for increasing the probability of crew survival from a launch system that originally was not designed for high reliability. Several times, NASA considered the possibility of launching a crew on spacecraft riding atop available expendable launch vehicles. As calculated by the Johnson Space Center Safety and Mission Assurance Office in early 2004, the raw reliability of these expendable launch systems ranges from .77 to .96. From these calculations, it can be shown that, when these expendable launch vehicles are used, a launch abort system with even a .9 probability of success can raise the expectation of crew survival into the range of .91 to .996.

Table 7.4 Probability of Failure After Operation

Hours of Operation	Probability of Failure (MTBF = 500 h)	Probability of Failure (MTBF = 1000 h)
250	.394	.221
500	.632	.394
1,000	.865	.632
5,000	.99996	.993
10,000	1	.99996

MTBF = Mean time between failures.

At first blush, it might sound as though a launch escape system with a .9 reliability should not be too difficult to develop. However, in many phases of flight, abort systems cannot function due to combinations of speed, altitude, and dynamic pressure. These are called *black zones*. They severely limit the use of abort and escape systems, and they especially are limiting when ejection systems are used. For the *Gemini* and early *Space Shuttle*, ejection seats were provided for crew escape. These are limited considerably as to the phase of flight in which they are effective. During ascent for the early Space Shuttle flights, Mission Control would issue a call to the crew with the words *negative seats*. This indicated that speed and altitude had reached limits where ejection was no longer possible. It is worth noting that, even in military jet aircraft, ejection seats achieve crew survival for only 90% of all ejections and in only 95% of cases where ejection occurs within the certified ejection envelope for speed and altitude. It is noteworthy that ejection seat survival statistics presently exceed 91% (Campbell 2003). The modern ejection seat survival rate reflects 60 a of development effort and the lessons learned through thousands of operational uses. Modern space rescue systems have neither the luxury of this design heritage nor the operational experience; therefore, a goal of .9 reliability for a rescue system can be one that is difficult to achieve.

Spaceflight is a very risky form of flight. Rescue and escape systems of reasonable reliability can have a major effect on the expectation for crew survival; however, even reasonable reliability numbers can be difficult to achieve in the space launch environment.

7.1.4 Hazards in the Different Phases of Flight

Human spaceflight has seven primary phases (Table 7.5), each associated with a number of potentially catastrophic hazards (Table 7.6). The phase of flight and the associated hazards determine the types of rescue and escape systems that might be required to ensure survival of the crew should an incident occur.

Many of the entries in Table 7.6 are the same for all flight phases. For example, malfunction in life or mission critical systems can occur during any phase, and it can be catastrophic. Aerospace systems engineers developed techniques, such as systems redundancy, to avoid this type of predicament. Similarly, structural failure can occur during any phase of flight. Again, aerospace structural engineers developed techniques, such as defining a design limit for a load and preserving a factor of safety against that load, to prevent failures under anticipated design conditions. In many ways, the job of a rescue system designer is to consider design solutions for those scenarios that cannot be anticipated by these design techniques. To achieve a practical design solution in terms of weight and performance, the design engineer for the primary system must play the odds. For example, it is often impossible to design a structure capable of withstanding the worst-case meteoroid impact or that can protect the crew and life critical systems for the worst-case solar flare. It is even impossible to develop an in-space crew medical facility capable of handling every ailment or injury that can arise during flight for an otherwise healthy crew. For these risks, which are very hard to evaluate and control, a space rescue system often provides the degree of assurance necessary to proceed to flight.

Table 7.5 Primary Phases of Spaceflight

Phase	Definition
Prelaunch	Crew ingress to liftoff
Ascent	Liftoff to achieving orbit
Orbit	Achieving orbit to initiating an orbital change that results in entry back into the atmosphere. This phase actually could include periods of time in a transfer from Earth orbit to orbit about another body such as the Moon, an asteroid, or another planet.
Rendezvous, docking, and departure to or from another spacecraft	From the start of maneuvers to bring two spacecraft together, through docking or undocking, and departure from the other spacecraft
Descent and ascent from a non-terrestrial surface	From initiating an orbital change that results in a non-terrestrial landing to the time back in a stable orbit above a non-terrestrial location. The non-terrestrial location could be the Moon, an asteroid, or another planet.
Extravehicular activity	Time spent outside the spacecraft
Entry	From initiating an orbital change that results in atmospheric entry until landing

7.1.5 Historic Distribution of Failures

David Shayler's *Disasters and Accidents in Manned Spaceflight* (2000) presents an excellent chronology of spacecraft accidents and near accidents. In reviewing that chronology, it is possible to separate major events into the seven flight phases (Table 7.7).

In reviewing Shayler's chronology, the incident tally can be characterized by phase (Table 7.8). This information leads to a conclusion that, although the risk of incident is approximately uniform throughout flight, the risk of a fatal accident is largest during the dynamic phases of flight, ascent and entry. This is not inconsistent with conventionally held wisdom within the aerospace industry that the dynamic phases of flight represent the greatest hazard and can be summarized by the adage, "the farther the hardware is from the launch site, the safer it is." The experience of the industry indicates that, once space hardware is within the quiescent state for which it was designed, such as the space environment, it generally is less likely to succumb to critical failure. It is interesting to note that, even when very dramatic failures have occurred in the space environment, such as the Gemini 8 thruster failure, the Apollo 13 explosion, the Mir fire, and the Mir collision, each represents a situation where the crew and ground control were able to stabilize a precarious situation and bring the crew home alive.

Table 7.6 Hazards Associated with Primary Spaceflight Phases

Phase	Primary Hazards
Prelaunch	Fire or explosion due to systems failure, loss of structural integrity, natural environment induced failure, or propulsion related failure
Ascent	Systems malfunction, loss of control, loss of structural integrity, natural environment induced failure, or propulsion related failure
Orbit	Systems failure (explosion, loss of attitude control, loss of critical function, toxic material release), natural environmental hazard (solar radiation, micrometeoroid orbital debris), or health issue for the crew
Rendezvous, docking, and departure to or from another spacecraft	Collision with another spacecraft, systems failure (explosion, loss of attitude control, loss of critical function, toxic material release), natural environmental hazard (solar radiation, micrometeoroid, orbital debris), health issue for the crew, or improper targeting or trajectory (off course)
Descent and ascent from a non-terrestrial surface	Takeoff or landing related accident due to systems malfunction, propulsion malfunction, or natural environment induced failure; improper targeting or trajectory (off course); or surface impact
Extravehicular activity	Suit systems malfunction, hole in suit, crew health issue, loss of crew connection to spacecraft (crewmember adrift, tether protocol lost)
Entry	Systems or structural failure, natural environment induced failure, or loss of control

Dynamic flight phase incidents generally do not afford the luxury of time. Rescue and escape mechanisms must be designed, implemented, and ready for use at a moment's notice, for there is usually no time to improvise when an incident occurs.

7.1.6 Historic Rescue Systems

Given the conventional wisdom and the reality of historical incidents, the development of rescue and escape systems reflects an approach to control risk during the dynamic phase of flight (Table 7.9). Most of the rescue systems for the more quiescent phases of flight in the space environment are conceptual, and few have proceeded into any hardware development stage.

Table 7.7 Events Associated with Launch Phases

Phase	Major Historic Incidents
Prelaunch	Apollo 1 fire, Soyuz T-10 abort, Gemini 6 prelaunch abort, prelaunch aborts following engine start on multiple Space Shuttle flights
Ascent	Apollo 12 lightning strike, Soyuz 18–1 loss of control during ascent, Challenger explosion, Columbia debris damage, STS-51F engine shutdown and abort to orbit, STS-93 electrical short
Orbit	Gemini 8 thruster fails and loses control, Apollo 13 oxygen tank explodes en route to the Moon, fuel cell failures on STS-2, fire onboard *Mir*, and medical conditions on multiple Mir flights
Rendezvous, docking, and departure to or from another spacecraft	Collision between *Mir* and Progress resupply vehicle
Descent and ascent from a non-terrestrial surface	Apollo 10 ascent loss of control during practice landing mission
Extravehicular activity	Crew helmets fogging (*Gemini*, *Mir*), crew exhaustion (Apollo lunar surface)
Entry	Soyuz 1 parachute failure, Soyuz 11 decompression, *Columbia*, Soyuz 23 landing on a frozen lake

Table 7.8 Incidents Categorized by Launch Phase

Phase	Major Historical Incidents
Prelaunch/ascent	12 (2 fatal)
Orbit/extravehicular activity/rendezvous	13 (0 fatal)
Entry	10 (3 fatal)

Prelaunch and Ascent Escape

Prelaunch escape options generally are of two modes. The first is a ground egress mode that basically consists of disconnecting from the spacecraft, opening the hatch, and departing from the launch pad area as fast as possible, usually by means of some sort of slide wire. The Space Shuttle slide wire system shown in Figure 7.1 involves the crew getting into baskets and sliding down the wire to the entry point of an underground bunker that offers protection from explosion. The second prelaunch mode is similar to in flight ascent abort modes and involves a flyaway concept.

7.1 Space Rescue

Table 7.9 Rescue Systems Used During Primary Flight Phases

Phase	Historical Controls
Prelaunch	Rapid pad egress systems (slide wires), launch abort rocket systems, and ejection seats
Ascent	Launch abort rocket systems, ejection seats, intact abort modes such as return to launch site, abort to orbit, and bailout systems
Orbit	Return vehicle for *Skylab, Salyut, Mir, International Space Station* (various concepts)
Rendezvous, docking, and departure to or from another spacecraft	None other than orbit rescue concepts
Descent or ascent from a non-terrestrial surface	None; some ability for spacecraft on orbit above the non-terrestrial surface to maneuver to rescue the lander vehicle if it is in low orbit
Extravehicular activity	Self rescue (secondary oxygen pack) and the simplified aid for extravehicular activity rescue for emergency return of an astronaut adrift
Entry	Ejection seat or bailout

FIGURE 7.1 Space Shuttle ground escape baskets and slide wire (Courtesy NASA).

236 CHAPTER 7 Emergency Systems

Prelaunch and ascent escape flyaway capabilities have been dominated by escape rockets that lift the entire crew module away from the launch vehicle stack. The requirements for these systems are driven by two estimates, the warning time for an imminent explosion and the blast danger radius. The large blast danger area associated with most launch vehicles, combined with short to nonexistent warning times, forces launch escape rockets to have very high thrust with a rapid buildup rate and short firing times. These characteristics, along with the desire for storability and low complexity, tend to force the selection of solid rocket motors for these tasks. These high thrust and rapid buildup characteristics tend to make escape rocket systems into propulsion units that are of little use in the flight profile except for performing the escape function. As such, they normally are jettisoned during flight after they are no longer required for escape. This improves launch efficiency by decreasing the mass to orbit.

Several features of the Apollo launch escape system shown in Figure 7.2 are noteworthy. First, the escape system consists of a launch escape motor, a tower, and a boost protective cover. The cover provides protection for the crew module during nominal ascent and aborts. The launch escape motor actually consists of three distinct motors: the pitch control motor; the tower jettison motor; and the launch escape motor. The launch escape motor is the largest. It is the one fired to implement an abort scenario. The pitch control motor imparts a large pitching moment that is intended to increase the range capability for pad aborts. It also is used to increase the lateral separation of the

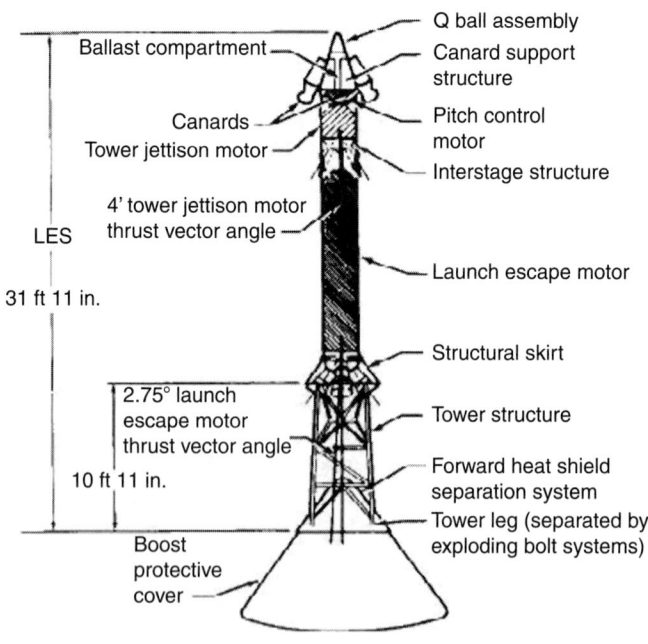

FIGURE 7.2 Apollo launch escape system (Townsend 1973) (Courtesy of NASA).

command module from the launch vehicle during in-flight aborts below 100,000 ft. This motor is designed to be disabled 42 s into flight. The tower jettison motor is a smaller motor used to pull the tower and boost protective cover from the command module during nominal flight or after the firing of the launch escape motor (Townsend 1973).

The Apollo abort system possesses many of the features typical of a crew escape system. In particular, it has several modes of operation (Figure 7.3). Switching between the modes to implement crew escape is critical, and considerable preflight analysis is required to determine their boundaries. As can be seen from Figure 7.3, the mode boundaries are either altitude or velocity, that is, dynamic pressure based, and each mode of operation utilizes different aspects of the propulsion capabilities of the system (Hyle, Fogatt, and Weber 1972).

A launch escape system is not limited to escape rocket systems. The Gemini spacecraft and the early Space Shuttle flights utilized ejection seats (Figure 7.4), which are desirable because of their low weight and similarity to ejection seats used on aircraft. However, due to the environment to which they expose an escaping crewmember, systems of this type are of limited use on spacecraft. The possibility of a relatively unprotected crewmember passing through rocket plumes or launch vehicle debris makes these systems less acceptable for use. Notable in this mode is a need to stabilize a crewmember during high altitude escapes. For *Gemini*, a ballute was used as a drag device for this purpose.

As for rocket based launch escape systems, there are switchover points between adjacent modes as increasing altitude and speed make the ejection seats progressively less useful until the only option remaining is to use the propulsion systems of the spacecraft (Figure 7.5). As for the Apollo system, the mode boundaries are either altitude or velocity based.

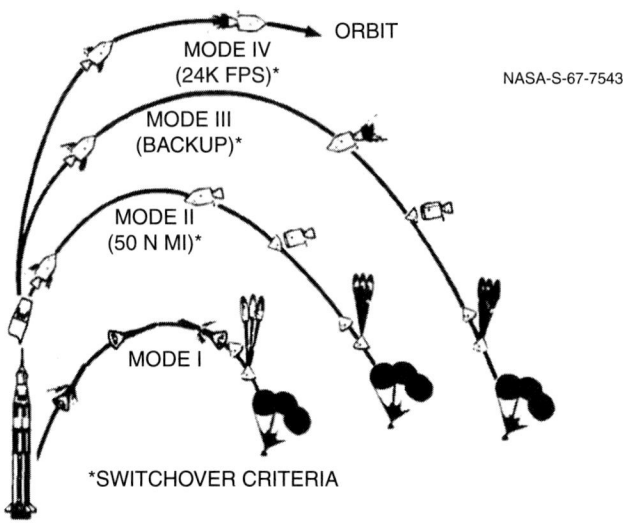

FIGURE 7.3 Apollo launch abort modes (Hyle, Foggatt, and Weber 1972) (Courtesy of NASA).

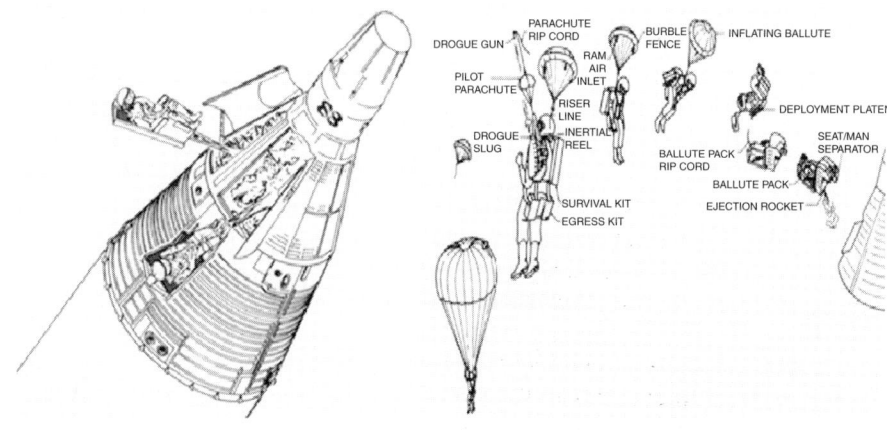

FIGURE 7.4 Gemini ejection seats (Grimwood et al. 1969) (Courtesy of NASA).

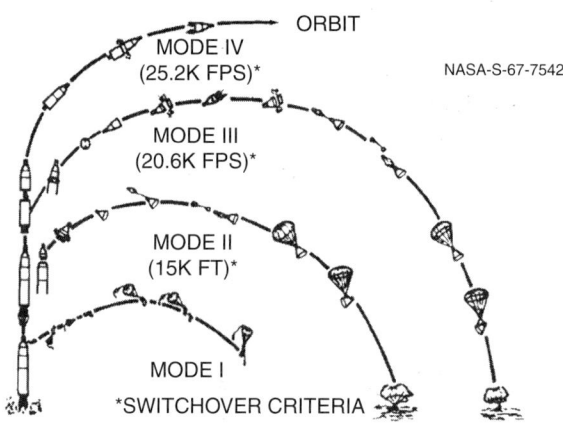

FIGURE 7.5 Gemini escape modes (Hyle et al. 1972) (Courtesy of NASA).

A more difficult problem for ascent abort systems is deciding when to initiate them. Whether activated automatically or by crew intervention, it is difficult to determine whether the criteria and instrumentation information are sufficient to decide that it is time to undertake an irrecoverable and potentially hazardous operation. According to Hyle (1968), during the Mercury, Gemini, and Apollo missions, three different malfunction detection systems were used, all operating under similar philosophies. For these systems, vehicle attitude rates and thrust chamber pressures were the primary cues for initiating automatic aborts. The flight crew could initiate an abort after observing excessive attitude rates, attitude errors, total attitude of dispersions, or angles of attack. Furthermore, the

Gemini launch vehicle malfunction detection system utilized fuel tank pressure, oxidizer tank pressure, guidance and control failure signals, and stage separation signals as cues to initiate aborts (North and Cassidy 1963).

Although the *Space Shuttle* has no crew escape rocket system, the winged vehicle design permits new options for self rescue that were not possible for previous launch systems. This type of self rescue, called an *intact abort* (Figure 7.6), represents those missions that do not achieve the planned orbit, yet result in landing the *Space Shuttle* and its crew on a runway.

Several classes of intact aborts are defined. The first of these types is an abort to orbit. Here, a propulsion malfunction results in reduced engine performance or engine shutdown. Both anomalies occurred on Space Shuttle mission 51-F, each resulting from a malfunction in a different engine. Depending on the nature of a malfunction, the situation is likely to be such that it is no longer possible to reach a specified orbit with the available propellant, so the vehicle is steered to the best orbit achievable. Once an orbit is achieved, a plan can be devised to use the onboard propellant to achieve the mission, then to deorbit and land. Alternatively, the decision can be made to abort the mission and directly deorbit and land the spacecraft. Another type of intact abort is the abort once around, a demanding maneuver used for system malfunctions that severely limit the ability of a crew to survive. It is used in cases such as a total loss of Space Shuttle cooling or depressurization and loss of cabin

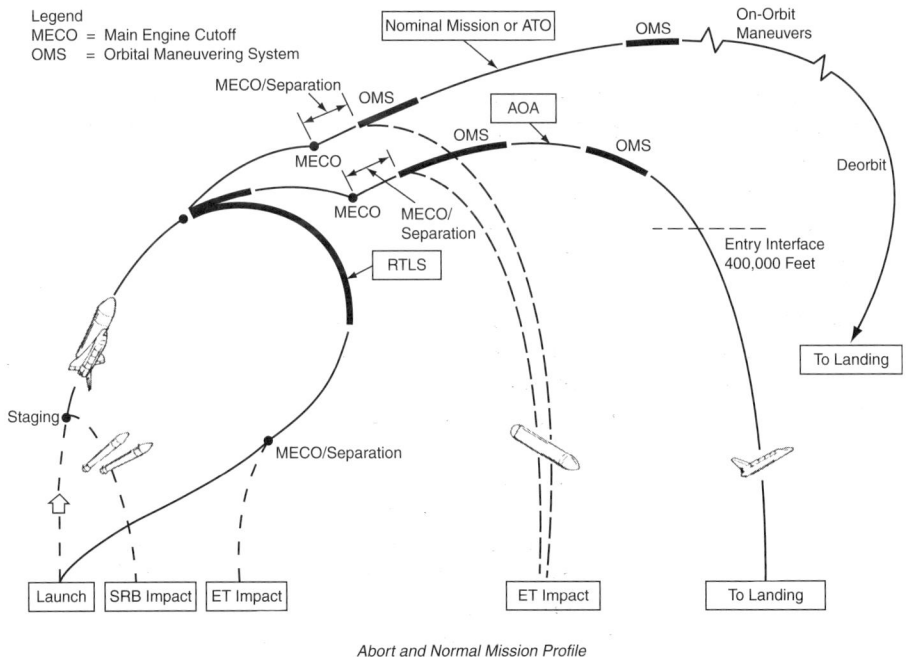

Abort and Normal Mission Profile

FIGURE 7.6 Space Shuttle intact abort modes (Courtesy NASA).

atmosphere during ascent. For such cases, the crew presses on to orbit, then immediately deorbits to land at a West Coast landing site such as Edwards Air Force Base or in New Mexico at White Sands Space Harbor.

There are two other dramatic intact abort scenarios for which a landing is attempted within minutes after launch. The first nominally is designed for the case where a Space Shuttle main engine shuts down just as the *Space Shuttle* leaves the pad. As a result, it is no longer possible to achieve orbit. Despite the loss of the Space Shuttle main engine, however, it is impossible for the crew to take any action until the two solid rocket boosters have expended. The vehicle continues in flight until the solid rocket booster separation occurs. The crew then flies the *Space Shuttle* in a powered trajectory back to the launch site. As the propellant is exhausted, the main engines shut down and the external tank is jettisoned into the Atlantic Ocean. Finally, the *Space Shuttle* glides back to the launch site runway. This intact abort mode, called *return to landing site*, is one of the highest stress cases for the *Space Shuttle* and many of its components. It has never been attempted during an actual flight.

The second intact abort, resulting in a landing before achieving an orbit, is a transatlantic landing or an East Coast abort landing. These intact aborts are designed for the case when a propulsion malfunction occurs after the *Space Shuttle* leaves the launch pad and has been propelled far enough downrange that a return to landing site is no longer possible. Because there is insufficient energy for the *Space Shuttle* to achieve orbit, the crew exercises a transatlantic landing abort and flies to a landing site in Europe or Africa. Alternatively, for an East Coast abort landing, the crew attempts to land at a runway on the eastern coast of the United States.

Should a propulsion malfunction occur for which it is impossible for the *Space Shuttle* to reach a runway, the crew can perform a contingency abort. For this mode, the crew flies a trajectory to get as close as possible to land. Once solid rocket booster separation occurs, the Space Shuttle main engines are shut down and the external tank jettisoned. The *Space Shuttle* is then flown as a glider toward the nearest land mass. At a specified altitude, the crew then engages an automated routine to pilot the vehicle in a straight path while the crew opens the side hatch and prepares to bail out. The crew then deploys the escape pole from the side hatch (Figure 7.7). Attaching themselves to a slide on the pole, crewmembers bail out of the *Space Shuttle*, sliding along the pole, and are released at its end, which is positioned to minimize the possibility of recontact with the spacecraft.

Contingency abort capability mainly is intended to deal with cases where the only alternative is an ocean ditching. Although the Space Shuttle shape has been studied and found to have good ditching characteristics, the trunions and fittings in the payload bay cannot withstand the deceleration forces resulting from water impact, and the vehicle rapidly will be torn apart by the loose cargo in the payload bay. This bailout option allows the crew to leave the *Space Shuttle* before it reaches the ocean and is broken up.

The orange advanced crew escape suit worn by Space Shuttle crews (Figure 7.8) provides physiological support for a high altitude bailout. The advanced crew escape suit, along with the small life rafts and survival kits that are similar to those used by military fliers, enable the crew to survive in the ocean until rescue forces arrive.

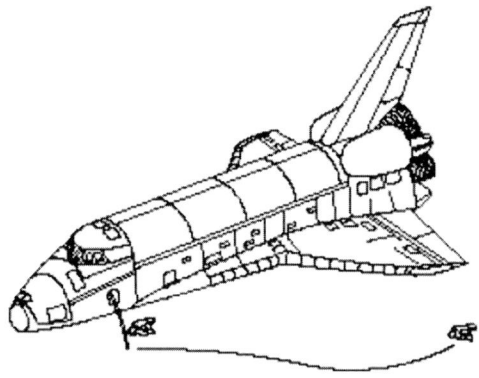

FIGURE 7.7 Space Shuttle escape pole system (NASA 2004) (Courtesy of NASA).

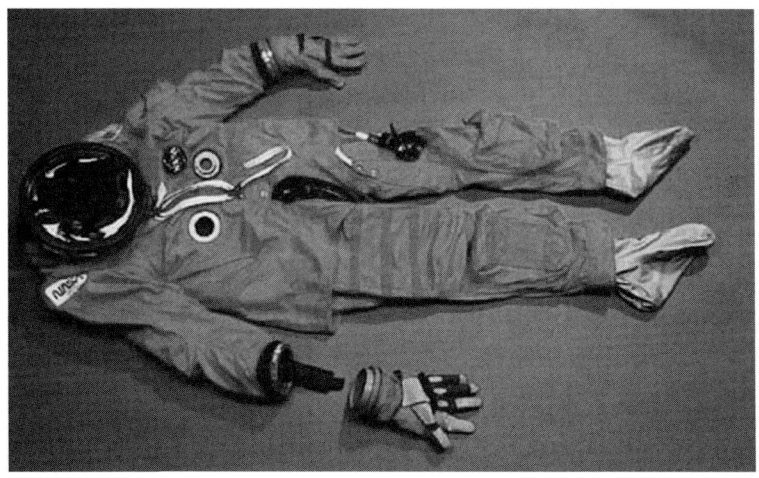

FIGURE 7.8 Space Shuttle advanced crew escape suit (Courtesy of NASA).

Another Space Shuttle escape mode is available for use post-landing. The *Space Shuttle* is equipped with an inflatable crew rescue slide that allows the crew to egress the side hatch and slide to the ground. The slide is necessary because the side hatch is at a considerable distance in the air when the *Space Shuttle* is on its landing gear on the runway. The inflatable slide is similar to those used by commercial airliners (Figure 7.9). The escape pole, survival suits and equipment, and the escape slide were added after the loss of the *Challenger*.

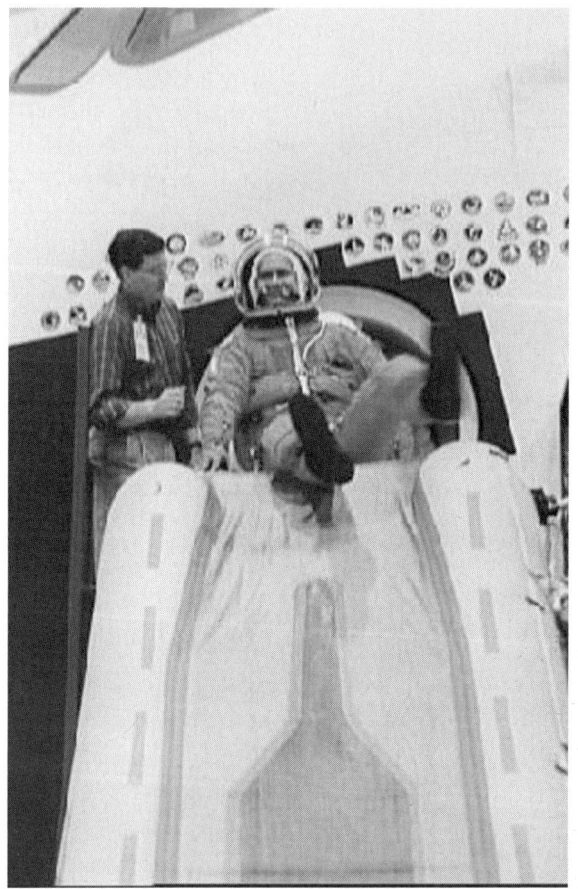

FIGURE 7.9 Space Shuttle escape slide during mockup training (Courtesy of NASA).

Orbit Rescue

The subject of rescue from a vehicle that is stranded on orbit received a lot of concept attention over the last five decades. This subject is treated in some detail on Mark Wade's Web site (Wade 2007). On this site, Wade provides the details of 35 concepts for rescue from Earth orbit. These concepts generally arise from two paradigms. Orbit rescue concepts based on the lifeboat paradigm generally employ a small spacecraft that can serve as a lifeboat to bring the crew home. In this paradigm, it is a struggle to pack crewmembers and spacecraft systems into a small functional vehicle. Another grouping of rescue concepts are based on a parachute paradigm. Such concepts generally are intended for use by an individual crewmember, involve deployable systems made from inflatable or foam structures, and usually involve a parachute as the final stage for crew recovery.

The parachute paradigm systems struggle with implementing structures that are sufficiently robust to maintain the integrity of the system and protect the crewmember during reentry from aerodynamic loads and heating. Parachute paradigm systems also must possess full capability to perform a deorbit maneuver and sustain life during entry. Computations performed by the Johnson Space Center Aerosciences and Flight Mechanics Division evaluated the entry environment for the parachute paradigm personal rescue enclosures (Cerimele 2002).

Systems of this type present an interesting design challenge. To reduce heating, they require a relatively large diameter. In turn, however, the large diameter necessitates additional weight, and this results in increased heating. For a notional 2000-lb personal rescue enclosure having a 15-ft deployed diameter, the aerodynamic load during entry would range from 20 lb/ft^2 in a controlled lifting entry to 50 lb/ft^2 for a purely ballistic entry. The maximum temperature anticipated for this type of device would be 2200°F for a lifting entry and 2500°F for a purely ballistic entry. The device also would need to withstand 2.5 g of deceleration in a lifting entry and a significant 7 g of deceleration during a ballistic entry. These figures represent considerable design challenges that preclude most simple foam or inflatable devices.

One exception to the on-orbit rescue concept has been the implementation of the simplified aid for extravehicular activity rescue, which is a simple rescue device that allows an astronaut performing an extravehicular activity or spacewalk to return to the *Space Shuttle* or *International Space Station* if the tether becomes disconnected or broken. The simplified aid for extravehicular activity rescue is a small compressed nitrogen system that can be worn at the bottom of an astronaut's life support backpack (Figure 7.10). This system has thrusters and controls to enable the crewmember to maintain attitude control and travel back to the *Space Shuttle* or *International Space Station*. Often referred to as an *extravehicular activity parachute*, it mitigates one of the major risks of extravehicular activity operations, that is, becoming separated from the spacecraft.

7.1.7 Space Rescue Is Primarily Self Rescue

Many factors make rescue of one spacecraft by another very difficult given the current state of human spaceflight technology. This forces space rescue to be primarily a self rescue type of operation, although rescue by other vehicles is possible given certain circumstances.

The problem of rescuing one spacecraft by using another is defined by the characteristics of both the spacecraft in distress and the rescue spacecraft. The rescue task is made increasingly easier and has a greater chance for success if the spacecraft in distress can sustain and maintain human life for a period of time (the longer the better). The ease and success of a rescue task is related directly to the speed with which a rescue craft can be prepared for launch (the sooner the better). Finally the rescue task is improved by minimizing the limitations for launch of a rescue vehicle, such as weather, systems, orbital mechanics. Given these conditions, a scenario can be optimized so that a spacecraft in distress can dock with another vehicle that will provide life support during the wait for the arrival of a rescue vehicle. Another optimization can be considered where a rescue vehicle rapidly can be prepared and launched into the same orbit as the spacecraft in distress.

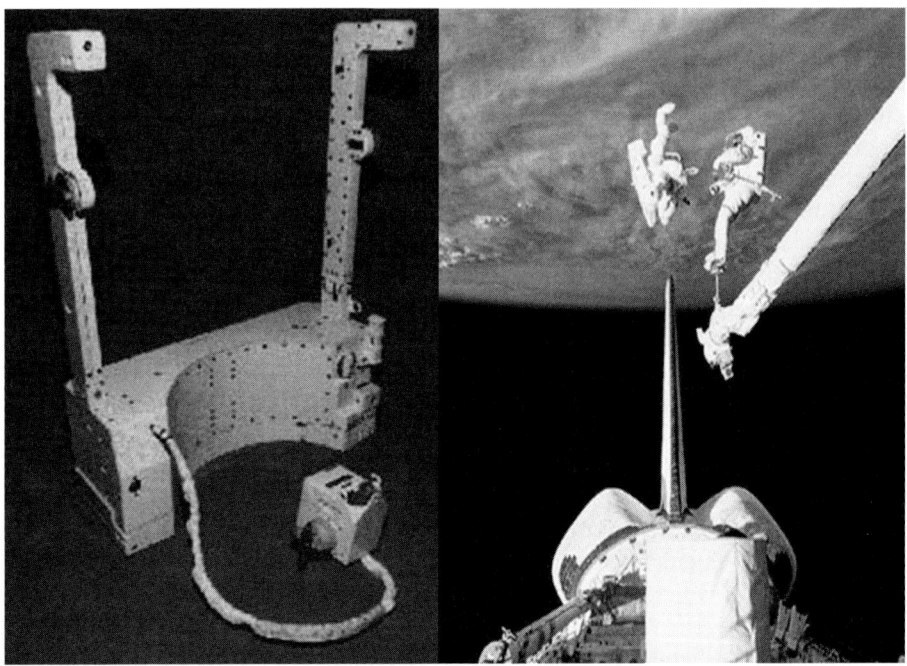

FIGURE 7.10 Simplified aid for extravehicular activity rescue: backpack and flight testing (Courtesy of NASA).

The successful response to the famous Apollo 13 incident was made possible by the fact that the Apollo *Command Service Module*, which suffered the explosion, was docked to the *Lunar Module* (another spacecraft that was able to support life). This allowed the crew to use the limited capabilities of the *Command Module* to return to Earth. Even though two vehicles were involved, this clearly can be seen as a case of self rescue.

For current low Earth orbit operations, docking a damaged spacecraft to the *International Space Station* can extend the duration of crew survival during a contingency. This is possible only when the spacecraft in distress can maneuver into a position to rendezvous and dock to the *International Space Station*. Orbital mechanics can limit the ability of the spacecraft in distress to perform this maneuver, so if the spacecraft in distress is not already in an orbit planned for rendezvous with the *International Space Station*, such a docking is unlikely to be possible.

One of the questions asked after the loss of the *Space Shuttle Columbia* is whether the crew could have found safe haven aboard the *International Space Station*. Reviewing this case is instructional with regard to the difficulties of space rescue. *Columbia*, on this mission, did not carry a docking port for the *International Space Station*, and it did not carry full spacesuits for each crewmember. Solutions could have been developed for each of these problems by station keeping the *Columbia* near the *International Space Station* and transferring the crewmembers using the robotic arms of the *Space Shuttle* and the *International Space Station*. The advanced crew escape suits have some limited capability in vacuum, and crewmembers in these suits could have been carried between the *Space Shuttle* and the *International Space Station* by astronauts wearing regular spacesuits.

As difficult as these solutions are, they are simple to solve in comparison to the problems associated with *Columbia* actually being able to rendezvous with the *International Space Station* during its last flight. To maximize the scientific payload mass, its last flight was launched into an orbit at a 39° inclination to the equator, whereas the *International Space Station* possesses an orbit inclined 51.6° to the equator. To shift the orbit by only 2° in inclination would require almost all the orbital maneuvering fuel onboard the *Columbia*. So, even if practical solutions existed for the transfer of crew between vehicles, there was no way for *Columbia* to have maneuvered anywhere near the orbit of the *International Space Station*.

An optimal rescue vehicle is one that can be prepared rapidly for launch and has few limitations imposed by the conditions for launch. The Soyuz vehicle is by this metric a very good rescue vehicle. These vehicles and their launchers are being manufactured on a production line and are, therefore, usually available for use. In addition, the launchers are capable of performing in a wide range of weather conditions. The *Space Shuttle* is less than optimal by this metric because the turnaround time varies considerably based on anomalies found during post-launch inspections, and the weather constraints for a Space Shuttle launch are more restrictive than for *Soyuz*.

There are other matters to consider, however. *Soyuz* normally is launched with at least one crewmember and is capable of returning three. The *Space Shuttle* can be launched with three or four crewmembers and nominally returns seven. An even larger crew can be returned in a contingency using the special add-on seats developed after the loss of *Columbia*. Indeed, it would require three Soyuz flights to return to Earth the number of astronauts that can be carried within a single *Space Shuttle* during a contingency.

The production line for Soyuz vehicle has a substantial lead time, and the number of rescue vehicles that could be available is limited to those in an advanced state of production. Custom seat liners are required in the Soyuz vehicle to protect the crewmembers against the loads that would occur if the landing retrorocket system does not fire during the final parachute touchdown. These seat liners must be produced on the ground from body molds of the intended crewmember. If a crewmember for whom there is no existing seat liner must be rescued, a very hard landing is possible. Typically, the *Soyuz* is operated with crewmembers suited as a safety precaution against accidental depressurization. Another problem is presented if the crewmembers being rescued do not have *Soyuz* compatible suits or are injured to the extent that they cannot be suited. As well, the *Soyuz* exhibits more restrictive size limits for crewmembers than the *Space Shuttle*, so there is a potential for a Space Shuttle crewmember to not fit into a Soyuz rescue vehicle. The *Soyuz* has no medical capability for use to tend to an injured crewmember during descent. Any medical treatment, therefore, would have to be suspended until after landing. Furthermore, it might not be at all possible to fit a badly injured crewmember into a *Soyuz* given the seating position required in the vehicle. When docked to the *International Space Station*, the on-orbit *Soyuz* is not kept ready for immediate departure.

Considerable work is required to prepare it for independent flight. It, therefore, is likely to be difficult to utilize the *Soyuz* in a rapid evacuation situation from the *International Space Station*. Finally, the *Soyuz* is limited to a 6-mo on-orbit life due to the degradation of the hydrogen peroxide utilized in its attitude control system for reentry. This necessitates the exchange of the *Soyuz* on the *International Space Station* every 6 mo. Although having stated all these limitations, it is worthwhile to point out that the *Soyuz* has delivered very reliable service as a crew return vehicle for both *Mir* and the *International Space Station*. Through operational means and the prudent management of the vehicle, the *Soyuz* can be used as a crew return vehicle.

Current launch rates, vehicle limitations, and a limited ability to launch on time make it difficult to implement a rescue launch to a spacecraft with a short mission life, except where measures are taken to prepare the rescue mission prior to the launch of the first or primary flight. For example, the United States has taken great measures following the Columbia incident to maintain a rescue flight capability in case a *Space Shuttle* is stranded on orbit. Essentially, it is necessary to prepare for two missions for a scheduled launch, the primary mission and the rescue mission. As a practical matter, the rescue mission usually is not prepared completely before launch of the primary mission. Techniques have been developed to allow the rescue mission to be assembled only partially before the primary mission is launched.

In particular, Space Shuttle rescue capability is based on the stage of preparation for the next regularly scheduled mission. The *Space Shuttle* is considered ready when the number of days required to complete launch preparations, as based on a highly expedited schedule, for the rescue mission is less than the maximum number of days that the *International Space Station* can support life for both the resident crew and the *Space Shuttle* crew using the station as a safe haven. In this way, the *International Space Station* serves as a place for the crew of a damaged *Space Shuttle* to dwell while waiting for a rescue launch.

It is normal practice during a *Soyuz* exchange for two Soyuz vehicles to be docked to the *International Space Station*. Although there are two ports on the *International Space Station* capable of Space Shuttle docking, mechanical interference prevents two of these vehicles from being docked simultaneously. Therefore, if a Space Shuttle crew is seeking safe haven in the *International Space Station*, the vehicle within which they arrived must be jettisoned while it still has enough electrical power and fuel to undock successfully and maneuver to a safe distance from the *International Space Station*. This automated departure procedure requires special technique development to allow the *Space Shuttle* to be undocked while the entire crew is aboard the *International Space Station*.

The Russian Soyuz vehicle is more suitable as a rescue vehicle for an *International Space Station* based crew than the *Space Shuttle* for several reasons. Once a Soyuz vehicle is on the *International Space Station*, it can dwell there for some time. The basic Soyuz vehicle has an on-orbit lifetime of 6 mo versus the 14- to 19-d lifetime of the *Space Shuttle*. It is, therefore, possible to keep a *Soyuz* at the *International Space Station* continuously by launching a new *Soyuz* every 6 mo and, at the same time, using the older *Soyuz* to rotate a crew to Earth. While the *Soyuz* is at the station, it is available for activation for an emergency return of the crew.

The major limitation for the *Soyuz* as a rescue vehicle is crew size. As a return vehicle for an *International Space Station* crew of three, the *Soyuz* is very near an optimal solution. However, if a *Space Shuttle* with seven crew members on board were stranded at the *International Space Station*, it would take four Soyuz flights (each launched with only one crewmember) to return all seven of the Space Shuttle crewmembers to Earth. A tremendous amount of hardware would need to be readied and launched to accomplish such a rescue mission.

It is worthwhile to note that, when multiple vehicles are required to rescue an entire crew, the probability for full survival is defined by Equation (7.4):

$$P_{\text{entire crew survival}} = 1.0 - (1.0 - P_{\text{primary success}})\left[1.0 - (P_{\text{rescue success}})^{\text{flights required}}\right] \quad (7.4)$$

So, as the number of rescue flights required to bring the entire crew home rises, so must the reliability of each rescue flight to maintain the probability for entire crew surviving. This translates to a considerable increase in complexity for each rescue vehicle and is the reason the U.S. *Crew Return Vehicle*, which was proposed for use on the *International Space Station*, was designed to be a single vehicle fully equipped to return the entire 7-person crew of the *International Space Station*.

7.1.8 Limitations of Ground Based Rescue

The fundamental issue with a ground based rescue is the limitations associated with its use. A rescue mission must be able to be launched within the available life support capability of the stranded spacecraft. The time required to prepare a mission, as well as the possibility of a mission scrub due to systems failure or weather, appreciably reduces the effectiveness of this type of rescue option. These limitations can be major, as in the case of the *Space Shuttle*, or minor, as in the case of the *Soyuz*.

Launch sites for a rescue mission must be located so that the rescue vehicle is able to reach the orbit of the stranded spacecraft. The minimum inclination that can be reached by a launch site is determined by the latitude at which the launch site is located. This limitation is a function of basic orbital mechanics, which require the center of the Earth to be located at the center of an orbit. Because of these physics, a rescue vehicle can never be launched into an orbit of lower inclination than the latitude of the launch site, thus limiting rescue options. For example, when NASA launches a Hubble Space Telescope maintenance mission to a 28.5° inclination orbit, it is impossible for Russian launch vehicles to be of assistance because the latitude of the Russian launch sites are so far north that their vehicles can never achieve a 28.5° inclination orbit.

It also should be said that orbital mechanics place severe limitations on the number of launch opportunities available to perform a rescue on any particular day. This is because an orbital ground track of an object in low Earth orbit, such as the *International Space Station*, moves westward approximately 22° of longitude per orbit due to the rotation of the Earth. Given this, only two opportunities per day occur when the orbiting vehicle is within range of the launch site. These launch windows tend to be very small, on the order of 5 to 9 min, for a Space Shuttle rendezvous mission with the *International Space Station*.

A more difficult question is whether the cause for the failure that strands one spacecraft on orbit also is likely to occur in an identical rescue spacecraft. If the answer is yes (and likely it is), then launching the rescue flight can have the possible result of additional crewmembers being stranded on orbit if the same failure occurs and the rescue spacecraft is disabled. This is a major question and cause for concern when considering the use of Space Shuttle vehicles to rescue another *Space Shuttle* that has been stranded at the *International Space Station* because of ascent debris damage. If ascent debris has disabled the first vehicle, now docked for safe haven at the *International Space Station*, then the possibility exists that a rescue flight could be disabled by a similar debris strike.

This issue is not limited to spacecraft systems, but concerns the launcher as well. It is interesting to note that the Soyuz launch vehicle is used for both human and cargo missions. There have been cases where a malfunction during a cargo launch caused temporary grounding of the Soyuz launcher, temporarily arresting human as well as cargo flights.

Given the short time that the *Space Shuttle* can dwell at the *International Space Station*, this common cause failure conundrum places NASA managers in a difficult posture. During the 14- to 19-d on-orbit life of the *Space Shuttle* when docked to the *International Space Station*, they must determine if any serious damage to the vehicle occurred during launch and ascent. If so, they must then evaluate the risk of similar debris damage for the rescue flight. Based on these assessments, the managers must decide either to commit to returning the crew in the damaged or repaired vehicle or to jettison the damaged *Space Shuttle* while it still has sufficient power and propulsion to clear the way for a rescue flight.

Another major problem with ground based rescue is that of docking with a tumbling spacecraft. The United States has demonstrated the ability to rendezvous and capture spacecraft spinning about a single axis on a number of occasions, such as the solar maximum satellite, the Westar and Palapa rescues, and the Intelsat Syncom rescue; however, docking with a tumbling spacecraft is a much more difficult task.

Simulations conducted during the *X-38/Crew Rescue Vehicle* project showed that it is much easier to depart from a tumbling spacecraft in a crew escape vehicle than for a ground based rescue vehicle to approach and dock to a tumbling spacecraft. In particular, large appendages, such as solar arrays, radiators, and antennae, sweep a large area when a spacecraft is tumbling. This area becomes a keep out zone for any rescuing vehicles attempting to approach and dock to the tumbling spacecraft. As the rescuing spacecraft maneuvers toward a tumbling spacecraft with its complex geometries, it becomes very difficult to determine if the rescue vehicle is even in a location where a safe approach path exists.

In contrast, to depart using an escape vehicle already docked to the tumbling spacecraft is a much easier problem. The geometry is fixed. The escape spacecraft is positioned at a known orientation with regard to the body coordinates of the tumbling spacecraft. Given the appendages that are part of the tumbling spacecraft, it is possible to compute a trajectory that leads to clearance from the appendages, usually by three propulsive maneuvers of achievable magnitude. In a large number of simulations with a mix of pilot and non-pilot astronauts, it was demonstrated repeatedly that escaping from a docking port on the bottom of the *International Space Station* was relatively easy using the *Crew Return Vehicle*, even with the *International Space Station* tumbling at rates of up to 5° per second. This rate of tumble was established as a reasonable loss of control limit, because this is the amount of rotation that would occur should a single propellant tank on the *International Space Station* expel all of its contents in a single propulsive impulse.

Because of the inherent difficulties of ground based rescue operation, as explained in this chapter, the International Space Station Program office chose to implement a space based self rescue system, the *Crew Return Vehicle*. This capability was planned to be provided initially by the Russian *Soyuz*, then implemented with the U.S.-developed *Crew Return Vehicle*. As history has developed, however, the Crew Return Vehicle project was cancelled, and the Russian *Soyuz* now fills this role for the *International Space Station*.

7.1.9 The *Crew Return Vehicle* as a Study in Space Rescue

Because the *Crew Return Vehicle* was the first custom built space rescue vehicle, it is worthwhile to examine its development and its driving requirements to understand the desirable characteristics and rationale for any future space rescue system.

After the loss of *Challenger* in 1986, the Space Station Freedom Program (predecessor to the *International Space Station*) baselined the *Assured Crew Return Vehicle* (Aldrich and Rusnak 2002). This vehicle was to accommodate three basic missions:

- To return the crew if a crewmember became ill or injured while the *Space Shuttle* was not at *Space Station Freedom*.
- To return the crew in the event that a catastrophic failure of *Space Station Freedom* made it unable to support life and the *Space Shuttle* is neither at *Space Station Freedom* nor is it able to reach *Space Station Freedom* in the required time.
- To return the crew in the event that a problem with the *Space Shuttle* made it unavailable to resupply *Space Station Freedom* or change out crew in a required time frame.

Prior to baselining the *Assured Crew Return Vehicle*, the *Space Station Freedom* had committed to an onboard health facility to deal with astronaut medical emergencies and a safe haven capability to deal with failures aboard the station. Major architectural modifications were incorporated into the *Space Station Freedom* design to make safe haven a possibility. For example, the modules of the *Space Station Freedom* were arranged in an elliptical racetrack pattern. This precludes the possibility of a failure in one module making other modules inaccessible, as well as ensuring two exits from every module. The subsystems of the *Space Station Freedom* also were distributed about the vehicle so that loss of any of the racetrack modules did not disrupt critical services.

Baselining the *Assured Crew Return Vehicle* occurred with considerable trepidation. The concern was that the addition of another fully equipped spacecraft to the *Space Station Freedom* would greatly increase the development and cost risk to the program. It is important to note that the requirement was baselined with no increase in the expected cost to complete the *Space Station Freedom* program, that is, the budget was not increased to accommodate the cost for the *Assured Crew Return Vehicle*.

When the *Space Station Freedom* evolved into the *International Space Station*, the United States, in an international memorandum of understanding, committed itself to produce the *Assured Crew Return Vehicle*, whereas Russia committed itself to providing crew return vehicle capability with its Soyuz vehicle during initial International Space Station operations. Once again, however, no resources were committed to the development of an *Assured Crew Return Vehicle*, and it was viewed as a threat to the International Space Station budget.

From 1986 to 1995, 12 attempts were made to determine an Assured Crew Return Vehicle configuration. Despite considerable work to refine the requirements for the vehicle, limited progress was made toward actually starting a full development effort. The options studied during this time ranged widely from Apollo-type capsules, biconic vehicles, and lifting bodies to a mini-shuttle. Consideration was even given to building an entire additional *Space Shuttle* so that one could be located continuously at the *International Space Station*. For each of these concepts, test and verification were a major cost to the International Space Station Program.

During 1995, the final and most long lived of the Crew Return Vehicle projects was started. The X-38/Crew Return Vehicle project ran from 1995 until 2002. It was based on using a lifting body with a detachable propulsion unit. The propulsion unit would maneuver the craft away from the *International Space Station* and provide for a deorbit maneuver. The propulsion unit would then be jettisoned and the lifting body would fly back to Earth much in the same manner as a *Space Shuttle*. At the end of the flight, a large parafoil would be deployed to allow landing at low velocities (35 kn) and slow descent rates (25 ft/s) on a wide variety of desert type terrain.

The *X-38* project conducted over 40 development tests of the large parafoil using conventional air drop techniques. The project also conducted eight free flight tests of the lifting body parafoil combination by dropping the lifting body from the wing of the NASA B-52 bomber (Figure 7.11). On the final flight of the *X-38* project, the 25,000-lb, 24-ft lifting body intercepted the calculated trajectory of a Crew Return Vehicle returning from space at approximately 50,000 ft and flew the trajectory successfully to large parafoil deployment at 15,000 ft. The largest parafoil in the world (7500 ft^2) then deployed and flew the vehicle to a precision touchdown within 150 ft of the planned target. This accuracy was obtained while dropping the *X-38* from the B-52 almost 10 mi from the intended landing point. Inertial global positioning system navigation and controls via aerosurface actuators during lifting body mode and steering winches connected to parafoil risers during parafoil flight provided precision approach and landing.

A space test *Crew Return Vehicle*, which was to be carried aloft by the *Space Shuttle*, was over 75% completed at the time of project cancellation. The space test vehicle and the space test flight profile are pictured in Figure 7.12. Structural and systems tests conducted after the project cancellation showed that the design would have passed NASA requirements to fly in the Space Shuttle bay, and its systems were capable of completing a vehicle return mission.

At the time the *X-38* project was cancelled, it was projected that a fleet of three vehicles could be built and qualified fully for approximately $1.5 billion. This represented a dramatic decrease in the cost to produce the *Crew Return Vehicle*, which according to

7.1 Space Rescue

Dryden Flight Research Center EC99 45080-25 Photographed 09JUL1999 The X-38 Ship 2, Crew Return Vehicle is released from the B-52 Mothership. NASA/Dryden Carla Thomas

EC99-44923-157 Dryden Flight Research Center 05MAR99 X-38 v132 free flight NASA photo by Bill Isbell

FIGURE 7.11 *X-38* lifting body and parafoil test (Courtesy of NASA).

previous NASA cost models would have cost over $3 billion to develop alone. They key to the reduced cost to develop this vehicle was in its verification approach, which used a buildup method similar to modern aircraft testing for atmospheric flight phases. The basic lifting body shape was a modified version of the *X-23/X-24* shape that the U.S. Air Force had flown in the atmosphere and to space during the 1970s. This vehicle served as the basis for the X-38/Crew Return Vehicle design; and although modified for the Crew Return Vehicle role, the new design was in many ways pretested before the start of the program. Parafoil only flights were conducted to qualify the parafoil system. For reference, the parafoil system was the largest square parachute in the world, with a deployed wing area 50% bigger than that of a Boeing 747. Flights utilizing the lifting body and parafoil combination in the atmosphere of the Earth were conducted. Components of the *X-38/ Crew Return Vehicle* were tested independently on other vehicles. For example, the inertial global positioning system navigation system was tested aboard the *Space Shuttle* and the electromechanical flap actuator was tested aboard a NASA F-15 to qualify these components for flight. Afterward, a full set of ground tests and a spaceflight test were to be conducted. This buildup approach was an important factor for reducing the number of tests and the associated cost for the *Crew Return Vehicle*.

The X-38/Crew Return Vehicle project included several technology firsts. In addition to the largest parafoil ever and the first new lifting body shape flown in 20 a, the *X-38* project developed and incorporated the following new technologies:

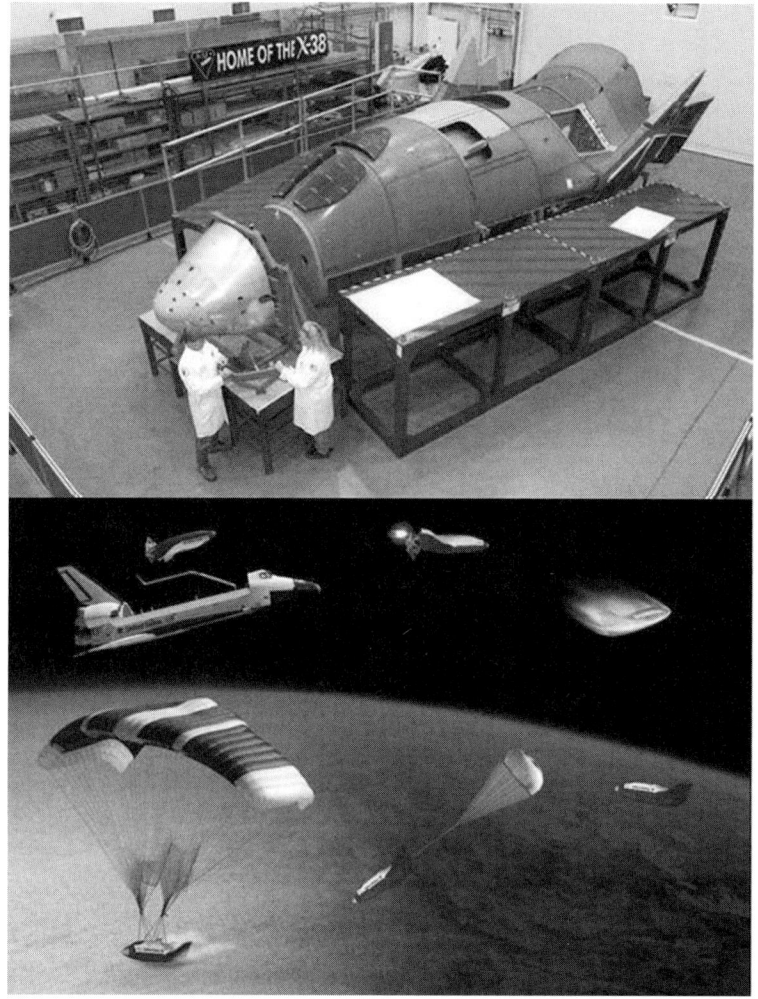

FIGURE 7.12 *X-38* space test vehicle and space test profile (Courtesy of NASA).

- Flush air data system for angle of attack and sideslip tied directly into the flight control system.
- Neural network for computing angle of attack and sideslip from flush air data system.
- Laser initiated pyrotechnics.
- Inertial platform alignment and attitude determination by the global positioning system on orbit without other attitude references.
- Electromechanical actuators for full authority flight control surfaces.

- Hot structures for nose cap and body flaps.
- Dynamic inversion flight control laws.
- Electromagnetic international berthing docking mechanism.
- Large space qualified lithium-ion batteries.

The X-38/Crew Return Vehicle represents the first attempt in space technology to build a complete spacecraft dedicated to the rescue and escape role. Future space rescue vehicles might or might not incorporate the technologies pioneered in this project. However, the requirements developed by the project to drive this design are more instructive than the actual technologies involved, because they represent the first clear statement for a rescue spacecraft designed from a clean sheet of paper.

Thirteen major requirements were identified for the *Crew Return Vehicle*. As further definition occurred, the medical community did an extensive review of what would be necessary for a successful space ambulance and identified additional requirements. These 13 major requirements are listed in Table 7.10.

On orbit, in the microgravity environment of space, deconditioning of the crew starts to happen almost immediately, although notable symptoms begin to occur only as the mission duration increases. Because of this time dependency, current constraints place a maximum duration for Space Shuttle flight at 19 d to minimize the effects of deconditioning on the crew.

It is interesting to examine how the medical community expanded the requirements to turn the *Crew Return Vehicle* into the first space ambulance. They identified a specific cadre of equipment to be carried onboard and a set of vehicle capabilities unique to the crew return vehicle role of rescue ambulance:

- "The [*Crew Return Vehicle*] shall accommodate medical life support for one ill or injured crewmember, including, but not limited to ventilation, physiological monitoring with defibrillation, intravenous fluid therapy, and pharmacotherapy services.
- Crew Return Vehicle emergency medical kit (16 in × 15 in × 10 in, or 2400 in^3 at 30 lbs)
  1. Bandages, equipment, and medications (1080 in^3 at 20 lbs)
  2. Pulse oximeter (50 in^3 at 1.2 lbs)
  3. Intravenous fluids pump (200 in^3 at 3.0 lbs)
  4. Suction device (14 in^3 at 0.5 lbs)
  5. Two liters of intravenous fluids (200 in^3 at 4.4 lbs)
  6. Ambu bag (250 in^3 at 0.9 lbs)
- Crew Return Vehicle hard mounted medical equipment (1200 in^3 at 24 lbs)
  1. Automated electronic defibrillator/monitor (768 in^3 at 17 lbs)
  2. Autovent ventilator/respiratory support pack (432 in^3 at 7 lbs)
- Totals 3600 in^3 at 54 lbs."

Most interesting to the vehicle designer are the implications of these capabilities on vehicle design. The X-38/Crew Return Vehicle team members spent a considerable time assuring themselves that defibrillating someone in a metal spacecraft cabin would not result in hazardous currents to any other crewmember. Physiological monitoring, ventilation, and intravenous fluids all required special accommodations in the vehicle design.

Table 7.10 Major Requirements for a Crew Rescue or Escape Vehicle

Requirement	Rationale
Operate in a shirtsleeve environment	Crew might have neither time nor physical ability to don a spacesuit
Dual fault tolerance in critical systems	Necessary for human rating
Landing within a 5 nautical mile radius	Minimize search and rescue time
Dry land touchdown	Deconditioned crew members can have trouble in an ocean survival situation; avoid the need for ocean recovery forces
Supports medical mission	To be used as an ambulance, an injured crew member must be able to ingress and egress the *Crew Return Vehicle*; it must contain medical equipment and get injured crew member to care within 24 h
Less than 4-g sustained load in flight	Support deconditioned injured crew member
7 crew members of 95% American men	Bring the entire crew down in a single flight
All attitude separation	Able to perform function even if the *International Space Station* has lost attitude control
Separation in less than 3 min	Able to perform function rapidly in case of a growing catastrophe onboard the *International Space Station*
Two-way communications with the ground	Ability to coordinate rescue
Autonomous capability, pilot not required	Surviving crewmembers might not be pilot trained
Operation in English	U.S. provided system
3-a life with 95% availability	Minimize costs associated with launch of replacement units

The medical community also identified a minimum survival kit for in-flight and postlanding survival and specified that the Crew Return Vehicle system must provide a minimum of 24 h of crew survival equipment and consumables, including but not limited to

- Survival kits for seven crew members, including clothing, shelter, hygiene, and rations.
- Potable water

    10.5 L (7 × 1.5 L for fluid loading).
    7.0 L (7 × 1.0 L for 24-h survival).
    (17.5 L = 17.5 kg = 38.5 lb).

Fluid loading refers to the procedure of having crewmembers drink large amounts of water prior to entry to prevent a sudden drop in blood pressure when they are back in an Earth gravity environment.

The medical team also placed requirements on the design that required the crew to be in a prone position at landing. As well, they levied challenging requirements on the life support system designer. To accomplish the ambulance function, the medical community required the capability to place an injured crewmember on 100% oxygen (4 lb/h) during the entire Crew Return Vehicle flight. This was a problem for the life support designer because each exhalation of a human breathing 100% oxygen contains considerable amounts of oxygen in the exhaled gas. In a small vehicle, this can lead rapidly to an atmosphere that is highly conducive to flammability. To accommodate this requirement, the *Crew Return Vehicle* was designed to be equipped with nitrogen tanks and computer controlled valves to dump part of the atmosphere overboard and subsequently repressurize with nitrogen to keep the inside of the cabin within flammability limits. Given the small volume of the *Crew Return Vehicle*, this limit otherwise would be reached quickly when a crewmember is breathing 100% oxygen.

The second challenging requirement initiated by the medical community was to purge the Crew Return Vehicle atmosphere. The medical community pointed out that a main scenario for Crew Return Vehicle use is the case where there is a fire aboard the *International Space Station*. In this situation, the Crew Return Vehicle atmosphere might become contaminated with smoke and toxic combustion products before the hatch is closed. To address the requirement, the life support system was provided with an ability to purge and replace the atmosphere, and the design included activated carbon filters to scrub the atmosphere.

7.1.10 Safe Haven

The final subject to be covered in space rescue is that of safe haven. When a crew in space is in trouble, the longer its members can survive represents more time to prepare a rescue mission, thus increasing the probability for success. On long duration missions to the Moon and Mars, safe haven technology is likely to be the only rescue capability available to the crew.

After the loss of *Columbia*, NASA decided to place a contingency shuttle crew support capability onboard the *International Space Station*. Prior to every Space Shuttle launch, a computation is made to determine how long the *International Space Station* could support its crew along with the Space Shuttle crew. To have rescue capability, the rescue vehicle would need to be readied for flight on an expedited basis in fewer days than the maximum capability of the *International Space Station* to support the lives of the station and the shuttle crew together. The experience of NASA with regard to this is instructive in that it can seem a simple task to compute the maximum life support capacity of the *International Space Station*; however, several questions immediately arise to define the bounding assumptions:

- What assumptions should be made about equipment failure rate?
- What assumptions should be made about equipment lifetime?

- Should hardware installed on the *International Space Station* but not yet exercised be counted as part of the capability?
- Should food, water, and exercise (oxygen) consumption be as planned, or should a reduced consumption rate be assumed?
- Should resupply of the *International Space Station* on normal schedules with non-Space Shuttle assets (Russian *Progress/Soyuz*) be assumed?
- Should it be assumed that three crewmembers would be evacuated immediately from the *International Space Station* using the *Soyuz* to decrease the overall life support requirement?

In the end, NASA decided to compute three separate estimates. One estimate is based on the entire *International Space Station* with the Space Shuttle crew remaining aboard the *International Space Station* being resupplied and consuming at nominal rates. When examining this number, perfect hardware and resupply performance is assumed; however, some reduction in the requirement also can be made by reduced food, water, and oxygen (exercise) consumption. A second estimate, called the *engineering estimate*, assumes a failure rate of hardware consistent with recent experience. For example, when the water electrolysis device, which generates oxygen on the *International Space Station*, had a high failure rate, engineering estimates included it as an already failed component. A third estimate, called the *worst case failure*, assumes the worst possible failures resulting in the minimum life support capability for the *International Space Station*. All three estimates are briefed at the flight readiness review, and the engineering estimate is compared to the number of days required to prepare a rescue vehicle before the launch of a primary Space Shuttle mission begins.

7.1.11 Conclusions

Space rescue is in its infancy as a technical capability. Given that human spaceflight remains the most risky of human flight endeavors for the foreseeable future, much more work needs to be done in this field. The attempts to put together space rescue capabilities from operational capabilities as well as attempts to design a custom space rescue system have been reviewed.

7.2 PERSONAL PROTECTIVE EQUIPMENT

7.2.1 Purpose of Personal Protective Equipment

The space environment poses numerous hazards and obstacles not encountered in a terrestrial environment. These hazards and obstacles generate unique challenges to the development of safety programs for human space systems. The development of an effective human space safety program begins with the identification of health hazards within the human space system. This hazard assessment results in the formation of a comprehensive list of hazards associated with the system. Five methods commonly are used to control and mitigate these hazards:

- Eliminate the hazard through design, for example, design a nontoxic substance to replace a toxic substance.
- Implement engineering controls, for example, require additional vessels to contain toxic material.
- Implement administrative controls, for example, require training prior to handling toxic material.
- Develop operational procedures, for example, develop specific handling procedures for toxic material.
- Prescribe the use of personal protective equipment, for example, require gloves when handling toxic material.

This section focuses on the use of personal protective equipment in human space systems and describes the various types of personal protective equipment and their usage.

7.2.2 Types of Personal Protective Equipment

The Occupational Safety and Health Administration defines *personal protective equipment* as "equipment that protects employees from serious injury or illness resulting from contact with chemical, radiological, physical, electrical, mechanical, or other hazards" (OSHA 2003). Any piece of equipment that eliminates or reduces a hazard can be considered personal protective equipment. Common forms of personal protective equipment include gloves, eye shields, and respirators. An analysis of hazards encountered in a human space program yields two broad classifications of hazards, physiologic and environmental. Therefore, personal protective equipment can be classified as physiological or environmental, depending on the type of hazard for which it is used.

Physiologic Personal Protective Equipment

As described in Chapter 3, "Overview of Bioastronautics," weightlessness causes a wide variety of physiologic effects on the human body that, over time, have negative physiologic consequences. The two consequences of most importance are the loss of bone density and the loss of muscle mass. These losses considerably reduce the ability of the body to function in a non-weightless environment and impair the performance of such tasks as piloting during atmospheric reentry and egress on landing. To mitigate these effects, physiologic personal protective equipment must be employed, which takes the forms of resistive and cardiovascular exercise devices.

Resistive Exercise Devices In a terrestrial environment, muscle and bone counteract the downward force of gravity. Because a human standing on Earth is using his or her muscles and bones constantly, they do not atrophy. In space, however, those same muscles and bones are not counteracting gravity, and therefore they do tend to atrophy, the result being a reduction in muscle mass and the loss of bone density. A strength training program can be implemented in space to encourage bone and muscle use, thus reducing the degree of muscle and bone loss.

Traditional strength training programs, consisting of leg and arm presses and leg and arm pulls using conventional weights do not work in a weightless environment. In place of weights, bungee cords are used to provide the resistance necessary to load bone and muscle actively (Figure 7.13). To exercise, one end of the cord is fixed while the other end is tied to a bar. This bar is pulled or pushed, thus stretching the cord, providing resistance. A variety of cords is used, each with a characteristic resistance. By adjusting the number and stiffness of these cords, varying levels of resistance can be achieved that correlate to varying weights in a traditional terrestrial strength training program.

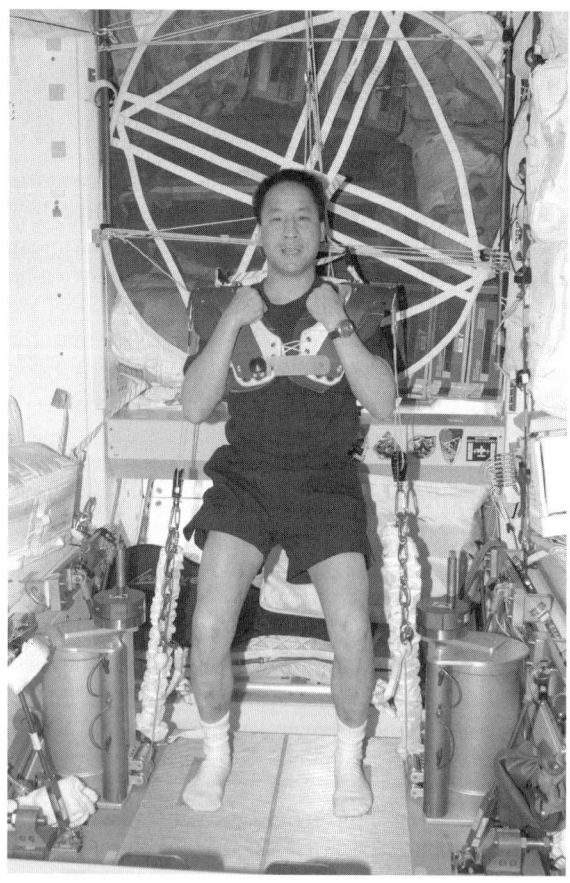

FIGURE 7.13 Example of a resistive exercise device. Expedition 7 science officer Ed Lu performs knee bends (Courtesy of NASA).

Cardiovascular Exercise Devices Cardiovascular exercise devices, such as a treadmill and a cycle ergometer, constitute another class of personal protective equipment that combats the deleterious physiologic effects of weightlessness. These exercise devices typically combine a cardiovascular exercise, such as jogging or cycling, with resistive exercise. Treadmills in space, in addition to providing a stationary platform for in-place jogging, have been fitted with bungee cords that can be strapped to the waist and shoulders of a user, thereby pulling the astronaut toward the treads on the treadmill (Figure 7.14). The bungee cords statically load the astronaut so that the physiologic effects of running in space are similar to those of running on Earth.

The cycle ergometer is the horizontal equivalent of a standard stationary bicycle in that the user is in a recumbent rather than an upright position. A cycle ergometer requires less

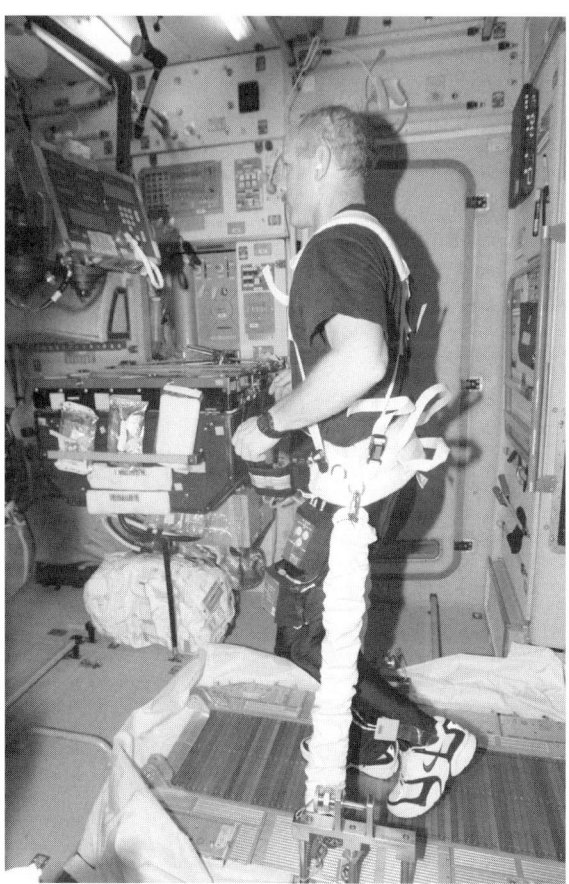

FIGURE 7.14 Example of a treadmill in space. Expedition 6 mission commander Kenneth Bowersox, wearing a body harness and earplugs, runs on the treadmill (Courtesy of NASA).

modification than a treadmill to provide cardiovascular training and resistive training. The cycling activity provides cardiovascular training while the resistance of the pedals provides resistive exercise. The resistance can be varied to meet the desired resistance of the person exercising.

Environmental Personal Protective Equipment

Human rated space vehicles pose numerous environmental hazards ranging from the extreme, such as electrocution or fire, to more moderate hazards, such as sharp edges. These hazards and others like them require mitigation in the form of environmental personal protective equipment. A unified and accepted code of environmental regulations for space does not exist, but the Occupational Safety and Health Administration guidelines, the American National Standards Institute standards, and the U.S. Code of Federal Regulations provide guidance in describing the types of personal protective equipment needed for an application. Occupational Safety and Health Administration publication 3151 (OSHA 2003) describes personal protective equipment for all parts of the human body. Table 7.11 provides a listing of the U.S. Code of Federal Regulations (CFR) and ANSI standards that define specific personal protective equipment requirements.

Head, Leg, Foot, and Body Personal Protective Equipment Injury to the head, leg, foot, and body can be caused by tripping, contact with sharp objects, and being struck by falling or rolling objects. In general, the greater the velocity at which a hazard of this type occurs, the more severe are the results. To combat such hazards in a terrestrial environment, personal protective equipment such as hard hats, steel toed boots, and shin guards are employed. In space, however, the likelihood of these hazards causing a major injury to the head, leg, foot, or body is low, because it is extremely difficult to achieve the velocity

Table 7.11 Governing ANSI Standards and CFR Regulations for Personal Protective Equipment

Code of Federal Regulation (CFR) for Personal Protective Equipment	
General requirements	29 CFR 1910.132
Eye and face protection	29 CFR 1910.133
Head protection	29 CFR 1910.135
Foot protection	29 CFR 1910.136
Electrical protective equipment	29 CFR 1910.137
Hand protection	29 CFR 1910.138
American National Standards Institute (ANSI) Standards	
Head protection	ANSI Z89.1-1996
Foot and leg protection	ANSI Z41-1991
Eye and face protection	ANSI Z87.1-1989

required for consequential injury. Therefore, the need for this type of personal protective equipment is low. For additional information regarding head, leg, foot, and body personal protective equipment, refer to Table 7.11 or Occupational Safety and Health Administration publication 3151 (OSHA 2003).

Eye and Face Personal Protective Equipment Space exacerbates hazards associated with ocular and facial damage. Exposure to contaminated particles, caustic liquids, or harmful gases and vapors can cause permanent ocular and facial injury on Earth, but the potential for exposure is much greater in space. On Earth, if a component fractures into numerous smaller pieces, such as a window being shattered, the shards are accelerated to the ground by gravity, thus eliminating the risk for ocular and facial injury (assuming that the face and eyes are not in the path to ground). In space, however, the shards would not accelerate to the ground but rather float in the environment, increasing the likelihood of coming in contact with the face or eyes. Air circulation fans generate currents and eventually do remove this type of material, but until it is removed, the material remains airborne and generates a credible risk for ocular or facial damage that must be mitigated by personal protective equipment. A wide variety of face and eye protection devices exist (OSHA 2003), including

- **Safety spectacles**, which are protective eyeglasses constructed of a metal or plastic frame and transparent impact resistant plastic lenses. They are worn over the eyes; however, along the top, bottom, and sides of these spectacles, there is commonly a gap through which particulate matter can travel and strike the face or eye. Side shields (flexible transparent pieces of plastic) attach to the spectacles to reduce the gap size along the sides, thus lowering the likelihood of particles reaching the eye.
- **Safety goggles** are tight fitting eye protection that completely covers the eyes, eye sockets, and the facial area immediately surrounding the eyes. They typically are constructed of a soft, flexible, transparent plastic and have a distinct advantage over safety spectacles because the seal created by the safety goggles considerably reduces the likelihood of particles reaching the eye. However, the material used to construct safety goggles is not as impact resistant as that used for spectacles.
- **Face shields** are transparent sheets of impact resistant plastic that extend from the eyebrow to below the chin and span the entire width of the head. Face shields share a weakness with spectacles, in that they do not form a tight seal around the face, so it is possible that contaminants still can reach the eye and face.

Face and eye personal protective equipment selected for use in space should

- Protect against specific known or anticipated hazards.
- Fit properly.
- Be reasonably comfortable.
- Provide unrestricted vision and movement.
- Be durable and easily cleaned.
- Allow unrestricted function with any other personal protective equipment.

Hand Personal Protective Equipment Astronauts use their hands to perform many tasks in space, thus exposing them to a number of hazards. Numerous types of commercially available gloves can protect against a wide variety of hazards. Potential applications for common materials used in the manufacture of these gloves are listed in Table 7.12. The types of gloves and their applications are categorized (OSHA 2003) as follows:

- **Leather, canvas, or metal mesh gloves** are sturdy and provide protection against cuts and burns.
- **Fabric and coated fabric gloves** provide varying degrees of protection against abrasion and chafing. The plastic based coating added to some gloves provides additional protection against abrasion and chafing, while also providing increased slip resistance.
- **Rubber and plastic gloves** are made from various types of rubber, such as natural, butyl, neoprene, nitrile, and fluorocarbon (Viton®), or various kinds of plastic, such as polyvinyl chloride, polyvinyl alcohol, and polyethylene (Silver Shield® and 4H®), and provide protection against hazardous chemicals and liquids.

The selection of a particular glove depends on numerous factors, including

- The type of chemicals handled.
- The nature of contact, such as total immersion or splash.
- The duration of contact.
- The grip and dexterity requirements.
- The need for thermal protection.
- Size and comfort.

Table 7.12 Applications for Common Glove Materials (Source: HSE 2000)

Chemical Group	Natural Rubber	Nitrile Rubber	Neoprene®	Polyvinylchloride	Butyl	Viton®
Water miscible substances, weak acids/alkalis	√	√	√	√		
Oils		√				
Chlorinated hydrocarbons						√
Aromatic solvents						√
Aliphatic solvents		√				√
Strong acids					√	
Strong alkalis			√			
PCBs						√

As a rule, the thickness of the glove material increases the protective properties of the glove but impairs grip and dexterity. All these factors must be evaluated to ensure a successful application.

Hearing Protection Sound propagates through the air via longitudinal pressure waves. These are created when an object, such as stereo speakers or the vocal cords, vibrate and create a localized disturbance in the density of the air. The disturbance is propagated by the molecules of the air vibrating in the direction of propagation (Tipler 1991). The human body perceives sound when pressure waves travel through the ear canal and reach the tympanic membrane, commonly called the *eardrum*, causing it to vibrate. These vibrations subsequently are converted to electrical impulses along the auditory nerve, which are interpreted by the brain as sound.

Hearing protection reduces the intensity of sound waves that reach the tympanic membrane by impeding the forward propagation of sound pressure waves. The noise reduction rating, expressed in decibels, is a measure of the effectiveness of hearing protection. A larger noise reduction rating correlates to greater hearing protection. Several of the most common types of hearing protection (OSHA 2003) follow:

- **Single use earplugs** typically are made of waxed cotton, foam, silicone, or fiberglass wool. They are compressed just prior to insertion into the ear canal and expand to their maximum volume once inserted. Achieving the advertised noise reduction rating values of single use earplugs depends heavily on correct insertion into the ear canal. This often requires training. Incorrect installation reduces the overall noise reduction rating of the earplugs.
- **Molded earplugs** typically are made of a flexible silicon or acrylic material molded to the ear canal of an individual by a professional. This type of earplug offers increased effectiveness over single use earplugs because they conform to the unique features of the individual, thus providing increased resistance to sound propagation through the ear canal. Additionally, achieving the advertised noise reduction rating values is more reliable because installation is much easier.
- **Earmuffs** are protective devices that completely enclose the ears, thus blocking the ear canal and reducing the amount of sound waves that reach the tympanic membrane. These are constructed of a variety of materials, and they typically employ a hard plastic outer shell with a flexible foam interior that seals around the ear. To be maximally effective, earmuffs require a perfect seal around the ear. Glasses, facial hair, long hair, and facial movements, all of which disrupt this seal, can reduce the effectiveness of earmuffs.

To select appropriate hearing protection, the loudness of the noise, as measured in decibels, and the duration of noise exposure must be considered. The louder the noise, the shorter the exposure time before hearing protection is required. Occupational Safety and Health Administration publication 3074 (OSHA 2002) and 29 CFR 1910.95 provide specific noise exposure limits, a summary of which can be found in Table 7.13 (OSHA 2003; CFR 2006).

Respiratory Protection Occupational Safety and Health Administration publication 3079 defines a *respirator* as a device designed to protect the wearer from inhaling harmful

Table 7.13 Permissible Noise Exposure

Sound level (dBA)	Duration of Noise Exposure per 24 h
90	8 h
92	6 h
95	4 h
97	3 h
100	2 h
102	1.5 h
105	1 h
110	0.5 h
115	0.25 h or less

dusts, fumes, vapors, or gases (OSHA 2002). Respirators have been developed by the military and private industry and therefore come in a wide range of types and sizes. They can be loose or tight fitting. Tight fitting respirators adhere closely to the skin and can cover just the mouth and nose or the entire face from the hairline to below the chin. Loose fitting respirators are typically hoods and helmets that cover the head completely and are often worn as a complement to a full body protection device. Respirators can be categorized as either air purifying or air supplied (OSHA 2002):

- **Air purifying respirators** use filters to remove harmful contaminants from inhaled air, and they range from simple dust masks to complex masks that filter a wide variety of harmful chemical contaminants. These respirators do not provide supplemental oxygen, and they are not recommended for use in oxygen deficient atmospheres or other atmospheres that are immediately dangerous to life or health. Examples of air purifying respirators are gas masks, chemical canister masks, and chemical cartridge masks.
- **Air supplied respirators** use an alternate source of breathable air to supplant the contaminated air. These range from supplied air respirators, that is, masks connected to flexible hoses that provide positive pressure breathable air, to self contained breathing apparatuses, that is, masks connected to a portable storage tank that provides positive pressure breathable air.

The selection of a respirator depends heavily on the environment in which the respirator is to be used, and the length of time for which the respirator is to be needed. Like all other forms of personal protective equipment, respirators often have targeted environments, where they operate most effectively. Particulate size, particulate concentration, and specific chemical compounds and vapors are just a few of the environmental characteristics that

Table 7.14 Respirator Selection Criteria

Hazard	Respiratory Options
Immediately Dangerous to Life or Health	
Oxygen deficiency	Full face piece, self-contained breathing apparatus certified for a minimum service life of 30 min
	Combination full face piece supplied air respirator with auxiliary self-contained air supply
Harmful gas or vapor contaminants or other highly toxic air contaminants	Full face piece, self-contained breathing apparatus certified for a minimum service life of 30 min
	Combination full face piece supplied air respirator with auxiliary self-contained air supply
	Gas mask
Not Immediately Dangerous to Life or Health	
Gas and vapor contaminants	Self-contained breathing apparatus
	Supplied air respirator
	Gas mask
	Chemical cartridge or canister respirator
Particulate contaminants	Any air purifying respirator with a suitable particulate filter for size and type particulates
Smoke and other fire related contaminants	Self-contained breathing apparatus

dictate the choice of respirator. Usage time also dictates respirator selection. For example, supplied air respirators with their bigger breathable gas sources can provide relatively long periods of breathable air, while self contained breathing apparatuses and cartridge respirators provide shorter periods of respiratory protection, because they are limited by breathable air source capacity and cartridge life, respectively (OSHA 2002). Table 7.14 provides a simplified version of the characteristics and factors used to select a respirator.

REFERENCES

Aldrich, A., and K. Rusnak. (2002) Oral history 2 transcript. Oral History Project. National Aeronautics and Apace Administration, Johnson Space Center.

Caidin, M. (1964) *Marooned*. New York: Dutton.

Campbell, R. (2003) Ejection summary. *Flying Safety* 59, nos. 1–2.

Cerimele C. (2002) Personal communication. National Aeronautics and Space Administration, Johnson Space Center.

Code of the Federal Regulations. (2006) *Occupational Safety and Health Standards: Occupational Noise Exposure*. CFR 1910.95, available from http://osha.gov/oshaweb/owadisp.show_document?p_table=-standard&p_id9735 (cited June 15, 2007).

Grimwood, J. M., B. C. Hacker, and P. J. Vorzimmer. (1969) *Project Gemini: Technology and Operations, a Chronology*. Special Publication SP-4002. Houston, TX: National Aeronautics and Space Administration, Manned Spaceflight Center.

Health and Safety Executive. (2000) *Selecting Protective Gloves for Work with Chemicals*. Health and Safety Executive publication, available from www.hse.gov.uk/pubns/indg330.pdf (cited June 15, 2007).

Hyle, C. T. (1968) *Launch Abort Philosophy for Manned Space Flights*. Internal Note 68-FM-104. Houston, TX: National Aeronautics and Space Administration, Manned Spaceflight Center.

Hyle, C. T., C. E. Foggatt, and B. D. Weber. (1972) *Apollo Experience Report—Abort Planning*. Technical Note TN D-6847. Houston, TX: National Aeronautics and Space Administration, Manned Spaceflight Center.

National Aeronautics and Space Administration. (2004) *Space Shuttle Crew Escape System*. Report NSTS-60518. Houston, TX: National Aeronautics and Space Administration, Johnson Space Center.

North, W. G., and W. B. Cassidy. (1963) Gemini launch escape. The Second Manned Spaceflight Meeting. National Aeronautics and Space Administration, Manned Spaceflight Center.

Occupational Safety and Health Administration. (2002) Publication 3079, available from www.osha.gov/Publications/osha3079.pdf (cited June 15, 2007).

Occupational Safety and Health Administration. (2003) Publication 3151-12R, available from www.osha.gov/Publications/osha3151.pdf (cited June 15, 2007).

Shayler, D. (2000) *Disasters and Accidents in Manned Spaceflight*. Chicester, UK: Springer-Praxis.

Tipler, P. A. (1991) *Physics for Scientists and Engineers*. New York: Worth Publishers.

Townsend, N. (1973) *Apollo Experience Report—Launch Escape Propulsion*. Technical Note TN D-7083. Houston, TX: National Aeronautics and Space Administration, Johnson Space Center.

Wade, M. (2007) Rescue page, available from www.astronautix.com/craftfam/rescue.htm (cited May 23, 2007).

CHAPTER 8

Collision Avoidance Systems

Michael J. Eiden, Dipl. Ing.
Senior Advisor for Multidisciplinary Mechanical Systems, European Space Research and Technology Center, European Space Agency, Noordwijk, the Netherlands

CONTENTS

8.1. Docking Systems and Operations 268
  By Dr. Andrey V. Yaskevich
8.2. Descent and Landing Systems 280
  By Michael J. Eiden, Dipl. Ing.

Collision avoidance systems for space applications cover a wide range of safety critical functions and key enabling elements for space mission scenarios. They comprise on-orbit rendezvous and docking systems, essential for crewed and non-crewed space missions, as well as entry, descent, and landing systems. The latter are the keys to successful recovery of capsules with crew or payloads from orbital and suborbital flights, such as sounding rockets, and from missions to the surface of other planetary bodies within our solar system. From a technological point of view, collision avoidance systems can be subdivided into on-orbit systems and descent and landing systems, which are employed for deceleration and touchdown on the surface of the Earth or other planetary bodies. Because collision avoidance systems are clearly potential single point failures that can result in the loss of life or the complete space mission, the reliability and safety of these systems are of utmost importance for both crewed and uncrewed space missions.

This chapter presents a discussion of docking systems and parachute descent and landing systems. On-orbit rendezvous systems are employed in the early phases of dual or multiple spacecraft interactions, with the goal to prepare adequate approach conditions for subsequent docking operations. However, this is a subject beyond the scope of this text.

8.1 DOCKING SYSTEMS AND OPERATIONS

8.1.1 Docking Systems as a Means for Spacecraft Orbital Mating

Docking is a controlled mechanical process for spacecraft orbital mating. It begins at the moment of spacecraft first contact after the final approach. It is performed with a docking system that includes docking assemblies and control avionics. Methods for the design, simulation, and testing of different docking systems systematically were reviewed first by Syromiatnikov (1984). The docking assemblies themselves actually perform the mechanical process of spacecraft mating, whereas the initial conditions for docking are provided by a rendezvous system. This process can be split into four main stages:

- The first stage begins with first contact and ends with an acquisition of a primary mechanical connection or capture of the docking assemblies.
- During the second stage, the major portion of the kinetic energy of spacecraft relative motion is absorbed and dissipated (attenuated).
- The third stage begins with the issuance of commands to align the relative positions of the spacecraft and ends with retraction of the docking assembly interfaces to full contact.
- At phase four, a stiff and solid mating of the docking interfaces is ensured.

The docking assemblies on each spacecraft have guides specially designed to provide for mutual alignment during the contact interaction. Elimination of any misalignment between the guides is essential to facilitate capture.

Several mechanisms are used in docking assemblies to apply consecutive restrictions on the relative displacement of the joining spacecraft. Normally, only one of the two docking assemblies actively functions during the docking sequence. Termed the *active docking assembly*, it is located on the approaching spacecraft. This implementation is important because it permits the designers and operators to concentrate the major docking control functions in just one spacecraft.

Docking assemblies consist of a docking assembly body, mechanisms, and other devices. The docking mechanism is installed along the longitudinal axis of the active assembly body, and it performs the following functions:

- Elimination of misalignments between the guides for acquiring capture.
- Attenuation of the kinetic energy of spacecraft's relative motion, and restriction of its relative displacements.
- Alignment and full contact of the docking interfaces.

To compensate for spatial misalignments, the mechanism provides for several degrees of freedom. It contains special function devices, such as springs and dampers, that serve to attenuate kinetic energy. The docking mechanism includes a mechanical drive that is used to extend the docking mechanism to the initial position for docking, and it provides mechanisms for alignment and establishing full contact of the docking interfaces. Finally, guides are provided to affect a gradual decrease of any misalignment between the active docking mechanism and the passive assembly once capture occurs.

Capture mechanisms are installed on the docking mechanism. Their links can move under the action of mechanical drives, that is, an active mechanism, or under the action of external contact forces by overcoming spring mechanical resistance, that is, a passive mechanism. The spring return links of the passive mechanism are called *latches*. They engage with the mechanical stops on the mating docking assembly body. Mechanical drives on both active and passive docking mechanisms set the latches to their initial position during preparation for docking and return them to that position after a stiff interface closure is acquired.

Docking interface closure mechanisms are part of the active docking assembly body. These mechanisms can have either an individual or a common actuator. Regardless, the mechanism is activated after the docking interfaces have achieved full contact. Docking interface closure mechanisms provide for a rigid, strong, leakproof link by tightly fastening the interface elements. These elements subsequently are released by the drive during preparation for undocking. Depending on the requirements applied to the design of a docking system, some of these mechanisms can be eliminated or replaced by components that function based on different principles of physics. For instance, magnetic devices can be used instead of capture mechanisms.

Sensors installed within the mechanisms of both active and passive docking systems provide insight into typical docking events. Using sensor data, the avionics of the active assembly assesses the status of the process and performs automatic control. The signals, first contact, capture, and so forth are transmitted to spacecraft control systems and provide the basis for the vehicles to change their motion.

Docking systems are based on the probe and cone (Bloom 1970; Syromiatnikov 1984; Rivera, Motaghedi, and Hays 2005) as well as peripheral or even hybrid designs. A peculiarity of the probe and cone docking system is its strict separation of functions. Only one of the docking assemblies can be active, whereas the other only passive. The docking mechanism probe and the receiving cone of the mating assembly act as guides. The capture mechanism latches are located at the end of the probe, whereas the mechanical stops of the mating assembly are located at the apex of the receiving cone. The probe and cone system provides capture once the end of the probe extends into the apex of the receiving cone. In other docking assembly designs utilized for micro-satellites, a flexible probe, post, and receptacle function as attenuation and alignment devices (Rivera et al. 2005).

The active side of a peripheral docking system contains a ring shaped docking mechanism (Syromiatnikov 1984; Burns, Prise, and Buchanan 1988; Gonzalez-Vallejo, Fehse, and Tobias 1992; Urgoiti and Belikov 1999; IBDM 2005) connected to the assembly body by six multilink mechanisms possessing energy absorbing and other control devices. The end joints of these mechanisms are located around the perimeter of the ring and the docking assembly body.

The fixed mechanical rings that are part of the active and passive docking assemblies are termed *docking rings*. Several guides are installed on each docking ring, each guide having a conical petal shape. If the geometry of the design satisfies the principle of inverse symmetry (Syromiatnikov 1984) and both docking units have identical mechanisms, either can be active or passive; that is, they are androgynous. These are components of the capture mechanisms normally installed on the docking rings.

Peripheral androgynous docking systems require more precise guidance than the probe and cone system. The requirement for greater precision results from the use of identical guides on both docking adapters, the dimensions of which (in plane) cannot exceed one half radius of the docking ring. The capture condition for a peripheral docking system is subject to a proximity of no less than three pairs of characteristic points that correspond to locations on the capture mechanisms of the docking mechanism ring and mechanical stops on the passive docking assembly body. To achieve this condition, almost full closing is required of the docking interface planes. Therefore, provided the same initial conditions apply, a larger margin of kinetic energy is needed for capture with peripheral docking assemblies versus probe and cone systems. These drawbacks are more than compensated for by the greater capability of peripheral docking systems to attenuate relative motion kinetic energy and its ability to restrict relative displacements. Only this type of system can be used to enable docking for spacecraft like the *Space Shuttle*, which is characterized by a large mass and a center of mass eccentrically positioned with respect to the docking assembly axis.

Hybrid docking systems are based on a combination of the two previously presented types of docking assembly designs. For example, the Gemini space vehicle docked with the Agena rocket stage by using a docking unit having a movable cone installed on a peripheral mechanism (Meyer 1966). A system combining a compliant probe and peripheral guides can be developed for berthing robotic operations.

8.1.2 Design Approaches Ensuring Docking Safety and Reliability

Docking operations safety for a specific flight and type of spacecraft is achieved by implementing a design with no possibility of failure or contingency. The reliability of a docking system is defined by the product of the stability of its qualitative and quantitative characteristics and its suitability for a specific operation. An acceptable docking system design must meet the following requirements:

- One failure must not prevent the execution of the required system functions.
- Two failures must not lead to a catastrophic situation leading to loss of crew or spacecraft.

An off nominal situation can occur at any docking stage, the consequences for flight safety being different at each. If capture is not achieved, uncontrolled relative spacecraft motion can cause the spacecraft to impact catastrophically. Allowable interface loads can be exceeded or the spacecraft can impact during attenuation of the relative motion kinetic energy. Motion can occur during alignment, and contact of docking interfaces can be interrupted due to a failure of a drive, the avionics, or because of some external mechanical obstacle. Failure of closure mechanisms can result in the docking interface opening or lead to a situation where a nominal undocking is impossible.

Docking safety and reliability are provided by the following:

- Consistency with general design principles for docking systems.
- Design solutions based on in-depth studies of docking features, particularly by using mathematical simulations.

- Structure stress analysis.
- Detailed multistage testing.

The following general design principles, which ensure docking process safety and docking system reliability, remain valid for different design solutions:

- Geometrical and inertia parameters for docked spacecraft, geometric and kinematic rendezvous system errors, and recovery capabilities must be made coherent (the first prerequisite for docking system design).
- Kinetic energy and control system capability of the active spacecraft must be used for increasing capture reliability.
- Guide geometry must ensure alignment and exclude seizure.
- Energy attenuation device type, kinematics, and control of mechanisms must be adequate to the environment, and the number of idler wheels and links should be minimized.
- Mechanism drives must be redundant.
- Design of mechanisms and docking structures must ensure minimal distance between a load source and counteracted devices.
- Functional redundancy for sensors and data transmission must be provided at the avionics design level.
- Sensor data should provide monitoring of all typical docking events, particularly capture, completion of mechanism motion, and initial or final position.
- Number of ready to dock preparation operations must be minimized.
- Untimely operations of high criticality, such as undocking, must be inhibited.
- Docking interruption must be foreseen at any operation stage in an off nominal case.

Design solutions largely affect the quality, safety, and reliability of operations. The design is driven by mechanical processes featured at each docking phase. These critical processes can be described adequately using classic theoretical mechanics, impact theory, strength theory, and modern theory of rigid and flexible multi-body systems.

Parameters used for mating spacecraft, assessing approach errors, and determining an appropriate docking system type preliminarily are coordinated by comparison of the maximum lateral and angular misalignments of the assemblies at first contact, the geometry of their guides, relative motion kinetic energy, and the characteristics of docking mechanism attenuators. The type and associated characteristics of attenuators, the kinematical chains for the docking and capture mechanisms, and the methods for their control are selected based on accumulated design heritage to ensure maximum docking safety and reliability. Detailed mathematical simulation is used to compare any potential alternative solutions that account for the effect of spacecraft control systems on the docking process and evaluate the operational docking and undocking capabilities in off nominal cases.

Selection of Attenuation Devices and Mechanisms Kinematics

Attenuation devices should provide a low mechanical resistance for the docking mechanism during contact before capture. The mechanism should be self centering after unloading so that attenuation capability is recovered before the next impact. However, these devices must be able to attenuate a large amount of kinetic energy effectively after capture. These essential requirements easily are met using spring mechanisms and controlled dampers.

A spring mechanism represents an assembly of serially preloaded springs with limited strokes. Stroke limitation for each preloaded spring, as well as for the entire mechanism, is provided by mechanical stops. Spring mechanisms with symmetric load characteristics are used to self center the docking mechanism.

A damper can be a hydraulic, electrical, or friction device. However, highly efficient hydraulic dampers are sensitive to the vacuum and temperature differentials of the space environment. The friction clutches of such devices have a short lifetime and exhibit steeply rising resistance characteristics with time.

One type of damper is a rotational electromechanical device comprised of permanent stator magnets and a hollow rotor, that is, a so-called electromagnetic brake. It presents a simple design, requires no electrical power supply, and therefore is characterized by high reliability. The electromechanical brake is connected to the kinematic chains of a docking mechanism by a controlled feeble current electric clutch, which is activated after capture. This type of damper possesses kinetic energy attenuation capacity considerably greater than that exhibited by the probe and cone system. However, electromagnetic brakes do not provide for self centering of the mechanism; therefore, they can be used only to complement spring mechanisms.

An axial attenuator can utilize a friction clutch to provide limitation of axial interface loads. To provide for an axial displacement resource, the docking mechanism should be extended to its initial position. The maximum axial displacement is calculated, and the estimate is verified by mathematical simulation and testing.

The selection of a particular rotational electromechanical brake as a damper can introduce rotational spring mechanisms into the design. In this case, translation is transformed into rotation by ball-screw gears. The high gear ratio of this electromechanical brake permits the use of small, low power, high speed devices. However, the high gear ratio, apart from these advantages, considerably increases the total inertia of the mechanism. Additionally, ball-screw gears have rather low efficiency, in that they exhibit high friction losses that further increase at negative temperatures. These two factors can decrease the mobility of the mechanism prior to capture and increase the loads experienced during the kinetic energy attenuation phase of docking.

Despite these issues, high quality attenuation characteristics for a docking mechanism can be achieved using a hybrid system that combines rotational electromagnetic dampers and translational spring mechanisms in the design.

The kinematics of docking and capture mechanisms depend on the type of docking system selected. For example, the probe and cone docking mechanism must have a rotational joint at its axis to ensure axial and lateral mobility of the probe end. Accordingly, separate axial and lateral attenuators for relative motion kinetic energy must be utilized. Latches located at the probe end ensure capture. Because of the size limitation, the capture latches can be only spring loaded. They are located distantly from the kinetic energy attenuator, an unavoidable situation that leads inevitably to deformation of the probe and hence limitations on interface load levels.

For peripheral docking systems, multilink mechanisms that connect the ring to the docking assembly body have different kinematics that should eliminate docking mechanism workspace irregularities, such as loss of controllability. The capture mechanisms of peripheral systems should be located as close as possible to the end joint mechanisms of the ring through which contact forces are conducted from the attenuation devices to docking assembly body. This design reduces ring deformation and thus the probability of a capture loss under high interface load levels in the event of off nominal initial docking conditions.

Limitations on lateral and relative angular displacements are driven by the geometries of the docking interface and spacecraft body, which determine the maximum allowable strokes for the spring mechanisms of the attenuator. Load limitations for the mechanical stops in the spring mechanisms are ensured at the design stage by coordinating the value of attenuated kinetic energy with the attenuation capability of the docking mechanism.

Because latches or other mechanisms are moved by the docking mechanism during capture, the kinematic scheme of these devices can allow only simple motion with one degree of freedom. Docking interface closure mechanisms work at practically zero tolerance for misalignment at the initial point of engagement for guide pins and sockets as well as for electrical and fluid connectors. The narrow alignment tolerance allows for the use of simple kinematical schemes having one degree of freedom. The required closure force is provided by the high gear ratio of the drive, and thus movement of the mechanism links is essentially quasistatic. Considering any contradictory criteria, such as the number of redundant components and the cost, such closure mechanisms and their drives can be designed independently from other active docking subsystems.

Redundancy within the kinematic chains increases the inertia of the docking mechanism and the friction within its joints. Should a failure occur in even one of the redundant chains, mobility would be restricted, symmetrical operation would be compromised, and the function of the entire mechanism would be degraded. Redundancy, therefore, can be used only for independently moving capture and interface closure mechanisms having only one degree of freedom. Redundancy, however, is provided for sensors located within the kinematic chains of every mechanism of the docking assembly and interface.

Selection of Mechanism Control Methods for Different Docking Stages During docking, the first physical contact must initiate the docking sequence. Once initiated, the process continues regardless of any change in the contact between the two interfaces to ensure capture.

For more severe capture conditions, other design solutions that involve use of peripheral mechanisms can be utilized to facilitate mating. However, the advantages for these design solutions are not apparent. For example, controlled gripping levers have been suggested for use with peripheral docking systems (Gonzalez-Vallejo et al. 1992; Urgoiti and Belikov 1999); however, these increase not only the inertia but also the design complexity for the system. Moreover, additional avionics is required to activate the gripping levers, further reducing the overall reliability of the docking system. Another design approach is an active peripheral docking mechanism that ensures capture by utilizing specific algorithms to control a series of linear actuators connecting the ring with the docking assembly body (IBDM 2005). This approach relies on actively controlled drives rather than springs and electromagnetic brakes; however, such a system consumes a large amount of electrical power, and the potential for avionics failures reduce its reliability.

The process for attenuation of spacecraft relative motion kinetic energy is influenced strongly by the initial docking conditions. As compared to mechanical latches, electromagnetic capture devices exhibit a lower load capacity during the relative motion kinetic energy attenuation stage of docking. Additionally, an active control system used during this stage requires a large amount of power. Another important consideration is that the avionics required to control an active docking mechanism are complicated, and because the system is engaged fully during docking, the probability of failure is increased. Therefore, the most effective, simple, and accordingly reliable docking system design is one that utilizes an attenuator with electromagnetic brakes, where the most simple means of control provides for activation and deactivation of its clutches during docking.

During final alignment and closure of the docking interfaces, considerable torques and forces are required for preliminary engagement of electrical and fluid connectors. During this stage, actively controlled drives possess the same deficiencies as those used during kinetic energy attenuation. The most simple and reliable solution is the use of an uncontrolled electrical drive having a high gear ratio and special mechanical devices for alignment.

Requirements for Mating Spacecraft Control Deactivation of active spacecraft attitude control systems and the subsequent activation of thrusters that generate axial post-contact forces increase capture reliability even under extreme initial docking conditions. The thruster firing timeline for casual contacts nominally is fixed and activated on the first contact signal from the avionics of the active docking assembly. The crew manages the process during manual docking control for the spacecraft.

After capture, post-contact forces can remain for some time because of various characteristics of the docking system design and certain features of the active spacecraft control system. Lingering interface loads due to resonance phenomena can be diminished by deactivation of the passive spacecraft attitude control system once the capture sensors of the passive docking assembly are triggered.

Possible Methods of Docking Interruption in Off Nominal Situations Methods for interrupting the docking process are dictated by the stage within which an anomaly occurs. For example, if a predefined capture time limit is exceeded, then the active spacecraft control system can fire its thrusters for retreat. This process is used to avoid collisions, thus ensuring safety in case of emergency.

In the case of a docking mechanism failure that occurs before interface closure, the spacecraft can separate after using the capture mechanism drives to open the latches. Should the spacecraft then be unable to separate, pyrocartridges, located at the attachment points of the docking mechanism to the active docking assembly body, are fired. Similarly, for the case of an interface closure failure, separation is performed by firing pyrocartridges located at the attachment points to the fastening elements.

Accounting for Operation Environment Factors

Vibration, acoustic, and thermal loads are induced on docking assemblies and avionics during launch. Vacuum, solar radiation, mechanical, and thermal cycling loads are the most severe factors encountered during orbital flight. To account for these, the designer must consider, in particular, selection of materials compatible with the harsh space environment, component locking methods, inclusion of tolerances for temperature deformation compensation, and other factors. Docking system operational conditions are well known and provide a sound basis on which the adequacy of design can be assessed by analyses and testing of the various components and assemblies of the mechanism and avionics.

8.1.3 Design Features Ensuring the Safety and Reliability of Russian Docking Systems

The building and operation of the *International Space Station* are supported by two Russian docking systems, including the space station berthing mechanism (Foster et al. 2004). The Russian Soyuz and Progress vehicles and modules dock to the *International Space Station* using the probe and cone system. Space Shuttle vehicles dock using the androgynous peripheral docking system. All active spacecraft generate a post-contact force to accelerate capture. Once capture is acquired, the International Space Station attitude control system is deactivated.

Probe and Cone System

In the Russian probe and cone system (Figure 8.1), the ball-screw gear transforms the axial translation of the probe screw into rotation of the spring and axial electromagnetic brakes. The axial attenuator resistance force then is restricted by a friction clutch. Attenuation of the lateral probe motion is accomplished by a combination of translational spring mechanisms and rotational electromagnetic brakes. The geometry and size of the probe and cone units allow for considerable uncontrolled relative displacements after capture.

The attenuation capability of the docking mechanism is relatively low; therefore, the probe and cone system is used for spacecraft with low eccentricity for the center of mass with respect to the docking axis. Docking interface alignment is ensured by the contact with the receiving cone and the levers extended from the docking mechanism structure. On the probe head, four kinematically connected contact sensors detect the cone surface and two contact sensors detect the socket bottom. Four capture latches on the probe head interlock with four stops in the cone socket. The capture signal is initiated by redundant sensors on the probe head and in the socket of the passive assembly. If it becomes impossible to complete the interface closure, the entire docking mechanism can be jettisoned by activation of pyrocartridges.

Androgynous Peripheral Docking System

Ball-screw gears and differential mechanisms transform the linear and angular displacements from the peripheral docking mechanism ring (Figure 8.2) to rotation of the friction clutch and to two groups of spring mechanisms and electromagnetic brakes. One of these groups attenuates energy from the relative lateral motion and roll rotation, whereas the other, having higher attenuation capability, absorbs and dissipates the energy from axial motion, as well as from pitch and yaw rotations. The resistance force of axial docking mechanism is determined by the friction clutch.

FIGURE 8.1 Probe and cone docking assemblies (Courtesy of S. P. Korolev Rocket Space Corporation, Energia).

Capture is accelerated by the separation of the relative motion of the docking assemblies through the use of differential mechanisms within the docking system. These minimize contact loads by permitting rotation of the docking ring. Such rotation of the docking rings allows for optimal selection of attenuator parameters, thus reducing the relative motion of the ring and spacecraft. Because uncontrollable relative displacement of the docking assemblies after capture are negligible with this system, use of the androgynous peripheral docking system sensors are the safest means for docking spacecraft having high eccentricity of the center of mass with respect to the docking assembly axis.

Following the process of attenuation, docking interfaces are aligned by extending the docking mechanism to its forward position up to the stops in the screw mechanisms. This position is maintained with mechanical locks during the process of retraction and up to full contact of docking interfaces.

Contact of docking rings is controlled by use of relative displacement sensors. Three pairs of (redundant) latches and their associated drives are located on three guide petals.

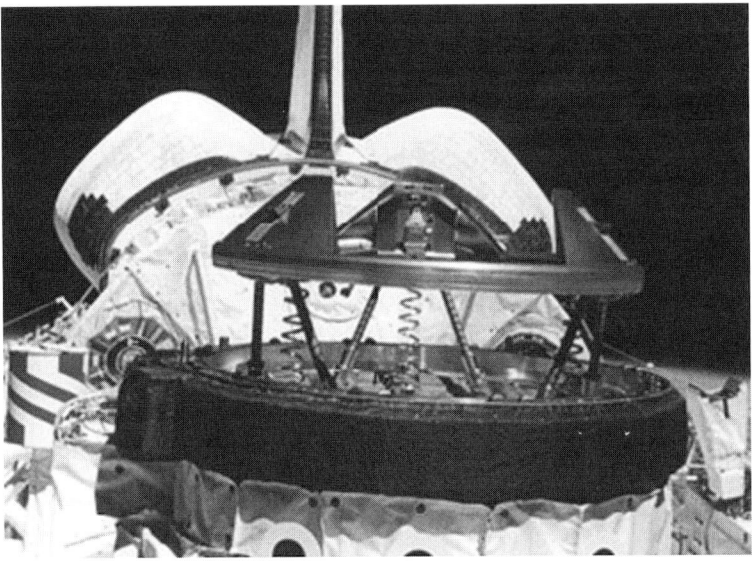

FIGURE 8.2 Androgynous peripheral docking assembly on a *Space Shuttle* (Courtesy of NASA).

The capture signal is generated when three of these sensors, which are located on each docking ring and connected in series, operate simultaneously.

For both types of docking systems, the docking mechanism is moved either to the initial (fully extended) or final (fully retracted) position by redundant drives, a process controlled by contact sensors. Retracting the docking interfaces presses spring return pushers to separate the spacecraft during undocking.

Closure of Docking Interfaces

Docking interface closure for the probe and cone and the androgynous peripheral docking systems is performed using hook mechanisms located in the structure latches of the docking assembly body. Safety and reliability of the hook mechanisms are ensured by

- Minimizing the number of strength critical parts and increasing their strength margin by analysis and tests.
- Use of the hook shape and kinematics to transform an interface tension force to a hook tightening torque.
- Redundancy for the hook closure drive.
- Redundancy for the sets of active and passive closing hooks.
- Use of inhibitors in the hook drive activation procedure.
- Monitoring the active hook status via telemetry onboard the spacecraft and in the Mission Control center.

Off nominal separation should a hook mechanism failure occur is performed by firing pyrotechnics, which results in active and passive hook opening and subsequent operation of the pushers.

8.1.4 Analyses and Tests Performed for Verification of Safety and Reliability of Russian Docking Systems

Before actual use, design solutions must be verified by strength analyses and tests. Strength standards define all types of analysis and tests to verify static and dynamic strength, vibration strength, and durability of a structure. Static, dynamic, and cycle load capability are calculated and proven by testing, while taking into account a safety factor that implies a certain uncertainty factor for load levels. In addition, a margin factor is taken into account during the design phase, the value of margin depending on the criticality of the parts or unit. Structural parameters for most critical elements, interface hooks, are determined by considering the highest value for the yielding margin. Strength analysis is conducted using formulas from strength theory or finite element software. The capacity of the structure to work under ultimate operational loads and strength margins are verified during strength tests performed as part of the qualification test. For newly designed docking systems, preliminary tests can be performed to verify the design solutions applied.

The Qualification Test

The qualification test for a docking system normally is run in two stages. The first stage verifies the ability of the system to work within operational conditions as defined in the specification, such as the vacuum, mechanical, thermal, and electrical environments, with a specified number of corresponding cycles. During the second stage, a margin of capacity to work within a wider range of operational conditions is defined. The docking assembly and its avionics are tested separately because of different operational requirements. The docking assembly is controlled with autonomous control panels. The functionality of avionics is verified using specific test facilities.

The first stage of the docking assembly qualification testing includes static, vibration, acoustic, thermal vacuum, shock, and dynamic strength tests. Static tests are performed on the entire assembly, while taking into consideration a safety factor. Maximum limit load levels are defined by analysis and verified by detailed mathematical simulation. The load spectrum for acoustic and vibration tests is defined by analyzing the launch process. Each type of test is completed with a functional check of the assembly.

Mechanical docking under vacuum and thermal environments is verified using a simplified test facility that reproduces a limited number of initial docking conditions. This facility can be placed into a thermal vacuum chamber reproducing different combinations of flight conditions: temperature, vacuum, and kinematics at first contact.

A dynamic test on a six degrees of freedom test facility is performed at the end of the first qualification stage. It provides functional verification of docking assemblies under a full range of spatial initial conditions, as well as under positive and negative temperatures.

The first qualification stage for the docking system avionics includes thermal, vibration, thermal vacuum cycling in a vacuum chamber, and a functional check under maximum and minimum voltages. A functional check of the avionics is performed after each type of test, taking into account failure tolerance. The success criteria for each type of test of the first qualification stage are the retention of the capacity of the system to function and support of the specified dynamic load levels.

At the second qualification stage, the range and the number of cycles for mechanical, thermal, and electrical loads are extended. Short duration load peaks also are reproduced. Mechanical tests are performed for the capture and hook mechanisms and docking assembly structures. Mechanical cycling loading parameters for docking interfaces are determined based on operational conditions for spacecraft docked on orbit. Docking assembly structures fastened by hooks are tested for durability, taking into account maximum interior atmospheric pressure. The hardware can undergo intentional destruction during testing. Functional check of docking system avionics is performed at voltages exceeding allowable limits.

Acceptance Test

The acceptance test validates the flight prototypes of the docking system. The relevant vibration, dynamic, thermal vacuum, and leak tests are performed under lower load levels and in a lower scope with regard to qualification. This stage is completed with functional checks of all mechanisms and sensors, calibration of relative displacement sensors, and checks of compatibility for mechanical and electrical interfaces. If the docking system has passed the acceptance test, it can operate in real flight.

End-to-End Test of the Docking System

For newly designed docking systems, an autonomous end-to-end functional test of its assemblies and avionics can be performed after the completion of the qualification test. This test would be conducted using an autonomous test facility, without reproducing mechanical, thermal, and other space environment conditions.

If flight avionics is supposed to communicate with a flight computer, then the docking system is tested in a frame for ground testing flight software. Here, a check of the communication between the docking system and other spacecraft systems is performed on an end-to-end test facility that reproduces the functionality of the entire flight hardware. Similar end-to-end tests are performed after prelaunch integration of the spacecraft.

Flight Test of the Docking System

Flight safety for the docking system is provided by a technical support team in the Mission Control center. Basic data to be analyzed in real time are video and telemetry from the sensors of the docking assemblies and their mechanisms. The data are displayed on video monitors and stored in the database. Post-flight analysis is done to reproduce docking kinematics and dynamics. A detailed post-flight analysis report is produced, capturing all the peculiarities of the docking operation.

At the time of this writing, more than 230 dockings have been performed with the probe and cone system and more than 30 with the androgynous peripheral docking system. All were successful.

ACKNOWLEDGMENT

I acknowledge the efficient docking safety discussions held with my colleagues V. Pavlov, E. Bobrov, S. Temnov, A. Subchev, and L. Ostroukhov.

8.2 DESCENT AND LANDING SYSTEMS

Human space exploration critically depends on mission operations involving Earth reentry, atmospheric descent, and landing for safe recovery of astronauts and payloads (Eiden 1989). Recent exploration developments for Moon and Mars are turning from reusable winged entry and descent vehicles like the *Space Shuttle* to embrace the use of capsule type vehicles. By doing so, the strong position of parachute landing systems as a key factor in mission feasibility is confirmed. In addition to Earth reentry and descent, parachutes are the basic landing system used for placement of robotic payloads on planetary bodies. Human rated launch vehicles similarly employ parachute systems for crew escape and launch abort emergency systems applicable to a range of situations that could occur prior to reaching orbit. Even winged reentry vehicles like the *Space Shuttle* employ parachutes as the final speed retardation device after touchdown (Figure 8.3).

Descent and landing systems can be classified roughly into systems that employ onboard propulsion systems for entry and descent speed retardation and those using

FIGURE 8.3 STS-121 mission: *Space Shuttle Discovery* landing with drag chute (Courtesy of NASA).

aerodynamic decelerators. The latter beneficially employ atmospheric friction for energy efficient descent and touchdown. In contrast to entry, descent, and landing operations on planetary bodies having atmosphere, similar operations for crewed spacecraft or robotic vehicles on planetary bodies devoid of atmosphere, such as Moon or Mercury, must depend entirely on the use of onboard propulsion systems or retrorockets. However, use of a propulsive braking system greatly limits the achievable mass of the landing vehicle reaching the surface, thus placing strong limitations on the range and scope of possible missions that can be conducted.

Earth and the other planetary bodies of our solar system that have an atmosphere, such as Mars, Venus, and Titan, provide conditions for use of the most volume and weight efficient descent and landing systems existing, aerodynamic decelerators. Such systems are based on parachutes inflated by dynamic pressure or related decelerator devices, such as the ballute, parafoil, or deployable parawing.

In view of design and performance differences involved in use of various aerodynamic descent and landing devices, it should be noted that the material in this section is limited to a discussion of parachute systems and concentrates on their safety related design and testing issues.

Although propulsive breaking systems have and will again be employed for lunar landings, parachutes today are employed in a majority of descent and landing operations and are to be used extensively in the NASA Constellation program. It is important to mention that the use of parachute landing systems for space applications generally requires the employment of deorbit modules and aerodynamic heat shield decelerators. These reduce the entry speed through aerodynamic breaking to a velocity of which safe deployment of a specific parachute type is possible. A typical entry and landing operation that employs parachutes is shown in Figure 8.4, which depicts the uncrewed landing sequence of the ESA *Huygens* mission to Titan (Achterman, Kapp, and Lehra 1986).

Space missions with human rated descent and landing systems that utilize parachutes as the primary breaking system are limited. To date, parachute descent and landing systems have found application on the Russian *Soyuz* capsule, the U.S. *Apollo* command module, the *Mercury* and *Gemini* spacecraft, and more recently the Chinese *Shenzhou* capsules. It is important to note that the Constellation program of NASA is going to utilize only parachute systems for Earth reentry of the human rated *Orion* capsule.

8.2.1 Parachute Systems

To establish a reliable and safe design for a parachute descent and landing system, it is essential to develop a set of comprehensive system requirements specific to the intended space mission. Key factors influencing the size of a parachute system must be established clearly, or if specific planetary conditions for the mission are known insufficiently, the atmospheric conditions must be estimated but with inclusion of considerable engineering margins. Key aspects to be covered include

- Planet type, with associated gravity, atmospheric composition and pressure, temperature, and density profile over altitude and lateral wind speed at touchdown.

FIGURE 8.4 Entry, descent, and landing scenario of the ESA *Huygens* probe to Titan (Image: ESA).

- Altitude for initiation of payload operations (this can be very demanding, as in the case of *Huygens*: 160 km) and achievable entry speed and altitude conditions (supersonic to subsonic speed) for reliable initiation of parachute deployment.
- Entry body geometry, thermal conditions, dynamic state, and stability at the time of parachute deployment.
- Functional performance requirements, such as recoverable payload mass, maximum allowable g level during descent for payload or crew, allowable touchdown speed, stability requirements at touchdown, and consideration of land or water landing.
- Storage volume and geometric constraints for parachute system implementation.
- Reliability, redundancy, and launch safety requirements.
- Environmental conditions and loads for the launch vehicle and on-orbit or interplanetary trajectory.

Based on these key requirements, the choice and complexity of parachute system design is determined. For example, the requirements can lead to either a single or multiple sequential parachute deployment with different sizes and types of parachutes. Furthermore, these requirements can be used to determine other components of parachute design, such as reefing or dis-reefing the parachute canopy, length requirements for parachute suspension lines and risers, and equipment and devices required for initiation of parachute deployment and control of autonomous sequencing of parachute staging operations.

The wide range of performance requirements can influence strongly the choice of parachute type and sequencing operations. Initial parachute deployment conditions could require the use of a supersonic parachute; however, because performance of this parachute in the subsonic regime might not be sufficient to retard the velocity of the vehicle to the maximum allowable for touchdown, a second subsonic parachute having high drag performance can be needed. In many cases, the final design choice for a parachute system can be determined only after many iterative advanced simulations for entry and descent operation. The result of these analysis processes is a satisfactory mission profile that takes into account any parachute design and performance limitations.

If, for reasons of mass and volume constraints, this parachute set also is to be used for launch abort and crew escape in a contingency situation, the complexity of the parachute system likely is to be increased greatly. The engineer is faced with the difficulty of finding design solutions adequate to the full range of flight and contingency conditions over the entire mission profile. The analytical and test effort required to ascertain a final design solution is proportional to the complexity of the mission and its demands.

Parachute Types and Associated Recovery and Landing System Devices

To fulfill a full range of mission and performance requirements for safe deceleration and touchdown for various space missions, a parachute system must be composed of many mechanisms and devices. The parachute system is necessarily complex to provide for and achieve successfully sequencing for all aspects of its operation from deployment initiation through touchdown. The various components that constitute the parachute and its rigging are the mechanisms for release and deployment initiation, for sequencing and descent control, and sensors and actuation devices along with their associated control electronics.

A parachute breaking system requires a precise and rather complex interaction of actuation and control mechanisms. Mission success largely depends on the flawless functioning of each device in an autonomous manner. Automated sequencing is essential because no real-time interaction or manual control generally is possible due to the great distances from Earth involved for planetary entry missions.

Because of the timing required and the speed at which sequencing events occur during a parachute deployment, automated sequencing is used to avoid any exceedance of design limitations or aerodynamic performance. Numerous examples of complex parachute sequencing performed successfully and with high reliability have proven the feasibility of safe parachute recovery systems.

Taking a planetary entry vehicle for a Mars or Titan mission as a generic example (Achterman et al. 1986), a number of key operations and associated mechanisms generally are involved in sequencing:

- A parachute deployment scheme that defines the intermediate sequencing and control of parachute stages is initiated. This process includes use of a timer, G-switch, a baroswitch, an altitude sensing device, or some combination of these devices to ensure that the parachute design loads are not exceeded, parachute deployment starts at the intended altitude, and when needed, the parachute is released at touchdown to avoid the vehicle or payload being dragged by shear winds.

- The stowed parachute compartment cover or reentry body aftercover employed to protect the parachute compartment from excessive entry heating is deployed to free a path for extraction of the drogue chute. The release mechanism utilizing a mortar or gun type pyrotechnic device ejects the cover with sufficient speed into the airflow. Note that the accelerated cover often is used to pull the drogue chute along with its containment bag from its storage location, extract the drogue suspension lines and riser, pull away the drogue chute bag, and extend the canopy into the undisturbed airflow.

- The drogue chute stowage bag and the drogue chute with its associated dereefing or reefing devices is used either to limit the initial deployment loads to the drogue canopy and suspension lines or to change the descent speed of the payload in accordance with a particular descent profile. Sometimes a parachute is pressure packed to save volume. The primary task of the drogue chute is to stabilize the vehicle and reduce descent speed to a velocity at which a main parachute of larger size can be deployed safely. The drogue chute can be either a supersonic or a subsonic parachute, depending on initial deployment and descent conditions. Because stability typically is poor in the transonic regime for ballistic entry vehicles and for capsules like *Apollo* when attitude control and guidance are not available, supersonic or transonic parachutes quite commonly are used as the initial device.

- A pyrotechnic cutter is employed to cut the drogue riser lines, releasing the drogue chute. Drogue chute release frequently is combined with the function of main parachute or main parachute pilot chute extraction and deployment.

- Mortar deployment devices for the main parachute (and the related pilot chute if one is used for main parachute deployment), the main parachute stowage bag, and associated dereefing or reefing devices with pyrotechnic reefing line cutters are used either to limit initial deployment loads on the main parachute canopy and the suspension lines or to change descent speed of the payload in accordance with a particular descent profile. If multiple main parachutes are used, they generally are deployed by individual pyrotechnic mortars to achieve highest reliability for the deployment function. As for the drogue chute, extraction of the main chute, its suspension lines, and riser from the descent vehicle, this usually is performed by the pilot chute.

- A parachute swivel device is employed if decoupling of the relative rotation of the deployed parachute and the vehicle is required.

- Finally, the parachute release device is initiated at touchdown to avoid dragging the landed capsule over the surface or possible drowning in case of a water landing.

The parachute types employed in space mission operations are represented by a rather limited range of identifiable parachute designs (Table 8.1). Without addressing any special designs, such as parafoils, parawings, and ballutes, parachutes roughly can be classified into supersonic and subsonic types. In these two categories, one can find a range of

Table 8.1 Parachute Types and Characteristics

Characteristics	Parachute System				
	Conical Ribbon	Disk-Gap-Band	Modified Ringsail	Cross	Ribless Guide Surface
Mach range, wind tunnel, free flight	$0.1 < M < 2.0$ up to $M \sim 3.0$ (WT tests)	$M < 0.5$ (WT tests) up to $M \sim 2.6$ (FF tests)	$M < 0.5$ up to $M \sim 1.4$ (FF tests)	up to $M \sim 1.64$ (FF tests)	
Drag coefficient	0.5 to 0.55	0.52 to 0.58	0.52 to 0.8	0.6 to 0.78	0.3 to 0.34
Opening load factor	~ 1.05 to ~ 1.3	~ 1.3	~ 0.1	~ 1.2	~ 1.4
Average angle of oscillation	0 to $\pm 3°$	± 3 to $6°$ (WT tests)	$\sim \pm 7°$	0 to $\pm 3°$	0 to $\pm 3°$
Canopy stability	Beginning of severe pulsation and ribbon flutter at $M > 1.5$	These parachutes were characterized by partial collapse and fluctuations of the canopy immediately after the first inflation peak at Mach numbers $M > 1.4$. The partial collapse was most severe for the disk-gap-band configuration and least severe for the modified ringsail system			No data available
			Stable for $M < 1.4$	Never stable in $1.1 < M < 1.64$	

CD = drag coefficient.
FF = free flight.
M = Mach range.
WT = wind tunnel.

concepts, each associated with a particular set of performance characteristics, for example, drag coefficient, dynamic stability, mass, porosity, deployment range, deployment loads, and the like, that are relevant for consideration in the choice of parachute system design. In view of the extensive effort necessary to validate a parachute design for a space mission application, design concepts often are reused in new mission applications with a delta qualification for the specific size and flight conditions anticipated. Therefore, the selection of a reliable and safe parachute system places great importance on the individual experience of the personnel involved, and the background of a manufacturer working with similar applications.

Parachute System Design

Because weight is a key factor for both spacecraft design and the feasibility of a space mission, it is very important to consider a complete systems engineering approach from entry to touchdown in any design of parachute descent and landing devices. Such an approach is essential to arrive at acceptable design solutions for parachute sizing and optimize the design for the individual components. The sizing of parachute system components must be performed with due consideration of functional performance characteristics, material limitations, and validated experience with the particular parachute type proposed. Compared to the mechanical sizing of spacecraft structures, the sizing of parachute systems is highly complex because it involves an interaction of aerodynamic, thermal, and structural analyses for which integrated tools generally are not available. Furthermore, the aerodynamic flow regime strongly interacts with parachute geometry, as characterized by the behavior of the inflated canopy and the suspension lines. The parachute behaves like a flexible body, whose shape depends on dynamic pressure and aerodynamic flow conditions. It can be exposed to a wide range of flight regimes, covering possibly supersonic, transonic, and subsonic descent in its deployed shape.

Also to be considered are the complex deployment dynamics of an entry vehicle with limited stability. The parachute, along with its riser or suspension lines, is either ejected into the rear airflow where it is affected by the entry forebody wake or laterally into the airstream. For the ejection phase, a purely analytical, quantitative determination of parachute behavior and loads barely is possible with available analytical tools. As well, this analysis is very difficult to support economically, considering the typical funding limitations for recovery systems. To have a highly reliable parachute system design, the engineer must rely on a combination of analytic tools, such as descent trajectory analysis, computational fluid dynamics, structural and thermal analysis, and multi-body dynamics tools, which are used in an iterative manner with parametric variations of minimum and maximum atmospheric conditions. Further iterations are likely to be required by the analysis because of nuances within the descent interrelationship with sizing, that is, mass and performance limitations, of the parachute components. Once completed, a detailed sizing prediction for components and their associated load assessment is established using a range of empirically derived equations by incorporating data obtained through extensive wind tunnel and flight testing experience.

The various types of parachutes employed for space missions are chosen either because of extensive experience from past space missions or there is sufficient evidence that they have gone through a vigorous development and validation program commensurate with the required reliability for use with a spacecraft or payload. From the performance perspective, some parachute types can be used in the low supersonic as well as in the subsonic regime; however, stability problems and drag performance often place constraints on the range of velocity for which they can be used. In view of such performance constraints, parachutes known to perform best in the supersonic regime typically are used for the initial retardation sequence at high speed. When a satisfactorily low level of subsonic velocity (Mach < 0.8) is achieved using the initial parachute, deployment of a subsonic main parachute is initiated for the final descent of the landing capsule. An entry and descent operation begins with the deorbiting of an entry capsule by either onboard propulsion or selection of a satisfactory entry angle, as is the case for a ballistic entry vehicle arriving at a planet from along a particular interplanetary trajectory.

For a heat shield protected vehicle, aerothermal braking generally is used to reduce the interplanetary or deorbit entry speed (on the order of 12.8 km/s) to a velocity for which successful deployment of a sequence of parachutes is feasible. Because the stability of a limited diameter entry vehicle is often critical in the transonic regime, unless it is controlled by an onboard attitude control system, a small diameter stabilization and retardation parachute, often called a *drogue chute*, is deployed at the beginning of the descent operation. The drogue chute exhibits high drag and stability performance in the supersonic and transonic regimes and sometimes is used to free the aftercover protection from the rear vehicle structure, where the main parachute is housed.

To deploy the drogue chute into the flow field, a small pilot chute, which is protected from entry heating by the capsule aerothermal structure, is ejected through a pyrotechnically operated mortar or deployment gun. The deployment either separates a cover plate or ruptures breakout patches located in the cover of the entry capsule. The aim is to eject the drogue chute with its containment bag either rearward or sideways with a typical speed of 5 to 10 m/s and as far as possible into the undisturbed flow field behind the entry capsule. The small pilot chute then pulls the drogue chute and its bag from the descending vehicle. Finally, the pilot chute unfolds, pulling the packed riser and suspension lines and the parachute canopy, which is then inflated by aerodynamic pressure from the containment bag.

The entry capsule, acting as the forebody of the deploying parachute, generates a substantial wake effect. This turbulence is likely to have a substantial impact on the stability and drag performance of a drogue chute and can be detrimental to a successful deployment. Therefore, parachute suspension line and riser lengths need to be of sufficient length to be independent of the flow regime, forebody size, and the chosen diameter of the parachute drag surface.

The choice of suspension line and riser lengths represents a complex two-body problem with a number of interacting performance parameters. A practical solution can be derived through computational fluid dynamic analyses using extensive wind tunnel and free flight test data. Alternatively, the selection can be made by limiting the choice of lengths to those utilized in existing validated parachute and forebody combinations and supported by information obtained from the flight experiences of earlier missions. As a general guideline, separation distances of 8 to 9 times the forebody diameter are required to achieve high reliability in parachute deployment and avoid substantial drag performance reduction for the drogue chute.

Important factors in the choice of parachute type are its stability, drag performance, inflation time history, associated peak loads, and the porosity of the canopy. All of these are derived either from tests or previous flight experience. Based on empirical data, parameter based sizing equations have been derived (Knacke 1992) that can be used by experienced parachute designers to produce approximations. These are sufficiently accurate to derive a satisfactory initial design for sizing a parachute landing system. A parachute system designed in this way sequentially undergoes detailed validation in a wind tunnel and during flight tests representative of the actual application.

In many cases, the peak loads experienced on inflation of a drogue or main parachute are found to be too severe for the intended full deployment of the parachute canopy. Deployment, therefore, is initiated using a reefed parachute diameter, that is, the mouth opening of the parachute canopy is restrained by a circumferential reefing line. After sufficient velocity retardation is achieved by deploying the reefed parachute, the reefing line is cut by a timer operated pyrotechnical device, thus allowing for full deployment of the unrestrained canopy. Descent with further speed reduction then continues with the full parachute diameter.

Once the velocity of the descending vehicle has been reduced sufficiently by the high speed drogue chute so that the transonic regime has been traversed, a second, generally larger parachute, called the *main chute*, is deployed within the subsonic flight regime. The function of the main chute is to reduce further the speed of the descending vehicle to a specified terminal descent velocity. It also provides stability and attitude control for the vehicle during the landing.

The main chute staging sequence is initiated when the velocity of the vehicle has been reduced sufficiently for deployment of the large main canopy. Initially, the drogue chute is released by pyrotechnic cutters that sever its load carrying attachment to the capsule. The main chute then is injected into the flow field. Deployment of the main chute canopy can be performed by simply releasing the drogue chute itself and having it pull the main chute along with its riser and suspension lines from containment located on the capsule. Alternatively, a similar procedure using mortar ejection for initial pilot chute deployment is applied as described earlier. During the drogue release operation and until the vehicle is stabilized by the main chute, the attitude of the capsule can change substantially. The designer must pay particular attention to any change in capsule attitude so as not to compromise the success of main chute deployment and inflation. As described for the drogue chute, the reefed main chute inflation is often the process used to limit the loads for mass efficient sizing of parachute components. Two-stage reefing often is employed to further limit inflation loads. In this way, descent can be performed using a single main chute optimized for minimum mass and stowage volume.

Depending on mission operations or the atmospheric density profile with respect to altitude, reefing of the first main chute could be required after a prescribed descent time to permit the vehicle or payload to reach the surface within a specified period of time. Time based active reefing devices or pyrotechnical operated reefing cutters are employed to perform these functions. Their proper performance within tight time constraints and the actual cutting of the flexible rope reefing lines require a specific design validation selected for the actual application.

Safety Critical Components and Functions

From the description of parachute systems employed for space missions, it is quite clear that a series of safety critical functions and components are involved in the achievement of a successful descent and landing mission. Safety and reliability, when presented with detrimental effects that range from exposure to launch vibration, space (vacuum, thermal, radiation), and the entry environments associated with a space mission, must be ensured by design and selection of materials. Rigorous testing to validate functionalities during and after exposure to these environments is mandatory. For application on crewed missions and on the ground handling purposes, relevant safety requirements are prescribed by the pertinent safety regulations of the launch authorities or the site from which launch of the space mission is planned.

Compliance with specified safety regulations is mandatory for the design and operational procedures of safety critical items. This typically applies to all equipment used for the release and ejection of manacle clamps, mortars, guns, separation pistons used for covers and pilot chutes, the electroexplosive or pyrotechnic devices, batteries, pressurized containers, control valves, and staging line and reefing cutters. Hazard classification for all safety items, including the related procedures for operations and hazardous commands, must be performed. Technical personnel, handling and installing procedures, and pyrotechnic device testing must be certified according to applicable national regulations. The system architecture and design for each safety critical component has to be accepted by the relevant launch authority responsible for safety. An example of particular design requirements applied to safety critical items such as the electroexplosive devices used in the deployment and staging of a parachute system follows (Widal et al. 2006):

- All electroexplosive devices must meet the 1-A/1-W no-fire requirement and be qualified with a 500 VDC megohm meter test from bridgewire to case for 5 s and from bridgewire to bridgewire if dual bridgewires are used, unless exemption is granted by the safety board of the launch authority.

- The electrical wiring and power source must be completely independent and isolated from all other systems. They must not share common cables, terminals, power sources, tie points, or connectors with any other system.

- The system initiator must be isolated electrically by switches in both the power and return legs.

- All electrical circuit wiring must be twisted, shielded, and independent of all other systems. When it is not physically possible to maintain the shield throughout the entire electrical circuit, at a minimum, the wiring must be twisted and shielded from the system initiator to the point of the first short circuit condition. This requirement is applicable both before and after installation of safe and arm type connectors. The use of single wire firing lines having their shield as the return is prohibited.

- Shielding must provide a minimum 20-dB safety margin below the minimum rated function current of the system initiator, that is, the maximum no-fire current for electroexplosive devices, and provide a minimum of 85% optical coverage. Note that a solid shield rather than a mesh would provide 100% optical coverage.

- Shielding must be continuous and terminated to the shell of connectors and components. The shield must be joined electrically to the shell of the connector or component around the full 360° of the shield. The shell of connectors or components must provide attenuation at least equal to that of the shield.

- The electrical circuit to which the system electroexplosive device is connected must be isolated from vehicle ground by no less than 10 kΩ.

- All circuits must be designed with a minimum of two independent safety devices. Any time personnel are exposed to a hazardous system, a minimum of two independent safety devices are required to be in place.

- The system electroexplosive device must be protected by an electrical short until its programmed actuation unless an exemption is granted by the safety board of the launch authority. This requirement does not negate the use of solid state switches.

- Any electrical relay or switch electrically adjacent to the system initiator, either in the power or return leg of the electrical circuit, must not have voltage applied to the switching coil or the enable or disable circuit for solid state relays and switches until the programmed initiation event.

Associated with the previously described parachute sequencing operation are a substantial number of control and initiation devices essential to the successful staging of different parachutes. Their timely actuation in an autonomous mode during descent within tight tolerance limits typically is required to ensure a successful mission. In view of their importance, redundancy generally is implemented for all control devices of the parachute system, such as timers, baroswitches, *g*-sensors, global positioning system or retrieval beacon devices, pyrotechnic devices for deployment and dereefing, and on all electronic circuits controlling the sequencing process.

Solutions to Avoid Failure

Analysis In view of the complexity of the sequential staging operations involved in a typical parachute based payload recovery mission, the achievement of a successful parachute descent and landing operation strongly depends on the implementation of appropriate solutions for failure avoidance. Foremost, the sequencing scenario should be assessed rigorously with respect to any potential for single point failures as well as for multiple parallel and series failures. Solutions for avoidance of single point failures should be assessed and implemented, unless these design solutions introduce unacceptable complexity, which would otherwise promote an increased risk of failure or introduce additional single point failures into the system. Such assessments should cover the aerodynamic stability and attitude motion of the descent vehicle, the definition of single or multiple parachute systems, and the aerodynamic interaction between or among parachutes, as well as the forebody wake. Parachute failures related to the adequacy of design margins for the specified velocity range, sequencing failures, sensor and related component failures and tolerance levels, pyrotechnically actuated devices failures, and delayed initiation anomalies should be the principal elements of this assessment. Moreover, a wide range of potential failures or performance deviations involved in manual manufacturing operations for parachute canopies, suspension lines and risers, and the subassembly integration and assembly for the complete parachute descent and landing system should not be overlooked.

An example of a rigorous probabilistic approach to the assessment of failure modes considered for the development of the *Apollo* crew recovery parachutes is chronicled by Knacke (1968). Based on the identification of potential failure modes for a parachute system, the means for their avoidance must be identified. Design solutions could be either the implementation of redundant devices or adequate design criteria and margins to ensure that the identified failure mode is unlikely or noncredible. Whereas uncrewed planetary entry and descent missions generally do not utilize redundancy concepts in the parachutes themselves, crewed spacecraft are flown with both redundant parachute devices and single parachutes. A typical example of a vehicle that relies on redundant parachute systems is the Apollo *Command Module*, where triple parachutes are used for the descent to touchdown and water landing. Two out of the three main chutes would have been sufficient to achieve landing speed retardation to a level compliant with the allowable g loads for the crew. Although multiple parachutes can enhance the reliability of the mission, the increase in the number of components and mechanisms and the possible interactions among them also can have detrimental effects on the overall reliability of the system. The decrease in reliability results from the interaction of the parachutes during deployment and aerodynamic descent with an associated increased number of failure modes. An actual Apollo capsule landing with one failed main parachute out of three deployed is reported and shown in Figure 8.5. Redundancy of the parachute system also has a substantial impact on the additional mass required and increases the effort and complexity of test verification.

FIGURE 8.5 Apollo *Command Module* descent with one failed main parachute out of three (Courtesy of NASA).

At the design stage, analytical assessment of the parachute staging operation and associated sizing loads for the components depend largely on a detailed parametric simulation by analysis. The complexity of the parachute to descent vehicle dynamic interaction typically requires the use of multibody dynamics and therefore an aerodynamic descent analysis tool. Six degrees of freedom coverage can be sufficient for the determination of a basic sizing load using an appropriate atmospheric model with temperature and density parameter variation. Such a model allows derivation of entry and descent trajectory (altitude over time) independent of mass and body drag surfaces, Mach speed, steady state decelerations, and descent and terminal velocity. It is important to note that the actual drag performance of the parachute substantially varies over the velocity range and must be known in advance to ensure proper sequencing and sizing.

For sizing a parachute, specialized tools that permit analysis sometimes reaching 18 degrees of freedom are employed to assess the interaction of parachute and suspension lines, rotary swivels, stabilizer rigging, and descent vehicle coupling, where specific requirements such as high descent vehicle stability or specific spin motion are required. In these simulations, it is essential to vary the parameters sufficiently to encompass maximum to minimum atmospheric density profiles, thus deriving the maximum range of possible initiation conditions for deployment, such as altitude, time, velocity, and aerodynamic pressure, for each parachute evaluated in sequence. Having determined the overall staging point range for the descent sequence by means of this basic analytical simulation, more specialized tools are employed to determine the sizing loads and performance of each individual parachute extraction and deployment. The process for deployment follows a staging operation characterized by certain events, such as the initial stretch of suspension lines, avoidance of possible malfunctions due to line sail effects, the initial canopy air mass influx and subsequent canopy expansion to the first inflated stage, and overinflation and resultant canopy depression by the momentum of the surrounding air mass.

Rather than to apply a fully analytic solution to this complex problem of aerodynamic interaction, where the riser, parachute suspension lines, and the canopy are extracted by a mortar or other means, a combination of empirical parameter based equations, along with certain analytical formulations, is used to determine the sizing loads and the required performance of the parachute. These empirical parameters are derived from previous experience with the type of parachute being evaluated and often need to be interpolated to determine the actual size of parachute needed. To complete the analysis, potential failure scenarios and the likely tolerances for the controlling sensors and actuation devices must be considered. In this way, a superposition of their influence on the trajectory profile can be evaluated to derive cases for sizing the final design of the parachute descent and landing system components.

In addition to functional performance assessments for dimensioning, the packed parachute system and associated equipment must be analyzed to account for compliant sizing associated with the dynamic environment of the launcher and the thermal environment encountered during the on-orbit transfer process. It is mandatory to conduct a thorough failure mode and effects analysis during the design of the equipment and each component of the layout as well as for the function of every component of the system. Essential guidance as to the approach to be followed is provided by the European Corporation on Space Standardization standard, ECSS-Q-30-02A (ECSS 2001).

Design Loads As can be appreciated from the approximations (and their associated limitations) employed in the analytical sizing assessment, a prediction of loads and performances with a margin for error of 5 to 10% is extremely difficult to achieve, if not impossible. Therefore, for parachute systems, failure avoidance relies on the application of appropriate safety factors applied to the verification of expected loads and the strength of the individual mechanical parts. Safety factors of 1.5 for uncrewed missions and of 1.35 for crewed missions (Knacke 1968) have been applied successfully and are considered acceptable. In addition, load derating factors are applied for joint efficiency at critical locations, such as canopy stitching and suspension line joints and knots, to account for degradation due to vacuum stripping, a process that reduces friction within woven textiles, and operation at extreme temperatures. An aging factor of 10% for long-term interplanetary flight conditions, as on the *Huygens* probe to Titan (~7 a) also is taken into account (Underwood 1995). These factors generally are derived from a history of testing.

Aerodynamic performance related failure modes can be avoided by the application of established experience based engineering margins to the limiting flow conditions for a specific parachute and any identifiable zones of possible instability that can be determined only by wind tunnel or free flight tests of the parachute system and a representative forebody descent vehicle.

Unlike materials for a typical spacecraft structure, the main parts of the parachute system, that is, the canopy, riser, and suspension lines, are composed of woven textile materials whose performances are known generally from civil or military parachute applications in the environment of the lower Earth atmosphere. These high strength materials are made from polyamid (Nylon®), polyester (Dacron®, Trevira®, and Diolen®), or aramid (Kevlar®) and more recently Dyneema®. Considering the compliance of these materials when exposed to the space environment, that is, temperature extremes, vacuum, radiation, and ultraviolet light, and their compatibility with the harsh and sometime hostile environment of planetary atmospheres, a substantial effort is required to confirm that the material performance characteristics are appropriate to the specific application and to ensure that the material data used in the sizing determination are valid. The choice of material could further be constrained if the parachute materials are required to be radio frequency transparent to ensure continuity of a radio link to an orbiting spacecraft. In addition, the components of the parachute system and associated subassembly loads need to be verified because their composition is based on stitching, knotting, or otherwise joining together various sections of textiles. Derating of manufactured textile joints is utilized to ensure accounting for the weakening effects of the joining operation. The derating factors are derived by proof load testing for a series of joints and subsequently applying safety factors of between 1.3 and 2 to the results, depending on the application. Redundancy generally is implemented to avoid the consequence of failure for every control device, such as timers, baroswitches, *g*-sensors, global positioning system or retrieval beacon devices, pyrotechnic devices for deployment and reefing or dereefing, and electronic circuits controlling sequencing. Even so, the sizing and reliability of parachute system components must comply typically with applicable standards for spacecraft systems, which cover mechanical, functional, electrical, electronics, and material sizing requirements, along with their related specific safety factors and all required test verifications.

Development and Flight Testing The adequacy of the materials chosen along with associated environmental performance data for the various components of the parachute system need to be confirmed by detailed materials testing. Such a materials testing program should cover the space environment, temperature range, atmosphere composition effects, long-term storage in a densely packed vacuum condition, aging and creep effects, and possibly radio frequency transparency. Determination of material properties for representative conditions is of utmost importance because application of the aforementioned safety factors does not provide for an uncertainty of several percent to be applied to performance data. A rigorous test and validation program for nondegraded functionality after exposure of the controlling components to these environments is mandatory for descent and safety critical operations. Particularly, environmental testing should be performed on the descent controlling sensors and active devices in all operational conditions, because they not only are exposed to launch vibration in a passive condition but also must operate during exposure to dynamic loads during the various phases of descent. It should be noted that peak loads during actual deployment could exceed those imparted by the launch environment. These devices, therefore, must be verified to the absolute worst-case peak loads expected on the vehicle during parachute deployment and vehicle deceleration to ensure proper function during actual operational conditions. It should be noted that certain devices, such as reefing cutters, can be located within the canopy of the parachute, and thus they are exposed to the environmental conditions of the atmosphere during flight.

As explained earlier, the design approach typically is based on approximated performance data for the specific parachute type and size, the actual configuration of the required number and size of suspension lines, and the riser configuration dictated by descending vehicle geometry. Confirmation of the sizing loads and performance by wind tunnel or free flight testing with instrumented payload simulators early in the development cycle is essential to confirm the adequacy of parachute choice and associated loading.

The complexity involved in representing a specific flight environment for the planetary entry conditions of a specific mission usually requires a set of tests in which particular flight phases can be represented properly in terms of atmospheric pressure, velocity, density, and sequencing. A recommended approach initially would be to confirm the specific parachute type and performance in the wind tunnel. Separation of covers or breakout patches then is verified at the subsystem level before drop testing. Having obtained a satisfactory level of confidence with the basic choice of the parachute, the sequencing operation utilizing autonomous control devices must be confirmed by a set of drop tests that encompass the key critical sequences for parachute deployment and inflation that represent the actual application. Depending on the initiation conditions at deployment, drop testing for a parachute can be performed from a helicopter, an airplane, or even a high altitude balloon. Matching representative deployment and descent conditions, particularly for missions to other planetary bodies, is sometimes extremely difficult to achieve, and therefore the worst possible cases must be employed during testing to account for variation in descent velocity, dispersion in the sensor signals, and the like. Very high altitude deployments above the surface of planets can lead to extremely low dynamic pressure deployment conditions for which no ground experience might exist. Related test validation programs are likely to be very demanding and therefore are substantial cost drivers for the validation of a reliable parachute design.

Manufacturing and Quality Assurance The parachute canopy and its riser and suspension lines are produced from lightweight textile materials available in various weavings and sizes. In view of the complex shape of the inflated canopy and the importance of the porosity, which substantially affects the deployment and inflation characteristics specified by the design of the parachute system, the canopy typically is composed of many textile sections, that is, a meshwork of ribbons, cut according to the deployed shape. They are assembled into a complete canopy by specific stitching operations generally used by parachute manufacturers for military and civil parachute applications. Stitching is essentially a manual operation that has a corresponding spread in the quality of achieved performances. The riser and suspension lines are selected from a range of woven cord and ribbon whose load performances are established for the ambient environment in Earth atmosphere conditions.

The assembly process for the suspension lines and riser into the final product is performed by stitching and interconnecting processes. Specific stitching patterns using localized reinforcement textiles are required to achieve the necessary strength at the joining locations. Product assurance applicable to commercial stitching fabrication generally is acceptable; however, some exceptions, such as cutting by laser tools to avoid edge fraying of cut weavings, are required for high reliability space applications. To confirm the selected design choice for stitching and materials, key load transfer joints are subjected to proof load testing of samples. This similarly applies to the stitching of riser joints located at the structural attachments to the descent vehicle.

Installation of the overall assembly of the manufactured parachute, including its riser and suspension lines, into a parachute bag or other appropriate stowage container onboard the descent vehicle is a critical task. The process by which the parachute assembly is packed can lead to many sources of functional anomalies or even mission loss. The process is a manual operation that should be performed with quality assurance supervision and control in accordance with a validated and approved procedure already in use for the qualification tests and flight of the parachute system.

Often pressure packing is involved to control package density because the available volume for the recovery system is constrained considerably on space missions. Should pressurized packing for the parachute be part of the system design, the degradation of material performance due to high density packaging must be determined as an integral part of the testing program and the data appropriately considered in assessing the functional performance of the parachute.

8.2.2 Known Parachute Anomalies and Lessons Learned

As discussed, the manufacturing of parachutes depends substantially on the particular experience of the companies and the design and manufacturing staff involved. This experience is derived mostly from civil or military applications of parachutes used in the environment of Earth atmosphere. Compared to these Earth based applications, experience related to space mission is limited. The encounter of various anomalies in the development of a parachute system for a designated space mission in which environmental conditions are likely to be substantially different from Earth application is not surprising. Indeed, a number of parachute system failures on actual space missions and during their associated development programs have occurred. A failure that occurred during 2004 is reported for the NASA *Genesis* space mission, when the drogue parachute of the reentry capsule, which was supposed to be deployed at an altitude of 33 km, did not deploy because of a design flaw in a deceleration sensor. The sensor did not initiate the deployment sequence, clearly identifying a case of single point failure in a control device that almost led to complete loss of the mission (Wikipedia 2007). To further inform the reader about problems encountered and related lessons learned in space mission applications, the following nonexhaustive set of *Apollo* related parachute anomalies were summarized by Wade (1999):

- **June 28, 1963: Apollo Pioneer triconical solid parachutes cancelled; Program: Apollo.** A cluster of two Pioneer triconical solid parachutes was tested. Both parachutes failed. Because of this unsatisfactory performance, the Pioneer solid parachute program officially was cancelled on July 15, 1963.

- **September 6, 1963: Apollo *Command Module* boilerplate destroyed during tests; Program: Apollo.** At El Centro, California, command module Boilerplate-3, a parachute test vehicle, was destroyed during tests simulating the new Boilerplate-6 configuration without strakes or apex cover. Drogue parachute descent, disconnect, and pilot mortar fire appeared normal. However, one pilot parachute was cut by contact with the vehicle, and its main parachute did not deploy. Because of harness damage, the remaining two main parachutes failed while reefed. Subsequent investigation of the Boilerplate-3 failure resulted in rigging and design changes on Boilerplate-6 and Boilerplate-19.

- **December 8, 1964: Apollo main parachute drop tested; Program: Apollo.** A single main parachute was drop tested at El Centro, California to verify its ultimate strength. The parachute was designed for a disreef load of 11,703 kg (25,800 lb) and a 1.35 safety factor. The test conditions were to achieve a disreef load of 15,876 kg (35,000 lb). Preliminary information indicated the parachute deployed normally to the reefed shape (78,017 kg, or 17,200 lb force), disreefed after the programmed 3 s, and achieved an inflated load of 16,193 kg (35,700 lb), after which the canopy failed.

- **August 5, 1965: Apollo Boilerplate-6A sustained considerable damage; Program: Apollo.** During tests of the Apollo Earth landing system at El Centro,

California, Boilerplate-6A sustained considerable damage in a drop that was to have demonstrated Earth landing system performance during a simulated apex forward pad abort. Oscillating severely at the time, the auxiliary brake parachute opened. The spacecraft severed two of the electrical lines that were to have released that device. Although the Earth landing system sequence took place as planned, the still attached brake prevented proper operation of the drogues and full inflation of the mains. As a result, Boilerplate-6A landed at a speed of about 50 f/s.

- **October 30, 1967: Apollo drop test failure 84–1; Program: Apollo; Spacecraft: Apollo *Command Service Module*.** A parachute test (Apollo Drop Test 84-1) failed at El Centro, California. The parachute test vehicle was dropped from a C-133A aircraft at an altitude of 9144 m to test a new 5-m drogue chute and investigate late deployment of one of the three main chutes. Launch and drogue chute deployment occurred as planned, but about 1.5 s later, both drogue chutes prematurely disconnected from the parachute test vehicle. A backup emergency drogue chute, which had been installed in the test vehicle and was designed to be deployed by ground command in the event of drogue chute failure, also failed to operate. The parachute test vehicle fell for about 43 s before the main chutes were deployed. Dynamic pressure at the time of chute deployment was estimated at about 1.2 N/cm^2 (1.7 lb/in^2). All parachutes failed at or shortly after main parachute line stretch. The parachute test vehicle struck the ground in the drop zone and was buried to a depth of about 1.5 m. An accident investigation board was formed at El Centro to survey mechanical components and structures, fabric components, and electrical and sequential systems. It was determined that two primary failures had occurred: failure of both drogue parachute reefing systems immediately after deployment and failure of the ground radio commanded emergency programmer parachute system. On November 3, a preliminary analysis of the drop test failure was made at Downey, California, with representatives of NASA, North American Rockwell, and Northrop participating. The failure of the drogue, which was being tested for the first time, was determined to be a result of the failure of the reefing ring attachment to the canopy skirt. The reason the ring attachment failed seemed to be the lack of a good preflight load analysis and an error in the assumption used to determine the load capacity of the attachment. The failure of the deployment of the emergency system was still being investigated at that time.

- **December 8, 1967: Apollo drop test failed; Program: Apollo.** An Apollo drop test failed at El Centro, California. The two-drogue verification test had been planned to provide confidence in the drogue chute design (using a weighted bomb) before repeating the parachute test. Preliminary information indicated that, in the test, one drogue entangled with the other during deployment and only one drogue inflated. The failure appeared to be related to the test deployment method rather than to drogue design. The test vehicle successfully was recovered intact and reusable by an Air Force recovery parachute.

- **January 11, 1968: Apollo parachute test vehicle failed; Program: Apollo; Spacecraft: Apollo *Command Service Module*.** A test of the parachute test vehicle failed at El Centro, California. The parachute test vehicle was released from a B-52 aircraft at 15,240-m altitude, and the drogue chute programmer was actuated by a static line connected to the aircraft. One drogue chute appeared to fail on deployment, followed by failure of the second drogue 7 s later. Disreefing of these drogues normally occurred 8 s after deployment with disconnect at deployment at plus 18 s. The main chute programmer deployed and was effective for only 14 out of the expected 40-s duration. This action was followed by normal deployment of one main parachute, which failed, followed by the second main parachute as programmed after 0.4 s, which then failed. The main chute failure was observed from the ground, and the emergency parachute system deployment was commanded. However, it also failed because of high dynamic pressure, allowing the parachute test vehicle to impact and be destroyed. Investigation was under way, and NASA Marshall Space Flight Center personnel traveled to El Centro and Northrop-Ventura to determine the cause and effect a solution.

- **March 23, 1968: Apollo drogue chute test failure 99–5; Program: Apollo.** Apollo Drogue Chute Test 99–5 failed at the El Centro, California, parachute facility. The drop was conducted to demonstrate the slight change made in the reefed area and the 10-s reefing cutter at ultimate load conditions. The 5897-kg vehicle was launched from a B-52 aircraft at 10,668 m. Programmed chute operation and timing appeared normal. At drogue deployment following mortar activation, one drogue appeared to separate from the vehicle. This chute was not recovered, but ground observers indicated the failure seemed to occur in either the riser or vehicle attachment. The second drogue remained on the vehicle but seemed to slip in the reefed state. This chute was recovered and inspection confirmed the canopy failure. The Air Force parachute system, which was to recover the vehicle, also failed in the reefed state.

- **July 26, 1971: *Apollo 15*; Program: Apollo.** *Command Module and Service Module* separation, parachute deployment, and other reentry events went as planned; however, one of the three main parachutes failed, causing a hard but safe landing. Splashdown at 4:47 PM EDT August 7, after 12 d, 7 h, and 12 min from launch was 530 km north of Hawaii and 10 km from the recovery ship USS *Okinawa*. The astronauts were carried to the ship by helicopter, and the command module was retrieved and placed onboard. All primary mission objectives had been achieved.

- **July 15, 1975: *Apollo* (*Apollo-Soyuz* Test Project).** A failure in switchology caused the automatic landing sequence not to be armed at the same time the reaction control system was still active. When the *Apollo* had not initiated the parachute deployment sequence by 7000-m altitude, Brand hit the manual switches for the apex cover and the drogues. The manual deployment of the drogue chutes caused the command module to sway, and the reaction control system thrusters worked vigorously to counteract that motion. When the crew finally armed the automatic Earth landing system 30 s later, the thruster action terminated. During that 30 s, the cabin was flooded with a mixture of toxic unignited propellants from the thrusters.

Prior to drogue deployment, the cabin pressure relief valve had opened automatically, and in addition to drawing in fresh air, it also brought in unwanted gases being expelled from the roll thrusters located about 0.6 m from the relief valve. Brand manually deployed the main parachutes at about 2700 m despite the gas fumes in the cabin. By the time of splashdown, the crew was nearly unconscious from the fumes. Stafford managed to get an oxygen mask over Brand's face, and he began to come around. When the command module was upright in the water, Stafford opened the vent valve, and with the in-rush of air, the remaining fumes disappeared. The crew ended up with a 2-wk hospital stay in Honolulu. For Slayton, it also led to the discovery of a small lesion on his left lung, and an exploratory operation that indicated it was a nonmalignant tumor.

ACKNOWLEDGMENT

I acknowledge the great contributions of parachute experiences described in the identified references. Without these contributions, the provided summary of safety aspects to be considered in the development and flight of successful parachute descent and landing systems would not have been possible.

REFERENCES

Achterman, E., R. Kapp, and H. Lehra. (1986) *Parachute Characteristics of Titan Descent Modules*. Report CR(P) 2438, Noordwijk, the Netherlands: European Space Agency.

Bloom, K. A. (1970) The Apollo docking system. *Proceedings of the Fifth Aerospace Mechanism Symposium*. Houston, TX: National Aeronautics and Space Administration, Johnson Space Center.

Burns, G. C., H. A. Prise, and D. B. Buchanan. (1988) *Space Station Full-Scale Docking/Berthing Mechanism Development*. Huntington Beach, CA: McDonnell Douglas Astronautics Company.

European Corporation on Space Standardization. (2001) *Failure Modes, Effects, and Criticality Analysis [FMECA]*. Standard: ECSS-Q-30-02A. Noordwijk, the Netherlands: European Space Agency.

Eiden, M. J. (1989) Aerodynamic decelerators for future European space missions. *Proceedings of the American Institute of Aeronautics and Astronautics 10th Aerodynamic Decelerator Systems Technology Conference*. Cocoa Beach, FL: American Institute of Aeronautics and Astronautics.

Foster, R. M., J. G. Cook, P. R. Smudde, and M. A. Henry. (2004) Space station berthing mechanisms: Attaching large structures on-orbit that were never mated on the ground. *Proceedings of the 37th Aerospace Mechanism Symposium*. Greenbelt, MD: National Aeronautics and Space Administration, Goddard Space Flight Center.

Gonzalez-Vallejo, J. J., W. Fehse, and A. Tobias. (1992) A multipurpose model of Hermes-Columbus docking mechanism. *Proceedings of the 26th Aerospace Mechanism Symposium*. Greenbelt, MD: National Aeronautics and Space Administration, Goddard Space Flight Center.

IBDM. (2005) International berthing and docking mechanism. *Vlaamse Ruimtevaart Industrielen Newsletter* (April): 3.

Knacke, T. W. (1968) *The Apollo Parachute Landing System*. NASA-CR-131200. Houston, TX: National Aeronautics and Space Administration, Johnson Space Center.

Knacke, T. W. (1992) *Parachute Recovery Systems Design Manual*. Santa Barbara, CA: Para Publishing.

Meyer, P. H. (1966) Gemini/Agena docking mechanism. *Proceedings of the First Aerospace Mechanism Symposium*. Houston, TX: National Aeronautics and Space Administration, Johnson Space Center.

Rivera, D. E., P. Motaghedi, and A. Hays. (2005) Modeling and simulation of the Michigan Aerospace autonomous satellite docking system I. Paper 5799. *Proceedings of the International Society for Optical Engineering*. Orlando, FL: International Society for Optical Engineering.

Syromiatnikov, V. S. (1984) *Spacecraft Docking Assemblies*. Moscow: Mashinostroenie.

Underwood, J. C. (1995) Development testing of disk-gap-band parachutes for the Huygens Probe. Paper presented at the 13th AIAA Aerodynamic Decelerator Systems Technology Conference, Clearwater, FL.

Urgoiti, E., and E. M. Belikov. (1999) Docking and berthing systems for the International Space Station. A proposal for a CRV docking and berthing system. *Proceedings of the 50th International Astronautical Congress*, Amsterdam, the Netherlands.

Wade, M. (1999) *CSM Parachute*. Available at www.friends-partners.org/partners/mwade/craft/csmchute.htm (cited: December 1999).

Widal, O., S. Kemi, O. Norberg, and C.-G. Borg. (2006) *ESRANGE Safety Manual*. EUA00-E538. Kiruna, Sweden: SSC Esrange.

Wikipedia. (2007) Genesis (Spacecraft), available at http://en.wikipedia.org/wiki/Genesis (spacecraft) (cited: July 17, 2007).

Robotic Systems Safety

CHAPTER 9

Victor Chang
Safety and Mission Assurance and Configuration Management, Canadian Space Agency, St. Hubert, Canada

Lindsay Evans
Senior Robotics Instructor, Canadian Space Agency, St. Hubert, Canada

CONTENTS

- 9.1. Generic Robotic Systems 301
- 9.2. Space Robotics Overview 303
- 9.3. Identification of Hazards and Their Causes 305
- 9.4. Hazard Mitigation in Design 308
- 9.5. Hazard Mitigation Through Training 310
- 9.6. Hazard Mitigation for Operations 312
- 9.7. Case Study: Understanding Canadarm2 and Space Safety 313
- 9.8. Summary 317

9.1 GENERIC ROBOTIC SYSTEMS

If you do a quick search of the Internet for the definition of a *robot*, you can find many variations; however, all have the same undertones, in that it is a machine built to perform automatic tasks (Figure 9.1).

Merriam-Webster's Online Dictionary (Webster 2007) provides the following definition: "an anthropomorphic machine, guided by automated controls that performs various complex acts of a human being." A similar definition is from the *Cambridge Online Dictionary* (Cambridge 2007), which defines a robot as "a machine used to perform jobs automatically, which is controlled by a robot." One final definition from the International Standards Organization (ISO 2007) is "an automatically controlled, reprogrammable, multipurpose, manipulator programmable in three or more axes, which can be either fixed in place or mobile for use in industrial automation applications." To begin this chapter, the various components that make up any robotic system are explored.

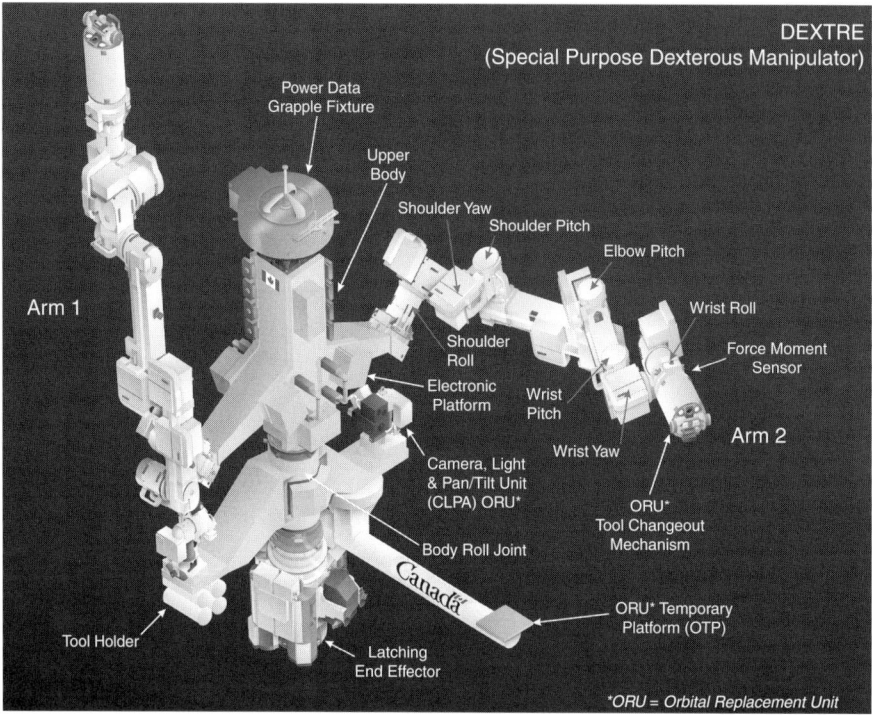

FIGURE 9.1 Robotic system diagram (Courtesy of CSA).

9.1.1 Controller and Operator Interface

A controller, that is, a computer, functions as the brain of a robot by passing commands to lower level processors, then reading back telemetry from these processors to ensure the robot is healthy and operating in a coordinated fashion. The controller is run by software programs that provide the robot with varying levels of artificial intelligence. The operator interface can either be in close proximity to the robot or at a remote location.

9.1.2 Arms and Joints

The arm of the robot defines its reach limit. Having joints that provide for shoulder, elbow, and wrist motion, most robotic arms look similar to those of a human. Each joint typically provides the robot with one degree of freedom. To position the end of the arm anywhere within its reach limit, that is, workspace, six degrees of freedom are required, one each for the X, Y, and Z positions and one each for pitch, yaw, and roll orientation. Most robotic arms today possess a six degrees of freedom construction with two joints at the shoulder, one at the elbow, and three at the wrist. These joints can be offset providing a greater range of motion, that is, the upper and lower portions of the arm do not collide with one another when rotated through 360°, or they can be in-line joints that provide a limited range of motion similar that of the human elbow.

9.1.3 Drive System

The drive system is the engine that moves the joints of the robot into the specified position. Pneumatics, hydraulics, or electric motors, some of which obtain electricity from solar cells or batteries, can power this engine. Because of the complex and extreme nature of the space environment, all space robots presently use electric motors to drive their joints. To assist with control of the arm once it is in motion, mechanical brakes are an essential part of the drive system as well.

9.1.4 Sensors

Sensors are the perceptual tools of the robot and provide essential feedback to the controllers, indicating the status of parameters such as temperatures, forces, and moments exerted on the tip of the arm, as well as the joint angles during installation tasks. Typical sensors found on space robots include optical and electrical position sensors on the joints, current and voltage sensors within the drive mechanisms, strain gauges on the end effector, and cameras mounted on the arm to aid the operator with situational awareness, that is, where the arm is positioned relative to other structures.

9.1.5 End Effector

The end effector is the hand of the robot. It typically does not resemble the human hand; however, it can be used in conjunction with various tools to perform the same functions, such as grasp, move, fasten, remove, replace, and install. In addition, the end effector has the ability to establish power, fluid, and mechanical connections by mating the connectors located at its tip to those on the grapple fixture of the object or payload of interest.

9.2 SPACE ROBOTICS OVERVIEW

Robotic arms onboard the *International Space Station* are essential for assisting with construction and maintenance of the vehicle, supporting externally located International Space Station experiments, capturing free flying vehicles and modules, supporting extravehicular activities, and aiding in other scientific activities. To date, the space agencies of Canada, Japan, and Europe each designed and built robotic arms for installation and use on the *International Space Station*. Each of these robots possesses a unique design and task specific characteristics. The following paragraphs summarize each of the robotic systems from these International Space Station international partners.

FIGURE 9.2 Canadarm2 supporting extravehicular activities (Courtesy of NASA).

Canadarm2, otherwise known as the *space station remote manipulator system*, is a Canadian built robot used to assemble the *International Space Station* (Figure 9.2). It is 16.9 m long and possesses a seven degrees of freedom manipulator with a relocatable base. Launched during April 2001, it was assembled on orbit and used to install the U.S. Airlock, *Quest*, a month later. During March 2008, Dextre, a smaller two-arm robot, was launched to the *International Space Station*. Dextre possesses a one degree of freedom body with two seven degrees of freedom arms attached to it. The body measures 3.7 m, and each arm is 3.3 m in length. When grappled at the end of Canadarm2, Dextre provides the capability to perform the more dexterous, intense maintenance tasks of servicing International Space Station external orbital replaceable units.

The Japanese experiment module remote manipulator system is a contribution by Japan to the *International Space Station* (Figure 9.3). The main arm, a 9.9-m, six degrees of freedom manipulator, is attached permanently to the pressurized Japanese experiment module. Unlike Canadarm2, the Japanese experiment module remote manipulator system is not symmetric about the elbow; in fact, it has a third, smaller boom that joins the wrist roll joint to the end effector. The Japanese experiment module remote manipulator system is used to support experiments and maintenance tasks on the Japanese experiment module exposed facility and module.

Tasks demanding more dexterity than can be provided by the Japanese experiment module remote manipulator system alone are accomplished using the small, fine arm, also a six degrees of freedom manipulator that can be operated when grappled at the end of

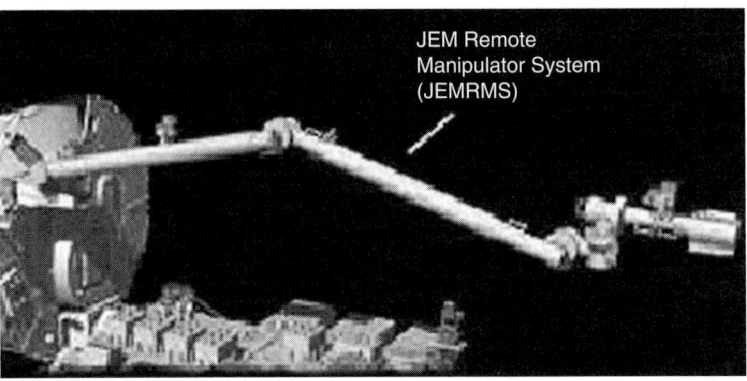

FIGURE 9.3 Japanese experiment module remote manipulator system (Courtesy of JAXA).

the main arm. The small, fine arm measures 2.2 m and is used primarily for capture and release operations as well as for orbital replaceable unit transfer and maintenance.

The European robotic arm is an 11-m, seven degrees of freedom manipulator having a relocatable base (Figure 9.4). The European robotic arm is used for the installation of solar panels on the Russian solar power platform, handling various Russian payloads, and carrying out maintenance tasks. Unique to this manipulator is its ability to be operated by a crewmember from either inside or outside the *International Space Station*. The European robotic arm is symmetrical about the elbow, and all joints are operationally interchangeable.

Canadarm, or the Space Shuttle remote manipulator system, is a Canadian built space robot whose shoulder is fixed permanently to the Space Shuttle cargo bay. It is a 15.2-m long, six degrees of freedom arm with in-line joints. Because its workspace is limited to an area just above the Space Shuttle cargo bay, it most often is used to conduct camera surveys of the vehicle, providing camera views for other International Space Station robotic arm operations, and transferring payloads from the cargo bay to Canadarm2.

9.3 IDENTIFICATION OF HAZARDS AND THEIR CAUSES

Robotic systems operating in a crewed space environment are discussed in this chapter. Payloads manipulated by robotics include experiments, orbital replaceable units, modules, vehicles, and extravehicular activity. Hazard criticality for these differs from that of uncrewed systems; however, the identification, causes, and controls can be similar.

A catastrophic hazard is an occurrence that can result in the potential for a disabling or fatal injury to one or more crewmembers or loss of the spacecraft or space station. Two catastrophic hazards can be identified for robotic systems: uncommanded motion and uncommanded payload release.

FIGURE 9.4 Artist impression of the European robotic arm being used to transport an astronaut during a spacewalk (Courtesy of ESA).

Uncommanded motion of a robotic arm, with or without a payload attached, could result in physical contact with the *International Space Station* or other spacecraft. Such contact could cause penetration of a pressurized module, the result being rapid decompression of the module and subsequent fatal or disabling injury to the crew inside. Contact also could damage critical systems on the spacecraft, such as the Space Shuttle cargo bay doors or its wings, and prevent safe reentry. If the robotic system is maneuvering or transporting a crewmember in extravehicular activity, a failure could occur whereby the crewmember collides with the structure with sufficient energy to inflict fatal or disabling injuries. Another consequence of physical contact or collision of the robotic system is the generation of free flying debris. These materials could cause injury to crew and equipment as just outlined. Uncommanded payload release essentially results in one large mass of free flying debris, and the consequences are likely to be considerable if a rogue payload drifts into a structure.

A variety of causes can give rise to the hazard conditions of uncommanded motion or uncommanded payload release. These can include electrical and electromechanical malfunctions, such as brake slippage, mechanical and structural failures, failure in the control path, and operator error.

9.3.1 Electrical and Electromechanical Malfunctions

Electrical malfunctions result from failures within the electronic units that control individual joint or end effector motion and the coordination of the motion of the various joints in the arm. A failure within the display or the control panel, including its switches and the hand controllers, as well as electronic malfunctions within power and data cables or the portable computer system circuitry are possible sources of electrical malfunctions that could lead to a catastrophic hazard.

Electromechanical malfunctions as well can lead to the hazards of uncontrolled motion or release. Malfunctions of this type can result from failures in electromagnetic motor windings, brake windings, motor resolver windings, and joint angle resolvers.

9.3.2 Mechanical and Structural Failures

Mechanical failures within robotic systems include failure of the end effector latching mechanism, joint motors, gearboxes, and other moving parts. They are caused by galling, binding, jamming, high friction, debris, or even improper selection of materials or manufacturing processes. Structural failures of the boom, end effector latches, snare cables, and other safety critical structural components, such as bolts, can be caused by incorrect assembly, over-torquing, or exceedance of external loads. Failure of the braking systems can include slippage due to brake pin failure, friction surface failure, or excessively high external load impulses. If the robotic system is manipulating a massive payload and an emergency motion arrest occurs, high loads could result in end brake slippage of the effector mechanism and compromise the integrity of the interface between the end effector and the payload; that is, the payload is released inadvertently.

9.3.3 Failure in the Control Path

Failure can occur in the data path between computer units and computer software configuration items or hardware configuration items. Such failures result in loss of control of the robotic arm, therefore presenting catastrophic hazard conditions.

9.3.4 Operator Error

Operator errors include the inadvertent actuation of a critical switch or the issuance of a deviant or incorrect command. Operator errors are difficult to control and are a legitimate cause of catastrophic hazards.

9.3.5 Other Hazards

Other robotic systems hazards that must be addressed include pinch points, entrapment, extravehicular activity contact hazards, electrical shock, arcing and sparking, venting, and containment. Each of these hazards is a potential cause for damage to the extravehicular activity suit. Any damage to the extravehicular activity suit presents a catastrophic hazard because it is the only life support available to the extravehicular activity crewmember. Extravehicular activity contact hazards include collision with the robotic system, sharp edges, holes, pinch points, entrapments, snagging points, high or low touch temperature extremes, failure of handholds or restraints, release of stored strain energy, and shock. Deployment and assembly operations represent the highest potential for contact hazards, because the various system components awaiting assembly on orbit likely are to have sharp edges at their interfaces. Bolts or screws also can have sharp edges, and pinch points can be present at the physical interface during assembly. Detailed information regarding extravehicular activity contact hazards can be found in Chapter 22, "Extravehicular Activity Safety."

9.4 HAZARD MITIGATION IN DESIGN

9.4.1 Electrical and Mechanical Design and Redundancy

Clearly, for such critical systems, a stringent product assurance process must be implemented to ensure high quality during the design, manufacture, assembly, testing, and integration phases for robotic hardware. Adequate fault tolerance is required to ensure proper control of hazards. The design should be reviewed to ensure that redundancy is incorporated in key areas, such as power, data, and commanding systems. Such reviews should include the display as well as control switches and hand controllers. Component selection must be stringent, with adequate derating and qualification considered. Both the space station remote manipulator system and European robotic arm have fully redundant power and data channels, meaning that they employ two physically independent electrical channels to ensure continued safe operation should a failure occur.

During the design and manufacture of a robotic system, an effective program must be implemented to control materials and processes. Detailed requirements and procedures with respect to the selection, application, specification, qualification, and traceability of the materials and process used in the development help minimize mechanical failures. Structural load margins are calculated to define flight planning load limits. Prior to robotic operations, load cases are run and optimization of the joint angles and rates are defined to ensure that loads do not exceed the flight planning load limits.

9.4.2 Operator Error

Ready-arm-fire protocols can be implemented for all commands that initiate arm movement or release motions for the end effector mechanism. These special protocols can prevent uncommanded motion in the event of one or two operator errors. Even so, training remains the prime mitigator for operator error.

9.4.3 System Health Checks

System health checks should be implemented to monitor for command and control of hazardous functions. Health checks could include command protocol validation, command and status sequence validation, communication checks, and command and status health checks.

Command protocol validation ensures that every command sent to a spacecraft arrives at its intended destination, is acted on, and the results of the command execution are provided in a timely manner. Any consecutive command rejects or command aborts should result in a warning or other action to prevent a hazardous outcome. Command and status health checks ensure that commands are sent in the expected order and extra or missing commands are not sent. Communications checks ensure that messages sent from processor to processor are error free. Command and status health checks ensure that the reported state of a computer unit corresponds to its commanded state. These health checks catch unintentional changes in the systems software configuration or software controlled hardware. Any single or consecutive errors should result in issuance of a warning and initiation of an automatic action to prevent a hazardous operation.

9.4.4 Emergency Motion Arrest

One method for controlling the hazard of uncommanded motion is to ensure the safety of the robotic system when an error in command, communication, or function is detected. Safing for the robotic system is implemented by the application of brakes on all joints and end effector mechanisms and the inhibition of the joint and end effector motor drive amplifiers, thus preventing voltage to the joint motors. This control should be fail-safe, such that, if power is removed, the brakes engage. With this safing scheme, a finite time is required to detect a failure or error, activate the brakes, inhibit the motor drive amplifiers, and overcome the inertia of the motion to obtain a full stop of the robotic arm end effector mechanism. Any delay or increase in the stopping distance can result in a collision with a structure. As part of the control for this hazard, an operational constraint to maintain a minimum distance to structure during robotic arm maneuvers can be implemented.

9.4.5 Proximity Operations

Typically, robotic systems need to operate in close proximity to structure; therefore, minimization of the impact energy becomes part of the hazard control. Reduction in the rate of translation reduces the energy transfer in the event of an impact as well as the time and distance required for this system to come to a full stop. Determination of the impact energy at this lower rate should be calculated and the possibility of damage determined on a case by case basis. Because it is a zero-fault tolerant case, controls for this type of proximity operation are expected to include the reduction of the probability of occurrence and the effects of collisions. Reducing the probability of occurrence can be achieved by defining keep out zones and planning trajectories that minimize the amount of time the system operates within a keep out zone. Checkout procedures should be performed prior to performing proximity operations as well. Furthermore, reducing the effect of collision could include the addition of guides, bumper plates, or scuff plates designed to survive nominal contact loads on both the end effector and any location to be grappled.

9.4.6 Built in Test

Built in tests provide for detection of electronic or electrical failures. A built in test should monitor critical hardware, such as the joint motor drive amplifiers and the motor windings and resolvers, and compare the position of the motor resolver to that of the joint resolver to ensure consistency. As well, built in tests are used to monitor the hardware that ensures communication between the display and control panel, the workstation, and other subsystems.

9.4.7 Safety Algorithms

Algorithms can be imbedded into the control software used to assess the location of booms and joints relative to each other and to any structure within the robotic workspace. Arm position can be calculated by joint angle data, and location of the structure relative to the arm can be obtained from structural models. The European robotic arm contains an algorithm that detects its proximity to a structure and stops the robot before a collision can occur. This algorithm depends on having an accurate, up-to-date *International Space Station* model. The space station remote manipulator system contains a self-collision algorithm. Because the space station remote manipulator system has offset joints, it is physically possible for the wrist or shoulder cluster joint angles to be set to values that can drive the end effector into the boom. The self-collision algorithm of the space station remote manipulator system software monitors the joint angles to avoid this situation.

9.5 HAZARD MITIGATION THROUGH TRAINING

The process of ensuring crew safety and mission success is critical for human spaceflight. In large part, crew safety and mission success rely on the specialized skills of crewmembers to perform their on-orbit tasks. To operate a robot in space safely and efficiently, it is imperative that astronaut operators participate in a structured and rigorous robotics training and evaluation program. Here, they can acquire the specific skills, knowledge, and attitudes necessary for the job. A successful training program ensures that the crew has a clear understanding of all potential hazards associated with robotic operations, the potential impacts if any such hazard is realized, and how, through their actions, they can eliminate or minimize operational risks, that is, loss of life, payload, or mission effectiveness.

At the end of each robotics training program, it is imperative that there is a stringent evaluation to ensure that learner's performance complies with a predefined standard. Depending on the measure of success achieved during the evaluation, appropriate certification is granted to the crewmember. For Canadarm2, a specialist, operator, or user certification is granted (see case study later for details).

It is to be expected, however, that without practice, knowledge and skills degrade with time, and some form of intervention is required for crewmembers to remain proficient. During the assembly of the *International Space Station*, crewmembers lived and worked onboard for a period of up to 6 mo per increment. During their stay on orbit, there is an ever increasing risk of skill degradation and potential operational errors. To minimize this increasing potential for risk, appropriate proficiency sessions are scheduled for the on-orbit crew.

The process of maintaining proficiency onboard the *International Space Station* poses concerns because crew time is very limited. Still, International Space Station crewmembers are expected to maintain their robotics proficiency by performing scheduled robotics operations or conducting practice sessions at regular intervals. Currently, no training simulator is onboard to support proficiency training; therefore, crewmembers are required to train by manipulating the actual robotic system in free space.

Proficiency training using the actual flight system is hardly advantageous for a number of reasons. From the hardware perspective, it causes greater wear and tear on the system, thereby affecting its maintenance cycle. From a training perspective, only a limited number of teaching points can be exercised because of safety considerations. Knowledge, skills, and attitudes associated with clearance monitoring, application of flight rules, and certain other tasks, such as payload maneuvering and extravehicular activity support, cannot be refreshed effectively during training with the flight system. For these reasons, alternate methods of proficiency training have been developed, such as drill and practice computer based training and static procedure rehearsal. If, by some chance, it is found that robotics proficiency has been lost, it is imperative to have a training program in place that addresses the issue and allows the operator to regain proficiency in a safe, guided, and timely fashion.

Two other training factors that contributed greatly to hazard mitigation must be considered. The first is the generation of safety guidelines. These are identified during training development and enforced from the beginning of the course to ensure that the crewmember has acquired all necessary safe operation habits. These guidelines can range from verifying system feedback prior to executing the next command to informing Mission Control of hot switch status. The safety guidelines for the Canadarm2 are listed in Table 9.1.

The second factor to consider is the learning interference. Eventually, several robotic systems are to be onboard the *International Space Station*, such as the Canadarm2, the Japanese experiment module remote manipulator system, the European robotic arm. Only one crewmember is required to receive training, become certified, and operate each of them. Each manipulator is similar to another in a number of ways. However, a sufficient number of subtle differences exist among them for training interference to occur as the crewmember learns to operate multiple systems. Training interference leads to poor retention. This leads to confusion, resulting in potential operational errors. To minimize this risk, safety related differences among all International Space Station robotic arms have been identified, and an effort is currently in place to generate quick reference cards to remind operators of the key differences among the robots. For example, the cards would indicate that, for one robotic arm, an active collision avoidance algorithm exists to prevent collisions with structure; however, for another robotic arm, this piece of software does not exist and cannot be relied on to stop motion if the arm gets too close.

Table 9.1 Canadarm2 Safety Guidelines

Verify system feedback before issuing next command.
Verbally identify hot hand controllers or switches on the robotic workstation.
Ensure hand does not bump hand controllers or switches.
Apply brakes or safety procedures when leaving the robotic workstation unattended.
Monitor clearances between structure and Canadarm2 and payload with camera views.
Maintain a 2-ft clearance from structure.
Use vernier rates when within 5 ft of a structure.
Use rate hold only when greater than 10 ft from a structure and moving away.
Verify target values on the personal computer system and overlay before all automatic trajectories.
Verify frame of reference and display frame before frame of reference operator command and sequence and autotrajectories.
Keep hand over pause/proceed switch at the beginning of an autotrajectory until expected motion is confirmed.
Be prepared to initiate safety procedures if uncommanded latching end effector mechanism motion is observed.
Review end effector cue card before payload operations.
Monitor tension on effector overlay when maneuvering a payload.

9.6 HAZARD MITIGATION FOR OPERATIONS

Another means to minimize hazards and thus increase robotics safety is to ensure that operations are performed in a manner consistent with the training received. For example, all operations are now performed on the *International Space Station* with two crewmembers (designated M1 and M2) at the robotics workstation or one crewmember, M1, at the robotics workstation and the second certified operator assisting from the control center on the ground, M2. Each operator has specific roles and responsibilities that are known, defined, and rehearsed. The intent is to share the operational overhead and prevent events that otherwise might have gone undetected; that is, some step in a procedure is skipped due to crew fatigue.

The procedures also provide the robotics operators with suggested camera views for optimal situational awareness. The cadre of essential views is known and understood sufficiently well that operations do not proceed if one of these essential views is lost.

In addition, the operations community generated a list of instructions, housed at the robotics workstation, that includes steps to perform if a payload ever comes free from the manipulator. This cue card is learned, discussed, and implemented during training. It must be reviewed consistently and constantly while in orbit prior to payload operations.

The operations community developed recovery procedures for implementation should a failure occur within a robotic system, be it either electrical or mechanical. These procedures specify the steps that should be taken to regain enough arm functionality to complete the operation. If the failure is deemed time critical, application of these procedures permits the crew to recover in the timeliest fashion possible. Finally, flight rules generated from applicable hazard reports are written to provide the operations community a rapid means for making good, safe decisions when situations arise. These rules are generated by the flight control team and reviewed by operations, training, and crew offices.

9.7 CASE STUDY: UNDERSTANDING CANADARM2 AND SPACE SAFETY

9.7.1 The Canadarm2

The primary contribution of Canada to the *International Space Station* is the mobile servicing system, which consists of

- Two robotic arms (Canadarm2 and Dextre).
- The mobile remote servicer base system.
- The robotics workstation from which the onboard crew can operate the mobile remote servicer base system.

Canadarm2, designed and built by McDonald Douglas Aerospace Space Missions in Toronto, Ontario, for CSA, is a 16.9-m long, seven degrees of freedom robot consisting of three joints at the shoulder (pitch, yaw, and roll), one joint at the elbow (pitch), and three joints at the wrist (pitch, yaw, and roll) (Figure 9.5). It has a tip positioning accuracy of 6.5 cm and 0.7°, a maximum stopping distance of 61 cm and 3°, and a maximum handling capability of 116,000 kg, that is, the mass of a fully loaded *Space Shuttle*. The arm is symmetrical about the elbow giving it the capability to relocate by moving end over end and changing base to tip. Each joint is offset from the next, and each provides 540° of motion, which adds complexity to its operation because the large range of motion means that the arm position with respect to a structure and the arm/payload itself must be monitored closely.

9.7.2 Cameras

Canadarm2 has a vision system composed of two camera pan and tilt units, that is, optical sensors located on the booms, and a fixed boresight camera on each end effector. These cameras provide a situational awareness aid to the robotics operator during maneuvering or grappling and when performing inspection tasks. This is important because the crew has no ability to view the arm directly from a window. Without proper camera views,

FIGURE 9.5 Canadian mobile servicing system showing Dextre, Canadarm2, and the mobile remote servicer base system (Courtesy of CSA).

the operator loses important situational awareness cues and risks flying the arm in too close to structure.

9.7.3 Force Moment Sensor

Canadarm2 has a force moment sensor located at each end effector to measure the forces and moments exerted at the tip during payload installation tasks. The force moment sensor feeds data back to the main controller, which applies additional commands to the arm to null out any detected forces and moments. To ensure that the arm does not become unstable when force and moment sensing are active, it is important to ensure that the operation has been simulated properly with a force moment sensor on the ground prior to it being used on orbit (Figure 9.6).

9.7 Case Study: Understanding Canadarm2 and Space Safety

FIGURE 9.6 Robotic workstation simulator at CSA (Courtesy of CSA).

9.7.4 Training

Currently, on-orbit proficiency training for Canadarm2 is required every 30 to 45 d, depending on the intricacy of the task. Proficiency training consists of exercises in psychomotor skills for hand controller operations, situational awareness skills using camera views to appreciate the location of the robot relative to a structure, and procedure execution skills to ensure steps are not skipped. Training is sufficiently intensive that there is no lapse in any cognitive component of robotic operations.

Since the deployment of Canadarm2 in April 2001, the skills required to perform procedure execution tasks have been observed to degrade faster than those required for psychomotor tasks, such as using the hand controllers to position the tip of the arm in space. To present an example, approximately 4 mo into a stay on orbit, a robotics specialist commanded the wrist roll joint in single mode to an incorrect joint angle (−4°, instead of +4°). In this case, proficiency training on single joint maneuvers had occurred just 10 d prior to this event; however, at that time, no M2 support was available to assist the M1 specialist. It was from this event that the need for two operators at all times became a standard and integrated in the Canadarm2 training program. Space robotics tasks are not simple, considering that the operational environment is very complex and often consists of narrow corridors, varying light conditions, absence of out of the window views, and reliance on cameras for teleoperations. For this reason, mobile servicing system operational tasks now are partitioned between two crewmembers, with M1 controlling the arm and M2 controlling the cameras and monitoring displays. This is analogous to the roles of pilot and copilot in aviation. Given the intricate nature of the operational tasks, the role of the M2 is to assist in mitigating errors.

Both the M1 and M2 receive Canadarm2 training at CSA as part of an 18-mo International Space Station training flow that leads to crew qualification in accordance with clearly defined performance standards as a user, operator, or specialist in robotic operations. This qualification dictates the specific tasks that a crewmember is permitted to perform during an increment; that is, only a specialist can maneuver an astronaut performing an extravehicular activity or robotically install an International Space Station hardware component.

In an effort to reduce learning interference and consolidate the safety related differences among all International Space Station robotic arms, effort is proceeding to ensure that the crewmembers are trained properly on a specific system. This training includes those areas of information crossover that might lead to negative training. For example, Table 9.2 illustrates some major safety related differences among Canadarm2, the Japanese experiment module remote manipulator system, and the European robotic arm.

In a further effort to increase the level of safety during operations, a number of changes to the operational philosophy and procedure development have come about over the years as experience has been gained. Some of these follow:

- Despite being a seven degrees of freedom robot, the Canadarm2 is operated using only a six degrees of freedom control algorithm; that is, one shoulder joint always is locked, thereby making motion more deterministic.
- Manual flying of the arm using hand controllers is limited to shorter trajectories when grappling and releasing payloads and during extravehicular activity support operations.
- When no M2 is available on orbit (due to other crewmembers performing extravehicular activity tasks outside the *International Space Station* or crewmembers having never attained or having lost their robotics qualification), the M2 functions are performed by a qualified robotics operator located at the mission control center in Houston, Texas.
- Generic procedures are embedded within each task specific procedure.
- Single joint tables have been modified for clarity.
- Uplink daily summaries provide a high level overview of the planned operations to prepare crew for the day ahead.

9.7.5 Hazard Concerns and Associated Hazard Mitigation

The seventh joint provides Canadarm2 with another degree of freedom and provides the robot with the capability for keeping the tip and base fixed while swinging the two booms and elbow joint; that is, like a human arm, it can hold a cup of coffee and swing the elbow without spilling the coffee. It should be noted, however, that this increased capability comes with a price, that is, additional safety concerns that must be managed. The seventh joint means that, for any position and orientation in space, the plane defined by the three pitch joints is no longer unique. This can result in a number of hazards if the arm booms and the elbow begins to swing in toward a structure as the tip is being commanded in the proper direction. For this reason, the arm is now operated in the six degrees of freedom

Table 9.2 Robotic Arm Differences: Canadarm2, Japanese Experiment Module Remote Manipulator System, and European Robotic Arm

Canadarm2	Japanese Experiment Module Remote Manipulator System	European Robotic Arm
Arm to payload collisions are not detected	Arm to payload collisions are detected	Arm to payload collisions are detected
Releasing brakes puts arm into standby mode, where hand controller inputs are nulled	Releasing brakes puts arm into previous mode, where hand controller inputs cannot be nulled	Releasing brakes takes arm from standby mode into dynamic hold mode
In a joint autosequence, all joints start and stop at the same time	In a joint autosequence, joint motion is executed one joint at a time from base to tip	In a joint autosequence, all joints start and stop at the same time

mode as much as possible, with one of the base joints locked in place. This minimizes any motion of the elbow perpendicular to the pitch plane.

Canadarm2 typically is operated from the robotic workstation inside the U.S. laboratory module, unlike the European robotic arm, whose workstation is located outside the *International Space Station* and on the base of the arm itself. However, more and more Canadarm2 operations are being commanded from the ground. This frees the onboard crew to perform other pressing operations or conduct scientific experiments; however, it also results in the emergence of another safety concern during operations, loss of signal.

The *International Space Station* must maintain line of site communication with the tracking and data relay satellites, which in turn must maintain line of site communication with the ground tracking station to permit communication between the *International Space Station* and the ground. Depending on the orbit of the *International Space Station*, in some periods of time, the required line of site is unobtainable and a loss of signal results. Loss of communication is a new failure mode that would allow the hazard of uncommanded motion. The hazard can be manifested by the ground operator losing the ability to command or ensure the safety of the mobile servicing system during motion. The ground operator also can lose the ability to monitor the motion of the mobile servicing system or command the next step of a sequence. Any of these situations can leave the mobile servicing system in an unsafe condition. To mitigate these hazards, ground commands can be limited to autosequences, where the arm moves automatically over very short distances in a preplanned, preverified trajectory.

9.8 SUMMARY

Robotic arms onboard the *International Space Station* are essential for the following operations: construction and maintenance of the *International Space Station*, supporting external International Space Station experiments, capture of free flying vehicles or modules, supporting extravehicular activity, and aiding in other scientific activities. These robotic systems can perform simple, mundane tasks, such as camera positioning, to daunting complex tasks, such as manipulating and installing 100,000-kg modules. Safety controls and training are key areas required to enable the successful use of robotic systems in the crewed space arena.

REFERENCES

Cambridge Online Dictionary. (2007) Available at http://dictionary.cambridge.org (cited April 23, 2007).

International Standards Organization. (2007) Available at www.iso.org/iso/en/ISOOnline.frontpage (cited April 23, 2007).

Merriam-Webster's Online Dictionary. (2007) Available at www.m-w.com (cited April 23, 2007).

Meteoroid and Debris Protection

CHAPTER 10

Heiner Klinkrad, Ph.D.
Head, Space Debris Office, European Space Operations Center, European Space Agency, Darmstadt, Germany

CONTENTS

10.1. Risk Control Measures 319
    By Heiner Klinkrad, Ph.D.

10.2. Emergency Repair Considerations for Spacecraft Pressure Wall Damage 332
    By Kornel Nagy, Ph.D., and Russell Graves

10.1 RISK CONTROL MEASURES

Space systems exposed to human made space debris and the natural meteoroid environment can avoid an impact induced mission termination in two ways: by maneuvering to avoid conjunctions of high risk with cataloged, tracked objects or by deploying shields to mitigate the effect of impacts from unobservable objects.

10.1.1 Maneuvering

Orbit Prediction Accuracy

The most comprehensive source of information for space objects of sufficient size to be tracked is the U.S. Space Surveillance Network catalog. During the year 2007, it comprised about 10,000 on-orbit objects. Because of the sensitivity limitations of the Space Surveillance Network, the size threshold for detection ranges from 5 to 10 cm in low Earth orbit and about 1 m in geostationary Earth orbit. Of 10,000 tracked objects, approximately 700, or 7%, are operational payloads, 46% of which are in low Earth orbit, 43% are in geostationary Earth orbit, and 11% are distributed across medium Earth orbits and highly eccentric Earth orbits.

Space agencies and spacecraft operators have implemented collision avoidance procedures for their operational payloads. These procedures use the orbit information from the Space Surveillance Network catalog to determine close conjunctions of their spacecraft with other tracked objects. The orbit data within the Space Surveillance Network catalog are provided in a two-line element format, using doubly averaged orbital elements that are defined in the True Equator and Mean Equinox coordinate system. Two-line element data must be propagated in time and converted into osculating elements by means of the simplified general perturbations theory, SGP-4, which is applicable for orbit periods $T \leq 225$ min or the simplified deep space perturbations theory, SDP-4, which is applicable for orbit periods $T > 225$ min.

The orbit prediction errors from using two-line element data in combination with the SGP-4 or SDP-4 propagators mainly are due to the omission errors of the simplified orbit theory. To a lesser extent, these errors result from the undisclosed measurement and orbit determination errors of the Space Surveillance Network system. The two-line element providers do not recommend using their data for precise analyses. All detailed collision risk assessments for U.S. payloads, therefore, are based on osculating ephemerides and error covariance information obtained from Space Surveillance Network measurement data.

For those payload operators who have no access to accurate orbit data, fits to two-line element based SGP-4 generated osculating orbit arcs with a high resolution orbit determination program can provide an indication of the accuracy of two-line element based orbits and the associated error covariance in position and velocity. Alternatively, for payloads with an accurately known orbit, direct comparison with a two-line element based arc can provide an even more reliable estimate of the error covariance for the Space Surveillance Network catalog orbit. As an example, a two-line element based Sun synchronous low Earth orbit at 800-km altitude and 98.5° inclination has typical 1σ position errors in radial (u), transversal (v), and out-of-plane direction (w) of $\Delta r_u = 130$ m, $\Delta r_v = 440$ m, and $\Delta r_w = 160$ m. The corresponding 1σ velocity errors are on the order of $\Delta v_u = 430$ mm/s, $\Delta v_v = 140$ mm/s, and $\Delta v_w = 190$ mm/s. If high quality tracking data are available, then the accuracy of an operational orbit determination process, in most cases, is two orders of magnitude better. Apart from uncertainties in the orbit state, air drag related effects also contribute, particularly at low altitudes. The combined 1σ error due to air density and ballistic parameter uncertainties can be expected to be on the order of 20% or more, relative to nominal conditions.

Collision Risk Estimation

The risk of collision between two space objects that can be tracked always is associated with a close conjunction and orbit position uncertainties for which a given flyby distance and collision cross section can lead to some finite probability of collision that exceeds an accepted threshold level. To quantify the risk, one needs to determine

- The conjunction distance and geometry.
- The position error covariance at conjunction epoch.
- The effective collision cross section of the two objects.

Different methods have been devised to determine close conjunction events between pairs of objects during a predefined time interval. Hoots, Crawford, and Roehrich (1984) developed a theory based on geometric considerations. The conjunction detection in this case is performed by the successive application of an orbit altitude filter, an orbit plane or geometry filter, and an orbit position or phase filter. This approach is useful if the two orbits are available through an analytic orbit theory. Even so, it uses complex filter algorithms that have computationally intensive iterative root finders.

With the advent of increasing computer capabilities, numerical algorithms for conjunction event detections have become an alternative. These more robust sieve algorithms (Alarcon 2002) consist of an initial altitude filter and successively applied range filters to detect a zero transition of the range rate as the indicator for a conjunction. This is accepted as a close approach if it is located within a certain volume, centered on the target object. For the *International Space Station*, the initial threshold volume is a box of ±10 km × ±40 km × ±40 km in the radial, along track, and out-of-plane directions.

Each event that passes the criteria of the conjunction event sieve must be assessed for its collision probability with regard to the target object. Several authors developed methods to achieve this objective (Foster 1992; Khutorovsky, Boikov, and Kamensky 1993; Bérend 1997; Alfriend et al. 1999; Patera 2001; Alarcon 2002). All of these approaches possess the following assumptions:

- During the conjunction, target and risk object move along straight lines at constant velocities.
- Uncertainties in the velocities can be neglected.
- Target and risk object position uncertainties are not correlated.
- Position error covariances during the encounter are constant, corresponding to those at the time of closest approach.
- Position uncertainties can be combined in a common three-dimensional Gaussian distribution.

Most collision probability assessment schemes are based on formulations by Alfriend et al. (1999), who use as inputs the relative position. $\Delta \vec{r}_{tca}$ and relative velocity, $\Delta \vec{v}_{tca}$ of the conjunction risk object with respect to the target at the time t_{tca} of the closest approach (Figure 10.1). At the conjunction time t_{tca} and in accordance with the assumptions made, the propagated and uncorrelated 6 × 6 error covariance matrices of the target and risk object state vectors can be added [Equation (10.1)] to retain the upper left 3 × 3 combined position error covariance matrix $C(t_{tca})$:

$$C = C(t_{tca}) = C_t(t_{tca}) + C_r(t_{tca}) \tag{10.1}$$

This 3 × 3 error ellipsoid C can be mapped onto a 2 × 2 position error ellipse C_B in the B plane. The B plane contains the miss vector $\Delta \vec{r}_{tca}$ and is perpendicular to the approach velocity vector $\Delta \vec{v}_{tca}$ (Figure 10.1). Due to this transformation into a two-dimensional problem, Equations (10.2) and (10.3) are used to determine the resulting collision probability P_c:

$$P_c = \frac{1}{2\pi\sqrt{\det(C_B)}} \int_{-R_c}^{+R_c} \int_{\sqrt{R_c^2 - x_B^2}}^{+\sqrt{R_c^2 - x_B^2}} \exp[-A_B] dy_B \, dx_B \tag{10.2}$$

$$A_B = \frac{1}{2}(\Delta \vec{r}_B - \Delta \vec{r}_{tca})^T C_B^{-1} (\Delta \vec{r}_B - \Delta \vec{r}_{tca}) \tag{10.3}$$

Here, a circular collision cross section of radius R_c is assumed, which surrounds the equally circular cross sections of the risk object and the target (Figure 10.1, upper left

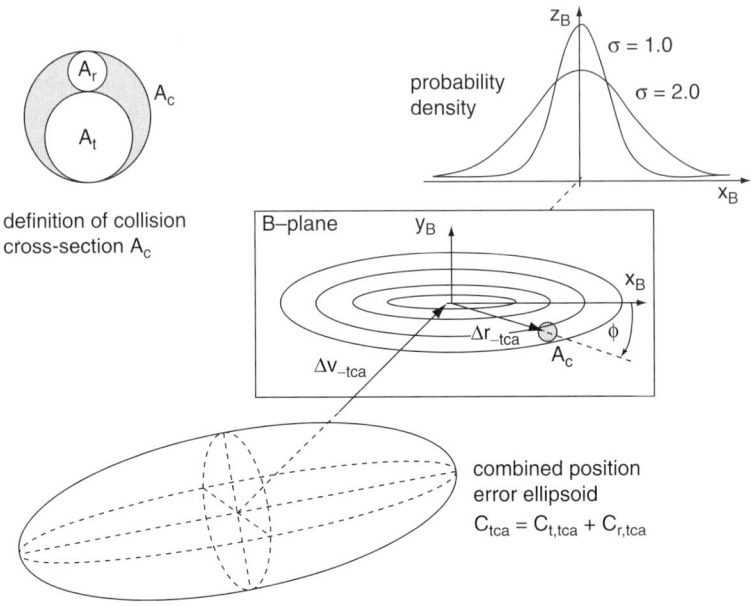

FIGURE 10.1 Two-dimensional mapping onto the B plane of the three-dimensional combined position uncertainty of the target and risk object at the time of closest approach. The collision probability is determined by the size of the combined collision cross section A_c and the conjunction location within the two-dimensional probability density distribution. (Graphic reprinted from Klinkrad 2006: *Space Debris: Models and Risk Analysis*. Springer-Praxis Books in Aeronautical Engineering. Berlin–Heidelberg–New York: Springer Verlag, Copyright: Praxis Publishing, Ltd., Chichester, UK, 2006 with kind permission of Springer Science+Business Media).

diagram). One can show that, for a given conjunction distance and a fixed shape and orientation of the combined position error ellipse within the B plane, the maximum achievable collision probability is given by Equation (10.4):

$$P_{c,\max} = \frac{A_c}{p\,e\sqrt{\det(C_B)\Delta\vec{r}_{tca}^T C_B^{-1} \Delta\vec{r}_{tca}}} \tag{10.4}$$

where $e = \exp(1)$. More refined analysis methods allow the transformation of Equation (10.2) into a single contour integral over a general shape body (Patera 2005) and permit nonlinear relative motion during the conjunction event. Such options are particularly meaningful in the analysis of close conjunctions of geostationary Earth orbit objects at low velocities and with long dwell times.

Collision Avoidance Maneuvers

If, for a given conjunction event, an accepted collision risk level $P_{c,\mathrm{acc}}$ is exceeded, then a collision avoidance maneuver should be considered. To avoid unnecessary maneuvers, one should acquire independent tracking data for the generally poorly known risk object and perform orbit determinations that are more reliable than the a priori estimates. These determinations should be a similar accuracy as the operational orbit determinations for the target spacecraft. Such tracking and orbit determination campaigns can be simulated by scaling the a priori risk object covariance C_r with a factor $k_{\sigma,r}$, as in Equation (10.5):

$$\widehat{C}_r = k_{\sigma,r}^2 C_r \tag{10.5}$$

The decision factors required to initiate a collision avoidance maneuver are

- The accepted probability of collision $P_{c,acc}$.
- The achievable reduction $k_{\sigma,r}$ in the orbit determination error of the risk object.
- The time to go Δt_{tca} until the close approach.

The following sensitivity analysis illustrates the effect of these parameters for the ESA *Envisat* satellite, which operates on a near circular Sun-synchronous orbit at 784-km mean altitude and 98.52° inclination. Its circular cross section is $A_t = 530$ m², which for most risk objects with $A_r \ll A_t$ corresponds to $A_c \approx A_t$. If the *Envisat* operators were prepared to accept any collision risk $P_{c,acc} \leq 1$, then no maneuvers would need to be performed ($\dot{N}_{c,man} = 0$), and the residual collision rate would be equal to the collision rate from the Space Surveillance Network catalog objects ($\dot{N}_{c,res} = \dot{N}_{c,cat} = 0.00727 y^{-1}$). In this case, the risk is invariant with pre-event notification times and orbit prediction uncertainties. Because no maneuvers are performed, the probability of false alarms P_{fa} is zero. If maneuvers are performed, the false alarm rate grows with increasing uncertainty in the orbit determination (expressed in terms of $k_{\sigma,r}$), with decreasing levels of accepted collision probability $P_{c,acc}$, and (in general) with increasing time to go Δt_{tca}. Under all circumstances, the execution of a maneuver leads to a reduction of the statistical collision risk. This can be expressed by the efficiency parameter $\eta_c = \dot{N}_c/\dot{N}_{c,cat}$, which indicates the fraction of the conjunctions of high risk by catalog objects that can be avoided. The effectiveness of the maneuver increases with decreasing time to go ($\eta_c = 61.87\%$, 82.24%, and 93.15% for $\Delta t_{tca} = 48$ h, 24 h, and 8 h, where $P_{c,acc} = 10^{-4}$ and $k_{\sigma,r} = 1.0$), with increasing orbit determination accuracy of the risk object ($\eta_c = 93.15\%$, 97.82%, and 98.83% for $k_{\sigma,r} = 1.0$, 0.1, and 0.01, where $P_{c,acc} = 10^{-4}$ and $\Delta t_{tca} = 8$ h), and with a decreasing level of accepted collision probability (η_c 57.14% and 93.15%, for $P_{c,acc} = 10^{-3}$ and 10^{-4}, where $k_{\sigma,r} = 1.0$ and $\Delta t_{tca} = 8$ h). Hence, to achieve the maximum risk reduction for a small number of avoidance maneuvers, the orbit of the risk object should be improved to the same accuracy level as the well-known target orbit (typically, $k_{\sigma,r} \approx k_{\sigma,t} \approx 0.01$). The maneuver decision then should be taken at the latest possible time compatible with operational requirements (typically, $\Delta t_{tca} \approx 8$ h). Orbit accuracies of risk objects corresponding to $k_{\sigma,r} \approx 0.01$ reduce the two-line element based estimate of the *Envisat* avoidance maneuver rate by more than one order of magnitude, from $\dot{N}_{c,man} \geq 9$ per year to $\dot{N}_{c,man} \leq 1$ in 4 a, for $P_{c,acc} = 10^{-4}$, with a resulting risk reduction of 98.83%. Accepting a larger collision probability of $P_{c,acc} = 10^{-3}$ causes only a minor deterioration of the risk reduction to 98.17%. Note that this applies only for small $k_{\sigma,r}$.

Covariance and correlation information for orbit prediction errors of the target and chaser objects are the dominant factors for a meaningful assessment of collision risk and dependable planning for evasive maneuvers while having acceptable false alarm rates

(Leleux et al. 2002; Jenkin 2004). Time polynomials for the main axes of the 1σ or 3σ position error ellipsoids often are used to provide analytical approximations for predicted maneuver rates as a function of the prediction time span and orbit accuracy.

When collision avoidance maneuvers are prepared, two main strategies can be taken:

- Increasing the radial clearance.
- Increasing the along track clearance.

Increasing the radial clearance is the most commonly used concept because it increases the clearance along the steepest gradient of the combined position error ellipsoid, thus achieving the largest possible risk reduction per range unit. The range increase directly maps onto an increase of the semi-major axis of the orbit that governs orbital energy. Hence, such maneuvers for orbit raising or lowering are ΔV demanding. For low Earth orbit satellites and small impulsive maneuvers (up to ± 10 m/s), Equation (10.6) provides the ΔV required for a single burn orbit raise by ΔH opposite to the maneuver location from an initial circular orbit altitude of H_o (Klinkrad 2003):

$$\frac{\Delta V}{[\text{m/s}]} = \frac{\Delta H}{[\text{km}]} \left(3.54 + 0.00047 \frac{H_o}{[\text{km}]}\right)^{-1} \quad (10.6)$$

The maneuver should be performed $n + \frac{1}{2}$ orbits before the conjunction location, where $n \geq 0$.

An alternative collision avoidance concept aims for the time delayed or time advanced arrival of a target object at the conjunction location. This along track clearance option is suited mainly for collision scenarios that occur under oblique approach angles of the orbits and have sufficient warning times. In this case, a relatively small ΔV changes the orbital period such that the along track separation at the conjunction epoch is increased by ΔS within n orbital revolutions preceding the time of closest approach [Equation (10.7)] (Klinkrad 2003):

$$\frac{\Delta V}{[\text{m/s}]} = \frac{1}{3\pi n} \frac{\Delta S}{[\text{km}]} \left(3.54 + 0.00047 \frac{H_o}{[\text{km}]}\right)^{-1} \quad (10.7)$$

Often, a restituting maneuver is required to restore the operational orbit after the avoidance operation.

10.1.2 Shielding

Spacecraft, particularly those that operate at densely populated low Earth orbit altitudes, experience a continuous flux of human made space debris and natural meteoroids. This flux level increases considerably with decreasing particle size. The consequences of the resulting impacts on an unprotected spacecraft can range from small pits due to micrometer size objects, via clear hole penetrations for millimeter size objects, to mission terminating hits by centimeter size objects that possess a kinetic energy equivalent to that of an exploding hand grenade.

At the most frequently used orbital altitudes, the dominant contributors to risk relevant impacts are space debris. Depending on the orbit of the target spacecraft, the most probable impact velocities can vary considerably. They are on the order of 10.5 km/s for the *International Space Station* at 400-km altitude with an orbit inclination of 51.5°, 14.5 km/s for a typical remote sensing satellite at 800-km altitude with an orbit inclination of 98.5°, and 0.8 km/s for a geostationary satellite at 35,786-km altitude with an orbit inclination of 0°. The survivability of a spacecraft can be increased considerably if it is equipped with shields to defeat debris and meteoroids in the most abundant size regimes. The effectiveness of such shields can be verified in laboratory tests with hypervelocity accelerators, such as light gas guns, shaped charges, electromagnetic rail guns, and electrostatic guns. For projectile velocities and masses outside their performance range, numerical hydro-codes can be used to simulate the impact response for a given shield design. Physical conditions during an impact are similar to the pressure and temperature conditions at the center of the Earth ($\sim$365 GPa and $\sim$6000 K).

Single Wall Damage Equations

The consequences of hypervelocity impacts commonly are described by means of damage equations. These are used mainly to determine whether particles exceed the ballistic limit for a target, therefore providing a means for assessing the probability of penetration. Care must be taken when using damage equations to understand their underlying assumptions and physical limitations. For instance, all equations listed hereafter have been derived for spherical projectiles, ignoring particle shape effects. For a given mass, such a spherical projectile is likely to cause less damage than an elongated object. It also should be noted that damage equations generally have been validated experimentally only in a confined range of velocities and projectile masses, as well as for certain shield and projectile materials. Accordingly, these empirical relationships should be used only for preliminary assessments. The validation of a final shield design should, in any case, be performed by hypervelocity impact tests and numerical hydro-code simulations. Recurring parameters used in the following damage equations are listed in Table 10.1. The damage equations quoted hereafter are based on information compiled in Klinkrad (2006).

The diameter of an impact crater on a semi-infinite single wall of monolithic material is characterized by Equation (10.8), with the coefficients listed in Table 10.2:

$$d_c = K_1 d_p^\lambda \rho_p^\beta \rho_t^\kappa v_p^\gamma (\cos \alpha_p)^\xi \tag{10.8}$$

For a fully perforated target, the impact damage is described by the following hole diameter Equation (10.9), with coefficients listed in Table 10.3:

$$d_c = d_p \left[K_1 \left(\frac{t_t}{d_p} \right)^\lambda \rho_p^\beta \rho_t^\kappa v_p^\gamma (\cos \alpha_p)^\xi + K_0 \right] \tag{10.9}$$

In between the thick target case of Equation (10.8) and the thin target case of Equation (10.9) are intermediate damage scenarios, whereby the capability of a single wall shield to withstand a perforation is described by the ballistic limit. This corresponds to the minimum target thickness $t_{t,\text{lim}}$, which can defeat an incoming object for given projectile and shield properties and a given impact velocity and impact incident angle [Equation (10.10)]:

$$t_t \geq t_{t,\text{lim}} = K_1 d_p^\lambda \rho_p^\beta \rho_t^\kappa v_p^\gamma (\cos \alpha_p)^\xi \tag{10.10}$$

Table 10.1 Definition of Quantities Used in Hypervelocity Impact Damage Equations (Table from Klinkrad 2006: *Space Debris: Models and Risk Analysis*. Springer-Praxis Books in Aeronautical Engineering. Berlin–Heidelberg–New York: Springer Verlag, Copyright: Praxis Publishing, Ltd., Chichester, UK, 2006 with kind permission of Springer Science+Business Media)

Symbol	Units	Description
t_t, t_w, t_s	cm	Thickness of target, back wall, and shield (total)
$t_{t,lim}$	cm	Ballistic limit (minimum required single wall thickness)
$t_{w,lim}$	cm	Ballistic limit (minimum required back wall thickness)
C_w	—	Ballistic limit parameter
K_1, K_0	—	Calibration constants
$\lambda, \beta, \gamma, \xi, \kappa, \delta, \mu, \upsilon, \upsilon_1, \upsilon_2$	—	Calibration exponents
d_c	cm	Crater or hole diameter
D_c	cm	Crater depth
d_p	cm	Particle (projectile) diameter
$d_{p,lim}$	cm	Maximum defeatable projectile diameter
S	cm	Spacing between bumper shield and back wall
$\rho_t, \rho_p, \rho_s, \rho_w$	g/cm^3	Density of target, particle, shield and back wall
V_p, V_n	km/s	Particle impact velocity and its normal component
α_p	deg.	Particle impact angle with respect to the local vertical
τ	Pa	Yield stress of back wall
τ_1^*	Pa	Reference yield stress = 276 × 106 Pa
τ_2^*	Pa	Reference yield stress = 483 × 106 Pa

The coefficients for Equation (10.10) are listed in Table 10.4. This equation can be rearranged as Equation (10.1) to indicate the maximum defeatable (spherical) projectile diameter $d_{p,lim}$:

$$d_p \geq d_{p,lim} = \left[\frac{t_t}{K_1 \rho_p^\beta \rho_t^\kappa \upsilon_p^\gamma (\cos \alpha_p)^\xi} \right]^{\frac{1}{\lambda}} \tag{10.11}$$

According to Equation (10.11), the shieldable projectile diameter decreases with increasing impact velocity according to $d_{p,lim} \propto v^{-2/3}$.

Multiple Wall Damage Equations

Single wall shields do not offer efficient protection against hypervelocity impacts. For a given structural mass per shield area, multiwall protection systems turn out to be much

Table 10.2 Calibration Parameters for Impact Crater Equations of Ductile and Brittle Materials According to Different Authors (Table from Klinkrad 2006: *Space Debris: Models and Risk Analysis.* Springer-Praxis Books in Aeronautical Engineering. Berlin–Heidelberg–New York: Springer Verlag, Copyright: Praxis Publishing, Ltd., Chichester, UK, 2006 with kind permission of Springer Science+Business Media)

Equation	Material	K_1	λ	β	γ	ξ	κ
ESA	Ductile	0.8 to 1.32	1.056	0.519	2/3	2/3	0
Gault	Brittle	1.08	1.071	0.524	0.714	0.714	−0.5
McHugh et al.	Brittle	1.28	1.2	0	2/3	2/3	0.5
Cour-Palais	Brittle	1.06	1.06	0.5	2/3	2/3	0

Table 10.3 Calibration Parameters for Clear Hole Impact Equations According to Different Authors (Table from Klinkrad 2006: *Space Debris: Models and Risk Analysis.* Springer-Praxis Books in Aeronautical Engineering. Berlin–Heidelberg–New York: Springer Verlag, Copyright: Praxis Publishing, Ltd., Chichester, UK, 2006 with kind permission of Springer Science+Business Media)

Equation	K_1	λ	β	γ	ξ	υ	K_0
Maiden	0.88	2/3	0	1	1	0	0.9
Nysmith	0.88	0.45	0.5	0.5	0.5	0	0
Sawle	0.209	2/3	0.2	0.2	0.2	−0.2	1
Fechtig	5.24×10^{-5}	0	1/3	2/3	2/3	0	0

more effective. In such Whipple shields, a thin outer shield exploits the kinetic energy of the projectile to break it up and a back wall defeats the dispersed fragments. As the projectile hits the bumper, compression and tensile waves are generated that cause complete disintegration of the object, provided that $t_s/d_p \geq 0.1$ and $v_p \geq 7$ km/s. Apart from a few surface ejecta particles, a cloud of mainly liquid projectile and shield material forms. It disperses laterally from the impact hole and progresses longitudinally at almost unaltered speed before hitting the back wall. Due to the lateral dispersion and the time delayed arrival of the different cloud particles, surface loading on the back wall is reduced considerably as compared to the point load caused by the intact projectile at the time of impact. Hence, the back wall can defeat the cloud of projectile fragments. For a classical Whipple shield made of aluminum 7075-T6 alloy, the minimum thickness of the back wall can be estimated from Equation (10.12):

$$t_w \geq t_{w,\lim} = \frac{C_w \rho_p d_p^3 v_p}{S^2} \quad (10.12)$$

Table 10.4 Calibration Parameters for Single Wall Ballistic Limit Equations According to Different Authors (Table from Klinkrad 2006: *Space Debris: Models and Risk Analysis*. Springer-Praxis Books in Aeronautical Engineering. Berlin–Heidelberg–New York: Springer Verlag, Copyright: Praxis Publishing, Ltd., Chichester, UK, 2006 with kind permission of Springer Science+Business Media)

Equation	Target	K_1	λ	β	γ	ξ	κ
ESA	Thick plate	0.36 to 0.99	1.056	0.519	2/3	2/3	0
ESA	Thin plate	0.26 to 0.64	1.056	0.519	0.875	0.875	0
Pailer and Gruen	Any	0.77	1.212	0.737	0.875	0.875	−0.5
Frost	Any	0.43	1.056	0.519	0.875	0.875	0
Naumann et al.	Any	0.65	1.056	0.5	0.875	0.875	−0.5
Naumann	Any	0.326	1.056	0.499	2/3	2/3	0
McHugh et al.	Thick glass	1.18 to 4.48	1.2	0	2/3	2/3	0.5
Cour-Palais	Thick glass	0.98 to 3.17	1.06	0.5	2/3	2/3	0

where $C_w = 21.7 \pm 7.3$ for $\sigma_{0.2}$ yield conditions, and $C_w = 4.3 \pm 0.7$ for fracture conditions. For a given back wall thickness t_w (typically $t_w \geq 0.2 d_p$), this equation also can be used to estimate the necessary spacing S between the bumper and back wall (typically $S \geq 30 d_p$). Figure 10.2 shows the ballistic limit curve in terms of maximum defeatable projectile diameter d_p for a Whipple shield with back wall thickness $t_w = 4$ mm, standoff distance $S = 100$ mm, and varying bumper shield thickness t_s. At velocities below 3 km/s, in the ballistic region, the material strength of the projectile exceeds the dynamic pressure at impact, and so the projectile is fragmented only poorly by the bumper. At velocities above 3 km/s and below 7 km/s, in the shatter or transition region, the projectile breaks up into a finite number of pieces, which can be either solid or liquid. At velocities beyond 7 km/s, in the true hypervelocity range, shock induced pressure and tensile strain by far exceed the material strength of the projectile, thus causing it to break up completely into a cloud of numerous liquid particles and vaporized material. The ballistic limit of a multiple wall shield can be expressed in terms of the minimum required back wall thickness $t_{w,\text{lim}}$ [Equation (10.13)] or in terms of the maximum defeatable projectile diameter $d_{p,\text{lim}}$ [Equation (10.14)]:

$$t_w \geq t_{w,\text{lim}} = K_1 d_p^\lambda \rho_p^\beta \rho_w^\kappa \rho_s^{\nu_1} v_p^\gamma (\cos \alpha_p)^\xi s^\delta - K_2 t_s^\mu \rho_s^{\nu_2} \tag{10.13}$$

$$d_p \geq d_{p,\text{lim}} = \left[\frac{t_w + K_2 t_s^\mu \rho_s^{\nu_2}}{K_1 \rho_p^\beta \rho_w^\kappa \rho_s^{\nu_1} v_p^\gamma (\cos \alpha_p)^\xi s^\delta} \right]^{\frac{1}{\lambda}} \tag{10.14}$$

These ballistic limits are defined for the two velocity ranges $0 \leq v_n \leq v_1 = 3$ km/s and 7 km/s $= v_2 \leq v_n \leq \infty$, where v_n is the normal velocity component. Corresponding model coefficients are listed in Table 10.5. In the intermediate velocity regime at $v_1 \leq v_n \leq v_2$,

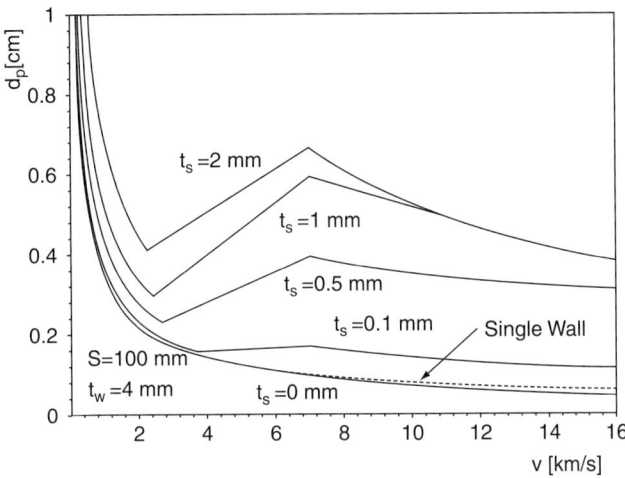

FIGURE 10.2 Ballistic limit curves for an aluminum Whipple shield with a wall thickness of $t_w = 4$ mm and a separation distance of $S = 100$ mm, as a function of projectile diameter d_p, shield thickness t_s, and normal impact velocity v_n (Graphic from Klinkrad 2006: *Space Debris: Models and Risk Analysis*. Springer-Praxis Books in Aeronautical Engineering. Berlin–Heidelberg–New York: Springer Verlag, Copyright: Praxis Publishing, Ltd., Chichester, UK, 2006 with the kind permission of Springer Science+Business Media).

results are interpolated linearly between the left and right boundaries according to Equations (10.15) and (10.16):

$$t_{w,\lim}(v_n) = t_{w,\lim}(v_1) \cdot f_1(v_n) + t_{w,\lim}(v_2) \cdot f_2(v_n) \tag{10.15}$$

$$d_{p,\lim}(v_n) = d_{p,\lim}(v_1) \cdot f_1(v_n) + d_{p,\lim}(v_2) \cdot f_2(v_n) \tag{10.16}$$

where $f_1(v_n) = (v_2 - v_n)/(v_2 - v_1)$ and $f_2(v_n) = (v_n - v_1)/(v_2 - v_1)$. The multiple wall ballistic limit equations can be used as well to estimate the behavior of honeycomb structures, which normally consist of aluminum cells enclosed in face sheets made of aluminum or carbon fiber reinforced plastic. Such structures provide excellent material stiffness with a small mass per area ratio. However, because the honeycomb cells are aligned parallel to the surface normal, an impact of a hypervelocity projectile results in a channeled fragment cloud with the corresponding concentration of the pressure load on the back wall. To reduce this problem, multiple layers of honeycombs can be used, with horizontal displacements by half the cell width between different layers. For aluminum-aluminum single and double honeycomb panels, one can use the ESA triple wall equation for near vertical impacts, the coefficients of which are listed in Table 10.5. In this case the dependency on the incident angle is adjusted via $\xi = 8/3$ (instead of $\xi = 5/3$) to represent channeling effects. Beyond impact velocities of 7, one can introduce an equivalent back wall thickness that mimics the combined behavior of the heterogenic structure.

Table 10.5 Calibration Parameters for Multiple Wall Ballistic Limit Equations According to Different Sources (with $\tau_1' < \sqrt{\tau_1^*/\tau}$, $\tau_2' < \sqrt{\tau_2^*/\tau}$, $\gamma = 2/3$, $\xi = 5/3$, $\mu = 1$ for $v_n < 3$ km/s and $\gamma = 1$, $\xi = 1$, $\mu = 0$ for $v_n < 6(7)$ km/s) (Table from Klinkrad 2006: Space Debris: Models and Risk Analysis. Springer-Praxis Books in Aeronautical Engineering. Berlin-Heidelberg-New York: Springer Verlag, Copyright: Praxis Publishing, Ltd., Chichester, UK, 2006 with kind permission of Springer Science+Business Media)

Equation	v_n [km/s]	K_1	K_2	λ	β	κ	δ	v_1, v_2
ESA (triple wall equation)	$v_n < 3$	$0.312\,\tau_1'$	$1.667 K_1$	1.056	0.5	0	0	0, 0
	$v_n < 7$	$0.107\,\tau_1'$	0	1.5	0.5	0	−0.5	0.167, 0
NASA (modified Cour-Palais)	$v_n < 3$	$0.6\,\tau_1'$	$1.667\,K_1$	1.056	0.5	0	0	0, 0
	$v_n < 7$	$0.129\,\tau_1'$	0	1.5	0.5	0	−0.5	0.167, 0
NASA (shock equation)	$v_n < 3$	$0.3\,\tau_1'$	$1.233\,K_1$	1.056	0.5	0	0	0, 1.0
	$v_n < 6$	$22.545\,\tau_1'$	0	3.0	1.0	−1.0	−2.0	0, 0
ESA (triple wall equation)	$v_n < 3$	$0.4\,\tau_1'$	$0.925\,K_1$	1.056	0.5	0	0	0, 1.0
	$v_n < 6$	$18.224\,\tau_1'$	0	3.0	1.0	−1.0	−2.0	0, 0

Protection Through Shielding and Design

The effectiveness of Whipple shields can be improved in different ways. One approach is to increase the number of bumper shields, which in succession cause further shock waves to disrupt and disperse the projectile and raise its thermal state, causing it to melt or vaporize within a short penetration distance. Another option often used in combination with multiple shields is to deploy different bumper materials. For the outer bumper, which is exposed to the harsh space environment, sufficiently inert materials are preferred, such as aluminum alloys, corrugated aluminum, metal matrix composites, carbon fiber reinforced plastic composites, or Kevlar®. For the inner bumper layers, Nextel® fabrics can be applied. For instance, the crewed *Columbus* module of the *International Space Station* uses a stuffed Whipple shield with an Al-6061-T6 outer bumper of 2.5-mm thickness, followed at 71.1-mm distance by a 4-mm sheet of four Nextel 312-AF62 fabric layers on top of a 6-mm Kevlar 129–812 layer in an epoxy resin 914. This internal bumper is separated by 42 mm from an Al-2219-T851 back wall of 4.8-mm thickness. The shield is designed to withstand impacts at orbital speeds by objects of up to 1 cm in size.

The protection of crewed space systems is a high priority for several reasons. The penetration of high velocity impact projectiles can cause the immediate injury or death of crewmembers due to the kinetic energy of their fragments, heat, light flash, and overpressurization. Time delayed consequences can include hypoxia induced crew unconsciousness and suffocation, especially if the impact hole is large and the depressurization rate is rapid. Because crewed modules are pressurized, special precautions must be taken in their structural design to avoid propagating crack formations after local failures (holes) in the pressure shell. Moreover, the air escaping from the leak can cause perturbing torques that can exceed the control capabilities of the spacecraft and lead to a loss of altitude, power degradation, and for large objects like the *International Space Station*, the possibility of structural damage due to centrifugal forces.

For uncrewed spacecraft, the acceptable risk level due to hypervelocity impacts is higher than for crewed systems. However, with minor adaptations in the spacecraft design, survivability can be enhanced considerably. One such example is the Canadian *Radarsat*, which operates at 790-km altitude in a Sun-synchronous orbit. Because *Radarsat* is three-axis stabilized with constant yaw, roll, and pitch relative to an orbit related coordinate system, the deployment of debris impact protection could be concentrated on surfaces close to the ram direction. As the result of high velocity impact tests, a layer of Nextel was added to the multilayer insulation blankets that cover the external electronic boxes and harnesses. Moreover, critical boxes were given thicker walls or moved away from the debris approach direction and located behind less critical hardware. With a total mass penalty of 0.6% (17 kg) for debris protection measures, the *Radarsat* designers achieved an improvement in system survivability from 50 to 87% over a planned 5-a mission lifetime.

The survivability of a space object is a function of its size and geometry, its structure, its orbit and attitude, and its exposure time to the space debris and meteoroid environment. A measure of survivability is the probability of no failure, which oftentimes is denoted synonymously as *probability of no penetration*. Let an individual system component have a ballistic limit of $d_{p,\text{lim}}$, in terms of defeatable projectile diameter. If the most probable impact velocities are high, such as $v_p \approx 10$ km/s for the *International Space Station* or $v_p \approx 14.5$ km/s for Sun-synchronous orbits, then one can assume a constant mean value of $d_{p,\text{lim}}$ across the relevant velocity range with the highest flux contribution. Based on the concepts of Poisson statistics, the probability of no penetration can be determined via Equation (10.17) from the impact flux $F(d \geq d_{p\text{lim}})$ of debris and meteoroid objects that exceed the ballistic limit, from the collision cross section A_c of the target, and from the time Δt of exposure to the space debris and meteoroid environment:

$$P_{\text{PNP}} = \exp[-F(d \geq d_{p,\text{lim}}) \cdot A_c \cdot \Delta t] \qquad (10.17)$$

The corresponding probability of one or more failures due to shield penetration is expressed by Equation (10.18):

$$P_{n \geq 1} = 1 - P_{\text{PNP}} \qquad (10.18)$$

For the *International Space Station*, design requirements call for a probability of critical failures of less than 0.5% per year.

10.2 EMERGENCY REPAIR CONSIDERATIONS FOR SPACECRAFT PRESSURE WALL DAMAGE

10.2.1 Balanced Mitigation of Program Risks

Many atmospheric leak scenarios requiring emergency repair of a spacecraft pressure wall during a mission are of low probability. The scenarios tend to be very diverse in nature, such that prior certification of the comprehensive repair methods needed for each possibility is prohibitive. However, the possibly of catastrophic consequences from such events dictate that a potential event cannot be ignored. To address this, general methods and processes can be developed, recognizing that some scenarios require unique applications of those methods and processes.

General Space and Space Vehicle Constraints and Challenges

Most spacecraft have no autonomous leak detection, location, and repair (self healing) capabilities. The result is that on-orbit repair of spacecraft typically requires crew actions to do the work. Damage to a subsystem component, such as electronics and life support systems, can be mitigated with the inclusion of inherent fault detection, isolation, and recovery features in the original design of a spacecraft.

The space environments that most directly affect the design and development of repair techniques are zero gravity, vacuum, temperature variations, and the limited amount of atmosphere for crew operations during the repair of an active leak. Design constraints common to all spacecraft designs include weight, power, and heat rejection capability, which limit the design options for available repair techniques.

Spacecraft instrumentation for an automatic leak location system should be included during the initial design and development phase. Integration into the spacecraft and into the caution and warning system can be accomplished as part of the initial assembly. The retrofit of a system after the spacecraft already has been assembled poses severe difficulties in access for the location of the instrumentation and modification of already developed software. For a retrofit system, a wireless instrumentation system might be the primary option available.

The best physical phenomenon to be utilized and measured when locating a leak has not been established firmly as of this writing. For the detection of atmospheric leaks from a spacecraft compartment, airflow induced structural borne ultrasonic acoustic emissions are a leading candidate for allowing triangulation of the leak location. For interior measurements, airborne ultrasonic and audible noises are also candidates. Depending on access to the surface of a spacecraft, infrared thermal imaging is a possibility. Externally, the light flash associated with micrometeoroid and orbital debris impacts or gas density is a means that could be utilized to locate a leak.

Spacecraft Damage

Damage to a spacecraft can result in different severity of problems: catastrophic loss, that is loss of spacecraft or crew, mission loss, component damage, and surface degradation. A subset of catastrophic loss applicable only to human spaceflight is penetration of the pressure wall of a crew module. Damage to a crew module can compromise the pressure and structural integrity of the pressure wall and cause damage to or loss of internally or externally mounted equipment items that are part of other subsystems. Depending on the function of any damaged equipment, additional hazards can be created.

Spacecraft are subject to damage from various sources. In low Earth orbit, the spacecraft is subjected to the meteoroid environment as well as human made debris. Very high relative velocity particles can impact the spacecraft causing a penetration of the pressure wall and damage to externally mounted equipment. Spacecraft to be used on lunar and Mars missions will be exposed to the meteoroid environment during transit as well as when on these planets. These spacecraft also are subject to damage resulting from secondary ejecta caused by large meteor impacts to the surface. Such spacecraft also can be damaged by collision with other spacecraft, inherent failure due to insufficient structural life, and corrosion of a pressure wall caused by various environmental effects. Additionally leaks can occur due to seal damage or failure, leaking or stuck valves, a breach in the pressure wall at a utility feedthrough, or a leak in the feedthrough itself. An example of micrometeoroid and orbital debris induced damage to a pressure wall is shown in Figure 10.3.

Damage severe enough to penetrate a pressure wall results in a leak of spacecraft crew compartment atmosphere to vacuum. Depending on the area of the opening created by the damage event on the pressure wall and the size of the crew compartment pressurized volume, the amount of time to vent the cabin atmosphere to vacuum can vary. The duration of venting is a key parameter for evaluating the choices of operational controls available for this catastrophic hazard. If the atmospheric leak is slow, there can be sufficient time for the crew to locate the damage and affect a repair. The limiting factor for the amount of crew time available to affect a repair is the time from the initiation of the leak to the time when the pressure has lowered to a level inducing hypoxia for the crew. This amount of time might be extended by the use of oxygen masks. However, if the leak rate is

334 CHAPTER 10 Meteoroid and Debris Protection

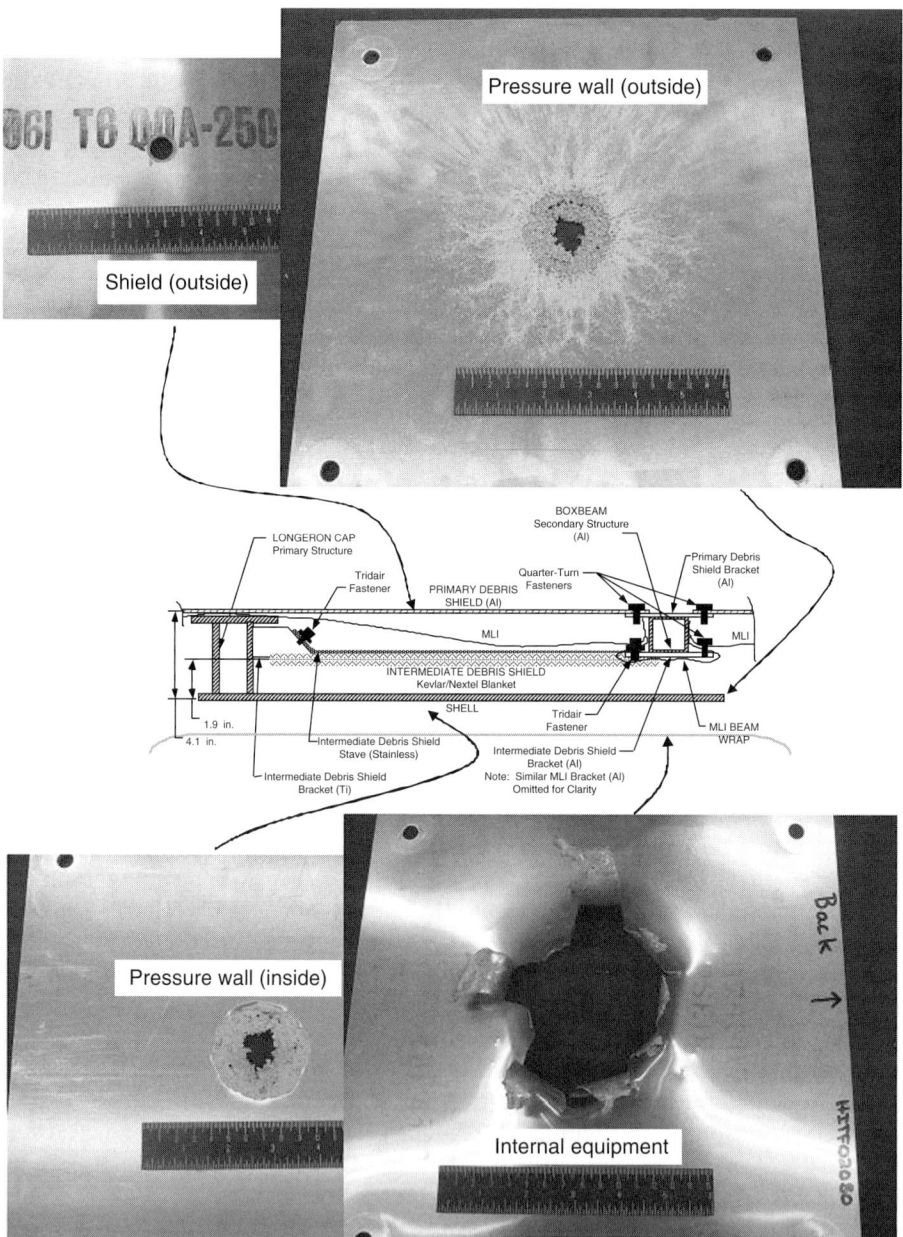

FIGURE 10.3 An example of micrometeoroid and orbital debris induced damage to a pressure wall (Courtesy of NASA).

high, then the crew must evacuate immediately and isolate the compartment from the rest of the spacecraft. The repair is therefore postponed to a later time.

Any damage to the spacecraft pressure wall has to be evaluated based on the severity of the damage. If the structural strength and structural life of the pressure wall has been reduced such that negative margins exist, then the structural and pressure integrity of the pressure wall must be restored prior to repressurization and subsequent use of the compartment. However, if the damage to the pressure wall does not result in negative margins for structural strength and life, then only the pressure integrity needs to be restored.

Damage Repair

Repairing damage to a crew module pressure wall introduces a number of requirements for implementation:

- Knowledge of the loss of pressure integrity. Generally, these data are available from the crew module internal pressure sensors. Software alarms are included to alert the crew to a drop in internal pressure.
- The time to repair the damage before module isolation, evacuation, or crew loss. Available crew time to react to the pressure drop is calculated based on the rate of pressure drop and the internal volume of the modules.
- Ability to access the pressure wall combined with volume to store the removed equipment without compromising wall access or crew mobility.
- Equipment available to affect the repair.
- Equipment available to certify the repair following its completion.

Any restoration of the integrity of the pressure wall must employ a method whereby the repair restores the original leak rate requirement for the damaged compartment or results in a repaired leak rate that can be supported by the atmosphere resupply capability. For the case of a very slow leak and if the allowable time to depressurization permits, the pressure wall repair can be achieved by repair devices that are applied by the crew from inside the pressurized module. However, if there is insufficient time for the crew to perform the procedure before depressurization, then the pressure wall repair must be performed on a depressurized module isolated from the other spacecraft compartments. Such a repair is performed presumably by crewmembers in spacesuits from outside of the module.

The on-orbit restoration of spacecraft pressurized compartment pressure integrity requires a repair method whereby the damaged area is sealed. The method for sealing must withstand the internal pressure of the spacecraft and either restore the original leak rate specified by its design or be no greater than that which can be supported by atmosphere resupply capability. The method for sealing must be able to close the opening for the duration of the mission for that particular spacecraft.

Methods are available for implementing these types of repair, such as using tape, pressure domes with seals, and epoxy compounds cured in place after using appropriate fixturing and material preparation. Some of these repair techniques include methodology whereby they are applied from inside the pressurized module, while others are applied from the outside by a pressure suited crewmember.

On-orbit restoration of the structural integrity of the pressure wall of a spacecraft pressurized compartment is a formidable technical and operational challenge. The general methodology employed in the aircraft industry for ground repair of pressurized airplane fuselages is an option to consider for adaptation to on-orbit repair. These techniques include the application of riveted or bonded structural doublers over the damaged area. Additionally, welded structural doublers can be considered as a repair technique that is adapted and used on spacecraft pressure walls constructed of aluminum. Spacecraft pressure walls made from composite structures require bonded structural doublers for repair of the damaged pressure wall.

These repair techniques are well developed for ground repair of airplanes; however their adaptation for on-orbit application requires extensive development efforts. The basic challenge is development of an effective methodology whereby the spacecraft crew can apply these repair techniques successfully to affect a pressure wall structural repair under on-orbit conditions. The zero-g effects, the probability that the repair might need to be performed while the crewmembers are wearing pressure suits, the temperature extremes experienced on orbit, and the vacuum environment are all factors that have to be considered and included in repair technique development. The use of robotics to conduct or support a repair is possible; however, this would require a considerable development effort to establish and verify the necessary techniques.

Because of the myriad of subsystems that could be damaged, the approach for developing an emergency repair kit is limited to restoration of either pressure integrity or both pressure and structural integrity of the pressure wall of a spacecraft crew compartment. Damage to internally or externally mounted equipment that is part of other subsystems must be addressed on a case by case basis. Due to the diversity of spacecraft equipment, the repair of any damage usually requires the availability of specific spares for each item.

Component Damage

Spacecraft components can be damaged as well by impact from micrometeoroids and orbital debris. A spacecraft must possess adequate component redundancy and separation of redundant paths to meet the reliability and the probability of mission success requirements of the program. For an orbital platform, damaged components could be designed to accommodate on-orbit replacement. For a lunar, planetary, or an uncrewed mission, reliance on sparing is not likely an initially viable solution. For these missions, the designer must consider that, by the time components have been determined to possess adequate robustness to meet launch loads, they must be sufficiently robust to provide an acceptable level of component vulnerability.

For those cases where mechanisms are deployed on orbit and for critical surfaces, this does not hold true. Many deployed mechanisms can be packaged to survive launch loads but have an unacceptable tolerance to impact. For this situation, on-orbit impacts become the design driver for the components of the mechanism. Certain components possessing critical surfaces, such as radiators, solar arrays, optics, and antennas, cannot be designed in a considerably different configuration or architecture nor can they be shielded. Therefore, they must be designed to accommodate an expected amount of surface damage while still providing the required performance. A residual risk remains and mission ending damage can occur, but a design for graceful degradation should support adequately the probability of mission success requirements.

10.2.2 Leak Location System and Operational Design Considerations

Perhaps the key consideration in spacecraft repair is awareness and knowledge of any damage to the vehicle or penetration of the pressure wall. Atmospheric venting through large holes, that is, 0.25-in. diameter or larger, clearly should cause audible and ultrasonic noise. In contrast, holes 0.025-in. diameter and smaller can generate limited or no discernable audible or ultrasonic signal, depending on the physical parameter measured.

For a human rated space vehicle, there must be instrumentation to provide data indicating the pressure status of the spacecraft. Such instrumentation includes internal pressure gauges, pressure change rate indicators, and hardware to assess the partial pressure of atmospheric constituent gases. In spacecraft applications, hardware of this type typically is part of the environmental control and life support system instrumentation.

Hardware integrated into a vehicle to monitor for structural leaks and structural integrity are part of what commonly is referred to as the *structural* or *integrated vehicle health monitoring system*.

The capability to identify the location of a leak source requires hardware and software to triangulate the leak within practical constraints. Triangulation must be accurate within a specified spatial resolution that typically corresponds to normal equipment. Any system used to identify a leak location would need to be tailored to the specific vehicle within which it is to be used; therefore, it would require a specific development effort.

Should module isolation and depressurization be necessary, a different approach would have to be taken to locate the leak site. After a crew compartment has vented to vacuum, the noise emitted by the leaking air or other air induced phenomena that could be used for sensing the leak location stops.

Alternate methods for leak site location were tried on the *Mir* following the collision, isolation of the Spectre module from the rest of that space station, and subsequent module depressurization. As an example of one such approach, low pressure gas containing a fluorescing compound was introduced into the depressurized compartment. It was hoped that observation of the escaping gas from outside of the module would indicate the leak site location. However, the precise location of the leak, in fact, was not detected.

As with any design, operational requirements must be included during the initial design. The effectiveness of a repair method for control of a catastrophic hazard during a time critical repair operation includes its expeditious application by the crew. Operational considerations for a repair method include the emergency kit storage location, emergency warning method, and process path decisions based on situation severity (evacuation, isolation, or repair of the damaged compartment and crew and vehicle safety).

10.2.3 Ability to Access the Damaged Area

The on-orbit repair of a damaged area of the pressure wall of a spacecraft compartment requires access by the crew to affect the repair. The crew can be required to reach a leak source from the inside of the pressure wall or from the outside. Accordingly, access can be achieved only if spacecraft subsystem components located adjacent to the pressure wall can be removed by the crew; alternately, there must be sufficient clearance between these components and the pressure wall to provide access for repair. This objective can be met only if the original design documentation of the spacecraft is detailed sufficiently to have incorporated this requirement. As an example, on the *International Space Station*, the design solutions are such that the internal stowage racks are rotatable. This allows the crew to gain access to the pressure wall. Cabling and utility lines are installed using a 1-in. standoff from the pressure wall. Implementation of this requirement poses severe technical challenges in terms of the weight and space limitations typical of spacecraft design.

10.2.4 Kit Design and Certification Considerations (1 is too many; 100 are not enough)

The design of repair kits must utilize the smallest mass and volume possible. This is a requirement, because on-orbit stowage mass and volume allocation typically are limited severely for all spacecraft. As well, there are a minimum number of separate repair kits that must be stowed aboard the spacecraft. Depending on the spacecraft pressurized compartment configuration, repair kits are stowed at easily accessible locations. The choices of location and the number of kits required are supported by safe haven studies unique to the specific spacecraft.

For orbital or near Earth missions, a number of spare kits also should be certified and stored on the ground. Depending on the number of missions planned for a spacecraft, the number of spare kits can vary widely. For the *International Space Station*, the number of spare kits required is expected to be considerably different from that of a lunar mission spacecraft system. The repair kits can contain components that degrade rapidly with time, such as two-part adhesives; therefore, the sparing process must account for resupply and replacement of all such limited life items within the kits.

The design of the repair kits must facilitate their use by the crew under emergency conditions. This requires that the packaging, labeling, and installation procedures are simple to perceive and easy to use.

10.2.5 Recertification of the Repaired Pressure Compartment for Use by the Crew

Any completed repair of a spacecraft pressure wall must be certified prior to use of the pressurized compartment by the crew.

Restoring Pressure Integrity

A number of specified processes and verification procedures are required to obtain certifications for any repair that restores the pressure integrity of a pressure wall:

- **Structural analysis.** Structural analysis must be performed for the pressure wall of a damaged module to assure that adequate structural life remains, given the type and location of the damage.

- **Pressure integrity qualification testing.** Prior to its actual flight and any use during an emergency, adequate qualification testing of a repair technique must be conducted. The qualification testing has to address all aspects of its anticipated use during a repair, such as leak rate, structural integrity, service life, and vacuum compatibility. In essence, the qualification process must assure not only that it is safe to use but also its functionality in sealing a leak path. The qualification testing is performed on the ground prior to the first flight of the kit.

- **Pressure integrity acceptance testing.** Acceptance testing of an actual repair must be conducted on orbit after the repair is completed on a spacecraft pressure wall and before reoccupation or use of the pressurized volume by the crew. Pressure integrity acceptance testing consists of a leak check of the completed repair, which is conducted in situ by the crew.

Restoring Structural Integrity

A number of specified processes and verification procedures are used to restore the structural integrity of a pressure wall. For these, certifications must be obtained before their use.

- **Structural analysis.** Structural analysis is required for a repaired pressure wall of a module to assure that adequate structural life remains, given the specific repair method employed and the type and location of the damage.

- **Structural integrity qualification testing.** Prior to its actual flight and use during an emergency, adequate qualification testing of the repair technique must be conducted. The qualification testing has to address all aspects of its anticipated use during a repair, such as leak rate, structural integrity, service life, and vacuum compatibility. In essence, the qualification process has to assure not only that it is safe to use but also its functionality in sealing a leak path. The qualification testing is performed on the ground prior to the first flight of the kit.

- **Structural integrity acceptance testing.** Acceptance testing of an actual repair must be conducted on orbit after the repair is completed on a spacecraft pressure wall and before reoccupation or use of the pressurized volume by the crew. Pressure integrity acceptance testing consists of a leak check of the completed repair, conducted in situ by the crew. This test must be a nondestructive evaluation of the repair. Proof pressure testing of the repaired compartment is not an option, because internal components and the subsystems of the compartment usually are not certified to operate above the maximum dynamic pressure of the spacecraft.

REFERENCES

Alarcon, J. (2002) *Development of a Collision Risk Analysis Tool*. Final Report, ESA Contract 14801/00/D/HK. Darmstadt, Germany: European Space Agency.

Alfriend, K., M. Akella, D. Lee, J. Frisbee, and J. Foster. (1999) Probability of collision error analysis. *Space Debris* 1, no. 1: 21–35.

Bérend, N. (1997) *Étude de la probabilité de collision entre un satellite et des débris spatiaux*. Technical Report RT 38/3605 SY. Paris: Onera.

Foster, J. (1992) *A Parametric Analysis of Orbital Debris Collision Probability and Maneuver Rate for Space Vehicles*. Technical Report JSC-25898. Houston, TX: National Aeronautics and Space Administration, Johnson Space Center.

Hoots, F., L. Crawford, and R. Roehrich. (1984) An analytical method to determine future close approaches between satellites. *Celestial Mechanics* 33.

Jenkin, A. (2004) Effect of orbital data quality on the feasibility of collision risk management. *Journal of Spacecraft and Rockets* 41: 677-683.

Khutorovsky, Z., V. Boikov, and S. Kamensky. (1993) Direct method for the analysis of collision probability of artificial space objects in LEO—techniques, results, and applications. SD-01. Proceedings of the First European Conference on Space Debris. Darmstadt, Germany: European Space Agency, pp. 491-508.

Klinkrad, H. (ed.). (2003) *ESA Space Debris Mitigation Handbook*. Darmstadt, Germany: European Space Agency.

Klinkrad, H. (2006) *Space Debris—Models and Risk Analysis*. Berlin-Heidelberg-New York: Springer-Praxis.

Leleux, D., R. Spencer, P. Zimmermann, C. Propst, W. Heilman, J. Frisbee, and M. Wortham. (2002) Probability based Space Shuttle collision avoidance. 02-R-T3-50-1. *Proceedings of the SpaceOps Conference*. Houston, TX: National Aeronautics and Space Administration, Johnson Space Center.

Patera, R. (2001) General method for calculating satellite collision probability. *Journal of Guidance, Control and Dynamics* 24: 716-722.

Patera, R. (2005) Calculating collision probability for arbitrary space vehicle shapes via numerical quadrature. *Journal of Guidance, Control and Dynamics* 28: 1326-1328.

CHAPTER 11

Noise Control Design

Jerry R. Goodman, Ph.D.
Chairman (Former), Acoustics Working Group and ISS Acoustics Lead, Johnson Space Center, National Aeronautics and Space Administration, Houston, Texas

Ferdinand W. Grosveld, D.E.
Consultant, Hampton, Virginia

CONTENTS

11.1.	Introduction	341
11.2.	Noise Control Plan	341
11.3.	Noise Control Design Applications	345
11.4.	Conclusions and Recommendations	355

11.1 INTRODUCTION

The acoustics environment during space operations is characterized in Section 2.4, "Acoustics," in Chapter 2 of this book. Limiting the acoustic exposure levels in the crew compartment and habitat to the defined requirements is deemed essential to achieve a safe, functional, effective, and comfortable acoustic environment for the crew. A noise control plan is necessary to define and lay out the plans and efforts required to achieve compliance with the acoustic requirements. The status and progress of the noise control plan needs to be monitored actively to ensure good communications on efforts to limit noise, identify any areas of emphasis and concerns early in the design process, and allow timely remedial actions to be taken. A detailed discussion of the noise control plan and its major components are presented, followed by various applications of successful noise control design in habitable space environments.

11.2 NOISE CONTROL PLAN

A noise control plan, at a minimum, should include

- The overall noise control strategy.
- The supporting acoustic analysis approach.
- The testing and verification procedures for the system and hardware components.

11.2.1 Noise Control Strategy

A sound source radiates energy that is perceived at the receiver location as a pressure deviation from the local ambient pressure. The source is characterized by the sound energy per unit time, or sound power, and the pressure deviations at the receiver are measured as sound pressure levels. The sound energy emitted from the source follows various paths into the crew compartment. The acceptability of the resultant acoustic levels at the crew receiver location is defined by the requirements for the habitable environment.

Unwanted sound is defined as *noise*. Noise control is the application of designs and technology necessary to limit the noise at the source, along its path, and at the receiver location to acceptable levels (Beranek 1988).

Noise Sources

It is important to identify and control noise sources, for they provide acoustic energy to the crew compartment or habitat of a spacecraft. Sources need to be classified as to whether they are continuous or intermittent, because environmental limits in space operations are specified in this manner. Fans, pumps, motors, and compressors are usually the dominant continuous noise sources.

There are two basic alternatives to noise source control:

- Select or develop noise sources that are quiet by design while considering acoustic emission as well as other characteristics in the choice of this hardware.
- Focus on development activities to quiet the selected design or hardware to the extent required.

Sound sources should be characterized by their sound power output level. This information is provided by either the designer or the supplier and measured in accordance to applicable international standards (ISO 2003).

Noise Paths

Three basic sound paths need to be addressed:

- Airborne.
- Structure borne.
- Enclosure radiated.

Airborne sound comes from the inlets and exhausts of air ducts, directly from exposed equipment, or from sound leaking through air passageways or gaps. This type of sound can be controlled using mufflers or silencers for broadband noise, resonators for narrow band noise, active acoustic control systems inside the duct, applications of sound absorbing materials in the duct lining, and use of appropriate materials to seal the gaps or otherwise block the noise.

Structure borne noise is transmitted by structural vibrations and the resultant energy transfer at mountings, connections, and from surfaces. This noise can be reduced by the use of vibration isolators, active vibration control systems, applications of passive or active damping materials, and decoupling of lines to preclude the transfer of vibration.

Enclosure radiated sound is radiated from or transmitted through structural enclosures, panels, shelves, and other types of closeout materials. This noise contribution can be lowered by material changes, the addition of barrier or stiffening materials to reduce transmission, the addition of damping or viscoelastic materials to minimize radiation, addition of absorbent materials inside the enclosure to absorb acoustic energy, or the use of active structural acoustic control.

Noise at the Receiver Location

Acoustic requirements for the various types of limits to be met at the location of the receiver or the ear of a crewmember were discussed in Section 2.4, "Acoustics," in Chapter 2. The acoustic environment in the receiving space is affected by the volume, surface area, the dimensions relative to the acoustic wavelength, the ratio of the dimensions, the reverberation time, and the absorption properties of the crew compartment. At higher frequencies, where the sound pressure in the reverberant field can be assumed constant, the noise in the receiving space is best controlled by increasing the absorption coefficient of the bounding surface areas.

The application of these absorption materials to the interior surfaces of the crew habitat can be limited in use because of flammability, offgassing, wear and tear resistance, and other properties of the material. Although porous acoustic materials often have good sound absorbing properties, they might not be suitable for use within the crew compartment if they either particulate or collect moisture, dirt, or other contaminants detrimental to the health and well-being of the inhabitants. If their use is necessary, these materials need to be covered or contained such that the concerns are remedied and good absorption properties are maintained.

At the lower frequencies, a noise control strategy can be based on active acoustic control, if the application can be made practical using reliable hardware and robust control software. The design should address redundancy and mitigation measures relating to a possible failure of the active control system. The acoustic environment in the crew compartment or habitat should be controlled at all potential receiver locations or, at least, at established crew operation positions. At crew receiver locations, other approaches for reducing the sound pressure levels or changing the effects of the factors described are limited. Options at the receiver are to enclose the receiver, move the receiver, or require that the receiver wear hearing protection. If the receiver acoustic levels are too high because the predicted or measured levels have been underestimated or not understood adequately, the remedial alternatives lead back to reducing emissions from the noise sources or along the paths to the receiver. This is all the more reason why the crew compartment with the basic systems installed should be tested early to ensure problems can be found, quantified, and appropriate remedial actions implemented in a timely fashion. When this assessment is postponed until late in the flow schedule, any noncompliance discovered at that time more severely impacts the design and delivery schedules. Remedial action then proves to be more difficult and costly. The noise control plan and design flow schedules should include time for this valuable effort, and it should be conducted as early in the program as possible.

The option of moving the receiver is practical only if the crew can be relocated to areas not affected by the higher noise levels. By providing separate sleeping quarters, the crew can be isolated from noise that otherwise would disturb their rest or sleep cycles. Controlling the noise directly at the ear of the receiver usually is not acceptable, because the levels would be tolerable only with the use of hearing protection. Exceptions can be made for short duration events, such as cabin depressurization, the launch sequence, or some segments during the descent of the space vehicle.

11.2.2 Acoustic Analysis

An acoustic analysis is an important part of the noise control plan because its predictions provide an estimate for the resultant noise levels in the crew compartment habitat throughout the design phase. The acoustic analysis should be based on a semiempirical approach, in which possibly inaccurate assumptions, calculations, and procedures in the analysis can be replaced by validated test results. The analyses should be performed at the component or assembly level of the contributing sources and along their paths to the receiver location. The purpose of the semiempirical acoustic analysis is to have a continuously updated and documented assessment of the acoustic environment as it relates to compliance with the requirements and to provide insight and understanding of the underlying acoustic principles. This allows efficient and effective implementation of noise control.

The first step in estimating the noise environment is to quantify the sound power of the noise sources to determine which measures need to be implemented along the pathways to the receiver location and establish priorities for noise control efforts. Analysis and testing should be maximized to provide updated information on source, path, and receiver information. Breadboard or piggyback testing on major noise source subsystems should be used to expose acoustic effects, and the actual noise levels should be used to update the analysis.

A variety of tools are available for the acoustic analyses, each of which has advantages and disadvantages, depending on the frequency range of interest, the computational and financial resources available, the accuracy required, the type of source, the nature of the noise paths, and the characterization of the receiving space. Tools include the use of analytic formulas, geometric computer aided design models, statistical energy analysis programs, finite element and boundary element codes, acoustic ray tracing programs (Pilkinton and Denham 2005), technical and mathematical computing languages, and the traditional programming languages.

11.2.3 Testing and Verification

The sound power and directivity of noise sources need to be measured, and the results should be used to determine possible quieting approaches. Simple mockups or prototypes can be employed to determine inexpensively the effectiveness of mufflers or other noise reduction devices. Designs and design approaches should be tested as much as possible prior to formal verification testing to minimize unforeseen results, provide time for remedial actions if required, and supply a basis for updating the analysis to reflect test results. Acoustic measurements should be included in the breadboard testing of systems, like, for example, the environmental control system.

It is important to operate each equipment item individually to determine its noise contribution and frequency content relative to the total noise levels. This provides information for ranking the contributing sound sources in selected frequency bands and helps establish priorities for the work to be done. Testing setup, conditions, instrumentation, procedures, and results should be included or referenced in the noise control plan and implemented accordingly. As noted previously, it is recommended to allow for testing early in the final checkout, so that time is available for remedial action with minimized impact. Verification is very important, in that it defines how and what needs to be done to prove that the requirements have been met. Verification needs to address the testing, demonstrations, analysis, and the equipment and programs used in the verification process.

11.3 NOISE CONTROL DESIGN APPLICATIONS

The noise control plan should define the approaches to be used and the efforts that can be made at the source, path, and receiver levels to control the noise for compliance with the established requirements. The Space Shuttle Program uses an approach in which all continuous noise sources are identified, the source to listener paths are determined, the combined systems noise in the flight deck and middeck are estimated, the contribution of each source relative to the total noise is established, and the applicable noise criteria are specified (Hill 1992, 1994). This approach is illustrated by the flowchart in Figure 11.1 (Hill 1992). The figure also shows that the typical fan powered source radiation in the *Space Shuttle* consists of contributions from the aerodynamic noise emanating from the inlet and exhaust, contributions from the structure borne vibration at the mounting interface, and noise from equipment enclosure radiation. In addition, the flowchart illustrates that the noise emitted at the duct outlet already has been reduced by losses within the duct due to absorption and by the bends and branches of the duct. The structure borne vibration is affected by structural losses, joints, and the mass and damping of the structural element. Finally, the source surface radiation is dependent on the enclosure losses, the transmission loss, and the mass, stiffness, and damping of the enclosure. Typical noise paths aboard the *Space Shuttle* are shown in Figure 11.2 (Hill 1992).

The *Space Shuttle* and the *International Space Station* control the noise permitted in the habitable environment by budgeting allocations to the equipment sources and noise pathways. European modules of the *International Space Station* use a somewhat different approach. Budgets are established for the allowable sound power of hardware systems. The sound power contributions of these sources are then determined, and any necessary pathway reduction efforts using testing or a database of prior testing are implemented. The test results are used to ensure compliance (Destafanis and Marucchi-Chierro 2002).

The sound power present in the crew compartment is the result of controlling the noise source power being channeled through the various radiation and transmission paths, while taking into account any insertion loss through panels and materials. The sound power at the receiver is then converted into sound pressure levels by using the room equation and constants (Beranek 1988). Although concentrating more on predictions than budgeting

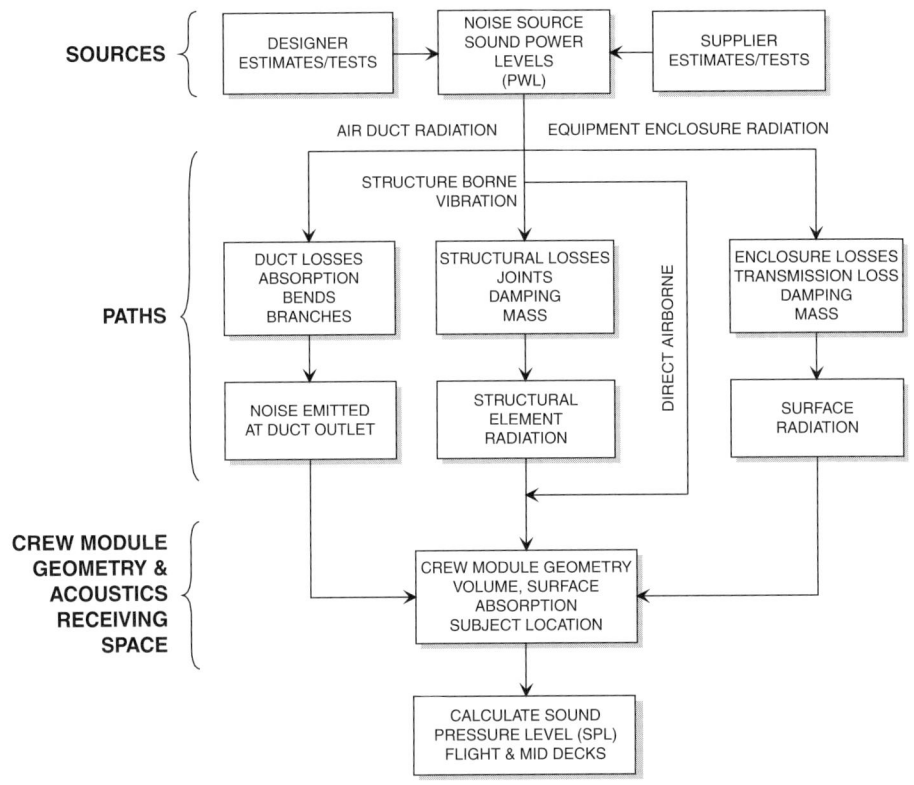

FIGURE 11.1 Space Shuttle approach to estimating continuous noise in the crew compartment (Courtesy of NASA).

and control, the approach used for the International Space Station *U.S. Laboratory Module, Destiny*, similarly focused on the sound power and resultant effects of the design (Denham and Kidd 1996). The International Space Station Program also developed a good noise control plan for use with International Space Station payload racks, as defined in Appendix H of SSP 57010B (NASA 2000).

11.3.1 Noise Control at the Source

Fans are the dominant noise sources within the Space Shuttle flight deck and middeck and in the *International Space Station*. Because of Apollo mission acoustic concerns, an effort was made to develop a new, quiet fan under NASA research funding. Originally, this type of fan was part of the Space Shuttle design baseline but was later dropped with some debate because of cost and schedule. Late in the design, mufflers were added to offset the fan noise.

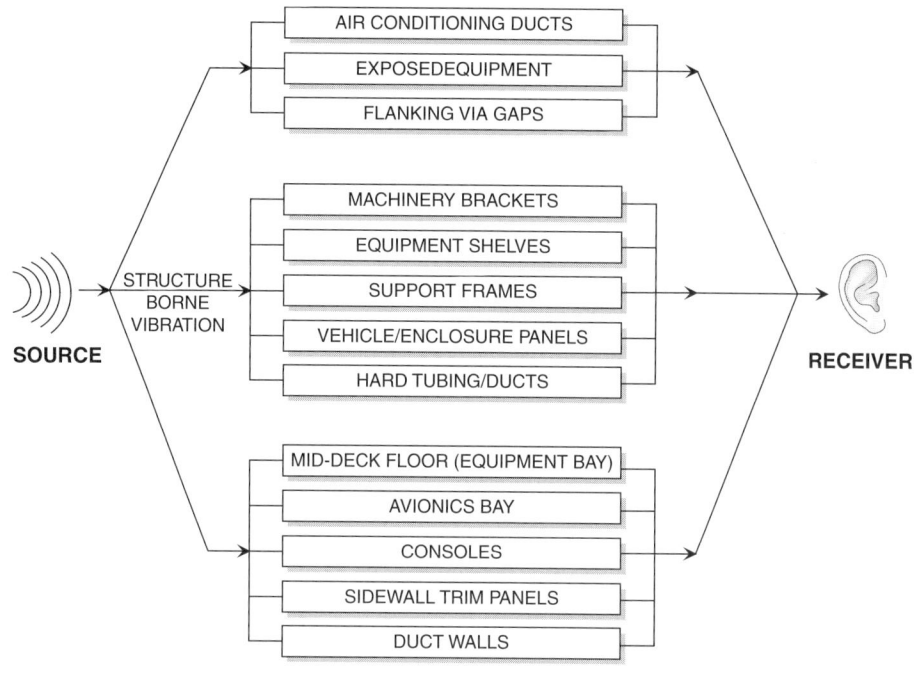

FIGURE 11.2 Space Shuttle noise paths (Courtesy of NASA).

Fans have been tested for flow, flow resistance, and acoustics and have been cataloged to help fan selection for International Space Station payloads. Quiet fans now are being developed for the International Space Station *Service Module*, because acoustic pathway improvements have not been able to reduce the noise sufficiently and the *Service Module* contains so many internal noise sources.

The NASA Constellation program began multi-center efforts to focus on acoustics in the design of larger environmental and thermal control fans. Fan design to meet acoustic requirements is a trade-off involving many factors, and elements that must be matched include fan source noise, power versus frequency requirements, and size as a function of speed (O'Conner 1995). Fan balance, blade shape, bearings, and motor design are some areas where improvements can be made to lower the noise. Reducing the fan speeds or voltage has been used where feasible to lower the noise emission levels.

Although fans draw most of the attention, pumps, compressors, and other notable noise sources need to be attended to in the same manner. Technology and expertise exist for this hardware as for fans. In the case of the *Service Module*, considerable design and development efforts, funding, and costly on-orbit time has been spent on mitigation methods to remedy noise problems. Applying resources and technology early in a program to obtain quiet noise sources is recommended.

11.3.2 Path Noise Control

Noisy fans generate loud airborne noise in air duct inlets and exhausts that is transmitted into the crew compartment. Inlet and outlet mufflers are commonplace accessories used on the *International Space Station* to lower noise produced by fans. A muffler or silencer used at the intermodule ventilation fan inlet and outlet in the U.S. segment is shown in Figure 11.3. It is lined inside with a feltmetal (a micron-size fiber sinter bonded into continuous felt) screen covering applied over absorbent foam material. The European modules use similar mufflers.

Considerable noise concerns existed for the *Space Shuttle* before its first flight, and government furnished equipment mufflers were developed to quiet the effects of the most dominant noise source, the inertial measurement unit fans (Figure 11.4). The acoustic benefits for the use of the government furnished equipment foam lined reactive and dissipative muffler designs is shown in Figure 11.5 (Hill 1992). These government furnished equipment mufflers subsequently were changed from the four individual mufflers (three inlets and one outlet) to one unified muffler.

For the International Space Station *Functional Cargo Block*, NASA developed a unique muffler (Figure 11.6), incorporating improved flow, noise barrier, absorption, and Helmholtz resonator concepts that reduced both broadband and narrowband noise (Grosveld and Goodman 2003). Reserving an envelope and provisioning for future addition of mufflers (scaring) should be considered in the design of space systems so that, if needed, mufflers can be added later without major impacts. Air duct noise can be attenuated by improving the design of the ducts, bends, absorbent liners, and the diffusers or grills that draw air in or let it out. The airflow passageways to and from fans can produce noise because of restrictions and turbulent flow. They, therefore, can raise the total fan related noise. Space Shuttle airborne noise ducting losses are shown in Figure 11.7 (Hill 1992). Acoustically treated devices, termed *splitters*, with Helmholtz resonators tuned to attenuate fan inlet or outlet

FIGURE 11.3 U.S. Laboratory intermodule ventilation fan muffler (Courtesy of NASA).

11.3 Noise Control Design Applications

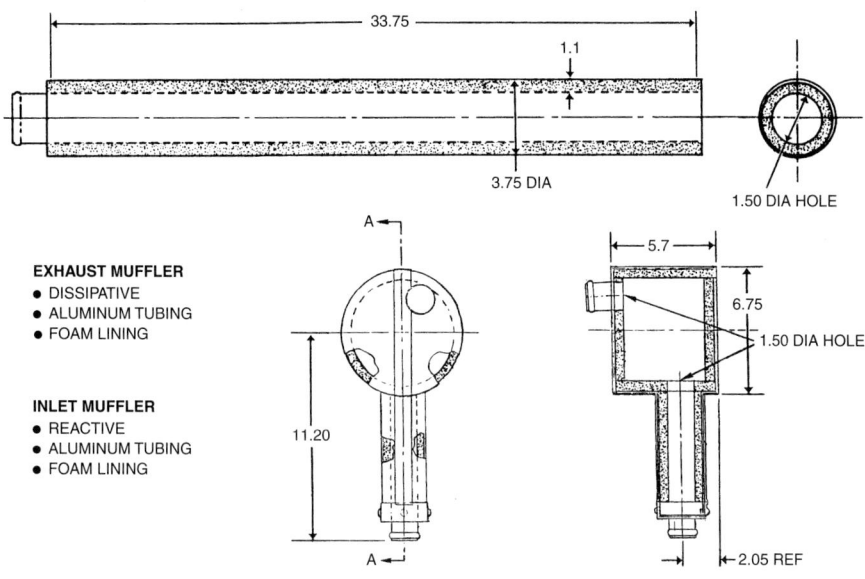

FIGURE 11.4 Space Shuttle inertial measurement unit cooling fan mufflers (Courtesy of NASA).

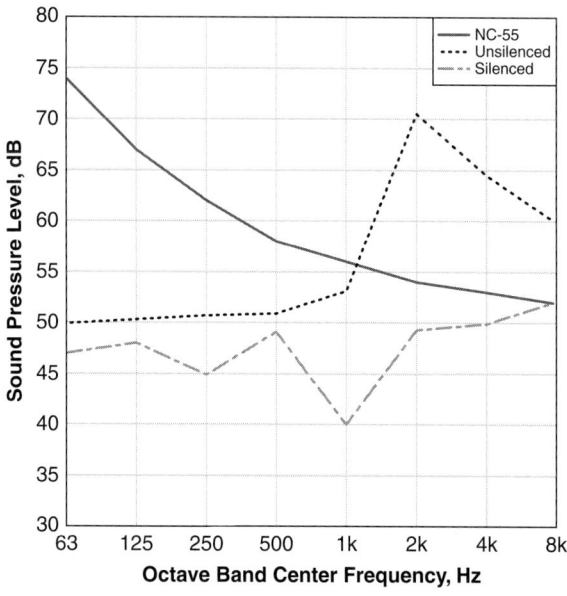

FIGURE 11.5 Space Shuttle inertial measurement unit muffler attenuation (Hill 1992) (Courtesy of NASA).

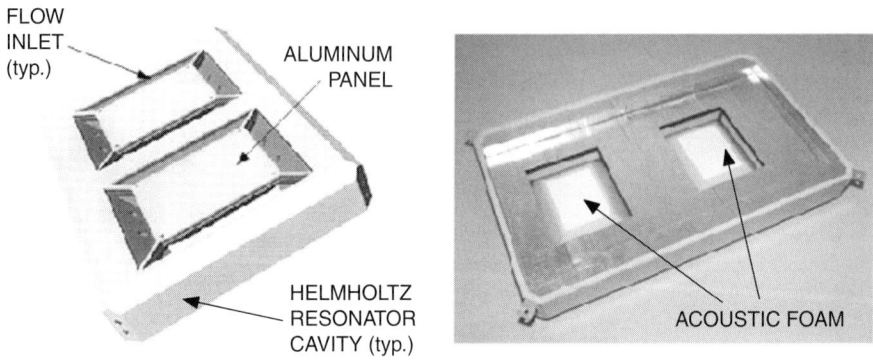

FIGURE 11.6 NASA muffler for the Functional Cargo Block (Courtesy of NASA).

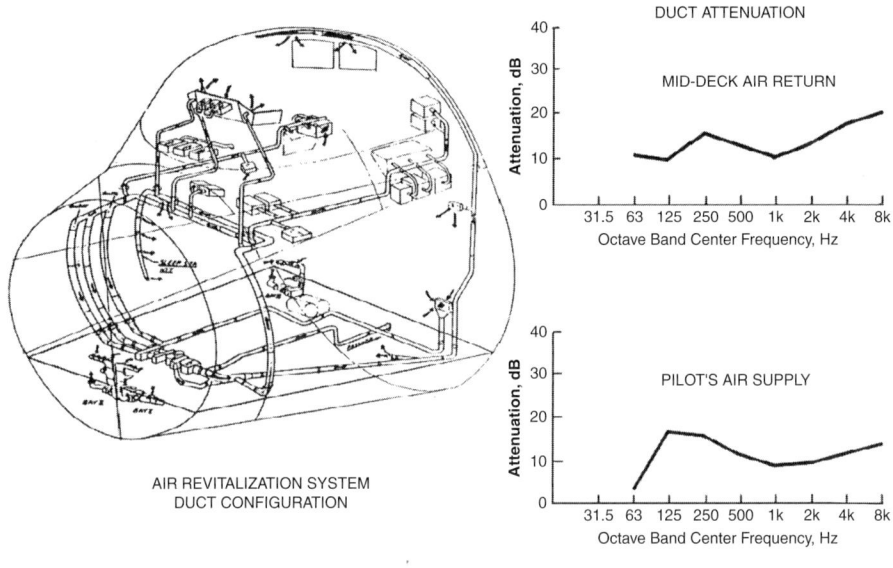

FIGURE 11.7 Space Shuttle airborne noise path air duct attenuation (Courtesy of NASA).

noise were added in a number of places in the International Space Station U.S. Laboratory Module ducting to attenuate duct noise (Denham and Kidd 1996). As well, International Space Station air inlet and outlet registers have been designed or later modified to lower the noise in the design of the outlets.

If a source, such as a fan, cannot be quieted by design, then strong consideration should be given to the use of a unified package that attenuates airborne emissions by using mufflers, attenuating case radiated noise by barrier applications, and reducing

11.3 Noise Control Design Applications

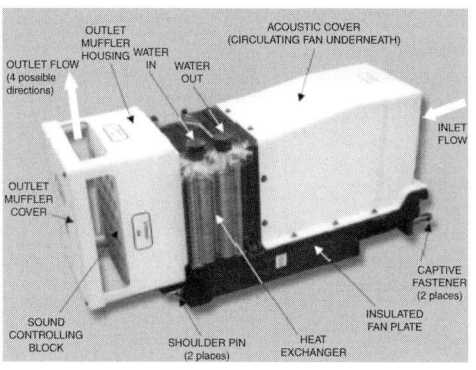 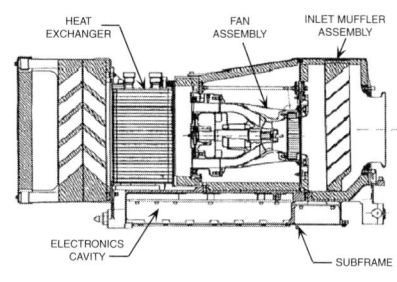

FIGURE 11.8 International Space Station avionics air assembly fan and packaging (Courtesy of NASA).

structure borne noise by the implementation of isolation or anti-vibration mounts. A good example of a system for which most of these features have been implemented is shown in an avionics air assembly fan package used in the *U.S. Laboratory* (Figure 11.8).

Another U.S. Laboratory fan, the intermodule ventilation fan, illustrates several control measures (isolators and acoustic barriers) that can be implemented on fans and other noise sources, as shown in Figure 11.9. Use of vibration isolation is recommended strongly to control *structure borne* noise by mechanically isolating fans, motors, pumps, compressors, other major noise sources, as well as the attachment of ducting and lines to them. Vibration paths in ducting to ducting or fan to ducting connections can be reduced by using rubber type booties for connections. Vibration isolators are used widely in the *Space Shuttle* and the *International Space Station*.

Vibration isolators were not used to mount the pump package assembly in the International Space Station *U.S. Laboratory*. One pump package assembly is used for two separate thermal cooling loops, each located in a separate rack. The operating pump package assembly produces high level noises and excites the structure of the rack within which it is mounted because of its hard mounting (Figure 11.10) and its high mass and energy emission. The dual pump package assembly units operating within the *U.S. Laboratory* produced the highest continuous noise level of any source. Sound pressure levels on orbit were measured to be very high in locations near the rack. Later, it was found that single pump package assembly operations were feasible if the one pump loop worked at a higher rate. Even so, the resultant single pump package assembly operation still produces the highest broadband noise and narrow band tone of all the prime movers in the *U.S. Laboratory*. A pump package assembly quieting kit has been developed to silence this hardware by improving its structural isolation and encasing it in barrier material.

NASA successfully quieted a very loud depressurization pump in the *U.S. Airlock* primarily by the addition of four inexpensive, off-the-shelf, commercial isolators (Grosveld, Goodman, and Pilkinton 2003). This pump, the pump package assembly, and the fans noted previously are good examples of where vibration isolation should be applied.

352 CHAPTER 11 Noise Control Design

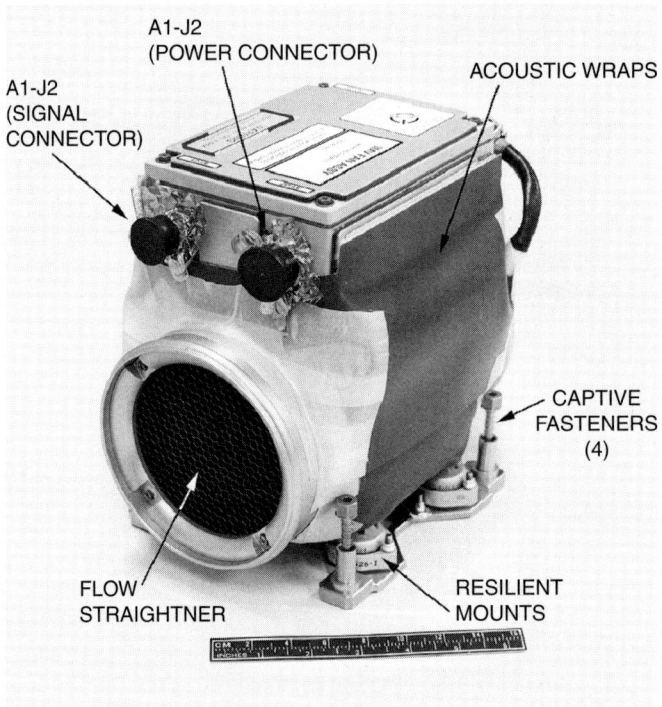

FIGURE 11.9 International Space Station intermodule ventilation fan (Courtesy of NASA).

In structural borne noise situations, it is important to reduce the radiating surface area of the vibrating parts to minimize the noise emissions. Rubber pads have been used successfully for isolation in other International Space Station applications, where there is insufficient room for an isolator or to isolate ducts or tubing at their mounting to a structure.

To reduce enclosure radiation, acoustic foam has been used effectively inside a large number of International Space Station module and payload racks to absorb, and thus lower, noise levels inside the racks. Figure 11.10 shows foam added to the pump package assembly rack interior door and the underside of the pump package assembly mounting shelf, as well as the damping material added to the inside face of the rack door to reduce vibrations.

Barriers have been used on enclosures or as wraps around ducting to reduce radiated noise. These applications were used in quieting the ducting in the −80°C laboratory freezer payload rack (Figure 11.11) (Tang, Goodman, and Allen 2003).

Various types of materials and material lay-ups have been employed to reduce emissions through rack front faces, structural closeouts, or simply as closeouts. Materials are very important in acoustic applications, and it is essential to have space qualified materials with good acoustic properties available.

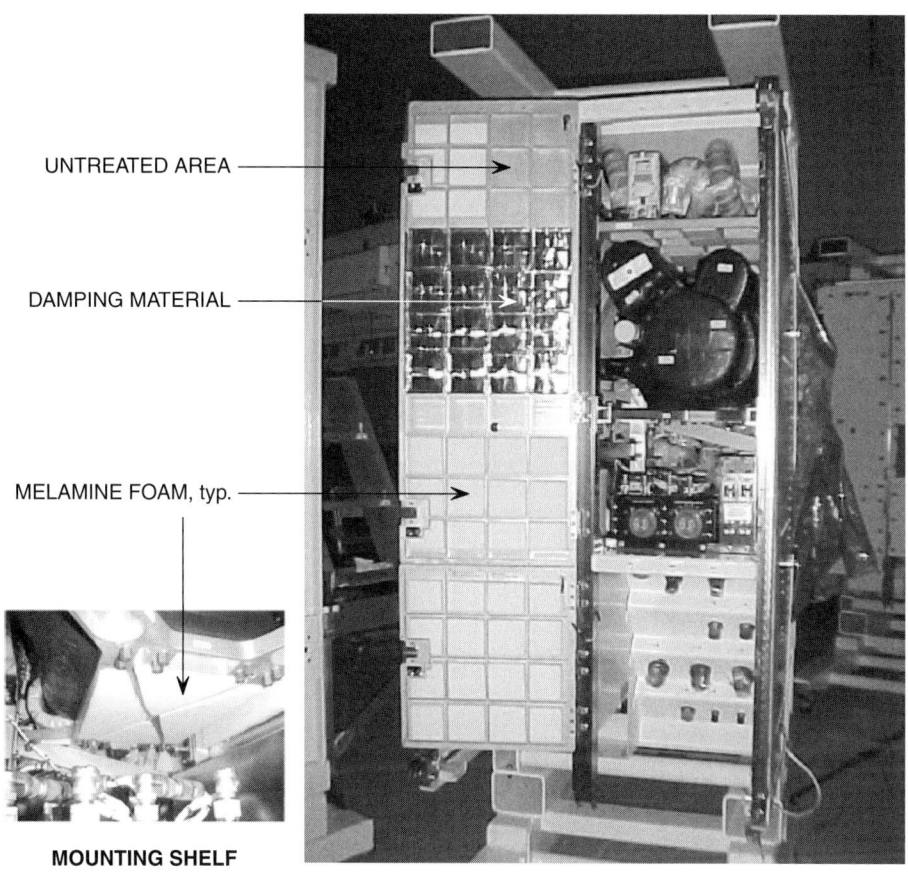

FIGURE 11.10 International Space Station rack with pump package assembly (Courtesy of NASA).

11.3.3 Noise Control in the Receiving Space

Applications of foam end cone cushions were considered for use in the International Space Station *U.S. Laboratory* as a way to help lower acoustic levels by changing the absorption properties of the module and the related room coefficient (Beranek 1988). The results are shown in Figure 11.12. This approach, although beneficial, was not used because of concerns with the cushions being damaged and contents coming out during on-orbit operations. This area is worthy of further consideration to improve the surface absorption, if surfaces can be made durable and reliable.

Another way to provide acceptable sound pressure levels at the receiver is to provide special isolating enclosures, like sleep stations, for use by the crew during periods of rest and sleep. This approach was used in the *Space Shuttle* and the *International Space Station*. Such enclosures, generally designed into the crew compartment or added later as a kit, accommodate the need for lower noise levels for rest and sleep.

FIGURE 11.11 Duct wrapping of the −80°C laboratory freezer (Courtesy of NASA).

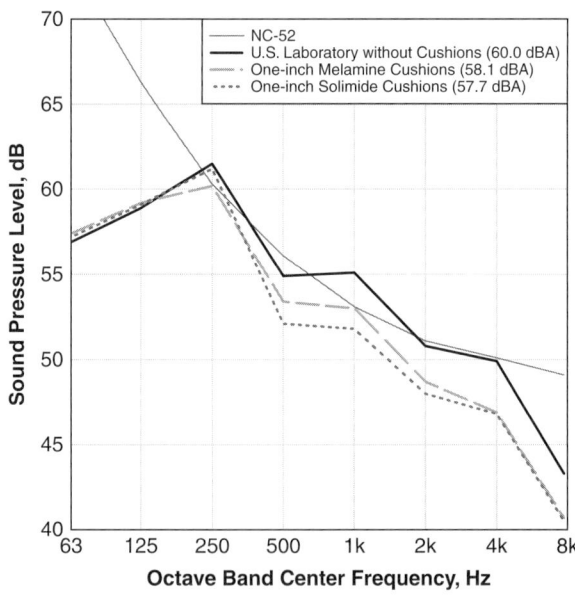

FIGURE 11.12 U.S. Laboratory Melamine® and Solimide® absorbing cushion applications (Courtesy of NASA).

The provision of special, closed off areas for exercise is an approach to lower the noise exposure to crewmembers who are not exercising. In other cases, systems can be turned off or flows can be diminished if such adjustments are acceptable. An example of this was within the Apollo *Lunar Module*, where fans were turned off to solve the noise interference with crew communications. The use of hearing protection devices for launch, entry, and during limited applications also is an acceptable way to control levels at the receiver locations, but only for relatively short durations. They have been used in the Apollo, Space Shuttle, International Space Station, and other space programs. As can be seen from these examples, options for reducing noise at the receiver are limited, which is why efforts need to be focused and expended on effective source and path measures.

11.3.4 Post-Design Noise Mitigation

Noise control is most effective when it is implemented as part of a normal design effort, and it should be approached in that manner. Many examples of successful noise control efforts are designed into International Space Station modules and payloads. Most International Space Station modules were successful in meeting their acoustic requirements or being within an acceptable deviation from them. International Space Station payloads implemented a comprehensive noise control plan and, for the most part, were successful in obtaining compliance. A good example of this is the human research facility payload quieting efforts described by Phillips and Tang (2003).

When mitigation efforts are required to remedy an unacceptable noise situation after design completion, there is risk of considerable impacts on development, costs, and schedules. Also, late mitigation might be only partially effective, because the design or impacts preclude a more effective remedy. A successful mitigation effort to limit noise along numerous pathways was implemented late in the flight assembly process for the −80°C laboratory freezer payload (Tang et al. 2003). This effort, however, was possible only because the design allowed such modifications. It also took considerable technical consultation, design efforts, travel, materials support, testing efforts, and impacts to pull it off.

As discussed before, the International Space Station *Service Module* mitigation effort has taken considerable time and been costly in terms of funding and mission timeline impacts. Remedial pathway actions have been extensive but insufficient to bring the module to specification levels without further work at the noise sources. The International Space Station *Functional Cargo Block* is another example where mitigation efforts added a lot of additional hardware after the first flights but with mixed results. These and other experiences show that acoustics should be considered and designed into the crew compartments and habitats early in their development phases.

11.4 CONCLUSIONS AND RECOMMENDATIONS

A noise control plan is essential to define and lay out all the basic efforts required to achieve acoustic compliance. Included in the plan should be the overall noise control strategy and acoustic analysis approach, testing and verification plans, and focused efforts to use or develop reasonably quiet noise sources and otherwise deal with pathway treatments. The noise control plan needs to be monitored actively. To implement effective noise control in the design it is necessary to understand the principles of acoustics, have noise control experience, and be able to apply these to making the acoustics in the compartment acceptable. Such background and capabilities are needed by those responsible for a safe, functional, and comfortable acoustic environment in the crew compartment. It is imperative that the program management be supportive of the need to comply with established requirements and use the noise control efforts required to achieve compliance. Support of this nature is necessary for acoustics successfully to be designed into the crew compartments and habitats.

RECOMMENDED READING

Bies, D. A., and C. H. Hansen. (2003) *Engineering Noise Control—Theory and Practice*. New York: Spon Press.

Department of Health and Human Services. (1978) *Industrial Noise Control Manual*. Publication No. 79-117. Washington, DC: DHHS, National Institute of Occupational Safety and Health. Available at www.cdc.gov/niosh/79-117pd.html (cited March 14, 2007).

Harris, C. M. (ed.). (1998) *Handbook of Acoustical Measurements and Noise Control*. Acoustical Society of America. New York: McGraw Hill.

Kelso, D., and A. Perez, 1983. *Noise control terms made somewhat easier*. Minneapolis, MN: Minnesota Pollution Control Agency. Available at www.nonoise.org/library/diction/soundict.htm (cited March 14, 2007).

Minnesota Pollution Control Agency. 1999. *A Guide to Noise Control in Minnesota—Acoustical Properties Measurement, Analysis, and Regulation*. Saint Paul, MN: Minnesota Pollution Control Agency. Available at www.nonoise.org/library/sndbasic/sndbasic.htm (cited March 14, 2007).

Occupational Safety and Health Administration. (1980) *Noise Control—A Guide for Workers and Employers*. Washington, DC: United States Department of Labor, Occupational Safety and Health Administration. Available at www.nonoise.org/hearing/noisecon//noisecon.htm (cited March 14, 2007).

Orr, W. G. (ed.). (1981) *Handbook for Industrial Noise Control*. Special Publication SP-5108. The Bionetics Corporation. Hampton, VA: National Aeronautics and Space Administration, Langley Research Center.

Ver, I. L. M., and L. L. Beranek (eds.). (2006) *Noise and Vibration Control Engineering—Principles and Applications*. Hoboken, NJ: Wiley and Sons.

REFERENCES

Beranek, L. L. (ed.). (1988) *Noise and Vibration Control*. New York: McGraw-Hill.

Denham, S. A., and Kidd, G. (1996) *US Laboratory Architectural Control Document*, vol. 14. *Acoustics*. D683-149-147-1-14. Houston, TX: National Aeronautics and Space Administration, Johnson Space Center.

References

Destafanis, S., and P. C. Marucchi-Chierro. (2002) *Node 3 Audible Noise/Human Vibration Environments Analysis and Budget Report*. Report N3-RP-AI-0014. Turino, Italy: Alenia Aerospazio, Space Division.

Grosveld, F. W., and J. R. Goodman. (2003) Design of an acoustic muffler prototype for an air filtration system inlet on the International Space Station. Proceedings of NOISE-CON 2003. Washington, DC: U.S. Institute of Noise Control Engineering.

Grosveld, F. W., J. R. Goodman, and G. D. Pilkinton (2003) International Space Station acoustic noise control—Case studies. Proceedings of NOISE-CON 2003. Washington, DC: U.S. Institute of Noise Control Engineering.

Hill, R. E. (1992) *Space Shuttle Crew Module Prior Noise Reduction Efforts*. The Acoustical Noise Working Group. Houston, TX: National Aeronautics and Space Administration, Johnson Space Center.

Hill, R. E. (1994) *Space Shuttle Orbiter Crew Compartment Acoustic Noise—Environments and Control Considerations*. Report 94SSV154970. Houston, TX: Rockwell International.

International Standards Organization. (2003) *Acoustics—Determination of Sound Power Levels of Noise Sources Using Sound Pressure—Precision Methods for Anechoic and Hemi-anechoic Rooms*. ISO Standard 3745:2003E. Geneva, Switzerland: International Standards Organization.

National Aeronautics and Space Administration. (2000) Acoustic noise control plan for ISS payloads. SSP-57010B Draft, Appendix H. Houston, TX: National Aeronautics and Space Administration, Johnson Space Center.

O'Conner, E. W. (1995) Space vehicle fan package acoustic characteristics. Technical Paper 951647. Society of Automotive Engineering.

Phillips, E. N., and P. Tang. (2003) ISS Human Research Facility (HRF) acoustics. *Proceedings of NOISE-CON 2003*. Washington, DC: U.S. Institute of Noise Control Engineering.

Pilkinton, G. D., and S. A. Denham. (2005) Accuracy of the International Space Station acoustic modeling. *Proceedings of NOISE-CON 2005*. Washington, DC: U.S. Institute of Noise Control Engineering.

Tang, P., Goodman, J., and C. S. Allen. (2003) Testing, evaluation, and design support of the minus eighty degree laboratory freezer (MELFI) payload rack. *Proceedings of NOISE-CON 2003*. Washington, DC: U.S. Institute of Noise Control Engineering.

CHAPTER 12

Materials Safety

Michael D. Pedley, Ph.D.
Materials and Processes Branch, Johnson Space Center, National Aeronautics and
Space Administration, Houston, Texas

CONTENTS

12.1.	Toxic Offgassing	360
	By Tony Brown	
12.2.	Stress-Corrosion Cracking	363
	By Giancarlo Bussu, Ph.D.	
12.3.	Conclusions	373

It goes without saying that all flight hardware is fabricated from materials, and safe and reliable materials and materials processes are essential to safety in space. Materials safety covers a very wide range of material and process technical issues, which simply cannot be covered in a single chapter of a book. For many years, the safety community focused on a limited subset of material and process requirements as key elements to assure safety:

- Materials flammability (covered in Chapter 27, "Fire Safety").
- Toxic offgassing (covered in this chapter).
- Propellant compatibility (covered in Chapter 20, "Propellant Systems Safety").
- Oxygen compatibility (covered in Chapter 13, "Oxygen Systems Safety").
- Stress-corrosion cracking (covered in this chapter).

Each of these items is important, but many other items are equally important to materials safety. Materials used in the fabrication and processing of flight hardware need to be selected by considering the worst-case operational requirements for the particular application and the design engineering properties of the candidate materials. For example, the operational requirements include, but are not limited to, operational temperature limits, loads, contamination, life expectancy, moisture or other fluid media exposure, and vehicle related induced and natural space environments. Other properties that need to be considered in material selection include mechanical properties, fracture toughness, corrosion, thermal and mechanical fatigue properties, glass transition temperature, coefficient of thermal expansion mismatch, vacuum offgassing, fluids compatibility, microbial resistance, moisture resistance, fretting, galling, susceptibility to electrostatic discharge, and contamination. Conditions that could contribute to deterioration of hardware while in service need to receive special consideration. Other important materials safety requirements include

- **Materials process control.** The proper qualification and execution of manufacturing processes, such as structural adhesive bonding, welding, and forging are crucial to materials safety. Limitation of discussions on materials process control to an arbitrary subset known as critical processes is a mistake because any process where the quality of the product cannot be verified by inspection potentially could result in unsafe hardware.

- **Materials structural design allowables.** The development of valid design allowables when none exists in the literature requires extensive testing of multiple lots of materials, an extremely expensive proposition. However, without valid allowables, structural analysis cannot verify structural safety.

- **Contamination control during ground processing.** This is critical to manufacturing processes such as adhesive bonding and to eliminate any foreign object debris that are hazardous to the crew of crewed spacecraft in microgravity environments.

- **Fasteners.** The design, use of verifiable secondary locking features, and proper verification of fastener torque are essential for proper structural performance.

Because the scope of this chapter is limited to materials safety, the interested reader is encouraged to review NASA-STD-(I)-6016 (NASA 2007a), which provides a full set of material and process requirements for the safety and reliability of materials used for flight hardware. As well, NASA-HDBK-6009 (to be published by NASA in 2009), which provides guidelines on materials and processes selection for manufacturing flight hardware, should be reviewed.

12.1 TOXIC OFFGASSING

Volatile chemicals become trapped within other solid or liquid materials through a variety of mechanisms. They usually come from intermediate steps in the material manufacturing process (residual solvents are very common offgassed products). Over time, these chemicals gradually are released. When these gases are released into a habitable area, the process is known as *offgassing*. In the enclosed environment of a spacecraft, the buildup of offgassed products can result in an atmosphere that is hazardous to the crew, depending on the quantity and toxicity of the released compounds.

Offgassing is generally at its highest level soon after a material is manufactured or cured and gradually decreases with time. A commonplace example of materials offgassing is the new car smell, which is caused by offgassing from automobile materials of construction and gradually fades over a period of 6 to 12 mo.

Toxic offgassing controls are implemented to assure the protection of the flight crew while confined within an enclosed space. Selection of materials is one of three controls for reducing the amount of toxic products in the enclosed atmosphere. Other controls are a trace contaminant removal system, which is an essential part of the environmental control and life support system design, and offgassing tests conducted on assembled and outfitted modules prior to launch.

12.1.1 Materials Offgassing Controls

Materials are tested as characterized by NASA-STD-(I)-6001A Test 7 (NASA 2008) to determine their offgassing constituents. The quantity of each offgassed product is compared with the spacecraft maximum allowable concentration for that product.

Spacecraft Maximum Allowable Concentrations

The NASA Johnson Space Center Toxicology Group establishes the guidelines for safe and acceptable levels of individual chemical contaminants of spacecraft air in collaboration with the National Research Council Committee on Toxicology. These safe and acceptable levels are known as the *spacecraft maximum allowable concentrations*. The rationales for these values are documented in books published by the National Academy Press (NRC 1994, 1996a, 1996b, 2000). A table listing all official NASA spacecraft maximum allowable concentration values is published in JSC 20584 (NASA 1999). Unofficial spacecraft maximum allowable concentration values, as well as those designated as interim, temporary, or tentative, are listed in the NASA Materials and Processes Technical Information System (MAPTIS) database.

It is understood that spacecraft maximum allowable concentration values vary with the duration of exposure. For most chemical contaminants for which such official values are defined, the spacecraft maximum allowable concentration values are specified for continuous 1-h, 24-h, 7-d, 30-d, and 180-d exposures. The spacecraft maximum allowable concentration values usually are lower than the industrial American Conference of Governmental Industrial Hygienists threshold limit values or the Occupational Safety and Health Administration permissible exposure limits, because they are for continuous exposure rather than an 8-h workday.

Toxic Hazard Index Values

The standard method for dealing with exposures to a group of offgassing constituents is to calculate a toxic hazard index (T) by adding the ratios of the concentration of each (C_n) to the spacecraft maximum allowable concentration of each constituent for the time of exposure ($SMAC_n$) as per Equation (12.1):

$$T = C_1/\text{SMAC}_1 + C_2/\text{SMAC}_2 + \ldots + C_n/\text{SMAC}_n \tag{12.1}$$

In general, the partial T values are additive for constituents in the same toxicological category but not for constituents with different toxicological effects. However, this distinction normally is ignored when calculating T values for offgassing from materials or assembled articles, and all partial T values are treated as additive.

12.1.2 Materials Testing

All nonmetallic materials used in habitable flight compartments, except for ceramics and metal oxides, are required to meet the offgassing requirements of NASA-STD-6001 Test 7 (NASA 2007c) using one of the following methodologies:

- Bulk and other materials not inside a container are evaluated individually using the ratings in the MAPTIS database. The maximum quantity and associated rating is specified for each material code.
- For offgassing tested as an assembled article, the summation of the T values for all offgassed constituent products is required to be less than 0.5.
- For hardware components evaluated on a materials basis, the individual materials used to make up a component are evaluated based on the actual or estimated mass of the material used in the hardware component. The total T value for all materials used to make up the component is required to be less than 0.5.
- If a single hardware component is tested or evaluated for toxicity but more than one is to be flown, the T value obtained for one unit times the number of flight units is required to be less than 0.5.

Evaluation of hardware on a materials basis is difficult to perform because it is hard to estimate the weight of the individual materials within the component, and it is common to find that not all materials in the component have existing offgassing data. Evaluation of commercial off-the-shelf hardware on a materials basis is almost impossible, because it is difficult to identify all the materials present in adequate detail. For these reasons, testing an assembled article is always the best option.

Materials used in large quantities for spacecraft construction usually are evaluated on a bulk materials basis using the data in MAPTIS (NASA 2007b). In contrast, payload experiments nearly always are evaluated by conducting an assembled article offgassing test.

The analytical technique used to identify and quantify offgassed products by the NASA-STD-6001 standard test is not specified. However, the technique normally consists of gas analysis by a gas chromatograph mass spectrometer to identify the offgassed products, followed by gas chromatography to quantity the individual products. Traditional techniques have been insensitive to formaldehyde, which is a common and relatively toxic spacecraft contaminant. The recently released NASA-STD-(I)-6001A requires the analysis to be capable of detecting formaldehyde concentrations of 0.1 ppm.

When the T value for an assembled article exceeds 0.5, a bakeout process can be implemented to reduce the concentrations of the offgassing constituents. The bakeout should be performed in an oven with active ventilation to ensure adequate removal of the offgassing constituents. The recommended time and temperature is 48 h at 120°F, which can be broken into separate periods. Once the bakeout is completed, a new test is needed to verify that the T value now meets requirements. If the results continue to be above 0.5, the process can be repeated until the desired result is achieved. Once the effectiveness of the bakeout process is verified, subsequent identical components can be processed in the same way without conducting the offgassing test after the bakeout.

12.1.3 Spacecraft Module Testing

When a large module, such as *Spacelab*, *Spacehab*, or an International Space Station segment is planned to fly for the first time, offgas testing of the whole module is performed by a special procedure. Ideally, the module is to be outfitted fully with all cargo onboard before the test. In practice, this is rarely feasible, and a correction has to be made based on the weight of nonmetallic materials that are not present. The module is sealed up, and samples of the atmosphere are taken at multiple locations over a period of several days. The specifics of the test are customized for the application.

12.2 STRESS-CORROSION CRACKING

In many textbooks, *stress-corrosion cracking* is defined as a failure phenomenon associated typically with industrial applications where materials are exposed to saltwater or highly corrosive substances, such as the oil and chemical industry. It is less known that stress-corrosion cracking is an insidious failure phenomenon that can occur in a less dirty environment such as that of an aircraft or a spacecraft. Many cases where structural failure of spacecraft is caused by stress-corrosion cracking can be found in literature. These failures have occurred in both the American and European space programs.

The first documented case of stress-corrosion failure of a propellant tank can be traced back to 1965 and was the failure of a pressurized titanium alloy tank containing liquid nitrogen tetroxide at Bell Aerospace, USA (Jackson and Boyd 1966). Although considered a successful launcher, the European *Ariane 4* suffered from stress-corrosion failures during the early phases of its deployment. On one of these occasions, during final integration at the Guyana Spaceport in Kourou, leaks were detected in a water tank located in the second stage of the rocket. This tank was made by welding AA-7020 alloy, an aluminum-zinc alloy that, if not specifically heat treated, can exhibit low stress-corrosion resistance. Laboratory examination carried out at the ESA European Space Research and Technology Center indicated that the failure was caused by stress-corrosion cracking of welds that had not been heat treated suitably to relieve weld residual stresses and improve the alloy resistance to stress-corrosion (Dunn 1997). Another documented case of stress-corrosion failure was observed on the auxiliary power units of the Space Shuttle *Columbia*. The injector tubes of auxiliary power units made from Hastelloy-B material failed following exposure under stress to ammonium hydroxide vapors (Korb et al. 1985).

These examples show that stress-corrosion failures can occur in spacecraft hardware when consideration has not been given to their susceptibility to this phenomenon during materials selection. Stress-corrosion failures can be caused by residual stress in materials and can happen in the absence of external loads, such as during storage, transport, or before launch. Humidity and water condensation present in pressurized living quarters provide an environment that can promote stress-corrosion failures of spacecraft hardware, such as payloads. For these reasons, stress-corrosion cracking should be considered as an insidious failure phenomenon that severely can affect the integrity and safety of crewed spacecraft.

12.2.1 What Is Stress-Corrosion Cracking?

Stress-corrosion cracking affects engineering materials, typically metals but also ceramics and polymers. For stress-corrosion to occur the following three conditions must be present:

- A sustained tensile stress (which can be residual tensile stress as well as stress from external loads).
- A corrosive environment.
- A material that is susceptible to stress-corrosion.

The damage consists of environmentally induced crack propagation. Crack initiation and propagation are the result of the combined synergetic interaction of mechanical stress and metal corrosion (Jones and Ricker 1993). Environmentally assisted cracking is a wider definition that also includes corrosion fatigue, a failure phenomenon induced by dynamic loading in the presence of a corrosive environment. Water condensation from either the atmosphere or produced within inhabited modules, from exposure to a coastal environment as typically found at launch sites, or resulting from the presence of chemical substances, such as cleaning fluids or propellants, can promote stress-corrosion cracking in susceptible materials. Exposure to these environments can occur during hardware storage, transportation, and service.

Stress-corrosion cracking is a complex phenomenon, and in many cases, its mechanisms have not been understood. In metals, stress-corrosion resistance is lower along the short transverse grain direction. It is influenced by many factors, such as the chemical composition and heat treatment of an alloy, the alloy grain microstructure, the stress level, and the chemistry and temperature of the environment. Magnesium alloys particularly are susceptible to stress corrosion, and the factors influencing this behavior are varied (Winzer et al. 2005). Aluminum alloys might or might not suffer from stress-corrosion cracking, depending on the specific heat treatment to which they are subjected. Figure 12.1 shows an example of intergranular stress-corrosion cracking in an AA-7075-T651 aluminum alloy specimen after being tested for stress-corrosion susceptibility (Bussu and Dunn 2002). The photo shows a crack that propagated from the surface with no visible sign of corrosion. In these alloys, according to the electrochemical theory, the localized decomposition of solid solution along the grain boundaries makes them anodic to the surrounding regions, thus promoting intergranular stress-corrosion cracking in the presence of a sustained tensile stress (Jones and Ricker 1993).

The fact that corrosion is involved in the damage mechanism can lead to the assumption that alloys possessing good corrosion resistance are immune to stress-corrosion cracking. However, this assumption is not always correct and, therefore, must be substantiated by experimental evidence. Stress-corrosion tests conducted on several alloys showed that materials with good corrosion resistance can exhibit high susceptibility to stress-corrosion cracking (Bussu and Dunn 2002).

12.2.2 Prevention of Stress-Corrosion Cracking

As previously stated, stress-corrosion cracking is influenced by many factors, such as the alloy chemical composition and heat treatment, the alloy grain microstructure, the stress

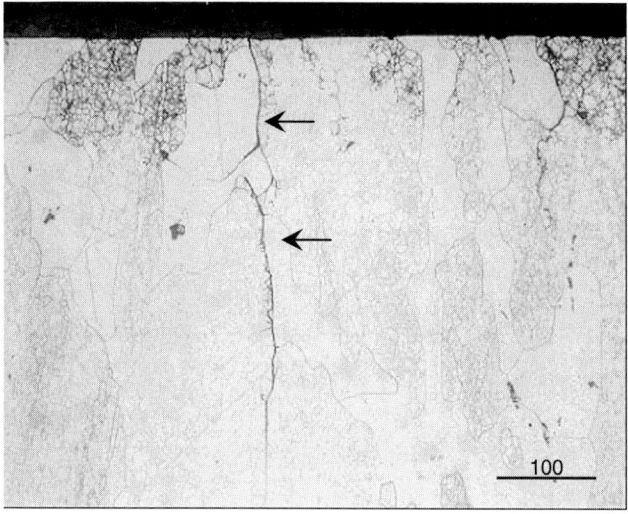

FIGURE 12.1 Example of intergranular stress-corrosion cracking in an AA-7075-T651 aluminum alloy specimen. The sample was tested for stress-corrosion susceptibility. The crack is indicated by arrows (Bussu and Dunn 2002) (Courtesy of ESA).

level, and the chemistry and temperature of the environment. It is, in addition, a time dependent phenomenon. Time to failure in a stress-corrosion susceptible alloy subjected to constant tensile stress decreases rapidly with an increasing stress level. Figure 12.2 shows the typical effect of stress on time to failure for a structural aluminum alloy exposed to a solution of sodium chloride and water. The threshold stress value is defined as the maximum stress that the material can sustain in a particular environment without failure occurring. Although the definition of such a threshold is useful as an appreciation of the level of susceptibility of materials to stress-corrosion cracking, predicting the time to failure of real components with an acceptable level of confidence is difficult and cannot be used in design.

The current approach consists of preventing rather than controlling stress-corrosion cracking. In other words, it is based on a safe life design philosophy by which only stress-corrosion resistant alloys are selected for spacecraft structural applications. The approach differs from that applied to prevent failures by fatigue or monotonic load fracture, for which reliable crack initiation and growth models can be used to predict whether, under the foreseen service loads, an initial defect present in a structure might or might not reach a critical crack size.

In this light, the safety implications of this approach are very important. The determination of the stress-corrosion susceptibility and the use of appropriate criteria for the selection of materials and the verification of structures during the preliminary stages of the design are of paramount importance in preventing failures in spacecraft hardware.

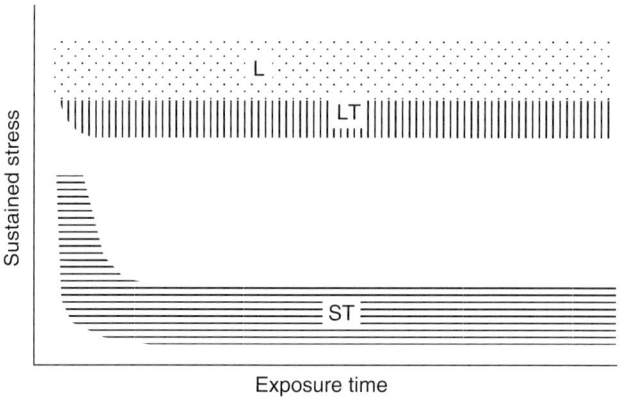

FIGURE 12.2 Typical scatter bands of stress versus time-to-failure for a structural aluminum alloy exposed to a solution of sodium chloride and water. Time-to-failure in a stress-corrosion susceptible alloy subjected to constant tensile stress decreases rapidly with increasing stress level. L = longitudinal; LT = long transverse; ST = short transverse.

12.2.3 Testing Materials for Stress-Corrosion Cracking

When testing a material for stress-corrosion cracking, smooth tensile specimens are loaded at a constant stress level and exposed to a standard 3.5% sodium chloride-water solution for a fixed period of time. The majority of experiment data published in literature refers to a 30-d alternate immersion test under constant load to achieve 75% of the yield stress of the parent material. Typically, the tests are carried out in triplicate as a minimum, the specimens being loaded in tension by calibrated springs or hydraulic loading devices. Because the stress-corrosion resistance of metals is lowest in the short transverse grain direction, specimens are produced to have this orientation parallel to their main axis.

Exposure by alternate immersion consists of immersing the specimens in the solution for 10 min followed by a 50-min drying phase. The cycle constantly is repeated for the full duration of the test. Unstressed control specimens are exposed simultaneously with the stressed specimens to provide a basis for comparison.

Prior to stress-corrosion testing, tensile tests are carried out on similar specimens to obtain the 0.2% proof stress for ultimate stress and elongation of the material. These values are then used to establish the stress-corrosion test stress level, typically calculated as 75% of the 0.2% proof stress for the material. They also provide the reference for comparison with specimens to be tensile tested if the material survives the 30-d duration of the stress-corrosion test.

The susceptibility to stress-corrosion is assessed by tensile tests to compare the residual strength of the specimens exposed while being stressed and unstressed and by metallographic examination of microsections obtained from the stressed and control specimens. Metallographic examination allows the investigator to distinguish failures caused by stress-corrosion that would occur from the damage caused by severe corrosion, independent of stress.

Detailed procedures for testing materials for stress-corrosion cracking in sodium chloride–aqueous environments can be found in European standard ECSS-Q-70-37A (ESA 1998b) and in NASA MSFC-STD-3029A (NASA 2005). The European standard requires tensile specimens to be tested by the constant load test method. The advantage of this method is that it overcomes the reproducibility problems associated with constant strain stress-corrosion tests, which can suffer from relaxation of the stressing jig. Particularly for small specimens, following the onset of stress corrosion, there is a concomitant reduction in specimen stiffness, plastic strain, and creep with an attendant reduction in the initial stress level (Dunn 1997). This can lead to an underestimation of the level of susceptibility of the tested materials.

In MSFC-STD-3029A, the recommended test method involves testing round tensile specimens stressed to 50%, 75%, and 90% of the yield strength, which have been exposed to 3.5% sodium chloride-water solution by alternate immersion for 30-d (NASA 2005). Other test configurations and exposure environments are considered complementary. The specimens are loaded in frames to achieve a desired initial stress level under constant strain. The initial stress level remains constant if there is no stress relaxation, either in the sample or in the frame, for the entire duration of the test. This approach simulates well, for example, the case of a bolt considered to be loaded by constant strain, and in addition, it has the advantage of being simple.

The interpretation of the test results includes the reduction in load carrying ability and the metallographic examination. Based on this, tested alloys are classified as Class or Table I, II, or III materials, depending on their respective high, moderate, or poor resistance to stress-corrosion cracking. Multiple batch testing, in-service experience, and literature data are used to include an alloy in Table I, II, or III of the standards.

Damage during exposure is generally the result of two factors, stress-corrosion and metal corrosion, typically in the form of surface pitting or other forms. Although some alloys are not stress-corrosion sensitive, they exhibit poor corrosion resistance. This is the case for the MIG welded AA-7020-T6 aluminum alloy, which has been post-weld solution heat treated and precipitation hardened. This alloy was tested using the constant load method at ESA (Bussu and Dunn 2002). Metallographic analysis indicated no evidence of stress-corrosion in the stressed specimens (Figure 12.3), and both stressed and control specimens showed the same damage extent regardless of the presence of sustained tensile stresses in the former ones. It was concluded that the weldments exhibited good resistance to stress-corrosion cracking and the damage was caused essentially by pitting corrosion in the transition region between the fused metal and heat affected zone.

The alternate immersion exposure to 3.5% sodium chloride-water solution provides a means of accelerating the stress-corrosion phenomenon. This exposure method is not necessarily representative of a specific corrosive environment, that is, exposure to saltwater or a coastal environment. In fact, the implicit assumption of this method is that an alternate immersion exposure to 3.5% sodium chloride-water for 30-d is representative in terms of stress-corrosion damage of a long-term exposure in a generic corrosive environment. This is the fundamental engineering assumption of the testing method, and its validity and

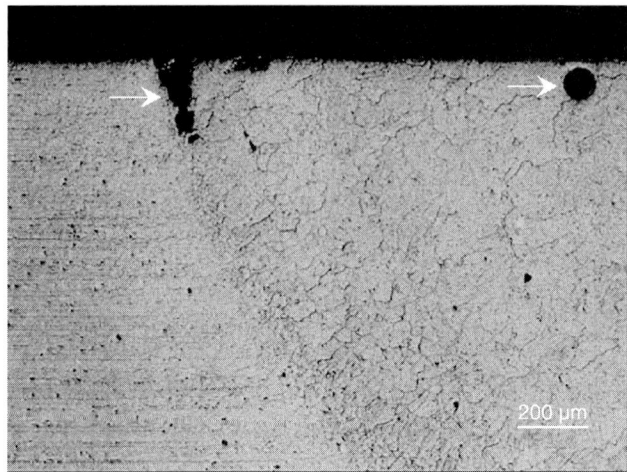

FIGURE 12.3 Cross section of a welded specimen subjected to stress-corrosion test. The photo shows that the damage is caused essentially by pitting corrosion in the transition region between the fused metal and parent plate (Bussu and Dunn 2002) (Courtesy of ESA).

importance should always be discussed in the light of the specific in-service exposure conditions. For example, the suitability of an alloy to be used in the fabrication of a propellant tank would need to be assessed by a dedicated stress-corrosion test program, including the exposure of representative specimens to the specific propellant environment.

12.2.4 Design for Stress-Corrosion Cracking

It has been mentioned that stress-corrosion cracking occurs in the presence of sustained tensile stresses. In a component, these stresses are the result of externally applied loads experienced during storage, transportation, and service and the residual stresses introduced during manufacture and assembly. Whereas tensile stresses associated with external loads can be determined by design with a certain degree of confidence, residual stresses introduced during manufacture and assembly are more difficult to predict and therefore often are underestimated or neglected. Residual stresses are generally of a sustained nature and are present in the absence of externally applied loads. For this reason, they can promote stress-corrosion cracking. Residual stresses always should be considered when assessing the risk of failure by stress-corrosion cracking.

The selection of manufacturing processes that minimize residual stress can lower the risk of stress-corrosion failures by reducing tensile stresses acting in the short transverse direction of the material. The application of a specific design to reduce assembly stresses resulting from mismatch or excessive clearance between elements in mechanical joints can further increase safety. Figure 12.4 shows a typical example of how sustained stresses are introduced into a mechanical connection. The excessive clearance between lugs

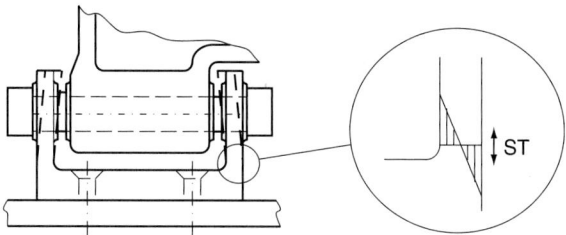

FIGURE 12.4 A typical example of how sustained stresses can be introduced into a mechanical connection. The shaft axial preload and the excessive clearance between lugs results in normal stresses parallel to the short transverse (ST) direction in the material (Bussu and Dunn 2002) (Courtesy of ESA).

results in normal stresses parallel to the short transverse direction in the material. A heat treated alloy that is highly resistant to stress-corrosion should be selected for this application. In addition, a manufacturing process should be selected to change the grain orientation with respect to the direction of sustained tensile stress. Clearance between lugs should also be reduced.

Components such as tanks and other pressurized elements typically are manufactured by welding, and they are exposed during service to highly corrosive substances. Weld residual stresses can approach the yield strength of an alloy, and unless suitably relieved by means of stretching and heat treatment, such stresses can promote stress-corrosion cracking (Hill 1961).

Surface protection treatments against corrosion, such as chemical conversion coating, are not particularly effective in preventing stress-corrosion failures. The effect of chemical conversion coating on stress-corrosion cracking was assessed by ESA (Bussu and Dunn 2002). Samples of stress-corrosion sensitive AA-7010-T7451 aluminum alloy were chemical conversion coated and tested for stress corrosion together with unprotected samples of the same alloy using the constant load test method of ECSS-Q-70-37A (ESA 1998b). Metallographic analysis showed evidence of stress-corrosion cracking in both bare and conversion coated samples.

12.2.5 Requirements for Spacecraft Hardware

Materials for spacecraft structural applications should be selected from those known to possess a high resistance to stress-corrosion cracking. Standards ECSS-Q-70-36A (ESA 1998a) and MSFC-STD-3029A (NASA 2005) describe the material selection criteria for stress-corrosion. Heat treated alloys that have been tested are classified in these standards with respect to their stress-corrosion resistance. In the standards, these alloys are listed in three tables. Materials listed in Table I possess high resistance to stress-corrosion cracking, and they are preferred.

Table II materials, which are moderately resistant to stress-corrosion cracking, are used only in cases where a suitable alloy cannot be found in Table I. To prevent occurrence of in-service stress-corrosion failures, the suitability for use of a Table II material is assessed on a case by case basis. The designer formally is asked to provide justification for the selection of a Table II material. The evaluation is carried out by estimating sustained tensile stresses (including residual stress) and their orientation with respect to the short transverse direction in the material. The environment to which the component is to be exposed, the fabrication processes, and any finishes applied for corrosion protection also are considered. The effect of failure by stress-corrosion of the component on the overall integrity and safety of the major assembly or mission is assessed as well.

Table III materials, being highly susceptible to stress-corrosion cracking, are considered for use only in nonstructural applications or applications where it can be demonstrated conclusively that the probability for stress-corrosion failure is remote. This would be because of low sustained tensile stress, an innocuous environment, or the use of suitable protective measures. The evaluation procedure is similar to that carried out to assess the use of Table II materials.

Special applications can require the use of heat treated alloys that are not listed in ECSS-Q-70-36A and MSFC-STD-3029A or involve exposure to specific environments. The suitability of these materials is assessed by means of tests conducted in an environment representative of the specific application. Direct comparison with similar heat treated alloys for which stress-corrosion susceptibility is known to be low also provides an indication of the expected level of susceptibility. This practice is essential to guarantee the integrity and safety of spacecraft hardware.

Information from any previous service application is of great value when assessing the suitability for use of an alloy not listed in Table I. However, it should be noted that cases where the same heat treated alloy was applied in service are relevant only if there is documented evidence that no stress-corrosion failures occurred. In such cases, it should be verified that the environment and the sustained stress levels were of the same or of a higher degree of severity than those foreseen for the application in question.

Assessing literature data is usually the first step in investigating the stress-corrosion behavior of an alloy. Many test methods can be employed for characterizing the stress-corrosion behavior of a material, and plenty of test data can be found in open literature. Unfortunately, because stress-corrosion is complex and depends on many factors, different stress-corrosion test conditions often produce results that are not necessarily comparable (Jones and Ricker 1993). For example, the slow strain rate test per ASTM G 129 (ASTM 2006) involves the application of a slow strain rate to tensile samples while they are exposed to a corrosive environment. This laboratory test method is essentially an accelerated way to produce a stress-corrosion failure in a specimen of a susceptible material. The assessment is based on the comparison between results obtained from applying the same method in air. However, the results depend on strain rate values, and because the test conditions are difficult to compare to those of a real application, generally, it does not provide data that can be used in design.

For these reasons, the alloys listed in the tables of ECSS-Q-70-36A were evaluated using the constant load test method with a stress level of 75% of the 0.2% proof stress of the material (ESA 1998a). This guarantees homogeneity and comparability among the test results. To ensure coherency between the tables of ECSS-Q-70-36A and those of MSFC-STD-3029A, the final rating of new alloys is discussed during ESA-NASA dedicated meetings. The introduction of new heat treated alloys into the table system is agreed based on mutual exchange of test results. This process also guarantees that more than one batch of the same heat treated alloy is evaluated. Multiple batch testing accounts for batch to batch variability in the stress-corrosion behavior that a heat treated alloy can exhibit.

12.2.6 Stress-Corrosion Cracking in Propulsion Systems

The possibility of stress-corrosion cracking occurring in a propellant or oxidizer tank directly affects the safety and reliability of a spacecraft. *Rosetta* is an ESA scientific satellite for deep space exploration (comet observation and sampling). Originally scheduled to launch in January 2003, its launch was put on hold after *Ariane 5* experienced a defective orbit injection in December 2002. During this standby period, the satellite was stored with its Ti-6Al-4V alloy tanks filled with dinitrogen tetroxide and from 0.7 to 1.0% (wt) nitric oxide. For safety reasons, it was decided that the tanks should be emptied if the new launch date was to be postponed beyond March 2004. On March 2, 2004, *Rosetta* was launched successfully on *Ariane 5*. However, the Rosetta launch postponement highlighted particular problems. Notably, if the tanks had to be emptied, inevitable chemical modifications of the residual propellant would have exposed the titanium alloy walls to the risk of stress-corrosion cracking. In addition, insufficient reliable test data were available to allow the realistic assessment of such a risk (Bussu, Stramaccioni, and Kaelsch 2004).

Propellants and oxidizers used in spacecraft propulsion systems are extremely reactive and aggressive substances. Materials used in components of propulsion systems in contact with such substances are selected based on their capacity to retain their initial mechanical and chemical properties. Such materials compatibility issues include resistance to stress-corrosion cracking. Titanium alloys and stainless steels are among the most widely used materials because of their combination of good mechanical properties and chemical stability.

Although they possess good chemical and physical properties, titanium alloys can be susceptible to stress-corrosion cracking in certain chemical environments. Much literature data are available on the subject. However, these data are often of limited applicability because test conditions under which the results were obtained are not specified.

Typically based on fracture mechanics, the test methods for materials used in propellant systems differ from those usually used for testing materials in the sodium chloride-water environments discussed so far. Fatigue precracked test specimens are exposed to the test environment while constantly loaded to reach a given stress intensity factor value, K. The test is carried out for a predefined duration that would allow the initial crack to propagate under the combined effect of stress and environment. Evidence of any crack growth can be obtained either by fractographic examination of the specimen at the end of the test or crack growth monitoring techniques and changes in specimen dimension during exposure. The latter methods have the advantage of immediate detection of the onset of crack propagation. By using different specimens loaded at different K values, or sometimes by loading the same specimen in step increments, a threshold stress intensity factor value for stress-corrosion cracking (K_{ISCC}) is determined. It is essential that exposure duration is equal to or longer than the minimum duration to produce stress-corrosion crack propagation. An appropriate exposure duration should be chosen based on literature data or determined by testing. It should also be mentioned that, in large test programs, test duration has a considerable impact on cost and schedule.

The use of precracked specimens acknowledges the fact that defects introduced during manufacturing and service are always present in real structures. For example, flat tensile specimens with thumbnail surface cracks are used to represent a typical surface defect in a tank wall. Furthermore, some materials, for example titanium alloys, in the presence of surface defects, exhibit a susceptibility to stress-corrosion cracking that is not evident from tests conducted with smooth specimens (Jones and Ricker 1993).

Alloys should be tested in the specific chemical environment representative of planned service conditions to show that no failure or unacceptable damage occurs during operation. To prevent stress-corrosion cracking of titanium alloys, from 0.7 to 1.0% (wt) nitric oxide is added to the oxidizer, dinitrogen tetroxide, to obtain MON-1, which widely is used in spacecraft propulsion systems. In a MON-1 tank, local concentrations of nitric oxide can vary. During tank offloading, there is the possibility of the formation of nitric oxide lean nitrogen tetroxide. This is caused by the greater volatility of nitric oxide than dinitrogen tetroxide and nitric acid (caused by the inevitable contamination with water during nitrogen flushing). For *Rosetta*, the effects of these changes on the susceptibility to stress-corrosion of the tank material had to be assessed by testing. This involved carrying out tests in a complex test setup to simulate tank offloading, draining, and refilling (Bussu et al. 2004). For the *Space Shuttle*, the problem with nitric oxide lean dinitrogen tetroxide is eliminated by using MON-3, which contains nominally 3% (wt) nitric oxide. This is sufficient to prevent depletion below 0.7%.

The interpretation of test results is one of the most critical phases of an investigation. Typically, stress-corrosion test results are characterized by high scatter, and their application in design can be difficult. Test results obtained in a specific environment always should be compared with those obtained in air under the same test conditions. Titanium alloys are susceptible to sustained load cracking in air due to local hydrogen absorption and embrittlement. In fracture mechanics terms, this means that the crack propagation threshold under sustained loading in air (K_{Ith}) is lower than the materials toughness (K_{IC}). It also implies that threshold stress intensity factor values for stress-corrosion cracking (i_{ISCC}) more correctly should be compared to K_{Ith} than to fracture toughness values. This behavior should be considered when carrying out the fractographic examination of test specimens. Fracture surface features indicating stress-corrosion cracking on exposed samples also can be present on specimens tested in air and therefore are not suggestive of a degradation effect of the test environment.

For the design and damage tolerance verification of a tank, the predicted stress intensity factor values, calculated by assuming certain initial crack sizes and shapes, should always be lower than the K_{ISCC}. In the absence of reliable test data for the selected material in the specific service environment, this approach cannot be applied with confidence in the design and verification of tanks. The current approach mainly involves the use of safety factors to account for stress-corrosion in space propulsion components. Although it has been used successfully in space programs, this approach has two fundamental limitations. The safety factors are derived from industrial practice and based on assumptions concerning stress-corrosion cracking that mostly have not been verified by systematic testing. In addition, because no testing is done to determine K_{ISCC}, there is no appreciation of the effective margins of each specific design.

Cost effective manufacturing solutions and new materials are being used to satisfy the high structural efficiency requirements of the latest propulsion components. In this light, further structural optimization can be achieved by a more refined approach to the prevention of stress-corrosion. In the opinion of the author, this approach should be based on fracture mechanics, as discussed previously.

12.3 CONCLUSIONS

Stress-corrosion cracking is an insidious failure phenomenon that occurs in susceptible materials exposed to saltwater or corrosive substances in the presence of a sustained tensile stress. Stress-corrosion cracking affects the safety and reliability of spacecraft because it can happen with no warning, even when a structure is not subjected to service loads. Because time-to-failure predictions for real structures are unreliable, the current approach is based on preventing failures by selecting heat treated alloys possessing high resistance to stress-corrosion cracking in the service environment. Appropriate design and fabrication processes to minimize residual stresses can be used to improve the resistance to stress-corrosion cracking of structural components. The standards MSFC-STD-3029A, ECSS-Q-70-36A, and ECSS-Q-70-37A detail the materials selection criteria for stress-corrosion cracking and the test methods for classifying materials based on their stress-corrosion susceptibility. Due to the aggressive nature of propellants and oxidizers, stress-corrosion cracking is a highly important issue in propulsion components. No standard design and verification methods exist for such components, and the current industrial practice is based on the use of safety factors to account for stress-corrosion.

REFERENCES

American Society for Testing and Materials. 2006. Standard for Slow Strain Rate Testing to Evaluate the Susceptibility of Metallic Materials to Environmentally Assisted Cracking. ASTM G 129–00 (Reapproved 2006). West Conshohocken, PA: American Society for Testing and Materials.

Bussu, G., and B. D. Dunn. (2002) ESA approach to the prevention of stress-corrosion cracking in spacecraft hardware. Joint ESA-NASA Space Flight Safety Conference. Noordwijk, the Netherlands: European Space Agency, European Space Research and Technology Center.

Bussu, G., D. Stramaccioni, and I. Kaelsch. (2004) Experimental assessment of the susceptibility of stress-corrosion cracking of Ti-6Al-4V alloy exposed to MON-1 propellant tank environment—background and test design. International Propulsion Conference. Chia, Sardinia: The European Space Agency.

Dunn, B. D. (1997) Metallurgical Assessment of Spacecraft Parts, Materials and Processes. Chichester, UK: Praxis.

European Space Agency. (1998a) Determination of the Susceptibility of Metals to Stress-Corrosion Cracking. ECSS-Q-70-36A. Noordwijk, the Netherlands: The European Space Agency, European Space Research and Technology Center.

European Space Agency. (1998b) Materials Selection for Controlling Stress-Corrosion Cracking. ECSS-Q-70-37A. Noordwijk, the Netherlands: European Space Agency, European Space Research and Technology Center.

Hill, H. N. (1961) Residual welding stresses in aluminum alloys. Metal Progress: 92–96.

Jackson, J. D., and W. K. Boyd. (1966) Corrosion of Titanium. Memorandum 218. Columbus, OH: Battelle Memorial Institute, Defense Materials Information Center.

Jones, R. H., and R. E. Ricker (eds.). (1993) Stress-Corrosion Cracking: Materials Performance End Evaluation. West Conshohocken, PA: American Society for Testing and Materials International, pp. 1–40.

Korb, L. J., D. C. Augustine, W. L. Castner, and C. D. Brownfield. (1985) A metallurgical analysis of the failure of two APU's of the spacecraft Columbia. *Proceedings of the Space Technology Conference*, Anaheim, CA.

National Aeronautics and Space Administration. (1998) Flammability, Odor, Offgassing, and Compatibility Requirements and Test Procedures for Materials in Environments That Support Combustion. Standard NASA-STD-6001, Huntsville, AL: National Aeronautics and Space Administration, Marshall Space Flight Center.

National Aeronautics and Space Administration. (1999) Spacecraft Maximum Allowable Concentrations for Airborne Contaminants. JSC 20584. Houston, TX: National Aeronautics and Space Administration, Johnson Space Center.

National Aeronautics and Space Administration. (2005) Guidelines for the Selection of Metallic Materials For Stress-Corrosion Cracking Resistance in Sodium Chloride Environments. MSFC-STD-3029A. Huntsville, AL: National Aeronautics and Space Administration, Marshall Space Flight Center.

National Aeronautics and Space Administration. (2007a) Standard Materials and Processes Requirements for Spacecraft. Interim Technical Standard NASA-STD-(I)-6016. Washington, DC: National Aeronautics and Space Administration, Headquarters.

National Aeronautics and Space Administration. (2007b) Materials and Processes Technical Information System-II (MAPTIS-II). Available at http://maptis.nasa.gov (cited August 13, 2007).

National Research Council. (1994) Spacecraft Maximum Allowable Concentrations for Selected Airborne Contaminants, vol. 1. Washington, DC: National Research Council, Board on Environmental Studies and Toxicology, Committee on Toxicology, Subcommittee on Spacecraft Maximum Allowable Concentrations.

National Research Council. (1996a) Spacecraft Maximum Allowable Concentrations for Selected Airborne Contaminants, vol. 2. Washington, DC: National Research Council, Board on Environmental Studies and Toxicology, Committee on Toxicology, Subcommittee on Spacecraft Maximum Allowable Concentrations.

National Research Council. (1996b) Spacecraft Maximum Allowable Concentrations for Selected Airborne Contaminants, vol. 3. Washington, DC: National Research Council, Board on Environmental Studies and Toxicology, Committee on Toxicology, Subcommittee on Spacecraft Maximum Allowable Concentrations.

National Research Council. (2000) Spacecraft Maximum Allowable Concentrations for Selected Airborne Contaminants, vol. 4. Washington, DC: National Research Council, Board on Environmental Studies and Toxicology, Committee on Toxicology, Subcommittee on Spacecraft Maximum Allowable Concentrations.

Winzer, N., A. Atrens, G. Song, E. Ghali, W. Dietzel, K. U. Kainer, N. Hort, and C. Blawert. (2005) A critical review of the stress-corrosion cracking (SCC) of magnesium alloys. *Advanced Engineering Materials* 7, no. 8.

CHAPTER 13

Oxygen Systems Safety

Michael D. Pedley, Ph.D.
Materials and Processes Branch, Johnson Space Center, National Aeronautics and Space Administration, Houston, Texas

CONTENTS

13.1. Oxygen Pressure System Design 375
By Sarah R. Smith and Joel M. Stoltzfus
13.2. Oxygen Generators 392
    13.2.1. Electrochemical Systems for Oxygen Production 392
By Stephen S. Woods
    13.2.2. Solid Fuel Oxygen Generators (Oxygen Candles) 398
By Jon P. Haas

In this chapter, three types of spacecraft oxygen systems are addressed:

- Conventional gaseous and liquid oxygen storage systems and the hazards associated with high pressure oxygen.
- Electrochemical systems for oxygen generation by water electrolysis.
- Solid fuel oxygen generators from which oxygen is generated by thermal decomposition of highly oxygenated chemicals.

The last two systems commonly are used for generating oxygen in situ in crewed spacecraft. All three systems are potentially hazardous because of the high reactivity of oxygen and the high flammability of common materials in an enriched oxygen environment.

13.1 OXYGEN PRESSURE SYSTEM DESIGN

13.1.1 Introduction

Oxygen is a powerful oxidizer in both gaseous and liquid states, and its use involves a degree of risk that should never be overlooked. Oxygen is reactive at ambient conditions, and its reactivity increases with increasing pressure, temperature, and concentration. The successful design, development, and operation of oxygen systems require special knowledge and understanding of ignition mechanisms, material properties, design practices, and test data.

Many materials that do not burn in air, burn in oxygen enriched atmospheres and do so with lower ignition energies and faster burn rates than in air. Most nonmetals are flammable in 100% oxygen at ambient pressure, and most metals are flammable in oxygen at increased pressure. Catastrophic fires have occurred in both low pressure and high pressure oxygen systems, in gaseous oxygen and liquid oxygen systems, and even in oxygen enriched systems operating with less than 100% oxygen. Appreciably destructive fire events have occurred in every industry or venue in which oxygen enriched atmospheres are used.

For example, patients who smoke while using medical oxygen experience fires that are ignited with surprising ease and quick spreading. During surgery, particularly surgery of the head and neck, in which an oxygen enriched environment exists in the surgical area, fires are sometimes initiated by electrosurgical scalpels or lasers. The oxygen production and distribution industry experienced catastrophic fire events in which carbon steel and even stainless steel components ignited and burned, leading to costly destruction of equipment and casualties to human operators and bystanders. The increased use of oxygen enriched pressurized air for underwater diving has led to an increase in the number of fires in dive shops when bottles are filled. Fires occurred in cutting and welding, industrial processing, laboratory, and deep sea operations; all involved pressurized, oxygen enriched atmospheres. Dramatic fires occurred in military, private, and commercial aircraft systems, often leading to the costly and total destruction of the aircraft.

Of particular interest are the fire events that occurred in the space industry. A Titan missile in a silo was being defueled when leaking liquid oxygen infiltrated an adjacent equipment silo. An unknown ignition source, perhaps a spark from operating machinery, ignited materials in the oxygen enriched atmosphere causing a fire that spread rapidly and ultimately initiated an explosion of the missile fuel. Fortunately, all the workers were evacuated safely prior to the explosion, but the loss estimate exceeded $7 million.

In January 1967, astronauts Roger Chaffee, Gus Grissom, and Ed White were burned fatally in a fire that occurred in the pure oxygen environment of the Apollo *Command Module*. In spite of an extensive investigation, the origin of the fire was not determined conclusively, due to the destruction from the fire. The most probable source of ignition was an electrical spark in faulty wiring. The fire propagated rapidly through spacecraft materials that were not highly flammable in air but were very flammable in the pure oxygen atmosphere in the spacecraft.

The Apollo 13 lunar mission was interrupted by a dramatic fire that caused the rupture of two oxygen storage vessels, crippling the spacecraft and threatening the lives of the astronauts. The heroic efforts of the crew, flight controllers, and support engineers enabled the safe return of the crewmembers. The fire was started as a result of a damaged electrical contact in the stirring motor of a liquid oxygen storage vessel. The damage to the contact had occurred during a ground based checkout procedure and was undetected prior to its use during the flight. Fortunately, the Apollo 13 crew survived this oxygen enriched fire; however, the entire mission was lost.

A Space Shuttle extravehicular mobility unit was destroyed by a flash fire during a functional test in a Johnson Space Center laboratory (Figure 13.1). A technician, who was standing next to the suit, received second degree burns on his upper body in the mishap. It was

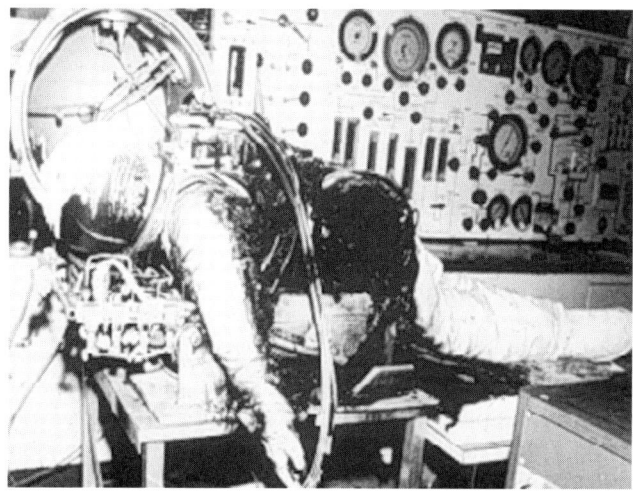

FIGURE 13.1 Fire damage in a NASA space suit and backpack (Courtesy of NASA).

determined that the fire originated in the aluminum bodied regulator valve assembly when the valve was opened and high pressure oxygen flowed into the regulator. The cost of the fire included the destruction of the $3.1-million suit and backpack. Over $20-million was spent investigating the fire and developing the suit and backpack currently used in the Space Shuttle Program.

It is clear that fire hazards in oxygen enriched environments far exceed the fire hazards in air. As well, fires occur in systems that are thought to be designed well and safe. These fires are costly because lives are lost, equipment is damaged, and missions are lost or interrupted. Because of the likelihood and consequences of fires in oxygen enriched atmospheres, special care must be taken in the design of such systems to avoid these fire hazards. This section presents an approach to oxygen system design and material selection that controls these hazards.

13.1.2 Design Approach

The design approach to managing the risks associated with oxygen systems should focus on limiting the amount of oxygen available, using ignition and burn resistant materials where practical, limiting the amount of heat generated within oxygen systems, and limiting the exposure of personnel and equipment.

Oxygen pressure and concentration can have sizeable effects on material flammability and ignitability. In general, materials are easier to ignite and burn more readily as oxygen pressure or concentration increases; therefore, oxygen systems should be operated at the lowest possible pressure and oxygen concentration. Likewise, poor material choice greatly can increase the likelihood of a fire occurring in an oxygen system. Some materials are harder to ignite than are others, and when ignited, they are resistant to sustained burning. Materials also vary in the amount of energy released when they burn. Therefore, careful selection of materials can enhance the ignition and burn resistance of a system and limit the amount of damage resulting from a fire. Even though heat sources can be inherent to an oxygen system or its surroundings, design elements can limit the amount or dissipate altogether the heat generated within an oxygen system.

Materials Selection

The fire hazards inherent in oxygen systems make materials selection a crucial step in designing and maintaining a safe system. To ensure the safety of any oxygen system, the designer must have an understanding of the numerous factors relating to the selection of suitable materials for oxygen service. These would include the material properties related to the design and operating conditions, compatibility with the operating environment, ignition, and combustion behavior, property changes that occur at cryogenic temperatures, and ease of fabrication, assembly, and cleaning. The focus of this subsection is materials selection related to flammability, ignition, and combustion.

Materials selection for oxygen systems used by NASA is regulated by NASA-STD-(I)-6001(A) (NASA 2008). The approach outlined in this document begins with preselection of materials based on flammability and combustion test data, followed by a flammability assessment. If the materials are determined to be nonflammable in their use configuration and environment and no additional material control is required, then they can be used. If the materials are determined to be flammable, an oxygen compatibility assessment must be performed per NASA/TM-2007-213740 (NASA 2007). The oxygen compatibility assessment is a process to determine whether the material can be used, or whether there is a need to perform supplemental material, configuration, or component testing. A further description of the oxygen compatibility assessment process can be found in a subsequent subsection of this section.

Metals Metals generally are the bulk of the materials used in construction of oxygen systems, and most metals require very high concentrations of oxygen to support combustion. Bulk metals are generally less susceptible to ignition than nonmetals. However, metals, including those that normally exhibit high resistance to ignition, are more flammable in oxygen when they have thin cross sections, such as in thin walled tubing, or when they are finely divided, such as in wire mesh or sintered filters. With the exception of commercially pure nickel, all common materials are flammable in oxygen at near ambient pressures when configured as thin cross sections or finely divided configurations. Therefore, special care should be taken to avoid ignition sources in locations where thin cross sections or finely divided metals are used. Once ignited, burning metals can cause more damage than burning nonmetals because of their higher flame temperatures and their usual production of liquid combustion products that are more likely to spread fires.

Nickel and copper based alloys, such as Monel®, brass, and bronze, are among the least ignitable alloys commonly used as structural materials. When configured as ⅛-in. diameter rods, these materials do not support self sustained combustion in upward flammability tests in oxygen at pressures as high as 10,000 psia. In bulk form, these materials are very resistant to ignition and combustion; therefore, they are suitable for use in oxygen systems at all pressures. Monels, brasses, and bronzes are recommended for manually operated systems and systems where the consequences of a fire are high. Aluminum bronzes containing greater than 5% aluminum, although containing a high amount of copper, are not recommended for use in oxygen systems because the presence of aluminum increases their flammability and ignitability (Werley et al. 1993).

Stainless steels have been used extensively in high pressure oxygen systems but are known to be more flammable and more easily ignited than nickel and copper based alloys. Stainless steels are considered flammable in relatively low pressure oxygen, and they release more heat than nickel and copper based alloys when ignited. Few problems have been experienced with the use of stainless steel for storage tanks and lines, but ignitions have occurred when stainless steels were used in dynamic locations of high velocity, high pressure, or high flow rates, such as in valves and regulators. Therefore, care should be taken to minimize ignition hazards when using stainless steel in dynamic locations or thin cross sections.

Aluminum and its alloys, when configured as ⅛-in. diameter rods, support self sustained combustion in upward flammability tests conducted in oxygen at near ambient pressures. In addition, aluminum and its alloys easily can be ignited by some ignition mechanisms, and they release a large amount of heat once ignited. Therefore, caution should be exercised in using alloys containing even small percentages of aluminum. Aluminum alloys are attractive candidate materials for pressure vessels and other applications where no credible ignition hazards exist because of their high strength-to-weight ratios. However, the use of aluminum alloys in lines, valves, and other dynamic components should be avoided whenever possible because of their poor ignition and combustion characteristics.

The use of certain metals including titanium, cadmium, beryllium, magnesium, and mercury must be restricted in oxygen systems. When configured as ⅛-in. diameter rods, titanium and magnesium support self sustained combustion in upward flammability tests in oxygen at absolute pressures as low as 1 psi. The toxicity and vapor pressure of cadmium restrict its use, and systems containing breathing oxygen must not include cadmium if temperatures can exceed 120°F at any time. Beryllium and its oxides are highly toxic and must not be used in oxygen systems or near oxygen systems where they could be consumed in a fire. Mercury is toxic and must not be used in breathing oxygen systems in any form, including amalgamations.

Many other metals and alloys have mechanical properties suited to applications in high pressure oxygen systems. New alloys are being developed continually. Some are being designed to resist ignition and not support self sustained combustion in high pressure oxygen systems. The ignitability of these metals and alloys in high pressure oxygen and their ability to propagate fire after ignition must be compared to the flammability properties of the common structural materials previously described before determining how suitable they are for use in high pressure oxygen systems. Before a new alloy is used in an oxygen system, its flammability and resistance to the ignition mechanisms present in the proposed application must be determined based on applicable test data.

Nonmetals The use of nonmetals, such as polymers and plastics, is often necessary in oxygen systems for purposes such as valve seats and seals. Most nonmetals are flammable in 100% oxygen at ambient pressure. In addition, nonmetals generally are easier to ignite, and they generally ignite at lower temperatures and pressures than metals. Therefore, the use of nonmetals should be limited, and their quantity and exposure to oxygen should be minimized. Some damage that might result from the ignition of nonmetals includes propagation of a fire to metallic components, loss of function arising from system leaks, and toxic combustion products entering breathing oxygen systems.

The nonmetals used in oxygen service are usually polymers (including plastics and elastomers), composites, and lubricants. The use of ceramics and glasses in oxygen systems is increasing. These materials are considered to be inert in use, and they are not discussed further in this book. Nonmetals that are preferred for use in oxygen systems are those that require a high temperature for autoignition, release relatively low amounts of heat when burned, and require high concentrations of oxygen to be flammable. Fluorinated materials typically meet these criteria and generally are preferred for use in oxygen systems. It should be noted that oxygen compatibility is not the only factor in the selection of nonmetals, and the functional performance of materials is critical. For instance, a material with high oxygen compatibility but poor mechanical performance can introduce fire hazards that would not exist if a material with lesser oxygen compatibility but better mechanical performance had been used.

Elastomers typically are used for O-rings and diaphragms because of their flexibility. Fluorinated elastomers, such as polyhexafluoropropylene-co-vinylidene fluoride (Viton® and Fluorel®), generally are preferred in terms of oxygen compatibility. Silicone rubbers have been used in oxygen systems because of their extremely low glass transition temperatures; however, they are not as ignition resistant as fully fluorinated compounds. Therefore, when using silicone rubbers, extra care should be taken to minimize ignition sources, especially in high pressure systems. In addition, extra care must be taken to minimize ignition mechanisms when using Buna-N®, neoprene rubber, polyurethane rubbers, and ethylene-propylene rubbers because of their poor ignition and combustion characteristics. If ignited in oxygen, these hydrocarbon based materials burn energetically and more easily can kindle ignition to surrounding materials. Furthermore, several catastrophic fires have resulted from the use of these hydrocarbon based elastomers instead of fluorine based compounds.

Plastics are used typically for seat and seal applications. The most frequently used plastics in oxygen systems include polyimides, such as Vespel®, and fluorinated materials, such as polytetrafluoroethylene (PTFE Teflon®), fluorinated ethylenepropylene (FEP Teflon), and polychlorotrifluoroethylene (CTFE Neoflon®). Commonly, polytetrafluoroethylene is used in oxygen systems because of its good oxygen compatibility. Unfortunately, this material has poor creep resistance; therefore, it is sometimes necessary to replace it with polymers that are less compatible with oxygen. Nylon® has been used in oxygen systems when its superior mechanical properties are needed; however, caution should be used when using Nylon because its ignition and combustion characteristics are not as favorable as the fully fluorinated materials.

Composites include plastics and elastomers that have nonpolymer reinforcement, such as glass and graphite. Caution should be exercised when incorporating a reinforcement material into a polymer, because the addition of the reinforcement can lower the ignition resistance of the material. For example, glass filled Teflon is more vulnerable to ignition by certain ignition mechanisms than unfilled Teflon.

Lubricants and greases used in oxygen systems are based mainly on polychlorotrifluoroethylene, polytetrafluoroethylene, or fluorinated ethylenepropylene. These include fluorinated or halogenated polychlorotrifluoroethylene fluids thickened with higher molecular weight polychlorotrifluoroethylene materials, such as Fluorolube®, and perfluoroalkyl-ether fluids thickened with polychlorotrifluoroethylene or fluorinated ethylenepropylene telomers (short chain polymers), such as Braycote® and Krytox®. These materials are preferred as the result of their good oxygen compatibility. Some polychlorotrifluoroethylene based products use additives to increase lubricity, but the oxygen compatibility of these products can be compromised as a result of the additives. Lubricants such as Krytox, Braycote, and polychlorotrifluoroethylene fluids thickened with silicon oxide have been found to allow moisture to penetrate the oil film and cause severe corrosion. Therefore, care should be taken when using them in systems where moisture is present.

In breathing gas systems, the toxicity of the combustion products of the nonmetal components can be an issue when selecting materials. In general, fluorinated nonmetals generate the most toxic combustion products if they burn; however, they also have a much greater resistance to ignition and burning than the alternative materials for these applications. Furthermore, the fluorinated materials, if ignited, are less likely to lead to a burnout of the component because of their low heats of combustion. For these reasons, fluorinated components commonly are selected for breathing oxygen systems as providing the best overall safety for personnel, despite their high combustion product toxicity.

Minimize Ignition Mechanisms

Ignition mechanisms in oxygen systems are simply sources of heat that can lead to ignition of the materials of construction or contaminants. Oxygen systems must be designed purposefully to minimize ignition mechanisms, and the designer must consider the relative importance of the various ignition mechanisms when designing new or modified hardware. For instance, certain designs can be more vulnerable to specific ignition mechanisms than others simply by their function, that is, components that produce high velocities, or because of the size and exposure of soft goods.

The following is a list of some potential ignition mechanisms for oxygen systems. This list is not intended to be representative of all possible ignition mechanisms; however, it should be considered as a starting point for identifying sources of heat in oxygen systems:

- Particle impact.
- Rapid pressurization.
- Flow friction.
- Mechanical impact.
- Friction.
- Static discharge.
- Electrical arc.
- External heat.

Most designs can be optimized to minimize ignition if emphasis is placed on minimizing or removing the characteristic elements of a particular ignition mechanism inherent in the design. For any ignition mechanism to be active, certain characteristic elements must be present. These characteristic elements are unique for each ignition mechanism, and they represent the best understanding of the elements typically required for ignition to occur. Specific ignition mechanisms and design guidelines are presented.

Particle Impact The particle impact ignition mechanism is the heat generated when particles strike a material with sufficient velocity to ignite the particles or the material (Figure 13.2). Particle impact is a very effective ignition mechanism for metals. Nonmetals are considered to be less susceptible to ignition by particle impact than metals, but limited data exist. The characteristic elements necessary for ignition by particle impact are

- Particles that can be entrained in flowing oxygen.
- High gas velocities, typically greater than ~30 m/s (100 ft/s) (Williams, Benz, and McIlroy 1988).
- An impact point ranging from 45° to perpendicular to the path of the particle (Benz 1988).

Test data show that, in most cases, the particulate must be flammable to ignite the target material. However, some highly reactive materials, such as aluminum and titanium, can be ignited when impacted by inert particles. In general, copper and nickel based alloys are resistant to ignition by particle impact. Hard polymers have been ignited in particle impact tests, but limited data exist (Forsyth, Gallus, and Stoltzfus 2000).

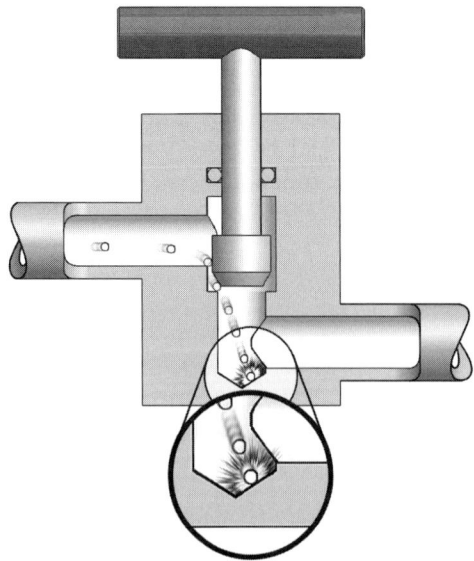

FIGURE 13.2 Particle impact ignition could occur as a result of assembly generated particles accelerated across the orifice of a valve and striking the flammable body just downstream of the orifice (Graphic by WSTF Publications Department; courtesy of NASA).

An ideal design to eliminate particle impact ignition according to the characteristic elements limits fluid velocities, minimizes contamination through design and filtering, and reduces the potential for particle impacts on blunt surfaces. A best-case example of a design minimizing particle impact ignition is filtered, low velocity flow through a straight section of piping. Component designs purposely should minimize particulate generation through the normal operation of valve stems, pistons, and other moving parts. However, some components, such as compressors, pumps, check valves, rotating stem valves, regulators, and quick disconnect fittings, generate particulates simply by their function. The locations and effects of the operationally generated particulates from these components should be considered.

Even small pressure differentials across components can generate gas velocities in excess of those recommended for various metals in oxygen service. In areas where high velocities can be present, such as internal to and immediately upstream and downstream of valves, designs should avoid blunt particle impingement and use materials that are resistant to ignition by particle impact. In addition, filters can be used to limit particulate immediately upstream of high velocity areas.

Rapid Pressurization The rapid pressurization ignition mechanism, also known as *heat of compression* or *adiabatic compression*, is the heat generated when a gas is compressed from a low pressure to a high pressure (Figure 13.3). Rapid pressurization is the most efficient igniter of nonmetals, but generally it is not capable of igniting bulk metals. The characteristic elements for rapid pressurization are

- Rapid pressurization of oxygen (generally less than 1 s for small diameter, higher pressure systems, and generally on the order of a few seconds for those of larger diameter).
- An exposed nonmetal close to a rapidly pressurized dead end.
- Pressure ratio that causes the maximum temperature from compression to exceed the situational autoignition temperature of the nonmetal.

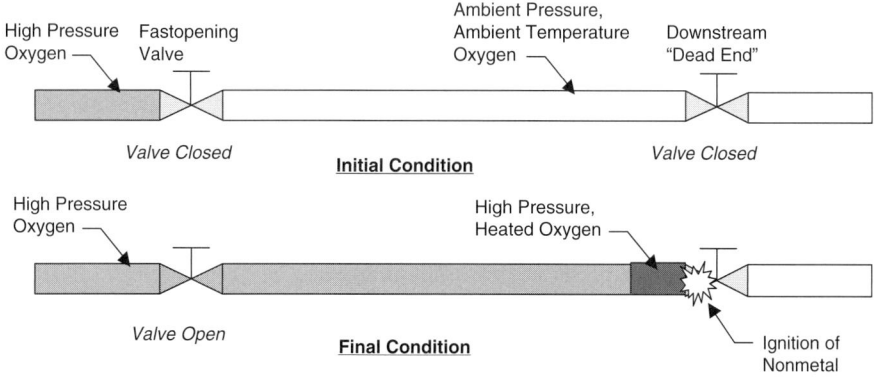

FIGURE 13.3 Rapid pressurization ignition could occur as a result of rapidly compressing oxygen against the nonmetallic seat of a valve (Graphic by the instructors of the NASA and ASTM oxygen training courses; courtesy of NASA).

Although theoretically possible at pressures below 500 psia, rapid pressurization ignition has not been observed below 500 psia because the temperatures reached in real systems are too low to produce ignition. Rapid pressurization ignition is considered by NASA to be noncredible at pressures below 265 psia (NASA 2006).

Ideal designs to eliminate rapid pressurization ignition according to its characteristic elements limit pressurization rates, minimize the amount of soft goods and contaminants, use metallic parts to protect soft goods from fluid flow, and do not compress oxygen against soft goods, such as exposed valve seats, lubricants, and seals. In some applications, flow metering devices such as orifices can be used to limit pressurization rates downstream of high flow components. Metal to metal seals also can be used in some applications to limit the amount of soft goods. Rotating valve stems and sealing configurations that require rotation on assembly should be avoided because they can result in damage to the soft goods, thus making them more susceptible to ignition by heat of compression.

Flow Friction The flow friction ignition mechanism is the heat generated when oxygen flows across or impinges on a polymer and produces erosion, friction, or vibration. Flow friction is a poorly understood ignition mechanism, but current theory indicates that the characteristic elements for flow friction ignition are

- Oxygen at elevated pressures, generally greater than 3.4 MPa (500 psi).
- A nonmetal exposed to the flow.
- Flow or leaking that produces erosion, friction, or vibration of the nonmetal.

Surfaces of highly fibrous nonmetals being chafed, abraded, eroded, or plastically deformed can render flow friction heating effects more severe. An ideal design to eliminate flow friction ignition uses redundant seals to prevent leaking, limits the amount and size of soft goods, and is purposefully designed to prevent damaging the soft goods during assembly, operation, and maintenance. Rotating valve stems and sealing configurations that require rotation on assembly, rotation of seals, and rotation against seals should be avoided because such configurations can damage soft goods. After assembly or maintenance, leak checks should be performed using dry, oil free, filtered, inert gas. Any leaks across polymers should be repaired promptly.

Mechanical Impact The mechanical impact ignition mechanism is the heat generated as a result of single or repeated impacts on a material (Figure 13.4). Most metals cannot be ignited by mechanical impact; however, nonmetals are susceptible to ignition by this means. The characteristic elements for mechanical impact ignition are

- A single large impact or repeated impacts.
- A nonmetal or reactive metal at the point of impact.

Data have shown that reactive metals such as aluminum, magnesium, titanium, and lithium based alloys, as well as some lead containing solders can be ignited by mechanical impact. The presence of liquid oxygen can cause some porous materials to become dramatically more sensitive to mechanical impact. Some components, such as check valves, regulators, and relief valves, can become unstable and chatter during use. Chattering can result in multiple impacts in rapid succession on polymer parts within these

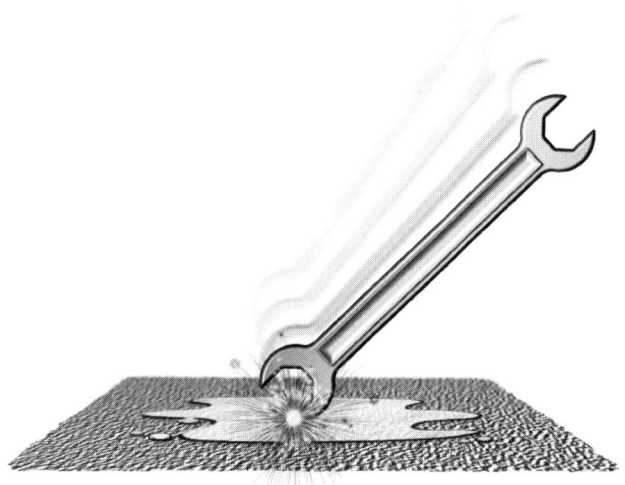

FIGURE 13.4 Mechanical impact ignition can occur as a result of a wrench dropping onto a porous hydrocarbon, such as asphalt, soaked with liquid oxygen (Graphic by WSTF Publications Department; courtesy of NASA).

components, thus creating a mechanical impact ignition hazard. Inert gas flow checks of regulators, relief valves, and check valves in their use configuration and environment should be performed to ensure chatter does not occur in the range of flow rates and pressures to which the component is exposed during operation.

Friction As two or more parts are rubbed together, heat can be generated because of friction and galling at the rubbing interface (Figure 13.5). Data indicate that metals, not polymers, are most susceptible to ignition by friction and galling in the friction heating tests currently available. Current research indicates that polymers and composites also can be susceptible to ignition under certain conditions. The characteristic elements for friction ignition are

- Two or more rubbing surfaces, generally metal to metal.
- Rapid relative motion.
- High normal loading between rubbing surfaces.

Rubbing of metallic parts should be avoided unless the design has been analyzed carefully. Rotational or translational sliding contact between two parts has the potential to generate enough heat to ignite parts at the interface in gaseous oxygen as well as in liquid oxygen. Any easily ignited material near the heated region, such as lubrication or particulates generated by seal wear, also can be ignited. Common configurations where frictional heating

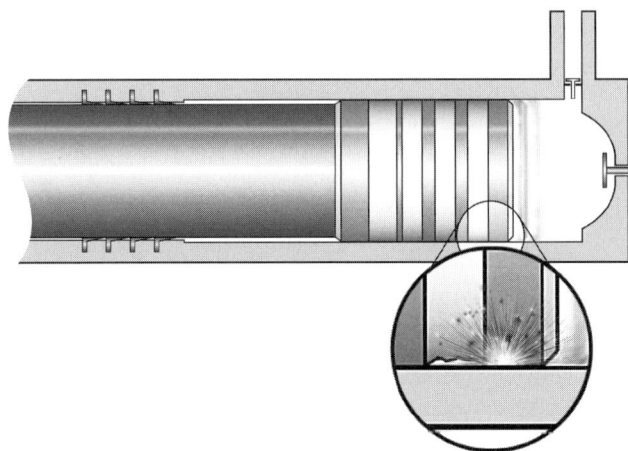

FIGURE 13.5 Frictional ignition can occur due to damaged or worn soft goods resulting in metal to metal rubbing between the piston and cylinder of a reciprocating compressor (Graphic by WSTF Publications Department; courtesy of NASA).

might be observed are components with bearings and pistons. Frictional ignition hazards can be reduced by careful control of surface finishes, coefficients of friction, alignment, and flow induced cooling. Burn resistant materials should be used if frictional heating cannot be eliminated or limited sufficiently. Rotating machinery should be designed with adequate clearances that can be verified.

Others Potential ignition sources to consider initially should include any heat sources, such as electrical arc, static discharge, lightning, explosive charges, personnel smoking, open flames, fragments from bursting vessels, welding, and exhaust from combustion engines.

Importance of Cleaning

Scrupulous cleaning is the most fundamental fire safety measure that can be applied to oxygen systems. The presence of contaminants in otherwise robust oxygen systems can lead to catastrophic fires. To reduce the hazard of ignition, components used in oxygen systems always should be reasonably clean before initial assembly to ensure the removal of contaminants, such as particulates, and hydrocarbon oils and greases that potentially could cause mechanical malfunctions, system failures, fires, or explosions. Visual cleanliness is not always a sufficient criterion when dealing with oxygen systems because of the hazards associated with contaminants that cannot be detected with the naked eye.

13.1.3 Oxygen Compatibility Assessment Process

The oxygen compatibility assessment approach, which is defined in NASA/TM-2007-213740, is recommended as a risk management tool that can be used to evaluate the fire risks associated with materials and components intended for use in oxygen systems

(NASA 2007). The focus of the oxygen compatibility assessment process is fire hazards, and large emphasis is placed on evaluating ignition mechanisms and applying materials test data. This process is a systematic approach that can be used as both a design guide and an approval process for materials, components, and systems. The necessity for conducting an oxygen compatibility assessment is tied directly to minimizing the risk of fire and the potential effects of fire on personnel safety and the system. All oxygen systems should be reviewed by a person, or preferably a group, trained in oxygen system fire hazards, design principles, and materials selection.

The oxygen compatibility assessment process is designed to be applied to individual components. To analyze an entire system, the process can be applied to each component in a system, or techniques can be applied quickly to evaluate the severity of system components and piping so that the most severe components are identified and analyzed by this method (Forsyth et al. 2003). The suggested oxygen compatibility assessment procedure is (in sequence)

- Determine the worst-case operating conditions.
- Assess the flammability of the oxygen wetted materials at the use conditions.
- Evaluate the presence and probability of ignition mechanisms.
- Evaluate the kindling chain, which is the potential for a fire to breach the system.
- Determine the reaction effect, which is the potential loss of life, mission, and system functionality as the result of a fire.
- Document the results of the assessment.

Determine Worst-Case Operating Conditions

An increasing oxygen concentration, temperature, pressure, and flow rate, as well as the presence of contamination, can intensify flammability and ignition risks. Therefore, it is necessary to quantify the worst-case operating conditions before analyzing each component. A system flow schematic, process flow diagram, or piping and instrumentation diagram generally are required for determining the worst-case operating conditions for each component. In general, the analyst should determine the conditions that can exist as a result of single point failures and minimize reliance on procedural controls to regulate the conditions within the oxygen system. In addition to environmental factors, such as oxygen concentration, temperature, and pressure, the analyst should determine the worst-case cleanliness level for each component.

Assess Flammability

As noted in the subsection on materials selection, the flammability of materials greatly is dependent on their configuration. Therefore, when assessing flammability, it is important to reference a cross-sectional view of each component that shows the configuration of the materials of construction. An example of a cross-sectional view is shown in Figure 13.6.

Once there is an understanding of the configuration of the materials of construction, the analyst should reference test methods and data to determine whether the materials are flammable. Such data can be found in the most current version of ASTM Manual 36 (Beeson, Stewart, and Woods 2007), which is the NASA oxygen safety standard for oxygen systems.

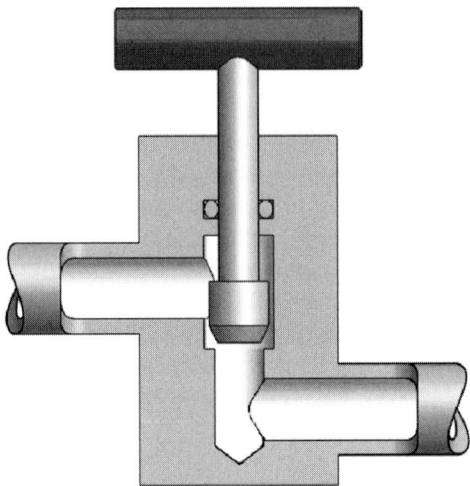

FIGURE 13.6 Example of a cross-sectional view (Courtesy of NASA WSTF Publications Department).

If the flammability of the materials is unknown or the materials of construction have not been selected, then the materials should be considered flammable. Material flammability is affected by many factors, and absolute flammability thresholds are difficult to establish without testing the actual use configuration. Therefore, much of the oxygen compatibility assessment process focuses on the presence and probability of ignition mechanisms.

Evaluate Ignition Mechanisms

The most effective way to analyze the ignition risk in a component is to perform a systematic analysis of known ignition mechanisms. To assess ignition mechanisms, the analyst should focus on evaluating the presence of the characteristic elements and applying materials test data related to the ignition mechanisms, which can be found in ASTM Manual 36 (Beeson, Stewart, and Woods 2007). Knowledge of the system layout, system flow schematic, process flow diagram, or piping and instrumentation diagram generally are required to perform this assessment.

The analyst should assign a subjective probability rating for each ignition mechanism. This rating is based on the assessment of the characteristic elements and the flammability of the materials of construction. These ratings provide a basis for determining which ignition mechanisms are most prevalent in each component. An example of a probability rating logic that can be used is shown in Table 13.1.

When the ignition mechanism assessment indicates that fire hazards are present, the analyst should make recommendations for their mitigation. These recommendations assist the system owner, user, and approval authority in making the system safe to use. The recommendations can encompass topics such as changes in materials, replacement of components, and the implementation of procedural controls.

Table 13.1 Ignition Mechanism Probability Rating Logic

Rating	Code	Criteria	
		Characteristic Elements	Material Flammability
Not possible	0	Not all present	Nonflammable or flammable
Remotely possible	1	All present (some weak)	Nonflammable or flammable
Possible	2	All present	Flammable
Probable	3	All present (some severe)	Flammable
Highly probable	4	All present (all severe)	Flammable

Determine Kindling Chain

A kindling chain is defined as the ability of ignition to propagate within a component or system, leading to burnout. A kindling chain reaction can occur if the heat of combustion and specific configuration of the ignited materials are sufficient to ignite or melt the surrounding materials, leading to a burnout. The analyst should assess the kindling chain based on the presence of ignition mechanisms and the ability of the materials of construction to contain a fire. If ignition of one material can promote ignition to surrounding materials, a kindling chain is present.

Determine Reaction Effect

The reaction effect assessment is performed to determine the effects of a fire on personnel, mission, and system functionality. The analyst should assign a reaction effect rating for each component based on the presence of a kindling chain and the potential consequences of a fire. The potential consequences of a fire are based on the extent of fire propagation in the materials that surround the component, which in turn determines the effect on personnel safety and mission and system functionality. Because it is difficult to conceive of all possible fire scenarios that could result in injury or damage, reaction effect ratings should be applied conservatively; that is, the worst-case scenario should drive the reaction effect assessment. Reaction effect ratings provide a basis for determining those components that have the potential for causing the greatest damage and injury. An example of a reaction effect rating logic that can be used is shown in Table 13.2, extracted from G63–99 (ASTM 1999) and G94–05 (ASTM 2005b).

Document Results

It strongly is recommended that the results of the oxygen compatibility assessment be documented in a written report. This report can facilitate communication and dissemination of results to interested parties and serves as a record of the findings for future reference. The report should include system schematics, drawings for each component, references to data, and notes to document the rationale used in determining the various ratings. In addition, the report should identify the components with the highest probability of fire and recommend changes to design, materials, and procedures that mitigate the

Table 13.2 Reaction Effect Raing Logic, Based on ASTM G63-99 (ASTM 1999) and G94-05 (ASTM 2005b) (Courtesy of NASA)

Rating	Code	Effect on Personnel Safety	Effect on System Objectives	Effect on Functional Capability
Negligible	A	No injury to personnel	No unacceptable effect on production, storage, transportation, distribution, or use as applicable	No unacceptable damage to the system
Marginal	B	Personnel injuring factors can be controlled by automatic devices, warning devices, or special operating procedures	Production, storage, transportation, distribution, or use as applicable is possible by utilizing available redundant operational options	No more than one component or subsystem damaged. This condition is either repairable or replaceable within an acceptable time frame on site
Critical	C	Personnel can be injured operating the system, maintaining the system, or being in the vicinity of the system	Production, storage, transportation, distribution, or use as applicable impaired seriously	Two or more major subsystems are damaged, a condition that requires extensive maintenance
Catastrophic	D	Personnel suffer death or multiple injuries	Production, storage, transportation, distribution, or use as applicable rendered impossible; a major unit is lost	No portion of system can be salvaged; it is a total loss

Note: Because it is difficult to conceive of all possible fire scenarios that could result in injury and damage, reaction effect ratings should be applied conservatively; that is, the worst-case scenario should drive the reaction effect assessment.

fire hazards identified. For large systems, reports should include a concise listing of the most severe hazards and suggested mitigations for those hazards.

Using the Oxygen Compatibility Assessment Process to Select Materials

The oxygen compatibility assessment process can be used in selecting materials for use in oxygen systems (Figure 13.7). The material selection process begins by defining the application for the material, followed by assessing the oxygen compatibility and locating or generating test data relevant to the credible ignition mechanisms. In defining the application for the material, designers should ensure that the materials selected have the proper

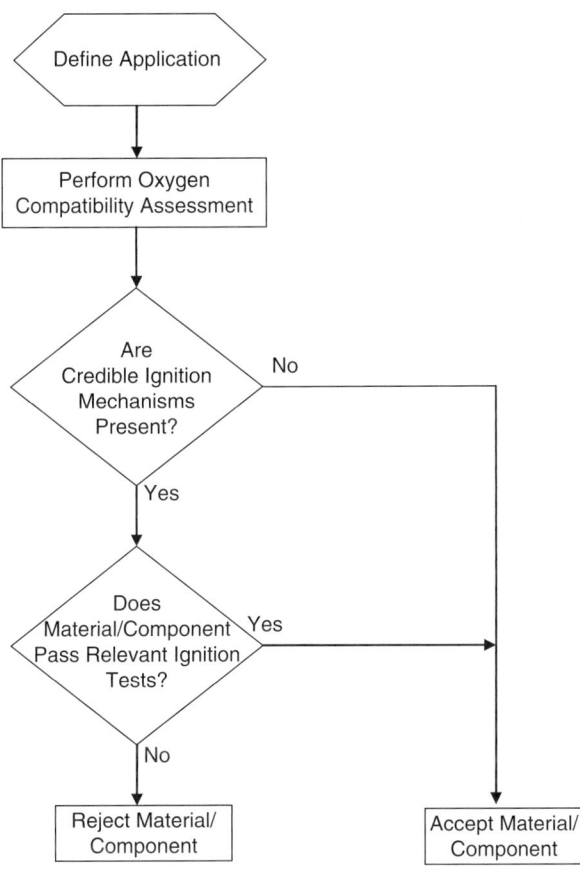

FIGURE 13.7 Material selection process (Courtesy of NASA).

material properties, such as strength, ductility, and hardness, to operate safely under all use conditions. Further, it is important to consider the ability of each material to undergo specific cleaning procedures to remove contaminants, particulates, and combustible materials without damage. In addition to the material requirements for gaseous oxygen service, materials used for liquid oxygen service should have satisfactory physical properties, such as strength and ductility at low operating temperatures.

The oxygen compatibility assessment process allows the designer to identify credible ignition mechanisms, then focus on locating or generating relevant data. Whenever possible, materials should be used below their ignition thresholds for the applicable ignition mechanisms. For up to 20 MPa (3000 psi), a large base of experience and test data can be used when selecting materials. However, limited experience exists at pressures greater than 20 MPa (3000 psi). When selecting materials where little use experience exists, application specific material tests and configuration tests should be performed.

Additional information relating to oxygen pressure system design, material selection, and cleanliness can be found in ASTM Manual 36 (Beeson, Stewart, and Woods 2007), ASTM G63-99 (ASTM 1999), ASTM G88-05 (ASTM 2005a), ASTM G93-03e1 (ASTM 2003), and ASTM G94-05 (ASTM 2005b).

13.2 OXYGEN GENERATORS

13.2.1 Electrochemical Systems for Oxygen Production

Electrolysis is a process by which electricity is used to decompose water into oxygen and hydrogen. By discharging the hydrogen overboard as waste, this can form a basis for oxygen production onboard spacecraft. Decomposition of 1 kg of water can supply 25 L of oxygen per hour at 1 atm for 24 h. This is enough for one crewmember.

Introduction

The primary safety issue for electrolysis systems is the potential for fire and explosion associated with the formation of accidental mixtures of hydrogen and oxygen. Additional concerns are for the reduction in strength of certain metals by hydrogen embrittlement and, for the case of alkaline systems, the release of the caustic electrolyte.

This section describes important features of aerospace electrolysis systems, addresses hydrogen and oxygen combustion hazards, and discusses basic remediation of these hazards. It should be noted that fuel cells, also used in aerospace application, operate by the reverse process to produce electricity from hydrogen and oxygen. The hazards of fuel cell systems, discussed in a later section of this chapter, are very similar to those of electrolyzers. Other issues with oxygen safety apply, and these must be considered.

Electrolytic Processes

At present, aerospace systems use technologies based on two electrolytes: an alkaline liquid electrolyte formed from a solution of potassium hydroxide and water and a solid electrolyte made from an ionomer that forms the core of the proton exchange membrane system.

Alkaline Process A 30% solution of potassium hydroxide in water develops maximum conductivity as an electrolyte. When contained in a vessel and subjected to a current, the following reactions proceed:

- Negative hydroxyl ions, migrate to the positive anode and release electrons, causing the production of oxygen and formation of water: $[4\ OH^- + 4e^- \rightarrow 2H_2O + O_2]$.
- At the negative cathode, water molecules split to form negative hydroxyl ions and release hydrogen: $[2H_2O + 2e^- \rightarrow 2OH^- + H_2]$.

The Russian built Elektron system used aboard *Mir* and the *International Space Station* relies on the alkaline process.

Proton Exchange Membrane Process In this process, the solid electrolyte is an ionomer that acts as a semipermeable membrane. In the proton exchange membrane, it allows conduction of ionic hydrogen as part of a battery-like circuit. Because the ionomer is impermeable to reactant gases, such as molecular hydrogen and oxygen, the gases are kept isolated from each other.

- Water is supplied to the anode, where molecular oxygen is produced and electrons are released: $[2H_2O \rightarrow 4H^+ + 4e^- + O_2]$.
- Hydrogen ions flow to the cathode, where electrons are supplied to form molecular hydrogen: $[4H^+ + 4e^- \rightarrow 2H_2]$.

In both systems, water and electricity are supplied by the spacecraft. Oxygen is released into the crew cabin, while hydrogen is discarded overboard as waste. Future systems will use the waste hydrogen in the Sabatier process to recover oxygen in the form of water from metabolic carbon dioxide and methane. The methane then will be discarded overboard as waste. The oxygen generator assembly developed by Hamilton-Sundstrand for use on the *International Space Station* employs Nafion®, an ionomer with a composition related to Teflon.

General Combustion Issues

Hydrogen is a waste byproduct of the electrolysis of water. When mixed with air, oxygen, or other oxidizers, hydrogen can form very energetic mixtures that are very sensitive to ignition. Therefore, care must be taken for its disposal. In terrestrial settings, the primary approach to hydrogen disposal is to promote its dispersion in concentrations below the level of flammability in air, or where there are much larger quantities, it is combusted under controlled conditions in a flare stack. Confinement of hydrogen in pipes, ducts, or enclosures of flammable mixtures under any circumstances dramatically increases its hazard, both in frequency of occurrence and consequences. In the face of these realities, hydrogen use within spacecraft systems presents the challenge of avoiding any leakage of the gas within the spacecraft until it is ejected outside into the vacuum of space.

Hydrogen Flammability Hazards arising from the failure of oxygen producing equipment onboard a spacecraft entail consideration of several environments to evaluate the formation of flammable mixtures, as shown in Table 13.3.

To assess the hazards associated with an electrolysis system, it is important to understand the energetic mixtures that might form. Flammability and explosion data for

Table 13.3 Hazard Environments

Environment Description	Type of Mixture Formed
Within the system (including oxygen exhaust)	Hydrogen and oxygen or hydrogen and air
Within the cabin	Hydrogen and air
Between the system and exhaust to vacuum	Hydrogen and air or hydrogen and oxygen at subambient pressure

Table 13.4 Selected Data (Benz et al. 1988) for Hydrogen Combustion Hazards at Ambient Temperature and Pressure

Parameter	H_2-Air	H_2-Oxygen
Flammability limits	3.9 to 75.0%	3.9 to 95.8%
Flame temperatures at stoichiometry	2387 K	3080 K
Quenching gap*	0.064 cm	0.023 cm
Criteria for flame acceleration		
Concentration	≥12%	NA, but less
Dimensions	Length/width > 8	NA, but less
Approximate detonability limits (for weak ignition sources)	18.3 to 59.0%	15.0 to 90.0%
Critical dimension1 for detonation propagation for a stoichiometric mixture	0.51 cm	0.04 cm
Minimum ignition energy	0.017 mJ	0.008 mJ
Maximum overpressure		
Confined reaction	8 X	11 X
Deflagration	12 X	NA
Detonation	15 X	20 X
Detonation reflection off flat surface	28 X	36 X

*Dimensions increase for less sensitive mixtures.
NA = Data not available.
X = Represents initial mixture pressure in system. The numerical values, 8, 12, 15, and 28, are based on a complete reaction of a stoichiometric mixture.

hydrogen in ambient air and oxygen (Benz, Bishop, and Pedley 1988) are shown in Table 13.4. Combustion of hydrogen and oxygen produces water in the form of steam, infrared and ultraviolet radiation, and high pressure and shock waves when under confinement. Note that a reaction of hydrogen and oxygen in a sealed system leads to pressures considerably less than the starting pressure after cooling and condensation, because the volume of liquid water is negligible compared with that of hydrogen and oxygen gases. Hydrogen is flammable over a wide range of mixtures in air and an even wider range if mixed in oxygen.

Modes of Combustion The process of combustion varies with the conditions of mixing, concentrations of the reactants, and the effects of confining surfaces. All processes produce high flame temperatures, and confinement can lead to high pressures and flame acceleration.

- **Fire.** A stationary flame with the flammable mixture being fed into the reaction zone is one way to define a *fire*. As long as conditions support combustion and the source of hydrogen persists, combustion continues. In the microgravity environment, more exotic flame structures are possible. For example, flame balls where diffusion of the reactive species across the surface of a sphere with hydrogen inside can persist for some time. Such structures typically require quiescent air and are destroyed by eddies or the presence of surfaces. With total confinement under adiabatic conditions, the pressure for stoichiometric mixtures can reach 8 times the initial pressure for hydrogen-air mixtures and 11 times the initial pressure for hydrogen-oxygen mixtures.

- **Deflagration.** A flame can move through a flammable mixture at a velocity that is subsonic with respect to the unburned mixture. Propagation is weak until mixtures reach concentrations exceeding a volume of 12% hydrogen. Without confinement by surrounding surfaces, a stoichiometric hydrogen-air flame moves at approximately 2.7 m/s. When surfaces (consider pipes or ducts) confine expanding product gases, the flame front is propelled in the open direction. This movement can, in turn, increase mixing within the flame front and accelerate the rate of combustion. Such a feedback loop can accelerate flames to the speed of sound with respect to the unburned media, that is, 400 to 800 m/s, if the geometry of the confinement continues to reinforce the acceleration. This is possible in volumes with a length-to-diameter ratio greater than 8. This behavior is supported by the diffusion of reactive species and fluid dynamic processes. Deflagration waves reflected off walls or by elbows in a pipe or duct can produce higher pressures, where superposition occurs. Deflagration pressures can approach those of detonations.

- **Detonation.** A combustion process in which an exothermic chemical reaction is coupled to a shock wave, and propagation occurs at a velocity supersonic with respect to unburned gases is the definition of *detonation*. Accelerated flames can transition from diffusion based combustion to a shock wave driven process if turbulent flow or turbulence enhancing baffles in the flow path produce sufficiently strong shock waves. This is called *deflagration to detonation transition*. Very large electrical discharges or explosives can initiate detonation directly. Detonation waves travel at velocities greater than 1500 m/s and subject their surroundings to high pressures, that is, 15 times the initial pressure for a stoichiometric hydrogen-air mixture and 20 times the initial pressure for a stoichiometric hydrogen-oxygen mixture and even higher pressures if the shock wave is reflected back on itself (Table 13.4). In addition, transient overpressures during the deflagration to detonation transition can exceed the overpressure considerably in stable detonation waves. Experimental tests have shown that, for a near stoichiometric mixture confined to a 1.5-cm smooth stainless steel tube, ignition at a confined end with a candle flame can result in a transition to detonation within a little more than a half meter of travel (Nettleton 1987). Pressure relief systems cannot protect against supersonic events.

Strategies for Hazard Mitigation

Strategies for hazard mitigation are listed first in a general context; specific mitigation strategies that work for hydrogen follow. General strategies for the mitigation of hydrogen hazards are suggested in the priority order of consideration:

- Minimize opportunities for promoted ignition:
  - Fireproofing protection should be provided for components external to the system that might be exposed to a leak.
  - Selected seals should be capable of withstanding heat and resisting ignition.
  - Filters to maintain water purity should be constructed of nonflammable materials. Care must be taken with filters that can capture and accumulate catalytic fines shed by the cell stack. Interaction with escaped gas, as might occur with a component failure, poses a fire or explosion hazard.
- Minimize ignition sources:
  - Energized subcomponent power levels should be reduced to that only necessary for the job.
  - Design, machining, and cleaning should be planned to minimize any potential for conductive debris. Insulate or otherwise isolate conducting surfaces that, in a microgravity environment, can be spanned by conductive debris that might float into a position and cause a short.
- Ensure that the component design strength exceeds the safety factor for maximum internal hazard pressure. Protect against worst-case overpressures from fire, deflagrations, and detonations.

Some specific strategies found to help mitigate hydrogen combustion hazards are

- Do not let hydrogen and oxidizers mix. Purges using inert gas or evacuation to space vacuum are approaches used to isolate hydrogen from air or oxidizer. Care must be taken with evacuated lines to ensure that leakage of cabin air into a line does not create flammable mixtures. Catalysts have been used to scrub small amounts of hydrogen from oxygen released to the cabin; however, controls must be in place to prevent excess hydrogen from entering and creating excessive heat.
- Hazard assessment should focus on hydrogen wetted volumes as regions having the most probable hazard. A secondary level of concern is for regions exposed following accidental hydrogen release.
- The sensitivity of energetic fluids used or generated by a process must be minimized to the extent possible. Near stoichiometric mixtures of hydrogen and air or oxygen are extremely sensitive to ignition by very small sources of energy: sparks, friction, and the like. Diluents, such as helium, nitrogen, carbon dioxide, and water, can reduce the sensitivity of mixtures to ignition. Keep mixtures out of the flammable range where possible.
- Flame processes are damped when heat is drawn out by the proximity of material surfaces. The quenching gap is a measure of this phenomenon. Narrow gaps in cell stacks and small passages in flame arrestors can prevent flames or detonation cells

from forming. Take advantage of confinements that incorporate physical combustion limits.
- Use dimensions less than those of the quenching gap or less than the critical cell size for detonation to be possible:
  - Use this where it is not possible to reduce flow paths.
  - Flow constrictions that reduce openings below the quenching gap dimension or critical cell size for detonation can be considered to minimize the flow path and interfere with flame acceleration and deflagration to detonation transition.
  - Minimize accumulations of energetic fluids in process loops to only that which is needed.
- Use hydrogen detection and active controls to maintain the system within safe operating limits. Control formation of hydrogen in air so that it never exceeds a concentration a volume of 1%.

Hydrogen and Materials Considerations for Oxygen Production Systems

A variety of materials have been used successfully for hydrogen service. Areas where care must be taken are noted.

Hydrogen is compatible with most elastomers, however, the constituent molecules are very small and have less viscosity than helium. As a result, hydrogen leaks through small passages where other gases cannot and readily permeates many materials (Chernicoff et al. 2005).

Some care must be taken with metals used for hydrogen service. Hydrogen undergoes a complex surface reaction with some metals, entering into the metal crystalline matrix as a hydrogen ion. Grain boundaries, voids, and cracks can provide collection points for hydrogen. If the metal in question is under stress, hydrogen can weaken bonds and cause embrittlement. This is a complex topic, so designers should seek expert metallurgical help to resolve these issues.

Mitigation of these effects is accomplished by proper selection of materials. Note that hydrogen dissolved in water can migrate into containment materials and cause embrittlement. An example of a commonly overlooked application where embrittlement effects should be considered is the metal used in instrument diaphragms. Because of the resulting embrittlement, pressure gauges with improper metal diaphragms can fail rapidly in a hydrogen environment.

The most general advice is to use durable materials that can withstand a hydrogen fire. Stainless steel lines with welded connections are preferred over plastic lines. Avoid combustible seal materials. Do not use soldered connections that can melt from the heat of an ignited hydrogen leak.

Technology Specific Issues

Several technology specific issues are covered for alkaline and proton exchange membrane systems.

- **Alkaline systems.** Potassium hydroxide is a caustic material, and any leak poses a severe hazard in a microgravity environment. Previous incidents have involved excessive heating of the cell stack (Oberg 2005) and ignition of hydrogen-oxygen mixtures (DoE 2007) that overpressurized system seals. In potassium hydroxide systems, impurities can accumulate as deposits in flow passages to the cell stack, thus reducing the electrolyte flow and causing overheating (DoE 2007). In addition, impurities that built up in the gas separators impede gas flow from the electrolyte. This condition led to hydrogen-oxygen mixtures forming in the oxygen separator and an explosion. It is desirable to use detectors to monitor for adverse mixture formation within gas separators. Catalytic getters are employed in oxygen outlet lines to remove residual hydrogen. This arrangement can lead to a potential hazard should excess hydrogen flow across the getter, producing an unexpected amount of heat.

- **Proton exchange membrane systems.** All proton exchange membrane systems present the possibility of a single point failure should the electrolyte membrane develop a pinhole. Oxygen and hydrogen then can mix across the defect, and the presence of catalyst or charged plates promotes a reaction. Pinholes form from wear or penetration by some metal shard overlooked during inspection. Initially, the reaction can be so small as not to be detected by system performance checks. With time, the reaction enlarges the pinhole. The threat is that unreacted hydrogen or oxygen can travel elsewhere and accumulate to levels that pose a future explosion hazard. The system must monitor for these conditions, and provide for a safe shutdown if warranted.

Summary

The primary hazard facing both alkaline and proton exchange membrane electrolysis systems in spacecraft is the potential for formation and ignition of flammable mixtures of hydrogen and oxygen or hydrogen and cabin air. A secondary issue is that hydrogen embrittlement must be considered in the selection of metals for system service. Mitigation of these hazards is accomplished by using established hydrogen practice in the design and operation of electrolysis systems.

13.2.2 Solid Fuel Oxygen Generators (Oxygen Candles)

"*Pozhar!*" or "Fire!" in Russian, is not an exclamation one wants to hear on a spacecraft. On a Sunday in February 1997, Mir flight engineer Aleksandr Lazutkin raised that very alarm as the solid fuel oxygen generator candle he lit began spouting sparks, smoke, and flame (Burrough 1998). Though not proven, it was suggested that contamination, perhaps introduced during manufacture, ignited in the pure oxygen environment and kindled ignition of the metallic components of the unit. Metal fires in microgravity are particularly dangerous.

Solid fuel oxygen generator systems are fairly common. They are known as *oxygen candles*, *self contained oxygen generators*, or *chemical oxygen generators*. The oxygen masks we hope will not deploy from the compartment above us on commercial airliners are another example of solid fuel oxygen generator usage. The tug to the mask starting the flow of oxygen releases a spring loaded firing pin, which, striking a percussion cap, ignites the solid fuel, thus generating oxygen. Solid fuel oxygen generators have been used on airlines for emergency use and submarines during various operations for many decades. In addition to having been used on *Mir*, solid fuel oxygen generators are onboard the *International Space Station* to provide a backup supply of oxygen should the primary system fail or become inadequate.

The environment inside a solid fuel oxygen generator is harsh; the devices contain nearly pure oxygen and hot molten salt. Contaminants can ignite and kindle further ignition of the components of this hardware. Fires in oxygen environments are hard to extinguish. This necessitates sound process controls, judicious materials selection, and clean environments.

How much oxygen does one need? The metabolic requirement for oxygen is approximately 0.8 kg/person/d. A sodium chlorate candle is about 30% oxygen by weight, so the consumable material for one day for one person has a mass of 2.7 kg. This is not substantially different from the 30 to 50% mass efficiency of a pressurized tank system with its associated plumbing. An important solid fuel oxygen generator advantage is found in the volume of the stored oxygen. The solid fuel oxygen generator consumables occupy roughly 10% of the volume required for an equal mass of oxygen stored at 3000 psi. Additional mass and volume are associated with solid fuel oxygen generator safety and handling equipment, however.

In this section, only solid phase chemical oxygen generation (candles) is covered, although oxygen can be liberated by means of gas and liquid phase processes. Oxygen also can be reclaimed from air via superoxide reactions with carbon dioxide and moisture. This section summarizes the general considerations faced by the aerospace systems designer. An excellent example of the challenges encountered while implementing a solid fuel oxygen generator system can be found in Graf et al. (2000).

Systems

Fundamental similarities lie between various solid fuel oxygen generator systems. The oxygen is chemically bound as an oxyanion, such as a chlorate or perchlorate. The decomposition reaction liberates oxygen and is initiated by means of an igniter, often a shock sensitive primer or percussion cap like those used in ammunition to initiate a kindling chain. However, any heat source can be used in principle. Thermite has been studied for one application (Graf et al. 2000). In practice, small amounts of other materials can be added to fuel formulations, either to scavenge fugitive emissions of toxic or corrosive gases, such as chlorine, or to produce additional heat through oxidation to sustain or regulate the decomposition reaction.

Two common forms of the solid fuel reaction are

$$\text{(Sodium chlorate)} \quad 2NaClO_3 \rightarrow 2NaCl + 3O_2 + \text{heat}$$

$$\text{(Lithium perchlorate)} \quad LiClO_4 \rightarrow LiCl + 2O_2 + \text{heat}$$

Sodium chlorate decomposes at around 500°C, whereas lithium perchlorate decomposition proceeds about 100°C cooler. Additives, packaging, and the physical form of the fuel modify the reaction temperatures and rates somewhat.

Storage

Solid fuel oxygen generators are used primarily as backup or supplemental systems for life support, so they need to be reliable and storable. Safety and reliability trade-offs must be considered when evaluating solid fuel oxygen generators for use. Properly stored solid fuel oxygen generators are relatively safe. Some models store igniters separately for added safety. Recall that the fatal crash of ValuJet flight 596 was attributed to improperly stored, illegally shipped, and unlabeled airline solid fuel oxygen generators that were thought to have been activated during flight (NTSB 1997). Compare this to a high pressure gas tank that can be susceptible to bursting from damage, heat, corrosion, or age deterioration. There is one reported case of a submarine solid fuel oxygen generator apparently exploding, resulting in the death of two sailors (Pemberton 2007). The incident is under investigation as of this writing. Most solid fuel oxygen generators are fairly stable, in that they possess high resistance to shock and vibration loads, can be stored in a range of pressures and temperatures, and age predictably. However, they must be kept internally dry, because most solid fuel oxygen generator chemicals are hydroscopic, and excess absorbed water slows the production of oxygen or causes failure to ignite. Historically, solid fuel oxygen generators have reliability records that vary by design, so care must be taken to understand reliability during development and testing repeated throughout operations cycles. A failed solid fuel oxygen generator in space is an expensive piece of hazardous waste.

Heat Generation

In use, solid fuel oxygen generators become hot enough to require some form of cooling to moderate the reaction, protect against hot surfaces, and allow the molten salt byproduct to solidify quickly. Normally, this happens by buoyant convection, which is absent in the spacecraft microgravity environment. Additionally, in some situations, forced convection can be unavailable as well. Surfaces can become hot enough to cause severe burns or become ignition sources for combustible materials, hence, the need to be isolated from accidental contact. Measured surface temperatures can exceed $400°C$ under conditions of buoyant convective cooling. Even safely dissipated, the additional heat load to the spacecraft environment needs to be absorbed by the thermal control system. Once ignited, solid fuel oxygen generators cannot be turned off, and their emissions can contain trace compounds, such as those offgassed from hot materials or oxides of nitrogen, that are considered harmful in an enclosed, recirculated environment. Oxygen candles emit nearly 100% oxygen, which can require mixing to avoid creating a region of oxygen enrichment and increased fire risk.

Application in Space Systems

Both space system unique designs and modified commercial off-the-shelf solid fuel oxygen generator systems have been produced for spacecraft applications. Thermal control, atmospheric mixing, and the action of decomposition are different from the better characterized terrestrial behaviors. As with all oxygen systems and particularly because of their high operating temperature, care must be taken in materials selection for spacecraft solid fuel oxygen generator components. Additionally, contamination control is critical in manufacture and processing. Metals used in terrestrial solid fuel oxygen generator units, normally considered nonflammable, can be unacceptable for microgravity applications. Nearly pure oxygen, operating pressure, temperature (components and effluent), convection, fugitive emissions, materials, and failure modes must all be accounted for in the design risk assessment. Human factors in the space environment and off-nominal operating conditions require procedural controls. Contamination and damage are mitigated with proper checks and process controls during manufacture and handling. Comprehensive testing and evaluation of a solid fuel oxygen generator concept prior to deployment is prudent. Additionally, life cycle analysis and repeat testing of aging systems can reveal problems and reduce accidents. With adequate design analysis and testing, solid fuel oxygen generators can be used as part of a safe, reliable spacecraft life support system.

REFERENCES

American Society for Testing and Materials. (1999) *Standard Guide for Evaluating Nonmetallic Materials for Oxygen Service*. G63-99. West Conshohocken, PA: American Society for Testing and Materials.

American Society for Testing and Materials. (2003) *Standard Practice for Cleaning Methods and Cleanliness Levels for Material and Equipment Used in Oxygen Enriched Environments*. G88-05. West Conshohocken, PA: American Society for Testing and Materials.

American Society for Testing and Materials. (2005a) *Standard Guide for Designing Systems for Oxygen Service*. G88-05. West Conshohocken, PA: American Society for Testing and Materials.

American Society for Testing and Materials. (2005b) *Standard Guide for Evaluating Metals for Oxygen Service*. G94-05. West Conshohocken, PA: American Society for Testing and Materials.

Beeson, H. D., W. F. Stewart, and S. S. Woods (eds.). (2007) *Safe Use of Oxygen and Oxygen Systems: Guidelines for Oxygen System Design, Materials Selection, Operations, Storage, and Transportation*. Manual 36. Philadelphia, PA: American Society for Testing and Materials.

Benz, F. (1988) *Summary of Testing on Metals and Alloys in Oxygen at the NASA White Sands Test Facility (WSTF) during the Last Six Months*. WSTF Metals Work Memo RF/DLPippen:kp:09/14/88:5722. Los Cruces, NM: National Aeronautics and Space Administration, Johnson Space Center, White Sands Test Facility.

Benz, F., C. Bishop, and M. Pedley. (1988) *Ignition and Thermal Hazards of Selected Aerospace Fluids*. Research Document RD-WSTF-0001. Los Cruces, NM: National Aeronautics and Space Administration, Johnson Space Center, White Sands Test Facility.

Burrough, B. (1998) *Dragonfly, An Epic Adventure of Survival in Outer Space*. New York: HarperCollins, pp. 123-132.

Chernicoff, W., L. Engblom, W. Houf, R. Schefer, and C. San Marchi. (2005) *Characterization of Leaks From Compressed Hydrogen Systems and Related Components*. DOT-T-05-01. Washington, DC: Department of Transportation, Research and Special Programs Administration, Office of Innovation Research and Education.

Department of Energy. (2007) Water electrolysis system explosion. Incident Report. Department of Energy 2H Incidents, *Hydrogen Reporting Tool*. Available at www.h2incidents.org (cited May 2007).

Forsyth, E. T., T. D. Gallus, and J. M. Stoltzfus. (2000) Ignition resistance of polymeric materials to particle impact in high pressure oxygen. In T. A. Steinberg, B. E. Newton, and H. D. Beeson (eds.), *Flammability and Sensitivity of Materials in Oxygen Enriched Atmospheres*, vol. 9. STP 1395. West Conshohocken, PA: American Society for Testing and Materials.

Forsyth, E. T., B. E. Newton, J. Rantala, and T. Hirschfield. (2003) Using ASTM standard guide G 88 to identify and rank system level hazards in large scale oxygen systems. In T. A. Steinberg, H. D. Beeson, and B. E. Newton (eds.), *Flammability and Sensitivity of Materials in Oxygen Enriched Atmospheres*, vol. 10. STP 1454. West Conshohocken, PA: American Society for Testing and Materials, pp. 211-229.

Graf, J., C. Dunlap, J. Haas, M. Weislogel, J. Lewis, K. Meyers, and A. McKernan. (2000) Development of a solid chlorate backup oxygen delivery system for the International Space Station. Technical Paper Series 2000-01-2348. Toulouse, France: 30th International Conference on Environmental Systems.

National Aeronautics and Space Administration. (2006) *Standard Materials and Processes Requirements for Spacecraft*. Technical Standard NASA-STD-(I)-6016. Washington, DC: National Aeronautics and Space Administration, Headquarters.

National Aeronautics and Space Administration. (2007) *Guide for Oxygen Compatibility Assessments on Oxygen Components and Systems*. Technical Memorandum NASA-TM-2007-213740. Los Cruces, NM: National Aeronautics and Space Administration, Johnson Space Center, White Sands Test Facility.

National Aeronautics and Space Administration. (2008) *Flammability, Odor, Offgassing, and Compatibility Requirements and Test Procedures for Materials in Environments That Support Combustion*. Technical Standard NASA-STD-(I)-6001(A). Huntsville, AL: National Aeronautics and Space Administration, Marshall Space Flight Center.

National Transportation Safety Board. (1997) *In-Flight Fire and Impact with Terrain ValuJet Airlines Flight 592 DC-9-32, N904VJ Everglades, near Miami, Florida. Aircraft Accident Investigation Report PB97-910406*. NTSB/AAR-97/06; DCA96MA054. Washington, DC: National Transportation Safety Board.

Nettleton, M. A. (1987) *Gaseous Detonations, Their Nature and Control*. New York: Chapman and Hall.

Oberg, J. (2005) Oxygen has its ups and downs on space station. MSNBC (January 7).

Pemberton, M. (2007) Accident kills two British submariners. Associated Press (March 21).

Werley, B. L., H. Barthélémy, R. Gates, J. W. Slusser, K. B. Wilson, and R. Zawierucha. (1993) A critical review of flammability data for aluminum. In D. D. Janoff and J. M. Stoltzfus (eds.), *Flammability and Sensitivity of Materials in Oxygen Enriched Atmospheres*, vol. 6. STP 1197. Philadelphia, PA: American Society for Testing and Materials, pp. 300-345.

Williams, R. E., F. J. Benz, and K. McIlroy. (1988) Ignition of steel by impact of low velocity iron/inert particles in gaseous oxygen. In D. W. Schroll (ed.), *Flammability and Sensitivity of Materials in Oxygen Enriched Atmospheres*, vol. 3. STP 986. Philadelphia: American Society for Testing and Materials, pp. 72-84.

CHAPTER 14

Avionics Safety

David E. Tadlock
Manager, Operational Space Systems Support Office (Retired), Johnson Space Center,
National Aeronautics and Space Administration, Houston, Texas

CONTENTS

14.1.	Introduction to Avionics Safety	403
14.2.	Electrical Grounding and Electrical Bonding	404
	By Robert C. Scully, MSEE, PE, NCE	
14.3.	Safety Critical Computer Control	411
	By David E. Tadlock	
14.4.	Circuit Protection: Fusing	414
14.5.	Electrostatic Discharge Control	417
	By Robert C. Scully, MSEE, PE, NCE	
14.6.	Arc Tracking	428
	By David E. Tadlock	
14.7.	Corona Control in High Voltage Systems	434
	By Robert C. Scully, MSEE, PE, NCE	
14.8.	Extravehicular Activity Considerations	437
	By David E. Tadlock	
14.9.	Spacecraft Electromagnetic Interference and Electromagnetic Compatibility Control	442
	By Antonio Ciccolella, Ph.D.	
14.10.	Design and Testing of Safety Critical Circuits	450
	By Antonio Ciccolella, Ph.D.	
14.11.	Electrical Hazards	461
	By Johannes Wolf, Dr.-Ing.	
14.12.	Avionics Lessons Learned	469
	By David E. Tadlock	

14.1 INTRODUCTION TO AVIONICS SAFETY

Avionics safety is the area of safety engineering that uses electrical and electronic systems for the control or management of hazards. Major types of avionics hardware include command and data handling systems, displays and controls, communication and tracking systems, instrumentation, and health management systems. This chapter concentrates on the particular areas of grounding and bonding, fusing of power distribution systems, static electricity, arc tracking, high voltage, extravehicular activity considerations, electrical hazards, electromagnetic interference, electromagnetic compatibility, the use of computers to control hazards, and some lessons learned from the U.S. space programs and addresses most areas of avionic safety engineering.

The phases of avionics safety engineering include design, construction, analysis, and testing of the electronic systems used to manage hazards inherent to space systems. As stated elsewhere in this text, the preference is to design space systems in such a way that no hazardous operations are associated with them. When this is not possible, effective measures must be provided to manage hazardous functions.

Space systems avionics can be categorized into two major types of circuits for the control of hazards:

- Those that must function.
- Those that must not function inadvertently.

Some circuits absolutely must function to manage associated hazards. A simple example of a must function design is the use of fuses or circuit breakers to protect power distribution circuitry from overheating during an overload. Some circuits must not function inadvertently if a hazard is to be avoided. For example, pyrotechnic deployment devices for landing drag chutes on aircraft must not be deployed inadvertently during takeoff or launch. These same drag chutes must function during landing and hence are a combination of must function and must not function designs.

The use of avionics to manage hazards is applied across each of the various phases of design. During the conceptual design phase, the engineer has an interest in making sure that all hazards associated with a particular design have been identified adequately. As the project matures into a preliminary design, the engineer considers whether the avionics design incorporates effective and adequate measures to provide the needed safety. This is the time when verification and end to end testing of an avionics hazard management design is developed at a conceptual level. As the project progresses into the critical design phase, the emphasis now is to ensure that the design does in fact incorporate all appropriate hazard controls for each identified hazard, that these controls are effective, and that an acceptable verification can be achieved for each.

14.2 ELECTRICAL GROUNDING AND ELECTRICAL BONDING

Electrical grounding and electrical bonding are easily the most misunderstood aspects of space vehicle electrical system design, and improper applications of either or both quite often are found to be the root cause of various system flaws or failures. Poor or improper electrical grounding and bonding can contribute directly to personnel or equipment hazards from the development of unintended high potentials. To name but a few of the possible ramifications, failure to provide proper electrical grounding and bonding can result in electrical shock injury or damage to equipment from dielectric breakdown and associated electrical discharge, accelerated structural metal corrosion, loss of the useful range of radio frequency communications systems, undesirable accumulation of electrical charge leading to damaging static electrical discharges, nuisance tripping in protective circuits, unexplained noise related problems during full scale operations, and loss of major equipment resulting from lightning storm activity.

It is important to realize that electrical grounding and electrical bonding are not terms that can be interchanged directly. An electrical ground is a physical connection, typically exhibiting low electrical impedance, to the Earth or an equivalent large equipotential region, an electrical source, or in some cases both, whose primary purpose is to reduce or eliminate potential differences between various physical points of an equipment installation by allowing for the free movement of electrical charge between the interconnected points.

Electrical bonding, in contrast, refers to the process of joining two or more pieces of material together to create a path exhibiting low electrical impedance of some predetermined quality. The specification for electrical bonding necessarily includes details concerning characteristics such as surface preparation and finishing and the prevention of galvanic or dissimilar metal corrosion. All electrical ground connections must be completed through proper application of at least two electrical bonds, the quality of which is driven by the needs of application. In addition, some electrical bonds are not intended to complete electrical ground connections but instead are used to control voltages and currents from the generation of or exposure to radio frequency energy, provide a means of control over the buildup of electrostatic charge, or provide low impedance to the flow of lightning energy across the outer mold line of a space vehicle.

14.2.1 Defining Characteristics of an Electrical Ground Connection

All proper electrical ground connections share the characteristic that they are not meant to carry return current, either signal or power, during normal operation. However, an electrical ground connection can function as a return path under certain circumstances. Electrical ground connections are intended to carry electrical currents related to or resulting from large transient events, in particular those that can result respectively from unintended or abnormal events, such as a lightning attachment or a wiring short to an equipment enclosure.

Proper electrical system design utilizes protective devices, such as gas discharge tubes, metal oxide varistors, and other transient absorption devices to redirect excessive transient energy that could appear at the input or output of an operational circuit into an electrical ground connection. Equipment short circuits can raise the electric potential on an enclosure to an unsafe level for human exposure. An electrical ground connection acts to mitigate such a potential rise. Similarly, such faults can raise the current from the source to a level that exceeds its capability and is, therefore, unsafe for it to provide. The additional current in the electrical ground path adds to the normal current. The sum of these currents enables any properly designed protective devices to sense the increased current flow and open the circuit, thus protecting the human operators and the affected circuitry.

Electrical ground connections also provide a means whereby subsystem elements can be maintained at nearly the same reference potential, so that noise and operational margins of circuits can be maintained over long distances or between signals promulgated from two electrical sources that might or might not share a common reference at their origins. An example that combines the effects of both long distances and different electrical sources is an analog DC input signal from an engine pressure transducer routed into an analog to digital converter circuit of an avionics device located within a spacecraft cockpit. For the converter to provide the correct output signal, the input and output signals necessarily must share a common reference. This is provided by a common ground connection.

14.2.2 Control of Electric Current

Anyone who has ever seen a lightning bolt or, after walking across a carpet on a dry day, has touched a metal doorknob is familiar with the movement of electrical charge from regions of higher potential to regions of lower potential. The transfer of charge from a point of higher potential to a point of lower potential results in a closer equilibrium of charge and thus a smaller potential difference between the two points. If a low impedance path exists between two points such that an electrical charge is free to move back and forth, no great difference of potential can exist. This is the basis for an electrical ground connection that establishes a common reference between two points in a system. It is the reason why, under normal circumstances, no current is expected to flow in the interconnecting path.

14.2.3 Electrical Grounds Can Be Signal Return Paths

Note that it is possible for a given connection or path to function as an electrical ground connection and as a power or signal return path. An example of this is a space vehicle having a metallic fuselage within which both critical and non-critical signals are passed back and forth among various equipment installations. Critical signals should always be routed utilizing one wire for the signal and another wire for the signal return. Most often, the two wires are twisted together to provide enhancement of magnetic field immunity and shielded with a woven metallic overbraid of at least 98% coverage to provide electric field immunity. For critical signals, the body of the space vehicle then serves as a path for fault currents to flow in the event of a lightning attachment or a wiring short to an equipment enclosure. In either case, no current related to the signal flows in the fuselage during normal operation. In contrast, non-critical signals can utilize the fuselage for signal return to avoid the additional mass and volume of return cabling in the design. Therefore, for critical signals, the body of the space vehicle serves as an electrical ground, whereas for non-critical signals, it serves as the signal return.

14.2.4 Where and How Electrical Grounds Should Be Connected

Although it is a given that every circuit design is unique, there are some fundamental rules for where and how an electrical ground connection should be made. To begin, as based on common and widely accepted safety standards, any potential greater than or equal to

30 V rms (42.4 V peak) or 60 VDC is considered hazardous to humans. Military and aerospace applications typically place the limit for all applications at 30 V rms and VDC for the sake of simplicity. Thus, if any electrical or electronic device interfaces with or internally generates voltages in excess of these values, an electrical ground connection should be made between the device enclosure and the energy source of interfacing voltages. For internally generated voltages, the electrical ground connections should be made between the enclosure and local Earth or an equivalent large equipotential region.

In typical older military and aerospace applications, all electrical and electronic equipment having external metallic regions are bonded electrically to the metallic frame of the vehicle as a matter of course for several reasons, one being protection from case faults. This works well in aerospace vehicles having metallic frames, because the negative side of the common electrical power source usually is connected to the metallic frame for safety, reference control, and power return. Note that, in this typical case, no additional electrical ground connection is necessary for equipment that is bonded electrically to the vehicle frame.

This practice, however, is not the case for equipment that is not bonded electrically to the metallic frame of the vehicle, that possesses a non-electrically conductive enclosure, or for newer design aerospace vehicles made from large amounts of non-electrically conductive composite materials. For these applications, the ground connection for normal operation should be made externally to the case of the device and must never be made to the case via a pin of a connector. By doing so, a magnetic loop antenna internal to the device is created, thereby destroying the shielding integrity of the enclosure. This is considered bad engineering practice today. Instead, an external electrical ground connection (bus) should be routed closely with the power return. As well, the designer should avoid using more than one ground bus; otherwise, unnecessary weight and volume would be added to the wiring harness. Perhaps more importantly, multiple ground connections can create parasitic ground loops that can become excited by ambient radio frequency energy, reradiate strong local magnetic field interference, and conductively couple the interfering electrical energy into the power circuit wiring harness.

An external electrical ground connection provides protection from transient events by providing a path to local Earth or the vehicle frame for energy that can couple conductively or radiatively to the interconnecting wiring harness and attempt to enter an electrical or electronic device. Conversely, this is not the case for internally generated transients that conductively or radiatively couple to the interconnecting wiring harness and attempt to exit the device. An example of such a transient event would be the appearance of electric potentials at the pins of connectors resulting from the coupling of magnetic fields from a lightning strike to a shielded wiring harness. Equipment that is designed properly with transient protective devices and mounted in a conductive enclosure and electrically bonded to a metallic aerospace vehicle frame often is protected well from such transients. Such equipment generally does not require a separate external electrical ground connection. As before, though, for the case of non-electrically conductive enclosures or vehicle frames, a separate external electrical ground connection in the interconnecting wiring harness is needed.

In addition to those made to protect crewmembers from exposure to hazardous voltages and transient events, electrical ground connections to an internal signal reference can be made ohmically to enclosure cases to help mitigate radio frequency interference in the megahertz range and above. This type connection should be made internally at connector penetration points to mitigate both susceptibility to external radio frequency fields and any unintended emission of radio frequency interference. Note that, if interference from relatively low radio frequency radiation in the kilohertz range is of concern, no two pieces of equipment that are interconnected should share ohmic connections to the enclosure. Rather, every other piece of equipment so interconnected should employ a capacitive connection to the vehicle frame. This breaks any possible ground loops that otherwise would exhibit effects similar to those of a long pigtail at higher frequencies.

14.2.5 Defining Characteristics of an Electrical Bond

All proper electrical bonds share the characteristic that they always are intended to carry current, sometimes during normal operation and sometimes during abnormal operation. In some cases, they carry a lot of current. The primary purpose of an electrical bond is to facilitate the passage of electrical current with minimum impedance to the flow of that current as based on the needs of the particular application. As indicated already, an electrical bond can be part of an electrical ground path, a radio frequency operational or protective circuit, the design of the outer mold line of a space vehicle that protects from lightning attachment or electrostatic charge accumulation, and part of a power or signal return path.

14.2.6 Types of Electrical Bonds

Electrical bonds typically are separated into six classes, designated respectfully by the alphabetical characters A, C, H, L, R, and S. A-class electrical bonds are used for antenna applications wherein a low impedance path is required for proper operations. This class of electrical bonds is appropriate for use when a counterpoise is necessary for a monopole or when radio frequency currents must be passed between an antenna base and a coaxial shield connection. These bonds typically are required to be on the order of 2.5 mΩ or less to be considered effective; however, they can be any low impedance value deemed appropriate by the engineering design for a particular application.

C-class electrical bonds are used for intentional current return paths and most often are employed for connections between power return leads and a power bus or the metallic body of a space vehicle. This classification of electrical bonds is used as well for intentional current paths that pass through areas likely to contain explosive vapors, such as a fuel tank. The value of C-class bonds is driven by the allowable voltage drop between the source and the point of regulation and, for the case of explosive vapor areas, by ignition energies associated with fault currents. These particular bonds can have values as low as 0.01 mΩ, depending on the application.

H-class electrical bonds are used for protection against possible personnel shock hazards that can result from large transients, such as lightning attachments or system switching events, or from equipment faults, such as wiring shorts to equipment enclosures. These bonds typically are required to be on the order of 100 mΩ or less to be considered effective.

L-class electrical bonds are used for protection against lightning currents and generally are applicable to any possible entry point into the body of the space vehicle. Examples of entry points include navigation lights, antenna mounts, cargo bay access panels, and externally mounted sensors and transducers. The value of these bonds is driven by a number of variables, including the constitutive properties of space vehicle skin material, the allowable voltage drops across discontinuities, and the allowable magnitude of associated electromagnetic fields that can penetrate into the body of the space vehicle and couple to internal wiring harness and structure. L-class bonds typically are required to be on the order of 2.5 mΩ or less to be considered effective but can be any low impedance value deemed appropriate by the engineering design for a particular location or application.

R-class electrical bonds are used to provide low impedance paths for the control of radio frequency currents and generally are applicable to all equipment capable of producing electromagnetic energy and all external or outer mold line permanent skin panels, removable access panels, and structural components. These bonds are required to be on the order of 2.5 mΩ or less to be considered effective.

Finally, S-class electrical bonds are used to provide protection against the discharge of electrical potentials that can accumulate on external surfaces of the body of the space vehicle or internal plumbing that carries fluid or gas in motion. These bonds are required to be on the order of 1.0 Ω or less when dry to be considered effective.

14.2.7 Electrical Bond Considerations for Dissimilar Metals

Electrical bonds always are made between conductive materials but not necessarily the same materials. Differences in material electrochemical composition can lead to galvanic corrosion, the prevention of which can present great difficulty. As part of the electrical bond process, it is necessary to determine the electrochemical potential difference between the materials that are to be joined together. In all cases, the materials being electrically bonded form an electrochemical cell or battery, with one material acting as the cathode and the other acting as the anode. The anode always corrodes, the amount of corrosion being proportional to the electrochemical difference between it and the cathode. If the potential difference is small enough, typically less than 0.15 V for aerospace applications, it might not be necessary to do more than clean the surfaces that are to be joined and fasten them together. If the potential is too large, that is, greater than 0.15 V, it likely is possible to interpose other materials whose electrochemical potentials with the initial materials are within the acceptable range. For example, copper or brass can be interposed between aluminum and carbon graphite. The surface of the anodic material must be thoroughly cleaned of any surface oxidation or other contamination that might be present before any such treatment. In some applications, the interposition technique cannot be used, and alternate materials must be utilized to solve the incompatibility.

Typical of aerospace applications, electrical bonds, particularly those intended to be semipermanent or permanent, must be sealed to mitigate the intrusion of electrolyte into the interstitial space between the two materials.

14.2.8 Electrical Ground and Bond Connections for Shields

A common aerospace application that involves both electrical grounding and electrical bonding is wiring harness shielding. If a wiring harness having a gross overshield is used to interconnect equipment possessing metallic enclosures, a convenient and very effective external connection approximating a Faraday shield for the harness can be made by terminating the gross overshield using a 360° backshell termination at each equipment connector. The termination is completed using an R-class electrical bond. This type of connection provides maximum shielding protection from external electromagnetic fields. The shielding effectiveness of such a protected harness is not dependent on whether the enclosures have electrical ground connections but is dependent completely on the integrity of the 360° bond. If a gross overshield is not used, external connections can be made for individually shielded singlets, pairs, triplets, and so forth, using very short connections, that is, pigtails, to the enclosure. As well, these connections should be completed using an R-class bond.

Pigtails are simply short wires extending from the common connection of the individual shields at a single point to the enclosure, generally and preferably located at the connector mounting location or as near to it as possible. Pigtails effectively are small magnetic loops, so if the wiring harness is exposed to very high radio frequency signals, they can act as efficient magnetic loop pickups, coupling the radio frequency energy directly onto the wiring harness and obviating any shielding the external shield connection was intended to provide. For this reason, pigtails should be kept as short as practically possible.

One of the most misunderstood issues is whether a particular shield on a wiring harness should be somehow electrically connected to ground. In most cases, the answer to this question is to determine the frequency range of most concern. Much like the internal connections for signal reference, if radio frequency interference in the megahertz range and above is the concern, the shield should be connected electrically to ground at both ends. If the enclosures of interconnected equipment are bonded electrically to a metallic frame or external electrical ground connections have been made, then it is necessary to establish an R-class electrical bond between the shields and the enclosures only at either end. If interference from relatively low radio frequency radiation in the kilohertz range is of concern, one end of the shield should be left floating, thus providing a capacitive couple for any interfering signals and avoiding possible magnetic loops.

RECOMMENDED READING

Many very good references on the topics of electrical grounding and bonding exist in the literature. These include military standards, texts specifically on topic, texts of a more general nature dealing with electromagnetic compatibility engineering, and of course many technical papers written for and published by organizations such as the National Fire Protection Association, the Institute of Electrical and Electronics Engineers, and the International Electrotechnical Commission. Rather than attempt to provide a comprehensive listing, the following is a good basic subset of reference material for both the interested reader and the serious student that can serve as foundational material as well as a good source leading to additional advanced reading and study.

Air Force System Command. (1984) *Electromagnetic Compatibility.* Air Force System Command. Washington, DC.

Denny, H. W. (1989) *Grounding for the Control of EMI.* Gainesville, VA: Don White Consultants.

Department of Defense. (1964) *Bonding, Electrical, and Lightning Protection, for Aerospace Systems.* Technical Standard MIL-B-5087B. Washington, DC: Department of Defense.

Kaiser, K. L., 2004. *Electromagnetic Compatibility Handbook.* Boca Raton, FL: CRC Press.

Keiser, B. (1987) *Principles of Electromagnetic Compatibility.* Norwood, MA: Artech House.

Mardiguian, M. (1988) *A Handbook Series on Electromagnetic Interference and Compatibility*, vol. 2, *Grounding and Bonding.* Gainesville, VA: Don White Consultants.

Mardiguian, M. (1999) *EMI Troubleshooting Techniques.* New York: McGraw-Hill.

Mills, J. P. (1993) *Electromagnetic Interference Reduction in Electronic Systems.* Englewood Cliffs, NJ: Prentice-Hall.

Morrison, R. (1998) *Grounding and Shielding Techniques.* New York: Wiley and Sons.

Violette, J. L. N., D. R. J. White, and M. F. Violette. (1987) *Electromagnetic Compatibility Handbook.* New York: Van Nostrand Reinhold.

14.3 SAFETY CRITICAL COMPUTER CONTROL

In the exploration of space, computers increasingly are needed to control potentially hazardous hardware or functions. Computers can reduce the workload of the crew by assisting with complex and time dependent control of hardware. The use of computers to control hazardous payload functions safely often falls within three categories of computer usage: timers, partial control, and fail-safe control.

The distinction between payload functions and critical spacecraft functions is considerable. There are requirements for the various safety critical computer control functions of a spacecraft far beyond what this text is able to address. Only those applications of computer control for payload hardware on a spacecraft are treated in this section.

For those applications in which computers are used as timers, the content of the commanding and the enabling of the commands are under the control of humans. Computers are able to time precisely the execution of commands. In other situations, commands are to be executed at some later time. In either application, the computer provides only a timing function.

For applications in which computers are used to provide partial control of a process, the computer can interact extensively with potentially hazardous functions. When it is performing a safety critical function, however, the computer is able to exert only partial control over hazardous functions or processes. For example, a safety critical function is hazardous only if it occurs at an unexpected or unmonitored time. In such situations, a manual lockout or override is available that prevents the function until the crew intends for it to operate.

For some potentially hazardous processes, functions, and operations, the computer is in total control. For the computer to be able to manage these hazards, the hardware, its systems, and its subsystems must be designed to fail in a way so as not to create a hazard, that is, it must fail safe.

The design of hazard control systems that utilize computers as part of the system is addressed in two parts. The controlled part of the system is the recipient of the standard electronic design activity already described. The controlling part of the system receives the additional review and design considerations described next.

For all three categories of computer application, designers need to ensure that the computer hardware has been certified safe in the applicable worst-case natural and induced environments defined for the intended mission.

14.3.1 Partial Computer Control

There are two types of partial computer control systems, based on their complexity. The more simple system uses one computer to provide a single required control. Here, normal mission success requirements for the project, coupled with space qualified computer hardware are considered adequate to ensure acceptable safety. The other, more complex type utilizes multiple computers. In this case, the computerized part of the system is to be considered more than zero-fault tolerant. Said a different way, the computerized part of the system manages more than one of the required controls for the hazardous function. Such a system must have a design that incorporates robust system hardware, and it, as a system, must be robust as well.

For the more complex type of system, the computers must have independent possessing functions, such as independent power, clocks, and input circuitry, unless the system can be verified to be safe during power or timing anomalies. These verifications would be based on actual testing and not on analyses or simply being deemed similar to other hardware.

Software and firmware for these computers must be unique in functionality and implementation for each computer. Descriptions of how the computers are to perform their distinct functions and how these functions would be implemented with uniqueness should be documented thoroughly. Having multiple programmers develop software is not adequate within itself to ensure uniqueness.

For the non-computer method for hazard control to be effective, there must be adequate response time to control a hazardous system in a safe manner. The total response time includes the time necessary for human response to messages as well as is the response time characteristic of the hardware. A design also should address any required hazard control that would need to occur while the crew is sleeping or during telemetry outages.

The non-computer method also requires independent data on which to base decisions for the safe control of hazardous functions. Data on which these control decisions are made, even though based on data collected, provided, and displayed by a computer, do not satisfy the requirement to have a hazard control that is independent of computers.

14.3.2 Total Computer Control: Fail Safe

For total computer control systems, computers are in complete control of hazardous functions. Usually, the choice to utilize total computer control is made because the task is either very complex or requires a more rapid response to control parameters than can be achieved by non-computerized control. The most important characteristic of this category of computerized system is that it is a fail safe computer based control system.

The use of fail safe control systems is limited to those applications in which a computer based control system can be interrupted after a failure occurs without resulting in an impending hazard to spacecraft or crew. The intent of the fail safe concept is to apply validated computer based control system designs that exhibit multiple functionally unique computers (and/or firmware controllers) that reliably can detect the first failure and make the transition of the system to a safe state when failure is detected. A control system that is deemed fail safe possesses all 10 of the following characteristics:

- **A formal development process.** The hardware and software are developed under a formal development process that ensures all system and safety requirements are met throughout the life cycle of the system.

- **Fault containment.** A fail safe computer based control system is designed to have an architecture that prevents the propagation of faults (which affect function) from one computer to another. A failure occurring within one computer or its interfaces should not prevent other computers from performing their intended safety functions. All expected (normal) computer to computer interactions must be verified safe during developmental testing and analysis. Design features should be in place that provide detection for each unexpected interaction, and the quantity and complexity of computer to computer interactions should be minimized.

- **Failure and error detection.** A function must be implemented that monitors the status of the hardware and software components within the computers, detects any failures or error and provides error message notification. The failure and error detection program actively should monitor the system during hazard control operations. During power-up or restart, the system must initiate self test functions to ensure that the computers are healthy and ready for operation. Integrity checks must be performed when data or commands are retrieved from memory and when data or commands are exchanged among entities such as computers, transmission and reception lines, and other devices.

- **Controlled system failure detection.** The system must contain the capability for real time detection of failures in the controlled system. When a detector is used for closed loop control, a different detector must be used to satisfy the requirement for real time detection of failures in the controlled system.

- **Failure response.** When a failure is detected, a system failure must be declared and system activity halted. The remaining computers must assist in the immediate issuance of appropriate safety assuring actions independent of the failed computer or hazard control. After a failure, system operations must be suspended until the failure is resolved or substitute components brought online. A computer cannot be solely responsible for detecting and assuring the safety of its own hardware failures or software corruptions that have potential to affect the safe operation of a payload.

- **Independence.** Each computer, at a minimum, should have independent power and independent clocks unless the system can be verified to be safe during power or timing anomalies. Each computer should be unique in functionality and implementation of the software and firmware. A single computer must be incapable of satisfying all the requirements for the initiation of a hazardous event without concurrence from another computer. Additionally, a single computer should not be able to control more than one of the system hazard controls for a specific hazard without concurrence from another computer.

- **Prerequisite checks.** The system must verify prerequisites prior to command issuance to ensure that each command is valid and in the proper sequence.

- **Procedural fault tolerance.** The operator interface should be designed such that deliberate actions consistent with the hazard level are required to initiate a hazardous event.

- **Reconfiguration for safe return.** Fail safe control system designs, which present no immediate hazard after a failure but must be reconfigured to permit the safe return of a spacecraft, should have design features that permit the assurance of safety for return. The design features that permit safety assurance for return must be independent of the failed control system and provide a level of fault tolerance appropriate to the hazard potential associated with spacecraft return.

- **Hazard detection and Safing.** The need for hazard detection and safing by a computer based control system to control time critical hazards should be minimized and implemented only when an alternate means of reduction or control of hazardous conditions is not available. Hazard detection and safety assurance can be utilized to support control of hazardous functions, provided that adequate system response time is available and demonstrated by test.

14.4 CIRCUIT PROTECTION: FUSING

14.4.1 Circuit Protection Methods

The power distribution circuitry of either a space vehicle or a piece of avionics is a critical component. Without electrical power, avionics quite obviously cannot function. This section addresses various methods, such as physical, electrical, and electronic, that can be employed to protect this critical circuitry. Strengths, weaknesses, and implementation guidance for each method is presented. The term *fusing* is used to describe electrical and electronic methods of circuit protection to distinguish them from physical protection.

Need for Fusing

There are several conditions in which an electrical or electronic means for fusing is required to provide the amount and type of protection necessary to meet safety requirements. The conditions can be summarized as the need for protection of power sources, prevention of failure propagation, prevention of toxic fume generation, and protection of distribution circuitry.

Fusing can protect the power sources of avionics from overloads that otherwise would lower the voltage of the overall power grid to unacceptable levels. These overloads oftentimes are caused by short circuits or component failures. Whatever the source, in-line circuit protection essentially disconnects the overload from the power grid and allows nominally functioning equipment on the power grid to continue operation.

One robust power source can serve a function normally reserved for multiple or redundant power sources if it is fused properly. In this case, the robust power source is considered equivalent to multiple sources because the fusing scheme isolates overloads that might otherwise compromise the grid.

Fusing can interrupt the propagation of damage from an overload, thus preventing it from compromising other avionics or circuitry. Proper fusing in this venue is important because damage propagation is unacceptable for multiple use crewed space vehicles and other vehicle designs where the loss of all or part of one instrument or subsystem cannot be permitted to compromise the function and effectiveness of other critical portions of their avionics.

Critical protection of power distribution circuitry can be provided by proper fusing to ensure that the operating temperature of wiring and other components does not elevate to temperatures that would otherwise generate toxic fumes from heated insulation and other materials. A related failure is the unexpected reconfiguration or rewiring of circuits that can occur when power wiring is heated to the point of insulation failure. Once this happens, aberrant circuit paths can be established that can result in unsafe conditions. As examples, the reconfigured circuits could bypass more than one control or inhibit a safety critical or hazardous function when power is bundled with the control line or even result in the manifestation of a hazard previously unforeseen.

Fusing Alternatives

For some spacecraft or avionics systems, alternative design approaches can be used to achieve acceptable safety. These can provide physical protection that renders the likelihood of a circuit overload a non-credible event. This approach to design considers the robustness of the insulation, cable supports, edge protection, routing paths, and the like The manufacturing process is modified to provide any additional insight necessary to ensure that the implementation of the design realizes the intent of the design. An engineering inspection of the final flight configuration is included in the process. This design approach must consider all potential damage to the hardware that can occur and the physical protection required during shipping, handling, and installation. Such damage is often hidden, and it can compromise the physical protection.

Crewed Vehicle Fusing

Payload electrical power distribution circuitry that is designed for use on a crewed space vehicle must protect the wiring of the vehicle from overload conditions as well as protect the vehicle itself and its crew (Gaston 1991). This is especially true if any payload avionics are to be energized by the vehicle itself.

When a payload derives redundant safety critical power from a single source on a vehicle, fusing must be provided to prevent the condition whereby a fault in one redundant safety critical circuit causes the loss of the power source to other redundant safety critical circuits.

14.4.2 Circuit Protectors

Qualification of Circuit Protectors

All fusing used on spacecraft requires the use of space qualified fusing devices. If the design calls for devices that are not available commercially, the manufacturing organization for the space system must be able to conduct all testing necessary to establish the credibility of the fusing device and any associated hardware. In other words, the fusing device must be qualified to function as intended in the system and the environment within which it is to be used.

Important Hardware Characteristics

The functional characteristics of circuit protection or fusing devices used in spacecraft are critical elements of a safe design. One important characteristic is the sure fire or trip level of the fusing device when operated in the intended environment. Another important characteristic is the no fire level, which is the maximum level of current that the device can sustain reliably in its intended operational environment.

To assure the intended operation of the hardware within which the device is installed, parameters such as these must be derated accordingly. The two derating factors of importance are

- Space derating of 50% is used when a fusing devise is applied in a low gravity situation where normal convective cooling is unavailable, that is, a 10-A fuse would be considered to be a 5-A fuse to provide reliable power.
- Bundle derating of 50% is used when a failure could expose more than 50% of the wires in a bundle to an overload current. In cases where only a few of the bundled wires would be subjected to the overload, bundle derating is not applicable because conductive cooling by the nearby wires within the bundle offsets the loss of convection cooling in this environment.

14.4.3 Design Guidance

Robust Design: Smart Shorts

When analyzing a power distribution design, an excellent design screen is the use of a smart short. The concept is one that subjects a power distribution system to a worst-case overload stress. One assumes that a smart short subjects power circuits to a current equal to the maximum current capability, that is, sure fire, of the circuit protection device. If this overload current does not elevate the wire and its insulation to a level in excess of the qualification temperature for the wire, then the system has a sufficiently robust wire and fusing design and therefore is appropriate for use in managing a hazardous function.

Protection Device Selection Criteria
Properly selected circuit protection devices have operating characteristics such that the recommended operating temperature limit for the wire insulation is not exceeded for any possible loading or fault condition of the circuit under worst-case environmental conditions. An example of design guidance that correlates wire sizes and circuit protectors can be found in interpretation letters available from NASA Johnson Space Center (Lambert 1993).

Flammability Concerns
Compliance with circuit protection criteria is not considered an adequate hazard control when reviewing a payload design for compliance with flammability requirements (NASA 1989). Circuit protective devices limit the energy delivered to a fault or failed component only when the current is sufficient to cause the protective devices to open. The energy limiting action of circuit protection devices likely is not adequate to eliminate electrical ignition sources for certain materials configurations; therefore, proper selection of materials (NASA 1991) is the primary control method for a flammability hazard.

14.5 ELECTROSTATIC DISCHARGE CONTROL

An uncontrolled electrostatic discharge can result in catastrophic loss of facilities, vehicles, and personnel. The electrostatic discharge phenomenon results from an electrification of materials and covers a spectrum of scale ranging from electronic components to full size aerospace vehicles. In general, three mechanisms can lead to an electrification of materials: contact electrification (sometimes referred to as *triboelectrification*), exogenous charging, and ionic charging. Contact electrification is the process in which dissimilar substances are brought into intimate physical contact, then separated. A transfer of charge results, typically but not always, with the material having the greater dielectric constant accumulating a net positive charge. This is the process that most often occurs during manufacturing or material handling and processing. Exogenous charging is the process by which charge separation occurs on a material or body by virtue of being immersed in a strong, externally generated electromagnetic field. An example of this process is the charge separation that occurs on aircraft when they fly near or through clouds. Ionic charging is the process whereby a material or body accumulates an electrical charge through bombardment of charged particles, such as those contained in an engine exhaust plume or encountered during interaction with plasma in low Earth orbit. This also is referred to as *spacecraft charging*. Contact electrification is of the most concern and most likely of the three to be encountered in practice.

14.5.1 Fundamentals

Material Classifications

All materials can be charged through contact electrification. *Insulators* are defined as materials with a surface or volume resistance of 1×10^{11} Ω or greater. Electrons cannot move about freely on the surface of these materials, and so insulators can have both net positive and net negative charges on their surface in different locations at the same time. These isolated areas of charge remain on the surface of the material until they are neutralized through interaction with oppositely polarized charges in near proximity. The charges cannot be removed from the material by attaching any sort of conductive wire or strap.

Conductors are defined as materials with a surface or volume resistance of 1×10^{4} Ω or less. Electrons are able to move freely about on the surface of these materials, so any charge deposited on a conductor spreads completely over the surface, creating an equal charge density. If the conductor is isolated, these charges can remain on the surface of the material for a long time, again depending on surrounding conditions. As for the insulator, these charges can be neutralized through interaction with oppositely polarized charges in near proximity. Unlike the insulator, these charges can be removed by attaching a conductive wire or strap between the charged conductor and another uncharged or oppositely charged conductor.

Dissipative materials are defined as having a surface or volume resistance between 1×10^{4} Ω and 1×10^{11} Ω. Electrons cannot move about on the surface of these materials as freely as they can on conductors; however, they can move. Isolated charged areas having both net positive and net negative charges can exist on their surface in different locations at the same time but only for a relatively short time. Charge can be removed from dissipative materials by using a conductive wire or strap, but it can take a long time for this to occur, depending on the surface or volume resistance of the material. A conductor in contact with a dissipative material can drain its charge through the dissipative material slowly to another uncharged or oppositely charged conductor.

Controlling Factors

Charge transfers between materials are based on several controlling factors:

- Area of surface contact.
- Speed of contact and separation.
- Relative values of the dielectric constants of materials.
- Ambient temperature.
- Material temperature.
- Relative humidity.
- Surface or volume resistance of materials.
- Work function of materials, that is, the minimum energy needed to remove an electron from the material to a point immediately outside the material surface.

The larger the area of surface contact between objects, the greater the amount of charge that can transfer from one to the next. Rubbing two objects together with a brisk motion exposes larger surface areas in a shorter period of time. Although the action of rubbing two objects together seems to be the means of transferring the charge, actually the relevant motion simply is bringing larger surface areas into contact. Similar reasoning leads to the conclusion that, if more charge can be transferred through larger areas, then exposing more surface area in a shorter period of time also increases the amount of charge transferred.

Conversely, charge can move only so fast on a given material surface. If the material is a conductor, charge can move about readily and most if not all of any stored charge can move from a conductor to another object very quickly. Anyone who has ever walked across a carpet on a cool dry day and grabbed a brass doorknob can attest to just how quickly that stored charge can move across a conductor. So, when a conductor is one of two objects involved in a charge transfer, the contact and separation can be very rapid and still transfer a high amount of charge. Charge moves more slowly across dissipative materials, depending on the surface resistance of the material. In the case where two dissipative materials are involved, the contact and separation time ultimately makes a difference in how much charge actually is transferred. Finally, if one or both of the materials is an insulator, charge might or might not transfer. A net charge can result simply from the interaction of nearby, oppositely polarized charges. Contact and separation time are important for insulators but must be comparatively longer because charges are immobile on these types of materials.

The dielectric constant, or relative permittivity, of a material is the ratio of the stored electrical energy in the material in the presence of an electric field to the electrical energy present in the same volume in the absence of the material. Therefore, a material having a higher relative permittivity can store more electrical energy than the equivalent volume of free space or a material of equivalent volume having a lower relative permittivity. The term *dielectric constant* is used most often for electrostatic conditions. The term *relative permittivity* is the more general form. Permittivity itself can be thought of as the measure of the ability of a material to become polarized. *Polarization* in this case refers to the separation of positive and negative charges on the atomic or molecular level within the material. Conceptually, in an electric field, the electrons in a given atom statistically more often are located on one side of the atom than on the other, the location being relative to the direction of the ambient electric field. In a material having a high permittivity, a large number of the atoms in the material respond by becoming polarized. This results in an internal electric field that has an opposite polarity than the external ambient electric field. Dissipative and insulative materials generally have higher values of relative permittivity than conductors, and consequently, they tend to polarize much more easily in electric fields than conductors. Electrons in the atoms found in materials with high relative permittivity or dielectric constants tend to be bound less tightly to the nucleus, increasing the ease of removal of electrons and allowing for the atoms in such materials to become ionized. On contact of two materials, electrons can transfer away from the surface with the higher dielectric constant (where electrons easily more are removed) to that with the lower dielectric constant, thus leaving the former material positively charged.

Temperature is a factor in the sense that, as the temperature rises, more energy is available to the electrons in the atoms of a material. As electron motion increases in response to the greater energy level, it becomes easier to liberate electrons from parent atoms. Therefore, higher temperatures typically yield higher charge transfers.

When the relative humidity is greater than 50%, electrostatic discharge is not normally a concern because moisture in the air acts as a high resistance bleeder that allows accumulated charges on the surface of a material to move about and drain off before a discharge occurs. A few materials, such as Teflon® and vinyl, do not absorb moisture, and therefore charge does not bleed off readily even in environments above 50% relative humidity. These and similar materials should be avoided where electrostatic discharge is a concern. Environments below 50% relative humidity can require special attention for the selection and use of tapes, plastic films, and electrostatic flooring material. Operations below 30% relative humidity should be assessed carefully and avoided whenever possible. At levels below 30% relative humidity, additional precautions need to be employed, such as the use of air ionizers and local humidifiers.

Surface or volume resistance directly affects electron mobility over the surface of or through a material. Less electron mobility means the material can store a very high amount of charge and, consequently, release a substantial amount of charge in a given discharge event. Charge released by an insulator can lead to explosions because of the high energy content. Conductors can store charge as well, but the electron mobility is high and the accumulated charge can drain quickly if a conductive path is provided. Because a lot of charge can move easily from a conductor, if the drain path has a higher resistance than the conductor, heat buildup in the bleed path can occur and the temperature of the material can become elevated to an ignition potential in a flammable or explosive environment.

Often, a so-called triboelectric series, based on the relative values of either dielectric constants or work functions, is used to determine the potential for charge transfer between two materials. According to a relationship once known as Coehn's law, which for a variety of reasons has been shown to fail in many cases, two materials brought into contact are charged either positively or negatively, depending on the material possessing the higher dielectric constant. A work function is the property of the ability of a material to hold onto the electrons orbiting the outer most shell of its atoms. The greater the work function of the material, the less likely it is to give up electrons during contact. The weaker the work function, the more likely is the material to acquire a more positive charge by giving up or losing some of its electrons. In general, materials with higher work functions tend to appropriate electrons from materials with lower work functions. Many versions of such tables are available, each based to some greater or lesser degree on variations of the preceding factors; therefore, caution should be exercised when employing any triboelectric series, particularly those found to be based on Coehn's law. The best approach is to use this information as a guide, unless one is completely confident of the variation of factors used in the creation of the table. An example of a triboelectric series is shown in Table 14.1 for illustration.

14.5.2 Various Levels of Electrostatic Discharge Concern

Manufacturing

At the manufacturing level, electrostatic discharge immunity is of paramount concern in the handling and processing of electrical and electronic components and circuitry. Uncontrolled electrostatic discharge at this level can range from the loss of a single integrated circuit to

14.5 Electrostatic Discharge Control

Table 14.1 Typical Triboelectric Series

Tends to Charge Positively (Lower Work Function)
Nylon
Wool
Lead
Silk
Aluminum
Cotton
Steel
Natural rubber
Copper
Silver
Polyethylene
Teflon
Tends to charge negatively (higher work function)

damage resulting in a cascade that renders an entire manufacturing lot of printed circuit boards nonfunctional. This dramatically affects delivery schedules and could cost thousands in terms of material dollars and labor costs.

Damage due to electrostatic discharge can be more subtle, in that it can create partial opens or shorts inside devices that, even months later, can result in failure under moderate to heavy operational stress. Because of the threat of this latent damage alone in the manufacturing environment, components should be required to meet some predetermined level of immunity, as demonstrated by test, before they are selected for use by a given project or program. At this level, electrostatic discharge testing involves injection of test waveforms to component pins and the application of calibrated electrostatic discharges to component surfaces. Devices and components can have a wide range of damage sensitivity levels that range from as low as 50 V to greater than 8000 V. These levels are not DC values but rather refer to peak values normally associated with test waveforms having very fast rise and slightly slower fall times. Typical electrostatic discharge test waveforms are specified in terms of current provided by a source generator through a specified resistance-capacitance (RC) network across a short circuit. A typical electrostatic discharge current test waveform is shown in Figure 14.1.

Because of the potential severity of damage, the aerospace manufacturing processes of both the government and vendors are required to have in place an electrostatic discharge control plan. This plan should detail every aspect of product handling, including the proper attire and the use of wrist straps and heel grounders for assembly and test personnel; the use of

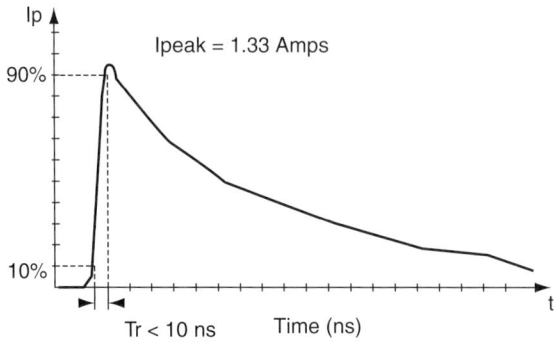

FIGURE 14.1 Typical electrostatic discharge current test waveform from MIL-STD-883F (DoD 2004) (Courtesy of U.S. Department of Defense).

electrostatic discharge safe tooling, lighting, and workbenches; and the proper use of electrostatic discharge safe containers for devices, components, and finished products up to and including completely assembled electrical and electronic equipment. The ANSI standard, ANSI/ESD S20.20 (ANSI 1999), has been adopted by NASA as the de facto standard for electrostatic discharge control at the manufacturing level for all new development. Previously, MIL-STD-1686 (DoD 1995) and MIL-HDBK-263 (DoD 1994) were used for this purpose and are still referenced by many vendors in their in-house documentation.

Equipment and Subsystems

As indicated earlier, an electrostatic discharge can damage assembled printed circuit boards, whether or not installed into equipment enclosures. As assemblies are built up, electrostatic discharge requirements change in their application and severity. At the level of a printed circuit board, the handling requirements and stress levels are very similar to those for individual devices or components. Once the printed circuit boards are assembled and installed into an enclosure, the handling requirements and stress levels begin to change in accordance with the expected entry points for an electrostatic discharge, such as through connector pins and backshells as well as panel mounted switches, dials, and displays. This level can be thought of as a transitional area, where electrostatic discharge moves from an electronic, electrical, and electromechanical parts, which is of concern for systems.

Before equipment is ready for storage and subsequent installation, it must go through the certification processes, which include electrostatic discharge testing. At this level, electrostatic discharge testing involves the injection of test waveforms to connector pins and calibrated electrostatic discharges to enclosure surfaces as well as to any devices mounted on enclosure surfaces that are accessible during operation of the equipment. At this level, equipment can be entering or in the middle of the certification processes. It also can be certified and ready for installation into its service application. Because of the handling processes typical for this level, equipment generally should be enclosed in so-called antistatic bags and any exposed connectors covered with antistatic dust caps. Precautions of this kind provide good protection against casual contact from either personnel or surfaces carrying an electrical charge of sufficient magnitude to damage the equipment. The electrostatic discharge control plan discussed in the previous section should extend the definition and enforcement of protection and handling procedures to this level.

System Level

At the system level, prevention of ignition or explosion involving flammable materials is of the highest concern. Surface discharges between isolated differentially charged areas can lead to high levels of radio frequency interference with communications and tracking systems, loss of structural protective coatings resulting in structure surface degradation, damage to solar arrays causing loss of electrical power, and in crewed space applications such as extravehicular activity personnel shock hazards. At this level, electrostatic discharge is controlled primarily by electrical bonding methods. An S-class electrical bond with a value of 1 Ω is used for static controlled connections. In aerospace applications, this class of electrical bond is used for all isolated conducting items, except antennas, that have any linear dimension greater than 3 in. (7.62 cm) subject to contact or frictional charging, carry fluids in motion, or is external to the vehicle. Additional techniques, such as material selection, material surface treatments, reduction or elimination of relative motion between materials in contact, control of the environment to avoid low relative humidity, and leak or intrusion control, are used to preclude the presence of fluids, fumes, or gases having relatively low ignition potential thresholds. Spacecraft charging and its unique problems and possible corrective or preventive measures are discussed in another section.

Ignition or explosion involving flammable materials results from the sudden release of sufficient energy in an electrostatic discharge event. For a stoichiometric mixture of hydrogen and air, the minimum ignition energy reported in the literature is 19 µJ for a 29% by volume concentration of hydrogen in air at sea level standard pressure, 101 kPa. This value increases as the pressure decreases, reaching a value of 500 µJ at 20 kPa. For hydrogen mixed with oxygen, the value is much lower, the minimum being 1.2 µJ.

To determine whether a given material configuration presents an electrostatic discharge hazard in a flammable or explosive environment, it is first necessary to determine if the configuration can store sufficient energy to equal or exceed the minimum ignition energy of the environment. For a flammable fuel, fume, or gaseous environment, the best way to do this is to perform incendivity testing of the material configuration under controlled conditions. In this type of testing, a charge is placed on the material configuration by any of several methods. Note: The chosen process should be representative of a credible charge mechanism for the configuration under test. In general and except under unusual circumstances, a contact electrification process is the most common mechanism used in an aerospace application and should be employed for this test. Once the material under test has been charged, a grounded wand or probe, which is fitted with a nozzle to provide a continuous flow of the flammable environment mixture across the material surface in near proximity to the wand or probe tip, is moved close to the material under test.

The test operator then attempts to provoke a discharge event between the wand or probe tip and the material under test. If the discharge has sufficient energy to ignite the flammable environment, the material under test fails.

A less satisfactory but more conservative method is to perform an analytical assessment. Analysis is effective generally for isolated conductors for which a capacitance to another conductor, the airframe, for example, can be found or assumed. However, an analytic assessment becomes increasingly difficult, and thus increasingly less accurate, if dissipative or insulating materials are involved. Many factors are involved in an analytical approach, but a good starting point is to establish the lowest breakdown potential or voltage in the expected environment. Breakdown voltage in a gaseous environment is a function of the environment pressure and the shapes and separation distance of the materials between which the discharge is expected to occur. These characteristics and the associated breakdown potentials for various gaseous environments have been collected and are shown in the literature as Paschen curves. As an example, the minimum breakdown potential of air at 1 atm for planar electrodes separated by 5 to 6 μm is between 330 and 340 V. If expanded to show sufficiently low values of the product of pressure and the material or electrode distance, all these curves show a minimum breakdown potential. In some cases, the minimum breakdown potential occurs over a very narrow range of pressure-distance values, the curve quickly rising again to either side of the minimum; whereas in others, the minimum is very broad, stretching over a wide range of pressure-distance values. One common characteristic of this is that, as pressure decreases for a given gap distance or the gap distance decreases for a given pressure, the breakdown potential increases, sometimes dramatically. Once a breakdown voltage has been identified, it can be combined with the known minimum ignition energy of the environment to determine the acceptable amount of charge that can be stored by the material configuration or the acceptable value of capacitance the material configuration can have and remain safe. This determination can be made by using the energy relationship for a capacitance as in Equation (14.1):

$$W = \frac{1}{2}QV = \frac{1}{2}CV^2 \tag{14.1}$$

If capacitance is found, then one can focus on reducing its value as a means of mitigating the threat. This usually involves either reducing the common surface area of the two materials of interest or separating them by a greater distance. The latter is preferred, because this not only reduces the capacitance but also increases the breakdown voltage. If the capacitance cannot be altered, then the amount of charge must be controlled. This involves determining if there is a credible charging mechanism, such as a lot of particulate matter that can be picked up and blown about by the movement of the gaseous environment coming into contact with and subsequently separating from the surface of the material. Clearly, if no credible charging mechanism exists, the threat is mitigated. If a credible charging mechanism does exist, it is necessary to determine if there is also a credible means for discharge that can mitigate the hazard. Charge can be drained away from a conductive or dissipative material through an electrical bond strap, but it has to be neutralized on an insulating material. Electrical charge that is neutralized is said to decay over time, and the amount of time required differs for different materials and environments.

Essentially, charge decay occurs as the bound charge on the insulative material is neutralized slowly by the presence of oppositely polarized charges in a conductive fluid, such as a pool of mercury or water or a stream of ionized air. Clearly using a mercury bath would not be a very useful method for most practical situations encountered, but the use of water in hand wipes or sprays could be effective. Air ionizers widely are available and a very effective means of neutralizing charges on insulative materials. Caution should be employed when using ionizers, because they contain electric motors that can present a hazard when used in a flammable or explosive environment. In addition, by definition, ionizers are a source of electrical noise, so if the environment is sensitive to the increased noise produced by an ionizer, that solution probably is not acceptable.

Most often, at the system level, the best solutions incorporate more than one approach, thereby reducing the probability of a hazardous electrostatic discharge. Conductive or dissipative materials can be connected electrically to a large, common reference such as a metallic vehicle chassis or airframe. Electrical bonding as a means of charge accumulation control for insulative materials is not effective. These materials require either surface treatments to introduce some small amount of surface conductivity, making them at least dissipative, or some means of neutralizing any accumulated charge.

Spacecraft Charging Effects

The design of spacecraft must take into consideration the effects of the severe environment in which operations take place. That environment is comprised to a large extent of any electromagnetic or gravitational fields that can be encountered, ionizing radiation of varying types and intensities, natural space plasma, meteoroids and other orbital or space debris, and thermal environmental effects. All these environmental constituents interact directly with a spacecraft immersed in them and have an array of effects on the vehicle that can in one or more ways affect normal operations.

The interaction of a spacecraft with plasma, in particular, is responsible for the phenomenon of spacecraft charging, that is, the accumulation of charged particles on its exposed surfaces. Plasma consists of electrically charged particles, electrically neutral particles, and electric and magnetic fields stretching between the particles, which loosely contain them while allowing them to move about freely. Particle motion within the plasma constitutes an electrical current and, as such, generates additional magnetic fields that further interact with the other particles. Plasma, sometimes referred to as the *fourth state of matter*, in fact is its most common form. Plasma exists throughout the solar system and the interstellar and intergalactic environments.

Spacecraft charging effects present serious concerns and challenges to spacecraft engineering design. Contaminants attracted to charged surfaces can mitigate the charge, but by so doing, they result in increased surface contamination that can, for example, lead to surface erosion and degradation of thermal protective materials. Inbound ions, accelerated by the high fields, also can cause sputtering from surfaces on which they impact. Instrumentation electrically referenced to the structure of a spacecraft can suffer interference or damage from increased electric and magnetic fields related to surface charging. Electrical discharges between differentially charged areas dramatically can impact spacecraft operations, resulting in interference and damage to scientific instrumentation, communications and tracking systems, electrical power systems, and spacecraft structural surface.

Currents collected by electrically biased surfaces easily can alter the potentials of different parts of the spacecraft. Because of their large mass and low mobility, ions collected by negatively biased surfaces result in a relatively small plasma current density. Solar arrays or other surfaces whose charge is biased positively with respect to the plasma collect electrons from the plasma and can result in a parasitic loss to the power system. Because the mass of an electron is much less than an ion, the magnitude of current density is much greater for surfaces with positive bias. At bias potentials in the 200-V range, sheath formation and secondary electron emission from the surface causes the entire surrounding surface, normally an insulator, to behave as if it were a conductor. This effect, called *snapover*, results in large current collection from even a very small exposed area. As the spacecraft moves through the plasma, it creates a compressed shock wave preceding its motion and a large volume depleted of ions behind. Such ram and wake effects further complicate the charging process. The worst situations occur when the spacecraft power system uses a negative ground, so that large surfaces are negative and collect slow moving ions to balance the current from electron collection. In this arrangement, parts of the spacecraft are biased with respect to the surrounding plasma. If solar arrays are employed, potentials can rise to a level very near the maximum voltage generated by the arrays.

No one combination of techniques completely is effective for all possible spacecraft locations in space. What works in geostationary Earth orbit does not work well in low Earth orbit. Many very good references have been published over the years by NASA concerning protective techniques. Several of them are listed in the Recommended Reading section, which follows. The reader is encouraged to seek out these references for much more detailed information about this unique form of electrostatic discharge.

RECOMMENDED READING

Many very good references on the topics of electrostatics and electrostatic discharge exist in the literature. These include military standards, texts specifically on topic, texts of a more general nature dealing with electromagnetic compatibility engineering, and of course many technical papers written for and published by organizations such as the ElectroStatic Discharge Association (ESDA), the Institute of Electrical and Electronics Engineers, and the International Electrotechnical Commission. Rather than attempt to provide a comprehensive listing, the following and the references cited in the text are a good basic subset of references for both the interested reader and the serious student that can serve as foundational material, as well as a good source leading to additional, advanced reading and study.

Air Force Systems Command. (1984) *Electromagnetic Compatibility.* Air Force Systems Command.

American National Standards Institute. (1993) *American National Standard Guide for Electrostatic Discharge Test Methodologies and Criteria for Electronic Equipment.* ANSI C63.16. New York: American National Standards Institute.

Boxleitner, W. (1989) *Electrostatic Discharge and Electronic Equipment.* Piscataway, NJ: IEEE Press.

Cobine, J. D. (1941) *Gaseous Conductors.* New York: McGraw-Hill.

Dangelmayer, G. T. (1999) *ESD Program Management: A Realistic Approach to Continuous Measurable Improvement in Static Control.* Norwell, MA: Kluwer Academic Publishers.

Department of Defense. (1964) *Bonding, Electrical, and Lightning Protection for Aerospace Systems.* Technical STANDARD MIL-B-5087B. Washington, DC: Department of Defense.

Electrostatic Discharge Association. (2001) *ESD Association Technical Report for the Development of an Electrostatic Discharge Control Program for Protection of Electrical and Electronic Parts, Assemblies and Equipment.* Technical Report ESD TR20.20. Rome, NY: Electrostatic Discharge Association.

Ferguson, D. C., and G. B. Hillard. (2003) *Low Earth Orbit Spacecraft Charging Design Guidelines.* Technical Paper 2003-21228. Cleveland, OH: National Aeronautics and Space Administration, Glenn Research Center.

Harper, W. R. (1967) *Contact and Frictional Electrification.* London: Oxford University Press.

Herr, J. L., and M. B. McCollum. (1994) *Spacecraft Interactions: Protecting against the Effects of Spacecraft Charging.* Research Paper 1354. Huntsville, AL: National Aeronautics and Space Administration, Marshall Space Flight Center.

International Electrotechnical Commission. (2001) *Electromagnetic Compatibility (EMC),* Part 4-2, *Testing and Measurement Techniques—Electrostatic Discharge Immunity Test.* IEC 61000-4-2. Geneva, Switzerland: International Electrotechnical Commission.

Institute of Electrical and Electronic Engineers. (1993) *IEEE Guide on Electrostatic Discharge (ESD): Characterization of the ESD Environment.* IEEE STD C62.47. Piscataway, NJ: Institute of Electrical and Electronic Engineers.

Institute of Electrical and Electronic Engineers. (1994) *IEEE Guide on Electrostatic Discharge (ESD): ESD Withstand Capability Evaluation Methods (for Electronic Equipment Subassemblies).* IEEE STD C62.38. Piscataway, NJ: Institute of Electrical and Electronic Engineers.

Kaiser, K. L. (2004) *Electromagnetic Compatibility Handbook.* Boca Raton, FL: CRC Press.

Leach, R. D., and R. B. Alexander. (1995) *Failures and Anomalies Attributed to Spacecraft Charging.* Research Paper 1375. Huntsville, AL: National Aeronautics and Space Administration, Marshall Space Flight Center.

Loeb, L. B. (1958) *Static Electrification.* Berlin: Springer-Verlag.

Moore, A. D. (1968) *Electrostatics.* New York: Anchor Books, Doubleday.

National Aeronautics and Space Administration. (1999) *Avoiding Problems Caused by Spacecraft on Orbit Internal Charging Effects.* Handbook HDBK-4002. Washington, DC: National Aeronautics and Space Administration, Headquarters.

Penning, F. M. (1957) *Electrical Discharges in Gases.* New York: Macmillan.

Purvis, C. K., H. B. Garrett, and N. J. Stevens. (1984) *Design Guidelines for Assessing and Controlling Spacecraft Charging Effects.* Technical Paper 2361. Cleveland, OH: National Aeronautics and Space Administration, Lewis Research Center.

Raizer, Y. P. (1997) *Gas Discharge Physics.* Berlin: Springer-Verlag.

Violette, J. L. N., D. R. J. White, and M. F. Violette. (1987) *Electromagnetic Compatibility Handbook.* New York: Van Nostrand Reinhold.

von Engel, A. (1965) *Ionized Gases.* London: Oxford University Press.

14.6 ARC TRACKING

14.6.1 A New Failure Mode

Within the Space Shuttle program, a new failure mode of the electrical power distribution system was observed. This failure mode was peculiar to the Kapton® polyimide insulation of this particular wiring. The insulation, which had shown excellent toughness and resistance to heat, had a previously unknown characteristic that could support low current arcing. Once started, an arc could progress a considerable distance in the direction of the power source. The failure mode is called *arc tracking* because of this characteristic. This section describes the characteristic and the experience related to the phenomenon during Space Shuttle missions and methods for avoiding it and any related damage to power wiring.

The Initial Event

The arc tracking phenomenon essentially was unknown during early U.S. spacecraft experience. In the late 1980s, however, a technician accidentally stepped on a 28-V battery charger power cable, and the wiring started to arc. The ground support equipment cabling did not extinguish immediately but burned several inches of the 20 AWG wiring before the power could be removed from the cable.

Wire insulation testing was conducted according to the standard at the time. The test procedures assessed for flammability by subjecting a single wire to a flame to see if the insulation would continue to burn after being ignited and determine if the ignited insulation would self extinguish. The insulation of the battery charging power cable had been tested and deemed suitable for power wiring because no evidence was found that it could be ignited and subsequently support flame propagation. The insulation was a polyimide, Kapton.

Laboratory Testing

Careful re-creation of that particular failure showed that, if two polyimide insulated wires were shorted together, a surprising phenomenon occurred. The arc from the initial short was observed to propagate toward the source of power. An initial current pulse of very short duration and nearly 60 A would occur. The current would then quickly drop to an average level of 5 to 10 A. Under some test conditions, a traveling arc was produced that consumed several feet of wire before it intersected a physical barrier or the power was removed from the wire. The low average current of the series of arcs was such that the circuit protecting fuse would not always melt open and stop the arcing.

This event, which was so surprising at that time, has now been reproduced in many test laboratories. When the aromatic polyimide insulation material, Kapton, which is used extensively in the *Space Shuttle*, is exposed to 28-VDC power, it becomes sensitive particularly to the arc tracking phenomenon. The failure mode begins when twisted power and return wires become shorted together; that is, the insulation on both wires is compromised so that the actual wires touch. This contact produces an arc that causes the insulation to transition from an insulator into a resistive material. The current through the now resistive portion of the insulation causes it to heat and burn clear, allowing the now exposed wires in the direction of the power source to come into contact. This contact, in turn, produces a new arc and sparking short that burns another portion of the contacting wire. These events recur as the arc progresses or tracks from the initial short toward the power source.

The data from testing showed that twisted Kapton insulated wiring of around 20 AWG carrying 28-VDC power easily is triggered into an arc tracking event. However, another popular insulation material, Teflon, would not track with 28-VDC power. When the voltage was increased to 115 V, as used on the *International Space Station*, the Teflon, however, was found to support arc tracking. Depending on the chosen voltage for a new spacecraft design, the insulation selected likely is to be vulnerable to this failure mode.

Flight Experience

Initially, the Space Shuttle Program assumed that there was no real likelihood of this failure mode affecting the flight harnessing. Analysis indicated that arc tracking occurred only to wiring that had been damaged in a way that bare wires were exposed. Because the wire insulation was so tough and the flight harnesses were so well treated, the Space Shuttle Program was convinced that this failure would be confined to harnessing used with ground support equipment.

Unfortunately, this was not to be the case. It was discovered that some arc tracking failures had been experienced during flight. One such event occurred inside a wire bundle on Space Shuttle mission STS-6 (NASA 2003a, 2003b). This particular wire bundle inadvertently and partially had been crushed during the replacement of a nearby avionics box. There was no obvious damage to the wire, and no shorts or arcing were observed prior to launch. Late in the first day of the mission, during a quiescent part of the flight, a short and subsequent arc tracking event occurred and eventually tripped a circuit breaker. Later examination showed that a total of six wires positioned deeply inside a wire bundle had become shorted together. This case, however, was one in which the wires were energized by a 115-V, three-phase, 400-Hz AC power source.

Several disturbing observations were made about this failure:

- The initial damage to the wiring was inside a wire bundle, and therefore, visual inspection could not detect the latent damage that had occurred during the maintenance operation.
- The arcing did not occur during dynamic flight, thus the time frame in which such a failure can occur is very long indeed.
- The initial failure between two insulated wires spread to four other wires and defeated their insulation before the circuit breaker opened and removed power to the wire harness.

Previous experience with this particular wiring insulation convinced the engineering team that it had to be damaged severely for a short circuit to occur. It was assumed that, to inflict such damage, the insulation would have required exposure to sharp edges, which most certainly would produce very visible damage. Neither this engineering team nor any other had experience with the type of damage mechanism whereby there is no noticeable injury to the outside of the wiring harness while severe damage is caused to the insulation of wires buried within. This was, however, the mechanism of the wire failure on STS-6.

All previous experience was such that wire harness short circuits were considered to be caused at the time of an impact or abrasion. It had been assumed that, if the wiring survived an impact without power loss to equipment or circuit breakers being tripped, then it was not necessary to worry about the integrity of the power source until some future event that might cause additional trauma to the wire harness.

Statistical techniques used this information to reduce the probability of a relevant failure to an acceptably small number based on the ratio of time in dynamic flight (ascent, launch, docking, and the like) to that spent in quiescent flight (attitude control and coasting). For this statistical exercise, it was concluded that the damage occurred during ground processing, before ascent. However, the short circuit and arc tracking event occurred after ascent, after launching a large heavy satellite, and after several more hours in quiescent flight.

Again, previous experience with this type of wire insulation led the engineering team to assume that a short circuit would compromise only the two conductors involved with the initial damage. Heretofore, the engineering team had not considered it necessary to route redundant power wires in bundles physically separated from the primary power wiring. It was not expected that the damage to the wire harness spread from the initial two wires to four additional wires within the bundle.

As well, it was thought that a current limited power source is an effective step toward controlling damage or propagation of wire damage in a short circuit situation. Testing would later demonstrate that a current limited power supply would deny the short circuit the amount of current required for it to burn clear. Instead, the current limited power source would provide sufficient current for the arc to initiate, then to progress in a low current mode for a duration sufficiently long for the damage to spread to adjacent power wires.

A second arc tracking event occurred during Space Shuttle mission STS-28 (NASA 2003a), and as well, it involved polyimide insulated power harnessing. Investigation of this anomaly found that the 28-VDC power cable was a polyimide insulated cable having the power supply and return or ground wires twisted tightly together. The particular strain relief used for this harness was one with an uncushioned tang to restrain the wires. An overwrapped Teflon sleeve had been added to the power cable during manufacture to extend its life. The power cable wires and their Teflon sleeve were secured to the tang. Installation and removal of this cable over several flights subjected the wire insulation to a very tight radius bend at the strain relief tang. The white Teflon overwrap precluded any ability to inspect the wires for damage. Had the wires been inspected at the stage where the polyimide insulation had begun cracking, the in-flight anomaly could have been avoided. As it was, the insulation cracked, the conductors were exposed to each other inside the overwrapping sleeve, and arc tracking was initiated. The tracking continued from the tip of the tang strain relief toward the power source. Because the power source was on the other side of the connector, the arc tracked inside the connector backshell to a point where the two conductors separated to attach to their individual connector pins. At that point, the wires were physically too far apart for the arc to sustain. In-line circuit protection was not effective in stopping this event.

A third arc tracking event occurred during Space Shuttle mission STS-93 (NASA 2003b) that involved polyimide insulated power harnessing. Investigation of this failure showed that a 115-V, 400-Hz AC power cable was damaged by a burr on a screw head. The screw head was holding a portion of the cable tray within which this wire was located. The screw head had abraded the wire during every previous mission of that vehicle. X-ray examination of the wire revealed corrosion consistent with the conductor being exposed for 5 a. During that time, this *Space Shuttle* flew five missions. During the fifth mission, contact with the screw produced a short circuit of sufficiently long duration to open the circuit breaker and shut down one of two redundant engine controllers on two of the three main engines during ascent. This arcing failure was unique, in that it occurred between a powered wire and a grounded screw. Because the screw was stationary and the power supply was current limited, the arcing could not track away from its initial location even though it reoccurred periodically.

14.6.2 Characteristics of Arc Tracking

The in-flight anomalies experienced and subsequent laboratory testing yielded the following characteristics of the arc tracking signature. All arc tracking begins with a short duration, high current short circuit and resultant arc. This is followed by either a high power mode in which current in excess of 10 A exists and can produce considerable damage in a short period of time or a low power mode in which a current of less than 10 A, while dissipating less power, still can consume the sample but over a period of minutes.

The characteristic feature of low power arc tracking is that this mode requires a series resistance of approximately 3 to 4 Ω. In one series of tests, when the series resistance was increased to values above 3 Ω, a single low power mode was available. In this case, the arc voltage could be reduced to as low as 6 V. Because the current remained low, such as from 3 to 5 A, this event oftentimes could occur without causing the lower rated circuit protection to open. The power in the arc was determined to be only 30 to 40 W, and it was observed to move slowly along the sample, taking up to 2 min to consume 5 in. of wire. Because the arc was of such low power, only two wires in a bundle were destroyed, and without damage to any of the other wires.

In a vacuum, DC arc track events tend to produce longer damage lengths than those in atmosphere. In multiwire tests, when several wires became involved in the event, the current flowing in the individual wires was reduced, although the total current increased. Because effective wire protection uses individual fusing for each leg of the circuit, this tends to slow the opening of the circuit protection device. Such low power arc tracking often does not trip the circuit protection device.

14.6.3 Likelihood of an Arc Tracking Event

Early assessments by the Space Shuttle Program were conducted without benefit of the insights gained from investigating these three in-flight anomalies. They concluded that wiring short circuits were unlikely failure modes of Kapton insulated power harnesses. These assessments were conducted without knowledge of the amount of deterioration of the Kapton insulation that would occur on a reusable spacecraft over an extended period of time. Presently available data indicate that Kapton insulated wiring has a progressive tendency to develop cracks at those areas where it has been bent sharply or stressed by ground handling.

Factors That Increase the Likelihood of Arc Tracking

Test and flight data indicate that the repeated bending of wiring at the strain reliefs of connectors can cause cracks in Kapton wiring insulation. Similarly, impact loads to wiring harnesses caused by inadvertent rough contact when electronic boxes were removed and reinstalled between flights can damage this insulation, as can repeated contact with sharp edges. All three mechanisms eventually precipitate arc tracking during the in-flight portion of a mission.

The Kapton insulation is so tough and resistant to damage when it is new that it had given technician and ground personnel a false sense that the insulation was almost damage proof. Indeed, several pictures of early Space Shuttle processing showed technicians routinely using cable bundles as handholds and standing on the bundles as they walked across the payload bay.

Testing for Precursor Conditions

The current state of the art for harness testing cannot detect insulation damage within harnessing that does not have a measurable short circuit; therefore, any testing for precursor conditions for arc tracking is of very limited utility. Short circuits manifest in differing resistances. Because wires are cylindrical and any contact between cylinders represents a point contact, the initial surface area of wire to wire contact, that is, a short circuit, is usually quite small. In addition, because all wires have finite resistivity, short circuits are also finite.

Time domain reflectometers are often able to detect subtle wiring damage in a quiet laboratory environment. The application of this technology to a vehicle having many different items in test has had limited success. At any given time, the varying electromagnetic interference and radio frequency fields tend to confound the attempts to detect wiring damage, even when attempting to detect known damage.

14.6.4 Prevention of Arc Tracking

Prevention of arc tracking is a challenge. The methods in use today are a combination of circuit protection and fusing, physical wiring protection, inspection, and physical separation or isolation of circuits. These are addressed in sequence.

Damage to circuits and harnessing caused by short circuits is limited effectively by circuit protection or fusing techniques. These techniques usually are used to manage hazards other than arc tracking. Even so, some experiences demonstrate that proper fusing and circuit protection devices can limit the duration of arc tracking. However, as is exemplified by the aforementioned STS-28 anomaly that progressed for 1.7 s before being terminated by physical separation of the conductors at the connector, this is not always the case. During this anomaly, the protecting fuse did not open the circuit.

The physical protection for the wiring, which is implemented to increase the reliability of harnessing, tends to prevent tight bends and other related conditions that can lead to cracking of the polyimide insulation. Such protection includes strain reliefs, edge protection devices, and cable clamps.

Any physical protection used for harnesses must be inspected to ensure that the design has been implemented during the buildup process as originally intended. An important process during the buildup is having an engineering inspection conducted as opposed to a quality inspection. Many quality inspections are designed to ensure that the physical protection material is applied according to design documentation and drawings. In contrast, the engineering inspection ensures that the actual hardware build process has placed edge protection and cable clamps at locations that best protect the as-built cabling. Because the actual length and diameter of cabling can vary during the buildup process, the engineering inspection is critically important.

The final means of prevention is physical separation and isolation of the wiring and harnesses. This aspect of the design can assure that a single event is unlikely to damage more than one of the redundant control circuits. Separation also assures that, if the wiring starts arc tracking, the damage is limited to those circuits bundled with the wires that suffered the initial short circuit. By so doing, even if we do not prevent the arc tracking, the damage that such a failure can cause to safety critical circuits can be managed.

14.6.5 Verification of Protection and Management of Hazards

Circuit protection for wiring is verified by checking the drawings against the as-built hardware. Verification of physical protection for wiring requires an engineering inspection to ensure that credible methods have been applied and is no different from that required if adequate circuit protection has not been imposed. To verify the physical separation of wire paths requires a review of physical design and the inspection of as-built hardware to ensure that the wiring paths actually are separated for their entire lengths.

14.6.6 Summary

The arc tracking phenomenon is caused by a unique type of short circuit that, once initiated, does not progress to a hard short or draw a very high current. The arc tracking process can be described as a walking series of brief duration shorts. The average current drawn is low, and it is difficult to stop with typically used fuses or circuit breakers. The arc tracking threat can be mitigated by the application of adequate physical protection methods, including the separation of redundant power circuits. All physical protection techniques used to mitigate arc tracking require an engineering inspection to validate the implementation of any protective measures applied.

14.7 CORONA CONTROL IN HIGH VOLTAGE SYSTEMS

The use of high voltages in space applications can result in the manifestation of an undesirable electrostatic effect known as *corona*. Corona is a form of glow discharge that occurs in a strongly nonuniform electric field near electrically charged or excited objects having small to very small radii, such as wires or pointed electrodes immersed in a gaseous or fluidic environment. A corona discharge occurs at potentials less than those required to produce a spark discharge. The term is used to describe the breakdown process that proceeds from ionization of the surrounding gaseous environment. Corona might or might not be visible.

The onset of corona is a function of several factors, among them the constituent components of the gaseous environment: cleanliness, pressure, temperature, humidity, and electrical charge or potential difference; and the phenomenon can occur at any voltage level. Corona damage occurs as the result of insulation degradation caused by the collision of high speed electrons generated in the air surrounding the electrodes, located within voids in the insulation, or from electron leakage resulting from the carbonization of insulation surfaces adjacent to the electrodes in air, all at high voltages. Corona damage also can result from the recombination of ionized species into acidic or basic compounds, which then attack wiring insulation and nearby metallic or dielectric components. Successful control of corona generally is effected by careful material selection, attention to cleanliness, and most important, adherence to design practices and good workmanship. These practices mitigate conditions conducive to the formation or proliferation of corona discharge.

14.7.1 Associated Environments

Electrical and electronic equipment located within a spacecraft is expected to operate with varying degrees of contamination, under many changes in pressure, and in varied gaseous environments. The equipment itself can contain various gaseous constituents that can interact with the surrounding environment as well as under off nominal conditions. Orbital environments present a number of challenges to the design of high voltage hardware. The design constraints to be considered include mission duration, orbital altitude, equipment operating times and durations, and the atmospheric pressure within the spacecraft. On the spacecraft surface, the peak and duration of pressure and voltage transients are all elements of concern in the proper design and operation of high voltage equipment. The designer must remain cognizant that any of these elements can be expected to change in a dynamic sense from one mission to the next or even during a single mission.

Contamination

Contamination in the environment can lead to reduced breakdown potentials. Sources of contamination include particulates ranging from foreign gases and dust particles to corrosion products, such as various oxides or salts deposited during handling and processing. The onset of corona can be affected as well by the presence of charged particle radiation, thus necessitating protection schemes for high voltage equipment exposed to the high fluence rates of charged particles.

Pressurization

At standard temperature, data collected for parallel plate electrodes with 400-Hz AC excitation indicate that gases such as carbon monoxide, nitrogen, propane, and xenon exhibit ionization potentials similar to or less than that of air at the same pressure. These data form what are known as Paschen curves, named after Frederick Paschen, who first established the relationship between temperature, pressure, gap spacing, and breakdown voltage for various gaseous environments and electrode types. Argon and helium have much lower ionization potentials than air at standard temperature and the same pressure. Thus, equipment immersed in a gaseous environment containing sufficient concentrations of argon or helium can be susceptible to corona onset at voltages approximately 50% lower than if they were immersed in air. Because of this drastic reduction in corona onset voltage, equipment expected to operate in such environments specifically must be designed for those environments and demonstrate they can operate safely in them during the certification process.

Temperature

Based on their relative position on the Paschen curve for the subject gaseous environment, it can be seen that temperature changes can affect corona onset voltages dramatically, thus causing previously safe conditions to become hazardous. It is, therefore, very important to understand the operating temperature profiles for equipment in various gaseous environments to be able to avoid operation in unsafe temperature ranges.

Other effects of temperature that can affect corona control are those related to material stress caused by thermal cycling or sustained low or high temperatures, which eventually can lead to problems such as softening, melting, plastic flow, chemical decomposition, embrittlement, and cracking. Effects such as these on system electrical and electronic components, and particularly on system wiring, can result in insulation failure, thus allowing corona to form in areas previously controlled and safe. Once a corona discharge forms, further damage is inflicted on the affected insulation by the discharge process. This can lead to complete system failure.

14.7.2 Design Criteria

Design guidance is based on good workmanship and the application of common sense and lessons learned. Fundamental considerations applicable at any operating voltage include

- The prevention of contamination on conductor and dielectric surfaces.
- Employment of insulating materials having high resistivity and dielectric strength and low permittivity.
- Selection of materials that exhibit outgassing properties that do not result in the introduction of undesirable gaseous constituents into the operating environment.
- Ensure that encapsulation of electrical and electronic equipment, including wiring, is void free.
- Pressurized hermetically sealed equipment using a dry gas having a high dielectric strength, and ensure that the equipment thoroughly is dry before pressurization.

- Design for an operating margin to accommodate abnormal transient and fault conditions.
- Maintaining good separation between equipment and wiring that operate at widely separated voltages.
- Use of rounded corners and edges on conductors.
- Venting of all unsealed volumes containing critical parts, because pressure in any unvented volume decreases gradually and results in corona or arcing. Dielectric inserts should be slotted to vent the interior volume.

Additional detailed guidance can be found in the references included in the Recommended Reading.

14.7.3 Verification and Testing

Verification and testing are performed to demonstrate that equipment performs its tasks and remains safe from corona for the duration of its intended use. Inspection, analysis, and testing or a combination of some or all three can be used for this demonstration.

Many factors determine whether corona can occur, including temperature, humidity, ambient pressure, test specimen shape, rate of voltage change, and previous history of the applied voltage. Test methods such as ASTM D 1868, ASTM D 149, and MIL-STD-202G must be used with caution due to interpretability of results and with great care because of the personnel hazards involved.

Inspection

Design and installation drawings should be inspected to verify that good workmanship guidance applicable to the operating conditions and voltages of the affected equipment has been incorporated. If doubt exists as to the incorporation of proper workmanship guidance or the design or drawings show that proper workmanship guidance was not incorporated, then analysis might be used to determine if the equipment functions safely in its intended operating environment.

Analysis

Analysis should include not only an examination of the materials and workmanship issues that affect the immunity of equipment to corona but also such criteria as operation only under very high vacuum conditions, no operation during ascent or descent, time of exposure to pressure transient values conducive to corona, no operation with possible exposures to high fluence rates of charged particle radiation, and whether the design contains all high voltages within a sealed chassis or sealed component. If a sealed chassis or component is used for corona control, the equipment should be tracked to assure that sealing is maintained at design levels over its operational life.

Testing

If equipment cannot be verified by inspection or analysis as just described, the equipment must be tested for compliance with corona design requirements. Such tests are conducted under service conditions using worst-case voltages over the full operational range of pressure and environments. Testing usually is done on energized subassemblies and lowest replaceable units and should include, to the maximum extent possible, any associated cable runs and peripheral items and equipment, such as bus bars and enclosures. Test equipment that applies an electrical charge to the surfaces of hardware under test must enable the isolation and identification of corona sources and surface creepage. Tests are considered successful when they demonstrate that equipment safely can be operated without corona discharge for conditions conducive to the formation of corona.

RECOMMENDED READING

References on the topic of high voltage design and the control of electrostatic discharge phenomena related to high voltage applications exist in the literature. However, this is a relatively specialized area, and reference materials are scarcer than for other similar topics. References include portions of military standards, texts specifically on this topic, texts of a more general nature dealing with electrostatics and applications of electrostatics in design and engineering, and of course technical papers written for and published by organizations such as the Institute of Electrical and Electronics Engineers. Rather than attempt to provide a comprehensive listing, the following is a basic subset of references for both the interested reader and the serious student. These can serve as foundational material, as well as a good source leading to additional, advanced reading and study.

Cobine, J. D. (1941) *Gaseous Conductors*. New York: McGraw-Hill.

Dunbar, W. G. (1983) *High Voltage Design Guide: Aircraft*. AFWAL-TR-82-2057. Wright Patterson Air Force Base: Dayton, OH, vols. 1-5.

Latham, R. V. (1995) *High Voltage Vacuum Insulation: Basic Concepts and Technological Practice*. New York: Academic Press.

Malik, N. H., A. A. Al-Arainy, and M. I. Qureshi. (1998) *Electrical Insulation in Power Systems*. New York: Marcel Dekker.

Moore, A. D. (1968) *Electrostatics*. New York: Anchor Books, Doubleday.

NASA. (1978) *High Voltage Design Criteria*. Technical Standard MSFC-STD-531. Huntsville, AL: National Aeronautics and Space Administration, Marshall Space Flight Center.

Peek, F. (1915) *Dielectric Phenomena in Engineering*. New York: McGraw-Hill.

Penning, F.M. (1957) *Electrical Discharges in Gases*. New York: Macmillan.

Raizer, Y. P. (1997) *Gas Discharge Physics*. Berlin: Springer-Verlag.

Ushakov, V. I. (2004) *Insulation of High Voltage Equipment*. Berlin: Springer-Verlag.

von Engel, A. (1965) *Ionized Gases*. London: Oxford University Press.

14.8 EXTRAVEHICULAR ACTIVITY CONSIDERATIONS

In consideration of extravehicular activity, the designer faces several unique avionic challenges. These challenges can be categorized broadly into the areas of protecting a crewmember from exposure to hazardous electromagnetic radiation, exposure to molten metal that would damage the spacesuit, and use of displays under intense illumination or total darkness. Other sections of this text address the specific radiation limits acceptable for crew exposure.

14.8.1 Displays and Indicators Used in Space

During the flight of a payload on the *Space Shuttle*, the extravehicular activity crew was unable to determine the state of the payload restraint devices. Light emitting diodes had been chosen to provide the status indication for the restraint device. During the design and the testing of the restraint prior to flight, the illumination conditions considered were those of laboratories. Under these test conditions, the light emitting diodes functioned quite adequately as status indicators. When the equipment was used in space, however, the combination of the unimpeded illumination of the Sun and the light shields of the helmet rendered the light emitting diodes unusable, and the crew was unable to determine the status of the release latches.

This unfortunate design prevented the Space Shuttle crew from using the indicator as an aid during deployment of the device. The designers had failed to consider the shortcomings of light emitting diodes used as indicators when in the intensely bright direct sunlight in space. The guidance offered here is that the designer should seek indicators that do not become washed out by direct sunlight and can function in the total darkness of the space night. Often mechanical flags or other displays are more appropriate for these conditions. Testing of indicators in conditions of extremely bright light and total darkness helps reveal any weakness of a proposed indicator. The designer is encouraged to consider that the temperature extremes between dark and sunlight in space can be destructive to certain indicators as well. Some liquid crystal displays, for instance, cannot tolerate the cold temperatures to which equipment is exposed when pointed at deep space.

Another area of challenge is to provide status indicators to a crew at the site of a particular hazard. If there are electromagnetic hazardous radiation sites, the extravehicular activity crew needs to know when the transmitter is energized. Indicators for this purpose should be ones that catch the eye. Obviously, the spacesuits and the vacuum of space preclude audible alerts. Motion based alerts, varying illumination, such as flashing lights, and the like can be effective means to alert a crew that the area in which they are being used possesses a potential hazard requiring their attention.

14.8.2 Mating and Demating of Powered Connectors

The intent of this section is to present the safety policy regarding the design provisions required when electrical connectors must be mated or demated during extravehicular activity. The specific approach is to eliminate potentially hazardous energy levels at the connector interface during mating and demating operations by limiting the energy of the power source or isolating power sources from the connector. The design also must prevent generation of molten metal and damage to safety critical circuits.

Low Power Connectors

The mating and demating of low power connectors is permissible without upstream inhibits or special connector design features. *Low power connections* are defined as those used with a design having a power supply capacity or appropriate upstream circuit protection that limits maximum continuous current to 3 A or less with an open circuit voltage no greater than 32 VAC rms or 32 VDC.

Test data associated with a 22-AWG connector indicate that the first arcs occur with the application of 1.5 to 3.8 A (average is 3 A) at 33 V. The smallest pin size considered was 22 AWG. These test criteria should not be used for smaller pins. The low power connection is based on upstream hardware design features that limit voltage and current to the values specified for each contact in the connector. Typically, circuit protection devices that satisfy the maximum continuous current criteria are rated at 2 A. The downstream design is not a factor in this determination. Sustained arcs are the major concern. The risk of momentary exceedances normally is acceptable based on the speed of the circuit protection device.

Medium Power Connectors

Each powered circuit must have at least one verifiable upstream inhibit. The design must provide for verification of the inhibit status at the time it is inserted. An additional upstream inhibit is required when the short circuit current is greater than 65 A.

In this case, any molten metal generation concern is controlled by an upstream inhibit. A downstream break in the circuit, that is, a downstream inhibit, or the reduction of the load presented to the power source also is acceptable, provided any concerns associated with a short circuit at the connector are addressed. A reduction of the upstream load is addressed by the low power criteria. Because connector testing has shown that 67 A at 33 V is the threshold for causing considerable damage to sockets, a current of 65 A was chosen as the limit for connector shells. Therefore, a more stringent requirement is imposed for circuits exceeding this value.

14.8.3 Single Strand Melting Points

If the payload has a power supply capacity or possesses upstream circuit protection that limits the short circuit current to less than the single wire strand melting current, reduction of current draw to less than 3 A on the downstream side can be used instead of an upstream inhibit. If the melting current value is approached, however, the power supply or upstream circuit protection must remove all power from the connector within 5 s. The single wire strand melting current value is

- 5.1 A for 22-AWG wire or pin.
- 7.2 A for 20-AWG wire or pin.
- 10.2 A for 18-AWG wire or pin.
- 12.3 A for 16-AWG and larger wire or pin sizes.

The amperage listed for the different wire and pin sizes is based on the fusing or melting current for one strand of the wire. Because of the various means and associated conditions by which heat can be removed from a heated strand, such as heat sinks, initial wire temperature, or the presence of forced air cooling, these amperage values are considered to be only in the ballpark for each of the different pin sizes. If a strand of wire became separated from the main wire and shorted, the main wire, functioning as a heat sink, would remove heat from it. In addition, if the fusing current were reached, some time would pass before the shorted strand heated to its melting temperature. Considering these data and any possible modes envisioned, the 5-s requirement to remove power is based on the engineering judgment that more time would be required to cause molten metal, for which the concern is damage to an extravehicular activity suit. Any circuits above this current threshold must have an upstream inhibit. These criteria are to be applied to all connectors, including those used on battery cables.

Connector Design Considerations

Connectors must employ design features that completely enclose or shroud the pins and sockets during any making or breaking of electrical contact. The primary design feature that keeps a person from being injured by molten metal is the connector housing. It should be designed so that the pins and sockets separate before the shell is opened.

The design of a connector should provide for protection of the powered side from debris and inadvertent shorting when unmated or when mating or demating; for example, the powered side should be terminated in sockets rather than pins. It also should minimize the risk of a bent pin causing a short during mating or demating operations.

As mentioned earlier, each powered circuit must contain at least two inhibits. At least one of these inhibits must be upstream and function to remove voltage from the connector. The other design feature must provide either an additional inhibit upstream of the connector or a reduction of power or current draws to the lesser of 180 W or 3 A. Payloads should contain design features that limit the voltage across the connector to less than 200 V.

Any input electromagnetic interference filters upstream of these switching devices that remove the downstream load can cause transient exceedance of the 3-A limit until the capacitors in the electromagnetic interference filters become charged. This type of design is acceptable if the input filter energy storage capability is no greater than that allowed in the energy storage calculation chart (Table 14.2) for the corresponding connector pin gauge.

The design must provide for verification of at least one of the upstream inhibits at the time that it is inserted. The theory associated with this subject is that the potential to arc is a function of available power and the sharpness of the pins. Although other tests have been performed with inconsistent results and the phenomenon remains not understood fully (consistent sharpness among test results is difficult to establish), a series of tests managed by the NASA Engineering Directorate showed that contacts begin pitting at 1.5 A and 123 V for 22-AWG pins or at a power draw of 184.5 W. Based on these data, the 180-W limit was chosen as a conservative value for this interpretation. Test data associated with 22-AWG pins also have shown that minimal damage occurs from 1.5 to 3.8 A (average is 3 A) at 33 V. For higher voltages, the limit is based on power and for lower voltages, current.

These criteria provide an adequate margin of safety for connectors because the limits are set based on the initiation of pitting or contact damage rather than the contact failure

Table 14.2 Energy Storage Calculation Per Connector Pin Gauge

Connector Pin Gauge	Allowable Electromagnetic Interference Filter Stored Energy
4	49.0
8	20.5
10	13.0
12	8.0
14	4.9
16	3.0
18	2.0
20	1.3
22	0.8

threshold. Additionally, the limits are set based on the smallest pin size, which rarely is used for extravehicular activity applications. Because connector testing associated with mating and demating of powered connectors has been performed only up to 173 V, a more stringent requirement is imposed for connectors with an open circuit voltage above 200 V based on extrapolation of existing test data. Concerns about corona in proximity to the suit were considered and dismissed because the worst-case voltage buildup is below the corona pressure threshold.

Energy Storage Calculation

Energy storage is calculated using Equation (14.2):

$$E = \frac{1}{2}CV^2 \tag{14.2}$$

where E is stored energy in Joules, C is the input line to line capacitance, and V is the line voltage maximum.

14.8.4 Battery Removal and Installation

The NASA safety review organizations developed appropriate extravehicular activity battery change out activities in support of the Hubble Space Telescope payload. The safety review determined that electrical shock was not a hazard to the crew while in the extravehicular activity spacesuits because there was no conductive path to the crewmember. The overriding concern was molten metal generation as the result of an arc, because the hot debris can compromise the integrity of an extravehicular activity spacesuit or potentially ignite any materials in the suit exposed to 100% oxygen.

14.8.5 Computer or Operational Control of Inhibits

In the low power case, computer or operational control of the upstream circuit protection device is not allowed; rather, it must be controlled by hardware design. Even though a disconnected cable satisfies the intent of limiting the upstream power capacity, this is not an acceptable operational solution.

14.9 SPACECRAFT ELECTROMAGNETIC INTERFERENCE AND ELECTROMAGNETIC COMPATIBILITY CONTROL

All spacecraft and associated equipment must meet acceptable standards for electromagnetic compatibility. The verification of electromagnetic compatibility is always the subject of a dedicated test campaign for acceptance by the procuring body. This fact establishes the implicit contractual importance of the discipline.

It is essential that electromagnetic compatibility engineering ensures that space vehicles as well as their systems and subsystems do not exhibit or produce electromagnetic interference throughout the life cycle of a program. This must be attained through built-in design compatibility, which is the real indicator of success, instead of installed as an after the fact remedial measure.

The rapid development of technology increased not only the number of onboard electrical equipment items but also their complexity. Hence, the effectiveness of performing any single basic function presently depends on the efficient performance of many other functions. Faster and more sensitive electronic technologies for space applications and the use of wider bandwidths in the equipment design drive new and challenging requirements to the flight hardware and increasing pressure to accommodate it in ever smaller and more crowded spaces. As a result, the probability of performance degradation by undesired electromagnetic interaction is increased.

Depending on the nature of a mission, electromagnetic interference can upset spacecraft in a variety of fashions. These range from DC effects, such as electrostatic charging and magnetization, to AC and transient effects, both conducted and radiated, that span the frequency range. The effects include interference hazards at the intrasystem level of spacecraft avionics and payload elements and at the intersystem level between the launch vehicle and spacecraft.

Interaction with the elements of the natural space environment, such as radiation, plasma, and cosmic rays, or the occurrence of geomagnetic substorms that can cause potentially disruptive effects through electrostatic discharge are other issues for which there must be an accounting. Therefore, assurance of electromagnetic compatibility entails the following topics:

- Knowledge and characterization of the composite noise environment, that is, human made and natural electromagnetic noise.
- The effect of noise on system performance, including analysis and verification.
- The scientific and technological basis of noise and interference control.
- Control of spectrum use and wireless communications.

14.9 Spacecraft Electromagnetic Interference

Although potentially catastrophic, the consequences of electromagnetic compatibility related disturbances on spacecraft essentially are nuisances that jeopardize the correct performance of a mission and reduce the efficiency of some functions. Such consequences can translate into irreversible loss of some operational capability with any relevant impact assessed in terms of scientific and programmatic yields, cost overruns, and schedule impacts. Effects of electromagnetic compatibility related disturbances can be temporary interruption of telemetry, noisy data from onboard science, permanent damage to power supplies, accidental tripping of circuit protection devices, false commanding, and instability of the power distribution subsystem, to mention a few. It is apparent that assurance of electromagnetic compatibility involves risk management and implies a working knowledge of the spacecraft subsystems.

Because the funding allocated for a space program is limited, the electromagnetic compatibility process includes cost effective considerations throughout every phase in the system life cycle. In fact, electromagnetic compatibility control decisions particularly are susceptible to the influence of cost effective trade offs. Analysis, testing, and corrections involve considerable program expense. Still, every effort should be made to apply a commensurate level of effort and provisions to ensure an acceptable level of electromagnetic compatibility so that mission objectives are achieved and reliability safeguarded. For this reason, electromagnetic compatibility engineering also encompasses programmatic responsibilities.

The field of electromagnetic compatibility per se is multidisciplinary, although it is a matter for specialists. The control of noise and interference involves generic applications spanning many engineering domains. The collective wisdom of each of these disciplines is critical for successful electromagnetic compatibility implementation. Full cognizance of electromagnetic compatibility requires a thorough understanding of electromagnetic theory and modeling techniques, as well as familiarity with the sophisticated practices of manufacturing and testing. All this must be considered with an attentive eye turned to the cost and schedule of the program.

At the spacecraft level, it is important to establish an electromagnetic compatibility control program early to avoid costly fixes for the integrated system. The approach varies from project to project and follows no predetermined paradigm. Currently, only heuristic first approximation considerations and tailored military standards constitute the basic step to establish an electromagnetic compatibility control program.

14.9.1 Electromagnetic Compatibility Needs for Space Applications

The ability to design for and achieve electromagnetic compatibility is becoming more demanding for space applications. On the one hand, the rapid development of faster and more sensitive electronic technologies for space applications drives new and challenging requirements for the flight hardware and increases the pressure to accommodate equipment in ever smaller, more crowded spaces. These demands greatly increase the probability of electromagnetic interference. On the other hand, tighter space program budgets encourage the procurement of commercial off-the-shelf subsystems or the reuse of hardware manufactured for previous missions.

The space industry has the difficult task of harmonizing these two tendencies while reducing the costs of the spacecraft electromagnetic compatibility program, especially when emerging technologies are involved, and safeguarding its reliability. This difficulty is due, in part, to present day capabilities to analyze, predict, and verify the electromagnetic status of a system throughout its design stages and the many inevitable modifications to the hardware during the development stage until the spacecraft is assembled. Nowadays, electromagnetic compatibility customarily is reached through a balanced combination of guidelines, standards, heritage of previous projects, partial modeling, and testing throughout all the design and development phases of a spacecraft. The limitations of presently available electromagnetic models, especially in the intrasystem area, prevent procuring bodies from reducing the extent of expensive test campaigns. This leads to considerable cost-benefit requirements.

The fulfillment of specified equipment level requirements does not imply necessarily that electromagnetic compatibility is achieved for the integrated system. Such requirements mainly address procurement activities. When anomalies occur at the system level, empirical troubleshooting usually allows the discovery of a workaround solution. Even so, an exhaustive identification of the interference mechanism is seldom achieved due to the expense to facility costs and schedule. The converse also is true, in that an out-of-specification condition detected during equipment level electromagnetic compatibility testing cannot lead to any consequence when the unit is integrated into its operational environment. The principal objective of electromagnetic compatibility for space systems is the attainment of built-in design compatibility instead of applying after the fact remedial measures.

14.9.2 Basic Electromagnetic Compatibility Interactions and a Safety Margin

Achievement of electromagnetic compatibility implies the quantitative control of four classes of measurable parameters, covering the detection of all the coupling mechanisms for both emission and susceptibility in conducted and radiated mode:

- Conducted emission.
- Conducted susceptibility.
- Radiated emission.
- Radiated susceptibility.

Relevant requirements are specified in electromagnetic compatibility standards, such as MIL-STD-461 for equipment procurement activities, but usually they are tailored to specific demands of the space program. Historically, electromagnetic compatibility acceptance for systems relied on the notion of a safety margin, which is the ratio of the susceptibility threshold of the receptor to the level of the interfering emission evaluated at a single frequency. Safety margins are specified quantitatively in accordance with the worst-case potential criticality for the effects of interference induced anomalies. The following categories are used widely:

- Category I includes serious injury or loss of life, damage to property, or major loss or delay of mission capability; that is, safety margin ≥ 12 dB, usually 20 dB.

- Category II represents degradation of mission capability, including any loss of autonomous operational capability; that is, safety margin ≥ 6 dB, typically 6 dB is considered. This is the classical case concerning intrasystem compatibility.
- Category III represents loss of functions that are not essential to a mission; that is, safety margin 0 dB.

Demonstration of the safety margin verifies the attainment of electromagnetic compatibility at the system level, even from a contractual standpoint. This is achieved by a combination of test and analysis. Handbooks of electromagnetic interference for space applications readily are available, such as the one by Clark et al. (1995), which details many useful design guidelines for space that go beyond the scope of this text.

14.9.3 Mission Driven Electromagnetic Interference Design: The Case for Grounding

As an example of how missions can drive the electromagnetic interference design, different grounding solutions as they have been implemented successfully by major space agencies and industries are presented. Grounding is the process that maintains all parts of the system at the same reference potential by providing a low impedance path at all frequencies for current returns throughout the system. Designing for electromagnetic compatibility inherently aims to avoid any electromagnetic field, voltage, or current generated or utilized at one point in the system being shared through common path impedance with other units and thus interfering with their operation. The grounding approach chosen and its implementation play a fundamental role in achieving system electromagnetic compatibility.

Each of the grounding concepts possesses pros and cons, and these include reliability considerations that are the subject of trade studies that must be conducted after the conceptual phase of spacecraft definition. The concept appropriate for the application must be determined before the electromagnetic compatibility design for the equipment takes place.

The main power distribution system of first generation Russian spacecraft was isolated from the chassis. More precisely, the return power line was connected to the chassis through a bleed resistor shunted by a capacitor close to the power source. Typically, these were a few thousand ohms and a few hundred nanofarads, respectively. The resistor tied the power bus to the DC chassis potential and simultaneously limited the current flowing onto the structure should an isolation fault occur in the main power bus. This grounding system had an intrinsic single failure tolerance against any short circuiting of the primary bus that could cause the loss of a mission. However, it also caused an increase in common mode noise at user interfaces and consequently an increased radiated emission from the power harness. Hence, users of the main power bus in these spacecraft had to have made provisions for common mode immunity.

Reliability considerations suggested using this particular grounding architecture on the spacecraft of critical interplanetary missions, especially those using radioisotope thermoelectric generators, such as *Huygens-Cassini*, *Voyager*, and *Galileo*. In the longer term, however, nuclear radiation could alter the properties of insulation materials used inside the generator container. This could lead to current leakage or short circuits. Some launch vehicles utilized this grounding architecture as well. When it was used, hardware along the primary power bus was designed in a way that the isolated grounding concept was not violated. The success of many space missions demonstrates the validity and sometimes the necessity of this design approach.

For reliability reasons, pyrotechnic initiator units are used widely in the isolated grounding architecture, even when the rest of the spacecraft utilizes different architecture at the system level. When an electroexplosive device is detonated, conductive hot plasma generated by the powder charge can close the loop from the positive power line to the chassis, causing a persistent short circuit. In a case such as this, currents of 10 A or so can flow onto the chassis, giving rise to magnetic coupling into nearby circuits until the available power source might be exhausted. An isolated power bus having an appropriate bleed resistor is a simple remedy, albeit not the only one, that can limit the consequences of this phenomenon.

The single point grounding concept ensures that no ground loops are created within the electrical network of a spacecraft. All circuit grounds are returned to a single common point, called the *star point*, usually located on the spacecraft chassis. This practice is applicable to small spacecraft, when the length of the grounding wires is very small compared to the wavelength of interest. The true single point ground seldom is used, and it is discouraged when the system largely is distributed. However, use of the single point ground is mandatory in the case of accurate magnetometric missions.

A widely used architecture in spacecraft is the distributed single point ground. The basic principle is to isolate power networks in the system through DC to DC converters. For this architecture, the power circuits can be referred independently to the same ground, that is, spacecraft structure, without creating ground loops. The star point is the entire spacecraft chassis, if it is metallic. The separation of the grounding network minimizes mutual interaction, and it particularly is suitable for large systems, allowing flexibility for grounding individual items of equipment.

14.9.4 Electromagnetic Compatibility Program for Spacecraft

The final objective of an electromagnetic compatibility program applied throughout the development of a space system is to ensure that, during the spacecraft lifetime, from equipment integration until spacecraft decommissioning and launch, the system is compatible with itself. It must neither cause disturbances to other systems nor suffer loss of performance due to other systems or any external environment that it might encounter during its life cycle. This goal is pursued by a balanced combination of standards, guidelines, heritage of previous projects, modeling, and testing that spans from the earliest stage of the program to the final integration of the space system. The skeleton of a procedure and relevant documentation aimed at ensuring the attainment of electromagnetic compatibility at the system level is presented. The main documents managed by electromagnetic compatibility engineers are the electromagnetic compatibility

- Specification, which contains requirements at the system, subsystem, and unit levels.

- Control plan, which describes the methods, means, and the rules to follow throughout the project to guarantee compliance with the requirements as defined in the electromagnetic compatibility specification.
- Test and verification plan and procedure, which present the test setup and procedures to verify the specifications.
- Test and verification report, which presents all test results and any relevant nonconformances.
- Analysis, which contains all ancillary analyses conducted in support of design and testing activities, such as predictions of intrasystem electromagnetic interference and electromagnetic compatibility based on equipment electromagnetic interference characteristics to assess design solutions such as filtering, grounding, and shielding.

These documents periodically are updated throughout the development phases of the system. Several milestones can be used to trace the design cycle for a spacecraft. Failure to complete any one of them precludes the possibility to progress to the next. A nonexhaustive list of data and deliverables that must be completed for each milestone and reported in the appropriate electromagnetic compatibility document as follows:

- Request for proposal or response to an invitation to tender:
  - Study and define known operational environments.
  - Identify functional criticality for all equipment and subsystems and allocate the pertinent category.
  - Define a safety margin for critical functions and electroexplosive devices to account for lifetime, degradation of circuit, and circuit protection. Typically, 20 dB is allocated for the electroexplosive device and 6 dB for signal, power, and control lines.
  - Provide some general guidelines, such as separate signal and primary power buses, select DC to DC frequency outside signal bands, and twist and shield the harness with the appropriate twist rate.
- System readiness review or requirements definition review:
  - Define requirements at system, subsystem, and unit levels in the electromagnetic compatibility specification. Typical requirements include grounding, bonding, inrush current, conducted (continuous wave and transient) and radiated emissions, susceptibility, and electrostatic discharge. These requirements ensure the technical reliability of the equipment procurement.
  - Define margin verification methods at the system level in the electromagnetic compatibility control plan. De facto, the verification of the compatibility of a system is achieved by imposing and demonstrating a safety margin between the susceptibility threshold of the units and the actual noise at the system level during worst-case conditions.
  - Consolidate the guidelines along with any necessary special precautions for critical cases. For magnetically sensitive spacecraft, such as *Cluster*, several preventive

measures have matured along the years, and nowadays very reliable engineering practices are known and implemented.

- Preliminary design review:
  - Identify the most critical electromagnetic interference aspects, and consolidate appropriate countermeasures in the electromagnetic compatibility control plan and through electromagnetic compatibility analyses.
  - Analyze any electromagnetic compatibility impact from commercial off-the-shelf hardware on the system, and assess their electromagnetic interference and electromagnetic compatibility performances.
  - Generate an exhaustive grounding diagram of the primary and secondary power systems, identifying all units, shielding, and principal interfaces; and include it in the electromagnetic compatibility control plan.
  - Consolidate the model philosophy and all relevant verification methods in the electromagnetic compatibility control plan.

- Critical design review:
  - Between the preliminary design review and critical design review, all the units and subsystems are designed, assembled and qualified.
  - The system electromagnetic compatibility control plan is updated with the unit and subsystem results from the electromagnetic compatibility test report, any potential criticality identified, and any relevant countermeasures decided.
  - The electromagnetic compatibility managers evaluate all requests for waivers and noncompliance reports and disclose any relevant actions to be completed prior to the final system level test.
  - Based on test results from the unit and subsystem levels, the system level test, which is to be performed before the flight acceptance review, is defined fully.

- Flight acceptance review:
  - Only limited tests are performed at the system level. Most often, the conducted emissions and susceptibility tests are confined to those areas (essentially power and signal lines) that have shown a certain degree of marginality at the subsystem level or constitute the core power distribution points for the spacecraft. The radiated emissions and susceptibility tests are performed to prove the system margin, that is, 6 dB, and particular attention is devoted to the measurement of specific niches, such as telecommand, launcher, and sensitive payloads.

The principal spacecraft models used during the preliminary design review to perform this assessment are the

- Avionics verification model.
- Engineering model or electrical qualification model.
- Protoflight model.
- Flight model.

14.9 Spacecraft Electromagnetic Interference

The verification methods employed for this assessment during the preliminary design review are

- Analysis.
- Review of design, such as correct use of shielded twisted wires, shield grounding, power isolation by review of drawings.
- Inspection to verify the conformance of hardware to drawings, and the use of proper parts and materials, such as harness separation or correct routing.
- Tests conducted to demonstrate compliance with the requirements during different stages of a project, that is, development, qualification, and acceptance.
- The application of similarity to equipment and subsystems that previously were qualified to the same or a more severe environment.

To conclude, a flowchart that depicts the various stages of the electromagnetic compatibility program used to control electromagnetic interference and that leads to safety margin verification, that is, the achievement of electromagnetic compatibility at the system level, is presented (Figure 14.2). For this flowchart, the spacecraft is subdivided into the service module (the platform) and the payload module and the interfaces where the payloads are

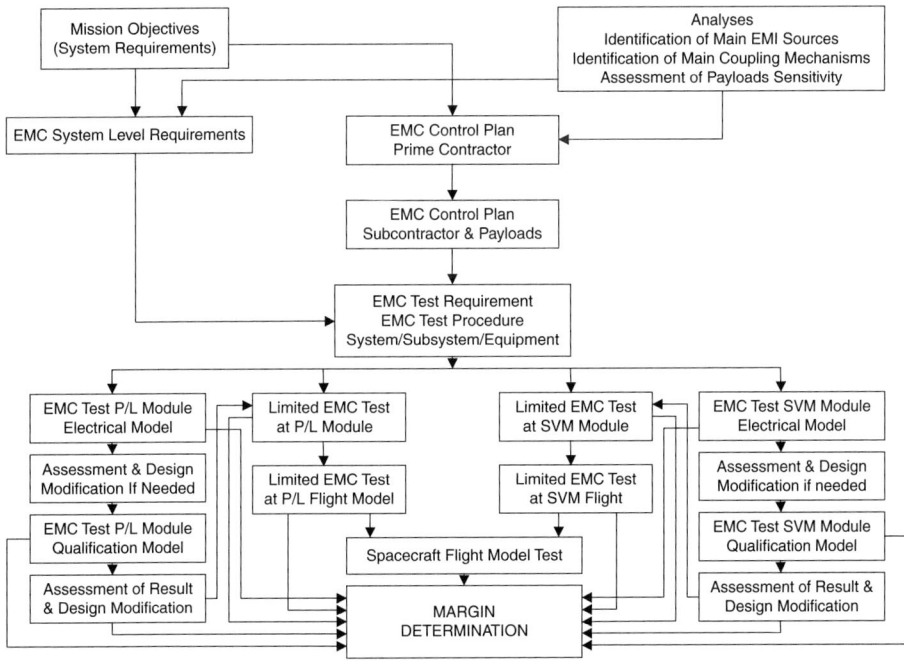

FIGURE 14.2 Electromagnetic interference and electromagnetic compatibility engineering flow diagram for a spacecraft project.

located. Verification of the intrasystem compatibility safety margin is achieved throughout testing and analysis is supported by testing at the interface between the service and payload modules.

14.10 DESIGN AND TESTING OF SAFETY CRITICAL CIRCUITS

14.10.1 Safety Critical Circuits: Conducted Mode

Conducted emissions seldom present critical safety hazards. However, some topics involving the conducted area can be safety critical if not analyzed carefully. Two case studies are presented in this section. One concerns power stability for the docking of an orbital infrastructure; the other regards design provisions employed to avoid failure propagation.

Ensuring Power Stability for Docking

The *Automated Transfer Vehicle* is an automatic, uncrewed space transport vehicle being developed by the ESA to carry cargo and supplies from Earth to the *International Space Station*. The *Automated Transfer Vehicle* is to be docked to the Russian segment of the *International Space Station*, from which it draws part of the power it requires.

The Russian service module provides a regulated 28-V power bus that is floating from the structure potential. Hence, particular attention is required to design an interface for the *Automated Transfer Vehicle* that possesses four separate switching mode power supplies referred to structure to regulate the internal Automated Transfer Vehicle buses. The negative resistance input characteristics of the converters, in conjunction with the impedance interaction between the electromagnetic interference filters and the power source, can cause the power system to become unstable and oscillate. The result likely is to have consequences for reliability and safety.

Design and verification of the Automated Transfer Vehicle power system interface to the Russian *Service Module* required crucial information about the source impedance seen by the *Automated Transfer Vehicle* when it is docked to the Russian *Service Module*. Verification at the system level was performed through use of an equivalent network. Although a complete hardware emulator of the Russian *Service Module* is available in Korolev (Russia) at RSC Energia, the size of the *Automated Transfer Vehicle* precludes transporting the spacecraft to Russia to test the interfaces. This constraint dictated the experimental characterization of the source impedance of the Russian *Service Module*, which then allowed ESA to derive the information necessary for the design of a stable power interface.

This approach, widely used in the framework of the International Space Station *Program*, is based on the underlying assumption that the interacting subsystems are linear, with small signal stability being the objective to achieve. That is, a source block with forward gain, F_S, and a load block with forward gain, F_L, are considered (Figure 14.3); Z_S is considered to be the source impedance and Z_L the load impedance.

14.10 Design and Testing of Safety Critical Circuits

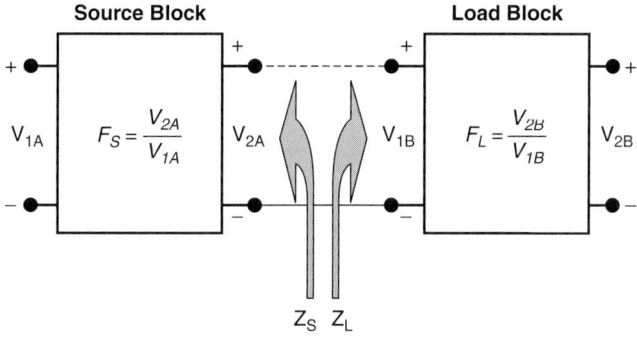

FIGURE 14.3 Source block cascaded with a load block.

Working with the g-parameters (hybrid), it becomes possible to verify that the input to output transfer function, F_{SL}, of the two cascaded blocks results; that is,

$$F_{SL} = \frac{F_S \cdot F_C}{1 + H_m} \quad (14.3)$$

$$H_m = \frac{Z_S}{Z_L} \quad (14.4)$$

The impedance ratio H_m can be considered the open loop gain of the integrated system. When $Z_S \ll Z_L$ for all the frequencies of interest, such as $f \leq 100$ kHz, the stability of both the individual source and load blocks guarantees stability for the integrated system. In general, this condition is not realistic. For some frequency intervals, we can expect $Z_S > Z_L$. Although this case does not imply a stability problem, additional analysis is required for the open loop gain H_m, such as by applying the Nyquist criterion to determine whether system stability is achieved.

Consequently, a measurement campaign was planned to measure the Russian *Service Module* emulator source impedance, Z_S, to provide the power control and distribution unit manufacturer substance to build a stable system, that is, to design for an appropriate Z_L. Figures 14.4 and 14.5 depict typical measurements of common mode and differential mode, respectively.

With a novel network synthesis technique, a single circuit reproducing both the common and differential mode impedance can be reconstructed having impedances as shown in Figures 14.6 and 14.7 for the differential and common modes, respectively.

The requirement levied on the industry was to design the interface to achieve an impedance separation of at least 3 dB and, in case of crossover, a phase separation of at least 45°.

Electrical ground support equipment was built having an output impedance identical to that of the Russian *Service Module*, and the Automated Transfer Vehicle power interface was connected to it through a harness representative of the actual installation. No instability was detected, so the design was validated. Further details on the methods for

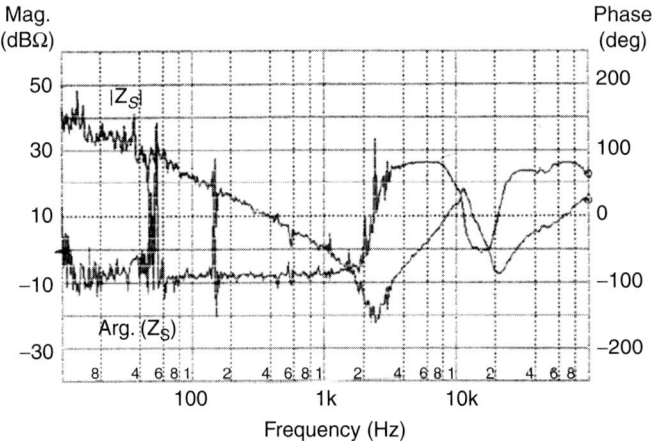

FIGURE 14.4 Typical measurement for the Russian *Service Module*: Common mode impedance.

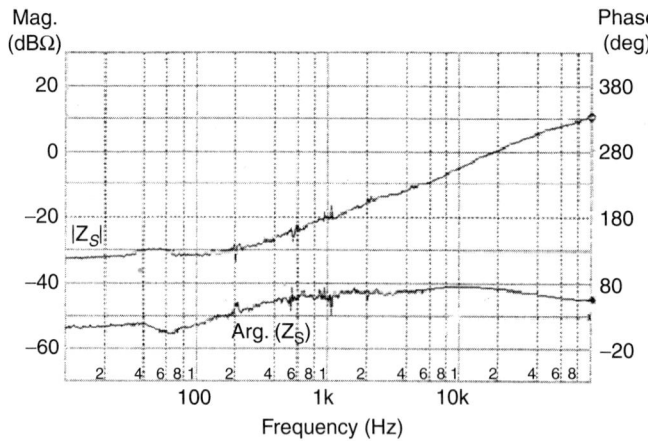

FIGURE 14.5 Typical measurement for the Russian *Service Module*: Differential impedance.

network synthesis and the test setup for these measurements can be found in Ciccolella and Blancquaert (2002).

Failure Propagation Control in Safety Circuits The management of power faults is related to safety and is essential for a crewed orbital infrastructure. The International Space Station modules take power from two parallel DC to DC converters, which are provided by NASA. Solid state power controllers are in place for overload protection purposes in the unit that distributes power inside the *Columbus Attached Pressurized Module*.

14.10 Design and Testing of Safety Critical Circuits

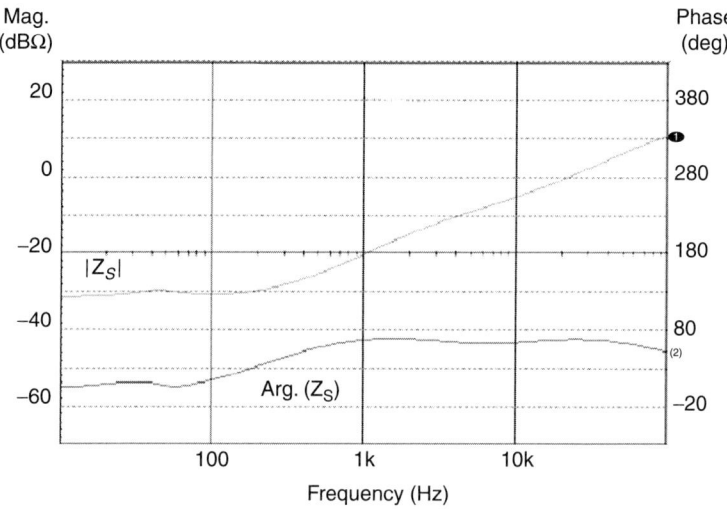

FIGURE 14.6 Results of the simulation of the equivalent network for differential measurements.

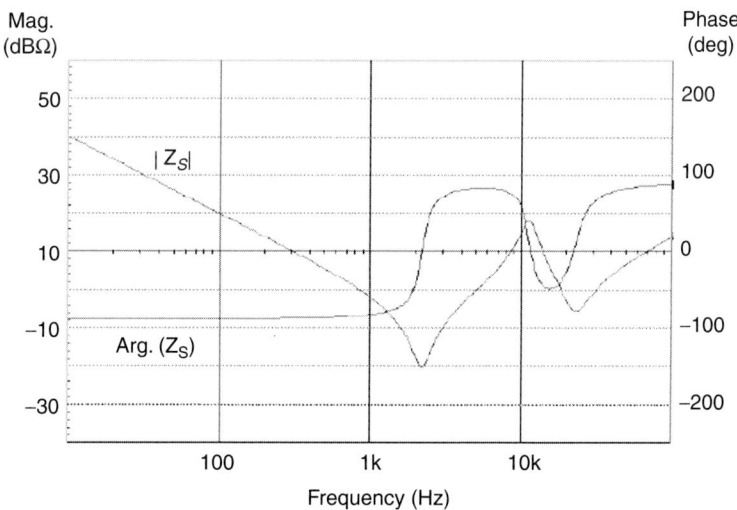

FIGURE 14.7 Results of the simulation of the equivalent network for common mode measurements.

A sequence of tests was planned to verify the preliminary compatibility of the source with boards containing one or two solid state power controllers, especially focusing on critical operational scenarios, such as induced faults. The test configuration is shown in Figure 14.8.

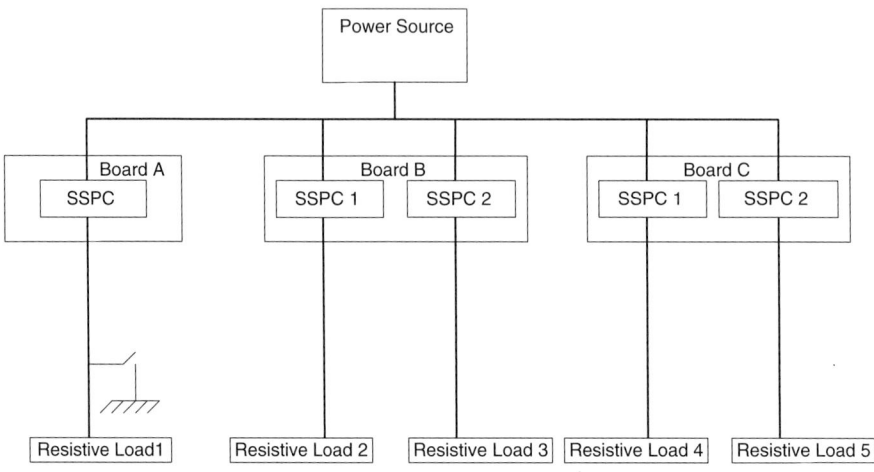

FIGURE 14.8 Description of the test configuration (SSPC = Solid State Power Controller).

Should a fault occur on any power line, only the affected solid-state power controller will trip, isolating the faulted channel (pass-fail criterion). All other channels should remain on. Faults induced during testing consisted of both hot to return and hot to chassis short circuits in various channels. During the hot to chassis fault on the A channel, all the solid state power controller hybrids tripped at once, shutting down the power to all the resistive loads in the test setup. Other induced faults, however, resulted in nominally expected behavior. This anomaly is typical of common mode transient phenomena.

It was postulated that the individual circuitry of the solid state power controllers was susceptible to a common triggering event. Subsequent measurements of the bus common mode transient voltage at various solid state power controller inputs while the hot to chassis fault was reproduced proved this assertion. Negative transient peaks as high as -140 V were detected between the bus return line and the chassis at the input to adjacent channels (Figure 14.9).

A common mode voltage transient is generated by the reaction of the inductance of the return power cable to a sudden current interruption as per Equation (14.5):

$$V_{cm} = L\frac{di}{dt} \tag{14.5}$$

Any common mode voltage transient is seen by the auxiliary circuitry of the solid state power controllers, which are connected in parallel on the power bus. This specific type of fault, which is associated with long cables, generates transient levels that exceed the susceptibility threshold of the auxiliary circuitry of the solid state power controller. When this type of fault occurs, the circuitry of the solid state power controller is switched off and the device transitions to OFF status.

A Tranzorb® is a combination of three groups of two unidirectional transient voltage suppressor diodes mounted back to back and in parallel. This ensures at least one failure

14.10 Design and Testing of Safety Critical Circuits

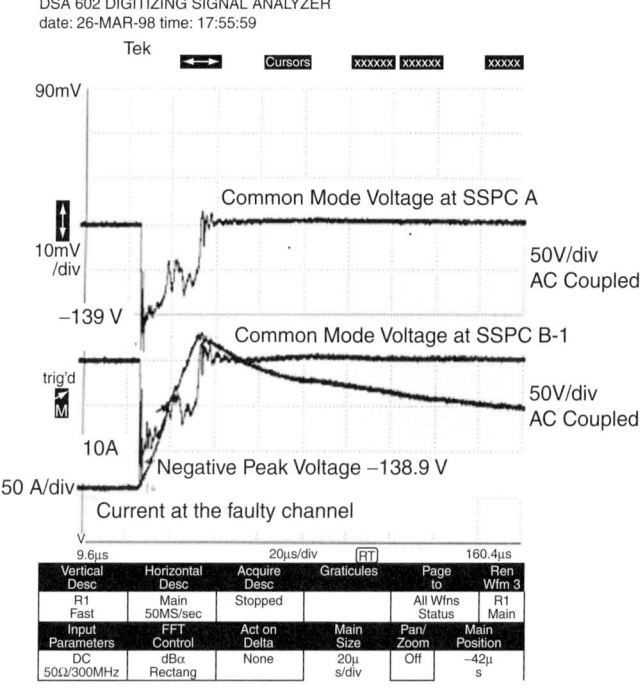

FIGURE 14.9 Common mode transient on two adjacent channels during the fault test (Courtesy of ESA).

tolerance. The Tranzorb clamping voltage is between 20 and 30 V, and the device can withstand very high currents for short periods. During a test campaign, a Tranzorb was inserted between the return line close to a solid state power controller and ground as a temporary fix. The fault again was applied to Channel A and the other channels. The problem did not manifest itself again; only the faulted channel tripped, while the adjacent channels remained on.

As the result of this test campaign, Tranzorbs were installed in both the breadboard and the flight hardware. Fault tests then were passed successfully. The clamping effect of the Tranzorb is shown in Figure 14.10.

In conclusion, the performance of the solid state power controller during a fault depends greatly on

- The nature of the overload.
- The inductance of the operational environment, both upstream and downstream.
- The interaction of the auxiliary circuitry with the transient phenomena associated with the fault.

Hence, design requirements must account for the expected system environment rather than simply relying on the rules considered general practice.

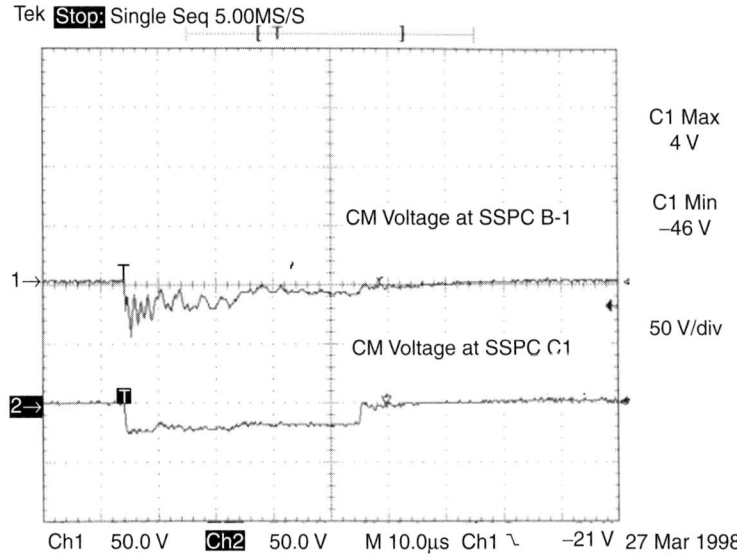

FIGURE 14.10 Clamping effect of a Tranzorb after insertion (Courtesy of ESA).

14.10.2 Safety Critical Circuits: Radiated Mode

Safety impairment due to radiated electromagnetic interference depends on either the purpose of the space system under consideration or the nature of the resulting effect. The classic example of detrimental consequences is the interaction of strong electromagnetic fields with spacecraft and rockets, causing unwanted ignition of electroexplosive devices. Further, critical communications, such as during search and rescue or the operation of hazardous payloads, can be disrupted by radiated electromagnetic interference, thus endangering the life of people who depend on these relevant services.

In general, radiated electromagnetic interaction within the pressurized environment of the crewed orbital infrastructure does not imply any hazard to personnel or crew, because standard design techniques and suitable frequency plans prevent electromagnetic interference from jeopardizing critical functions, such as ultrahigh frequency audio communication.

A realistic estimation of the human made noise intensity for a space vehicle operating at an altitude of 100 miles with orbit inclinations of 57° and 90° is given by Apirian and Baummer (2006). Electromagnetic fields as much as 200 V/m^2 for frequencies ranging from 2 to 3 GHz can be expected for that orbit. However, by scaling the field at the International Space Station orbit and considering the shielding effectiveness of the pressurized modules (say, 20 dB), the field penetrating the *International Space Station* can be considered not to be endangering any critical functions that concern crew safety. This has been confirmed by use.

The effect of electromagnetic interference on the self destruction subsystem of a rocket potentially is an element of concern for safety. However, the destruction signal is transmitted in spread spectrum mode and can be deciphered only by an ad hoc receiver. This virtually rules out any possibility of hazardous interference. Hence, the discussion on radiated electromagnetic interference hazard is limited to the interaction of strong field electromagnetic interference with electroexplosive devices and critical communication issues.

Radio Frequency Interaction with Electroexplosive Devices

The presence of strong human made electromagnetic field sources at launch sites or in other environments where the space vehicles or rockets must operate safely throughout their life cycle presents the problem of the electromagnetic interference control for electroexplosive devices. These potentially hazardous devices are used for several purposes, including rocket motor initiation, rocket stage separation, and solar array deployment in spacecraft.

The response of electroexplosive devices to intense radio frequency fields can lead to two unwanted effects. The first is activation of the electroexplosive device either by electromagnetic energy coupled directly into the device or by enabling the controlling electronics to send a firing signal to it. In this case, inadvertent electroexplosive device initiation can have catastrophic effects on safety, such as premature rocket motor ignition along with the associated consequences of causalities and property damage. Even if an electroexplosive device is activated unwittingly without immediate safety implications, it loses the capability of performing its intended function, resulting in severe reliability and mission success impacts.

A second effect is desensitization of the electroexplosive device when it repeatedly is subjected to energy induced by radio frequency fields below the minimum level required to fire the device. Repeated exposure to radio frequency fields of this nature changes the physical characteristics of the electrostatic discharge components. This phenomenon results mainly in degradation of performance, thus hampering the ability to intentionally activate the device. In principle, desensitization is not catastrophic per se, but it increases the probability for the occurrence of related hazardous events, should the inability to fire-disable properly some safety function. However, reliability rather than safety consequences usually are expected for this situation.

Factors Driving Electromagnetic Interactions with Electroexplosive Devices

Most electroexplosive devices use a small resistive element, called a *bridgewire*, embedded within the explosive charge to detonate the device. When an electroexplosive device is activated intentionally, a current is injected into the bridgewire, heating it until the explosive charge ignites. It is known that electromagnetic interference can force a false ignition command to the control circuitry for the electroexplosive device or directly heat the bridgewire. In either case, radio frequency coupling mostly occurs via the interconnecting harness and common impedance paths, such as cables. These common impedance pathways often are referred respectively as pin to pin, that is, differential mode, and pin to chassis, that is, common mode.

Assuming that the bridgewire resistance is typically about 1 Ω ±10% and the characteristic impedance of the interconnecting cables is around 50 Ω, at a first glance, one would expect that most radio frequency radiation is reflected at the bridgewire junction and therefore little radio frequency power passes through it. This would seem to prevent the device from detonating accidentally. In reality, electroexplosive devices tend to detonate when exposed to intense radio frequency fields, and simple analytical techniques, such as interaction of an electromagnetic field with a transmission line alone, can provide only a coarse estimate that safety is achieved. The basic parameters that should be used to provide a preliminary safety assessment are

- The physical layout of the electroexplosive device activation subsystem and equipment and their location in the intended installation.
- The detailed procedure used to fire the electroexplosive device.
- The electromagnetic environment that the system is expected to survive.
- The type of electroexplosive device, firing sensitivity, and bridgewire resistance.
- The electroexplosive device response time.

Characteristic Parameters of Electroexplosive Devices

Several types of electroexplosive devices are available, each having specific applications and characteristics. Firing sensitivity is a fundamental characteristic of an electroexplosive device. Typically, it is determined by the device manufacturer, although third parties can perform an independent assessment. Firing sensitivity is measured by examining the threshold of the ignition point when the electroexplosive device is energized with electrical stimuli. It is described in terms of maximum no-fire stimulus.

This parameter essentially provides a measure of the susceptibility of the device and thus a reference for which a safety margin (usually 20 dB) can be established. The preferred procedure is to establish the maximum no-fire stimulus statistically, by determining the greatest firing stimulus that does not cause activation within 300 s for more than 0.1% of all electroexplosive devices of a given design at a confidence level of 0.95. Alternatively, the 1-A/1-W rating method can be used, which states that the electroexplosive device should not explode within 300 s when a current of 1 A per bridge with an associated 1 W per bridge is applied. More details on this issue can be found in MIL-STD-1576.

The duration of this characterization implies that the response time of an electroexplosive device is a fundamental element in safety evaluation for these devices. The temperature response of a bridgewire to a current step can be approximated by an exponential function; that is, it evolves in time in proportion to $1 - e^{-t/\tau}$. The parameter τ is the thermal time constant of the bridgewire. Time constants are seldom determined as a standard practice, and they depend on the type of electroexplosive device being considered. Typical values for commonly used bridgewire range from 5 to 100 ms. However, electroexplosive devices of the carbon bridge type have a time constant shorter than 1 μs and therefore have a rapid temperature response to the current step.

Knowledge of the thermal time constant enables calculations for assessing responses to pulsed fields once the electroexplosive device response to a continuous wave field is known. This is evaluated by calculating the multiplying factor, MF, as in Equation (14.6):

$$MF = \frac{1 - e^{-\frac{T_2}{\tau}}}{1 - e^{-\frac{T_1}{\tau}}} \quad (14.6)$$

where T_1 is the pulse width, T_2 is the pulse interval, and τ is the thermal time constant.

A numerical example is given to show how to use this formula. If an electroexplosive device with a thermal time constant of 5 ms has a maximum no-fire power of 1 W continuous wave at the operating frequency of a pulsed field with pulse repetition frequency of 1 kHz and a 4% duty cycle, that is, $T_2 = 1$ ms and $T_1 = 50$ μs, then MF = 18.2. Therefore, the maximum no-fire level of the electroexplosive device for peak pulse power is 18.2 W. Similarly, if the electroexplosive device can withstand a continuous wave power density of 2 W/m², then it can tolerate 36.4 W/m² for that particular field. The calculation of equivalent voltages or current requires using the square root of the multiplying factor.

Radio Frequency Protective Methods Implemented on Electroexplosive Device Circuits

The firing system of an electroexplosive device consists of a power source, transmission line, and all switching devices, such as safe and arm switches that permit energy to be transferred intentionally to it. Various electromagnetic interference filters, shielding enclosures, means of circuit isolation, harness segregations, and any combination of these can be used to protect the electroexplosive device firing system from unwanted activation. Typical design provisions are imposed by the following requirements:

- Electroexplosive device circuits must be isolated from other electrical circuits and each other by a minimum resistance of 1 MΩ.
- Each firing circuit must be returned to the firing power source and should be isolated from ground by a minimum resistance of 100 kΩ.
- In the firing circuit, static bleed resistors of 100 kΩ must be placed from line to line and line to ground for electrostatic protection.
- Ignition lines, that is, those downstream of electrical safety barriers must be twisted and shielded; the shields should be connected to ground at each end.
- The firing circuit harness must have dedicated connectors that should not be used by other lines.
- Shields terminating at the connector must be connected electrically with no gaps around the full 360° circumference of the shield.
- When a filter is inserted into the system, it must be installed close to the initiator. The filter enclosure must be grounded to chassis and the harness between the filter and the initiator shielded.

- The electrical circuits of the firing system must be designed to limit the current induced on the ignition circuit to at least 20 dB below the maximum no-fire current when they are exposed to the maximum allowable electromagnetic field over the frequency range of 50 kHz to 40 GHz.

Margin Verification Techniques

MIL-STD-1576 provides guidance on the use of electroexplosive devices in orbiting spacecraft and their launch vehicles and describes methods to perform an analysis of acceptable margins based on the radio frequency threshold determination of the maximum no-fire stimulus. If a purely analytical, worst-case assessment indicates that radio frequency induced energy on electroexplosive device firing lines or in electronic circuits associated with safety critical functions is low enough to warrant the specified safety margin in the expected electromagnetic environment, a practical test, preferably on an instrumented electroexplosive device, should be conducted. However, in conventional space systems, this is seldom necessary.

Accurate simulation of the firing circuitry, including its geometry, cable enclosure, filters, and safeguard devices in principle can be established. However, this would be at the expense of a huge modeling effort for which uncertainties in the results remain and would be valid only up to certain frequencies. However, with good approximation, numerical methods allow detection of the resonant frequencies for which the system is most vulnerable. For this process, direct comparison of the results with the full spectrum of the intended electromagnetic environment leads one to decide whether the risks to safety are ruled out or additional investigations need to be undertaken.

Search and Rescue Payload

Some satellite telecommunications systems provide a number of critical services, such as reception of emergency transmissions in the ultrahigh frequency band via a search and rescue payload. The band, 406.0 to 406.1 MHz, is dedicated for use by only low power satellite emergency position indicating radio beacons (Earth to space). The protection requirement can be as low as -210 dBW/Hz onboard the spacecraft, a level so stringent that even a very small signal generated by a satellite at that frequency can saturate the receiver and interfere with the service. Despite the very narrow bandwidth of interest, interference has been detected during the ground testing of a spacecraft, and a great deal of troubleshooting effort has been expended.

The radio frequency field radiated by a spacecraft is generated by currents originating from electronic equipment, subsystems, and systems being induced onto a structure that is not perfectly conductive. To prevent the occurrence of a disturbing current, it is necessary to exercise stringent control of the frequencies generated by the spacecraft equipment, such as clock and local oscillator harmonics, the rise time of switching mode power supplies, and the enforced provision for harness shielding and grounding. The avoidance of nonlinear material that can generate intermodulation also should be strictly observed.

Testing on the ground is mandatory, and its successful accomplishment is a necessary condition for the spacecraft delivery. Although this phenomenon is essentially a matter of reliability, it becomes indirectly an element of safety for potential users.

14.11 ELECTRICAL HAZARDS

14.11.1 Introduction

The presence of electrical equipment and devices in modern spacecraft systems implies potential electrical hazards as soon as there is any interaction with crew or personnel. This is the case for on-the-ground integration and testing of uncrewed as well as crewed spacecraft, which also pose these hazards during the utilization phase. In this section, design requirements to prevent crewmembers or personnel from encountering electrical hazards are discussed and related design considerations presented. It should be noted that the protection of hardware from electrical hazards is a different topic and is not covered here.

14.11.2 Electrical Shock

An electrical shock occurs when any human comes directly into contact with an energized circuit in a way that allows a current to flow through the body. To enable the flow of current, it is necessary to have a second connection to the body to return the flow of current to the energy source.

On the ground, power networks usually are referenced to Earth. If a person in contact with a live circuit is standing on a floor that possesses a certain resistance to Earth, the second or return current path is established. In space, power supply systems are referenced to the spacecraft structure. Here, the condition for a return path is that the crewmember is in contact with the spacecraft structure. This can be via handrails, footrests, and the like. The current that flows through the body in case of an electrical shock (Figure 14.11) is dependent on various factors:

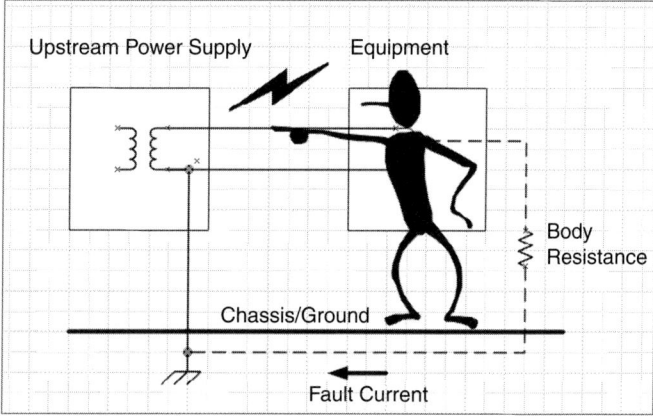

FIGURE 14.11 Current path for electric shock.

- Impedance of the power source.
- Type of power source, that is, AC or DC (AC current is waveform and frequency dependent).
- Resistance between power source and the point touched by the human.
- Skin resistance at the entry location, which itself is dependent on physiological factors such as skin humidity and thus it differs from one person to another.
- Body resistance, which mainly is dependent on the path length for the current flowing through the human body.
- Skin resistance and other transition resistances at the exit location, such as, for the feet, the additional resistance of socks and shoes.
- Resistance of the return path to the power source.

The current path is established more easily for AC, because even for the case of galvanic insulation, any capacitances present can represent an impedance small enough to allow a dangerous current to flow through the body. As an example, socks and shoes with rubber soles are considered to provide good insulation for DC; however, depending on the thickness of the soles, their AC impedance can be small enough to allow harmful currents to flow.

The reaction of the human body to an electrical shock depends on the voltage and the type of current flowing through the body and ranges from perception to muscular inhibition. In the worst-case, it can lead to ventricular fibrillation and death.

14.11.3 Physiological Considerations

The resistance of the body, including the resistance of the skin, determines to a major degree the amount of the current that can flow through the human body. Skin resistance depends on several factors:

- Consistency of the skin, that is, whether it is dry or moist.
- Contact pressure; that is, increasing pressure reduces skin resistance.
- Contact area; that is, the larger is the area, the lower the skin resistance.

It also has been reported that cuts or punctures can reduce skin resistance to values for wet skin or even below. Because of its variable nature, a conservative and safe approach is to consider skin resistance negligible.

Body impedance is an important factor, but it appears to be nonlinear with voltage. In addition, it depends on the physics and constitution of the considered person, that is, strong musculature creates less resistance than weak musculature, whereas long body parts create more resistance than short ones. Therefore, only approximate (statistical) values can be used.

The major factor in the determination of internal body resistance is attributed to joints, for which approximate values of resistance reasonably are established (Table 14.3):

- Body resistance from hand to hand:

$$2 \times (R_{\text{wrist}} + R_{\text{elbow}} + R_{\text{shoulder}}) = 1000 \; \Omega$$

- Body resistance from hand to foot:

$$R_{wrist} + R_{elbow} + R_{shoulder} + R_{torso} + R_{hip} + R_{knee} + R_{ankle} = 1000\ \Omega$$

Table 14.3 Typical Resistance Values for the Joints of the Human Body

Body Part	Resistance
Wrist	250 Ω
Elbow	150 Ω
Shoulder	100 Ω
Neck	50 Ω
Torso	100 Ω
Hip	50 Ω
Knee	100 Ω
Ankle	250 Ω

Note: As a general rule of thumb, the resistance of the human body can be considered to be approximately 1000 Ω.

14.11.4 Electrical Hazard Classification

The physiological effects of electric current flowing through the human body follow in the order of current level and severity of effect (from low to high):

- Perception: a tickling sensation that does not hurt and causes no harm.
- Surprise: a stronger feeling in which one actually is surprised, startled, or perplexed, but still causing neither harm nor pain.
- Reflex action: an immediate, subconscious reflex usually resulting in a movement in a direction away from the energized surface that has been touched.
- Muscular inhibition: the inability to move the hand or arm away from the energized surface without external help because control of the muscles is blocked, and the muscles are contracted strongly.
- Respiratory arrest: paralysis of the intercostal muscles and diaphragm and inhibition of the central respiratory centers.
- Ventricular fibrillation (and usually death): disruption of the cardiac conductive cycle causing uncoordinated, erratic contractions of the heart that completely are ineffective.

To classify electrical hazards into the two main categories, critical and catastrophic, and to define a threshold, a clear criterion is needed. By taking into account that neither respiratory arrest nor ventricular fibrillation necessarily leads to a fatal injury if the current flow is stopped quickly, it is clear that the major criterion for hazard classification is whether the shock victim is able to retreat from the source autonomously, interrupting the flow of current through the body. For this case, the maximum amount of current that can be permitted to flow through the body is defined as the let-go current, which is the current threshold above which a person is unable to release his or her grip on an electrically energized surface because of involuntary muscle contractions (NASA 1995). This value is used to classify hazards as catastrophic.

Keeping in mind the previous physiological considerations, it is clear that let-go threshold values can be defined only on a statistical basis. Usually the 99.5 percentile ranking is used to specify limits for the let-go current. In NASA-STD-3000 (NASA 1995), these limits are provided for men and women, and they are stated for DC, sinusoidal AC for different frequencies, and complex waveform AC (sine with various DC offsets and half and full wave rectified). The important values are

- DC: 60 mA for men, 40 mA for women.
- AC at the power line frequency: 9 mA rms for men, 6 mA rms for women.

14.11.5 Leakage Current

In AC powered equipment or equipment that contains AC signals, so-called leakage paths are formed by capacitances between the AC circuitry and the equipment housing. These capacitances can be physical capacitors, such as components of filters, or capacitances formed by the proximity of insulated conductors to the housing. For AC, these capacitances present a value of impedance depending on the frequency. Leakage currents along these leakage paths can flow back to the source through contact with the human body. Therefore, the let-go limits for AC currents apply.

14.11.6 Bioinstrumentation

Bioinstrumentation is used in space mainly to monitor the health status of the crew and for research purposes. With respect to electrical safety, bioinstrumentation forms a special case, because here it is necessary and intended to expose the human body to very small electrical currents that are typically far below 0.1 mA. The bioinstrumentation used in space is of two basic types: indwelling catheters and surface electrodes. For each, different limits for nominal, critical, and catastrophic currents are defined. These limits are frequency dependent. Table 14.4 shows the main values.

In most cases, bioinstrumentation is hardware that previously has been qualified for medical use on the ground. For space applications, it is assumed that the human interface to any bioinstrumentation used is safe, and the safety focus for qualification is merely on the hazard controls related to the galvanic insulation between it and the power supply system of the spacecraft. Typically, a minimum of three levels of galvanic insulation must be located between the power level in contact with the human body and the main power level.

Table 14.4 Current Limits for Bioinstrumentation (Courtesy of NASA)

Equipment Type	Limits/Frequency Band	
	DC $< f \leq$ 1 kHz	$f >$ 1 kHz
Indwelling catheters		
Nominal current limit (I_{nom})	10 µA	$I_{nom} = f/\text{kHz} \times 10 \text{ µA/kHz} \leq 1000 \text{ µA}$
Catastrophic current limit (I_{cat})	20 µA	$I_{cat} = f/\text{kHz} \times 20 \text{ µA/kHz} \leq 1000 \text{ µA}$
Surface electrodes		
Nominal current limit (I_{nom})	100 µA	$I_{nom} = f/\text{kHz} \times 100 \text{ µA/kHz} \leq 5000 \text{ µA}$
Critical current limit (I_{crt})	500 µA	$I_{crt} = f/\text{kHz} \times 500 \text{ µA/kHz} \leq 5000 \text{ µA}$
Catastrophic current limit (I_{cat})	1000 µA	$I_{cat} = f/\text{kHz} \times 1000 \text{ µA/kHz} \leq 5000 \text{ µA}$

14.11.7 Electrical Hazard Controls

All hazard controls to prevent electrical shock must be independent, that is, no single failure or event can disable the control and no single control failure can affect more than one control. Typical methods of implementing hazard controls follow:

- Safety green wire.
- Bonding.
- Insulation.
- Barriers to energized surfaces.
- Ground fault current interrupter.

Safety Green Wire

This method is well known from terrestrial applications connected to the AC power network. It is used in conjunction with ground fault current interrupter devices. The green wire is an integral part of the power cable and provides a connection to ground or Earth to the device. It usually is connected to the metallic housing of the equipment. Safety green wires are used mostly for portable equipment where no other bonding connection to ground is feasible. Should a fault from one power line (high side, positive, or phase for AC) to the equipment housing occur, the green wire provides a connection to ground. Because the power source or network is referenced to ground, a short path is established via the green wire back to the power source that

- Limits the voltage from which the housing can become energized.
- Causes the upstream fuse to blow.
- Causes the ground fault current interrupter, if one is present, to trip.

Because the equipment housing has a connection to ground, that is, the return line for the power source, it cannot be energized to voltages that would present an electrical hazard.

For this method to be efficient, it is of vital importance to keep the impedance of the entire path between the equipment and the power source via the green wire very low. At a minimum, the resistance of the green wire, including all connections toward the power source, needs to be low enough to carry the current necessary to trip the upstream fuse. In the case where a ground fault current interrupter is used as a second control, the green wire need not take the full current because the ground fault current interrupter will trip before the upstream fuse blows.

Bonding

Bonding is an alternative method to establish a reliable low impedance connection to chassis or ground. The functional principle is the same as discussed for the safety green wire. Bonding is also important for achieving electromagnetic compatibility by establishing a low impedance conductive path to chassis or structure.

Insulation

Here, *insulation* means that there must be no galvanic connection from a power circuit to any conductive surface or other power source. This can be achieved in two ways:

- Using batteries as independent power sources (usually for portable equipment).
- Using galvanic insulated DC to DC converters. Galvanic insulation is achieved by means of a transformer.

The principle employed for this type of control is that in no case can a current flow to or via chassis or structure be established should either the hot or the return wire of the insulated power source be touched by a human. The person even might exhibit a good connection to the chassis or ground, thus establishing a conductive path to chassis or ground through the body, but no current can flow because there is no return path to close the circuit (Figure 14.12).

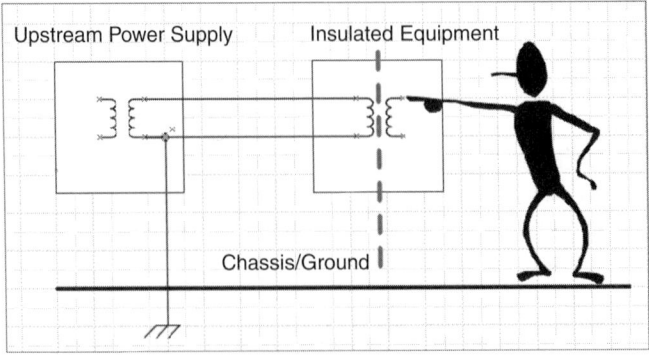

FIGURE 14.12 Insulation as an electrical hazard control.

In the case where insulated power is used within a conductive enclosure or housing, even a single insulation fault connecting one of the power terminals to the housing cannot create a hazard. This is because there is no return path to the power source, and therefore no current can flow through a person who is in some way connected to the chassis or ground that touches the case.

Barriers to Energized Surfaces

Barriers prevent a person from touching an energized surface or circuit. Barriers can be made of nonconductive, such as a plastic housing, or conductive materials, which in this case must be bonded to chassis or ground. From an electromagnetic compatibility perspective, a nonconductive case or housing is not the preferred solution because it offers no shielding effect. Nevertheless, for simple equipment that neither emits electromagnetic noise nor is sensitive to electromagnetic disturbances, an insulating case might be a good solution.

Barriers are used mainly where a sensor operating at hazardous high voltages (several sensor types require high voltages in the range of kilovolts) is used to monitor the environment outside the housing. In some cases, even very thin metallic or nonmetallic plates used to cover a sensor dramatically reduce its sensitivity or completely prevent it from functioning. For these, a mesh or grid can be used to prevent a person from touching the energized sensor surface. If made of conductive material, these barriers must be connected to chassis or ground and bonded. The main criterion for sizing a mesh or grid is that it must prevent a human finger from being able to touch the energized surface. This is achieved by sizing the grid so that the smallest finger cannot penetrate it. If a wider grid is required to ensure sensor performance, the grid must be installed at a distance longer than a finger from the energized surface.

Ground Fault Current Interrupters

The principle of a ground fault current interrupter is to switch off the power to the load as soon as the difference between the load current on the positive line and that on the return line exceeds a certain limit. This is usually in the range 8 to 15 mA. Should a fault current, that is, a current drawn by a device but not flowing back on the return line, occur but flowing back to the power source via the chassis or ground, the current difference between the positive and the return lines is detected by the ground fault current interrupter, and this trips the device (Figure 14.13).

A ground fault current interrupter can be used as a hazard control only when the device to be protected possesses a conductive housing bonded to a chassis or ground. This usually can be achieved by a safety green wire. As explained before, the requirements for the safety green wire are less stringent in terms of current carrying capability if a ground fault current interrupter is used, because currents above 15 mA trip the device, interrupting the power.

The main application for use of ground fault current interrupters in crewed spaceflight is in portable equipment that cannot be bonded directly to the spacecraft structure. Here, most electrical devices are equipped with DC to DC converters and have power input filters to reduce common mode noise. If powered via a ground fault current interrupter, currents filtering into the safety green wire have to be considered carefully.

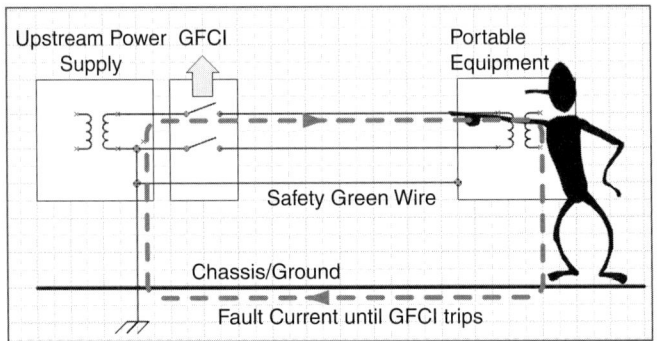

FIGURE 14.13 Ground fault current interrupter (GFCI) protection.

14.11.8 Verification of Electrical Hazard Controls

As part of the safety review process, verification of electrical hazard controls has to be confirmed. Depending on the type of hazard control employed, verification can be performed by review of design, inspection, or testing. Except for barriers, all electrical hazard controls normally are verified by test. This is not necessarily a test performed only for safety purposes; in most cases, tests performed in the framework of electrical interface verification, such as bonding or insulation measurements, can be used to verify electrical hazard controls. It is important that verification is considered part of a design; for example, inspection points must be accessible and test points can be foreseen to support testing and verification.

14.11.9 Electrical Safety Design Considerations

Failure Tolerant Design and Design for Minimum Risk

Two main safety design approaches are applicable to electrical design:

- Design for failure tolerance.
- Design for minimum risk.

Failure tolerance means that, depending on the particular hazard level, should a minimum number of credible failures or operator errors occur, the payload still must be safe. For a catastrophic hazard, a two-failure tolerant design is required; whereas for critical hazards, a single failure tolerant design is considered sufficient. On the other hand, design for minimum risk demands that special requirements be fulfilled to ensure that the control in place functions as intended. This can be considered to be an approach that reduces the number of credible failures to near zero. Compared to a failure tolerant design, this often-times demands more stringent requirements related to materials, parts, and processes. As well, the hazard controls employed are more critical and require extremely careful verification. A typical example is bonding, for which a failure tolerant design requires that a second, independent bonding path be implemented. This is not difficult to do, and it is easy to verify. When following a design for minimum risk approach, there is to be only one bonding path, but a number of requirements regarding contact area, surface cleanliness, absence of oxidation and anodic films, and use of materials must be fulfilled, and verification by testing is required.

Commercial Off-the-Shelf Hardware

Regarding the use of commercial off-the-shelf hardware, that is, electronic parts or equipment, for use in space, it is important to distinguish between the systems and equipment used for life support and maintenance of vital functions in a crewed spacecraft and the small equipment and payloads used for scientific experiments. The latter are more and more often assembled using commercial parts and components that are not space qualified. If the commercial off-the-shelf hardware does not impose a critical or catastrophic hazard and it can be ensured that no propagation of failures can occur toward other equipment and systems or result in hazards to the crew, from an electrical point of view there is no fundamental problem with its use.

It is a certainty that commercial off-the-shelf parts and equipment can possess a very high level of quality, such as parts produced in large series that are considered adequate for use in space. In case of bioinstrumentation, medical equipment qualified for public use in hospitals can be considered safe for use in space as well. In most cases, additional qualification and testing are necessary, such as electromagnetic interference and electromagnetic compatibility, vibration, offgassing, and thermal vacuum testing. In general, it can be stated that using commercial off-the-shelf hardware and parts for small experiments does not impose electrical hazards automatically. It is very important that, during the design phase, aspects of electrical safety and error propagation be taken diligently into account. For example, the addition of properly designed interface adapter circuits that provide insulation and protection can ensure that using equipment based on commercial off-the-shelf hardware is safe.

Nevertheless, it must be noted that a good quality level needs to be ensured during the design and manufacturing phases of commercial off-the-shelf based equipment. Equipment that has been developed in a very short time frame and costs very little but is found to be dead on arrival in space is of no use at all.

14.12 AVIONICS LESSONS LEARNED

Throughout the years of the U.S. Space Shuttle Program, several things have happened that provide insights of use to engineers. A collection of these insights follows.

14.12.1 Electronic Design

Off-Nominal Voltage Operation

Flight hardware generally is not subjected intentionally to off-nominal voltage. Such exposure can overstress the hardware; therefore, a procedure of this type would be permitted only for a qualification article. However, equipment powered from batteries almost is certain to be exposed to off-nominal low voltages as the battery becomes exhausted. Battery powered safety critical hardware is tested to determine its limits of safe operation and the characteristics of its operation as the supply voltage decays. The test results provide valuable insight into its operation during a mission.

- **Battery powered instrumentation.** A number of battery powered items of instrumentation have failed because of launch delays due to weather and from failure of other Space Shuttle systems that resulted in prolonged launch delays for corrective measures. Because no provision was made to deactivate the keep-alive power to this equipment, the battery charge became exhausted during the delays.

- **Sneak circuits.** In preparation for one particular mission, a quiescent vehicle structure had been anticipated by the testing program. However, the vehicle structure was found to have developed a voltage differential of 0.75 V, and resistance testing became erroneous and misleading. It was discovered that the anomaly was caused by the operation of an unrelated hydraulic pump.

- **On-orbit testing.** Susceptibility to false indications. As mentioned in the comments on sneak circuits, in-flight or even some ground testing can produce unexpected results. Time domain reflectometers are used widely in the telephone industry to determine the location of damaged wiring. However, their use on spacecraft has been problematic. The usefulness of these and other devices, such as digital ohmmeters, with their very high input impedances can be compromised severely by unexpected voltages when testing powered circuits.

- **Jam tolerant electronic controls.** Although not all hardware can be built to operate in every potential environment of space, sometimes equipment can be designed and built to exhibit a graceful shutdown and recovery if exposed to some hostile environment within which its operation was never intended. The user of commercial laptop computers in space often has to cope with lockup, causing them to be rebooted as many as 12 times in a day's operation.

14.12.2 Physical Design

Wire Bundling

Wire bundles can contain the power wires of a number of independent circuits. The proximity of these circuits to each other oftentimes makes it expedient to include these wires in the bundle when building a harness. Although sensitive critical circuits can be affected by this practice, electronic designers often do not specify the bundling of the wire harness external to the circuit cards or the electronic box. If critical circuits are involved, the designer should address the requirements for wire bundling and harness buildup external to the electronics as part of the design.

Schematics Versus Circuit Boards

The analysis of circuits normally is done based on schematics and not on the as-built circuit boards and wire harnesses. The design engineer must consider the probable or actual location of the circuitry in the final build when determining likely failure modes. Sometimes, a power source on the bottom of a schematic can be very close to circuitry at the top of the schematic when the circuit paths are established on the circuit board and internal wiring harnesses are fabricated. And, sometimes, electromagnetic field generators inadvertently are placed near susceptible components.

Limitation of Damage from Crew Movement in Zero-g

Due to actual movement during crew operations, the crew can impose push and pull loads beyond those anticipated during development of the hardware. This is especially true of commercial off-the-shelf hardware. Small signal cables can be exposed to loads in access of 100 lb. If so, the manufacturer is not likely to have test data for this situation. For this reason, design engineers need to test samples of the hardware to determine if it is robust enough for space operations.

Launch and Impact Loads: Co-Aligned Hardware

If the hardware contains relays or other mechanical devices, the independence of the hardware can be nullified when launch or docking loads are sufficient to overdrive the device. In such situations, reorienting the components during the design phase can avoid this susceptibility.

14.12.3 Materials and Sources

Limited Life Hardware

Equipment built and then stored for several years developed problems because of epoxy embrittlement. The use of epoxy as a mounting method for cable clamps and supports had been effective at the time of manufacture. However, when the equipment was removed from storage and prepared for flight, many cable supports were found to have failed. A major rework of the hardware was necessary before it could be flown.

Source of Relevant Experience for Robust Hardware Designs

Although much data have been generated by each of the space programs, often relevant knowledge and hardware experience resides in industries unrelated to space. Sometimes engineers in those fields are willing to share their sources of hardware and experience in building robust hardware. The following list describes fields forced to develop various types of robust hardware and software:

- Off road racing for durable electronics.
- Auto industry for electronics with margins for use in hostile environments.
- Military equipment for hostile environment and operator tolerant operation.
- Mountain climbing sports for safeguards against inadvertent operation, confusing displays, and operator fatigue.

14.12.4 Damage Avoidance

Shipping Damage

Sensitive electronic equipment is shipped to the launch facility or other locations where it can be exposed inadvertently to electrostatic discharge stress. This type of stress immediately is not obvious after the event. When electrostatic discharge sensitive equipment is to be shipped, the shipping and packing methods should be selected carefully; it should be anticipated that the people doing receiving and inventory control are not likely to be educated in the appropriate handling of electrostatic discharge sensitive equipment.

Damage by the Cleaning Crew

In many organizations, the cleaning crew likely is not trained in the proper handling of sensitive equipment. The storage of such equipment should provide protection during after hours when trained technicians are not present to ensure the equipment is handled properly.

14.12.5 System Aspects

Hazard Analysis Versus Failure Modes and Effects Analysis

Space hardware must undergo a safety analysis based on failure modes and their effect on the safety of the hardware and associated personnel. Hazard analyses are recommended instead of the much more time consuming and expensive failure modes and effects analysis. Hazard analysis is used to determine all potential hazards for the equipment and evaluate the components and circuitry to determine the hazard causes and the most appropriate means for their control.

Concurrent Engineering

Concurrent engineering is a term used in engineering organizations to describe a team activity. The team pulls members from the various aspects of the life cycle of the hardware or software. The diversity of team disciplines, such as design, fabrication, quality inspection, and testing, enriches team activities and often helps avoid mistakes.

Quality Inspections

Design engineers benefit from discussions with the quality inspection teams. The quality inspectors can often provide valuable insights into the limitations and challenges of building hardware of which the design team might be unaware, thus ensuring that the product is of high quality.

Repair Practices

If a technician accidentally or mistakenly damages hardware or software, under certain circumstances, it is permissible to correct the problem without starting over. Note that, if the normal practices of the technician can compromise the integrity of the design, this condition needs to be factored into the design or build instructions. Information of this type can be reviewed during concurrent engineering reviews.

REFERENCES

Various subsections of this chapter have few text citations, if any at all. The authors of these sections felt that text citations were unnecessary because the majority of the material presented is taken principally from various NASA safety documents, which are the accepted standards presently in use by NASA, JAXA, and ESA in the design and development of spaceflight hardware. References for these documents follow. Although RSA safety documents are somewhat different, the basic principles and practices included in them essentially are similar to those contained expressed by NASA documentation. Several of the subsections of this chapter provide a rather extensive Recommended Reading list at their end for perusal by the student. This material provides considerable insight into the realm of avionics safety, and the student seriously should consider studying this material.

American National Standards Institute. (1999) *ESD Association Standard for the Development of an Electrostatic Discharge Control Program for Protection of Electrical and Electronic Parts, Assemblies and Equipment (excluding Electrically Initiated Explosive Devices)*. ANSI/ESD S20.20. Rome, NY: Electrostatic Discharge Association.

Apirian, L., and P. Baummer. (2006) *Space Vehicle RF Environments*. DCA100-00-C-4012, and Contractor Report JSC-CR-06-070. Huntsville, AL: Department of Defense and National Aeronautics and Space Administration, Johnson Space Center.

Ciccolella, A., and T. Blancquaert. (2002) Power interface characterization between the Russian Service Module and the ESA automated transfer vehicle. Proceedings of the 6th European Space Power Conference. Oporto, Portugal: Institute of Electrical and Electronic Engineers.

Clark, T. L., M. B. McCollum, D. H. Trout, and K. Javor. (1995) *Space Flight Center Electromagnetic Compatibility Design and Interference Control (MEDIC) Handbook*. CDDF Final Report, Project No. 93-15. Reference Publication 1368. Huntsville, AL: National Aeronautics and Space Administration, Marshall Space Flight Center.

Department of Defense. (1964) *Bonding, Electrical, and Lightning Protection, for Aerospace Systems*. Technical Standard MIL-B-5087B. Washington, DC: Department of Defense.

Department of Defense. (1994) *Electrostatic Discharge Control Handbook for Protection of Electrical and Electronic Parts, Assemblies, and Equipment (excluding Electrically Initiated Explosive Devices)*. Technical Handbook MIL-HDBK-263. Washington, DC: Department of Defense.

Department of Defense. (1995) *Electrostatic Discharge Control Program for Protection of Electrical and Electronic Parts, Assemblies and Equipment (excluding Electrically Initiated Explosive Devices)*. Technical Standard MIL-STD-1686. Washington, DC: Department of Defense.

Department of Defense. (2004) *Test Methods and Procedures for Microelectronics*. Technical Handbook MIL-STD-883F. Washington, DC: Department of Defense.

Gaston, D. M. (1991) *Selection of Wires and Circuit Protective Devices for STS Orbiter Vehicle Payload Electrical Circuits*. Technical Memorandum TM-102179. Houston, TX: National Aeronautics and Space Administration, Johnson Space Center.

Lambert, H. C., Jr. (1993) *Protection of Payload Electrical Power Circuits*. NASA Interpretation Letter TA-92-038. Available at http://psrp-pub.jsc.nasa.gov (cited June 22, 2007).

National Aeronautics and Space Administration. (1989) *Safety Policy and Requirements for Payloads Using the Space Transportation System*. Space Shuttle Document NSTS-1700.7B. Houston, TX: National Aeronautics and Space Administration, Johnson Space Center.

National Aeronautics and Space Administration. (1991) *Flammability, Odor, Offgassing, and Compatibility Requirements and Test Procedures for Materials in Environments That Support Combustion*. NASA Handbook NHB 8060.1C. Houston, TX: National Aeronautics and Space Administration, Johnson Space Center.

National Aeronautics and Space Administration. (1995) *Man-Systems Integration Standards*, vol. 1. Standard STD-3000. Houston, TX: National Aeronautics and Space Administration, Johnson Space Center.

National Aeronautics and Space Administration. (2003a) *Teleprinter Cable Short Circuit*. IFA STS-28-V-11. Available at www.jsc.nasa.gov/news/columbia/anomaly/STS28.pdf (cited June 20, 2007).

National Aeronautics and Space Administration. (2003b) *AC1 Phase A Short*. IFA STS-93-V-01. Available atwww.jsc.nasa.gov/news/columbia/anomaly/STS93.pdf (cited June 20, 2007).

National Aeronautics and Space Administration. (2006) *Humidity Separator "B" Circuit Breakers Opened*. IFA STS-6-V-08. Available at www.jsc.nasa.gov/news/columbia/anomaly/STS6.pdf (cited June 20, 2007).

CHAPTER 15

Software System Safety

Professor Nancy G. Leveson
Aeronautics and Astronautics/Engineering Systems, Massachusetts Institute of Technology, Boston, Massachusetts

Kathryn Anne Weiss, Ph.D.
NASA Jet Propulsion Laboratory, California Institute of Technology, Flight Software and Data Systems Section, Pasadena, California

CONTENTS

15.1. Introduction	475
15.2. The Software Safety Problem	476
15.3. Current Practice	486
15.4. Best Practice	489
15.5. Summary	503

15.1 INTRODUCTION

Software is quickly becoming a major part of and a major concern in space applications. Whereas software always has played a role in the design and control of spacecraft, the functionality being assigned to software is quickly increasing; and conservative design, which minimizes the role and complexity of software components, is rapidly decreasing. Serious software errors have occurred during each of the recent NASA Mars missions, for example, with varying impacts depending on the phase of the mission involved. If the error does not have an immediately fatal effect and occurs during a time when the spacecraft can be put into a sleep or other safe state, e.g., *Pathfinder* and *Spirit/Opportunity*, there is often time to find the error and correct it before it causes damage. However, if the error occurs during orbit insertion or entry, descent, and landing, when such time does not exist, then the consequences are and have been disastrous for the spacecraft, for example the Mars *Polar Lander* (MPL) and the Mars *Climate Orbiter* (MCO).

This chapter describes the roots of the software safety problem, why they exist, and some approaches that can be used to mitigate them. The increasing number of incidents and losses related to software, despite great care in its development and testing, show the difficulty involved in spacecraft software engineering and the need for more attention and rigor.

The next section describes the software safety problem, and why it is special and needs to be treated differently than hardware. Then, the state of current practice in spacecraft software safety is described. Finally, best practices specifically related to software (and systems) are outlined.

15.2 THE SOFTWARE SAFETY PROBLEM

Although it might seem that spacecraft software can be treated from a safety perspective in the same way as the physical components, it is important to understand why software is such a special problem and needs additional attention. This section describes the unique aspects of software and the new safety problems that it introduces into spacecraft design and operations. Some past spacecraft losses related to software are used to illustrate the problems. For more information about the role of software in past spacecraft losses see Leveson (2004).

15.2.1 System Accidents

The introduction of embedded software to control physical systems has led to a new safety concern—system accidents. Traditionally, accidents have been treated as the result of single or multiple component failures, usually with an assumption that these failures exhibit some random behavior. Consequently, the logical approach to preventing *component failure accidents* is to provide redundancy or enhanced component integrity, thereby reducing the probability of component failure or the impact of the failure on the overall system.

With the proliferation of software control of physical systems and system components, a different type of accident is taking on increasing importance. In these accidents, labeled *system accidents*, losses arise from dysfunctional interactions among system components in which no components have failed. The MPL loss provides an example. In this case, it is believed that the software turned off the spacecraft descent engines prematurely, 40 m above the Mars surface (Euler, Jolly, and Curtis 2001). The software did not "fail" in the traditional sense; it satisfied all requirements as provided to the software designers. However, the designers were not expressly informed about the possibility of the landing leg sensors prematurely emitting signals before the spacecraft had reached the surface. Although this description is somewhat oversimplified, the example demonstrates the basic characteristics of a system accident where each component operates as specified, i.e., does not fail, but the combined behavior leads to a system loss.

System accidents arise in the interactions among components and are related to interactive complexity and tight coupling (Perrow 1984), both of which are usually increased when software is used to control spacecraft (or systems in general). Software allows the construction of systems with much greater complexity than permitted by analog components alone, which, ironically, is a major reason it is used. Whereas a simple system might be defined as having a small number of unknowns in its interactions within the system and with its environment, a system becomes *interactively complex* when the level of interactions reaches the point where they cannot be thoroughly anticipated, planned, understood, and guarded against. The MPL loss resulted from an unexpected interaction between the Hall effect sensors in the landing legs and the descent engine control software that was not handled correctly by the software logic (Jet Propulsion Laboratory Special Review Board 2000). Although system accidents can occur in systems without software, the interactions among components in these systems are usually simple and few enough that all the interactions can be understood, anticipated, and tested.

A second common factor in system accidents is tight coupling. A *tightly coupled* system is one that is highly interactive, with each part linked to many other parts. Failure or unplanned behavior in one can rapidly affect the status of others. Often the processes are time dependent and cannot wait, i.e., there is little slack in the system, and sequences are invariant so that there is only one way to achieve a goal. System accidents arise from unplanned interactions among the coupled components. Coupling creates an increased number of interfaces and potential interactions, which again raises the difficulty of handling them during test and operations.

There are, of course, many other types of complexity such as structural complexity (the structural decomposition is not consistent with the functional decomposition), nonlinear complexity (cause and effect are not related in an obvious way), and dynamic complexity (the system and environment change over time).

The potential for building systems with all these types of complexity is enhanced by using software and digital components and leads to systems that are intellectually unmanageable and hence to accidents where the cause is not a component failure but a design flaw. Intellectually unmanageable complexity is, of course, not required in system design; however, eliminating it requires discipline on the part of the system designer. This fact leads to the first basic solution for enhancing safety of software-intensive space applications—reduce the complexity of the system design. This solution, however, is not readily accepted by spacecraft designers, particularly because one of the reasons for introducing software is to allow functionality not easily achievable by using analog components alone.

15.2.2 The Power and Limitations of Abstraction from Physical Design

A second factor in software system safety is what might be termed the *curse of flexibility*. One reason software has become such an important part in the embedded control of physical systems is that computers provide a general purpose machine that can be changed into a special purpose machine simply by adding software, that is, adding the *design* of a special purpose machine. Thus, the same computer hardware might function as an autopilot when the design or logical functionality of an autopilot is loaded, or an engine controller when different logic is loaded. In addition, the design of the special purpose machine can easily be changed without retooling or manufacturing new hardware components. The power of this concept, i.e., that the design of a machine can be abstracted from its physical realization, has revolutionized engineering. Machines that in the past were physically impossible or impractical to build become feasible to create and use. The designer can concentrate on the logical steps to be realized without worrying about how those steps will be realized physically.

This power creates both the advantages and the disadvantages of using computers and is both its blessing and its curse. Software is so powerful and useful because it has eliminated many of the physical constraints of previous machines. Of course, the physical constraints of the components being controlled by software are not changed, but the physical constraints on the controllers have been for the most part eliminated. On the positive side, designers are no longer limited by the need to physically realize the functional logic of their designs, and thus much greater complexity is allowed. On the negative side, designers no longer have physical laws to limit the complexity of their designs: The physical constraints of analog systems enforced discipline on system design, construction, and modification and tended to limit the overall system complexity.

Complacency, misunderstood risks, and ill considered trades often lead to software that contains more functions than absolutely necessary or required. Both the *Ariane 5* (Lions et al. 1996) and *Titan IV/Milstar* (Pavlovich 1999) losses involved software that was not needed. The more features included in software and the greater the resulting complexity (both software and system complexity), the more difficult and expensive it is to test, provide assurance through reviews and analysis, maintain, and reuse in the future. Software functions are not free.

The flexibility afforded by computers has some other unfortunate side effects. Because software is easily changed, it often becomes the resting place of afterthoughts. *Feature creep* and late addition of functionality is common in software projects, and software often becomes the place where late discovered limitations or problems in the physical system design are rectified. Consequently, software complexity, even when originally carefully controlled, quickly increases.

In addition, the flexibility of software leads to a tendency to start creating code before the overall architecture has been carefully planned. This approach to creating software usually leads to unnecessary complexity and difficulty in debugging and getting the software correct, as well as making modification and maintenance very costly and difficult.

Specification and careful design are arguably more important for software than for hardware, not only because of the complexity of the software artifact being created, but also because software development is usually done by experts in software construction and engineering, rather than experts in the application. The control engineer, for example, usually provides requirements (of varying completeness and clarity) to the software engineer, who actually creates the detailed functional logic design. In essence, the software engineer is doing system design, but may not (and probably will not) be as expert in that design as the hardware or controls engineer. It is not surprising, therefore, that most errors in operational software can be traced to flawed requirements, not to flaws in the implementation of those requirements. Considering MPL again, the system requirement to not use the sensor touchdown data above 12 m over the planetary surface was never traced down and included in the software requirements. Both ambiguity and incompleteness in the requirements specification are common factors in software-related accidents. The tendency of software engineers to promote development methodologies that omit the requirements specification phase or reduce the attention paid to it simply exacerbate the problems (Martin 2002; Beck and Andres 2004).

A final consequence of software being a pure abstraction is that the failure modes are different than for hardware. Indeed, it is not clear that the term *failure* even applies. Software rarely just stops working, i.e., "breaks," but in most cases does exactly what it is told to do. The problems occur from operation of the software, not its nonoperation. In almost all software-related accidents, the software was doing exactly what the software engineers intended it to do: The problem arose not because the software failed to operate as designed, but rather that it did operate the way it was designed, i.e., from a system design error.

15.2.3 Reliability Versus Safety for Software

Reliability is important in spacecraft. Furthermore, for hardware components and component failure accidents, increasing component reliability is an important approach to reducing the probability of an accident and increasing safety. In system accidents and software-intensive systems, however, component reliability is only indirectly related to safety, and increasing component reliability is not adequate to prevent losses. Most system accidents involve "correct" operation of the individual components as discussed above. Even for some hardware components, reliability is not equivalent to safety, and increasing reliability can even decrease safety, such as increasing the burst-pressure to working-pressure ratio of a tank decreases tank failures but it may introduce new dangers of an explosion or chemical reaction in the event of a rupture. Increasing component reliability in ways that increase system complexity, as in some types of redundant design, can also reduce safety by increasing the potential for system accidents.

For hardware and software, safety has a broader scope than failures in that it includes dysfunctional interactions among system components. Safety also has a narrower scope in that failures may not compromise safety. In addition, accidents may be caused by equipment operation outside the parameters and time limits upon which the reliability analyses were based. Therefore, a system may have very high reliability and still have catastrophic accidents. The calculated MTBF (mean time between failures) for Chernobyl style nuclear power plants is 10,000 years. The Therac-25, a medical device that massively overdosed several patients because of software flaws, worked safely tens of thousands of times before the peculiar conditions arose that triggered the software flaw.

In addition, reliability engineering approaches to safety do not apply to software, although many engineers are vainly attempting to use them. In general, reliability approaches to increasing safety rely on preventing functional failures through redundancy, increasing component reliability, and reuse of designs and components. The following three subsections address these three reliability approaches to increasing safety and why they break down for software-intensive systems.

Redundancy

Redundant logic that contains the same design flaw will not increase safety, as exemplified by the *Ariane 5* loss. The maiden flight of the *Ariane 5* ended in failure when about 40 s after initiation of the flight sequence at an altitude of 2700 m, the launcher veered off its flight path, broke up, and exploded. The accident report described the primary cause as the complete loss of guidance and altitude information 37 s after the main engine ignition sequence when both the primary and backup inertial-reference system computers, which were running the same software, turned themselves off due to an unexpectedly large and unhandled horizontal velocity value read from the strapdown inertial guidance platform (Lions et al. 1996). The accident report makes it clear that the engineers did not understand the limitations of redundancy for preventing software failures. In fact, for system accidents that are caused by unhandled complexity, redundancy simply adds more potential interactions and thus increases interactive complexity. A NASA study of an experimental aircraft with two versions of the software control system found that all of the software problems occurring during flight-testing resulted from errors in the redundancy management system and not in the control software itself, which worked perfectly (Mackall 1998). The redundancy management system software was much more complicated than the control software.

In addition, the majority of software-related losses, as discussed above, are caused by requirements flaws, which will not be fixed by multiple implementations or multiple uses of one implementation from the same flawed requirements specification.

Redundancy provides little help even for requirements implementation errors. Software errors are not caused by random wear out failures—they are caused by misunderstandings or mistakes made by the developers of the software. The assumption that independently developed versions of the same software functionality will fail in a statistically independent manner has been shown repeatedly to be false in practice and in careful experimentation (for example, Knight and Leveson 1986). Humans tend to forget or mishandle the same unusual or difficult-to-handle inputs, resulting in common-cause software logic errors. Furthermore, multiple-version designs usually involve adding to system complexity (as noted in the NASA aircraft example above), which can result in failures itself.

In summary, any solutions that involve adding complexity, as redundancy inevitably does, will not solve problems that stem from intellectual unmanageability and interactive complexity. And assuming that designers make random mistakes contradicts almost everything known about human psychology and cognitive abilities.

Increasing Component (Software) Reliability

If specified (i.e., required) behavior is unsafe, increasing the likelihood of the software satisfying the specification, i.e., making it more reliable will not make it safer. Almost all software-related accidents can be traced to flaws in the requirements provided to the software engineers, not coding or implementation problems. Safety involves more than simply getting the software "correct."

As an example, consider a simple controller that issues a signal when pressure reaches a particular level. To know what the pressure is, the system design will probably include at least one pressure sensor, and probably more to account for possible sensor failures. If the signal emitted by the software controller is used by the system to increase safety, and failing to issue a signal is worse than issuing it at the wrong time, then the software logic should require that the signal be produced if any of the redundant sensors report that the threshold pressure has been reached. On the other hand, if the signal is safety reducing, and it would be disastrous if a signal was issued when not required, then safety concerns would probably lead to software logic that requires all three sensors to report that the threshold has been reached before issuing the signal. The point is that simply implementing the logic is not the issue; safety requires that the content of the logic be safe within the particular system context, and not just that it work as designed. In other words, safety is an emergent property of the system design and operation in a particular environment as opposed to a "correct" implementation.

Reuse

A common approach to increasing safety in hardware is to use standard components and component designs that have been thoroughly proven through extensive use in many different systems. Again, this approach does not work for software. In fact, a large percentage of past software-related spacecraft and other losses have involved reuse of software constructed for other systems. Examples include the *Ariane 5*, the *MCO*, and the *Titan IV/Centaur/Milstar* losses and the *SOHO* (*SOlar Heliospheric Observatory*) mission interruption (Lions et al. 1996; U.S. NASA 1998; Pavlovich 1999; Euler et al. 2001; Leveson 2004).

The belief that software that has operated safely in a different system will be safe in other systems is related to the confusion between reliability and safety for software. Software, as a controller, always includes assumptions about the controlled system and its environment. Accidents occur when those assumptions are incorrect. Reusing software in another system or environment often violates at least some of those assumptions. For example, the trajectory of the *Ariane 5* differed from the trajectory of the *Ariane 4* for which the attitude control software was originally developed. The software operated safely when the trajectory of the launcher remained in the envelope of the *Ariane 4* design for which it was originally developed, but a software design flaw was exposed when the trajectory of the *Ariane 5* exceeded that envelope.

The fact that software has been used safely in another environment provides no information about its safety in the current one. In fact, reused software is probably less safe because the original decisions about the required software behavior were made for a different system design and were therefore based on different environmental assumptions. Changing the environment in which the software operates makes all previous usage experience with the software irrelevant for determining safety.

This difficulty does not mean that software cannot be reused, but it does mean that reuse by itself will not only *not* increase safety but, in fact, can decrease safety unless great care and extensive effort are applied to ensuring that the software is safe in the new or modified system or environment. For some software, building a new version may be safer and less costly than trying to verify the safety of the reused software.

15.2.4 Inadequate System Engineering

With the increasing use of software in system designs, system engineering has been changed, although some system engineers are not fully aware of the changes in practice that are needed. Others are aware, but their training has not prepared them for this revolution in engineering. The implications of software and the risks associated with it are often underestimated or even completely ignored in early feasibility and conceptual design stages. There are two particular areas of concern in the system engineering of software-intensive systems:

- Incorporating software in early system trade studies.
- Software system safety analysis.

Software in Early Trade Studies

Currently, little emphasis is placed on the fact that software-related cost and risk are often two of the main cost and risk drivers in spacecraft system design. Traditionally, costing is usually the only, if any, analysis performed on software during the early concept development and trade study stages of spacecraft development. Furthermore, information produced by software costing techniques often is not provided to system engineers, or when information is provided, it often is not used by the system engineers as a data point for trade studies. Unfortunately, the lack of attention to software during these early development stages is a contributing factor in many of the cost and schedule overrun problems encountered during downstream development activities. These cost and schedule overruns force management to make cuts in certain engineering development efforts, often in requirements, documentation, and verification and validation, all of which are areas that have been highlighted throughout this chapter as important in ensuring the safety of the software system.

Currently, there are a variety of techniques that can be used to estimate the cost of software development. These software costing techniques rely heavily on expert opinion, previous company experience, historical data, estimated lines of code or function points, or a combination of these techniques, such as parametric models like COCOMO II. For large, unprecedented systems, which characterize most spacecraft, these techniques break down due to a lack of historical data and experience, as well as the unreliability of estimates based on lines of code or function points. Furthermore, the information required to generate the parameters required for these models is usually not available at the time system engineers are selecting an overall system architecture and need the cost and programmatic risk information. For some systems, particularly those in the aerospace industry, this problem is compounded because similar systems have never been built before and therefore historical information is lacking.

Consequently, the impact of the system architecture on the cost and risk associated with developing software to support that architecture is not taken into account. As previously mentioned, the fact that software does not play an integral role in early system trades may lead system engineers to allocate functionality to software that might be more easily or safely implemented in hardware or operations.

To ensure that software is taken into consideration during early trade studies and that unforeseen cost and schedule overruns are avoided, thereby ensuring proper resource allocation to requirements and documentation, software-related cost and risk must become a data point in the large trade space of early spacecraft design. Instead of determining absolute software cost or risk, which is virtually impossible for spacecraft software, relative cost and risk rankings of the software required to support the various system architectures in the trade space can be utilized to ensure software is adequately considered. Weiss, Leveson, and Francis (2007) provide one example of how software can be included in these early design stages. In this approach, various parameters are used to assign values to cost and risk. The cost and risk values are summed to obtain total cost and risk rankings for the various evaluated system architectures. The system architectures are then ranked relative to one another and with respect to a nominal case, or average, system architecture. Finally, discounts are applied to development cost and risk based on how the development timeline allows for reuse.

Software System Safety Analysis

Software is not by itself unsafe. It is an abstraction and abstractions cannot inflict damage. Software only comes into the safety picture when it is controlling potentially unsafe physical processes. Therefore, the term *software safety* actually makes no sense, and system safety engineers usually qualify the term, as in the title of this chapter, as software system safety, i.e., the implications of software on system safety.

Indeed, although some hardware components can themselves be unsafe, such as they may have sharp edges or may be capable of delivering an electrical shock, for the most part safety is an emergent system property. That is, it emerges when the components interact together as a system within a particular environment. As such, it can only be analyzed in the context of the whole, and system safety is properly a top-down system engineering activity.

Most software-related spacecraft losses have involved inadequate system engineering, including poor specification practices, unnecessary complexity and software functionality, software reuse without appropriate system safety analysis, and violation of basic system safety engineering design practices in the system and the software (Leveson 2004).

Overconfidence in and complacency about software is rampant in the spacecraft realm; spacecraft engineers often underestimate the difficulty of software engineering. The *Ariane 5* accident report notes that the software was assumed to be correct until it was shown to be faulty (Lions et al. 1996). Before the *Titan/Centaur IV/Milstar* loss, resources were directed elsewhere because the software was thought to be mature, stable, and had not experienced problems in the past (Pavlovich 1999).

Complacency can arise from a misunderstanding of the risks associated with software. Most accident reports (and indeed many engineering textbooks) emphasize failures as the cause of accidents and redundancy as the solution. The *Ariane 5* accident report, for example, notes that according to the culture of the Ariane program, only random failures were addressed, and these failures were primarily handled with redundancy. Not only are software design errors not random, but also the complexity of most software precludes examining all the ways it can misbehave.

Finally, software is often treated as just another system component, and software development is assigned to an isolated group with little visibility at the system level into the software development process. The MPL accident report states that the Jet Propulsion Laboratory (JPL) technical experts had little oversight or involvement with the software subsystem development at the contractor and suggests that authority, accountability, and lack of participation of the right people at meetings and reviews were significant factors in the loss (JPL Review Board 2000). Roles and responsibilities were not clearly allocated, resulting in the reviews not uncovering mission-critical issues and concerns early in the program. System engineering participation in and oversight of the software development process is particularly important as software usually controls the interactions among system components and, in essence, encompasses the overall system design.

15.2.5 Characteristics of Embedded Software

Another factor that makes software engineering for spacecraft applications difficult is the fact that unlike most software applications, spacecraft software is not the end deliverable. The spacecraft itself is the end deliverable, and the software is merely embedded within the spacecraft as one medium for implementing functionality. The difference between software as an end deliverable and software as an implementation mechanism has significant impact on how that software is engineered. System and software engineers must take into account the characteristics of the system and the system environment in which the software is embedded.

The spacecraft and space environments impose several constraints on the form that spacecraft software takes. As described in Dvorak et al. (1999), resource limited robots such as unmanned spacecraft and Mars rovers have only enough resources, plus a small margin, to carry out their mission objectives due to several reasons, including the cost of launching mass into space and the cost of radiation-hardened avionics. Consequently, mission activities must be designed with these stringent limitations in mind: Turning on a camera to take pictures uses power and storing those pictures consumes part of a very limited amount of non-volatile memory.

Not only are the computer's processing speed amount of memory significantly less on board spacecraft due to the aforementioned cost and weight constraints, but the physical side effects of these limitations are also nonnegligible and therefore must be managed. As examples, monitoring the effects of a heater on power consumption, the effect of temperature on a sensor measurement, and the health of a science instrument all become paramount. These functions are almost always allocated to implementation in software, further increasing its complexity.

Moreover, traditional spacecraft software architectures follow a similar pattern: The spacecraft is decomposed into a series of subsystems and from that subsystem into a series of components, until the actual hardware is reached, depicted in Figure 15.1. As stated in Dvorak et al. (1999),

> *The main problem in applying a device/subsystem architecture to resource-limited systems is that the architecture provides no leverage in dealing with the many nonnegligible inter-subsystem couplings. Each such coupling has to be handled as a*

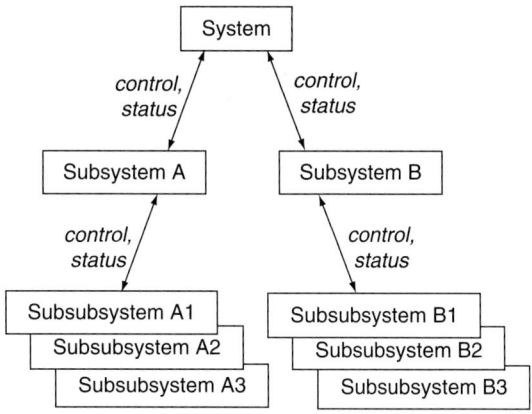

FIGURE 15.1 Subsystem-based decomposition of spacecraft control software.

special case, leading to a tangle of subsystem-to-subsystem interactions hidden behind the façade of modular decomposition. In effect, the original architecture becomes an appealing fiction.

This "tangle of subsystem-to-subsystem interactions" also occurs in software. In spacecraft, the subsystem interfaces are often handled through the command and data handling computer, which consists of computer hardware, or avionics, and software. The command and data handling computer is responsible for orchestrating the interactions between subsystems, estimating and controlling the spacecraft state, and providing fault detection, isolation and recovery for off nominal operations. The architecture of the software, therefore, has significant impact on the complexity inherent in that software. The impact of architecture on spacecraft software is discussed in greater detail in Section 15.4.6.

Finally, in terms of command and control, the real-time requirements of spacecraft software coupled with the asynchronous environment of space further increases the software complexity. Spacecraft must be commanded and controlled using the limited amount of available resources within the context of an unpredictable environment where violation of timing constraints can lead to a mission loss. Human intervention is minimal, especially for deep space missions in which the time delay between the Earth and the spacecraft precludes direct human control. Consequently, closed-loop, onboard control becomes a necessity. However, it is impossible to determine all the possible interactions that the spacecraft can have in the space environment and, moreover, that the software embedded in that spacecraft will have within that context. Any use cases or scenarios that software engineers utilize to help elucidate requirements, design and test modules, and perform verification and validation activities become a trivial set of possible operational scenarios in the space environment. New techniques such as Model-Based Engineering (discussed in Section 15.4.5) and new control architectures (discussed in Section 15.4.6) are needed to deal with these special characteristics.

15.3 CURRENT PRACTICE

How is software safety handled today? The practices differ in Europe and the U.S. and between those who treat safety as a component reliability problem and those who treat it as an emergent system property.

First, it is important to note the dichotomy in the definition of *safety* throughout the aerospace industry, even within NASA itself. NASA centers and aerospace projects that focus on the human exploration of space tend to adopt the definition of safety used throughout the System Safety community, i.e., safety refers to a loss event, whether that loss refers to humans or equipment. The distinction used in this definition of *safety* to separate the loss of human life and the loss of equipment is the classification of the hazard that leads to the loss. In other words, a hazard that can lead to a loss of human life would have a much greater assigned severity than that of a hazard with lesser consequences. However, at the JPL and other organizations that focus on the robotic exploration of space, *safety* only refers to human safety. In these organizations, the safety team often focuses on assembly, test, and launch operations, where there are individuals who can be adversely impacted by the system entering a hazardous state. All other hazards fall under the category of mission critical, where the result of a hazardous state can lead to loss of equipment or loss of mission.

This section first describes the safety-related approaches and tools currently used in industry, including the SPICE (Software Process Improvement and Capability dEtermination) project, CMMI (Capability Maturity Model Integration), dependability/safety cases, and ARINC 653. Next a discussion of the implementation of System Safety as a discipline at NASA and the NASA Software Safety Standard is presented.

The SPICE project and CMMI are both approaches to standardizing process improvement for software. The SPICE project is an initiative to support the development of an international standard for software process assessment and is supported by the International Committee on Software Engineering Standards through its Working Group on Software Process Assessment (Dorling et al. 1995). The project has three principal goals:

1. To develop a working draft for a standard for software process assessment.
2. To conduct industry trials of the emerging standard.
3. To promote the technology transfer of software process assessment into the software industry worldwide.

CMMI is a process improvement approach from the Software Engineering Institute that identifies the essential elements of effective processes for both systems and software engineering. It is used to guide process improvement across a project, a division, or an entire organization (Software Engineering Institute 2007).

Both SPICE and CMMI focus on the standard evaluation of an organization with respect to its software engineering processes. The objective is to help ensure software quality through rigorous adherence to standard software process. Unfortunately, organizations often focus on the maturity levels themselves as outlined in SPICE and CMMI instead of their true process capability (Humphrey et al. 2007). This misplaced focus can result in high maturity ratings that do not lead to better organizational performance or product quality.

Another approach to safety analysis, used primarily in Europe, is called *safety cases*, often referred to in the United States as *dependability cases* by the Software Engineering Institute. In fact, safety cases are considered a subset of the overall dependability or quality case. A safety case is a demonstration of software product safety. The demonstration typically takes the form of a presentation of evidence with argumentation linking the evidence to the claim of software safety. Safety cases have a postdesign focus, which is the opposite of the system safety approach discussed below. The emphasis of a safety case is on proving that the current design is safe; therefore, considerable emphasis is placed on creating a proof of system safety whether the design is safe or not. In the system safety approach, safety is designed into the system from the beginning making the safety case trivial to provide.

Finally, there are several attempts at creating avionics and software standards, as well as reusable avionics interfaces, that enforce some safe design principles, the most notable of which is the ARINC 653 specification for system partitioning and scheduling. This standard is often required in safety- and mission-critical systems, particularly in the commercial aviation industry. ARINC 653 defines an Application Executive (APEX) for space and time partitioning that can be used wherever multiple applications need to share a single processor and memory in order to guarantee that one application cannot bring down another in the event of application failure. Each partition in an ARINC 653 system represents a separate application and makes use of memory space that is dedicated to it. Similarly, the APEX allots a dedicated time slice to each, thus creating time partitioning. Each ARINC 653 partition supports multitasking (ARINC 2003).

15.3.1 System Safety

As argued above, treating safety as a component reliability problem and, for software, attempting simply to get the software correct has limitations with respect to safety. Traditionally in NASA and the U.S. defense industry, safety has been treated as an emergent system property, and special approaches labeled *system safety* have been used. These same approaches apply to software, although the application to software is not always done well.

Although many of the basic concepts of system safety, such as anticipating hazards and accidents and building in safety, predate the post-World War II period, much of the early development of system safety as a separate discipline began with flight engineers immediately after the war and was developed into a mature discipline in the early ballistic missile programs of the 1950s and 1960s. It was developed in response to the same changes that are being seen more widely today, i.e., increased complexity and use of software and computers that led to accidents and potentially devastating near misses in very expensive and inherently dangerous systems.

The American space program was the second major application area to apply system safety approaches in a disciplined manner. After the AS 204 (*Apollo 1*) fire in 1967, NASA hired Jerome Lederer, the head of the Flight Safety Foundation, to head manned spaceflight safety, and later, all NASA safety activities. Through his leadership, an extensive program of system safety was established for space projects, much of it patterned after the Air Force and Department of Defense programs.

In contrast to the reliability engineering focus on preventing failures, system safety is concerned primarily with the management of hazards: their identification, evaluation, elimination, and control through analysis, design, and management procedures. As a subdiscipline of system engineering, it deals with systems as a whole rather than with subsystems or components. In system safety, safety is treated as an emergent property that arises at the system level when components are operating together. System safety not only considers the possibility of accidents related to component failures, but also the potential damage that could result from successful operation of the individual components.

Another unique feature of system safety is its inclusion of non-technical aspects of systems. Jerome Lederer wrote in 1968:

> *System safety covers the entire spectrum of risk management. It goes* beyond the hardware *and associated procedures to system safety* engineering. *It involves: attitudes and motivation of designers and production people, employee/management rapport, the relation of industrial associations among themselves and with government, human factors in supervision and quality control, documentation on the interfaces of industrial and public safety with design and operations, the interest and attitudes of top management, the effects of the legal system on accident investigations and exchange of information, the certification of critical workers, political considerations, resources, public sentiment and many other non-technical but vital influences on the attainment of an acceptable level of risk control. These non-technical aspects of system safety cannot be ignored.* (Lederer 1986)

While, like most everyone else, NASA first attempted to treat system safety for software and software intensive systems in the same way as it treated it in hardware systems, the agency has recognized the differences between software and hardware and has recently adopted a revised NASA Software Safety Standard that takes standard system safety practices and adapts them for software (NASA 2004).

The activities in the software safety standard start in the concept phase and prior to the start, or in the early stages, of the acquisition or planning for the software. These activities implement a systematic approach to software safety as an integral part of the project's overall system safety program, software development, and software assurance processes. The goal is to design safety into the software and maintain safety throughout the software and system life cycle.

The first step in the NASA standard is to perform system and software safety analyses to determine if the software is safety critical, and how it can impact the safety of the system. Software is classified as safety critical if it:

- Can cause or contribute to a system hazard.
- Provides control or mitigation for hazards.
- Controls safety-critical functions.
- Processes safety-critical commands or data.
- Detects and reports, or takes corrective action, if the system reaches a specific hazardous state.
- Mitigates damage if a hazard occurs.

- Resides on the same processor as safety-critical software (in which case methods of separating or partitioning the critical from the noncritical components may be required).
- Processes data or analyzes trends that lead directly to safety-related decisions.
- Provides full or partial verification or validation of safety-critical systems, including hardware or software subsystems.

System safety and software safety analyses are used to identify potentially hazardous software behavior, and specific software requirements and design constraints are imposed on the software to eliminate or mitigate any such hazardous behaviors, e.g., the software must not turn off the descent engines before the lander reaches the surface, or the software must ignore any inputs from the landing leg sensors while the lander is over 12 m above the planet's surface. The standard also specifies requirements for tracing the flow of safety requirements and constraints from the system hazards down to the software requirements and from there down to specific software design features that implement or resolve them and to test other verification activities.

This standard differs from some other approaches to software system safety by focusing (in specification, design, verification, and assurance) only on the potentially hazardous behavior of the software. The most popular alternative approaches identify safety-critical software and then put more effort into the standard software engineering activities used to implement and verify all the software requirements for those software components deemed safety critical, and not just the safety-critical behaviors. These alternatives, thus, take a reliability engineering approach to the problem.

System safety approaches to software safety (and the new NASA standard) also involve integrating the software system safety analysis with system safety analyses and activities to ensure that the software does not compromise any safety controls or processes and that it maintains the system in a safe state during all modes of operation. The activities continue throughout the operational use of the software to detect any unsafe behavior before it causes an accident (and to perform a root cause analysis if a loss or near miss does occur) and to ensure that routine upgrades and reconfigurations do not compromise safety. Change analysis is used to evaluate whether a proposed software or system change could invoke a hazardous state, affect a hazard control, increase the likelihood of a hazardous state, adversely affect safety-critical software, or change the safety criticality of an existing software element.

Procedures and methods for implementing the new standard are still being developed and validated, but now are starting to be used on NASA projects.

15.4 BEST PRACTICE

So far, this chapter has described what is currently done for software safety. This section outlines what the authors believe are the best practices in the domain of spacecraft software safety that should be followed to provide the greatest degree of safety possible and practical with current knowledge. These practices fall into the categories of managing software-intensive projects, system engineering practices, specifications, requirements analysis, model-based software engineering and safe reuse of software, architectures that support software safety, software design, human factors (human-computer interaction, software reviews, and verification and assurance. Whereas only a brief description can be provided here, more detailed information is available in Leveson (1995 and 2007). The practices described here are those that implement a system safety approach to software system safety rather than a reliability engineering approach.

15.4.1 Management of Software-Intensive, Safety-Critical Projects

In a classic system safety engineering program, the system hazard analyses necessary to identify risks are continually performed and those risks are communicated to all segments of the project team and institutional management. System trades are vigorously worked that eliminate or control the risks in order to maximize the likelihood of mission success, and the progress of the risk mitigation plans and trades are regularly communicated to project, program, and institutional management.

One of the benefits of using a system safety engineering process is simply that someone becomes responsible for ensuring that particular hazardous behaviors are eliminated if possible or their likelihood reduced and their effects mitigated in the design. Almost all attention during development is normally focused on what the system and software are supposed to do. System safety and software system safety engineers are responsible for ensuring that adequate attention is also paid to what the system and software are *not* supposed to do as well as verifying that hazardous behavior will not occur. It is this unique focus that has made the difference in systems where safety engineering successfully identified problems not found by the other engineering processes.

During a program or project planning phase, a number of policies, procedures, etc. need to be developed to identify how safety is to be handled. A system safety plan should be created to detail the safety goals and activities. In addition, a safety management structure should be created that includes assignment of authority, accountability, and responsibility for safety as well as communication channels for safety-related information flow between system engineerers, software developers, system component design engineers, management, and all other system safety stakeholders. The defense industry uses system safety and software system safety working groups very successfully to enhance communication between and within the various levels of the program or project structure. Someone within software engineering should be responsible for interacting with the system safety engineers and with ensuring that the software is safe.

System and software system hazard logs and hazard tracking systems, with links between them, need to be established with the goal of getting information from system safety engineers to software system designers (and vice versa) in a timely manner. System safety management also includes auditing and overseeing activity, reviewing, testing, and assurance procedures.

Researchers have found that the second most important factor in the success of any safety program (after top management concerns) is the quality of the hazard information system. Both collection of critical information, as well as dissemination of that information to the appropriate people for action, is required. After accidents, these activities are usually seen to have been haphazard at best for the projects involved. The most important problem usually is lack of dissemination of information about hazardous behavior to be avoided early enough in the design process to have an impact on the design. Later, during development and in operations, new hazards and concerns will be identified, and there must be a simple and non-onerous way for software (and other) engineers, as well as operations personnel, to raise safety concerns and get the information disseminated to the appropriate groups and questions answered in a timely fashion. It is common to find that a lack of discipline in reporting problems and insufficient follow up are important causal factors in accidents (Leveson 2004).

Establishing proper communication channels is critical. Most software-related spacecraft losses involve inadequate information dissemination between the software developers and the other engineers or between the developers and the operations team.

15.4.2 Basic System Safety Engineering Practices and Their Implications for Software Intensive Systems

Preventing system accidents falls into the province of system engineering—those building individual components have little control over events arising from dysfunctional interactions among system components. As systems become more complex, with much of that complexity being made possible by the use of software, system engineering plays an increasingly important role in the engineering effort. In turn, system engineers need to use modeling and analysis tools that can handle the complexity inherent in these systems. Considering only part of the system, because the tools are unable to handle the entire system, usually results in ineffective, or worse, misleading analysis results. Because most of the spacecraft systems include hardware, software, and humans (if only in mission control and launch operations), appropriate modeling and analysis methodologies must include all of these components.

Safety must be specified and designed into the system and the software from the beginning. It has been estimated that from 70 to 90% of the safety-related decisions in an engineering project are made during the early concept development stage (Leveson 1995). When hazard analyses are not performed, are done only after the fact (for example, as part of quality or mission assurance of a completed design), or are performed but the information is never integrated into the system design environment, they can have no effect on the most important safety-related decisions, and the safety effort is reduced to a cosmetic and perfunctory role.

Different activities are appropriate for the different system and software development stages. Performing a preliminary hazard analysis during concept development facilitates the consideration of risk in the early system trade studies (an example preliminary hazard analysis used in early system architecture trade studies for Project Constellation can be found in Dulac and Leveson (2005). The preliminary hazard analysis identifies the system hazards and provides some preliminary analysis of causality at a very high level. As the design process continues, the identified hazards are used to create designs that eliminate or control them. Hazard analysis can be applied to the design alternatives, if necessary, to determine if and how the system could get into a hazardous state. Once some basic architectural decisions have been made and functions are allocated to components, then the system hazards that cannot be resolved, i.e., eliminated or completely controlled, at the system level must be traced down to the system components (software, hardware, and human) and specified as requirements and design constraints for the component designers to take into account in their subsystem designs. For software, this means tracing the system safety requirements and design constraints to the software interfaces and software functions.

During system implementation, system engineers should be responsible for coordinating the component-level safety activities and reviewing the subsystem designs to ensure that hazards are being resolved and that those designs do not create new hazards. Any changes, either during development or operations, must be evaluated for their impact on system safety and the system hazards. Also, during operations, system engineering is responsible for incident and accident analysis, performance monitoring, and periodic safety audits to ensure that the original assumptions underlying the hazard analyses are correct and still hold.

The use of software in the system has two important implications for this standard system safety process. The first is the difficulty it introduces into the early risk assessment stage, usually implemented by creating a risk matrix that combines likelihood and severity of system hazards. Likelihood in software-intensive systems, however, is almost impossible to predict. Estimating likelihood is possible for physical systems in which particular hazardous behaviors have been observed over a large amount of time. But software is newly created for every system, and the likelihood of the software exhibiting a specific unsafe behavior is unknowable. In addition, as argued above, experience with the same software used in a different system provides no information about its safety in any new environment. In almost all practical cases, severity is usually enough to determine priority of treating hazards, which is the goal of generating the risk matrix at this stage in development, for complex systems. Alternatively, mitigation potential (Dulac and Leveson 2005) and other factors (Weiss 2006) can be used that are more relevant than likelihood to determine relative risk associated with system hazards. Although the software risk factors identified in this chapter are applied to software cost prediction, they are equally applicable to estimating safety risk.

The second impact of a software-intensive system design is that the traditional hazard analysis techniques developed for electromechanical systems, such as failure mode and effects analysis and fault tree analysis, based upon assumptions about how and why failures occur and assumptions about accidents occurring as a result of component failures, do not apply to software (and to human error). The contribution of software to accidents, as noted previously, is different than that of purely mechanical or electronic components. In particular, software does not fail in the sense assumed by these techniques. Therefore, the resulting analyses often are less than useful, e.g., a box is inserted in a fault tree that says "software failure" or "software error." Some attempts have been made to create software versions of these techniques, such as Software Failure Modes and Effects Analysis (SFMEA), but simply looking at all the possible failure modes of software and determining whether they are potentially safety critical is a hopeless task; there are millions of erroneous behaviors that software can exhibit. Such bottom-up, reliability-based engineering techniques need to be replaced or augmented with top-down approaches that work down from the system hazards, because these approaches consider dysfunctional interactions among components as well as component failures. A new hazard analysis technique called STPA (STAMP-Based Hazard Analysis), based on an expanded model of accident causality, has been developed to handle software-intensive systems (Leveson 2007).

15.4.3 Specifications

Some specification practices are vital in the development of safety-critical systems. Traceability and specification (documentation) of safety-related design decisions, design rationale, and other safety-related information such as assumptions about the environment are just a few examples of specification practices that have a significant impact on the design for safety process. These practices not only are needed during the original development process, but they are also necessary during operations to allow reuse and safety analysis of impact when changes are proposed. Assumptions related to safety, such as assumptions about how the humans will interact with or in the system, and assumptions about human behavior that are safety related, must also be documented so they can be checked during operations. Although this process may seem expensive, the cost savings when changes are needed and safety analysis is required will more than compensate for the up-front costs.

Lutz (1992) examined software errors uncovered during integration and system testing of the *Voyager* and *Galileo* spacecraft and concluded that the software errors identified as potentially hazardous tended to be produced by different error mechanisms than the non-safety-related software errors. She showed that for these two spacecraft, the safety-related software errors arose most commonly from discrepancies between the documented requirements specifications and the requirements needed for correct functioning of the system and from misunderstandings about the software's interface with the rest of the system. Specification and thorough review and analysis can help prevent these safety-critical errors.

Inadequate documentation and design rationale to allow effective review of design decisions is a very common problem in system and software specifications. The *Ariane 5* accident report mentions poor specification practices in several places and notes that the structure of the documentation obscured the ability to review the critical design decisions and their underlying rationale. The Ariane report recommends that justification documents be given the same attention as code and that techniques for keeping code and its justification consistent be improved. Intent specifications are one way to accomplish this goal in a cost effective way (Leveson 2000, 2007).

Software specifications often describe nominal behavior well but are very incomplete with respect to required software behavior under off nominal conditions and rarely describe what the software is not supposed to do. This problem is often exacerbated by standards and industry practices that forbid such negative requirements specifications. Most safety-related requirements and design constraints, however, are best described using negative statements.

Complete and understandable specifications are not only necessary for development, but they are critical for operations and the handoff between developers, maintainers, and operators. Software is some of the most complex engineering artifacts in existence and without appropriate specification cannot be produced and maintained without introducing errors. Just having specifications is not enough. These documents must be easily readable and reviewable in order to be reviewed in-depth by domain experts who might not be software engineering experts or familiar with arcane software specification languages.

15.4.4 Requirements Analysis

Most software-related losses are caused by flawed software requirements related to

- Incomplete or wrong assumptions either about the operation of the controlled system or the required operation of the computer.
- Unhandled, controlled system states and environmental conditions.

Therefore, merely trying to get the software "correct" (whatever that means) or to correctly implement the requirements will not necessarily make it safer. In fact, software may be highly reliable and correct (in terms of implementing its requirements) and still be unsafe because

- The software logic correctly implements the functional requirements, but the specified behavior is unsafe from a system perspective.
- The requirements do not specify some particular behavior required for system safety (incompleteness).
- The implemented software has unintended (and unsafe) behavior beyond what is specified in the requirements.

Consequently, requirements analysis and getting the requirements correct and complete with respect to safe and unsafe behavior is of paramount importance in providing software that does not contribute to accidents. Note that specifying complete requirements from all aspects is impractical and probably impossible. However, for safety, only the safety-related requirements and constraints need be complete. This validation process is performed by ensuring that the software safety requirements and constraints include all the safety-related system requirements and constraints. This match should be simple if, as outlined above, the software safety requirements and constraints are derived from those at the system level. To provide further assurance, additional validation activities can be used to examine the software to determine how its nominal performance, operational degradation, functional failure (including no or erroneous outputs), unintended function, and inadvertent function (proper function but at the wrong time or in the wrong order) could contribute to system hazards. Such analysis is difficult and requires sophisticated system modeling.

Additional assurance can be obtained by examining the software requirements for common types of incompleteness that have led to accidents in the past or violate basic engineering principles that are known to be related to unsafe behavior. A set of criteria detailing these types of incompleteness can be found in Leveson (1995).

15.4.5 Model-Based Software Engineering and Software Reuse

Earlier, the dangers of application software reuse were discussed. The conclusion should not be that software cannot be reused, but that a safety analysis of its operation in the new system context is mandatory. Testing alone is not adequate to accomplish this goal. For complex designs, the safety analysis required for safe reuse stretches the limits of current technology. For such analysis to be technically and financially feasible, the reused software must contain only the features necessary to perform critical functions, yet another reason to avoid unnecessary functionality.

Commercial off-the-shelf software is often constructed with as many features as possible to make it commercially useful in a variety of systems. Thus, there is tension between using commercial off-the-shelf software and being able to perform a safety analysis and have confidence in the safety of the system. This tension must be resolved in management decisions about acceptable risk in specific projects. Ignoring the potential safety issues associated with commercial off-the-shelf software can lead to accidents and losses that are greater than the additional cost of designing and building new components instead of buying them.

If the commercial off-the-shelf or reused software makes any assumptions about the computing environment, then these assumptions need to be checked, and software/system modeling and analysis used to determine that the software does not result in hazardous system behavior. If the commercial off-the-shelf software is simply software used to operate the internal functions of the computer (and not related to the application), such as operating systems and computer utilities, then it can usually be safely reused because the assumptions involved are only about the computing environment (the computer hardware) and not the application or controlled system environment.

One solution to the reuse dilemma is to reuse the earlier system engineering and software engineering efforts and artifacts rather than the code itself. Instead of component-based *software* engineering, where reuse occurs at the code level, an engineering team can perform component-based *system* engineering in which each component contains system engineering information, including generic specifications of the component's externally visible behavior as well as formal models of that behavior (Leveson and Weiss 2004). Using executable specification languages allows requirements specifications to be executed and analyzed before any code is ever written or any hardware implementation is completed. Changes in these components as well as the integration of multiple components can subsequently be documented easily and tested before design and implementation. Details specific to the spacecraft and its mission can be elided from the reusable specifications and then can be easily added to the generic components, making the specification suitable for the wide range of spacecraft applications.

Because coding is such a small part of the software engineering process, particularly for real-time embedded control software, and coding for these applications can even be partially automated from requirements specifications, the cost of repeating the coding step is insignificant compared to the potential cost of revalidating the reused code. In addition, safely changing code is easier when the changes are made at an earlier development phase. Finally, accidents involving computers are usually the result of flaws in the software requirements, not coding errors, as previously mentioned. Consequently, applying systems engineering principles during these early development stages aids engineers in recognizing software interconnections at every stage of the life cycle, including requirements specification, thereby increasing the quality of requirements specifications by uncovering problems early in the life cycle and also decreasing the cost of correcting these mistakes.

15.4.6 Software Architecture

The software architecture is arguably one of the most important artifacts produced during the software development effort. The ability of the software to meet functional requirements while adhering to stakeholder properties of concern, often referred to as *quality attributes* (one of which is safety) is largely decided by the time the software architecture has been formulated. Consequently, developing a software architecture that is appropriate given the requirements and characteristics of the system is paramount. This section discusses some of the best practices in software architecture with respect to ensuring system safety.

First, it is important to note that safety is a quality attribute, i.e., it is an indirectly measurable system property that must be designed into the system from the beginning. Often, quality attributes are difficult to quantify and characterize, and therefore nonfunctional, or quality attribute–related requirements, are overlooked or not given the appropriate amount of attention during project formulation phases in which the architecture is defined. Unfortunately, the ability of a system to meet quality attribute requirements is largely decided once the architecture has been chosen. Clearly, more emphasis and rigor needs to be placed on architecture definition and evaluation to ensure that quality attribute requirements are being addressed, and more specifically, that safety concerns are handled throughout the architecting process. There are two main ways in which safety can be addressed throughout the architecting process, each of which is discussed below:

- The use of appropriate architecture styles.
- The integration of fault protection into the architecture.

Given a set of high-level requirements, both functional and non-functional, architecture styles can be identified that will help in constructing an overall software architecture that caters to safety. Architectural styles are idiomatic patterns of system organization that are developed to reflect the value of specific organizational principles and structures for certain classes of software. Styles define a vocabulary of component and connector types and a set of constraints on how they can be combined. When selecting a certain architecture style with its corresponding engineering principles and materials for a given project, architects are guided by the problem to be solved, as well as the larger context in which the problem occurs. Consequently, certain architectural styles are more appropriate for use on a particular project than others; the styles either support or detract from achieving the requirements and quality attributes desired by the stakeholders (Weiss 2006).

The current architecture of most spacecraft is the subsystem-based decomposition described in Section 15.2.5. As previously stated, this architecture leads to unnecessary coupling and complexity, and it does not address the impacts of the physical space environment on the spacecraft. Another problem with the as-built, subsystem-based architecture of spacecraft is the command and control portion of the software; unmanned spacecraft follow a series of time-stamped commands, frequently uploaded in real-time from ground controllers. To assure that the mission objectives are achieved, the abilities of the spacecraft are limited to this predefined operational model. The occurrence of unpredictable events outside nominal variations is dealt with by high-level fault protection software, which is often added at the end of development as a separate module following the subsystem-based decomposition. This protection software may be inadequate if time or resources are constrained and recovery actions interfere with satisfying mission objectives. In these situations, the spacecraft enters a safe mode in which all systems are shut down except for those needed to communicate with Earth. The spacecraft then waits for instructions from the ground controller.

There are several important architectural principles that enhance an architecture's analyzability with respect to safety as defined in Weiss (2006), including reviewability, traceability, isolation of critical functions, and assurance of completeness and determinism. New architectures are being proposed (such as the Mission Data System, MDS) from JPL to address not only the general criteria for creating architectures that are analyzable with respect to safety, but also address the physical constraints of spacecraft, while reducing coupling and complexity (Ingham et al. 2005). MDS incorporates a broader and more detailed set of architectural principles for addressing these issues, such as migrating functionality from ground to flight, closed-loop control, real-time resource management, and integral fault protection (Dvorak et al. 1999).

15.4.7 Software Design

Figure 15.2 shows the system safety design precedence (Leveson 1995). As depicted in the figure, the highest precedence is accorded to eliminating hazards from the design entirely. If hazard elimination is not feasible, then (in order of precedence) the designer

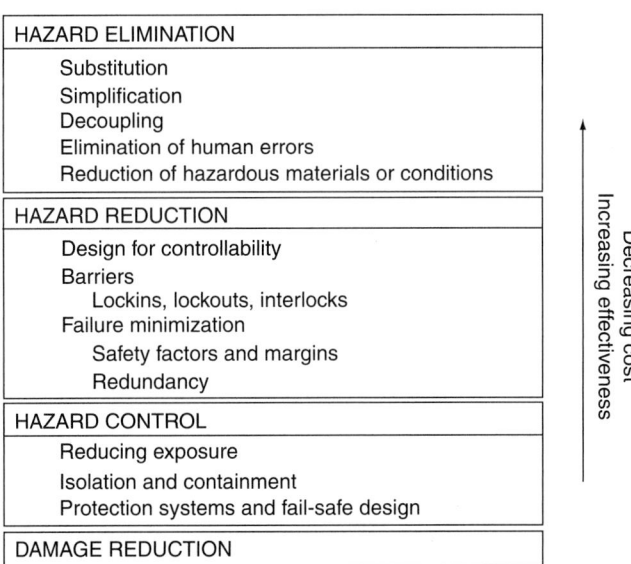

FIGURE 15.2 System safety design precedence.

should attempt to reduce the occurrence of hazards, control hazards if they occur, and finally reduce damage. Design does not necessarily involve only one of these; for example, efforts to reduce hazards may be unsuccessful. Therefore, the design should also include ways to control the hazards if the hazard does occur despite efforts to reduce their occurrence. If all else fails, damage reduction is a last resort.

Consider the Space Shuttle External Tank foam insulation impacting the orbiter thermal protection system as an example hazard. Eliminating this hazard could involve an alternative method for insulating the External Tank that does not utilize a material that can damage other parts of the spacecraft. Reducing the hazard may involve designing to reduce the likelihood of the insulation being shed. Controlling the hazard might involve ways of preventing the insulation from impacting the Space Shuttle (e.g., using barriers) or hardening the thermal insulation such that this type of impact does not create significant damage. Finally, damage reduction (if an impact does occur and the thermal tiles are significantly damaged) might involve an astronaut escape system. Note that the higher the precedence, the greater effectiveness and the lower the cost (including mission-related losses) involved.

Design of software for safety involves enforcing the software system safety requirements and design constraints identified by the system engineers and traced down to the software. Most of these requirements and constraints will involve standard software engineering techniques but will focus on safety-critical software behavior. Software frequently implements some of the system safety design features (described in the previous paragraph), and therefore the software engineers need to understand these features and the role the software is playing in their enforcement. In addition, many of the system safety design approaches are applicable to the software design as well as to the system hardware (Leveson 1995).

Starting from the highest precedence (see Figure 15.2), hazard elimination may be accomplished through substituting safe or safer materials or designs. If there are well understood and effective hardware designs that eliminate hazards through passive means, it may be safer to use these than to introduce an active software control system to control the device. There is no technological imperative to use computers to control dangerous devices when simpler and safer analog means are available.

Because system accidents are related to interactive complexity and coupling, hazards and accidents can potentially be eliminated through the use of simplification and decoupling in both the hardware and the software. Simplicity of design is unfortunately rarely taught in programming and software engineering classes and, indeed, some of the latest software design methodologies actually increase software complexity and functional coupling. In general, criteria for a simple software design include the following:

- The design is testable; that is, the number of software states is limited. Examples of how testability can be enhanced include using deterministic vs. nondeterministic designs, single tasking vs. multitasking, and polling vs. interrupts.
- The design is easily understood and readable from the code listing.
- Interactions between components are limited and straightforward, and functional coupling between components is minimized.
- The code includes only the minimum features and capability required by the system.
- Worst-case timing is determinable by looking at the code.

Unlike most hardware (although not all), it is easier to create a complex software design than a simple one: Constructing a simple software design requires discipline, creativity, restraint, and time.

The principles of decoupling are usually determined by the modularization of the software and which functions are assigned to the various modules. Other aspects of decoupling involve firewalls and barriers between critical and noncritical code, read-only and restricted-write memories, eliminating the effects of common computer hardware and sensor failures on the software behavior, and functional cohesion within modules.

Other aspects of eliminating hazards related to software include better human-computer interaction design (see Section 15.4.8) and using programming languages that reduce programming errors. Furthermore, the programming language itself should not only be simple (masterable), but it should also encourage the production of intellectually manageable and understandable programs. This requirement, plus the goal of functional decoupling (and other reasons), argue against the use of object-oriented design for safety-critical embedded software, although objects and abstract data types are obviously important and safe in the design of the individual software modules.

At a lower precedence, but still important, are techniques to prevent or reduce the likelihood of hazardous software behavior. The difference between software designed using the higher precedence design approaches and software using design approaches at lower precedence levels is that the lower precedence level approaches do not attempt to eliminate software design flaws, but rather to detect them and prevent any resulting hazardous software behavior. For example, using incremental control and various types of fallback and intermediate states allows recovery from some types of software design errors. Monitoring and insertion of checks into the software to detect errors are important, but it is often difficult, if not impossible, to make monitors independent. It is also difficult to write effective self-checking software, because the amount of checking is usually limited by time and memory constraints. System hazard analysis, and the resulting software safety requirements and constraints, can be used to determine which variables need to be protected and which checks are the most important to make.

A third category of software design techniques to reduce hazardous software behavior uses the standard system safety design approach of applying barriers, i.e., lockins, lockouts, and interlocks. For example, software synchronization techniques and critical sections are a way of implementing interlocks in software.

A lower level of hazard reduction design approaches is the use of failure minimization. Again, these approaches arise from hardware design and are much less effective, or do not apply at all, for software because of the different nature of software (described in Section 15.2). Safety factors and margins are used to cope with uncertainties in engineering that arise from inaccurate calculations or models; limitations in knowledge; and variations in the strength of a specific material due to differences in composition, manufacturing, assembly, handling, environment, or usage. These factors do not apply to software. The use of safety margins and safety factors is most appropriate for continuous and non-action systems.

Redundancy is a second way failures are reduced in hardware. As mentioned above, there are a very large number of both theoretical and experimental research results that show this approach has limited effectiveness in software. In addition, the costs are extremely high and usually cuts must be made in other parts of the software development and assurance efforts. There is no data to show that putting resources into software redundancy rather than other software engineering techniques provides better results and some data even show the results will be worse (Leveson 1995).

Finally, as repeatedly noted, most software-related accidents can be traced to flawed requirements and misunderstanding about the proper behavior of the software. Running multiple versions of software that does the wrong or dangerous thing does not increase safety. Software as a pure abstraction does not exhibit the random wear-out failures that redundancy in hardware protects against. There are much more effective and less costly approaches to protect against software design errors than redundancy.

The third level of system safety design precedence is to control hazards once they occur. Because the hazard in software is unsafe behavior, the protections here usually involve system designs that protect the controlled system against inadvertent unsafe software behavior (and not protections within the software itself). Standard system hazard control design approaches include limiting exposure of the hazardous system state, isolation and containment, and protection systems and fail-safe design. Some implications for software design are discussed in Leveson (1995).

15.4.8 Design of Human-Computer Interaction

One goal of introducing automation in control systems is to reduce human errors. Unfortunately, this new automation, while eliminating some types of human errors, particularly errors of omission, has at the same time created or contributed to new operator errors (Sarter and Woods 1995). The new errors, often errors of omission, are frequently more difficult to mitigate than the previous errors that were eliminated. There are other problems found in human-computer interaction, such as the human not getting appropriate feedback to monitor the system and the automation and to make safe decisions about necessary and safe human control actions. While human factors experts have focused most of their attention on design of the physical interface between the human and the machine and on trying to change human behavior and reduce errors through training and operational procedures, another necessary part of the solution must focus on the software design and the design of the functional interaction between humans and computers. Unfortunately, creating safe human-computer interaction has proven to be extremely difficult. Some principles are known, however, and can be used in the system design (Leveson 1995).

Figure 15.3 shows a human-machine interaction (HMI) design process that adapts the standard system safety engineering process to the design of human-computer interaction. The same hazard analyses and system safety engineering processes can provide the information necessary to enhance the design of HMI. In this process, the system hazards are traced down to the human system components, and requirements and constraints are specified on their behavior. Then the HMI is designed and tested with these requirements and constraints in mind and training procedures developed. Like the other parts of the system, operational feedback and monitoring needs to be established and audited during operational use of the system to ensure that the assumptions about human behavior used in the hazard analysis are still correct and that new hazards have not arisen as human behavior inevitably changes over time.

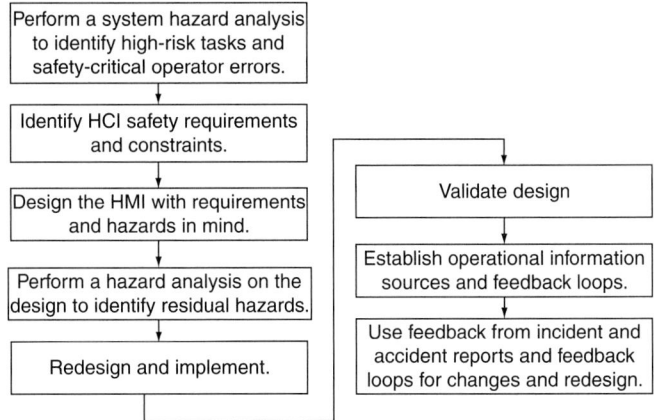

FIGURE 15.3 The human-machine interaction design process.

High tech automation is changing the cognitive demands on operators: They are often supervising rather than directly controlling; they are engaged in more cognitively complex decision making; they must interact with complicated, mode rich systems; and they are under increased demands for cooperation and communication with other humans and with the automation itself. Human-factors experts are complaining that engineers are creating technology-centered automation where they focus on the technical issues rather than supporting operator tasks. In the end, this leads to clumsy automation (Weiner and Curry 1980) and to confusing, error-prone interface designs and interactions. Many accidents and incidents in aircraft are now being blamed on mode confusion and other sophisticated design issues (Leveson 1995; Sarter and Woods 1995; Billings 1996). More attention needs to be paid to the human aspects of automation design.

15.4.9 Software Reviews

Because of its inherent complexity, software is difficult to review and the success of any such effort is greatly dependent on the quality of the specifications. However, identifying potentially unsafe software behavior, i.e., the things the software should not do, through the system hazard analysis and concentrating on that behavior for at least part of the review process, helps to focus the review and to ensure that critical issues are adequately considered.

The potentially unsafe or mission-critical software behavior should be identified in the system engineering hazard analysis process before software development begins and should be documented and identified as such in the specified software requirements and design constraints. This behavior then can be the focus of such a safety review.

As noted earlier, almost all software-related accidents have involved incomplete requirements specification and unhandled or mishandled system states or conditions. System and software hazard analysis and requirements analysis techniques and tools exist to assist in finding these types of specification incompleteness. To make such a review feasible, the requirements should include only the externally visible (blackbox) behavior. All implementation-specific information should be put into a separate software design specification, which can be subjected to a later software design review by a different set of reviewers. The only information relevant for a software requirements review is the software behavior that is visible outside the computer. Specifying only blackbox behavior allows the reviewers (usually including system and component engineers who do not specialize in software engineering) to concentrate on the information of importance to them without being overwhelmed by internal design information that has no impact on the required, externally observable behavior.

The language used to specify the software behavior is critical to the success of such a review. The best way to find errors in the software requirements is to include a wide range of disciplines and expertise in the review process. The reviewers must be able to read and understand the specifications without extensive training and, ideally, the notation should not differ significantly from standard engineering notations. While formal and executable specification languages have great potential for enhancing our ability to understand the implications of complex software behavior and to provide correct and complete requirements, most of the languages created by computer scientists require too much reviewer training to be practical for engineers to use them to find specification errors. A high priority on readability and learnability has not been placed on the development of such languages.

15.4.10 Verification and Assurance

It is not feasible to verify safety or assure safety, particularly for software; instead it must be built in from the beginning. But safety-related requirements and constraints must be thoroughly tested and all test results reviewed for potential safety issues. Any newly identified hazards must be traced back to the system level. Safety-related problems should never get to the test phase if the software engineering is done well, so finding something during test should trigger a root-cause analysis to determine why the flaw got through the software engineering process, how to prevent a reoccurrence, and if other potential flaws could also have gotten through.

Most software-related accidents have involved situations that were not considered during development or were assumed to be impossible and therefore not necessary for the software to handle. Thus, the likelihood that testing will find these errors is remote, especially given that testers usually use the same flawed assumptions and specifications as the engineers that designed and implemented the system. Testing of unexpected, off nominal, and stress situations (inputs that violate the specifications and operational assumptions) are more likely to detect safety problems than testing only expected use cases or inputs that satisfy the specification.

Whereas assurance is often augmented with probabilistic methods, such as probabilistic risk assessment (PRA), such analyses do not apply to a deterministic engineering abstraction such as software, which contains only design errors and does not fail in a random way. In fact, one could argue that it cannot fail at all in the standard engineering sense. In addition, even if one could get some measure of software reliability, such a general reliability measure is useless for a PRA as most of the errors (incorrect behavior) included in that measure are irrelevant to safety. The only quantity that would be applicable would be the probability the software will exhibit a *particular* unsafe behavior due to a software implementation error. In addition, most accidents (as repeatedly stressed throughout this chapter) result from requirements errors, not software implementation errors. Consequently, the relevant quantity is the probability that the software requirements are flawed in a safety-critical way, and again not just that they are flawed. Such probabilities are not obtainable. Indeed, if there was enough information to get these probabilities, then there would be enough information to fix the problem and measurement would be unnecessary.

15.4.11 Operations

During operations, all proposed software changes (upgrades and routine maintenance), must be thoroughly tested and analyzed for their potential effect on system safety and system hazards. Such change analysis will not be feasible or affordable unless special steps are taken during development to document the information needed.

The environment in which the system and software are operating also changes over time, partly as a result of the introduction of the automation or system itself. Basic assumptions made in the original hazard analysis process must be recorded and periodically evaluated to ensure they are not being violated in practice. Incident and accident analysis are also important, as well as performance monitoring and periodic operational process audits.

15.5 SUMMARY

Complacency, misunderstanding software and its risks, and inadequate system safety engineering programs applied to software have been at the root of most software-related spacecraft losses. Software presents tremendous potential for increasing our engineering capabilities. At the same time, it introduces new causal factors for accidents and requires changes in the techniques used to prevent the old factors. There is no magic in software—it requires hard work and is difficult and expensive to do well—but the results are worth the effort.

REFERENCES

ARINC. (2003) *ARINC 653 Avionics Application Standard Software Interface*. Available at www.arinc.com/cf/store/catalog_detail.cfm?item_id=632.

Beck, K., and C. Andres. (2004) *Extreme Programming Explained: Embrace Change*. Boston: Addison-Wesley.

Billings, C. E. (1996) *Aviation Automation: The Search for a Human Centered Approach*. Boca Raton, FL: CRC Press.

Dorling, A., H. Barker, J.-N. Drouin, M. Paulk, M. Konrad, D. Kitson, and T. Rout. (1995) *SPICE Software Process Assessment*, Parts 1–9. Vol. 1.00. Available at www.sqi.gu.edu.au/spice/suite.

Dulac, N., and N. Leveson. (2005) Incorporating safety into early system architecture trade studies. Conference of the System Safety Society. San Diego, CA: International System Safety Society.

Dvorak, D. (2000) Challenging encapsulation in the design of high risk control systems. 17th Annual ACM Conference on Object Oriented Programming, Systems, Languages, and Applications, Seattle, WA.

Dvorak, D., R. Rasmussen, G. Reeves, and S. A. Sacks. (1999) Software architecture themes in JPL's mission data system. AIAA-99-4553. Proceedings of the AIAA Guidance, Navigation, and Control Conference. New Orleans: American Institute of Aeronautics and Astronautics.

Euler, E. A., S. D. Jolly, and H. H. Curtis. (2001) The failures of the Mars climate orbiter and Mars Polar Lander: A perspective from the people involved. AAS 01-074. Proceedings of Guidance and Control. Pasadena, CA: American Astronautical Society.

Gomez, S. (2005) *Three Years of Global Positioning System Experience on International Space Station*. NASA Technical Report TM-2005-213715. Houston, TX: National Aeronautics and Space Administration, Johnson Space Center.

Humphrey, W. S., M. D. Konrad, J. W. Over, and W. C. Peterson. (2007) Future directions in process improvement. *CrossTalk, the Journal of Defense Software Engineering* 20, no. 2.

Ingham, M. D., R. D. Rasmussen, M. B. Bennett, and A. C. Moncada. (2005) Engineering complex embedded systems with state analysis and the mission data system. *Journal of Aerospace Computing, Information, and Communication* 2, no. 12: 507–536.

Jet Propulsion Laboratory Special Review Board (2000) *Report on the Loss of the Mars Polar Lander and Deep Space 2 Missions*. NASA Jet Propulsion Laboratory (March 22).

Knight, J., and N. Leveson. (1986) An experimental evaluation of the assumption of independence in multi-version programming. *IEEE Transactions on Software Engineering* SE-12, no. 1: 96–109.

Lederer, J. (1986) How far have we come? A look back at the leading edge of system safety eighteen years ago. *Hazard Prevention* (May/June): 8–10.

Leveson, N. G. (1995) *Safeware: System Safety and Computers*. Reading, MA: Addison-Wesley.

Leveson, N. G. (2000) Intent specifications: An approach to building human-centered specifications. *IEEE Transactions on Software Engineering* SE-26, no. 1: 15–35.

Leveson, N. G. (2004) The role of software in spacecraft accidents. *Journal of Spacecraft and Rockets* 41, no. 4.

Leveson, N. (2007) System Safety Engineering: Back to the Future (draft). Available at http://sunnyday.mit.edu/book2.html (accessed: July 12, 2007).

Leveson, N. G., and K. A. Weiss. (2004) Making embedded software reuse practical and safe. Proceedings of Foundations of Software Engineering. Toulouse, France: Foundations of Software Engineering.

Lions, J. L., L. L'Beck, J. L. Fauquembergue, G. Kahn, W. Kubbat, S. Levedag, L. Mazzini, D. M. Thomson, and C. O'Halloran. (1996) *Ariane 5 Flight 501 Failure: Report by the Inquiry Board*. Paris: Europen Space Agency, Centre National d'Etudes Spatiales.

Lutz, R. R. (1992) Analyzing software requirements errors in safety-critical embedded systems. Proceedings of the International Conference on Software Requirements, Baltimore, MD: Institute of Electrical and Electronic Engineers, pp. 53–65.

Mackall, D. A. (1988) *Development and Flight Test Experiences with a Flight-Crucial Digital Control System*. NASA Technical Paper TP-2857. National Aeronautics and Space Administration, Dryden Flight Research Center. Los Angeles County, CA.

Martin, R.C. (2002) *Agile Software Development*. Boston: Prentice Hall.

National Aeronautics and Space Administration. 1998. *SOHO Mission Interruption Joint NASA/ESA Investigation Board*. Washington, DC: European Space Agency and National Aeronautics and Space Administration.

National Aeronautics and Space Administration. 2004. *NASA Software Safety Standard*. NASA-STD-8719.13B. Washington, DC: National Aeronautics and Space Administration.

Pavlovich, J. G. (1999) *Final Report of the 30 April Titan IV B/Centaur TC-14/Milstar-3 (B-32) Space Launch Mishap*. United States Air Force, Space Command. Colorado Springs, CO.

Perrow, C, 1984. *Normal Accidents: Living with High-Risk Technologies*. Cambridge, MA: Basic Books.

Sarter, N., and D. Woods. 1995. Mode error and awareness in supervisory control. *Human Factors* 37: 5019.

Software Engineering Institute. (2007) *CMMI*. Available at www.sei.cmu.edu/cmmi (accessed July 12, 2007).

Weiss, K. A. (2006) Incorporating modern development techniques into the creation of large scale, spacecraft control software. Cambridge, MA: Massachusetts Institute of Technology, Ph.D. dissertation.

Weiss, K. A., N. Leveson, and J. Francis. (2007) Incorporating software cost and risk assessment into early system development trade studies. Proceedings of INCOSE 2007, San Diego, CA: International Council on Software Engineering.

Wiener, E. L. and R. E. Curry. (1980) *Flight Deck Automation: Promises and Problems*. NASA Technical Memorandum 81206, Moffett, CA: National Aeronautics and Space Administration, Ames Research Center.

Battery Safety

16

Judith A. Jeevarajan, Ph.D.
Senior Scientist, Power Systems Branch, Johnson Space Center, National Aeronautics and Space Administration, Houston, Texas

CONTENTS

16.1. Introduction	507
16.2. General Design and Safety Guidelines	508
16.3. Battery Types	508
16.4. Battery Models	509
16.5. Hazard and Toxicity Categorization	509
16.6. Battery Chemistry	509
16.7. Storage, Transportation, and Handling	544

16.1 INTRODUCTION

The basic unit of a battery is an electrochemical cell. A cell consists of four main components: the cathode or positive electrode, the anode or negative electrode, the separator that physically separates the two, and the electrolyte that provides the medium and facilitates ion conduction. There are many different types of electrochemical cells. The variation in their characteristics allows them to be used in different applications, generic or unique. A battery can be a single cell, a combination of cells in series or parallel, or both. For convenience, all single cell batteries are referred to as *cells*, and designs containing more than one is referred to as a *battery* throughout this chapter.

Typically, the quantity of electric charge is measured in coulombs, and the flow of electrons through a metal conductor or wire is stated in coulombs per second. A current, I, of 1 coulomb per second is equal to 1 ampere, A. It is well known that all materials that conduct electricity resist the flow of electrons, and the unit of this electrical resistance, R, is an ohm, Ω. The potential needed to cause 1 coulomb per second, that is, a current of 1 A, to flow through a conductor having a resistance of 1 Ω is 1 volt, V. In summary, current is directly proportional to voltage and inversely proportional to resistance. Ohm's law gives the relation between voltage, resistance, and current:

$$V = I \times R \tag{16.1}$$

One of the most common electrical measurements used is the watt, W, a unit of electrical power. Power, P, is the product of current and voltage:

$$P = I \times V \tag{16.2}$$

A few textbooks provide an in-depth treatment of the electrochemistry of cells and batteries (Bockris and Reddy 1970; Bard and Faulkner 2001; Linden and Reddy 2002).

16.2 GENERAL DESIGN AND SAFETY GUIDELINES

Cells and batteries need to be designed for performance and safety from the initial phases of any hardware development requiring stand-alone power. Although millions of batteries are available in the commercial market, not all are safe for use in a crewed space vehicle or on-orbit environment. Batteries are a source of voltage and current. If not handled carefully, they can be, at the very least, a source of electrical shock or, in many cases, cause burns to personnel and damage to equipment. Batteries are also a toxic hazard, with the toxicity level varying with the battery chemistry, which can range from a skin irritant to one of a lethal nature. It is very important to understand that, when designing a battery, several factors need to be considered, including such factors as mass, volume, voltage, power, gravimetric and volumetric energy density, pulse capability, and safety. In general, the choice of battery chemistry and the manufacturer can be a preference if there are no restrictions on mass and volume. However, in many cases, a specific battery chemistry is likely to be required because of the power requirements. It is recommended that the hardware owner work with the relevant battery experts from the initial phase of a project to design for optimum performance and safety.

This chapter provides a basic knowledge of the different battery chemistries currently used in space applications and an understanding of the hazards and their controls one can encounter. All test data, unless referenced, is based on testing performed by the author in collaboration with other members of the Johnson Space Center battery team or with battery experts at other NASA centers.

16.3 BATTERY TYPES

Cells and batteries fall under two main categories, primary and secondary. Primary cells and batteries are those that cannot be recharged. They can be used until their energy is depleted and have to be disposed of at that time. Secondary cells and batteries are rechargeable and hence can be used for many cycles depending on the chemistry and manufacturer. Reserve batteries are sometimes considered a third category, wherein the active components of the cell are combined just before use. These, after becoming active, can fall under either the primary or the secondary category.

16.4 BATTERY MODELS

Cells come in different shapes and sizes. The usual common types available in the market are button cells, A, AA, AAA, AAAA, 2/3A, 4/3A, 4/5A, C, D, DD, 9V, and so forth. Lithium-ion cells do not conform to this standardized categorization, in that their model number indicates their dimensions. For example, an 18650 cell has the dimensions of 18 mm diameter and 65 mm length. A prismatic cell with the model number 353452 has the dimensions of 35 mm thickness by 34 mm width by 52 mm length. Polymer lithium-ion cells use the same format for model numbers as the prismatic ones.

16.5 HAZARD AND TOXICITY CATEGORIZATION

The hazard and toxicity categorizations are combined to determine the level of testing and number of protective safety controls a battery system is required to have. Batteries flown for space applications by NASA in a human tended environment are subject to toxicity categorization because electrolyte leakage and gaseous vapors result from cell or battery failure. The level of toxicity is based on the nature of the electrolyte and calculated from its permissible levels and toxic effects. Per the NASA Johnson Space Center standards (Lam, Wong, and Coleman 1991), a toxicity rating of 2 or above is considered to be of the catastrophic type, for any electrolyte leakage from these could result in hazards that range from permanent injury to the crew to those of a lethal nature. For batteries that fall under the categories of catastrophic and above, two-fault tolerance to all failures must be provided. Those that do not cause permanent injury or damage likely fall within the critical category, that is, toxicity level 1, and those that do not pose a toxic hazard at all are designated a toxicity level of 0.

16.6 BATTERY CHEMISTRY

The battery chemistries used most commonly for space applications are discussed here. These batteries are either custom designed or commercial off-the-shelf varieties. The chemistries of primary batteries are described first, followed by those for the secondary or rechargeable types.

16.6.1 Alkaline Batteries

Alkaline cells and batteries are the most commonly available commercial off-the-shelf product. The basic electrochemical processes taking place in these alkaline cells are

Anode: $Zn + 2OH^- \rightarrow Zn(OH)_2 + 2e^-$
Cathode: $2MnO_2 + H_2O + 2e^- \rightarrow 2MnOOH + 2OH^-$
Overall reaction: $Zn + 2MnO_2 + 2H_2O \rightarrow Zn(OH)_2 + 2MnOOH$

The anode is made up of zinc powder and the cathode of manganese dioxide powder. The electrolyte is 35% potassium hydroxide. The nominal voltage of alkaline cells ranges from

1.5 to 1.6 V, and most of their capacity has been consumed when their voltage drops to about 1.0 V. The internal resistance of the cell varies with the depth of discharge and ranges from 800 mΩ at the start to about 100 mΩ at the end of discharge. Alkaline cells maintain up to 85% of their capacity after a shelf life of 4 years. A typical discharge curve for an alkaline cell is given in Figure 16.1.

Due to improvements in active materials, the capacity of alkaline cells has increased considerably during the past decade. For example, a D-size cell that was rated at 13 Ah in 1990 currently possesses greater than 20 Ah capacity. With increases in capacity and volumetric energy density, it is important to keep in mind that the hazards faced by the user also have increased.

Recently, an update to the test database on alkaline cells of varying capacities from a few cell manufacturers was conducted at the NASA Johnson Space Center. The results of the test program indicate that alkaline cells can be hazardous under certain abusive conditions. Figure 16.2 shows the results of external short circuit testing, where the maximum temperature observed was 110°C. Variations in active material can influence the behavior of a cell under abusive conditions. For instance, a 4.5-V alkaline battery manufactured with a slight variation to the cathode material resulted in the cell exploding under an external short circuit of 50 mΩ. Tests conducted on cells in the NASA Johnson Space Center inventory did not show any explosion under the same test conditions. For this reason, lot testing is recommended even for alkaline cells.

Irrespective of age, alkaline cells have a very high tendency to leak that can be attributed to their cell construction and exposure to extreme thermal environments. Alkaline cells have a unique cell construction because they were used initially as a replacement for Leclanche batteries. The construction of alkaline cells is such that the cell can is positive with a button terminal, whereas the flat cover, crimp, and vent are at the negative end. The presence of the vent at the negative terminal, that is, the cell bottom, can contribute appreciably to the increased number of leakage incidences. Equipment using alkaline cells should be examined frequently, for the electrolyte is extremely corrosive and can

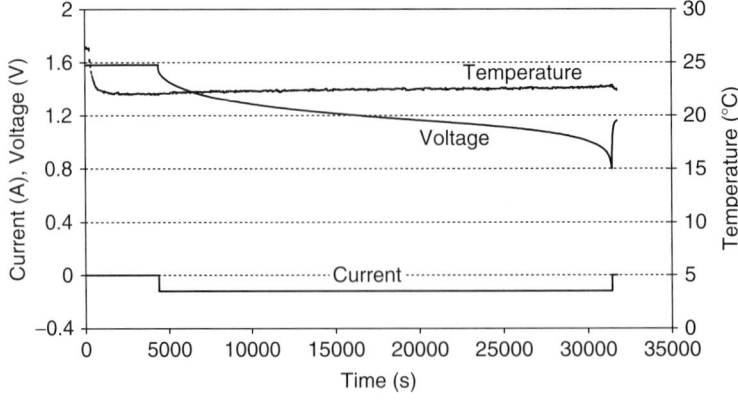

FIGURE 16.1 Typical voltage profile for alkaline cells (Courtesy of NASA).

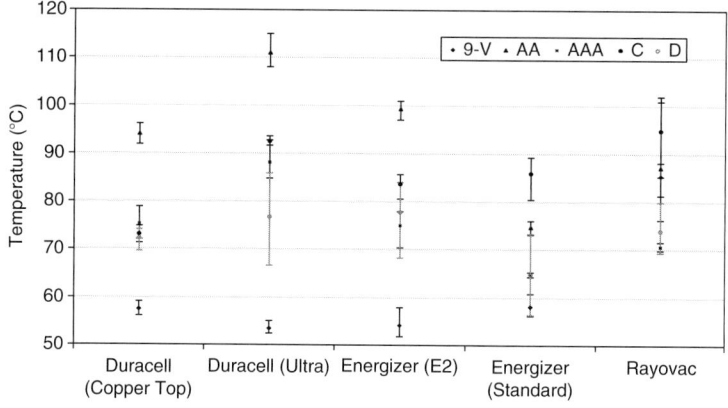

FIGURE 16.2 External short circuit test on commercial off-the-shelf alkaline cells of different sizes (Courtesy of NASA).

cause extensive damage to equipment and injury to personnel. Whenever possible, adequate quantities of absorbent or wicking material should be used in a battery compartment, and the cells should not be enclosed in gastight containers.

For battery designs, including those of alkaline cells, the safety circuitry should contain a fuse and resistor in each battery string for short circuit protection and a pair of blocking diodes if there are any charging sources in the circuitry. For high voltage batteries, typically greater than 12 V, and for those constructed with D-size cells, bypass diodes should be provided for protection against voltage reversal. Figure 16.3 shows typical design protection for primary batteries of all chemistries.

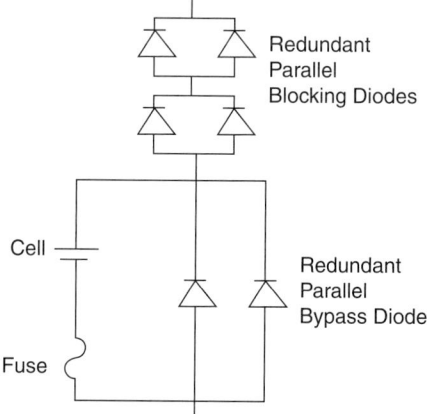

FIGURE 16.3 Design guidelines for protection of primary batteries against inadvertent charge or overdischarge into reversal and external short circuit (Courtesy of NASA).

16.6.2 Lithium Batteries

Lithium batteries of the primary or non-rechargeable type are of several different chemistries. In most cases, primary lithium cells have a lithium metal anode. Cells of this chemistry have a nonaqueous electrolyte that is either organic or inorganic in nature. Depending on the cathode present, the nominal voltage of the cell differs. The most commonly used lithium primary batteries for space applications are lithium manganese dioxide ($LiMnO_2$), lithium iron disulfide ($LiFeS_2$), lithium polycarbonmonofluoride ($LiCF_x$), lithium thionyl chloride ($LiSOCl_2$), and lithium bromine chloride complex ($LiBC_x$). The less commonly used ones are lithium sulfur dioxide ($LiSO_2$) and lithium sulfuryl chloride ($LiSO_2Cl_2$).

At first, the safety characteristics common to all primary lithium batteries are discussed; subsequently, those characteristics specific to each chemistry are addressed. All lithium primary batteries should be protected against the hazards of inadvertent charge, overdischarge into reversal, external and internal short circuit, and high temperatures. The resultant heat and the accumulation of energy inside sealed cells are considered to be the major causes for cell and battery related fire and explosions. The U.S. Department of Transportation regulations for shipping and transportation apply to all primary lithium batteries (DoT 2007).

Hunger and Christopulos (1975) point out that the heat produced internally due to high rate discharges, external and internal short circuits, or produced externally due to high thermal environments increases the temperature of the internal cell components. This leads to a change in the physical and chemical properties of the formerly stable components, resulting in a hazardous condition. At elevated temperatures, the vapor pressure of the solvents inside the cell can become high enough to rupture the gasket seal or even the metal casing. Hot electrolyte is then released into the atmosphere along with the vapors of the flammable solvents, which ignite in the presence of air at temperatures above their flash point. In this instance, a spark is required to ignite the vapors (ASTM 2007). However, if the pressure causes their metallic containers to rupture, the resulting frictional forces are sufficient to cause localized high temperatures sufficient to ignite the flammable solvents (ASTM 2005).

The lithium anode can fail in many ways. Lithium metal has a melting point of 182°C, and if a cell reaches this temperature, explosion results. It would be good to keep in mind that the internal cell temperatures range from 5 to 10°F (or more in some cases) different from the external temperature of the cell, and this depends on the size and design configuration, such as AA versus D. As well, the lithium anode can ignite in the event of forced contact with the cathode due to an internal short circuit, caused for example by a damaged separator, external impact, or penetration. If there is sufficient heat, a lithium anode can ignite in the presence of air when the cell seal or can is ruptured (DoT 2007).

Lithium primary batteries lack the capability to be recharged. This is because the electrochemical reaction and equilibrium proceed in one direction only. The components are consumed during the discharge process and cannot be regenerated or converted back to the parent material. Hence, any attempt to recharge these cells can be catastrophic, resulting in violent venting or explosion. Note that lithium metal along with inorganic liquid electrolytes can provide an explosive atmosphere with a TNT equivalency that is discussed later.

General design guidelines for safety include fuses, resistors, blocking and bypass diodes, thermal fuses, and external and internal positive temperature coefficient devices. Figure 16.3 provides some methods for the incorporation of fuses and diodes. Each cell string should have a pair of blocking diodes in series for protection against inadvertent charge, a fuse and a resistor or two fuses of different ratings for protection against external shorts, and two bypass diodes in parallel to each cell for overdischarge into reversal. If the cells do not have internal fuses or positive temperature coefficients, a fuse must be inserted in series with each parallel bypass diode (not depicted in Figure 16.3) to protect against external shorts due to failure of the diodes, that is, they fail closed. Cells should be tested under vacuum conditions for a few hours or at elevated temperature, that is, 70°C, for extended periods of up to 3 d to detect instances of leaky cell seals or any structural damage that might have occurred during the cell assembly process. Such damage could be caused by spot welding or other chemical incompatibilities (Hunger and Christopulos 1975). Design considerations should take into account any limitations of the protective devices. When diodes are used, their leakage current and voltage limitations should be understood thoroughly. Devices with a positive temperature coefficient have a voltage limitation dependent on their components, thermal environment, and cell internal construction and composition that should be characterized.

At this time, the basic electrochemical processes, cell composition and components, and the hazards associated with each lithium battery system are to be discussed independently. All lithium primary batteries discussed under this section have lithium metal as their anode.

Lithium Manganese Dioxide (LiMnO$_2$)

The half reactions for the LiMnO$_2$ cell chemistry are

$$\text{Anode:} \quad \text{Li} \rightarrow \text{Li}^+ + e^-$$
$$\text{Cathode:} \quad \text{Mn}^{\text{IV}}\text{O}_2 + \text{Li}^+ + e^- \rightarrow \text{Mn}^{\text{III}}\text{O}_2(\text{Li}^+)$$
$$\text{Overall reaction:} \quad \text{Li} + \text{Mn}^{\text{IV}}\text{O}_2 \rightarrow \text{Mn}^{\text{III}}\text{O}_2(\text{Li}^+)$$

The cathode is made up of solid heat treated MnO$_2$ and the electrolyte is a mixture of propylene carbonate or butylene carbonate and dimethoxyethane, along with a salt, such as lithium perchlorate or lithium trifluoroethylacetate (triflate). The open circuit voltage of this cell chemistry is 3.3 V. The nominal or operating voltage of the cell is 3.0 V. Caution should be exercised when using cells with an open circuit voltage greater than 3.3 V. Higher voltages accelerate corrosion and lead to considerable reduction in the calendar life of the battery. The internal impedance of the cell varies from 2 to 6 Ω for cylindrical spiral wound cells, from 5 to 13 Ω for cylindrical bobbin type cells, and from 8 to 18 Ω for low rate coin cells.

LiMnO$_2$ cells of various sizes have been used in space applications. Typically, the cells are of a spiral wound design. Because of their high internal resistance, they are not as hazardous as those cells with an inorganic electrolyte. However, at the end of a discharge process, if excess lithium metal is present, the products of the decomposition of the electrolyte can react with the lithium metal to produce excessive gassing and venting, which is likely to result in smoke and fire. Figures 16.4 and 16.5 depict the variation in safety for cells obtained during different time periods, or in other words, cells from different lots, wherein one lot showed a tolerance to overdischarge and another did not (Dimpault-Darcy 2007). Hence, lot sample qualification testing is of paramount importance in establishing the safety of any new cell lot. Spiral wound cylindrical $LiMnO_2$ cells typically are fitted with a positive temperature coefficient device, and thus their voltage limitations need to be taken into consideration during the buildup of batteries because positive temperature coefficient devices tend to ignite if activated at high voltages.

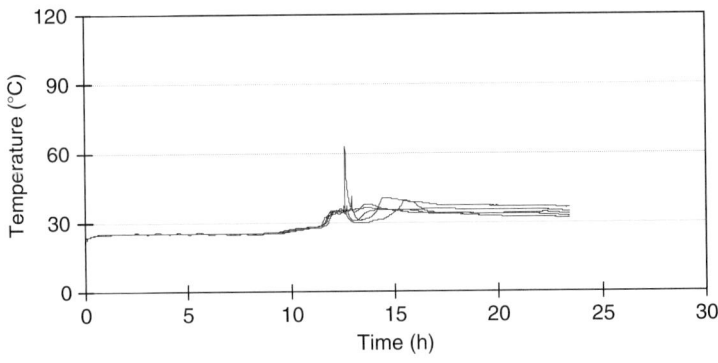

FIGURE 16.4 Cell temperatures recorded during the overdischarge into reversal test on the first sample lot of the 2/3-A $LiMnO_2$ cells with a load of 130 mA (no venting) (Courtesy of NASA).

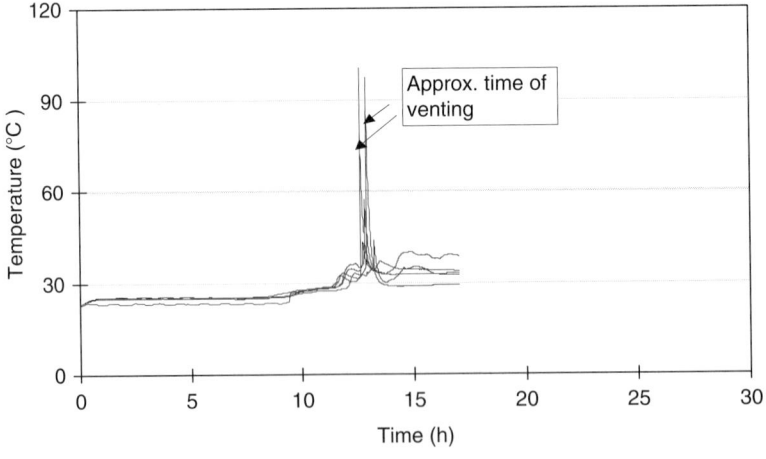

FIGURE 16.5 Cell temperatures recorded during the overdischarge into reversal test on subsequent lots of 2/3-A $LiMnO_2$ cells with a load of 130 mA (Courtesy of NASA).

Lithium Polycarbonmonofluoride [Li(CF)$_x$ or LiCF$_x$]

The half reactions are

Anode: $xLi \rightarrow xLi^+ + xe^-$
Cathode: $(CF)x + xe^- \rightarrow xC + xF^-$
Overall reaction: $xLi + (CF)x \rightarrow xLiF + xC$

The cathode is solid polycarbonmonofluoride and the electrolyte is a mixture of gamma-butyrolactone combined with a salt, such as lithium tetrafluoroborate. In coin cells, dimethoxyethane also is added to the solvent mixture. The open circuit voltage is 3.2 V and the operating voltage is between 2.5 and 2.6 V, which drops down further under higher loads (Figure 16.6). Voltage delays are very common with this chemistry especially at low temperatures, and predischarge is performed typically to remove this feature. The shelf life for this chemistry is 8 to 10 a.

LiCF$_x$ cells were used more commonly during the 1980s and 1990s, but their use in many applications has decreased as cells like lithium thionyl chloride and LiBC$_x$, which provide higher discharge rates, were designed to be safer. The cathode of these cells is stable thermally up to 400°C, and any unsafe condition that can cause the internal cell temperature to reach this value results in venting, smoke, and fire. Inadvertent charging of D cells with high (approximately 1 A) and low currents have caused the cells to smoke considerably. Venting and flames occurred only when the cells were charged with the higher current independent of the depth of discharge (Davis et al. 1983). The combustion gas analysis showed a variety of low molecular weight hydrocarbons, such as acetylene, ethylene, and propylene, as well as fluorocarbons, carbon dioxide, oxygen, carbon monoxide, hydrogen, and low levels of benzene (Davis et al. 1983). Under short circuit and

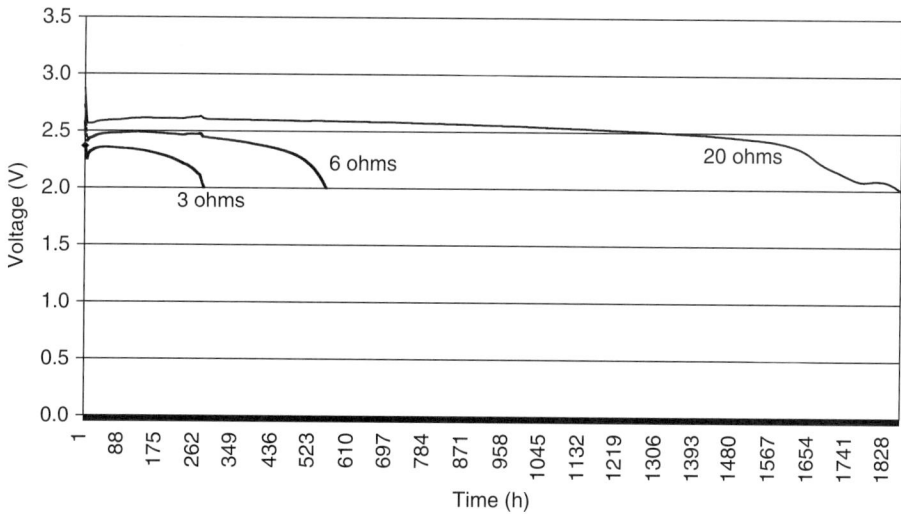

FIGURE 16.6 Discharge profiles for LiCFx aged DD cells under different loads (Courtesy of NASA).

overdischarge conditions, sufficient heating to melt the separator is attained, and the cell dies, with electrolyte leakage occurring after some time, that is, hours or in some cases days later.

As with other solid cathode lithium primaries, the internal resistance for coin and low capacity cells is high and does not result in high energy hazards under abusive conditions, such as short circuits and voltage reversal. However, the higher capacity D and DD cells are capable of explosive venting if subjected to high rates of discharge, external shorts, and voltage reversal.

Lithium Iron Disulfide (LiFeS$_2$)

The half reactions are

Anode: $4Li \rightarrow 4Li^+ + 4e^-$
Cathode: $FeS_2 + 4e^- \rightarrow Fe + 2S^{--}$
Overall reaction: $4Li + FeS_2 \rightarrow Fe + 2Li_2S$

The cathode is solid iron disulfide (FeS$_2$), and the electrolyte is a mixture of organic solvents such as dimethoxyethane and dioxalane and a salt, such as lithium iodide. The open circuit voltage of the cells is as high as 1.9 V, and their nominal voltage is 1.5 V. The operating voltage range is from about 1.6 to 1.0 V (Figure 16.7), and the cells are capable of providing up to a C rate discharge. The internal impedance of the cells is approximately 180 mΩ. The shelf life for this chemistry is about 15 a.

LiFeS$_2$ cells have been tested extensively under conditions of inadvertent charge, overdischarge into reversal, external short circuit, simulated internal short circuit, and high temperature exposure. The cells, when overcharged with a current for 6 h, showed a maximum rise in temperature (Figure 16.8) of 20°C and did not exhibit venting or fire (Jeevarajan 2005a, 2005b). When the cells were overdischarged into reversal, the maximum temperature recorded was 35°C. Under conditions of an external short circuit

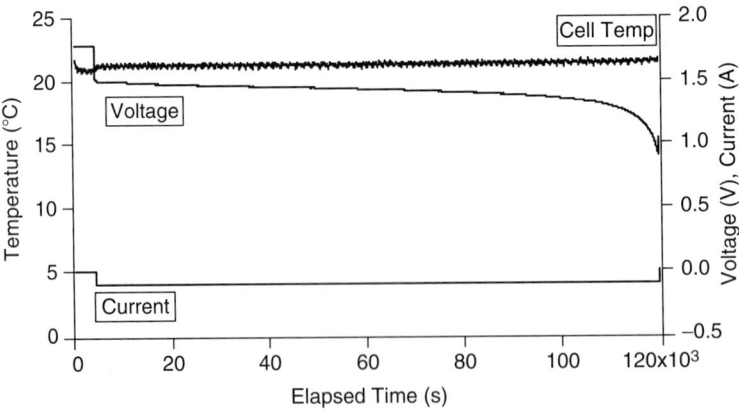

FIGURE 16.7 Typical discharge profile for a LiFeS$_2$ AA cell under a 100 mA load (Courtesy of NASA).

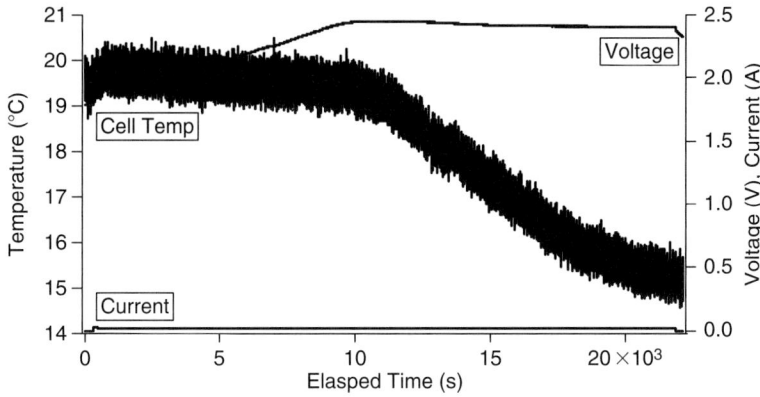

FIGURE 16.8 Charging test on LiFeS$_2$ primary lithium cell for 6 h with a current of 20 mA (Courtesy of NASA).

using a load of 50 mΩ, the positive temperature coefficient device inside the cell activated, dropping the current to approximately 1 A until the cell had discharged completely. Cells exposed to high temperatures at a ramp rate of 3°F per minute and up to 200°C were charred; however, they did not explode (Figure 16.9). No venting or flame was observed under simulated internal short circuit tests using a crush method. Although the cells exhibit tolerance to most abuse conditions, they tend to vent violently with smoke and fire when impacted, as in bullet shot tests. Due to the presence of a lithium metal anode and an organic electrolyte, they should be designed with adequate protection as with other solid cathode primary lithium cells.

Liquid Cathode Lithium Primary

The lithium types of LiSOCl$_2$, LiBCX, LiSO$_2$, and LiSO$_2$Cl$_2$ have a liquid cathode that also functions as the electrolyte. This material, therefore, is termed a *catholyte*. The cells are sealed

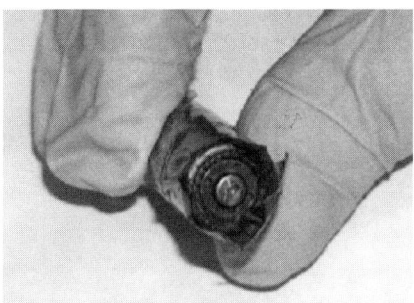

FIGURE 16.9 A LiFeS$_2$ AA cell after a heat to vent test indicating an intact cell with no disassembly or shrapnel (Courtesy of NASA).

hermetically and fitted typically with a glass to metal seal that provides a leak before burst design feature. Venting would occur through the glass seal that would crack and the leak propagate well before the cell container bursts.

Lithium metal along with inorganic solvents can provide an explosive atmosphere having a TNT equivalency. Theoretical calculations have shown that a pound of this type of cell has the same explosive potential as a pound of TNT. During heat to vent testing at NASA Johnson Space Center (Figure 16.10) on LiBCX C cells, pressures as high as 120 psig were recorded and the cells exploded, producing shrapnel (Figure 16.11). In contrast, the $LiFeS_2$ AA cells produce only 4 psig of pressure under the same test conditions (Jeevarajan 2005a, 2005b).

With respect to performance characteristics, close attention should be paid to voltage delays or depression common to primary lithium cells (Figure 16.12). This characteristic is due typically to passivation and more commonly is found in cells with a liquid cathode. A thin passivation layer is formed on the surface of the anode as soon as the electrolyte is added to the cell (Figure 16.13). This layer is formed by the interaction of the metallic lithium anode with the oxyhalide electrolyte. The passivation is important because it provides a very long shelf life for the cell by protecting the anode from reactions.

The passivation layer can be removed in different ways, the simplest of which is to subject the cells to short periods of higher discharge current rates than are used normally but still at a rate that the cell can withstand. If a passivated cell is subjected to a very high rate of continuous or pulse discharge without regard to an undervoltage limit, the already increased internal resistance can cause the cells to go into an overdischarged or, in some cases, a voltage reversal condition. This can result in venting, fire, or explosion. Voltage reversal does not mean that the cell is being charged; rather, it typically indicates the cell is being demanded to give more than its rated capacity and results in the cell being taken into negative voltages. This condition results in an electrochemical reduction of internal cell components. It is important to note that voltage reversal should never be confused with the inadvertent charging of a primary cell.

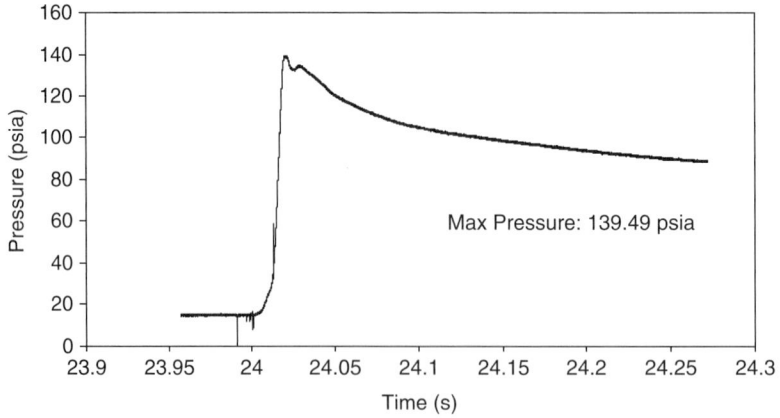

FIGURE 16.10 Pressure recorded during the heat to vent testing on a LiBCX C cell (Courtesy of NASA).

16.6 Battery Chemistry 519

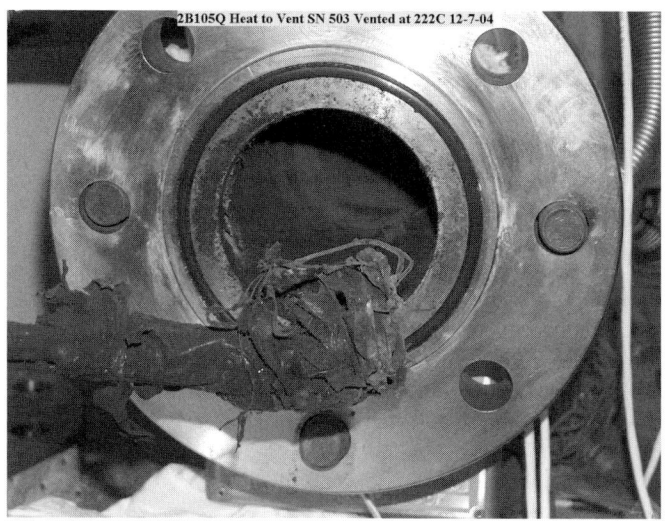

FIGURE 16.11 The remains of a heat to vent test on a LiBCX C cell (Courtesy of NASA).

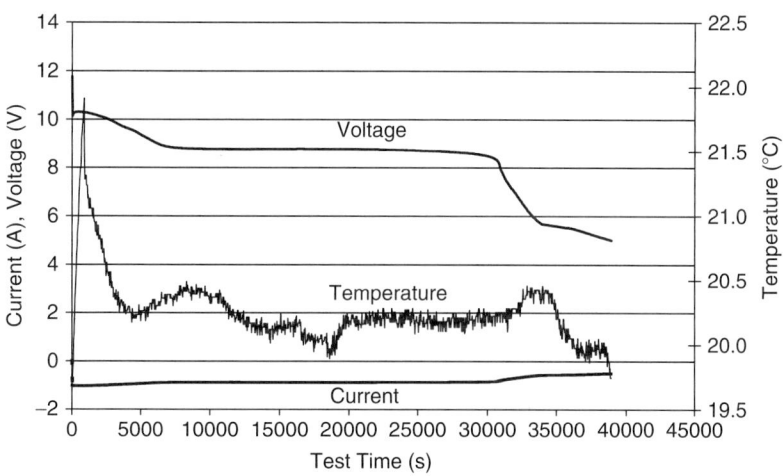

FIGURE 16.12 Internal wireless instrumentation system (12-V) LiBCX battery discharge characteristics after storage at ambient temperatures for 6 mo (load: 180 mA) (Courtesy of NASA).

The half reactions for the $LiSOCl_2$ cell are

$$
\begin{aligned}
\text{Anode:} &\quad Li \rightarrow Li^+ + e^- \\
\text{Cathode:} &\quad 2SOCl_2 + 2e^- \rightarrow SO_2 + S + 4Cl^- \\
\text{Overall reaction:} &\quad 4Li + 2SOCl_2 \rightarrow 4LiCl + SO_2 + S
\end{aligned}
$$

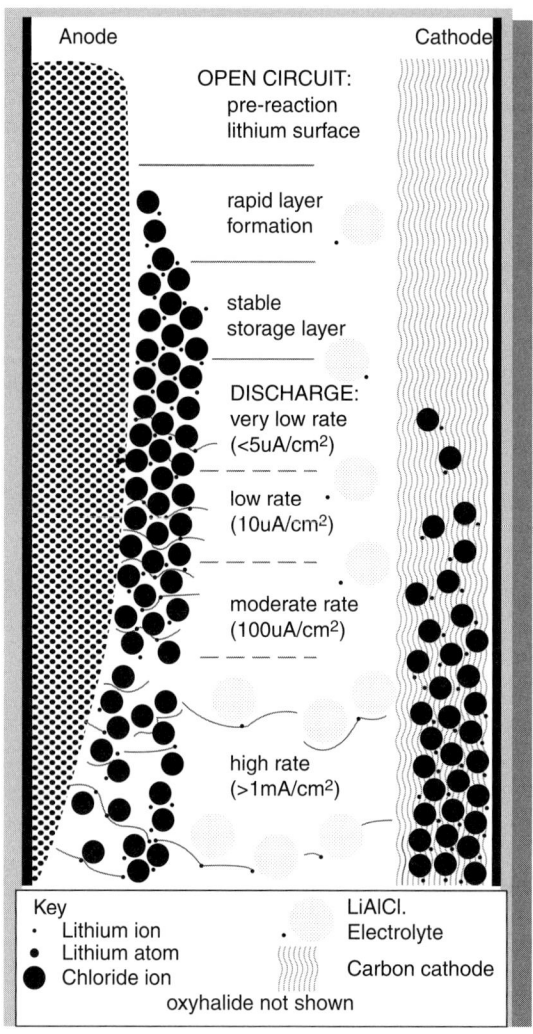

FIGURE 16.13 Passivation observed in a lithium primary cell with a liquid electrolyte (Reproduced by permission of Electrochem).

The catholyte is a mixture of thionyl chloride and lithium aluminum tetrachloride on high surface area carbon. The open circuit voltage is 3.7 V, and the nominal operating voltage for the former is 3.5 V. Typical discharge profiles for $LiSOCl_2$ cells are as shown in Figure 16.14 (Frerker, Jeevarajan, and Bragg 2000).

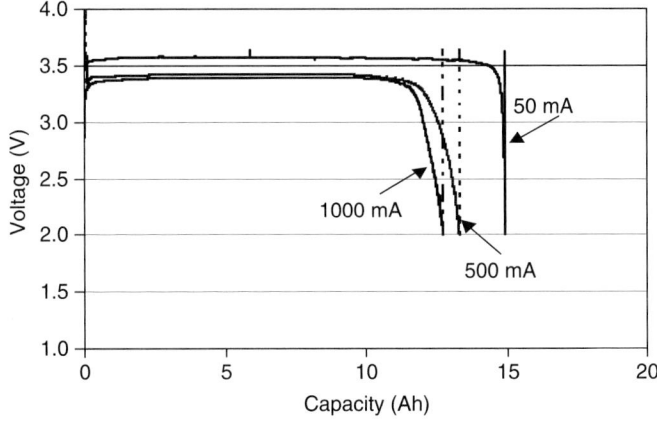

FIGURE 16.14 Typical voltage profiles for the LiSOCl$_2$ cell under different discharge currents (Courtesy of NASA).

The half reactions for the LiBCX cell are

Anode: Li $\rightarrow$ Li$^+$ + e$^-$
Cathode: 2SOCl$_2$BrCl + 2e$^-$ $\rightarrow$ SO$_2$ + S + 3Cl$_2$ + Br$_2$
Overall reaction: 6Li + 2SOCl$_2$BrCl $\rightarrow$ 4LiCl + SO$_2$ + S + 2LiBr

The catholyte in LiBCX cells is a mixture of thionyl chloride (SOCl$_2$), lithium aluminum tetrachloride, and bromine chloride (BrCl), with the SOCl$_2$ to BrCl ratio being 6:1. The open circuit voltage for LiBCX is approximately 3.8 to 3.9 V with an operating voltage of 3.6 V that varies very little, even at low temperatures (Figures 16.15 and 16.16).

Lithium thionyl chloride cells have been used for many decades for government, aerospace, and military applications. Brooks (1974), as early as 1974, performed a series of tests on cells from three manufacturers, the work being performed in collaboration with them. External and internal short circuit, increasing load, hot plate, deformation (simulated internal short), dynamic environment, and case rupture tests were conducted at the cell level. Other tests, including a discharge into reversal test at the battery level also were conducted. During the short circuit, fast discharge, and hot plate tests, the cells vented with no fire or explosion. In the deformation test, the cells vented due to internal shorts, but in some cases, they vented prematurely, before sufficient crush had been applied. During the case rupture and drilling tests conducted at room temperature, the cells did not catch fire, but when the drilling was performed after the cells were heated, fire was observed.

The battery level tests showed venting under conditions of short circuit, hot plate, immersion, and discharge into reversal with a C/5 load. When the rate was increased to C/2.5 for the discharge into reversal test, there were incidences of cell explosion. This prompted the manufacturers to change their cell design to accommodate a slightly larger vent that would permit the cell to vent rather than explode.

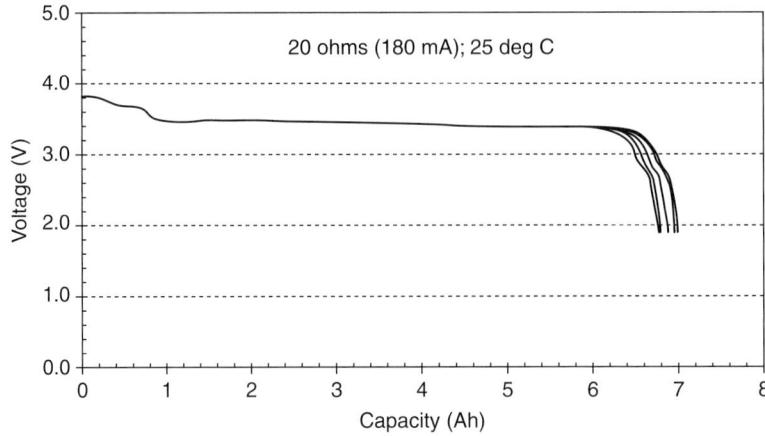

FIGURE 16.15 LiBCX voltage profile under a 180-mA load at room temperature (Courtesy of NASA).

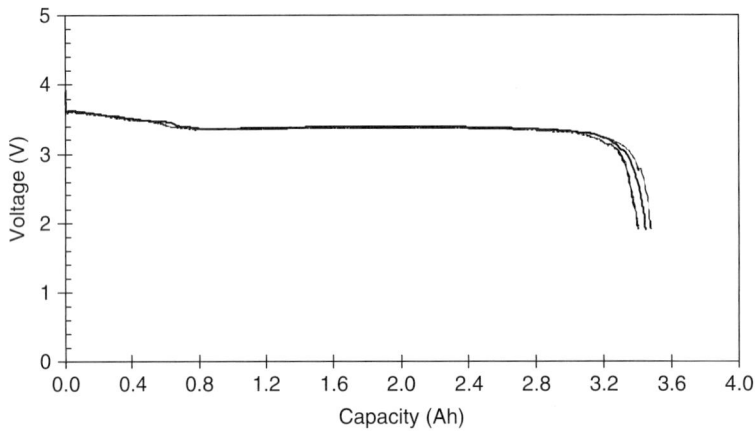

FIGURE 16.16 LiBCX voltage profile under a 500-mA load at 85°C (Courtesy of NASA).

At NASA Johnson Space Center, collaborative research with Wilson Greatbatch, Ltd., resulted in the production of LiBCX cells with lower electrolyte molarity and thicker can lids (Bragg and McDonald 1992). Despite all of the design improvements, the thionyl chloride and BCX cells have a tendency to vent and explode because of the production of intermediates during the discharge process.

Carter et al. (1985) explained that this likely is due to an increased concentration of metastable compounds to a point that their decomposition causes venting or explosion. As well, the discharge process could be producing sufficient heat to trigger a reaction between the lithium metal anode and the liquid thionyl chloride cathode. Although several intermediates have been proposed, among which are SO, $(SO)_2$, S_2O, and S_2Cl_2, only some of them are confirmed to have been produced. The compounds S_2O, SO_2, and SO_2Cl_2 have been identified by infrared spectroscopy (Istone and Brode 1982; Carter et al. 1985) and S_2Cl_2 by gas chromatography (Blomgren et al. 1978). Electron spin resonance spectroscopy (Carter et al. 1985) indicated the presence of a dimer of OClS and sulfur. The compound OClS is similar to the hazardous compound ClO_2, which might be one of the contributors to the venting and explosion of these cells.

The $LiSO_2$ and $LiSO_2Cl_2$ types are not used commonly in space applications, especially in a human tended environment and, hence, are not discussed here. As well, the two chemistries discussed in this section are recommended for use only if there is adequate hazardous gas detection and cleanup capability, because the toxicity level values for hydrogen sulfide and sulphur dioxide, the byproducts of the sulfur containing electrolytes, is as low as 5 mg/m^3 (Lam et al. 1991).

16.6.3 Silver Zinc Batteries

Silver zinc batteries found in the market today are typically of the rechargeable category. The first silver zinc cells were built in the 1940s by Professor Andre in collaboration with Michel Yardney (Karpinski et al. 1999). The cells initially were primary or non-rechargeable in nature, and later with the improvements in separator and electrode components, the cyclability of these cells became possible. Not many manufacturers of silver zinc batteries can be found around the world because of the availability of other battery chemistries that have long shelf and cycle life and are less expensive. However, this chemistry is used in many applications by government agencies; therefore, it is discussed in this chapter.

The half reactions are

Anode: $\quad Zn \leftrightarrow Zn^{++} + 2e^-$
$\quad\quad\quad\quad Zn^{++} + 2OH^- \leftrightarrow Zn(OH)_2$
Cathode: $\quad 2AgO + H_2O + 2e^- \leftrightarrow Ag_2O + 2OH^-$
$\quad\quad\quad\quad Ag_2O + 2H_2O + 2e^- \leftrightarrow 2Ag + 2OH^-$
Overall reaction: $Ag_2O + Zn + H_2O \leftrightarrow Zn(OH)_2 + Ag$

The anode is a zinc electrode supported by a silver or copper current collector. The cathode is silver oxide on a silver grid that serves as the current collector. Silver is in its pure metallic form during discharge and is converted to silver (II) oxide (AgO) at the end of a full charge. An intermediate product formed is silver (I) oxide or monovalent silver oxide (Ag_2O) (Figures 16.17 and 16.18). The electrolyte is potassium hydroxide (KOH) at an optimum concentration of 45%, which is corrosive and can cause burns on the skin and damage to equipment.

The zinc anode is one of the major contributors to the capacity degradation of these cells (Himy 1986). In these cells, the zinc oxide formed during discharge undergoes dissolution due to its high solubility (5 to 8%) in the electrolyte. Typically, at the end of charge or in an overcharge condition, some of the zinc oxide that went into solution replates onto the zinc electrodes in a different location, inevitably on the bottom. This results in

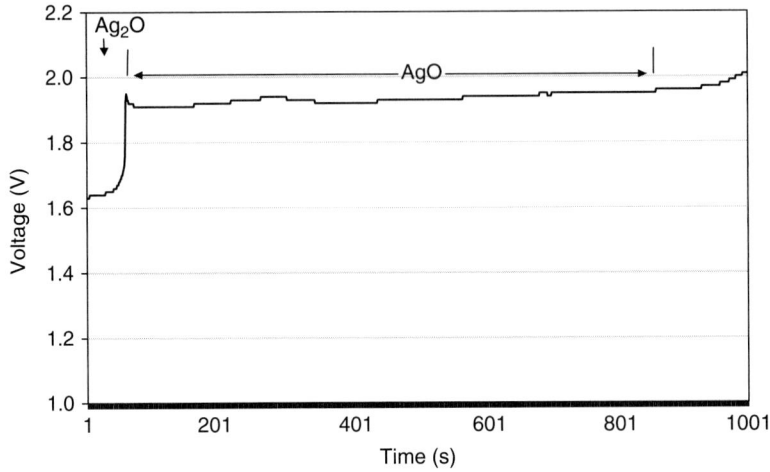

FIGURE 16.17 Voltage profile during the charging of a silver zinc cell depicting the two plateaus indicating the Ag_2O and AgO phases (Courtesy of NASA).

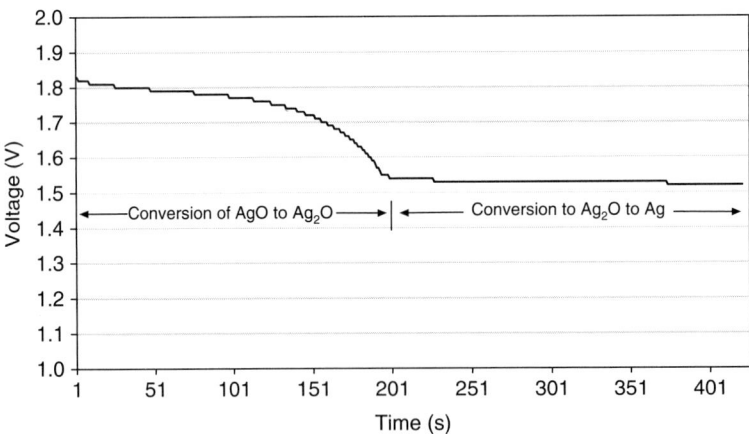

FIGURE 16.18 Voltage profile during the discharge of a silver zinc cell depicting the two plateaus indicating the conversions of the Ag_2O and AgO phases (Courtesy of NASA).

depletion of the negative active material from the exposed parts of the electrode. Depletion at the top and exposed sides of the electrode leads to a loss in cell capacity.

Internal shorts are the second cause for capacity fading. The replated zinc forms needlelike dendrites capable of piercing the separator, causing internal shorts between the two active electrodes. Because of its solubility, silver oxide goes into solution, and along with it, silver also goes into solution and remains as a suspension. Both diffuse slowly through the separator membrane by a phenomenon called *silver penetration* or *migration* and eventually form a silver trail between the two active electrodes and thus an internal short. This phenomenon can be reduced by the use of several separator membranes.

A third contributor to the capacity loss is the cellulose separator, which is made of a material similar to cellophane or a sausage casing, and an absorbent material that is very porous. Because water readily attacks cellophane, its degradation becomes marked when the concentration of potassium hydroxide drops, thus shortening the cell life. Cellophane also reacts with potassium hydroxide in such a way that the potassium ions become attached to the cellulose molecule, causing a decrease in the concentration of potassium hydroxide, thus reducing the active ions within the cell.

Another hazard inherent to silver zinc chemistry is the formation of oxalate crystals. Their formation begins soon after the addition of electrolyte to the dry cell. If the cell is left to stand fully charged for prolonged periods without use, the oxalate crystals increase in size to the point where they can pierce the separator causing an internal short. The addition of electrolyte can dissolve the oxalate crystals; however, this method is not very practical.

The typical method for charging silver zinc cells and batteries is a constant voltage protocol with a current limit. The open circuit voltage for a silver zinc cell is typically 1.86 V at room temperature. This voltage remarkably is stable on a short-term basis, that is, days or weeks, and can be used as a good indicator for internal shorts during manufacturing. A drop in voltage of 50 mV over a 3-d stand indicates that the cell has a slow short. This slow short can turn into a fast or hot short later, resulting in a thermal runaway with the electrolyte temperature reaching its boiling point. In the worst case, this can result in a fire.

Silver zinc cells are built with resettable vents that typically activate between 5 and 10 psi for low capacity cells of up to 10 Ah. Higher capacity cells are equipped with vents rated at higher pressures of up to 50 psi. The potassium hydroxide electrolyte in silver zinc cells is highly hygroscopic and tends to creep and crystallize if it leaks from the cell. The low vent pressure setting and the electrolyte flooded design result in a high possibility of electrolyte release during venting. Absorbent material, therefore, should always be used in this battery design.

Silver zinc cells have a very low cycle life, that is, 30 to 35 cycles, due to the failures described earlier. An increase in the robustness of the separator by the use of many layers decreases the rate capability of the cells. The cells are good for a maximum of 3 a after activation by the addition of electrolyte; however, if the cells have to undergo many continuous cycles, they are not likely to last for 3 a. Conversely, if the cell has a shelf life of 3 a, it probably has not undergone many cycles.

16.6.4 Lead Acid Batteries

Lead acid batteries have been in existence since 1860, when Plante built the first practical lead acid battery. Lead acid batteries are not used frequently for space applications due to the lack of small cell designs typically required for portable applications, and they offer no weight and volume savings for higher power applications.

The half reactions are

Anode: $Pb \leftrightarrow Pb^{++} + 2e^-$

$Pb^{++} + SO_4^{--} \leftrightarrow PbSO_4$

Cathode: $PbO_2 + 4H^+ + 2e^- \leftrightarrow Pb^{++} + 2H_2O$

$Pb^{++} + SO_4^{--} \leftrightarrow PbSO_4$

Overall reaction: $Pb + PbO_2 + H_2SO_4 \leftrightarrow 2PbSO_4 + 2H_2O$

The anode in lead acid cells is made up of a spongelike metallic lead, and the cathode is lead dioxide (PbO_2) in both α and β forms. The α form contributes to a higher cycle life, although it is slightly less electrochemically active and lower in gravimetric energy density. The electrolyte is sulfuric acid of about 37% by weight (approximately 2 M H_2SO_4) and has a specific gravity of between 1.26 and 1.28 when the cell is charged fully.

The open circuit voltage of a lead acid cell can vary from 2.125 to 2.05 V, depending on the electrolyte concentration, the specific gravity of the electrolyte being 1.28 in the former and 1.21 in the latter. Typically, the operating voltage of the cell is 2.0 to 1.75 V; however, the cell can be discharged to 1.0 V. Lead acid cells are charged using a constant voltage protocol to 2.39 V, and because most of these are used for uninterruptible power sources, they should be kept fully charged at a float voltage between 2.17 and 2.25 V. The float voltage is dependant on the nature of the anode and the specific gravity of the electrolyte. A shelf life of 7 a can be obtained if the cells are maintained at a float voltage when not in use.

The reaction of the lead acid cell during charge and discharge, termed the *double sulfate reaction*, involves dissolutions and precipitations (Mantell 1983). Sulfuric acid, the capacity limiting material, is consumed in the discharge reaction, producing water and causing a change in specific gravity.

During the charging process, lead sulfate is converted to lead and PbO_2. Above 2.39 V, overcharging results, causing the production of hydrogen and oxygen gas from the electrolysis of water. In sealed lead acid cells, the oxygen evolved typically recombines with the negative lead plate. However, when the hydrogen in air exceeds 4% by volume, that is, its lower explosive limit, explosions result, especially in the presence of large quantities of oxygen, as would be in an overcharged cell. Hydrogen detectors that provide a warning when within 20 to 25% of the lower explosive limit should be used whenever large lead acid batteries are used in enclosed areas. For this reason, lead acid batteries should always be used and stored in well ventilated areas.

Because the thermodynamically stable state dictates the direction of the discharge, an unused lead acid cell has a high rate of self-discharge. This can be lowered by the addition of small quantities of antimony to the lead in the negative plate. Overdischarging lead acid cells and batteries can have a profound effect on the pore structure of the cell due to the reduction in the electrolyte concentration.

High temperatures can have a considerably negative effect on the life of a lead acid battery, because the heat increases corrosion, the solubility of the metal components, and the rate of self-discharge. Hence, batteries that are to be used in high thermal environments should use electrolytes with a lower initial specific gravity. For long-term storage, cold temperatures are recommended.

Other safety hazards associated with lead acid cells and batteries include leakage of the corrosive sulfuric acid, which can burn skin and cause permanent eye damage. In addition, toxic gases, such as stibine and arsine, which are colorless and odorless, are generated. These can cause serious illness and death. Good ventilation that keeps the level of hydrogen below the recommended 20% of the lower explosive limit also keeps the arsine and stibine below their toxic limits.

16.6.5 Nickel Cadmium Batteries

Nickel cadmium (NiCd) batteries are the first choice in the battery market for use in high power tools and especially for low temperature applications. The NiCd batteries have been used widely for portable power space applications, and they are used to power the Russian module of the *International Space Station*.

The half reactions are

Anode: $Cd + 2OH^- \leftrightarrow Cd(OH)_2 + 2e^-$
Cathode: $NiO_2 + 2H_2O + 2e^- \leftrightarrow Ni(OH)_2 + 2OH^-$
Overall reaction: $Cd + NiO_2 + 2H_2O \leftrightarrow Cd(OH)_2 + Ni(OH)_2$

The cathode in the NiCd cell is made up of nickel oxyhydroxide (NiOOH) and the anode of cadmium hydroxide [$Cd(OH)_2$]. The electrolyte is 45% by weight potassium hydroxide, which is approximately a 6 M concentration. The open circuit voltage of NiCd cells can be as high as 1.55 V, their nominal operating voltage is 1.2 V, and they typically are discharged down to 1.0 V per cell. Their service life can vary between 8 and 25 a, depending on the design, application, and operating conditions. The internal resistance of NiCd cells is very low and can be less than 1 mΩ for high capacity cells of about 100 Ah.

NiCd cells suffer from a memory effect that requires frequent maintenance by reconditioning. The memory effect is most pronounced if the NiCd battery is left on the charger for days or recharged repeatedly without a periodic full discharge. In this condition, there is a buildup of large crystals of an intermetallic compound of nickel and cadmium. Fresh NiCd cells have particle cross sections of 1 μ, whereas those of the intermetallic compound are between 50 and 100 μ. These crystals tie up some of the much needed cadmium, thus creating extra resistance in the cell (Buchmann 2001). In the event the memory effect has been established, continuous cycling with discharges down to 0.6 V can break down the crystals to 3- to 5-μ particle sizes. This considerably reverses capacity loss.

Typically, NiCd cells are designed with a nickel limited electrode; however, if it is cadmium limited, oxygen evolution occurs due to the drop in potential of the cadmium electrode. When the nickel oxyhydroxide is depleted, cell reversal occurs that is dependent on the rate of discharge. The hydrogen evolution reaction has been studied in the past by many research groups (Ritterman 1976; Badcock and Martinelli 1979; Zimmerman and Effa 1982). Zimmerman and Effa (1982) concluded from their experiments that reduction of cadmium hydroxide to cadmium metal at the nickel electrode competes effectively with the hydrogen evolution when the discharge rate is low during the reversal process. Prolonged periods of reversal result in the growth of cadmium metal dendrites from the nickel

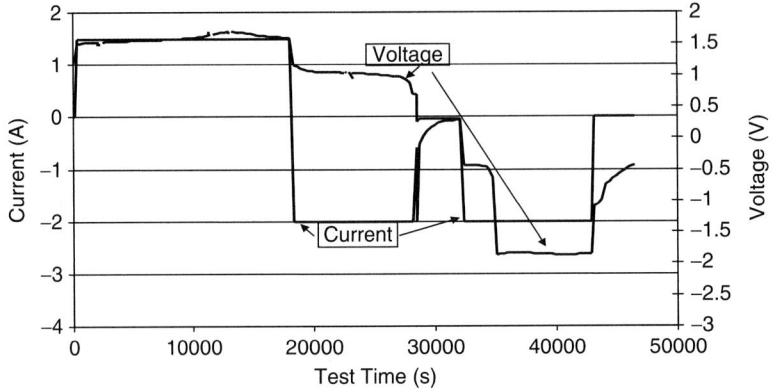

FIGURE 16.19 Overdischarge into reversal of a NiCd D cell (Courtesy of NASA).

electrode to the cadmium electrode, forming internal short circuits. However, after this metal formation, no evolution of hydrogen is observed on continued reversal. This type of short circuiting is reversible because of the oxidation of the cadmium metal during the recharge process. However, in low capacity cells used for commercial applications, they were found to be unusable after a reversal test (Figure 16.19).

Charging of NiCd cells typically is performed by using a constant charge protocol. Under overcharge and overvoltage conditions, irreversible evolution of hydrogen occurs. This eventually results in cell venting or bursting. For this reason, Pipoli et al. (1998) studied the safe voltage to which NiCd cells can be charged during continuous charge–discharge cycles in space, as in satellite applications. They observed that, with changes in taper voltage, the amount of gaseous hydrogen evolved varied. This also changed with cycle life. The study concluded that the safe voltage for charge was a maximum of 1.55 V. Voltage and temperature control typically are used as a cutoff limit during the charging process.

External short circuit testing of NiCd D cells resulted in cell venting as observed by the drop in voltage. However, no flames or other catastrophic events were observed (Figure 16.20). The internal reactions that occur are similar to the overdischarge condition, resulting in the formation of dendrites if the short is maintained for a long time (Fiala et al. 1985). High temperature exposures, above 200°C, can cause considerable venting and electrolyte leakage, resulting in loss of performance.

16.6.6 Nickel Metal Hydride Batteries

Nickel metal hydride batteries commonly are used in portable power applications, especially those such as power tools that require high rate and pulse capability. This chemistry was first discovered by Stanford Ovshinsky in the early 1980s as a replacement for the NiCd.

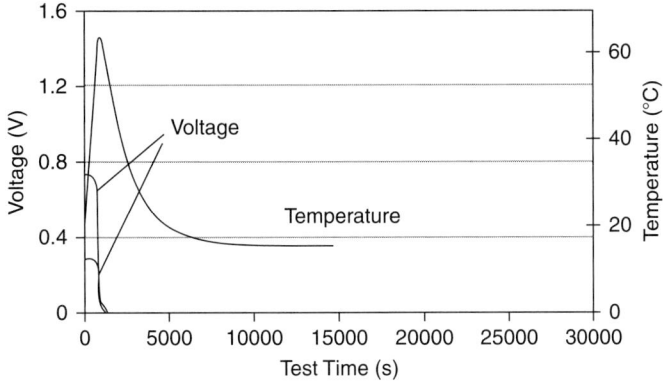

FIGURE 16.20 External short circuit test on commercial NiCd D cells (Courtesy of NASA).

The half reactions are

Anode: $MH + OH^- \leftrightarrow M + H_2O + e^-$ (M = metal alloy)
Cathode: $NiOOH + H_2O + e^- \leftrightarrow Ni(OH)_2 + OH^-$
Overall reaction: $NiOOH + MH \leftrightarrow Ni(OH)_2 + M$

The cathode is nickel oxyhydroxide, and the anode is typically one of two types of metallic alloys. One such alloy is the Misch metal type, made up of rare earth alloys of lanthanum nickel ($LaNi_5$), known as the AB_5 class, and another is made up of titanium and zirconium, known as the AB_2 class. Substitutions of the Misch metal alloys can be made with cesium, neodymium, praseodymium, gadolinium, and ytterbium in the place of lanthanum. In the case of AB_2 types, major substitutions by the use of vanadium, titanium, and zirconium provide improved hydrogen storage. Other metal substitutions that would result in longer life are made in both cases to suppress corrosion. The electrolyte is typically 30% by weight of potassium hydroxide. The open circuit voltage for the nickel metal hydride is similar to the NiCd and is approximately 1.55 to 1.6 V. The nominal or operating voltage of the cell is 1.2 V, and the cell is discharged down to 1.0 V. The internal resistance of typical cells utilized in portable electronic equipment is about 20 mΩ.

The nature of the metal alloys makes them thermodynamically more stable as oxides than as hydrides, and with the alkaline electrolyte, oxidation is unavoidable. However, it is most essential that the anode possess high oxidation as well as corrosion resistance to avoid using up the electrolyte. This also prevents the evolution of oxygen at the nickel hydroxide electrode. Extensive research has been carried out in the development of the anodes for nickel metal hydride cells (Fetcenko and Venkatesan 1990; Corrigan and Srinivasan 1992).

Nickel metal hydride batteries can be charged in more than one way. A typical charge method is a constant current at a low rate, such as a C/10 rate, for a period of 16 h (Figure 16.21), with a temperature limit of 45°C and a negative delta voltage change of 10 mV. At the end of charge, the drop in voltage and increase in temperature are not as

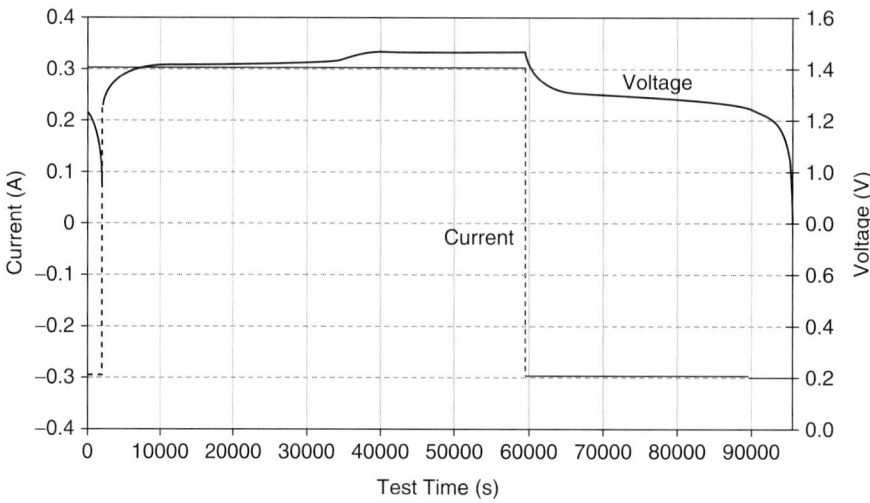

FIGURE 16.21 Typical profile for a constant current charge-discharge cycle for a nickel metal hydride subC cell (Charge: 300 mA for 16 h; discharge: 300 mA to 0.8 V) (Courtesy of NASA).

pronounced for nickel metal hydride as they are for NiCd batteries, thus a time limit along with the temperature limit should be used. Another well studied and used method is the burp charge method (Dimpault-Darcy 1999). In this method, reverse (discharge) pulses are levied during the charging process. This is an effort to remove gas bubbles formed on the electrodes during the charging process to expose more active material for the chemical reaction. A test program on D-size cells from Panasonic® was conducted, wherein both charge protocols were compared for 500 cycles. The results showed no difference in performance for the cycles studied. The capacity lost after 500 cycles under either method was about 3%, making it difficult to determine the best charge method for moderate cycle life batteries (Jeevarajan and Cook 2004). The small capacity loss is no small wonder because this cell is used in the first generation hybrid electric vehicles, the Toyota Prius® and Honda Insight®.

Overcharging nickel metal hydride cells should be avoided, because the cells can become extremely hot. They can vent explosive gases, that is, hydrogen and oxygen; and therefore, they lose the active gaseous materials required for the recombination reactions. The pressure inside the cell during an overcharge condition has been found to be proportional to the charge rate (Venkatesan et al. 1989). This pressure decays as the cell is discharged or when it goes into an open circuit condition because the gases either recombine or are reabsorbed into the metal anode. It was observed that, at the C/10 rate, the pressure inside the cell is as high as 50 psig but remains at this value even if the cell is charged continuously in a float mode (Figure 16.22). The nickel metal hydride cells are fitted with a resealable vent to release excessive pressure that builds up inside the cells from an overcharge or other abusive condition, such as external short, overdischarge, or exposure to high temperatures.

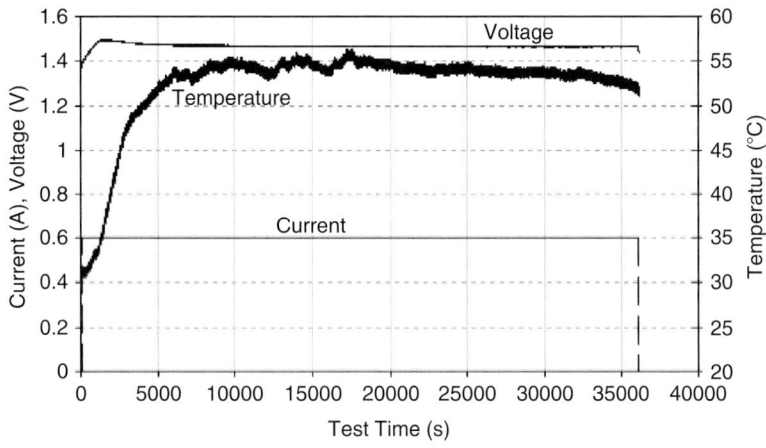

FIGURE 16.22 Overcharge test on nickel metal hydride subC cell using a current of 600 mA for 21 h (Courtesy of NASA).

Theoretically, overdischarge into reversal of nickel metal hydride cells leads to gassing from hydrogen production, increased internal pressures, cell venting, and deterioration of performance due to the release of the gases through the resealable vent. At low rates of discharge, nickel metal hydride cells are capable of being reversed indefinitely without resulting in a catastrophic event or notable loss of performance (Figure 16.23), because the hydrogen produced rapidly is reabsorbed by the negative electrode (Venkatesan et al. 1989).

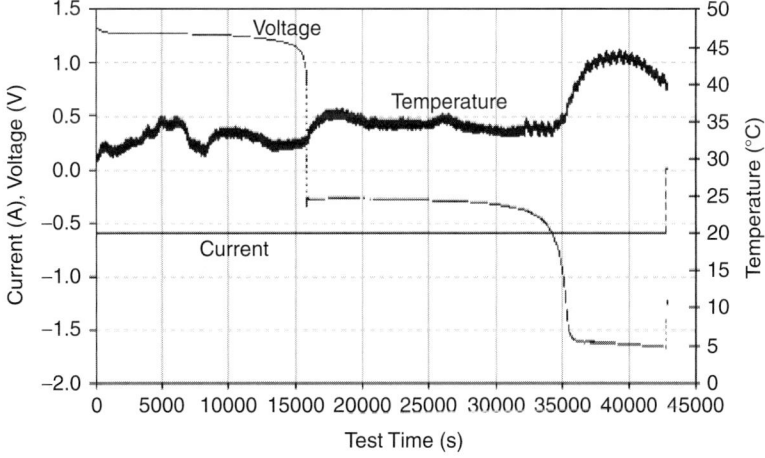

FIGURE 16.23 Overdischarge into reversal of nickel metal hydride subC cell using a current of 0.6 A for an additional 150% of original capacity removal (Courtesy of NASA).

Although nickel metal hydride cells do not have the same memory effect seen in NiCd cells, that is, the formation of intermetallic crystals of nickel and cadmium, they do have considerable passivation if the cells are not maintained well. Storing nickel metal hydride batteries on the shelf at ambient temperatures for long periods leads to passivation, which can be manifested as a voltage depression or incomplete subsequent charge due to a high internal resistance in the cell. This results in high cell temperatures. Despite inadequate literature to describe the exact reactions that occur during this passivation, it is clear that both physical and chemical changes occur on the inactive portions of the electrodes. This causes a change in the internal resistance of the cells (Linden and Reddy 2002). Some have described this phenomenon as the deposition of potassium hydroxide crystals on the surface of the electrodes. These crystals then go back into solution during reconditioning cycles when the cells are charged and discharged a few times. As with the NiCd, nickel metal hydride cells can be cycled a few times with discharges to voltages as low as 0.6 V per cell to remove the passivation and improve their capacity.

Nickel metal hydrides are quite tolerant of external shorts. However, current spikes as high as 50 A were observed with a 30-mΩ load. Under these conditions, permanent vent activation can occur, as indicated with the drop in voltage shown in Figure 16.24. No remarkable increase in temperature was observed during this test or with lower ohmic load short circuits. It was also observed that the cells vent irreversibly when exposed to temperatures above 250°C (Figure 16.25).

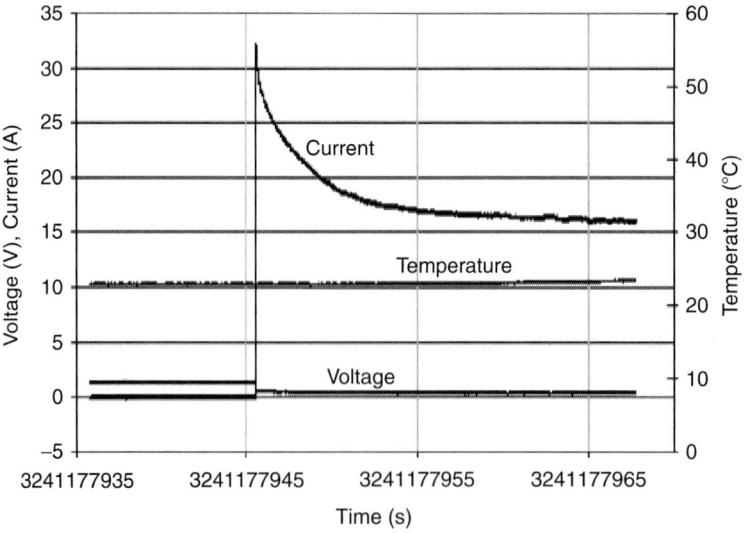

FIGURE 16.24 External short circuit test on nickel metal hydride subC Cell with a 30-Ω load (Courtesy of NASA).

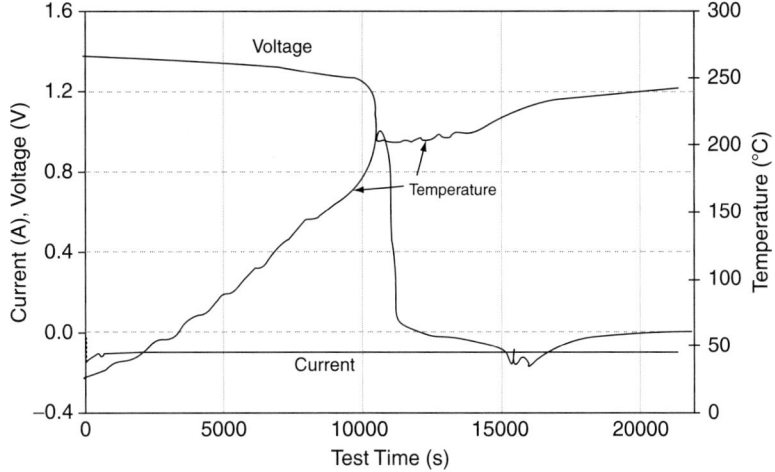

FIGURE 16.25 Heat to vent test on nickel metal hydride subC cell (Courtesy of NASA).

16.6.7 Nickel Hydrogen Batteries

Nickel hydrogen (NiH_2) batteries widely are used in aerospace applications, specifically in satellite systems for both low Earth orbit and geosynchronous Earth orbit. Although satellite systems are switching over to lithium-ion batteries because of weight and cost savings, a majority of satellites still use NiH_2 batteries. No chemistry available today has a proven long-term and high cycle life capability as NiH_2, especially for aerospace applications. The first aerospace application using NiH_2 batteries was *IntellSat V*, Flight 6, in 1983. The first NASA application that used NiH_2 batteries was the *Hubble Space Telescope* in 1990, which was developed by the NASA Goddard Space Flight Center.

The half reactions are

$$\text{Anode:} \quad 1/2\, H_2 + OH^- \leftrightarrow H_2O + e^-$$
$$\text{Cathode:} \quad NiOOH + H_2O + e^- \leftrightarrow Ni(OH)_2 + OH^-$$
$$\text{Overall reaction:} \quad NiOOH + 1/2\, H_2 \leftrightarrow Ni(OH)_2$$

The anode in NiH_2 cells is the hydrogen electrode, which is hydrogen gas in a compressed state and a Teflon® bonded platinum black catalyst supported on a photoetched nickel substrate. The cathode is similar to the other nickel chemistries, composed of nickel hydroxide impregnated on a porous nickel sintered plaque. The electrolyte is concentrated potassium hydroxide, but it varies in concentration depending on the application. The concentration of potassium hydroxide in a fully discharged cell used for low Earth orbit applications ranges from 26 to 31% and from 31 to 38% in those used for geosynchronous Earth orbit applications.

Because NiH_2 batteries contain hydrogen under pressure, they are true pressure vessels and should be treated as such. Batteries used for critical applications such as in powering the *International Space Station* are fitted with strain gauges for monitoring safety and performance. Three main designs are in existence for the pressure vessels used in NiH_2 cell construction: the *COMSAT*, the Air Force, and the hybrid Mantech (Dunlop, Rao, and Yi 1993; Linden and Reddy 2002). The cells also can be of an individual pressure vessel or common pressure vessel design (Linden and Reddy 2002).

In an overcharge condition, the following reactions occur at the two electrodes (Dunlop et al. 1993):

1. Nickel electrode: $2OH^- \rightarrow 2e^- + 1/2\,O_2 + H_2O$
2. Hydrogen electrode: $1/2\,O_2 + H_2O + 2e^- \rightarrow 2OH^-$
3. Hydrogen electrode: $H_2O + e^- \rightarrow OH^- + 1/2\,H_2$
4. Chemical recombination: $1/2\,O_2 + H_2 \rightarrow H_2O$

Typically, neither the combination of reactions 1 and 2 nor the combination of reactions 1, 3, and 4 produces any net. However, reaction 4 can cause popping if there is a localized mixture of hydrogen and oxygen gases in combustible concentrations. Due to the efficient recombination processes that occur after the evolution of oxygen under an overcharge condition, the cell is capable of withstanding continuous overcharge as long as the high temperatures created are dissipated, effectively avoiding a thermal runaway situation.

Similarly, in an overdischarge into reversal condition, the following reactions occur (Dunlop et al. 1993):

5. Nickel electrode: $H_2O + e^- \rightarrow OH^- + 1/2\,H_2$
6. Hydrogen electrode: $1/2\,H_2 + OH^- \rightarrow H_2O + e^-$
7. Nickel electrode: $2NiOOH + 2H_2O + 2e^- \rightarrow 2Ni(OH)_2 + 2OH^-$
8. Hydrogen electrode: $2(OH)^- \rightarrow 2e^- + 1/2\,O_2 + H_2O$
9. Net reaction: $2NiOOH + 2H_2O \rightarrow 2Ni(OH)_2 + 1/2\,O_2$

Reactions 5 and 6 occur in a negative or hydrogen precharge cell, and reactions 7 and 8 occur in the positive precharge cell. In a positive precharge cell, at full discharge to 0 V, all the hydrogen is consumed. If this cell is forced into reversal, oxygen gas is generated at the hydrogen electrode per reaction 8 until nickel hydroxide electrode is discharged fully as in reaction 7. The oxygen gas produced can be consumed during the recharge process, or it can combine with hydrogen gas in localized areas causing an explosive atmosphere and ignite (popping) if in combustible concentrations. In the presence of an alkaline environment, the oxygen can dissolve the platinum metal on the negative electrode, forming soluble platinum hydroxide that can diffuse through the whole cell. In a hydrogen precharge cell at 0 V, the positive electrode is discharged fully, and when the cell is forced into reversal, the hydrogen gas generated at the positive electrode per reaction 5 is consumed at the negative electrode per reaction 6 at the same rate. This allows for the continuous cell reversal in a hydrogen precharge cell without the accumulation of gases or a net change in electrolyte concentration. This is an advantage of this cell design.

An interesting concept in NiH_2 cells is the charge method. Under long cycle life requirements, the cells are neither charged nor discharged fully. This allows for a change and extension of the voltage limits for charge and discharge to obtain the required capacity. Two commonly used methods for charge are the voltage limited taper current charge protocol and the use of a fixed charge-discharge ratio protocol to terminate charge. The preferred charge method for geosynchronous Earth orbit satellites is the latter, with the charge being to 105 to 115% of the discharged capacity, charging at a high rate followed by a low trickle charge rate for the remainder of the eclipse period. For low Earth orbit applications, the battery has to be charged during the period of sunlight and discharged during the period of eclipse and not become overcharged or overdischarged during these cycles. For the Hubble Space Telescope battery, a temperature compensated voltage limited charging method is used because this method originally was used to charge the NiCd batteries. In this case, after a high rate charge, the charge protocol was switched to a very low rate (C/100) trickle charge and a cell voltage limit of 1.513 V at 0°C (Nawrocki et al. 1990; Lowery et al. 1990).

Safety tests indicate that short circuiting an 85-Ah aerospace cell with a 0.65-Ω load produced an initial current draw of 700 to 1000 A, which then stabilized to 400 to 500 A during the first 10 min of the discharge. The current then dropped down to 10 A through the rest of the test (Trancinski and Applewhite 1997). The peak temperatures recorded during the test were 193°C initially, which then fell to about 75°C and finally down to 37°C after approximately an hour into the test. The cell voltage fell to almost 0.0 V during the short circuit test but recovered to approximately 1.15 V after the short was removed. No venting, smoke, or flame was observed, even with the high temperatures recorded.

In general, design and operational faults such as overcharge, overdischarge, shorting (external and internal), and temperature faults (high, that is, above 30°C, and low, that is, below −25°C) should be avoided or mitigated. Because these cells are pressure vessels, the cells should be proof tested to 1.5 times their maximum operating pressure. The cells also should have a 4:1 ratio for the cell burst to maximum operating pressure inside the cell.

Cells and batteries should be charged and used only in well ventilated areas to provide for any dissipation of hydrogen gas due to inadvertent abuse. Electrolyte leakage often accompanies gas leakage, and this can be identified by visual observation of the whitish gray crystals of potassium hydroxide deposited near the cell seals. If any evidence of such leakage is observed, these cells immediately should be replaced. Rapid drops in cell voltage, that is, high self-discharge, is a good indication of an internally shorted cell.

Cells and batteries exposed to temperatures above 30°C have been shown to undergo permanent loss of capacity. Cells should not be allowed to go below −25°C because the potassium hydroxide electrolyte freezes at about −30°C (Falk and Salkind 1969).

16.6.8 Lithium-Ion Batteries

Lithium-ion batteries are of the liquid electrolyte or polymer electrolyte types. At present, lithium-ion batteries are used most commonly for portable electronic applications, having replaced the nickel based systems because of their higher voltage (3.6 V versus 1.2 V) and higher energy density. This chemistry also is being used extensively for aerospace applications, including satellite systems. Improvements in cell materials have made them capable of providing high rate discharges. The cells are of the cylindrical, prismatic, or pouch design.

The half reactions are

Cathode: $LiMO_2 \leftrightarrow Li_{1-x}MO_2 + xLi^+ + xe$ (M = metal)
Anode: $C + xLi^+ + xe^- \leftrightarrow Li_xC$
Overall reaction: $LiMO_2 + C \leftrightarrow Li_xC + Li_{1-x}MO_2$

Lithium-ion cells have a metal oxide cathode with an aluminum current collector, a carbon anode with a copper current collector, and an organic electrolyte with an inorganic salt. In a lithium-ion cell, the charge and discharge reactions are described as simple intercalation and deintercalation processes, wherein the lithium ions are inserted between layers of the carbon matrix and remain affixed there by van der Waals forces during the charging or intercalation process. During the discharge or deintercalation process, they return to their parent matrix. Six carbons are associated with each Li^+ ion that is intercalated. The typical operating voltage for a lithium-ion cell is 3.6 to 3.7 V when using a traditional transition metal oxide cathode, such as $LiCoO_2$, $LiNiO_2$, $LiMn_2O_4$, or when mixtures of these metal oxides are used, such as in $LiCo_xMn_yNi_zO_2$, $LiCo_xMn_yAl_zO_2$, and the like. In recent years, other cathodes (Ritchie and Wilmont 2006; Paulsen, Park, and Kwon 2006) have become available that have a lower operating voltage, such as $LiFePO_4$, which has an operating voltage of 3.3 V (Kim et al. 2006).

The most common anode used in lithium-ion cells is carbon in natural or synthetic forms (such as hard carbon, coke, graphite), mesophase carbon microbeads, carbon fibers, and carbon nanorods. Advances in materials for anodes have indicated that the use of a nanotitanate might provide additional safety, but it has a lower operating voltage of 2.65 V. The electrolyte used in lithium-ion cells is a combination of organic carbonates with a lithium salt, such as lithium hexafluorophosphate, $LiPF_6$. The salt facilitates ionic conductivity; it is used because organic carbonates are very poor ionic conductors. A traditional lithium-ion cell with the transition metal oxides can be charged between 4.1 and 4.2 V using a constant current, constant voltage method (Figure 16.26) and discharged to

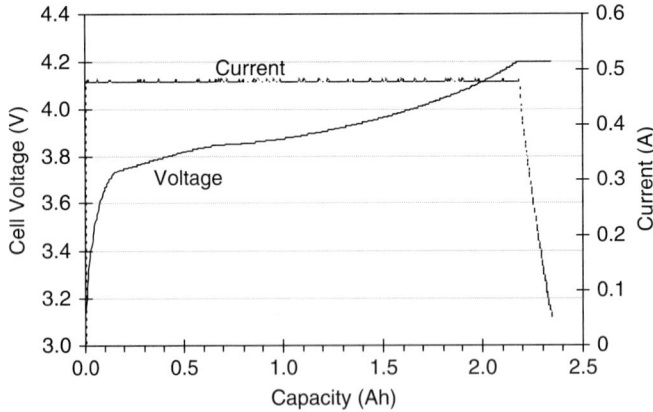

FIGURE 16.26 Typical constant current, constant voltage charge profile for a lithium-ion cell of 2.4-Ah capacity to 4.2 V at 0.48 A current to a 50-mA taper current (Courtesy of NASA).

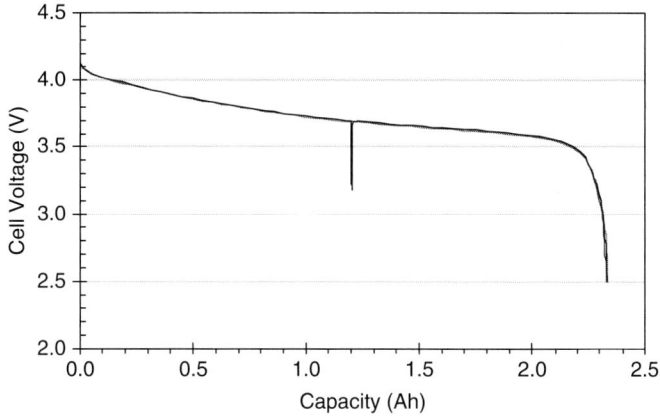

FIGURE 16.27 Typical constant current discharge profile for a lithium-ion cell with a discharge current of approximately 1.0 A (Courtesy of NASA).

voltages between 2.5 and 3.0 V (Figure 16.27), the end voltages depending on the nature of electrode materials. The internal resistance of standard 18650-size lithium-ion cells is between 90 and 110 mΩ, this value decreasing with increasing size to a value of approximately 0.6 mΩ for 200-Ah capacity cells (Jeevarajan, Zhang, and Irlbeck 2002).

Commercial lithium-ion batteries have been used extensively by NASA to provide portable power for space applications since 1998 (Jeevarajan and Bragg 1999; Jeevarajan et al. 2000; Jeevarajan, Cook, and Collins 2003a). The cells range in capacity from 0.75 Ah to about 2.0 Ah, with a maximum battery capacity of 4.0 Ah and a battery voltage of about 10.8 V. Until recently, lithium-ion batteries were capable of discharging at medium rates only (1C rate), but with improvements in cell design, electrode materials, and electrolyte compositions, cells capable of high rates (Jeevarajan and Inoue 2006) have become available in the market. The most recent innovation is the use of a lithium iron phosphate cell in a 36-V Dewalt® impact wrench. Testing of the cells that are used in this tool indicated that the 2.3-Ah cells are capable of providing pulses of up to 130 A (Figure 16.28) (Jeevarajan, Varela, and Nelson 2007).

All commercial battery packs are fitted with smart circuit boards that provide for cell level voltage monitoring and string level voltage balancing, as well as protection against catastrophic failures such as overcurrent due to an external short, overvoltage due to overcharge, and undervoltage caused by overdischarge (Jeevarajan, Cook, and Collins 2004). These are accomplished by the use of hard blow fuses, thermal fuses, and overvoltage and undervoltage limiters. The smart circuit boards used in commercial batteries have integrated circuit chips that continuously are monitoring voltages and currents. These integrated circuits also provide performance management, as well as safety protection. Charge management is provided by cell balancing using different methods, such as high voltage cell bypass with continued charge of low voltage cells or resistive loads on high voltage cells.

Management during discharge is provided in a similar manner by bypassing cells that are at lower voltages than those at higher voltages. Commercial batteries that require

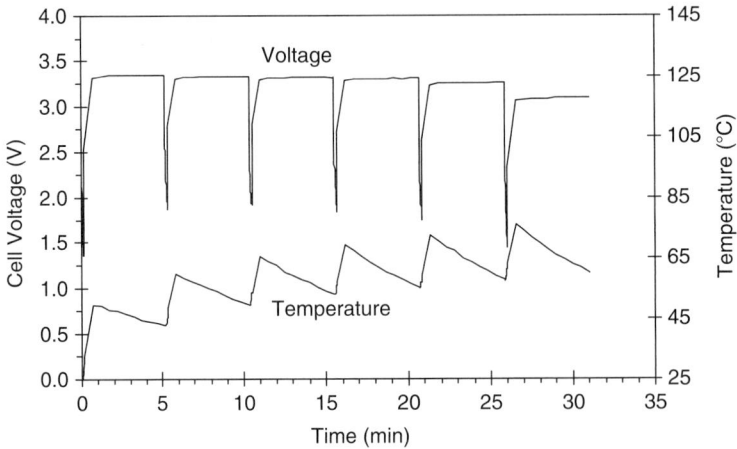

FIGURE 16.28 Pulse power capability of olivine lithium-ion cells under 130-A 10-s loads every 5 min with a nominal discharge of 2.3 A (Courtesy of NASA).

multiple cells to be connected in series and in parallel are configured by connecting banks of cells in parallel before these banks are connected in series. This helps reduce the complexity of current and voltage monitoring (Jeevarajan and Bragg 1999, Jeevarajan et al. 2000, 2003a).

The conditions of overcharge and high temperatures are the most catastrophic failures in most lithium-ion cells and batteries. Under overcharge conditions, instability of the cathode and decomposition of the electrolyte occur. For the case of the cobaltate cathode (an octahedral structure) (Goodenough and Mizushima 1980a, 1980b), above 4.8 V, the cathode releases its oxygen readily, and hence oxygen gas accumulates in the cell. In the case of the spinel cathodes, such as $LiMn_2O_4$ (a tetrahedral structure), the oxygen is not given up readily due to the inherent structure of the cathode. With the invention of new cathode materials, such as $LiFePO_4$ (an olivine structure), the oxygen atoms no longer are attached directly to the metal and therefore are not released as readily. This provides higher stability up to almost 1000°C, at which point the cathode decomposes. In the latter case, the voltage of the cathode stabilizes at the end of charge and does not get into a voltage regime that causes decomposition of the electrolyte.

External short circuit protection at the cell level in cylindrical 18650 and 26650 designs is provided by the positive temperature coefficient and a shutdown separator (Venugopal 2001; Balakrishna, Ramesh, and Prem Kumar 2006). In larger or prismatic cells, short circuit protection is provided by other devices, such as fusible links. Although the shutdown separator has not always proven to be a safety feature, the positive temperature coefficient provides adequate protection at the cell level.

The positive temperature coefficient is a doughnut shaped device, made up of a polymeric material sandwiched between two stainless steel plates. The polymeric material increases in resistance due to the high currents and temperatures resulting from expansion and breaking of the polymeric chains, thus decreasing the number of paths for current to flow. The increased resistance typically drops the current going into the cell to approximately 1 A or lower, which the cell can always tolerate. Once the cause of heating or short circuiting is removed, the positive temperature coefficient can regain about 98% of its original resistance, and the cell returns to its normal state. However, the positive temperature coefficient also has a voltage limitation. Depending on the nature of the materials used, its voltage tolerance can vary from 30 to more than 60 V. Laboratory scale testing on a positive temperature coefficient removed from a commercial cell indicated that these spontaneously ignite at voltages above 30 V. Test programs in the past (Cowles et al. 2002; Jeevarajan et al. 2003b), which indicated this limitation, initiated tests on single as well as series strings of varying lengths to understand this phenomenon (Trancinski and Jeevarajan 2006; Jeevarajan and Patel 2007). Short circuit tests on single cells (Figure 16.29) and a 4-cell series string showed that the positive temperature coefficient functioned very well as required. However, for a string of 14 cells, the cells vented with smoke and high temperatures. Closer study of the voltages and temperatures on each cell (Figures 16.30 and 16.31) indicates that the positive temperature coefficient devices try to balance the current and voltage seen on each cell but do not tolerate it for long due to their withstanding voltage limitation. As the positive temperature coefficient devices ignite at high voltages, the cells could have some softening of seal material. This causes a short circuit in the cell container, resulting in the cell shorting with excessive gas production and venting with flame due to the ignition of the flammable electrolyte and its vapors. The reaction is cascading, and it occurs within a matter of seconds.

Under overcharge conditions, at voltages above 4.5 V, electrochemical decomposition of the organic electrolyte occurs, producing large quantities of carbon monoxide that gets

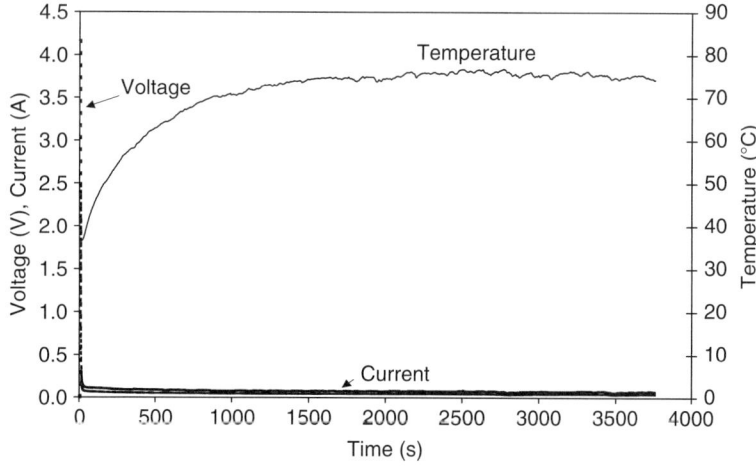

FIGURE 16.29 External short circuit test on a single lithium-ion cell using a 50-mΩ load (Courtesy of NASA).

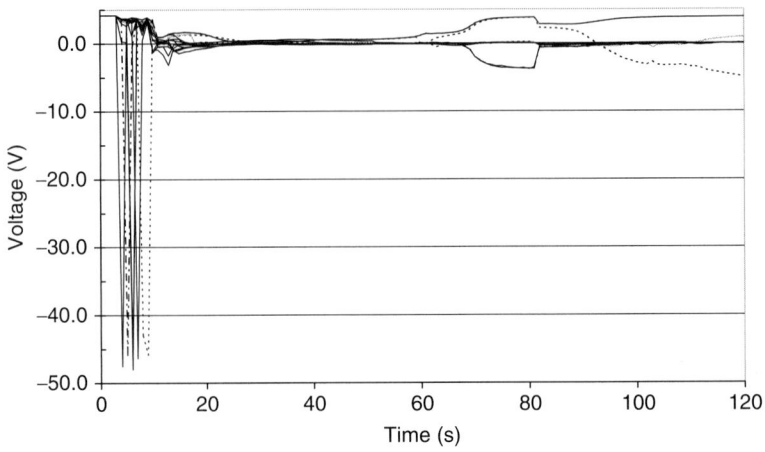

FIGURE 16.30 External short circuit voltage data for a 14-cell lithium-ion cell string (Courtesy of NASA).

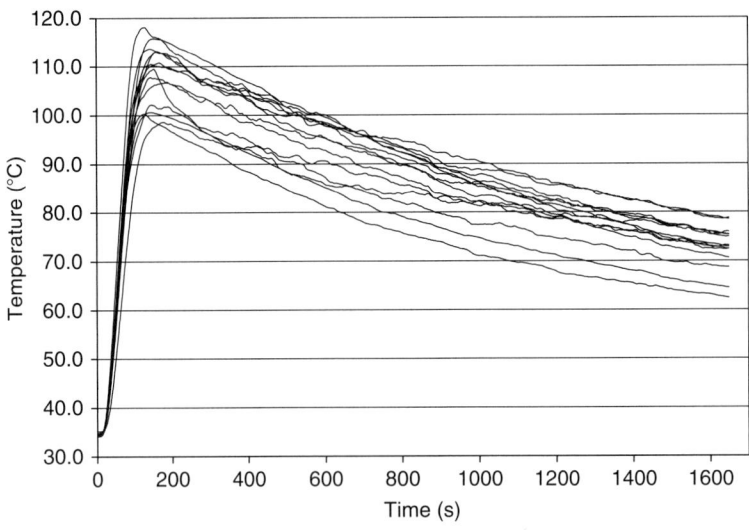

FIGURE 16.31 Temperatures recorded during the 14-cell short circuit test on lithium-ion 18650 cells (Courtesy of NASA).

converted to carbon dioxide with continued voltage increase. Many additives (Venugopal 2001) have been researched to reduce the hazards associated with overcharging, but for the 18650 cell design, an internally located current interrupt device, which activates under pressure and is typically non-resettable, causes the cell to remain open (unusable) after activation.

Observations of the current interrupt device not being protective in high voltage cell strings led to testing the single, 4-cell and 14-cell series strings under overcharge conditions (Jeevarajan and Patel 2007). The current interrupt device activated in all single cells above 5.0 V (Figure 16.32) even though the time for activation varied from cell to cell. The 4-cell string became inactive after the current interrupt device in one cell activated, resulting in a high resistance in the string. In the 14-cell string (Figures 16.33 and 16.34), a close analysis of oscilloscope data, as well as fast rate data collection, indicated that the first current interrupt device activation (cell 36, at 11 s after the 50-min hold period) was incomplete. In this test, a current spike of 2.5 A was observed that lowered to about 1.6 A, then oscillated between 1.6 and 1.5 A for a period of about 35 s, after which it fell to 0 A. During this period, a second one (cell 31) spiked to about 24 V and fell to about 8 V. For the next 200 s, the voltages of at least two cells increased but remained below 10 V, even though the current to the cell string was at 0 A. This indicates some communication between the cells, although the high resistance at the total string level allows no charging currents.

A closer look at the cell voltages indicated that at least three cells dropped in voltage below 0 V, indicating positive temperature coefficient failures. The increased vapor pressure of the flammable electrolytes, the presence of oxygen released from the cathode at high voltages, and the presence of the positive temperature coefficient ignition source resulted in the cells venting and thermal runaway, resulting in cell disassembly and violent reactions. Another possibility is the creation of sharp edges during the opening of the current interrupt device that could cause the incidence of a flame in the presence of hot flammable gases. For designs that rely on the cell level positive temperature coefficient

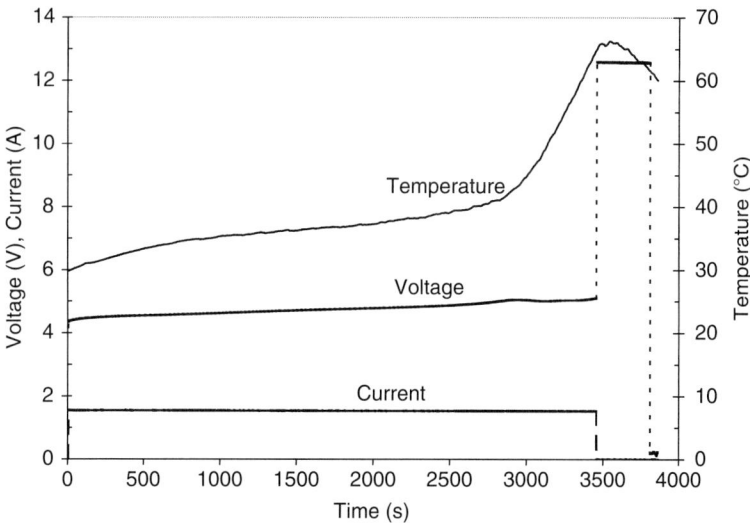

FIGURE 16.32 Overcharge test on a single lithium-ion cell using a current of 1.5 A and voltage limit of 12 V (Courtesy of NASA).

542 CHAPTER 16 Battery Safety

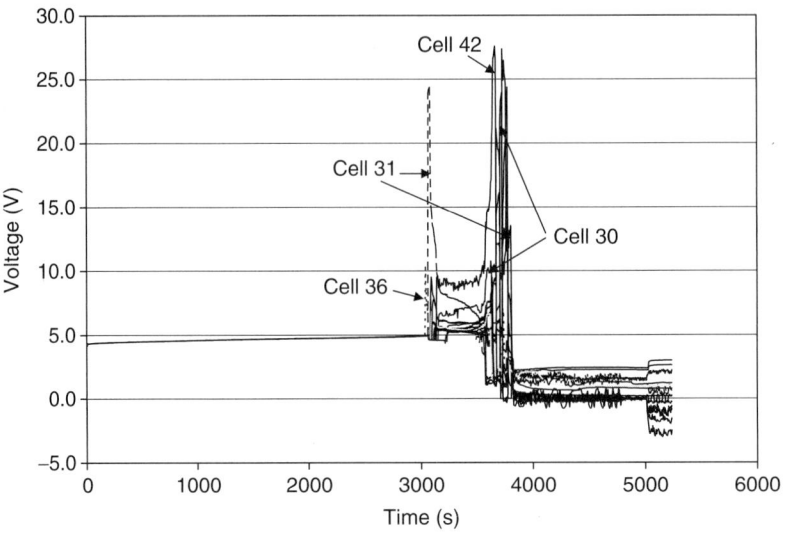

FIGURE 16.33 Voltage pattern during an overcharge test on a lithium-ion 14-cell string (Courtesy of NASA).

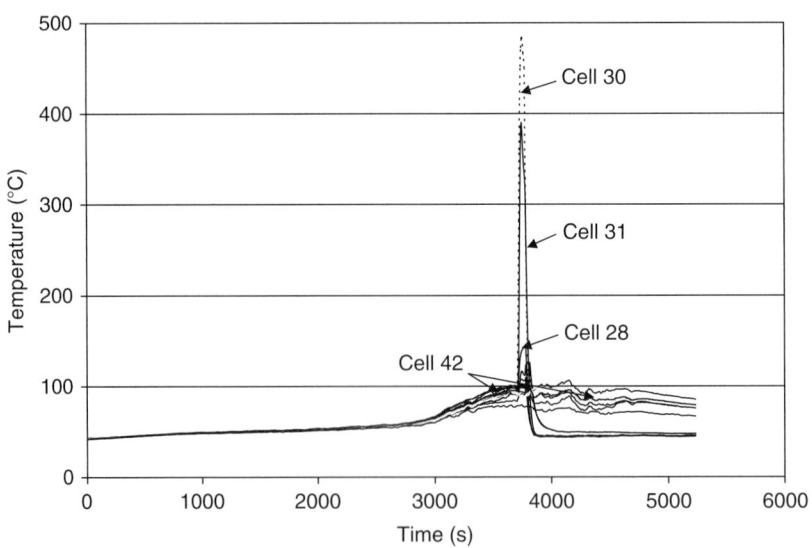

FIGURE 16.34 Temperature readings recorded during an overcharge test of a 14-cell lithium-ion series string (Courtesy of NASA).

and current interrupt device as controls, these features should be characterized and their tolerance and limitations well understood. In prismatic cells and cell designs that do not incorporate a current interrupt device, overcharge has resulted in the breakage of the cell header weld or pouch, releasing flammable gases and electrolyte. In some cases, this resulted in disassembly of the cell. For this reason, cell vents or rupture disks should be redundant and work as expected. Adequate samples from every new production lot should be tested to confirm that the vents work adequately and as designed for protection.

All lithium-ion cells have a three-layer polymeric separator material of polyethylene sandwiched in between two layers of polypropylene. At ambient temperatures, the separator is permeable to the passage of ions. When the temperature of the cell increases to about 130°C, the polyethylene layer melts and makes the separator opaque to the passage of ions. Scanning electron microscopy of the polymeric material shows that the pores in a good separator are blocked in the activated separator. Although this property of the separator is in place to protect the cells under internal and external short circuit and high thermal environment conditions, the temperature at which the separator melts is too close to the thermal runaway temperature of the lithium-ion cells. Separators that activate at lower temperatures of between 80 and 110°C have been studied (Jeevarajan and Sun 2006). A refinement in the manufacturing process and the incorporation into actual lithium-ion cells would help provide the much needed protection offered by the shutdown separator.

Overdischarge into reversal for lithium-ion cells traditionally has not resulted in a catastrophic hazard (Figure 16.35). Below 2.5 V, electrochemical dissolution of copper (the negative current collector) occurs, and the copper ions migrate through the separator to be deposited on the cathode. This can cause copper deposition on the separator as well. The result is a typically benign internal short that occurs due to the deposition of copper on the separator and cathode (Vaidyanathan and Jeevarajan 2006), thus rendering

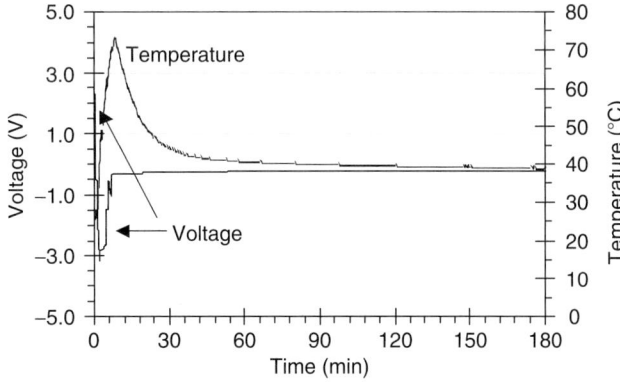

FIGURE 16.35 Overdischarge into reversal for a typical lithium-ion cell with a C rate current (Courtesy of NASA).

the cell inactive. In critical applications, the loss of a cell can result in loss of power, and hence cells should not be subjected to an overdischarge condition.

Internal shorts in lithium-ion cells often result in disassembly of the cell. This can be prevented only by excellent manufacturing control and very high quality. The recent lithium-ion battery recalls due to venting with flames have been related typically to poor quality control during the manufacturing process. Destructive analysis of cells has indicated that burrs from improperly cut current interrupt devices, flakes of anodized materials from the interior of the cell can, and active material impurities adhering to the electrodes have been the major causes of these internal short circuits. The author, with consultation and collaboration of the structures group at NASA Johnson Space Center, used a vibration screening method (Jeevarajan and Darcy 2006) to screen all flight batteries for internal shorts and for battery chemistries that are intolerant to internal shorts. This method of vibration screening has been used since the first lithium-ion battery was certified for flight in space in 1998 as well as for other battery chemistries that have been tested and shown to have an intolerance to internal shorts due to internal cell defects.

In short, adequate controls using hardware, software, or a combination of both should be in place for protection against overcharge, overdischarge, and external shorts. Because external controls cannot be used for internal short failure protection, a good method of screening for internal shorts, such as vibration, should be used. The thermal environment within which the lithium-ion batteries are used should not exceed 100°C to provide adequate margin from causing a thermal runaway in the batteries (Figure 16.36).

16.7 STORAGE, TRANSPORTATION, AND HANDLING

All cells and batteries should be stored at the manufacturer's recommended temperatures. Long-term storage should be at low temperatures to reduce the irreversible self-discharge

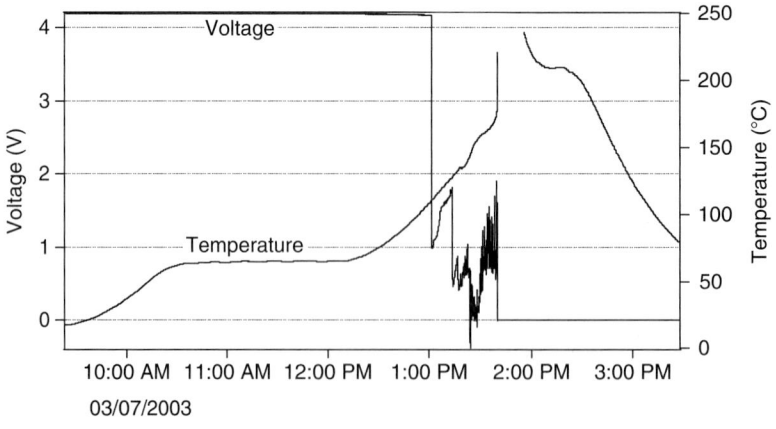

FIGURE 16.36 Typical heat to vent test result on a $LiCoO_2$ cell (Courtesy of NASA).

caused by decomposition of electrolyte and other undesirable side reactions. The U.S. Department of Transportation has regulations for transportation of batteries, especially the primary lithium and rechargeable lithium-ion.

Packaging of batteries for transportation should take the toxicity of the electrolyte into consideration (Lam et al. 1991). Absorbent material usage helps absorb any leaked electrolytes. Packaging in clear plastic bags allows the user and the crew to observe warning signs, such as electrolyte leakage or venting, before they are exposed inadvertently to toxic or corrosive substances due to a defective battery. Batteries should be handled with care through all phases of research, development, design, and usage as they are energetic devices and contain components that are of different toxicity ratings.

REFERENCES

American Society for Testing and Materials. (2005) *Standard Test Method for Autoignition Temperature of Liquid Chemicals*. Standard E659-78. West Conshohocken, PA: American Society for Testing and Materials.

American Society for Testing and Materials. (2007) *Standard Test Method for Flash Point and Fire Point of Liquids by Tag Open-Cup Apparatus*. D 1310-01. West Conshohocken, PA: American Society for Testing and Materials, pp. 484-497.

Badcock, C., and M. R. Martinelli. (1979) Reversal of nickel cadmium cells. Proceedings of the 1979 GSFC Battery Workshop. Greenbelt, MD: National Aeronautics and Space Administration, Goddard Space Flight Center, p. 355.

Balakrishna, P. G., R. Ramesh, and T. Prem Kumar. (2006) Safety mechanisms in lithium-ion batteries. *Journal of Power Sources* 155: 401.

Bard, A. J., and L. R. Faulkner. (2001) *Electrochemical Methods: Fundamentals and Applications*. New York: Wiley and Sons.

Blomgren, G. E., V. Z. Leger, T. Kalnoki-Kio, M. L. Kronenberg, and R. J. Brodd. (1978) Projected mechanism for thionyl chloride and sulfuryl chloride cathode reactions. *Journal of Power Sources* 7: 583.

Bockris, J. O'M., and A. K. N. Reddy. (1970) *Modern Electrochemistry*. New York: Plenum.

Bragg, B. J., and R. C. McDonald. (1992) Development of internal/external short circuit protection for lithium D cells. Proceedings of the 1991 NASA Aerospace Battery Workshop. Huntsville, AL: National Aeronautics and Space Administration, Marshall Space Flight Center, pp. 215-227.

Brooks, E. S. (1974) Evaluation of designs for safe operation of lithium batteries. *Proceedings of the 26th Annual Power Sources Conference*. Atlantic City, NJ: Department of Defense, United States Army CECOM, United States Army DECOM, and United States Army Research Laboratory.

Buchmann, I. (2001) *Batteries in a Portable World: A Handbook on Rechargeable Batteries for Non-Engineers*. Richmond, BC, Canada: Cadex Electronics.

Carter, B. J., R. M. Williams, F. D. Tsay, A. Rodriguez, S. Kim, M. M. Evans, and H. Frank. (1985) Mechanistic studies related to the safety of Li/SOCl2 cells. *Journal of the Electrochemical Society* 132, no. 3: 525.

Corrigan, D. A. and S. Srinivasan (eds.). (1992) *Hydrogen Storage Materials, Batteries and Electrochemistry*, vol. 92-5. Pennington, NJ: Electrochemical Society.

Cowles, P. R., E. C. Darcy, F. J. Davies, and J. A. Jeevarajan. (2002) Safety performance of small lithium-ion cells in high voltage batteries. Proceedings of the 2002 NASA Aerospace Battery Workshop. Huntsville, AL: National Aeronautics and Space Administration, Marshall Space Flight Center.

Davis, P. B., R. F. Bis, J. A. Barnes, S. F. Buchholz, F. C. Debold, and L. A. Kowachik. (1983) *Safety Evaluation of an Electronic Totalizer Containing a Lithium/Poly (Carbon Monofluoride) (CFx) Cell.* NSWC TR 83508. Silver Spring, MD: National Technical Information Service, Naval Surface Warfare Center.

Dimpault-Darcy, E. C. (1999) Investigation of the response of NiMH cells to bump charging. Ph.D. dissertation, University of Houston.

Dimpault-Darcy, E. C. (2007) Personal communication. Houston, TX: National Aeronautics and Space Administration, Johnson Space Center.

Department of Transportation. (2007) *Title 49—Transportation: Subchapter C—Hazardous Materials Regulations.* 49 CFR Parts 171, 172, 173 and 175. Available at http://ecfr.gpoaccess.gov/cgi/t/text/text-idx?sid=585c275ee19254ba07625d8c92fe925f&c=ecfr&tpl=/ecfrbrowse/Title49/49cfrv2_02.tpl (cited July 6, 2007).

Dunlop, J. D., G. M. Rao, and T. M. Yi. (1993) *NASA Handbook for NiH2 Batteries.* Research Publication 1314. Greenbelt, MD: National Aeronautics and Space Administration, Goddard Space Flight Center.

Falk, S. U., and A. J. Salkind. (1969) *Alkaline Storage Batteries.* New York: Wiley and Sons.

Fetcenko, M. A., and S. Venkatesan. (1990) Metal hydride materials for rechargeable NiMH batteries. *Progress in Batteries and Solar Cells* 9: 259.

Fiala, V., J. Mrha, J. Jindra, and M. Musilova. (1985) Protection of sealed Ni-Cd cells from cell voltage reversal: I. Experimental conditions for the formation of cadmium bridges. *Journal of Power Sources* 14: 285.

Frerker, R., J. A. Jeevarajan, and B. J. Bragg. (2000) Performance and abuse testing of low rate and medium rate lithium thionyl chloride cells. *Proceedings of the 2000 NASA Aerospace Battery Workshop.* Huntsville, AL: National Aeronautics and Space Administration, Marshall Space Flight Center.

Goodenough, J. B., and K. Mizushima. (1980a) Electrochemical cell with new fast ion conductors. U.S. Patent 44,302,518. Washington, DC: U.S. Patent and Trademark Office.

Goodenough, J. B., and K. Mizushima. (1980b) Fast ion conductors. U.S. Patent 4,357,215. Washington, DC: U.S. Patent and Trademark Office.

Himy, A. (1986) *Silver-Zinc Battery: Phenomena and Design Principles.* New York: Vantage Press.

Hunger, H. F., and J. A. Christopulos. (1975) *Preliminary Safety Analysis of Lithium Batteries.* Technical Report ECOM-4292, U.S. NTIS ADIA-006149. Forth Monmouth, NJ: Department of Defense, United States Army Electronics Command.

Istone, W. K., and R. J. Brodd. (1982) On the mechanism of thionyl chloride reduction. *Journal of the Electrochemical Society* 129: 1853.

Jeevarajan, J. A. (2005a) Comparison of safety of two primary lithium batteries for the orbiter wing leading edge impact sensors. *Proceedings the 208th Electrochemical Society Meeting,* San Diego, CA: The Electrochemical Society.

Jeevarajan, J. A. (2005b) Comparison of safety of two primary lithium batteries for the orbiter wing leading edge impact sensors. *Proceedings of the 1st IAASS Conference,* Nice, France: International Association for the Advancement of Space Safety.

Jeevarajan, J. A., and B. J. Bragg. (1999) Engineering and abuse testing of Panasonic lithium-ion battery and cells. *Proceedings of the 1999 NASA Aerospace Battery Workshop,* Huntsville, AL: National Aeronautics and Space Administration, Marshall Space Flight Center.

Jeevarajan, J. A., and J. Cook. (2004) Comparison of charge methods for a high capacity Panasonic NiMH commercial cell. *Proceedings of the 203rd Electrochemical Society Meeting.* San Antonio, TX: The Electrochemical Society.

Jeevarajan, J. A., J. S. Cook, and J. Collins. (2003a) Engineering evaluation of performance and safety of Moli spinel cells. *Proceedings of the 2003 NASA Aerospace Workshop*, Huntsville, AL: National Aeronautics and Space Administration, Marshall Space Flight Center.

Jeevarajan, J. A., J. S. Cook, and J. Collins. (2004) Safety evaluation of two commercial lithium-ion batteries for space applications. *Proceedings of the 41st Power Sources Symposium*, Philadelphia: Department of Defense, United States Army CECOM, United States Army DECOM, and United States Army Research Laboratory.

Jeevarajan, J. A., and E. C. Darcy. (2006) *Crewed Space Vehicle Battery Safety Requirements*. Document JSC 20793. Houston, TX: National Aeronautics and Space Administration, Johnson Space Center.

Jeevarajan, J. A., and T. Inoue. (2006) A novel Li-ion pouch cell tested for performance and safety. *Proceedings of the 42nd Power Sources Symposium*, Philadelphia: Department of Defense, United States Army CECOM, United States Army DECOM, and United States Army Research Laboratory.

Jeevarajan, J. A., and P. Patel. (2007) Comparison of safety of 18650 lithium-ion cells from three different manufacturers. *Proceedings of the 2007 Space Power Workshop*. Los Angeles: The Aerospace Corporation.

Jeevarajan, J. A., and L.-Y. Sun. (2006) A variable temperature shut down separator for safer lithium-ion cells. *Proceedings of the 2006 Space Power Workshop*, Manhattan Beach, CA: The Aerospace Corporation.

Jeevarajan, J. A., G. Varela, and T. Nelson. (2007) High rate olivine lithium-ion cells for power tool applications. The Olivine Workshop. Pasadena, CA: California Institute of Technology, Jet Propulsion Laboratory.

Jeevarajan, J. A., W. Zhang, and B. W. Irlbeck. (2002) Cycle life testing of high capacity lithium-ion cells for electric auxiliary power unit. Proceedings of the 40th Power Sources Symposium. Cherry Hill, NJ: Department of Defense, United States Army CECOM, United States Army DECOM, and United States Army Research Laboratory.

Jeevarajan, J. A., E. C. Darcy, F. J. Davies, and P. Cowles. (2003b) Lithium-ion cell PTC limitations and solutions for high voltage battery applications. *Proceedings of the 203rd Meeting of the Electrochemical Society*, Paris: The Electrochemical Society.

Jeevarajan, J. A., F. J. Davies, B. J. Bragg, and S. M. Lazaroff. (2000) Safety evaluation of two commercial lithium-ion batteries. *Proceedings of the 39th Power Sources Symposium*, Cherry Hill, NJ: Department of Defense, United States Army CECOM, United States Army DECOM, and United States Army Research Laboratory.

Karpinski, A. P., B. Makovetski, S. J. Russell, J. R. Serenyi, and D. C. Williams. (1999) Silver-zinc: Status of technology and applications. *Journal of Power Sources* 80, nos. 1-2: 53-60.

Kim, D.-K., H.-M. Park, S.-J. Jung, Y. U. Jeong, J.-H. Lee, and J.-J. Kim. (2006) Effect of synthesis conditions on the properties of LiFePO4 for secondary lithium batteries. *Journal of Power Sources* 159: 237.

Lam, C. W., K. T. Wong, and M. Coleman. (1991) *Toxicologic Hazard Assessments on Batteries Used in Space Shuttle Missions*. NASA-JSC 25159. Houston, TX: National Aeronautics and Space Administration, Johnson Space Center.

Linden, D., and T. B. Reddy. (2002) *Handbook of Batteries*. New York: McGraw-Hill.

Lowery, J. E., J. R. Lanier Jr., C. I. Hall, and T. H. Whitt. (1990) Ongoing nickel-hydrogen energy storage device testing at George C. Marshall Space Flight Center. *Proceedings of the 25th International Energy Conversion Engineering Conference*. Reno, NV: Institute of Electrical and Electronic Engineers, pp. 28-32.

Mantell, C. (1983) *Batteries and Energy Systems*. New York: McGraw-Hill.

Nawrocki, D. E., J. D. Armantrout, D. J. Standlee, R. C. Baker, and J. R. Lanier. (1990) The Hubble Space Telescope nickel-hydrogen battery design. *Proceedings of the 25th Intersociety Energy Conversion Engineering Conference*. New York: American Institute of Chemical Engineers.

Paulsen, J. M., H.-K. Park, and Y. H. Kwon. (2006) Ni-based lithium transition metal oxide. U.S. Patent 2006/0233696 A1 20061019. Washington, DC: U.S. Patent Office.

Pipoli, T., B. Hendel, G. Dedley, and M. Schautz. (1998) Evaluation of maximum safe voltage for nickel cadmium cells. *Proceedings of the 5th European Space Power Conference*, Tarragona, Spain: European Space Agency, p. 735.

Ritchie, A., and H. Wilmont. (2006) Recent developments and likely advances in lithium-ion batteries. *Journal of Power Sources* 162: 809.

Ritterman, P. (1976) Hydrogen recombination in sealed nickel cadmium cells. *Annual Proceedings of the 27th Power Sources Conference*, Atlantic City, NJ: Department of Defense, United States Army CECOM, United States Army DECOM, and United States Army Research Laboratory.

Trancinski, W. A., and A. Z. Applewhite. (1997) Short circuit testing of a nickel hydrogen cell for compliance with range safety requirements. *Proceedings of the Annual Battery Conference on Applications and Advances*, Long Beach, CA: Institute of Electrical and Electronic Engineers, pp. 61–62.

Trancinski, W. A. and J. A. Jeevarajan. (2006) Safety considerations in lithium-ion series strings. *Proceedings of the 2006 NASA Battery Workshop*, Huntsville, AL: National Aeronautics and Space Administration, Marshall Space Flight Center.

Vaidyanathan, H., and J. A. Jeevarajan. (2005) Destructive physical analysis of components of cycled and abused lithium-ion cells. *Proceedings of the 2005 NASA Aerospace Battery Workshop*. Huntsville, AL: National Aeronautics and Space Administration, Marshall Space Flight Center.

Venkatesan, S., M. Fetcenko, B. Reichman, and K. C. Hong. (1989) Development of ovonic rechargeable metal hydride batteries. *Proceedings of the Intersociety Energy Conversion Engineering Conference*, New York: Institute of Electrical and Electronic Engineers, p. 1659.

Venugopal, G. (2001) Characteristics of thermal cut-off mechanisms in prismatic lithium-ion batteries. *Journal of Power Sources* 101: 231.

Zimmerman, A. H., and P. K. Effa. (1982) *Short Circuit Formation during NiCd Cell Reversal*. Technical Report SD-TR-82-26. Los Angeles: The Aerospace Corporation, National Technical Information Service.

CHAPTER 17

Mechanical Systems Safety

Dean W. Moreland
Executive Officer, Payload Safety Review Panel, Johnson Space Center, National Aeronautics and Space Administration, Houston, Texas

CONTENTS

- 17.1. Safety Factors 549
  By Dean W. Moreland
- 17.2. Spacecraft Structures 551
  By Constantinos Stavrinidis
- 17.3. Fracture Control 567
  By Scott C. Forth, Ph.D.
- 17.4. Pressure Vessels, Lines, and Fittings 568
  By William D. Manha and John D. Albright
- 17.5. Composite Overwrapped Pressure Vessels 576
  By Nathanael J. Greene
- 17.6. Structural Design of Glass and Ceramic Components for Space System Safety 581
  By Karen S. Bernstein
- 17.7. Safety Critical Mechanisms 591
  By Brandan R. Robertson

The structural layout of a spacecraft is in many ways analogous to the structure of the human body. Both have a primary load carrying structure, numerous secondary structures, and depending on the spacecraft design, mechanisms and pressurized fluid systems. The primary purpose of these structural systems is to provide physical support for various mission functionalities: sensors, power systems, command and data handling, and communication systems. This chapter attempts to provide the reader with some useful insight into this primary aspect of spacecraft design.

17.1 SAFETY FACTORS

The use of safety factors to provide ensurance that a structure can perform its intended function, where the failure of that structure would result in negative consequences, is common in all industries. In its most simple form, a structural safety factor is the ratio of the strength of the structure to the maximum expected load. Safety factors as used in the aerospace industry and most other industries are simply design factors imposed as requirements to assure that risk of failure for safety or mission success is reduced to an acceptable level:

$$\text{Safety factor} = \text{Actual capability}/\text{Required capability} \tag{17.1}$$

Safety factors might be better described as ignorance or fudge factors. If the design engineer could determine the exact or, at least with great accuracy, the maximum stress of a structure and knew with similar accuracy the strength of the structural component, the part would need to be only slightly stronger than required, and failure would not be possible. Many factors make it virtually impossible to determine with the required accuracy either the maximum stress or the load capability of the structure. One need only review any of a number of space vehicle failures in recent decades to gain some appreciation of the consequences resulting from the failure to understand the operating environment and the performance of the structure under these conditions.

17.1.1 Types of Safety Factors

Design safety factors can be either deterministic or probabilistic. The deterministic approach generally is used in the space industry for payloads on human rated systems. A deterministic safety factor is essentially a consensus of the structural community's experience of what constitutes a conservative engineering approach to account for all uncertainties that can arise on the path from concept to operations in space. The attractiveness, at least from a programmatic standpoint, of using safety factors is that they easily are understandable using conventional verification techniques, such as stress analysis and structural testing. They also provide project managers with a line in the sand by which to judge whether a given design is safe or unsafe. When a structure is made of conventional aerospace materials such as aluminum, has relatively simple loads and load paths, and can be tested to verify the stress analysis that several decades of flight experience has demonstrated, the classical safety factor approach generally is conservative. For a more complex structural design and loading environment, the deterministic approach to safety factors is less conservative, simply by consideration of the potential for unknown variations in the strength of the structure and the environment within which it must operate.

The use of deterministic safety factors provides insight into neither the actual risk of structural failure nor a situation that can result in a catastrophic structural failure; or for the opposite case, a structure that is considerably heavier than would otherwise be required to meet the safety and mission success goals of the project. The effects of overdesigning a structure from a strength point of view are not apparent always, but in many cases they are not detrimental when the weight of a payload is not constrained severely.

To use a probabilistic approach for structural design, it is necessary to establish a structural reliability requirement for the design that takes into consideration the probability of variation in the load environment and the actual strength of the structure. A detailed treatment of the mathematics of this probabilistic approach to a safety factor can be found in the structures and mechanisms text listed in the reference materials (Sarafin and Larson 1995).

17.1.2 Safety Factors Typical of Human Rated Space Programs

The design safety factors used by payloads for the *Space Shuttle* and International Space Station Programs are described in their basic forms in *Safety Policy and Requirements for Payloads Using the Space Transportation System* (NASA 1989) and the ISS addendum to this document (NASA 1995). More detailed structural design and test factors of safety are described in NASA Technical Standard 5001 (NASA 1996). Note that the safety factors described in this document exclude certain system requirements, such as pressure vessels, ground support equipment, rocket engines, and various areas of interest where other NASA standards are used to guide application of appropriate design constraints.

The intent of NASA Standard 5001 is to provide guidance for the design and verification of the structural features of the payload using methods typical of current practice in the aerospace industry. The designs covered by the standard are assumed to use materials of construction and methods of manufacture that are well understood for the intended use environment and the loads imposed on the structure are well defined. The standard describes in some detail the safety factor requirements for ultimate, yield, qualification, and acceptance testing, as well as test and no-test options. The information is not included here, and the reader should refer to the standard for the details. Also note that the safety factors used by the *Space Shuttle* and International Space Station Programs do not apply necessarily to other launch vehicles or to on-orbit transfer vehicles.

17.1.3 Things That Influence the Choice of Safety Factors

When determining the appropriate safety factor to apply, the design constraints to consider must take into account the expected environment, the materials of construction, and the manufacturing process for the structure. The more uncertainty there is in understanding the true nature of these three areas, the higher the required safety factor. Some of the relevant constraints that drive the choice of a specific safety factor are described in more detail in other chapters.

17.2 SPACECRAFT STRUCTURES

Structures form the skeleton of all spacecraft, and thus, the means of support for key spacecraft components in locations where various constraints need to be considered, such as thermal control, fields of view for antenna and sensors, and the length and weight of cables. For payload hardware, the stowed configuration must fit within the payload envelope of the launch vehicle; yet, the structural design must provide access for installing and maintaining components.

Structures need to provide adequate protection for spacecraft components from dynamic environments during ground operations, launch, on-orbit operations, reentry, and landing. Successful deployment and mission operations are a prime consideration during structural design. As an example, structures should be designed to ensure deployment of antennas and sensors and provide adequate structural stability during their operation.

Structures must, of course, be sufficiently light to maximize the utilization of available payload mass.

The stiffness of a structure must comply with the frequency constraints of the launcher so as to not interfere with the dynamics of the launch vehicle during flight. Similarly, the structural frequency characteristics of the spacecraft in its deployed configuration must not interfere with its own control system.

The materials used must survive ground, launch, and on-orbit environments, such as time, varying applied forces, pressure, humidity, radiation, contamination, thermal cycling, and atomic particles. It must do so without rupturing, collapsing, excessively distorting, or contaminating any critical components. Additionally, structural materials need to support thermal control and, in some cases, electrical conductivity.

The manufacturing, handling, and storage processes must be selected or conceived so as not to introduce any life limiting factors to the structure.

There are three categories of structure (Figure 17.1):

- The primary structure is the backbone, or the major load path, between the components of the spacecraft and the launch vehicle.
- Secondary structures include support beams, booms, trusses, and solar panels.
- Tertiary structures refer to the smallest structures, such as boxes that house electronics and brackets that support electrical cables.

Typical characteristics of these structure categories are

- The primary structure usually is designed for stiffness or natural frequency and to survive the steady state accelerations and transient loading during launch. A major characteristic of the primary structure is to interface adequately with the launcher.
- The design of secondary structures is influenced appreciably by steady state accelerations and transient loading during launch; however, loads generated by mission operations, on-orbit thermal cycling, and acoustic pressure during launch often constitutes a more severe environment for deployable appendages.
- Stiffness and structural stability, including thermal cycling over the lifetime, are other driving requirements for primary, secondary, and tertiary structures.

17.2.1 Mechanical Requirements

Mechanical requirements are defined and typical sources for requirements are identified for spacecraft structures in this section.

Strength is the amount of load applied once to a structure that must not cause rupture, collapse, or deformation sufficient to jeopardize a mission. This basic structural requirement for all space programs applies to all life cycle events. Strength inadequacy can lead to spectacular failures.

Structural life is the number of loading cycles a structure can withstand before its materials fatigue, such as a rupture from cycling loading. Structural life determined by the duration of sustained loads a structure can withstand before its materials creep or

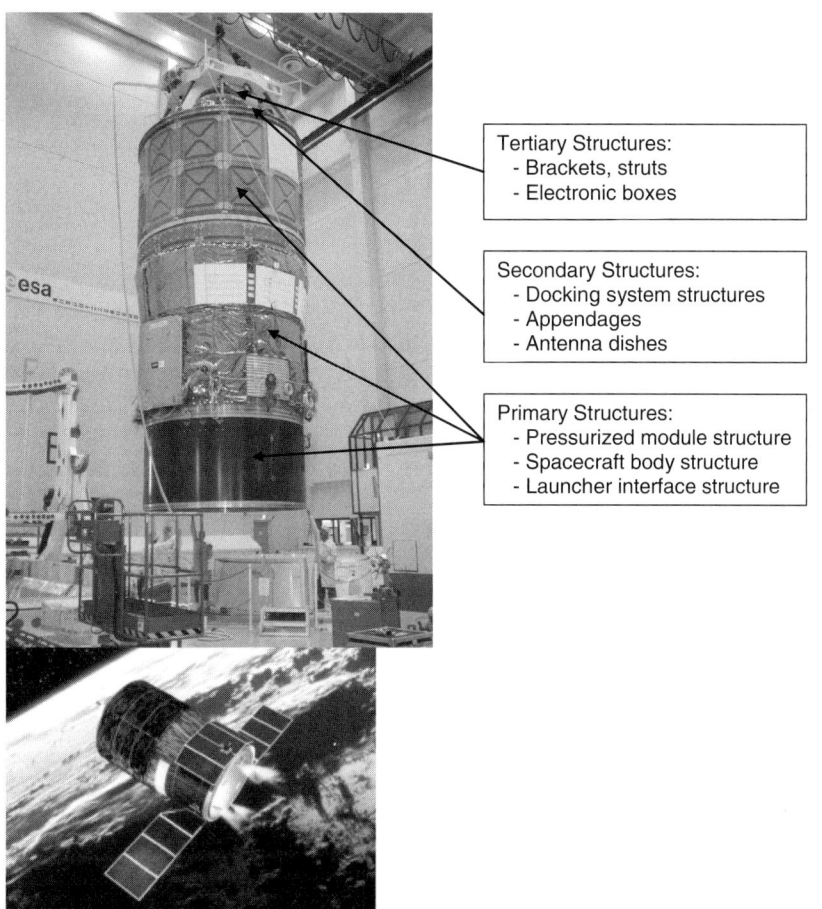

FIGURE 17.1 Categories of structures (ESA *Automated Transfer Vehicle*) (Courtesy of ESA).

deform excessively from sustained loading or before they crack from stress corrosion. This applies to all life cycle events.

Structural response attenuation is the magnitude of the vibration in response to external loads that must be contained to avoid damage to critical components or the launch vehicle.

Natural frequency is the frequency at which a structure vibrates when excited by a transient load or when left undisturbed. It depends on mass properties and stiffness. Each structure has an infinite number of natural frequencies, each corresponding to different mode shapes of vibration. Natural frequencies usually must be above a given value or outside of a specified range. Of particular concern is the lowest fundamental frequency of a structure, which corresponds to its first mode of operation. In the stowed configuration, the natural frequencies of the combined spacecraft-launch vehicle system must not interfere with the control system of the launch vehicle by causing excessive loads. Likewise, for the on-orbit configuration, vibrations must not affect the control system of the spacecraft.

Stiffness is often prescribed for substructures to achieve the required natural frequency for a larger assembly. As well, stiffness is used to provide the necessary structural stability for a sensor or antenna.

Damping is the dissipation of energy during vibration. This is a structural characteristic that limits the magnitude and duration of response to input forces. It is used to control loads and ensure that any vibration decays before influencing the control system of the spacecraft.

Mass properties include mass, center of mass, mass moments, and products of inertia. These are imposed by the launch vehicle and allocated to all substructures to achieve the required natural frequencies of a larger assembly.

Dynamic envelope is the physical space within which a spacecraft or substructure must remain while deflecting under loads to avoid contact between the spacecraft and the payload fairing of the launch vehicle or between parts of the spacecraft.

Structural stability is the ability to maintain location or orientation within a certain range. Typical concerns are thermoelastic distortions, material yielding, and shifting of mechanical joints. The intention is to ensure that critical instruments, such as antennas and sensors, can find their targets because inadequacy in this area results in performance degradation.

A mechanical interface includes features such as flatness, stiffness, and locations of bolt holes that define how structures and components attach. It is derived from designs of mating structures to ensure fit and avoid excessive loads and deformations.

17.2.2 Space Mission Environment and Mechanical Loads

Environments on Earth, during launch, and in space are design drivers for spacecraft structures. When applicable, on-orbit performance of spacecraft structures, such as pointing accuracy and structural stability, are further important design drivers. Activities related to space mission environments and mechanical loads are identified:

- Structures must not only survive the environments to which the spacecraft is subjected but also protect its nonstructural components and allow them to function.
- The selected materials, nonstructural and structural, must not degrade either before or during a mission.
- Ground testing must envelop mission environments with margins. As a result, test environments need to be defined at an early stage and structures designed adequately to cover them.
- Mechanical loads can be static or dynamic. Mechanical loads can be external, such as engine thrust, sound pressure, gusts of wind during launch, or self contained, as for the mass loading of a vibrating satellite during environmental testing or in space after the force that caused the excitation is removed.

The design of a spacecraft and its subsystems must cater to all loads to be experienced. Typical loading events are

- Testing tailored to cover adequately flight, on-orbit, and ground transportation environments.
- Ground handling and transportation.
- Liftoff.
- Stage separation.
- Stage ignition.
- Stage or main engine cutoff.
- Maximum aerodynamic pressure.
- Spin up and deployment.
- Attitude control system firings.
- Reentry.
- Emergency landing (as in the case for the *Space Shuttle*).

These events induce acceleration, shock, and vibration to structure. Careful attention is needed to cover these effects adequately. In general, the limiting factors in a structural design are set by dynamic effects rather than by steady state acceleration. The primary source of loads occurs during the launch phase. All types of launchers apply different load levels according to their design. For example,

- Solid rocket engines have combustion chambers that run the length of the fuel column that burn continuously and cannot be throttled.
- Liquid fuel rockets have a combustion chamber fed by separate fuel tanks that can be, but rarely are, throttled to reduce thrust loads at critical aerodynamic phases of the flight and give a more controlled trajectory. The tanks of liquid fuel act to some extent as a damper, attenuating vibration from the engines.

The launcher agency issues standard guidelines for the design and qualification of spacecraft. These apply to the mounting interfaces at its base and are concerned with both steady and dynamic forces. They are stated as flight limit loads, that is, loads that one would not expect to be exceeded during 99% of launches. The launcher agency does, however, require the spacecraft designer to demonstrate by a combination of test and analysis that the design can withstand these loads with considerable margins. Differing from those of aircraft systems, margins typically specified are

- Flight limit loads: 1.0.
- Flight acceptance: 1.1.
- Design qualification: 1.25 to 1.4.

Test levels for flight acceptance, therefore, are set for structures and equipment (in prototype form) that previously passed testing at the design qualification level. The implication here is that, if a one-model program is followed, that is, a protoflight approach in which the actual prototype is flown, then the model must pass the higher qualification test. Separate factors for the materials also can be specified by the launch agency.

Typically, these are 1.1 at yield, 1.25 ultimate stress for metals, and up to 2.0 for composites. Therefore, a structure that passes acceptance test levels and has sufficient material safety factors possesses a healthy margin over actual flight loads.

In assessing the margin of safety, terms used in strength analysis follow:

- The *load factor*, *n*, is a dimensionless multiple of *g* that represents inertial force. This term is often used to define limit load.
- The *limit load* (or design limit load) is the maximum acceleration, force, or moment expected during a mission or for a given event at a statistical probability defined by the selected design criteria, usually between .99 and .9987.
- The *limit stress* is the predicted stress level corresponding to limit load.

The allowable loads and factors of safety are determined by the

- *Yield failure*, a permanent deformation.
- *Ultimate failure*, a rupture or collapse.
- *Allowable load* or *stress*, the minimum strength, that is, load or stress, of a material or a structure at a statistical probability defined by the selected design criteria, usually .99 probability at 95% confidence level.
- *Allowable yield load* or *stress*, the highest load or stress that, based on statistical probability, cannot cause yield failure.
- *Allowable ultimate load* or *stress*, the highest load or stress that, based on statistical probability, cannot cause ultimate failure.
- *Yield factor of safety*, FS_y, a factor applied to the limit load or stress to decrease the chance of detrimental deformation. Usually, this value is between 1.0 and 2.0 for flight structures, depending on whether personnel safety is at risk and how sensitive the mission is to small deformations.
- *Ultimate factor of safety*, FS_u, a factor applied to the limit load or stress to decrease the chance of ultimate failure. This value is usually between 1.25 and 3.0 for flight structures, depending on the test option and whether people are risk.
- *Design yield load*, the limit load multiplied by the yield factor of safety. This value must be no greater than the allowable yield load.
- *Design yield stress*, the predicted stress caused by the design yield load. This value must not exceed the allowable yield stress.
- *Design ultimate load*, the limit load multiplied by the ultimate factor of safety. This value must be no greater than the allowable ultimate load.
- *Design ultimate stress*, the predicted stress caused by the design ultimate load. This value must not exceed the allowable ultimate stress.

- *Yield margin of safety*, MS_y, is defined as the relation

$$MSy = (\text{Allowable yield load or stress/Design yield load or stress}) - 1 \quad (17.2)$$

- *Ultimate margin of safety*, MS_u, is defined as the relation

$$MSu = (\text{Allowable ultimate yield load or stress/Design ultimate yield load or stress}) - 1 \quad (17.3)$$

17.2.3 Project Overview: Successive Designs and Iterative Verification of Structural Requirements

The assessment of load distribution within a spacecraft structure largely is an iterative process, starting with generalized launcher predictions. These are used as a basis for the design of an initial spacecraft concept and, therefore, provide subsystem target specifications according to their location.

The first steps in structural design are to convert the mission requirements into a spacecraft concept and to specify the underlying parameters with as much detail as can be expected at the concept stage. These processes can be quite missions specific, as exemplified by contrasting the thermal dissipation of a communication satellite with the precision requirements of large antennas and telescopes.

In general, the following requirements need to be covered from an early stage:

- Overall configuration for meeting mission objectives.
- Accommodation for the meeting mission objectives.
- Ability to withstand launch loads.
- Stiffness.
- Provision of environmental protection.
- Alignment.
- Thermal and electrical paths.
- Accessibility.

At this stage, the configuration effort dominates as the distribution of masses becomes established. The concept of a load path from the interface with the launch vehicle, from which all accelerations are imparted, through the spacecraft structure to the mounting points of individual systems or units is followed. At each mounting point, individual sets of interface requirements, that is, alignment, thermal, field of view, screening, connections, and accessibility, are generated. This, therefore, sets some of the constraints to be applied to the design of the structure. Subsequently, the problem of providing a structure that meets the specification can be outlined quite well. However, the need for both high mass efficiency and reliability presents challenges. The selection of materials often is dominated by stiffness and structural dynamics rather than stress levels. Testing as a means of validation is an inherent part of the design. Validation is accomplished by full testing or by limited testing supported with analysis and modeling. The final design stage requires a coupled loads analysis, which combines the characteristics of the full assembly of the spacecraft with those of the launch rocket.

In the case for large structures such as the *International Space Station*, reconfigurations well beyond the initial concept are going to occur on orbit during its 25-a lifetime. The addition of modules and long beams change its very low frequency dynamic characteristics during attitude maneuvers and orbit boosting, thus presenting an additional design criterion. The design methodology is summarized in Figure 17.2.

To exclude the failure of space vehicle structures caused by propagation of flaws and delaminations, damage tolerance design principles are often applied. To the extent possible, the structure is designed to be fail-safe, meaning that redundancy is built into the structure by introducing redundant load paths into the structure and its interfaces. For critical structures where multiple load paths are not possible, the structure is designed to be

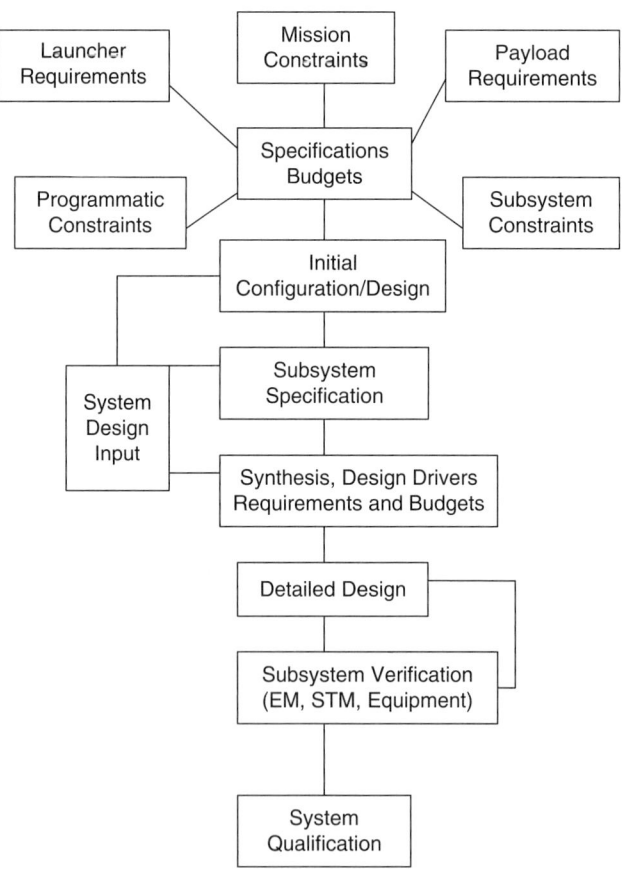

FIGURE 17.2 Design process (Courtesy of ESA).

safe-life. In this case it is verified that the flaws inherent in the structure cannot propagate to a critical size during the life cycle of the structure. For launchers and crewed systems, damage tolerance principles are applied to all structural parts, whereas for uncrewed satellite structures, only the most critical structural parts are subject to such design and verification rules. Examples of this are single point fasteners, primary structures known to contain flaws such as castings and forgings, and primary structures made from composite materials.

17.2.4 Analytical Evaluations

With the development of modern fast access large memory computers, it is now possible with a readily available commercial finite element code to model in detail satellite mechanical systems and examine their behavior under various static or time dependent load conditions. Commercial code packages are available, such as NASTRAN®, ABAQUS®, SAMCEF®, and PERMAS®, that enable a spacecraft structure to be evaluated under static, dynamic, and thermal loads. The natural frequencies and modes of vibration are determined as the first step in a dynamic evaluation to assess the characteristics of the structure. These are used to evaluate the compatibility of the spacecraft with launcher requirements and for modal response analysis of the structure excited by external loads to determine the spacecraft response at different locations. The finite element model of the spacecraft system is employed in launcher coupled loads dynamic analysis to examine the launch assembly configuration and determine representative text excitation, that is, tailoring of the test profile and level. Figure 17.3 shows a model of a highly three-dimensional launch vehicle for which these evaluations have been conducted.

In such analytical evaluations, it particularly is important to assess qualitatively and quantitatively the analytical predictions. A number of qualitative checks are presented here that can assist practicing engineers to identify mathematical errors. Quantitative checks are covered in the following subsection.

The underlying physics of a problem always should be respected in analytical modeling and used to assist in qualitative checks of the finite element mathematical model. For example,

- No stress or loads should be generated due to any rigid body translations and rotations of a structure.
- The mass matrix of the structure, often simplified for dynamic analysis, should represent the total mass and inertial properties of the structure.
- No stress or loads should be generated in a homogeneous structure that is free to expand when subjected to a uniform temperature field.

17.2.5 Structural Test Verification

The primary objectives of a space system verification program are to

- Qualify the design.
- Ensure that the product is in agreement with the qualified design and free from workmanship defects and therefore acceptable for use.

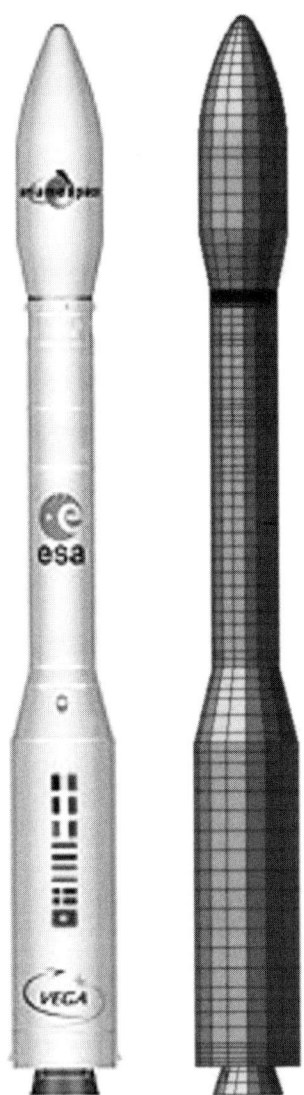

FIGURE 17.3 The *VEGA* launcher and its three-dimensional finite element model (Courtesy of ESA).

- Certify that the space system, including tools, procedures, and resources, is able to fulfill mission requirements.
- Confirm product integrity and performance after particular steps of the project life cycle, such as prelaunch, on orbit, and post-landing.

As an integral part of a test verification program, the structure is subjected to a range of environmental tests, the objectives of which are to obtain confidence in the analytical predictions performed during the design phase and ultimately support the qualification and flight acceptance of the space system.

A static test is achieved by subjecting the structure to discrete loads or centrifuge testing to simulate inertia loading. Although it is used mainly to qualify the strength adequacy of the primary structure and critical structural interfaces, such as spacecraft to launcher and spacecraft to payload, the static test provides insight to the validity of the stiffness matrix. In some cases, the static test can be replaced by a sinusoidal shaker vibration test at low frequency that is intended to excite the structure with an equivalent quasi-static load.

A modal survey test is achieved by using small exciters to excite the structure locally to determine the natural frequencies and modes of vibration of a spacecraft. Because the application of excitation forces at resonance is compensated only by damping, large responses can be achieved from small excitation forces. Modal survey tests identify the spacecraft natural frequencies, modes of vibration, and damping, all of which are necessary to determine its dynamic compatibility. These tests are used in the loads cycle assessment of the launch vehicle-to-spacecraft coupled loads dynamic analysis as well as to tailor the vibration test level.

A shaker vibration sinusoidal test supports the verification of the mathematical model used in forced frequency response predictions and is useful particularly to determine the amplification of the excitation input through the structure, such as from the launcher interface to various elements of the spacecraft. Here, the amplification factor Q = Output/Input. The main purpose of the shaker vibration sine test is to qualify the adequacy of secondary structures when subjected to a dynamic environment and verify the adequacy of the spacecraft system by performing functional tests after the spacecraft system qualification and flight acceptance shaker tests.

A random shaker vibration test supports the verification of spacecraft units subjected to a random dynamic environment representative of that which might be experienced during flight.

A vibroacoustic test is performed for larger structures that are susceptible to direct excitation induced by the acoustic dynamic pressure. Such structures include launcher fairings, large spacecraft, and spacecraft appendages such as antenna structures and solar arrays.

A shock test supports the verification and qualification of the space system structure and instruments subjected to a shock resulting from pyrotechnic firing and latching loads, such as separation of launcher stages, release of the launcher interface clamp band, and the release of booms, solar panels, antennas, and the like.

17.2.6 Spacecraft Structural Model Philosophy

Classical Approach
In the classical test verification approach, structural and, in some cases, engineering and flight models are tested environmentally prior to space system flight. The purpose of these models and the tests are summarized in Table 17.1.

Protoflight Approach
In a number of cases, specifically for spacecraft and payloads, a protoflight approach can be used. The major difference between the protoflight approach and the classical

Table 17.1 Classical Structural Environmental Testing of Spacecraft

Model	Testing
Structural model (representative of space system flight structure, units might be representative mechanically or, with electronics, represented as mass dummies)	Static or centrifuge test at qualification load level to check and qualify the structural strength, in particular, primary structure and critical interfaces.
	Modal survey test to identify vibration natural frequencies, natural modes, and damping to support verification of mathematical model used in loads cycle and coupled loads dynamic analysis of the spacecraft with the launcher.
	Shaker sine vibration qualification test input level to qualify spacecraft secondary amplification from the spacecraft-launcher interface to various parts of the spacecraft.
	Vibroacoustic test to derive structural response to the dynamic pressure loads, and for derivation of random vibration environments for subsystems and equipment.
	Shock or separation tests to derive the shock vibration transmission through the structure for defining shock vibration environments for subsystems and equipment.
Flight model	Modal survey test in general is not performed with the flight model unless considerable design changes have been introduced or there are built-in standard differences with the structural model.
	Shaker sine vibration test is conducted at flight acceptance level.
	Acoustics test is conducted at flight acceptance level Shock test to confirm compatibility of the space system with shock environment in particular payload units.
Engineering model	This model usually is not representative structurally and is employed for electrical purposes. Sometimes, the structural model is employed as the engineering model after the mechanical qualification tests.

approach is that the flight hardware is subjected to qualification loads. The hardware model and associated testing are summarized in Table 17.2.

17.2.7 Materials and Processes

The selection of appropriate materials for space vehicle structures applications requires knowledge of the way in which each material property can best be used and where various limitations must be recognized adequately. Selection criteria encompass the specific strength, specific stiffness, stress-corrosion resistance, thermal parameters, sublimation and erosion, and ease of manufacture and modification.

Table 17.2 Structural Environmental Testing of Protoflight Spacecraft

Model	Testing
Structural model (this is refurbished after qualification test, as the flight model units might be representative mechanically or with electronics represented as mass dummies)	Static or centrifuge test at qualification load level Modal survey test Shaker sine vibration test at qualification test input level Acoustic test at qualification level Shock test
Flight model (structural model refurbished to flight model standard after qualification test)	Shaker sine vibration test at flight level Acoustic test at flight level (qualification level acceptance duration) Shock test

Ground and launch environments have to be considered when selecting materials for handling, moisture absorption or desorption, shocks, and quasi-static and dynamic loads. The space environment, in particular, influences the selection of materials. Among the most important factors are

- Atomic oxygen affects mostly polymeric materials, an important exception being silver.
- Charged particles require conductive materials to prevent local electrical charging (indium tin oxide conductive coating).
- Thermal radiation and uneven temperatures lead to thermal distortions, particularly in multilayer composite materials.
- Vacuum facilitates outgassing, an important selection criterion.
- Micrometeoroids and orbital debris.

Analysis of the possible impact of an environment aspect and the verification of suitability or acceptability of selected materials are performed using a structured and documented approach. In addition, costs, availability of data, previous experience, and ability to be repaired easily can affect the choice of materials.

The most frequently used classes of materials for space vehicle structures are metal alloys and composites. Metals are homogeneous and isotropic. Composites are inhomogeneous and orthotropic. Typical applications of metals and composites are

- Aluminum, used for skins, shells, truss members, face sheets, and the core of sandwich structures.
- Titanium, used for attachments, fittings, fasteners, and pressure vessels.
- High strength steel, used for support structures, and solid rocket motor cases.

Composites are used in mass and stiffness driven applications, such as primary structures and adapters, as well as for appendages, such as sun shields, antenna reflectors, and solar array structures. Generally, thermosetting resin materials are used in combination with high stiffness carbon fibers because of their superior performance characteristics.

For structures requiring very high thermoelastic stability, ceramic materials such as silicon-carbon increasingly are used. For hot structures, such as high temperature applications in reentry vehicles, ceramics made from carbon-carbon and carbon-silicon-carbon have been developed and qualified to operate at temperatures well above 1200°C. However, prevention of these materials from oxidizing under severe heat fluxes requires the development and use of adequate coatings.

Jointing is important, and it is basically of three types: mechanically fastened, bonded, and welded. In general, mechanical joints lead to stress concentrations that require special attention for load introduction. This, in particular, is a concern for composite and ceramic materials that are susceptible to such concentrations. Bonding is the most elegant approach, for it guarantees a better continuity of load transfer. However, this approach requires proper shaping of the parts to be assembled so that there is a progressive load transfer to minimize thickness stresses. For composites, attention must be given to avoid or limit peeling forces. The tolerance between parts and cleanliness of the surfaces also is very important. Welding is reserved for metals, and provides excellent tightness. Fusion associated with the welding process introduces weak zones in the joint, particularly when the heat affected zone is not small, as is typical for the tungsten–inert gas process. High performance welded joints are required for pressure vessels, and these can be obtained by electron beam welding, laser welding, and friction stir welding.

17.2.8 Manufacturing of Spacecraft Structures

The manufacturing process for metallic and composites parts is different fundamentally. Because the processing of composites begins with raw materials and not with partly finished components, as is the case for metals, it is possible to produce fairly complex parts as a single piece, that is, unitization, thus saving manufacturing and assembly costs (Figure 17.4).

Metals

Machining is the process of removing material with cutting or grinding tools. Many machining operations are automated to some extent, although for low volume production, such as spacecraft manufacturing, manual processes supported by CAD/CAM tools are often cost effective.

FIGURE 17.4 Spacecraft to launcher adapter with integrated interface rings (Courtesy of ESA).

Forming is one of the most economical methods of fabrication. The most limiting aspect of designing formed parts is the bend radius, which must be made quite large to limit the amount of plastic strain in the material. Super plastic forming and diffusion bonding, at temperatures up to 1000°C, can produce complex components; however, this process can be used only for titanium alloys because others are prone to surface oxidization that inhibits the diffusion bonding. Spin forming has been applied successfully for manufacturing of aluminum pressure vessels.

Forging, a process in which structural shapes are produced by pressure, is well suited for massive parts, such as load introduction elements. Where large sheets have to be manufactured, such as the skin of main thrust cylinders, chemical milling frequently is used to reduce the thickness of the sheet where possible. This process is more reliable than machining when processing very thin elements; however, tolerances must be sufficient to account for various inaccuracies, such as thickness variations in the original piece of material.

Casting can be used to produce parts of complex shapes. However, the quality of a casting is difficult to control because, as the material solidifies, gas bubbles can form. This results in porosity. Material strength and ductility are not as high as with most other processes.

Composites

Impregnated tapes are the most widely used precursor for composite manufacturing. Fibers can be woven to form fabrics. The key parameters when specifying a fabric or tape are the type of fibers and the resin as well as the content, tack, and drape.

The fiber controls the major mechanical properties of the part, such as strength and stiffness. The resin that binds the fibers together determines the maximum temperature under which the part can be used safely. It also influences moisture absorption and desorption, with possible effects on geometrical distortions under thermal and moisture cycling.

Tack is a measure of how much the preimpregnated material (prepreg) sticks to itself and to other layers. Prepregs that have too much tack can be difficult to handle because a misplace layer is difficult to reposition without disrupting the resin or fiber direction. Prepregs with little tack are difficult to keep in place as more plies are applied. Lack of tack often indicates the prepreg resin has cured beyond an acceptable limit. Should this be the case, the composite part cannot cure properly.

Drape is the ability of the prepeg to form around contours and complex shapes, and it is influenced by the fiber material and the diameter of the filaments as well as the cross section of the tow. It also is conditioned by the weaving pattern. For flat panels, a low tack and drape are acceptable, but for complex shapes and cavities, tack and drape are important to keep the composite in place during the process of laying-up and subsequent preparation for curing.

Manual lay-up is a costly technique adapted quite well to the production of single parts of very small series. When the number of identical parts to be produced grows, filament winding, resin transfer molding, and braiding are cost-effective alternatives.

Filament winding consists of wrapping bands of continuous fiber or strands, that is, rovings, over a mandrel in a single machine controlled operation. A number of layers of the same or different patterns are placed on the mandrel. The fibers can be impregnated with the resin before winding (wet winding), preimpregnated (dry winding), or post-impregnated.

The first two winding sequences are analogous to wet or dry lay-up in the reinforced plastic fabrication methods. The process is completed by curing the resin binder and removing the mandrel.

Resin-transfer molding is a closed mold, low pressure process. The fiber reinforcement is placed into a tool cavity, which is then closed. The dry reinforcement and the resin are combined within the mold to form the composite part. This process allows the fabrication of composites ranging in complexity from simple, low performance small parts to complex, large elements.

In a braiding operation, a mandrel is fed through the center of a braiding machine at a uniform rate, and the fibers or yarns from the carriers are braided around the mandrel at a controlled angle. The machine operates like a maypole, the carriers working in pairs to accomplish the over and under braiding sequence. Parameters in the braiding operation include strand tension, mandrel feed rate, braider rotational speed, number of strands, and the width and perimeter being braided. Interlaced fibers result in stronger joints. Applications include lightweight ducts for aerospace applications.

RECOMMENDED READING

Agarwal, B. D., and L. J. Broutman. (1990) *Analysis and Performance of Fiber Composites*. New York: Wiley.

Bruhn, E. F. (1965) *Analysis and Design of Flight Vehicle Structures*. Cincinnati, OH: Tri-State Offset.

Curtis, H. D. (1997) *Fundamentals of Aircraft Structural Analysis*. Los Angeles: Times-Mirror HE Group.

Dowel, E. H. (1995) *A Modern Course in Aerolasticity*. Dordrecht, the Netherlands: Kluwer Academic Press.

Foreman, R. G., V. E. Kearney, and R. M. Engle. (1967) Numerical analysis of crack propagation in cyclically loaded structures. ASME Transactions. *Journal of Basic Engineering* 89, no. D: 459–464.

Hull, D. (1981) *An Introduction to Composite Materials*. Cambridge, UK: Cambridge University Press.

Kush, P. (1956) *Stresses in Aircraft Shell Structures*. New York: McGraw-Hill.

Megson, T. H. (1997) *Aircraft Shell Structures for Engineering Students*. London: Edward Arnold.

Miner, M. A. (1954) Cumulative damage in fatigue. ASME Transactions, *Journal of Applied Mechanics* 67: A159.

Niu, M. C. Y. (1988) *Airframe Structural Design: Practical Information and Data on Aircraft Structures*. Tabernash, CO: Aircraft Technical Book Company.

Niu, M. C. Y. (1992) *Composite Airframe Structures: Practical Design Information and Data*. Hong Kong: Adaso Adastra Engineering Center.

Paris, P. C., R. J. Bucci, E. T. Wessel, W. G. Clark, and T. R. Mager. (1972) *Extensive Study of Low Fatigue Crack Growth Rates in A533 and A508 Steels*. STP 513. Philadelphia: American Society for Testing and Materials, pp. 141-176.

Rivello, R. M. (1969) *Theory and Analysis of Aircraft Structures*. New York: McGraw-Hill.

Rooke, D. P., and D. J. Cartwright. (1976) *Compendium of Stress Intensity Factors*. London: Her Majesty's Stationary Office.

Timoshenko, S., and J. M. Gere. (1961) *Theory of Elastic Stability*. New York: McGraw-Hill.

17.3 FRACTURE CONTROL

Fracture control is implemented to reduce appreciably the risk of a catastrophic failure caused by propagation of undetected preexisting cracklike defects or flaws for a prescribed service period. Fracture control includes the engineering disciplines of fracture mechanics for the assessment of damage tolerance (safe operating life), nondestructive evaluation to screen for damage, quality and process controls, material lot traceability, and proof testing to assure the highest quality part. The combination of these disciplines into a viable fracture control plan ensures that parts whose failure could lead to loss of life or vehicle have adequate safety.

17.3.1 Basic Requirements

The basic assumptions that underlie implementation of Fracture control include

- All individual parts contain flaws or cracklike defects. The minimum life of a part can be determined by assuming that a flaw in the most critical area of the part and in the most unfavorable orientation propagates to failure.
- The use of nondestructive evaluation techniques does not negate the assumption that flaws are inherent to a part. The techniques of nondestructive evaluation establish a probable upper bound on the size of an assumed initial flaw at a specified confidence level. If no flaws are detected during inspection, a flaw size at least as large as the probable upper bound flaw size established by the appropriate nondestructive evaluation technique is used for analysis.
- All spaceflight hardware is of a good design, certified for the application, acceptance tested as required, and manufactured and assembled using high quality aerospace processes.

17.3.2 Implementation

To meet safety requirements for crewed spaceflight systems, all payload and experiment hardware flown on the *Space Shuttle* or *International Space Station* must be assessed using fracture control. The actual implementation of fracture control utilizes specific classifications to identify the criticality of parts: exempt, nonfracture critical, and fracture critical. Exempt hardware typically includes nonstructural items, such as flexible insulation blankets, enclosed electrical circuit components or boards, electrical connectors, wire bundles, and seals. Nonfracture critical hardware includes low released or contained masses, redundant structure, non-hazardous leak-before-burst pressurized components, low speed and low momentum rotating machinery, and parts with high margins on cycle life and strength. Fracture critical hardware includes pressure vessels, high energy or high momentum rotating equipment, hazardous fluid containers, habitable modules, and any remaining hardware that does not fit the categories of exempt or nonfracture critical.

Any assessment of hardware criticality examines the different phases of application, including transportation, launch, on orbit, interplanetary, or lunar travel including surface operations and return to ground (including contingencies) to determine the applicability and extent of fracture control. For example, a part might not be fracture critical during the launch phase but could be fracture critical for on-orbit service. In this case, the fracture control assessments must address the on-orbit phase as well as other phases and their potential effects on the on-orbit performance.

In general, relatively few parts or components in payloads and experiments are truly fracture critical. Some units or assemblies might have no fracture critical parts. The fracture critical designation requires special considerations and treatment for that part. This is absolutely necessary for parts whose failure is catastrophic, but it can become resource consuming and potentially schedule impacting if parts are classified fracture critical in a casual manner.

17.3.3 Summary

Because Fracture control deals with what might happen should crack propagation lead to structural failure, reasonableness and credibility must prevail. Many bad things can be imagined as a result of chained, unlikely events. Consequently, those who do fracture control and those who judge it must put some restraint on their imaginations, tempering them with the likelihood that the events under consideration have a reasonable chance of occurring.

17.4 PRESSURE VESSELS, LINES, AND FITTINGS

17.4.1 Pressure Vessels

Pressure vessels are used on space vehicles to store a variety of consumable commodities in liquid or gaseous form over a wide range of pressures and temperatures. The stored commodities can be used as pressurants, propellants, pneumatic gases, hydraulic fluids, power reactants, coolants, purge gases, or sources of breathable atmosphere. The pressure vessels themselves can be designed with any geometry; however, due to packaging efficiency and manufacturing considerations, they are typically spherical or cylindrical in shape. For space applications, classification as a pressure vessel is based on the following:

- The pressure level for a liquid or gas, typically 100 psi or greater.
- The total energy content for a gas, which is a function of pressure, volume, and gas expansion properties, typically 14,240 ft-lbf or greater.
- The potential for catastrophic hazards associated with fluid release at pressures as low as 15 psi, such as toxicity, asphyxiation, chemical or thermal incompatibility with other nearby materials, and overpressurization of surrounding compartments.

Pressure vessels also can be categorized broadly as all metallic; composite overwrapped with a metallic liner, that is, a composite overwrapped pressure vessel; or all composite.

The Apollo and Space Shuttle vehicles used a combination of all metallic and composite overwrapped pressure vessels (Kevlar® and graphite epoxy) to accomplish their mission objectives safely. Important problems associated with these vessels included manufacturing flaws, metallurgical defects, and material incompatibility with the fluids used for fabrication, testing, or flight service. The reusability and long service life of the *Space Shuttle* presented additional challenges for the pressure vessels relative to cycle life, time at pressure, material aging, and maintenance.

Lightweight space pressure vessel safety was preceded by the historic development of safety practices and codes for pressure vessels used as boilers. These historic origins are described by Canonico (2000) and are the basis for the safety requirements and practices in use today for pressure vessels used on space systems and in other industrial applications.

Origins of the American Society of Mechanical Engineers Boiler and Pressure Vessel Code

The American Society of Mechanical Engineers (ASME) boiler and pressure vessel code was conceived in 1911 to protect public safety and provide uniformity in the manufacturing of boilers. This action was considered necessary after many boiler explosions in the United States and Europe throughout the 1800s resulted in numerous injuries and deaths. It also was intended to address the lack of uniform legal codes governing the design, fabrication, packaging, transport, installation, operation, and maintenance of industrial boilers. A consistent approach in addressing these topics was recognized to have both safety and economic benefits. Today, all the provinces in Canada, 48 of the 50 U.S. states, and numerous regulatory agencies around the world have adopted, by law or regulation, various sections of the ASME boiler and pressure vessel code.

Pressure vessels complying with ASME codes have relatively high structural safety factors, that is, ~4.0 or more, on internal or external pressure loads as compared to spacecraft pressure vessels, which can have ultimate safety factors as low as 1.5. Pressure vessels designed to the ASME code, except for very small units in specific applications, seldom are used for spaceflight because of the high weight penalty associated with such generous safety factors.

Spacecraft Pressure Vessel Design and Operation

Due to the reduced margins and lack of fault tolerance in spacecraft pressure vessels, a design for minimum risk approach must be taken to control critical and catastrophic hazards. Critical hazards include damage to equipment, facilities, and non-disabling personnel injuries. Catastrophic hazards include complete destruction of equipment, facilities, and disabling or fatal personnel injuries. Due to these hazards, the design, fabrication, installation, and operation of spacecraft pressure vessels require special considerations and data products to ensure system safety. Notable examples based on past experience and lessons learned from the development and operation of Apollo and Space Shuttle pressure vessels include

- Establishment of a single technical authority having expertise in materials selection and compatibility, metallurgy, manufacturing techniques, stress analysis, and fracture mechanics to oversee the development and usage of all pressure vessels on the spacecraft. The intent is to ensure appropriateness and uniformity of design requirements, material selection, test requirements (in process, acceptance, and qualification), fabrication and analysis methods, verification approaches, test data interpretation, issue tracking and resolution, handling and storage requirements, inspection intervals and techniques, repair thresholds and techniques, and operational placards.

- Creating detailed specifications for each pressure vessel application.
- Conducting design reviews with pressure vessel suppliers.
- Employing fracture control principles as applicable during the design process.
- Imposing stringent controls on material selection.
- Performing thorough inspections and acceptance tests on each production vessel to ensure materials, manufacturing processes, and workmanship meet specifications. As a minimum, acceptance test procedures should include visual inspections, weight measurements, external dimension measurements, leak testing, nondestructive inspection of fusion joints, proof pressure testing, and internal volume measurements before and after proof.
- Performance of rigorous qualification tests on a sample of production vessels in flight representative environments. As a minimum, qualification test procedures should include pressure and temperature exposure at flight-like levels, durations, rise rates, and cycle counts using the actual service fluid or a carefully chosen referee fluid, vibration, shock, acceleration, vacuum exposure as applicable, and burst pressure testing to verify adequate performance of the design over the projected service life of the hardware. Note that a proof pressure test on each production unit along with a fatigue analysis showing a minimum of 10 design lifetimes can sometimes be used in lieu of a dedicated burst test.
- Maintaining a data package for each production vessel with information on its entire manufacturing and processing history. Key items include material and process specifications; weld parameters and repair history; chemical composition; mechanical property verification of materials including weld wire; pressurization and fluid exposure history including levels, durations, and fluid types; discrepancy reports and resolutions; and nondestructive test and acceptance test procedure certifications and results. This data pack always should accompany the pressure vessel and be included in the vehicle data file for which the pressure vessel is installed.
- Application of a quantitative approach to inspection and repair based on stress analysis and fracture mechanics, such that an appropriate inspection interval, repair schedule, and repair criteria can be established and documented to govern pressure vessel usage.

If a catastrophic hazard can result from structural failure of a pressure vessel due to the initiation and propagation of flaws or crack-like defects, the pressure vessel is considered fracture critical. In that case, Fracture control techniques must be employed in the design, fabrication, and testing of the hardware to prevent such failures in service. To address these requirements, several important aspects of ensuring pressure vessel safety are discussed in the following sections.

Maximum Design Pressure

The first step in assessing a pressure vessel for safe operation is to establish the maximum design pressure of the hardware. This pressure, which also can be referred to in the literature as *maximum operating pressure* or *maximum expected operating pressure*, frequently is associated with the setting of a pressure relief device in the system, such as a relief valve or burst disk, which is sometimes used in series or parallel combinations depending on the application. Additional factors influencing maximum design pressure include thermal inputs, such as compression heating from tank loading, line or tank heater operation, or incident radiation from solar flux, and transient pressure responses within the system that greatly can exceed static pressures under nominal operating conditions, such as pressure pulses from water hammer in systems containing liquid or slam starts in systems containing gas. In some cases, credible hardware or software failures need to be considered when determining maximum design pressure, such as relief valve failed closed, heater failed on, procedural errors, or ground support equipment failures producing an excess pressurization rate.

Ultimate Safety Factors

Once maximum design pressure is established, an ultimate safety factor is used as a multiplier to establish the minimum design burst requirement for the pressure vessel. This safety factor, typically 1.5 to 2.0 for spacecraft applications, provides the primary means for ensuring pressure vessel safety by precluding rupture at pressures below the design burst value, assuming no flaws or defects exist. The basic intent of the safety factor is to account for variability in material properties and, to a lesser extent, any uncertainty and statistical variation in the stress producing pressure environments to which the vessel is exposed. However, the small safety factors on spacecraft pressure vessels do not provide much margin to accommodate both sources of error, making an accurate prediction of material properties and loading environments more critical. To ensure the vessel can meet the design burst pressure requirement, classical stress analysis can be used for simplistic designs along with finite element methods for complex geometries or features. Details of these techniques are provided in other textbooks and journal articles.

As a direct means for verifying the adequacy of a pressure vessel design, burst pressure testing can be performed on a representative sample of the production units, such as the first article or a random sample, as part of the qualification program. The intent simply is to verify that the as-built units do not fail, that is, rupture or leak excessively, below the design burst pressure. Ideally, burst testing should be performed at the maximum predicted service temperature of the vessel. However, it usually is performed at ambient temperature with the test pressure increased as required to account for the difference in material properties between test conditions and service conditions. In some cases, the vessel is taken only to the design burst pressure, while in other cases, the unit is taken beyond that point until physical failure occurs. In both cases, the design burst test is considered destructive, and the test vessel is not used for any other purpose afterward. Due to the hazard potential associated with burst testing, water or another incompressible liquid normally is chosen as the test fluid. Compressible gases also can be used, but sufficient blast protection must be provided for test personnel and equipment. Although the test vessels are not used again after burst testing, short term compatibility of the test fluid with pressure vessel materials must still be confirmed via existing compatibility data or special material exposure testing to ensure validity of the test results.

In addition to the internal pressure discussed previously, other important mechanical loads are experienced by a pressure vessel in service. Sources include vibration, shock, acceleration, thermal gradients, differential expansion and contraction, installation, packaging, transportation, and handling. These loads are not accounted for directly by the design burst safety factor. Instead, they are covered by a separate structural safety factor requirement of 1.4 minimum on worst-case combined loads.

Flaw Screening

To help screen a pressure vessel for flaws and defects, which exist in all manufactured parts but are not addressed explicitly by stress or fatigue analysis and factors of safety, a combination of nondestructive and proof pressure testing can be performed during and after vessel fabrication. Examples of manufacturing flaws and defects experienced on Apollo and Space Shuttle pressure vessels include scratches, dents, machining errors, weld mismatches, weld porosity, weld filler inconsistencies, dimensional anomalies, and heat treat errors. Various nondestructive test methods are used to help identify these conditions in pressure vessel welds, membranes, and bosses, including radiographic inspection (X-ray, gamma ray, and neutron), ultrasonic inspection, eddy current inspection, liquid penetrant inspection, magnetic particle inspection, thermographic inspection, and computer aided tomography. However, each of these techniques has limitations related to the type, orientation, and size of flaws that reliably can be detected in different materials. Also, some rare metallurgical conditions, such as embrittled alpha phase in titanium and titanium hydride formation at welds, simply are not detectable with available nondestructive test techniques.

To address nondestructive testing limitations and screen for other undesirable conditions, a proof pressure test normally is performed on each pressure vessel as part of acceptance testing. The purpose of this test is to verify satisfactory workmanship and material quality to withstand all subsequent service conditions by ensuring that no detrimental yielding or deformation occurs at pressures above maximum design pressure, typically by a factor of 1.1 to 1.3. It is used as well to screen for flaws that might have escaped detection during prior in-process nondestructive testing and visual inspections. Note that proof testing should be performed at the maximum predicted service temperature of the vessel, or else the test pressure must be increased to account for the difference in material properties at the actual test temperature. As learned on *Space Shuttle* composite overwrapped pressure vessels and in addition to qualitative visual inspections, testing also should include an accurate measurement of the external dimensions and internal volume of the vessel to provide a quantitative indicator of permanent deformation or growth.

Safe-Life and Leak-Before-Burst Via Fracture Mechanics

The proof test and nondestructive test results just discussed can be used to establish an initial flaw or defect size that could have passed inspection and survived a single cycle to proof stress without propagation to failure. With this information, the analytical concept of linear elastic fracture mechanics can be used to predict flaw growth and fracture behavior of vessel materials during subsequent exposure to ground test or flight loads and environments. Four primary factors govern the susceptibility of a pressure vessel to fracture:

- Fracture toughness properties of the material.
- Size, shape, and location of the initial flaw or defect.
- Tensile stress level, including the effect of residual stresses.
- Fluid environment to which the pressure vessel material is exposed.

Fracture mechanics relates these factors to the stress field at the tip of a crack using a parameter called the *stress intensity factor, K*. Unstable fracture occurs when the stress intensity factor at the crack tip reaches a critical value, K_c, representing the fracture toughness of the material. Fracture toughness is a function of material thickness, temperature, fluid exposure environment, and rate of load application (normally determined by test). At that point, sudden unstable fracture can occur, resulting in vessel rupture.

Due to the hazard potential of a rupture event, a fracture critical vessel must be designed for safe-life. This means that any undetected flaw or defect cannot grow to the point of instability or break through under the combined static and cyclic loads encountered in at least four complete service lifetimes of the vessel. The factor of 4 is intended to account for typical scatter in fracture mechanics properties of materials. Crack growth analysis is based on an initial flaw or defect size determined from the proof pressure test or nondestructive test results combined with an assumed worst-case location and shape. Alternatively, testing can be performed in an appropriate pressure, temperature, or fluid exposure environment using pre-cracked specimens of representative vessel materials to demonstrate controlled and stable crack growth. However, it can be very difficult to envelop all possible dimensions, tolerances, material properties, and crack shapes or locations with a test, so analysis typically is used for this purpose. In either case, the safe-life of a pressure vessel ultimately can be determined for the predicted operating conditions to ensure that it fully envelops the intended service life.

In the case where safe life cannot be demonstrated by fracture analysis, a leak-before-burst analysis can be performed to confirm that the failure mode of a pressure vessel under cyclic stress is a benign event involving leakage alone, not a violent rupture producing shrapnel. In fact, leak-before-burst occurs when a crack-like flaw grows slowly through the thickness of a pressurized membrane before becoming critical and causing vessel rupture. This means that the flaw grows through the wall thickness (t) before the stress intensity (K) exceeds the critical value (K_c). The leak-before-burst analysis can be based on an initial flaw size and location established by nondestructive testing. Alternatively, it can be shown that a flaw having a depth equal to the wall thickness (t) and, by convention, a length equal to $10t$ cannot become a critical flaw at the maximum vessel stress. The latter approach is attractive because it eliminates the need for nondestructive testing, which can be an expensive and time consuming process when multiple production units are involved. Note that leak before burst as described here differs from the ultimate failure mode of a vessel at its actual burst pressure. Of course, the leak-before-burst approach is acceptable only if leakage or release of the contained fluid does not constitute a catastrophic hazard.

17.4.2 Lines and Fittings

For most spacecraft fluid systems, containment must be designed for a minimum risk of bursting and leaking for three reasons:

- The fluid contained can be reactive and either burn or cause reactions with hot surfaces or fire. These can result in a cascade of failures toward a catastrophic end.
- Many of the storable fluids are hazardous and, if leaked, can cause flammable, toxic, or carcinogenic hazards to spacecraft or personnel.
- If the fluids contained are leaked, the function that requires the fluid is lost, potentially jeopardizing an essential capability.

Rigid lines and welded joints are preferred for containment integrity, but there must be a trade evaluation between leak tightness and all the risks associated with leaks and the easy replacement of (failed) components. Cutting a line and again welding it easily introduces particulate and other contamination into the system. Flexible hose should be used only where vibration, relative motion, and contraction or expansion properties are severe enough to warrant their use. Rigid tube wall thickness and weld thickness requirements should be verified by nondestructive tests. Even though X-ray testing requires extensive personnel training and is expensive, it is best used for determining weld penetration and wall thickness.

It is important that all pressure lines be clamped properly and supported to minimize flexing, chafing, abrasion, and strain that could lead to line leakage or rupture. System lines should be supported by firm structure rather than by connecting to tubing. All lines should be clamped and supported independently. Clamping and supporting precautions must be taken to allow for thermal expansion and contraction that could damage the lines and fittings. All rigid lines should be supported as close as possible to each bend in the line. Flexible hoses should have a maximum slack of 5% of the total length. Flexible hoses should be clamped or supported by a structure at the connecting ends or have hose restraints across the hose connections.

Flexible hose applications should be assessed to determine whether a protective coating is required to preclude damage from abrasion or chafing. Redundant pressure lines should be separated as far as practical. Where lines and flexible hoses are accessible to personnel, they should be shielded, located, or otherwise protected to preclude being used as handholds or footholds. Vent lines and pressure relief lines should be located so that escaping gas is not hazardous to personnel and does not create thrust that would impart a force to the spacecraft.

Staggered fittings, different diameters, and clocking of fittings are advisable to prevent cross connecting lines. Positive sealing capability should be provided at quick disconnects to prevent leaks or drips when the lines are disconnected. Connectors and fittings that are to be disconnected during normal operations should be provided with tethered end plates, caps, plugs, or covers to protect the system from contamination or damage when not in use. The integrity of mechanical fittings used for connecting components, subassemblies, and systems must be qualified for use in the environment to which the propulsion system is exposed, including acoustic and vibration.

Redundant mechanical fittings and back off prevention should be used on all mechanical fittings. The fluid system should be designed so a completely assembled system can be internal to external pressure or leak tested at the maximum expected operating pressure or greater. Specifically, leak testing should be performed for all the welds, fittings, and connections.

17.4.3 Space Pressure Systems Standards

One of the commonly used American standards for spacecraft pressure systems, including pressure vessels, is MIL-STD-1522A (DoD 1984). This standard was last reviewed and declared active for military procurement purposes in September 1992. No changes have been made since that time.

Other, more recent standards for spacecraft pressure systems used in the United States include ANSI/AIAA S-080-1998, *Space Systems—Metallic Pressure Vessels, Pressurized Structures and Pressure Components* (ANSI 1998), and ANSI/AIAA S-081A-2006, *Space Systems—Composite Overwrapped Pressure Vessels (COPVs)* (ANSI 2006).

17.4.4 Summary

The NASA Apollo and Space Shuttle Programs demonstrated that, for high performance flight, lightweight pressure vessels with safety factors as low as 1.5 can be designed, tested, and utilized safely. The design analyses, acceptance and qualification testing, and quality assurance provisions are extremely demanding and make these types of pressure vessels very expensive. Therefore, in the early stages of a program, trade studies must be performed to evaluate the use of lower cost and heavier pressure vessels with safety factors of 2.0 and above as compared to the expensive pressure vessels with lower safety factors and very demanding quality assurance. The International Space Station Program, which came after the Apollo and Space Shuttle Programs, requires pressure vessels with safety factors of 2.0 or greater. It appears the pressure vessels with the higher safety factors, that is, 2.0 or greater, have a greater life and are more damage tolerant. With increasing orbital debris and concomitantly increasing probability for penetration, it is advisable to use pressure vessels with safety factors higher than those used on *Apollo* and the *Space Shuttle* and to provide appropriate protection or shielding to reduce the risk of ruptures in space. Present composite overwrapped pressure vessel technology, discussed in the next section of this chapter, does not appear to be fully mature, so there is likely to be additional possibilities for reducing weight and increasing safety from this technology in the future.

17.5 COMPOSITE OVERWRAPPED PRESSURE VESSELS

Many decades ago, NASA identified a need for low mass pressure vessels to carry various fluids aboard rockets, spacecraft, and satellites. The composite overwrapped pressure vessel was developed to provide weight savings without a loss in safety margins over traditional single material pressure vessels. The composite overwrapped pressure vessel design consists of a thin liner material, typically a metal, overwrapped with a continuous fiber yarn impregnated with epoxy. Most designs allow the overwrapped fiber to carry a majority of the load at normal operating pressure. The weight advantage for a composite overwrapped pressure vessel versus a traditional single material pressure vessel contributes to widespread use of composite overwrapped pressure vessels by NASA, the military, and industry. Initial Kevlar overwrapped pressure vessel designs of the 1970s realized a weight savings of approximately 25% over metal vessels. More recent improvements in fiber and resin overwrap technologies demonstrated weight savings well beyond 25% (Ecord 1977). The promise of new applications for composite overwrapped pressure vessels in cryogenic pressurant storage and fluid densification represents additional weight savings and suggests a strong future for composite overwrapped pressure vessels (Greene and Ray 1977). The following sections provide a discussion of characteristics of composite overwrapped pressure vessels that affect spacecraft system design for safe use.

17.5.1 The Composite Overwrapped Pressure Vessel System

A lined composite overwrapped pressure vessel is a system of two distinctly different load carrying materials. The properties of the materials work together to provide advantages over a vessel manufactured from either of these materials alone. Traditional composite overwrapped pressure vessels consist of a composite overwrap and a metallic liner. Composite overwraps used in space systems today utilize carbon fiber, Kevlar, Zylon®, and glass fiber in an epoxy matrix. Due to availability and a high tensile modulus-to-mass ratio, carbon is the most popular fiber material. Table 17.3 provides a summary of fiber properties. Groups of 8,000 or 12,000 fibers, called *tows*, are wound onto a spool. The fiber from the spool is either impregnated with epoxy by drawing it

Table 17.3 Comparison of Composite Overwrapped Pressure Vessel Fiber Properties (MatWeb 2007)

Fiber	Modulus, GPa	Ultimate tensile strength, MPa	Elongation, %	Density, g/cm^3
Carbon IM7	276	5450	2	1.78
Kevlar 49	112	3000	2.4	1.44
Zylon HM	280	5800	2.5	1.56
S-glass	85.5 – 86.9	4585	5.4	2.4 – 2.49

through a bath just before it is filament wound onto a liner or the fiber is pre-impregnated prior to winding. The process for filament winding is determined by how the epoxy is impregnated into the fiber tow. Filament winding using fiber that is dry before drawing through an epoxy bath is called *wet winding*. The process is characterized by liquid resin flow during filament winding. Filament winding with pre-impregnated fiber is characterized by a dry wind without the addition of resin. A cure process is performed that allows the epoxy to flow and crosslink.

Most composite overwrapped pressure vessel designs use liners of a monolithic material such as metals or polymers. Some composite overwrapped pressure vessel designs for liquid storage are without liners. These designs are filament wound using a mandrel that is dissolved after the composite overwrap is cured. Liners are used in most composite overwrapped pressure vessel designs for the following reasons:

- Liners provide a fluid permeation barrier.
- In some designs, liners carry a considerable amount of the pressure load.
- Liners provide a surface for initiating a filament winding pattern.
- Liners provide a connection port to the pressure system.

The behavior of a composite overwrapped pressure vessel is a synergy of the composite and the liner after manufacturing. An autofrettage or seizing cycle experienced by most aerospace composite overwrapped pressure vessels influences this behavior and sets the stress state of the vessel.

17.5.2 Monolithic Metallic Pressure Vessel Failure Modes

A metallic pressure vessel has two distinct failure modes: leakage and burst. Leakage, with the exception of hazardous fluids, is considered a benign failure. A crack develops in the vessel and propagates to a length where pressure is relieved, and the crack arrests. The most serious failure mode from a safety standpoint is burst failure. Burst failure is catastrophic, resulting in a sudden and unstable rupture with vessel wall fragmentation. A considerable amount of energy is released during a burst failure. This would include a step change in the pressure within pressurized gas vessels. Table 17.4 identifies failure mechanisms and controls for avoiding the burst failure mode in monolithic pressure vessels.

A crack can begin to propagate in the wall of a metallic pressure vessel through several mechanisms. One such mechanism is corrosion due to material incompatibility or exposure to a corrosive environment. Another mechanism involves inherent flaws from

Table 17.4 Monolithic Pressure Vessel Failure Mode: Mechanisms and Controls

Failure Mode	Failure Mechanism	Control
Burst failure	Fatigue	Design requirement
	Unstable crack growth	Leak before burst
	Corrosion	Fluid use restriction
	Unstable crack growth	Fluid specification

manufacturing or damage during processing, handling, or installation. Each of these processes must be controlled adequately to ensure the quality of the system. Control of the corrosion mechanism is undertaken through fluid compatibility and exposure requirements separate from leak-before-burst requirements. The inadvertent introduction of flaws or cracks is controlled via well qualified manufacturing processes and damage control plans.

Cycling of vessels can result in fatigue driven crack development while in service, and it must be shown that such a crack will develop a stable leak before the vessel can be considered leak-before-burst. Indeed, the criterion for leak-before-burst is a stable crack, that is, the stress intensity at the crack is less than the material toughness, whose length, $2c$, is 10 times the vessel wall thickness, t. More discussion of the leak-before-burst criterion is described in Gregg and Lambert (1994). Simplified formulation of the leak-before-burst criteria is given by

$$2c \leq 10 \cdot t \tag{17.4}$$

The stable crack growth requirement applies only to fatigue induced crack growth in a monolithic vessel wall and has been identified as the leak-before-burst criteria. The term *leak-before-burst* has been a point of confusion. The term has been interpreted to mean that the vessel can burst even though it first leaks. The correct interpretation of *leak-before-burst* is that the vessel leaks the pressurizing fluid due to fatigue crack growth precluding a burst failure. Burst-before-leak failure of a monolithic pressure vessel can be controlled through design, by ensuring that cracks are stable and run through the vessel wall. Because of concern about ensuring safe use of pressure vessels, metal pressure vessels are required to meet leak-before-burst criterion as specified in MIL-STD-1522A (DoD 1984) and (ANSI 2006). Figures 17.5 and 17.6 demonstrate stable and unstable crack growth characteristics that lead to leak or burst failure for monolithic pressure vessels.

17.5.3 Composite Overwrapped Pressure Vessel Failure Modes

The leak-before-burst failure mode is characterized by development of a stable leak path that enables the pressurizing fluid to leak from the vessel without fragmentation of the vessel wall, thereby ensuring that the pressure vessel cannot burst. An unstable catastrophic burst failure is averted, considerably improving system safety.

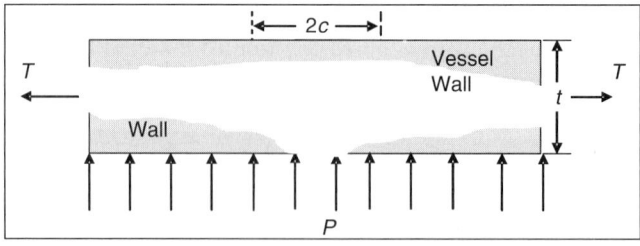

FIGURE 17.5 Leak-before-burst behavior in a monolithic pressure vessel wall (stable crack growth), where $2c$ is the crack length, t is vessel wall thickness, T is tension in the vessel wall, and P is the applied pressure due to the pressurizing fluid (Courtesy of NASA).

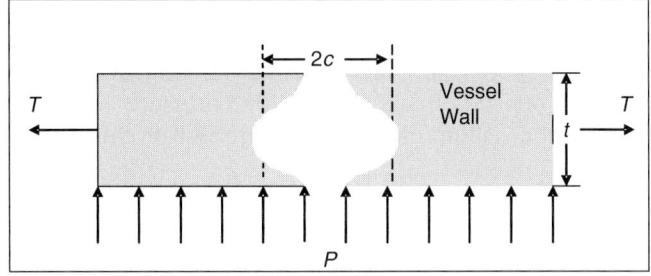

FIGURE 17.6 Burst-before-leak behavior in a monolithic pressure vessel wall (unstable crack growth, not through wall), where $2c$ is the crack length, t is vessel wall thickness, T is tension in the vessel wall, and P is the applied pressure due to the pressurizing fluid (Courtesy of NASA).

17.5.4 Composite Overwrapped Pressure Vessel Impact Sensitivity

The leak-before-burst requirement has been applied to composite overwrapped pressure vessels to assure that crack development and growth in the liner results in benign leakage, as done successfully for monolithic metal pressure vessels. However, the leak-before-burst requirement has been misinterpreted by some to mean that the composite overwrapped pressure vessel system is not susceptible to burst failure should the liner be leak-before-burst with respect to the fatigue crack growth mechanism. The reality is that steps must be taken to protect the composite and overwrap from fatigue and other mechanisms that can result the burst failure mode. Figure 17.7 demonstrates the components of the composite overwrapped pressure vessel system.

The composite overwrap has more tensile strength than the liner and undergoes a more brittle failure. Once the liner fails, it can increase the stresses in the composite to the point that brittle failure occurs. Therefore, it is not enough to simply model a liner as leak-before-burst, then presume the system will exhibit only leak failure.

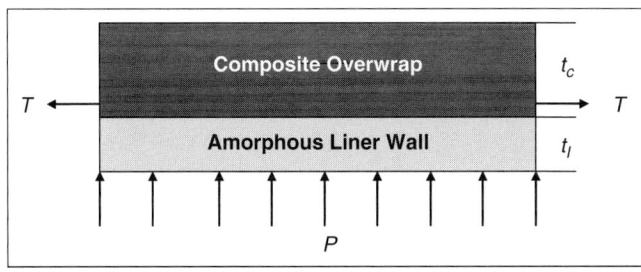

FIGURE 17.7 Composite overwrapped pressure vessel wall layout, where t_c is the composite thickness, t_l is the liner thickness, T is tension in the vessel wall, and P is the applied pressure due to the pressurizing fluid (Courtesy of NASA).

Because the composite typically carries a majority of the pressure load, controls to preclude the propagation of composite failure mechanisms are especially important in preventing the burst failure mode. More failure mechanisms for a composite overwrapped pressure vessel can result in burst behavior than for monolithic pressure vessels (Table 17.5). As an example showing that other failure mechanisms can result in burst behavior, Figure 17.8

Table 17.5 Composite Overwrapped Pressure Vessel Failure Mode, Mechanisms, and Controls

Failure Mode	Failure Mechanism	Control
Burst failure	Liner fatigue and unstable crack growth	Leak-before-burst
	Liner corrosion and unstable crack growth	Fluid use restriction
	Overwrap corrosion and stress corrosion	Fluid contact restriction
	Overwrap damage and damage propagation	Impact control plan
	Time, temperature, and pressure stress rupture	Stress rupture assessment
	Overstress and damage propagation	Maximum operating pressure, proof pressure requirements
	Manufacturing defect and damage propagation	Nondestructive evaluation

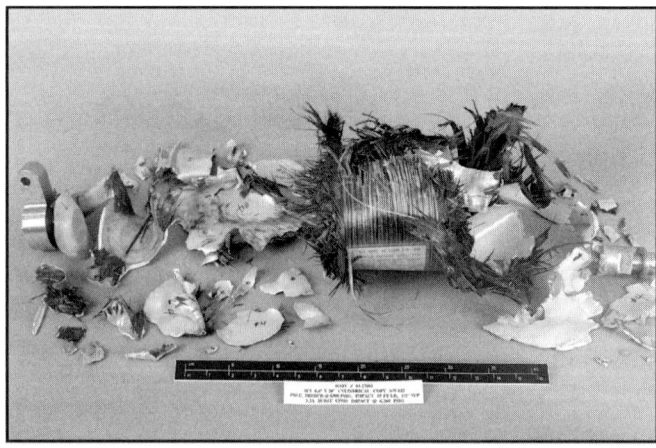

FIGURE 17.8 Burst failure of a 4.40 × 20″ cylinder on impact while pressurized to maximum expected operating pressure with nitrogen gas at ambient temperature (WSTF 1295-2874 — COPVs/n 033) (Courtesy of NASA).

shows a burst vessel from impact damage to a composite overwrapped pressure vessel that had a leak-before-burst liner (Beeson 2002). A more detailed discussion of the progressive failure nature of composite overwrapped pressure vessels can be found in Greene et al. (2007b).

A proposed solution to the leak-before-burst design for composite overwrapped pressure vessels is to design the composite overwrap so that its load carrying capability has a margin to absorb the dynamic load transferred to the overwrap from unstable crack growth in the liner. An overwrap designed to absorb unstable liner failure could remove the need for the leak-before-burst fatigue requirement for a liner. This would be useful in the effort to reduce composite overwrapped pressure vessel mass by allowing for considerably reduced liner thicknesses. Overwrap designed to absorb unstable crack growth in the liner follows the energy absorption relationship given in Equation (17.5), where $E_{OW,\,fail}$ is the energy absorption capability of the overwrap at failure, $E_{OW,\,MEOP}$ is the energy carrying capability of the overwrap at maximum expected operating pressure, and $E_{L,\,MEOP}$ is the energy carrying capability of the liner at maximum expected operating pressure. Therefore, any leakage shown during qualification testing of the composite overwrapped pressure vessel would be a function of the system as a whole, instead of just the liner (Greene et al. 2007a).

$$E_{OW,\,fail} > E_{OW,\,MEOP} + E_{L,\,MEOP} \quad (17.5)$$

An understanding of the structural response of the vessel is required in the determination of the amount of load carried by the liner and the amount of load carried by the overwrap. Guidance for the approach is given in Thesken et al. (2005).

17.5.5 Summary

All mechanisms that affect composite overwrapped pressure vessel failure must be included in a use assessment. Controls can be implemented to force a leak failure in a monolithic pressure vessel or composite overwrapped pressure vessel. It has been shown here that these controls are not the same. The leak-before-burst fracture control requirement for monolithic vessels is not sufficient to control the failure mode of a composite overwrapped pressure vessel. A proposed energy requirement would demonstrate a system level leak-before-burst failure mode for composite overwrapped pressure vessels and possibly result in lighter, safer vessels.

17.6 STRUCTURAL DESIGN OF GLASS AND CERAMIC COMPONENTS FOR SPACE SYSTEM SAFETY

It is without question that crewed spaceflight programs always will have windows as part of the structural shell of the crew compartment. Astronauts and cosmonauts need to and enjoy looking out of the spacecraft windows at Earth, approaching vehicles, scientific objectives, and the stars. With few exceptions, spacecraft windows have been made of glass, and the lessons learned over 40 years of crewed spaceflight have resulted in a well defined approach for using this brittle, unforgiving material within NASA vehicles, in windows and in other structural applications. This section outlines the best practices developed at NASA for designing, verifying, and accepting glass and ceramic windows and other components for safe and reliable use in any space system.

17.6.1 Strength Characteristics of Glass and Ceramics

Glass is a brittle material. Structural design with glass is governed by fracture mechanics and static fatigue analysis. Every glass material has characteristic properties associated with fracture and static fatigue that must be known by the designer and analyst to meet the strength, life, and safety requirements specified by the spacecraft or payload developer. The following subsections are a brief summary of fracture and static fatigue. For more detailed information about these topics, the reader is directed to any textbook about the fracture of brittle materials.

Fracture of Glass

As originally described by A. A. Griffith (1920), brittle materials like glass fail in tension as the result of tiny flaws in the surfaces of the part that were created during manufacturing or handling. When these cracks are placed in a tensile stress field they grow; and when a crack reaches the critical stage, the glass fails. The fracture strength of a piece of glass is related inversely to the size of the surface flaw:

$$\sigma_F = \frac{YK_{IC}}{\sqrt{\pi a}} \tag{17.6}$$

where Y is a factor related to crack and part geometry, a is the crack depth from the surface, and K_{IC} is the critical stress intensity, discussed later. The critical stress intensity is also called the *fracture toughness* of the material.

Because most of the initial flaws are too tiny to see, failure of glass can come with no forewarning. It is important to note that surface flaws are the controlling feature of the strength of any glass product. Glass parts fail from surface flaws for four reasons. Manufactured glass has few internal flaws, and those that do exist usually are smooth in nature, like bubbles, and do not concentrate stresses. For most loading conditions, the maximum tensile stress is on the surface. The surface also is subject to flaws induced by the manufacturing process, such as polishing, and other contact events, both intentional and accidental. Finally, surface cracks are exposed to the environment and subject to subcritical crack growth or static fatigue, which is discussed in the next subsection (Varner 1996).

To determine the design strength of a glass part, it is necessary to know K_{IC}, the critical stress intensity, and the size of the flaws present in the final part.

Static Fatigue of Glass

Unlike metals, glass and ceramics experience static fatigue, something similar to stress-corrosion, where the strength of any part decreases over time at load when water molecules are present in the operational environment. Cyclic loading normally is not detrimental to glass parts, but the total time at load and the humidity of the operating environment are critical. The rate of static fatigue or subcritical crack growth in glass is described by special crack growth parameters determined by test.

17.6 Structural Design of Glass and Ceramic Components

Some compositions of glass, like soda lime, experience subcritical crack growth only when the tensile stress exceeds a certain threshold. In the space program, the glass most commonly used is fused silica, for which no threshold stress has ever been determined. Therefore, even at the lowest operating stresses, subcritical crack growth can be expected in fused silica parts. A good understanding of the crack growth properties, the flaw population, and the operating stresses is extremely important for accurately predicting the structural life of this kind of hardware.

Fatigue and Fracture Parameters

Both fracture toughness and the crack growth parameters are determined by test programs involving many samples. The ASTM International and other standards agencies published several test standards in recent years describing programs to determine these parameters for glass and ceramics. Some of the relevant standards and which properties are determined by them are included in Table 17.6.

Notes About Modeling Static Fatigue

The model of crack growth noted in Table 17.6, a power law relationship between the crack velocity and the crack tip stress, is the most common approach to modeling subcritical crack growth. This model often is called the *Paris equation*:

$$v = A(K_I/K_{IC})^n \qquad (17.7)$$

Table 17.6 Test Standards for Fracture and Fatigue Properties

Standard Title	Standard Number	Parameters Tested
Standard test methods for strength of glass by flexure: determination of modulus of rupture	ASTM C 158-02	Modulus of rupture
Testing of glass and ceramics: determination of bending strength	DIN 52-292, part 1	Modulus of rupture
Standard test method for determination of slow crack growth parameters of advanced ceramics by constant stress rate flexural testing at ambient temperature	ASTM C 1368-06	Slow crack growth parameters n and A for a model of crack growth velocity represented by the power law form shown in Equation (17.6)
Standard test method for determination of slow crack growth parameters of advanced ceramics by constant stress flexural testing (stress rupture) at ambient temperature	ASTM C 1576-05	
Standard test methods for determination of fracture toughness of advanced ceramics at ambient temperature	ASTM C 1421-01b	K_{IC}

It is based empirically and relatively simple to apply. However, an exponential relationship also has been used, and this has more basis in the physics of crack growth:

$$v = V_0 e^{n(K_I/K_{IC})} \tag{17.8}$$

The parameters in the exponential form can be developed using a curve fit process somewhat more complicated than for the power law. No published standards have been published for this numerical analysis. Some versions of the exponential formulation account for variations in humidity and temperature, which typically is not done using the power formulation. Some versions of the power law model account for different regions of crack growth, which is not done using the exponential model.

For low stresses or long times to failure, the two formulations diverge. Figure 17.9 illustrates static fatigue data for borosilicate glass and the two fatigue models fit to the data (NASGRO 2005). Figure 17.10 shows the divergence of the time to failure predictions for

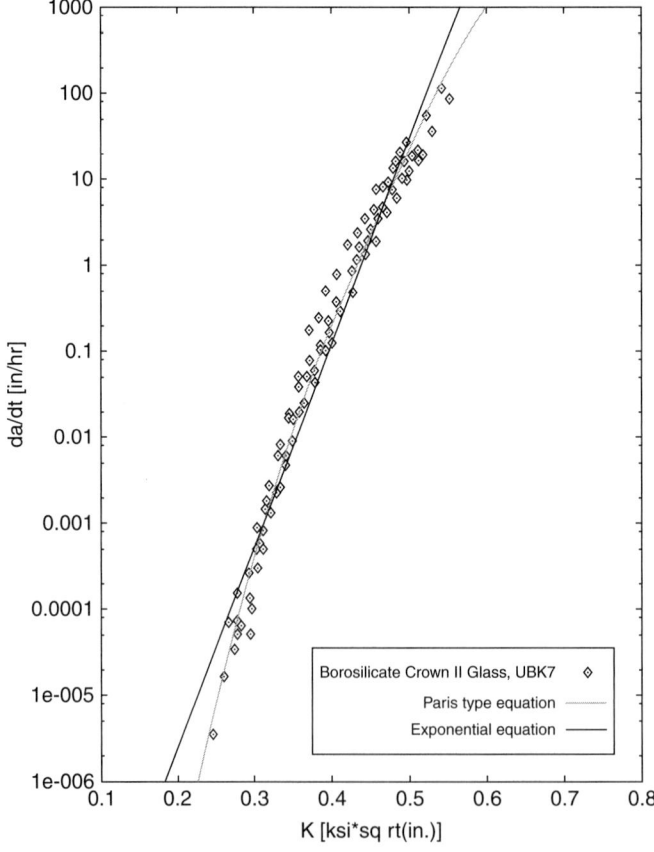

FIGURE 17.9 BK7 fatigue data (Courtesy of NIST).

17.6 Structural Design of Glass and Ceramic Components

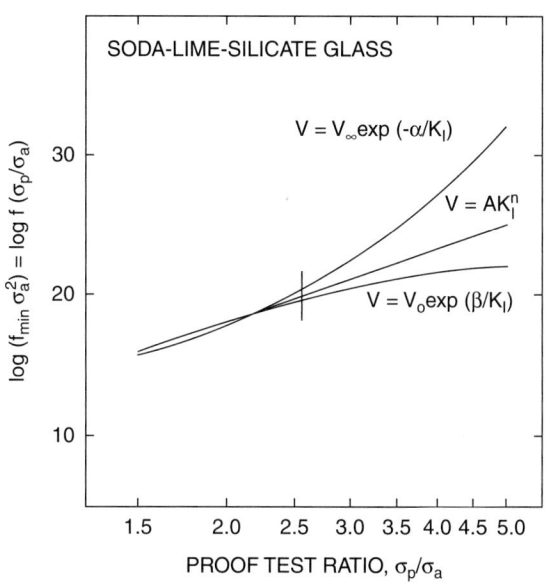

FIGURE 17.10 Time to failure for soda lime silicate glass (Wiederhorn 1977) (Courtesy of NIST).

each model. At a proof test ratio of 3.0, which is typical for human rated spaceflight hardware, the time to failure for soda lime silicate glass modeled with the power law formulation is almost 60 times longer than the predicted life using the exponential form.

For most applications in the payload community, the power law relationship is appropriate. For long life parts like International Space Station windows, the exponential relationship was specified because it gives a more conservative result. Industrial users of glass, such as optical fiber manufacturing and cabling, implement modifications of the power law with good results (Baker 2001). Other researchers prefer the power law formulation and show that results are more accurate for both long and intermediate life parts (NASGRO 2005). NASGRO, a widely used numerical analysis tool for predicting fatigue life, offers both models.

Alternative Approaches to Assessing Strength and Reliability

For some applications of glass in space hardware, the operating stresses are very low compared to the advertised strength of the material. NASA permits an alternative verification path that requires the hardware provider to demonstrate that the part is at least five times stronger than the applied load. This approach derives from the fact that, at such low stresses, the critical crack size likely exceeds the nominal dimensions of the part and the fracture analysis becomes invalid.

To pursue this path, the hardware provider must show a glass rupture strength value developed from controlled test data with the appropriate statistical analysis (using the

B0.15 in a Weibull curve fit, which indicates the 0.15 percentile failure strength). If this can be done, no life analysis or acceptance proof testing is required, and a simple stress analysis of the glass part is sufficient.

An example of this application is for a small payload window pressurized to 0.1 MPa. The maximum stress in the window is 0.6 MPa. The hardware provider has no test verified strength data for this window material as polished by the glass vendor, so 20 samples are tested in a biaxial ring fixture according to the DIN 52-292 standard (DIN 1984). Figure 17.11 shows the results with a Weibull fit and that the B0.15 strength value of 50 MPa is more than sufficient to permit this hardware provider to forego proof testing and further life analysis.

17.6.2 Defining Loads and Environments

All structural designers need to know the loads and environments within which their hardware must perform. Where glass and ceramics are concerned, the definition of these environments is critical with respect to understanding the operating conditions like vacuum or humid air that could interact with the flaw growth properties of a material.

The designers should be aware of any environments that lead to surface damage, like atmospheric or spaceflight impacts or crew contact. Operating temperature is less of a concern, for the material strength and properties of glasses and ceramics typically do not vary over

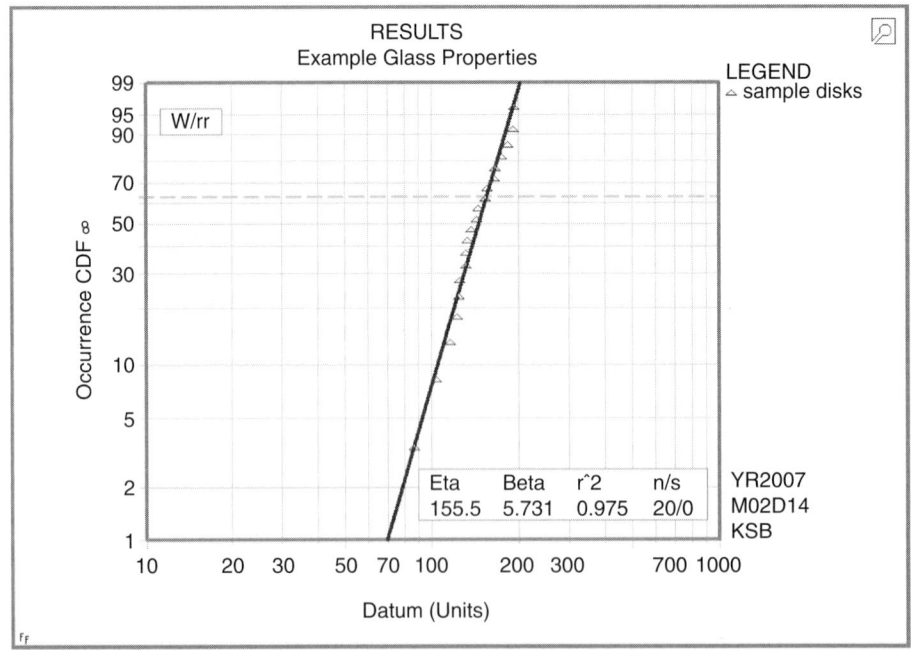

FIGURE 17.11 Weibull plot of glass rupture strength (Courtesy of NASA).

17.6 Structural Design of Glass and Ceramic Components

temperature to any great degree. For this reason, low expansion glass is an excellent choice for the external pane of a reentry vehicle window as compared to a transparent polycarbonate, which is much tougher than glass but intolerant of high temperatures. For example, the Space Shuttle windshield glass can exceed 650°C during reentry, but the fused silica panes installed in the windshield have an annealing temperature of 1042°C and a softening point considerably higher than that (Corning 2003). These fused silica panes survive multiple reentry events without being affected by the temperature or plasma environment.

Applied Environments
Applied environments should include thermal shock and structural loads due to thermal strain, pressure, vibration, and acceleration. Space vehicles must launch, ascend, orbit, descend, and land; and the environment of each operational phase must be considered. The specific hardware design might include vacuum or pressure cycling, and these must be considered.

Stress Distribution Verification Validating the stress distribution in glass and ceramic structures must be done by testing of flight or flight-like hardware. Experience from several major programs and vehicles has demonstrated that 10% or more of the stresses in any window come from the secondary effects of warping and deformation in the supporting structure. In testing a window installation, the glass panes typically are replaced by aluminum simulators, and strain gauges are applied to provide sufficient data to confirm the stress distribution. In other structures, the design team can put strain gauges directly on the glass components; however, it is important to control the process carefully to prevent glass failure caused by flaws induced or magnified by the strain gauge. It should be noted that the strain gauges cannot be removed without introducing new damage to the glass.

Stress Analysis Stress analysis of glass or ceramic parts follows a typical process. Only the ultimate load case is examined, and the strength value used here is the statistically valid rupture strength found during materials characterization testing. An alternative is to use the initial flaw size verified by proof test and a calculated initial strength from Equation (17.6). This calculation should be performed using $K_{IC} - 3\sigma$ from the materials characterization tests to ensure a conservative assessment of fracture strength.

Inadvertent Contact
Inadvertent contact during hardware processing or flight is one of the most difficult environments to manage. The design team must be cognizant that inadvertent contact is going to occur, and the team must specify what kind of inspections must be performed to detect the inevitable damage. These people must also be able to calculate the strength and structural life loss caused by this contact and either manage that effect during the mission of the item or be prepared to replace damaged hardware. Some examples of inadvertent contact that damaged glass components in NASA programs include dropped tools, grit contamination on a gloved hand, scratches due to cameras used on-orbit, and tool scratches from ground processing on equipment adjacent to the glass parts. When the initial design of the hardware specifies an acceptable flaw size of 0.0018 in. in depth, even a minor scratch can render a part unusable if it is in the wrong place.

Inadvertent contact should be prevented by the use of protective covers whenever practical. Ground processing of windows or other glass hardware should be reviewed so adequate protection can be provided for these items. During flight, transparent protective panes, covers, or grills can be used to keep crew contact with the glass minimized.

Glass to Metal Contact

Glass to metal contact should be prevented in the design of any structural glass component. Glass is extremely brittle and it fractures readily if even a small point load is applied. If the assembly includes a glass component supported by a metallic structure, designers should provide a pliable interface of some kind between the two parts.

Seals and Cushions in Assemblies with Glass

The seals and cushions used in assemblies containing glass must be selected with the temperature extremes of the hardware in mind. These materials must remain pliable at the coldest expected temperature. Most elastomers have a glassy transition point where they become quite hard, and this transition point should be avoided when selecting appropriate materials for the required design environment.

Special Considerations for Coatings

Coatings have been shown to propagate preexisting surface flaws in glass when the coated surface is under tension. If coatings are to be used, it is best to apply them to the surface that is to be in compression.

17.6.3 Design Factors

As with all structural design efforts, factors of safety, uncertainty factors, and other factors always are specified. For glass and ceramic components, these can have one value for the beginning of life and a different value for the end of life in an attempt to address static fatigue concerns. Because glasses and ceramics are brittle materials, no yield factor needs to be considered, because no yielding ever occurs. It is necessary only to assess the ultimate load condition for strength and the limit load condition for life.

An uncertainty factor can be applied to the operating stress for the life analysis. In the International Space Station Program, NASA specified different uncertainty factors for parts with differing life requirements. Short life components have higher uncertainty factors, and the windows that are required to perform for the full 15-a life of the *International Space Station* had the lowest uncertainty factor at 1.1. This was intended to address the discrepancies in the models used to predict crack growth. See Section 17.6.1, subsection "Notes About Modeling Static Fatigue," for more discussion on this topic.

Factors of Safety for Annealed Glass

The minimum ultimate factor of safety at the beginning of life for annealed glass used in structural applications in NASA programs has been 3.0 for the International Space Station and payload hardware. Other programs specified lower design factors, but they have been overridden by acceptance proof test requirements, which typically drive the factor higher than 3.0. This is discussed later in the chapter. The end of life factor of safety for most NASA programs has been 1.4 or less.

Factors of Safety for Tempered (Strengthened) Glass

Usually, the magnitude of the surface compression in chemically tempered glass is quite high compared to the expected operating stress, and NASA programs specified that the surface must not go into tension at twice the operating stress. However, there is also a factor of safety requirement, and the most recent glass design requirements document (NASA 2006a) specifies an ultimate safety factor of 3.0 at the beginning and end of life. No static fatigue strength degradation is ever expected for tempered panes because the surface should never be in tension, and therefore static fatigue crack growth is prohibited.

Other Factors

Unique hardware designs can require other factors. For example, glass windows in NASA spacecraft always are made redundant, so that a single pane failure is not catastrophic. Design requirements for these window systems specify a dynamic factor to be applied to the load on the redundant pane because in a failure event nothing is static. Payload providers might face similar issues and carefully should consider all the operational scenarios their hardware must survive.

17.6.4 Meeting Life Requirements with Glass and Ceramics

Structural components made of glass or ceramics are considered fracture critical, and adequate life is required to be demonstrated analytically. Generally, this means that the hardware provider must determine the maximum initial flaw size present in the final part. The provider also must perform a numerical analysis of flaw growth by using commercially available software or another proven method for applying accepted models of crack growth in glass or ceramics. Alternative approaches to life certification are available in special circumstances; one was described in Section 17.6, subsection "Alternative Approaches to Assessing Strength and Reliability."

Scatter Factor

Typically, a scatter factor of 4 is required to demonstrate adequate life. This means that, if the hardware has a required mission life of 1 a, the analysis must show that at the end of 4 a the part has residual strength adequate to meet the limit load times a specified end of life factor of safety.

Proof Test

Each piece of glass or ceramic structure delivered for the flight hardware must have acceptance proof testing performed unless a special approach, like the one described in Section 17.6, subsection "Alternative Approaches to Assessing Strength and Reliability," is approved by NASA. The acceptance proof test demonstrates that the maximum initial flaw size present in the hardware cannot propagate to failure during the life of the part.

Designing the acceptance test involves controlling the environment and determining the necessary delta pressure or other external load to achieve the appropriate screening stress. Typically, the glass vendor determines the polishing process, which defines the population of flaws in the surface. The maximum initial flaw size in the final polished item should be approximately three times the size of the final grit. It is necessary for the proof test to reach a pressure that causes failure for a part where the flaw sizes exceed the specification of the vendor. Items that fail the acceptance proof test do so destructively, assuring that an item delivered for flight meets the requirements.

To calculate the proof stress and envelop the rather large scatter inherent in glass properties, use the K_{IC} resulting from material characterization tests and add a single standard deviation. This provides some conservatism in the screening process:

$$\sigma_{proof} = \frac{(K_{IC} + 1 \cdot \sigma)}{Y\sqrt{\pi a_{initial}}} \qquad (17.9)$$

The proof load can be applied as slow or as fast as desired, but once the proof stress is reached, the test must be ended as quickly as possible. Flaws propagate during the proof test, so it is imperative to limit the time of the test following the proof stress to as little as is possible.

It also is important to keep moisture out of the proof test environment. To this end, NASA windows are heated in an oven for several hours before a proof test. When the dew point reaches −35°C, the test is begun. In a pressure test of a NASA window, only dry nitrogen gas is used.

Life Analysis

Life analysis is performed typically using numerical analysis codes. The most widely used is NASGRO, which calculates subcritical flaw growth in glass or ceramic materials and outputs a flaw size at the end of the analysis. This code compares stress intensity values against the critical stress intensity for the material.

If the NASGRO analysis is completed successfully, an end of life flaw size is reported. Using the end of life flaw size, the end of life strength can be computed using the relationship between flaw size and stress intensity:

$$\sigma_{end\text{-}of\text{-}life} = \frac{K_{IC}}{Y\sqrt{\pi \cdot a_{end\text{-}of\text{-}life}}} \qquad (17.10)$$

From this strength value and using the required end of life factor of safety, the hardware provider calculates the end of life margin of safety. A negative result indicates that the part can fail at the end of its life and a redesign is necessary to lower the operational stresses. In the cases where this issue has arisen, usually, a single high stress event in the design life of the part caused the majority of the crack propagation. Judicious redesign can focus on those few high stress cases.

17.7 SAFETY CRITICAL MECHANISMS

Spaceflight mechanisms have a reputation for being difficult to develop and operate successfully. This reputation is earned well. Many circumstances conspire to make this so:

- The environments in which mechanisms are used are extremely severe.
- Usually limited or no maintenance opportunity is available during operation due to this environment.
- The environments are difficult to replicate accurately on the ground.
- The expense of the mechanism development makes it impractical to build and test many units for long periods of time before use.
- Mechanisms tend to be highly specialized and not prone to interchangeability or off-the-shelf use.
- The mechanisms can generate and store a lot of energy.
- The nature of mechanisms themselves, as a combination of structures, electronics, and the like, that are designed to accomplish specific dynamic performance makes them very complex and subject to many types of unpredictable interactions.

In addition to their complexity, mechanisms often are counted on to provide critical vehicle functions that can result in catastrophic events should their functions not be performed. For this reason, mechanisms frequently are subjected to special scrutiny in safety processes. However, a failure tolerant approach, along with good design and development practices and detailed design reviews, can be developed to allow such notoriously troublesome mechanisms to be utilized confidently in safety critical applications.

17.7.1 Designing for Failure Tolerance

The essence of a failure tolerant approach is to identify potential hazards and ensure that proper controls or redundancies are in place to prevent undesirable incidents from occurring. For the present, the time that it can take for a detrimental incident actually to manifest itself after the hazard is created (often called the *time to effect*) can be neglected, and it can be assumed that the incident happens at the instant the hazard is created. From a safety perspective, the basic goal in the design of any safety critical mechanism is to prevent the creation of any hazard that could result in an undesirable incident. There are two approaches to this:

- Designing for reliability.
- Designing for failure tolerance.

Reliability in Space Mechanisms

In an ideal world with unlimited schedules and budgets, designing mechanisms for reliability is the best approach. If systems can operate reliably without failure, not only are hazards avoided but also the operations problems associated with using backup systems. To create a mechanism with the kind of reliability typically required for use on human rated spacecraft, the design must go through many expensive and time consuming iterations involving development design life testing. The initial design iteration must go through a life test, after which any failures or unacceptable performance degradations observed must be addressed with design modifications followed by retesting. Once a satisfactory life test has been completed on one unit, more units of that design must be constructed and run through their life tests to gain statistical confidence in the design and work out any problems that could be a result of the assembly process. Only when enough units to generate meaningful statistical data have demonstrated sufficient life can the mechanism be assigned a credible reliability number for use in a probabilistic risk assessment.

Even when a reliability approach can be used, requirements usually are instituted such that the mechanisms fail into a safe configuration. In practice, this often, but not always, equates to failure tolerance. But most space hardware development programs lack the luxuries of budgets or schedules sufficiently large to build and test the dozens or even hundreds of units needed to pursue a reliability approach. Reliability cannot be demonstrated sufficiently on a qualification unit that is the first of its kind. When only a small number of units can be built, as is almost always the case with space hardware, a failure tolerant approach must be followed. For that reason, the rest of this section deals mainly with the aspects of the failure tolerance approach.

Failure Tolerance Assessments and Recognizing Failure Modes

The first steps in designing a failure tolerant mechanism are to perform a proper failure tolerance assessment and identify the operations or constituent mechanisms that require failure tolerance. The key to these steps is to make sure that all possible failure modes actually are assessed. This can be more difficult than it sounds. The most obvious and easily identified failure modes generally are a failure to function when needed and inadvertent operation at all other times. However, often the failure of a mechanism in mid-travel presents a unique failure mode that can have safety implications.

For example, imagine you are determining the mechanism safety implications of the following scenario. The *Space Shuttle* is to rendezvous with a very large orbiting satellite. The robotic arm is to grapple the satellite and berth it to the *Space Shuttle* near a new piece of equipment that is to be attached to the satellite and in an orientation such that the long axis of the satellite is perpendicular to the axis of the payload bay. To cut down on the mass of the attachment mechanisms, the new equipment is retained on the *Space Shuttle* with an attachment system that latches the equipment onto the satellite at the same time that it is released from the payload bay. Once attached, the new equipment undergoes system checks. If the new equipment operates correctly, the satellite is unberthed and released back into orbit with its upgrade. However, if the new equipment does not work properly, then it is to be removed from the satellite and returned to Earth to be repaired, and the satellite is released back into orbit without the upgrade.

If the attachment mechanism fails to operate in the first place, the mission can be aborted and the satellite released back into orbit without its new equipment. This is a failed mission, which has its own consequences, but the *Space Shuttle* and crew are safe. Now consider a case in which the mechanism works correctly when attaching to the satellite, but the new equipment fails its checkout. The mechanism is commanded to release from the satellite and reattach to the *Space Shuttle*, but the mechanism fails. In this case, the satellite can once again be released, albeit with useless equipment attached to it. This results in another mission failure. It is worse than the failure of the previous example because a new mission with new operations and support hardware must be created to repair the failed mechanism, launch a second upgrade, and swap the failed equipment. Even so, everyone still is safe, and no catastrophic problems are encountered.

Assuming that inadvertent actuation is guarded against properly, at this point a cursory look indicates that no catastrophic failure modes are associated with the release mechanism. Now let us say a closer look is taken at the operation of the latching mechanism. From this, it is learned that a clever single mechanism is used to move the system through the following series of states during its operation:

1. Prior to operation, there is a structural connection to the *Space Shuttle*.

2. On initial actuation, the structural connection is released, but the mechanism remains soft captured to the *Space Shuttle*. The mechanism achieves capture of the satellite prior to fully releasing the *Space Shuttle*, so that the mechanism cannot float away accidentally.

3. The mechanism moves further, releasing the *Space Shuttle* and captures only the satellite.

4. The mechanism creates a structural attachment with the satellite.

Failures in states 1, 3, or 4 have the same consequences as those initially assessed before looking into the mechanism. But, a new problem is hidden in state 2. If the mechanism was to seize in an unrecoverable way while in the process of capturing both the satellite and the *Space Shuttle*, the satellite can no longer be released, and it is stuck on the *Space Shuttle* in a position that is not structurally sound and prevents the payload bay doors from closing, a prerequisite for returning to Earth. The mechanism can pass through this critical range of motion quickly, but that makes no difference under a fault tolerance approach. A catastrophic failure mode has been uncovered that probably requires a design change to meet failure tolerance requirements.

Another error that can be made when determining failure modes is to miss mechanisms completely. This may seem unlikely, but the types of mechanisms most often missed are missed not because they were forgotten but because they were not considered mechanisms.

One class of mechanisms peculiar to human spaceflight frequently disregarded is that of threaded fasteners. Bolts and other fasteners commonly are used both internal and external to habitable volumes for assembly and latching. When properly installed and left undisturbed during flight, a threaded fastener is considered structure. However, as soon as a fastener configuration intentionally is altered in any way during flight, it becomes a mechanism, because new failure modes are introduced that are mechanical rather than structural in nature. For example, bolt threads can gall during installation or jam due to thread damage or debris.

Maintenance and contingency actions frequently are overlooked in failure tolerance assessments as well because they fall outside of the normal operations concepts. For example, one panel of a structure might be able to be removed during flight to allow for failure investigation or maintenance of the system it encloses. But if this panel is necessary to maintain structural integrity during subsequent loading events, then the inability to reinstall the panel is a safety critical hazard, and the fasteners or latching devices required to attach the panel must be considered safety critical mechanisms.

Failure Tolerance Assessments During the Design Phase

As easily can be imagined, figuring out that a mechanism design does not meet failure tolerance requirements can be very expensive if design changes are required once the hardware reaches the manufacturing stage and beyond. Like any other piece of hardware, the more rigor put into a mechanism during the design phase, the fewer problems experienced later. A failure tolerance analysis is an important part of the rigor that is too often overlooked. Recognizing all failure modes early can lead to elegant design solutions with little or no mass penalty.

17.7.2 Design and Verification of Safety Critical Mechanisms

Once a mechanism is recognized as safety critical and its failure modes are understood, the next step is to perform the detailed design and verification of the mechanism so that it can operate robustly, fail gracefully if failure does occur, accurately represent its status, and provide confidence that its performance in all these modes is as intended. The developer should pay particular attention to several important areas:

- Positive indication of status.
- Torque and force margin.
- Debris shielding.
- Lubrication.
- Thermal tolerance analysis.
- Qualification and acceptance testing, including design life testing and run-in.

Each of these is discussed briefly.

Positive Indication of Status

Positive indication of status is perhaps the most important feature of any safety critical mechanism. Put another way, positive indication of status means the ability to measure directly and reliably the actual state that constitutes the potential hazard. Indication of status is so important because, in most cases, for failure tolerance provisions to be put into effect, someone or something must know there is a failure in the first place. Every effort must be made to measure this condition with direct, not indirect, means.

Consider the case of a door mechanism on the belly of an entry vehicle. This door must be closed fully to prevent hot gases from burning a hole in the vehicle structure during atmospheric entry. One way to provide an indication of the status of the mechanism would be to place a limit switch near the shaft of the actuator that powers the door hinge so that, when the door is closed fully, a cam correspondingly positioned on the shaft strikes a set of redundant limit switches to indicate closure. The problem with this approach is that it is not the position of the shaft that causes the hazard but rather the position of the door. If a piece debris lodged itself in the door cavity or damaged the mechanism or door itself on ascent, the door might not be able to seat fully, but the inherent flexibility or kinematics of the door and mechanism still could allow the actuator shaft to reach the measured position. In this case, with no other indications available, the crew would assume all is well and initiate a potentially catastrophic deorbit burn. A better solution is to place a limit switch where it is actuated by contact with the door itself.

Torque or Force Margin

For linear devices, torque or force margin is a measure of the amount of force or torque a mechanism has in reserve, that is, the amount beyond the anticipated need for its predicted function. To operate, the available torque or force (for simplification, *torque* is used for both terms in the remainder of this section) in the mechanism must exceed the sum of all of the resisting torques, which can come from a number of sources: friction, bending of cables, material deflections, inertial motion, and the like. However, it often is difficult to predict all the potential sources. In addition, occasionally, circumstances can result in an unforeseen drop in available torque. As a result, having extra torque in reserve is crucial when designing dependable mechanisms to operate as intended.

In general, torque margin can be defined as

$$\text{TM} = \frac{\text{Available torque}}{\text{Resisting torque}} - 1 \qquad (17.11)$$

The proper torque margin to use in a mechanism generally depends on the amount of uncertainty in the design conditions. Still, most programs have specific requirements for the minimum required torque margin. Even with state-of-the-art analysis and testing techniques, it is very difficult to predict how a mechanism is going to function in a space environment after having been exposed to the environment of a launch. Therefore, with the exception of some special cases, the completed hardware always should possess a torque margin. Based on experience with past programs, the torque margin always should be at least 1.0 under worst-case conditions, unless specified otherwise by the program. To account for increased uncertainty in early design phases, it is recommended that the design torque margin should start higher than 1.0, be decreased as the design matures, and end with a minimum measured margin of 1.0 imposed at hardware qualification and acceptance. Table 17.7 illustrates a

Table 17.7 Recommended Minimum Static Torque or Force Margin

Development Phase	Torque or Force Margin
Conceptual design	1.75
Preliminary design review	1.50
Critical design review	1.25
Hardware acceptance	1.00

commonly utilized torque margin management approach, consistent with that recommended in AIAA S-114-2005, *Moving Mechanical Assemblies for Space and Launch Vehicles* (AIAA 2005).

These recommendations are for static conditions, so the kinetic energy from the motion of the mechanism should not be considered in the calculation of available torque. This ensures that, even if the motion of the mechanism should stop for some reason, enough torque is available to resume motion. These recommendations also represent minimum values. In general, mechanisms should try to achieve the highest margin possible within the design constraints, such as load or acceleration limits.

Debris Shielding

Debris is a constant concern for spacecraft and aircraft. Although the concern for aircraft is primarily that of engine damage, debris can cause a variety of problems for spacecraft due to the microgravity environment encountered during spaceflight. It can damage structure or instruments during liftoff, ascent, or landing. It can contaminate scientific instruments, rendering them useless, or it can find its way into mechanisms and prevent them from operating. For these reasons, it is important to shield mechanisms from debris to the largest extent practical.

Note that debris can come in two types:

- Debris from external sources not associated with the mechanism (commonly called *foreign object debris*).
- Debris generated by the mechanism itself.

It is easy to imagine foreign object debris; it is what most people think of when they consider debris shielding. Almost anything can become foreign object debris, a washer dropped during ground processing, a tool left behind, paint chips from a nearby surface, orbital debris from previous spacecraft, and even insects. These are just a few examples of foreign object debris that have been encountered. This type of debris can be guarded against rather easily. For example, closeout panels and flexible boots can be added to most mechanisms to protect their interior from intrusive debris.

Self generated debris can be a little harder to catch or protect against. For example, misaligned mechanisms or improper gear meshing can create metal shavings that either cause galling or can build up and act as debris. Dry film or solid film lubricants that are not burnished properly can generate a similar buildup. Internal features, such as wiper seals, can be useful in controlling self generated debris. In addition the usefulness of closeouts to prevent self generated debris from becoming foreign object debris for another system should not be overlooked.

Lubrication

Nearly all space mechanisms contain surfaces that move relative to each other. The use of a proper lubricant on these surfaces can extend the life of the mechanism and help them meet their service life requirements.

Tribology is a very intricate and complicated subject prone to frequent advances in technology. For this reason and the wide array of possible applications to mechanisms, it is not practical to include a comprehensive discussion of lubrication within the confines of this book. It is, therefore, important to ensure that lubrication experts are consulted during the determination of lubrication strategies. However, a few often overlooked lubrication pitfalls are addressed.

One class of subtle lubrication problems is migration. Due to the microgravity, thermal, and vacuum conditions, liquid or grease lubricants can travel, through various methods, from place to place within a mechanism. This can present three problems:

- In the components of some mechanisms, such as brakes, friction is crucial, and lubricant contamination can cause severe mechanism malfunction.
- Lubricants can contaminate surfaces that are sensitive for other reasons, such as optical surfaces.
- Lubricants can travel to locations utilizing materials or even other lubricants incompatible with the migrating lubricant and alter the chemical properties of the material or result in effects ranging from degraded lubricant performance, to corrosion of the mechanism, to the destruction of seals.

It is, therefore, important to ensure that lubricants are contained properly within a mechanism. When possible, material choices should be made so that incompatible materials are not used in the same mechanism, even if physically separated. It is also important not to overlook ground operations. A liquid lubricant that is perfect for an on-orbit application can pool, leak, or evaporate from a mechanism when operated or stored on the ground.

Another subtle issue is the proper specification of lubricant quantity. Drawing notes often specify the application of a lubricant to a surface without specifying a quantity or thickness. Whereas this obviously can cause problems if an insufficient quantity of lubricant is used, what is not so obvious is that different problems can be created if too much lubricant is used. Overfill of lubrication is a recognized concern in bearings, but several examples of past problems outside of this category are available. One example involves dry film lubricant applied too thickly to thread surfaces causing a bolt to jam on insertion. Another had excess grease applied to a mechanism. The excess grease collected, froze in the cold thermal environment, broke away, and subsequently jammed another sensitive portion of the mechanism.

Thermal Tolerance Analysis

A proper tolerance analysis is one of the most important design and verification steps that can be performed on a mechanism. Occasionally, circumstances arise wherein a mechanism function cannot be tested adequately in the proper environment before flight. In this case, a thermal tolerance analysis can provide the sole verification that a mechanism can function reliably in its design environment. A 2004 survey of the root causes of International Space Station mechanism failures and anomalies revealed that half of all tracked events had tolerancing problems as a sole or contributing cause (McCann 2004).

When performing a tolerance analysis, it is crucial that thermal effects be included. This is true particularly for mechanisms located external to the vehicle. Even though such external influences are often the driving conditions for the thermal environment, heat generated by the mechanism itself can cause notable distortions that must be considered. For this reason, thermal effects should not be neglected, even in thermally controlled environments.

Thermal effects can manifest themselves in two ways. First, materials with different coefficients of thermal expansion change size at different rates when exposed to the same uniform temperature. As an arbitrary example, an aluminum component that had sufficient clearance with a steel component at room temperature can have an interference at elevated or reduced temperatures. The second way for a thermal effect to manifest itself is if two components brought together are at different temperatures. For components of an integrated system, this is not often a problem, because the two components are in thermal equilibrium by virtue of their close proximity and contact. However, situations can arise where this effect becomes important. For example, consider a component removed from the *International Space Station* for return to Earth in the payload bay of the *Space Shuttle*. Say that the removed component has been exposed to the Sun for a long period of time and is transferred to flight support equipment that is shaded within the payload bay. Though the materials might be identical, they are at a different temperature; an interface that fit together when the two were at the same temperature no longer fits. A related situation occurs when a large mechanism such as a docking mechanism is shaded on one side, and exposed to the Sun on another. The temperature gradient across the system can cause a thermal distortion that brings the mechanism out of its natural shape and thus results in interface problems with the mating half. Though this type of effect can be controlled operationally by allowing the two components to come into thermal equilibrium prior to installation, this is not always convenient operationally or even possible.

Thermal tolerance stack ups can be complicated extremely, especially if the mechanism contains many parts with many different materials. The use of a geometric dimension and tolerance standard such as ASME Y14.5M-1994 (ASME 1994) can make the analysis much easier to perform and eliminate ambiguities in the interpretation of drawings by providing a well defined vector analysis method, thereby reducing or eliminating common sources of error in the analyses.

The designer should be cautious to not discount thermal effects in an analysis simply because the same or similar material is used throughout a mechanism and the system does not experience temperature differences between parts. Indeed, there is a difference between similar materials and identical materials when speaking of their coefficients of thermal expansion, and the difference can be important. For example, Custom 455, a common aerospace stainless steel, has a coefficient of thermal expansion between 78 and 200°F of 5.9 in./in./°F, whereas the coefficient of thermal expansion of A-286, another common aerospace stainless steel, is considerably higher, at 9.2 in./in./°F (DoT 2003). Though the magnitudes of the differences vary, their existence is common to all metal alloys.

Mechanism Testing

Of all considerations involved in the development of safety critical mechanisms, undoubtedly the most important is the performance of adequate testing. In the end, the suitability of a design and its ability to meet performance requirements can be confirmed in the most straightforward manner by demonstration of the hardware operation in the specified design environment.

In general, mechanism testing falls into one of three categories. The first of these is development testing. Development tests are very useful and often performed at the component level. These tests use non-flight designs to accomplish a variety of objectives, including characterizing specific performance parameters or sensitivities, trading or proving various design concepts, or meeting the incremental success gates of a larger development. Although the purpose of individual tests vary, development tests only provide aid in the development of a final design and are not used as the final verification of requirements. Rather, they help establish confidence in the design as the development progresses and can demonstrate an understanding of the design constraints.

The next test category is qualification testing. Qualification tests prove that the design of the mechanism meets the requirements imposed on it. Such tests are performed at levels and in environments above and beyond the design environments to demonstrate robustness and margin in the design. A wide variety of qualification tests are available for any given system, but for mechanisms, the most important are performance tests, vibration tests, thermal vacuum tests, and design life tests—all of which mimic the design environment. Because the qualification test levels exceed those expected during service, they are performed on flight or flight-like units. These are dedicated specifically for use in testing, so that any units potentially experiencing damage during testing are not used in service. Sometimes, if a system design has only very minor differences from a system that has already passed qualification testing, it can be considered qualified by similarity. This qualification should be used with great caution, however, for small design changes can have unintended and unexpected effects.

Certain inadequacies in qualification test plans often are encountered. Vibration and thermal testing usually are planned appropriately, but certain design life test inadequacies are encountered frequently. The factor to be applied to design life testing can vary depending on the program and the consequences of failure. It is, therefore, important that the correct factor is used and applied to the sum of all design cycles. However, the required number of design cycles to be used in the test often is calculated incorrectly; therefore, it must be done carefully. Proper design life testing includes not only the duty cycles during flight but also all cycles incurred during acceptance tests, functional tests, troubleshooting, maintenance, and other ground operations. Oftentimes, scenarios arise where cycles that were never planned are accumulated by a mechanism. One situation where this is prone to occur is during the planned testing of associated mechanisms. Take for example the Space Shuttle payload bay door actuators. These doors have cycle requirements imposed on them to account for their ground testing. However, several mechanisms within the payload bay cannot be operated with the payload bay doors closed. Unless the full range of flight operations and ground operations is understood, the fact that the doors have to open and incur cycles just so that other mechanisms can undergo their planned testing could be missed in the calculation of the required number of door cycles. This type of unforeseen cycle accumulation often is difficult to predict. For this reason, it is recommended that some reserve number of cycles be added to the cycle calculation prior to applying the required factor.

The last category of testing is that of acceptance. Whereas qualification testing serves as proof that the mechanism design can operate as intended during service, it does not prove that a subsequent individual unit manufactured per that design is up to the task. This is the purpose of acceptance tests. Acceptance tests are performed on every manufactured unit to screen out workmanship flaws and prove that each unit is capable of performing as designed. Acceptance testing usually is performed on qualification units before qualification testing so that workmanship issues can be excluded as possible causes of any qualification test failures. Important acceptance tests for mechanisms include run-in or wear-in tests, random vibration tests, thermal vacuum tests, and sometimes benchmarking tests that can aid in the later diagnosis of problems.

Attempts oftentimes are made to waive various acceptance tests. Acceptance vibration test waiver submissions usually and mistakenly use a low expected flight vibration level as rationale. However, this is not the purpose of an acceptance vibration test, which simply uses a random vibration environment as a disturbance to test workmanship issues such as soldering and fastener installation. This type of testing is not associated with the expected random vibration levels during flight. Developers often ask for relief from thermal tests, citing a completed thermal tolerance analysis as sufficient. However, as discussed previously, thermal tolerance analyses are notoriously prone to error, so in all but the simplest of mechanisms this is not sufficient rationale. Run-in tests are of the most frequently neglected tests. The purpose of the run-in test is twofold. It serves as a workmanship test to identify improper component assembly issues that manifest themselves quickly when placed in service, and second, it serves as sort of a burnishing process, wearing down initial rough spots and smoothing out any transient behavior of new components. Run-in testing is important to ensure that a mechanism operates within its steady state regime during acceptance testing and service and to avoid spurious test failures that are functions of transient behavior rather than a hardware or design defect. For this reason, it is important that the run-in test be performed before all other mechanical acceptance testing. Developers often cite initial functional tests to serve in lieu of the run-in tests. However, there are minimum cycle requirements for run-in tests designed to accomplish these two objectives. Finally, note that, although a design can be qualified by similarity, by definition, a mechanism cannot be accepted by similarity, and an attempt to do so must never be permitted.

On occasion, when the cost of a hardware unit is very expensive or for some other reason, such as the creation of a dedicated qualification article, is impractical, another testing technique called *protoflight* testing is used. Protoflight testing combines qualification and acceptance testing on a flight unit. For this type of testing, environmental levels generally are less than those used for a dedicated qualification unit due to the possibility of damaging the test article but are higher than those used for acceptance. Many unique pitfalls are associated with protoflight testing, so the decision to use this testing regimen needs to be carefully thought out.

Although deserving special scrutiny in protoflight situations, environments used in mechanism testing in general require careful attention. No test is worth much if inadequate environments are used. If resources allow, testing beyond certification limits is very valuable. Often the environment encountered during flight is found to be not quite what was predicted, circumstances arise that make operation beyond certification limits desirable, or a life extension is needed. In each of these cases, a decision must be made regarding the risk involved with operating a mechanism outside of its certification. This decision can be very difficult, and must rely on engineering judgment based on available data. Having such data, and having it well documented, can pay for itself many times over during the course of a program. Without these data, evaluation extends beyond engineering judgment into the realm of engineering intuition, which greatly increases risk.

Other Considerations

Although the areas just mentioned tend to play the most important roles in good design and verification, numerous other factors must be addressed that occasionally present notable problems.

Fasteners Fasteners are so common an element in structural and mechanical design that they often can be overlooked during the review process. However, they can be the root of a host of problems. Design practices that should be verified include the use of qualified, verifiable secondary locking features, specification of preload torques as being above running torque on drawings, and required measurement and documentation of all running torques. Additionally, liquid locking compounds generally should not be used as a secondary locking feature. A few problems are associated with liquid locking compounds. First, their locking effect is not verifiable, for it requires a broken bond to verify the lock, and the liquid locking compound no longer serves to lock the thread once the bond is broken. Second, the quality of the application, and thus the locking effect, is very process dependent and can be highly variable. As well, liquid locking compounds have a tendency to migrate in a vacuum and can end up locking together unintended surfaces. This obviously can be catastrophic for mechanisms and has been identified as the root cause of testing failures in the past. And last, when the liquid locking compounds do work, the strong bond can make removal difficult and can sometimes damage the hardware if removal becomes necessary.

As mentioned previously, threads that are operated during flight possess mechanical failure modes, and they should be treated as mechanisms. This applies, as well, to fasteners that can be used during maintenance operations. Such fastener applications always should be examined in the light of failure tolerance considerations.

Design for Assembly and Maintenance Sometimes a mechanism is doomed to fail before it ever leaves the ground because it was installed improperly during manufacture or maintenance. Although human error can never be eliminated entirely, it is possible in some instances to design a mechanism to preclude certain types of human error. Mechanisms always should be designed either to preclude installation in an incorrect orientation or clearly labeled in a manner that indicates proper installation orientation. Space program failures in the past have been traced to parts designed to be operated in only one direction and installed backwards. Yet, they had an interface that allowed the improper installation without providing a clear indication that something was wrong.

Strength Designing a mechanism to have adequate strength seems obvious, but the difficulty can lie in the details, particularly in the assumptions used in performing the analysis and deriving loads. One critical aspect that the designer fully should understand is the mechanism boundary conditions. Mechanisms usually are mounted to structure, and this mounting is assumed to behave rigidly. However, due to the lightweight nature of most aerospace structures, this mounting oftentimes has an inherent flexibility that can cause a change in both the external loads transferred to the mechanism and the behavior of the mechanism due to its own induced loads. For example, a motor mounted to a flexible structure can produce a rotational motion about its mount, generating moments or angular displacements on shaft couplings that are not predicted by a rigid mount.

Another problem that can be difficult to detect is an improperly understood load path. This is often manifested in one of two ways. The first is poor or neglected free body diagrams, which frequently are manifested in the effect that an offset force, that is, a force that generates a moment on the support structure, can have on a component. The moments created often are assumed to be of low enough magnitude that they do not contribute or they are omitted altogether. Surprisingly, small moments can be effective in binding surfaces designed to slide on one another. Precautions always should be taken to accommodate such moments fully and deal with the misalignments and friction they generate. The second manifestation is the existence of unintentional load paths. Oftentimes, mechanisms are designed such that two separate parts of the mechanisms are meant to share load equally, but natural tolerances and variations in assembly can produce uneven load sharing if the proper degrees of freedom are not included or the parts are not shimmed adequately. This is a consistently recognized design issue; however, the opposite situation also can arise, that is, mechanisms designed to withstand load in only one way can find this load being shared by components not designed to handle such loads. This condition is due to manufacturing tolerances and assembly clearances as well as material flexibility. Often, assembly models and drawings do not represent the true configuration the hardware is to take once subjected to the actual preloads and constraints it is to see once physically constructed. This can be a frustrating problem, for these issues can be very difficult to detect without an extremely thorough review of the design, which would include drawings, models, tolerance stack ups, and analysis.

Finally, strength problems can turn up in failure scenarios. Mechanisms and the structures they move should be designed so that they meet all necessary structural requirements, that is, strength and Fracture control, for redistributed loads after a mechanism failure commensurate with the hazard level. Operational procedures can be used to restore the load path or limit the subsequent loads; however, this approach should be developed and accepted prior to flight.

17.7.3 Reduced Failure Tolerance

Situations can arise where failure tolerance actually can lower overall system reliability or simply cannot be implemented in a feasible or practical way. In these situations, mechanisms must be subjected to a thorough review to eliminate as many potential problems as possible to minimize the risk incurred with such reduced failure tolerance. Although increased scrutiny certainly can be applied, realistically there should be little that can be done to the design process beyond normal design practice.

The first step in implementing reduced failure tolerance is to determine whether this reduction is necessary. Care must be taken not to provide a reduced fault tolerance approach as simply an alternative to full failure tolerance. There must be a solid reason for needing to increase the risk. Occasionally, the driving reason is that the nature of a particular mechanism is such that certain levels of failure tolerance actually reduce overall system reliability; but more often, the issue is just that it is highly impractical to implement a particular level of failure tolerance without considerable impacts. The rationale for the reduction typically includes reliability analyses comparing the full and reduced failure tolerant systems and trade studies detailing the impacts of implementing full failure tolerance. This would include cost, schedule, and vehicle packaging and performance impacts; the nature and severity of the hazards mitigated by the failure tolerance; and a detailed failure mode and effects analysis that includes the influence of the failed system on other systems, potential real time workarounds, and the like.

In most cases where reduced failure tolerance is warranted, the reduction is from two-failure tolerance to single failure tolerance, in which there remains in place a level of physical redundancy. At the cost of a more rigorous development and testing program, this reduction oftentimes can be achieved without an excessive increase in risk. Even so, reductions of this type should be limited to systems that utilize mechanisms, such as pyrotechnic devices, having a demonstrated history of high reliability and result in hazards having high potential for operational intervention prior to the occurrence of a catastrophic event. Here, design and testing should be subjected to periodic independent reviews at various life cycle stages: the conceptual design stage, preliminary design stage, critical design stage, and acceptance stage. The results of these reviews must be documented very closely. As well, the reviews should be conducted by a dedicated group of independent, that is, not associated with the project, mechanisms specialists to a greater depth of penetration than usually is associated with standard milestone design reviews. The review group should have access to the designers and all design and analysis information and should check against a well defined set of evaluation criteria, such as AIAA-S-114-2005 (AIAA 2005), NASA-STD-5017 (NASA 2006b), or a similar set of guidelines and lessons learned.

In a small number of cases, it is necessary or even highly desirable to operate a mechanism possessing a catastrophic failure mode and do so with no failure tolerance. Granting zero-failure tolerance to a catastrophic mechanism is equivalent to saying that there is no possible way that a given mechanism is going to fail. Great care should be taken in making this decision. For obvious reasons, use of zero-failure tolerance should be granted only under very special circumstances. The types of mechanisms that legitimately can be considered for this reduction are of the simplest nature, so simple in fact that there can be dispute as to whether these devices are even mechanisms, such as hinges, dovetails in slots, and the insertion of pins into holes. As with the single level reduction, such applications should be limited to systems that utilize mechanisms having a demonstrated history of very high reliability and result in hazards having high potential for operational intervention prior to the occurrence of the catastrophic event. A similar type of rigor as that used for the single level reduction should be employed in the independent review for this case, with increased attention paid to thorough testing and positive indication of status. No exceptions or waivers to any mechanism development or verification requirements can be permitted.

17.7.4 Review of Safety Critical Mechanisms

Thorough review of design and verification is of great importance and value, not only for mechanisms with reduced failure tolerance but for all safety critical mechanisms in general. It highly is recommended that a team of experienced independent mechanisms specialists be formed with the responsibility for reviewing and approving the designs of all safety critical mechanisms.

Independence is very important. For one thing, it provides a new set of eyes unfamiliar with the design and the history behind it. This team would be able to see with a fresh perspective and can challenge assumptions. Design practices and analysis assumptions are often subject to a sort of creep, wherein the evolution of a design can drag assumptions and techniques that were once appropriate into inappropriate regimes without it becoming obvious to those involved with the development. Independence also frees the reviewers to make unbiased assessments without the schedule and budgetary pressures that can be at work among those of the design team.

Experience with mechanical systems is crucial. Having a high degree of mechanisms experience resident in a review team provides a large pool of lessons learned that can be applied to new designs. Often the experiences of different people can supply different types of lessons, thus producing a very complete review, particularly when the group consists of specialists in different subfields, such as tribology, bearings, testing, and motors. As more and more systems are reviewed, more and more experience can be obtained by the group, so having a consistent team that can conduct such reviews is very useful and quickly can build expertise that can benefit the overall organization.

The review should be made as detailed as possible within the constraints of the project. The first step should be to conduct a technical meeting with the hardware developers to understand the requirements, environments, function, and operational scenarios of the mechanisms. Next, with the aid of the hardware developers, all possible failure modes should be mapped out. In addition to being useful for safety personnel creating hazard reports later in the design process, this mapping provides the group with clear direction regarding the failure modes that are the most critical and mechanisms or components that are candidates for reduced failure tolerance. Once the operation and the failure effects are established, the detailed review can begin. This includes reviews of the drawings, tolerance stack ups, materials, dynamic and strength analyses, verification and test plans, and any development hardware. It is helpful if the review team has available to them a set of requirements or checklist that can be used to assess the mechanisms systematically. Standards such as AIAA-S-114-2005 or NASA-STD-5017 have been used successfully in the past. In concert with this detailed review, it is important that the team document the compliance with each of the items in the standard along with the supporting data and rationale. This is invaluable for reference later. It is useful to have such compliance, along with the failure tolerance mapping and a general discussion of the mechanism and its operation documented in a single report that can be called up when issues arise. When properly implemented, such a review process can add considerable and important value by aiding in increasing the reliability of safety critical mechanisms.

REFERENCES

American Institute of Aeronautics and Astronautics. (1998) *Space systems—Metallic Pressure Vessels, Pressurized Structures and Pressure Components*. American National Standard ANSI/AIAA-S-080-1998. Reston, VA: American Institute of Aeronautics and Astronautics.

American Institute of Aeronautics and Astronautics. (2005) *Moving Mechanical Assemblies for Space and Launch Vehicles*. AIAA-S-114-2005. Reston, VA: American Institute of Aeronautics and Astronautics.

American Institute of Aeronautics and Astronautics. (2006) *Space Systems Composite Overwrapped Pressure Vessels (COPVs)*. American National Standard ANSI/AIAA-081A-2006. Reston, VA: American Institute of Aeronautics and Astronautics.

American Society of Mechanical Engineers. (1994) *Dimensioning and Tolerancing*. Standard Y14.5M-1994. Fairfield, NJ: American Society of Mechanical Engineers.

Baker, L. K. (2001) *Comparison of Reliability Models for Optical Fibers*. Corning, Incorporated, White Paper, WP5049. Available at www.corning.com/docs/opticalfiber/wp5049_06-01.pdf (cited February 11, 2006).

Beeson, H. (2002) *Composite Overwrapped Pressure Vessels*. Technical Paper TP-2002-210769. Las Cruces, NM: National Aeronautics and Space Administration, White Sands Test Facility.

Canonico, D. A. (2000) The origins of ASME's boiler and pressure vessel code. *Mechanical Engineering Magazine*.

Corning. (2003) *HPFS® Fused Silica Standard Grade*. Data sheet published September 30. Corning, NY: Corning, Incorporated.

Deutsches Institut fur Normung. (1984) *Testing of Glass and Glass Ceramics: Determination of Bending Strength*. DIN 52-292. Berlin: Deutsches Institut fur Normung.

Department of Defense. (1984) *Standard Requirements for Safe Design and Operation of Pressurized Missile and Space Systems*. Military Standard MIL-STD-1522A. Washington, DC: U.S. Department of Defense, U.S. Air Force.

Department of Transportation. (2003) *Metallic Materials Properties Development and Standardization (MMPDS)*. Scientific Report DOT/FAA/AR-MMPDS-01. Washington, DC: U.S. Department of Transportation, Federal Aviation Administration, Office of Aviation Research.

Ecord, G. (1977) Composite pressure vessels for the Space Shuttle Orbiter. Composites in pressure vessels and piping. *Proceedings of the Energy Technology Conference*. Houston, TX: American Society of Mechanical Engineers, pp. 129-140.

Greene, N., and D. Ray. (2007) The synergy of composite overwrapped pressure vessels (COPVs) with cryogenic fluid storage and propellant densification. *Proceedings of the Aging Aircraft*. Palm Springs, CA: Joint Council on Aging Aircraft.

Greene, N., D. Cone, R. Saulsberry, S. Forth, and G. Ecord. (2007a) Failure modes for composite overwrapped pressure vessels (COPVs). *Proceedings of the Aging Aircraft*. Palm Springs, CA: Joint Council on Aging Aircraft.

Greene, N., R. Saulsberry, T. Yoder, B. Forsyth, S. Thornton, and R. Wincheski. (2007b) Progressive failure indicators in composite overwrapped pressure vessels (COPVs). *Proceedings of the 54th Joint Army Navy NASA Air Force Conference*, Denver, CO.

Gregg, W., and D. Lambert. (1994) *Elastic Plastic Fracture Mechanics Methodology for Surface Cracks*. Contractor Report CR-196811. Atlanta, GA: Georgia Institute of Technology; Huntsville, AL: National Aeronautics and Space Administration, Marshall Space Flight Center.

Griffith, A. A. (1920) The phenomena of rupture and flow in solids. *Philosophical Transactions of the Royal Society* A221: 163.

MatWeb. (2007) *MatWeb, Your Source for Materials Information*. Available at www.matweb.com/index. asp?ckck=1 (cited: August 28, 2007).

McCann, D. (2004) Review of International Space Station mechanical system anomalies. Contractor Report CP-2004-212073 Proceedings of the 37th Aerospace Mechanisms Symposium. Galveston, TX: National Aeronautics and Space Administration, Johnson Space Center, p. 291.

National Aeronautics and Space Administration. (1989) *Safety Policy and Requirements for Payloads Using the Space Transportation System*. Space Shuttle Document NSTS-1700.7B. Houston, TX: National Aeronautics and Space Administration, Johnson Space Center.

National Aeronautics and Space Administration. (1995) *Safety Policy and Requirements for Payloads Using the International Space Station*. Space Shuttle Document NSTS-1700.7B, ISS Addendum. Houston, TX: National Aeronautics and Space Administration, Johnson Space Center.

National Aeronautics and Space Administration. (1996) *Structural Design and Test Factors of Safety for Spaceflight Hardware*. Standard NASA-STD-5001. Huntsville, AL: National Aeronautics and Space Administration, Marshall Space Flight Center.

National Aeronautics and Space Administration. (2006a) *Strength Design and Verification Criteria for Glass, Ceramics, and Windows in Human Spaceflight Applications*. Houston, TX: National Aeronautics and Space Administration, Johnson Space Center.

National Aeronautics and Space Administration. (2006b) *Design and Development Requirements for Mechanisms*. Technical Standard NASA-STD-5017. Washington, DC: National Aeronautics and Space Administration, Headquarters.

NASGRO. (2005) *Fracture Mechanics and Fatigue Crack Growth Analysis Software, NASGRO 5.0, Reference Manual*. Available at www.nasgro.swri.org (cited August 28, 2007).

Sarafin, T. P., and W. J. Larson (eds.). (1995) *Spacecraft Structures and Mechanisms from Concept to Launch*. Torrance, CA: Microcosm; Boston, MA: Kluwer Academic Publishers.

Thesken, J., N. Greene, P. L. N. Murthy, S. L. Phoenix, J. Palko, J. Eldridge, J. Sutter, R. Saulsberry, and H. Beeson. (2005) A theoretical investigation of composite overwrapped pressure vessel (COPV) mechanics applied to NASA full scale tests. *Proceedings of the 20th Technical Conference of the American Society for Composites*, Dearborn, MI.

Varner, J. R. (1996) Fatigue and fracture behavior of glasses. In *Fatigue and Fracture. ASM Handbook*, vol. 19. Russell Township, OH: American Society for Materials.

Wiederhorn, S. M. (1977) Dependence of lifetime predictions on the form of the crack propagation equation. *Proceedings of the 4th International Conference on Fracture*. Waterloo, ON, Canada: The International Conference on Fracture, pp. 893–901.

CHAPTER 18

Containment of Hazardous Materials

Summer L. Rose
System Safety Engineer for the ISS Program, Boeing, Houston, Texas

Rod Kujala
Consultant Engineer, AOES Netherlands B.V., Noordwijk, the Netherlands

CONTENTS

- 18.1. Toxic Materials 610
  By John T. James, Ph.D.
- 18.2. Biohazardous Materials 621
  By Duane L. Pierson, Ph.D.
- 18.3. Shatterable Materials 631
  By Karen S. Bernstein
- 18.4. Containment Design Approach 639
  By Summer L. Rose and Rod Kujala
- 18.5. Containment Design Methods 640
  By Summer L. Rose and Rod Kujala
- 18.6. Safety Controls 643
  By Summer L. Rose and Rod Kujala
- 18.7. Safety Verifications 644
  By Summer L. Rose and Rod Kujala
- 18.8. Conclusions 648
  By Summer L. Rose and Rod Kujala

The use of hazardous materials is a necessary part of human research activities in space. Space systems and experiments often consist of materials that pose a safety risk to humans due to toxicity, biohazards, or shatterable materials. The most common types of hazardous materials in human spaceflight experiments are toxic fluids, usually fixatives, for biological research. The safe use of hazardous materials dictates a verifiable design to contain these materials and their isolation from the environment within which humans can become exposed. Scientific objectives and operations shape the containment system design and have a large effect on the possible verifications that can be performed.

The designer of containment systems for human spaceflight, therefore, faces a unique and difficult challenge for every new hardware design.

Containment of hazardous materials is a very general expression used to describe a wide range of hardware designs for containing many different types of materials with a variety of hazards. The expression containment of hazardous materials contains three words that can be further defined to provide insight into its practical meaning. The word *containment* indicates that some verifiable method is used to prevent the release of a hazardous material. This containment must be valid for all environmental effects seen during the lifetime of the hardware. In particular, these effects include pressure loads, temperature, inertial loads, and cycle life. The word *hazardous* obviously indicates a condition that must be avoided according to the safety requirements; however, many types of hazards can arise because of the use of certain materials. Some examples of these hazards are illness due to toxicity, infection due to biohazards, injury due to broken glass, and burns due to flammability. Finally, *materials* is a generic term to describe the different states of matter, usually liquids, gases, and solids or particles. With these definitions, the phrase containment of hazardous materials is shown to be a very broad subject about preventing the release of potentially hazardous substances.

The safety authority requires containment of hazardous materials. This normally is dependent on a specific evaluation of each type of material. The hazard rating of a material determines the robustness of design verifications. Safety requirements purposely are open ended to allow flexibility of hardware design, while still requiring the necessary controls and verifications to ensure safety. The design engineer must rely on clarifications from the safety authority for a specific design during safety reviews. This makes the design and verification planning difficult.

This chapter provides guidelines and examples of hardware designs used to prevent release of hazardous materials to aid design engineers in meeting safety requirements. As an example, it describes the containment of materials within a metallic box with seals at all feedthrough locations. The guidelines include structural verification by test and analysis. The examples consist of the verification approaches for soft seals such as O-rings by leak testing and the qualification of on-the-ground assembly procedures for flight hardware.

Hazardous materials discussed in this chapter are grouped into three main categories:

- Toxic materials.
- Biohazardous materials.
- Shatterable materials.

Each type of material must be evaluated by the applicable safety authority to determine the level of hazard associated with the release of that material. Specific procedures, processes, and results are described here for each of these material types.

In designing adequate containment for hazardous materials, a number of pertinent definitions are used. Some of these relate to the design of pressurized systems but include key points necessary for the proper design of hazardous fluids containment. Specific definitions for designing systems to prevent hazardous materials release are contained in the requirements for fracture control and safety critical structures. Other general definitions, such as those pertaining to materials selection and quality assurance, although not listed here, remain applicable.

For pressurized hardware, the following definitions are particularly important to the containment of hazardous materials. The definition of *maximum design pressure* is applicable to these containers:

- *Pressurized hardware* is defined (AIAA 1999) as "those hardware items [that] contain primarily internal pressure. Included are pressure vessels, pressurized structures, special pressurized equipment, and pressure components."

- A *pressure component* is defined (AIAA 1999) as "a component in a pressurized system, other than a pressure vessel, pressurized structure, or special pressurized equipment that is designed largely by the internal pressure. Examples are lines, fittings, gauges, valves, bellows, and hoses."

- *Containers* are any single, independent container, component, or housing that has a stored energy of less than 14,240 ft-lb (19,310 J) that is not part of a pressurized system.

- The NASA safety requirements document for payloads (NASA 2006) defines *maximum design pressure* as "the highest pressure defined by maximum relief pressure, maximum regulator pressure, or maximum temperature. Transient pressures [must] be considered. Design factors of safety [must] apply to maximum design pressure. Where pressure regulators, relief devices, and/or a thermal control system (e.g., heaters) are used to control pressure, collectively they must be two-fault tolerant from causing the pressure to exceed the maximum design pressure of the system."

- A *hazardous fluid container* is defined (NASA 1996) as "any single, independent (not part of a pressurized system) container, or housing that contains a fluid whose release would cause a catastrophic hazard, and has stored energy of less than 14,240 foot pounds (19,310 Joules) with an internal pressure of less than 100 psia (689.5 kPa)."

- A *level of containment* is defined as a safety control used for the containment of hazardous materials. Levels of containment are usually physical barriers such as metallic boxes sealed with O-rings at all interfaces. Other examples of levels of containment are neutralizing and absorbent filters, monitored and maintained negative pressures, and sealed nonmetallic enclosures.

- A *hazardous fluid* is defined (NASA 2005) as "[for] fracture control, a fluid whose release would create a catastrophic hazard. Hazardous fluids include liquid chemical propellants and highly toxic liquids or gases. A fluid is also hazardous if its release would create a hazardous environment such as a danger of fire or explosion, unacceptable dilution of breathing oxygen, an increase of oxygen above flammability limits, over pressurization of a compartment, loss of a safety critical system, etc."

18.1 TOXIC MATERIALS

18.1.1 Fundamentals of Toxicology

Dose Makes the Poison

The basic premise in all of toxicology is that the dose to a living system determines the degree of poisoning. Compounds vary over many orders of magnitude in their inherent toxicity, so estimating the dose of a specific compound delivered or potentially delivered to a person is essential. Toxicologists refine this principal by estimating the dose of biologically active compound delivered directly to the target organs of an exposed person. This requires knowledge of how a compound is absorbed, distributed, metabolized, and eliminated from the body.

The dose of an inhaled toxicant typically is classified in terms of the concentration in the respired air and the length of time it remains in the air. The actual dose delivered deep into the lung depends on the respiration rate and whether the person is nose or mouth breathing. If the toxicant is a dust, then particles above 10 μ in diameter rarely reach the deep lung where the greatest damage can occur. For ordinary irritancy effects, the time of exposure is much less important than the airborne concentration of the toxicant. Irritants elicit their adverse effects almost immediately, and persons often adapt to their presence after a few minutes. The goal for irritating compounds is to keep their initial airborne concentrations below the irritation threshold and, if possible, the odor threshold as well. Many of these compounds have good warning properties, that is, they can be smelled in the air before any harm is done to the person.

Species Differences in Response to Toxicants

The differences in responses to identical airborne concentrations of a toxicant can be remarkable. Stories are told in the toxicology lore of the days when canaries were taken into mines to warn the miners that the atmosphere was becoming dangerous. If the canary died, it was time to leave. These days, species differences in toxic responses are an important problem for toxicologists. Before an animal exposure is undertaken, an investigator must address the question of whether the species to be exposed is an adequate model for estimating human responses. For example, the human response to some thyroid, liver, and renal toxicants is modeled by rats very poorly (James and Gardner 1996). When the results of an animal study are being used to estimate human toxicity, the first question to ask is whether the animal model is an appropriate surrogate for humans for each toxic response observed. Often, this question cannot be answered with certainty, in which case, the animal response must be assumed to be at least a qualitative indicator of human response.

Routes of Astronaut Exposure in Spacecraft

Airborne compounds inside spacecraft present a contact threat to the eyes, skin, and respiratory system. Oral ingestion is not considered a credible route of exposure. The nature of a toxic threat can vary greatly. Generally, the eyes are much more sensitive than the skin to toxic injury. If a pollutant suddenly enters a space habitat, it is often sufficient to protect the eyes and respiratory system and not be concerned about skin effects. For example, during the

NASA-Mir project in the 1990s, ethylene glycol often leaked from the thermal control system into the atmosphere. On one occasion, the face of a crewmember encountered a large floating blob of this liquid material, and severe eye irritation was experienced. At other times, the ethylene glycol vapor caused mild respiratory irritation to the crew. Another material that episodically has caused respiratory irritation in Space Shuttle crews is the lithium hydroxide dust from carbon dioxide scrubbing filters. This material forms a strong base when dissolved in water, so a small amount in the eyes can elicit discomfort.

Volatile toxicants and gases can enter the air and become absorbed by the respiratory system to cause effects beyond the lungs. One example of this is the gas carbon monoxide, a common toxic product of fire. Carbon monoxide enters the bloodstream from the lung and binds tenaciously to hemoglobin, forming carboxyhemoglobin. As this occurs, the blood begins to lose its ability to bind oxygen in the lungs and release oxygen to the tissues. During several hours of exposure to a few hundred parts per million of this gas, carboxyhemoglobin accumulates, and after several hours, the victim of exposure begins to experience headaches (Stewart et al. 1970). This adverse effect was reported aboard *Mir* several hours after pyrolysis of a paper filter inside the regenerable trace contaminant removal system. Monitoring of the air showed a peak concentration of about 500 ppm in the air, and this declined slowly during the next day and a half.

Metabolism and Disposition of Inhaled Pollutants

Once a pollutant is absorbed by the respiratory system, it can undergo metabolism to compounds that can be either more toxic or less toxic. Metabolism of pollutants can occur in the lung during absorption, in the liver as blood carries the toxicant through that large organ, or at the target organ level where tissue injury occurs. Enzyme systems typically metabolize toxicants; however, as the concentration of a toxicant increases, the basic enzyme system metabolizing the compound can be overwhelmed, and a new system then begins to metabolize the compound. Often the basic metabolic system is a detoxification pathway; however, once it is saturated, the toxicant can begin to injure tissue or be metabolized to another compound that injures tissue or causes an adverse effect. An interesting example of this is ethanol, which itself is a central nervous system depressant. Ethanol always is found in spacecraft air, but never at concentrations that would cause central nervous system depression. Alcohol dehydrogenase oxidizes ethanol to acetaldehyde, reducing any potential central nervous system effects; however, acetaldehyde has very unpleasant effects if it accumulates. Aldehyde dehydrogenase converts acetaldehyde to less toxic compounds (acetate), but in some individuals, especially those of oriental heritage, acetaldehyde dehydrogenase has much reduced activity levels. Therefore, during ethanol exposure, acetaldehyde accumulates, causing the flushing response evident in these people after ingestion of even small amounts of alcohol (Shibuya, Yasunami, and Yoshida 1989).

When dust is inhaled, it is removed from the lung via several mechanisms. If the particles are soluble, then they can dissolve, releasing their compounds into the bloodstream. They also can be ingested by alveolar macrophages, which are mobile scavenger cells in the lungs, and transported up the mucociliary clearance pathway into the upper airways. Once they reach the esophagus, they can be swallowed. As well, particles can be ingested by macrophages and transported to the lymph nodes draining the lung. Finally, particles, especially those of a fibrous nature, can be impossible to process and remove from the lung. These remain, and establish a chronic inflammatory response leading to scarring in the lung. This problem is important because NASA plans to return to the Moon and eventually travel to Mars. Dust was a major problem during the Apollo flights. The processes that form lunar dust produce a reactive dust with a large surface area and a fine particle size distribution, suggesting that much of the dust can be inhaled readily (James 2006).

Interindividual Differences in Response to Toxicants

Many factors can make the response of individual humans to toxic exposures different from the typical response; however, many of these can be disregarded for the purposes of spaceflight. For example, the population of astronauts is considered healthy, not pregnant, middle-aged persons in reasonably good physical condition. We simply do not have to protect elderly, unhealthy persons, persons with asthma, or unborn children. As NASA sets exposure standards in cooperation with the National Research Council Committee on Toxicology, the consensus decision was reached whereby a safety factor for interindividual variation is unnecessary unless well defined differences in the healthy population are known to exist. Often interindividual differences in response are due to well defined metabolic differences, as described for ethanol in the preceding subsection. Other differences can result from application of a model of toxicity. For example, the linearized multistage model used to assess cancer risk contains a variable for the age of the person exposed. This is to compensate for the fact that a given exposure is more likely to cause cancer if it occurs when the individual is 20 years old rather than 60 (NRC 1992). This recognizes the fact that more time is available for a young person to develop cancer than for someone who is within a few decades of the end of life. In practice this difference is set aside in establishing exposure guidelines, and a value of from 25 to 30 years is used for this parameter.

Adaptive Versus Adverse Responses to Toxicants

A fundamental problem in toxicology is deciding if effects seen in animals or humans are adaptive or adverse. For example, the liver makes several adaptive responses to the toxicants it processes. These include enzyme induction and increased deposition of fat. If the toxic insult is removed, the liver reverts to its previous condition. At some point, however, the liver is altered irreversibly, and the response it gives to toxic insult then becomes adverse. In addition, the effect itself might not be adverse, but it can make the liver more susceptible to a subsequent toxic insult by the toxicant itself or another liver toxicant. This opens a debate as to whether the original toxic insult was truly adverse. The decision about whether an effect is adverse or adaptive has important implications when setting air quality guidelines for human exposures in many settings, including that of spaceflight.

Response to Combined Exposures

All actual exposures to airborne compounds invariably involve exposures to several different compounds. Sometimes, the exposure to one compound is dominant, and the others can be ignored for purposes of predicting potential adverse effects. In spacecraft, we do not enjoy that situation. Typically, 15 to 20 compounds have been identified that are found in significant concentrations in samples taken from normal spacecraft air. These typically include alcohols, ketones, aldehydes, aromatic compounds, halogenated compounds, siloxanes, carbon monoxide, and carbon dioxide. Methane and hydrogen also tend to accumulate in spacecraft air, but these have no direct toxicological consequences. The typical method for dealing with exposures to a group of toxicants is to calculate a toxicity index (T) by adding the ratios of the concentration of each (C_n) and the limit of each for the time of exposure (L_n):

$$T = C_1/L_1 + C_2/L_2 + \ldots C_n/L_n \tag{18.1}$$

If the T value is <1, the air is considered safe. If it is ≥1, then the compounds contributing to the value are separated into toxicological groups, such as neurotoxicants, irritants, and carcinogens, and then a T value for each group (T_{group}) is calculated. The air is considered safe if each T_{group} is <1. This approach requires that the toxicologist understand the physiology of the target organs and the effects caused by each toxicant found in the air. This oftentimes is not an easy requirement to meet.

Effect of Spaceflight Stressors on Toxic Exposures

Spaceflight evokes many stressors on crewmembers. Some of these are important to consider when judging whether a compound could be toxic to a crew. Under some conditions, spaceflight is known to be arrhythmogenic to the heart. Halon 1301® (bromotrifluoromethane) is used as the fire extinguishant in the *Space Shuttle*. Indeed, NASA has set limits for exposure to this compound to protect against its arrhythmogenic effects should it be released into the cabin. These limits incorporate a safety factor of 5 to account for the possibility that crewmembers are unusually susceptible to the arrhythmogenic effects during spaceflight (Lam 1996). Other spaceflight factors that are considered important when assessing the toxic potential of compounds are immune system effects, loss of red blood cell mass, and radiation effects on bone marrow. The science has not progressed beyond assigning factors for these effects. Typically, the toxic and spaceflight mechanisms are not understood sufficiently to know how they could interact at the molecular level.

18.1.2 Toxicological Risks to Air Quality in Spacecraft

Quantity and Behavior of Escaped Compounds

Before beginning a risk assessment for a compound contained in an apparatus or system, the toxicologist must determine the amount of material that potentially could escape and the way in which the material will behave once it is no longer contained. The quantity of releasable material typically is estimated by the amount that could enter the environment if all containment were to disappear. Clearly, this can lead to extreme estimates, so if the apparatus or system builders can make a credible case for a smaller amount, then that is considered the amount at risk. For example, if the compound in question is contained in a paste form, then it can be unreasonable to do the risk assessment as if all the material escaped containment. For example it might be reasonable to assume that only 10% of such a material can escape. Another factor to consider is the rate at which a material can escape from containment. Whether the material suddenly can be propelled from containment or if it simply leaks slowly over several hours or days are important considerations.

Once a compound has escaped, the behavior of that compound depends on many factors that must be judged by the toxicologist because specific data typically are not available. This judgment must be made with an understanding of the air scrubbing mechanisms in use on the specific spacecraft. A prime example of this is the differences in particle scrubbing capability aboard a Space Shuttle compared to that of the U.S. segment of the *International Space Station*. The *International Space Station* has extremely efficient air filters, capable of removing even bacteria sized particles (<1 µ); however, the *Space Shuttle* has much coarser filters that lose filtering efficiency for particles sized below 40 µ. If the compound at risk is a liquid in a pressurized system, then it generally is assumed that any escaped material would be in aerosol form. The volatility of the material determines whether it is spread rapidly throughout the vehicle or much more slowly. For example, carbon monoxide gas spreads rapidly and distributes uniformly when released into a habitat, such as by a fire, whereas, ethylene glycol, because of its low volatility, spreads over a period counted in tens of days before it reaches a steady state concentration within the vehicle.

Another factor that must be considered is the volume into which a potentially toxic compound can be dispersed. Obviously, if a given amount of compound escaped into the small volume of the Soyuz capsule, it would be much more hazardous than if the same amount escaped into the *Space Shuttle* or International Space Station volume. This issue can become contentious if the time of highest risk of accidental release is when a crewmember is working with the hardware or system containing this material. In this situation, when the crewmember is known to be in close proximity to the point of any potential accidental release, the volume of dispersion is considered to be very small if the compound has immediate effects. One example of this is the use of large lithium thionyl chloride batteries in a heart monitoring experiment called Cardiocog. Thionyl chloride is a toxic, volatile compound that readily decomposes into sulfur dioxide and hydrogen chloride in the presence of water vapor. In this experiment, the crewmember actually wears the Cardiocog apparatus. If a battery were to rupture from a short circuit while conducting this experiment, the crewmember would have to remove the apparatus before being able to escape from the vicinity of the release. It would be inappropriate to consider using a large volume of dispersion, such as a whole module, in a situation like this.

The toxic hazard classifications are shown in Table 18.1 along with color coding for labels to communicate the hazard to the crew in the event of a leak. Subsequent sections consider specific issues for different classes of materials.

Particulates

Particles present an unusual hazard in spaceflight because they float in the air regardless of their size. One of the early problems identified on the *International Space Station* occurred when the crew first entered a new module that had been attached to the parent vehicle. Regardless of how clean the ground crew kept the module as it was processed, when it reached a microgravity condition many particles began to float in the air. When

Table 18.1 Criteria for Toxic Hazard Classification of Compounds (JSC 26895) (Courtesy of NASA)

Hazard Level	Irritancy	Systemic Effects	Contamination/Decontamination
0. Non-hazard (green)	Slight irritation that lasts <30 min and requires no therapy	None	Gas, solid, or liquid might or might not be containable
1. Critical (blue)	Slight to moderate irritation that lasts >30 min and requires therapy	Minimal effects, no potential for lasting internal tissue damage	Gas, solid, or liquid might or might not be containable; however, the crew is protected from liquids and solids by surgical masks, gloves, and goggles
2. Catastrophic (yellow)	Moderate to severe irritation that has the potential for long-term performance decrement and requires therapy Eye hazards: Might cause permanent damage	None	A solid or nonvolatile liquid that can be contained by a cleanup procedure and disposal. The crew is protected by 5-μ surgical masks, gloves, and goggles
3. Catastrophic (orange)	Irritancy alone does not constitute a level-3 hazard	Appreciable effects on coordination, perception, memory, and the like or potential for long-term (delayed) serious injury, such as cancer, or internal tissue damage	A solid or nonvolatile liquid that can be contained by a cleanup crew and disposal. Surgical masks and gloves do not protect the crew. Either quick-don masks or SEBs and gloves required
4. Catastrophic (red)	Moderate to severe irritancy with the potential for long-term crew performance decrement (for eye only hazards, risk of permanent damage). Requires therapy if the crew is exposed	Appreciable effects on coordination, perception, memory, and the like or potential for long-term (delayed) serious injury, such as cancer, or internal tissue damage	Gas, volatile liquid, or fumes that are not containable. The ARS is used to decontaminate. Either the quick-don masks or the SEBs are required, or the contaminated module is evacuated

crewmembers sought to enter for the first time on orbit, they were faced with much floating debris. Operational procedures now call for crewmembers to wear eye protection when they first open the hatch to a new module. In a similar situation, the Apollo astronauts reported that lunar dust became an important problem when the *Lunar Module* reacquired microgravity after it was launched into space to rendezvous with the command module orbiting the Moon. The problem of floating lunar dust was never solved fully during the Apollo flights. Such dust could present a difficult problem to manage as humans return to the surface of the Moon and remain there for months at a time (James 2006). Particulates from shattering of hard materials are considered in another section.

Liquids

Many systems and payloads contain liquid compounds or solutions that have the potential to escape containment and enter the habitable volume. These include strong bases and acids, fixatives, ionic solutions, organic liquids, and complex mixtures. The toxic hazard associated with strong acids and bases is based on the pH of the solution. For example, inorganic acids with a pH below 2.0 and bases with a pH above 11.5 are considered toxic hazard level 2, or THL-2, because they are corrosive to the eye. Formaldehyde and glutaraldehyde are fixatives commonly used in biological experiments. Concentrations of these fixatives at or above 1% are considered a THL-2 for eye irritation. Formaldehyde vapor at or above 10 ppm is considered a THL-4 because of eye irritation and the difficulty of removing it from the air. Nonreactive ionic solutions, such as sodium chloride, above a 1 M concentration, are considered THL-1, whereas reactive solutions are dealt with on a case by case basis. For example, liquid thionyl chloride is a THL-2 for the eyes, but its vapor hazard depends on the amount that could be released. Releases of this reactive compound causing a concentration above 5 ppm are considered a THL-4 to the respiratory system. Volatile organic liquids also must be considered for their liquid and vapor hazard. Complex mixtures are assessed based on the most toxic components present in a solution. It is rare that specific toxicity data are available on a complex mixture itself.

Gases

Materials in the gaseous state can present an especially dangerous hazard because they are oftentimes difficult to remove from the atmosphere; however, the accepted practice in spaceflight is to select nontoxic materials whenever possible. There are several good examples of these choices. Freon 218® (octafluoropropane) has been used in the International Space Station Service Module air conditioner. The amounts used are so small and the gas is so nontoxic that, if it all escaped, no harm to health would be done even though the time required to scrub this gas from the atmosphere is measured in weeks. Likewise, Halon 1301, the U.S. fire extinguishant on the *Space Shuttle* is nearly nontoxic, and weeks can be required to scrub this material from the air. Another low toxicity compound like this is sulfur hexafluoride, sometimes present in the cabin air from experiments. One gas that requires special consideration is carbon dioxide. Typically the International Space Station atmosphere contains a partial pressure of several millimeters of mercury of carbon dioxide from anthropogenic sources. Additional sources that require assessment include the carbon dioxide fire extinguishant used on the U.S. side of the *International Space Station* and experiments that release relatively small amounts of carbon dioxide into the breathing air. The toxicologist must always keep in mind that the running level of carbon dioxide can be close to the acceptable limits for human exposure. To deal with the case of possible release of carbon dioxide from fire extinguishers, an air scrubber is available for use after a sudden release of this gas.

Reactive Compounds

Compounds that react with systems when they are released accidentally from containment into the habitable environment can present a difficult assessment. One concern is that the material could react with the environmental control systems and damage them. Indirectly, the loss of capability of these systems could affect crew health as pollutants accumulate. On the other hand, the systems actually can convert a relatively nontoxic compound to one that is more toxic. There are examples of trichloroethylene being converted to the highly toxic dichloroacetylene and methanol being converted to formaldehyde by air revitalization systems (Graf et al. 2000). The toxicologist must be prepared to work with environmental engineers when the possibility of damage to a system or of increasing toxicity could occur as a result of interactions with these systems. For example, sulfurous compounds tend to poison the catalysts used in some of the International Space Station air revitalization systems. These assessments are also vehicle specific. Obviously, vehicles with different environmental systems must be considered on a case by case basis, even when upgrades are added.

Predictable Sources of Pollution

Predictable sources of pollution are those that involve release of compounds in an unchanged form. Because the compound is not changed when it is released, we can predict with some certainty how it behaves, how toxic it can be, whether it can be monitored, and how quickly it can be removed by air filtration. Typically such compounds are contained according to the toxicity hazard they present. Compounds with a THL-0 rating require no containment, although almost all are contained in some way. Those compounds with a rating of THL-1 require a containment system that is single-fault tolerant, that is, it presents two levels of containment. This means that two independent and unlikely failures would have to occur before the compound reached the air. Compounds having toxic hazard levels from 2 to 4 are required to have containment that is two-fault tolerant. This means that three independent and unlikely failures would have to occur before compounds in these categories reach the air. Historically, this approach has worked well. To present knowledge, no one has been injured to any extent by the escape of toxic compounds that are contained according to this paradigm. The design of robust containment is discussed later in this chapter.

Unpredictable Sources of Pollution

In one sense, fires are unpredictable sources of pollution. In another, decades of experience with spaceflight show that fires can be predicted, or more correctly, pyrolysis events do occur episodically. These events typically occur in association with wiring shorts, overheating of circuits, or overheated motors. Experience has shown that the most toxic products from such fires include carbon monoxide, hydrogen cyanide, hydrogen chloride, and hydrogen fluoride.

Therefore, a monitor for these compounds, not including hydrogen fluoride, routinely is flown to monitor the atmosphere in the event of a fire. Sensors for hydrogen fluoride have not been developed to the point where they have sufficient reliability for spaceflight applications.

The space toxicologist always must be prepared to deal with air pollution problems that could not have been foreseen. One such event aboard *International Space Station* occurred when some filters used during extravehicular activity were regenerated using heat (James 2003). The filters had been left partially open to the International Space Station air for 6 mo, during which time they had absorbed many pollutants. When regeneration was attempted, the heat drove many compounds from a charcoal bed located within the filters and the U.S. segment was filled with noxious pollutants. The regeneration process was stopped, and the crew took refuge in the Russian segment of the *International Space Station* for 30 h while the air revitalization system in the U.S. segment restored the air quality. In hindsight, it was obvious that the pollutants originated from the charcoal filter because it had been absorbing nominal vehicle pollutants for 6 mo. However, there is no reasonable way that this event could have been expected. The original design of the regeneration apparatus had the effluent being discharged overboard; however, to save money, the design was altered to allow the effluent to be discharged into the atmosphere of the vehicle.

The basic strategy for dealing with unpredictable pollution is to have a broad spectrum analyzer in place to monitor for compounds accidentally released into the air. This function currently is filled by the volatile organics analyzer, which employs dual gas chromatography ion mobility spectrometry (Limero et al. 1998). Recently, this instrument was able to detect some aromatic compounds released in the Russian segment of the *International Space Station* even though the instrument itself is located in the U.S. segment. This technology might soon be replaced by more recently developed instruments, including ones that use infrared spectroscopy, gas chromatography/mass spectrometry, or gas chromatography/differential mobility spectrometry (Mudgett, Packham, and Jan 2005). For exploration missions, these instruments essentially must be no larger than a small shoebox.

18.1.3 Risk Management Strategies

Replacement of Risky Compounds
The first way to manage compounds that are toxicologically risky is to avoid the risk by finding a less toxic substitute. When a compound is proposed for use in a vehicle, system, or payload and it is known to be toxic, the first declaration from the toxicologist should be to select a less toxic compound that is suitable for use in the intended application. This assertive approach resulted in several success stories. For example, glutaraldehyde was proposed for use in an aqueous loop of *International Space Station* to retard bacterial growth. It soon became clear that the concentration of glutaraldehyde needed to kill bacteria was also a risk to crew health if it were to leak from the system. Eventually, this led the developers to consider use of ortho-phthalaldehyde in place of glutaraldehyde after a toxicity assessment showed that the former compound was substantially less toxic than the latter (Coleman 2005). Another substitution occurred after the multiple problems with leakage of an ethylene glycol solution aboard

the Mir space station. In place of this compound, Russian engineers have used a solution called *triol*, an aqueous solution containing 35% glycerin and <5% additives to retard microbial growth. Testing of this material in U.S. laboratories demonstrated that it was nearly devoid of toxicity, and it was assigned a THL-0 rating.

Another success story involves the use of lithium thionyl chloride batteries aboard the *International Space Station*. The perception by the crew is that the containment afforded by these batteries and the hardware in which they are used is insufficient to control the possibility of a catastrophic (THL-4) release of thionyl chloride. These batteries offer exceptional power for the mass required, so they are preferred by hardware builders. However, the concerns over toxicity resulting from an accidental release have precluded use of larger batteries of this type, and currently an effort is under way to preclude the use of this type of battery regardless of its size. If one supposes that a crewmember could be using or near a piece of hardware when this type of battery shorts, then the argument can be made that the volume of dispersion is small before the ejected material reaches the crewmember. Because less hazardous substitutes are available, there is no robust counterargument to this appeal.

Containment of Hazardous Compounds

As described previously, the toxic hazard level and the expected behavior of a compound when released determine the amount of containment necessary for a payload or system to be accepted for use aboard the *International Space Station* or *Space Shuttle*. The risk associated with a compound stems from the degree of harm it could cause if released and the probability of its release. Our approach basically is to trade these two parameters against each other so that the level of risk stays below some maximum acceptable level. The difficulty oftentimes is to determine what constitutes a legitimate level of containment and if this containment is adequate at critical times, such as while the material in question is being used or an experiment is operated. These conditions lead to some flexibility in decisions made with respect to hazard and containment trade offs.

Monitoring Residual Risks

On occasion, the containment of a material and its toxic hazard level do not result in the perception of adequate protection against the threat. In this case, the view is that the residual risk needs a strategy for management. A prime example of this is anhydrous ammonia used for the International Space Station external thermal loops. It is now rated THL-4 because, in remote scenarios, the ammonia could be released from the external loop into the internal loop and from here into the International Space Station atmosphere. Sufficient concern has arisen over this risk that ammonia monitors have been placed onboard and flight rules have been developed to deal with this remote but serious risk.

Another risk that led to monitoring was the perception that some toxic propellants used in thrusters on the exterior of the *International Space Station* could result in contamination of the extravehicular activity suit and the toxic compound then could be brought into the air lock. The concern was that, once the crewmember doffed the helmet, the astronaut or cosmonaut would be exposed to the vapors as the propellant that had impregnated the suit began to evaporate into the atmosphere of the air lock. Despite flight rule stipulations for bakeout and for brushing the suit if contamination is suspected, propellant monitors were developed for use in the air lock to manage the residual risk that was perceived to be unacceptable without monitors.

Scrubbing Compounds from the Atmosphere

Air revitalization systems are designed to remove nominal levels of air pollutants from the air with a reasonable safety margin. The design might not be adequate to deal with plausible contingency events. On occasion, it is necessary to develop contingency strategies to manage a specific type of toxic release in a specific vehicle with limited air revitalization capability. One example of this is the flight rule governing the change out of lithium hydroxide canisters in the *Space Shuttle* after a fire event. Filters are changed out according to the level of pollutants (carbon monoxide, hydrogen cyanide, and hydrogen chloride) found in the atmosphere after a fire. Filters that efficiently remove carbon monoxide easily are poisoned by the acid gases. Until these gases are scrubbed sufficiently, the carbon monoxide removal filters are not moved into the airstream. In another example, a carbon dioxide removal kit is available aboard the *International Space Station* to remove this unwanted gas if it were to be released from the U.S. fire extinguishers (NASA 2004).

Crew Isolation from an Unhealthy Atmosphere

The last line of defense for the crew before abandoning a spacecraft or taking refuge in an uncontaminated module of a large compartmentalized vehicle is to protect their eyes and respiratory systems with masks and goggles (NASA 2004). Personal protective equipment consists of a variety of breathing masks and respirators. The breathing masks are self-contained and totally isolate the respiratory system of the wearer from the atmosphere. The respirators are able to remove specific compounds from the air before it is respired. The breathing masks have use times as short as 5 min, so it is essential to know how quickly a contaminated atmosphere can be scrubbed to a safe level. So-called zones of transition from one level of safety to another have been defined for ammonia, Halon 1301, Freon 218, and mercury (NASA 2004). Given a specific amount of compound released, one can predict how much time the air revitalization system requires to scrub the atmosphere to a safe level. If the time available on the breathing masks or filters does not extend past the time required for scrubbing to a safe level, then there is a risk that the crew cannot protect themselves from a large leak of material for a time sufficient for them to remain in the vehicle. If personal breathing masks are insufficient, then the final level of protection is to abandon the vehicle. Although there have been close calls in space and ground based incidents, to present knowledge, no vehicle has been abandoned because of a fire or other air pollution problem once it had reached space.

Ultimate Fate of Compounds

The toxicologist, in cooperation with environmental engineers, must always consider the ultimate fate of compounds released into the air. Reactive compounds and those with limited volatility can react with or adhere to surfaces. Water soluble compounds in the air are likely to find their way into the water recovery system through the formation of humidity condensate. Typical compounds found in the humidity condensate that originate primarily from the air include aldehydes, alcohols, and amines (Schultz et al. 2006). Under nominal conditions, the water recovery system is able to remove these compounds to a safe level; however, during an accidental large release into the air, the water recovery system might have to be deactivated. Condensation into the water also can be used to deal with a toxic release. One of the strategies in the event of a release of a water soluble compound aboard the *Space Shuttle* is to turn the thermostat to full cooling. This action maximizes the capture of soluble compounds in the condensate, which is routinely dumped overboard without purification. Obviously, this strategy is inappropriate for the *International Space Station*, where the humidity condensate is retained and purified for use as drinking water.

Sometimes the ultimate fate of a compound is challenging to predict. One of the approaches to thermal shield tile repair on the *Space Shuttle* was to fly materials that polymerize when combined so as to form a robust repair. One proposal was to use a system in which A-1100 silane (mostly 3-aminopropyltriethoxysilane) was employed in a thick paste. At first there was little concern about its hazard because it was thought that it would evaporate very slowly from the thick matrix if it were released. Ground workers discovered, however, that A-1100 silane settles out as a liquid, and if the liquid were to escape, it could have immediate toxic effects. The toxicity database on this liquid was sparse, so a study was undertaken to determine its effects on laboratory animals. During efforts to generate this material it was discovered that it reacts so quickly with water vapor, that only a small portion of the silane evaporates into the air before a crust forms over the material. Thus, it was determined that the fate of this material in a space vehicle would be to form a crust of material on the surfaces it impacts, and therefore, it has a relatively low toxicity due to limited volatility (Lam 2005).

18.2 BIOHAZARDOUS MATERIALS

18.2.1 Microbiological Risks Associated with Spaceflight

Microbiology is the science of microorganisms, which are exceedingly small organisms that generally cannot be seen by the unaided eye. Microorganisms include bacteria, fungi (mold), protozoa, and viruses. Their members span two kingdoms, *Monera* and *Protista*, and include both eukaryotic and prokaryotic organisms. Most microbial species are not harmful but instead are highly beneficial to humans and other forms of life. Microorganisms play essential roles in maintaining a habitable biosphere and are important participants in the carbon, nitrogen, sulfur, and oxygen cycles. Other roles include the food and beverage industry, such as bread and wine, agriculture, the absorption and synthesis of vitamins, such as vitamin D in the gastrointestinal tract, and many other applications to everyday life (Alcamo 2001). Many microbiology textbooks are available, but it is advisable to consult the microbiology department of a university for the latest recommendations.

This section focuses on the microbiological risks to astronauts and cosmonauts and their spacecraft or space habitats. Microbiological risks specific to spaceflight are identified along with approaches to control or mitigate any resulting adverse effects. Generally, infectious diseases are thought of as the major concern, but other adverse

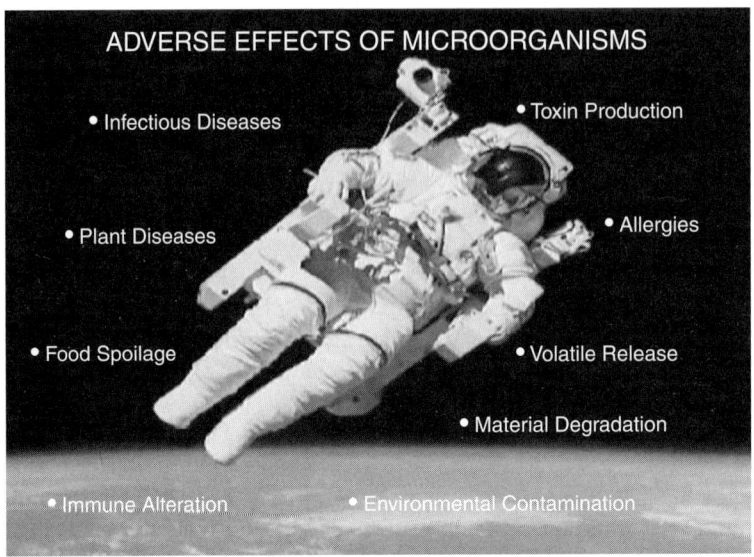

FIGURE 18.1 Adverse effects of microorganisms in space environment (Courtesy of NASA).

effects (Figure 18.1) also can affect the safety, health, and performance of humans in space. In addition to infectious diseases, allergies, microbial toxins, and production of noxious volatiles can cause crew discomfort or reduced productivity. Plant pathogens can endanger food supplies, microbial contamination can result in food spoilage and degraded water quality, and severe microbial buildup can lead to performance degradation of critical spacecraft systems, such as the life support system. Ironically, microbes have evolved with an amazing capability to degrade complex organic materials into basic molecules, and these building blocks can be used to synthesize the macromolecules essential for life. These very capabilities must be harnessed and applied to solid waste remediation, water and air purification, and food production for humans to be able to succeed in the exploration of the solar system (Pierson 2000).

More immediately, we must learn to prevent or reduce the risks associated with microbes in the closed environment of a spacecraft. The microbiological risks facing astronauts is not dissimilar to the risks ordinary people on Earth encounter. Major sources of microbiological risks to astronauts include breathing air, internal spacecraft surfaces, water and food, other crewmembers, and payloads and investigations. Each is considered here.

18.2.2 Risk Mitigation Approaches

Establishment of Acceptability Limits

Spaceflight acceptability limits for microbial contamination levels of the breathing air, internal spacecraft surfaces, drinking water, and food have been established to preserve the health and productivity of the astronauts (Table 18.2). The approach used to establish these

Table 18.2 Microbiology Acceptability Limits for the *International Space Station*

Parameter	Preflight	In-Flight
Air	Total bacteria: 300 CFU/m^3 Total fungi: 50 CFU/m^3	Total bacteria: 1000 CFU/m Total fungi: 100 CFU/m
Surfaces	Total bacteria: 500 CFU/100 cm^2 Total fungi: 10 CFU/100 cm^2	Total bacteria: 10,000 CFU/100 cm^2 Total fungi: 100 CFU/100 cm^2
Water	Total count: 50 CFU/1 mL Total coliforms: Non-detectable/100 mL	
Food	Total aerobic count: $\leq$20,000 CFU/g *Escherichia coli*: $\leq$1 CFU/g Coagulase positive *Staphylococci*: $\leq$1 CFU/5 *Salmonella*: $\leq$1 CFU/25 g *Clostridium perfringens*: <100 CFU/g Yeasts and molds: <100 CFU/g	

Note: CFU (colony forming unit) is used to quantify numbers of bacteria and fungi.

spaceflight specific limits utilizes industry standards, such as U.S. Environmental Protection Agency guidelines for drinking water, and expert panels. A robust microbiological monitoring program has been implemented to ensure conformance with acceptability limits. Monitoring of air, hardware, water, and surfaces is conducted on spacecraft and the modules of the *International Space Station* prior to flight. Monitoring of the International Space Station breathing air, internal surfaces, and drinking water is conducted on orbit. Specific monitoring approaches are identified in the subsections that follow.

18.2.3 Major Spaceflight Specific Microbiological Risks

Breathing Air

Numerous diseases are disseminated through the air. Many respiratory viruses (such as influenza, *varicella-zoster* virus, other respiratory viruses), bacterial diseases including tuberculosis, and fungal diseases such as aspergillosis commonly are spread by airborne routes. Gravity is an effective means for limiting the spread of airborne infectious diseases because larger droplets fall rapidly to Earth. On Earth, in normal gravity aerosol particles of 40 µ and larger settle to the floor within 60 s (Pierson, McGinnis, and Viktorov 1994). The more time airborne infectious agents persist in the breathing air, the greater the risk of infecting a crewmember. In the reduced gravity environment of spaceflight, generation of bioaerosols, that is, aerosols of microbes or microbial products, is problematic because aerosolized droplets are generated easily and remain suspended in the air until they either collide with a surface or evaporate.

Monitoring of spacecraft air ensures that acceptability limits for bacteria and fungi are maintained (Figure 18.2). Adherence to spaceflight limitations and constraints requires in-flight monitoring through the use of small, portable battery powered devices that are easy

FIGURE 18.2 In-flight monitoring of the International Space Station breathing air (Courtesy of NASA).

to operate, require low maintenance, and can be easily calibrated. Accuracy and reliability are additional essential factors. Low levels of airborne bacteria and fungi are found onboard the *International Space Station*. The most commonly recovered bacterial genera from the air on spacecraft are *Staphylococcus*, *Micrococcus*, and *Bacillus*. *Aspergillus* and *Penicillium* are the most prevalently observed fungal, that is, mold and yeast, genera. The bacterial genera found on spacecraft are those commonly associated with humans, except that *Bacillus* is a common environmental bacterium. The usual fungal genera observed on spacecraft are common environmental molds. Levels of bacteria and mold aboard the *International Space Station* have been consistently below acceptability limits and far below levels found in typical homes and offices. Additional information can be obtained from the Johnson Space Center Microbiology Web site, http://microbiology.jsc.nasa.gov.

Surfaces

Many diseases, such as influenza and tuberculosis, can be transmitted to others through human contact or contact with inanimate objects known as *fomites*, such as doorknobs. Accumulation of microorganisms within a spacecraft can lead to other undesirable effects (Figure 18.1). Degradation of the performance of critical spacecraft systems, such as the environmental control system, can result from heavy microbial growth. Acceptability limits for microbial levels on internal surfaces of spacecraft have been established (Table 18.2).

The use of swabs for collection of bacteria and fungi on surfaces to be sampled is a simple, cost effective approach for routine monitoring. The samples collected can be analyzed on-orbit in a manner similar to air samples to provide crewmembers with data on current environmental conditions. Results from the *International Space Station* indicate that bacteria of human origin, that is, *Staphylococcus* and *Bacillus*, are the most commonly recovered bacterial genera (Castro et al. 2004). *Penicillium*, *Aspergillus*, and

Cladosporium are the prevalent genera of mold found in these samples. Surface contamination levels on the *International Space Station* are consistently low and below acceptability limits. Over 7 a of data collected from the *International Space Station* verify the effectiveness of a rigorous housekeeping and monitoring schedule, which receives constant vigilance. However, on occasion, excessive fungal growth has occurred in most spacecraft, most recently aboard the *International Space Station* (Figure 18.3).

Water and Food

More than 200 diseases are transmitted through food (Bryan 1982). The Centers for Disease Control and Prevention estimate that 76 million cases of food borne illnesses and 5000 deaths occur in the United States yearly (Mead et al. 1999). Norwalk like viruses, *Campylobacter*, and *Salmonella*, are major causes of food borne illnesses (Mead et al. 1999).

The risk of infection from food is real, and every step practical is used to prevent contamination by pathogens in foods consumed by astronauts. These steps include strict adherence to regulations on food preparation and packaging. Three basic food supplies are available on spacecraft: commercial or locally prepared foods; finished foods, that is, foods in final flight packaging; and a very limited supply of fresh foods. The supply of

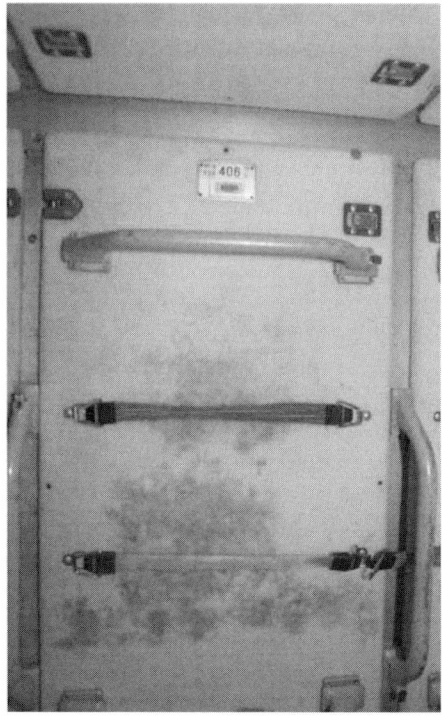

FIGURE 18.3 Fungal contamination on an *International Space Station* panel (Courtesy of NASA).

fresh foods is limited by shelf life and ability to resupply from Earth. Foods are analyzed for bacterial and fungal pathogens and must meet the acceptability limits (Table 18.2) established for foods destined for space use. All microbiological analyses of food are conducted prior to flight. Foods that fail to meet the established limits are replaced by acceptable substitutes. Further information on waterborne and food borne diseases can be found on the Centers for Disease Control and Prevention Web site (www.cdc.gov) and the Federal Drug Administration Web site (www.fda.gov/default.htm).

In response to the Safe Drinking Water Act Amendments of 1996, the Environmental Protection Agency and the Centers for Disease Control and Prevention completed a national workshop designed to assess the magnitude of endemic waterborne acute gastrointestinal illness associated with consumption of public drinking water. A joint report on the results of these studies is available at www.epa.gov/nheer/articles/2006/waterborne_disease.html. Most water quality experts believe the number of acute gastrointestinal illness cases in the United States is underreported. Estimates range from 4 million to 32 million cases of acute gastrointestinal illness per year. Clearly, drinking water is a valid risk factor in our health, and the same is true for astronauts.

The high cost of transporting drinking water to the *International Space Station* requires the reclamation and recycling of humidity condensate on the *International Space Station* to reduce the volume and mass of potable water that must be resupplied from the ground. The reclaimed and processed humidity condensate is supplemented by water provided by the *Space Shuttle* and ground supplied water. Potable water aboard the *International Space Station* is available from three water ports, including a dispenser of ground or Space Shuttle supplied water and two ports from the humidity condensate recovery system. A detailed description of these systems (Koenig et al. 1995) and analysis hardware has been described previously (Samsonov et al. 2002).

Postflight analysis of water samples identified the predominant genera recovered included *Sphingomonas*, *Ralstonia*, *Pseudomonas*, and *Methylobacterium* species. Although not uncommon in water, it is important to note that the opportunistic pathogens *Stenotrophomonas maltophilia* and *Pseudomonas aeruginosa* were recovered from the potable water systems (Castro et al. 2004; Ott, Bruce, and Pierson 2004). All three International Space Station potable water sources have been analyzed for coliform bacteria, a common indicator of fecal contamination and the potential for disease causing microorganisms, such as hepatitis B and other gastrointestinal bacterial pathogens (Figure 18.4). However, the closed International Space Station environment and occupation by exceptionally healthy astronauts precludes most pathogens associated with waterborne diseases. The major contaminants are the bacteria associated with the air, surfaces, and the normal bacterial flora associated with the crew. Therefore, no medically important bacterial contamination of the potable water system aboard the *International Space Station* has occurred.

Crewmembers

The two major priorities in human spaceflight are

- The safety and health of the astronauts.
- Achieving the mission objectives.

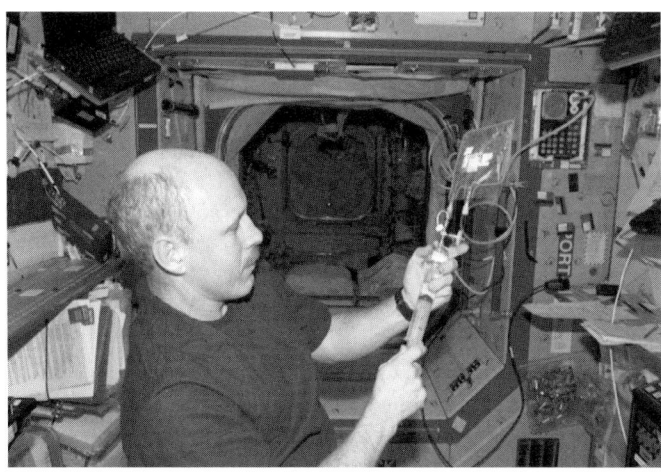

FIGURE 18.4 In-flight monitoring of potable water system aboard the *International Space Station* (Courtesy of NASA).

Even with the major advances, such as antibiotics, over the past 60 years, infectious diseases remain the second leading cause of death in the world and by far the leading cause of premature death and disability. To combat this threat to the astronauts, the health stabilization program was implemented after the Apollo 13 mission to reduce the incidence of astronaut illnesses. This program has been very effective in reducing illnesses both before and during flight. The health stabilization program focuses on reducing the exposure of astronauts from three large risk groups: people who are ill, large crowds, and small children.

Spaceflight can increase risk of infectious disease through alterations in the immune response resulting from stress and other factors. This interesting physiological effect of spaceflight has been reviewed (Pierson et al. 1996; Sonnenfeld, Taylor, and Kinney 2001). If longer duration missions result in medically significant reductions in the ability to resist infectious agents and tumor formation, a major reconsideration of the current approach would be essential.

Prevention being the goal, the limiting of risk factors to the crew before flight and conducting careful medical examinations prior to launch are the most effective countermeasures to illnesses with microbial etiologies. As on Earth, proper nutrition, adequate sleep, and stress reduction can bolster immunity and further reduce the risk of infections. Sending sound, healthy individuals into an environment with very limited opportunities to introduce microbial pathogens into their work and living environment considerably reduces the risk of infectious diseases among astronauts. Spaceflight conditions isolate astronauts from typical public health infectious diseases, such as hepatitis B and C viruses and tuberculosis, because there is no credible opportunity for such agents to enter the spacecraft.

The potential for contracting tuberculosis during a spaceflight mission is implausible, but normal bacteria (flora) associated with healthy individuals can result in disease. For example, *Staphylococcus aureus*, typically found in the nares of 40 to 50% of healthy individuals and astronauts (Pierson et al. 1996) and 80 to 90% of hospital workers can cause life threatening infections. Humans generally experience few infections while sharing a world with a near infinite number of microorganisms of amazing diversity and remain healthy because of a sophisticated immune system, good hygiene, and other factors. However, any factor that results in diminished immunity increases the risk of infectious disease. Such factors include aging, stress, poor nutrition, organ transplantation, cancer chemotherapy, infections (e.g., human immune deficiency virus), and many others. Other factors associated with spaceflight repeatedly have been shown to affect human immunity. However, the medical importance of these changes is not known (Pierson et al. 2006).

The potential for decreased immunity in astronauts over long-term stays in space and recent results demonstrating increased virulence in an important human pathogen, *Salmonella enterica* serotype *Typhimurium*, in microgravity and in modeled microgravity conditions on Earth are areas that require additional investigation (Nickerson et al. 2000; Pierson et al. 2006). Whether this phenomenon occurs with other human pathogens, currently is not known. If the immune response is diminished concurrent with increased virulence of microbial pathogens, the infectious disease risk could rise considerably among space explorers.

Microbiological data from astronauts participating in Space Shuttle flights and the Mir, International Space Station, Skylab, and Apollo missions are consistent with data from healthy populations with normal bacterial flora. This is not surprising, because astronauts are exceedingly healthy and physically fit. However, autoinfection from their own normal flora and human to human transfer of microorganisms remains the greatest risk of infections among crewmembers (Pierson et al. 1996; Pierson 2000). Staphylococcal and streptococcal infections can be life threatening, and both species are part of the normal flora of humans. For example, *Staphylococcus aureus* and *Streptococcus* species are leading causes of septicemia in humans. *Escherichia coli*, a common inhabitant of the human gastrointestinal tract, is the most usual etiology of urinary tract infections. In the United States, urinary tract infections account for about 4 million ambulatory care visits each year, representing about 1% of all outpatient visits. They usually respond to antibiotic treatment, but severe cases can result in death by sepsis. A urinary tract infection occurrence during the Apollo 13 mission was severe. Other normal flora agents of concern are the latent *herpes* viruses, such as *varicella-zoster* virus, which is the etiological agent of chickenpox and shingles (Pierson et al. 1996), and *Candida*, which is a yeast. For further reading, go to the Centers for Disease Control and Prevention Web site (www.cdc.gov).

Even if extended spaceflight does not affect immunity and microorganisms do not increase in pathogenicity, the risk of infections from microbes comprising the normal human flora does increase. Long duration spaceflight beyond low Earth orbit having no options for a rapid return to Earth, such as a Mars mission, will utilize countermeasures such as extensive medical evaluations in the selection of astronauts, increased medical training for the onboard physician, and the use of a quarantine period prior to flight.

Payloads

Astronauts are exposed to many biological materials, a small number of which can be hazardous. Most risks associated with biohazardous materials are microbiological, and these are assessed specifically for each spacecraft, space station, or space habitat, depending on a number of factors. These include the specific microorganism identified, the infectious dose, pathogenicity, the disease associated with the agent, the total number of microbes used in the investigation (allowing for growth during the mission), the biosafety level, availability of vaccine, and treatment options.

Flight payloads are proposed by payload organizations from countries around the world. All payloads undergo a rigorous safety evaluation by the payload safety review panel at NASA Johnson Space Center. The safety evaluation requires the preparation and submission of a safety data package by the payload organization.

One aspect of the overall safety review is biosafety. The Johnson Space Center Biosafety Review Board reviews all payloads that contain biological materials. If biohazardous materials are present, the board conducts an evaluation. Figure 18.5 shows the basic process followed for biosafety evaluation of payloads. The forms for biosafety assessment of payloads can be accessed at the Johnson Space Center microbiology Web site, http://microbiology.jsc.nasa.gov.

Some payloads might not be classified inherently as biohazardous, but they can harbor hazardous microorganisms. For instance, animals included in a payload for investigational purposes can harbor microorganisms that are hazardous. Animals must meet the requirements on microbial agents defined by the Johnson Space Center Committee for the Protection of Human Subjects document located at http://stic.jsc.gov/dbase/iso9000/master/master.cgi. Other payloads similarly can harbor microbes requiring evaluation. For example, plants, soil, or soil simulants can harbor fungi and bacteria that can present

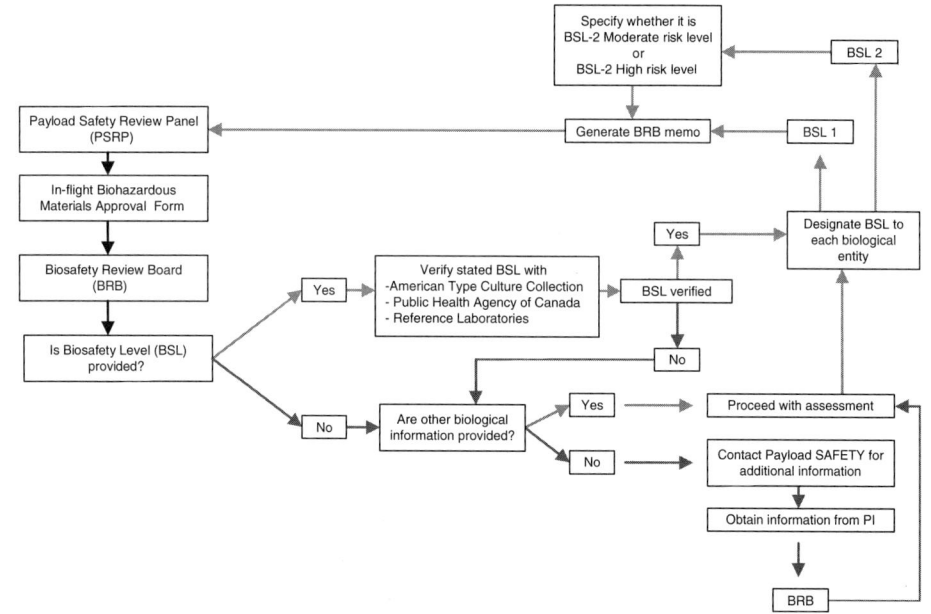

FIGURE 18.5 Biosafety assessment of payload (Courtesy of NASA).

a hazard to the crew or the spacecraft. In essentially all cases, the microbial hazards can be contained by hardware suitably designed with suitable levels of containment.

Risk Mitigation Countermeasures

Risk reduction should begin in the design phase of spacecraft and space habitats. Experience has shown that early identification of risks, followed by design and implementation of effective risk mitigation countermeasures is the most cost effective approach. Many microbiological risks can be reduced or eliminated by this approach. For example, the risk of airborne infectious or allergenic agents and nuisance particulates greatly can be reduced by placing safeguards into the air conditioning and distribution system. Inclusion of HEPA filters into the air circulation system has proven to be very effective in maintaining air with very low concentrations of bacteria, fungi, viruses, and particulates aboard the *International Space Station*. Various allergens, including fungal spores, pollen, and dust mites, are removed effectively as well. HEPA filters are 99.997% effective in removing particulates greater than 0.3 μ in diameter. HEPA filtration is the most effective proven technology to provide breathing air with very low levels of microbial contaminants. When possible, engineering and design solutions should be sought to eliminate environmental contaminants associated with adverse health effects or contamination of essential systems, such as the life support system. This is much more cost effective and often eliminates the problem instead of pursuing only monitoring approaches.

The selection of materials used to construct and outfit spacecraft is highly important in discouraging inappropriate microbial growth. Nonporous surfaces are cleaned and disinfected more easily than porous materials such as fabric. Materials containing antimicrobial substances should be considered. Vigilance with emphasis on water leaks, spills, and condensate is very effective in early detection of conditions that eventually lead to microbial growth. Generally, routine cleaning of exposed surfaces with cleaners containing surfactants to loosen and emulsify contaminants is sufficient to meet established acceptability limits for bacteria and fungi on spacecraft internal surfaces. However, disinfectant wipes are available for use when indicated. All cleaning and disinfecting substances must be compatible for use in closed environments that recycle air and water for crew consumption.

Maintaining environmental conditions unfavorable for sustained growth of microbial contaminants is essential to prevent or control environmental microbial contaminants. Spacecraft characteristically provide a shirtsleeve environment for crew comfort. Unfortunately, such temperatures promote microbial growth. However, controlling the availability of water, such as humidity and surface condensate, is essential in controlling microbial growth. Prevention of water condensation on surfaces, water leaks and spills, and holding relative humidity to 60% and below are effective controls.

Previous experience has demonstrated that spacecraft and space habitats can be built that provide a safe environment for human habitation. The lessons learned from previous human space habitats were applied to the *International Space Station*. Eight years of experience onboard the *International Space Station* has demonstrated that technology can provide a microbiologically safe and comfortable environment with adequate supplies of potable water, breathing air, and food.

18.3 SHATTERABLE MATERIALS

Glass fragments in a habitable zero-g environment can be very hazardous to the crew, having a potential for causing eye injuries or inhalation mishaps. With the advent of the Space Shuttle Program, NASA had to manage many shatterable items in the habitable compartment that were brought to the spacecraft by payload developers, NASA experiments, crew equipment, and the avionics and other instrumentation of the vehicle itself. This process has continued for the International Space Station Program, and it can be expected to be a part of every future crewed space effort.

18.3.1 Shatterable Materials in a Habitable Compartment

When the Space Shuttle vehicles were built, all but a few of the cockpit instruments were made with plastic covers over gauges to protect the glass from the crew and the crew from the glass. The safety philosophy that drove this decision in the cockpit flowed down to other hardware and equipment developers to provide a safe environment for the crew in the rest of the habitable module. Codifying this in the basic Johnson Space Center design standard, JSC-8080 included requirement G-9 that read, "material that can shatter shall not be used in the habitable compartment unless positive protection is provided to prevent fragments from entering the cabin environment" (Boeing 1976).

18.3.2 Program Implementation

Implementation in the Space Shuttle Program

Positive protection for the Space Shuttle project was described in the contractor implementation document, SD73-SH-0297 (Boeing 1976), which stated that (reformatted by the editor) "positive protection shall consist of one or more of the following:

- Substitution of non-shatterable material wherever possible.
- Use of guards to prevent damage to shatterable materials.
- Use of containment devices.
- Procedural controls to minimize activity around exposed glass until such glass is removed from the vehicle or protected."

All hardware installed in the Space Shuttle crew module met this requirement, with the exceptions of event indicators, heads up display units, elapsed time indicators, and structural glass in the windows. NASA permitted the first three pieces of equipment to be exceptions from G-9 and wrote the end-item specification for the Space Shuttle vehicle with those items listed as acceptable without covers. The windows, because they are structural elements of the vehicle, are controlled by other requirements in Johnson Space Center documents, as well as in the Space Shuttle specification; therefore, no internal safety covers were required or provided.

Implementation in the International Space Station

The International Space Station Program deals with the exact same issues as the Space Shuttle Program. Many sources of shatterable materials are seen in every module and the equipment used onboard. Each item designed for the *International Space Station* was procured to specifications written from requirements documents, like the glass structural design requirements in SSP 30560 (NASA 1997), the general structural design requirements in SSP 30559 (NASA 2000a), and the crewed systems integration requirements in SSP 50005 (NASA 1999).

Containment of one source of shatterable materials is addressed directly in SSP 50005 (NASA 1999), paragraph 8.13.3.4 (D), which requires light fixtures to be designed so that lens or lamp failures must not release fragments. This requirement is repeated in SSP 50021 (NASA 1995), the International Space Station safety requirements document. Other pieces of equipment derived a containment requirement from the hazard control process intended to address all hazards to the crew of a vehicle.

A considerable effort was focused on the various and numerous display screens used on International Space Station computer and video monitors. A walk through the International Space Station mockup at Johnson Space Center in the early 1990s revealed the number of display screens in or near the translation path. All these would be exposed to the large pieces of equipment the crew would be moving through the modules, such as racks and experiment boxes. Following this hazard review, all screens procured possessed films covering the glass panels.

Figure 18.6 is a photograph downlinked from *International Space Station* showing damage to a monitor screen. No specific event is known to have caused these scuff marks and the tear in the protective film, but the damage happened during orbital activities. The engineering of this particular monitor included successful impact tests to demonstrate the damage tolerance of the film and glass assembly.

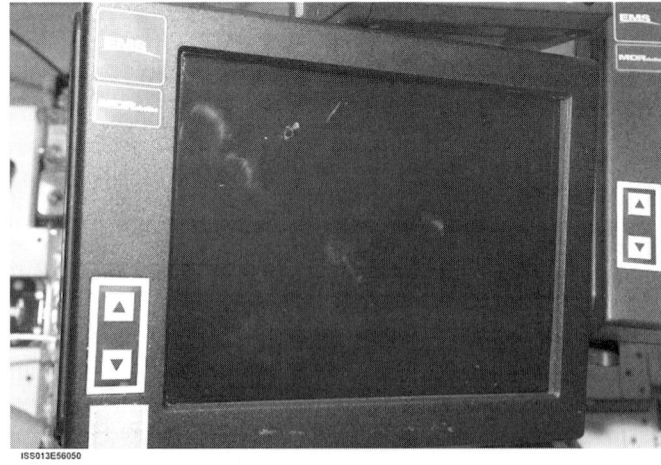

FIGURE 18.6 International Space Station monitor screen damaged by orbital activities (Courtesy of NASA).

Implementation in the Payload Community

Payloads are a primary source of glass components in the habitable volume. To address the payload community, the payload verification requirements, NSTS-14046, includes a section dedicated to verification issues for glass and ceramic structures (NASA 2000b). The first two sentences of this section read:

> *Uncontained ceramic and glass parts are always safety critical when located in a habitable area because of the inherent hazard to the crew. Therefore, glass in a habitable area must be shown safe from breakage or proven contained.*

Proving a glass component safe from breakage is discussed in Section 17.6 of this book. Proving an item is contained is the subject of the following sections.

NASA Payload Safety Reviews When the safety community is reviewing a proposed experiment or piece of equipment for manifesting, the hardware provider must show that the hazards from potential glass fragments are contained during all phases of flight. The hardware provider must document the containment method and concept in the safety data packages provided to NASA.

The payload provider has a choice of describing the shatterable materials on the standardized hazard control report form, JSC-1230, or writing a unique hazard report for the shatterable materials in the payload. Usually the complexity of the equipment drives this decision. If the glass components are nonstructural, it is likely that these controls can be described on the JSC-1230. Experiments with many nonstructural glass components requiring detailed descriptions of the hazard controls probably needs a unique hazard report. Structural glass usually requires a unique hazard report as well as strength and fracture analysis (discussed elsewhere) unless it is completely contained by a simple design or failure of the structural glass is of no consequence to the crew or vehicle.

Standardized Hazard Control Options On form JSC-1230, hazards from shatterable materials are listed in block 5. The hardware provider can select among three options for describing the hazard controls:

- All materials are contained.
- Optical glass (that is, lenses, filters, and the like) components of crew cabin experiment hardware that are non-stressed (no delta pressure) and have passed both a vibration test at flight levels and a post-test visual inspection.
- Payload bay hardware shatterable material components that weigh less than 0.25 lb and are non-stressed or nonstructural.

18.3.3 Containment Concepts for Internal Equipment

The approach to controlling hazards from shatterable materials in a habitable environment starts with the four criteria listed in Section 18.3.2, subsection "Implementation in the Space Shuttle Program." Containment of glass components typically has been achieved by plastic films and bags or transparent polycarbonate or acrylic covers or laminations.

From launch to orbit, many experiments are stowed in Space Shuttle middeck lockers or bags or in similar stowage compartments in the International Space Station or SpaceHab® racks. These stowage compartments, however, are not containment devices because the crew opens them once it is time to unstow the hardware. If the equipment contains glass components, the hardware provider is advised to put the equipment into a sealed plastic bag before it is stowed in a locker or rack before launch. Once on orbit, when the crew unstows the equipment, a simple inspection of the bag can be performed to determine if any glass fragments are present. If the crew finds fragments, the equipment can be restowed safely without risking a release of the fragments into the habitable environment.

During the time when the equipment is being used, the hardware ideally should contain all glass components. However, in many instances, this is not feasible. Cameras, for example, usually cannot have plastic covers over their lenses when photographs are being taken. Experiment hardware with intricate internal optics cannot always be completely contained for a variety of reasons, ranging from operational needs to cooling requirements. For these cases, special approaches have been developed.

Cameras

Still photo and moving image cameras are pieces of optical equipment that virtually every adult has handled in everyday life. There is almost universal familiarity with the techniques necessary to manage a camera and take a picture. Astronauts are trained further in how to make good photographs using the specific cameras provided for flight. Because of the high level of familiarity a crew has with cameras, the hazards from potential glass fragments are managed in a special way.

Cameras and lenses with lens covers attached are wrapped in plastic bags with zippered closures (Figure 18.7). These are packed in foam and stowed in middeck lockers or similar locations. Prior to use, the bags are taken out of the lockers, and the contents are inspected to ensure no glass fragments are present. When the photography task is

FIGURE 18.7 Camera packed for flight (Courtesy of NASA).

complete, the cameras are put back into plastic bags and restowed in the locker from which they were removed originally. At no time are cameras to be left loose in the habitable compartment.

Most cameras are subject to ground vibration tests to establish that they will function as required after exposure to the launch and ascent environment. Given these procedures and tests, the safety community and technical experts agree that the hazards from shatterable materials released from digital and film cameras are managed appropriately.

Experiment Boxes with Internal Optics

A common payload experiment configuration is a box built to fit into an International Space Station rack or replace a Space Shuttle middeck locker. An experiment box might have internal mirrors or lenses made from shatterable materials. Occasionally, an experiment box might contain a glass window and camera. The basic safety requirement for these boxes is that all shatterable material must be contained.

The payload provider should plan to make the contents of the box completely contained. Openings can be closed with transparent polycarbonate covers that permit excellent viewing and also provide containment.

Some payload developers wish to approach hazard control for optics internal to boxes using the same philosophy as described for handheld cameras. This presents difficulties, because the basis for accepting fewer positive controls with handheld cameras is due largely to the idea that the crew is familiar with this hardware and can handle it appropriately. Cameras and optics internal to boxes are not a direct part of the handling of the experiment hardware by the crew, and the crew even might not be aware that there are glass components inside the equipment.

In these cases, the payload safety engineers have required that experiment hardware go through an acceptance vibration test that envelops the launch and ascent environment. Following the test, the hardware is inspected as thoroughly as possible. Functionality also is tested to ensure that all optics are working as expected. To be acceptable, the shatterable components must not be accessible by the crew, cannot be damaged by handling, and must be demonstrably resilient to the environment.

Many payload providers have been able to provide containment up until the time the experiment is initiated. The crew can inspect the equipment and even look through the optics before the containment device or material is removed, thus, establishing that no shatterable material release is imminent.

Acceptable Released Fragment Size

If the experiment requires cooling, the box is likely to include a ventilator fan and air vents. These vents are open paths for fragments of broken glass to escape, so complete containment usually is not possible for these configurations. If an internal glass component fractures, fragments would be released into the habitable compartment through the ventilation system.

When considering this issue for the first time, NASA flight surgeons were asked to determine the maximum size for a fragment that would be acceptable. The result of their work was an upper limit of 50 μ for fragment characteristic length. Therefore, where appropriate, experiment boxes that cannot contain glass fragments completely must have a 50-µ screen on any open path so that hazardous fragments are not released into the habitable environment. If this can be achieved, safety engineers consider the hazard controlled.

18.3.4 Containment Concepts for Exterior Equipment

Even if payload or flight hardware is never to be a part of the habitable environment, the hardware provider is not relieved of all hazard control responsibility. For example, an item that launches in the Space Shuttle payload bay must not release fragments that could become loose object hazards to the vehicle. The hardware provider must establish that no piece weighing more than 0.25 lb can be released by the equipment. Fragments smaller than 0.25 lb also can present hazards to mechanisms; therefore, they are likely to require evaluation by the mechanical systems working group at NASA Johnson Space Center. Containment is recommended, but this requirement can be met by test and analysis as well. Exterior equipment also can present hazards to the crew through a contamination event that brings shards into the habitable compartment.

Extravehicular Activity Hazard

If a piece of equipment is in or near an extravehicular activity translation zone or worksite, any shatterable material exposed to the extravehicular activity crewmember is at risk for damage by errant tools. If a tool strikes a piece of glass during an extravehicular activity and shards are released, the extravehicular activity crew faces potential sharp edge hazards and contamination hazards for the intravehicular activity crew once the extravehicular activity is complete and the contaminated suits are doffed and stowed in the air lock. To address this risk, NASA defined (NASA 2005b) the tool impact environment (Table 18.3). Hardware providers are required to show that fragments larger than 50 µ are not released by a tool impact event. This requirement also can be found in the *International Space Station Flight Crew Integration Standards*, SSP 50005 (NASA 1999), paragraph 14.1.3.N.

Although this is not strictly a containment issue, the ultimate hazard is the same, that is, contamination of the habitable environment by fragments of shatterable materials. Controls available to the hardware provider include designing the equipment so that the shatterable materials are protected from tool impact by using grids or screens, designating a mass limit on the tools and equipment brought near the hardware during the

Table 18.3 Tool Impact Requirement for Shatterable Materials on the Spacecraft Exterior

Tool impact	60 lbm	Concentrated mass traveling at 1 fps on a 0.08-in. diameter circular area	Any direction	Windows and exposed glass surfaces or other shatterable materials	Applicable to all shatterable materials within translation paths or worksites

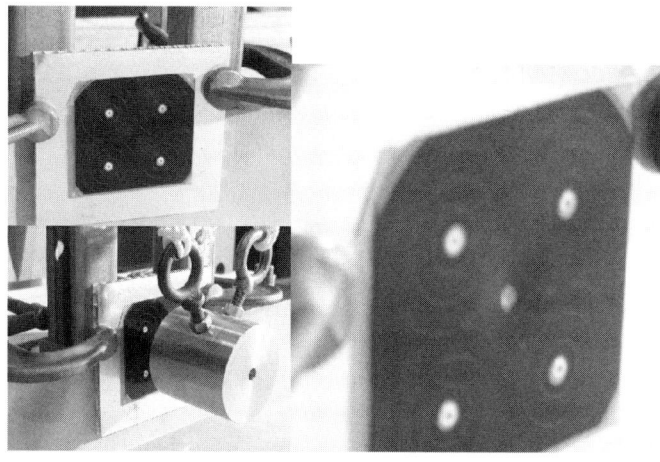

FIGURE 18.8 Impact test of solar panel for International Space Station external hardware (Courtesy of NASA).

extravehicular activity, or specifying a keep out zone for the hardware. Verification of the tool impact capability of a piece of shatterable material usually is achieved by an impact test. Figure 18.8 shows several photos from a test of a solar panel for an external device on the *International Space Station*. Figure 18.9 shows an external light with a grille designed to keep tools away from the shatterable materials.

FIGURE 18.9 Luminaire lamp with grille (Courtesy of NASA).

18.3.5 General Comments About Working with Shatterable Materials

Good Design Practices for Shatterable Components

The following suggestions should help any hardware designer successfully complete the design and verification cycle:

- Design glass friendly mounting systems:
    - Do not use rigid mounting systems.
    - Simply support the glass and protect the glass edges from damage.
    - Use soft seal mounts, such as silicone.
- Prevent glass to metal contact by using nonmetallic bumpers to prevent contact.
- Do not use tight tolerances between the mount and the glass.
- Consider thermal effects of the mount on the glass.
- Minimize the loads that are transmitted into the glass.

Crew Handling of Shatterable Components

One of the greatest risks to a shatterable component is crew handling. Handling loads are very difficult to predict, and it is best to design hardware that does not require crew handling. If crew handling is required, the worst-case handling loads must be considered in the design and verification of the shatterable components. Detailed crew procedures must be developed to control these loads as much as possible. Tests with crewmembers are necessary to envelop all the conditions the hardware can experience.

Failure Propagation and Shatterable Materials

Glass components inside an experiment box should be protected from other items in the experiment. Even if a failure inside an experiment box presents no direct hazards to the crew or mission, it is still important to consider whether this failure could propagate and damage the glass components. For example, internal fasteners might need to have redundant locking features to ensure that loose fasteners cannot rattle around an experiment box and cause impact damage to glass components.

Containment Materials and Devices

Lexan® is a common polycarbonate material chosen for containment in the zero-g environment. It is a tough material with excellent optics, and it provides a very good solution for the majority of experiments. Another brand name polycarbonate that has been used successfully is Makrolon®.

For small pieces of glass, Kapton® tape frequently has been used. The tape simply is applied over the glass item, and it contains any fragments should the component be damaged. This is the most common approach to containing the tiny glass covers of LCDs and LEDs. It should be noted, however, that today, many liquid crystal diodes and light emitting diodes are manufactured without glass covers and do not require the Kapton tape.

Zippered plastic bags are a very effective means of enclosure for components of many different sizes. Cameras and experiment boxes have been stowed in plastic bags, as well as many other items with components made with shatterable materials.

For a detailed review of containment issues involving a series of glass failures during an orbital experiment, read "Microgravity Furnaces and Sample Containment—A Reflective Perspective," by Shaefer and Fiske (2004).

18.4 CONTAINMENT DESIGN APPROACH

The containment of hazardous materials can be performed with either a fault tolerant or a design for minimum risk approach. The fault tolerant approach requires separate and independent containers to prevent release of the hazardous materials. For example, critical hazards require single-fault tolerance, meaning that THL-1 fluids must be enclosed in two independent leak tight containers. The design for minimum risk approach permits use of a single container, but one that has a very robust design and verification process.

18.4.1 Fault Tolerance

The classic fault tolerant design for the containment of hazardous materials is the use of two or three separate levels of containment, such as a sealed box completely enclosed within another sealed box. Each box is qualified independently and acceptance tested as leak tight and is compatible with associated materials and environments. This method is preferred for containing hazardous materials and should be used whenever possible.

Safety authorities commonly approve some designs that are not strictly fault tolerant, such as when the hardware developer proves a robust design and verification. An example design is a single metallic box that is dual sealed at all interfaces. The verifications of this design include high margins of safety for ultimate strength, proof testing, seal design qualification, and final acceptance leak testing. This type of design requires specific approval from the applicable safety authority on a case by case basis.

18.4.2 Design for Minimum Risk

Hardware designs following the design for minimum risk approach for containment of hazardous materials must meet the fracture control requirements for hazardous fluid containers. Normally, hazardous fluid containers are used to contain catastrophically hazardous materials and exceed the requirements for containing critically hazardous fluids. A typical design is a metallic box with welds at all interfaces.

Implementing a design that meets all requirements for hazardous fluid containers is often impractical for on-orbit operations. For example, an experiment using THL-1 fluids would need to be observed on orbit with diagnostic systems located outside the container. This is impossible when the experiment is enclosed fully within a welded metallic box. These types of experiments typically have soft seals, view ports, or nonmetallic parts that cannot be verified as a hazardous fluid container.

18.5 CONTAINMENT DESIGN METHODS

18.5.1 Containment Environments

Containers must be designed to be leak tight for all environments expected during the lifetime of the hardware. In particular these effects include structural loads, pressure loads, temperature extremes, cycle life, and exposed materials. Other environments can be relevant for particular hardware. Pressure loads typically are the primary design environment for containers of hazardous materials. Most of the design and verification guidelines that follow are driven by the need to withstand pressure loads without rupture or leakage.

An important design consideration is the vacuum environment in space. Sealed hardware mounted external to a pressurized environment is going to experience a delta pressure during normal operations. In addition, the International Space Station safety authority requires internal hardware to be designed to withstand a depressurization within the International Space Station; therefore, the design engineer must consider normal operating conditions, worst-case thermal environments, and two failures in addition to a vacuum environment. Containment of the fluids is required generally for the duration of the depressurization to prevent International Space Station crew exposure to potentially released hazardous materials after recovery and restart of normal operations. Depressurization effects that might reduce the perceived hazard, such as release of the fluids into a vacuum or dilution during repressurization, generally are unacceptable controls for the applicable safety authority, because they cannot be verified definitively and potentially could deposit hazardous materials on surfaces accessible to the crew or might evaporate after module repressurization.

18.5.2 Design of Containment Systems

Hardware designed to contain hazardous materials can be described at the system level and at the component level. Overall, a containment system or a set of containers must be designed to prevent release of hazardous materials. This is the system level design. The actual hardware design and usage procedure at the individual component level is sometimes described as a level of containment. A containment system usually is constructed with two or three means of containment. As described previously, pressurized systems also can be used to contain hazardous materials.

The primary justification for the applicable safety authority to accept a level of containment is qualified design and acceptance testing. At a minimum, this should include a container that is designed with safety margins, proof and leak tested, made from aerospace quality materials, and compatible with all exposed materials.

Metallic Containers

Metallic containers are the most common type of hardware designed to contain hazardous materials because requirements and design practices governing their verification are established firmly. Design engineers can use either design approach, fault tolerance or design for minimum risk, for metallic containers; however, the verifications can differ between the two approaches. The design of the metallic container should be inherently robust, and it should be designed as such from the beginning. The engineer should design the container to withstand all applicable environments, such as pressure, thermal, and inertial loads. Typically, the most important environment is due to pressure loads and maximum design pressure. The metallic container must be designed to withstand the maximum design pressure multiplied by the applicable safety factor. The minimum safety factor on ultimate strength for human spaceflight is typically 1.5, although higher safety factors are applicable for pressure components, lines, and fittings. Pressure vessels can be designed with a lower safety factor with special approval from the safety authority. The metallic container also must be designed with a safety factor on yield strength to ensure failure does not occur during proof testing of the flight hardware.

The obvious geometry for a metallic container primarily under pressure loads is a sphere or other typical pressure vessel shape. A properly designed pressure vessel is recommended whenever practical. However, pressure vessels are often impractical for most experiment hardware, and therefore, a milled metallic box with thick reinforcing ribs is recommended. The surface areas of flat plates should be as small as possible and the plate stiffness as high as possible to avoid excessive bending stresses under pressure loads. The designer should select aerospace quality materials from the beginning that are compatible with the hazardous materials and highly resistant to stress-corrosion cracking.

Plastic Bags

The use of plastic bags for levels of containment generally is discouraged. The appropriate use of plastic bags is as one level of containment, particularly for short periods of time. The bag must be verified with qualification and acceptance pressure testing, including an external vacuum test when applicable. Particular attention should be placed on the verification of bag seals. Hazardous materials must be compatible with the plastic bag materials, and the hardware within the bag must not be capable of damaging or puncturing the plastic.

Soft Seals

A container of hazardous materials must be sealed at all interfaces to achieve leak tightness. Typically, containers are sealed with soft seals such as O-rings or gaskets at all interfaces. The term *soft seal* indicates the seal material is nonmetallic. Common soft seal materials are Viton®, silicon, ethylene propylenediene monomer rubber, or another polymer based substance compatible with a wide range of materials.

Each verifiable soft seal is valid as one safety control to prevent leakage. Two soft seals are needed for each leak path to contain critically hazardous materials, and three seals are required for catastrophically hazardous materials. The design engineer must select soft seals to be compatible with the hazardous materials, cleaning fluids, and other materials with which it might come into contact, considering the life of the hardware and associated thermal environments.

Most containers include a soft seal at the interface for fluid fill and drain ports, power and data connectors, and between parts of the assembled container. The designer must remember to consider all potential leak paths and ensure two or three independent seals along each path. As an example, fill and drain ports are fluid connectors that pass through a metallic container. Two seals must be present to prevent a critical hazard for the leak path external to the fluid connector, usually with a type of O-ring sealed flange on either side of the metallic wall. Additionally, the fill and drain valve must be dual sealed for the internal leak path, such as with an internal valve seal and an external sealed cap.

Each type of seal design must be qualified as a valid seal to prevent leakage of the hazardous material. Whenever possible the design engineer should select aerospace or military qualified seal designs, dimensions, and applications. The design engineer must qualify unique seal designs or demonstrate compliance with qualified sealed designs following the specifications of the manufacturer. In any case, each specific seal application must be acceptance leak tested. The recommended pressure for this leak test is maximum design pressure. The preferred seal designs are static seals and face seals rather than seals on moving surfaces or circumferential seals to avoid potential wearing during installation or operation.

Metallic Seals

Metallic seals are a common feature of pressurized systems, especially within fluid line fittings. The typical design is a thin metallic disk that is deformed when the fitting is tightened. In some cases, the metallic seal becomes a permanent connection between the fitting and line due to this deformation. These types of metallic seals are well qualified and have a long and safe history when used according to specified assembly procedures and within qualified environments. Metallic seals of this sort usually are considered equivalent to two levels of containment, meaning a single metallic seal can be used instead of two soft O-ring seals at fluid line connections containing critically hazardous materials. The use of metallic seals as two levels of containment must be approved by the applicable safety authority on a case by case basis, however.

Welds

Containers that include welded interfaces must meet the general requirements for safety critical structures. The welds should be of aerospace quality and must undergo nondestructive inspections. The designer can be required to perform nondestructive inspections both before and after proof testing of a container. Furthermore, nondestructive inspections should be performed using dye penetrant inspections or the equivalent for surface cracks and X-ray should be used to assess for internal defects. According to NASA, welds should be avoided wherever possible because of the stringent requirements for space qualified welds in terms of qualification of processes, weldment design, tests, and inspection. However, if properly verified, welded interfaces can be used in containers, and the applicable safety authority can accept them as the equivalent of at least two levels of containment (NASA 2002).

Electrical Connectors

The design engineer should select aerospace or military qualified hermetic connectors for all electrical interfaces through the container. The external leak path of the electrical connector itself should be sealed, such as around the connector housing at the interface to the metallic container. The internal leak path, such as along the pins, should be hermetically sealed. This hermetic sealing usually is performed by aerospace or military qualified potting materials considering the container design environments.

18.6 SAFETY CONTROLS

This section contains a brief overview of the main safety controls for containment of hazardous materials. These controls should appear on a hazard report titled, *Release of Hazardous Materials*. The hazard report must be marked with a hazard level, critical or catastrophic, in accordance with the evaluation from the safety authority. Specific controls for particular operations or designs are not addressed.

The release of hazardous materials can occur during rupture or leakage of the containment system. Leakage is associated with the slow release of a fluid from a container through a seal or crack. The following controls focus on preventing leakage. Specific controls for the release of the hazardous materials due to incorrect operational design or crew procedures are not included.

18.6.1 Proper Design

The primary control for preventing the release of hazardous materials due to hardware design or operations is containment. For example, physical containment is possible using a metallic box with seals at all interfaces. Specific recommendations for different elements of containment design were described previously. The design engineer must demonstrate to the applicable safety authority that each containment design is robust for the defined environments.

18.6.2 Materials Selection

All materials used in the construction of a container must be selected in accordance with the material requirements. Selection is performed usually with the aid of a database or material standards. The design engineer should select materials from these sources whenever possible. When not possible, however, the design engineer must present a request for approval or a materials usage agreement and receive approval by the appropriate agency. Among other selection criteria, the metals should be selected to be highly resistant to stress-corrosion cracking.

18.6.3 Materials Compatibility

The designer must demonstrate the compatibility of hardware materials with all exposed materials, cleaning products and fluids, and in particular the hazardous materials. This materials selection process usually includes considerations for materials compatibility. The only additional concern for materials compatibility is to ensure that specific fluids, other than normal atmospheric or humid conditions, are compatible with the container materials. The particular concern here is the materials compatibility of soft seals with contained hazardous materials. A specific lifetime often is required by the applicable safety authority for soft seals as limited life items for the safety certification. Soft seal life limits often determine the operational limits for safety.

18.6.4 Proper Workmanship

The construction and assembly of containers, including all subparts, must be controlled with a quality assurance program. The hardware developer must designate someone responsible for performing inspections and other quality assurance activities, including the approval of drawings, procedures, processes, and parts lists. Any nonconformances must be documented and accepted by the appropriate agency. The primary purpose of quality assurance for safety is to ensure the as built flight hardware is designed and verified according to the approved hazard reports.

18.6.5 Proper Loading or Filling

The hazardous materials must be verified for proper loading or filling within the hardware and container. Proper loading or filling includes a qualified procedure to ensure both the proper material quantity and quality. A quality assurance representative should witness the loading or filling of the materials into the flight hardware.

18.6.6 Fracture Control

Fracture control must be applied to any items identified as potentially fracture critical. The payload designer must provide a fracture control plan at the beginning of the project and finally a fracture control report for acceptance of the flight hardware. The plan and report are reviewed and approved by the safety authority to ensure compliance with the applicable fracture control requirements. Typically, fracture control is implemented for verification of metallic containers. This includes proof testing, crack growth analysis, and fail safe analysis. As well, nondestructive inspections and qualification of welds can be derived from a fracture control program.

18.7 SAFETY VERIFICATIONS

Each of the previous safety controls must be verified through qualification and acceptance testing. Specific verifications for containers of hazardous materials, which focus on strength analysis and tests, seal qualification, leak tests and procedures, are included here. The qualification and acceptance tests, analyses, or procedures required for safety appear as verifications in the hazard reports. As stated before, common safety verifications, such as materials selection or quality assurance, are not addressed.

The verifications described here are based on the prototype verification approach with a separate development of a qualification model and flight model. Qualification level tests are performed on the qualification model to demonstrate compliance of the design with the safety controls. Acceptance level tests are performed on the flight model, although they usually are fewer and less severe than qualification tests. Many design engineers choose the protoflight model approach instead of the qualification model or flight model. Notes and suggestions for protoflight model verification are included when appropriate.

18.7.1 Strength Analysis

The purpose of strength analysis is to demonstrate that the container cannot rupture because of the worst-case structural loads. These primarily are assumed to be pressure loads. The container should be designed to withstand a safety factor on ultimate strength with the maximum design pressure as the load. When possible, the analysis should apply a safety factor on yield strength to prevent damage during the proof test. The verification method should be structural analysis, typically using finite element model analysis. The finite element model analysis should be performed according to the structural requirements. Positive margins of safety must be demonstrated using applicable safety factors.

Strength testing is acceptable if a structural analysis is not possible. The test must be performed up to the ultimate strength pressure, including the safety factor. Material yielding, permanent deformation, and possibly even leakage are acceptable during the strength test as long as structural integrity is demonstrated without bursting. Strength testing should be performed on the qualification model. Strength testing can be performed on the protoflight model if the designer can demonstrate high margins of safety by engineering judgment, and the designer accepts the risk of possibly destroying the protoflight model during the tests.

18.7.2 Qualification Tests

The purpose of qualification testing is to demonstrate the functional design of the hardware. Qualification tests are performed on non-flight hardware, such as the qualification model. If the protoflight model verification approach is used, the qualification tests should be performed on the protoflight model. The following qualification tests are recommended as safety verifications for containers of hazardous materials.

Qualification tests must be performed separately for each soft seal to demonstrate that each seal design application in the container functions in the design environment. Typically, qualification tests for soft seals are pressure leak tests and should be performed at least to maximum design pressure. Additional testing for other environments, such as thermal or cycle life, must be qualified as deemed appropriate. Each seal must be qualified separately; however, typical aerospace or military qualified seal designs can be accepted without qualification tests if previous qualification tests are valid for the flight model design, environment, and life. Most design engineers perform soft seal qualification tests on the qualification model by placing only one soft seal at each interface and performing a leak test. The designer then removes the tested soft seal and places the other soft seal in the qualification model for a second leak test. In this way each soft seal is qualified separately. This approach can be used for the protoflight model. The acceptance criteria to demonstrate leak tightness depends on the type of leak test performed and the type of materials to be contained. A typical acceptance criterion for leak testing of fluids is less than 10^{-3} sccs (standard cubic centimeters per second) and for gases is less than 10^{-6} sccs.

Metallic seals are qualified similar to soft seals as just described. The only difference is that metallic seals can be considered as two levels of containment with approval from the safety authority, and therefore, only one leak test to maximum design pressure is needed because no secondary seal exists.

Qualification testing of welds are required only as needed by the welding procedure, process, or specific design considerations. Refer to the applicable welding requirements for more information.

Electrical connectors must be qualification tested using the qualification model similar to the tests described for soft seals. In most cases, the external mounting of the electrical connector is sealed with soft seals. Qualification testing performed for aerospace or military standard designs can cover the connector hermetic seal through the internal leak path, such as along the pins. The designer must demonstrate the connector is qualified for use in the specific container design, environment, and life.

Qualification tests of plastic bags consist of a pressure test to at least maximum design pressure on several samples from the same lot of bags as the flight model. The pressure test is performed usually in a vacuum chamber to reach the required delta pressure. The designer should give particular attention to the amount of air in the bag during testing to ensure the pressure test applies a pressure load to the bag, such as by inflating the bag as much as possible before the test.

18.7.3 Acceptance Tests

The purpose of acceptance testing is to accept the workmanship and functionality of each piece of flight hardware. All acceptance tests are performed on flight units, such as the flight model or protoflight model. The following acceptance tests are recommended as safety verifications for containers of hazardous materials.

Acceptance tests are required for safety to demonstrate each flight model is leak tight. The type and purpose of acceptance tests vary for each design and loading or filling procedure. The design engineer must demonstrate to the applicable safety authority that the acceptance tests performed on the flight model are adequate for safety. This should be accomplished with two types of acceptance leak tests of the flight model. The first type of leak test demonstrates the as built hardware is capable of sealing as designed. This test is not in the final flight configuration but verifies the workmanship and functionality of the flight model. The second is a leak test of the flight model in the final flight condition. For clarity, the first type of leak test is referred to as the *functional acceptance test*, and the second is the *flight condition acceptance test*. These test approaches also can be used for the protoflight model.

As an example, the acceptance tests for containers with two soft seals at each interface should consist of two functional acceptance tests and one flight condition acceptance test. The two functional acceptance tests should be performed on the flight model, one for each soft seal, at maximum design pressure. If successful, these leak tests demonstrate the flight hardware is built as designed and each seal is functioning. The flight condition acceptance test should be performed after final filling of the container and at maximum design pressure. This leak test demonstrates the flight hardware to be leak tight as finally assembled and filled. The acceptance criteria for leak tests are the same as the qualification leak tests and depend on the type of leak test performed.

Hardware design and mission success requirements often do not allow for all three of these recommended leak tests to be performed. The design engineer should propose acceptance leak test verifications to the applicable safety authority that provide equivalent safety and do not harm the functionality of the hardware. The typical rationale is to assemble the flight model in accordance with qualified procedures as described previously and perform at least one leak test to maximum design pressure.

Acceptance tests for metallic seals, welds, and electrical connectors should follow the same philosophy as described in the previous two paragraphs for soft seals. Each flight model plastic bag should be leak tested to maximum design pressure if possible. In addition, the qualification test defined previously should be performed for each lot of plastic bags. Welds should undergo nondestructive inspection both before and after proof testing.

18.7.4 Proof Tests

Proof testing of metallic containers is a safety verification to ensure structural integrity. It is applicable only as a safety verification for the flight hardware, that is, the flight model or protoflight model. The design engineer can choose to perform a pressure test at the proof level on the qualification model to demonstrate functionality of the design; however, this is not required for safety. Proof testing of flight hardware to at least 1.5 times maximum design pressure is encouraged strongly for containers of hazardous materials. Proof testing is one of the main reasons the applicable safety authority commonly approves container designs. Acceptance criteria for proof testing are that there is no bursting, plastic deformation, leakage of the container, or similar indications in the post-proof nondestructive inspection for welds. Repeated proof testing on the same flight hardware should be avoided if possible to prevent cyclic pressure loads and inducing crack growth during repeated proof pressure tests.

Proof testing flight model containers and demonstrating high margins of safety for strength also satisfy fracture control requirements. As stated in the fracture control requirements (ECSS 1999), "Safe life analysis is not required if the item is proof tested to a level of 1.5 or more times the limit load, including [maximum design pressure] and vehicle accelerations."

18.7.5 Qualification of Procedures

As described previously, two types of acceptance leak tests should be performed on the flight model. Operational constraints often limit the types of leak tests that can be performed on it. For example, a flight condition acceptance test likely is not possible for some hardware because it disturbs the contained materials and destroys the functionality. In situations like these, the designer can rely on the specific qualification of an assembly procedure to demonstrate confidence in the hardware design and assembly. Qualification of filling or assembly procedures are meant to substitute for acceptance leak tests that cannot be performed as recommended. The applicable safety authority must approve each application of qualified procedures to substitute leak tests on the flight model.

Qualification of procedures typically involves quality assurance personnel and practices. Specific quality assurance guidelines are not included in this description; however, the general steps a designer should use to qualify procedures are as follows. At a minimum, the qualification should include an assembly or filling procedure that can be followed for the flight model. The designer must demonstrate that implementing this procedure results in a leak tight flight model with repeatability. The assembly or filling procedure usually is performed several times on the qualification model followed by a leak test. Qualification of assembly or filling procedures is often necessary to ensure the hardware is not harmed; however, this should be avoided whenever possible.

18.8 CONCLUSIONS

Toxicological assessments of compounds to be used in spacecraft and space habitats must never be done in isolation, away from knowledge of the apparatus and the vehicle in which they are to be used. Ways in which the compound can escape from the apparatus must be considered in cooperation with the hardware builders. The vehicle environmental engineers must be consulted when detailed understanding of removal capabilities have a bearing on the hazard assessment. The toxicologist must avoid arbitrarily invoking too many worst-case scenarios when assessing compounds. This requires a balanced level of situational awareness, as the toxicologist finalizes risk assessments. There is always an obligation to the hardware builders to not overstate the risks; however, at the same time there is an obligation to the crew to protect them from exposure to toxic compounds from accidental release.

Microbiological agents can affect adversely the health, safety, and performance of astronauts. In addition to direct effects on crewmembers, microorganisms can degrade the environment and performance of critical spacecraft systems, ultimately jeopardizing mission objectives. Microbiological risks must be identified, and levels of acceptable risks must be defined. These risks can be mitigated by early development and implementation of effective countermeasures beginning with the spacecraft design phase. Specifically, selection of antimicrobial materials, use of air filtration, control of humidity and condensate, establishment of acceptability limits, verification monitoring, and remediation technologies all play important roles in the successful human exploration of space.

As NASA goes forward with long exploration missions, beginning with a return to the Moon and eventually exploring Mars, the risks from microbiological agents are going to increase as mission duration increases. Lessons learned from the Space Shuttle, International Space Station, and earlier programs can be applied to future longer duration missions and prevent many of the microbiological problems previously experienced.

A related issue is the need to determine the medically significant effects (if any) of various factors associated with spaceflight on human immunity and microbial virulence. Decreased immunity or increased virulence resulting from spaceflight would alter current plans for safeguarding the crew on long exploration missions. A recommended approach is to send exceptionally healthy and physically fit astronauts into a well controlled and microbiologically safe environment.

Forty years of crewed spaceflight have taught designers and safety engineers many things about the hazards of a zero-*g* habitable environment. Even with robust controls in place, some glass components have broken during spaceflight, but due to the redundant layers of controls, no person has been injured. NASA intends to keep that safe record intact through good design specifications and vigorous safety reviews. The information in this chapter has outlined the important points for the reader and future hardware providers.

REFERENCES

Alcamo, I. E. (2001) *Fundamental of Microbiology*. Sudbury, MA: Jones and Bartlett.

American Institute of Astronautics and Aeronautics. (1999) *Space Systems—Metallic Pressure Vessels, Pressurized Structures, and Pressure Components*. ANSI/AIAA Standard S-080-1998. Reston, VA: American Institute of Astronautics and Aeronautics.

Boeing Corporation. (1976) *Implementation Report of the Manned Spacecraft Criteria and Standards JSCM 8080 for the Space Shuttle Orbiter*. Document SD73-SH-0297. The Boeing Corporation.

Bryan, F. L. (1982) *Diseases Transmitted by Food*. Atlanta, GA: U.S. Department of Health Services, Centers for Disease Control.

Castro, V. A., A. N. Thrasher, M. Healy, C. M. Ott, and D. L. Pierson. (2004) Microbial diversity aboard spacecraft: Evaluation of the International Space Station. *Microbial Ecology* 47: 119–126.

Coleman, M. E. (2005) *Toxicological Assessment of Internal-Active-Thermal-Control-System Coolant Biocides*. Memorandum #632. Houston, TX: National Aeronautics and Space Administration, Johnson Space Center.

European Cooperation on Space Standardization. (1999) *Space Engineering—Fracture Control*. Standard ECSS-E-30-01A. Noordwijk, the Netherlands: European Space Agency, European Space Research and Technology Center, European Cooperation on Space Standardization.

Graf, J., J. Perry, J. Wright, and J. Bahr. (2000) Process upsets involving trace contaminant control systems. Technical Paper 2000-01-2429. International Conference on Environmental Systems. Toulouse, France: Society of Automotive Engineers.

James, J. T. (2003) Toxicological assessment of the International Space Station atmosphere with emphasis on Metox canister regeneration. Technical Paper 2003-01-2647. International Conference on Environmental Systems. Vancouver, BC, Canada: Society of Automotive Engineers.

James, J. T. (2006) Airborne dust in space vehicles and habitats. Technical Paper 2006-01-2152. International Conference on Environmental Systems. Norfolk, VA: Society of Automotive Engineers.

James, J. T., and D. E. Gardner. (1996) Exposure limits for airborne contaminants in spacecraft atmospheres. *Applied Occupational and Environmental Hygiene* 11: 1424–1432.

Koenig, D. W., D. M. Bell-Robinson, S. M. Johnson, S. K. Mishra, R. L. Sauer, and D. L. Pierson. (1995) Microbial analysis of water in space. Technical Paper 951683. The 25th International Conference on Environmental Systems. San Diego, CA: Society of Automotive Engineers.

Lam, C-W. (1996) Bromotrifluoromethane. In *Spacecraft Maximum Allowable Concentrations for Selected Airborne Contaminants*. Washington, DC: National Academy Press.

Lam, C-W. (2005) *Toxicological Assessment of STA-54 Type 1, Part A & B Materials*. Memorandum JSC Tox-Lam-050105-CIPA to R. Dasgupta/JSC/ES. Houston, TX: National Aeronautics and Space Administration, Johnson Space Center.

Limero, T., J. Trowbridge, S. Taraszewski, J. Foster, and J. James. (1998) Results of the risk mitigation experiment for the volatile organics analyzer. Technical Paper 981745. International Conference on Environmental Systems. Danvers, MA: Society of Automotive Engineers.

Mead, P. S., L. Slutsker, V. Dietz, L. F. McCaig, J. S. Bessee, C. Shapiro, P. M. Griffin, and R. V. Tauxe. (1999) Food related illness and death in the United States. *Emerging Infectious Diseases* 5:607–625.

Mudgett, P. D., N. J. Packham, and D. L. Jan. (2005) An environmental sensor technology selection process for exploration. Technical Paper 2005-01-2872. International Conference on Environmental Systems. Rome, Italy: Society of Automotive Engineers.

National Aeronautics and Space Administration. (1995) *Safety Requirements Document: International Space Station Program*. Space Station Program Document SSP-50021. Houston, TX: National Aeronautics and Space Administration, Johnson Space Center.

National Aeronautics and Space Administration. (1996) *Fracture Control Requirements for Payloads Using the Space Shuttle*. Technical Standard NASA-STD-5003. Houston, TX: National Aeronautics and Space Administration, Headquarters.

National Aeronautics and Space Administration. (1997) *Glass, Window and Ceramic Structural Design and Verification Requirements: International Space Station*. Space Station Program Document SSP-30560. Houston, TX: National Aeronautics and Space Administration, Headquarters.

National Aeronautics and Space Administration. (1999) *International Space Station Flight Crew Integration Standards*. NASA-STD-3000/T, Space Station Program Document SSP-50005. Houston, TX: National Aeronautics and Space Administration, Headquarters.

National Aeronautics and Space Administration. (2000a) *Structural Design and Verification Requirements: International Space Station*. Space Station Program Document SSP-30559. Houston, TX: National Aeronautics and Space Administration, Headquarters.

National Aeronautics and Space Administration. (2000b) *Payload Verification Requirements: Space Shuttle Program*. Space Shuttle Document NSTS-14046. Houston, TX: National Aeronautics and Space Administration, Johnson Space Center.

National Aeronautics and Space Administration. (2002) *Payload Flight Equipment Requirements and Guidelines for Safety-Critical Structures*. Space Station Program Document SSP 52005C. Houston, TX: National Aeronautics and Space Administration, Johnson Space Center.

National Aeronautics and Space Administration. (2004) *Basic Provisions on Crew Actions in the Event of a Toxic Release on the International Space Station*. Space Station Program Document SSP 50653-01. Houston, TX: National Aeronautics and Space Administration, Johnson Space Center.

National Aeronautics and Space Administration. (2005a) *Fracture Control Implementation Handbook for Payloads, Experiments, and Similar Hardware*. Handbook NASA-HDBK-5010. Houston, TX: National Aeronautics and Space Administration, Johnson Space Center.

National Aeronautics and Space Administration. (2005b) *EVA Design Requirements and Considerations*. Document JSC-28918. Houston, TX: National Aeronautics and Space Administration, Johnson Space Center.

National Aeronautics and Space Administration. (2006) *Safety Policy and Requirements for Payloads Using the International Space Station*. Space Station Program Document NSTS-1700.7B. Houston, TX: National Aeronautics and Space Administration, Johnson Space Center.

National Research Council. 1992. *Guidelines for Developing Spacecraft Maximum Allowable Concentrations for Station Contaminants*. National Research Council, Committee on Toxicology. Washington, DC: National Academy Press.

Nickerson, C. A., C. M. Ott, S. J. Mister, B. J. Morrow, L. Burns-Keliher, and D. L. Pierson. (2000) Microgravity as a novel environmental signal affecting *Salmonella enterica* serovar *typhimurium* virulence. *Infection and Immunity* 68, no. 6: 3147-3152.

Ott, C. M., R. J. Bruce, and D. L. Pierson. (2004) Microbial characterization of free floating condensate aboard the *Mir* space station. *Microbial Ecology* 47: 133-136.

Pierson, D. L. (2000) Microbial contamination of spacecraft. American Society for Gravitational and Space Biology. *Montreal Bulletin* 14, no. 2: 1-5.

Pierson, D. L., M. R. McGinnis, and A. N. Viktorov. (1994) Microbiological contamination. In: F. M. Sulzman and A. M. Genin (eds.), *Foundations of Space Biology and Medicine*, vol. 2, *Life Support and Habitability*. Washington, DC: American Institute of Aeronautics and Astronautics, pp. 77-93.

Pierson, D. L., M. Chidambarum, J. D. Heath, L. Mallary, S. K. Mishra, B. Sharma, and G. M. Weinstock. (1996) Epidemiology of *Staphylococcus aureus* in the Space Shuttle. *FEMS Immunology and Medical Microbiology* 16: 273-281.

Samsonov, N. M., L. S. Bobe, L. I. Gavrilov, V. P. Korolev, V. M. Novikov, N. S. Farafonov, V. A. Soloukhin, S. J. Romanov, P. O. Andrechuk, N. N. Protasov, A. M. Rjabkin, A. A. Telegin, J. E. Sinjak, and V. M. Skuratov. (2002) Technical Paper 2002-01-2358. The 32nd International Conference on Environmental Systems. San Antonio, TX: Society of Automotive Engineers.

Schultz, J. R., D. K. Plumlee, and P. D. Mudgett. (2006) Chemical characterization of US Laboratory humidity condensate. Technical Paper 2006-01-2016. International Conference on Environmental Systems. Norfolk, VA: Society of Automotive Engineers.

Shaefer, D. A., and M. R. Fiske. (2004) Microgravity furnaces and sample containment—a reflective perspective. *Proceedings of the 42nd AIAA Aerospace Sciences Meeting and Exhibit*. Reno, NV: American Institute of Aeronautics and Astronautics.

Shibuya, A., M. Yasunami, and A. Yoshida. (1989) Genotypes of alcohol dehydrogenase and aldehyde dehydrogenase loci in Japanese alcohol flushers and non-flushers. *Human Genetics* 82: 14-16.

Sonnenfeld, G., G. R. Taylor, and K. S. Kinney. (2001) Acute and chronic effects of space flight on immune functions. In R. Ader, D. L. Felten, and N. Cohen (eds.), *Psychoneuroimmunology*, vol. 2. San Diego, CA: Academic Press, pp. 279-289.

Stewart, R. D., J. E. Peterson, R. T. Baretta, R. T. Bachand, M. J. Hosko, and A. A. Herrmann. (1970) Experimental human exposures to carbon monoxide. *Archives of Environmental Health* 21: 154-162.

Sulzman, F. M., D. L. Pierson, S. K. Mehta, and R. P. Stowe. (2006) Effects of space flight-associated stress and environmental factors on reactivation of latent *Herpes* viruses. In *Psychoneuroimmunology*, vol. 2. San Diego, CA: Academic Press, pp. 851-868.

Failure Tolerance Design

Gregg John Baumer
Chairman, International Space Station Safety Review Panel (Retired),
Johnson Space Center, National Aeronautics and Space Administration,
Houston, Texas

CONTENTS

19.1. Safe 653
19.2. Hazard 655
19.3. Hazardous Functions 658
19.4. Design for Minimum Risk 659
19.5. Conclusions 660

19.1 SAFE

Safe is a general term, denoting an acceptable level of risk, relative freedom from and low probability of personal injury, fatality, damage to property, or loss of critical equipment function (NASA 1989). It should be the goal of all space hardware designers to develop hardware that is safe. When there is a breakdown of safety in design, accidents occur.

An accident is an unplanned event or series of events that result in death or major injury to personnel or damage to the launch vehicle, experiments, or spacecraft (DoD 1979). A major injury is a severe injury to personnel that might require admission to a hospital, such as bone fracture, second or third degree burns, severe lacerations, internal injury, severe radiation exposure, exposure to toxic agents, or unconsciousness (DoD 1979). Damage includes breakage or destruction of hardware that could cause loss of critical functions or require repair or replacement. Acceptance of credible risk is based on the magnitude of risk compared with the impact of compensating for it.

19.1.1 Order of Precedence

Risks are not totally controllable by design action because of cost, performance, or schedule. They are to be dealt with at the highest feasible order of precedence. All corrective action taken in response to the risks of the design must follow the order of precedence specified in MIL-STD-1574A (DoD 1979).

Design for Minimum Hazard

Assurance of the inherent safety of a system through the use of appropriate design features and qualified components includes design features to permit detection of impending hazardous conditions in sufficient time to complete automatic or manual control actions (DoD 1979). This is the category of corrective action most used by space hardware designers and is required by most safety approval boards. It includes failure tolerant design and design for minimum risk. The use of appropriate design features to assure inherent safety and control of hazards should be a major goal of engineers throughout the development phase of their hardware.

Safety Devices

Hazards that cannot be eliminated through design selection must be reduced and made controllable through the use of appropriate safety devices, such as mechanical internal barriers or inhibiting mechanisms as part of the system, subsystem, or equipment (DoD 1979). An example of such a safety device would be the safe and arm device used in pyrotechnic systems, which interrupts the pyrotechnic chain between the initiator and the explosive charge. Deployable spacecraft using the *Space Shuttle* to reach low Earth orbit are required by NASA to utilize a safe and arm device in the pyrotechnic chain used to ignite the solid propellant rocket motors.

Protective Systems

For designs in which accident risk cannot be eliminated totally, appropriate protective systems, such as fire suppression, radiation shielding, or explosion shielding, should be used (DoD 1979). As an example, spacecraft developers utilize the general practice of conducting extensive materials testing to identify flammable materials, and they implement stringent materials selection criteria to minimize flame propagation paths on crewed spacecraft. However, it is standard policy as well to install fire detection and suppression systems to further protect the flight crew. Another example is the system for tracking the use of toxic materials on crewed spacecraft and the development of cleanup procedures for use in the event of a toxic spill. These processes are in addition to the appropriate materials selection criteria used to avoid materials with high offgassing rates and the design features employed to assure that these materials properly are contained if they are used.

Warning Devices

When it is not practical to preclude the existence or occurrence of known hazards or to use automatic safety devices to control a hazard, devices must be employed for the timely detection of a hazardous condition and the generation of an adequate warning signal. A detection and warning system should be coupled with the use of emergency controls to allow for operating personnel to take appropriate corrective action, in which the affected subsystem is brought to a safe state or shut down altogether. Such devices and their application must be designed to minimize the probability of incorrect warning signals or improper reaction by operating personnel to the signal (NASA 1989). These controls commonly are used in ground processing facilities for spacecraft where hazardous operations, such as the loading of propellant or the use of toxic cleaning fluids, routinely are conducted. On the *International Space Station*, warning devices are used as well to alert the crew to the hazards of depressurization, fire, or any spillage or release of toxic chemicals. Procedural controls have been developed for the emergency evacuation of a crew to some safe area and to isolate the affected modules. If isolation of the hazard is not possible, the crew has available the capability to evacuate the *International Space Station* using the Russian *Progress* as a Crew Return Vehicle.

Special Procedures

In situations where it is not possible to reduce the magnitude of an existing or potential hazard through either design or the use of safety and warning devices, special procedures must be developed to minimize any associated hazardous conditions to enhance personnel safety.

19.2 HAZARD

A hazard is the presence of a potential risk situation caused by an unsafe act or condition. It is a condition or changing set of circumstances that presents a potential for adverse or harmful consequences, or it can be an inherent characteristic of an activity, condition, or circumstance that can produce adverse or harmful consequences (NASA 1989). A common misconception is that, if the design of a system, subsystem, or hardware item includes controls to prevent the occurrence of hazardous consequences, then there are no hazards. A primary task of the safety engineer is to identify all hazards and associated hazard controls in a design. The safety engineer then has the responsibility to verify that those hazard controls actually are incorporated into the subsystem, system, or hardware and they are effective in preventing the hazard consequence. The analytical process used is referred to as *hazard analysis*.

The potential consequence of a hazard determines its severity ranking. This is an important concept of safety, because it allows the designer to identify those hazards having the greatest consequences and that present the highest risk to the program. This knowledge then is used to assist management in making design trades while having a sound understanding of the safety impact. Table 19.1 contains the hazard severity categories, catastrophic, critical, and marginal, that the partners of the *International Space Station* use to rate hazards.

19.2.1 Hazard Controls

When designers are investigating hazard controls to incorporate into their designs, they must consider the acceptable level of risk tolerance of the program for which the hardware is being built. The level of acceptable risk for a non-crewed spacecraft is considerably different from that of a crewed spacecraft. That which is safe for one program likely can be unacceptable in another. This should not be a hindrance to the hardware designer because one of the first development steps taken by any program is to conduct

Table 19.1 Hazard Severity Category (NASA 1995) (Courtesy of NASA)

Description, Category	Mishap Definition
Catastrophic, I	Any condition that can cause a disabling or fatal personnel injury or loss of one of the following: *Space Shuttle, International Space Station*, or major ground facility. Loss of the *International Space Station* is to be limited to those conditions resulting from failure or damage to elements in the critical path of the *International Space Station* that render it unusable for further operations, even with contingency repair or replacement of hardware, or render it in a condition that prevents further rendezvous and docking operations with other launch elements.
Critical, II	Any condition that can cause a non-disabling personnel injury; severe occupational illness; loss of an *International Space Station* element, on-orbit life sustaining function, or emergency system; or involves damage to the *Space Shuttle* or a major ground facility. For safety failure tolerance considerations, critical hazards include loss of International Space Station elements that are not in the critical path for its survival or damage to an element in the critical path that can be restored through contingency repair.
Marginal, III	Any condition that can cause major damage to an emergency system or an element in a non-critical path or minor personnel injury or occupational illness.

a program requirements review to establish the requirements to which the hardware must be built. For crewed space programs, safety requirements generally focus on two concepts:

- Design to tolerate failures.
- Design for minimum risk.

19.2.2 Design to Tolerate Failures

Failure tolerance is the primary safety process that designers should use to control identified hazards. Many safety specialists and safety requirements documents use the terms *failure* and *fault* as if they are interchangeable. In reliability circles, there is a distinct difference. A failure is the inability of a system, subsystem, component, or part to perform its required function under specified conditions for a specified duration. The focus is clearly on the hardware. Fault, on the other hand, is an undesired system state and, therefore, is nonspecific about the cause. In this chapter, *fault tolerance* and *failure tolerance* are considered synonymous. The concept of fault or failure tolerance means that the design must endure a minimum number of credible failures or operator errors without resulting in a hazardous consequence. Credible failures are those that can occur or are reasonably likely to occur. The term *operator error* is not meant to include all possible crew actions or deliberate malicious acts but rather addresses mistakes in procedure, such as an out of sequence step, an inadvertent single switch throw, or other action that has not been addressed specifically during proficiency training. The severity of the hazardous consequence determines the minimum number of failures or operator errors that the designer must protect against. The failure tolerance design concept applies to the control of hazardous functions. A hazardous function is an operational event or service, such as an appendage deployment, firing of a liquid propellant motor, control of a habitable atmosphere, or control of the thermal environment, whose inadvertent operation or loss can result in a hazardous consequence. The concept includes operations that are single-failure tolerant, two-failure tolerant, fail operational/fail operational (fail op/fail op), fail operational/fail safe (fail op/fail safe), fail safe/fail safe, and fail safe.

Single-Failure Tolerant

For a design to be single-failure tolerant, a minimum of two failures are required for the hazardous consequence to occur. This concept typically is applied to hazards with a severity category of critical (Table 19.1).

Two-Failure Tolerant

A two-failure tolerant design requires a minimum of three failures for the hazardous consequence to occur. This concept typically is applied to hazards with a severity category of catastrophic (Table 19.1).

The Fail Op/Fail Op Concept

The fail op/fail op concept is one in which the system or function to which it is being applied maintains functionality after the first and the second failure. This concept is applied to systems that are especially critical. Here, the goal is to maintain safety and achieve mission success with the greatest degree of operational flexibility. In building a spacecraft computer control system, an implementation of this concept is a system of four or five voting computers to control spacecraft functions, where as least two computers must be operational and online to provide safe control of the spacecraft.

The Fail Op/Fail Safe Concept

For designs utilizing a fail op/fail safe concept, the system or function to which it is being applied maintains functionality after the first failure but not after the second. However, the system remains safe even after two failures have occurred. An implementation of this concept might be a spacecraft with a fully redundant design to maintain performance and mission objectives and that possesses a separate and independent watchdog system to monitor primary and secondary systems and take control when they fail. This watchdog system reconfigures the vehicle to a survival mode in the event both primary and secondary systems fail. This watchdog system action also prevents the failure conditions from propagating, possibly affording ground controllers an opportunity to take corrective action.

The Fail Safe/Fail Safe Concept

Designs incorporating the fail safe/fail safe concept do not maintain functionality after any failure but ensure that no hazardous consequence will result after even two failures.

Designers should apply this concept to hardware where the impact of mission failure does not warrant the cost of adding redundancy to the design. A typical application of this concept would be for low priority payloads where the cost of developing a highly reliable redundant design is too high, considering the payload priority. Another interesting aspect of this concept is that it might be adopted by the host vehicle as the minimum requirement to be certified for potential payloads. In this case, however, the payload is free to develop and incorporate as much reliability and redundancy as is affordable without incurring the cost of certifying that reliability and redundancy to the host vehicle.

19.3 HAZARDOUS FUNCTIONS

All functions deemed hazardous are operational events or services whose inadvertent operation or failure to operate results in a hazardous consequence. The deployment of an antenna on a satellite is a planned event. However it would be considered a hazardous function if early deployment could cause the satellite to become entangled with its launch vehicle, causing loss of the satellite or perhaps damage to the launch vehicle. Similarly, the firing of a pyrotechnic separation system that severs the attachment bolts between a satellite and its launch vehicle is a planned event but would be considered a hazardous function if failure to separate would cause loss of the satellite. Designers must be aware of hazardous functions so that appropriate failure tolerance can be applied to the function of concern.

19.3.1 Must Not Work Hazardous Function

A must not work hazardous function is one in which its inadvertent operation results in a hazardous consequence. To meet safety failure tolerance requirements, the designer must implement serial design features to prevent the inadvertent operation of the function. These serial design features, referred to as *inhibits*, provide a physical interruption between an energy source and a particular function. An electrical inhibit can be a transistor or relay between a power source, such as a battery or power distribution bus, and an electrically operated function, such as a motor or pyrotechnic device used to deploy an antenna. An example of a mechanical inhibit would be an isolation valve in a propellant line located between a propellant tank and a satellite thruster. It is common aerospace practice to require that a function whose inadvertent operation can result in a catastrophic consequence must be controlled by a minimum of three independent inhibits whenever the hazard potential exists. For critical hazardous effects, the safety requirement is for two independent inhibits to the function. Two or more inhibits are considered independent if a single event, environment, or credible failure cannot defeat more than one of them. It is important to recognize that inhibits are in the path between the energy source and the function. Numerous devices in the circuits can cause a change of state for an inhibit. However, these devices are referred to as *controls*, and they do not count with respect to meeting the inhibit or failure tolerance requirements for hazardous functions.

19.3.2 Must Work Hazardous Function

A hazardous function designated as must work is one for which failure to operate would result in a hazardous consequence. To meet safety failure tolerance requirements, a designer must implement parallel design features to assure operation of the function. Where loss of a function would result in a catastrophic hazard, the design must be two-failure tolerant to assure operation of the function. For critical hazardous effects, the design must be single-failure tolerant. This overall failure tolerance can be met by one, two, or three independent methods, depending on the failure tolerance of each method to assure functionality.

19.4 DESIGN FOR MINIMUM RISK

Instead of designing to tolerate failures, another option available to the designer is to use proven hardware design concepts where the safety related properties and characteristics of the design have been accepted by the program as sufficient to control hazards and reduce any risk to an acceptable level. For this design approach, the safety criteria are built into the specifications in the form of minimum qualification and acceptance requirements that address environments, testing at the component and system levels, mandatory inspection points, material and parts selection and traceability, redundancy, reliability, and the like. The selection of this approach is not simply an alternative to failure tolerance because of cost, schedule, or noncompliance with failure tolerance but rather a deliberate decision to select a proven design concept that has been demonstrated through experience, the characteristic of eliminating or reducing credible failures.

Structures is one of those systems where it is standard practice to use a design for minimum risk approach. Using material properties for yield and ultimate strength, margins of safety, factors of safety, stress-corrosion considerations, fracture control criteria, and a comprehensive structural verification plan, designers are able to build complex aerospace hardware without risk of structural failure. Other examples are pressure vessels, pressurized line and fittings, pyrotechnic devices, mechanisms in critical applications, materials compatibility, and flammability. For each of these systems, established aerospace specifications define the criteria to control hazards in the system. Hazard controls related to these areas of design are extremely critical and warrant careful attention to the details of verification of compliance. The verifications include quality control, procedures, independent analyses, inspection, demonstration, test, or combinations of these activities. Minimum supporting data requirements for these areas of design typically are required by a procuring organization to assure and verify compliance to applicable specifications.

Use of the design for minimum risk approach, however, does not provide complete relief from the use of failure tolerance. Most design for minimum risk specifications contain some form of failure tolerance in the details of the specification. Using fail safe structural load paths is an alternative to implementing fracture control. Design specifications for safety critical mechanisms typically require redundant sliding surfaces to control binding and jamming. In addition, designers still must apply failure tolerance criteria to these designs as necessary to assure for systems that interface with the design and exhibit credible failures do not invalidate the safety properties of the design. For example, a pressure vessel can be certified safe based on its inherent properties to withstand pressure loading that have been verified by analysis and qualification and acceptance testing. Failure tolerance still must be imposed on external systems that can affect the pressure vessel, such as a tank heater, to assure that failures of the heater do not cause the pressure to exceed the maximum design pressure of the pressure vessel.

19.5 CONCLUSIONS

When exploring for design solutions to control hazards, it is important to follow the safety design order of precedence to minimize risk. Failure tolerance design is a key element of the design order of precedence in the control of critical and catastrophic hazards. When considering the failure tolerance to be implemented to control hazardous effects, one must also consider the reliability or mission success effects of failure. The greater the failure tolerance, the more capability exists to continue operations in a safe manner after experiencing failures. In lieu of failure tolerance, the designer must also consider using design for minimum risk solutions for hazard control. This trade is not just a switch in hazard control approaches but a deliberate decision to utilize proven hazard control design solutions that meet the risk tolerance of the program.

REFERENCES

Department of Defense. (1979) *System Safety Program for Space and Missile Systems*. MIL-STD-1574A. Washington, DC: Department of Defense.

National Aeronautics and Space Administration. (1989) *Safety Policy and Requirements for Payloads Using the Space Transportation System*. Space Shuttle Document NSTS 1700.7B. Houston, TX: National Aeronautics and Space Administration, Johnson Space Center.

National Aeronautics and Space Administration. (1995) *Safety Review Process: International Space Station Alpha Program*. Space Station Document SSP 30599A. Houston, TX: National Aeronautics and Space Administration, Johnson Space Center.

Propellant Systems Safety

CHAPTER 20

William D. Manha
Senior Specialist, Propulsion Pressure Systems, Jacobs Engineering, Houston, Texas

CONTENTS

20.1.	Solid Propellant Propulsion Systems Safety	662
	By H. F. R. Schöyer	
20.2.	Liquid Propellant Propulsion Systems Safety	673
	By William D. Manha	
20.3.	Hypergolic Propellants	683
20.4.	Propellant Fire	686
	By David L. Baker and Miguel J. Maes	

Success in space demands perfection. Many of the brilliant achievements made in this vast, austere environment seem almost miraculous. Behind each apparent miracle, however stands the flawless performance of numerous highly complex systems. All are important. The failure of only one portion of a launch vehicle of spacecraft can cause failure of an entire mission. But the first to feel this awesome imperative for perfection are the propulsion systems, especially the engines. Unless they operate flawlessly first, none of the other systems will get a chance to perform in space.

Perfection begins in the design of space hardware. This book emphasizes quality and reliability in the design of propulsion and engine systems. It draws deeply from the vast know how and experience which have been the essence of several well designed, reliable systems of the past and present. And, with a thoroughness and completeness not previously available, it tells how the present high state of reliability gained through years of research and testing can be maintained, and perhaps improved, in engines of the future.

As man ventures deeper into space to explore the planets, the search for perfection in the design of propulsion systems will continue. This book will aid materially in achieving this goal.

—*Wernher von Braun (Huzel and Huang 1992)*

20.1 SOLID PROPELLANT PROPULSION SYSTEMS SAFETY

In the Western world, solid rockets have been used since the Middle Ages, when Jeanne d'Arc defended Orléans against the English. Although these were very primitive gunpowder based rockets, nowadays the experience with solid propellants is large and based on a heritage of more than seven centuries. In military rocket applications, solid propellants are used predominantly, adding to the overall experience. It is not surprising, therefore, that the demonstrated reliability record of solid propellant space systems is at least 1% better than for liquid propellant systems, that is, ~99% versus ~98%. This fact notwithstanding, every user should be aware that the difference between solid propellants and explosives is marginal. Indeed, many solid propellants make excellent explosives. For these devices, a large amount of energy is stored in a small volume. Unprofessional or careless handling of solid propellants can cause serious injuries and damage, if not worse.

The design of solid propellant propulsion systems must take this into account if inherently safe and reliable systems are to be produced. The designer continuously must be aware that solid propellant propulsion systems are composed of potentially dangerous materials that have a high energy density. By taking these aspects into account, paying attention to details, and being aware of all potential dangers, perfectly safe and extremely reliable systems can be designed and manufactured. As well, the facts that solid propellant propulsion systems are low in cost and can be stored for long periods before use makes them very attractive for many applications, including space.

20.1.1 Solid Propellants

There are two types of solid propellants:

- Homogeneous or double base propellants, where fuel and oxidizer are combined at molecular level.
- Composite propellants that are a mixture of oxidizer and fuel.

Double base propellants were first made by Unge in Sweden in 1896 (von Braun and Ordway 1976). Composite propellants were invented by Malina and von Kàrmàn during World War II (von Kàrmàn and Edson 1967).

Double Base Propellants

Double base propellants are a mixture of glycerol trinitrate and cellulose nitrate. Because these propellants are relatively low in performance and generally are more dangerous to transport and store than composite propellants (see ammonium perchlorate based composite propellants), they seldom are used in space applications. They do find application in some older sounding rockets. Usually, double base propellants are classified Class 1.1 (UN 2005). An advantage is that double base propellants are not hygroscopic.

Ammonium Perchlorate Based Composite Propellants

Composite propellants based on ammonium perchlorate consist of 65 to 89% ammonium perchlorate, a binder composed usually of 11 to 18% hydroxyl-terminated polybutadiene, up to 20% aluminum powder, and various additives, such as a curing agent, a burn rate modifier, and plasticizers. Because this type of propellant is hygroscopic, it should not be exposed to more than 60% relative humidity. If the humidity is too high, the performance and ignition characteristics can be affected, as can its structural integrity, which could lead to an operational failure, thereby causing a hazard. Therefore, ammonium perchlorate based composite propellant motors usually are hermetically sealed and filled with dry nitrogen at a slight overpressure (0.3 MPa).

Other Composite Propellants

Instead of ammonium perchlorate, other materials (ammonium nitrate, ammonium dinitramide, or hydrazinium nitroformate) can be used as the oxidizer in solid composite propellant systems. Although these presently do not find use in space applications, each has specific characteristics that must be considered for its application, storage, handling, and transportation requirements.

Applications

Today, ammonium perchlorate based composite propellants represent the majority of solid propellants used in the Western world for space applications. They find such widespread and common use because of their excellent mechanical properties and the flexible and rubbery nature of their binder, hydroxyl-terminated polybutadiene in most cases, which allows the propellant to be cast directly into the motor case, that is, a case bonded design. This process leads to a light strong motor design in which the propellant itself provides good thermal protection of the chamber walls.

In contrast, double base propellants characteristically are rather rigid and do not allow for the expansion and contraction afforded by composite propellants. Therefore, the double base propellant grain has to be cast in advance and mounted into the motor later. A different, usually heavier motor construction is required, as is a more extensive internal thermal protection system.

Double base propellants usually are classified as Class 1.1 in the *United Nations Recommendations on the Transport of Dangerous Goods* (UN 2005) because they present a mass explosion hazard. However, ammonium perchlorate based composite propellants usually are classified as Class 1.3, meaning that they present a fire hazard and either a minor blast hazard or a minor projection hazard or both. They do not present a mass explosion hazard.

Except for those based on ammonium nitrate, the other solid propellants, to which there is a reference in Section 20.1.1, subsection "Other Composite Propellants," are still under development. They offer substantially higher performance than the classical composite propellants. However, a safety classification for them is yet to be known because no operational systems have used them to date.

It must be emphasized that the safety classification assigned depends on the propellant-container combination. For instance, rocket motors for military systems have been produced using double base propellants so that, in a fire, the case will lose strength rapidly and a mass explosion, therefore, cannot occur. Such systems using double base propellants are classified as Class 1.3.

20.1.2 Solid Propellant Systems for Space Applications

Applications

For space applications, solid propellant systems are used mainly in the following areas:

- Solid boosters for launchers, such as for the *Space Shuttle* or *Ariane V*,
- Solid apogee boost motors, such as the European MAGE series and the American STAR series,
- Sounding rockets, such as the *Skylark*, *Orion*, or *VSB-30*,
- Separation motors as used on the *Ariane V* to separate the solid boosters from the launcher after burnout,
- Deorbiting and reorbiting motors to ensure the return of a spacecraft into the atmosphere or the transfer of a spacecraft into a graveyard orbit,
- Landing motors for landing on Earth or some heavenly object.

Except for the last two applications, the time between manufacturing and the use of a solid propulsion system is short, less than 3 to 4 a, and often less than 12 mo. This is in contrast with military applications, where lifetimes can be as long as 25 a.

Design

The solid propellant systems used for space applications strongly rely on a heritage from military applications. There are no particular design requirements, except that the designer must be proficient in the field of solid rocket propulsion, propellants, internal ballistics, fluid dynamics, heat transfer, and mechanical engineering; and the designer must carefully pay attention to details. The design must account for all safety aspects as mentioned in Section 20.1.3, "Safety Hazards"; for example, the design must be such that electrical continuity can be assured and protection against lightning can be implemented. The main requirement for arriving at a safe design is a sound design analysis, including failure mode, effects, and criticality analysis.

It is important that proper estimates are made for the probabilities of the occurrence of failures, as well as for their effects. Experienced manufacturers have a large database on which these estimates can be established. Those manufacturers that are rather new in the field or have limited experience must invest sufficient time and effort in literature research to obtain this statistical information. Taking this information into account makes for a safe and reliable design. All this notwithstanding, most failures and accidents that occurred in practice were never identified in the failure mode, effects, and criticality analysis during the development period. A flawed design, in combination with use outside the intended operational envelope, such as temperature too low, and bad management, can lead to dramatic and catastrophic failures as illustrated by the Challenger incident.

20.1.3 Safety Hazards

In this section, the major safety hazards for solid propulsion systems are discussed. To take the proper safety measures during storage, handling, and transportation, it is important to be aware of these hazards.

The hazards of electromagnetic and electrostatic effects on solid propellant rockets, propellants, and explosives are illustrated by the following list from *Peace Magazine* (Babst and Axelrod 1990):

- On March 26, 1987, in Florida, a $160 million Atlas Centaur rocket blew up. Investigators said electromagnetic interference caused by lightning likely was responsible.
- A Soviet Union ammunition depot experienced a massive explosion during 1984, about 900 mi north of Moscow. This explosion was the largest of 12 very large detonations in the western USSR over a 7-mo period. The suspected cause was computer controlled experiments using over the horizon radar that likely concentrated energy in areas near the depots.
- On June 10, 1987, at Wallops Island, Virginia, three small NASA rockets were ignited by lightning, causing them to be launched.
- On January 11, 1985, a Pershing II missile being assembled unexpectedly exploded. Investigators attributed it to static electricity.
- On December 29, 1987, the solid fuel of an MX missile exploded, killing all workers in a Utah assembly building. The Air Force suspected static electricity.
- The April 19, 1989, accidental explosion on the battleship, *USS Iowa* that killed 47 sailors might have been caused by an electrostatic discharge. The HERO project made an investigation for the U.S. Senate and found that the gunpowder was stored in silk bags. The more silk is moved, the greater the buildup of static electricity. The gunpowder was packed in 1945, and its stabilizer could have been weak.

The last incident is not related to rockets but illustrates that electrostatic discharge can have similar effects on solid rockets. This was illustrated in 1964, when the third stage of a Delta rocket motor suddenly ignited and filled the workroom with searing hot gases. Eleven engineers and technicians were burned, three of them fatally. A spark of static electricity had probably set off the fuse that ignited the solid propellant (Newell 1980).

Electrostatic Discharge

Sparks resulting from electrostatic discharge can ignite solid rocket motors. Therefore, the occurrence of electrostatic discharge in the neighborhood of solid propulsion systems must be prevented. This is accomplished by ensuring that all parts of the motor are connected electrically, that the maximum resistance between any two parts does not exceed 1 Ω, preventing the creation of a critical potential difference. At all times, the motor must be grounded; when handling the motor, the personnel must be grounded as well. Handling the motor should take place only in an environment with relative humidity exceeding 30%.

Radio Frequency and Electromagnetic Fields

By themselves, radio frequency and electric fields constitute no safety hazard for solid motors. However, if an electric initiator is used, it can be activated by radio frequency energy or an electromagnetic field. Therefore, it must be ensured that once an initiator is present in the motor, electric wires associated with the device must be twisted and short circuited or properly shielded to prevent inadvertent firing. These devices also must be grounded or bonded during assembly if the motor is to fly on a spacecraft or launcher. If the device is not on the ground, all systems that otherwise would be grounded are connected electrically to the common platform, that is, the launcher or spacecraft, so that no potential differences can arise that could result in detonation.

At all times, wires leading to initiators must be prevented from acting as an antenna, and there can be no loops in which a rapidly changing electromagnetic field could introduce a difference in potential. Either situation could result in an inadvertent firing of an initiator.

Lightning

Solid propellant motors always must be stored in buildings that are protected thoroughly from lightning. On launch sites, multiple lightning conductors are placed in the near vicinity of the launcher. These precautions ensure that the probability of a direct hit on a solid propellant propulsion system by lightning is extremely small. During thunderstorms, launches are not permitted because wind shear can induce excessive loads on the launcher. Radio communications and visual tracking are jeopardized as well.

The probability that a solid propellant propulsion system receives a direct hit by lightning during launch is extremely small. By ensuring proper bonding between all parts of the launcher and spacecraft, the effects of a lightning strike are minimized. Nevertheless, the induced effects of lightning in the vicinity of the solid propulsion system should not be underestimated or ignored.

During lightning, the magnetic flux density, B, changes with extreme rapidity, and according to one of Maxwell's laws (or Faraday's law of induction), these changes introduce a potential difference, E, in a closed loop circuit:

$$\frac{\partial B}{\partial t} = -\nabla \times E \tag{20.1}$$

or

$$\oint_C E \cdot dl = -\frac{d}{dt} \int_S B \cdot dA \tag{20.2}$$

where C is the (closed) contour of an enclosed area, S.

If one cannot prevent loops in the electrical circuits, adequate safeguards must be built in to prevent excessive currents from occurring in the electrical circuits. These measures are extremely important because initiators that have an all-fire current of 0.3 A sometimes are used.

The hazards from the combination of lightning and solid motors are illustrated by an accident in 1987, when three NASA sounding rockets at Wallops Island, Virginia, ignited and were launched because of lightning. Ironically these sounding rockets were intended for lighting research. It also is highly likely that the initiators of ESA *Hipparcos MAGE* were fired because of lightning in Kourou during summer 1989. Because the safe and arm device (see Section 20.1.4, subsection "Handling") was in the safe position, the *MAGE* did not ignite on the ground; however, it later could not be ignited in space.

Impact and Shock

The impact sensitivity (Meyer, Köhler, and Hamburg 2002), that is, the impact (measured in Nm) at which an explosive or propellant ignites, usually is known for explosives. The propellant or rocket motor manufacturer must be aware of the impact sensitivity of the propellant. It is highly unlikely that a rocket motor could ever experience an impact that would ignite the propellant. However, if a solid rocket motor is dropped, cracks or high internal deformation stresses can be introduced that can lead to deflagration to detonation transition after the motor is ignited intentionally. In addition, the shock can cause pieces of propellant to break off, causing nozzle blockage, increased burning surface area, and debonding between the propellant and liner—all of which can lead to a failure or explosion of the motor.

Therefore, if a rocket motor has experienced a shock that exceeds specified limits, it should be destroyed. It can be argued that, by x-raying the motor or using ultrasonic tests, a determination can be made of whether cracks or debonding has occurred or the propellant has disintegrated. However, extremely small cracks and debonds cannot be discerned by these methods, and any induced stresses cannot be determined at all by such tests.

To avoid using a motor that has been exposed to excessive shock, all solid propellant rocket motors must be monitored continuously for shocks in three directions. The automatic recording equipment fixed to the motor for this purpose allows inspection of the shock history.

Friction

The friction sensitivity (Meyer et al. 2002) for explosives usually is known, and the propellant or rocket motor manufacturer must be aware of the friction sensitivity of the propellant being used. Friction sensitivity is the friction, measured in Newtons, at which the explosive or propellant ignites. Under normal use and even when inspecting the motor with a boroscope, it is unlikely that friction sensitivity can lead to hazardous situations in rocket motors that are in good condition. For normal propellants used in solid rocket motors, friction sensitivity does not create a hazard at the motor level.

Deflagration to Detonation Transition

Under certain circumstances, the rate of deflagration, which is the steady burning of a propellant, can begin to increase continuously, approaching shock velocities. At some point in this process, a sudden transfer from deflagration to detonation occurs (Cooper and Kurowski 1996). Whether this happens depends on the general characteristics of the propellant, that is, the porosity, particle size, density, and pressure. In a well designed rocket, the operating pressure is far away from the level where the deflagration to detonation transition would occur. However, if the propellant possesses many cracks, say, because the motor was subjected to severe shock, deflagration to detonation transition can take place, usually with catastrophic results. This highlights the necessity of X-ray or ultrasonic inspection of the motor before use and having on hand a proper record of the shock history of the motor.

Ruptures, Cracks, and Debonds

In a well designed rocket motor, ruptures and cracks can occur only as the result of mechanical overload, that is, impact or shock, or by cooling the propellant below its glass transition temperature. At the glass point, the modulus of elasticity changes dramatically, and the propellant becomes extremely brittle. For hydroxyl-terminated polybutadiene based propellants the glass point is −63°C (Schöyer et al. 1995) or lower. Double base propellants, depending on their composition can have glass transition points as high as −45°C or even higher. A shock or excessive load at low temperatures can have a damaging effect on the integrity of the propellant grain. Debonds can occur due to contamination during manufacturing or overloads. In any case, a solid rocket motor must be inspected by X-ray or ultrasonic inspection techniques to ensure that it is free of debonds, cracks, and ruptures.

Debonds can lead to uncontrolled burning or local overheating, the latter resulting in a mechanical failure. Cracks and ruptures can have similar effects but, in addition, are likely to cause chunks of propellant to break off and temporarily block the nozzle. Such a blockage can be sufficient to raise the chamber pressure well beyond the design limit and cause all the corresponding dramatic results. Therefore, motors must be inspected thoroughly before use and, if transportation between inspection and use takes place, shocks and accelerations to which the motor is subjected must be recorded (Section 20.1.3, subsection "Impact and Shock").

Overheating and Undercooling

The danger of overheating is small for solid motors used in space applications because military systems require solid motors to operate between −40 and +70°C. Most solid motors in space applications inherited their technology from military applications, but the operating temperatures usually are limited to a range of 0 to 40°C. In practice, this leads to large margins that contribute to safety.

Overheating of a solid rocket motor can be caused by solar radiation while in space or if it is stored or transported incorrectly. Overheating in space can be avoided by proper thermal control of the spacecraft or launcher. Adherence to the rules governing the proper conditions prevents overheating during transport or storage. Overheating can result in a loss of strength or integrity because the strength of composite materials, that is, the case, propellant grain, liners, and bonding agents, deteriorates.

Long duration storage at very elevated temperatures can lead to a chemical change in the propellant composition from migration of the ingredients or slow reactions occurring among the constituents in the propellant. As solid propulsion systems used in space applications normally are not stored for long periods, the danger of overheating is small.

Undercooling occurs if the solid propulsion system is cooled below its operational range. It is not that the binder becomes brittle, because the glass point of hydroxyl-terminated polybutadiene is quite low (Section 20.1.3, subsection "Ruptures, Cracks, and Debonds"). If the temperature becomes too low, the propellant can shrink excessively causing debonding or even a crack between the liner and the propellant or the liner and the case. Cracks in the propellant itself can be created, especially in locations of high residual stress. If composite cases are used at excessively low temperatures, the connection of the composite material to the polar bosses can fail as well. Also O-rings and seals may become brittle, so that the sealing function is no longer maintained. The means for thermal control of a propulsion system in space must ensure that the temperatures to which it will be subjected remain within the operational envelope of the system.

It is good practice to record the temperature of solid propellant propulsion systems continuously during transport and storage, as is done with shocks and accelerations. This practice provides knowledge of any unacceptable temperature deviation and allows corrective measures to be taken should they occur. Observing the instructions of the manufacturer for handling and use ensures that no hazardous situation occurs.

Floater

The floater allows a case bonded grain to expand and contract (Figure 20.1). A floater that does not open on ignition of the rocket motor within which it is installed constitutes a safety hazard. Therefore, during manufacturing, measures must be taken to ensure easy parting of the relief flap and thermal protection. This is accomplished by putting a piece of paper or combustible foil between the floater and the thermal protection during the manufacturing process.

Combustion Instability

Combustion instability can be a major problem during the development of new solid propellant systems. Although the mechanisms of combustion instability are understood reasonably well (Price and Culick 1968; Harrje and Reardon 1972; Schöyer 1993), it is very difficult, even with the use of advanced computer codes, to be sure that there are no unstable modes in the operational domain of a newly conceived solid propellant motor, because the dynamic combustion characteristics of a solid propellant have to be measured at conditions in agreement with the operational conditions of the envisaged motor. The implication is that extensive and expensive laboratory measurements have

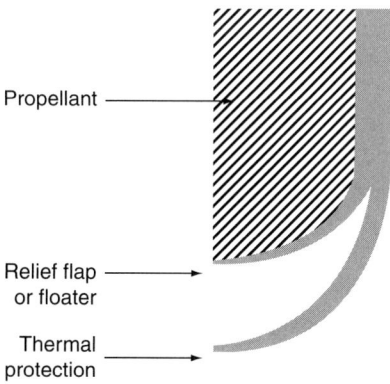

FIGURE 20.1 The floater.

to be made. In many cases, this is not done to save costs; however, if not done, problems can occur.

While the works by Schöyer et al. (1995) and Harrje and Reardon (1972) deal with combustion instability in liquid systems, much of this is applicable to solid propellant systems as well. The peculiarities of solid propellant combustion are dealt with by Price and Culick (1968).

It must be demonstrated that newly developed engines are perfectly stable, with sufficient margin in their operational envelope. However, the operational envelope can be extended only if it is a certainty that no combustion instability can occur.

Combustion instability has two main effects, the introduction of vibrations and oscillations into the system and a severely increased heat transfer. The effects of combustion instability can be disastrous but need not be. The motor can fail and disintegrate due to increased heat transfer and mechanical loads, but it also is possible that vibrations caused by combustion instability can damage the payload, instruments, or the guidance, navigation, and control systems. For a safe use of solid propellant systems, thorough development and qualification processes must be employed to ensure that no damaging combustion instability can occur within the operational domain.

Proof and Leak Test

During development, the motor case is burst tested. It is essential that every motor case is proof tested and leak tested as well. Proof testing allows measurement of deformation and the comparison of test results with requirements. The proof testing process thus enables the engineer to decide whether the design requirements of the case are met.

Leak testing is extremely important. For example, small leaks can allow the storage nitrogen (Section 20.1.1, subsection "Ammonium Perchlorate Based Composite Propellants") to escape and moisture to enter a tank. Large leaks constitute a safety risk, in that hot combustion gases (temperature ~3600 K) could escape and, within a short time, enlarge the leak. The result would be a fire in or damage to other parts of the launcher or spacecraft. A tragic example of the dramatic consequences a leak can have is the Challenger incident. Every solid motor therefore has to be leak tested under conditions reflecting the worst operational cases.

20.1.4 Handling, Transport, and Storage

Adherence to the *United Nations Recommendations on the Transport of Dangerous Goods* (UN 2005) during transport and storage is essential for the safety of personnel and goods.

Handling

During handling, all safety regulations for the site where the work is performed must be observed. It is imperative that these safety regulations stipulate that the motor and the personnel working on the motor are electrically grounded, all open fire and tobacco smoking in the vicinity of the motor is forbidden, limitations are established on the intensity and wavelength of radio frequency fields in the vicinity of the motor (including those from mobile or cellular phones and wireless computer transmissions), and the relative humidity and the temperature range of the environment is set. It must be verified that all these values agree with the requirements for the solid propellant system.

Anomalies during handling must be reported and recorded to allow appropriate corrective actions to be taken and facilitate tracing them back to the causes of anomalies in later stages.

Transport

During transport, the same precautions must be taken as during handling. In addition, where possible, the solid propellant system should be put into a pressurized container during transport. The temperature and acceleration, that is, shock, must be measured and recorded continuously as well.

Storage

To ensure safe storage of solid propellant systems, facilities must be temperature and humidity controlled. As well, the storage facility must provide for the motors to be grounded. Lightning protection is a condicio sine qua non (without which it could not be). At least one wall or the roof of the storage facility must be of the blowout type. Finally, the total amount of explosive material, that is, propellants, must never exceed that for which the storage facility has been certified.

If several rocket motors are to be stored in the same facility, cookoff tests can provide information about the behavior of the motors and the associated risks in case of accidental fire.

It is good practice to store the motor with its length axis in an approximate horizontal position, and once every 3 or 4 mo rotate the motor about 50°. In this way, the grain cannot deform unacceptably in one direction, and therefore, the motor remains safe for use.

20.1.5 Inadvertent Ignition

The inadvertent ignition of solid rocket motors is one of the major causes of accidents with these devices. Although rare in occurrence, the results can be disastrous.

Lightning

During lightning, peak currents of $J = 3$ to 500 kA have been observed with time rates of change measured as high as $dJ/dt = 10^{11}$ A/s. Because the induced potential difference in an electrical circuit is proportional to the rate of change of lightning current (dJ/dt), due to direct coupling with the magnetic field that penetrated through apertures, the peak voltage or peak potential difference is proportional to the maximum value of dJ/dt. This potential difference causes a current through a closed circuit. If this closed circuit contains an electrical initiator or an electrically activated switch, these can be activated.

Radio Signals

No open electrical connections to an initiator that can act as antennae are allowed. Any such connections must be closed, wires twisted and grounded or bonded to the spacecraft. In the vicinity of live ignition systems, radio communications, including cellular phones, usually are not allowed; and if they are, only under strict conditions.

Electrostatic Discharge

To avoid electrostatic discharge, the relative humidity should be kept above 30% wherever possible and all parts of a solid propulsion system grounded or bonded. Moreover, all parts of the propulsion system must be connected electrically and the maximum resistance between any two parts must be less than 1 Ω. When handling or transporting, system personnel and handling equipment must be grounded to the same common ground.

20.1.6 Safe Ignition Systems Design

Through proper design of ignition systems, the probability of inadvertent ignition can be made extremely small. The design of a safe ignition system usually includes a safe and arm device, which only can be armed remotely. However, it is possible to disarm the safe and arm device either manually or remotely. A safe and arm device contains two independent barriers of a different physical nature, such as an electrical barrier and a mechanical barrier or an optical barrier and a mechanical barrier. Both barriers are installed or removed simultaneously by the safe and arm device.

Electrical

For initiation, electroexplosive devices require an electrical current. The most common devices cannot ignite if they are fed with a current of 1 A or an electrical power of 1 W, even for 5 min. However, some electroexplosive devices have an all-fire current of 0.3 A or less. To minimize the risk of accidental ignition, practically all solid propellant space systems use a safe and arm device. The schematic of a typical safe and arm device is given in Figure 20.2. The S_1 switches, which are outside the device, are the ignition switches. They are the last to be switched, and they ignite the solid propellant system. The safe and arm device depicted in this schematic is in the safe position. The S_2 switches are open and thereby close a grounded loop so that no radio frequency induced potential difference can be created. The loop, C, itself cannot be made always so that the enclosed area, S (Section 20.1.3, subsection "Lightning"), is zero. This is because the loops are physically

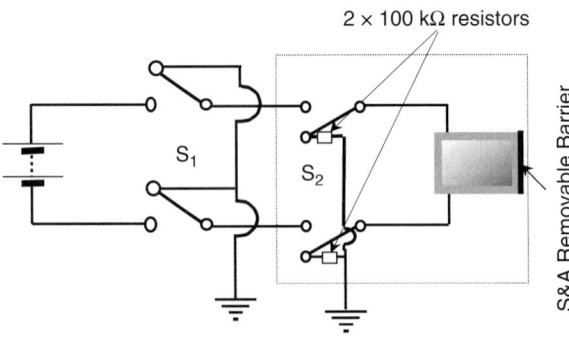

FIGURE 20.2 A safe and arm device.

located within the safe and arm device. Normally wires are twisted or shielded like coaxial cables to avoid creating a nonzero enclosed area, S.

Because it is possible that, during lightning, a rapidly changing magnetic field can pass through the loop, C, and induce a large potential difference, two large value resistors, such as 100 kΩ, have been introduced into the grounded loop. With these resistors present, even if a large potential difference in the loop is created, the actual current now is limited to very low values so as not to cause accidental ignition.

Also, a mechanical barrier shuts off the initiator from the downstream part of the ignition train, the solid propellant system igniter. This barrier is put in place when the safe and arm device is in the safe position by having switches S_1 and S_2 open so that no connection can be made between the power source and the electroexplosive device. Also, when switches S_1 and S_2 are open, they are grounded to avoid the circuit acting as an antenna and picking up radiated electromagnetic energy.

Optical

In addition to electroexplosive devices, optical explosive devices are being used. Here, a laser of a specified wavelength, intensity, and power initiates the optical explosive device. A safe and arm device can be used to block or deflect the laser beam or mechanically shut off the initiator from the downstream part of the ignition train.

Mechanical

Mechanical initiation is not common. In this case, a percussion mechanism is used to activate the initiator. It is sometimes used when a burned out lower stage that falls away is used to activate the stage above. In this situation, a mechanical safe and arm device either is armed by the same source used to activate the initiator or armed before launch. Such a safe and arm device cannot be armed manually.

20.1.7 Conclusions

A professional design for solid propellant rocket systems greatly affects the safety of the system, provided that it allows for grounding or bonding and the proper installation of safe ignition systems.

A major contributor to the safe operation of the solid propulsion system is using the system only within its operational envelope and ensuring strict adherence to storage, transport, and handling regulations. Proper knowledge of the history of the system, that is, load history, temperature history, and exposure to abnormal conditions, allows one to make a well founded decision as to whether or not to use it.

20.2 LIQUID PROPELLANT PROPULSION SYSTEMS SAFETY

Early in the 1900s, a Russian schoolteacher, Konstantin Tsiolkovsky wrote about basic theories for using liquid rocket propellants but did not experiment with any of his ideas. American physicist Robert Goddard began experimenting with gasoline and liquid oxygen, making the first liquid propellant rockets in 1923. The first of these rockets was launched on March 16, 1926 (von Braun and Ordway 1976). The German army began evaluating liquid propelled rockets during 1932, and the first successful flight of the *V-2* (*Vergeltungswaffe-2* or *Vengeance Weapon-2*) occurred during 1942 (Arnold 2003). Following World War II, some of the experienced German liquid propellant rocket scientists, engineers, and technicians, including Wernher von Braun, emigrated to the United States.

The United States had begun to lag in the space race with the Union of Soviet Socialist Republics, which orbited the first satellite and placed the first man into space. There was much pressure to accelerate American achievements in space, but the Navy Vanguard missile, which was first choice as a manned booster, continued to experience failures. It appears that von Braun's *V-2* experience and emphasis on perfection played a major roll in establishing the reliability of the Redstone missile and its adaptation to become human rated. Adaptation of the tactical missile was made through a series of design changes and modifications based on ground and flight tests. The guidelines for conversion of the Redstone design and operations to a crewed payload were safety, acceptable human factors, and no marginal performance. At the top of the list of engineering problems was the high explosive yield of propellants, followed by acceleration, noise and vibration environments, and safety for ground personnel and facilities. The human rating program development had three major guidelines:

- Safety during launch.
- Satisfactory operation within human factors.
- Adequate performance margins for mission reliability.

The pull away launch escape systems became a distinctive feature of crewed vehicles such as the *Mercury, Gemini, Apollo, Soyuz, Chang Zheng*-2F (*Shenzhou*), and the *Crew Exploration Vehicle*. Those responsible for implementing the design and operational integration of the human rated Mercury-Redstone project were so successful (Kuettener et al. 1964) that they became the organization responsible for the technical coordination of the long and highly accomplished Apollo and Saturn Programs.

Solid propulsion systems are simpler but cannot provide many of the liquid propulsion system capabilities required, such as propellant offload safing; stoppable, restartable, and throttable thrusting; and multiple thrusters pulsating for attitude control. Minimizing the risk of explosion for liquid propellant propulsion systems continues to be one of the most important top level requirements. This requirement is on the same level as that specifying the necessity for propulsion systems being designed as must-work, that is, they are engineered to have minimum risk of not functioning.

An exploding propulsion system, in addition to causing major disruptions in a space program, can cause a considerable amount of orbital debris, which is extremely hazardous to ongoing and subsequent spaceflight. Therefore, even for uncrewed space vehicles, minimizing the potential for an exploding propulsion system is a top level requirement for all launch and spacecraft propulsion systems designed to achieve orbit. The following sections provide top level guidelines for assuring successful and safe liquid propellant system designs. This section, however, is not a how to engineer document, such as *Modern*

Engineering for Design of Liquid Propellant Rocket Engines (Huzel and Huang 1992) and the commonly used *Rocket Propulsion Elements* (Sutton 2001).

20.2.1 Planning

The quality assurance, engineering management analysis, engineering, production, verification, and qualification required for a highly reliable, high quality, and very safe propulsion system and all its interfacing and supporting subsystems should be included in the original planning and design for a production, qualification, and operations program.

Top Level Requirements

The user of a liquid propellant propulsion system, in conjunction with potential suppliers, initially must establish its capabilities and coordinate all trade studies of top level propulsion system requirements, such as total impulse, velocity changes, attitude control and precision, life, propellant, and plume impingement restrictions. Reliability, redundancy, and safety factors should be established as top level requirements for not just the propulsion system but also for interfacing and supporting subsystems. If the system is human rated or can cause orbital debris, the requirements should be established for must-work redundancy and minimum risk of failing catastrophically, that is, exploding.

Iterative Requirements Coordination

The propulsion system is a subsystem of the entire launch vehicle or spacecraft, and its requirements must be coordinated iteratively with all interfacing subsystems, such as structures and thruster positioning and thermal management. This especially is necessary for the guidance, navigation, and control; thrust vector control; and gimbal actuators, because the integrated structures, propulsion, guidance, navigation, and control dynamics capabilities and limitations, and other supporting subsystems, such as software and electrical power control and distribution, must be considered. Transient pressures, engine firings, vibration, and acoustics greatly affect the propulsion system controls, and the guidance, navigation, and control end to end interactions. All interfacing subsystems requirements must be integrated with those of the propulsion system to ensure the reliability and safety required of propulsion system for the launch vehicle or spacecraft.

Often, conflicting interface requirements must be solved. The guidance, navigation, and control and the flight controls can have a requirement for very short duration, minimum impulse, precision attitude control thruster firings that might not be possible with a selected thruster because it cannot provide the minimum impulse or it might be very inefficient. In this case, special purpose vernier thrusters might be required. On the other hand, very small thrusters that are used for large velocity changes require very long duration burns and, therefore, can require a special thermal control design. As well, shock and vibration due to propulsion engine firings can have considerable impact on the designs or locations of other systems.

A propulsion system must function as an integrated system. During the preliminary design and development testing, there must be very close coordination with all the propulsion interfacing subsystems; structures; guidance, navigation, and control; thermal control; electrical power and control; instrumentation; and data systems to assure the propulsion system functions as an end to end integrated system. The test capability to verify the satisfactory end to end integrated propulsion system with all interfacing and supporting subsystems must be provided by the developer.

20.2.2 Containment Integrity

The containment of liquid propellants and supporting fluid systems must be designed for a minimum risk of bursting and leaking for three reasons:

- The liquid propellant contained often is very reactive and can cause reactions with hot surfaces, burn, and cause cascading failures, resulting in a catastrophic event.
- Many of the storable chemical propellants, such as dinitrogen tetroxide and nitrotriazolone, unsymmetrical dimethylhydrazine, monomethylhydrazine, and hydrazine, are very toxic or carcinogenic. Leakage creates an extreme hazard to personnel, so any risk of leakage must be minimized.
- If the propellants or pressurants are leaked, the capability of the propulsion system to provide the required total impulse is reduced and mission success and safety are jeopardized.

Welded joints are preferred for containment integrity, but there must be a trade evaluation between leak tightness and all the risks associated with leaks and the easy replacement of failed components. Cutting a line and again welding it, easily introduces particulate and other contamination into the fluid system of which this line is a part. The integrity of mechanical fittings for connecting components, subassemblies, and systems must be qualified for the environment to which the propulsion system is exposed, including acoustic and vibration. Redundant back off prevention, preferably a mechanical method such as a safety wire or tab, should be used on all mechanical fittings.

Design and Test Requirements

Propulsion system lines and fittings, components, and metallic pressure vessels should meet the design and test requirements of ANSI/AIAA S-080–1998 (ANSI 1998) and ANSI/AIAA-081A-2006 (ANSI 2006), or equivalent.

Transient pressures, such as those caused by pulsating thruster valves, pyrovalve system activation, or slam activation, that is, the quick opening of a valve that isolates a low pressure subsystem from one of high pressure and can be a multiple of the static pressure, must be considered when determining the maximum expected operating pressure. This pressure subsequently is used to determine the factors of safety.

Fluid system strength integrity must be verified by a proof pressure test, whereby the test pressure is greater than the maximum expected operating pressure. A subsequent inspection is required to show that no detrimental deformation of the fluid system lines, fittings, and components occurred. A leak test at the maximum expected operating pressure is required to show that no leaks have been caused by the proof test. A 1.5 proof factor commonly is used as specified for lines, fittings, components, and subassemblies within fluid subsystems. However, after the pressure vessels and components with lower factors of safety are installed, a system level proof test must be conducted with the test pressures reduced low enough to prevent damaging or reducing the life of equipment with low factors of safety.

Unnecessary repeated proof tests on equipment with low factors of safety should be discouraged, because it tends to reduce the life of the pressurized hardware. Special consideration must be made to provide high point bleeds to reduce trapped gas in a liquid system. Low point drains must be provided so the maximum amount of propellant can be drained for offload safing.

No Leak Verification

Helium mass spectrometer leak testing, with a sensitivity of about 1.0×10^{-7} sccs helium, is the preferred leak test method for liquid propulsion systems. This method can detect leaks from systems that would not leak fluids and is sensitive enough to detect flaws or cracks that would then require fracture control assessment. Because helium mass spectrometer leak testing is searching for no indicated leakage, it is very easy to go through the motions of helium mass spectrometer leak testing without actually having performed a leak test. Technicians operating the helium mass spectrometer, quality assurance representatives, and responsible engineers should be trained and certified as proficient in the use of a helium mass spectrometer for leak testing propulsion systems (Moore, Jackson, and Sherlock 1998).

End of Life Safety

In some cases, orbiting spacecraft have exploded after the useful life of the spacecraft. Some of these have been caused by propulsion systems in which residual high pressure leaked into a low pressure subsystem that was not relief protected for overpressure. Consequently, the low pressure subsystem exploded. Also, the residual propellants somehow could have caused the exploding pressures. The propulsion system must be analyzed for the afterlife potential to explode. It is preferable to deorbit a spacecraft that has the potential of an afterlife propulsion system explosion. If there is a potential that the propulsion system could explode and the spacecraft cannot be deorbited, the propulsion system must have a planned post-life depressurization or safety capability to prevent it from exploding and causing considerable orbital debris as the consequence of the event.

20.2.3 Thermal Control

Most spacecraft equipment have qualified operating temperature limits that are narrower than the extremely low or high temperatures that can be experienced in space. The thermal operating limits for propellants are probably the narrowest of thermal limits for the spacecraft, and special propulsion system thermal control is required to keep them within the specified range. If propellants freeze, they are not available for utilization, possibly rendering a must-work system inoperable until thawed. Even though propellants can decrease in volume on freezing, frozen propellants still have the potential to rupture the system. Freezing can form freeze blocks, and on thawing, the rising temperatures between freeze blocks can result in hydrostatic pressures that can exceed rupture pressures.

Experience has shown that it is preferable to have thermal control to prevent propellant freezing. Because freezing and the subsequent rupture and loss of containment is unacceptable, the thermal control system should be redundant and fault tolerant. Over temperature propellant is unacceptable as well, so there also must be fault tolerance protection from failed-on heaters. Thermal modeling used for thermal control subsystem design must be validated through testing because of the many interacting variables. The propulsion system thermal control must be qualified as an integrated subsystem.

20.2.4 Materials Compatibility

Propellants are reactive and not found commonly in combinations with the propulsion system materials. Despite a wealth of evaluations of propellants and propellant system materials, actual components testing should be accomplished because of potentially overlooked materials, such as residual cleaning fluids, that can have a considerable impact on compatibility. During the early Apollo hypergolic propulsion systems development, many lessons were learned about the incompatibilities between the chemical propellants and the new lightweight, high strength materials, such as titanium alloys. Acceptability for any new combinations must be proven by actual tests at the full range of all the environments to which they are exposed.

As an example of just how dangerous materials incompatibility can be, on March 18, 1980, a Vostok 8A92M booster, while on the pad being serviced, exploded and killed 48 people. The cause was a hydrogen peroxide filter assembled with tin based solder that had been replaced with one using a lead based solder. This caused runaway hydrogen peroxide decomposition and the subsequent explosion. The root cause was that apparently no double check had been done on the decision to change the material and the change was not qualified by analysis or tests. Most errors are not caused by a person but by a deficiency in the double checking required by high quality assurance or the process of total quality management (Savary and Crawford-Mason 2006).

20.2.5 Contamination Control

Liquid propellants are reactive, and contamination of a propulsion system with foreign materials should be prevented. Moisture, hydrocarbons, and particulates are the most common contaminates controlled. Most fluid propulsion system components are vulnerable to unacceptable internal leakage caused by particulate contamination. Experience has shown that it is preferable to start with precision clean components, assemble clean, and maintain subassembly cleanliness; or in other words, basically keep the system clean during assembly, rather than attempt to clean a contaminated system after it has been assembled with no contamination control. Despite starting clean, particulate contamination can be generated internally, such as by a pyrotechnic valve; and where that is possible, internal system filtering is necessary.

20.2.6 Environmental Considerations

The environment to which a propulsion system is exposed can have detrimental effects on the system. The salty, high humidity Florida atmosphere caused many problems with the pilot operated Space Shuttle primary reaction control thruster valves, even though they individually had been qualified acceptable for exposure to salt, humidity, and propellants. When all these factors came together for an extended period of time, valve leaks and failures began to occur as a result of the hygroscopic nature of the propellants and the formation of propellant reaction products.

The launch vibration environment is especially harsh on propulsion systems reliability. Combinations of thermal vacuum and end to end guidance, navigation, and control hot firing of the engines must be integrated to demonstrate the acceptability of the propulsion system to function in flight environments.

20.2.7 Engine and Thruster Firing Inhibits

Several major propulsion disasters were caused by inadvertent rocket engine firings. As a minimum, the engine firing commands should have an electrical safe and arm inhibit on the electrical power to the engine firing drivers and mechanical inhibits with safe and arm position monitoring so that, even if the monitoring signal is not functioning, it is detectable. Usually, the engine firing command requires redundant signals. There generally are two fluids inhibits to the engine, such as propellant tank isolation valves and manifold isolation valves in series with the engine valves. The isolation valves, that is, the inhibits, should have indicators for both valve positions to show closed, or safe, and open, or arm. Likewise, the engine valve drivers safe and arm electrical power controls should possess open, or safe, and closed, or arm, indicators for monitoring.

One of the worst inadvertent propulsion engine firing accidents was the Nedelin catastrophe. On October 24, 1960, the first attempt of the USSR to launch the R-16 intercontinental ballistic missile, a relatively small rocket, at the Baikonur Cosmodrome resulted in an explosion on the pad, killing 122 people, when a spurious radio signal fired the second stage rocket engine. The fluid system inhibiting valves were open and work was being done on the electrical control system when the accident occurred. The number of personnel and visitors in close proximity to the launch pad exceeded safe limits, given that the technicians were performing repairs on a fully fuelled rocket. At least 74 people died from the fireball and toxic gases, and approximately 50 died later from injuries received that day. Marshal of Artillery, Mitrofan Nedelin, who was personally in charge of the R-16 program and on the spot overseeing work, was at the launch pad. He was one of the first 74 killed. Even 40 years since the Nedelin disaster, system failures remain poised to strike whenever key casual factors interact, compound, or align. "Don't be seduced by success—the price of safety is constant vigilance" (Lloyd 2005). There are many other examples of such failures.

Propulsion system engines and thrusters are must-work systems and extremely dangerous if inadvertently fired. In reaction control systems having numerous thrusters, it is common to group control of more than one thruster into fewer safe and arm controls than the number of thrusters on the spacecraft. On *Gemini VIII*, the return vehicle had a thruster fail-on and the vehicle started spinning up. In a time consuming attempt to isolate and inhibit only the failed-on thruster, the vehicle continued to wind up to the point where the aeromedical doctors were concerned about the crew being on the verge of loosing consciousness. As well, concerns were being raised about adequate propellant to reenter, as well as how to stop the vehicle from spinning. Because other thrusters were attempting to compensate, the specific failed-on thruster could not be identified. This ultimately required powering down all thrusters, so that the failed thruster would be shut down along with the rest. Individual thruster activation was conducted to locate the failed-on thruster. Once isolated, the correct combination of functional attitude control thrusters needed to stop the spacecraft from spinning and configure it for entry was powered up and armed without arming the failed-on thruster (Carlton 2001).

At the time, NASA was not prepared for a failed-on thruster and failed to shut it down by inhibiting propellant flow through the process of closing a tank isolation or manifold valve. Such action would have limited the propulsion subsystems power down. Subsequent Apollo and Space Shuttle designs grouped fewer inhibits of must-work attitude control thrusters together into common safe and arm controls, and much attention was given to a process for identifying a specific failed-on thruster as well as to training for this failure mode. Incidentally, Neil Armstrong was the Gemini VIII commander, and the fact that he survived this close call gave him an opportunity to be the first man to set foot on the Moon.

20.2.8 Heightened Risk (Risk Creep)

Launch vehicles and spacecraft tend to be very complicated. In spite of Wernher von Braun's noble goal of perfection, the reality is that it is impossible to have all specifications met perfectly because of the complexity. The reality is that decisions must be made to accept some less than perfect hardware and software design or functional performance. These less than perfect, not per specification or drawing decisions are made using discrepancy reports, nonconformance reports, and waivers. It is most desirable to limit greatly the number of these less than perfect decisions because they heighten risk. Within complicated launch and spacecraft systems, much of the series and parallel redundancy is fail safe, that is, fail operational, for which heightened risk is more forgiving in the event of a failure. On the other hand, propulsion systems tend to have more single catastrophic failure points, which are designed for minimum risk of failure. These are extremely unforgiving if the heightened risk is excessive. Over an extended period of time, the tendency is to become more accepting of more and more nonconformances. This further heightens risk for design for minimum risk applications and increases the risk for potential catastrophic failure, especially in propulsion systems (Leveson and Cutcher-Gershenfeld 2004). Since the Columbia incident, NASA requires probabilistic risk assessment for hazards to more thoroughly analyze the risk of less than perfect nonconformances (Pelton et al. 2005).

20.2.9 Instrumentation and Telemetry Data

Because propulsion systems failures and explosions can leave so little physical evidence, recorded telemetry data must be available for failure analysis, as well as operational verification. There must be an adequate number of instrumentation measurements with enough sensitivity, adequate frequency response characteristics, a sufficiently wide range, and high sampling rates that telemetry data are adequate to determine the cause of a failure and the corrective action to be taken to prevent subsequent failures.

20.2.10 End to End Integrated Instrumentation, Controls, and Redundancy Verification

The instrumentation and the electrical and flight controls used to support the propulsion system must be designed with comparable redundancy or design for minimum risk reliability and safety, because the propulsion system, including all supporting systems, must work together and be designed for minimum risk of catastrophic failure. The propulsion system is the end effector of the flight control system. It cannot function by itself, and it must function as a part of an integrated flight control system.

Propulsion systems instrumentation and controls must be subjected to a rigorous verification program comparable to the scrutiny of the fluid system. Comprehensive system level testing of the integrated end to end propulsion system, supported by qualification testing, is the preferred verification method. It is essential that the integrated propulsion system be verified from the input stimuli to the end function. The end to end integrated verification should verify supporting subsystems for propulsion systems, such as electrical power, power distribution, power control, flight controlling software, and gimbal actuators, where the power source has redundancies, fault tolerance, and reliability comparable to the propulsion system. Test requirements, procedures, and test apparatus must be derived from intended functional requirements rather than from the design, and all items must be maintained under strict configuration control (Shaw 1994).

20.2.11 Qualification

Qualification, the demonstration that the production design complies with all the system requirements, usually is a combination of similarities, analyses, and tests.

Fluid propulsion systems are very expensive to qualify by test demonstration, and there is always much financial incentive to reduce costs and time by using previously obtained qualification data, thus building on heritage. The similarity, differences, and application of the previous qualification to the new system to be qualified must be scrutinized to ensure that qualification by similarity is valid. Some propulsion systems requirements cannot be qualified by ground testing and therefore require analyses to qualify the propulsion system until the flight test qualification can be obtained. Some qualification is done more satisfactorily by analysis than by testing. In such cases, the analyses or model accuracies must be validated with substantiating test data.

End to end complete production testing of a propulsion and guidance, navigation, and control system, including computer software, by an integrated system hot firing qualification test of the propulsion system over the full range of its operating conditions and in all the operating environments is most desirable. Some of the tests might have to be done serially in different facilities, like in a thermal vacuum chamber, acoustic and vibration test facilities, and on a propulsion system hot firing stand. Hot fire testing of a propulsion system for performance verification should use the flight-like feed and pressurization system software used in flight-like operations and high fidelity thermal and electrical interfaces. All design changes after propulsion system qualification must be requalified, preferably by testing. Sufficient qualification testing must be done to assure the repeatability of performance, reliability, and safety of the propulsion system. In the case for human rating, it is most desirable to have multiple uncrewed qualification flight tests before the first crewed flight.

20.2.12 Total Quality Management (ISO 9001 or Equivalent)

Reliability, including the failure mode and effects analysis and the critical item list, should be performed concurrently during design to determine where redundancies are required to make the system fail operational and identify any potential single point catastrophic failures. These determine where greater reliability and factors of safety are required so that the design can be for minimum risk of catastrophic failure. The failure mode and effects analysis becomes a valuable tool for understanding how the system operates and identifies alternate configuration options for mission success and safety (ISO 2000).

An important aspect of liquid propulsion systems and rocket engine development and qualification testing is to determine failure causes so that corrective designs can be implemented to avoid repeated failures (continuous improvement) and the corrected design can be qualified. The quality control must be adequate to assure minimum variation between the test, production, and qualification articles. There must be documentation authorization control to assure consistency in implementing engineering orders and shop work with enough peer review given to reduce the possibilities of mistakes being made (Yarnell 2006).

20.2.13 Preservicing Integrity Verification

Usually, the launch vehicle or spacecraft propulsion system production contractor performs acceptance, proof, leak, and function testing on the propulsion system to show compliance with the requirements for customer acceptance of the vehicle at the contractor facility. Then, the launch vehicle or spacecraft is delivered to the launch site. It is preferable that a different test team, that is, a test, checkout, and launch team versus the manufacturing final acceptance test team, repeat the acceptance proof, leak, and functional testing at the launch site. This redundant verification can find flaws missed in the factory acceptance test. The liquid containment integrity, components acceptable functionality, cleanliness, and end to end integrated control and instrumentation acceptability should be verified before servicing with the potentially highly explosive propellants.

20.2.14 Propellants Servicing

A liquid propulsion system is relatively benign before servicing begins. There must be much preparation for safety before the hazards of the propellants are introduced. The design of the liquid propulsion system and ground servicing system must be reviewed and verified capable of off-load safing before any propellants are loaded. The special equipment and personnel training for safety during and after servicing must be available and verified fully functional. Toxic chemical propellants require special equipment and training. Cryogenic propellants require special equipment and training. The servicing equipment, procedures, and personnel must be checked out and certified ready for propellant servicing before starting the servicing. During servicing and after being serviced with propellants, potential explosive or fire hazards for the launch vehicle or spacecraft must be recognized so that there is limited access, special safety controls are implemented, and emergency procedures have been proven and ready to minimize risk to personnel.

20.2.15 Conclusions

Liquid propellant systems have been demonstrated to be developed, produced, and operated safely in spite of the tremendous potential for high explosive yield of the propellants. It appears most engineering and design deficiencies are found during development and qualification testing, some of which can be catastrophic explosions. Even though operational liquid propellant systems have been qualified and operated safely, there have been numerous operational catastrophic explosive failures, usually because of deficiencies in the processes and the total quality management system. The propulsion systems developers and users have an extensive experience base, with many lessons learned, that must be heavily relied on and applied diligently for the safe design, development, production, and operation of liquid propellant propulsion systems. There is a need for consistent perfection in propulsion systems, as stated by Wernher von Braun (Huzel and Huang 1992), for hardware, software, and procedures to be within the allowed tolerances of the qualified design and specifications. Less than perfect, heightened risk designs should be avoided or qualified acceptable. As a last resort, less than perfect, heightened risk designs might be considered, but only with a thorough failure mode and effects analysis and probabilistic risk assessment before being allowed.

20.3 HYPERGOLIC PROPELLANTS

20.3.1 Materials Compatibility

This section considers the reaction of hypergolic propellants, such as hydrazine, monomethylhydrazine, and dinitrogen tetroxide, with materials of construction normally contacted in hypergolic propellant operating systems and the reaction of hypergolic propellants with other materials used in aerospace operations that inadvertently can come into contact with these propellants.

The chemical reaction of hypergolic propellants with materials potentially is detrimental for two reasons. The propellant can degrade the material sufficiently to impair its useful function, such as loss of integrity through corrosion or loss of ductility, and in the case of the hydrazine fuels, it can lead to the decomposition of the propellant by catalytic material surfaces, thus causing thermal runaway. It should be noted that dinitrogen tetroxide, however, does not decompose exothermically.

When assessing the compatibility of a material with a hypergolic propellant, it is important to carefully detail the conditions of the system and its environment. Principal considerations include the phases present, the concentration, and purity of the propellant in contact with the material, the surface area of the material, and the temperature of both the propellant and the material. The condition of the material surface also is important. This would include the presence of oxide coatings or contamination. The nature of the contact of the material with the propellant is crucial when assessing the hazard. A material can have adequate compatibility if it inadvertently comes into contact with hypergolic propellants but be unacceptable for routine and long-term use in operating systems. Factors driving this are shorter exposure times, lower temperatures, and acceptance of some degradation, because materials can be changed out more easily than they can be made compatible. With this information, catalytic processes can be identified, and an assessment can be made as to whether the process causes material degradation or a thermal runaway.

20.3.2 Material Degradation

The principal types of materials selection criteria are the compatibility rating under specified operating conditions, such as temperature and pressure, and when time or rate information exists, the length of time the material is exposed to the propellant. Very little information specific to monomethylhydrazine has been reported; however, similar information is available for hydrazine, and it can be used as a starting point for a monomethylhydrazine materials compatibility evaluation (Pedley et al. 1990).

Material Degradation Criteria

The gross compatibility of materials with liquid hypergolic propellants as a function of exposure time and temperature is determined by the immersion test, that is, Test 15 entitled the *Reactivity of Materials in Aerospace Fluids* located in NASA STD-6001 (NASA 1998). In Test 15, the total pressure and test temperature must simulate the worst-case use environment for degradation of material or fluid. The material is immersed in liquid propellant under its own vapor pressure plus an ambient nitrogen blanket, elevated to a minimum temperature of 344.1 K (160°F), and held for 48 h or until sufficient hydrazine or monomethylhydrazine decomposition occurs to end the test. In the particular Test 15 that utilizes dinitrogen tetroxide, decomposition of the liquid does not occur, although pressure can increase as the result of the degradation of the material.

The gross reactivity of materials exposed to hypergolic propellant can be described with a compatibility rating. General compatibility ratings for the degradation effects of propellant on materials are based on past experience in handling and are qualitative in nature. The qualitative nature of the ratings is a result of not having a standardized test method (Uney and Fester 1972). Often the results of tests performed in a given system do not agree with results of tests performed in another system. The compatibility data seldom separate the effects of monomethylhydrazine on materials from the effects of materials on the propellant (Pedley et al. 1990).

Material Degradation Effects

Temperature and contamination are the major parameters affecting materials compatibility. In general, increased temperature leads to increased incompatibility (Pedley et al. 1990). Many materials that are compatible at ambient temperature are incompatible at elevated temperatures. The presence of contaminants, both dissolved and suspended, has been shown to affect adversely hydrazine compatibility with materials (Pedley et al. 1990). Similar behavior is anticipated for monomethylhydrazine. Other sources of dinitrogen tetroxide contamination are extraction of impurities in hardware or soft goods such as cleaning solvents, plasticizers, lubricants, and polymeric materials.

20.3.3 Hypergolic Propellant Degradation

Degradation of hypergolic propellant infers the decomposition of the molecule into two or more products. In a closed system, that is, one with no exchange of mass with the surroundings, hydrazine fuel degradation results in a pressure buildup because of vapor and gaseous decomposition products. In an adiabatic system, the temperature increases and a thermal runaway can result. Again, dinitrogen tetroxide does not decompose exothermically, although it does dissociate to equilibrium at elevated temperatures.

The buildup of pressure in a closed system is an indication that degradation has occurred. However, in many propellant systems, a pressure pad of approximately 2.07 MPa (300 psi) is used, which makes it difficult to detect pressure buildup from decomposition. Clearly, when the liquid fuel system pressure exceeds the vapor pressure for a given temperature, degradation has occurred. Even in a non-adiabatic system, when hydrazine or monomethylhydrazine decomposition occurs, temperature buildup could result if the rate of thermal energy loss is less than the rate of thermal energy generation.

Reaction Rate Testing

The kinetic reaction rate is a function representing a specific mechanism, or more frequently, an empirically based function for the reacting system. The rate is a function of temperature, pressure, concentration, catalyst, and the like. To be most useful, the kinetic reaction rate regime must be isolated from all other rate affecting mechanisms in the experiment system; therefore, standard procedures need to be followed in each experiment. Two such methods have been developed and used: exothermic calorimetry (Benz 1980; Pedley 1987) and accelerated rate calorimetry (Wedlich, Davis, and Peters 1988). Two additional methods have been reported in the literature and applied using the isothermal rate of gas evolution (Chang and Gokcen 1976) and Fourier transform infrared spectroscopy (Martin et al. 1989).

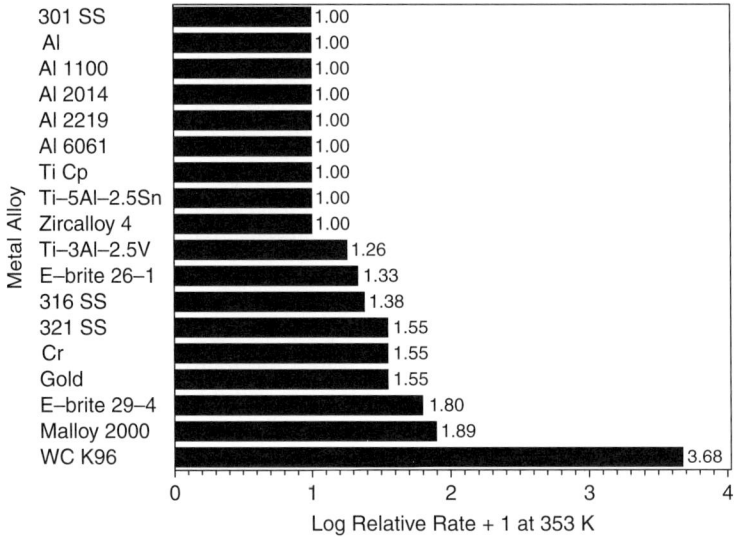

FIGURE 20.3 Reactivity of metals in monomethylhydrazine at 353 K (176°F) relative to the reactivity of titanium (test vessel made of commercially pure titanium) (Courtesy of NASA).

Exothermic data include the heat generation rate observed for the decomposition of hydrazine fuels per unit area of material as a function of temperature. Accelerated rate calorimetry testing yields the rate of temperature change at a given temperature, from which Arrhenius parameters can be determined. Hydrazine fuels decomposition proceeds on metal surfaces at a rate proportional to the metal surface area (Davis and Wedlich 1991). The heat generation rate can be used to rank materials by their relative compatibility with hydrazine fuels at a specific temperature, as shown in Figure 20.3. When heat generation rates are compared at the same temperature, all other factors being equal, materials that are most compatible with the fuel produce the lowest rates.

20.4 PROPELLANT FIRE

Fire is a rapid chemical reaction that produces heat and light (Lapedes 1974; Strehlow 1984). Fire normally requires a fuel, an ignition source, and an oxidizer. However, hydrazine is a monopropellant that decomposes exothermically; therefore, it does not require the presence of an oxidizer. Hydrazine and monomethylhydrazine fires also can begin without the usual ignition source. Hydrazine fuels are hypergolic with oxidizing reagents and propellant oxidizers such as dinitrogen tetroxide. On the other hand, dinitrogen tetroxide is not flammable, although it does support combustion.

A fire can occur with either vapor or gaseous hydrazine or monomethylhydrazine fuel, as well as in the mist, droplet, and spray forms. The fire hazard from the burning fuel is affected by system and environmental conditions, that is, the temperature, concentration, and pressure of the fuel vapor, and the type of ignition source involved.

For hydrazine and monomethylhydrazine vapor, the fire hazard can be quantified by considering the flammability limits, ignition source, flame speed, or flame velocity. The fire hazard in hydrazine and monomethylhydrazine mists, droplets, or sprays also can be measured by considering flash and fire points and burning rates. These factors vary with the purity of fuel vapor: neat or completely free of contaminants, mixed with air, mixed with an inert diluent, or mixed with an oxidizer or liquid.

20.4.1 Hydrazine and Monomethylhydrazine Vapor

The reactants in a hypergolic fuel fire usually are in the gaseous phase and must be present within a specific concentration range to burn. The fire hazard of hydrazine and monomethylhydrazine vapor can be assessed by considering flammability, ignition, flame speed, and flame velocity in neat vapor, fuel-air mixtures, fuel-diluent mixtures, and fuel-oxidizer mixtures.

Flammability

Flammability is a measure of the extent to which a vapor concentration of a fuel in a mixture ignites and propagates a flame. The limits of flammability for a gas or vapor are the minimum and maximum fuel concentrations that can support flame propagation. The upper flammability limit is defined as the concentration of the most concentrated mixture that is flammable and the lower flammability limit as the concentration of the most dilute fuel-air or fuel-diluent mixture that is flammable.

Flammability varies for neat fuel, fuel mixed with air, fuel mixed with an inert diluent, and fuel mixed with an oxidizer. The flammability limits for each mixture are affected by pressure, temperature, and other factors. The range of flammability is the range of concentrations between the lower and upper flammability limits. In general, for many flammable mixtures, the flammability range is widened by increasing temperature (Burgess and Wheeler 1911). Decreased pressure, that is, below ambient, narrows the flammability range by increasing the lower flammability limit and decreasing the upper flammability limit (Coward and Jones 1952). As the pressure decreases, the two limits approach each other. When the upper and lower limits are identical, the low pressure limit, that is, the minimum pressure required for ignition, is reached. The low pressure limit is affected by ignition energy in that, as the ignition energy is increased, the low pressure limit decreases (Benz, Bishop, and Pedley 1988).

Flammability limits also vary with the shape of the confining container and the direction of flame propagation. Generally, for many flammable mixtures, the flammability range widens with increasing tube diameter (Coward and Jones 1952) and narrows as the closed tube is lengthened (Mullins and Penner 1959). The direction of flame propagation affects flammability limits by widening the limits for upward propagation and narrowing the limits for downward propagation. The limits for horizontal propagation are between those for upward and downward propagation (Benz et al. 1988).

Ignition

Hydrazine and monomethylhydrazine vapor can be ignited with or without an ignition source. Ignition by a source requires that minimum ignition energy be present; autoignition, that is, ignition with no source, requires the autoignition temperature to be reached. In general, the minimum ignition energy decreases with increasing oxygen content (Lewis and von Elbe 1961), increasing temperature (Drell and Belles 1958), and increasing pressure (Benz et al. 1988). The standard method for determining the minimum ignition energy is ASTM E582-07 (ASTM 2007).

Ignition Sources

Various sources have been used to ignite hydrazine, including electric sparks (Overly 1976); a 15-kV, 60-mA transformer (Furno, Martindill, and Zabetakis 1962); and shock waves (Michel, Troe, and Wagner 1963). Other common ignition sources include flames, heated or fused wires, incendiaries, hot surfaces, and rapid compression. Many of these ignition sources, such as hot surfaces, can be affected by surface area, surface temperature, and type or volume of container or extent of confinement.

Autoignition

Autoignition occurs when a mixture of gases or vapors ignites spontaneously with no external ignition source and after reaching a certain temperature, the autoignition temperature. The autoignition temperature is not an intrinsic property of the gases or vapors (Kanury 1975) but is the lowest temperature in a system where the rate of heat evolved from the gases or vapors increases beyond the rate of heat loss to the surroundings, resulting in ignition. The autoignition temperature of a mixture of gases or vapors is affected by pressure, vessel shape and volume, surface activity, contaminants, flow rate, reaction rate, droplet and mist formation, gravity, and reactant concentration (Benz et al. 1988).

In general, decreased pressure leads to an increased autoignition temperature (Furno, Imhof, and Kuchta 1968; Benz and Pippen 1980; Bodurtha 1980), whereas increased vessel size leads to a decreased autoignition temperature (Setchkin 1954). For fuels in general, the autoignition temperature is not very sensitive to fuel concentration except at near limiting concentrations (Furno et al. 1968); however, some studies with hydrazine show that off-stoichiometric mixtures lead to increased autoignition temperatures (Miller and Schluter 1978). The effect of catalytic surfaces on autoignition temperature varies with the system. Some studies show that catalytic surfaces in some systems increase the autoignition temperature (Lewis and von Elbe 1961; Miller and Schluter 1978), whereas others indicate that the reaction between the fuel and the surface material leads to a decreased autoignition temperature (Scott, Burns, and Lewis 1949; Stevens and Benz 1978).

One study found that, as hydrazine is heated, decomposition increases, hydrazine concentration changes, and the final result is an increased autoignition temperature. In a flowing system where concentration is constant, the effect of materials appears to be determined by the catalytic properties of the materials (Miller and Schluter 1978).

Fuel vapor concentration and the presence of diluents affect the autoignition temperature. Mixtures near the flammability limits have a higher autoignition temperature than those of intermediate composition (Bodurtha 1980). Fuel-oxygen mixtures have a slightly lower autoignition temperature than similar concentrations of fuel-air mixtures (Bodurtha 1980). Mixing inert gases with fuel generally increases the autoignition temperature by diluting or altering the thermal conductivity, specific heat, or diffusivity of the mixture (Mullins and Penner 1959).

The standard methods for determining the autoignition temperature are detailed by ASTM E 659-78 (ASTM 2000), Mullins (1955), and Setchkin (1954). The autoignition temperature is recorded as the lowest temperature at which autoignition occurs for a fuel. The ASTM E 659-78 (ASTM 2000), notes that the method is not designed for evaluating materials capable of exothermic decomposition. For this reason, autoignition temperature data should be applied cautiously (Benz et al. 1988).

Flame Velocity

The movement of a flame is defined relative to a specific coordinate system. For coordinates centered in the flame, the velocity at which unburned gases move through the combustion zone in the direction normal to the flame front is defined as the flame velocity. For laboratory coordinates, the flame speed is the velocity of the flame through the unburned gases.

The flame velocity of hydrazine vapor can be calculated from the flame speed, and the equation used depends on the nature of the confinement. When a flame propagates throughout a tube, the flame velocity can be calculated using Equation (20.3) (Kanury 1975):

$$U_v = U_s - U_r \tag{20.3}$$

where

U_v = flame velocity, m/s (ft/s),
U_s = flame speed, m/s (ft/s),
U_r = fresh reactant mixture velocity, m/s (ft/s).

For a spherical flame, the flame velocity can be calculated by Equation (20.4) (Strehlow 1984):

$$U_v = \left[\frac{V_1}{V_2}\right] U_s = \left[\frac{r_1}{r_2}\right]^3 U_s \tag{20.4}$$

where

V_1 = initial volume of flame ball, m³ (in.³),
V_2 = final volume of flame ball, m³ (in.³),
r_1 = initial radius of flame ball, m (in.),
r_2 = final radius of flame ball, m (in.).

Combustion in a system can affect the boundary conditions of the combustion process, causing the flame velocities and flame speeds to vary. An average value can be used for the purpose of estimation.

In a flow system, the flame speed depends on whether the flow is laminar or turbulent. It is important to consider flame speed for determining the rate that a fire can move through the flammable mixture and subsequently distribute the fire throughout the system.

In evaluating the rate that heat is being generated throughout the system, the flame speed is used to calculate the mass rate of consumption of the fuel.

Flame Speed in Laminar Flow

The elementary combustion theory of Mallard and Le Chatelier (Glassman 1987) predicts that the laminar flow flame speed is proportional to the square root of the product of the thermal diffusivity and the reaction rate, Equation (20.5):

$$U_L \propto \sqrt{(\alpha \times \text{RR})} \qquad (20.5)$$

where

U_L = laminar flow flame speed, m/s (ft/s),
α = thermal diffusivity, m/s (ft/s),
RR = reaction rate, s^{-1}.

20.4.2 Liquid Hydrazine and Monomethylhydrazine

Liquid hydrazine and monomethylhydrazine support fire only if the temperature of these materials is above the fire point and an ignition source is present.

In general, liquid pool fires are affected by temperature and wind. As the temperature exceeds the fire point, the vaporization rate increases and the rate of flame spread across the liquid layer increases. At temperatures well above the fire point, the flame spread rate does not increase with increasing vaporization (Botteri et al. 1966; Kuchta 1973). A burning pool can be extinguished by a strong wind, because as the diameter of the pool increases, the velocity of the wind required to extinguish the fire also increases (Atkinson and Eklund 1971).

Flash and Fire Points

Flash and fire points of liquid hydrazine and monomethylhydrazine vary depending on whether the hydrazine vapor above the liquid surface is mixed with air, an inert gas, or another oxidizer.

The flash point of hydrazine or monomethylhydrazine is the lowest temperature at which the liquid gives off enough vapor to form an ignitable mixture with air at ambient pressure near the surface of the liquid (Sax 1975). The fire point, that is, the lowest temperature at which a liquid continuously can support a flame (Jensen 1976), is higher than the flash point. This can be determined by the open cup method.

Flash points of compounds and their mixtures normally are determined by the Tag closed cup or Cleveland open cup test method, with open cup flash points generally measuring higher than closed cup flash points. The Cleveland open cup method is described in the ASTM D 92-05a (ASTM 2005), whereas the Tag method used to determine closed cup flash points is detailed in ASTM D 56-05 (ASTM 2004). The methods used to determine fire points are detailed in Scott et al. (1949).

20.4.3 Hydrazine and Monomethylhydrazine Mists, Droplets, and Sprays

Mists can form when saturated vapor cools or liquid fuel is sprayed mechanically. If an ignition source persists, vaporization from the mist can result in ignition at lower temperatures than vapor ignition. In addition, the rapid vapor formation that occurs in hydrazine mists, droplets, or sprays can support fire if the amount of vapor that forms is equal to or greater than the amount of vapor that is consumed (Benz et al. 1988).

Comprehensive information on fire is found in the manuals *Fire, Explosion, Compatibility, and Safety Hazards of Hydrazine* (Pedley et al. 1990), *Fire, Explosion, Compatibility, and Safety Hazards of Monomethyl Hydrazine* (Woods et al. 1993), and *Fire, Explosion, Compatibility, and Safety Hazards of Nitrogen Tetroxide* (Davis et al. 1999).

REFERENCES

American Institute of Aeronautics and Astronautics. (1998) *Space Systems—Metallic Pressure Vessels, Pressurized Structures, and Pressure Components*. American National Standard ANSI/AIAA S-080-1998. Reston, VA: American Institute of Aeronautics and Astronautics.

American Institute of Aeronautics and Astronautics. (2006) *Space Systems Composite Overwrapped Pressure Vessels (COPVs)*. American National Standard ANSI/AIAA-081A-2006. Reston, VA: American Institute of Aeronautics and Astronautics.

Arnold, T. M. (2003) *Booster Systems Briefs*. Publication JSC-19041. Houston, TX: National Aeronautics and Space Administration, Johnson Space Center.

American Society for Testing and Materials. (2000) *Test Method for Autoignition Temperature of Liquid Chemicals*. Standard E 659-78. West Conshohocken, PA: American Society for Testing and Materials.

American Society for Testing and Materials. (2004) *Standard Test Method for Flash Point by Tag Closed Cup Tester*. Standard D 56-05. West Conshohocken, PA: American Society for Testing and Materials.

American Society for Testing and Materials. (2005) *Standard Test Method for Flash and Fire Points by Cleveland Open Cup Tester*. Standard D 92-05a. West Conshohocken, PA: American Society for Testing and Materials.

American Society for Testing and Materials. (2007) *Standard Test Method for Minimum Ignition Energy and Quenching Distance in Gaseous Mixtures*. Standard E582-07. West Conshohocken, PA: American Society for Testing and Materials.

Atkinson, A. J., and T. I. Eklund. (1971) *Crash Fire Hazard Evaluation of Jet Fuels*. FAA-RD-70-72. Philadelphia: Naval Air Propulsion Test Center.

Babst, D., and P. Axelrod. (1990) This HERO is a villain. *Peace Magazine* 6, no. 2: 23.

Benz, F. J. (1980) *The Exothermicity of Monomethyl Hydrazine*. Technical Report TR-271-001. Las Cruces, NM: National Aeronautics and Space Administration, Johnson Space Center, White Sands Test Facility.

Benz, F. J., C. V. Bishop, and M. D. Pedley. (1988) *Ignition and Thermal Hazards of Selected Aerospace Fluids: Overview, Data, and Procedures*. Research Document RD-WSTF-0001. Las Cruces, NM: National Aeronautics and Space Administration, Johnson Space Center, White Sands Test Facility.

Benz, F. J., and D. L. Pippen. (1980) Autoignition, flammability and explosion properties of hydrazine and monomethylhydrazine. In J. A. E. Hannum (ed.), *JANNAF Safety and Environmental Protection Specialist Session*. Laurel, MD: Chemical Propulsion Information Agency, pp. 603–616.

Bodurtha, F. A. (1980) *Industrial Explosion Prevention and Protection*. New York: McGraw-Hill.

Botteri, B. P., R. E. Cretcher, J. R. Fultz, and H. R. Lander. (1966) *A Review and Analysis of the Safety of Jet Fuel*. AFAPL-TR-66-9. Wright-Patterson AFB, OH: U.S. Air Force, Air Force Systems Command.

Burgess, M. J., and R. V. Wheeler. (1911) Lower limit of inflammation of mixtures of paraffin hydrocarbons with air. *Journal of the Chemical Society* 99: 2013-2030.

Carlton, R. L. (2001) Interview by Kevin M. Rusnak. Houston, TX: National Aeronautics and Space Administration, Johnson Space Center, Oral History Project.

Chang, E. T., and N. A. Gokcen. (1976) *Compatibility of Alloys with Hydrazine Containing Freon*. TR 76-34. El Segundo, CA: Aerospace Corporation.

Cooper, P. W., and S. R. Kurowski. (1996) *Introduction to the Technology of Explosives*. New York: Wiley-VCH, p. 84.

Coward, H. F., and G. W. Jones. (1952) *Limits of Flammability of Gases and Vapors*. Bulletin 503. Washington, DC: Bureau of Mines.

Davis, D. D., and R. C. Wedlich. (1991) Kinetics and mechanism of the thermal decomposition of monomethylhydrazine by accelerating rate calorimetry. *Thermochimica Acta* 175: 189-198.

Davis, D., D. L. Baker, L. A. Dee, B. Greene, C. H. Hart, and S. S. Woods. (1999) *Fire, Explosion, Compatibility, and Safety Hazards of Nitrogen Tetroxide*. Research Document RD-WSTF-0017. Las Cruces, NM: National Aeronautics and Space Administration, Johnson Space Center, White Sands Test Facility.

Drell, I. L., and F. E. Belles. (1958) *Survey of Hydrogen Combustion Properties*. Report 1383. Cleveland, OH: National Advisory Committee on Aviation, Lewis Flight Propulsion Laboratory.

Furno, A. L., A. C. Imhof, and J. M. Kuchta. (1968) Effect of pressure and oxidant concentration on auto-ignition temperatures of selected combustibles in various oxygen and dinitrogen tetroxide atmospheres. *Journal of Chemical Engineering Data* 13: 243-249.

Furno, A. L., G. H. Martindill, and M. G. Zabetakis. (1962) Limits of flammability of hydrazine-hydrocarbon vapor mixtures. *Journal of Chemical Engineering Data* 7:375-376.

Glassman, I. (1987) *Combustion*. Orlando, FL: Academic Press.

Harrje, D. T., and F. H. Reardon (eds.). (1972) *Liquid Propellant Rocket Combustion Instability*. Special Publication SP-194, Washington, DC: National Aeronautics and Space Administration, Headquarters.

Huzel, D. K., and D. H. Huang. (1992) *Modern Engineering for Design of Liquid-Propellant Rocket Engines*. Washington, DC: American Institute of Aeronautics and Astronautics.

Jensen, A. V. (ed.). (1976) *Chemical Rocket/Propellant Hazards*, vol. 1, *General Safety Engineering Design Criteria*. Silver Spring, MD: Johns Hopkins University.

Kanury, A. M. (1975) *Introduction to Combustion Phenomena*. New York: Gordon and Breach.

Kuchta, J. M. (1973) *Fire and Explosion Manual for Aircraft Accident Investigations*. AFAPL-TR-73-74. Wright-Patterson AFB, OH: U.S. Air Force, Air Force Systems Command.

Kuettener, J. P., F. E. Miller, J. L. Cassidy, J. C. Leveye, and R. I. Johnson. (1964) *The Mercury-Redstone Project*. Publication TMX 53107. Huntsville, AL: National Aeronautics and Space Administration, Marshall Space Flight Center.

Lapedes, D. N. (1974) *Dictionary of Scientific and Technical Terms*. New York: McGraw-Hill.

Leveson, N. G., and J. Cutcher-Gershenfeld. (2004) What system safety engineering can learn from the *Columbia* accident. Providence, RI: International System Safety Conference.

Lewis, B., and G. von Elbe. (1961) *Combustion, Flames, and Explosions of Gases*. New York: Academic Press.

Lloyd, J. (2005) *Death on the Steppes—The Nedelin Rocket Disaster*. Washington, DC: National Aeronautics and Space Administration, Headquarters.

Martin, N. B., D. D. Davis, J. E. Kilduff, and W. C. Mahone. (1989) *Environmental Rate of Hydrazines: AFESC Final Report*. Las Cruces, NM: National Aeronautics and Space Administration, Johnson Space Center, White Sands Test Facility.

Meyer, R., J. Köhler, and A. Homburg. (2002) *Explosives*. Weinberg, Germany: Wiley-VCH Verlag.

Michel, J. A., J. Troe, and H. G. Wagner. (1963) *Detonation and Shock-Tube Studies of Hydrazine and Nitrous Oxide*. Technical Report ARL 63-157. AD 419-097. Charlottesville, VA: United States Air Force, Aerospace Research Laboratory.

Miller, E. L., and L. A. Schluter. (1978) *Thermal Regeneration Temperatures of Materials Exposed to Hydrazine Vapor and Air Mixtures*. Technical Report TR-225-001. Las Cruces, NM: National Aeronautics and Space Administration, Johnson Space Center, White Sands Test Facility.

Moore, P. O., C. N. Jackson Jr., and C. N. Sherlock. (1998) *Nondestructive Testing Handbook*, vol. 1, *Leak Testing*. Columbus, OH: American Society for Nondestructive Testing.

Mullins, B. P. (1955) *Spontaneous Ignition of Liquid Fuels*. London: Butterworth.

Mullins, B. P., and S. S. Penner. (1959) *Explosions, Detonations Flammability and Ignition*. London: Pergammon Press.

National Aeronautics and Space Administration. (1998) *Flammability, Odor, Offgassing, and Compatibility Requirements and Test Procedures for Materials in Environments That Support Combustion*. Technical Standard NASA-STD-6001. Huntsville, AL: National Aeronautics and Space Administration, Marshall Space Flight Center.

Newell, H. E. (1980) *Beyond the Atmosphere: Early Years of Space Science*. Special Publication SP-4211. Washington, DC: National Aeronautics and Space Administration, Headquarters.

Overly, J. (1976) Establishing criteria for the ignition: An experimental and theoretical study of the effect of spark geometry on the minimum ignition energy of hydrazine. Ph.D. dissertation, Vanderbilt University.

Pedley, M. D. (1987) *Accelerated Rate Calorimetry Analysis of Various 300-Series Stainless Steels Found in the Rockwell RCS Low-Pressure Solenoid Valve*. Technical Report TR-515-001. Las Cruces, NM: National Aeronautics and Space Administration, Johnson Space Center, White Sands Test Facility.

Pedley, M. D., D. L. Baker, H. D. Beeson, R. C. Wedlich, F. J. Benz, R. L. Bunker, and N. B. Martin. (1990) *Fire, Explosion, Compatibility, and Safety Hazards of Hydrazine*. Research Document RD-WSTF-0002. Las Cruces, NM: National Aeronautics and Space Administration, Johnson Space Center, White Sands Test Facility.

Pelton, J. N., D. Smith, N. Helm, P. MacDoran, and P. Caughran. (2005) Space safety report—Vulnerabilities and RISK Reduction. In Peter Marshall (ed.), *U.S. Human Space Flight Programs*. Washington, DC: George Washington University.

Price, E. W., and F. E. C. Culick. (1968) *Combustion of Solid Rocket Propellants*. AIAA Professional Study Series. New York: American Institute of Aeronautics and Astronautics.

Savary, L. M., and C. Crawford-Mason. (2006) *The Nun and the Bureaucrat*. Washington, DC: CC-M Productions.

Sax, N. I. (1975) *Dangerous Properties of Industrial Materials*. New York: Van Nostrand Reinhold.

Schöyer, H. F. R. (ed.). (1993) *Combustion Instability in Liquid Rocket Engines*. WPP-062. Noordwijk, the Netherlands: European Space Agency, European Space Research and Technology Center.

Schöyer, H. F. R., A. J. Schnorhk, P. O. A. G. Korting, P. J. van Lith, J. M. Mul, G. M. H. J. L. Gadiot, and J. J. Meulenbrugge. (1995) High-performance propellants based on hydrazinium nitroformate. *Journal of Propulsion and Power* 11, no. 4: 856–869.

Scott, F. E., J. J. Burns, and B. Lewis. (1949) *Explosive Properties of Hydrazine*. Technical Report BM-RI-4460. Pittsburgh: Bureau of Mines.

Setchkin, N. P. (1954) Self-ignition temperatures of combustible liquids. *NBS Journal of Research* 53: 49–66.

Shaw, B. H. (1994) *Safety Policy for Detecting Payloads Design Errors*. Technical Assessment TA-94-018. Houston, TX: National Aeronautics and Space Administration, Johnson Space Center.

Stevens, B. D., and F. J. Benz. (1978) *Autoignition Characteristics of Monomethylhydrazine at Reduced Pressure*. Technical Report NASA-TR-205-003. Las Cruces, NM: National Aeronautics and Space Administration, Johnson Space Center, White Sands Test Facility.

Strehlow, R. A. (1984) *Combustion Fundamentals*. New York: McGraw-Hill.

Sutton, G. P. (2001) *Rocket Propulsion Element: An Introduction to the Engineering of Rockets*. New York: Wiley and Sons.

United Nations. (2005) *Recommendations on the Transport of Dangerous Goods: Model Regulations*. New York: United Nations Publications.

Uney, P. E., and D. A. Fester. (1972) *Material Compatibility with Space Storable Propellants: Design Guidebook*. Document NAS7-100. Pasadena, CA: California Institute of Technology, Jet Propulsion Laboratory.

von Braun, W., and F. I. Ordway. (1976) *The Rocket's Red Glare*. New York: Anchor Press.

von Kàrmàn, T., and L. Edson. (1967) *The Wind and Beyond*. Boston: Little, Brown and Company.

Wedlich, R. C., D. D. Davis, and T. Peters. (1988) *Evaluation of Thermal Hazards of Hydrazine Decomposition by Accelerating Rate Calorimetry*. Monterey, CA: Joint Army-Navy-NASA-Air Force Combustion Conference, Safety and Environmental Protection Subcommittee Meeting.

Woods, S. S., D. B. Wilson, R. L. Bunker, D. L. Baker, and N. B. Martin. (1993) *Fire, Explosion, Compatibility, and Safety Hazards of Monomethyl Hydrazine*. Research Document RD-WSTF-0003. Las Cruces, NM: National Aeronautics and Space Administration, Johnson Space Center, White Sands Test Facility.

Yarnell, N. (2006) *Design, Verification/Validation and Operations Principles for Flight Systems (Design Principles)*. DocID 43913. Pasadena, CA: California Institute of Technology, Jet Propulsion Laboratory.

CHAPTER 21

Pyrotechnic Safety

Keith E. Van Tassel
Pyrotechnics Project Manager and Explosives Safety Officer, Johnson Space Center, National Aeronautics and Space Administration, Houston, Texas

CONTENTS

21.1. Pyrotechnic Devices 695
21.2. Electroexplosive Devices 696

21.1 PYROTECHNIC DEVICES

During the era of the NASA Apollo Program, engineers who used explosives in the design of separation systems, emergency escape systems, and landing and deceleration systems chose to call these types of energetic devices *pyrotechnics* as opposed to *ordnance*.

The range of pyrotechnic devices encompasses all components and assemblies that contain or are operated or actuated by propellants or explosives. Such devices include items such as initiators, detonators, safe and arm devices, cartridges, disconnects, retractors, thrusters, transfer assemblies, through bulkhead initiators, shaped charges, military hardware worn for crew escape, mortars, circuit interrupters, dimple motors, oxygen candles, and igniters but specifically exclude large rocket motors.

The use of the term *pyrotechnics* has carried over for explosives and explosive devices used on human rated space vehicles. This convention, however, is not followed universally outside the human rated community; for engineers and technicians who design and build uncrewed launch vehicles nominally refer to explosive devices as *ordnance* or *energetics*. Some uncrewed space vehicle projects prefer to define pyrotechnics to be only those devices that contain explosive material, such as initiators, cartridges, detonators, squibs, and primers. As will be seen later in this chapter, it is important to include both the explosive and the device it operates when addressing the safe handling and operation of pyrotechnics.

21.1.1 Explosives

An explosive is a substance that undergoes a very rapid chemical transformation to other more stable products that are entirely or largely gaseous and whose combined volume is much greater than the original product when it is subjected to heat, impact, friction, or other suitable impulse (Hayes 1938).

Explosives can be broken into two basic types:

- Deflagrating.
- Detonating.

NASA-S-84-05656, defines *deflagration* as the propagation of a reaction front through a material at velocities less than the speed of sound for the unreacted mix. A deflagrating pyrotechnic produces a considerable amount of gas at high pressure that is used to do work, usually by pushing on a piston. Typical applications include thrusters, retractors, pin pullers, and mortars.

Detonation is defined as the propagation of the reaction front through a material at velocities greater than the speed of sound for the unreacted mix. A detonating pyrotechnic usually produces a very strong shock wave, which is used to break material rather than pushing on a piston, as does a deflagrating pyrotechnic. Typical applications include breaking frangible nuts and explosive bolts, and they find use in a range safety system where a linear shaped charge is used to cut through the casing of a solid rocket booster.

21.1.2 Initiators

The beginning of any explosive train is the initiator. Initiators are used to ignite both deflagrating and detonating pyrotechnics. Initiators can be designed to be ignited by a variety of methods, such as electronically, mechanically, or with laser energy. Although not uncommon in terrestrial applications, laser initiated pyrotechnics have not been used widely in NASA programs and are not addressed further in this chapter. An example of a mechanical initiator is the primer used in the Space Shuttle overhead window crew escape T-handle.

21.2 ELECTROEXPLOSIVE DEVICES

An electrically initiated device is known generically as an *electroexplosive device*. Most of these devices are ignited by passing a substantial electric current through a small bridgewire imbedded in the propellant charge. The hot bridgewire heats the propellant to autoignition temperature and causes the initiator to fire. Usually, the initiator is used to light at least one other explosive charge, which actually does the work. Occasionally, enough gas is produced by an electrical initiator that it is used by itself. However, such use of an initiator is not considered good design practice and should be considered an exception and not the rule.

21.2.1 Safe Handling of Electroexplosive Devices

Safe handling and operations practices are similar, whether an electroexplosive device is igniting a deflagrating pyrotechnic or a detonating pyrotechnic. The desire is to prevent the electroexplosive device from igniting prematurely. Training is an essential part of pyrotechnic safety. Attending formal explosive handler safety training that includes supervised hands-on training with an experienced explosives handler is considered mandatory.

For safe handling of pyrotechnics the cardinal rule of explosives operations is this: Expose the least number of people to the least amount of explosives for the least amount of time. Having the minimum number of operators present during live operations is based on this crucial rule and should be a part of the established operating procedures for any facility that tests pyrotechnics or installs pyrotechnics on spacecraft. Most facilities require use of the buddy system when performing pyrotechnic operations and installations. The safety benefits of the buddy system include the facts that there is an extra set of eyes to ensure the procedures are performed properly and that someone is available to go for help in the event of an accident.

When being handled, the electroexplosive device must be protected from electromagnetic radiation that could generate a sufficiently high voltage across the pins to cause initiation. Two methods generally are used to prevent this:

- Faraday caps.
- Shorting clips.

Shorting clips are placed across the pins of the initiator, causing a short circuit (Figure 21.1). A Faraday cap covers the entire connector end of the electroexplosive device forming a Faraday cage while simultaneously shorting the pins together (Figure 21.2).

FIGURE 21.1 NASA standard initiator with a shorting clip (Photo by F. Salazar; courtesy of NASA).

FIGURE 21.2 NASA standard initiator with a Faraday cap (Photo by F. Salazar; courtesy of NASA).

Static Electricity

When the Faraday cap or the shorting clip is removed so that the electrical connector from the firing circuit can be attached, the explosives handler must take all necessary action to ensure that no static electricity has accumulated on his or her body that could result in an electrostatic discharge to the electroexplosive device. The simplest approach is to use the first touch method, that is, the handler touches a large piece of nearby conductive metal just prior to handling the electroexplosive device. This allows any static electricity accumulation to dissipate from the handler. Although adequate, using this method alone usually is met with skepticism because it relies on the explosive handler being disciplined sufficiently to perform the procedure every time. Consequently, other means usually are employed to ensure that no static charge accumulates near the electroexplosive device. The most common is the use of a conductive wrist stat, a strap on device worn around the wrist of the explosives handler (Figure 21.3). The wrist stat is connected to facility ground by means of a grounding strap. Other methods include the use of leg stats, conductive shoes, conductive workbenches, and conductive mats grounded to facility ground, as well as floors painted with conductive paint. Explosives handlers also are required to wear appropriate personal protective equipment, which include safety glasses, cotton clothing, and in some cases, flame retardant smocks or coveralls. Periodic inspections and annual testing of facility ground are required to ensure proper grounding and bonding during explosive operations. Wrist stats, leg stats, and conductive shoes normally are tested daily and prior to each use.

Coding and Labeling

One of the standard adages for handling pyrotechnics is nearly identical to that for handling guns: Always treat a pyrotechnic as if it were live. Pyrotechnics should be color coded to indicate the status of the device. All inert explosives should be marked, labeled, stenciled, or

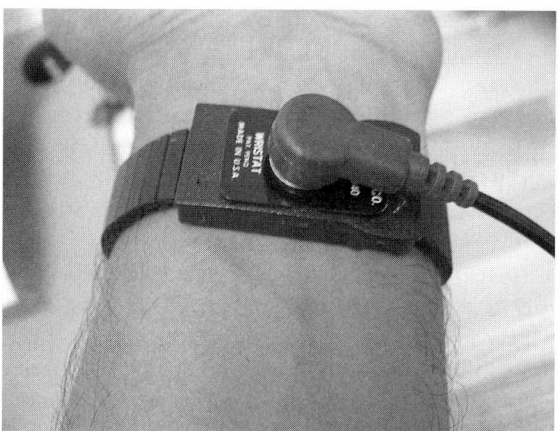

FIGURE 21.3 Wrist stat (Photo by F. Salazar; courtesy of NASA).

tagged as to their status. Explosives not identified as inert always should be assumed to be live. Usually, the color red designates live pyrotechnic devices, whereas the color blue indicates pyrotechnic devices that are inert. Note that this system of color coding is not the only one in use; for example, NASA flight pyrotechnic devices are never painted. Because color coding systems vary, training in color coding differences and the details of operating procedures must contain information pertaining to the color coding scheme in use.

Blast Shields

During electrostatic discharge ground testing, blast shields are required to provide protection between the test article and the technician performing the test. Requirements for blast shield design are derived from MIL-STD-398, *Shields, Operational for Ammunition Operations, Criteria for Design of and Test for Acceptance* (DoD 1976).

Transmitting Devices

During the installation process for pyrotechnics in a ground test fixture or a spacecraft, any broadcasting devices, including radios, walkie-talkies, and cell phones, must be turned off. This is usually a part the operating procedures for any facility.

Installation and Checkout

The two rules of thumb related to the installation process for pyrotechnics are

- Final hookup of pyrotechnics should take place as late as possible in the vehicle assembly flow.
- No other operations are allowed during final hookup.

Once installed in a spacecraft, an electroexplosive device, such as the NASA standard initiator, requires checkout before use. This is accomplished by measuring the bridgewire resistance.

The use of low current resistance measuring devices prevents inadvertent ignition during check out. There is a bridgewire resistance between 0.95Ω and 1.15Ω for the NASA standard initiator. Resistance measuring equipment used to determine bridgewire resistance must have current ratings much below the rated no-fire current of the electroexplosive device. The NASA standard initiator, for example, must not ignite when the bridgewire is subjected to a DC current of 1A for 5 min or a DC power of 1W for 5 min. The maximum allowed current for NASA standard initiator bridgewire resistance measuring equipment is 50mA.

One final area of safety operations is concerned with the post-flight checkout phase of a recovered spacecraft. Every electroexplosive device on the spacecraft must be checked for continuity to verify that firing actually has occurred. Any unfired electroexplosive device is considered live and dangerous and must be removed by certified explosives handlers. When disassembling pyrotechnic devices, such as thrusters and pin pullers, the operator must keep in mind that the pyrotechnic remains under pressure. To avoid injury during disassembly, the technician slowly must crack open the cartridge and let the gas escape before proceeding with the final disassembly.

21.2.2 Designing for Safe Electroexplosive Device Operation

As with the safe handling of pyrotechnics, the desire here is to design the pyrotechnic device and its firing system to prevent the electroexplosive device from igniting prematurely. A number of standard rules are used throughout the aerospace industry and the military. Many of the requirements come from *Electroexplosive Subsystem Safety Requirements and Test Methods for Space Systems*, MIL-STD-1576 (DoD 1984). In general, the three design objectives are

- To prevent inadvertent initiation from electromagnetic radiation, electrostatic discharge, or switch failures.
- To prevent inadvertent initiation of the pyrotechnic device from sudden impact, such as being mishandled or dropped.
- To protect the vehicle and personnel from shrapnel and hot gas should the pyrotechnic device discharge unexpectedly.

Shielding
Protection from electromagnetic radiation includes the incorporation of twisted, shielded pair wiring between the firing circuit and the electroexplosive device. The first line of defense here is the shielding. It must completely cover the wiring from initiator connector to the firing circuit and must be grounded to the spacecraft ground. The shielding substantially attenuates electromagnetic radiation from external sources, ranging from cell phones to radar antennas. Twisting the wires further serves to cancel out the effects of any electromagnetic fields that happen to penetrate the shielding (Figure 21.4).

Inhibits
The circuit used to send the firing current to an electroexplosive device is required to have three inhibits, all of which normally are included in the form of switches, relays, and transistors. Switches can be operated either manually, electronically, or in some combination,

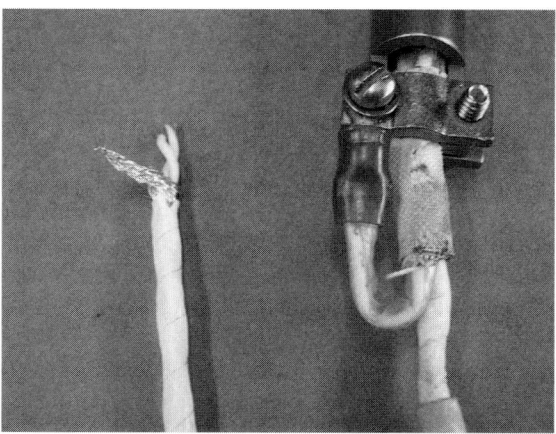

FIGURE 21.4 Twisted shielded pair (Photo by F. Salazar; courtesy of NASA).

depending on the requirement for a human operator to initiate a pyrotechnic function. One example of manually operated inhibits is the deployment of the pilot chute for the Space Shuttle drag chute. The first inhibit, or pre-arm, is a circuit breaker. The second inhibit, or arm, is a switch that powers up the circuit and charges the firing circuit capacitors. The third and final inhibit, or fire, is a switch that discharges the energy stored in the capacitor into the electroexplosive device. One very important safety design note is that all three inhibits switch both the high side and the return side of the firing circuit. This design prevents possible sneak paths in the firing circuitry from circumventing any of these inhibits.

Another design feature of a safe firing circuit is the use of bleed resistors when the firing circuit uses a capacitive discharge to fire the electroexplosive device. These resistors in the circuit allow any extraneous charge that might have accumulated in the capacitor bank to bleed off. They also bleed off any residual charge from the capacitor after it has been used to fire the electroexplosive device.

Impact Testing

To prevent inadvertent initiation of a pyrotechnic device from a sudden impact, such as being mishandled or dropped, pyrotechnic devices must be designed to withstand impact environments. The best way to verify that the design is adequate in this respect is to perform drop testing on a representative sample from the production lot of pyrotechnics. Two common tests are

- The 8-ft drop test.
- The 40-ft drop test.

For the 8-ft drop test, a pyrotechnic device is dropped from a height of at least 8 ft onto a steel plate. The pyrotechnic device usually is dropped in three orthogonal axes. The pyrotechnic device must not fire as the result of hitting the steel plate, and it must still function properly after the drop. For the 40-ft drop test, a pyrotechnic device is dropped from a height of at least 40 ft onto a steel plate. As with the 8-ft drop test, the pyrotechnic device usually is dropped in three orthogonal axes. The pyrotechnic device must not fire as a result of hitting the steel plate; however, it does not have to function properly after the drop.

Rupture

To protect the vehicle and personnel from shrapnel and hot gas should a pyrotechnic device be fired unexpectedly, it must be designed to withstand a locked shut condition. In a lock shut test, the moving article, usually a piston of some type, is constrained from movement while the device is fired. The pyrotechnic device must not rupture, create shrapnel, or allow hot gas to escape when fired in this manner.

21.2.3 Pyrotechnic Safety of Mechanically Initiated Explosive Devices

Historically, mechanically initiated explosive devices preceded development of the electroexplosive device by many centuries. However, that does not detract from their use in present day space vehicle applications. The use of a mechanically initiated explosive device is recommended for those applications where an electrical firing circuit would interfere with sensitive instruments; where actuation in case of an emergency is controlled directly by an astronaut, such as the T-handle for side hatch escape; and where electrical energy is low or unavailable, such as a reefing line cutter on a parachute. A mechanically initiated explosive device can be discharged by a number of means:

- Percussion.
- Use of an electroexplosive device precursor.
- Direct application of heat energy.

In space applications, a mechanically initiated explosive device nearly always is functioned by a spring loaded firing pin that strikes a common everyday percussion primer, such as the primers used in shotgun shells and gun cartridges. This percussion primer, or cap, is struck by a blow that causes deformation of the metal, which then pinches a small amount of a pressure sensitive pyrotechnic material between the deformation and an internal anvil. The sensitive powder then ignites, setting off the explosive train.

A few devices currently used in space vehicles meet the criteria of being activated strictly by mechanical means. These limited applications include the T-handles for overhead window severance, side hatch jettison, and vent severance on the *Space Shuttle*, the reefing line cutters for the drag chute, and the igniters for oxygen candles used on the *International Space Station*.

As stated in other sections of this chapter, every mechanically initiated explosive device must be handled with care and only by trained explosive handlers. Although the threats of electrostatic discharge and electromagnetic radiation are not as great as with a sealed device, loose or open powders are sensitive and can ignite when provided the proper stimulus. The standard personal protective equipment used in powder loading procedures and the handling of electroexplosive devices must be used when working with a mechanically initiated explosive device.

The design for a mechanically initiated explosive device that initiates detonations generally consists of a cup or shell in which a quantity of a secondary high explosive has been consolidated or pressed to near its theoretical density. This process is followed usually by additional increments of a consolidated high explosive until the desired quantity is reached. Next, a primary or particularly sensitive explosive is measured in and consolidated. At this point, the device can take two possible directions, depending on its desired end use. Initiation leading to the primary can take on the form of a delay or a direct initiation, each of which is described in some detail here. In space vehicles, the sensitive primary explosive is energized when the spring loaded firing pin hits the primer. The heat from the primer lights up the primary, which flashes to detonation. The heat and shock wave generated by the detonating primary then cause the secondary explosive to detonate.

Delay columns, as the name suggests, are a means of delaying the activation of the primary for a specified time interval, usually from between 5 ms to several seconds. The time delay can be controlled within a few milliseconds by the manufacturing process. Typically, the column consists of one to four increments of consolidated, specially blended pyrotechnic material having a well established and predictable burn rate. These blends are measured, loaded, and consolidated before being attached to the detonator cup. Intimate contact with the primary explosive is imperative. When the column is initiated, it burns in a manner similar to a fuse, supplying heat energy to flash the primary to detonation. The main difference between the delay column and a fuse is the tightly controlled delay time and the physical size of the column.

If no delay or only a few milliseconds delay is desired, a method can be used that is a combination of a direct ignition and a delay. This device, known as a *deflagration to detonation transition device*, uses the principle that certain rapidly deflagrating materials, when properly confined, ramp up in burn rate to a detonation that in turn is capable of driving a secondary explosive to detonation.

Another form of delayed initiation, though not by much, known as a *slapper*, uses the process of accelerating a small mass to a high velocity, which then impacts a sensitive primary. Although these devices normally are initiated electrically by means of an exploding bridgewire or foil, some have been developed to accelerate a mass across a gap to initiate a mechanically initiated explosive device. Note that the exploding bridgewire and exploding foil types are not used on the *Space Shuttle*, because the high electrical energy requirement could result in damage to avionics.

As with the electroexplosive device, a mechanically initiated explosive device is required to have three inhibits to preclude inadvertent firing. A T-handle, for example, could have the following three inhibits:

- A lid covering the entire T-handle that must be removed prior to firing.
- A release pin in the T-handle to prevent it from accidentally being pulled or vibrated out.
- A squeeze and pull handle to fire the T-handle.

A properly designed mechanically initiated explosive device must be able to protect against inadvertent initiation due to mishandling or dropping. It also must be designed to protect the vehicle and personnel from shrapnel and hot gas should the pyrotechnic device go off unexpectedly. These devices must be tested per the same drop test and lock shut criteria as for the electroexplosive device mentioned earlier in this chapter.

This chapter touched on several items crucial to the safe handling and operating of pyrotechnic devices used on space vehicles. When it comes to handling pyrotechnics, training with an experienced explosives handler is essential. When it comes to installing and operating pyrotechnics, several precautions need to be taken to protect against inadvertent initiation. Establishing and following safety procedures for both preflight and postflight checkouts are required for both personnel and flight vehicle safety. And the cardinal rule of the least number of people exposed to the least amount of explosive for the least amount of time should always be remembered.

REFERENCES

Department of Defense. (1976). *Shield, Operational for Ammunition Operations, Criteria for Design of and Test for Acceptance*. Military Standard MIL-STD-398. Washington, DC: Department of Defense.

Department of Defense. (1984). *Electroexplosive Subsystem Safety Requirements and Test Methods for Space Systems*. Military Standard MIL-STD-1576. Washington, DC: Department of Defense, United States Air Force.

Hayes, T. J. (1938) *Elements of Ordnance: A Textbook for Use of Cadets of the United States Military Academy*. New York: Wiley and Sons.

CHAPTER 22

Extravehicular Activity Safety

Christine E. Stewart
Senior Extravehicular Activities Operations Safety Engineer, Science and Applications International Corporation, Houston, Texas

CONTENTS

22.1. Extravehicular Activity Environment	705
22.2. Suit Hazards	712
22.3. Crew Hazards	716
22.4. Conclusions	722

22.1 EXTRAVEHICULAR ACTIVITY ENVIRONMENT

Designing hardware or experiments that are to be located or operated in an extravehicular environment requires understanding how materials and mechanisms react to the zero-g thermal vacuum environments. Even when experiment hardware is not designed to interface with a crewmember engaged in extravehicular activity, there are very credible scenarios in which incidental contact during this activity is possible. As the result of efforts to balance multiple competing requirements, the NASA extravehicular mobility unit (Figure 22.1) and the Russian Orlan space suits provide limited visibility, mobility, endurance, tactile feedback, and force application. These limitations negatively affect the ability of extravehicular activity crewmembers to react to adverse situations. This amplifies the need to preclude as many hazards as possible.

The extravehicular activity working environment is harsh, and the limitations presented by the extravehicular mobility unit are considerable. For example, the habitable environment is maintained at a pressure of 14.7 psi, but the extravehicular mobility unit is pressurized to only 4.3 psid (NASA 2005). Considering that, when the suit pressure drops to 2.8 psid, an extravehicular activity crewmember loses sight and at 2.0 psid consciousness is lost, it is apparent just how close to a catastrophic potential the extravehicular activity crews are operating.

FIGURE 22.1 The extravehicular mobility unit (Courtesy of NASA).

22.1.1 Definitions

Extravehicular activity is any activity performed by a pressure suited crewmember in an unpressurized or space environment. This can be on the *International Space Station*, in the Space Shuttle payload bay, or even on the surface of the Moon or Mars. An extravehicular activity generally begins with depressurization of the airlock or space module and ends with repressurization of the module or airlock after the crewmember ingresses.

There are three basic classes of extravehicular activity:

- Scheduled.
- Unscheduled.
- Contingency.

A scheduled extravehicular activity, also referred to as a *planned* or *nominal extravehicular activity*, is conducted to accomplish tasks included in the baseline scheduled timeline to support specified mission operations. An unscheduled extravehicular activity, also known as an *unplanned* or *off nominal extravehicular activity*, is conducted to accomplish tasks that are not included in the baseline timeline but are needed to achieve mission success, mission enhancement, or to repair or override failed systems. The final class, the contingency extravehicular activity, is conducted to affect the safety of the vehicle or crew or to ensure the safe return of the vehicle. A contingency extravehicular activity is considered a subset of unplanned extravehicular activities.

As well, tasks are categorized as basic, moderately difficult, or difficult, based on the complexity of the skills involved. Basic tasks are those that require the use of standard tools, restraints, or mobility aids; do not require special training; require little coordination between crewmembers; involve easily accessible work sites; and do not expose crewmembers to unique hazards. Moderately difficult tasks are those that require additional tools or equipment but are still procedurally simple. These tasks can require skills that although not necessarily unique are not used frequently. Some level of coordination between extravehicular activity crewmembers is required, modification of existing procedures and techniques can be necessary, and more extensive training likely is to be needed to accomplish the tasks. Difficult skills are those that require a notable extension of capabilities, such as the use of specialized tools or mobility aids, pose access or restraint problems, or require multiple related extravehicular activities for their accomplishment. These tasks require unique skills, can involve large orbital replacement units or extravehicular activity tools, and likely are to require intricate maneuvers. Time critical or intense coordination between the extravehicular activity crewmembers, the intravehicular activity crew, and the involvement of a remote manipulator system operator are required. The complex requirements necessary for completion of a difficult task are more hazardous for the extravehicular activity crew, and the necessary procedure development and verifications necessarily reflect the additional procedures, training, and crew safety considerations.

Uses of extravehicular activity can include the following operations:

- Mechanical support to major mission objectives, that is the capture, berthing, assembly, deployment, positioning, and mating or demating of large space structures or satellites. Associated activities conducted by the extravehicular activity crew include connecting and disconnecting utilities; removing launch restraints and covers; deploying antennas, sensors, cameras, and other appendages; and fastening (bolting or latching) together structural elements.
- Maintenance and support, that is preventive and corrective maintenance, such as inspecting and replacing equipment modules or orbital replacement units, activating or deactivating experiments, retrieving samples, resupplying propellant or fluids, and repairing meteoroid or other damage.
- Transfer, that is the movement of cargo, equipment, and personnel, including the transfer of disabled crewmembers. This can be done in conjunction with a robotic aide.
- Experimentation, that is conducting experiments with new hardware or techniques in an extravehicular environment.
- Inspection, that is the inspection or photographing of hardware.

22.1.2 Extravehicular Activity Space Suit

The current U.S. extravehicular activity suit is the extravehicular mobility unit. It is in essence a miniature spacecraft that provides the environmental protection, mobility, life support, and communication equipment necessary for a crewmember to perform an extravehicular activity. The Orlan, the Russian extravehicular activity suit, provides the same basic capabilities as the extravehicular mobility unit. It is critical that these suits fit well to maximize the potential of the crewmember performing an extravehicular activity. Without a good fit, discomfort and even injury is possible.

The actual use rates for consumables are a function of the metabolic rate of each crewmember, which in turn, depends on the individual workload and other physiological factors. Consumption of electrical power is relatively constant. The extravehicular mobility unit is designed to accommodate a maximum 7-h extravehicular activity that comprises 15 min for egress, 15 min for ingress, 6 h for useful extravehicular activity work, and 30 min of consumables reserve.

The backup life support provided by the secondary oxygen pack consists of open loop ventilation at a reduced suit pressure and is limited to 30-min duration. This capability represents the limit of the time available for a crewmember to return to the habitable environment or an umbilical should the cooling system of the suit fail.

The extravehicular mobility unit requires the crewmember to use force to overcome the pressure moment and friction forces inherent in the mobility joints. The joints are designed to approximate neutral stability throughout the full range of motion when pressurized to 4.3 psid (29.6 kPa). The extravehicular mobility unit suit joint neutral stability feature helps alleviate the necessity for a crewmember to apply a counteracting force to maintain a desired position.

The reach capability of a crewmember within the extravehicular mobility unit is evaluated by considering two aspects:

- The optimum work envelope of the various sized suited crewmembers (Figure 22.2).
- The omnidirectional reach of a suited crewmember.

The optimum area for one- or two-handed operation is centered about the upper chest and lower face area of the crewmember, a factor also to consider when positioning foot restraints. The work envelope is limited relative to a comparable shirtsleeve Earth based task (Figure 22.2). Reach limits are a function of the anthropometry of the subject and can be used to assess translation paths, handrail spacing, and temporary hardware retrieval or stowage.

The extravehicular activity crewmember in a microgravity environment assumes a position dependent on the space suit configuration. Because of the suit characteristics, this body position can be different from the neutral body posture of the shirtsleeve microgravity environment.

An unrestrained crewmember effectively can perform only low force, very short duration operations, such as actuation of toggle and rotary switches, surveillance of controls and displays, and visual inspections. Unrestrained, a crewmember can exert impulse loads of this type using one hand to hold onto structure and one hand to apply the load.

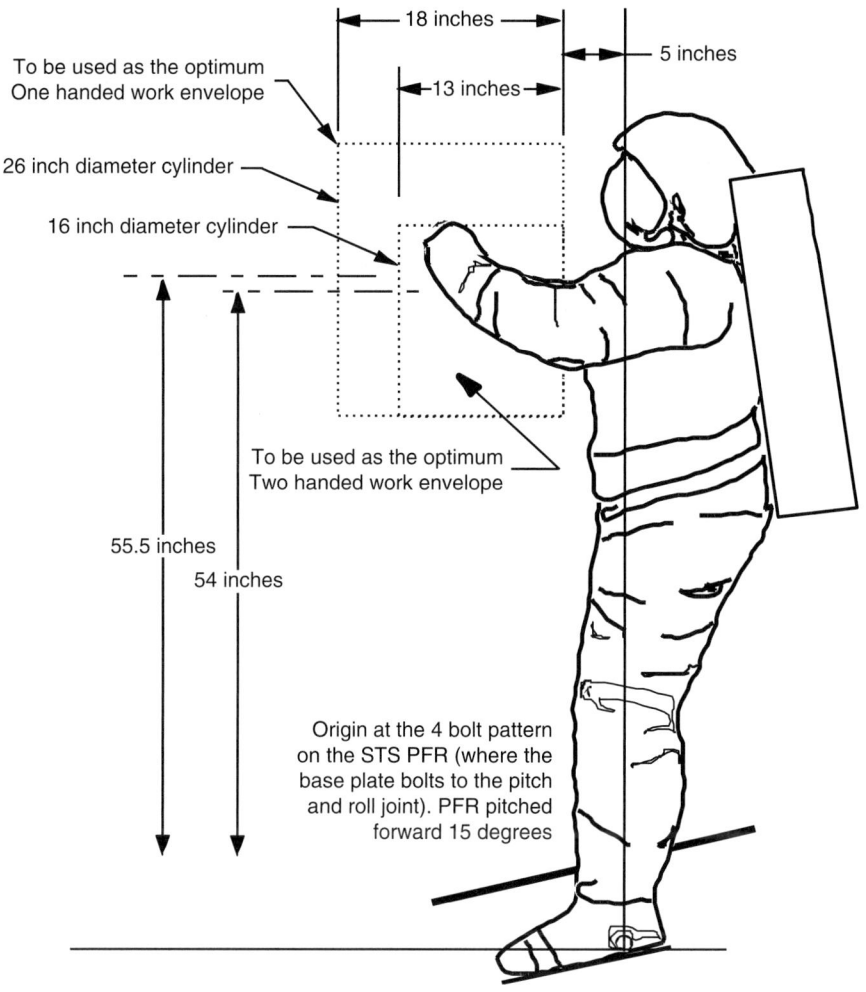

FIGURE 22.2 Crewmember optimum work envelope (Courtesy of NASA).

Use of a rigid tether, such as the body restraint tether, allows two-handed tasks and relatively low force activities to be conducted but with greater control than is possible with a free floating extravehicular activity crewmember. The body restraint tether can interface only with a dog bone handrail or equivalent.

Very high loads, similar to those for which the crewmember is capable in a shirtsleeve environment, can be applied using a foot restraint. The forces that can be applied are reduced considerably when the point of application is moved near the top of the reach envelope of the crewmember. This necessitates providing adequate restraint and proper body orientation for the extravehicular activity crewmember to optimize force output.

Foot restraints have proven to be the most effective means of stabilizing the crewmember and maximizing his or her capabilities. Even when using foot restraints, a crewmember might choose to use a handrail or other aide at the work site to provide additional stabilization, apply additional force, or as an aid while getting in or out of the foot restraint.

Foot restraints are limited in the magnitude of the loads that they can take. The remote manipulator systems of the *Space Shuttle* and the *International Space Station* move out of position under high loads for hard mounted foot restraints. On the *International Space Station*, load alleviators protect the structure from induced loads but can move the crewmember out of position if too much force is applied.

22.1.3 Sensory Degradation

Extravehicular activity crewmembers experience some sensory degradation because of the design of the space suit, which is intended to mitigate hazards of the extravehicular activity environment above and beyond the impacts of microgravity. The sensory degradation experienced by the crewmember is expressed by reduced tactile sensation and degraded visibility. The design of the visor is a compromise between multiple competing requirements, including visibility, impact protection, thermal protection, and solar protection. The result is a limited size for the visor window, which reduces the field of vision. The only important impact, however, is to the peripheral vision of the crewmember. Looking to the side requires the crewmember to turn the entire suit. For this reason, if care is not taken when locating equipment, the possibility exists that critical hardware cannot easily be seen. Also, in the environment of space, shaded areas are not illuminated by diffuse light as in an atmospheric environment. For this reason, the crew may not be able to see everything they can reach; therefore, they accidentally could impact on nearby hardware. The crew also has considerably less tactile sense due to the pressure differential across the suit and its required fabric thickness. Extravehicular activity gloves degrade tactile feedback relative to that during bare hand operations. Although dexterity can be compared to that when using heavy work gloves, crewmembers can operate some standard handles, knobs, toggle switches, and buttons while wearing extravehicular activity gloves. When tasks are sensitive to limits on dexterity, interface compatibility can sometimes be verified by ground based evaluations.

22.1.4 Maneuvering and Weightlessness

The crew has limited mobility in the extravehicular activity environment. Maneuvering is difficult, with or without any additional mass being carried. Translation, however, is practically effortless, and cargo transfer with robotic aids is nearly as easy. Even so, translating while transporting hardware can be much more difficult to accomplish. Torso and limb movements are partly a function of the agility of the crewmember. Roll and pitch control by the suited crewmember is minimal. Roll control is somewhat easier than pitch, although control of that aspect depends on handrail availability. Stopping with precision is difficult under the best of conditions. Yaw is the only direction over which the extravehicular activity crewmember has a manageable degree of control.

There are concerns about the momentum of not only the extravehicular mobility unit but also any hardware carried by the extravehicular activity crewmember. The extravehicular mobility unit plus a crewmember with tools and the simplified aid for extravehicular rescue attached has a total mass of approximately 700 lb. Momentum can be a disadvantage when trying to stop an object, particularly one of high mass or with an unusual center of gravity. An off nominal center of gravity also can add complications to the ability of the crewmember to control an object. In addition, weightlessness allows tools and other items to float away from the work site. To prevent this, rigorous tethering protocols are implemented to avoid loss of hardware and cluttering of space. One-g evaluations can be misleading in attempting to assess aids for the extravehicular activity crew to use for stability and access.

22.1.5 Glove Restrictions

The gloves on both the extravehicular mobility unit and the Orlan use special designs to maintain a semi-closed position to alleviate any pressure on the hand that might cause fatigue over the duration of an extravehicular activity. The suit gloves are similar to a balloon, in that the inflated glove has resistive force to closure. The gloves have a stiffness similar to a pressurized football. Tasks that require grasping or the use of fine motor control can easily fatigue the hands of the extravehicular activity crewmember. Combined with the lack of tactile feedback while wearing the glove, this results in the necessity to design hardware that does not require any considerable fine motor control to operate or service during an extravehicular activity.

22.1.6 Crew Fatigue

Crew fatigue is a concern during extravehicular activity. Performing an extravehicular activity task is very physical and mentally challenging. Both the length of translation and the duration of the extravehicular activity contribute to fatigue of the crew. Although the torques associated with space suit motion are relatively small, certain repetitive tasks requiring arm, wrist, or hand movements tend to fatigue the extravehicular activity crew as well. For hardware being designed for extravehicular activity servicing, tasks such as the manual removal or replacement of threaded fasteners, continuous force-torque applications, and extended gripping functions should be minimized. Hardware providers should simplify such tasks to reduce any unnecessary crew movement and provide for crew rest. Modular extravehicular activity tasks that have simple, straightforward interfaces are usually the least fatiguing. Operational controls such as avoidance of sharp edges, touch temperature exceedance warnings, or antenna and thruster keep out zones add to the mental stress of the crewmember during any extravehicular activity. These should be avoided where design controls are possible.

22.1.7 Thermal Environment

Because the extravehicular activity environment is a vacuum, the thermal environment is harsher than that of the habitable volume. Historically, the maximum thermal range of extravehicular activity is from $-180°F$ to $+300°F$ ($-118°C$ to $+149°C$). If evaluating an extravehicular activity in a new environment, such as the Moon or Mars, a thermal analysis should be performed to determine thermal extremes to which the crew and hardware are expected to be exposed. Because the hardware is exposed to sunlight in a cyclical fashion, its temperature generally transitions from cold to hot to cold, while exposed to the external environment. Additionally, hardware that is shadowed is considerably colder than hardware exposed to sunlight.

22.1.8 Extravehicular Activity Tools

A complement of standard extravehicular activity tools has been used during numerous extravehicular activity tasks. These tools were designed to withstand the extravehicular activity thermal extremes, have proper handling and gripping characteristics, and possess adequate tethering features. During an extravehicular activity task, the required tools are attached to the suit or tool caddies by means of tethers. The tools are not always visible to the crew and can be as far as 3 ft from them or the work site. Because of this, they can impact hardware near the extravehicular activity task for which they are being used. Hardware design must, therefore, account for potential tool impact. Any tool designed for a specific extravehicular activity operation, such as a new handling aid for an experiment, should adhere to the design requirements and good design practices that the extravehicular activity community developed and documented.

22.2 SUIT HAZARDS

The most common extravehicular activity hazards are associated with the space suit. Because the suit can be considered an independent space vehicle that provides life support to the crewmember, any damage to it can be life threatening.

22.2.1 Inadvertent Contact Hazards

Hazards due to inadvertent contact are the most difficult for the extravehicular activity crew to avoid. The extravehicular environment makes it difficult for the extravehicular activity crew to avoid certain areas when preparing for or performing their tasks. The best way to prevent these hazards is to design unpressurized hardware correctly.

Suit Puncture

One of the most common extravehicular activity hazards is puncture of the suit caused by sharp edges, burrs, corners, and the like. Although both the extravehicular mobility unit and Orlan have secondary oxygen supplies for use in the case of a leak, their design is based on a puncture of very small size. Therefore, any puncture to the suit greater than 0.25 in. most certainly would result in the death of the crewmember wearing it. Even a pinhole puncture makes it absolutely essential that the extravehicular activity task be aborted and the crewmember return immediately to the habitable environment. To preserve the integrity of the extravehicular mobility unit and its support equipment, requirements have been

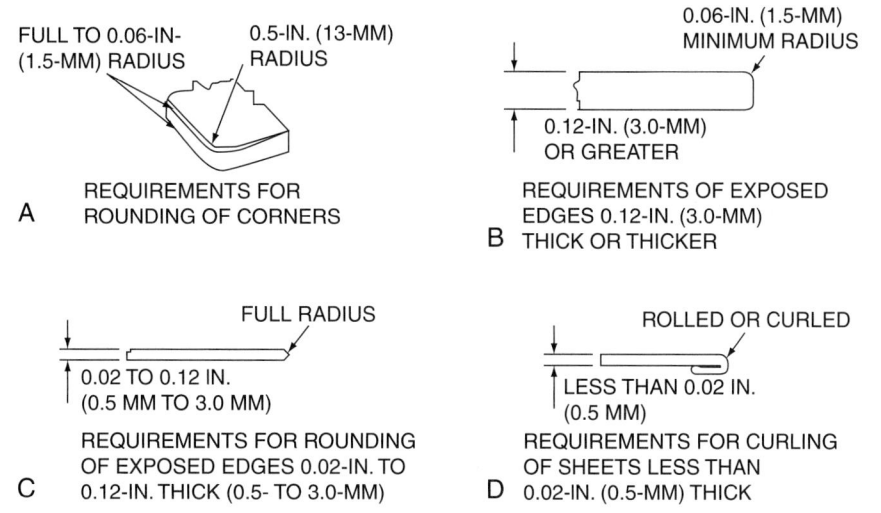

FIGURE 22.3 Exposed corners and edges (Courtesy of NASA).

established (NASA 2005b) that provide specifications for rounding corners and breaking the edges of hardware to which the crew can come into contact (Figure 22.3).

Designing hardware to these requirements unfortunately does not eliminate totally the hazard. During the manufacturing and assembly process, snags and burrs can be created. During respective post-flight inspections, sizeable cuts were observed on the extravehicular mobility unit gloves from extravehicular activities performed on Space Shuttle flights STS-72, STS-76, and STS-116. The STS-76 damage was caused by safety wire that had been exposed during maintenance activities prior to the flight. It is, therefore, very important to verify that all hardware is free of burrs and snags after it is manufactured. Hardware also should be inspected after it is integrated into the vehicle to assure that integration activities, such as tightening a screw, did not create a sharp edge (Figure 22.4). The sharp edge requirements are not comprehensive because the edge radius of certain parts, such as nut plates, screw threads, or spring clips, cannot be measured. Also, honing caused by the repeated cycling of certain hardware, such as the mating and demating of multi-pin connectors, can create sharp edges that are not present in the original design.

Suit Contamination

Certain chemicals present a hazard to the suit. Potential hazards include pressure barrier violations due to corrosive material and crazing of the polycarbonate materials in the helmet. Other considerations include staining, which could affect the thermal profile of the suit. During the return to flight activities for the Space Shuttle Program, NASA discovered that one of the materials proposed to repair the reinforced carbon-carbon material of the tiles stained the back of the glove and resulted in degradation of the thermal properties of the suit. Although the area of the suit actually affected was small, the alteration of the suits

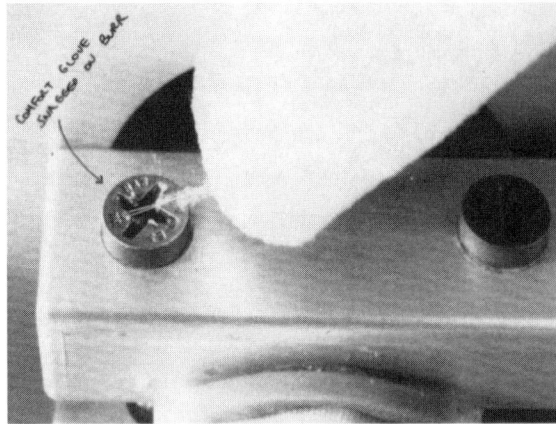

FIGURE 22.4 Example of a sharp edge due to a burr (Courtesy of NASA).

thermal properties give basis for concern during extravehicular activity tasks with a high temperature profile. Any material that can adhere to the suit visor and has the potential to affect visibility also is a concern that must be addressed. Material left on the gloves also can be transferred to and thus contaminate other hardware, such as hatch seals, preventing repressurization of the habitable environment or experiment hardware, causing loss of data.

Cleaning the suit is difficult to accomplish and verify in the extravehicular activity environment. The most effective approach to controlling this hazard is to contain the material appropriately. Any chemical used in the extravehicular activity environment, including lubricants, should be assessed to determine its compatibility with suit materials. Any material capable of causing corrosion sufficient to breach the pressure barrier, visor, or helmet of the extravehicular mobility unit should be contained using three levels of containment or equivalent controls. Any material capable of occluding the visor or affecting the thermal profile of the suit must have at a minimum two levels of containment. It is important to assess the hazard level early in the design, because adding levels of containment after the design is complete is costly and likely requires additional mass and volume. Operational controls to prevent exposure are difficult to implement in the extravehicular activity environment. Choosing to bake out the crew after exposure extends the duration of the extravehicular activity, affects consumable use, and increases crew fatigue.

Shatterable Materials

Many science experiments require the use of glass or another shatterable material to achieve the required results. Unfortunately, in the extravehicular activity environment, the use of such materials creates a potential hazard to the suit and possibly to the habitable environment post extravehicular activity. It is possible that either an extravehicular activity crewmember or tool can impact a shatterable material, such as window or some cover that directly is exposed to the extravehicular activity environment. To withstand such an inadvertent kick load, the material must be tested and verified to be able to survive a quasi-static load of 125 lbf over a circular area with a 0.5-in. diameter. Alternatively, the load can be applied as a dynamic half-second sine wave. An impact load from a tool is a concentrated force of 60 lbm traveling at 1 fps onto a circular area with a 0.08-in. diameter. Should the material shatter from crew contact or tool impact, the resulting fragments have potential to cut the suit. Small particles can become trapped within the folds of the suit, and, on return of the extravehicular activity crew to the habitable environment, these fragments can float free and, if inhaled, damage the lungs of the crew or become lodged in their eyes. Any fragment larger than 30μ can cause damage to eyes.

22.2.2 Area of Effect Hazards

The potential always exists for the need to perform an extravehicular activity in an unexpected area. Although many experiment developers do not plan for an extravehicular activity, experiment hardware failures do occur. When this happens, the spaceflight programs do everything possible to repair that hardware, up to and including the performance of extravehicular activity.

If experiment hardware with an identified hazard is located near another experiment that needs an extravehicular activity to service it or is adjacent to some translation path, the extravehicular activity crew might have to enter the area and be exposed to the hazard. For this reason, if extravehicular activity hardware cannot be designed to preclude hazards, then proper controls should be designed into the hardware to enable removing the hazards prior to performance of the extravehicular activity task.

Radiated Electromagnetic Emissions

Excessive radiated electromagnetic emissions can cause damage to the suit and are potentially harmful to the crew. Such radiated emissions include both radio frequency radiation and magnetic fields. Any magnetic field greater than 250 gauss at the surface of the extravehicular mobility unit (NASA 2005) shuts down the ventilation system. Radiated electromagnetic emissions in excess of those identified in ICD 2-19001, *Shuttle Orbiter/Cargo Standard Interface (Core ICD)*, have the potential to interfere with the suit electronics, including interference with suit to suit communications as well as communications between the suit and the *International Space Station* or *Space Shuttle* (Boeing 1998). Loss of the ventilation system or damage to suit electronics can lead to loss of the extravehicular activity crew or, at best, termination of the extravehicular activity task. When designing extravehicular hardware that is to be powered by or uses magnetic force, the hardware designer should keep the value of these parameters below designated limits or incorporate controls to permit removal of power to the circuits prior to an extravehicular activity task being conducted near the hardware. A minimum of three controls should be provided to ensure power is removed.

Miscellaneous Area of Effect Hazards

Other area of effect hazards generally controlled operationally, are as hazardous to the crew as they are to the suit. This miscellaneous area of effect hazards includes rotating structures, thrusters, and ionizing radiation. Rotating structures can snag the suit or collide with the crew. Thrusters can plume the extravehicular activity crew, leading to damage to the suit, as well as causing injury or death of the crewmember. Ideally, these hazards should be eliminated from the hardware design.

22.3 CREW HAZARDS

Hazards to the crew are just as momentous as hazards to the suit. There are several types of potential hazards to an extravehicular activity crew, although they are not found commonly in every piece of extravehicular hardware to which the crew has contact or exposure.

22.3.1 Contamination of the Habitable Environment

Even if a hazardous chemical does not damage the suit, once the suit has been contaminated, the chemical can be brought into the air lock. This presents a number of concerns relating to the crew being exposed to the uncontained hazardous substance in the habitable environment. After the extravehicular activity crewmember wearing the contaminated extravehicular mobility unit has returned to the air lock and it has been repressurized, the extravehicular activity crew is exposed to the contaminating substance while removing the suit. Because of the myriad of materials to which the crew could be exposed, detection of a particular chemical substance is not always possible. Even if detection was possible, often, the safe exposure level of material is beneath the sensitivity of the detector. It is very difficult to verify that cleaning the suit or allowing the contaminant to off gas in space by prolonging the extravehicular activity is effective. Therefore, the correct approach to addressing this hazard is to ensure that the chemical does not come into contact with the suit through proper design of the required containment of the chemical. All chemicals should be assessed for toxicity so that the appropriate level of containment can be established early in design and development of the hardware.

22.3.2 Thermal Extremes

In the extravehicular activity environment, thermal extremes are a primary consideration. Contact by an extravehicular activity crewmember with any external hardware that exceeds the touch temperature limits, either hot or cold, can result in damage to the suit and injury or even loss of crew. By coming into contact with hardware exceeding the temperature limits, the crew can overheat or freeze to an extent leading to illness or death for the worst case. At minimum thermal extremes, such contact can cause tissue damage in the form of a burn or frostbite. For hot surfaces, the pain threshold is usually reached prior to any physical injury to the crew. Cold extremes are more difficult to control operationally because the hands and feet become numb prior to any damage occurring from frostbite or other cold related injuries. In fact, at least one astronaut suffered frostbite because of holding onto cold hardware for too long. For this reason, a thermal analysis should be performed on every piece of hardware in the extravehicular activity environment.

Table 22.1 Passive Design Temperature Extremes (Courtesy of NASA)

Condition	Maximum Surface Optical Property Ratio	Continuous Contact with External Structure	Design Temperature Extremes °F (°C)
Nonoperating	0.9	Allowed	−180 to +300 (−118 to +149)
Operating, case 1	0.9	Allowed	−180 to +250 (−118 to +121)
Operating, case 2	0.9	Not allowed	−130 to +250 (−96 to +121)
Operating, case 3	0.45	Not allowed	−130 to +200 (−96 to +93)

Note: Verification by test should employ a range 20°F (11°C) in excess of the designated hot and cold extremes.

Table 22.1 lists passive design extravehicular activity temperature extremes enveloping the maximum thermal range, both historical and envisioned, that can be assumed without further thermal analysis. The values are for operating and nonoperating passive design cases, that is, those in which there is no internal heat generation, and with end-of-life surface optical property ratios of 0.9 or less.

Inadvertent Contact

The NASA extravehicular activity requirements currently state that hardware surfaces that can be touched must remain within the thermal range of −244°F to 320°F (−153°C to 160°C). Many hardware providers, however, do not consider that their hardware could ever require extravehicular activity access for any reason. This assumption is problematic in the sense that the hardware design does not address this hazard. Historically, on many occasions, an extravehicular activity crewmember, who was either translating or performing a task, inadvertently contacted hardware, which because no planned extravehicular activity task had ever been envisioned, had not been designed to meet these requirements.

Continuous Contact

For hardware that is known to require a planned extravehicular activity task and for which extended extravehicular activity crew contact is necessary, another set of thermal contact limits apply. Generally, continuous contact requirements are based on limits of 30 min. For glove palm contact, hardware that remains between −80°F to +150°F (−63°C to +65°C) does not damage the crew or suit. In rare cases, extravehicular activity operations or hardware design require long duration compression of the thermal micrometeoroid garment orthofabric, thereby thermally shorting the multilayer insulation and making thermal micrometeoroid garment layers vulnerable to thermal damage. Examples of these cases are hardware within or containing confined space that requires any portion of the extravehicular mobility unit to be compressed during extravehicular activity operations. For hardware that requires prolonged compression of the extravehicular mobility unit thermal micrometeoroid garment, thermal extremes outside the temperature range of −195°F to +240°F (−126°C to 116°C) during extended contact operations can harm the crew and even damage the suit.

22.3.3 Lasers

Laser radiation, while not likely to damage the suit, can injure the crew. The extravehicular mobility unit visor does not protect adequately in the visible light range. Any laser used in the extravehicular activity environment must be analyzed for hazard level. Any laser of the designation Class 3b or higher according to ANSI-LIA Z-136.1 (ANSI 2003) should be contained to preclude exposure of the crew. Alternatively, the nominal ocular hazard distance should be determined and sufficient inhibits to laser operation should be present to remove the hazard during an extravehicular activity task within that distance. During the assessment, it is important to take into account reflective surfaces.

22.3.4 Electrical Shock and Molten Metal

When dealing with powered hardware in the extravehicular activity environment, the hazards of both molten metal and electrical shock need to be considered. Although an electrical shock hazard is not applicable to crews using the Orlan suit for extravehicular activity, the extravehicular mobility unit does have a number of paths that result in crew susceptibility to electrical shock. Both, the extravehicular mobility unit and the Orlan suit, however, are at risk from molten metal due to arcing that can occur during a powered mate or demate operation for connectors.

Because the sweat soaked interior of the extravehicular mobility unit decreases the skin resistance of the crew wearing it, any powered hardware in the extravehicular activity environment has a potential to be an electrical shock hazard. It is, therefore, of concern that incidental contact with a powered chassis and the *International Space Station* structure can cause sufficient current, anything greater than 100 mA, to pass through the body of the crew and cause cardiac arrest. The adequate design of hardware to ensure proper bonding and grounding should mitigate this hazard. The designer should consider that it is not possible to verify grounding to the *International Space Station* on orbit; therefore, proper grounding design should be verified prior to flight. Recently, extravehicular activity operations near the solar array wing were necessary (Figure 22.5) and controls had to be determined to mitigate the hazard of contacting energized surfaces, thus preventing both electrical shock and molten metal.

When considering the hazard of molten metal, the amount of power on the pins of the connector during a mate or demate operation is the most important factor. Although generation of molten metal depends somewhat on the size of pin used in the connector, it is accepted conventionally that, at power levels below 3 A and 28 V, no appreciable molten metal is generated (Gerstenmaier and Greene 2000). For a mate or demate operation, a current flow greater than 3 A does generate molten metal and increasing amounts are

FIGURE 22.5 STS-116 solar array wing retraction (Courtesy of NASA).

generated as the current flow increases. The potential exists that, at sufficiently high voltage and current, not only is a considerable amount of molten metal generated, but the resulting force is enough to drive a person across a room in a 1-g environment. Obviously, the extravehicular activity crew would be highly endangered should that occur in a microgravity environment. To prevent this hazard from occurring, power should be removed prior to a mate or demate operation, with at least two inhibits to power being reapplied. At least one of those inhibits should be upstream of the connector being mated or demated, and the presence of this inhibit should be verifiable.

22.3.5 Entrapment

The possibility of entrapment, that is, the inability of the extravehicular activity crew to return to the habitable environment, must be considered when designing extravehicular activity hardware. Entrapment hazards can range from confined work areas to loose cables and hoses that can entangle the crewmember. This hazard also includes small openings and mechanisms that can snag portions of the extravehicular mobility unit or its attached tools. Latches, moving or rotating hardware, or similar devices can entrap the crewmember by snagging the suit or its tethers, thus preventing movement. This can occur without immediate damage to the suit or crewmember. Holes are also considered an entrapment hazard, because if they are sized improperly, a finger can become stuck in the opening.

22.3.6 Emergency Ingress

Due to the current design of the suits, certain failures can occur that make it necessary for the crewmember to return to the habitable environment within 30 min of the failure. The suits are designed with backup oxygen supplies that provide sufficient oxygen for this amount of time, should a failure such as the loss of ventilation occur. This is important for a hardware provider to consider in the design, because little time is available to safely configure hardware being installed or serviced when such a failure occurs. In addition, should hardware being serviced, installed, or operated impinge on designated translation corridors, the need to ensure a return time of 30 min must be addressed in its design.

22.3.7 Collision

The hazard of collision is that the crewmember can be impacted by a large mass or even a small mass at high velocity that results in injury or death. The potential means for this hazard to occur, which should be of concern to hardware designers, are due to mass handling concerns and structural failure of the hardware.

Mass Handling

Although items in space are weightless, the mass of the objects still must be considered. When handling hardware (Figure 22.6), the crew has to control their own mass, the mass of the space suit, and the mass of the object they are controlling. Any mass greater than 750 lbm is precluded because of the inertia it possesses. A large mass requires the crew to move very slowly to control it; and the larger the mass, the greater is the risk of damage, should the mass impact the extravehicular activity crew. Hardware that is to be transported by the extravehicular activity crew needs to have appropriate handling devices designed as an integral part of the hardware. Table 22.2 outlines the design features and constraints for transporting objects during extravehicular activity.

Structural Failure

Structural failure of hardware can lead to a collision of the mass with the crew or the International Space Station or Space Shuttle structure. When designing structures for

FIGURE 22.6 Extravehicular activity crewmember moving the crew and equipment translation aid cart (Courtesy of NASA).

Table 22.2 Constraints for Transporting Objects During Extravehicular Activity (Courtesy of NASA)

Object Size	Object Translation Mode	Object Handling Method	Design Features Required for Translation
0 to 20 lb, 0 to 2 ft^3 Max dimension: 2 ft	Hand over hand crew translation with object on a tether	Tether	Tether point
0 to 200 lb, 2 to 5 ft^3 Max dimension: 5 ft	Hand over hand crew translation with object on a body restraint tether	Body restraint tether	Handrail section[1]
0 to 750 lb, 5 to 100 ft^3 Max dimension: 8 ft	Object on a transfer device	Transfer device interface	Transfer device interface and two handrails[1,2,3]

[1] Handrail requirements also can be met using square or round scoops interfacing with square or round microfixtures on the object. The hardware provider must supply the square or round microfixture.
[2] For large items, tethering requirements are determined on a case by case basis.
[3] Transfer device interface is required for translation distances over 158 ft.

the extravehicular activity environment, it is important to consider all loads in the extravehicular activity environment. Specifically, inadvertent kick loads, tool impact loads, bump loads, and tether loads should be considered. Additional extravehicular activity induced loads can be found in JSC 28918, *Extravehicular Activity (EVA) Design Requirements and Considerations* (NASA 2005b). The structural life of the hardware in the space environment should be considered as well. Prolonged exposure to ultraviolet light and atomic oxygen can compromise the strength of hardware over time, resulting in an end of life structural failure hazard. The design needs to ensure the hardware remains structurally safe for the duration at which it is to be on orbit and not just the duration of the experiment.

22.3.8 Inadvertent Loss of Crew

When the extravehicular activity crew is working external to the vehicle in the microgravity environment, there is a concern with inadvertent loss of crew resulting from a member's detachment from the structure. Generally, the appropriate control for this hazard is the use of safety tethers (Figure 22.7) and either handholds or foot restraints to remain nominally in contact with structure at all times. For complex extravehicular activity tasks, the crew normally is secured by a foot restraint, because both hands are needed to perform a task. It is important to consider the design of the safety tether and assure that the hook mechanism that attaches to the structure has a locking mechanism that can withstand a failure while continuing to remain attached to the structure. If the vehicle on which the extravehicular activity crew is working is unable to maneuver quickly, a method of crew return, such as the simplified aid for extravehicular rescue is necessary to meet the two-fault tolerant requirement to mitigate this catastrophic hazard.

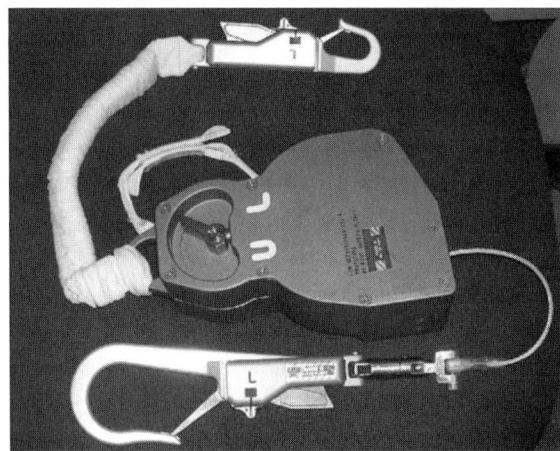

FIGURE 22.7 85-ft safety tether (Courtesy of NASA).

22.4 CONCLUSIONS

The list of hazards discussed within this chapter is not inclusive. The hazard of explosion should be assessed, particularly with regard to pressure vessels and batteries. Other hazards should be considered, such as micrometeoroid and orbital debris, battery explosion, explosion due to pressure systems or vessels, and other hazards mentioned in this book. When assessing the hardware for extravehicular activity hazards, consider what it would be like to work on a car in a swimming pool. The crew assumes the extravehicular activity environment is safe when they undertake extravehicular activity tasks. The hardware designer and manufacturer have a responsibility to ensure that the environment is safe.

Implementing the appropriate requirements for extravehicular activity hardware is paramount to performing successful experiments and tasks in the harsh environment of space. Understanding the basis and importance of these hazards is the first step in instituting safety minded design practices for future extravehicular activity.

REFERENCES

American National Standards Institute. (2003) *American National Standard for Safe Use of Lasers*. ANSI Standard Z136.1. Orlando, FL: American National Standards Institute, Laser Institute of America.

Boeing, Inc. (1998) *Shuttle Orbiter/Cargo Standard Interfaces (Core ICD)*. Space Shuttle Interface Control Document ICD 2-19001, Revision L. Seal Beach, CA: Boeing North American, Incorporated.

Gerstenmaier, W., and J. Greene. (2000) *Crew Mating/Demating of Powered Connectors*. Letter MA2-99-170. Houston, TX: National Aeronautics and Space Administration, Johnson Space Center.

National Aeronautics and Space Administration. (2005a) *NASA Extravehicular Mobility Unit Life Support System and Space Suit Assembly Data Book*. Winsor Locks, CT: Hamilton Sundstrand.

National Aeronautics and Space Administration. (2005b) *Extravehicular Activity (EVA) Design Requirements and Considerations*. Document JSC 28918. Houston, TX: National Aeronautics and Space Administration, Johnson Space Center.

CHAPTER

Emergency, Caution, and Warning System

23

David Hornyak
Lead System Interface Engineer, Payload Engineering Integration, Johnson Space Center, National Aeronautics and Space Administration, Houston, Texas

CONTENTS

23.1. System Overview 725
23.2. Historic NASA Emergency, Caution, and Warning Systems 726
23.3. Emergency, Caution, and Warning System Measures 727
23.4. Failure Isolation and Recovery 731

23.1 SYSTEM OVERVIEW

The emergency, caution, and warning system is used to notify on-orbit and ground crews when a hazard has been detected, and crew action is required. This system is designed to inform and assist the onboard crew to respond to and resolve hazardous conditions when situations arise that can endanger the resources of the spacecraft, the lives of the crew, or mission success. A notification by the emergency, caution, and warning system not only alerts the crew of a hazardous situation but does so in a timely manner, provides enough information to allow for immediate action without an investigation of the cause, and does this in a way that it is available immediately wherever the crew is located at the time.

The emergency, caution, and warning system must be able to assess and categorize the severity of hazardous events. Common classifications in use by the International Space Station Program (Surber and King 1998) include events designated as

- Emergency (Class I).
- Warning (Class II).
- Caution (Class III).
- Advisory (Class IV).

Each successive category indicates a lesser hazardous event and, therefore, requires different and usually less immediate action by the crew.

726 CHAPTER 23 Emergency, Caution, and Warning System

Because of the importance of the emergency, caution, and warning system, events that initiate its use are identified through hazard reports written for the design of the vehicle and its operation. Once these hazard reports are developed, a collective review of them must be conducted to ensure consistency in the use of the system. This review ensures that events identified in the individual hazard reports are given appropriate priority, with the more hazardous situations and those requiring more immediate action classified at higher levels.

The emergency, caution, and warning system is a data system that collects, distributes, transmits, and communicates information critical to controlling hazardous situations whenever they occur. It must be reliable, redundant, and distributed. To accomplish these needs, there must be multiple system interfaces to the crew, depending on the size of the space vehicle; and the processing system must be distributed to guard against loss of any part of the emergency, caution, and warning system caused by some disruptive consequence of a hazardous condition that has occurred.

23.2 HISTORIC NASA EMERGENCY, CAUTION, AND WARNING SYSTEMS

To learn from passed experience and understand the additional capabilities provided by advancing technology, it is beneficial to understand the emergency, caution, and warning systems used on historic space vehicles. Surber and King (1998) provide much of the background to these historical vehicles.

The Mercury and Gemini vehicles used hardwired emergency, caution, and warning systems that had no capability for on-orbit reconfiguration. Partly due to the limits of technology at the time and reflective of the nature of the Mercury and Gemini missions, which were shorter than those of today, their emergency, caution, and warning systems provided notification to the crew of hazardous events that were determined prior to a mission. These hardwired systems could not be adjusted to respond to any hazardous events that might develop during the course of a mission.

The emergency, caution, and warning systems used on *Apollo* and *Skylab* also were hardwired systems connected to the guidance computers of the spacecraft. As was the case for the Mercury and Gemini missions, these hardwired emergency, caution, and warning systems could not be reconfigured during a mission; however, the sensor limits or parameters that would initiate the alarm if exceeded could be modified by uplinking information to the vehicle during a mission, should that be determined necessary.

The Space Shuttle emergency, caution, and warning system possesses one emergency, caution, and warning annunciation system with on-orbit modifiable alarm limits implemented via the table maintenance block update application that writes new limits into computer memory and one hardwired caution and warning system that is modifiable by the onboard crew. The two systems complement and back up each other, depending on any specific event.

The International Space Station emergency, caution, and warning system is an integrated system of data management, annunciation, and display. It is based on a tiered data management structure, as opposed to a hardwired sensor and processor system. The integrated data system approach provides for increased robustness in the area of event classification changes, system redundancy, and long-term modification potential or upgrades. The International Space Station system includes hardware sensors, all reporting computational devices, the intermodule communication system, caution and warning panels, and the portable computer system displays. Integration of the caution and warning system ensures all these pieces work together as well as support flight procedures development, event classification activities, mission rules development, nomenclature standards, and failure detection, isolation, and recovery design.

23.3 EMERGENCY, CAUTION, AND WARNING SYSTEM MEASURES

The emergency, caution, and warning systems of today are composed of sensors that can identify hazardous conditions, a data system that manages the data from the sensors and evaluates the data to identify the condition, and the crew interfaces used to inform the on-orbit and ground crews when the system is responding to sensors detecting a hazardous condition. For the integrated emergency, caution, and warning system to operate as intended, all of these components must work together.

23.3.1 Event Classification Measures

When hazard reports identify any use of the emergency, caution, and warning system, an event classification must be determined. The emergency (Class I), warning (Class II), caution (Class III), and the advisory (Class IV) event classifications, which are in common use on the *International Space Station*, are for this purpose. These classifications have been documented by Surber and King (1998) and are presented in the following paragraphs.

The top level emergency, emergency (Class I), is intended to alert the crew to catastrophic hazards, those that could result in loss of life of a crewmember or loss of the vehicle. Emergency (Class I) events are "a life threatening condition requiring immediate attention. Predefined crew responses may be required prior to taking corrective action. Safe haven concept activation can become necessary. Included are the presence of fire and smoke, the presence of toxicity in the atmosphere, and the rapid loss of atmospheric pressure." Procedurally, emergency (Class I) events require all onboard crewmembers to react immediately.

The next level events are categorized as warning (Class II) events. These are "conditions that require immediate correction to avoid loss or major impact to mission, or potential loss of crew. Included are faults, failures, and out of tolerance conditions for functions critical to station survival and crew survival." Procedurally, warning (Class II) events require one crewmember to react immediately, that is, within approximately 5 min.

Caution (Class III) events are categorized as those "conditions of a less time critical nature, but with the potential for further degradation if crew attention is not given. Included are faults, failures, and out of tolerance conditions for functions critical to mission success." Caution (Class III) events require no immediate crew action.

Advisory (Class IV) events are used primarily for ground monitoring purposes, which is advantageous because of limited communication coverage and data recording. Advisories (Class IV) also are used for data items that most likely do not exist permanently in the telemetry list but should be time tagged and logged for failure isolation, trending, sustaining engineering, and the like. Procedurally, advisory (Class IV) events require no immediate crew action.

23.3.2 Sensor Measures

The first action of the emergency, caution, and warning system is to monitor the onboard sensors. This entails use of three important measures: the probability of event detection, the probability of a false alarm being initiated, and sensor redundancy.

The probability of detection, P_d, as the name implies, is a statistical measure of the likelihood that a sensor output unambiguously identifies the occurrence of the event it is intended to monitor. To determine the P_d, the designer must have a through understanding of

- The event being monitored.
- The basic engineering parameters of that event.
- How the sensor responds to these conditions.

Similar to the P_d, the probability of false alarm, P_f, is a statistical measure of the likelihood that a sensor output identifies the occurrence of the event it is intended to monitor when the event actually has not occurred. False alarms can occur for a number of reasons, including failure of a sensor. A design factor that can affect the P_f negatively, and one which must be considered by the designer, is the use of a sensor to measure the secondary effects of an event rather than the event itself. An example of this is a smoke detector used to detect fire. Smoke detectors are used to detect a secondary effect (smoke) generated by the fire and not fire itself. Without adequate maintenance to ensure the sensor and atmosphere are clean of particulates, the smoke detector also can detect particulates in the air other than smoke, the result being an indication that there is a fire when in fact one has not occurred. There are many other examples of sensors being used to measure the secondary effects of hazardous events. This practice remains a necessity due to technology and overall design concerns; however, its impact on P_f should be considered in the overall emergency, caution, and warning system design.

The P_d and P_f are not independent; rather, they are inversely codependent measures. A condition whereby a sensor can detect an event with a P_d of 1, meaning the event is always detected, and a P_f of 0, meaning there can be no false alarms, generally cannot be achieved. However, during system design and sensor selection, the goal is to move as close to these design measures as possible. The general relationship between P_d and P_f is shown in Figure 23.1.

When the importance or severity of the hazard dictates, multiple sensors measuring different phenomena, each of which would indicate the occurrence of an event, can be used. This can be more important as the hazard level, and thus emergency, caution, and warning level, increases. By using an integrated sensor scheme, it is possible to move this

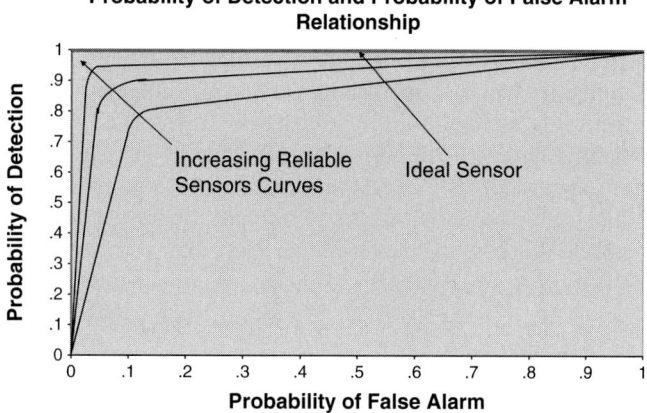

FIGURE 23.1 Probability of detection and probability of false alarm relationship (Courtesy of NASA).

relationship closer to the desired goals of the P_d being equal to 1 and P_f being equal to 0. However, such a design must be considered carefully; otherwise, the opposite result can occur. The designer must remember that, by integrating multiple sensors, the complexity and cost of a system also is increased; therefore, this design approach should be used only when necessary.

As an alternative to using multiple sensors to measure for an event, robust sensors can be made with built-in fault detection. These are useful, because when an integrated sensor system can detect its own failure modes and state of maintenance, its reliability increases.

23.3.3 Data System Measures

The emergency, caution, and warning data system includes the computer systems that monitor the sensors, the software on these computers that monitor and check the sensor outputs, the data ranges for the sensors, and the software that initiates the annunciation system and displays the data to the crew. The data system also has the capability to isolate (power off) or reconfigure systems to aid in the crew response when appropriate.

For system reliability, the computing system should be a distributed rather than having one centralized computer controlling the entire emergency, caution, and warning system. When distributed among two or more computers, each on a different power source, a single failure cannot bring down the system. During development of a distributed system, care must be taken to consider the hazard situations being monitored by the emergency, caution, and warning system to ensure that the occurrence of any of those events does not have the ability to shut it down.

When a sensor output indicates that a predefined hazardous situation has occurred, it is referred to as being *out of limits*. The computing system compares the monitored sensor outputs to predefined out of limits values. When the computer determines that a sensor output has exceeded a limit, a predefined annunciation, which depends on the classification level, is initiated. Along with the annunciation, information about the event is provided to the crew displays. For longer missions, where the crew and vehicle likely are to have to respond to developing events and account for previously unforeseen false alarms due to the sensitivity of a sensor being set too high, the out of limits values maintained in the computing system should be adjustable during the mission.

When a hazardous situation is identified by the emergency, caution, and warning system, its data system and computer have the capability to command the affected system or critical subsystems off as an initial means to control the hazardous event. Autonomous commanding of vehicle systems is referred to as *automated failure detection, isolation, and recovery.*

23.3.4 Annunciation Measures

Two methods of annunciation, audible and visual, are used to alert the crew by the emergency, caution, and warning system. Simultaneously the annunciation, data describing the event, and its location must be provided to the crew. For both audible and visual annunciation, the emergency, caution, and warning system must take into account the environment of the spacecraft to be sure that the alarm easily is distinguishable from all other events and activities in its vicinity.

Audible alarms generally are the primary annunciation method for higher level emergency, caution, and warning events, such as Class I. They generally are not used for lower level events such as Class IV. When determining if an audible alarm is required, the designer should consider the crew action required. Should the event classification require immediate action by the crew, an audible alarm likely is necessary. The alarm tone itself should be designed to provide a distinct tone and rhythm, so that it can be distinguished from other noises in the spacecraft. It should be sounded sufficiently loud to be noticed quickly above the background spacecraft noise level but not be so loud as to present a danger to the hearing or prevent the crew from communicating.

Visual alerts can accompany the audible alarm, to aid in providing an immediate understanding of the event that initiated it. Generally, visual alerts for emergency, caution, and warning events are colocated on an emergency, caution, and warning panel. An International Space Station emergency, caution, and warning panel is shown in Figure 23.2.

Once the annunciation method for each of the emergency, caution, and warning classifications has been determined, the system responds with the same annunciation to all events within that classification. It is important to understand this, because consistency in the classification annunciation must be maintained, so that the crew response remains consistent. There are no exceptions. This constraint should be considered in determining into which classification an event falls.

It is important that data describing the event and its location accompany the audible alarm. The format and content of the data provided must be clear, concise, and contain enough information for the crew to take appropriate action immediately and without further investigation. Two important considerations for the data system are the display of the information and location of the display. A visual display, such as a map of the spacecraft on a computer screen that shows the location and a unique symbol indicating the hazardous event generally are the most efficient means of conveying a lot of information to the crew

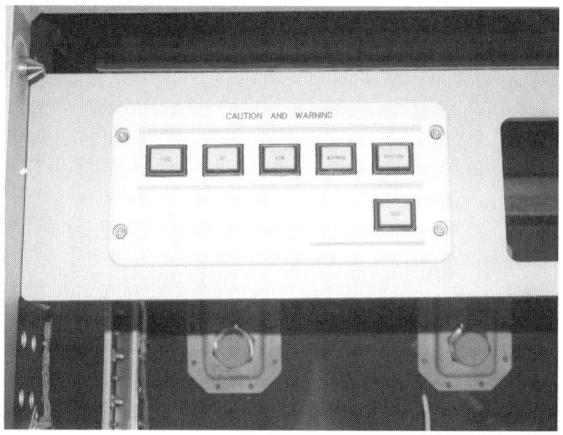

FIGURE 23.2 International Space Station emergency caution and warning panel (Courtesy of NASA).

quickly. When this method is used, the emergency, caution, and warning system display should override all other displays on the screen. By this action, the onboard crew can see the display immediately and without having to search through other displays in use. The second design consideration depends on the size of the spacecraft. On a larger spacecraft, such as a space station, the crew might not be near the location of a single centralized display at the time of an emergency, caution, and warning event; therefore, the system must be able to provide critical information to data displays located throughout the vehicle. By doing so, the crew has quick access to this critical information. On a smaller vehicle, a distributed emergency, caution, and warning system might not be necessary. The displays on the notional *International Space Station* are documented by Surber and King (1998). An example of these displays is shown in Figure 23.3.

Although designers do their best during the spacecraft design phase to determine all potential emergency, caution, and warning events and identify the sensor outputs that indicate the occurrence of a hazardous event, the system also must be able to react to unforeseen events during the progression of a mission. To account for these unforeseen situations, the annunciation system should have a crew override to silence an alarm. This affords the crew an ability to impede the annunciation system when an event has been sounded, so that, after the crew has determined that the event is not or is no longer a hazard to onboard personnel or the spacecraft, the alarm can be turned off. Having the ability to impede the annunciation system also provides a means for the crew to respond appropriately to any false alarms or preexisting conditions, such as associated system failures, that might occur.

23.4 FAILURE ISOLATION AND RECOVERY

As the name implies, failure detection, isolation, and recovery is a process by which a system is monitored for failure situations. When a failure is detected, the onboard computing system automatically isolates the event, usually by powering off the subsystem that

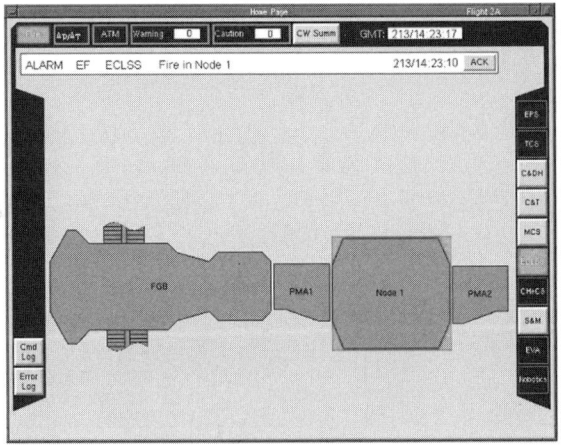

FIGURE 23.3 Notional International Space Station emergency, caution, and warning display in an alarm state (Courtesy of NASA).

experienced the fault. When the failure is located within or involves a critical system, the computing system can automatically command configuration changes to the system so that it can continue its operation. This usually is affected by automatically employing backup systems or subsystems. This recovery step of the process requires that the system being monitored is a redundant, robust system possessing backup capabilities; however, these capabilities increase the complexity and the size of the system, as well as the cost of development and testing. Due to the increased size and cost, recovery capabilities generally are reserved for use only on the most critical systems of the spacecraft.

The emergency, caution, and warning system detects system failures for the purpose of crew notification. However, to aid in the crew response to failure scenarios, the onboard computing systems also have the capability to isolate many of the failures at the time the emergency, caution, and warning alarm is initiated when such action has been determined appropriate. When the automatic isolation process is to be used, the design team must predetermine what actions would be necessary to isolate a failure that caused the hazardous situation. This can entail closing valves, powering off fans, or completely powering off systems. However, a thorough review must be conducted prior to implementing automatic isolation of these systems, and the results must show that these automated steps to isolate a failure neither present an additional hazard nor prevent the crew from addressing the original hazardous event.

REFERENCE

Surber, M., and G. King. (1998) *Caution and Warning System Description Document.* ISS Document D684-10299-01, Revision A. Houston, TX: Boeing and the National Aeronautics and Space Administration, Johnson Space Center.

Laser Safety

Joe M. Victor, Ph.D.
Laser Safety Officer, Johnson Space Center, National Aeronautics and Space Administration, Houston, Texas

CONTENTS

- 24.1. Background 733
- 24.2. Laser Characteristics 735
- 24.3. Laser Standards 738
- 24.4. Lasers Used in Space 740
- 24.5. Design Considerations for Laser Safety 742
- 24.6. Conclusions 744

There is a safety requirement at NASA for Class 3B and Class 4 lasers that applies to the operation of these devices in space. This requirement is called out in JPG 1700.1 (NASA 1999).

24.1 BACKGROUND

Before discussing lasers, a brief discussion of the optical spectrum, as well as the biological effects of light on the human body, are given as background.

24.1.1 Optical Spectrum

The optical spectrum is a small portion of the electromagnetic spectrum (Figure 24.1), ranging from 200 to 1400 nm. This is the portion of the spectrum that is of concern in this discussion of laser safety.

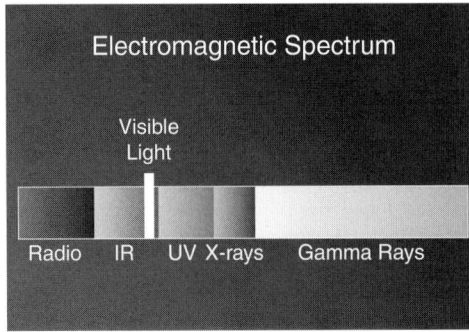

FIGURE 24.1 The electromagnetic spectrum (Courtesy of NASA).

24.1.2 Biological Effects

The Eye

The basic components of the eye are depicted in Figure 24.2. The human eye transmits and focuses visible and near infrared light very efficiently. The incident corneal irradiance, or the radiant exposure of the eye, is increased by approximately 100,000 times at the retina as the result of the focusing effects of the cornea and lens (EH&S 2007).

Ultraviolet light within the range of wavelengths from 200 to 315 nm is absorbed in the cornea, whereas near ultraviolet radiation of wavelengths between 315 and 400 nm is absorbed in the lens and can contribute to some forms of cataract.

Radiation within the visible spectrum, that is wavelengths between 400 and 780 nm, and the near infrared, between 780 and 1400 nm, is transmitted through the ocular media

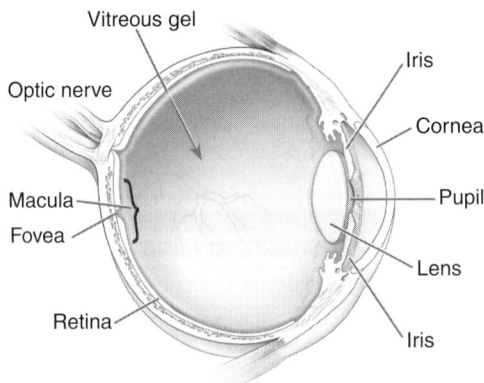

FIGURE 24.2 Structure of the eye (Courtesy of the National Eye Institute/National Institutes of Health).

with little loss. The radiation is focused by the cornea and lens to a spot on the retina only 10 to 20 in diameter. This can result in intensities sufficient to damage the retina. The range of wavelengths from 400 to 1400 nm, therefore, is termed the *ocular hazard region* of the electromagnetic spectrum.

Although far infrared radiation, that part of the spectrum having wavelengths between 3 μ and 1 mm, is absorbed by the front surface of the eye, some middle infrared radiation, that possessing wavelengths between 1.4 and 3.0 μ, penetrates somewhat deeper into the eye and can contribute to a condition known as *glass blower cataract*.

The Skin

From a safety perspective, skin effects usually have been considered of secondary importance. However, with the more widely spread use of lasers emitting in the ultraviolet spectral region and certain high power lasers, skin effects have assumed greater importance. For example, the 200 to 315 nm ultraviolet wavelengths are absorbed into the skin, producing erythema, or sunburn. The principal reactions of the skin to radiation in the infrared spectral region, 0.7 to 1.0 μ, are burns and excessively dry skin.

24.2 LASER CHARACTERISTICS

LASER is an acronym for light amplification by stimulated emission of radiation (ANSI 2007). A laser is a device that produces radiant energy predominantly by stimulated emission. All lasers have three basic characteristics:

- A material medium.
- Population inversion of electron states.
- A resonant structure.

The material medium can be either a solid, liquid, or gas. The medium is excited by a variety of methods to achieve an electronic population inversion, a phenomenon whereby more electrons are in an excited state than in a lower level state. Finally, the medium is placed in a resonant structure so that stimulated emission can make multiple passes through the excitation region. Figure 24.3 shows the basic laser requirements.

24.2.1 Laser Principles

The following are various laser characteristics.

Operating Material Medium

All lasers require an operating medium in which to achieve population inversion. This medium can be a gas, liquid, or solid material.

Common gas lasers include helium-neon and carbon dioxide lasers. These gas lasers incorporate other gases to improve their efficiency, because when a mixture of gases is used to improve excitation, population inversion and depletion of lower band levels are improved.

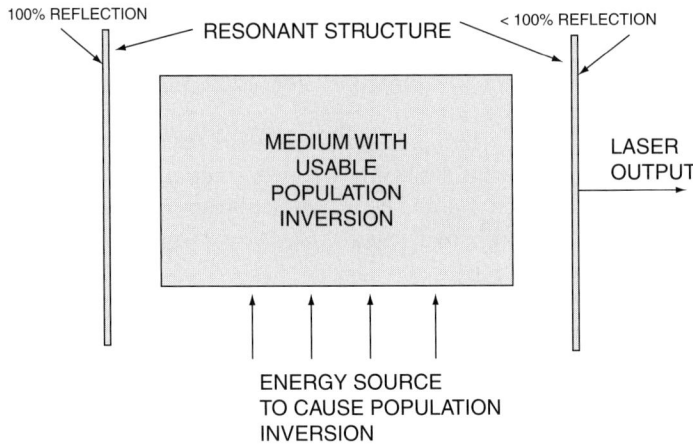

FIGURE 24.3 Basic characteristics of a laser.

Liquid lasers include the dye lasers, which have an ability to operate in a band of wavelengths that depends on the resonant structure and the liquids used. High power chemical lasers of this type utilize the mixing of highly reactive chemicals for excitation.

Solid lasers are of two basic types, solid state and semiconductor. An example of a solid state device is the ruby laser, one of the first developed; whereas the neodymium-doped yttrium-aluminum-garnet Nd:$Y_3Al_5O_{12}$, (Nd-Yag) laser, a high power laser used for a variety of machining operations, is an example of a semiconductor laser. Semiconductor lasers produce laser radiation within the PN junction. Depending on the semiconductor material used, a wide range of wavelengths is possible with these lasers.

Energy Excitation

The operating material medium can be excited by a number of methods, including light pumping, which is a method whereby a flash lamp shines on the operating medium. Ruby and Nd-Yag rods are excited using the light pumping method, although a semiconductor laser also can be used as an excitation source. Operating mediums such as helium-neon and carbon dioxide are excited by the high voltage breakdown of the gases. Lasers using the carbon dioxide medium can be excited by using radio frequency energy as well. For very high power lasers, excitation occurs by flowing together highly volatile chemicals, a technique used for systems sufficiently powerful to shoot down missiles in flight. Direct current is used to excite semiconductor lasers. All these methods are used to excite electrons at a lower state into higher states of excitation, thus causing the population inversion necessary for laser action.

Stimulated Emission

Excited state electrons normally decay to the ground state. The decay can be enhanced through stimulated emission. A photon traveling through a medium can induce excited electrons to go to the ground state at the same energy level. The result is two photons traveling in the same direction, at the same energy level and wavelength. Such photons are called *coherent*, and this process of photon multiplication is the basis of laser action.

Operating Modes

There are two basic operating modes:

- Continuous wave.
- Pulse.

A continuous wave laser emits a continuous steam of photons. A laser operating in continuous wave mode must be excited constantly. Lasers operating in this mode tend to have lower peak values than pulsed lasers, because their lower efficiency results in the medium overheating. Examples of continuous wave lasers are helium-neon and carbon dioxide gas lasers. Semiconductor lasers also can produce a continuous wave output.

A pulse laser can emit a burst of energy by two methods. The first is by periodic excitation of the medium. In this case, flash lamps periodically illuminate the medium. Electrons excited by the light pulse result in lasing action and a pulsed output. A second method to produce a pulsed output is by the periodic creation of a highly resonant cavity. Here, a highly resonant cavity that promotes laser action is formed by using a rotating mirror. Pulse lasers permit high peak output values without the medium having to be constantly energized. Because most lasers do not convert input energy to output photon energy efficiently, heat removal can be a problem during continuous operation.

24.2.2 Laser Types

The four basic medium types of lasers are

- Gas.
- Semiconductor.
- Solid state.
- Liquid.

Gas lasers use various gases as the lasing medium. The operating pressure of the gases usually is lower than atmospheric pressures. The medium can be excited by a high voltage discharge, as used in some helium-neon and carbon dioxide lasers. Additional gases can be added to the gas mixture to improve excitation and energy absorption and to deplete the lower energy level by colliding with the lasing medium, thus aiding the transfer of energy between species. Figure 24.4 shows a four energy level excitation method.

Solid state lasers use a solid medium to support the lasing. Ruby and Nd-Yag rods are common examples of solid state lasers. Rods of these mediums are placed in a resonant structure and excited by either flash lamps or semiconductor lamps.

Semiconductor lasers use the PN junction as the lasing medium. A PN junction is created by doping an intrinsic semiconductor with elements that have fewer active electrons than the host material to create a hole (P type) semiconductor or with elements that have more active electrons than the host material to create a free electron (N type)

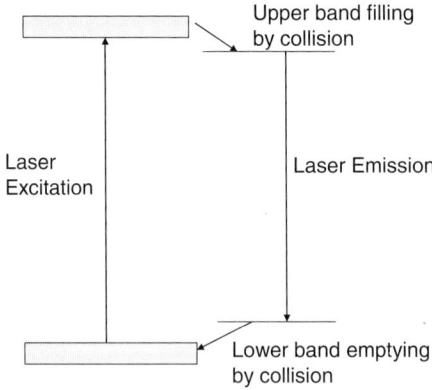

FIGURE 24.4 Four level laser band structure.

semiconductor. Fabrication of a junction of these two types of semiconductor form a PN junction that has diode-like effects. By flowing current through the junction in the forward direction, light can be generated in the junction area. With proper design, a laser action can be made to occur. Lasing wavelengths depend on the characteristics of the semiconductor junction material and the doping levels. A wide number of materials are used to construct lasers of various wavelengths extending into the infrared.

Liquid medium lasers are limited usually to dye lasers that can have different output wavelengths bands, depending on the dye concentration. These lasers usually are excited by other higher power lasers. Most of the liquid dyes are hazardous materials, being either poisonous or carcinogenic, and should be handled with care.

24.3 LASER STANDARDS

A number of standards concern the safe use of lasers. They call out safe levels of laser energy for the eyes, which are more sensitive, and for the skin.

24.3.1 NASA Johnson Space Center Requirements

The NASA Johnson Space Center laser safety hazard requirement is called out in JPG 1700.1 for Category II hazardous operations and operating Class 3B and Class 4 lasers or solar simulators (NASA 1999). The requirement specifies the issuance of a hazardous operations permit and a medical examination that includes a detailed examination of the inside of the eye. It basically references the ANSI Z136-1 standard (ANSI 2007) for engineering and administrative controls and refers to standards 21 CFR 1040.10 (FDA 2006a) and 21 CFR 1040.11 (FDA 2006b).

24.3.2 ANSI Standard Z136-1

The ANSI Standard Z136-1 is the generally accepted laser standard in the United States (ANSI 2007). It defines

- Laser Classes 1, 2, 3, and 4.
- The maximum permissible exposure for laser radiation.
- The nominal hazard zone.
- Safeguards.
- Signs.
- Examples of laser hazard calculations.

Laser Classification

Lasers and laser systems are classified into four levels of hazardous operations. Section 3.3 of ANSI Z136-1 defines the laser classes (ANSI 2007). The treatment of Class 3B and Class 4 lasers is unchanged from the last version of this document, which was released in 2000.

The lowest laser level is Class 1. These lasers are safe, either because the laser output is sufficiently low as to not cause damage to the eye or skin or because it is enclosed so that any radiation escaping to the environment is below hazardous levels. Note that the latest standard establishes a new Class 1M for cases involving the use of collecting optics.

Class 2 lasers are low level visible lasers that could be hazardous under extended exposure periods. In normal operation, they are not considered hazardous. Class 2M was established in the 2007 release of ANSI Z136-1 for cases involving the use of collecting optics.

Class 3 lasers are hazardous, especially to the eyes, and precautions must be taken when using them. These lasers present no particular hazard from scattered radiation. A new Class 3R is a laser system created for lasers that potentially are hazardous under direct and specular reflection viewing conditions, where the eye is focused and stable.

Class 4 lasers are the most hazardous types, having power levels greater than 0.5 W. Protection from direct and scattered reflections must be taken.

Maximum Permissible Exposure

The maximum permissible exposure is the level of laser radiation to which a person can be exposed without hazardous effect or adverse changes in the eye or skin. The criteria for maximum permissible exposure for the eye and skin are detailed in Section 8 of the ANSI standard (ANSI 2007). The maximum permissible exposure depends on the laser wavelength, exposure duration, and pulse repetition rate.

Nominal Hazard Zone

The nominal hazard zone is the space within which the level of direct, reflected, or scattered radiation during normal operation exceeds the applicable maximum permissible exposure. Exposure levels beyond the boundary of the nominal hazard zone are below the appropriate maximum permissible exposure level.

Safe Guards

Some safeguards can be used when it is possible for people to be within the nominal hazard zone. The following are the engineering and procedural safeguards:

- **Engineering safeguards.** Engineering safeguards are physical restrictions used to protect a person. These include protective housings, interlocks on removable protective housings, key control, viewing windows, beam path enclosures, beam blocks or attenuators, and warning signs and signals.

- **Administrative and procedural safeguards.** Administrative and procedural safeguards are methods or instructions that specify rules, work practices, or both to implement or supplement engineering control. These safeguards include the application of standard operating procedures, education, and training as well as specifying the use of personal protective equipment, which is available for specific wavelengths and power levels. Administrative and procedural safeguards apply only to Class 3B and Class 4 lasers and laser systems. In addition to the harmful effect of lasers on the eyes, skin and other body areas also can be damaged by these devices. Ultraviolet light can burn the body as well as the eyes. Lasers can create harmful by-products from their intended action. A listing of these by-products can be found in the ANSI standard Z136-1, Table F1B, Laser Generated Airborne Contaminants (ANSI 2007). Furthermore, some lasers utilize very high voltage, which can be a shock hazard. All these hazards must be addressed.

- **Signs.** The ANSI standard Z136-1 also defines the warning and hazard sign content (ANSI 2007).

- **Hazard zone calculation examples.** Appendix B of the ANSI standard provides 64 sample calculations to illustrate the use of the various charts and tables located in the standard.

24.3.3 Russian Standard

Other laser standards are in use in addition to ANSI standard Z136-1. The Russian standard is the *System of Occupational Safety Standards, Laser Safety, General Regulations*, GOST 12.1.040-83 (GOST 1983). This standard is not used in the United States but is mentioned here for completeness.

24.4 LASERS USED IN SPACE

Lasers have many uses in space. Lasers are used as optical light sources for photography and illumination for video camera viewing. Various types of laser radar have been used. Lasers also are used in a variety of sensor applications. A few examples of these devices are included in the following sections.

24.4 Lasers Used in Space

24.4.1 Radars

NASA currently uses ranging lasers in the handheld laser and the trajectory control sensor. The handheld laser is a wavelength modified police pulse laser ranger used by Space Shuttle crew to measure the range and the range rate between the *Space Shuttle* and a target of opportunity, such as the *International Space Station*. The handheld laser can measure these values for a variety of skin tracking surfaces. The wavelength of the handheld laser allows for sufficient transmission of the laser through the Space Shuttle windows. This device is a Class 1 laser, and it is safe unless used at close range with magnifying optics.

The trajectory control sensor uses two Class 3B semiconductor lasers, one pulse and the other modulated continuous wave. The trajectory control sensor lasers measure range and range rate as well as bearing and bearing rate between the *Space Shuttle* and the *International Space Station*, on which several retroreflectors are mounted. The pulse ranging laser provides automatic tracking from about 5000 ft to within 65 ft of the target. It is somewhat noisy. Use of the continuous wave ranger is limited to about 1000 ft to docking. It is more accurate in range and range rate than the pulse mode laser. Laser hazard analysis has shown that there is no hazard for *International Space Station* viewers looking through magnifying camera optics during its nominal operation.

Both the European and Japanese space agencies have laser rangers used for docking to the *International Space Station*. As for the trajectory control sensor, both employ the use of retroreflectors mounted on the *International Space Station*.

Other high powered lasers are used to measure ranges to targets. Some of these lasers are used to measure the distance of orbital debris from the ground. Other lasers of similar type have been used to measure the distance between the Earth and a retroreflector located on the Moon.

24.4.2 Illumination

Some lasers have been used simply as illumination sources in space. The orbiter boom sensor system is used to illuminate the Space Shuttle tiles for visual inspection using video cameras. Lasers of this type also are used as excitation sources for other lasers.

24.4.3 Sensors

Lasers have been used in the *Space Shuttle* and the *International Space Station* as light sources for a variety of enclosed sensors. Common computers have enclosed lasers in their compact disc read-write modules that could be hazardous if exposed. Another enclosed sensor is located in the modulated laser analyzer for combustion products and is used to determine the concentration of four gases (carbon dioxide, carbon monoxide, acetylene, and hydrogen cyanide) using a wavelength modulation spectroscopy technique and one near infrared diode laser.

24.5 DESIGN CONSIDERATIONS FOR LASER SAFETY

Space is a unique environment. Not only is there no atmosphere to cause scintillation of the laser beam, as there is on Earth, only limited possibilities for exposure to hazardous laser beams are found in space.

24.5.1 Ground Testing

Before a laser device is used in space, it must be tested on the ground. At the Johnson Space Center, all laser operation is under the supervision of the radiation safety officer. Lasers and laser operators must meet all pertinent requirements. The lasers themselves must be registered, using a form JSC-44b, which requires various characteristics of the lasers to be noted. In addition, each laser operator must complete a form JSC-1023, which details the qualifications for operating the laser. These forms are submitted to the radiation safety officer. All laser operators are required to have a complete medical examination including photographs of the inside of the eyes.

Individual buildings similarly have requirements associated with the use of lasers. These are usually under the control of the facility manager. Usually a test readiness review is performed before the laser can be used in a facility. This review is conducted by a panel that includes the user, safety, and facility personnel. Detailed usage procedures and personnel hazard protection are included in this review. A hazard analysis usually is written and reviewed by safety personnel.

24.5.2 Unique Space Environment

Space is a unique environment for operating a laser. Because of the vacuum of space, a limited number of observers might be present, and usually some type of optics lie between the laser and the viewer.

Laser safety requirements within the human habitat of the *Space Shuttle* and of the *International Space Station* are the same as on Earth. Normally, a Class 1, totally enclosed laser system is desired. Either engineering or administrative safeguards should be taken if the laser in use is other than Class 1 or if repair of the Class 1 system exposes a hazardous laser to onboard personnel. In addition to the eye and skin hazards posed by the laser, high voltage electrical shock, biological, and other hazards can be present. Therefore, a hazard analysis should be part of the documentation of any laser system.

In space, laser radiation can be from an outside source, such as ranger sensors on the *Space Shuttle* being used to expose the *International Space Station*. In such cases, the hazard analysis should take into account the reduced laser energy caused by transmission through the Space Shuttle and International Space Station windows and the increased hazard caused by the use of magnifying optics on the viewer scopes of cameras. Figure 24.5 depicts the transmission of light as a function of wavelength through the aft window of the *Space Shuttle*, whereas Figure 24.6 shows the transmission of light as a function of wavelength through the Space Shuttle side hatch photographic window (Boeing 1998). As yet, the author is not aware of any similar curves for the new Constellation spacecraft.

24.5 Design Considerations for Laser Safety

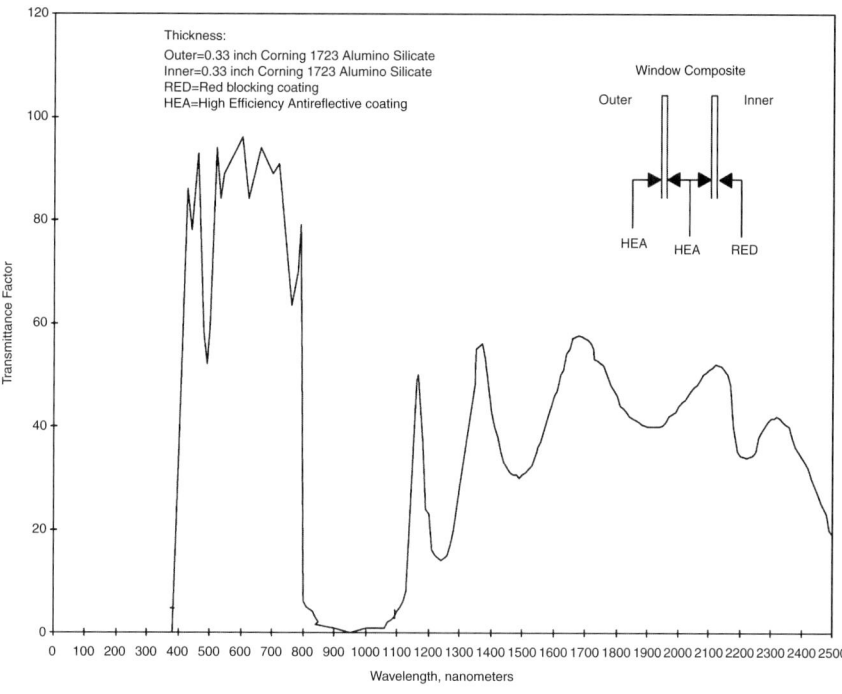

FIGURE 24.5 Space Shuttle aft window transmittance as a function of wavelength (Courtesy of NASA).

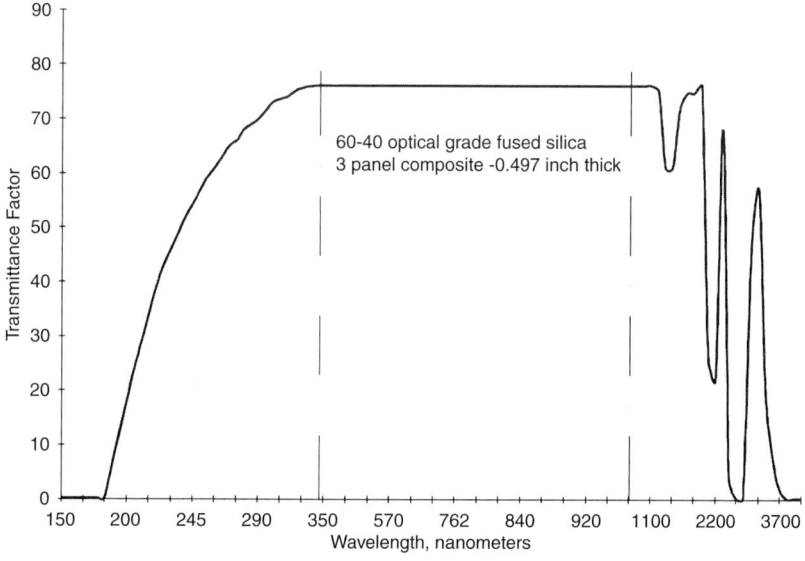

FIGURE 24.6 Space Shuttle side hatch photographic window transmittance as a function of wavelength (Courtesy of NASA).

In addition, astronauts have sunglasses available, which might attenuate a laser beam sufficiently, depending on the wavelength involved.

The *International Space Station* has at least 19 viewing ports. More are likely to be added in the near future as the European and Japanese modules are integrated onto the *International Space Station*. Some are open, whereas others are closed during docking. Hazards to viewers looking through these windows from lasers shooting toward the *International Space Station* must be considered. In some cases, cameras with magnifying lenses and direct vision viewfinders are used, especially for Space Shuttle tile inspection during the rendezvous pitch maneuver. Because no complete database exists, some effort is required to determine window characteristics and usage. This information is then used to establish appropriate hazard zones.

24.6 CONCLUSIONS

Within human habitats, laser safety requirements are similar to those used on Earth, and they should be treated in the same manner. With laser sources outside of the human habitat, the hazard analysis can take advantage of limited exposure due to flight constraints: minimum vehicle separation and limited viewing ports with their associated transmission.

REFERENCES

American National Standards Institute. (2007) *American National Standard for Safe Use of Lasers*. ANSI Standard Z136-1. Orlando, FL: American National Standards Institute, Laser Institute of America.

Boeing, Inc. (1998) *International Space Station Interface Definition Document*. Interface Definition Document NSTS-21000-IDD-ISS. Houston, TX: Boeing North America, Incorporated, Reusable Space Systems.

EH&S. (2007) *SMH Laser Safety For Medical Care Providers*. Available at www.safety.rochester.edu/lasers/smhlasertrainingpart1.pdf (cited: September 19, 2007).

Food and Drug Administration. (2006a) Performance Standards for light-emitting products: Laser products. *Code of Federal Regulations* 21 CFR1040.10. Washington, DC: U.S. Department of Health and Human Services, Food and Drug Administration.

Food and Drug Administration. (2006b) Performance standards for light-emitting products: Specific purpose laser products. *Code of Federal Regulations* 21 CFR1040.11. Washington, DC: United States Department of Health and Human Services, Food and Drug Administration.

GOST. (1983) *Occupational Safety Standards System: Laser Safety* [in Russian]. GOST 12.1.040-s83. Moscow.

National Aeronautics and Space Administration. (1996) *U.S. Laboratory Research Window Characterization Document*. ISS Document RMO-02. Houston, TX: National Aeronautics and Space Administration, Johnson Space Center.

National Aeronautics and Space Administration. (1999) *JSC Safety and Health Handbook*. Joint Program Guideline JPG-1700.1, Rev G. Houston, TX: National Aeronautics and Space Administration, Johnson Space Center.

CHAPTER 25

Crew Training Safety: An Integrated Process

Jean-Bruno Marciacq
Crew Safety Officer, European Space Agency, Safety Manager, European Astronaut Center, Cologne, Germany, Centre National d'Etudes Spatiales, Toulouse, France, Institute for Medicine and Physiology in Space, Toulouse, France

Loredana Bessone
Head of Instructional Technologies, Special Skills Training and Exploration Unit, European Astronaut Center, European Space Agency, Cologne, Germany

CONTENTS

- 25.1. Training the Crew for Safety 746
- 25.2. Safety During Training 770
- 25.3. Training Development and Validation Process 803
- 25.4. Conclusions 815

"[The] crew training program is designed to provide the systems familiarization and flight skills required to effectively, efficiently, and safely control and operate the [system] as well as carry out mission tasks" (NASA 2007).

The crew is an essential factor in the safety of any mission. As such, astronaut training, by its very essence, plays a major role in ensuring the safety of a flight. Even though training cannot be considered a space system per se, it is a function whose contribution to the overall safety of a space mission is as essential as a safe design or good flight procedures. Because it represents the ultimate encounter of the end users, that is, the astronauts, with the hardware before its use during a space mission, the training development, validation, and evaluation process naturally completes the hardware certification and the flight procedures validation processes. Further evidence that crew training is both directly and indirectly essential to safety is that the crewmembers themselves are certified through training, and this eventually enables their flight readiness.

This chapter represents an overview of how astronauts are trained on the general and specific safety aspects of the systems they use to accomplish their required tasks. The main principles, methodology, and requirements for ensuring the safety of the crew during training are presented. Last, a discussion of training validation and how it, together with comments from the crew, can provide invaluable feedback for the safe and efficient utilization of a space system is presented.

25.1 TRAINING THE CREW FOR SAFETY

Training represents the first practical utilization of space systems before a spaceflight, and it enables the crew to operate these systems safely. Training a crew entails training them for safety, efficacy, and efficiency. In the first sections of this chapter, the organization and generic content of crew training is outlined; subsequently, the specific safety aspects of training are discussed.

25.1.1 Typical Training Flow

Crew training has been provided by all space agencies, at different points in time, and for various space programs during the history of crewed spaceflight. Despite the variety of topics and objectives encompassed, crew training generally follows sequential phases, as shown by Figure 25.1.

The major phases of crew training are

- Basic training, which provides the candidate astronauts with general knowledge on space technology and science and on the basic skills related to their future operational tasks.

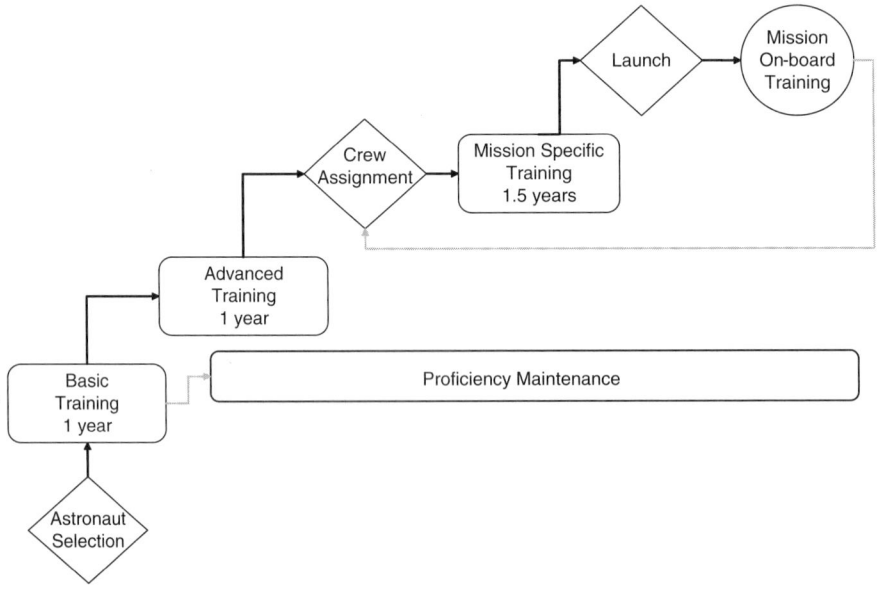

FIGURE 25.1 Typical training flow (Courtesy of NASA).

- Advanced training, which provides astronauts with the knowledge and skills related to the operation of specific space systems, payloads, transport vehicles, and their related interaction with the ground.

- Mission specific training, which provides the flight crew, and the backup crew if applicable, with the specific knowledge and skills required to perform the specific mission onboard tasks within their assigned roles.

- Onboard training, which is used to maintain the proficiency of the crew for critical skills while on-orbit.

- In parallel with these, proficiency maintenance training in areas such as extravehicular activity, robotics, language skills, diving, and air flight training is performed by crewmembers throughout their professional career.

Basic Training: General Crew Training

Once selected, astronaut candidates undergo a year of generic spaceflight familiarization training, consisting of introduction to spaceflight, space system engineering, space programs, vehicles, systems and operations, and space science (NASA 2004). Along with this introductory training, the astronaut candidates learn the relative organization of space systems and how to recognize and utilize the multiple acronyms widely used in space engineering. During this initial phase, they are exposed to spaceflight-like environmental conditions, such as microgravity, thermal changes, high altitude, and high accelerations. The astronaut candidates also are expected to develop the basic skills that will be required during their later training, such as medical, language, robotics, diving, piloting, and survival.

Specifically relevant to safety in this phase is the adaptation of the astronaut candidates to different environments, their capacity to perform under stress, their ability to resolve problems and make decisions, and their general attitude toward adherence to rules and procedures (Dietlein and Pestov 2004).

After successful completion of the basic training, the astronaut candidates become astronauts.

Advanced Training: Program Specific Crew Training or Vehicle, Module, and System Specific Training

During the advanced phase, astronauts are trained on the systems and the operations of the space vehicles and modules that are part of the spaceflight program to which they are assigned. Advanced training builds on basic training, but it is independent of a specific mission. The typical duration of the advanced training for an astronaut is 1 a.

Training during the advanced phase is job oriented, concentrating on the task and system knowledge associated with a single job. It can involve either single or multiple students. Astronauts receive training on operation of all systems, plus a subset of specialties, such as resource and data management, robotics, navigation, servicing, intravehicular and extravehicular activity, medical, and long-term on-orbit payload operations.

Mission Specific (Team) Training

Once assigned to a crew for a mission, astronauts train for the specific operations of the vehicles, systems, and payloads they will be operating on-orbit. Astronauts are assigned not only to the mission but to a role and a set of responsibilities within the mission as well: commander, onboard engineer for a particular system, extravehicular activity, or robotic operator (Figure 25.2). For each crewmember, the training is now specific to the vehicle and the configuration, role, tasks, and experiments of the mission.

The overall goal of mission specific training is to train the crew to be able to safely achieve mission objectives. Vehicle and mission specific survival training also are conducted during this phase, which typically lasts for 18 mo for Space Shuttle and International Space Station increment crews.

Distribution of Crew Roles and Associated Mission Training

For the earliest spaceflights, the number of crewmembers was limited to one or two individuals flying for short durations. Over time, however, longer missions with vehicles

FIGURE 25.2 STS-100 crew official picture. The commander (left) and pilot (right) traditionally are holding their helmets, whereas the extravehicular activity crewmembers are wearing white extravehicular mobility suits. All other crewmembers, including ESA astronaut and mission specialist Umberto Guidoni (center) are dressed in their orange pressurized launch and entry suits (Courtesy of NASA).

FIGURE 25.3 Andromede mission crew in front of the Soyuz-TM simulator in Star City, Russia. Like in the Soyuz capsule, the commander (Victor Afanassiev) sits in the middle, with flight engineer 1 (Claudie Haigneré) on his left and flight engineer 2 (Konstantin Kozeev) on his right. The ring bound books held by the crew hold the flight procedures (Courtesy of ESA).

capable of transporting larger crews have resulted in the specialization of the roles of the crewmembers.

The roles identified by both the U.S. and the Russian spaceflight programs are similar and consist of a commander, a mission specialist or board or flight engineer, and a payload specialist or cosmonaut researcher (Figure 25.3).

In the Russian as in the U.S. space program, the major responsibility for safety of the crew, integrity and control of the vehicle, and mission success lies with the commander and sometimes the pilot, whereas the responsibility for safe vehicle operation, robotics, and extravehicular activity resides with the flight engineer, board engineer, or mission specialist. The main responsibilities of the cosmonaut researchers or payload specialists are the safe operation of scientific equipment and the conduct of experiments (Dietlein and Pestov 2004).

In addition to the overall role in a mission, each individual astronaut or cosmonaut in a crew is assigned responsibility for system, task, or payload specific onboard operations, according to the requirements of the specific mission. For the *International Space Station*, those requirements are documented in a mission specific crew qualification responsibility matrix that identifies for each crewmember those operations that he or she is specifically responsible for carrying out onboard. A sample simplified extract of a generic crew qualification responsibility matrix is provided in Table 25.1.

Table 25.1 Sample Simplified Extract from a Generic International Space Station Crew Qualification Responsibility Matrix (Courtesy of ESA)

Systems/Operations	System Qualifications and Responsibilities			
	Segment	Crew A	Crew B	Crew C
Complex Operations				
Emergency operations	ISS	O	O	O
Activation and deactivation	USOS	S	O	S
Stowage and inventory	ISS	S	O	O
Command and Data Group (CDG)				
Operations local area network	ISS	S	O	O
Command and data handling	USOS	S	O	S
Onboard computer system	ROS	O	S	O
Data management system	EOS	O	U	S
Electrical Power Group (EPG)				
Electrical power system	USOS	S	O	S
Electrical supply system	ROS	O	S	O
Electrical power distribution system	EOS	O	U	S
Integrated Medical Group (IMG)				
Crew medical officer	ISS	S	S	U
Crew health care system	USOS	S	U	O
Medical support system	ROS	O	S	O

USOS = U.S. operating segment.
ROS = Russian operating segment.
EOS = European operating segment.

Simplified for clarity, for each system, task, or payload, three levels of operational responsibility are defined (NASA 1999b):

- User level: Ability to perform automatically all habitability related, emergency, and safing operations.
- Operator level: Ability to perform autonomously all nominal and frequent routine operations.
- Specialist level: Ability to perform, with ground support, off-nominal and infrequent maintenance operations.

Maintenance of Proficiency

Some of the complex astronautical skills learned during the various phases of the training require exercise according to a predefined schedule to maintain a minimum level of proficiency. Those skills include piloting, robotics, rendezvous and docking, diving, and extravehicular activity. As an example, aircraft piloting and robotics skills require complex tasks to be repeated every 45 d on a simulator, as well as operating the real hardware every 90 d.

During mission specific training, maintenance of proficiency is very important because the more time consuming remedial training for performance decrements are not compatible with a very tight mission preparation schedule.

Onboard Training

During flight, onboard training is used by the crewmembers to retain a high level of proficiency in core skills for tasks dealing with emergency, servicing, medical, and robotic operations.

Emergency and robotic skills are maintained on a regular basis, while rendezvous and docking and extravehicular activity skills are refreshed just before the scheduled onboard activity (NASA 1999a). Table 25.2 is a list of the types of onboard training and the frequency of such training.

Table 25.2 Sample Simplified Extract from a Generic International Space Station Onboard Training Plan (Courtesy of ESA)

Onboard Training Lessons	Duration (h)	Frequency
Rendezvous and Docking Nominal Training		
Manual docking	2	1 to 2 d prior to docking
Vehicle redocking	2	1 to 2 d prior to undocking
Emergency Drills		
Readiness for emergency egress drill	2	Once within 7 d after departure of previous crew
Emergency fire procedure drill	1	Every 2½ mo
Emergency depressurization procedure drill	1.5	Every 2½ mo
Emergency descent training	3	Every 3 mo
New module delta emergency procedure drill	2	No later than 7 to 10 d after arrival of new module
Emergency health maintenance system contingency drill	1	Launch plus 1 to 2 mo
Other Proficiency Training		
Crew medical officer proficiency	0.5	Monthly
Robotics proficiency	1.5	Once every 4 wk

25.1.2 Principles of Safety Training for the Different Training Phases

In this section, the basic principles of good training for safety, from safety oriented training (how to perform operations in the safest way) to safety specific training (what to do in case a hazardous situation occurs) are reviewed.

Basic Principles of Safety Training

To achieve the ultimate goal, which is the correct implementation by the crew of all safety rules and regulations, the best approach is to involve the subject as a responsible actor with respect to his or her own safety and the safety of crewmates. The crew fully must be made aware of potential safety issues and their potential consequences. To this end, it is essential that those who actually must implement a safety rule fully be aware of the rationale leading to its enforcement. This is why, in the text box of caution and warning labels and the procedural steps used on orbit and during training, it is necessary to mention not only any restriction or interdiction, such as do not connect or disconnect while powered, but also identify any risk (electrical shock and arching), its consequences (severe injuries, burns, short circuit, fire), and explicit ways to avoid the problem (ensure that power is actually off by checking that the switch is in the OFF position or the power status light is OFF). Flight procedure requirements, such as the onboard data file or the payload onboard data file, provide information as to the adequate and appropriate wording and formatting for use with these labels.

The next step is for the crew to become informed and safety proactive users of all equipment available for their use during a mission. This is achieved through regular training, by explaining safety issues as they appear in the course of training operations, and through safety dedicated training, during which the most notable space hazards, the means for their mitigation, and any necessary contingency actions are explained.

By choosing this approach, the crew becomes capable of managing and preventing potential hazards in the safest possible way and applying appropriate contingency measures should something happen. The trained crew must be able to elect the safest approach and prepare for unexpected events with respect to the occurrence of a known hazard.

Basic Safety Training

As part of basic training, astronaut candidates must become familiar with typical spaceflight hazards and any respective means for mitigation. The astronaut candidates also should be introduced to good safety practices, that is, the dos and don'ts; and safety awareness must be fostered by exposure to pertinent examples of mishaps and associated lessons learned. Through this process, the crew learns, for example, never to connect or disconnect a powered cable due to the risks of electrical arching, detachment of molten metal parts, circuit overload and tripping, and the like, or to always check a connector for bent pins before connecting it. In addition, it is important that the general principles of emergency management be provided as part of basic training.

Safety training is integrated into the basic training curriculum. Candidate astronauts first receive training on basic space emergencies (fire, depressurization, toxic releases) and associated caution and warning classifications. Then, crew environmental hazards (radiation or micrometeoroid and orbital debris) and their associated means of mitigation (shields, multilayer insulation) are introduced. Training of the astronaut candidates is complemented beneficially through the use of scenarios in which specific intravehicular activity hazards, such as shatterable materials release, are presented. Basic training scenarios also include the physiological effects of oxygen and carbon dioxide intoxication, microgravity and long stays in space, and the basic principles of responding to medical emergencies in space.

While performing air flight training, astronaut candidates are sensitized to the management of complex situations in a three-dimensional environment, a situation similar to that encountered during spaceflight. They practice approach procedures and patterns and become familiar with air traffic regulations and the use of checklists and standard communications protocols, such as the phonetic alphabet. This training especially is critical for launch and reentry. Parabolic flights are used to complement flight training by exposing astronaut candidates to the effects of weightlessness.

Diving training provides the astronaut candidates with an environment similar to that encountered during extravehicular activity. The trainees, however, must first learn the general rules of public open water scuba diving before entering the more specialized extravehicular activity training in a neutral buoyancy facility.

During basic training, astronaut candidates physically are acclimatized to specific aspects of the space environment to provide them with the basic knowledge and skills to perform safely in space. This is accomplished by using specific training facilities such as centrifuges, neutral buoyancy facilities, zero-g aircraft, or thermal or vacuum chambers. The principles of survival training also are provided at this stage.

Vehicle, Module, and Facility Specific Safety Training

During this phase, astronauts learn how to safely use the tools needed to operate effectively the modules and vehicles of a specific spaceflight program. Along with the development of space systems, an integrated safety review process is put in place to assess and review these systems against all space safety requirements. This process provides feedback not only to the designers but also to the operational community, which must verify how the equipment has been developed and determine how the hardware can be used in the safest possible way.

In turn, astronauts learn the functioning of the hardware and software they will use. The astronauts must understand the architecture of the hardware and be fully familiar with the interactions of the subsystems of which it is composed. This entails the acquisition of a working knowledge of the individual hazards of each subsystem, of any potential integrated hazards of the system as a whole, and any means by which these hazards are controlled by design or operations.

Mission Specific Safety Training

Successful performance of a particular mission is the ultimate goal of the crew. Because the term *mission specific* implies non-routine operations, particular emphasis is placed on the training for the associated procedures. To this end, mission specific training is the latest, that is, the closest to the mission, and the most specialized training for the space system and any experiment that is to be activated, operated, maintained, or repaired during the mission. Because it is the last stage of training before flight and specific roles have been assigned within the crew, this training phase is not only the most specialized but also the most loaded in safety training with respect to the specifics of the systems to be operated.

In this process, qualification levels that can be achieved by the crew are

- User level, that is, what to do and what not to do.
- Operator level, that is, how to use the equipment nominally.
- Specialist level, that is, how to conduct off nominal operations, maintenance, and the like.

Each level of qualification corresponds to an established allowed safe utilization of the equipment based on the degree of knowledge with respect to its function and hazards possessed by the crew. Specific safety critical operations and emergency procedures are trained or retrained at this stage. This usually occurs just before the flight to ensure that the crew remembers the main issues and how to cope with them proficiently once on orbit.

When getting closer to the flight, integrated simulations are conducted to provide a realistic environment not only for the crew but also the flight control teams, which are then better prepared to support safety critical phases of the operations and react appropriately in case of an unexpected development. During this training, realistic failure cases can be simulated and rehearsed by the crews.

It is interesting to note a major difference in the Russian and U.S. training philosophy with respect to the effect of the last training sessions before a mission. The last motion simulation in Space Shuttle mission training provides a nominal ascent, whereas in the case of Soyuz training, the last simulation includes an emergency escape.

Onboard Training

Safety is one of the main drivers for onboard training, because a certain level of proficiency must be maintained throughout the flight for complex and safety critical operations. Emergency drills are performed for the first time very soon after crew arrival to a station or vehicle. These drills typically are conducted just after docking to allow the crew to transfer and apply their learning to a specific environment. Thereafter, emergency drills are conducted on a regular basis.

Proficiency in the use of robotics is critical in the prevention of a collision of the robot with the structure of the habitable volume or vehicle; therefore, the crewmembers operating robotic arms are trained with a minimum recurrence of typically 45 d. Similarly, skills necessary for other critical tasks, such as rendezvous and docking and extravehicular activity, are refreshed a few days before the scheduled onboard task (NASA 1999a).

Some tasks might have to be performed on orbit for which the crew has had little or no prior training. For these tasks or situations, the onboard crew is then trained and rehearsed in the corresponding off nominal procedures prior to implementation, such as before returning a degraded vehicle to Earth. Should this situation arise, all necessary supporting information is uploaded to the spacecraft and briefing sessions for the crew planned with instructors on the ground.

Maintaining Safety with Proficiency

The safety practices and reflexes acquired by the crew can be lost in the absence of regular practice. Proficiency training reinforces these safety habits. By practicing technical, complex, and safety loaded operations, such as aircraft flying or scuba diving, the crew acquires and maintains good interdisciplinary safety awareness and analytical skills in demanding situations. The outcome of this training can be used profitably in safety critical orbital operations.

25.1.3 Specific Safety Training for Different Equipment Categories

The large variety of spaceflight specific equipment requires that specialized training be performed to ensure safe operation by the crew.

Launch and Return Vehicles

In addition to nominal launch operations, the crew is trained for aborted launch, such as the pyrotechnic extraction of the *Soyuz* from the top of its rocket or an early main engine cutoff for the *Space Shuttle*, and off nominal returns, such as landing at an alternate landing site or a ballistic instead of aerodynamically controlled reentry.

Since the Apollo 1 ground testing fire incident in 1967 that caused the tragic loss of the entire crew, the allowable oxygen partial pressure inside vehicles has been lowered. Precautionary measures relative to the utilization of oxygen in closed environments (monitoring of oxygen partial pressure, no grease or bare electrical contacts, no connecting or disconnecting powered cables) must be observed by the crew during all phases of the flight, especially during oxygen enriched phases such as extravehicular activity prebreathing.

Considering the inherent hazards of a launch vehicle and the potential risk of damage to the flight hardware within that vehicle, the actual testing of a flight vehicle by the crew is limited to the minimum considered necessary, such as the *premierka* or crew fit check inside the Soyuz capsule in Baikonur, conducted just before the flight. Spacecraft pilots are trained using realistic simulators. These can be either fixed based (the instruments and visual indications provided to the crew are simulated) or motion based (the accelerations of launch, reentry, and landing are simulated but to a lesser scale and duration) to provide the crew with a realistic imitation of the launch environment.

For capsules like the *Soyuz*, some reentry training is performed in cockpit replicas located at the end of a centrifuge boom. This type of simulator provides the crew with direct, realistic feedback with respect to the accelerations generated by the reentry path. Training of this nature is very exhausting for the crew, because it simultaneously pushes cognitive, psychomotor, and physical capabilities to their limits.

Any evacuation from a space vehicle, regardless of whether this occurs on orbit or at the launch pad or landing site, always is considered to be a critical operation

FIGURE 25.4 ESA astronaut Pedro Duque and NASA astronaut Scott Parazynski practice emergency evacuation of the *Space Shuttle* from its launch pad at the NASA Kennedy Space Center, using a slide wire basket from a 60 m (195 ft) height (Courtesy of NASA).

(Figure 25.4). It is an emergency mode conducted within a generally hostile environment. For example, the launch pad contains large amounts of explosives, flammable materials, and pyrotechnics; and a landing site is contaminated by toxic gases and trace materials generated by the vehicle propulsion system, pyrotechnics, or radioactive altitude detection systems.

For an evacuation from the *Space Shuttle*, the crew is trained extensively on the ground in the use of the space transportation system boom extension and implementation of the bailout procedure, as well as the splashdown and recovery operation. The crew dons standard pressurized orange garments, that is, the launch and entry suit, and practices the bailout procedure and the splashdown and recovery operation in a dedicated fixed mock-up of the *Space Shuttle* and in the neutral buoyancy laboratory, respectively.

Modules

For operations in International Space Station modules, the crew is trained in the proper management of the caution and warning system and the procedures for fault and leak detection and isolation using simulators and realistic fixed base mock-ups. The crew must be able to locate all required detection, protection, and curative equipment in an environment as close as possible to the one they will confront in space. Ideally, the only difference is the 1-g environment. This generally is not a problem, except in terms of three-dimensional realism; for in space, modules are oriented in all three axes, but on ground, they all are positioned horizontally on the floor.

Training for evacuation and compartmenting (hatch closing) is performed using realistic alarms and hatch mechanisms in the mock-ups. The crew also is trained on procedures for leak detection and subsequent isolation of the resources provided by the module, such as an ammonia cooling fluid leak.

Racks and Facilities

For racks and payload facilities (Figure 25.5), the crew is trained to perform an electrical shutdown of the facility and identify the exact location of the smoke detectors to be able to locate a fire. They also are trained to know the location of fire access ports so that they quickly can extinguish a fire inside a rack. As well, the crew is trained on the processes of leak detection, such as nitrogen leakage, and facility isolation.

Other aspects of training having potential consequences for flight safety are rack tilting or extraction and servicing. The proper process for connecting or disconnecting water, gas, and electrical lines is trained, as well as how to manipulate bulky pieces of hardware possessing high inertia in a fragile and restrained environment.

For those facilities dedicated to science, such as the ESA material science laboratory, the ESA fluid science laboratory, the ESA European physiology module, and the ESA biolab for biological experiments, the crew is trained on the specific hazards associated with the science (high temperatures, optical parts, and biological contamination) and any associated contingencies.

FIGURE 25.5 ESA astronauts Jean-Francois Clervoy and Frank de Winne during a training session on the ESA biolab training model at the ESA European Astronaut Center in Cologne, Germany (Courtesy of ESA).

Payloads and Experiments

The crew should be trained on all specific hazards presented by payloads and experiments. Such training should include hazard mitigation procedures and processes as well as how to react to off nominal situations and contingency cases. This instruction should be an integral part of specific payload and experiment training. Typical contingency cases for which the crew should be trained are recovery of leaks, spillages, or breakage of shatterable materials inside the habitable volume. These cases require the use of cleaning kits and donning personal protective equipment, such as gloves, goggles, or masks.

This kind of safety training should apply not only to internal payloads but as well to external payloads, such as SOLAR or the European technology exposure facility located external to the ESA Columbus module of the *International Space Station*. Some external payloads require processing by the crew inside a pressurized module for preparation before installation and after their retrieval and outside the module, where they are exposed to the space environment.

For the ESA Matroshka radiation exposure experiment, the crew must be trained not to contaminate the habitable environment while either preparing to deploy or after retrieval of the experiment through the air lock. This experiment utilizes extravehicular activity equipment, such as multilayer insulation, which contains a fabric made of beta cloth. The fibers of this material are coated with fiberglass, and if not handled properly, they can shatter, releasing sharp particulates into the habitable volume. Also, when this experiment is retrieved, residues of propellants might have accreted onto its surface during its exposure to the space environment and can be brought into, and thus contaminate, the habitable environment. Proper handling is trained through the use and disposal of containment bags or by cycling the air circulation system to filter out all contaminants and debris.

Robotics

The constant awareness of the robotic operator about the progression of the arm with respect to its surroundings is of paramount importance to safety (Figure 25.6). For this purpose, the operator must at all times use external cameras to have an overall view of the arm and adjacent structures, a view of the clearances, and a specific view of the task.

The safety specific parts of robotics training focuses on how to move within predefined envelopes, preventing collisions during movement and capture maneuvers, avoiding arm singularities, and careful positioning of extravehicular activity crewmembers when they are attached to an arm. The crew also learns how to manage the associated contingencies, such as engine overruns, deviations, and envelope run outs. Training for robotics operations focuses on sensitizing the crew to the importance of strict adherence to flight rules, procedures, and checklists.

Extravehicular Activity

The training for extravehicular activity (Figure 25.7) must address all potential hazards encountered by putting the crew in the hostile space environment, such as depressurization resulting from puncture of the extravehicular activity suit, which could be provoked by molten metal parts created when a powered connector is disconnected. As part of this

25.1 Training the Crew for Safety

FIGURE 25.6 ESA astronaut Paolo Nespoli training on the Space Shuttle Canadian robotic arm. Note the emergency stop pushbutton at the left of the window (Courtesy of ESA).

FIGURE 25.7 ESA instructors L. Bessone and H. Stevenin during an extravehicular activity run in the NASA neutral buoyancy laboratory. Note the extravehicular mobility unit space suits, the ground only backpacks and umbilicals, and the safety diver. To establish the proper buoyancy and balance control, adjustable floating and weighting devices are placed on the legs and in the backpack (Courtesy of NASA).

training, the crew should become familiar with the inherent dangers of the protective garment itself and its operation; for example, the gas composition breathed by the subject inside the suit must be controlled at all times, including during prebreathing and postbreathing protocols.

All extravehicular activity training should address the tethering protocols that oblige the crew and its tools to be tethered always to a structure. When properly trained, the crewmembers will neither let themselves nor their tools become separated from their vehicle and float freely in space. The crew must learn to avoid dangerous areas, designated keep out zones, which delineate a minimum safe distance from antennas, moving parts, hot and cold or glass surfaces, and solar panels. In addition, the crew should be trained to take extra care of fragile or sensitive parts of equipment, such as connectors and pins.

During training, it is fundamental that the crew be exposed to all relevant parameters of the actual space environment. For extravehicular activity, this is not possible using a single training facility. Training in overall procedures, including navigation and translation, therefore, is conducted on wet mock-ups in a neutral buoyancy facility. Here, low fidelity tools are employed, whereas individual operations are trained with flight-like tools in other training facilities.

Basic extravehicular activities, such as the tethering protocol and movements, can be trained at lower cost using low fidelity wet mock-ups, commercial scuba, and professional surface supplied diving system equipment when only the functionality is essential for training. Training of this nature is conducted in the ESA neutral buoyancy facility (Figure 25.8). A high fidelity training environment is provided at higher cost by using real space suits adapted for use in water. Training such as this is conducted using the extravehicular mobility unit at the NASA neutral buoyancy laboratory or the Russian Orlan suit in the hydrolab.

The extravehicular activity emergency procedures, such as incapacitated extravehicular activity crewmember recovery, are trained in water using extravehicular activity equipment and wet mock-ups. The mock-ups and equipment replicate as far as possible the exact environment the crew will experience during its assigned extravehicular activity. For this training, long structures, such as trusses and extensions, can be folded or laid horizontally instead of vertically to fit in existing pools.

Emergency Equipment and Procedures

The crew should be trained in the use of specific categories of emergency procedures and equipment:

- Alarms, cautions, and warnings.
- Fire and smoke, including the location of smoke detectors, gas masks, extinguishing equipment, and fire suppression ports, as well as the means for suppressing a fire, such as the dump command for purging the atmosphere of an isolated module.
- Toxic atmosphere, including ammonia.
- Depressurization, including the use of protective equipment and isolation of affected modules.

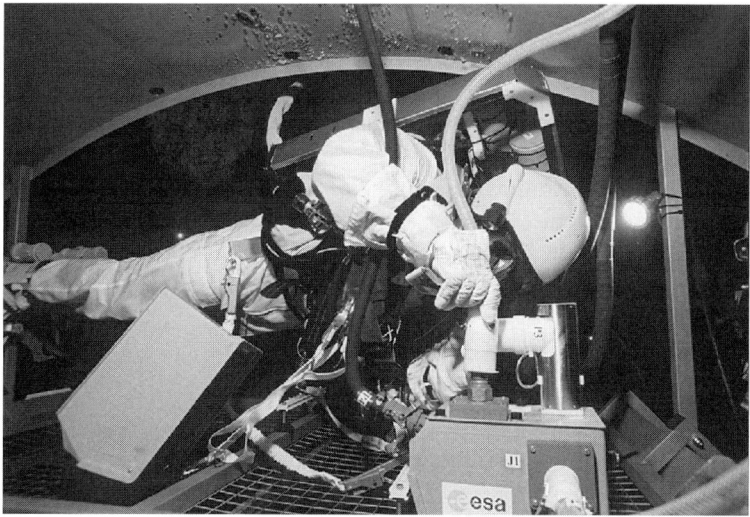

FIGURE 25.8 Extravehicular activity pre-familiarization training at the ESA neutral buoyancy facility. Note the extravehicular mobility unit gloves, the low fidelity flight-like connector, the tethers attaching the crew to the structure, and the payload box tethered to the crew. The helmet is not flight-like, although it provides both limitations of the field of view and protection (Courtesy of ESA).

- First aid, such as a crew health care system, including methods for cardiopulmonary resuscitation, the stabilized stretcher, and intravenous injection doses (see medical emergencies).
- Potable water collecting and sanitizing equipment, shelter, fire, radio, beacons, and the like.
- Environmental protection against wet or dry and cold or hot environments, such as sea, snow, or desert, including typical garments to be used for each purpose.
- Protection against wildlife and signalization.

25.1.4 Safety Training for Different Operations Categories

The levels of operational complexity range from basic good practices and nominal operations, where the operational hazard controls are embedded into procedures, to emergency procedures, whose purpose fully is dedicated to saving the crew. As the level of operational complexity increases, so does the associated level of safety.

Safety Training Syllabus

Safety order of precedence demands that specific procedures and training be developed to address the risks that cannot be eliminated by design. Here, the objective is preventive; that is, safety is embedded in the normal training. Additionally, the crew is to be taught how to react to failures and emergencies, including evacuation and off nominal landing.

Here, the objective is for the crew to react to contingencies, which are addressed by dedicated training (FAA 2000).

Based on this order of precedence, the safety specific training syllabus consists of

- System or payload task specific risk prevention and avoidance operations are trained within the standard training flow. This training sensitizes the crew to the importance of strict adherence to flight rules and procedures for risky operations. A notably important aspect of this training is to change the attitude of the crewmembers so that they adopt safe behaviors. This training is integrated within the standard training flows.

- Caution and warning training provides the crew with the knowledge and skills required to recognize and react to annunciations from the onboard caution and warning systems.

- Emergency training conditions the crew to respond automatically to emergencies. It is of utmost importance that all emergency training is delivered consistently and follow common and very clear procedures. This training is provided repetitively during the course of mission specific training and is repeated in flight through specific drills.

- Survival training provides the knowledge and skills required should a crew need to perform an emergency return to Earth at any phase of the flight and survive in the natural environment of the landing site.

- Human behavior and performance training instills the crew with an awareness of the impact of interpersonal interactions on the performance of team tasks and their implications toward safety. It provides methods to improve the safety and effectiveness of a team by means of best practices in areas like communication, teamwork, decision making, workload management, and conflict management. The crew also learns how to debrief problems and provide feedback to learn from mistakes.

Training for Nominal Operations

The training for nominal operations (color code: green) usually is done by having the crew execute the flight procedures on the training model and under the guidance and supervision of certified instructors. A briefing on the overall objectives, tools, and procedures to be used, where explicit care is taken to highlight specific safety cautions, is followed by a demonstration by the instructor. The crewmembers then are allowed to practice until they master the procedures. Finally, their proficiency is evaluated by the instructor against specific criteria.

The fact that operations are nominal, that is, running as planned, does not mean that no safety issues are associated with them. The safest operations are those that take into account all potential hazards, therefore, keeping the crew away from dangers or from making mistakes that jeopardize their safety. Here, training plays a great role. As an example, in nominal operations for the activation of oxygen generators/water separators, the procedures are written in such a way that the crew is prevented from inadvertently recombining oxygen with hydrogen. This is why, even for nominal operations, the crew should be made aware of the main potential hazards and how these are mitigated by design. Awareness of the hazards involved permits the crew to cope with them in the most efficacious manner should operations become off-nominal or contingent.

When the hazards cannot be mitigated by design, they must be mitigated by operational hazard controls, for which the crew receives specific training. An operational hazard control can take different forms in the training program.

The most common operational hazard control is to insert dedicated additional steps or instructions into a procedure. These would be in the form of cautions (a yellow box for hazards not directly affecting the safety of the crew but leading to degraded functioning), or warnings (a red box for health or life threatening hazards). While running the procedure during training, instructors guide the attention of the crew to any caution and warning. During this training, the instructors also provide background on the corresponding hazards and the means to mitigate them. This gives the crew sufficient safety awareness to prevent or correct mistakes and react to unexpected developments.

Some more critical hazards, or those requiring crew judgment or qualitative skills, cannot be addressed by caution and warning flags in the procedure. This includes the assessment of the proper attitude of an incoming docking vehicle or the return force of a joystick or physical trainer. The operational hazard control requires the crew to be trained specifically on this nominal but safety critical operation. In such cases, specific objectives are incorporated in the training and evaluation items against which the crew performance is assessed are incorporated into the tests. Through these means, potential hazards are mitigated by training.

Training for Off Nominal Operations

Once the crew has been familiarized with nominal operations, the next step is to acquaint them with non-safety critical failure modes and the appropriate way to respond to them.

As for nominal operations, anticipated off nominals (color code: amber) normally are described in flight procedures. Off nominal operations usually are trained by inserting failures during a simulation, and the crew is conditioned to identify signatures of the failure and call out the corresponding off nominal procedure. During the training, the crew is briefed initially on the recognition and correction of failures. Later in the process, failures are injected during nominal operations without prewarning. The ultimate level of difficulty is for the training instructor to combine several failures inserted into a simulation so that the crew has to perform an integrated failure analysis and call out the right off nominal procedures in exactly the right sequence to complete the exercise successfully.

Off nominal conditions and associated troubleshooting also can be trained during integrated simulations. These emulate the interaction between the ground controllers and crew for the resolution of a problem. Interesting failures, that is, the most credible ones, are inserted into the training operations either as errors in the simulators or by using the so-called green cards, which state the symptoms of a failure or advise the crew or the ground controllers of a degraded mode.

Training for Contingencies

After mastering nominal and off nominal operations, into which all preventive measures normally are embedded, the next step is to train the crew for contingency operations (color code: red). This type of training addresses cases where preventive measures were not sufficient or unexpected developments lower the level of safety with respect to requirements. Such situations require immediate crew reaction. Training usually is preventive and affected by informing the crew of the hazard. Should something unexpected happen, the crew is advised on the steps that need to be taken to regain control in the most efficient and appropriate way, that is, how to make it safe.

As typical examples, the loss of containment for a toxic liquid or the breakage of shatterable materials into the habitable volume are considered contingencies. The typical action sequence for the crew is to protect themselves and their crewmates by using personal protective equipment as required, isolate the source to prevent further contamination, clean to avoid dissemination of the existing contamination, and report the incident to fellow crewmates and the ground; that is, protect, isolate, clean, and report. The contingencies should be trained as a complement to nominal and off nominal operations.

Emergency Training

The last and least desirable step once all preventive, corrective, and recovery safety measures have been exhausted or if time does not allow for a sudden contingency to be controlled is to prepare the crew to cope with emergency cases, those requiring immediate action to ensure crew survival. Training for emergencies must be performed regularly and repeatedly, and it must be performed onboard to achieve automatic responses by the crew should an emergency occur.

In all emergency cases, the order of priority for the crew is always to

- Ensure their survival.
- Take care of the vehicle.
- Safe the systems.
- Safe the payloads or science.

Fire, depressurization, and toxic contamination typically are rated as the most critical emergencies possible on orbit. These correspond to catastrophic hazards in safety terminology, because they can lead to the loss of crew, station, or vehicle.

The first step of emergency training is to familiarize the crew with the audible alarms and visible caution and warning indications that correspond to these catastrophic hazards. For the three cases, the crew is taught how to locate and treat the problem by either suppressing it (extinguishing a fire or stopping a leak) or isolating the source (closing hatches).

Training for a reaction to fire or smoke starts by educating the crew as to how combustion develops in space. Further, the crew is familiarized with the location and utilization of protective gear, such as gas masks; Nomex® garments; extinguishing equipment, such as Halon®, carbon dioxide, or water based portable fire extinguishers (Figure 25.9); and fire ports for equipment that is not directly accessible. This type of emergency training also encompasses the last resort for the crew, which is to isolate the segment on fire, egress, and dump its atmosphere.

For depressurization cases, the crew should be trained to locate and, if necessary compartmentalize the leak by closing hatches to isolate modules or sections of the vehicle. The crew is trained to put on a breathing mask should the leak develop in some uncontrollable way.

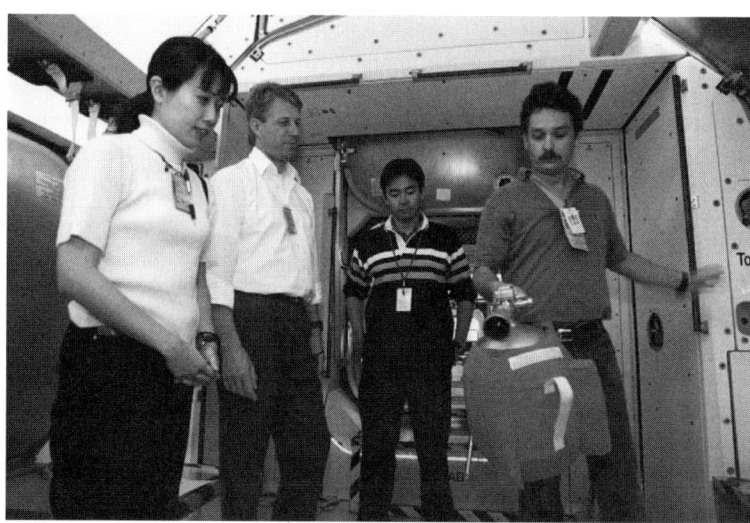

FIGURE 25.9 Portable fire extinguisher training inside NASA space station training facility (Courtesy of NASA).

For toxic contamination, the crew is taught how to locate and don all necessary personal protective equipment, such as gas masks, and how to clean up the leakage should there be a liquid spill. For the case of gaseous contamination, the crew is taught how to regenerate, filter, or if necessary dump the atmosphere of a module or vehicle.

Medical first aid is trained using the same basic principles as on Earth but adapted to the unique environment of weightlessness and, of course, including the lack of direct medical assistance from a physician. The crew is trained to use emergency medical equipment onboard the *International Space Station* and perform cardiopulmonary resuscitation and defibrillation, for example, by using a specially isolated stretcher that prevents the crewmember who is administering first aid from receiving an electrical shock when defibrillating the victim. The crew also should be trained to immobilize a victim, to secure themselves so as not to float away, and thus to be able to provide sufficient pressure when performing cardiac compressions. Learning to do intravenous injections, including on oneself, is also part of the training.

Evacuation and Egress

Should all means to recover from a critical situation onboard a spacecraft become exhausted, the crew must be trained to evacuate promptly and safely their space habitat or vehicle and return to Earth. The first step is to provide the crew with training on rapid safing. Should a partial or total evacuation be required, the crewmembers immediately must terminate their present activity and put any associated hardware or system into a safe mode of operation, so that an even more hazardous situation is not created. As part of this training, the crew learns how to clear the way for themselves and their crewmates along the designated evacuation path if time allows.

The crew then receives training for survival in a safe haven, that is, an isolated part of a vehicle or the *International Space Station* that possesses survival systems and gear that can be used until a recovery vehicle is sent for rescue. As well, the crew is trained for emergency undocking and return and bailout procedures should the *International Space Station* or vehicle need to be abandoned. Emergency landing or splashdown procedures also receive attention during this training.

Survival Training

Being back on Earth might be a relief to a crew that has had to abandon ship in an emergency situation, but it also can be rather hazardous depending on where they land. Deviations in the flight path or any unexpected events during the reentry can cause a capsule to land hundreds of kilometers away from its target. Since Alexei Leonov's offsite landing, which obliged him and his crewmate, Pavel Belyayev, to survive for 3 d in the Taiga before being located and recovered, all Soyuz return vehicles and crews are equipped with survival kits and the crews are prepared for such an emergency with dedicated survival training. Because of its criticality, its specificity with respect to a mission or vehicle, and the complex organization it requires to provide logistical and real emergency support to crew and instructors in the wilderness, survival training usually is conducted as part of the mission specific training.

During survival training, crews learn to secure water, shelter, and fire by any means at their disposal, from either their vehicle or the landing environment. Once the fundamental elements of survival are ensured, the crewmembers must learn how to protect themselves and adapt to the specific environment to which they are presented. To be prepared for a situation similar to that of Alexei Leonov, Soyuz crews usually spend a few days in the forest, regardless of the season. One common hazard in forest conditions are eye injuries caused by wood and other foreign material. These can be painful and incapacitating, and indeed, some training has had to be interrupted because of this problem.

Crews also must learn to signal their location to the recovery team, such as by spreading and displaying their colorful parachute. They must learn how to use flares without endangering their own safety (Figure 25.10). The use of standard localization tools, such as the beacon radio kit, is trained during these events.

Because winter conditions can be quite harsh in Russia, Soyuz crew are taught how to protect themselves against the cold using standard garments and specific techniques, such as building a shelter out of snow. During the summer, the crew trains for the hot and humid conditions of the Russian forest. Protection against wildlife is important whether the animals are big (bears, wolves, or warthogs) or much smaller albeit still aggressive (mosquitoes). To be prepared for all types of aggressive wildlife, each Soyuz capsule contains insect repellant and a three-canon pistol that can be loaded with three types of ammunition for use in hunting for food, signaling (cartridges), or for self defense (bullets). When fitted with an adapter, this pistol also can be used as an axe to cut wood or as a hammer for building a shelter. Any use of tools such as these requires specific training, because they present a number of inherent hazards, the most obvious being those of firearms, explosives, or sharp objects, as well as the extremely loud noise produced on discharge, which can cause appreciable hearing impairment.

FIGURE 25.10 ESA astronaut André Kuipers fires up a smoke signal as part of his winter survival training in a Russian forest during his preparation for the DELTA mission. Behind him is a makeshift tent built from a parachute (Courtesy of ESA).

Last, the training of the cosmonauts for splashdown requires that all three crewmembers doff their respective Sokol pressurized space suits in sequence and don an orange rubber watertight overall, called the Pinguin, which covers the whole body except for the face. The exchange of suits is done within the very confined space of the Soyuz reentry capsule. The next step is for the crew to jump into the water and wait to be recovered (Figure 25.11).

Medical Training and Baseline Data Collection

Training for human physiology experiments (Figure 25.12) in which the crew is the subject and the performance of baseline data collection for both preflight and postflight data collections requires intense coordination between safety and medical personnel. For the hardware used to conduct these experiments, safety engineering must ensure the compliance of equipment to applicable engineering safety requirements, whereas the medical authorities, that is, various medical boards, ensure compliance with applicable medical and ethical requirements and provide medical supervision as necessary. Coordination

FIGURE 25.11 ESA astronaut Frank de Winne jumping out of the Soyuz capsule into the Black Sea (Courtesy of ESA).

between the safety and medical entities is, therefore, of paramount importance, both in the preparation of a mission and during its execution.

Human Performance and Behavior: Crew Resource Management Training

Human performance errors are responsible for 34% of aerospace failures. Human errors are primarily those made during actual team operations, although they also are a factor in hardware design and procedure development.

In terms of design, human factors can intrude at different steps in the process and take various forms. All volumes, forms, switches, labels, light emitting diode colors, and so forth must adhere to human factors design and safety requirements. As an example, amber and red light emitting diodes usually are reserved for safety critical caution and warning indicators. For this reason, light emitting diode status indicators that are not safety critical must be green, white, or blue, colors that do not trigger unnecessary crew reactions.

Human factors training, including crew resource management, spaceflight resource management, and human behavior and performance training, is an essential complement to a well designed human interface. Such training strives to increase the efficiency and performance of the crew and prevent or mitigate the occurrence of human error during operations, thus contributing to safety. These ends are accomplished by training those key behavioral competencies that complement technical and procedural competencies.

As an illustration, behavioral competencies for long duration spaceflight include the capacity of the crew to manage their own well being and operate under stress, communicate effectively, be an effective member of a team, resolve conflicts, be aware situationally

25.1 Training the Crew for Safety

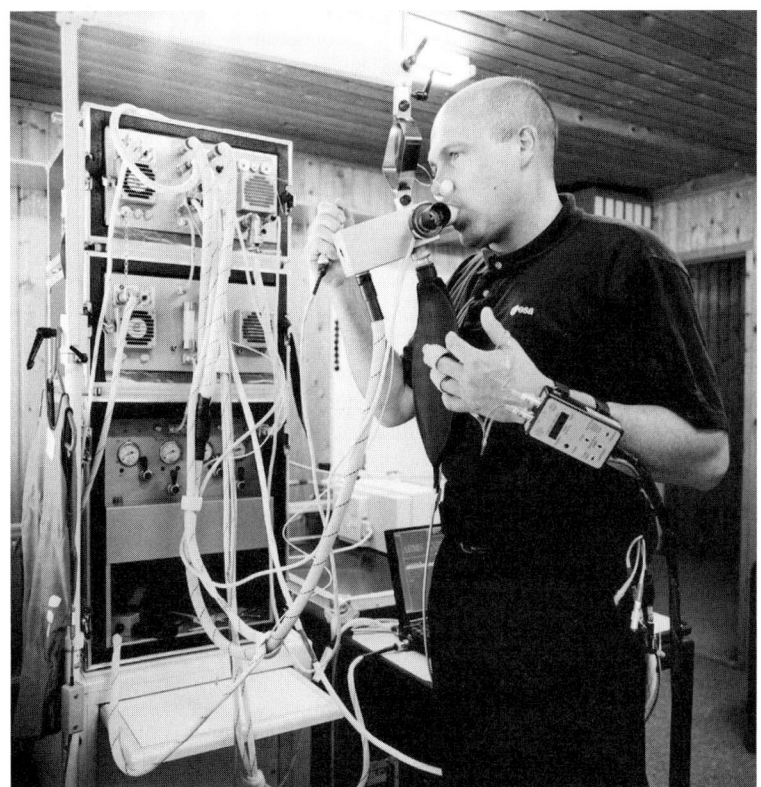

FIGURE 25.12 ESA astronaut André Kuipers during human physiology experiment training (Courtesy of ESA).

of system status and environment, make decisions and resolve problems, lead a team, and work effectively in multicultural teams. In addition, teams need to be able to give feedback to one another and debrief the outcome of operations so as to learn from their mistakes and adapt their performance and team interactions accordingly. A human behavior and performance competency model is illustrated in Figure 25.13.

Human behavior and performance training usually is conducted in environments where typical spaceflight stressors and similarly complex situations can be replicated. This allows the crew to exercise best practices under difficult conditions and transfer skills and techniques into their professional environment. Such human behavior and performance training can be applied to technical environments by providing a crew with feedback for behavioral competencies during complex simulations.

Additionally, human behavior and performance training is best performed in isolated or otherwise potentially hostile environments or in combination with survival training. As well as producing stressful conditions, the purpose of training in these environments is to detach the individuals from their usual and familiar surroundings and resources. This

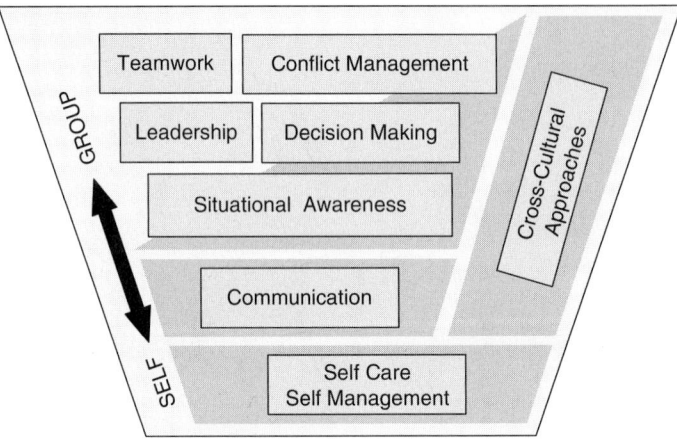

FIGURE 25.13 Human behavior and performance competency model.

forces them to use any and all limited resources provided by their environment, themselves, and their team.

25.2 SAFETY DURING TRAINING

Ensuring the safety of crew training or testing similarly is of high importance, because a mishap that occurs during the execution of training can jeopardize the ability of a crew to perform its spaceflight. The first rule is, therefore, not to endanger the crew during training, so that they can keep all their capabilities up to and during their flight. However, the training paradox then arises, that is, how the training sufficiently can be realistic for the crew to become familiarized with hazardous environments and difficult situations without being endangered during the training.

Space training is necessary to prepare a crew to handle complex and potentially dangerous systems in the hostile environment of space or emergency ground based situations, such as survival, while simultaneously striving to resolve this complex contradiction. Although this paradox especially is valid for emergency or survival training, it applies as well to all levels of the training chain.

25.2.1 Overview

As soon as the design for a space system is frozen, a set of processes parallel to the development and qualification of the spaceflight equipment is initiated. These encompass the development of not only flight procedures but the training curricula, flows, lessons, media, and facilities as well. This typically occurs at the time of the critical design review and corresponds to the phase III flight safety review, where all hazard controls and their means of verification are identified.

To ensure the safety of a training or test session, four root elements are considered:

- The training model and test equipment (the what).
- The test setup and environment (the where).
- The training procedures (the how).
- The crew, training, and ground support personnel (the who).

As a short mnemonic, one might want to remember the PEEP pneumonic standing for procedures, equipment, environment, and personnel.

To ease both the training and test validation process and the implementation of the training itself, these elements are summarized in one document, the training or test data package. This document is used

- To support the test and training readiness review.
- To perform an integrated safety assessment based on the four elements just mentioned.
- As a reference document for the development of training.

25.2.2 Training, Test, or Baseline Data Collection Model Versus Flight Model: Type, Fidelity, Source, Origin, and Category

A training model must provide a platform that allows the crew to exercise safely on the ground any operations that they are required to perform on orbit.

A certain relationship typically governs the development of the training model or test model. Within the assigned time and budget constraints, the training or test model must be as representative as possible of the safety and mission success critical functions of the flight model. It might appear that the simplest solution is to use the flight model as a training or test model. However, such a solution always is not possible. For example, using the *Space Shuttle* for Space Shuttle training might imply severe ground handling and 1-g adaptation constraints that could cause damage to the flight model while it is being used in a representative way during training on the ground. In addition, until it is integrated, the flight model usually follows very dense development and qualification processes. It is, therefore, difficult for crew and trainers to extract it from this process or even obtain access to it for training. The solution to using the flight model as training model, which might seem attractive at first, likely is to end up being very costly and even detrimental to mission success and safety because of the likelihood of damage to the vehicle.

A training model should allow the trainers to demonstrate and the crew to exercise functions and situations not easily represented by a flight model. A long docking approach or robotic deployment might need to be compressed to focus the attention of the crew on critical operations. The tilting of a heavy rack can be replicated in 1-g conditions by modifying the weight of the equipment or its anchoring. A training model also might allow for the disassembly of equipment if needed by the crew to comprehend the interfaces and behavior of components.

The preferred approach is to develop a training model that is dedicated fully to the purpose of training the crew with regard to all objectives of the training, including safety. A training model can be developed either from the flight model, that is, the space to ground approach, or by building the required training characteristics from scratch or commercial components, that is, the ground to space approach. Depending on the approach chosen, the training test data package either describes the differences between the flight model and their potential consequences for utilization in 1-g on-ground safety, or conversely, it follows a more classical occupational safety approach adapted to more stringent crew safety requirements.

Type

The first thing to identify is the type and purpose of the equipment to be used for training the crew (Table 25.3). The equipment can be a dedicated training model used only for training; a test model, whose purpose is to validate the design; or an operational concept with the crew or a crew surrogate. In addition, medical equipment can be used to perform baseline data collection or medical experiments using the crew as subjects on the ground and also for training the crew on the medical experiments they are to conduct in space.

Fidelity

From the most costly to the least expensive, the fidelity of the training model can range from being strictly identical (one to one), similar (looks the same but with no space rated components or materials), or representative (similar in form or function only) to the flight hardware or system being trained (Table 25.4). Also, some training equipment is designed for very specific environments such as the wet environment of neutral buoyancy facilities for utilization on deck or underwater. Last, ground support equipment can be part of the training equipment without being representative at all of any space system because of specific ground constraints, such as 1-g stands and power converters.

Source

Identifying the source of the test model (Table 25.5) is also of high importance, because it affects the way its safety assessment is realized. For example, the safety assessment of a flight model used on the ground for training or testing must rely on the flight safety data

Table 25.3 Training Model Types

Type	Description	Applicability
1	Training model	Training or familiarization purposes; that is, no medical use on human subjects
2	Test model	Tests, verification, and validation purposes
3	Medical equipment	Equipment used for medical training and baseline data collection (utilization on human subjects)

Table 25.4 Fidelity Types

Type	Identifier	Fidelity	Information to Be Provided
1	Id	Identical to flight model (one to one)	References, such as part number, serial number, flight, and the like
2	Sim	Similar (modified from flight model)	Differences from flight model, such as downgrading of components, materials, or coatings
3	Rep	Representative mock-up (dummy, similar in form only)	Physical representation only, such as volume, mass, indicators, displays
4	Func	Functional with different form (provides same functions)	Provides functions of flight model
5.1	Grnd	Ground only equipment or additional ground support equipment to the test model or training model	Functions provided by ground support equipment, such as mechanical or electrical
5.2	Wet	Water only equipment or additional wet support equipment to the test model or training model	Functions provided by wet support equipment, such as buoyancy, pressure, or umbilicals
5.3	Flight	Air flight only equipment or additional flight or parabolic flight support equipment to the test model or training model	Functions provided by flight support equipment, such as seat tracks, frame, nets, foam, or power adapters

package. This document is adapted for use under the conditions found on the ground, in the air, or underwater, that is, 1-g and dry for most of the training hardware, zero-g and air flight conditions for parabolic flight training hardware, and wet or inverted pressurized conditions for some downgraded space equipment, like extravehicular activity space suits, respectively. The safety assessment consists mainly of analyzing the impact of utilizing a spaceflight model under these different conditions on ground safety by comparing it with applicable occupational safety requirements.

Conversely, the safety assessment of commercial or prototype training on test equipment begins by evaluating its compliance with occupational safety requirements. This is based on existing ground safety certificates that are delivered with the hardware by its manufacturer or integrator.

Depending on their level of development and design similarity with the flight model, the development model (engineering model, qualification model, and test model) is assessed using a flight safety data package adapted to 1 g to account for ground safety requirements. Adapting the flight safety data package should take into account the differences between the flight model and the development model so that any potential impact on ground safety can be assessed as would be for a commercial or prototype model against standard ground and occupational safety requirements.

Table 25.5 Source Categories

Category	Identifier	Source	Information to Be Provided
1	Flight model	Flight model (flight spare)	Precautions to be taken with the flight model, such as gloves, antistatic wristband and table, accelerometers
2	Development model	Development model, engineering model, quality model, training model	Identification, such as engineering model, quality model, training model
3	Commercial off-the-shelf	Commercial off-the-shelf	Brand or model, description of commercial approval or utilization, references for instructions, and manual
3.1	Commercial off-the-shelf, assembly	Assembly of commercial off-the-shelf equipment	Description of the assembly with interfaces between the elements, and description of the integrated product, such as resources (inputs/outputs), power, functional scheme, drawings
3.2	Derived off-the-shelf	Commercial equipment derived from originally approved utilization	Description of the derived utilization from the initial authorized use of the commercial model
3.3	Modified off-the-shelf	Modified commercial equipment	Description of the modifications to the commercial model

Origin

The origin of the equipment provides important indications to training developers as well as to safety officers (Table 25.6). This information, which complements that related to fidelity and source, permits the equipment to be categorized and its provider identified. It is useful for trainers to know from where and in what form to expect information corresponding to a particular class of equipment and, if necessary, to be able to ask the provider for standard documentation that should accompany the equipment. As an example, commercial equipment must be accompanied at a minimum by a certificate of compliance to the applicable European safety and manufacturing requirements and an operating manual that includes all necessary safety instructions.

Specific Equipment Categories

Last, some pieces of equipment specifically must be identified because they must comply with specific requirements for their transportation, stowage, and utilization (Table 25.7). As an example, dangerous goods must be accompanied by their safety data sheet as well as transportation and stowage documents.

Table 25.6 Examples of Various Origins

Type	Description	Accompanying Documents
Class III	Spaceflight equipment downgraded or adapted for training, such as the extravehicular activity extravehicular mobility unit or Orlan space suits	Logbook
Government furnished equipment	Government furnished equipment certified by or for a space agency	Agency certificate
CE, OSHA	Commercial off-the-shelf equipment suitable for European community or U.S. public utilization	Instruction manual, including certificate of compliance and safety instructions
Military	Military equipment, compliant to military standards and norms, such as MIL-STD	Certificate of compliance or contract number
Aviation	Aviation equipment, compliant to aviation standards, such as joint aviation requirements (European Union) or federal aviation requirements (United States)	Logbook and certificates of compliance

25.2.3 Training Environments and Facilities

"Space crewmembers, like pilots, are considered operators who must function in unfamiliar environments" (Dietlein and Pestov 2004). During the early years of spaceflight, the crew received extensive exposure to endurance training. When it became clear that the environment was not affecting crewmembers adversely, a lot of the emphasis in training was shifted toward the use of simulators to develop operational skills.

Specific training environments (Table 25.8) are used to simulate spaceflight conditions. This allows crews to become familiar with and perform operations in such environments. These environments carry with them specific hazards, and the safety of crew training is guaranteed by very strict hazard prevention and control processes and adherence to strict operational and safety rules.

Specific Training Facilities

In this subsection, a synopsis of typical training and the respective facilities in which it is conducted is presented (Stromme 1998; Dietlein and Pestov 2004; NASA 2007).

Parabolic Flights The crews must be trained to function, move, and operate in microgravity. Such training is best done in the simulated zero-g environment afforded by parabolic flight. These flights are conducted using modified wide bodied commercial or military transport aircraft, such as the NASA KC-135, known as the *vomit comet*, the Gagarin cosmonaut training center, the Ilyushin Il-76 MDK, and the Novespace Airbus A-300 zero-g (Figure 25.14). These aircraft, which serve as training facilities, have been modified to permit flying parabolic arc patterns, that is, dive/pull-up/parabola/pull-up, and they are

Table 25.7 Examples of Specific Equipment Categories

Type	Description	Accompanying Documents
Electrical or high power	Power sources and management systems	Electrical certificates and inspection tags
Medical	Medical equipment	Medical or International Electric Code certificates and logbooks, including sterilization protocols if necessary
Automated robotics machinery	Robots and mechanisms	Instructions manual, including the working envelope
Light emitting devices	Lasers, hazardous, or high power light emitting devices, such as ultraviolet emitters	Certificate of compliance, such as ANSI
Pressurized	Pressurized vessels and associated equipment	Hydrostatic inspections logbook and tags
Hazardous substances and products	Dangerous goods, such as pyrotechnics, toxic materials or biocontaminants, flammables, explosives, corrosives, materials that develop toxic combustion products, high heat or deep cold generating materials, or materials having high volumetric expansion	Safety data sheets, transportation and stowage documents

equipped with soft padding, handholds, and nets for use by and to protect the crew during training.

Centrifuge Training Training on a large radius centrifuge that simulates the *g*-loading the crew experiences during launch and landing develops skills required to control spacecrafts during acceleration. The Gagarin cosmonaut training center centrifuge is used to train for Soyuz reentry and the Brooks Air Force Base centrifuge facility at the Armstrong laboratory for Space Shuttle launch and reentry simulations.

Vestibular Training Training to develop tolerance to vestibular stimulation is conducted using short radius centrifuges and rotating chairs (Figure 25.15).

Parachute Training Largely associated with the development of stress management and decision making skills, parachute training also requires an acute attention to safety because it replicates the risk factor of spaceflight. Parachute training also is indicated as part of bailout training.

High Altitude Training Using a barochamber as the test facility, high altitude training develops the skills required for the effective use of preventive measures to tolerate hypoxia caused by a reduction in atmospheric pressure. High altitude training also is used to practice donning space suits and for air lock operations.

Table 25.8 Typical Training Environments by Category

Category	Subcategory	Notes
Ground	1-g training facilities	Mock-ups
	Suspension/air cushion/pogo	Artificial zero-g facilities reproducing the absence of fixation by artificial suspension or airlift
	Hypobaric/hyperbaric chambers	Used for extravehicular activity suit tests and medical testing or recovery following nitrogen saturation sicknesses
	Medical environment	Centrifuges and baseline data collection facilities
	Launch site (for evacuation training)	Very hazardous or critical because of the presence of the launcher itself and relative isolation from emergency services
	Open air	Woods, ice, snow, or simulators for survival training
	Fitness training	Fitness rooms and facilities, such as stadium
Water	Subaquatic	Neutral buoyancy facilities, such as neutral buoyancy laboratory, neutral buoyancy facility, hydrolab, weightless environments
		Weightless environmental test system, neutral buoyancy training facility, and others; extravehicular activity training facilities, splashdown, diving (proficiency and fitness)
	Wet	Survival training in sea, lakes, or in neutral buoyancy facilities
Air	Parabolic flight	Flight intravehicular environments reproduced in wide bodied aircraft: the NASA KC-135, Ilyushin Il-76 MDK, Novespace AIRBUS A-300 zero-g, equipped with soft padding, handholds, and nets
	Air flight training	Various aircraft: military jet fighters, such as the MiG-25, used to train Buran reentries, or any aircraft in the inventory of military and test pilot astronauts, such as Panavia Tornado, Mirage 2000, or General Dynamics F-16; jet trainers, such as the Northrop T-38, Let L-39 Albatross or the Breguet-Dornier Alpha Jet; and general aviation modified aircraft, such as the NASA Grumman Gulftstream II with in-flight extendable thrust reversers and airbrakes to train Space Shuttle steep approaches; or business jets, such as Dassault Falcon 20 and propeller aircraft

Thermal Vacuum Chambers Replication of the real flight environment requires the use of thermal vacuum chambers. This facility (Figure 25.16) provides a realistic environment within which to train the crew in the use of actual flight tools for extravehicular activity.

All extravehicular activity crewmembers test their flight extravehicular activity suits in vacuum chambers that resemble the extravehicular activity air lock of the *Space Shuttle*. Here, they can practice their extravehicular activity preparation procedures and post-extravehicular activity tasks. More important, astronauts can feel and hear an

FIGURE 25.14 Novespace Airbus A-300 (Courtesy of Novespace/CNES).

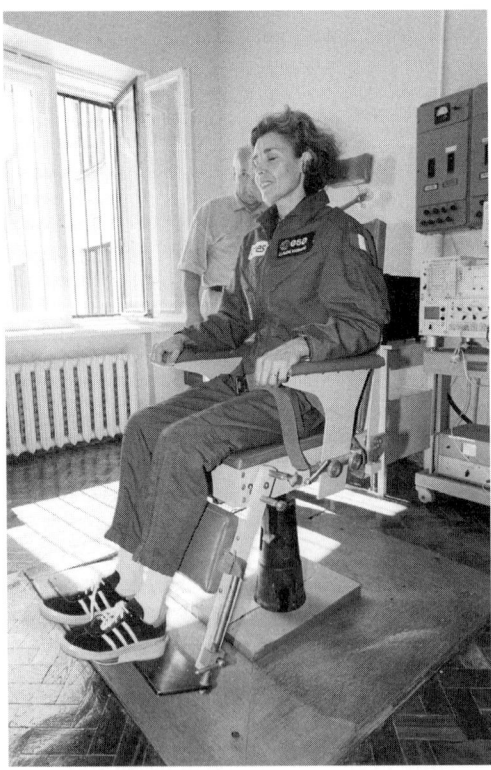

FIGURE 25.15 ESA astronaut Claudie Haigneré on a rotating chair during biomedical tests at the Gagarin cosmonaut training center (Courtesy of ESA).

FIGURE 25.16 NASA astronaut and extravehicular activity specialist Scott Parazynski within the extravehicular mobility unit in the NASA extravehicular training assembly (Courtesy of NASA).

actual suit purge, experience the difference in stiffness of their flight suit as compared to their pool suits, and conduct actual suit leak checks. The chamber runs are a definite confidence builder for extravehicular activity crews, demonstrating that their suits actually do work in a vacuum and will take care of them when they are called on to conduct an extravehicular activity.

The suits are exposed to a vacuum for about 1 h during this training. Extravehicular activity crewmembers must prebreathe 100% oxygen for 4 h prior to the exercise to reduce the nitrogen content of their bloodstream. Without this procedure, there is considerable risk of developing decompression sickness (Stromme 1998).

Neutral Buoyancy Training Facilities Microgravity conditions for operating extravehicular activity suits are simulated by the neutral buoyancy training facilities (Figure 25.17), which are simply large water tanks. Extravehicular activity training in these facilities allows the crew to develop spatial awareness and orientation in simulated microgravity conditions. In these facilities, crewmembers learn to move themselves and equipment efficiently and operate while restrained by tethers, the bulky space suits, and related equipment that constitute their life support and protection from the void of space.

Flight Training Stress management skills, decision making, communication, tolerance to the forces of acceleration, and control of the aircraft during acceleration are developed during flight training. To maintain and demonstrate proficiency, flight training must be performed on a regular and periodic basis; for example, a T-38 flight is required of the crew every 45 to 90 d.

FIGURE 25.17 The ESA European astronaut center neutral buoyancy facility located in Cologne, Germany. The overall dimensions of the water tank are 22 m × 17 m × 10 m deep, for a total volume of 3747 m^3 (Courtesy of ESA).

Escape and Survival Training The skills required to achieve a successful emergency escape from a spacecraft (Figure 25.18) and survive after an emergency landing using equipment and resources from the vehicle and the environment are developed through escape and survival training. This training addresses escapes and survival in water, on land, and under all weather conditions. Survival training also tests the psychological endurance of individuals and the functioning of the team. Survival environments are used for human behavior and performance training.

Spacecraft and Module Simulators Simulators provide for the training of operational skills in a highly representative environment (Figure 25.19 and Figure 25.20). The spacecraft and modules training models are divided into three categories:

- Mock-ups, such as passive mechanical devices, for the training of mechanical operations.
- Trainers, such as panels or laptops with active indications, instruments, or displays for functional simulations.
- Full fledged simulators combining both functions that provide the crew with a more realistic simulation of the flight.

Another example of the variety of training systems is shown in Table 25.9, which lists a sample of all facilities used to train the crew on the *Space Shuttle* (Stromme 1998).

FIGURE 25.18 Bailout from the Space Shuttle cockpit is trained at the Space Shuttle training facility (Courtesy of NASA).

25.2.4 Training Models, Test Models, and Safety Requirements

General Requirements

Every test model must be developed to comply with various requirements. In addition to being representative of flight operations and allowing a specific adaptation to enable training critical functions, these also must comply with occupational safety rules.

Because, on the Earth, training is performed on land, on sea, and in air, corresponding Occupational Health and Safety Administration requirements apply, and the design must be compliant from the onset with the local requirements, such as OSHA or the International Electrical Code for medical equipment. Should full compliance not be possible, a specific waiver must be granted by local safety authorities.

FIGURE 25.19 ESA astronaut Pedro Duque in the left hand seat (board engineer 1) of the Soyuz TMA simulator during a training session at the Gagarin cosmonaut training center. The commander (holding a pen) sits in the middle, and the flight engineer 2 (whose hand wearing a watch is visible) in the right hand seat (Courtesy of ESA).

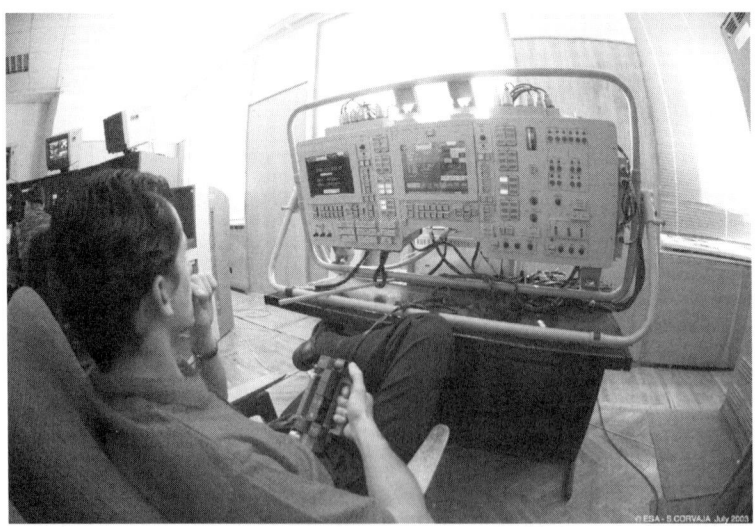

FIGURE 25.20 ESA astronaut Pedro Duque practices reentry on the manual controls of the Soyuz TMA trainer. Note that this trainer is an extraction of the main instruments panel of the Soyuz TMA but located in a more comfortable classroom environment (Courtesy of ESA).

Table 25.9 Sample Simulator and Mock-up Facilities for the Space Shuttle Program Used During the STS-90 Mission

Category	Name	Description
Static mock-ups	Full fuselage trainer	A full-scale mock-up of the *Space Shuttle* without wings, used for training crew escape procedures, in-cabin and payload bay photography, Spacelab ingress and tunnel operations, as well as stowage
	Crew compartment trainer	An accurate representation of the front end of the *Space Shuttle*, including the flight deck and middeck but without the payload bay
	Vertically tiltable mock-up	To train in the launch configuration
Simulators	Single system trainer	Medium fidelity simulators with very close representations of the Space Shuttle flight deck, used for basic Space Shuttle systems instruction and malfunction training. They are used early in the flight training flow to help refresh knowledge of each system and for some of the qualification lessons.
	Shuttle mission simulator	Motion based, high fidelity simulator used for training the dynamic phases of Space Shuttle flight. These are fixed based, fixed high fidelity simulators (without motion)
	Vertical motion simulator	Highest fidelity Space Shuttle landing simulator, the vertical motion simulator allows the crew to go all the way from flying down final approach to landing and rollout of the vehicle
Others	Shuttle training aircraft	Gulfstream II with modified cockpit and thrust reversers, airbrakes used for steep approach training
	External tank doors	1-g extravehicular activity simulator
	Ku-band antenna	1-g extravehicular activity manual retraction training facility
	Neutral buoyancy laboratory	Space Shuttle wet mock-up

Flight Models Safety Requirements for Utilization on the Ground

As a general principle, one should avoid using the flight model as the training model. If not otherwise possible and the flight model has to be used, the following precautions must be taken to ensure the integrity of the flight model and prevent future problems on-orbit:

- **Cleanliness.** The flight model should be manipulated in a clean room environment.
- **Gloves.** Latex or cotton gloves are required to prevent contamination of the flight model.
- **Antistatic wristband and table.** Appropriate electrostatic discharge equipment should be used to prevent electrostatic discharge from causing damage to the flight model.

- **Grounding.** The flight model should be grounded as it is in space; however, the grounding point must be of sufficient quality to not damage the flight model by return currents or create disturbances to its proper operation.
- **Power converters.** Secured and properly grounded power converters must be provided (avoid laboratory systems with rotating knobs and unprotected pins and cables).
- **Batteries and battery chargers.** Only space certified batteries are permitted for use in the flight model, and a specification data sheet for the charger must be provided.

General Constraints of the 1-g Environment

Gravity As obvious as it seems, the main constraint when training on the ground is gravity. Believe it or not, this factor very often is forgotten by training model providers and training developers, whose attention naturally is focused on the flight model and spaceflight operations. Consideration of the following factors is essential when the training model or flight model is being used in the 1-g training environment:

- A stable working area must be provided. Payloads, which are floating in space, must be displayed on a workbench or table, or suspended by cables if they are to be accessible from all sides.
- A scaffold should be used to elevate or extend the working surface (Figure 25.21, the depth of its surface must be at least 1 m (3 ft), and its edges must be delimited clearly by either a ramp or visible markings to prevent inadvertent falls and foot trapping.
- Loads greater than 5 kg must be secured, that is, tethered or fix mounted. Alternatively, they can be carried with the help of one or more assistants. Light volumes representative of structures or equipment, such as payload bags, can be filled with helium gas balloons to offset their weight.
- If the structure of a development model, that is, a qualification or engineering model, is used, it must be reinforced to accommodate any loads induced by its own weight, such as the extension of drawers, or crew and personnel repetitive induced loads (fatigue and overloads). If reinforcement is not possible, then structural limitations must be applied and indicated clearly.
- A structural qualification model that sustained an extensive launch loads qualification campaign is not qualified automatically for permanent utilization as a training model in a 1-g environment. This is because the direction of loads might be different (launch loads are applied longitudinally and in the direction of launch, whereas the modules are positioned horizontally on the ground) or the structural qualification model could be damaged and its structure weakened by the structural tests themselves.

FIGURE 25.21 ESA astronaut Frank De Winne during hands-on training with the ESA developed microgravity sciences glovebox experiment module at ESA ESTEC, the Netherlands. In the fully extended position, the rack requires a stable and secure fixture on the ground (visible on each side around the rack), and the scaffold (with its edge marked with yellow and black tape, to prevent inadvertent fall) on which the astronaut is standing provides a proper working position (Courtesy of ESA).

Power Power often receives less attention as a critical ground safety factor for the crew and support personnel during training. The fact that 340 VAC, 230 VAC, or 110 VAC is used instead of the standard space based 5, 12, 24, or 28 VDC presents greater electrical shock hazards and a greater risk of overheating and damaging components and the crew than would be expected on orbit. On the ground in the 1-g environment, the power source must be stabilized and reliable and proper bonding and grounding of the equipment established and verified. Safety considerations for the proper use of power sources in the 1-g environment are

- Appropriate power reduction must be ensured so as not to expose the crew to high voltages. Proper electrical protection for the circuit and the actual bonding and grounding of the equipment must be demonstrated by electrical schematics and resistance measurements on site.

- A low voltage power source is not a sufficient rationale to justify the safety of the crew and personnel, because even if the device is powered by a low voltage power source, the equipment could contain high voltages internally.

- End to end bonding and grounding should be verified systematically by using an ohmmeter to measure resistance between parts in contact with crew or personnel to the secured grounding point.

- A short in a powered circuit can result in a fire if the circuit is not isolated or protected properly. For this reason, proper isolation or protection of all cables and connectors must be verified.

- As a precautionary measure and especially for medical experiments, at least one red main power cutoff pushbutton on the training model or in the training facility clearly must be identifiable and immediately accessible to support personnel for the duration of any training, test, or baseline data collection.

- During any medical training, test, or baseline data collection procedures involving the use of electrical hardware worn by the crew, such as electroencephalogram cap or electrocardiogram electrodes, it is recommended to have an automated electronic defibrillator immediately available.

External Resources The proper functioning of a single rack or orbital replaceable unit deprived of its cooling or gas resources must be considered. In particular, all touch temperatures must stay below 45°C. Should fans be used to cool avionics or a facility, the total noise generated by the fans set to full power must stay within the occupational safety requirement for constant noise levels at a work site.

Mock-ups, Trainers, and Simulators

Access When designing 1-g mock-ups, the location and number of points of access and emergency exits is critical. As an example, modules preferably should have two escape routes. An emergency exit can be integrated into a fake stowage rack that opens to the outside of the mock-up. Also, for floor areas that are accessible in space but represent a well in 1 g, a proper means of access and escape must be provided.

All hatches and vertical drawers must be counterweighted so that they easily can be opened and closed and not act as guillotines. Opening and locking mechanisms must be accessible and easily operated to provide quick escape. Redundant locking mechanisms should be used so that hatches and drawers cannot be closed mistakenly, and ceiling and floor drawers should be secured properly to prevent inadvertent extension, fall over, or jump up.

The junction areas of the mock-up are considerable obstacles in 1-g because they are narrower than the rest of the structure, resulting in a raised floor and a lowered ceiling of the tunnels. Proper protections, such as rubber carpets and foam head protection, must be provided, as well as the means to facilitate access, such as stairs or ladders.

This similarly applies for conical, cylindrical, or inclined areas of the mock-up. Such geometries are not an issue in space but become a real ergonomic problem when trying to train in them in a 1-g environment. To ameliorate the difficulties, antislip devices and removable flat floors, stairs, or ladders must be provided.

For motion based simulators, the main access drawbridge always must be ready to be activated or a ladder deployed so that the crew and instructors can evacuate in case of fire in the cockpit.

Flammability Nonflammable materials must be selected to avoid the presence of highly flammable or toxic smoke generating materials in a closed environment. The mock-up builder or integrator must define an appropriate extinction means for a fire, should one occur. In all cases, carbon dioxide fire extinguishers should be avoided in closed mock-ups with personnel present.

Ventilation and Cooling Sufficient ventilation must be provided to the crew and support personnel working inside mock-ups, especially when hatches or accesses are closed. Proper cooling also must be ensured, especially within simulators containing a lot of avionics and monitors.

When fans are used to ventilate a mock-up or cool racks, the total noise generated by the fans set to full power must stay within the occupational safety requirement for constant noise levels at the work site.

Centrifuges and Disorientation Devices

Constant monitoring of the subject and the facility is required for centrifuges (Figure 25.22) and disorientation devices, such as equilibrium platforms and tilting rooms. This entails the use of medical telemetry to monitor permanently the condition of the subject with regard to the status of the facility. Furthermore, cameras can be used for closed circuit television monitoring and recording. Additionally, direct sighting should be used to maintain visual contact with the facilities and subject while operations are terminated, should the remote monitoring capabilities be lost.

All safety critical controls, such as power, speed, and brakes, must be redundant or designed for fail-safe operation. As for mechanisms, emergency stop pushbuttons not only must be provided for use by the operators but also by the subject. In case of a potential loss of consciousness, a deadman's switch is held in the closed position by the subject. On release, the switch opens and the device stops immediately.

The consequences of a voluntary or involuntary emergency stop must be assessed, both in terms of maximal deceleration and the stopping position of the device. The latter is important so that direct access to or direct egress by the subject is possible. As examples, a centrifuge preferably should be stopped in front of its main access point or a tilting room should stop in the position providing the access, that is, horizontal.

Should a malfunction cause an inadvertent stop of this equipment, accessibility to the subject and emergency egress must be provided regardless of position. This may entail the use of additional equipment such as ladders or ropes. All physical links, such as safety belts, tethers, oxygen tube, and communication and telemetry chords, between the subject and the gondola must be releasable by the subject in all positions. This capability typically is provided by the use of quick release or quick disconnect devices.

FIGURE 25.22 The small centrifuge of the Gagarin cosmonaut training center located in Zviezdny Gorodok (Star City) near Moscow (Courtesy of ESA).

The subject must be restrained by safety belts or tethers to avoid inadvertently being displaced in the device and suffering a collision with the surroundings during normal operations, emergency stops, and situations entailing a loss of consciousness (Figure 25.23). When the subject is required to be in a lying position, which is used to study space acceptance, shoulder belts must be provided to prevent the upward rotation of the upper torso during deceleration. Also, a central belt located between the legs must be provided to avoid the submarine effect, that is, an unconscious subject sliding down toward its feet. All devices surrounding the subject must be secured or soft padded to protect for any potential collision.

Proper venting and temperature control of the cabin must be provided and, if possible, directly to the subject. For a closed gondola, a supplemental air or oxygen supply must be provided, as in fighter aircraft cockpits. However, all precautionary measures relative to the utilization of oxygen in closed environments must be taken, such as monitoring to ensure that $ppO_2 < 24.5\%$, no grease or bare electrical contacts, and no connection or disconnection of powered cables.

Flight Training and Parabolic Flights

For flight training in general, air flight safety rules and air traffic regulations apply, such as federal and joint aviation regulations for civil aircraft and applicable military regulations for military aircraft.

It highly is recommended for the crew to wear flight approved protective equipment, such as Nomex fire protective suits or helmets, during all flight training (Figure 25.24). This precaution is to prevent the potentially lethal consequences of a crash, such as severe burns over the body resulting from a crash landing followed by a fire.

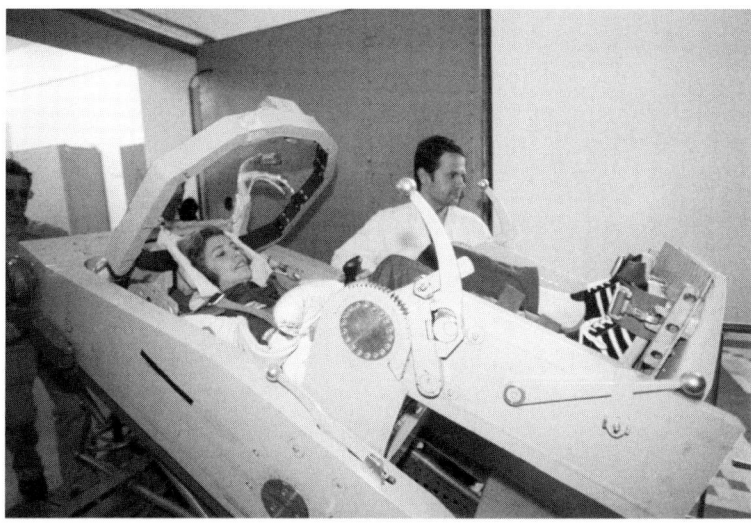

FIGURE 25.23 ESA astronaut Claudie Haigneré being prepared for centrifuge testing at the Gagarin cosmonaut training center. Note the foam around the head and the safety belts (Courtesy of ESA).

There must be no loose objects in the cockpit or cabin. Everything should be secured or tethered because of the potential collision damage with personnel and safety critical equipment, such as instruments or avionics, flight controls, electrical equipment, hydraulics, oxygen, bleed air, and fuel lines, that could occur during multidirectional maneuvers. The only exception to these requirements is made for parabolic flights during zero-g phases, when small items of equipment are used to demonstrate the effect of weightlessness. During use, they must be maintained under direct supervision of an operator and secured before the end of each zero-g phase.

In preparation for parabolic flights, soft padding or foam should be applied to all hard or sharp edges and hard accessible surfaces. For example, all frames must be wrapped in C-shaped foam tubes and all corners padded because of the risk that the personnel might collide with fixed equipment during flight maneuvers (Figure 25.25). Similarly, because hard shoes can cause injuries in case of inadvertent collision or squeezing after a zero-g phase, soft shoes are preferred for cabin personnel. Personnel who are blindfolded, such as for disorientation experiments, must be under constant supervision of at least one assistant.

The equipment itself must be certified both structurally and functionally for zero-g and pull up maneuvers. Safety margins corresponding to a slight exceedance of the nominal flight envelope are included. These typically are $-1\ g$ to $+5\ g$ vertically (g_z) and longitudinally (g_x) for resistance to crash, according to local requirements; for example, the attachment to seat tracks typically are 12-g resistant for civil aircraft. Additionally, all parts of equipment must be secured or under the direct and constant supervision of the subject or operator during the maneuver. There is to be no uncontrolled floating objects because of the risk of collision and foreign object damage.

FIGURE 25.24 ESA astronaut Christer Fuglesang in his Nomex suit, about to get onboard a T-38 jet for flight training at Ellington Field near Johnson Space Center in Houston, Texas (Courtesy of ESA).

Unprotected bare glass and potential sharp edges, either as part of the hardware design or following a structural breakage, must be avoided. Pressurized equipment must be protected and certified against puncture and kick loads, as well as against inadvertent pressure release and associated reaction or whipping effects. Convenient handling points, such as handrails, loops, or nets, also must be provided both inside the cabin and to the equipment itself to afford a good and comfortable position for all operators, subjects, and support personnel. Finally, the availability of an unobstructed escape path, emergency egress from the device, and rapid safing of the device also must be demonstrated, that is, typically within 30 s, and all items stowed and secured.

Pressurized, Depressurized, and Oxygen Enriched or Depleted Environments (Hyperbaric and Hypobaric Chambers, Thermal Vacuum Chambers)

Hyperbaric and hypobaric chambers as well as thermal vacuum chambers are specific training or research facilities to which a number of specific safety requirements correspond. As required for these facilities, design pressures must match the constraints induced by the environment. For example, specific gas dispensers must be designed to

FIGURE 25.25 ESA experiment performed in Airbus A-300 zero-g. Note the foam around the experiment frame (Courtesy of ESA).

supply the proper mixture and quantity of air and oxygen at all altitudes simulated by the chambers. The equipment also must be designed structurally to be resistant to all induced constraints, including thermal loads.

Some chambers are used for multiple purposes, that is, they can be used in both a hyperbaric mode by inflating them and in a hypobaric mode by using vacuum pumps to evacuate the air they contain. These chambers usually have double doors to hold the pressure from the outside (hypobaric) or inside (hyperbaric).

So as not to modify the internal environment whenever inserting or extracting objects, these chambers are equipped with air locks. To limit changes to the internal environment, these air locks are relatively small, thus limiting the size of objects that can be delivered to or extracted from the test environment.

Because of the exiguity and confinement inherent to these chambers, typically they are only 2 to 4 m in diameter, the use of elevated oxygen atmospheres, and the difficulty to egress, flammability is a greater concern than in space. Nonflammable materials must be selected for use in these chambers, and all grease or oils must be contained to minimize the risk of rapid oxidation or combustion of these substances in the enriched oxygen environment. Fire extinguishers and breathing masks also must be provided strategically inside the facility, because a rapid evacuation from the chamber is seldom possible on an immediate basis.

Rapid depressurization or repressurization can have serious adverse effects on the body. Because of the life threatening nature of these effects, emergency equipment and an adequate supply of medical oxygen always must be available immediately for use should an emergency egress be required from a hyperbaric or hypobaric chamber. Because of flammability constraints, however, the use of defibrillators is not advisable in oxygen enriched environments.

In an oxygen depleted environment, oxygen masks must be available for use by the subjects at all times (Figure 25.26). In addition, an assistant who is breathing normal oxygen must be present throughout the test to assist any subject having difficulties because of low oxygen or high carbon dioxide partial pressures. The availability of alternate air or oxygen sources is recommended should the main source fail.

Wet Environment

Structural and Mechanical Design Requirements The material used for the construction of wet mock-ups and training or test models must be selected so as not to allow corrosion and structural weaknesses to develop. Corrosion of materials is caused by either direct contact with the water or galvanic activity between different metallic materials in water (an electrolyte).

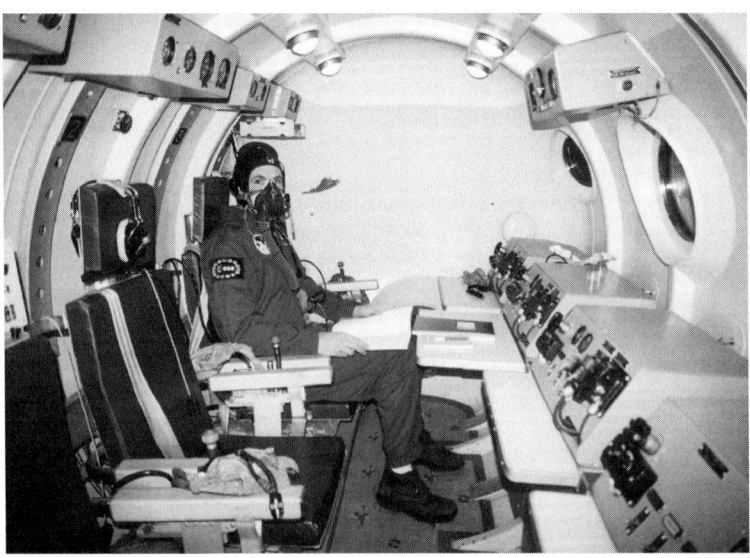

FIGURE 25.26 ESA astronaut André Kuipers undergoes training in a hypobaric altitude chamber at the Gagarin cosmonaut training center (Courtesy of ESA).

Galvanic action should be considered especially when selecting rivets, bolts, nuts, and screws and choosing welding electrodes or rods respectively for arc and oxyacetylene welders. The use of nonmetallic bolts and nuts, such as polyamide, macrolon, or ceramics, is recommended. The type of material from which these fasteners are made, as well as their strength, must match the demands of the design with a safety factor of 2.

Bolted assemblies can be disassembled for replacement or maintenance; therefore, they are easier to maintain than welded assemblies. Welded assemblies should be used only when absolutely necessary. For example, normal aluminum alloy rivets deteriorate within approximately 90 d of continued water exposure if not coated.

As for other structural considerations, all volumes that can be touched by the divers or test subjects must have a minimum radius of 1 cm, and any sharp edges must be filed or covered. Compliance with this requirement is necessary to avoid cuts to the personnel, damage to equipment, or other serious consequences like rupture of an umbilical.

Drainage holes should be provided in all sections of the mock-up to prevent any entrapment of water when the section is removed from the tank. Trapped water increases the weight of a mock-up section, and this must be taken into account when considering the capability of cranes and other lifting platforms. Vents, as well, should be designed to permit trapped air to escape as sections are submerged.

Electrical Requirements All electrical equipment to be used within 4 m of the water tank must be powered by a maximum of 30 VDC or 12 VAC. Only watertight or splash proof equipment should be authorized for use on the deck.

Water Contamination Equipment containing lubricants or fuel must provide proper containment of these materials, which should be verifiable at depths two times that of the facility. Double containment is recommended and, in some cases, is necessary, depending on the requirements imposed by the facility. As well, all equipment must be cleaned before placing it into the water. Furthermore, containment floats must be available on the deck to provide the capability to cope with any inadvertent spillage or leakage.

Neutral Buoyancy Apart from wet mock-ups that are heavy and stabilized at the bottom of the tank, equipment used in these facilities should be neutrally buoyant if possible. The center of mass for such equipment should be as close as possible to the center of hydrodynamic effects to avoid disturbance torques. For calculations, Archimedes' law applies.

To compensate for buoyancy and achieve mass equilibrium in the water, additional weights and buoyancy devices are used. Provision for accommodating these devices must be anticipated when designing a wet model. For example, it is preferred to use hard foams for buoyancy devices because those made from soft foams have a tendency to compress with depth, causing divergent buoyancy.

Underwater Equipment Commercial diving equipment must be certified and maintained in accordance with applicable commercial diving and professional diving requirements.

Oxygen enriched equipment, such as Nitrox®, must be maintained according to state-of-the-art rules, and special precautions must be taken to avoid contamination of the equipment by grease or oil. This minimizes the risk of oxidation, fire, or explosion.

All submersible equipment must be watertight at all depths. This should take into account both static and dynamic constraints.

Labels Yellow and white stripes or boxes must be used for emergency devices or modes, whereas red and white stripes are used to signify a keep out zone. Sufficient lighting must be provided in the pool to limit the effects of color absorption and increase visibility for underwater cameras.

Umbilicals and Cables All umbilicals and cables should be marked by brightly colored floating devices or stripes, pigtails, or small flags positioned along their length. Also, they should be adapted for buoyancy according to the specific environment within which they are to be used, such as a neutral buoyancy laboratory or neutral buoyancy facility. All but the last meters of cables and umbilicals, which correspond to the maximum operating depth of the facility, should be buoyant to prevent inadvertent entanglement of the divers. However, umbilicals used in open water environments are adjusted to have some degree of negative buoyancy to prevent them from floating in big arches and to keep them clear of the surface and any passing boats.

When several divers or devices require the use of an umbilical, distinctive color codes have been adopted for each to distinguish them on the surface and in water, such as one white, the second orange, the third black, and so forth. Greenish and bluish colors should be avoided because they become difficult to distinguish among themselves or from the background. Also red colors used for underwater devices are absorbed very fast by the water, and the color fades away after only a few meters distance.

Caution and Warning and Status Lights All lights used in the facility preferably must be white so as to be seen across the pool and from the surface. Different status modes are indicated by various blinking protocols, such as unpowered state (light off), powered but not active; light on, steady; active or failure mode state, light blinking slowly. Any combination of these light protocols can be used to signify the various states of the hardware or its operation.

Monitoring Because it is difficult to see through the water in the tank, especially with exhaust bubbles and the turmoil of moving equipment troubling the surface, it is an absolute necessity to provide constant monitoring of each subject and the facility. This can be accomplished by viewing directly through wide subsurface windows or using a closed circuit television system. For post-training as well as post-accident or incident analysis, all pictures must be recorded in real time from the beginning of the activity, that is, from the entry of the subjects and personnel into the tank until they exit from the water.

Because of the inherent criticality, it is recommended that all diving operations be monitored by a test safety officer. A test safety officer routinely is assigned to the NASA neutral buoyancy laboratory and the ESA neutral buoyancy facility. For suited operations in the tank, medical telemetry from the subjects is monitored by the test physician to provide evaluation of the condition of the subject, and to observe and assess any safety critical parameters, such as the partial pressures in oxygen and carbon dioxide.

Emergency Equipment for Wet Environments Emergency equipment such as rescue rings, ropes, extracting boards, and stretchers used to extract a diver from the water must be directly accessible. Oxygen must be on hand and in sufficient quantities for use until emergency medical personnel arrive. For unconscious divers, a forced ventilation system is used. This system must be equipped with an overpressure relief valve to avoid damage to the lungs of the victim when it is used. As with all oxygen systems, every related precautionary measure must be observed, such as no contact with grease or bare electrical contacts.

For the protection of support personnel, the use of defibrillators is not recommended in wet environments. If required to treat a cardiac arrest, the subject must be transported to a dry area, where it is safe to use the defibrillator.

In support of a diving operation, a hyperbaric chamber must be accessible within a short period, such as 1 h, from the time and location of any potential incidents. Should no hyperbaric chamber be accessible directly on-site, it is necessary to study and practice the transport of an injured subject to the nearest operating chamber. In addition, regular audits of these chambers must be conducted to ensure their availability during an emergency.

In some cases, transportation by helicopter could be required. If so, a dedicated landing zone free of obstacles must be available adjacent to the location of the diving activity, and personnel must be trained to interact properly with the helicopter crew. In case of decompression sickness, the helicopter must not fly higher than 300 m (1000 ft) above ground level.

Mechanisms

As for any mechanism used in occupational safety, whether autonomous or remotely controlled, accessible safety stops must be provided whose purpose it is to place the device in a safe mode by freezing its motion. A main power cutoff having red pushbuttons and a visible indication of status also must be provided.

25.2.5 Training Model, Test Model, and Baseline Data Collection Equipment Utilization Requirements

Technical specifications, as well as information on the setup, configuration, and operation of training equipment must be described and available to the subjects and personnel. In particular, ground support equipment delivery, assembly, and stowage processes must be documented. For small, passive payloads whose ground utilization requires no supplemental instructions, the training or test model can be accompanied only by the flight procedures. In all other cases, utilization instructions and requirements must be the subject of a dedicated operations manual that lists all ground specific procedures needed by support personnel and instructors to operate the training facility. These manuals are transparent to the crew but are required to configure properly the facility for optimal training. As an example, the operations manual tells support personnel how to power and configure the facility for a dedicated training session and how to perform the closeout of the facility and stowage. As a mandatory requirement, the operations manual must include all necessary safety instructions and restrictions for safe utilization of the training facility.

In the following section, sample instructions that must be included in the equipment operations manual are outlined as requirements imposed on the equipment developer, provider, and integrator.

Transportation and Delivery Plan

The equipment provider must provide documentation stating how a training, test, or baseline data collection model and its associated ground support equipment are to be conditioned for transportation. The provider must indicate the means by which it is to be delivered to and collected from the facility, such as truck, trailer, or mobile crane. In particular, a description including overall dimensions and maximum loaded weight per wheel must be provided.

Hoisting Plan

The equipment developer or provider must indicate how the training, test, and baseline data collection models and associated ground support equipment should be hoisted by cranes and by what means, such as chains and hook, rigid frame, or bands. A clear description of these and the data sheet of all furnished hoisting devices should be provided. In particular, the center of gravity of the training, test, and baseline data collection models, as well as all hoisting points, maximum overall dry and wet weights, and the water displacement must be identified clearly on the drawings provided.

Weight and Balance Buoyancy Plan

The equipment developer or integrator must indicate, as applicable, how the training, test, and baseline data collection models as well as any associated ground support equipment are to be balanced in both dry and wet environments. In particular, the center of gravity of the training, test, and baseline data collection models as well as any associated ground support equipment must be identified clearly on drawings. This applies to the center of lift as well. For wet equipment, maximum induced loads are to be indicated. This also should include forces and torques when the equipment is at the surface; when submerged partially in 1.0 m, 1.5 m, and 2.0 m of water; when submerged totally; and when at maximum depth. The natural buoyancy of the device, that is, the torques or forces when totally submerged, and the maximum induced momentum are to be indicated. This information is important should there be a loss of attachment. Last, should the equipment have a hydrodynamic form that can infer non-vertical displacement, a description of the induced hydrodynamic effects should be provided as well.

Handling

The equipment developer must state all handling requirements for the proper utilization of the test model, training model, or baseline data collection equipment. For example, if the training model is to be used as the flight model, gloves or electrostatic discharge wristbands are required when the hardware is manipulated in any way.

Power Supply

The equipment developer must document all power supply requirements for the proper utilization of the training model or baseline data collection equipment. This must include any grounding requirements, if applicable.

Reliability

The equipment developer must document the life limit of the equipment. The equipment developer also must indicate whether maintenance or specific stowage conditions are required to attain this life limit, such as whether batteries should be removed when the equipment is not to be used for a long period of time. Any life limitations for stowage items must be provided as well.

Marking

The equipment must be labeled "training model," "test model," or "baseline data collection" as applicable, using labels similar in size, material, and design to those applied to the flight model. This is described in the labeling reference document, which is provided along with the test data package. All markings must be in the same language as used for the flight model, such as English or Russian. As for flight equipment, all detachable parts must be labeled.

Packing

The equipment developer or provider must document any specific instructions necessary for packing the equipment and for any specific equipment that is to be used for packing if applicable. Equipment always should be packed in the original packing material within which it was delivered unless otherwise instructed by the equipment developer or provider. Instructions for transportation of the equipment must be in English as well as the local destination language, such as Russian, on the transport container of the training model.

Maintainability and Reparability

The equipment developer must document any requirements for the maintainability and reparability of the equipment. These requirements should include all allowed maintenance operations that can be performed at the baseline data collection, training, or test site by qualified personnel. If the equipment is not to be repaired on site, this constraint must be stated explicitly in the documentation.

Cleaning and Cleanliness

The equipment developer must document how the hardware should be cleaned and it should indicate clearly specific cleaning products that should or should not be used for cleaning. For baseline data collection or medical training equipment, any precautions to be taken for ensuring cleanliness and sterilization must be indicated. For wet equipment, the equipment provider should document how it was cleaned prior to entering the wet area. These instructions should address any procedures to be used for ensuring cleanliness throughout training operations to avoid water and wet area contamination.

Stowage

The equipment provider must indicate the proper stowage conditions for all hardware. For example, the equipment developer or provider must choose among and comply with the following statements, as applicable. To wit, the equipment must

- Be stored in its dedicated packaging within a temperature range of $-x°C$ to $+y°C$, a humidity range of $z\%$ to $w\%$, and pressure range of xx mmHg to yy mmHg.
- Be stored vertically or horizontally.
- Be or not be exposed to light during stowage.
- Be or not be exposed directly to the Sun.
- Be stowed in a vented area.

Transportation within the Training Location

Training, test, or baseline data collection equipment must be transported only in its dedicated transportation casing as described in the test data package or transportation safety data package as delivered. Additionally, training, test, or baseline data collection equipment can possess the following characteristics and must, therefore, follow special transport instructions as

- Fragile, that is, to be manipulated and transported with extra care. This designation is used if the transport container does not provide sufficient protection.
- Sensitive to shocks, such as composite materials. This hardware should be transported with active accelerometers. Any shock, kick load, or puncturing that occurs during transport must be reported.
- Sensitive to orientation, such as hardware containing liquids or orientation sensitive movable parts. This hardware should be transported according to a predetermined and properly documented orientation.
- Sensitive to temperature or pressure, such as pressurized vessels and live samples. For this hardware, temperature or pressure must not fall outside of a defined range during transportation. Temperature or pressure data loggers are required during transportation to monitor those parameters during shipment.

All instructions for transportation must be stated in both English and the local destination language, such as Russian, on the training model transportation container.

25.2.6 Qualification and Certification of Training Personnel

To ensure the safety and quality of training, it is necessary to ensure that all ground personnel are qualified to perform their tasks and roles. This typically applies to flight controllers and also to facility personnel and instructors.

Instructors are selected based on their ability to communicate, deliver information, and interact with people. They must be able to assimilate technical information and structure it into training material. Selected instructors must possess and maintain the following set of knowledge and qualifications (ESA 2002):

- Basic instructor qualification, that is, an ability to develop training plans according to structured instructional development processes, conduct lessons in classrooms and simulators, deliver constructive feedback, and debrief training sessions.
- Human spaceflight knowledge, that is, a thorough knowledge of the basic history of spaceflight, human spaceflight programs and vehicles, space agencies, and centers.
- Knowledge of the organizational rules, procedures, and working methodologies, including the safety policies of the respective space agency.
- Subject matter expertise, that is, an in-depth knowledge of space programs, vehicles, and systems, as well as their interfaces and their similarities and differences to any related systems on which the crew will be training.
- Operational knowledge, that is, an up to date knowledge of the operational use of the systems being used for training that apply to actual missions.
- Facility qualification, that is, an in-depth knowledge and ability to operate the training facilities, including startup, shutdown, and reconfiguration.
- Lesson certification, that is, a certification for the delivery of a specific lesson. This requires all previous qualifications, plus the proven ability to deliver the lesson to a representative crew effectively and answer questions from experienced crewmembers, subject matter experts, and operation personnel.

25.2.7 Training and Test Model Documentation

For each training or test assignment, it is recommended that a dedicated training or test data package be developed to support the training process, as well as for its validation and implementation. This material is to be used in combination with other documents by all participants, including hardware and procedure developers, integrators, instructors, support personnel, and safety engineers (ESA 2006).

The Training or Test Data Package

To be practical and usable, the training or test data package must contain four essential parts:

- A description of any differences between the training model and the flight model stated in terms of utilization, environment, and setup. It should include any inputs to flight or ground procedures, including safety instructions.
- A detailed description of the training and ground support equipment to be used during training.
- A safety assessment of the training process and equipment.

- A number of additional documents, such as the order of acceptance checklist, certificates of compliance and conformance, safety certificates, safety instructions, and all additionally needed information for the correct and safe utilization of the training equipment, setup, and operation by all parties.

To avoid negative training and ensure awareness of any discrepancies between the training and the actual flight, the training or test data package always should begin with a description of the differences between the training model and the flight model and it should address any potential consequences of these differences for the training.

The information contained in the crew procedures provides an overview of what is to be done, and in what sequence the crew is expected to perform a task or procedure on orbit. As deemed necessary, these procedures are complemented by specific ground procedures or an operations manual.

The detailed description of training or test equipment can be done in three parts:

- A descriptive part.
- The equipment summary list.
- The detailed equipment list.

The descriptive part briefly identifies and describes the equipment and includes pictures and drawings to provide for improved understanding. The equipment summary includes the big blocks of equipment to be used as well as provides information on their fidelity, source, and origin (Table 25.10). The legend for Table 25.10 is the subject of Table 25.11.

Finally, depending on the approach chosen, that is, developing the training model from the flight model (space to ground approach) or developing a training model from scratch or commercial components (ground to space approach), the safety assessment must either describe the potential consequences for safety of any differences between using the flight model in space and on the ground or follow a more classical occupational safety approach. In all cases, a safety instruction sheet must be used to summarize what is and is not permitted during use and any restriction of utilization needed to ensure the safety of ground personnel in a manner similar to what is done for commercial equipment in an instruction manual.

Training or Test Safety Data Package

Should training, test, or baseline data collection equipment be complex or possess safety critical features, a dedicated training or test safety data package must be developed if required by the training or test safety engineer or the facility safety manager. The training or test safety data package must be written by the product or quality assurance office and the safety manager of the training equipment developer or integrator. This document must be reviewed and approved by the training or test safety engineer, the facility safety manager for compliance with the local occupational safety requirements, and the crew safety officer to ensure the safety of the crewmembers during their training and during flight. Because it is a document developed and used by safety experts, the training or test safety data package uses a classical safety approach and analysis tools and goes deeper into safety analysis than a simple test data package.

Table 25.10 Sample Equipment Summary List, Examples Provided in the ESA-TrM-208 Template

#	Activity or System Full Name / Equipment Name	To Be Used for			Type of Equipment		References	Information Needed Depending on Category, Additional Comments
		Trng	Test	BDC	Fidelity	Source		
1	Portapress	X	X	X	Sim	FM+ COTS	CNES: *Andromede*	Powered via COTS 220 V/28 V AC/DC adapter on ground
2	Eye tracking device		X	X	Func	COTS	STS: Human research facility	Component quality is commercial
3	Tilt table		—	X	Gr	Lab	CE	Approved for utilization on public patients
4	Valsalva measuring device		—	X	Gr	MOTS Lab		Mouthpiece connected to COTS blood pressure measurement device via plastic tube
5	Rotating chair		—	X	Gr	MOTS		Translation movement of the chair added to the COTS; CE certified
6	SUIT		X	—	Func	DM	ESA: *Odissea*	Fabric is different from flight model. Powered by ground power
7	PROMISS TrM		X	—	Rep	DM+ Lab	ESA: *Odissea*	Opening of the cover is following a different axis
8	Body mass index		X	X	Sim	MOTS	ESA: *Odissea*	Modified commercial equipment. No pressure relief valve for the cuff
9	COSMIC		X	—	Rep	Lab	ESA: *Odissea*	Wooden mock-up with cables attached (not to be powered)

Table 25.11 Legend for Sample Equipment Summary List

To be used for	**TRNG–Training**, familiarization, demonstration (no medical use on subjects) **Test** verification, validation **Medical** training or baseline data collection (utilization on human subjects)
Type of equipment	**Fidelity** to the original or flight model (see following) **Source** of the equipment (see following)
References	Any preexisting certification(s): ESA, NASA, CE, IEC, or utilization in previous space missions, vehicles, or ground experiments: STS-###, LDM-##, *Automated Transfer Vehicle*, *Progress*, *Soyuz*, *Photon*, NEEMO
Fidelity	**Id–Identical** to the flight model (one to one, could fly in space) **Sim–Similar**, modified from the flight model **Rep–Representative** (dummy) mock-up (similar in form only) **Func–Functional** with different form (providing same functions) **Wet–Water** only equipment or additional wet support equipment **Grnd–Ground** only equipment, additional ground support equipment
Source	**FM–Flight model**, that is, flight spare **DM–Development model** (engineering model, quality model, test models) **COTS–Commercial off-the-shelf** **DOTS**–Commercial equipment **derived** from its originally approved use **MOTS–Modified** commercial equipment

As is the case for a flight safety data package, the training or test safety data package must first describe the flight equipment and operations for reference purposes, then the training equipment and all associated ground operations. Differences between the training model and the flight model, as well as the potential consequences of these for the safety of the crew and ground support personnel should be highlighted. All hazards associated with the utilization of the training equipment in the projected training environment and using appropriate training procedures should be identified in the training or test safety data package. For each identified hazard, its potential causes, consequences, mitigation process (hazard control), and verification must be stated. Each of the elements of the training or test safety data package is then translated into the form of instructions, labels, and corrective, contingency, or emergency ground utilization procedures by local safety officers.

A common mistake is for the developer or provider to submit the ground safety data package for the flight model in place of the ground certification for the training model or test model. Although some parts of the ground safety data package can be of common interest, these documents have very different purposes. The ground safety data package is relevant only for the manipulation and integration of spaceflight hardware into a launch vehicle. Its purpose is to ensure, on the one hand, the safety of ground personnel with respect to spaceflight hardware and, on the other hand, the integrity of the launch vehicle and payloads.

A dedicated training or test safety data package that addresses all potential hazards to the crew and support personnel associated with the equipment used for training, test, or baseline data collection must be developed. This document describes the equipment and its projected utilization on the ground and includes delivery, assembly, and stowage.

Attachments and Appendixes to the Data Packages

Order of Acceptance Checklist, Certificate of Compliance, and Certificate of Conformance The order of acceptance checklist (Table 25.12), the certificate of compliance (Figure 25.27), and the certificate of conformance (Figure 25.28) are used when performing the acceptance review, which occurs just before training takes place. The purpose of these documents is to avoid any change to or failure of the hardware before the arrival of the crew at the training facility.

Safety Certificates Safety certificates delivered by other safety authorities must be attached to both the test data package and training or test safety data package to ease and accelerate the safety review process.

Equipment and Safety Data Sheets For all commercial equipment, the data sheets and associated safety data sheets provided by the manufacturer must be attached to both the test data package and training or test safety data package. Copies or extracts from the instructions manuals are valuable information that should be provided along with the test data packages and training or test safety data packages when submitted for review and approval.

25.3 TRAINING DEVELOPMENT AND VALIDATION PROCESS

The training development and verification process represents an opportunity for the crew to question decisions made during the development reviews as deemed necessary from the evaluation of flight or representative hardware or software against the flight procedures.

Table 25.12 Order of Acceptance Checklist as Extracted from the EAC-NBF-208 Template (ESA 2006)

No.	Test	Applicability	Completion	Remarks
1	Completeness	X		
2	Visual inspection	X		
3	Physical inspection	X		
4	Marking check	X		
5	Functional test	X		
6	Weight and balance/buoyancy check	X		

> **Certificate of Compliance**
>
> We, the *Equipment Developer/Provider* hereby certify that the *ABC*-Program's:
>
> - ☐ Training Model (TrM) identified with the equipment code: *XXX*
> - ☐ Test Model (TM) identified with the equipment code: *YYY*
> - ☐ BDC Equipment identified with the equipment code: *ZZZ*
>
> is compliant with the applicable requirements as described in the Document NBF-User Handbook and meets the following generic requirements:
>
> *List here all applicable requirements such as:*
>
> - *NASA/Russian, EVA Requirements (Spaceflight derived H/W)*
> - *NSS/WS-1740.10 (NBL Diving Equipment)*
> - *CE, DIN, BGV, OSHA Certification (COTS Equipment)*
> - *IEC Certification (Medical Equipment)*
> - *Other specific requirements for utilisation in a saturated/wet/underwater environment,*
>
> and that their development, manufacturing and conditioning was done in full compliance with the following quality assurance and product assurance processes:
>
> *Refer here to all applicable quality assurance and product assurance labels such as*
>
> - *ISO Certification.*
>
> The delivered equipment is safe to be utilized for Training / Tests / Baseline Data Collection as described in Document NBF-ABC-208a
>
> Date:
>
> Name of the Equipment Developer/Provider PA/QA and Safety Responsible
>
> Signature:

FIGURE 25.27 Certificate of compliance as extracted from the EAC-NBF-208 template (ESA 2006) (Courtesy of ESA).

25.3 Training Development and Validation Process

Certificate of Conformance

We hereby certify that the *EXP*-Experiment:

- ☐ Training Model (TrM) identified with the equipment code: *EXP*-TrM
- ☐ Test Model (TM) identified with the equipment code: *EXP*-TM
- ☐ BDC Equipment identified with the equipment code: *EXP*-BDC
- ☐ and their respective documentation and setup

have been reviewed and found compliant with the applicable requirements as described in the Document LDM-200 (TS-EX),

have been inspected and found to conform to the description and requirements given in:

- Technical description of the Training Model (TrM) *or*

    Baseline Data Collection (BDC) Equipment (BDC) LDM-*EXP*–208,

and is accepted for:

- ☐ Ground/wet training activities following the Flight Procedures and in full compliance with applicable ground training and safety standards.
- ☐ Ground/wet test activities following the Flight Procedures and in full compliance with applicable ground test and safety standards.
- ☐ Baseline Data Collections (BDC-s) following the ESA-Medical Board approved medical protocol and under supervision of the ESA-Crew Surgeon.

Approved for ESA by:

EAC Training Manager: (for Training)

EAC Safety Manager:

EAC Infrastructures Manager:

EAC Physician/Delegate Physician: (for BDC-s)

FIGURE 25.28 Certificate of conformance as extracted from the EAC-NBF-208 template (Courtesy of ESA).

During this process, the training development teams and the crew have the prospect of detecting any safety relevant issues not apparent during previous paper reviews, thus providing valuable feedback to the safety community. This validation role is reinforced when training teams have been sensitized to both flight and ground safety or are supported directly by safety experts.

25.3.1 The Training Development Process

A systematic instructional system development process (Figure 25.29) is followed for the development of spaceflight training. Systematic development of training ensures that crews and flight control personnel be certified to perform their respective onboard and on the ground operations safely and efficiently. Certification of flight readiness is declared on the basis of successful performance as assessed through a formal evaluation.

The instructional system development process consists of five phases, of which the last step ensures the transfer of lessons learned into future training, as well as into systems operations (NASA 1997; ESA 2000a):

- **Analysis.** During this phase it is important to establish what is to be learned, by whom, and with what previously existing skills. The space systems and operations

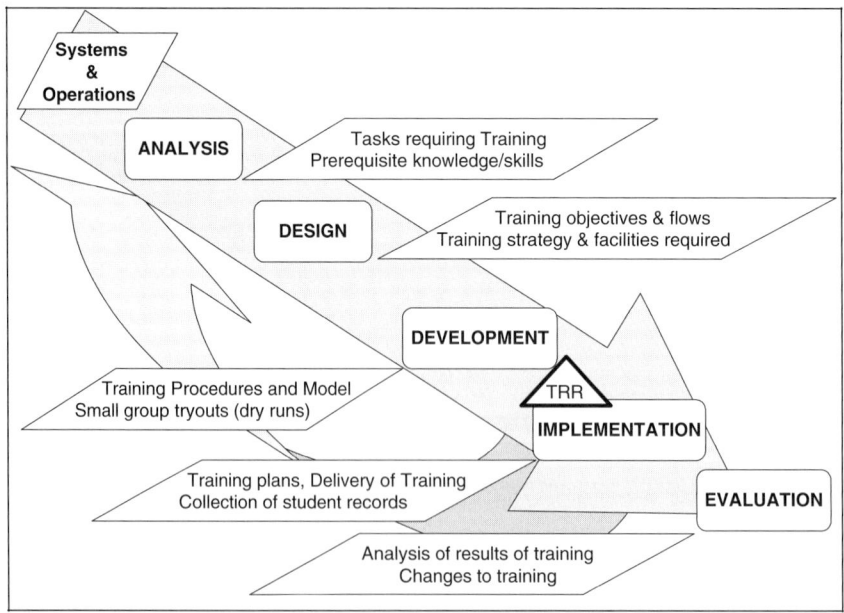

FIGURE 25.29 The instructional system development process (Courtesy of ESA).

are analyzed to identify those areas requiring training. The profiles of the trainees are analyzed to determine the skills they either have or must have prior to the start of training.

- **Design.** Once training requirements and learner characteristics have been identified, it is necessary to map out how the material to be learned can be best presented, practiced, and tested. Learning objectives are derived from the requirements, and strategies are developed with regard to how to best facilitate learning. Evaluation strategies are derived from the objectives to allow validation. Training media and facilities are selected. Training curricula containing catalogs of lessons and specific flows with associated qualifications for each job or position are created.

- **Development.** Only after the overall curriculum has been designed solidly can real training development begin. Lessons, training media, and facilities are developed. Instructors are certified. Small group tests, called *dry runs*, are performed to obtain feedback from system developers, operations personnel, and the crew. Training is reviewed and validated. Issues identified with the hardware and operations are fed back to the developers.

- **Implementation.** At this point, the training is ready to be delivered to participants. Plans and schedules are created. Performance of trainees is assessed. Training records are stored and reports created. Training models for hardware and software and training materials are maintained.

- **Evaluation.** Once training has been delivered, the actual performance of the trained personnel on tests and on the job, as well as any feedback received during training, is analyzed to evaluate the outcome and quality of training. Qualitative and quantitative data are collected, adequacy of training is assessed against trainee feedback, and actual job performance is analyzed. Changes to training systems and operations are recommended as required.

25.3.2 The Training Review Process

It is important to note that the instructional system development process just described is not simply a sequential process. Each step includes a set of formal reviews that can result in any previous steps being adjusted and revisited.

The review process is finalized by ensuring that, through training, the crew achieves the qualifications required to perform onboard operations. Therefore, it is important to involve in the review process all stakeholders, namely, crewmembers, system developers, safety personnel, ground control personnel, instructors, and managers, to ensure that all aspects of systems and operations are analyzed properly.

- **Requirements review.** At the end of the analysis phase, training requirements are reviewed. Sometimes the training analysis and review process can be integrated into a single step. In this case, a workshop consisting of all stakeholders is organized to capture and document training requirements.

- **Catalog review.** The overall catalog is compared with the requirements to verify that all are covered by an appropriate set of objectives and these objectives are grouped into structured courses and lessons.

- **Lesson plan review.** For each lesson, instructors, subject matter experts, and reference material and sources are identified. As well, training media and facilities are identified and verified as being ready. Furthermore, it is verified that these facilities provide a representative fidelity, that a lesson strategy and outline exist, and that an evaluation is planned to assess the achievement of the objectives at the end of the lesson.

- **Lesson flow review.** It is verified that flows are developed that lead to the qualification of individuals for each specific job or position. Each flow must contain a specific entry point and prerequisite qualification and lead to a specific qualification. A set of tests must be associated with the qualification, and the readiness to perform a set of tasks or operations on a given system verified.

- **Training material review.** Once lessons are developed, the training media, lesson slides, and handouts, as well as all training aids used in the training, are reviewed against associated training objectives, required standards, and instructional guidelines. Also, the material contained is verified by subject matter experts and experienced instructors.

- **Lesson certification.** Dry runs or small group tryouts are organized to assess the lesson in its delivery conditions using the actual facilities, procedures, and representative trainees. Subject matter experts participate to ensure the content of the lesson is presented correctly and questions are answered correctly. The lesson is certified formally only after all issues identified in all lesson related reviews and dry runs are resolved.

- **Instructor certification.** Each instructor must be certified for each lesson that is to be delivered to the crew. The instructor must demonstrate a strong lesson delivery and an ability to answer student questions accurately, as well as to control the training environment, including the handling of facilities, readiness of material, and leadership of the class.

- **Training readiness review.** The training readiness review is a formal step that is repeated every time training is delivered to a new crew. During the training readiness review, the readiness of everything required for delivering the training is verified. A schedule of the training must be available, all training facilities must be ready, training lessons and instructors must be certified, and all training material, including handouts, must be ready. Also, training feedback forms for the students and instructors must be available to ensure thorough record keeping.

It is important to note that some of these steps can be combined for efficiency, such as lesson and instructor certification can occur during the same dry run. However, it is fundamental that the outcome of all reviews is documented formally and that there is a follow up to ensure that the final product of training fulfills the operational and safety requirements (ESA 2000a, 2000b, 2000c).

25.3.3 The Role of Safety in the Training Development and Validation Processes

In this section, milestones linked to safety are described in more detail as illustrated by the instructional system development process diagram shown in Figure 25.30, which is complemented with typical safety milestones.

Training and Test Safety Organization and Roles

To support the test or training in the most efficient way, the safety roles and responsibilities usually are divided as follows.

For occupational health and safety in general, the manager of the hosting facility ultimately is responsible for the proper and safe implementation of the test or training operation. Nevertheless, this role usually is delegated to a facility safety manager, whose job it is to implement practically all preventive, corrective, and contingency measures. If the facility is divided into several areas of responsibility, the facility safety manager can be supported by specific facility safety officers and delegate the occupational safety responsibility to local safety officers.

For the test or training activity itself, a test director and a test conductor share the responsibility for the proper and safe implementation of the entire test. The role of the test director is to ensure the correct and safe performance of the test by all personnel

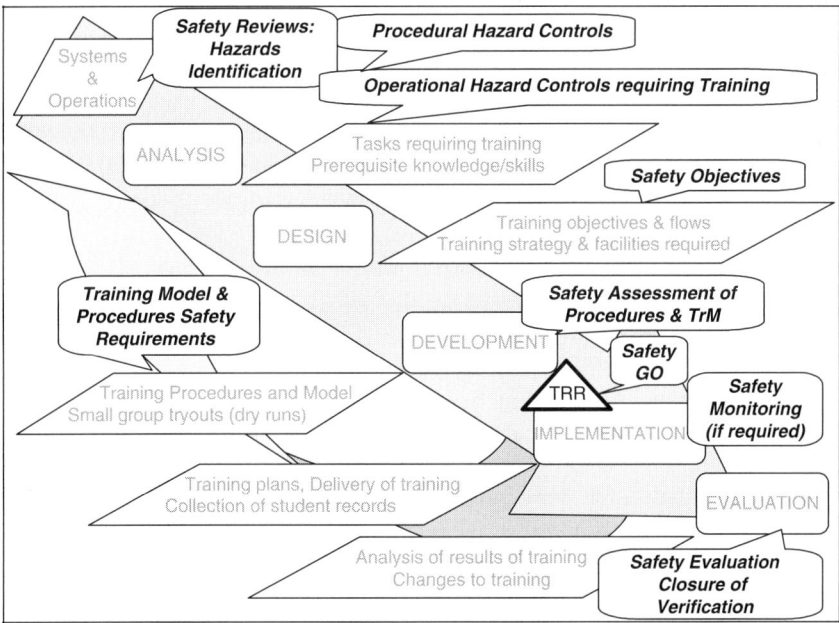

FIGURE 25.30 Safety inputs to the training development and validation process.

involved, whereas the test conductor focuses on the specific objectives and subjects of the test. The test conductor can act as an instructor to the crew as well.

The equipment developer, integrator, or provider always is responsible for delivering a safe product with regard to its intended conditions of utilization and must be able to demonstrate compliance to local requirements via appropriate documentation. To this end, facility requirements, an operations manual, and interface control documents between the facility and the equipment must be provided as required. Within the equipment developer team, the project assurance and quality assurance personnel and the safety officers are responsible for their respective specific safety aspects.

For test or training itself, the test safety engineer is responsible for ensuring the safety of test or training in its preparatory phase, whereas the test safety officer plays an active monitoring role during the implementation of the test or training itself. Those functions can be represented by the same person.

In the same way, the occupational physician has overall responsibility for the health of the personnel working at the facility as well as the state of all emergency medical devices. The medical officer would play an active role to ensure proper medical conditions are maintained before and during a test.

Last, if the crew is involved in a test, a crew safety officer ensures the conformity of the equipment and operation to both ground and flight safety standards.

Safety Assessment

In parallel to the training development process, the crew safety officer, the test safety officer, and representatives of the local occupational health and safety group and the facility or site are regrouped into a training or test safety review panel to perform a safety assessment based on documentation, such as drawings, test data package, or training or test safety data package, and any other additional information provided by the equipment developer or provider. This assessment is followed by request for additional information as needed and formal approval of the documentation reviewed.

Training, Test, or Baseline Data Collection Model Safety Review

Should the equipment to be used in training be complex or safety critical, the training or test safety data package would require and organize a formal training or test safety review. During the training or test safety review, the equipment developer or provider presents the equipment, describes its inherent hazards and the proposed hazard controls to the panel and any ad hoc specialists, such as a representative of the medical board in case of medical training or baseline data collection equipment. When completed and once all additional required information has been provided, the review is closed by formal approval of the minutes and the hazard reports.

Training Model Review, Inspection, and Acceptance

For the transfer of ownership of training equipment between the manufacturer, integrator, or provider and the receiving entity, an acceptance test must be conducted. This review is supported by crew and ground safety representatives. This process must be complemented by regular hands-on safety inspections during delivery, unpacking, assembly, and functional testing of the hardware and throughout its operational life to assess the safety impacts of upgrades or modifications, as well as aging or damage. Paper reviews and physical inspections as part of this process are sanctioned by safety certificates.

Validation of Human Factors Design Requirements

Noncompliance of the training model with human factors requirements usually is captured at this stage. Discovery of such a noncompliance can be an indication of related issues on the flight model.

Procedures Validation

The equipment developers or the training facility itself, with inputs provided by both operations and safety personnel, usually are responsible for writing the training procedures. The onboard data file panel is responsible for the compliance of procedures and their adherence to applicable procedure writing standards. The operations safety group is responsible for proper implementation of all required procedural hazard controls.

Before submitting procedures to the crew for the first time and to avoid negative training, it is recommended that an integrated procedures review be performed. This review is conducted with participation from procedure writers and representatives of the crew, the onboard data file panel, and the safety group.

Dry Runs

The proper application of all safety requirements and the validation of procedures with respect to any safety aspects against the hardware can be performed during dry runs with the participation of representatives from the crew and from safety.

Training or Test Readiness Review

At the end of this process, just before a training or test session, a training or test readiness review board is convened by the authority organizing the activity. The training or test readiness review board is composed of representatives from all involved entities, such as

- Training management.
- Training or test coordination.
- Instructors or test conductors.
- Crew.
- Facilities and infrastructure.
- Safety (training, test, or crew safety officer).
- Operations.

During the training or test readiness review, the current status of test or training preparation is presented in every detail, including the qualification of all participants to the activity. Deviations from applicable requirements and objectives are highlighted and issuance and approval of action items or waivers is decided. During the training or test readiness review, safety plays a considerable role because any unacceptable deviation from a safety requirement can lead to cancellation of the activity.

Safety Role During Training Implementation

If the training or test readiness review is concluded successfully, the test safety officer normally should not be required to attend the training. There are three exceptions to this rule:

- During training for flight safety critical operations, when it is necessary to supervise the correct performance of the training and answer safety specific questions.

- During mandatory safety training, when it is required by the safety review panel. The test safety officer must testify as to the proper training of the crew with respect to safety, to be able to close the verification of corresponding operational hazard controls.

- During ground safety critical testing or training, such as extravehicular activity training operations, which requires constant safety monitoring.

Changes to the Hardware After Lessons Learned from Training

After training, equipment or procedures can be changed as the result of unexpected events or required improvements that could not be detected earlier by the review and validation process.

As an illustration, following a slight injury to a crewmember during training, some screws securing the Automated Transfer Vehicle fluid management panel were found to be sharp and, therefore, noncompliant with flight safety requirements. Those screws were identical on the flight model and the training model. Once the problem was identified on the training model, they had to be modified on the flight model.

Similarly, during a thermal vacuum chamber extravehicular activity training simulation lasting 13 h in the space environment simulation laboratory at −112°C (−170°F) in preparation for the Hubble Space Telescope repair mission in 1993, Story Musgrave suffered frostbite on his forefingers. The outcome of the testing was the addition of thermal insulation to the gloves (Lenehan 2004).

25.3.4 Feedback to the Safety Community from the Training Development and Validation Processes

To ensure the safety of flight operations, it is of fundamental importance that all inputs from the safety community be incorporated into procedures and training (Figure 25.31). Similarly, it is very important and often underestimated that feedback provided by all involved parties during the training development and validation process be taken into account by the safety community.

Feedback from Procedure Developers

Noncompliance to general spaceflight safety requirements (such as disconnecting powered cables) or safety requirements specific to the experiment (operational hazard controls) must be captured by procedure writers when developing or reviewing their

25.3 Training Development and Validation Process

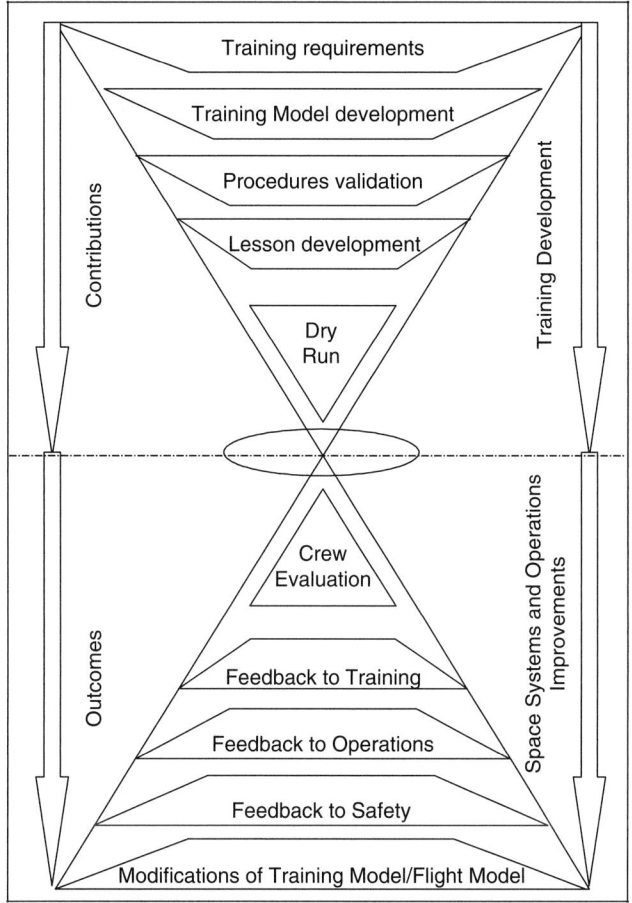

FIGURE 25.31 Feedback from training development and validation.

procedures. It is their duty to alert the safety community of their findings, especially when confronted with contradictory requirements or requirements that are not feasible operationally or make operations hazardous. The safety community must react proactively to this feedback by either confirming the safety approach or endorsing the required changes if necessary.

Feedback from Occupational or Site Safety

Ground safety issues, that is, noncompliances to occupational safety requirements that typically are detected at this stage, in some cases, can be relevant to flight safety. The case of the Automated Transfer Vehicle crews was revealed as a noncompliance to occupational safety requirements as well. For another example, on a rack training model, a male connector was powered with 230 VAC, a clear violation of occupational safety requirements. This nonconformance was investigated; fortunately it was not found on the flight model.

Feedback from the Instructors

The training development teams catch safety relevant issues that were not apparent during previous paper reviews. In doing so, they provide valuable feedback to the safety community. This validation role is reinforced when training teams have been sensitized to both flight and ground safety or are supported directly by safety experts.

A typical event of this stage, training is the first time that the users come together with the products of mission development, that is, the equipment and the procedures. Minor changes, such as sequence changes, can be discussed directly with the procedure writers using information provided to the safety community for approval, whereas more profound changes or questions that directly concern the safety of the flight must be routed back to the safety community for a more formal review and response.

Feedback from the Crew

Dry runs usually are organized with participation of crew representatives to validate the training before presenting it to trainees. The training sessions themselves represent a unique opportunity for the program to obtain feedback on how to improve the usability and safety of the equipment and procedures based on actual spaceflight experience from all astronauts. Thus, the training development and validation process (preparation) and training implementation (execution) naturally complete the processes of hardware and software certification and flight procedures validation. This brings an essential contribution to the safe utilization of space systems.

Feedback from Operations Simulations

Simulations and integrated simulations provide a realistic environment. Both play important roles in preparing the flight control teams, who are then better able to support safety critical phases of operations and react appropriately in case of unexpected developments. Realistic failure cases can be trained and rehearsed on these occasions and all the feedback collected and re-injected into the safety process.

Feedback Following the Flight

During the flight, operations support and monitoring provides direct feedback to the training development process. Feedback is provided on a daily basis by the crew via space to ground communications and by the flight control teams via operational reports. In case of in-flight mishaps, off nominal conditions, or contingencies, anomaly reports are generated. The causes of these can sometimes be traced to insufficient or improper training due to the lack of time, availability, or fidelity of the training model. Ultimately, post-flight debriefings or lessons learned from the crew are a great source of improvement, for both the training and the safety communities.

25.4 CONCLUSIONS

Through this comprehensive journey of the safety aspects of training, we hope to have demonstrated that training should be considered an integrated element of a spaceflight program and hence deserving of proper attention and resources from the system developers and the safety community. Doing so ensures the safety of the crew, both on the ground and during their spaceflight.

Although we tried to keep this chapter as informative as possible, the information provided cannot be exhaustive. Our approach should be taken as only an indication of the current status of techniques and processes.

REFERENCES

Dietlein, L. F., and I. D. Pestov (eds.). (2004) *Space Biology and Medicine*, vol. 4, *Health, Performance, and Safety of Space Crews*. Reston, VA: American Institute of Aeronautics and Astronautics.

European Space Agency. (2000a) *ESA Training Standards and Guidelines*. MSM-AT-MGT-GNL-RQ-001. Cologne, Germany: European Space Agency, European Astronaut Center.

European Space Agency. (2000b) *ESA Training Evaluation Process*. MSM-AT-MGT-GNL-PR-002. Cologne, Germany: European Space Agency, European Astronaut Center.

European Space Agency. (2000c) *ESA Training Courseware Evaluation Process*. MSM-AT-MGT-GNL-PR-001. Cologne, Germany: European Space Agency, European Astronaut Center.

European Space Agency. (2002) *ESA Space Flight Personnel Certification Plan*. MSM-AT-MGT-GNL-PL-007. Cologne, Germany: European Space Agency, European Astronaut Center.

European Space Agency. (2006) *Test Data Package Template for EAC NBF*. ESA-NBF-208a. Cologne, Germany: European Space Agency, European Astronaut Center.

Federal Aviation Administration. (2000) *FAA System Safety Handbook*. Washington, DC: U.S. Department of Transportation, Federal Aviation Administration, Chapter 3, "Principles of System Safety."

Lenehan, A. E. (2004) *Story: The Way of Water*. Hornsby Westfield, Australia: The Communications Agency, pp. 215–217.

National Aeronautics and Space Administration. (1997) *Station Program Implementation Plan (SPIP)*, vol. 7, *Training*. ISS Program Document SSP 5020002. Houston, TX: National Aeronautics and Space Administration, Johnson Space Center.

National Aeronautics and Space Administration. (1999b) *Station Program Implementation Plan (SPIP)*, vol. 9, *Real Time Operations*. ISS Program Document SSP 50200-09. Houston, TX: National Aeronautics and Space Administration, Johnson Space Center.

National Aeronautics and Space Administration. (2004) *Basic Training Requirements for the International Space Station Astronaut/Cosmonaut Candidate*. ISS Program Document SSP-50125. Houston, TX: National Aeronautics and Space Administration, Johnson Space Center.

National Aeronautics and Space Administration. (2007) *NASA Shuttle Crew Training Catalogue*. Available at https://mod2.jsc.nasa.gov/dt/dthome.htm (cited: March 15, 2007).

Stromme, K. (1998) STS-90 Training Facilities. Available at www.psu.edu/nasa (cited: March 15, 2007).

CHAPTER 26

Safety Considerations for the Ground Environment

Paul Kirkpatrick
Chairman, Ground Safety Review Panel, Kennedy Space Center, National Aeronautics and Space Administration, Kennedy Space Center, Florida

Thomas Palo
Aerospace Engineer, Kennedy Space Center, National Aeronautics and Space Administration, Kennedy Space Center, Florida

CONTENTS

26.1.	A Word About Ground Support Equipment	818
26.2.	Documentation and Reviews	819
26.3.	Roles and Responsibilities	819
26.4.	Contingency Planning	819
26.5.	Failure Tolerance	820
26.6.	Training	820
26.7.	Hazardous Operations	821
26.8.	Tools	822
26.9.	Human Factors	822
26.10.	Biological Systems and Materials	823
26.11.	Electrical	824
26.12.	Radiation	824
26.13.	Pressure Systems	825
26.14.	Ordinance	825
26.15.	Mechanical and Electromechanical Devices	826
26.16.	Propellants	826
26.17.	Cryogenics	826
26.18.	Oxygen	826
26.19.	Ground Handling	827
26.20.	Software Safety	827
26.21.	Summary	828

In the history of humankind, every great space adventure has begun on the ground. While this seems to be stating the obvious, mission and spacecraft designers who overlooked this fact have paid a high price, either in loss or damage to the spacecraft prelaunch or in mission failure or reduction. Spacecraft personnel risk not only their flight hardware but also their lives, the lives of their co-workers, and even those of the general public by not heeding safety on the ground. Their eyes might be on the stars, but their feet are on the ground!

One additional comment needs to be made. Although design requirements at times are very different for human rated as opposed to non-crewed flight hardware, while on the ground, the vehicle on which the flight hardware is to be launched and any related ground support equipment is irrelevant. On the ground, requirements additional to those for flight are levied to protect the workforce and the general public.

This chapter presents safety considerations during the performance of flight hardware prelaunch and post-flight operations in the ground environment. Although there are no ground safety requirements unique to or solely for the return of a spacecraft, personnel must apply the same care and precautions for post-flight as for prelaunch.

This chapter is not intended to cover launch vehicles. Whereas many, if not all of the principles to be discussed here can be applicable. Launch vehicles, by their very nature, operate in a consistent environment; to control costs, they depend very much on common configurations and stable processes. On the other hand, spacecraft tend to be unique, in that there is variation from spacecraft to spacecraft that usually depends on mission specifications.

The principles covered in this chapter are not intended to be all-inclusive but are general in nature. As well, they are to be considered generic. They are not meant to contradict specific requirements that might be applicable to any particular ground processing site. It is important to recognize that all flight hardware designers and operators are required to be cognizant of the requirements of the site from which they are operating.

26.1 A WORD ABOUT GROUND SUPPORT EQUIPMENT

In its broadest definition, ground support equipment is equipment related to the flight hardware, but which does not fly. Throughout the world, numerous names are associated with ground support equipment, such as *test equipment* or *factory equipment*, but all these terms mean the same.

All ground support equipment generally are designed and operated in accordance with the national laws of the country that produces it, subject to local processing site requirements. Despite great commonality across the world, ground support equipment is beyond the scope of this chapter. Personnel are urged to open lines of communications as early as possible with the appropriate authorities to ascertain the correct requirements for any ground support equipment that might be required.

26.2 DOCUMENTATION AND REVIEWS

Although the processing of flight hardware is the primary focus, safety documentation and reviews are an essential part of the safety process. Safety documentation and reviews provide the assurance that safety considerations have been identified, incorporated, and verified in the design and operation of hardware and in the use of ground facilities. Through this process, assessments are conducted to evaluate compliance to safety requirements; identify risks to personnel, resources, and mission; and steps are proposed and taken to mitigate the risks to an acceptable level.

The timing of safety reviews and the delivery of associated safety documentation typically is linked to major program milestones. The results of these reviews are presented to program management, processing and launch complex operators, and sponsoring agencies for approval. The complexity of the documentation and the required submittal dates and processes vary with each mission. Early contact with approving authorities is recommended to establish an understanding of applicable requirements and expectations.

26.3 ROLES AND RESPONSIBILITIES

Integration and eventual launch of flight hardware requires the coordination of numerous personnel from a variety of organizations who provide flight hardware elements, supporting equipment, and ground infrastructure. The efforts of the mission team can require work with hazardous materials or in a hazardous environment, either as part of the team's own activities or in conjunction with integrated activities.

The hardware provider is responsible for providing safe systems, equipment, facilities, and materials and for conducting operations in a manner that complies with established safety requirements. This responsibility extends to the preparation, coordination, and certification of documentation that provides assurance that their employees, other launch site personnel, and the general public are not exposed to unacceptable risk.

Employers are responsible for the safety of their personnel. Final authority and safety at the launch site rests with the operator at that site, who ensures that all pertinent operations at the range are reviewed, coordinated, and approved to assure range and public safety.

26.4 CONTINGENCY PLANNING

Despite the best design, implementation, and planning, an unexpected event can occur that challenges even the best engineering team. Consideration for contingencies early in the design aids in expediting a solution to ensure safety and to minimize any impact to hardware, schedule, and cost.

Planning for a contingency falls into two major areas:

- Programmatic plans are developed to respond to non-operational events, such as earthquakes, hurricanes, electrical power interruptions, or even labor disputes. These typically are beyond the control of a program and can result in lengthy launch delays.

- Contingency plans are developed for a response to anomalies or emergencies that can occur during ground operations, such as propellant leaks or spills, or an emergency power up or down of hardware. Hardware designs should support implementation of these plans by providing features such as accessibility to service points, pressurant and propellant service valves, calibration requirements, battery charging, installation and removal of protective and contamination covers, and access to safing plugs to render ordnance systems safe.

All operational plans should include back out steps to safe hardware if an unintended event occurs during processing. In addition, operational capability should be addressed for limited life items, such as battery reconditioning or replacement.

26.5 FAILURE TOLERANCE

The failure tolerance of flight hardware during ground operations depends greatly on the failure tolerance methods used to protect the hardware on orbit. Methods such as design for minimum risk, high reliability, and fault tolerance apply; however, the opportunity for human error is much greater on the ground if only because the flight hardware can be accessed physically. This scenario particularly is true during troubleshooting.

When planning spacecraft ground processing, designers must include in their analyses any interaction between the flight hardware, ground support equipment, ground facilities, and operators. During troubleshooting, inhibits and controls often have to be removed to uncover or resolve an anomaly. If the processing team fails to plan its troubleshooting steps or monitor a test properly, catastrophic results can occur because of the removal of too many controls. A suggested method for tracking inhibits or controls would be to place them in a matrix where they are visible and available to the test planners. This as well can serve as an aid when seeking approval for conducting the test from local approving authorities.

26.6 TRAINING

Employers are responsible to ensure that their employees receive adequate training for the activities they perform and for acquiring knowledge of the potential exposure to hazards. Certification can be required for certain specialized tasks, such as crane operations and propellant and ordnance handlers. Government, corporate, and local operating requirements can vary.

Personnel must be trained to identify and respond to any hazards they might encounter in the work area. This includes how to take appropriate precautions when working with hazardous materials present, use of required personal protective equipment, the location of emergency equipment and procedures, the reporting of emergencies and response to alarms, and the location and use of emergency equipment and first aid techniques.

Personnel training should include the use of safety processes, such as the lock out and tag out of equipment to prevent it from being energized inadvertently. Pathfinder operations conducted to enhance operational familiarization and response to emergencies should be included in operational training requirements.

26.7 HAZARDOUS OPERATIONS

When working in the ground environment, the operator must keep in mind the presence of personnel and other high value hardware and have a thorough knowledge of the facility. Spacecraft operations nominally conducted on orbit can have devastating effects when performed on the ground. For instance, the ignition of an upper stage or thruster has minimal external effect on orbit; however, on the ground, the same operation could be deadly. In some situations, the inherent fault tolerance of a system must be compromised, or very nearly so, to validate the operation of a system prior to launch.

Certain normal servicing operations, such as handling fluids or gases, might be required to be designated as hazardous. The hazardous designation of these operations is derived from related hazard analyses or in accordance with the requirements of the processing facility or area. In the case of hazard analyses, the designation of an operation as hazardous can be a hazard control. Processing facility requirements often are the culmination of years of experience, and many of the requirements are the result of previous accidents or near misses.

When conducting operations on the ground, whether hazardous or non-hazardous, the key to success is centered on written, step by step procedures as well as the existence of a structured process for their development. Well written, well coordinated procedures serve multiple purposes, which include assurance that the test team and support organizations are aware of the procedure, documentation of hazard controls and ensuring that they are in the correct location relative to the hazard, and providing a written record of the actual test in the event of problems later in processing. The importance of written procedures, indeed, cannot be understated.

When planning ground operations, it is not considered good practice to conduct concurrent hazardous operations. Concurrent operations usually involve overlapping control areas and can lead to confusion among the test team and the possibility of competing priorities.

A well run ground campaign recognizes the hazardous nature of preparing a spacecraft for launch and has in place the procedural processes to control any hazards so as to protect people, other flight hardware, and facilities.

26.8 TOOLS

Tools are the most essential and overlooked element of ground safety. Without proper tools, operational activities cannot be performed safely. A tool can be as simple as a screwdriver or as complex as computerized test equipment. Improperly selected or used on the job, they can lead to a hazardous condition.

For a hazardous operation, all tools are required to be identified in the hazardous technical operating procedure. The safety review of the procedure would include an assessment of the tools to ensure they are appropriate for an operation, that is, they are compatible with the flight hardware; ensure that another hazardous situation would not be caused by their use; or ensure that operational support or approval is required as a condition of their use.

Whereas tools obviously are selected to accomplish a task, other criteria can affect their selection and use. For example, working in the vicinity of sensitive instruments might require the use of nonmagnetic tools. Tools that generate high temperatures might require hot work permits, special shields or barriers, removal of combustible material, and fire protection equipment.

Electrical test equipment poses hazards in addition to those related to grounding concerns and potential exposure to energized electrical circuits. Electrical equipment must be used only in the environment for which it is designed. Operating areas where a potential exists for a propellant leak requires the equipment to be intrinsically safe, explosion proofed, or purged.

Pathfinder operations should be planned to demonstrate the adequacy and function of tools and equipment used for an operation. In such an assessment, special attention should be given to ease of use, accessibility, visibility, and any personal protective equipment that might be required for the operation and could affect performance.

A consideration equally important as the selection of appropriate tools is a means to account for them. Hand tools should be tethered at all times to prevent any of them from being lost, or accidentally dropped causing hardware damage or personnel injury. A tool control plan should be instituted to ensure that all tools are identified and an adequate accounting has been made. In addition, segregation of tools can be required to prevent undesired reactions. For example, the same tools used on a hydrazine system should not be used on an oxidizer system.

26.9 HUMAN FACTORS

Human factor considerations can affect the interface between ground personnel and the flight hardware. The spacecraft designer must be cognizant of these issues when considering the means by which the spacecraft is to be serviced on the ground. The placement of servicing panels and connections about the spacecraft is critical in avoiding errors leading to accidents. For instance, the placement of a battery connection that requires the technician to stand with his back to the connection point and reach over and behind his head to accomplish the task can be fertile ground for an accident (and it was). Some accidents such as this can be prevented by design (in this case, the use of scoop proof connectors) but not always.

The prevention of human error is essential, and it is important to design hardware and plan operations with this in mind. Design is the preferred solution; however, procedural controls can be used, if necessary. The labeling of equipment controls is critical. When developing procedures for real time operations or troubleshooting, the interface between personnel and the flight hardware must be considered. This especially is important if the location of the work is in an area previously not intended for access.

Exposure of personnel to hazardous materials must be avoided whenever possible through the implementation of adequate design features, such as redundant seals. In the event this is not possible or if there is active handling of hazardous materials, such as during fueling, personnel must be provided with appropriate protective equipment.

There should be an accounting for physical contact with the flight hardware, whether spacecraft or individual experiments, in the design. This should include contact with sharp surfaces or protrusions, rotating surfaces, and high or low temperatures. Adequate shielding, barriers, guards, or procedures for moving or removing the items must be used. Adjustments needed for electrically powered areas are best made with the power off; otherwise, shock protection must be provided.

The interaction of personnel with hardware must be accounted for at all times, whether for routine processing or troubleshooting.

26.10 BIOLOGICAL SYSTEMS AND MATERIALS

Biological systems cover the range from plant growth to human medical experiments. Because of the possibility of injury or potential harmful effects, hardware containing biological material requires special attention. This applies to both prelaunch and postlanding activities.

Although a biological experiment or sample has a low toxicity on orbit, this does not translate necessarily to the ground environment, especially in the area of sample preparation. For example, a low toxicity material like vinegar used on orbit is certainly of a higher toxicity when in the form of glacial acetic acid on the ground. The protection of personnel, both on the ground and in flight, is linked directly to the hazard presented by the material.

A special biological system that requires advance planning is trash that contains biological material. Special care must be taken when handling this material. Not only the potential presence of biologically active material needs to be considered but also the presence of contaminated physical material, such as needles and swabs. It is important that all personnel handling such material be made aware of the contents of the trash and utilize appropriate protective equipment and procedures to avoid inadvertent contamination or injury.

Whether in the form of a live virus, a human blood sample, or simply trash, it is the responsibility of the investigator or operator to ensure compliance with all programmatic and legal requirements. A close coordination with the appropriate authorities is required.

26.11 ELECTRICAL

Electrical ground support equipment and facilities should be designed to industry consensus standards. Equipment designs should ensure that a connection cannot be reversed or mated inadvertently, personnel are provided protection from accidental contact of energized components, and there is compliance with appropriate grounding or bonding schemes to ensure that equipment remains at ground potential at all times.

Special attention should be placed on battery charging and conditioning operations. Equipment design should incorporate protective devices such as fuses, diodes, and voltage and current limiters; and it should possess temperature and pressure monitoring ability as well. Continuous monitoring by personnel should occur during charging and conditioning operations.

Frequently, commercial off-the-shelf electrical equipment is utilized to support ground processing operations. This equipment should be used only in accordance with the intent of the manufacturer and in its intended environment. Any modification or integration of this equipment with other equipment should be assessed carefully.

In the event that troubleshooting, maintenance, or repair of electrical equipment is required, the activities should be performed in accordance with a documented process and accepted industrial practices. Special precautions, such as lock out and tag out of devices to prevent the accidental application of power to equipment undergoing service, are mandatory.

26.12 RADIATION

Radiation in the ground environment is classified into two categories: nonionizing and ionizing. Safety controls are required for radioactive materials, such as flight and ground calibration emitting sources, radiation producing equipment such as X-ray devices and radio frequency emitters, and lasers and optical emitters of high intensity light, infrared, and the like. Safety requirements for the ground environment provide specific engineering and operational controls for both types. All radiation sources and associated equipment must be designed to ensure that personnel exposure and any potential for release are as low as reasonably achievable while not exceeding applicable regulatory limits.

Flight radioactive sources should be installed as late in the countdown as is practical and handled only by approved personnel. Radioactive sources not in use should be secured against unauthorized access. Controls should be established to permit access only to authorized personnel, and all personnel exposure should be monitored.

Major radiological sources, such as radioisotope heater units and radioisotope thermoelectric generators, pose a greater risk and have additional requirements. Dedicated processing facilities are required, and more stringent controls are placed on the storage, access, use, and operation associated with major sources. Approval for flight requires increased coordination with the accompanying analyses, safety assessments, and contingency planning.

In addition to exposure limits, nonionizing sources and their controls should be assessed to preclude inadvertent operation. Integrated hazard assessment should be performed to ensure that radio frequency transmission inadvertently does not affect launch vehicle or spacecraft systems. Optical and laser systems require engineering controls, including beam stops, limit stops, interlocks, and shields. Materials used for targets or subject to exposure should be nonflammable and not emit toxic materials. Appropriate personal protective equipment should be utilized to protect against hazards associated with specific wavelengths, temperature extremes, and gases.

26.13 PRESSURE SYSTEMS

The design requirements of pressure systems are well defined in other locations within this book. Of greater concern is the operation of these systems on the ground.

An important datum to track is the pressure for which the system has been tested. Each new level of pressurization introduces different stresses to the system. As these pressures are increased, remote pressurization can be required to protect personnel.

Special care should be taken to ensure that any venting done by a pressure system, whether planned or unplanned, is done in such a way as to not create a hazardous condition. Also, it is good practice to design pressure system connections so that physically it is impossible to mix incompatible fluids, such as hypergolic fuels and oxidizers.

The use of composite overwrapped pressure vessels requires special mention. At all times, composite overwrapped pressure vessels are very sensitive to impact. This is even more of a consideration in the ground environment than on orbit because they can be accessed easily, making them susceptible to inadvertent contact or damage. It is, therefore, important that an impact protection plan be developed and adhered to on the ground and compliance with national or local standards is necessary for successful processing.

While on the ground, all pressure systems demand the utmost care and respect to avoid immediate or future damage to personnel, facilities, and equipment.

26.14 ORDINANCE

Ordinance devices can vary in size from safe, small handheld devices to larger devices that can initiate a chain of events causing injury or death. Initiation of ordinance devices can occur whether the ignition signal is intentional or inadvertent. Low energy inputs to these devices provide design advantages but result in potential safety hazards. Because these devices cannot be operated functionally, statistical reliability, production controls, and ground processing controls are imposed to ensure performance and safety.

All ordinance must be stored in a manner consistent with their hazard classification and compatibility. Faraday caps must be installed on ordinance until electrical connections are made. Ordinance test equipment must limit the energy input to the device.

Ordinance devices and systems are required to be designed to preclude inadvertent firing when subjected to environments, including shock, vibration, and static electricity, that can be encountered during ground processing. Ordinance circuits, hardware design, and accessibility must permit interrupts, such as safety plugs, placed as close to the device as possible and connection to be made as late as possible prior to launch.

26.15 MECHANICAL AND ELECTROMECHANICAL DEVICES

Flight hardware that contains deployment mechanisms must have all necessary controls in place to prevent inadvertent activation. These mechanisms include such items as solar arrays or sample gathering devices. Even if deployment is non-hazardous, controls highly are recommended to assure mission success. Special care must be taken during troubleshooting or repairs to prevent injury or damage.

26.16 PROPELLANTS

Propellants utilized in space vehicles vary in composition, form, and reactive properties. A single launch vehicle and spacecraft can include inert xenon gas, multiple solid rocket motors composed of homogeneous or composite propellants, Aerozine-50, dinitrogen tetroxide, rocket propellant-1, and liquid oxygen. Because each of these materials has distinct physical and reactive properties and a potential for a catastrophic hazard, the safety considerations during ground processing are extensive and result in closely monitored ground operations.

Considerations in the storage, transfer, and handling of propellants include material compatibility of system components, separation from reactive materials, capability to isolate system leaks, and electrostatic properties of materials. Emphasis is placed on protection of the personnel performing propellant operations or subsequent processing activities through the use of protective garments, toxic vapor detection, venting and scrubbing of vapors, explosion proofing of electrical equipment, and emergency planning.

26.17 CRYOGENICS

In general, cryogenic systems must comply with the same requirements as those that apply to propulsion systems. However, due to their unique physical properties, additional requirements often are levied. These additional requirements deal with prevention of the inadvertent conversion of liquid to gas and the subsequent pressure rise. The use of pressure relief devices or insulation is required in those parts of the system where this is a possibility. Liquefaction of air must be taken into account when designing cryogenic systems. Joints in cryogenic systems are recommended to be either butt-welded, flanged, bayonet, or hub type. In addition to these requirements, servicing flight hardware with cryogenics is subject to the requirements of the processing site.

26.18 OXYGEN

The safe design, development, and operation of oxygen systems requires special knowledge and understanding of design practices, materials, ignition mechanisms, manufacturing techniques, and operational controls. Materials that are highly reactive must be avoided, and those that are less reactive but still flammable should be protected from all ignition sources. Cleanliness of oxygen systems is important because particles could ignite or cause ignition when impacting other components of the system, and organic compounds such as hydrocarbon lubricants can ignite easily.

Oxygen systems should be analyzed to ensure leak prevention, adequate ventilation, suitable design of system components, and system cleanliness. Systems should be designed with sufficient redundancy to provide adequate fault tolerance for system integrity and personnel safety.

26.19 GROUND HANDLING

In general terms, the handling of flight hardware is enveloped by the need to sustain flight dynamics. This is a truism as long as the dynamics to be experienced on the ground fit the envelope.

Two special cases need to be considered during the design phase. The first involves the particular attachment points to be used during ground handling, and the second is an accounting of the potential for tip over. In the first case, if the flight attach points of a spacecraft are being used for handling, then further analysis is not required. If other points are to be used, then the designer must ensure that the expected loads are allowable for those points. This especially is important for those spacecraft being launched on a rocket where loads are transmitted through the base. In the second case, all flight hardware should have a center of gravity analysis performed to ensure that it does not tip, fall, slide, or allow for any type of sudden load shift while being handled on the ground. This should be of particular interest to those pieces of hardware that are lifted from below the center of gravity.

26.20 SOFTWARE SAFETY

A software safety assessment is required in the ground processing environment to ensure that flight or ground software does not contribute to or cause a hazardous condition. As well, the software must not alter the system configuration to the point where the potential risk of a hazard increases.

Embedded systems can exercise or provide real time control of a system with no direct user interface. Commands inadvertently could open valves, power transmitters, start sequence timers, allow power to relays, and remove other system or safety inhibits. The use of mechanical interrupts, such as safety plugs or safe and arm devices, provide a positive verifiable inhibit.

At the highest level, ground safety assessments should ensure that critical commands are identified and blocked adequately from execution. Sufficient independence should exist for software inhibits and be supplemented by controls such as watchdog timers and improper sequence detectors. Safety critical software should be configuration controlled closely to segregate ground test and flight software.

26.21 SUMMARY

As with life, the preparation for a mission in space is often fraught with as much peril as the journey itself. The preparation for the launch of a spacecraft is a dynamic event with numerous hurdles to overcome. The return of a spacecraft or experiment can be perilous because the mission team can let its guard down. After all, is the mission not over? The mission team that can focus on and clear these hurdles is well on the way to ensuring a successful mission.

CHAPTER 27

Fire Safety

Gary A. Ruff, Ph.D.
Advanced Capabilities Project Office, Glenn Research Center, National Aeronautics and Space Administration, Cleveland, Ohio

David L. Urban, Ph.D.
Chief, Combustion and Reacting Systems Branch, Space Processes and Experiments Division, Glenn Research Center, National Aeronautics and Space Administration, Cleveland, Ohio, USA

Michael D. Pedley, Ph.D.
Materials and Processes Branch, Johnson Space Center, National Aeronautics and Space Administration, Houston, Texas

Paul T. Johnson
Safety and Reliability, The Boeing Company, Huntsville, Alabama

CONTENTS

27.1. Characteristics of Fire in Space 830
   By David L. Urban, Ph.D., and Gary A. Ruff, Ph.D.
27.2. Design for Fire Prevention 847
   By Michael D. Pedley, Ph.D.; Tony Brown; and David E. Tadlock
27.3. Spacecraft Fire Detection 855
   By David L. Urban, Ph.D.; Gary A. Ruff, Ph.D.; and Paul T. Johnson
27.4. Spacecraft Fire Suppression 864
   By Gary A. Ruff, Ph.D., and David L. Urban, Ph.D.

Fire is a very serious threat in any confined volume but particularly is so when a crew is confined in a spacecraft far from Earth, where there is little or no possibility to receive aid or be rescued. Therefore, every measure must be taken to prevent a fire from happening. Not only must there be an effort to prevent a fire within the spacecraft, there must be a thorough understanding of what constitutes a fire in space, how to detect a fire event accurately, the best means to alert the crew, and the actions that should be taken, if any, should a fire event occur. These are the primary topics covered in this chapter.

Section 27.1 gives insight into the different phenomena of fire in space, including some surprising and unexpected results shown by various experiment investigations on prior crewed spaceflights. This is followed, in Section 27.2, by the two major considerations for preventing spacecraft fires: the choice of nonflammable materials and avoidance of ignition sources. Section 27.3 addresses the current philosophy for fire detectors, the detectors used on previous spacecraft, and an alternative means for fire detection using data parameter monitoring. Finally, in Section 27.4, fire suppression is discussed, and the strategies used on previous spacecraft are contrasted.

27.1 CHARACTERISTICS OF FIRE IN SPACE

27.1.1 Overview of Low Gravity Fire

Low gravity combustion research was first undertaken by Kumagi and Isoda (1957) using a short drop tower where droplet combustion phenomena were studied. Subsequently, considerably more interest developed for both spacecraft fire safety and fundamental research. Excellent reviews of the fundamental research are available in papers by Williams (1990); Law and Faeth (1994); and Friedman et al. (1999). Both fundamental low gravity combustion research and spacecraft fire safety are covered in some depth by Ross (2001), and Friedman and Ruff (2008) provide an annotated bibliography of the literature on spacecraft fire safety from 1956 to 2000. In the discussion that follows, the effects of the removal of gravity induced buoyant flow on fire behavior is discussed as it applies to spacecraft fire safety.

Many texts refer to zero-gravity and microgravity, but strictly speaking, these gravitation levels are achieved only in the NASA and Bremen drop towers (Ross 2001). For the purposes of future exploration missions by NASA, two reduced gravity regimes are of interest:

- The very low levels seen on orbiting spacecraft, typically on the order of $1 \times 10^{-4} \, g$ (Ross 2001).
- The levels seen on planetary surfaces, such as $0.376 \, g$ on Mars and $0.165 \, g$ on the Moon.

Hereafter in this chapter, the gravity levels seen in orbiting spacecraft, drop towers, and low gravity aircraft are referred to as *low gravity*, whereas the levels seen on the Moon and Mars are referred to as *reduced gravity*, and the Earth gravitation level is referred to as *normal gravity*. Overall low gravity research in material flammability generally has been limited to testing thin fuels, due to the short test items available in drop towers.

Space based material flammability experiments to date were conducted primarily using small samples of paper, polymethylmethacrylate, or plexiglass. Other tests have been conducted looking at fire detection as well as the fundamental properties of gaseous flames and flames over liquid fuels.

27.1.2 Fuel and Oxidizer Supply and Flame Behavior

Unlike flames in normal gravity, fires in low gravity are not supported by the strong pumping of oxidizer and fuel into the flame zone and the subsequent pumping of the reactants out of the reaction zone. For this reason, low gravity provides the only opportunity to observe combustion under truly quiescent conditions, that is, with no flow other than that caused by the processes of thermal expansion, production of gaseous species, and fuel surface evaporation or blowing. Under these conditions, molecular diffusion is usually the dominant mode of transport for the oxidizer to the fuel. In many cases, the diffusive transport of oxidizer is less than the flame spread rate, and flame spread then becomes the dominant oxidizer transport mode, creating a virtual wind as the flame moves into fresh fuel and oxidizer. Under these conditions, flame spread can occur in conditions that are not duplicated readily in normal gravity because of the large buoyant flows generated by the flame.

One matter of substantial speculation in low gravity material flammability is the ultimate fate of flames in a quiescent environment. Overall, marginally flammable materials have shown the tendency to self extinguish in quiescent low gravity, but uncertainty remains about the behavior of more flammable materials under quiescent conditions. These materials normally are not used for spacecraft construction but often are included as waivered materials subjected to extensive material controls. The extent of these necessary controls generally is affected by the common perception that most materials will self extinguish if the airflow is cut off. While low gravity tests of this hypothesis have been very limited, tests have been conducted in chambers where the volume is sufficiently large that it is not the limiting factor in the experiment. The tests to date include the work of Kimzey (1974) on *Skylab*, candle flame tests on *Mir* (Dietrich et al. 2000), and tests in the Skorost facility on *Mir* (Ivanov et al. 1999). The Skylab tests were conducted in a 41-L test chamber filled with the Skylab atmosphere: 65% oxygen at 35.8 kPa (5.2 psia) total pressure. The tested materials exhibited a variety of burning behaviors that were generally consistent with those of their 1-*g* flammability. Most notably, a test with Nylon® burned for 10 min, 43 s, until the crew eventually extinguished it by venting the chamber. A notable feature of the Nylon test was the self induced flow by the molten fuel ejecta from the flame, which enhanced the oxidizer transport to the flame. The Mir candle flame tests (Dietrich et al. 2000) also exhibited extensive burn times, with one candle burning to the point (45 min) where extinguishment is likely to have been due to exhaustion of the fuel rather than the oxidizer. This was a substantial increase over previous testing on the *Space Shuttle*, where candles typically extinguished within 60 s. The difference was attributed to the Mir tests being conducted in a larger test volume, which permitted the flame access to a greater reservoir of oxygen. The Skorost tests (Ivanov et al. 1999) with Delrin®, polymethylmethacrylate, and polyethylene were conducted in a 3.8-L flow duct, and the researchers found that these fuels generally extinguished immediately after flow termination, although with the notable exception of one test in which the sample was overheated erroneously during the ignition process. This sample continued to burn for 130 s after the flow was terminated, suggesting that the extinguishment condition was affected strongly by the temperature of the solid fuel. Interpretation of these results is affected by the very small chamber size, which likely limited the availability of oxygen. Although these results suggest that, while isolated samples of most materials do extinguish in quiescent low gravity, in cases of external heating or reduced heat loss, some materials can continue burning for an extended period without airflow, even in confined circumstances. This tendency to continue burning is enhanced by

- Increased chamber volume.
- Increased oxygen concentration.
- Increased ambient or fuel temperature.
- Enhanced fuel vaporization, such as the presence of wicking materials.

27.1.3 Fire Appearance and Signatures

Detection and response to fires requires an understanding of both the appearance and detection signature of a fire. The term *signature* encompasses all quantifiable features of the fire: its color or spectral emission, temperature, smoke particle emission, and gaseous species emission. Details of the signature are necessary for rapid fire detection, whereas the overall appearance is more important as a guide to understanding the controlling physical processes in low gravity fires. Combustion research to date has shown that low gravity fires exhibit markedly different appearances than their normal gravity counterparts. The most notable example is that of a candle flame, which was studied in low gravity by Dietrich et al. (2000). As seen in Figure 27.1, in the absence of buoyant flow, diffusion becomes the dominant mode of oxidizer transport. Without the preferred direction imposed by buoyancy, the flame assumes a quasi-spherical

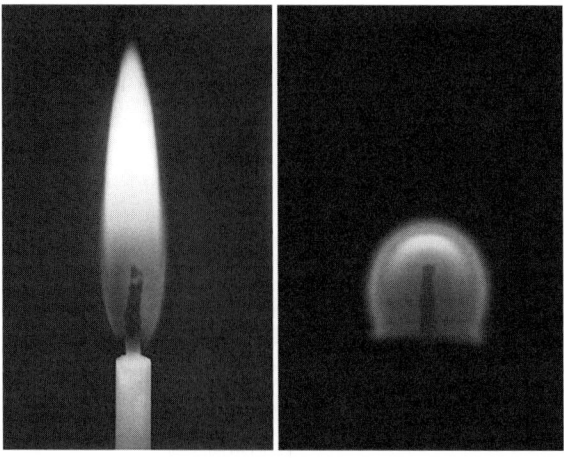

FIGURE 27.1 Candle flames in normal gravity and in low gravity. The low gravity images, taken on the Mir space station, illustrates the dominance of diffusion on the oxidizer transport to the flame front (Courtesy of NASA).

27.1 Characteristics of Fire in Space

shape that is smaller and bluer in color. The lack of yellow luminosity indicates the general absence of soot, which as a solid particle, emits light over a continuum of wavelengths, that is, white light, as compared to the narrow spectral emission of gaseous molecules that emit blue light. Under quiescent microgravity conditions, flow levels are below those achievable in normal gravity. This leads to very long residence or dwell times for fuel molecules as they pass from the fuel surface, in this case the wick, to the flame surface, where reaction occurs. One attribute of flames that often is misunderstood is that they essentially are transparent. The flame image we see is the integration of all of light from flame elements along the line of sight through the flame. So, although the normal gravity candle flame in Figure 27.1 appears to be bright uniformly throughout its volume, the flame is essentially hollow, with all the light being emitted by the outer shell. This hollow structure is more evident in the low gravity flame image in Figure 27.1.

Flame spread in slow convective environments in low gravity was studied by Olson et al. (2001) using the middeck glove box and drop towers. Figure 27.2 shows a typical spreading flame in low gravity. The fuel, here paper, was ignited in the middle of the sample, and the flame is spreading into the wind, rather than downwind, as occurs in normal gravity. This is a result of the transport of oxygen limiting the combustion process and thus controlling the flame spread rate. Flames spreading in low gravity across solid and liquid surfaces tend to follow the surface contour, maintain a smooth upper surface, and do not develop the buoyant flame tips and flicker seen in normal gravity flames. Any flame tips that develop are on the downstream end of the flame. Smoke emission can occur at this point as a result of incomplete combustion of the fuel when the flame quenches in the downstream portion of the flame. These tips can be seen in Figure 27.2 in the downstream portion of the flame, which exhibits flame tips driven by the effect of the incoming airflow.

The impact of buoyancy on the flame shape is shown in Figure 27.3, which contains images of flames spreading over a liquid pool in normal gravity and low gravity, as studied in the work of Ross and Miller (1996) and Miller et al. (2000). Overall, the visible appearance of low gravity fires tends to be bluer (less soot) than the corresponding normal

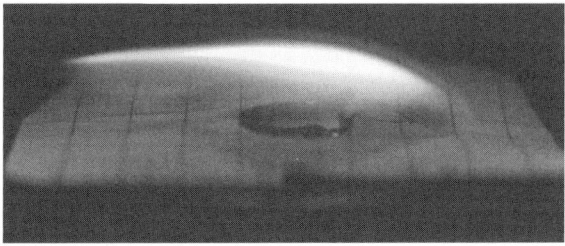

FIGURE 27.2 Flame spreading over a solid fuel (paper) in low gravity in the middeck glove box. Air is flowing from right to left, and the sample was ignited in the center of the sample via a radiative source (Courtesy of NASA).

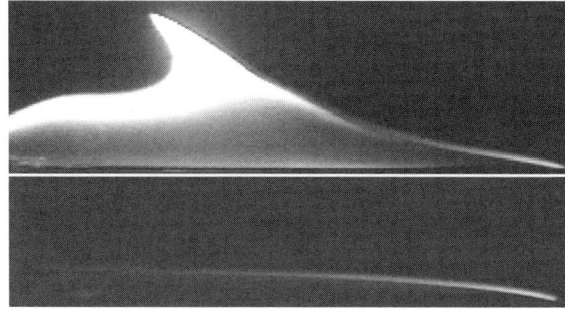

FIGURE 27.3 Flame spreading over a butanol pool in normal gravity (top) and low gravity (bottom). Air is flowing (30 cm/s) from right to left. The buoyant uplift is evident in the normal gravity flame, whereas the low gravity flame follows the fuel surface and the airflow (Miller et al. 2000; Ross and Miller 1996). (Reprinted by permission of The Combustion Institute).

gravity flame. The flame shape is changed dramatically in normal gravity by the buoyant forces lifting the hot gases above the fuel surface.

The fire signatures from low gravity fires received limited attention to date. Most of this work has been fundamental research that focused on changes in soot particle size and emission (Greenberg and Ku 1997; Urban et al. 1998, 2000). A limited set of experiments examined the smoke particulate from low gravity smoke sources (Paul et al. 1993; Srivastava, McKinnon, and Todd 1998; Urban et al. 1998, 2005). Overall, the fundamental soot studies demonstrated that both the primary and aggregate soot particles were substantially larger in low gravity than in normal gravity, the aggregate dimensions being as much as an order of magnitude larger in low gravity. This behavior was explained by the increase in flame residence time brought on by the elimination of buoyant acceleration. Further, the propensity to emit soot also increased in low gravity, with smoke point lengths being substantially smaller than the normal gravity counterparts. This behavior is due to the combination of the reduced oxygen flux to the flame zone and the increased residence time and consequent increase in radiative heat loss. This increased heat loss cools the flame to the point that soot oxidation is no longer supported, and unburned soot escapes from the flame tip. Generally, as the flame size increases while other conditions remain constant, hydrocarbon flames range from soot free to soot containing to soot emitting or smoking. For a given set of operating conditions, the smoke point measurement is an established indicator of sooting propensity and predicts soot emission and radiative emission for the majority of diffusion flame configurations. Studies with solid polyethylene in slow airstreams in low gravity (Greenberg et al. 1995) also showed a dramatic increase in soot emission, with the density of the emitted soot sufficiently high that the flame emitted a super-aggregate on the scale of the flame itself. Overall, these studies demonstrated that, for the same flame conditions, microgravity flames with typical flammable materials can be expected to release substantially more soot than their normal gravity counterparts. Because soot emission is an indicator of incomplete combustion and it correlates with carbon monoxide emission, it can be inferred that these flames emit substantial fractions of carbon monoxide.

All these results with enhanced soot emission occurred in cases in which there was some forced convection, either fuel jet or forced airflow. In cases where the air supply was quiescent, the oxygen transport limitation typically produced blue, soot free flames (Altenkirch et al. 1998; Dietrich et al. 2000). This has been attributed to reduced flame temperatures, resulting from reduced oxygen flux. This low temperature also has produced incomplete combustion with unreacted fuel emerging from the flame attachment point.

Soot is only one component of smoke, which is a general term that actually refers to any flame emitted condensed particulate. The smoke seen from flames can be composed of soot, ash, water vapor, and unreacted liquid or solid fuel components. Experiments examining the smoke production from typical spacecraft materials have seen increased particle sizes from polymers including Teflon® and Kapton® (Paul et al. 1993; Srivastava et al. 1998; Urban et al. 1998, 2005). These materials form smoke containing condensed polymer fragments that are solid at room temperature. This facilitates measurement of the particle sizes, because all the low gravity studies to date depended on transmission electron microscope analysis of captured smoke particles. Figure 27.4 contains transmission electron microscope images of smoke particles from three materials in low gravity: a candle flame, overheated Teflon, and overheated Kapton. The morphology of the smoke particles from each of these materials is unique. In fact, the work of Srivastava et al. (1998) demonstrated that the particulate morphology was affected by the color of wire insulation, even for wires of the same military specification.

The work of Urban et al. (1998, 2005) found evidence that the sizes of the smoke particulate from the combustion of materials that formed smoke composed of liquid particles were substantially larger than their normal gravity counterpart; however, these liquid droplets were not resolvable via transmission electron microscope analysis; consequently their increased size could be inferred only by other measurements.

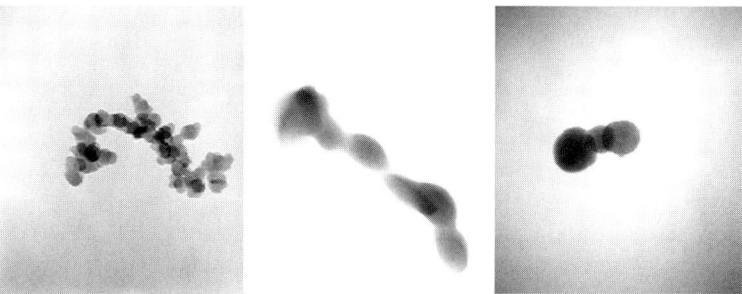

FIGURE 27.4 Transmission electron microscope images of smoke particulate from (L–R) candle soot, Teflon, and Kapton smoke (Urban et al. 1998). All images are at the same magnification. For scale reference, the candle soot primary particles are approximately 20 nm in diameter (Courtesy of NASA).

27.1.4 Flame Ignition and Spread

The previous sections examined the general behavior and appearance of flames in low gravity. However, flame behavior and appearance result from the complex interaction of heat transfer (conduction), forced convection, and radiation with convective mass transfer, thermal and mass diffusion, material properties, and chemical kinetics. When freed from the effects of buoyancy, low gravity flames generally exhibit a considerably increased dependence on thermal radiation, as well as thermal and mass diffusion. This dependence affects practically every characteristic of the flame, including flammability limits and flame speeds for laminar flames and ignition delay times and spread rates for flame spread over solid fuels. The following subsections highlight some of these differences, emphasizing those that have the greatest impact on spacecraft fire safety. The references should be consulted for a more comprehensive treatment of these topics.

Combustion of Gaseous and Particulate Clouds

For a quiescent gaseous fuel-oxidizer mixture, there is a mixture composition range for any initial pressure and temperature for which a sufficiently strong ignition source causes a premixed laminar flame to propagate throughout the entire mixture. The equivalence ratios for the mixtures that can propagate flames, called *flammability limits*, are important not only for understanding flame behavior but also in fire safety applications, because they determine the fuel to oxidizer ratio at which a mixture becomes flammable. The observed limits of propagation depend on a large number of variables, including the direction of flame propagation relative to the direction of gravity and even the method used to determine the limit itself. The dependence on gravity, of particular interest here, is a result of the flame speed in the gaseous mixture relative to the buoyant velocity and the associated curvature of the flame front. A detailed discussion of the impact of buoyancy, flame stretch, and burning velocity is beyond the scope of this text but can be found in the works of Ronney and Wachman (1985) and Strehlow, Noe, and Wherley (1986) and are summarized by Law and Faeth (1994) and Ronney (2001). As described in these reviews, the interaction between the physical and chemical phenomena leads to broader flammability limits, that is, the lean flammability limit is lower and the rich flammability limit is higher in low gravity, which has important implications for spacecraft fire and explosion hazards. Whereas additional work is required to understand this phenomenon fully, the work in these references demonstrates that it is incorrect to assume that flammability limits for gases at normal gravity are applicable to conditions at low gravity.

Furthermore, Ronney and coworkers (Ronney 1990; Ronney et al. 1994) demonstrated that, in a low gravity environment, stable, stationary flame balls can be formed in lean mixtures of flammable gases. These experiments were first conducted in drop towers and reduced gravity aircraft experiments and later confirmed by spaceflight experiments on the STS-83, STS-94, and STS-107 Space Shuttles (Ronney et al. 1998; Kwon et al. 2003). Without the introduction of instabilities caused by buoyancy, flame balls were found to burn and remain stable in a lean premixed gas mixture for up to 81 min. Even though these flames are very weak, they illustrate that, without buoyant convection, gaseous premixed flames in low gravity can exist for longer periods of time and at conditions in which they cannot exist in normal gravity.

Particle clouds also exhibit different behavior than their normal gravity counterparts, primarily because the lack of a settling velocity in low gravity allows larger particles to remain in suspension, thus producing a more homogenous distribution of particles. This single feature can increase the hazard of particle clouds in low gravity. Because the particulate does not settle, the particle clouds can persist for long times where they are formed unless dispersed by a convective flow. This would increase the likelihood that the particle cloud could encounter an ignition source and create a flame that propagates through the cloud. Studies of particle cloud combustion in low gravity have been conducted (Ballal 1983; Pu, Podfilipski, and Janosinski 1998; Hanai et al. 1999); yet, most have not reported flammability limits similar to those of premixed gases. Hanai et al. (1999) found that the flammability limits for normal gravity were difficult to resolve because of the difficulty in obtaining a uniform quiescent mixture in normal gravity. However, the results have shown that the temperatures of low-g flames, and therefore the pressures achieved in a closed vessel, remain constant in low gravity over a wide range of ignition delay times. In normal gravity, the pressures decrease with increased ignition delay because of particle settling.

Combustion of Solids in Low Gravity

Solid surface combustion in low gravity produces considerable variation of combustion properties. As discussed previously, convective and radiative heat transfer continue to play large roles, but material properties, including thermal diffusivity, conduction, pyrolysis temperature, and gases evolved, to name but a few, play a major role as well. One way to classify these phenomena is by the direction of flame spread relative to the flow of oxidizer and whether a convective flow is present in the ambient. These scenarios, along with some special situations, such as the smoldering of materials and a consideration of materials that can melt and drip in normal gravity, are discussed in the following subsections.

Flame Spread in a Quiescent Environment Flame spread over a solid can be classified as being either opposed or concurrent, based on the direction of flame spread relative to the convective flow. As shown in Figure 27.5, in opposed flow, the flame spreads into the incoming convective flow and the unburned fuel is upstream of the flame attachment point. In concurrent flame spread, the unburned material is downstream of the flame attachment point and the flame spreads in the direction of the convective flow. In normal gravity, flame spread occurs preferentially in a concurrent mode because buoyant convection induces an upward flow that brings fresh oxidizer to the flame and pushes the flame downstream. In a quiescent, low gravity environment, flames spread because of their own motion into the unburned material, a process that brings fresh oxidizer into the flame front. Therefore, flame behavior in quiescent low gravity is a special class of opposed flow. A more detailed discussion of opposed and concurrent flames in low gravity is presented in the following section.

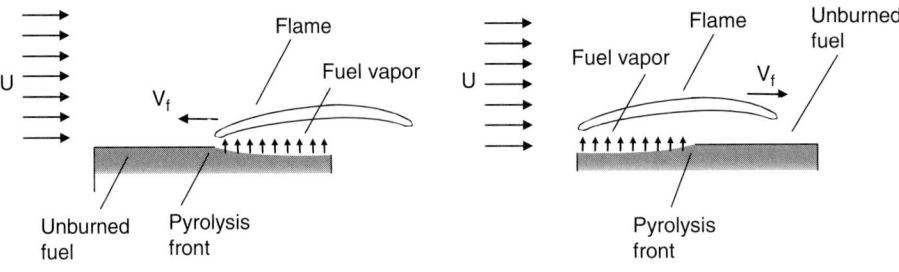

FIGURE 27.5 Opposed and concurrent flame spread. In this figure, U is the velocity of the incoming flow and V_f is the flame spread velocity. In opposed flow (left), the flame spreads into the incoming flow and the unburned fuel is upstream of the flame attachment point. In concurrent flow (right), the unburned material is downstream of flame attachment point and the flame spread is in the direction of the incoming flow (Courtesy of NASA).

Flame spread rates for sheet and foam materials in a quiescent ambiant have been found to be less than the corresponding rates in normal gravity. Strictly speaking, quiescent flame spread can occur only in the absence of buoyancy. Flames in normal gravity induce a natural convective flow with a velocity that depends on the rate of heat release and flame structure. However, low gravity flame spread rates typically are compared to normal gravity spread rates where there is no forced flow. Figure 27.6 shows the flame spread rate over thin filter paper as a function of oxygen mole fraction in both low and normal gravity. At high oxygen concentrations, the flame spread rates practically are identical between low and normal gravity. However, near the flame spread limit, at a constant oxygen concentration, the low gravity flame spread is less than that in normal gravity. This figure also shows that the oxygen concentration required to sustain low gravity flame spread in a quiescent low gravity environment is higher than that at normal gravity. This behavior suggests that, under quiescent low gravity conditions, a flame is considerably weaker than under similar conditions in normal gravity. T'ien et al. (2001) provide a detailed review of this literature and show that, for combustion in quiescent low gravity, the fuel mass loss rate is less, the flame temperature is lower, the rate of heat release is lower, and the soot production is reduced, as compared with the corresponding characteristics in normal gravity.

Flame Spread in a Convective Flow The presence of a forced convective flow in low gravity creates flame behavior that can be very different than that observed in quiescent conditions. With no flow, the flame characteristics are dictated by whether the flame motion and diffusion can provide sufficient oxygen to the flame zone to maintain the energy output required to vaporize fuel and overcome heat losses, both radiative and conductive. The addition of convective flow not only transports oxygen to the flame but also enhances heat transfer, both heat loss and heat gain, depending on the location. Of course, combustion in elevated oxygen levels is enhanced in both quiescent and convective flows. Figure 27.7 shows a computed extinction boundary as a function of oxygen

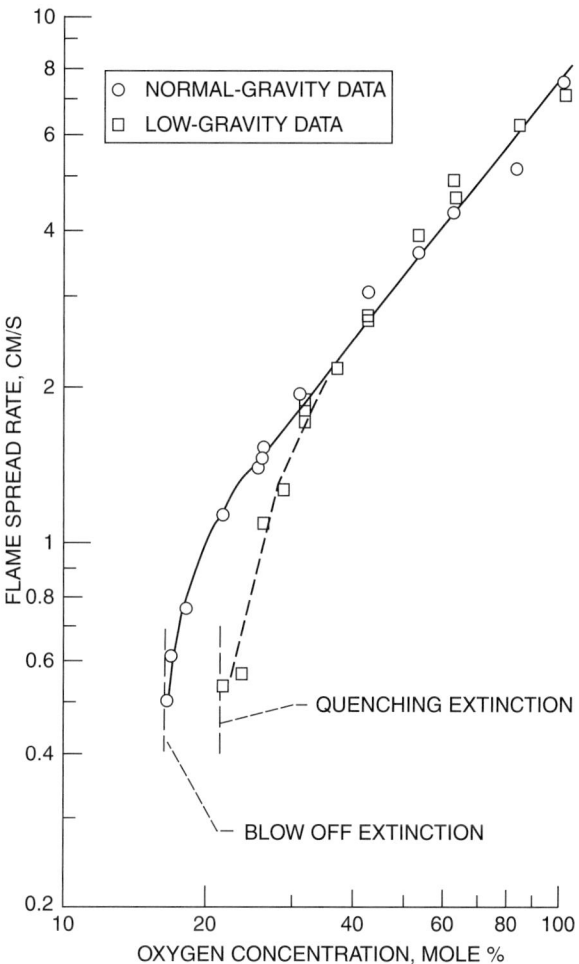

FIGURE 27.6 Flame spread rates over thin paper sheets in low gravity and reference normal gravity tests. Low gravity flames exhibited quenching extinction at the flammability limit, while normal gravity flames experienced blow off extinction because of the buoyant flow (Friedman and Olson 1989) (Courtesy of NASA).

mole fraction and free stream velocity for both an opposed flow and concurrent flow flame over a thin cellulose fuel sample (T'ien et al. 2003). Quenching and blow off modes of extinction are highlighted by this curve. Quenching extinction occurs when the flame is cooled by radiative losses to sufficiently low temperatures that combustion cannot be sustained. This occurs as the flow speed is reduced to near quiescent conditions. Blow off extinction occurs in higher speed flows when the residence time of the reactants in the flame zone is too short for the chemical reactions to proceed to completion.

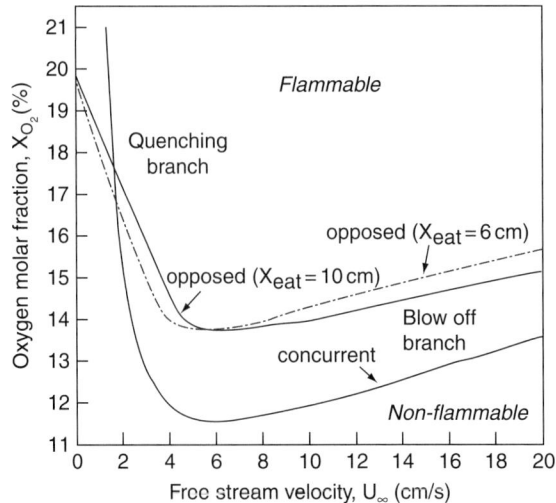

FIGURE 27.7 Computed extinction boundaries for opposed flow and concurrent flow flame spread over a thin sample. Conditions above the lines are flammable, whereas those below the lines are not (T'ien et al. 2003) (Courtesy of NASA).

Blow off extinction typically is observed in normal gravity because buoyancy introduces a flow of between 20 and 100 cm/s, which lies on the blow off branch of the extinction curve. Combustion in low gravity allows the quenching branch to be observed. This theoretical behavior has been observed experimentally for thin cellulose by Olson and co-workers (Olson, Ferkul, and T'ien 1988; Olson 1991), as shown in Figure 27.8. Unfortunately, the determination of this curve is difficult for many practical spacecraft materials. By definition, the determination of flammability limits requires the study of small, slow moving flames. To determine whether they propagate or extinguish under a specific set of conditions frequently requires a period of low gravity longer than can be obtained in ground based facilities. Of particular interest for spacecraft fire safety is the point where the quenching and blow off branches of the flammability boundary meet, a point that represents the minimum in the oxygen mole fraction required for a flame to propagate for any convective velocity. Maintaining a cabin oxygen concentration below the limiting oxygen mole fraction for spacecraft materials renders the material nonflammable in any convective environment, whether the flow was induced by cabin ventilation or as a result of natural convection, such as in a lunar or martian gravitational field.

Flammability tests conducted in space illustrate some additional features of solid combustion that are applicable to the determination of flammability limits just discussed. Hirsch, Beeson, and Friedman (2000) analyzed the data obtained by Ivanov et al. (1999) in the Skorost facility on *Mir* to determine the limiting velocities at which several polymers burn in a low-*g* convective flow. The results of this analysis are shown in Figure 27.9. The oxygen

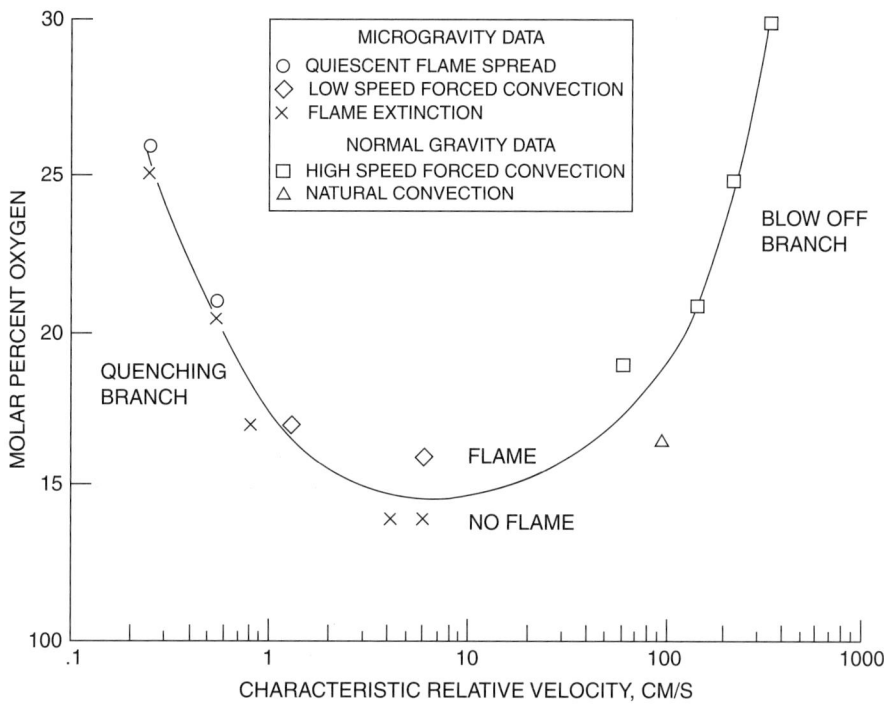

FIGURE 27.8 Extinction boundary for opposed flow flame spread over a thin fuel as a function of oxygen percent (by volume) and flow velocity (Olson et al. 1988) (Reprinted by permission of The Combustion Institute).

concentration and velocity combinations lying to the right of each line are flammable for that material, and those to the left are not. The data obtained on orbit certainly supports the more detailed flammability limits shown in Figure 27.8 and illustrates that the flow velocity required to realize the higher quiescent oxygen flammability limits is indeed very small.

Concurrent flow in low gravity has been studied considerably less than opposed flow because of the complicated interactions that govern the development and progression of the pyrolysis front, burnout front, and the flame tip. Because the entire flame is involved in preheating the unburned material, the flame characteristics play a significant role in the flame spreading behavior. Therefore, it takes much longer for the flame to reach equilibrium and, in general, longer than the time available in ground based low gravity facilities. Using the results of drop tower tests and numerical simulations, Grayson et al. (1994) showed that the flame spread rate for concurrent low speed flows is somewhat greater than the comparable low speed opposed flow and should continue to increase as the convective velocity increases. However, the characteristics of the flammability curve for concurrent flow was found to be similar to that for opposed flow shown in Figure 27.8. Also, their experiment results showed that, for thin cellulose

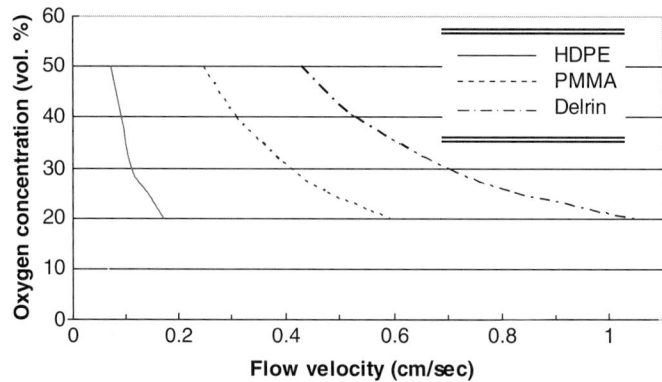

FIGURE 27.9 Limiting flow velocity in low gravity in various oxygen environments for high density polyethylene, polymethylmethacrylate, and Delrin (Hirsch et al. 2000) (Courtesy of NASA).

samples, the minimum oxygen concentration below which the flame cannot propagate occurred at approximately the same range of convective velocities, that is, from 5 to 10 cm/s, as in opposed flow. In a more detailed simulation, Kumar, Shih, and T'ien (2003) also showed that the velocity at the minimum oxygen concentration for flame propagation for opposed and concurrent flow was the same; however, the minimum oxygen concentration for concurrent flow was several percent less than that for opposed flow. This result has yet to be verified thoroughly by experiment and for materials other than thin cellulose.

The interaction between convective flow and an external heat flux causes unique ignition and flame spread characteristics in low gravity. Without buoyancy to induce a convective flow, materials exposed to an external heat source, such as from an adjacent hot surface or an overheated electrical conductor, can burn more readily. Ohlemiller and Villa (1991) conducted tests that followed the procedures of the NASA-STD-6001, Test 1 Upward Flame Propagation Test (NASA 2008), except that the configuration was modified to provide an external heat flux to the sample. They found that some materials, which would pass the standard Test 1, fail if simultaneously exposed to an external radiant flux. Of particular concern was the ignition delay time, that is, the time from the initiation of an external heat flux to when the material can be ignited by an external source. Shortened ignition delay times can be hazardous because materials could reach ignition conditions before an off-nominal condition were identified by other means, such as smoke detection, foul smell, or current overload.

Investigations of solid surface combustion under an external heat source in normal and low gravity have been conducted by Fernandez-Pello and coworkers (Cordova and Fernandez-Pello 2000; Fernandez-Pello et al. 2000; Roslon et al. 2001; Zhou et al. 2003). These researchers measured the ignition delay of fuel samples 73 mm long, 82 mm wide, and 12 mm thick in normal gravity and compared the results with data obtained in low gravity on the KC-135 reduced gravity aircraft. Two solid thermoplastic fuels were used in these tests: polymethylmethacrylate and blended polypropylene-fiberglass. The samples were mounted on the floor of a flow duct opposite a radiant heater, and the test parameters included the external heat flux and the flow velocity over the sample. Figures 27.10 and 27.11 show ignition delay data for the polypropylene-fiberglass and polymethylmethacrylate samples for both normal and low gravity, respectively. In normal gravity, the ignition delay is seen to decrease for both fuels as the convective velocity decreases. However, because of the flow induced by buoyancy, a minimum flow velocity can be obtained. With the elimination of a buoyant velocity in

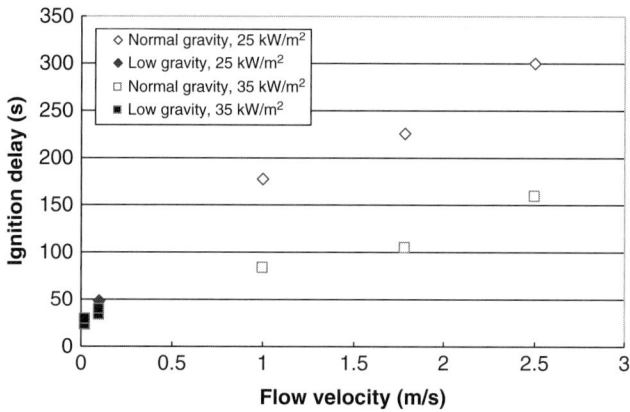

FIGURE 27.10 Ignition delay for blended polypropylene-fiberglass samples in normal and low gravity (Roslon et al. 2001) (Courtesy of NASA).

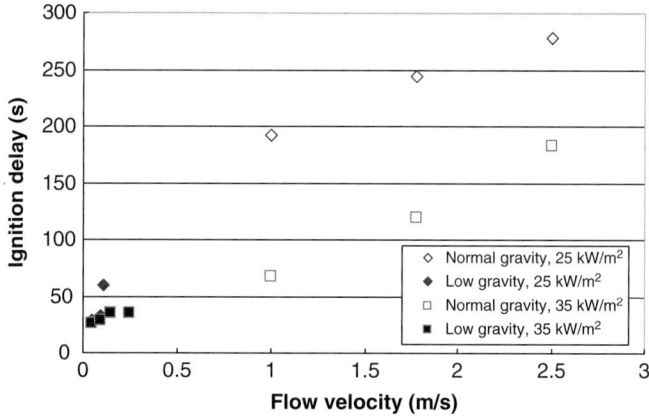

FIGURE 27.11 Ignition delay for polymethylmethacrylate samples in normal and low gravity (Roslon et al. 2001) (Courtesy of NASA).

low gravity, the ignition delay times continue to decrease and reach values significantly lower than those observed in normal gravity buoyant flows. These results show that, without buoyant convection, surface heat loss is reduced, which in turn allows materials to ignite more quickly in low-g than in normal gravity. Because of the reduced surface heat loss in low-g conditions, it follows that the critical heat flux for ignition also is reduced at these low flow velocities. Both findings have important ramifications for space facility fire safety and warrant additional study in long duration low gravity testing.

Smoldering Combustion Smoldering is a non-flaming exothermic combustion reaction that can occur in the interior of porous combustible materials. In contrast to flaming combustion, the oxidation of the fuel occurs in the solid phase rather than the gas phase. Smoldering combustion presents a unique fire risk because it can propagate slowly in the interior of the material and go undetected for long periods of time. During this period, it can release toxic products at a greater rate than a flaming fire, particularly heavy hydrocarbons and carbon monoxide. Under appropriate conditions, it can transition rapidly to a flaming fire. Although smoldering combustion has been studied extensively in normal gravity, only limited studies have been made of the phenomena in low gravity. The primary reason is because the timescales associated with the initiation and propagation of a smoldering reaction front are on the order of tens of minutes, that is, considerable longer than for a flaming fire. Only the long duration, low gravity periods available on a spaceflight provide sufficient time to study the low-g smoldering behavior of solids.

Experiments in extended low gravity have elucidated several important features of low gravity smoldering of materials (Walther, Fernandez-Pello, and Urban 1998). These tests were conducted using polyurethane foams having a 97.5% void fraction as fuel and airflow velocities up to 2 mm/s. In the quiescent experiments, conducted with oxygen concentrations of up to 40%, ignition did not lead to propagation of the smolder front without the continuous heat input from the igniter. For these conditions, heat losses prevented the initiation of a sustained smolder reaction. Similar conditions in normal gravity or with forced airflow through the sample lead to the initiation of smoldering throughout the material. Therefore, even though the lack of buoyancy reduced the heat loss by convection, the reduced transport of oxidizer to the reaction zone reduced the heat release from the oxidation process and the propensity for smoldering.

Smoldering combustion can occur in flow configurations similar to those identified in Figure 27.5, namely, opposed and concurrent propagation. During the smoldering process, heat released during the oxidation of the solid is transferred toward the unreacted material by conduction, convection, and radiation, thus supporting the propagation of the smolder front within the material. The oxidizer is transported to the reaction zone by diffusion and convection. The rate of propagation of the smolder front, as well as the limiting factors of the smolder process (ignition and extinction) and transition to flaming are determined by the balance between the heat released by oxidation, the energy required to heat the unreacted material and oxidizer to the smolder temperature, and the heat loss to the ambient environment. All these processes are affected by the presence or lack of gravity. Additional study is required to identify the limiting factors of these phenomena in low gravity.

Even though the findings to date have distinct implications for the requirements for smoldering combustion to occur in low gravity, it must be noted that the experiments were conducted using polyurethane foam with a 97.5% void fraction. Caution should be exercised in generalizing these results to other fuels, void fractions, and environmental conditions (T'ien et al. 2001).

Combustion of Dripping and Melting Materials In addition to the results of ignition, flame spread, and flammability limits previously discussed, other phenomena have been observed to play a role in spacecraft fire safety. These include materials that melt when heated and those that intumesce. Although a complete discussion of these phenomena is beyond the scope of this text, several primary characteristics are worthy of mention. Materials that can easily melt and vaporize or effervesce as they burn pose special problems in low gravity. In normal gravity, if a sufficient amount of material melts, the drip can fall away from the main burn, thereby removing energy and a portion of fuel. Also, in normal gravity, bubbles of vaporized material can form, rise through the melt, and be released into the ambient air. Both dripping and the release of hot gasses are appreciably different in low gravity. In low gravity, melted material does not drip away from the melt, thereby adding to the amount of fuel available to burn. This behavior has been observed with plastics (Ivanov et al. 1999) and in the combustion of metals (Steinberg, Wilson, and Benz 1992). Additional research on the combustion of metal rods in low gravity has been performed by Abbud-Madrid and coworkers (Abbud-Madrid, Amon, and McKinnon 1996; Abbud-Madrid, Branch, and Daily 2001) and Branch, Daily, and Abbud-Madrid (1996). These studies found that, for upward burning, the regression rate of the melt interface was significantly greater in low gravity than in normal gravity because the thermal energy of the melt remains in contact with the sample throughout the combustion process.

In low gravity, bubbles of vaporized material do not rise but continue to grow inside the material. The eventual release of these bubbles can be rather violent, expelling burning material called *firebrands* that could serve as potential ignition sources when they impact another object. The release of burning material by the expulsion of gas bubbles has been observed for Velcro® strips (Olson and Sotos 1987), polymethylmethacrylate cylinders (Ruff, Hicks, and Pettigrew 2004), and wire insulation (Greenberg *et al.* 1995). Combustion of the resin in burning composite also has been observed to result in the jetting of hot gases.

27.1.5 Summary of Low Gravity Fire Characteristics

The characteristics of flame structure and behavior discussed in this section and the references cited are but a fraction of the research performed on low gravity flame behavior. Whenever generalizations have been made about how flames behave in low gravity fire, they frequently have been proven incorrect by the next set of experimental results. In spite of this, it is useful to summarize the previous work and show our current assessment of flame and fire behavior in low gravity. A summary of the features discussed in this section, as well as those to be presented in subsequent sections, is shown in Table 27.1. Future researchers will undoubtedly continue to add to and quantify the observations listed in this table.

Table 27.1 Key Features of Flames in Low Gravity (Adapted from Friedman and Urban 2000)

Property	Trend	Remarks
Ignition	Promoted	Thermally stressed components can overheat rapidly. Particulate spills from aerosols that can persist for long times. Burning plastics eject hot material randomly and violently
Flame appearance	Altered	In quiescent environments, flames are often symmetrical and nearly invisible. Under imposed rates of low airflow, flames intensify and become bright and sooty
Flammability and flame spread rate (quiescent conditions)	Reduced or extinguished	Flames propagate slowly or extinguish due to the accumulation of combustion products
Flammability and flame spread rate (low flow conditions)	Increased, in some cases to match or exceed normal gravity levels	Low rate ventilating flows stimulate low gravity fires and greatly extend their flammability range and flame spread rates. Freely propagating flames tend to spread into the wind or into the oxygen source. Minimum oxygen mole fraction for combustion occurs at velocity lower than that induced by buoyancy in normal gravity
Flammability limits, gaseous	Broader limits, depending on fuel properties	Low-g flammability limits can be less than normal-g limits. Rich flammability limits might be higher, but tests have not been conducted
Detection signatures	Altered	Flames are often cooler and less radiant. Average size and range of soot particles are greater. Combustion product nature and quantities are altered
Fire suppression	Altered	Flame stability is increased in low gravity. Gaseous flames require higher suppression agent concentrations than in normal gravity. Suppression agent concentrations for flames on solids depend on convective velocity and the temperature of the solid

27.2 DESIGN FOR FIRE PREVENTION

This section describes spacecraft materials flammability control requirements, techniques used to mitigate materials flammability and techniques used to control ignition sources. Spacecraft fire safety is based on the assumption that ignition sources cannot be eliminated completely and the fuel must be eliminated by selecting materials that either do not ignite or are self extinguishing once ignited in the spacecraft atmosphere. However, ignition sources also must be controlled so that the number of credible ignition events is minimal.

27.2.1 Materials Flammability

Flammability is an oxidation process initiated by an ignition source. A material is ignited in the presence of oxygen, and the resulting heat of reaction supports or sustains the oxidation of the material. Oxygen, fuel, and heat, that is, an ignition source, combine to make what is called the *fire triangle*. When one of these three parts of the triangle is not present, fire does not occur.

Crewed spacecraft pressurized environments usually contain higher oxygen concentrations than air. As a result, many materials that are nonflammable or self extinguishing in air are flammable in spacecraft environments. Materials are tested for flammability by exposing them to a standard ignition source in an environment at the maximum oxygen concentration for a given space system design. The standard test and acceptance criteria are described in NASA-STD-(I)-6001A (NASA 2008) and ISO-14624-1 (ISO 2003). The test data are compiled by NASA in the Materials and Processes Technical Information System (MAPTIS) database for use in future designs. The MAPTIS materials selection database is available online at http://maptis.nasa.gov (registration required).

Materials Flammability Control

The first step in materials flammability control is the selection of materials that are inherently nonflammable or self extinguishing in the spacecraft environment. However, in practice, complete elimination of such materials is impossible. Spacecraft materials flammability control is designed to ensure that materials are self extinguishing in configuration. Thus, flammable materials can be used in small quantities, provided they are isolated from other flammable materials and flame propagation from one flammable material application to the next is precluded.

Many factors other than flammability control are considered in materials selection. The material must be able to perform its intended function reliably. This is something that might not be achievable every time with available nonflammable materials.

A safe design utilizes a configuration of nonflammable materials in conjunction with flammable materials that can maintain the flame propagation path within required limits. The nonflammable materials are used as firebreaks by interrupting potential fire propagation paths. Additional nonflammable materials might need to be incorporated into the design to contain, cover, or separate flammable materials and prevent them from providing continuous flame propagation paths. Requirements are developed around a defined amount of material to establish a limit for a flame propagation to be considered as controlled, therefore, mitigating a flammability safety hazard. For example, payload systems flown in the *Space Shuttle* are limited to 0.1 lb or 10 in.^{2} in the crew cabin and either 1 lb or 12 in. in the payload bay. In some other applications, flame propagation paths are limited to 6 linear in., which corresponds to the acceptable burn length for materials in the standard flammability test.

The minimum dimensions of firebreaks used to prevent fire propagation between flammable materials depend on the flammable material itself and the flammability of the material used as the firebreak. These dimensions are likely to have to be determined by test.

If fire propagation is possible, positive action must be taken to control or eliminate the hazard. Three types of positive action typically utilized for space hardware are containment, stowage, and covering the flammable materials with nonflammable materials. The primary methods used to reduce flammability hazards are the limitation of flammable materials by replacement with nonflammable materials and the restriction of propagation paths either by covering flammable materials with a nonflammable material or by separating them. Sample solutions are discussed in the following subsections.

Flammability Configuration Analysis A flammability configuration analysis that looks at all materials selected for the space system being designed is the best way to ensure flammability safety. Figure 27.12 shows an example of a flammability assessment logic diagram that is used to assess payloads flown on the Space Shuttle.

Fire Containment

Flight hardware is often in the form of a black box, a container with internal electronics, experiments, and the like. Such containers frequently are very effective at containing internal fires. Sealed containers have no vent openings and a verified maximum leak rate. Vented containers have active vents and associated cooling airflow. Intermediate containers, which have no active vents or cooling airflow, are not sealed physically to prevent air exchange. Many commercial off-the-shelf electronics items fall into this intermediate category. The fire containment capability of containers must be evaluated according to the amount of fuel involved, the container wall characteristics, and the presence of a combustion supporting environment.

Sealed Containers Hermetically sealed containers have a verified extremely low leak rate, and they can be filled with an inert gas such as nitrogen. Environmentally sealed containers also have a verified low leak rate, although higher than that of hermetically sealed containers. They normally contain air.

Fire propagation in a sealed container depends on the structural configuration of the container. If the sealed container does not contain oxygen or contains an inert gas, then it can be assumed that fire cannot be initiated. Further, it can be assumed that, for sealed metal containers that contain air or Space Shuttle or International Space Station atmospheres containing 30% oxygen or less, fire can be contained if the container wall is at least 60 mils thick. The same applies to nonmetallic containers, provided the container materials are nonflammable and are not melted away by an internal fire.

27.2 Design for Fire Prevention

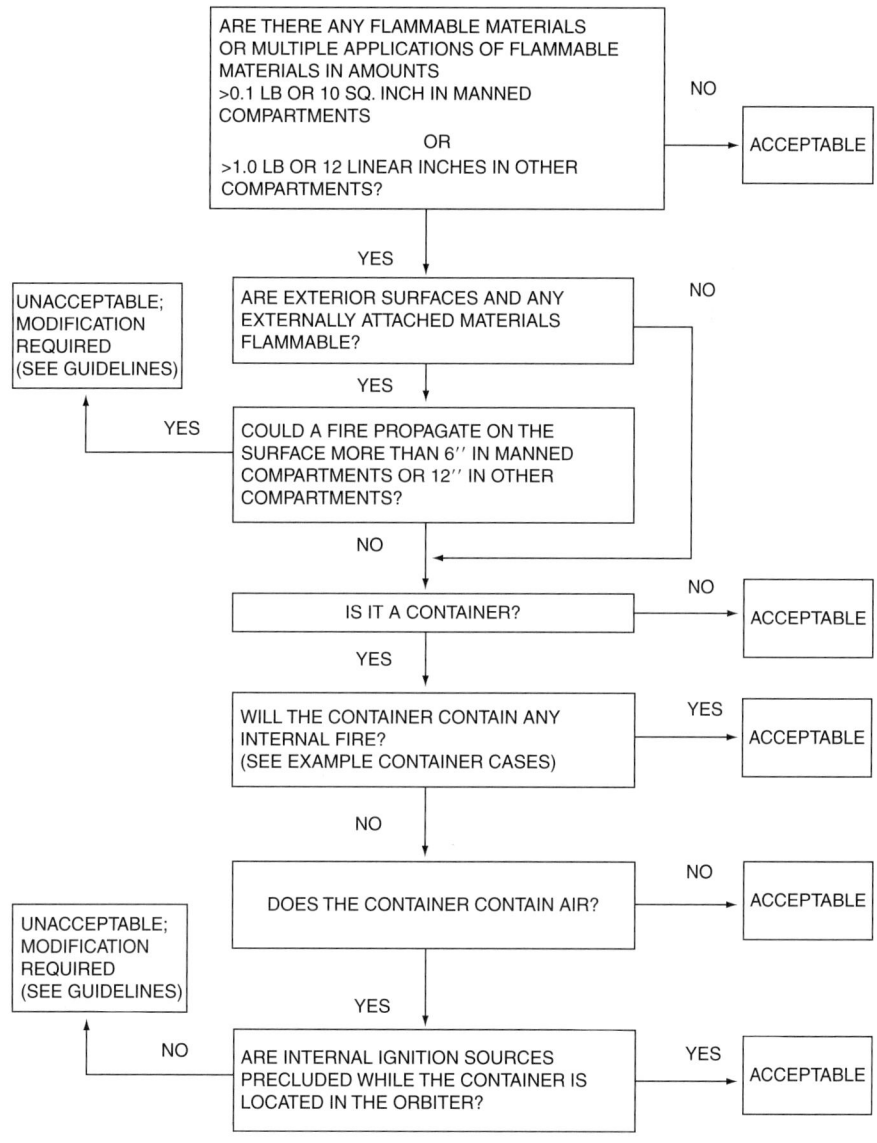

FIGURE 27.12 Flammability configuration analysis flow diagram.

Vented Containers Because oxygen is available to vented containers, it cannot be assumed that the container can contain a fire. However, tests have shown that it is possible for vented containers to contain fires if its vents are covered with a fine metal screen or the vent area is less than 1% of the total surface area.

The fire containment capability of vented containers must be evaluated carefully, because these containers allow replenishment of oxygen to support combustion of flammable materials. Definition of acceptable vented container configurations is very difficult, even with qualifications. In general, minimizing the number and size of vents and covering such vents with fine metal screens of fire resistant metals, such as stainless steel or nickel, rather than relatively flammable metals, such as aluminum, titanium, or magnesium, can reduce this hazard. Minimizing the free volume inside the container by adding nonflammable packaging materials, such as polyimide foam, also can help. An important consideration is that the use of flammable foams can increase the hazard.

The forced airflow velocity is also a major factor in the combustion of materials inside vented containers. If forced airflow is present, the relation between flow rate and flammability is determined best by configuration testing. If forced airflow is not required, it makes sense to cover all vents and assess the hardware as an intermediate container.

Intermediate Containers The intermediate container, which is not airtight but has no active vents or airflow, is a very common configuration. Examples include stowage lockers and bags and most commercial electronics items that do not contain a cooling fan.

Stowage lockers and stowage bags can be treated as containers that act as barriers to external fire. Flammable materials stowed in these containers do not constitute a fire risk while in the containers, provided they are not powered.

Many commercial electronics items can be addressed through the application of stowage constraints. However, items that do not comply with the stowage constraints can be acceptable if the case is shown to be nonflammable and capable of containing an internal fire. Many electronic items can be obtained commercially in metallic or nonflammable polycarbonate cases, and the internal components inevitably are packed sufficiently close that void space is not a concern.

In addition, many small commercial items are powered internally by alkaline batteries. Even in a hard short situation, such batteries are incapable of delivering sufficient energy to ignite solid flammable materials; however, the same cannot be said for rechargeable lithium batteries. Although potential ignition sources in spacecraft on a vehicle scale cannot be eliminated completely, it can be concluded that internal ignition is impossible for components powered by alkaline batteries. Therefore, the only potential for ignition is from external ignition sources, and the ignition potential can be eliminated by a nonflammable case or covering the case with a nonflammable material.

Stowed Hardware

Many small commercial off-the-shelf components are used in a spacecraft. Examples include cameras, power tools, compact and digital video discs, audio players, medical devices, medications, clothing, and personal hygiene items. Most items of this type are stowed in lockers or nonflammable stowage bags, taken out only as required, and returned after use. Regardless of the flammability of the hardware materials, stowed hardware of this type is acceptable for flammability provided it is relatively small, unstowed for only short periods, and controlled by the flight crew while in use.

Flammability Reduction Methods

This subsection briefly describes various methods for protecting flammable materials by covering them with nonflammable materials.

General Materials Protection Commercial items with flammable outer surfaces can be wrapped completely with a nonflammable tape. It is known that 3-mil aluminum tape protects most plastics, foam, and cardboard from external flame initiation. If aluminum tape cannot be used for electrical reasons, a nonflammable fiberglass tape with a silicone adhesive can be used. For long-term applications where tape aesthetically is unacceptable but for which fire protection is needed, the flammable surfaces can be coated with a nonflammable barrier material.

The most common nonmetallic case materials for commercial items used in space design are acrylonitrilebutadienestyrene, which is extremely flammable, and polycarbonates, which normally are acceptable at oxygen concentrations up to 30%. In many cases, the hardware organization is able to select a commercial item with a metallic or polycarbonate case, thus eliminating the need for wrapping or coating with a nonflammable material.

Electrically powered items containing internal flammable materials usually can be treated as a fire resistant container. In some cases, even a highly flammable case is an adequate fire barrier against propagation of an internal fire to the outside of the container.

Wire and Cable Most aerospace grade electrical wire insulation is nonflammable in space environments, and limitations usually are driven by other factors, such as flexibility and cut through resistance. Electrical wire insulation found in commercial off-the-shelf hardware typically is flammable in space environments. However, if commercial off-the-shelf hardware must be used, several protection methods exist, such as covering the wire within the device with a braided polytetrafluoroethylene sleeve, such as Gore-Tex®.

Wire and cable accessories such as cable markers, spacers, and cable ties should not contribute to fire propagation paths. Polyvinylidene fluoride or fluoroelastomeric cable markers generally are used. Acceptable lacing cords can be made from polytetrafluoroethylene, polytetrafluoroethylene-fiberglass, or Nomex®; and acceptable cable ties can be made from ethylene tetrafluoroethylene or ethylene chlorotrifluoroethylene fluoropolymers. When flammable cable tie wraps are used on nonflammable cables, they should be spaced at least 2 in. apart to prevent fire propagation.

Electrical Connectors In air and in moderately enriched oxygen environments of up to 40%, the shell of a metal shell type connector prevents fire propagation from the nonmetallic materials used inside the connector to other nonmetallic materials, regardless of the material inside the connector. The acceptability of nonmetallic shell connectors and the nonmetallic materials used inside the connector depends on the flammability of the shell material and its ability to act as a fire barrier.

Tubes and Hoses Tubes or hoses made from flammable materials can be replaced with a nonflammable material or covered with a fire barrier material. Clear polytetrafluoroethylene or fluorinated ethylene propylene tubes and hoses readily are available to replace flammable materials. If flammable tubes or hoses must be used, the exterior can be protected by a covering of a nonflammable fabric such as 7.2 oz/yd^2 natural Nomex HT-9040, polybenzimidazole, or Beta cloth®.

Hook and Loop Fasteners Although some hook and loop fastener materials are less flammable than others, all common types of hook and loop fasteners are flammable in spacecraft habitable areas. To prevent long flame propagation paths, the following usage limits are applied generally to hook and loop fasteners in habitable areas:

- Maximum size is 4 in.^{2}, either individually or in pieces.
- Maximum length is 4 in.
- Minimum separation distance is 2 in. in any direction from another piece.

Stowage Bags and Lockers Metal stowage lockers that contain no ignition sources are acceptable without reservation. Acceptable stowage bags can be constructed from 7.2 oz/yd^2 natural Nomex HT-9040, polybenzimidazole, Beta cloth, or other nonflammable fabric. Lighter weights of natural Nomex are acceptable in double layers. These containers, made of nonflammable nonmetallic materials, can have flammable items stowed inside them, provided they do not contain ignition sources and are not susceptible to spontaneous ignition or chemical reaction.

Most stowage bags and lockers used in space designs contain foam assemblies as part of the packaging. The most common foam packaging materials, polyurethane and polyethylene foams, are highly flammable. Several nonflammable foams exist, but none has been found completely to be satisfactory for spacecraft use. Flammable foam materials are acceptable if they remain inside the container, no ignition source is present, and the container is opened only briefly. If the container should be left open for any considerable period or if the foam likely is to be taken out of the container, it should be covered with a fire barrier material, such as a single layer of natural Nomex HT-9040.

Thermal Control Blankets Thermal control blankets are widely the most used potentially flammable external materials. These blankets typically contain 12 to 40 layers of film of between 0.0005 and 0.002 in. in thickness separated by some type of scrim cloth. Acceptable thermal control blankets are constructed typically as follows:

- The outer layer is made of nonflammable material such as polyimide film at least 1.5-mil thick, metal foil, silver Teflon, or Beta cloth.
- Internal layers can be a combination of flammable films or scrims.
- Edges are hemmed or suitably finished so that the inner flammable layers are protected.

27.2.2 Ignition Sources

Ignition sources are addressed as part of the overall control of combustion (fires) for spacecraft. Prior to the U.S. Apollo 204 disaster, the primary control of combustions hazards was by the control of ignition sources. Since that disaster, the safety concerns have expanded to recognize that, although ignition sources can be minimized, they never can completely be controlled and the safety of the vehicle and any crew onboard requires the control of ignition sources, flammable materials, and explosive environments. This section concentrates on the sources for ignition or explosion. In the discussion of specific ignition sources, the section addresses the sources in relation to the specific fuels of concern and methods of management of the hazard.

Overview of Sources

The ignition sources to be addressed are hot surfaces, sparks from mechanical impacts, electrical arcs and sparks, and electrostatic arcs. It should be noted that non-electrical ignition sources are very rare in practice and included only for completeness. Hot surfaces that exceed 400°C are considered potential ignition sources for flammable or explosive atmospheres. Mechanical impacts between some metals can cause sparks. These sparks can ignite flammable or explosive atmospheres but are rarely a credible source of ignition for flammable solid materials. Similarly, the presence of electrical shorts, arcing relay contacts, brushes of DC motors, and the like also are capable of igniting flammable or explosive atmospheres or insulation. Items of the final category of arc producing items are those that can acquire an electrostatic charge capable of discharging to a lower voltage structure. Those arcs are able to ignite flammable or explosive atmospheres as well.

The mission phases during which an explosive or flammable atmosphere is of particular concern are space vehicle entry, landing, and post-landing operations, whether planned or contingency. Normal equipment functions should not cause ignition of a flammable vehicle atmosphere that could have resulted from leakage or ingestion of fluids. Contingency landings, such as return to landing site or abort once around, although not normal, are considered to be of sufficient likelihood to require that they be considered as a normal phase for the purposes of design.

Electrical Sources

The preferred method for preventing electrical ignition of a flammable vehicle compartment atmosphere is for all electrical equipment to be non-powered during those mission phases that have the highest likelihood of containing an inadvertent flammable atmosphere. Those phases are launch and descent. If equipment must be powered during launch, it should be designed so that either all ignition sources are controlled or a method is provided for de-energizing all uncontrolled ignition sources. The method for de-energizing, of course, needs to comply with appropriate guidelines for such critical hardware. If overriding considerations dictate that the equipment must be powered during descent, it must be designed so that all ignition sources are controlled by appropriate means.

Electrical ignition sources can often be controlled by one of the following procedures, which are listed in the order of preference:

- Seal all relays, switches, motors, and other similar ignition sources to a leak rate of less than or equal to 1×10^{-4} scc of helium per second, at a pressure differential of 1 atm. These leak rates need to be verified by test.
- Perform the test stated in method 511.4, procedure 1 of MIL-STD-810F (DoD 2000) or method 109C of MIL-STD-202G (DoD 2002).

Smart Short Analyses The analysis method to establish that the power system has a robust design is called a *smart short analysis*. This analysis assumes that a short circuit occurs to a power system that is current limited. The particular value of the current limit is a value just slightly less than the current that trips the circuit protection device. For instance, a 10-A fuse can in some cases be able to sustain a current value of 15 A prior to opening. If that is a possibility for a particular fuse manufactured by a particular company, then the wire would be assumed to carry the 15 A until a steady state temperature is reached. That temperature would have to be below the qualification limit of the wire insulation. If the analysis shows that the current would produce temperatures above this limit, then the design would have to be changed to use fuses of lower rating or wire of larger gauge.

Temperatures above the qualified limit would expose equipment to unassessed conditions that cannot be assumed safe. If the insulation were to fail due to exposure beyond the qualification limits, the resultant new circuit paths would not have been recorded on design schematics. Without such schematics, the analyst would not have been able to complete the required hazard analysis. Because the safety analyst cannot ensure the design would be safe beyond the qualification limits of the insulation, the design needs to ensure that temperature stays within the qualification levels.

Commercial Off-the-Shelf Hardware Commercial off-the-shelf hardware is increasing in use for space applications for such devices as laptop computers, video cameras, printers, video recorders, and automated cameras. These units are valuable because they can adapt automatically to situations and record data for post-mission review. The automatic functions invariably are implemented using small DC motors with brushes and unsealed relays. These, of course, are potential sources of arcs. The commercial off-the-shelf equipment designers probably have not gone to the extra expense to make their devices safe to use in explosive atmospheres. For this reason, those pieces of equipment probably are not safe to be powered during the mission phases of concern.

Electrostatic Discharge The area of electrostatic charge control also is addressed in Chapter 14, "Avionics Safety." Generally speaking, this hazard is controlled normally by grounding items that could be exposed to static generation environments. The grounding system for static electricity control does not have to carry the same currents as grounding systems for other purposes owing to the slow buildup of static charge and the relatively low capacitance of thermal blankets and other insulating materials.

Other Electronic or Electrical Sources Other devices that can generate hot surfaces and must be considered potential ignition sources are lamps, heaters, and radioisotope thermal generators. With changing technologies, there likely are other devices that can produce hot surfaces. The designers need to be aware of the hazardous nature of any surface that reaches elevated temperatures nearing $400°C$.

Mechanical Sparks

The consideration of mechanical impact control is not always obvious. The sparks from such impacts can be ignition sources for some flammable or explosive atmospheres. An example of such a possible impact might be a fail-safe brake for quickly arresting the rotation of a centrifuge in the event of a need for rapid safing or loss of power. Another possible source might be the mechanical stops of a robotic device used to arrest the runaway motion of the robotic device or articulated antenna. The designer of such mechanisms should analyze or test the materials involved in such arresting devices to ensure that they are not subject to producing sparks under a worst-case operation.

27.3 SPACECRAFT FIRE DETECTION

Fire detectors operate by observing one of the properties or fingerprints of the combustion process: heat, light, or smoke, which can be either particles or gases (Bukowski 1987; Weiland 1994). The most common systems in use for terrestrial applications detect the particulate in the smoke or rely on thermal detectors. Both approaches are favored in normal gravity, because the buoyant airflow concentrates the heat and smoke near the ceiling, providing a reliable location to place detectors. Smoke detectors typically are ionization detectors, but light scattering systems are becoming more prevalent. Thermal detectors range from temperature sensors to fusible links on sprinklers and door closure devices. Gaseous species sensors also are used in terrestrial applications, and these have shown promise as a means to back up smoke particle detectors. By confirming their measurement, false alarms are reduced. Another type of fire detector employs a radiation sensor to detect another property of fire, light. These sensors are limited generally to line of sight applications and are most effective in systems where false alarms can be avoided. Examples of these cases include military applications where the detector has a high trigger level, such as munitions impact, or where the system looks for the characteristic flicker behavior seen in normal gravity fires. Development of spacecraft fire detection systems has been hampered by the lack of a natural smoke concentration mechanism like that provided by buoyancy and the limited data on the characteristics of low gravity fires.

27.3.1 Prior Spacecraft Systems

Prior to *Skylab*, NASA did not begin deploying fire detection systems. In the earlier spacecraft (*Mercury*, *Gemini*, and *Apollo*), given the combination of the short mission durations and the very small pressurized volumes, it was considered reasonable that the astronauts would detect any fire rapidly. The Skylab module initiated the combination of a large habitable volume (361 m^3) and extended duration missions. The Skylab spacecraft, therefore, included approximately 30 ultraviolet sensing fire detectors (Friedman 1992). These devices were limited to line of sight detection and were reported to have difficulties resulting from false alarms.

Space Shuttle Detector

The *Space Shuttle* is designed for extended missions of typically 2-wk duration. It exhibits a large pressurized volume (74 m^3), and operationally there are periods of time measured in hours when the entire crew can be asleep. In addition, during the launch sequence, many avionics systems cannot be accessed readily by the crew. Consequently, the vehicle is designed with a fire detection system and a centrally activated fire suppression system for the avionics systems. During the 1970s, when the *Space Shuttle* was being designed, ionization smoke detectors were becoming readily available for consumer home use, and there was extensive progress in the development of smoke detectors for terrestrial applications (Bukowski and Mulholland 1978). Ionization detectors, which used a radioactive source and whose stability was much greater than the expected lifetime of the electronics, readily were available, whereas photoelectric, that is, scattering or obscuration, detectors generally were unavailable due to the difficulty of producing stable light sources. No data were available regarding smoke particle size distributions in low gravity, and the database of normal gravity smoke characteristics was only a fraction of what is available today. Additionally, no data were available with regard to spacecraft dust particle size distributions, but the absence of gravitational settling suggested that the particles would be larger than those seen on Earth.

Given the state of knowledge at that time, it is quite reasonable that the Space Shuttle design (Figure 27.13) is an extension of the most common ground based approach, the use of ionization detectors. The *Space Shuttle* has nine particle ionization smoke detectors in the avionics cooling air return lines in the middeck and flight deck. *Spacelab* has six additional particle ionization smoke detectors in the avionics cooling air return lines (Martin and DaLee 1993). Although the design rationale for the Space Shuttle detector is not completely known, Celesco® (later Brunswick Defense®) based its design on data that

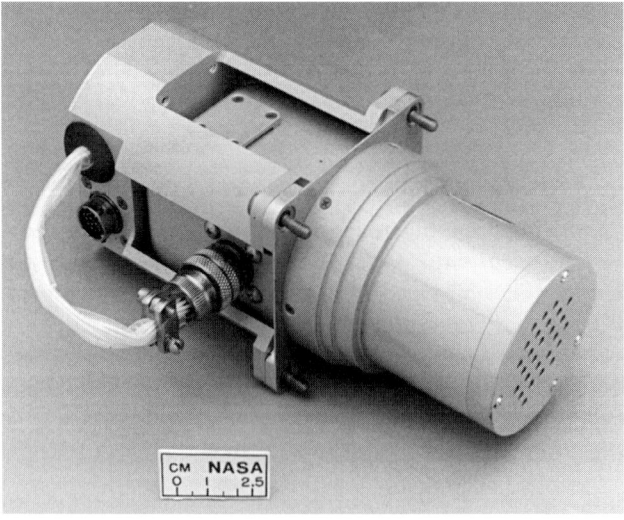

FIGURE 27.13 Brunswick Defense smoke detector used in the NASA Space Shuttle fleet. The inlet is on the right, and the gas is expelled out the small plate on the top (Courtesy of NASA).

suggested incipient fires could be discriminated by looking for particulate in the 0.4 to 0.7 µm range (Barr 1977). The device combined a dual chamber ionization detector with a small vane pump. The addition of a pump provided the opportunity to employ an inertial particle separator that rejected particulates larger than 1.0 µm because of their greater inertia. This focus on the smaller particulate was consistent with the understanding at the time that incipient smoke particles were smaller than 1.0 µm. Further, because the response from ionization chambers is affected by the ambient air velocity, the implementation of an ionization detector in a flow duct probably was facilitated by the use of an air pump to control the airflow through the smoke detector. This air pump increased the power requirements and reduced the operational life for the detectors.

International Space Station Smoke Detector

The designers of the environmental control and life support systems for the *International Space Station* took advantage of the fact that stable laser diode light sources had become readily available, and consequently, terrestrial smoke detectors using light scattering were becoming prevalent. Switching to a light scattering based instrument provided the opportunity to produce a detector that required substantially less power (1.5 W) than the Space Shuttle detector (9 W) (Steisslinger et al. 1993). Further, the International Space Station design had no moving parts and was quieter with a much longer operational life. Switching to light scattering was supported by the general understanding in the literature, which indicated that early smoldering fires produce larger particulates than established flaming fires (Bukowski and Mulholland 1978). The International Space Station detectors (Figure 27.14) are near infrared laser diode, forward scattering, smoke or particulate detectors. Consistent with light scattering theory, their sensitivity is greatest for particles larger than the laser wavelength of $\sim$1 µm, with their sensitivity extending down to 0.6 µm. The minimum reported sensitivity is 0.3 µm (Steisslinger et al. 1993). The current requirements for the *International Space Station* call for two detectors located in the open area of the module and detectors in racks that have cooling airflow (McKinnie 1997). The main sensing component of the International Space Station detectors is a two pass laser diode system that senses forward scattered light (30°). Also a 0° obscuration system is used as a measure of the beam strength. The detector is designed to release an alarm based on the magnitude of the scattered light signal referenced to the beam intensity. The current alarm value of 2 V is calibrated to 3.3% obscuration/meter (1% obscuration/foot) of visible light for punk type smoke in an Underwriters Laboratories smoke box. In this standard qualification method, the smoke detector is placed in a chamber filled with a level of smoke that is adjusted to achieve a target level of attenuation of visible light passing between a lamp and a photocell in the box (UL-268, UL 2006).

The International Space Station and the Space Shuttle detectors were designed two decades apart, using the best available data and technology at their respective times. The result is that the two detectors have particle sensitivities that nearly are mutually exclusive. Despite being based on the best available data, due to the complete absence of low-*g* data concerning the nature of particulate and radiant emission from incipient and fully developed low-*g* fires, different conclusions were drawn concerning the optimal design for spacecraft fire detection.

FIGURE 27.14 Allied Signal/Honeywell light scattering smoke detector used in the *International Space Station*. The near infrared laser beam emerges from the enclosure into the top assembly (A), is reflected by two mirrors (one visible at top right, B), and is reflected back to the sensors in the enclosure (C). One sensor detects the forward scattered light and is referenced by another sensor that looks directly at the incident beam (Courtesy of NASA).

27.3.2 Review of Low Gravity Smoke

Smoke is a nonspecific term that encompasses aerosol materials produced by a number of processes. In particular, it can include steam or dust from the fire suppression process; unburned, recondensed, original polymer or pyrolysis products that can be either liquid or solid; hydrocarbon soot; condensed water vapor; or ash particles. Ash and soot particles are the dominant particulate in established fires, whereas early ignition and fire establishment periods tend to produce higher concentrations of unburned pyrolysis products and recondensed polymer fragments. Early fire detection is always desirable, but given the constrained environment on any spacecraft, the target for any detection system necessarily must be the early ignition period and not after the fire has become established. Consequently, the primary target for detection is the pyrolysis products, not the soot and ash.

Other than the results reported here, the only combustion generated particulate samples that have been collected near the flame zone for well developed microgravity flames have been soot sampled from the laminar soot processes experiment (Urban et al. 1998).

All other data either came from drop tower tests, and therefore corresponded only to the early stages of a fire, or were collected far from the flame zone. The fuel sources in the drop tower tests were restricted to laminar gas jet diffusion flames (Ku et al. 1995) and very rapidly overheated wire insulation (Paul et al. 1993; Srivastava et al. 1998). The gas jet tests indicated, through thermophoretic sampling, that soot primaries and aggregates (groups of primary particles) in low gravity can be considerably larger than those in normal gravity. This raises new scientific questions about soot processes, as well as practical issues for particulate size sensitivity and detection and alarm threshold levels used in spacecraft smoke detectors. The overheated wire insulation tests conducted in the 2.2-s drop tower suggest that particulate generated by overheated wire insulation can be larger in low-g than in 1-g (Paul et al. 1993). One of the most surprising results was the determination by Srivastava et al. (1998) that the Teflon wire coloring agent had a very strong effect on the particle size distribution and morphology. Particles collected on transmission electron microscope grids downstream of the flame zone in the wire insulation flammability experiment (Greenberg et al. 1995), as well as visual observation of long string-like aggregates, further confirm this suggestion. Subsequently, the laminar soot processes experiment sampled soot from ethylene and propane gas jet diffusion flames in long term low gravity (Urban et al. 1998). The soot primary particles typically were twice the size of soot from similar normal gravity flames, and the aggregates were more than an order of magnitude larger than their 1-g counterparts. The combined impact of these limited results and theoretical predictions is that, as opposed to extrapolation from 1-g data, direct knowledge of low-g combustion particulate is needed for more confident design of smoke detectors for spacecraft.

27.3.3 Spacecraft Atmospheric Dust

Any effort to detect fires must be able to discriminate against the ambient background and nuisance signals. In the case of smoke detection, the background dust aerosol conditions must be considered. The only spacecraft background aerosol particulate measurements to date were made on STS-32 during 1990 (Liu et al. 1991) and on the *International Space Station* (Urban et al. 2005). The STS-32 measurements included the results from two cascade impactors and a light scattering device. The two impactors reported a bimodal particle size distribution with ~40% of the particles in each of the 2.5 to 10 μm and the >100 μm ranges. The other two ranges, <2.5 μm and 10 to 100 μm were populated very lightly (Table 27.2). Each impactor sampled approximately 15 m^3 of air over approximately 30 h. These results showed substantially higher concentrations than from typical indoor measurements. These data were supported by the light scattering instrument, which made 17 measurements in 12 locations, all of which reported mass concentrations ranging from 50 to 70 μg/m^3. The low particle levels in the <2.5 μm bin suggest a zone of opportunity for spacecraft fire detection, because typical normal gravity fires produce substantial particulates in this size range (Bukowski and Mulholland 1978; Bukowski et al. 2003), and the average sizes that were recorded by comparative soot diagnostics (later) are in this size range. However, more complete particle size statistics are needed.

Table 27.2 Airborne Particle Mass Concentration: Comparison of STS-32 to Indoor Environments (From Liu et al. 1991)

Sample Location	Particle Size Range					
	<2.5 m	Location	<2.5 m	Location	<2.5 m
Mass concentration μg/m^3						
Shuttle	2.3	Shuttle	2.3	Shuttle	2.3
Shuttle	2.1	Shuttle	2.1	Shuttle	2.1
Home	3.0	Home	3.0	Home	3.0
Office	3.4	Office	3.4	Office	3.4
Average mass concentration μg/m^3						
Shuttle	2.2	Shuttle	2.2	Shuttle	2.2
Indoor	3.1	Indoor	3.1	Indoor	3.1
Ratio of mass concentrations						
Shuttle/indoor	0.71	Shuttle/Indoor	0.71	Shuttle/Indoor	0.71

During testing on the *International Space Station*, Urban et al. (2005) found that the *International Space Station* environment was very clean compared to either the *Space Shuttle* or terrestrial laboratories. Contrary to the high particulate levels seen in the *Space Shuttle*, all three tests that sampled the undisturbed environment of the *International Space Station* showed very low levels, which averaged less that 0.005 mg/m^3 from the mass concentration device, and less than 15 particles/cm^3 from the particle counting device, with typical readings at the zero baseline for both instruments.

27.3.4 Sensors for Fire Detection

An excellent review of fire detection technology at the approximate time the *Space Shuttle* was under development is provided by Bukowski and Mulholland (1978), and a summary of the performance of current technology can be found in Bukowski et al. (2003). In general, well ventilated flames and established fires emit particulate with sizes tending toward the size range to which ionization detectors are more sensitive than optical detectors. On the other hand, smoldering fires and overheated or pyrolyzing materials produce larger particulates, owing to the large amount of condensed, un-oxidized fuel pyrolysis products and incomplete oxidation. For these larger particles, light scattering or obscuration detectors are more appropriate. Ionization detectors were favored in the 1970s through the mid-1990s due to the difficulty of producing light sources that would remain stable for several years. The advent of improvements in light emitting diode and diode laser technology has reduced the cost and increased the stability of photoelectric fire detectors.

The relative sensitivity of the two technologies is shown in Figure 27.15 (Mulholland and Liu 1980), which shows that below particle diameters of approximately 0.2 µm, ionization detectors are more sensitive. For particle sizes above this, light scattering is exponentially more sensitive. This further sensitivity difference is amplified by the fact that particle mass increases with diameter cubed, so for the same mass loading of smoke, the particle number count decreases as the diameter increases. Ionization detectors are, to the first order, proportional to the particle diameter, whereas light scattering is proportional to the diameter cubed. The net result is that, as particle diameter increases beyond the wavelength of the light, light scattering detectors become more sensitive than ionization detectors.

In an extensive comparative performance characterization in residential housing, Bukowski et al. (2003) tested ionization, photoelectric, and carbon monoxide detectors against a variety of smoke and nuisance sources. No detector was the clear winner. As expected, the ionization detectors performed better than the others for flaming fires but responded much more slowly than the scattering detector for the smoldering cases. The carbon monoxide detector performance was similar to but slightly slower than the scattering detector. In general, both smoke detector designs provided positive escape times if detectors were placed on every level and every bedroom. Ionization detectors were prone to nuisance alarms during early stages of cooking activities, even when no visible smoke was present. Although further work still is indicated to implement the results of the residential testing to date, the effectiveness of residential smoke detector technology and acceptable nuisance alarm levels have been demonstrated. Corresponding understanding for the spacecraft environment is needed to ensure adequate protection for future spacecraft.

Data Parameter Monitoring

The preferred method for fire detection in spacecraft is the use of an approved smoke detector that has sufficient airflow to carry the by-products of a fire. Because not all hardware or equipment possesses active airflow or has an adequate means for smoke to reach an approved fire detector, another method of fire detection was devised for use with these types of equipment on the *International Space Station*. To detect a fire in such equipment, specific hardware parameters are monitored for deviations that would be indicative of the occurrence of a fire event. This process, as defined in the *Safety Policy and Requirements for Payloads Using the International Space Station* (NASA 1995), commonly is referred to as *data parameter monitoring*. Typical data parameters used for this purpose include temperatures, pressures, and current. It is incumbent on the hardware developer to have thorough insight into the hardware to be able to establish a viable means for using data parameter monitoring for fire detection.

Properly understood, data parameter monitoring truly is indicative of a fire event. When using temperature as a data parameter, a thorough knowledge of the materials used in construction, including those of the electrical and electronics components, is required. Knowledge of the usage and operation of the hardware is necessary to understand and identify those areas of the hardware with a potential to become hot spots, and therefore, those that would require data monitoring through use of a strategically placed

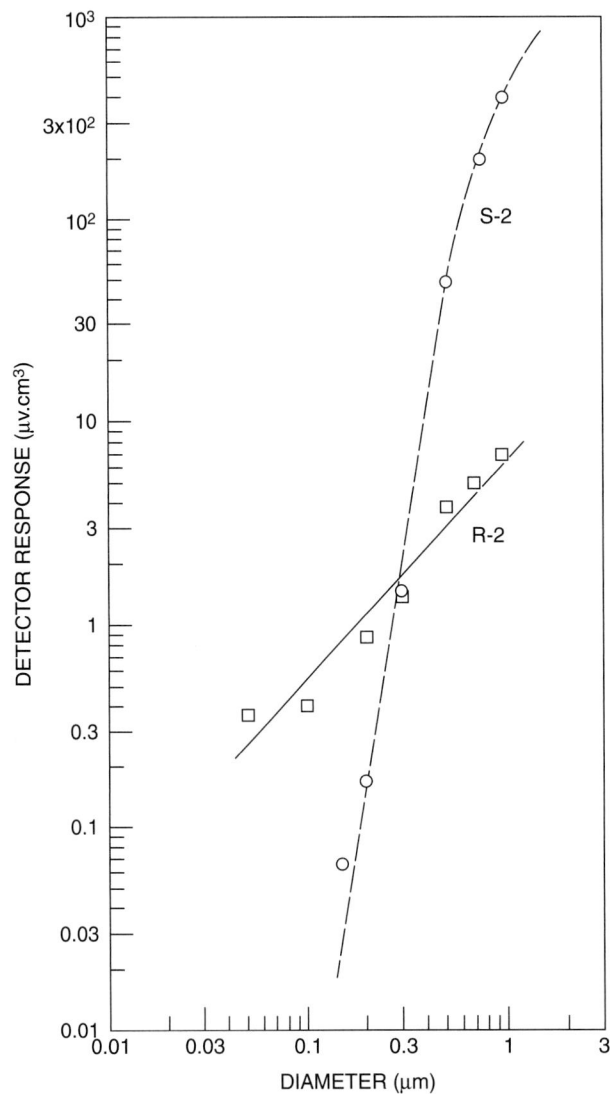

FIGURE 27.15 Ionization and scattering detector response as a function of particle size. The size response function, (detector output − background)/particle concentration, is plotted against the diameter for monodisperse aerosol for light scattering (S-2) and ionization (R-2) detectors (Mulholland and Liu 1980) (Courtesy of NIST).

temperature sensor. Even though monitored, it is not prudent to set an alarm temperature at or below the usage or temperature ratings of the components, because doing so would elicit an unacceptable number of false alarms. Most electronic components have a manufacturer's or Underwriters Laboratory rating in the range from 80 to 180°C, whereas most metals and polymeric materials are rated in hundreds of degrees Celsius. Therefore, it is anticipated that temperatures at or above 100°C are feasible as temperature set points for a fire event.

It is noted that the same sensors used for fire event parameter monitoring often are used to provide data relative to the health and status of the hardware or equipment. It needs to be made clear that, because the same health and status data can be used to power down or assure the safety of the hardware should an off-nominal condition occur and this is most likely to occur prior to the onset of a fire event, the data parameter monitoring needs to be powered from a source separate from that providing power to the hardware for this process to be able to continue to monitor for a fire event. The separate power feed to the data parameter monitoring is required unconditionally, because the hardware could retain enough stored heat or energy to ignite a fire even though the main power or heat generator has been powered down.

A classic example of an International Space Station payload facility using both an approved fire detector and data parameter monitoring is the microgravity sciences glovebox, which is comprised of various volumes with dedicated purposes. Its power distribution system has two major emphases; therefore, it is separated into separate feeds, which provide power for the operation of the facility itself and for use by experiments operating inside its work volume. During operation of the microgravity sciences glovebox, fans are used to provide airflow inside the experiment work volume and cooling air to its electrical and electronic components, and to the main fire detection device internal to the facility, but which is not part of the work volume. Various types of sensors for detecting pressure, temperature, and specific gases and numerous facility electronic devices, such as computer, video recorders, and data handling, are also in use. Because of a physical barrier between the experiment work area and the remainder of the facility, there is no transfer of air from the work volume to the glovebox facility smoke detector. The experiment work volume is a closed loop system in which no air can escape into the facility or crew habitable volume. To be able to detect a fire event inside the work volume, the air temperature is monitored at the intake to the fan filters. The logic used to support this method of fire detection is that, when all heat generating experiment hardware and all unnecessary microgravity sciences glovebox experiment support items, such as microscope, lights, and video recorders, inside the work volume have been powered down, should the air temperature being monitored inside the work volume continue to increase, the only reason for the increase is a fire event.

Another example to illustrate the use of data parameter monitoring is from the European drawer rack program, whereby the primary method of fire detection for the European drawer rack and any air cooled payloads it might contain during its use is by means of a smoke detector (EDR n.d.). However, for any contained and supported payloads that do not use air for cooling, data parameter monitoring must be used for fire detection.

When it is necessary to use data parameter monitoring, careful consideration of the parameters to be monitored; the type, number, and physical location of sensors to be used; their reliability, accuracy, and frequency of recalibration; and establishment of appropriate sensor set points for positive fire event detection is crucial for implementing a comprehensive, thorough, and accurate fire detection system. Even with all these considerations, ground based verification testing must be conducted prior to flight to ensure the data parameter monitoring system is operating correctly.

The concept of using data parameter monitoring for detecting fire events is not unique entirely to spacecraft. Forest fire management used Earth observation satellites in comprehensive studies of how to detect and track forest fires (Lynham et al. 2002; Vermillion et al. 1993). Although these are only two references of such studies, many related studies have been conducted on a global scale.

27.4 SPACECRAFT FIRE SUPPRESSION

27.4.1 Spacecraft Fire Suppression Methods

True fires in space have been very rare, as a result of the philosophy that fire prevention should be the first line of defense against unwanted fires. The success in preventing fires in spacecraft has been achieved by avoiding the use of flammable materials as much as possible. When the use of a flammable material is necessary, strict configuration control is followed to limit its use and to eliminate surrounding ignition sources. In spite of these preventative measures, fire suppression continues to be a critical issue, primarily because of the closed spacecraft environment, where escape is difficult for the crew and potentially just as dangerous as responding to and recovering from a fire. The spacecraft or space habitat, such as a structure on the Moon or Mars, presents a major additional challenge, in that the effects of reduced gravity on the flammability of materials and suppression processes are not understood fully.

Fire Suppression Systems on Previous Spacecraft

As discussed in the previous section, the selection of a spacecraft fire detector mirrored the development of advanced technologies for the detection of particulates. The selection of a fire suppressant on previous spacecraft relied to a lesser extent on the development of new technology and more on parameters such as mission duration, size of the vehicle, and number of crewmembers, as well as on compatibility with the environmental control and life support systems. Table 27.3 depicts a summary of the suppression method either used on or planned for previous spacecraft systems (Weiland 1994). For the Mercury and Gemini missions, fire detection and suppression were not well established in the environmental control and life support systems, and there were few, if any, formal requirements for its operation. As missions became longer and the consequences of a fire became more serious (prompted by the Apollo 204 fire), increased attention was given to detection and suppression systems onboard spacecraft.

One of the design trades inherent to a fire suppression system is whether the system should be fixed or portable. Prior to the *Space Shuttle*, spacecraft relied on portable extinguishers. The *Space Shuttle* used a combination of fixed and portable Halon 1301® systems

27.4 Spacecraft Fire Suppression

Table 27.3 Fire Suppression Methods and Requirements on Previous Space Vehicles

Space Habitat	Suppression Method	Requirements/Capabilities
Mercury	Water from food rehydration gun Depressurize habitat	No specific operational requirements
Gemini	Water from food rehydration gun Depressurize habitat	No specific operational requirements
Apollo	Portable aqueous gel extinguishers Water from food rehydration gun Depressurize habitat	Extinguishers capable of expelling 0.06 m^3 of foam in 30 s
Skylab	Portable aqueous gel extinguishers Depressurize habitat	Extinguishers capable of expelling 0.06 m^3 of foam in 30 s
Mir	Portable aqueous foam fire extinguisher	Unknown
Space Shuttle	Distributed rack or subfloor mounted Halon® bottles with distribution lines	Total flooding with 6% mean concentration of Halon in less than 10 s and a 10-s hold time
Spacelab	Portable Halon extinguishers Depressurize habitat	
Space Station Freedom	Centralized carbon dioxide distribution system Portable carbon dioxide extinguishers	Suppressant delivery rate of 1.0 lbm/s at normal pressure Oxygen dilution to 10.5% by volume in 33% of element's racks
International Space Station	Portable carbon dioxide extinguishers (U.S. modules) Aqueous foam fire extinguishers	Oxygen dilution to 10.5% by volume Unknown

with three fixed extinguisher cylinders protecting the instrument bays and three portable Halon extinguishers in the cabin (Friedman and Olson 1989). Although both fixed and portable carbon dioxide fire extinguishers were planned for the original U.S. prototype space station, *Space Station Freedom*, this design was not selected for the U.S. modules of the *International Space Station*. The results from system trade studies accounted for design and operational complexity, mass, weight, redundancy, and the like and led to the exclusive use of portable extinguishers on the *International Space Station*, which today uses only portable gaseous carbon dioxide fire extinguishers with access ports for their use provided in powered racks. The *Space Shuttle* was also unique in its use of Halon 1301. The choice of

Halon 1301 for suppression was based on its general effectiveness and relative inertness in small concentrations. Halon 1301 can be difficult to remove from the atmosphere by an onboard contaminant removal system, and the material produces toxic and corrosive hydrogen halides as a result of the fire extinguishing process. This is discussed in more detail later in this section. Also, for any of these systems, cabin depressurization was considered as an option for fire suppression. However, because of the penalties and increased risk associated with repressurization, this is considered to be a last resort response to a fire event.

Fire Classifications and Suppression Agents

Many methods to extinguish terrestrial fires have been developed, and the fire protection industry has various standards to select and evaluate suppressants for most types of fires. Terrestrial fires have been given alphabetical classifications based on the type of fuel being consumed, as shown in Table 27.4 (UL 2004; NFPA 2007). These classifications guide the

Table 27.4 Fire Classifications and Effective Extinguishing Agents for Portable Fire Extinguishers

Fire Classification	Fuel	Effective Extinguishing Agent
A	Wood, cloth, paper, rubber, many plastics	Water Foam (AFFF and FFFP) Halon 1211® Halocarbons Dry chemical (ammonium phosphate) Wet chemical
B	Flammable liquids, petroleum greases, tars, oils, oil based paints, solvents, lacquers, alcohols, and flammable gases	Foam (AFFF and FFFP) Carbon dioxide Halon 1301 Halon 1211 Halocarbons Dry chemical (ammonium phosphate) Wet chemical
C	Energized electrical equipment	Carbon dioxide Halon 1301 Halon 1211 Halocarbons Dry chemical (ammonium phosphate) Wet chemical
D	Combustible metals, including magnesium, titanium, zirconium, sodium, lithium, and potassium	Dry powder (metal fire agents)
K	Cooking fires (vegetable or animal oils and fats)	Wet chemical

selection of a fire suppressant, with effective suppression agents identified for each type of fire. Obviously, on a spacecraft, not all these fuels are of concern, and unlike terrestrial applications, where a unique extinguisher is required for each of the various types of burning material, the complexity of the response to a fire is increased. Therefore, design trades for spacecraft fire suppression systems generally attempt to identify the single best suppression agent for all anticipated types of fires.

The most common types of fires on spacecraft would be from energized electrical equipment (Class C) followed by the combustion of paper, cloth, and plastics (Class A) and most likely would be located in the open crew cabin or stowage. De-energizing an electrical circuit essentially causes a Class C fire to be redesignated Class A. The use of flammable liquids and gases (Class B) onboard a spacecraft is minimized with strict containment requirements. These hazards probably are best controlled by rapid detection and proper control system design, with the primary role of a fire suppression agent being to safeguard surrounding material to minimize damage. Class D fires certainly are possible because of the presence of pressurized oxygen and oxygen generation systems onboard spacecraft. Fires involving high pressure oxygen present a unique hazard because of the rapid and extensive damage they can cause to mechanical and electrical systems. In reality, it is unlikely that any fire suppressant would be useful for a high pressure oxygen fire in a closed cabin, and it actually can make fire recovery and cleanup more difficult. Therefore, this type of fire hazard is best addressed by the design and operation of the system. As with flammable gases, the primary goal of a fire suppression agent on a spacecraft is to reduce the damage to surrounding material.

The fire suppression agents most effective on Class A and C fires include water, foams, halocarbons (such as including Halon 1301 and Halon 1211), certain dry and wet chemical suppressants, and carbon dioxide. However, each has advantages and disadvantages for different types of fire and for use in a spacecraft system. These typically are evaluated in a trade study for a specific vehicle or mission design.

Two other fire suppression options are unique for a spacecraft. The first of these is to cease ventilation and power down any surrounding equipment, thereby reducing the flow of oxygen to the flame and the potential heat source. The second option is to depressurize the cabin to vacuum, thereby removing all the oxygen. The cabin would then be repressurized after it was verified that the fire had been extinguished and the components had cooled. Depressurization also would be an effective post-fire method to remove combustion products from the atmosphere. These are discussed later in this section.

27.4.2 Considerations for Spacecraft Fire Suppression

As previously stated, the selection of a fire suppression system on a spacecraft or habitat depends just as much on the mission requirements and environmental control and life support system compatibility as it does with the current state of the art in terrestrial suppression technology. These factors are considered, evaluated, and ranked in trade studies during the design phase of the vehicle and include

- Effectiveness on potential spacecraft fires.
- Reliability, that is, the fire suppression system must work after long periods of inactivity.

- Maintainability of the suppressant system, that is, the amount of maintenance required to ensure reliable operation.
- Weight.
- Required post-fire cleanup.
- Compatibility with spacecraft systems.
- Toxicity of the suppressant or post-suppression products.

The trade studies conducted for the *Space Station Freedom* and the *International Space Station* highlight these factors for the suppression methods identified in the previous section. For example, a study performed by Opfell (1985) evaluated carbon dioxide, nitrogen, water, and Halon as potential suppressants for use on *Space Station Freedom*. This study concluded that gaseous carbon dioxide should be the primary fire suppressant because of its effectiveness on a range of fires, compatibility with spacecraft systems, and ability to remove it from the cabin atmosphere by onboard systems. However, the authors also recommended that a water based fire extinguisher be included onboard because of the ineffectiveness of carbon dioxide on Class A, deep seated or smoldering fires. A similar study by Panzarella and Lewis (1990) used a similar criterion for carbon dioxide, nitrogen, helium, and water. They recommended gaseous nitrogen as the suppressant, although they acknowledged that the differences between nitrogen and carbon dioxide were small.

The experiment results of Sircar and Dees (1992) guided the selection of carbon dioxide as the suppressant on the *International Space Station*. These investigators conducted tests to determine the minimum concentrations of nitrogen, carbon dioxide, helium, and Halon 1301 required to extinguish self sustained combustion of various materials under various atmospheres. These tests were conducted in normal gravity and were based on suppressing a steady flame on a test sample using the configuration of the then NHB 8060.1C, Test 1 Upward Flammability Test. This test is now found in NASA-STD-6001, *Flammability, Odor, Offgassing, and Compatibility Requirements and Test Procedures for Materials in Environments That Support Combustion* (NASA 2008). The test system was designed so that the sample could be ignited in an 8.2-L chamber placed inside a 1400-L chamber. The test was prepared by installing the sample in the small ignition chamber and filling it with the desired oxygen-nitrogen ignition mixture. The larger chamber was filled with oxygen, nitrogen, and the extinguishing agent in the desired concentrations. The sample was ignited, and once combustion of the sample was self sustained, the walls of the ignition chamber were slid away from the sample, exposing it to the suppressant mixture. Their results indicated that Halon was by far the most effective fire suppressant, followed by carbon dioxide. Nitrogen and helium were equally less effective than Halon. Although they tested 24% oxygen concentration at 101.3 kPa (14.7 psia), they determined that an atmosphere with 30% oxygen concentration and 70.3 kPa (10.2 psia) was the most severe condition for extinguishing the flame. They also concluded that oxygen concentration was a much larger factor in determining suppressant effectiveness than ambient pressure.

Fire Suppression in the U.S. Modules of the International Space Station

Based on these studies and experience using carbon dioxide as a fire suppressant in ground based applications, as well as with several other operational constraints, a fire suppression system and associated procedures for the *International Space Station* were developed. The *U.S. Laboratory Module* contains two portable fire extinguishers, each containing 6 lb carbon dioxide at a charge pressure of 850 psi (Whitaker 2003). If a fire is detected either automatically or by the crew, the first action is to remove power to the affected racks to remove ignition sources. The ventilation system is then turned off to stop airflow within the module and any air exchange between modules. Atmospheric pressure control also is inhibited to prevent the introduction of oxygen and nitrogen into module. The crew can then use a portable fire extinguisher at their discretion.

Each portable fire extinguisher is supplied with two nozzles, a conical nozzle for use in open areas and a cylindrical nozzle for use in closed areas. In an open area, suppression of a fire is by dilution of the air with the carbon dioxide gas and the discharge velocity of the extinguishant gas, that is, 3 lb carbon dioxide discharged in the first 10 s of use. In a closed area, such as in a standard instrument rack on the *International Space Station*, the carbon dioxide concentration is increased to 50% within 60 s of discharge. In fact, the operational requirements for the portable fire extinguisher were determined by the ability to meet these specifications for an empty rack.

The trade studies for and the design of the fire extinguishers used on the *International Space Station* just discussed were based on the large amount of data on terrestrial fire suppression that led to the development of the standards used today by the fire protection industry. These data include information about when to use a fire suppression agent, the amount of agent required to suppress a fire, the best way to disperse an agent, and general fire response procedures.

Effect of Low Gravity on Fire Suppression

Low and partial gravity also play a role in the overall effectiveness of fire suppressants. Although there still is much to learn about fire suppression in low and reduced gravity, the following subsections summarize several key aspects of low gravity behavior of the suppression agents identified in previous sections.

Removal of Ventilation The removal of spacecraft ventilation flow is the first response to any confirmed spacecraft fire alarm. The basis for this action was discussed in Section 27.3. Without buoyant convection, forced convection and diffusion are the only modes of mass transfer available to carry oxidizer to the flame. As shown in Figure 27.8, the oxygen concentration required for a flame to exist increases as the flow velocity is decreased. The oxygen concentration at zero velocity is the concentration at which diffusion alone can sustain a flame. For many materials, this limit is higher than the oxygen concentration in the habitable volume (limited to 100%), indicating that most flames extinguish after a period of time. In contrast to this, Dietrich et al. (2000) conducted low-g experiments on *Mir* using a candle flame and demonstrated that a steady flame can exist in a quiescent atmosphere. Further, the analysis of the data taken by Ivanov et al. (1999) showed that the convective velocity required to sustain a flame on other polymeric materials can be on the order of millimeters per second (Hirsch et al. 2000), and this easily is attainable by the motion of the crew during a fire response, even with no cabin ventilation. Therefore, while the cessation of cabin ventilation is certainly the proper first response for a low-g fire, it is not an assured means of fire suppression.

Cabin Depressurization Cabin depressurization is a sure method to extinguish a fire and clear the cabin atmosphere of contaminants. However, it carries with it several physical and operational risks. Goldmeer (1996) studied the low-g suppression of a burning polymethylmethacrylate cylinder by depressurization and found that the pressure required for extinction depends greatly on the temperature of the solid. Figure 27.16 shows the ambient pressure required for extinction as a function of centerline temperature of the solid. These data were obtained for a 10 cm/s cross flow, the most difficult condition to extinguish in this study. We recommend a rapid depressurization to about 0.1 atm to suppress a fire, which would minimize additional heating of the fuel. Even though the studies of both Goldmeer (1996) and Kimzey (1974) show that rapid venting induces a convective flow that can intensify a flame, depressurization to 0.1 atm extinguishes the fire by allowing less time for the fire to heat the solid, spread, or damage surrounding equipment. Rapid venting, however, requires a large valve, thus increasing both the weight penalty and the risk associated with inadvertent valve opening and the ability to re-close it. The requirements of the *International Space Station* specify that a module must be able to vent from 1 to 0.3 atm within 10 min. With these specifications, fire suppression by venting is only a last resort, and the procedure would be more useful for post-fire cleanup after a serious fire had been extinguished.

Halon and Other Halocarbons The use of Halon 1301 has precedence on space vehicles as evidenced by its use on the *Space Shuttle*. However, since its selection for that application, its manufacture has been banned by the Montreal Protocol of 1987 because of its

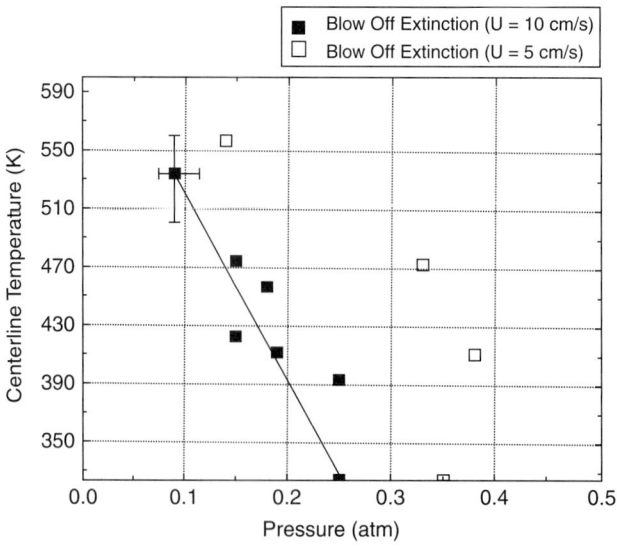

FIGURE 27.16 Dependence of extinction pressure as a function of temperature and cross flow velocity over a polymethylmethacrylate cylinder (Goldmeer 1996) (Graphic courtesy of NASA).

ozone depleting characteristics. Halon suppresses a fire by exothermic removal of hydrogen from the flame reaction zone. Because of the difference in structure and temperature between low and normal gravity flames, the behavior of Halon 1301 as a suppressant in low gravity can be expected to be different (VanDerWege et al. 1995). Limited tests of Halon 1301 effectiveness in low-g performed by Ronney (1985) on gaseous flames showed that the extinction concentrations in low-g were about the same as in normal gravity. In contrast, tests conducted by Haggard, as reported by Friedman and Dietrich (1991), showed that flame extinction on thin paper required about half the amount of agent in low gravity as required for normal-g extinction. These results were for 21% oxygen at 1 atm. Similar tests conducted at reduced pressure and higher oxygen concentrations gave nearly the same extinction concentration as seen in normal gravity.

In spite of its use, Halon also has undesirable characteristics for use in future space vehicles. For example, as the oxygen concentration increases, greater concentrations of Halon are required to suppress a fire (Kimzey 1967, 1968; Friedman and Dietrich 1991). Kimzey (1967) found that, at oxygen concentrations above 60%, Halon 1301, as well as carbon dioxide and most other suppression agents, relatively were ineffective. Water, however, was found to be most effective at these concentrations. Above 900°F, Halon 1301 can decompose into hydrogen fluoride and hydrogen bromide. These decomposition products are toxic even in low concentrations and would have to be monitored before declaring cabin air suitable for breathing by the crew. Also, even though the Halon fire extinguishers were not operated on early Space Shuttle flights, post-flight analysis detected trace quantities in the cabin atmosphere (Casserly and Russo 1990), indicating that it is likely that traces of Halon 1301 can make its way into the atmosphere during an extended mission. Although NASA made no plans to remove Halon 1301 from the Space Shuttle fleet, the agency actively was involved in a program to phase out Halon in ground and launch facilities through improved installation, leak prevention, and maintenance (Collins 1999).

Other halocarbons have been developed as Halon replacement fire suppressants that do not deplete the ozone, but they do still decompose into toxic products when exposed to high temperatures, as in a fire. Because Halon alternatives generally have a greater number of fluorine atoms per mole, they in fact can create more hydrogen fluoride per mole than Halon 1301 (Linteris and Gmurczyk 1995). Examples of these include HFC-23 (trifluoromethane, CHF_3), HFC-125 (pentafluoroethane, CHF_2CF_3), and FC-2-1-8 (perfluoropropane, $CF_3CF_2CF_3$). Although these have been demonstrated to be nontoxic to humans in the concentrations required for suppression, the long-term physiological effects remain to be determined.

Dry and Wet Chemicals Dry chemicals used for fire suppression include sodium bicarbonate, potassium bicarbonate, potassium chloride, and ammonia phosphate, to name a few. These agents are propelled by a compressed gas, such as nitrogen or carbon dioxide. Wet chemical agents are similar compounds in an aqueous solution and generally produce a foam that blankets the fire. Dry and wet chemical suppression agents have not been considered for use on spacecraft because they are rather dirty when used. Both types leave a chemical residue that must be cleaned thoroughly from all surfaces. Dispersion of a dry chemical would be problematic because, in low gravity, the powder would not settle on the fire source but instead would float in the cabin. Wet chemical agents could be directed onto a surface, but the residue would flake as the water evaporated. As with the introduction of any chemical into the spacecraft environment, the impact to other equipment, the operation of the environmental control and life support systems, and the long term effects to the crew must be considered.

Aqueous Foams As previously discussed, foam forming fire suppression agents have been used for fire suppression on *Apollo*, the Mir space station, and on the Russian modules of the *International Space Station*. This type of fire extinguisher has been used in space in response to the solid fuel oxygen generator fire aboard *Mir* in 1997 (Friedman and Ross 2001). In general, foam suppression agents act by forming a stable, homogeneous blanket over the combustible solid or liquid that cools the surface while excluding air, sealing liquids, and preventing flammable vapors from being emitted. Although it is unlikely that these extinction mechanisms are changed in low gravity, the dispersion mechanism and the ability of the foam to adhere to a surface is affected. Rygh (1995) conducted tests on the dispersion of a fire fighting foam with no fire in a reduced gravity aircraft. The researchers found that a slight increase in pressure was needed to produce foam similar to that produced in normal and low gravity. Also, they observed that, in low gravity, the foam did not curl over the sides of the target plate as it did in normal gravity, leaving the backside of the target uncoated. The low gravity behavior of any specific foam depends on the delivery technique and the foam properties, and it should be evaluated through low-g testing to verify the desired performance.

Water and Water Mist Water is a primary fire suppression agent, and as shown in Table 27.4, it also has been deployed on previous spacecraft. Water sprays extinguish a fire by directly wetting and cooling the combusting surface, cooling the air by vaporization (energy absorption), and dilution of the air with water vapor. Wetting is a characteristic of a water spray, whereas the cooling and dilution mechanisms are dominant for a water mist. As with foam fire suppression agents, these mechanisms are changed little by low gravity. What does change, however, is the dispersion of the water and its reaction on impact with a surface. In tests conducted on *Skylab*, Kimzey (1974) demonstrated the difficulty of controlling and directing water sprays in space. Rygh (1995) showed that, in low gravity, a water spray tended to rebound on impact with a surface, forming water droplets that were deflected away from the fire. The water mist fire suppression experiment, Mist, flew on STS-107 during January 2003 (Abbud-Madrid et al. 2004) with the objective being to study the effect of droplet size distribution and concentration on the burning velocity of a flame through a propane-air-mist mixture inside a cylindrical tube. These tests demonstrated that, in low-g, distributions of small droplets were more effective than larger size distributions in suppressing the propagation of lean flames. Although these tests were performed on a gaseous flame, normal gravity tests have shown that fine mist water suppression systems that use droplets on the order of 10 μm diameter do not wet the surface but instead act through cooling, which is made more efficient by the small droplet diameter, and oxygen dilution. However, the effectiveness of water mist on a cooler low gravity flame over a solid remains to be studied in detail.

Using water and water mist as a fire suppression agent on spacecraft offers distinct advantages, which include its effectiveness on many types of fires, zero toxicity, the ability to be removed from the atmosphere through the humidity control system, and the ability to replenish an extinguisher. The disadvantages of its use include electrical conductivity (the exception being ultrapure deionized water), reactivity with certain metals, and opacity of a dense water mist. In spite of these, evaluation and study for use of these agents on future spacecraft is warranted.

Inert Gas Agents The action of carbon dioxide and other inert gases is primarily by dilution of the oxygen concentration and the thermal effects, that is, absorption and diffusion of thermal energy from the flame zone and the burning surface. Although previous research indicated that carbon dioxide can have some chemical activity in suppression (Tapscott and Morehouse 1987), this contribution is small; therefore, carbon dioxide is considered inert for this discussion.

The behavior of inert gas fire suppression agents in low and partial gravity has been studied by several investigators including Ronney (1985); Hamins et al. (2001); Ruff et al. (2004); Katta, Takahashi, and Linteris (2004); Takahashi, Linteris, and Katta (2004, 2006a, 2006b, 2006c). These studies were conducted using various small scale geometries because of the limitations imposed by ground based low gravity facilities, that is, drop towers and reduced gravity aircraft. Additionally, these studies used different configurations and fuels, so application of the results requires a fairly detailed analysis of the methods and conclusions. However, because of the value of this limited amount of data, these works are summarized in the following sections.

Son and co-workers (Son and Ronney 2002; Son et al. 2006) conducted flame spread experiments over thermally thick polyurethane foams in the 2.2- and 5-s drop towers at NASA Glen Research Center. In these tests, they used the flame spread rate as a measure of the strength of the flame, with the extinction concentration defined as that which yielded a zero spread rate. The tests were conducted with no forced convection (normal gravity conditions included natural convection), ambient pressures from 0.5 to 2 atm, and oxygen concentrations from 21 to 50%. Tests were conducted with various concentrations of carbon dioxide, helium, and nitrogen to evaluate the impact on the flame spread rate. At ambient pressures greater than 1 atm, the spread rates showed little difference among the three suppressants in either normal or low gravity. At 0.75 atm and below, the results show that the low gravity flames are suppressed using less suppression agents, that is, lower mole fractions, than in normal gravity.

Interesting to note, the impact of low gravity at these pressures was to change the relative effectiveness between the agents. In normal gravity, the three agents required roughly the same concentrations to cease the flame spread. In low gravity, helium consistently required the lowest mole fraction, with carbon dioxide and nitrogen requiring higher, and somewhat similar, concentrations. Based on heat capacity arguments alone, carbon dioxide should have been consistently the most efficient suppressant. We attribute this behavior to the fact that energy transport is higher in oxygen-helium mixtures than in oxygen–carbon dioxide or oxygen-nitrogen mixtures, making the flame temperature lower and reducing the spread rate. Helium also has higher thermal conductivity and thermal diffusivity, which leads to a thicker flame and increased losses due to radiation. Last, unlike carbon dioxide, gaseous helium and nitrogen are radiatively nonparticipating; therefore, no reabsorption or reemission of the radiant energy accrues from the gas to the surface. This increases the heat loss from the surface for nitrogen and helium, so the surface cools more rapidly.

Ruff et al. (2004) studied the suppression of diffusion flames around a 19-mm diameter polymethylmethacrylate cylinder in a cross flow in low gravity on the NASA reduced gravity aircraft. This flow configuration is similar to that used by Goldmeer (1996) in his study of extinction by depressurization. Oxygen concentrations of 21 and 25% (balance nitrogen) were used in these experiments. The flame was stabilized on the cylinder and allowed to burn until the polymethylmethacrylate core temperature reached a predetermined threshold. On entering the low gravity period, the oxygen-nitrogen flow was switched to introduce an oxygen-nitrogen-suppressant stream (carbon dioxide, helium, or nitrogen) in which the mole fraction of suppressant was fixed. The fate of the flame in the oxidizer-suppressant cross flow was determined by observation. If the suppressant did not extinguish the flame before the end of the reduced gravity, the flame was extinguished by depressurization, the sample replaced with a fresh sample, and the test repeated. In these tests, both the concentration of the suppression agent and the cross flow velocity were varied.

The compilation of this data produced extinction maps showing the suppression agent-to-oxygen mole fraction ratio as a function of flow velocity. Similar to the studies of Goldmeer (1996) and Ivanov et al. (1999), this study found that the ability of a flame to be maintained on a solid depended strongly on the solid's temperature. This procedure allowed the solid's temperature to be fairly constant between tests, so that the effect of the suppression agent could be isolated. Helium and carbon dioxide were found to have similar extinction concentrations over the range of flow velocities, with nitrogen requiring slightly higher concentrations. We correlated the extinction boundary for these tests in terms of an effective suppressant mixture enthalpy, $\Delta H'$, introduced by Sheinson, Penner-Hahn, and Indritz (1989) and given by Equation (27.1):

$$\Delta H' = \sum_i \frac{X_i}{X_{O_2}} \int_{298}^{1600} Cp_i dT \qquad (27.1)$$

where X_i is the mole fraction and specific heat of the gases in the suppressant mixture (excluding oxygen), X_{O_2} is the mole fraction of oxygen, and T is the temperature. Figure 27.17 shows the extinction boundary for the three suppressants in terms of $\Delta H'$ as a function of flow velocity. For all suppressants, the amount required increased as the flow velocity increased, indicating that, in low gravity, the higher velocity flames were more difficult to extinguish. Interesting to note, the effective suppressant mixture enthalpies required to suppress these flames in normal gravity with only a buoyancy induced convective flow is approximately the same as that required in low gravity at 10 cm/s. Therefore, these results indicate that, when a forced convective flow, typical of cabin air velocities, is present to carry suppressant to the flame, normal gravity suppression limits for flames on thick solids appear to be conservative, as based on the corresponding limits in microgravity.

27.4 Spacecraft Fire Suppression

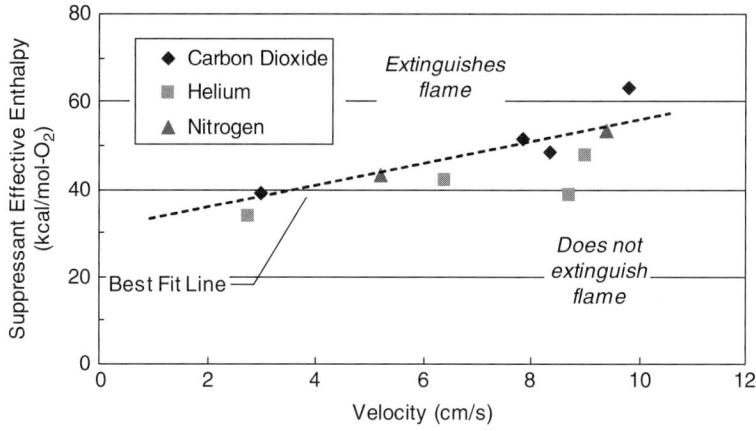

FIGURE 27.17 Suppressant mixture enthalpy required to extinguish a flame on a polymethylmethacrylate cylinder in cross flow in low gravity (Ruff et al. 2004) (Courtesy of NASA).

Takahashi and coworkers used cup burner geometry to determine the minimum extinction concentration for gaseous suppression agents in normal and low gravity. The cup burner, a widely used normal gravity standard in which to evaluate gaseous suppression agents (NFPA 2004, 2005; ISO 2006), consists of a circular fuel tube inside a coaxial chimney. The fuel, that is, gas jet, liquid pool, or solid, is introduced at the top of the fuel tube while the oxidizer flows coaxially around it through the chimney. Once ignited, the minimum concentration required to extinguish the co-flow diffusion flame is determined by gradually increasing the concentration of the suppression agent in the oxidizer stream. The low gravity tests were conducted on the NASA KC-135 reduced gravity aircraft, so this procedure was performed during the 20 s of low gravity on each parabola.

Figure 27.18 shows the minimum extinction concentrations for methane in normal gravity obtained by Takahashi et al. (2006c) for argon, nitrogen, helium, carbon dioxide, and trifluoromethane suppression agents. These data show that the chemically acting agent, trifluoromethane, was the most effective, followed by the inert agents carbon dioxide, nitrogen, helium, and argon, in that order, with carbon dioxide being the most effective and argon being the least. The concentrations of chemically acting agents are generally lower than inert agents that suppress a flame only by means of dilution and thermal effects. This figure also shows that the extinction concentrations from the cup burner facility relatively are independent of co-flow velocity. This is a simplifying feature of the cup burner and is not representative necessarily of every flow configuration.

Figure 27.19 shows experimental and calculated minimum extinction concentrations for carbon dioxide, nitrogen, and helium in low gravity (Takahashi et al. 2006b). Although carbon dioxide remained the most effective inert agent, helium only slightly was more effective than nitrogen for the cup burner in low gravity. In comparison, Ruff et al. (2004) found that the minimum effective concentration of carbon dioxide in low-g was 0.22 (at 8 cm/s) as compared to 0.237 from Takahashi et al. (2006b). Nitrogen also was

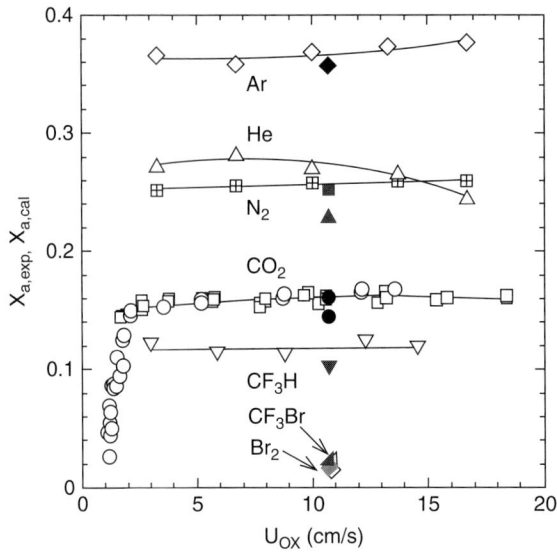

FIGURE 27.18 Measured and calculated inert suppression agent volume fractions at extinguishment in a methane flame in an air–suppression agent mixture in normal gravity (Takahashi et al. 2006c) (Reprinted by permission of The Combustion Institute).

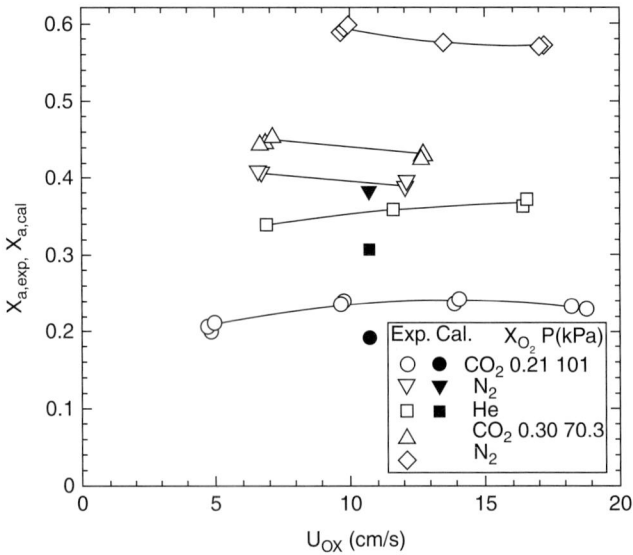

FIGURE 27.19 Measured and calculated inert suppression agent volume fractions at extinguishment for a methane flame in air–suppression agent mixture in low gravity (Takahashi et al. 2006b) (Reprinted by permission of The Combustion Institute).

higher, 0.22 (at 5.5 cm/s) (Ruff et al. 2004) as compared to 0.393 (Takahashi et al. 2006b). This difference is reasonable given that Ruff et al. (2004) studied flames formed over a solid phase, in which energy feedback from the flame to the surface is required to vaporize the solid and sustain the combustion process. Even though the extinction concentrations for the inert agents are different because of the fuel and flow configuration, the results of Takahashi et al. (2006c) also were well correlated by the specific heat of the oxidizer mixture, that is, oxygen, nitrogen, and suppression agent. Detailed numerical simulations of this configuration, made possible by the use of the cup burner and gaseous fuel, showed that extinction occurred at roughly the same flame temperature for the different agents (Katta et al. 2004; Takahashi et al. 2006a, 2006b, 2006c). For methane-air flames, the addition of various inert gas agents, such as carbon dioxide, nitrogen, helium, and argon, decreased the maximum flame temperature to a nearly constant value of 1590 K near extinguishment in low-g; the maximum temperature was approximately 1700 K in normal gravity. Similar results were found for ethane and propane, although the temperature at extinction was different and fuel dependent. This indicates that the dilution and thermal effects of these agents reduced the flame temperature, leading to destabilization of the reaction zone. The lack of a buoyant flow in low gravity delayed the onset of flame instabilities that allowed the flame to sustain to lower temperatures than in normal gravity.

This discussion demonstrates that, for any of the fire suppression agents that would be appropriate for the types of fires on spacecraft, notable differences are found between low and normal gravity behavior that could be a factor in the design of spacecraft suppression system. Whereas it certainly is preferable to obtain adequate data before proceeding with the design of any system, the difficulty in obtaining relevant low gravity suppression data could require another approach. One design approach could be to couple the breadth of the normal gravity data with the low gravity details of a few suppression agents, such as those previously discussed, to assist in the evaluation of suppressants and suppression concepts and determine the verification requirements. As exploration missions progress with increasing duration and complexity, additional data must be obtained at relevant scales and gravity levels to verify operation and performance of future spacecraft fire suppression systems.

REFERENCES

Abbud-Madrid, A., F. K. Amon, and J. T. McKinnon. (2004) The Mist experiment on STS-107: Fighting fire in microgravity. AIAA-2004-0288. *Proceedings of the 42nd AIAA Aerospace Sciences Meeting and Exhibit*, Reno, NV.

Abbud-Madrid, A., M. C. Branch, and J. W. Daily. (1996) Ignition and combustion of bulk titanium and magnesium at normal and reduced gravity. *Proceedings of the Combustion Institute* 26:1929-1936.

Abbud-Madrid, A., A. Modak, M. C. Branch, and J. W. Daily. (2001) Combustion of magnesium with carbon dioxide and carbon monoxide. *Journal of Propulsion and Power* 17, no. 4: 852-859.

Altenkirch, R. A., L. Teng, K. R. Sacksteder, S. Bhattacharjee, and M. A. Delichatsios. (1998) Quiescent flame spread over thick fuels in microgravity. *Proceedings of the Combustion Institute* 27: 2515-2524.

Ballal, D. R. (1983) Flame propagation through dust clouds of carbon, coal, aluminum, and magnesium in an environment of zero-gravity. *Proceedings of the Royal Society of London*, Series A, 385: 21–51.

Barr, L. G. (1977) Development of a quartz crystal incipient fire detector. In *Fire Detection for Life Safety*. Washington, DC: National Academy of Sciences.

Branch, M. C., J. W. Daily, and A. Abbud-Madrid. (1996) *Ignition and Combustion of Bulk Metals in a Microgravity Environment*. Contractor Report CR-202241. Cleveland, OH: National Aeronautics and Space Administration, Glenn Research Center.

Bukowski, R. W. (1987) Techniques for fire detection in spacecraft fire safety. In J. M. Margle (ed.), *Spacecraft Fire Safety*. Conference Publication CP-2476. Cleveland, OH: National Aeronautics and Space Administration, Glenn Research Center.

Bukowski, R. W., and G. W. Mulholland. (1978) *Smoke Detector Design and Smoke Properties*. Technical Note 973. Washington, DC: U.S. Department of Commerce, National Bureau of Standards, National Building Code.

Bukowski, R. W., R. D. Peacock, J. D. Averill, T. G. Cleary, N. P. Bryner, W. D. Walton, P. A. Reneke, and E. D. Kuligowski. (2003) *Performance of Home Smoke Alarms: Analysis of the Response of Several Available Technologies in Residential Fire Settings*. Technical Note 1455. Gaithersburg, MD: Department of Commerce, National Institute of Standard and Technologies.

Casserly, D. M., and D. M. Russo. (1990) *Identifying Atmospheric Monitoring Needs for Space Station Freedom*. Technical Paper 901383. Williamsburg, VA: International Conference on Environmental Systems, Society of Automotive Engineers.

Collins, M. M. (1999) NASA issues related to the use of Halon: Past, present, and future. In A. F. Whitaker, D. R. Cross, S. V. Caruso, and M. Clark-Ingram (eds.), *Proceedings of the Third Aerospace Environmental Technology Conference*. CP-1999-209258. Huntsville, AL: National Aeronautics and Space Administration, Marshall Space Flight Center, pp. 141–143.

Cordova, J. L., and A. C. Fernandez-Pello. (2000) Convection effects on the endothermic gasification and piloted ignition of a radiatively heated combustible solid. *Combustion Science and Technology* 156: 271–289.

Dietrich, D. L., H. D. Ross, Y. Shu, P. Chang, and J. S. T'ien. (2000) Candle flames in non-buoyant atmospheres. *Combustion Science and Technology* 156: 1–24.

Department of Defense. 2000. *Department of Defense Test Method Standard: Environmental Engineering Considerations and Laboratory Tests*. MIL-STD-810F. Washington, DC: Department of Defense.

Department of Defense. 2002. *Department of Defense Test Method Standard: Electronic and Electrical Component Parts*. MIL-STD-202G. Washington, DC: Department of Defense.

EDR. (no date) *European Drawer Rack: Multi-Discipline Flexible Experiment Carrier in Columbus*. Available at www.spaceflight.esa.int/users/downloads/factsheets/fs016_11_edr.pdf (cited: August 5, 2007).

Fernandez-Pello, A. C., D. C. Walther, J. L. Cordova, T. Steinhaus, J. L. Quintiere, and H. Ross. (2000) Test method for ranking materials flammability in reduced gravity. *Space Forum* 6: 237–243.

Friedman, R. (1992) Fire safety practices and needs in human crew spacecraft. *Journal of Applied Fire Safety* 2, no. 3: 243–259.

Friedman, R., and D. L. Dietrich. (1991) *Fire Suppression in Human-Crewed Spacecraft*. Technical Memorandum TM-104334. Cleveland, OH: National Aeronautics and Space Administration, Lewis Research Center.

Friedman, R., and S. L. Olson. (1989) *Fire Safety Applications for Spacecraft*. Technical Memorandum TM-101463. Cleveland, OH: National Aeronautics and Space Administration, Lewis Research Center.

Friedman, R., and H. D. Ross. (2001) Combustion technology and fire safety for human-crew space missions. In H. D. Ross (ed.) *Microgravity Combustion: Fire in Free Fall*. San Diego, CA: Academic Press, pp. 525-562.

Friedman, R., and G. A. Ruff. (2008) *Spacecraft Fire Safety 1956-1999: An Annotated Bibliography*. Technical Publication TP-2008- 211209. Cleveland, OH: National Aeronautics and Space Administration, Glenn Research Center.

Friedman, R., and D. L. Urban. (2000) *Progress in Fire Detection and Suppression Technology for Future Space Missions*. Technical Memorandum TM-2000-210377. Cleveland, OH: National Aeronautics and Space Administration, Glenn Research Center.

Friedman, R., A. Suleyman, S. G. Gokoglu, and D. L. Urban (eds.). (1999) *Microgravity Combustion Research: 1999 Program and Results*. Technical Memorandum TM-1999-209198. Cleveland, OH: National Aeronautics and Space Administration, Glenn Research Center.

Goldmeer, J. S. (1996) *Extinguishment of a Diffusion Flame Over a PMMA Cylinder by Depressurization in Reduced Gravity*. Contractor Report CR-198550. Cleveland, OH: National Aeronautics and Space Administration, Glenn Research Center.

Grayson, G., K. R. Sacksteder, P. V. Ferkul, and J. S. T'ien. (1994) Flame spreading over a thin solid in low-speed concurrent flow—drop tower experimental results and comparison with theory. *Microgravity Science and Technology* 2, no. 2: 187-195.

Greenberg, P. S., and J. C. Ku. (1997) Soot volume fraction imaging. *Applied Optics* 36, no. 22: 5514-5522.

Greenberg, P. S., K. R. Sacksteder, and T. Kashiwagi. (1995) The USML-1 wire insulation flammability glovebox experiment. Contractor Publication CP-10174. *Proceedings of the Third International Microgravity Combustion Workshop*. Cleveland, OH: National Aeronautics and Space Administration, Lewis Research Center.

Hamins, A., M. Bundy, I. K. Puri, K. McGrattan, and W. C. Park. (2001) Suppression of low strain rate nonpremixed flames by an agent. Contractor Publication CP-2001-210826. *Proceedings of the Sixth International Microgravity Combustion Workshop*. Cleveland, OH: National Aeronautics and Space Administration, Glenn Research Center, pp. 101-104.

Hanai, H., M. Ueki, K. Maruta, H. Kobayashi, S. Hasegawa, and T. Niioka. (1999) A lean flammability limit of polymethylmethacrylate particle-cloud in microgravity. *Combustion and Flame* 118: 359-369.

Hirsch, D. B., H. D. Beeson, and R. Friedman. (2000) *Microgravity Effects on Combustion of Polymers*. Technical Memorandum TM-2000-209900. Houston, TX: National Aeronautics and Space Administration, Johnson Space Center.

International Organization for Standardization. (2003) *Space Systems Safety and Compatibility of Materials, Part 1, Determination of Upward Flammability of Materials*. ISO 14624-1. Geneva, Switzerland: International Organization for Standardization.

International Organization for Standardization. (2006) *Gaseous-Fire-Extinguishing Systems—Physical Properties and System Design*. ISO 14520 Part I. Geneva, Switzerland: International Organization for Standardization.

Ivanov, A. V., Y. V. Balashov, T. V. Andreeva, and A. S. Melikhou. (1999) *Experimental Verification of Material Flammability in Space*. Contractor Report CR-1999-209405. Cleveland, OH: National Aeronautics and Space Administration, Glenn Research Center.

Katta, V. R., F. Takahashi, and G. T. Linteris. (2004) Suppression of cup-burner flames using carbon dioxide in microgravity. *Combustion and Flame* 137: 506-522.

Kimzey, J. H. (1967) *Freon 1301 as a Fire Fighting Medium in an Oxygen-Rich Atmosphere*. Manned Spacecraft Center Internal Note MSC-ES-R-67-11. Houston, TX: National Aeronautics and Space Administration, Manned Spaceflight Center.

Kimzey, J. H. (1968) Fighting Fires in an Oxygen-Rich Atmosphere. *Hospitals* 432: 91.

Kimzey, J. H. (1974) Skylab experiment M479: Zero gravity flammability. M-74-5. *Proceedings of the Third Space Processing Symposium.* Huntsville, AL: National Aeronautics and Space Administration, Marshall Space Flight Center, 1, pp. 115-130.

Ku, J. C., W. Devon, P. S. Greenberg, and J. Roma. (1995) Buoyancy-induced differences in soot morphology. *Combustion and Flame* 102: 216-218.

Kumagi, S., and H. Isoda. (1957) Combustion of fuel droplets in a falling chamber. *Proceedings of the Combustion Institute* 6: 726-731.

Kumar, A., H. Y. Shih, and J. S. T'ien. (2003) A comparison of extinction limits and spreading rates in opposed and concurrent spreading flames over thin solids. *Combustion and Flame* 132: 667-677.

Kwon, O. C., M. Abid, J. Porres, J. B. Liu, P. D. Ronney, P. M. Struk, and K. J. Weiland. (2003) Studies of premixed laminar and turbulent flames at microgravity. CP-2003-212376. *Proceedings of the Seventh International Workshop on Microgravity Combustion and Chemically-Reacting Systems.* Cleveland, OH: National Aeronautics and Space Administration, Glenn Research Center.

Law, C. K., and G. M. Faeth. (1994) Opportunities and challenges of combustion in microgravity. *Progress in Energy and Combustion Science* 20: 65-113.

Linteris, G., and G. Gmurczyk. (1995) Parametric study of hydrogen fluoride formation in suppressed fires. Halon Options Technical Working Conference, Albuquerque, NM.

Liu, B. Y. H., K. L. Rubow, P. H. McMurry, T. J. Kotz, and D. Russo. (1991) Airborne particulate matter and spacecraft internal environments. Technical Paper 911476. *Proceedings of the 21st International Conference on Environmental Systems.* San Francisco: Society of Automotive Engineers.

Lynham, T. J., C. W. Dull, and A. Singh. (2002) Requirements for space-based observations in fire management: A report by the wildland fire hazard team. *Proceedings of the 2002 International Geoscience and Remote Sensing Symposium.* New York: Institute of Electrical and Electronic Engineers, Committee on Earth Observation Satellites Disaster Management Group.

Martin, C. E., and R. C. DaLee. (1993). Spacecraft fire detection and suppression (FDS) systems: An overview and recommendations for future flights. Paper 932166. International Conference on Environmental Systems. Colorado Springs, CO: Society of Automotive Engineers.

McKinnie, J. M. (1997) Fire response aboard the International Space Station. Paper 972334. International Conference on Environmental Systems. Lake Tahoe, NV: Society of Automotive Engineers.

Miller, F. J., H. D. Ross, I. Kim, and W. A. Sirignano. (2000) Parametric investigations of pulsating flame spread across 1-butanol pools. *Proceedings of the Combustion Institute* 28: 2827-2834.

Mulholland, G. W., and B. Y. H. Liu. (1980) Response of smoke detectors to monodisperse aerosols. *Journal of Research of the National Bureau of Standards* 85: 223-238.

National Aeronautics and Space Administration. 1995. *Safety Policy and Requirements for Payloads Using the International Space Station.* International Space Station Document NSTS 1700.7B, ISS Addendum. Houston, TX: National Aeronautics and Space Administration, Johnson Space Center.

National Aeronautics and Space Administration. 2008. *Flammability, Odor, Offgassing and Compatibility Requirements and Test Procedures for Materials in Environments That Support Combustion.* Technical Standard NASA-STD-(I)-6001A. Washington, DC: National Aeronautics and Space Administration, Headquarters.

National Fire Protection Association. (2004) *Standard on Clean Agent Fire Extinguishing Systems.* Technical Report NFPA 2001. Quincy, MA: National Fire Protection Association.

National Fire Protection Association. (2005) *Standard on Carbon Dioxide Extinguishing Systems.* Technical Report NFPA 12. Quincy, MA: National Fire Protection Association.

National Fire Protection Association. (2007) *Standard for Portable Fire Extinguishers*. Technical Report NFPA 10. Quincy, MA: National Fire Protection Association.

Ohlemiller, T. J., and K. M. Villa. (1991) *Material Flammability Test Assessment for Space Station Freedom*. NISTIR 4591. Contractor Report CR-187115. Cleveland, OH: National Aeronautics and Space Administration, Lewis Research Center.

Olson, S. L. (1991) Mechanism of microgravity flame spread over a thin solid fuel: Oxygen and opposed flow effects. *Combustion Science and Technology* 76: 233-249.

Olson, S. L., P. V. Ferkul, and J. S. T'ien. (1988) Near-limit flame spread over a thin solid fuel in microgravity. *Proceedings of the Combustion Institute* 22: 1213-1222.

Olson, S. L., and R. G. Sotos. (1987) *Combustion of Velcro in Low Gravity*. Technical Memorandum TM-88970. Cleveland, OH: National Aeronautics and Space Administration, Lewis Research Center.

Olson, S. L., T. Kashiwagi, O. Fujita, M. Kikuchi, and K. Ito. (2001) Experimental observations of spot radiative ignition and subsequent three-dimensional flame spread over thin cellulose fuels. *Combustion and Flame* 125: 852-864.

Opfell, J. (1985) *Fire Detection and Fire Suppression Trade Study*. Report 85-22472, Data Item DR-19, Trade Study 73, Boeing Purchase Contract GS2804. Los Angeles: Allied-Signal Aerospace Company.

Panzarella, L. N., Jr., and P. Lewis. (1990) *Crew Lock/Hyperbaric Chamber FDS Fire Suppressant Selection Trade Study*. Memorandum A96-J753-STN-M-LP-900070. Houston, TX: McDonnell Douglas.

Paul, M., F. Issacci, G. E. Apostalakis, and I. Catton. (1993) Smoke and flammability of wires in microgravity. In S. S. Sadhal and A. Hashemi (eds.), *Heat Transfer in Microgravity Systems—1993*. American Society of Mechanical Engineers, Heat Transfer Division, vol. 235, pp. 59-66.

Pu, Y., J. Podfilipski, and J. Javosinski. (1998) Constant volume combustion of aluminum and cornstarch dust in microgravity. *Combustion Science and Technology* 135: 253-267.

Ronney, P. D. (1985) Effect of gravity on halocarbon flame retardant effectiveness. *Acta Astronautica* 12: 915-921.

Ronney, P. D. (1990) Near-limit flame structures at low Lewis number. *Combustion and Flame* 82: 1-14.

Ronney, P. D. (2001) Premixed gas flames. In H. D. Ross (ed.), *Microgravity Combustion: Fire in Free Fall*. San Diego, CA: Academic Press.

Ronney, P. D., and H. Y. Wachman. (1985) Effect of gravity on laminar premixed gas combustion. I, Flammability Limits and Burning Velocities. *Combustion and Flame* 62: 107-119.

Ronney, P. D., K. N. Whaling, A. Abbud-Madrid, J. L. Gatto, and V. L. Pisowicz. (1994) Stationary premixed flames in spherical and cylindrical geometries. *AIAA Journal* 32: 569-577.

Ronney, P. D., M. S. Wu, K. J. Weiland, and H. G. Pearlman. (1998) Structure of flame balls at low Lewis-number (SOFBALL): Preliminary results from the STS-83 space flight experiments. *AIAA Journal* 36: 1361-1368.

Roslon, M., S. Olenick, Y. Y. Zhou, D. C. Walther, J. L. Torero, A. C. Fernandez-Pello, and H. D. Ross. (2001) Microgravity ignition delay of solid fuels in low velocity flows. *AIAA Journal* 39, no. 12: 2336-2342.

Ross, H. D. (ed.). (2001) *Microgravity Combustion: Fire in Free Fall*. San Diego, CA: Academic Press.

Ross, H. D., and F. J. Miller. (1996) Detailed experiments of flame spread across deep butanol pools. *Proceedings of the Combustion Institute* 26: 1327-1334.

Ruff, G. A., M. Hicks, and R. Pettegrew. (2004) Assessment of CO_2, N_2, and He as suppressant agents in microgravity. Spring Meeting of the Western States Section. Davis, CA: The Combustion Institute.

Rygh, K. (1995) Fire safety research in microgravity: How to detect smoke and flames you cannot see. *Fire Technology* 31, no. 2: 175-185.

Sheinson, R. S., J. E. Penner-Hahn, and D. Indritz. (1989) The physical and chemical action of fire suppression. *Fire Safety Journal* 25: 437–450.

Sircar, S., and J. Dees. (1992) *Evaluation of Fire Extinguishants for Space Station Freedom*. Technical Report TR-650-001. Las Cruces, NM: National Aeronautics and Space Administration, White Sands Test Facility.

Son, Y., and P. D. Ronney. (2002) Radiation-driven flame spread over thermally thick fuels in quiescent microgravity environments. *Proceedings of the Combustion Institute* 29: 2587–2594.

Son, Y., G. Zeuein, S. A. Gokoglu, and P. D. Ronney. (2006) Comparison of carbon dioxide and helium as fire extinguishing agents for spacecraft. *Journal of American Society for Testing and Materials International* 3, no. 3.

Srivastava R., J. T. McKinnon, and P. Todd. (1998) Effect of pigmentation in particulate formation from fluoropolymer thermodegradation in microgravity. AIAA 98-0814. *Proceedings of the 36th Aerospace Sciences Meeting*. Reno, NV: American Institute of Aeronautics and Astronautics.

Steinberg, T. A., D. B. Wilson, and F. Benz. (1992) The burning of metals and alloys in microgravity. *Combustion and Flame* 88: 309.

Steisslinger, H. R., D. M. Hoy, J. A. McLin, and E. C. Thomas. (1993) Comparison testing of the Space Shuttle Orbiter and Space Station Freedom smoke detectors. Paper no. 932291. International Conference on Environmental Systems, Society of Automotive Engineers, Colorado Springs, CO.

Strehlow, R. A., K. A. Noe, and B. L. Wherley. (1986) The effect of gravity on premixed flame propagation and extinction in a vertical standard flammability tube. *Proceedings of the Combustion Institute*: pp. 899–908.

Takahashi, F., G. T. Linteris, and V. R. Katta. (2004) Suppression characteristics of cup-burner flames in low gravity. AIAA-2004-0957. *Proceedings of the 42nd AIAA Aerospace Sciences Meeting and Exhibit*. Reno, NV: American Institute of Aeronautics and Astronautics.

Takahashi, F., G. T. Linteris, and V. R. Katta. (2006a) Extinguishment mechanisms of cup-burner flames. AIAA-2006-0745. *Proceedings of the 44th AIAA Aerospace Sciences Meeting and Exhibit*. Reno, NV: American Institute of Aeronautics and Astronautics.

Takahashi, F., G. T. Linteris, and V. R. Katta. (2006b) Combustion and extinguishment characteristics of cup-burner flames. Technical Meeting of the Central States Section of the Combustion Institute, Cleveland, OH.

Takahashi, F., G. T. Linteris, and V. R. Katta. (2006c) Extinguishment mechanisms of co-flow diffusion flames in a cup-burner apparatus. *Proceedings of the Combustion Institute* 31: 2721–2729.

Tapscott, R. E., and E. T. Morehouse, Jr. (1987) *Next-Generation Fire Extinguishing Agent: Phase I-Suppression Concepts*. Air Force AF Report ESL-TR-87-03 (AD-A192279). Tyndall Air force Base, FL: U.S. Air Force, Engineering Services Division.

T'ien, J. S., H-Y Shih, C-B Jiang, H. Ross, F. L. Miller, A. C. Fernandez-Pello, J. L. Torero, and D. Walther. (2001) Mechanisms of flame spread and smolder wave propagation. In H. D. Ross (ed.), *Microgravity Combustion: Fire in Free Fall*. San Diego, CA: Academic Press.

T'ien, J., K. Sacksteder, P. Ferkul, R. Pettegrew, K. Street, A. Kumar, K. Tolejko, J. Kleinhenz, and N. Piltch. (2003) Solid inflammability boundary at low-speed (SIBAL). *Conference Proceedings CP-2003-212376 Rev 1*. Cleveland, OH: National Aeronautics and Space Administration, Glenn Research Center.

Underwriters Laboratories. (2004) *Rating and Fire Testing of Fire Extinguishers*. UL 711. Northbrook, IL: Underwriters Laboratories.

Underwriters Laboratories. (2006) *Standard for Smoke Detectors for Fire Alarm Signaling Systems*. UL 268. Northbrook, IL: Underwriters Laboratories.

Urban, D. L., D. W. Griffin, and M. Y. Gard. (1998) Comparative soot diagnostics: 1 year report. In *Third United States Microgravity Payload: One Year Report. Conference Proceedings CP-1998-207891*. Huntsville, AL: National Aeronautics and Space Administration, Marshall Space Flight Center.

Urban, D. L., Z-G Yuan, P. B. Sunderland, G. T. Linteris, J. E. Voss, K-C Lin, Z. Dai, K. Sun, and G. M. Faeth. (1998) Structure and soot properties of nonbuoyant ethylene/air laminar jet diffusion flames. *AIAA Journal* 36: 1346-1360.

Urban, D. L., Z-G Yuan, P. B. Sunderland, K-C Lin, Z. Dai, and G. M. Faeth. (2000) Smoke-point properties of nonbuoyant round laminar jet diffusion flames. *Proceedings of the Combustion Institute* 28: 1965-1972.

Urban, D., D. Griffin, G. Ruff, T. Cleary, J. Yang, G. Mulholland, and Z-G. Yuan. (2005) Detection of smoke from microgravity fires. Paper no. 2005-01-2930. 2005 International Conference on Environmental Systems. Rome: Society of Automotive Engineers.

VanDerWege, B. A., M. T. Bush, S. Hochgreb, and G. T. Linteris. (1995) Effect of CF3H and CF3Br on laminar diffusion flames in normal and microgravity. The Third International Microgravity Combustion Conference. Conference Proceedings CP-10174. Cleveland, OH: Lewis Research Center.

Vermillion, C., F. Stetina, P. Corondo, R. Mahoney, G. Shaffer, and A. Lunsford. (1993) A TIROS receiving and analysis system to automatically create image products for environmental monitoring and management. *Proceedings of the Geoscience and Remote Sensing Symposium*. Tokyo: Institute of Electrical and Electronic Engineers, pp. 1665-1667.

Walther, D. C., A. C. Fernandez-Pello, and D. L. Urban. (1998) Space Shuttle based microgravity smoldering combustion experiments. *Combustion and Flame* 116: 398-414.

Weiland, P. O. (1994) *Designing for Human Presence in Space: An Introduction to Environmental Control and Life Support Systems*. Reference Publication RP-1324. Huntsville, AL: National Aeronautics and Space Administration, Marshall Space Flight Center.

Whitaker, A. (2003) Overview of ISS US fire detection and suppression system. In G. A. Ruff (ed.), *Research Needs in Fire Safety for the Human Exploration and Utilization of Space*. Contractor Paper CP-2003-212103. Cleveland, OH: National Aeronautics and Space Administration, Lewis Research Center.

Williams, F. A. (1990) The combustion phenomena in relationship to microgravity research. *Microgravity Science and Technology* 3: 154-161.

Zhou, Y., D. Walther, C. Fernandez-Pello, J. Torero, and H. D. Ross. (2003) Theoretical prediction of microgravity ignition delay of polymeric fuels in low velocity flows. *Microgravity Science and Technology* 14: 44-50.

CHAPTER 28

Safe Without Services Design

Gregg John Baumer
Chairman, International Space Station Safety Review Panel (Retired),
Johnson Space Center, National Aeronautics and Space Administration,
Houston, Texas

The safety of the crewmembers and the spacecraft they inhabit must be the primary focus in the design of a crewed spacecraft. To assure the health of the crew and the survivability of the spacecraft, designers must allocate essential resources of electrical power, thermal control, and crew time to support essential vehicle systems, such as atmospheric control, life support, propulsion, communication, command and control, navigation, and descent and landing. To accomplish the goals of an overall space mission, spacecraft designers must include in the capacity of critical onboard support systems a resource budget for payloads and experiments.

Resources to be allocated to payloads and experiments are defined by payload integration plans and interface control documents. With proper planning and resource management during a nominal mission profile, the budget for the electrical power and thermal control systems of the spacecraft and for crew time can be adequate to supply all users their needed allocation of critical resources, and with margin.

Due to the constraints of spaceflight where on-orbit weight is at a premium, designers often must double book resources to address worst-case contingency scenarios. In other words, they must plan for vehicle operators to maintain the safety of the vehicle and crew continuously and remove electrical power users and heat generators from host vehicle services until the electrical and thermal balance issues of the host vehicle are resolved. This can be accomplished by using load shed tables developed preflight that provide the operators a priority listing of users or user hardware ranked according to their criticality to the mission. Because payloads and experiments are, in general, the least critical to the safety of the vehicle and crew and to the success of a mission, they are most likely to be high on the load shed tables.

From a safety perspective, one must determine whether payloads and experiments are operating within their allocated budget of electrical and thermal resources and what assurances exist that the payload or experiment is to be in a configuration where removal of those resources is safe. In fact, experience indicates that many payloads and experiments require that these resources are redundant and their operation must be monitored by flight and ground crews to maintain a safe condition. The concept of a safe without services design provides designers the capability to resolve this conflict and thus conserve the essential resources of electrical power, thermal control, and crew time during worst-case contingency shortfalls.

The concept of a safe without services design requires that a payload or experiment is compatible with a loss of services from the host vehicle, and it must maintain a configuration that defines its safe state without flight or ground crew intervention. This concept demands more than an absence of hazardous consequences. Indeed, the configuration that protects the crew and host vehicle from these consequences must be consistent with the safety requirements levied on the payload or experiment. This means that, after the removal of the electrical or thermal resources, the payload or experiment must maintain the safety margin, redundancy, and failure tolerance requirements defined and verified in its hazards analysis.

With respect to the use of host vehicle services, it is important for designers to understand that host vehicle systems are susceptible to failure. In conducting a hazard analysis, payload designers must understand and take into consideration the failure tolerance of the host vehicle resources being utilized. Despite the existence of redundancy and independence of controls on the payload side of the interface, these would be compromised if the payload controls are connected to a host resource having a single failure point. Therefore, payload developers must ensure that all payload hazards are assessed in an integrated perspective, and any analysis should include the host vehicle that provides hazard controls across the interface.

At first glance it might seem to be a contradiction that, on the one hand, the designers are allowed to use vehicle and crew resources to control hazards and, on the other, the designers must develop a payload or experiment that is safe without these same resources. The solution is found in the configuration management of the experiment or payload. While a payload is active and the host vehicle is operating in a non-contingency mode, the payload can use host vehicle resources within the constraints defined in the interface control document. However, designers must ensure that their payload possesses a configuration that is safe without the use of host vehicle resources. Some might call this a *survival mode configuration*.

Payload and experiment designers must remain aware of their use of vehicle and crew resources to control hazards, and they must ensure that these resources have been negotiated fully and documented with the host vehicle operators. This is especially important because a payload's safety critical designation of a resource during these negotiations factors into the priority of the payload in the load shed tables. The result is that, in the event of certain worst-case host vehicle contingencies, a payload is allowed to retain the use of host vehicle services only until the payload is reconfigured to a safe state that no longer requires these services to control hazards.

An example of this concept is a payload having a gimbaled telescope that cannot withstand certain mandatory loading conditions imparted by the host vehicle except when in a down and locked configuration. To meet the safety requirements for avoiding a structural failure on orbit, the designers have negotiated to use host vehicle triple-redundant electrical power to activate any one of three independent payload systems to drive the telescope into the locked position. Also negotiated is the operational window in the timeline where telescope operations can be allowed without execution of the hazardous loading event by the host vehicle.

During nominal operations all safety requirements are being met. The payload is conducting pointing operations during an acceptable loading phase of the mission and has triple-redundant capability to be reconfigured to a safe state prior to host vehicle initiating the hazardous load. If the host vehicle were to experience a contingency requiring the expedited conservation of electrical power, such as a loss of one of its primary power distribution systems, the host vehicle operators would implement a load shed scenario that would affect nominal payload operations. Given that the telescope is not in immediate danger of structural failure during a sudden loss of electrical power, the host vehicle operators would not be constrained from including the telescope in the load shed table of the host vehicle. In this unpowered state, the telescope, however, is in a hazardous configuration because it can fail structurally when the host vehicle eventually is required to execute the hazardous loading maneuver. By using a safe without services design concept, the designers have avoided a hazardous consequence. The host vehicle operators are aware of the safety critical need of the payload for reconfiguration power and are aware of the operational constraint to not perform the hazardous maneuver until the telescope is reconfigured. When contingency operations permit, the host vehicle operators are able to supply power temporarily to the telescope to allow reconfiguration. Once reconfigured, the payload then remains in a safe state without flight or ground crew intervention.

Another, more conservative example of a safe without services design would be a similar telescope to that of the first example. This time the designers have enhanced the structural capability of the telescope such that the telescope is not susceptible to structural failure for all host vehicle loading conditions with the telescope in an active scanning mode. Clearly, this solution is safer than the solution of the first example, because it was designed to eliminate the hazard and without the use of host vehicle resources. However, the design probably came at a price of increased weight, a situation that might not be an option available to all designers.

In some cases, designers likely are to be unable to comply with a safe without services design concept. In these cases, control of the affected hazards of the payload would become a risk to the host vehicle. If these payloads were allowed to be manifested, the host vehicle operators would be required to manage existing resources or develop new resources to control the affected hazards of the payload along with those of the host vehicle. An example of such a hazard might be the freeze-thaw cycling of the lines in a propellant system for a deployable satellite payload when the host vehicle is a reusable landing spacecraft. If the host vehicle had to make an emergency deorbit maneuver and land at a contingency landing site, the services of the host vehicle are lost after landing and the payload is then subjected to the freeze-thaw environmental conditions at the landing site. The hazard is the rupture of the propellant lines in the deployable spacecraft and the resulting leakage of propellant into the cargo bay of the host vehicle. Possible solutions to control this type of hazard are for the host vehicle to limit launch opportunities to avoid the potential for contingency landing sites that possess this environmental condition or preposition ground resources at potential contingency sites to control the payload hazard. Whether or not the host vehicle implements these additional hazard controls or elects to accept the increased risk due to the low likelihood of the whole hazard scenario is not the point. What is important is that the safety requirement for a safe without services design highlights this issue for disposition by host vehicle management.

Safe without services design is an important safety concept for payload developers. It recognizes the failure potential of host vehicle services and ensures that all payload hazards utilizing those resources for hazard control are addressed from a perspective integrated with the host vehicle. Most important, the safe without services design provides host vehicle operators the flexibility to respond to the worst-case contingencies without compromising the safety of payloads utilizing critical host vehicle resources.

Probabilistic Risk Assessment with Emphasis on Design

29

Michael G. Stamatelatos, Ph.D.
Office of Safety and Mission Assurance, Headquarters, National Aeronautics and Space Administration, Washington, DC

William E. Vesely, Ph.D.
Office of Safety and Mission Assurance, Headquarters, National Aeronautics and Space Administration, Washington, DC

Peter G. Prassinos
Office of Safety and Mission Assurance, Headquarters, National Aeronautics and Space Administration, Washington, DC

CONTENTS

29.1. Basic Elements of Probabilistic Risk Assessment	889
29.2. Construction of a Probabilistic Risk Assessment for Design Evaluations	894
29.3. Relative Risk Evaluations	898
29.4. Evaluations of the Relative Risks of Alternative Designs	904

29.1 BASIC ELEMENTS OF PROBABILISTIC RISK ASSESSMENT

This section presents the basic elements of a probabilistic risk assessment, which are based on material in the NASA *PRA Procedures Guide for Managers and Practitioners* (Stamatelatos et al. 2001). Basically, probabilistic risk assessment presents a set of scenarios, frequencies, and associated consequences developed in such a way as to inform decisions regarding the allocation of resources to safety, that is, accident prevention. A scenario contains an initiating event and one or more pivotal events leading to an end state (Figure 29.1). The end state is generally an undesired consequence, such as loss of mission or loss of crew. As modeled in most probabilistic risk assessments, an initiating event is an energetic event, failure, or other perturbation that requires response from one or more systems, operators, or pilots. The pivotal events generally include failures of these responses that enable the end state to occur when the initiating event occurs.

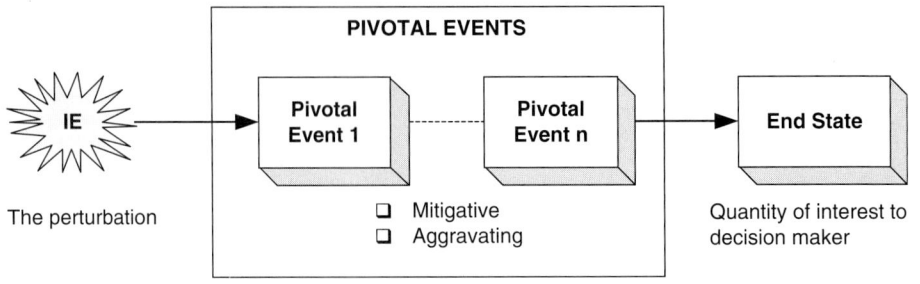

FIGURE 29.1 The concept of an accident scenario.

In certain cases, the initiating event itself can lead directly to the end state, such as an explosion of a propellant tank. The end states are formulated according to the decisions supported by the analysis. Scenarios are classified into end states according to the kind and severity of consequences, ranging from completely successful outcomes to losses of various kinds, such as

- Loss of life, injury, or illness of personnel.
- Damage to or loss of equipment or property (including software).
- Unexpected or collateral damage as a result of tests.
- Failure of the mission.
- Loss of system availability.
- Damage to the environment.

29.1.1 Identification of Initiating Events

Initiating events that potentially can lead to end states of interest can be identified in various ways. Documented hazard analysis can be used, as can past history of incidents. Brainstorming also can be a source. A systematic method often used is the master logic diagram. A master logic diagram (Figure 29.2) is a hierarchical, top down, logical decomposition of the general undesired end state, which is shown on the top of the tree, proceeding to increasingly detailed event descriptions at lower tiers and displaying basic initiating events as the leaves of the tree. The goal is not only to support identification of a comprehensive set of initiating events but to group them according to the challenges they pose, that is, responses required as a result of their occurrence. Initiating events that are completely equivalent in the challenges they pose, including their effects on subsequent pivotal events, are considered equivalent in the risk model.

A useful starting point for identification of initiating events is a specification of normal operation in terms of the nominal values of a suitably chosen set of physical variables and the envelope in this variable space outside which an initiating event would be deemed to have occurred. A comprehensive set of process deviations can thereby be identified and

29.1 Basic Elements of Probabilistic Risk Assessment

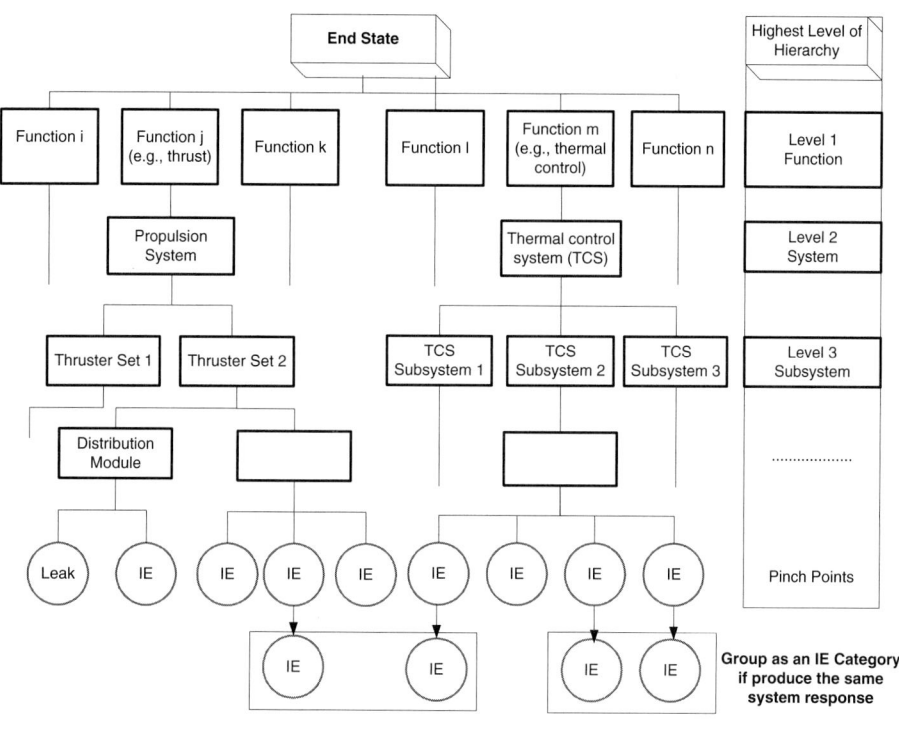

FIGURE 29.2 Typical structure of a master logic diagram.

causes for each of these can then be addressed in a systematic way. The goal not only is to identify a comprehensive set of initiating events but also to group them according to the challenges they pose and the responses required as a result of their occurrence. Those initiating events that completely are equivalent in the challenges that they pose, including their effects on subsequent pivotal events, are considered equivalent in the risk model.

29.1.2 Application of Event Sequence Diagrams and Event Trees

The scenarios that can be realized from a given initiating event are developed using an event sequence diagram or an event tree. An event sequence diagram is essentially a flowchart, with paths leading to different end states, where each path through this flowchart is a scenario. Along each path, pivotal events are identified as either occurring or not occurring (Figure 29.3). An event tree is similar to an event sequence diagram but lists the initiating event and pivotal events as the headings of the top row of the sheet. Paths are then drawn for each option of each pivotal event occurring or not occurring

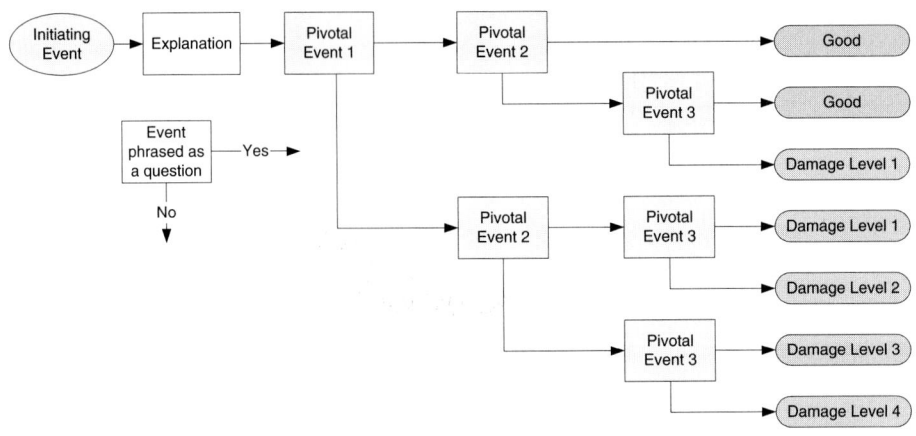

FIGURE 29.3 A conceptual event sequence diagram.

(Figure 29.4). An event sequence diagram is similar in logic to an event tree, and which of the two is selected depends on the type of scenario modeled. In situations that are well covered by operating procedures, the event sequence diagram flow can reflect these procedures, especially if the procedures branch according to the occurrence of pivotal events. Instrument readings that inform crew decisions can be indicated at the

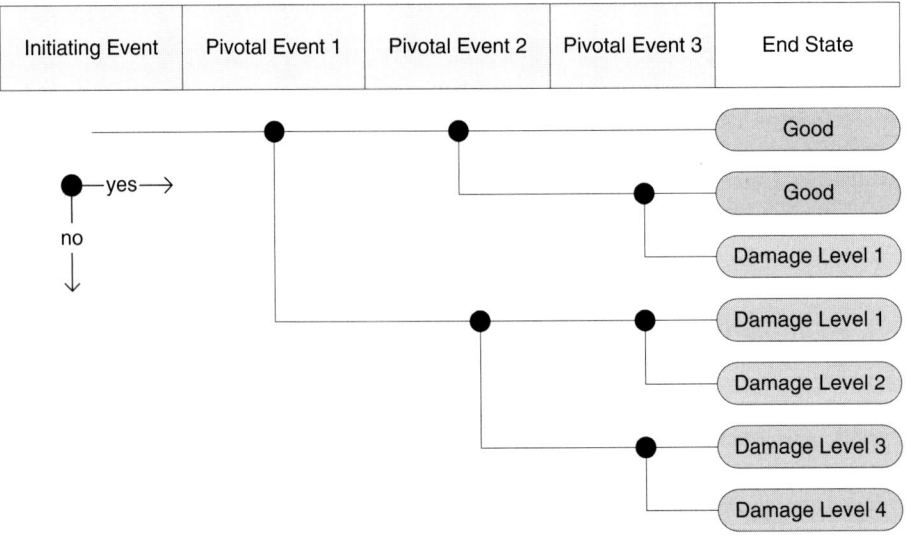

FIGURE 29.4 Event tree representation of the event sequence diagram shown in Figure 29.3.

appropriate pivotal event. A good deal of information, such as system level mission success criteria at each pivotal event, also can be displayed on the event sequence diagram. Even so, event trees are more structured and they more concisely can display scenarios involving many pivotal events. All the scenarios can be displayed with the same structure providing an effective organization of the scenarios. Most probabilistic risk assessment software can transform one type of display to the other.

29.1.3 Modeling of Pivotal Events

Pivotal events must be modeled in sufficient detail to support valid quantification of scenarios. The model must reach a level of detail at which data or expert estimates are available to support quantification of the model. The modeling of pivotal events also must be carried to sufficient depth to identify common basic events in different pivotal events and dependences among the pivotal events. Pivotal events that are simple can be quantified using simple models or direct estimation. Pivotal events corresponding to complex events or system failure generally are modeled using fault trees. Figure 29.5 shows fault tree models constructed for specific pivotal events involved in accident scenarios resulting from a hydrazine leak. The logic gates in the fault tree are OR and AND gates depicting the logic of the input events, that is, the events below a gate needed for the output event, which is the event above the gate.

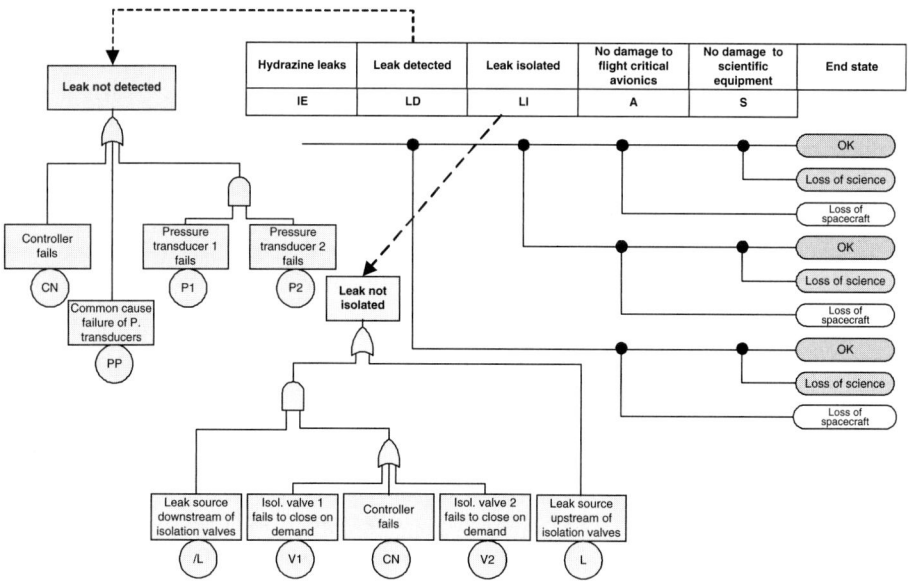

FIGURE 29.5 Fault trees for failure of leak detection and isolation, given detection, in an event tree.

29.1.4 Linkage and Quantification of Accident Scenarios

The fault trees are linked to the event sequence diagrams or event trees basically by substituting the fault trees for the pivotal events. This linkage is carried out by expressing the paths in the event sequence diagrams or event trees in terms of Boolean expressions. The fault trees also are expressed as Boolean expressions. The expressions for the top events in the fault trees are then substituted for the corresponding pivotal events. In this manner, the accident scenarios are expressed in terms of the basic events for which probabilities have been estimated based on data or expert opinion. Accident scenario probabilities are determined by manipulating the Boolean expressions containing the basic events with their probabilities. This manipulation is carried by reducing the Boolean expressions to a sum of products to produce the minimal cut sets, which are the smallest combinations of basic events resulting in the end state. The end state probability is then the probability of the logical sum of the minimal cut set probabilities, where each minimal cut set probability is the product of the basic event probabilities that also include dependencies among the basic events. Alternately, the top event probability is determined from the basic event probabilities using binary decision diagrams, which is another method of Boolean reduction that does not involve determination of the minimal cut sets.

Uncertainty quantification is an important element of the quantification of the probabilities and consequences of the end states in a probabilistic risk assessment. Uncertainties are quantified by assigning probability distributions to the failure rates and other basic data and propagating these through the probabilistic risk assessment using simulation, such as Monte Carlo techniques. Figure 29.6 illustrates the process using minimal cut sets to obtain the uncertainty distribution on the final probability that combines all the minimal cut set probabilities and their uncertainty distributions.

29.2 CONSTRUCTION OF A PROBABILISTIC RISK ASSESSMENT FOR DESIGN EVALUATIONS

29.2.1 Uses of Probabilistic Risk Assessment

This section describes the construction of a probabilistic risk assessment for design evaluations (Stamatelatos 2007). A specific application is described of a probabilistic risk assessment constructed to evaluate designs for the NASA crew return vehicle. Often a probabilistic risk assessment is constructed to evaluate a system or space vehicle already built. Such was the case of the probabilistic risk assessments for the *International Space Station* and *Space Shuttle*. However, a probabilistic risk assessment can be an extremely useful tool for evaluating a proposed design or conceptual design. At the initial stages of the design, the probabilistic risk assessment is top level, incorporating the function and system concepts and relationships. As the design evolves, the probabilistic risk

29.2 Construction of a Probabilistic Risk Assessment

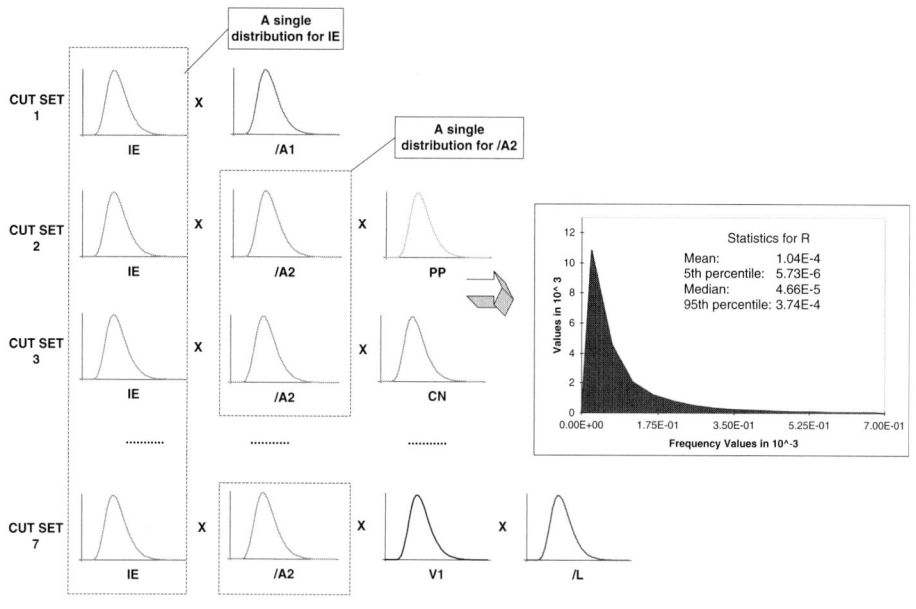

FIGURE 29.6 Uncertainty propagation in a probabilistic risk assessment (Courtesy of NASA).

assessment evolves and incorporates more detail on the structure and details of the functions and systems. At the initial stages, the probabilistic risk assessment is used to assist in determining risk allocations, identifying risk sensitivities and insensitivities, evaluating design trade-offs, and balancing risk control versus performance. As the design evolves and matures, these uses of probabilistic risk assessment are applied at successively more detailed levels.

To evaluate the risks of alternative designs for the crew exploration vehicle, NASA constructed a reference top level probabilistic risk assessment. A principal objective of the reference probabilistic risk assessment was to gain knowledge of the risk contributors and sensitivities associated with a design. The reference probabilistic risk assessment also was intended evaluate trade-offs between mission risk and mission performance. Furthermore, it could assist in selecting proposed designs and alternatives from its contractors. The following section describes the reference mission, the reference design, and the reference probabilistic risk assessment constructed. A preliminary design was used as the reference. Only sufficient detail of the design is provided to illustrate key points in developing the probabilistic risk assessment model. A subsequent section describes the use of the reference probabilistic risk assessment for evaluating the relative risks of different proposed designs for a commercial orbiter transport system.

29.2.2 Reference Mission

The reference mission was a lunar sortie, that is, a short duration mission to the Moon, and it was modeled to consist of eight phases. Figure 29.7 illustrates these different mission phases, which were defined to correspond to different functions being utilized and different requirements for the functions.

The phases are as follows:

- Phase 1. Two launches to low Earth orbit. The first launch is of the *Cargo Launch Vehicle* consisting of a lunar surface access module and an Earth departure stage. The second launch is of the *Crew Launch Vehicle*, which contains a crew module, a service module, and a launch abort system.

- Phase 2. Rendezvous of the *Cargo Launch Vehicle* and the *Crew Exploration Vehicle*. The Earth departure stage is used via a translunar injection burn and expended, leaving the *Crew Exploration Vehicle* and Lunar Surface Access Module.

- Phase 3. Lunar orbit insertion. The Lunar Surface Access Module performs the lunar orbit insertion for both the *Crew Exploration Vehicle* and lunar surface access module. All crewmembers of the *Crew Exploration Vehicle* transfer to the Lunar Surface Access Module, and it undocks from the *Crew Exploration Vehicle*.

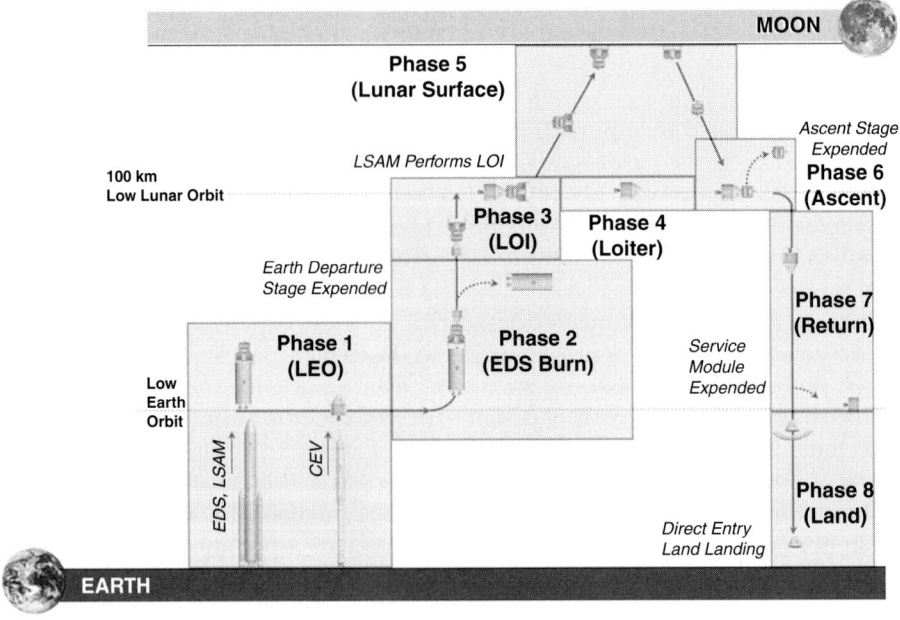

FIGURE 29.7 The different phases for a lunar sortie.

- Phase 4. *Crew Exploration Vehicle* in lunar orbit. The uncrewed *Crew Exploration Vehicle* remains in a lunar sortie but descends to a parking orbit during the short (up to 7 d) lunar mission.
- Phase 5. Lunar mission. The Lunar Surface Access Module is used to descend to and ascend from the lunar surface. A typical lunar mission lasts up to 7 d.
- Phase 6. Re-crew the *Crew Exploration Vehicle*. Following the Lunar Surface Access Module ascent (the descent stage is left on the lunar surface), the crew docks with and transfers back to the *Crew Exploration Vehicle*. The Lunar Surface Access Module (ascent stage) is expended back to the lunar surface.
- Phase 7. Return to Earth. The *Crew Exploration Vehicle* returns to Earth, where the service module is expended when no longer needed.
- Phase 8. Earth landing. The *Crew Exploration Vehicle* lands on the Earth with a direct entry and parachute assisted land touchdown.

Each of the phases was in turn divided into subphases to correspond to the different systems utilized, different system configurations, and different system success definitions. Table 29.1 describes the different subphases defined for each phase. The details associated with these subphases are not important here. What is important is that the subphases allowed fault trees to be constructed corresponding to the different requirements for each subphase. Because the systems architecture was for a preliminary design, the fault trees modeled the functional operation of the major systems of the *Crew Exploration Vehicle*, and they were represented by multi-string subsystems down to the major component level. The fault trees also modeled the interconnected nature of the various systems and subsystems both within a mission subphase and across mission phases. The basic event tree in Figure 29.7 was expanded (its top event headings) to include the subphases in each event. The generated, more detailed event trees consisted of failure or success of each subphase.

Two undesired end states were of interest: loss of crew and loss of mission. The fault tree models were linked finally to the event tree using standard probabilistic risk assessment methods and tools, in this case the SAPHIRE computer program. Figure 29.8 illustrates the type of relative results obtained using available data and expert opinion. Because the design was still preliminary, representative data were used with the intent of serving as a basis for relative evaluations, sensitivity and uncertainty evaluations, and trade studies. Examples of the type of trade studies that can be performed are

- Absence of one (of two) string of the active controlled thermal system.
- Absence of one (of three) string of the electrical power system.
- Absence of one (of three) string of the avionics system.
- Changes in landing parachutes success criteria.
- Absence of the launch abort system during phase 1.

Table 29.1 Subphase Events for Each Phase of the Lunar Mission (Abbreviated)

Phase	Subphase	Phase	Subphase
1 (120.9 h)	1. CaLV First stage launch 2. CaLV First stage ascent 3. CaLV SRB separation 4. CaLV Second stage ascent 5. CaLV Second stage separation 6. CaLV Third stage ignition 7. CaLV Third stage ascent 8. CaLV Third stage shutdown 9. CaLV EDS/LSAM loiter LEO 10. … 19. Rendezvous and docking w/ISS	4 (168 h) 5 (170.2 h) 6 (5.3 h)	28. CEV lunar loiter 29. Powered descent, lunar landing 30. Lunar surface mission 31. Ascent ignition, separation and ascent 32. Ascent stage LLO insertion 33. Ascent stage plane Δ burn 34. Ascent, rend, and docking/CEV 35. Crew transfer LSAM to CEV
2 (72.3 h)	20. TLI burn for LLO 21. TLI midcourse burn 22. Separation from LSAM/CEV	7 (108 h)	36. CEV separation from LSAM 37. CEV burn for Earth 38. Midcourse correction burn 39. CEV separation from ISS
3 (36.4 h)	23. Midcourse correction burn 24. Burn to slow near LLO 25. … 27. LSAM separation from CEV	8 (1.8 h)	40. Separation from service module 41. Entry 42. … 44. Crew recovery

29.3 RELATIVE RISK EVALUATIONS

This section describes relative risk assessments. The difference between relative risk assessments and absolute risk assessments, the role of relative risk assessments in design evaluations, and the modeling and quantifications carried out in relative risk assessments are described. Particular attention is given to the treatment of uncertainties in assessing relative risk. Proper treatment of uncertainties, including the proper accounting of correlations between the results compared, can result in considerable reduction in the uncertainties in the relative results as compared to the uncertainties in the absolute results. Descriptions of relative input data that can be used also are provided, bypassing the need to use absolute data, such as absolute equipment failure rates and human error rates. Modeling setups to take advantage of the relative risk evaluations are further described. An application is described in which relative risks are assessed on five proposed designs.

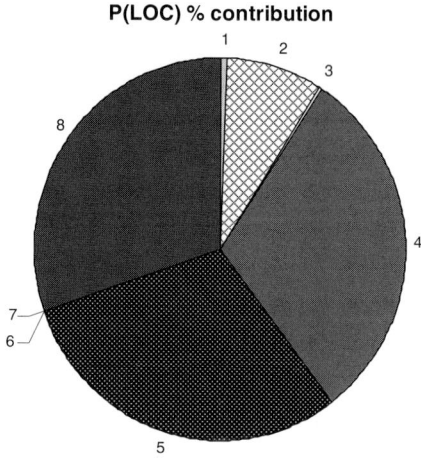

FIGURE 29.8 Example of the relative contributions to loss of mission from each phase (Courtesy of NASA).

The relative risks of the five designs are compared to a reference design to standardize the relative assessments. Of particular importance is the assessment of uncertainties, because the proposed designs have appreciably different heritages and legacies. The results and associated uncertainties illustrate the meaningful information obtainable from relative risk assessments when performed in the design stage.

29.3.1 Absolute Versus Relative Risk Assessments

Quantitative risk assessments can be categorized into two types: absolute and relative. An absolute risk assessment determines numerical risk values, such as probabilities and consequences, associated with accident scenarios. A relative risk assessment focuses on the relative risks, such as relative probabilities and relative consequences, among two or more alternatives, such as among two or more designs. Table 29.2 provides illustrations of absolute versus relative risk evaluations.

Absolute risk assessments can provide relative risk results by appropriate calculations. However, absolute data, such as equipment failure rates and human error rates, are entered. Absolute risk assessments produce relative results as a secondary result. Because the relative results are secondary, more input data are needed than if relative evaluations were the focus. Also, correlations among the entities compared often are not handled in a detailed manner, causing an overestimation of the uncertainties in the relative results.

Relative risk assessments, on the other hand, focus only on relative results. The relative results are generally in the form of ratios. For relative risk assessments, only relative data need to be entered, again generally in the form of ratios. When the relative evaluations are carried out, the correlations between the results compared accurately are taken into

Table 29.2 Examples of Absolute Versus Relative Risk Results

Absolute	Relative
Best estimate of the probability of loss of mission = 1/240	Ratio of probability of loss of mission for design 2 compared to design 1 = 1/5
Mean value of the probability of loss of crew (per mission) = 1/190	Relative contribution of avionics failure to loss of mission (mean value) = 12%
5 and 95% confidence bounds for probability of system failure = 1/780, 1/350	Relative increase in the probability of loss of mission when a redundant control train is removed = 0.3%

account, leading to uncertainties in the relative results that are often significantly smaller than the uncertainties in the absolute results.

There is thus a trade-off between absolute risk assessments and relative risk assessments. Absolute risk assessments provide absolute risk results but require more data, and the absolute risk results often have large uncertainties. Relative risk results also can be obtained, but often the uncertainties are not determined correctly. Relative risk assessments require less input data and can provide relative risk results with considerably smaller uncertainties than absolute risk assessments. However, relative risk assessments do not provide absolute risk values.

29.3.2 Roles of Relative Risk Assessments in Design Evaluations

The relative risk assessment of designs plays two major roles: evaluation of the relative risks of design changes and the relative risk differences between different designs. To evaluate the relative risk of design changes, the relative difference in the risk due to the design change is evaluated. Let

$$R_0 = \text{Risk of the original design} \tag{29.1}$$

$$R_1 = \text{Risk of the modified design} \tag{29.2}$$

Then,

$$\delta R = \frac{R_1 - R_0}{R_0} \tag{29.3}$$

which is the relative change in the risk.

Here, the risk can be any measure, such as the probability of loss of crew or the expected resources lost. These relations are basic. If the risk measure consists of a sum of contributors, such as the minimal cut sets for the accident sequences from a probabilistic risk assessment, then the preceding definition of the relative risk can be expressed as

$$\delta R = \frac{C_0}{R_0} \times \frac{(C_1 - C_0)}{C_0}. \tag{29.4}$$

The quantity C_0 is the original value of the contributors (minimal cut sets) affected by the change, and C_1 is the modified value with the modified design. Equation (29.4) can be stated in words as follows:

$$\text{Relative risk change} = \text{Relative importance of contributors affected} \times \text{Relative change in the affected contributors} \tag{29.5}$$

Therefore, only contributors affected by the design modification need to be evaluated for their relative change. The preceding relation and Equation (29.4) provide the basis for effective relative risk assessments of design changes. Useful bounds can be obtained based on considerations of the importance of the contributors and the maximum change that could result in the contributors. More detailed quantitative evaluations need to focus only on the relative importance of the contributors affected and the relative change in the contributors resulting from the design modification. Both data requirements and uncertainties in the results thereby can be reduced significantly.

Comparison of the risks of different designs can be expressed in a similar manner as the relative risk from a design modification, just discussed. Let

$$R_1 = \text{Risk associated with design 1} \tag{29.6}$$

$$R_2 = \text{Risk associated with design 2} \tag{29.7}$$

Again, the risk can be any specific measure. Then the relative risk ratio R_{21} of design 2 compared to design 1 is

$$R_{21} = \frac{R_2}{R_1} \tag{29.8}$$

Let the risk of each design be represented by a sum of contributions, such as from the accident sequences:

$$R_1 = \sum C_i^{(1)} \tag{29.9}$$

$$R_2 = \sum C_i^{(2)} \tag{29.10}$$

where $C_i^{(1)}$ and $C_i^{(2)}$ are the contributors for design 1 and design 2, respectively. The relative risk ratio R_{21} can be re-expressed as

$$R_{21} = \sum \frac{C_i^{(1)}}{R_1} \frac{C_i^{(2)}}{C_i^{(1)}} \tag{29.11}$$

or

$$R_{21} = \sum I_i^{(1)} \frac{C_i^{(2)}}{C_i^{(1)}} \tag{29.12}$$

where $I_i^{(1)}$ is the relative importance of the particular contribution to the total risk. Thus, the relative risk ratio R_{21} can be expressed as the relative importance of each contributor to the total risk for design 1 multiplied by the relative ratio of the contributor for design 2 over the contributor for design 1. In words,

$$\text{Risk ratio} = \sum \text{Contributor importance} \times \text{Ratio of contributor values} \qquad (29.13)$$

By matching contributors between the two designs, a number of the ratios of risk contributors (the second term in the preceding summation) cancels out. By accounting for these correlations, the uncertainties from these contributions can be reduced considerably.

The relative difference in risks between the two designs can be expressed in a similar manner to correlate the common contributions between the two designs. Let

$$\delta R_{21} = \frac{R_2 - R_1}{R_1} \qquad (29.14)$$

Then δR_{21} can be expressed as

$$\delta R_{21} = \sum \frac{C_i^{(1)}}{R_1} \frac{C_i^{(2)} - C_i^{(1)}}{C_i^{(1)}} \qquad (29.15)$$

or

$$\delta R_{21} = \sum I_i^{(1)} \delta C_i^{(21)} \qquad (29.16)$$

where $I_i^{(1)}$, again, is the importance of a particular contributor, and $\delta C_i^{(21)}$ is the relative difference in the contribution for design 2 compared to design 1. The summation in Equations (29.15) and (29.16) is now only over those contributors that are different in design 2 compared to design 1. Thus, the evaluations need be carried out only for those contributors that are different between the two designs and involve only relative calculations. This can result in considerably reduced calculations, less input data required, and fewer uncertainties in the relative risk difference calculated. If design 1 is used as reference design, all alternate designs can be compared in a relative fashion using these relationships.

29.3.3 Quantitative Evaluations

Once the contributors involved in the relative risk expressions are identified, the relative risk can be quantified using the appropriate input data. This generally involves assigning probability distributions to basic variables involved in the expressions and carrying out a Monte Carlo simulation. There are two basic approaches for constructing the input data required for the relative risk evaluations. The first is to estimate the relative values directly, which involves estimating the probability distributions for the basic relative variables, such as the ratios of failure rates or the relative differences in failure probabilities. The second approach involves estimating the probability distributions of the absolute variables involved in the relative expressions, such as estimating the probability distributions of the failure rates and failure probabilities.

For either approach, the correlations among the variables need to be incorporated in the uncertainty evaluations and Monte Carlo simulations. The principal method of doing this is to identify a common variable that appears in multiple locations in the relative risk expression. One probability distribution is then sampled and the random value assigned to the common variable in all locations. This is often called *coupling the common variable*. In this coupling approach, the correlation among the instances of the common variable is assumed to be 1.

A more general approach is to assign the correlation coefficient between related variables. In Monte Carlo simulations, this involves constructing bivariate or multivariate probability distributions for the correlated variables. For less complex expressions for the relative risk, the variance can alternatively be analytically determined. Sums of ratios of contributors generally appear in these expressions, such as in Equation (29.15). Let δR be a general relative risk measure. Then the variance, $V(\delta R)$, of δR is given by

$$V(\delta R) = \sum_i V(X_i) + \sum_{i \neq j} \rho_{ij} \sqrt{V(X_i) V(X_j)} \qquad (29.17)$$

where X_i is a general term in the expression for the relative risk, $V(X_i)$ is the variance of the term, and ρ_{ij} is the correlation between the two variables. For the relative risk evaluations performed, the correlation coefficients generally ranged from 0.5 to 1.0 when they were incorporated.

Often in quantitative risk assessments, lognormal distributions are used as the probability distributions to characterize the uncertainty in a variable. The preceding expression can be applied to a ratio by taking the log of the ratio. Then,

$$V\left[\ln\left(\frac{X_2}{X_1}\right)\right] V(\ln X_2) + V(\ln X_1) - 2\rho_{12} \sqrt{V(\ln X_1) V(\ln X_2)} \qquad (29.18)$$

where now ρ_{12} is the correlation between $\ln X_1$ and $\ln X_2$. When the variances of the logs are equal and the correlation coefficient is 1, such as for coupling the variable, the variance of the log of ratio, $V[\ln(X_2/X_1)]$, is 0. For highly correlated variables, the variance still is significantly reduced. This occurs, for example, if each term in the ratio consists of a product of factors, of which many are common to both the numerator and denominator. This occurs in applications when the ratio consists of accident sequence minimal cut set contributions, which have many common components in both the numerator and denominator.

As previously indicated, in various cases, direct estimation of the relative ratios involved in the relative risk expression is the most effective approach. This occurs when there are no data or little confidence in the data for the absolute variables, such as when new technologies or new processes are involved in the design. Also, in many instances, the system analyst can feel more confident in estimating relative differences between design contributors than absolute values. For these relative estimates, available relative comparison scales can be adapted for a particular application. One such scale that has been found to be applicable is an adaptation of the scale used in comparison in the relative evaluations made using the analytical hierarchical process. Such a scale is shown in Table 29.3. For ratios less than 1, the term *smaller* replaces the term *higher*. Uncertainties can be included with the ratio estimates. Also, a range can be estimated for the ratio when one description cannot be settled on. Even when absolute values are used in the evaluations, the analytical hierarchical process approach is useful for organizing the relative evaluations. The usefulness of the analytical hierarchical process approach is shown in the application to be described.

Table 29.3 Scale for Ratio Comparisons of Values

Value	Verbal Description
1	Basically equal
2	Somewhat higher
3	Moderately higher
5	Higher
7	Appreciably higher
9	Significantly higher

29.4 EVALUATIONS OF THE RELATIVE RISKS OF ALTERNATIVE DESIGNS

This section illustrates the evaluation of the relative risks of proposed designs using the reference design probabilistic risk assessment and the relative risk methods described in the previous sections. Specifically, the evaluation of the relative risks of five proposed designs for a proposed mission to the *International Space Station* is described (Vesely 2006). The risk of each design is compared to the risk of the reference design. The risk measure used is the probability of loss of crew in the mission. Certain designs involve new physical processes and new hardware concepts. The designs are of different complexity and involved different heritages and complexities. Therefore, treatment of uncertainties associated with each design is of high importance. The specific numbers presented are only illustrative, and unnecessary detail is not presented, to focus on application of the basic methodology described. The results illustrate the power of carrying out relative risk evaluations of proposed designs.

29.4.1 Overview of Probabilistic Risk Assessment Models Developed

A top level event tree is constructed for each design and the reference design for a proposed mission to the *International Space Station* for crew and cargo transport. A set of event trees is constructed for each alternative design and the reference design. The event tree headings are common for each set of event trees constructed and identify the principal functions that need to be performed. These principal functions are

- First stage launch and ascent.
- First stage separation.
- Second stage ignition and burn.
- Orbit insertion using upper stage or service propulsion.
- On-orbit operation utilizing active systems.
- Orbital rendezvous and docking.

29.4 Evaluations of the Relative Risks of Alternative Designs

- Vehicle docked at the *International Space Station* with dormant conditions maintained.
- International Space Station undocking and separation maneuvers.
- Deorbit burn for atmospheric entry and descent.

As seen from the single step downward in each event tree sequence representing failure of the principal function, a failure of any principal function is modeled as resulting in a loss of crew. Figure 29.9 illustrates the event tree of the basic function.

The preceding functions involving orbit insertion, on-orbit operation, deorbit burn, and entry and descent are further decomposed into principal risk important systems identified in the reference risk model assessment. These principal systems include the environmental control and life support systems, the attitude control system, the electrical power system, the thermal protective system, the orbital maneuvering system, and the reaction control system. A failure of any one of these principal systems is modeled as a failure of the associated function. Each function is, therefore, a logical OR of the principal systems.

29.4.2 Relative Risk Comparisons of the Alternative Designs

To compare the relative risks of the alternative designs, the event trees are used to assess the ratios of the probability of failure of each function for an alternative design to the reference design. The relative risk methodology previously described is used and in particular Equation (29.12), which is given here as Equation (29.19):

$$R_{21} = \sum I_i^{(1)} \frac{C_i^{(2)}}{C_i^{(1)}} \tag{29.19}$$

where R_{21} is the ratio of the risk of a given alternative design (subscript 2) to the reference design (subscript 1), $I_i^{(1)}$ is the importance of function i for the reference design (the Fussell-Vesely importance), $C_i^{(2)}$ is the failure probability of the function for the alternative design, and $C_i^{(1)}$ is the failure probability of the function for the reference design. To carry out these relative comparisons, the event trees are transformed to a hierarchy format as shown in Figure 29.10.

The leftmost box in the hierarchy represents the criteria for comparison of loss of crew. The level-1 criteria to the right define the principal functions, and the level-2 criteria to their right define the further breakdown of the functions according to the event tree. For the level-1 criteria in the hierarchy, first stage launch and ascent includes failure of abort, because this is the failure branch in the event tree. Similarly, second stage ignition and burn includes failure of abort corresponding to the failure branch in the event tree.

Figure 29.11 illustrates the importance weights used in one of the evaluations. When a level-1 criterion is directly used for level 2, the lowest level for the comparisons, then the weight is 1; otherwise, the weight is decomposed into the lower level criteria (subfunctions) whose sum of weights is 1. The net importance, denoted as $I_i^{(1)}$ in Equation (29.19), therefore, is the product of the level-1 weights and the level-2 weight for the given criterion.

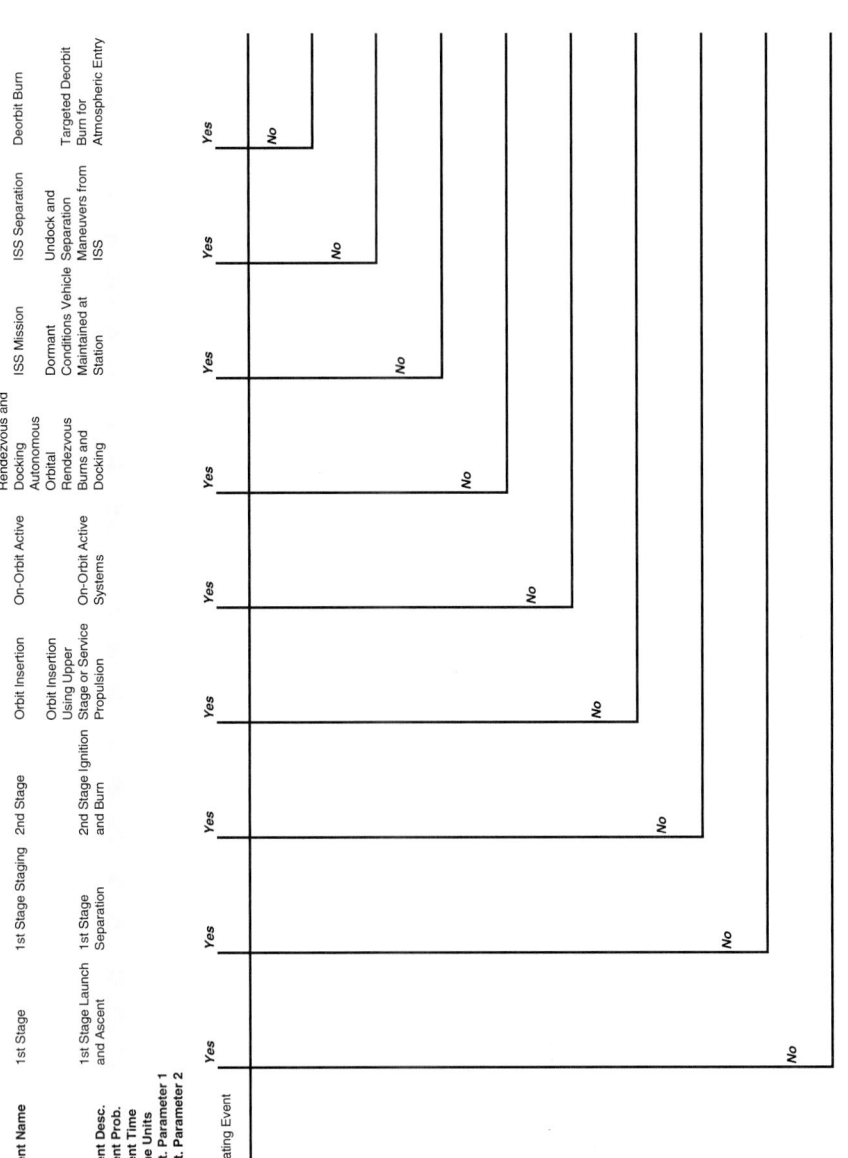

FIGURE 29.9 The event tree of the principal functions.

29.4 Evaluations of the Relative Risks of Alternative Designs

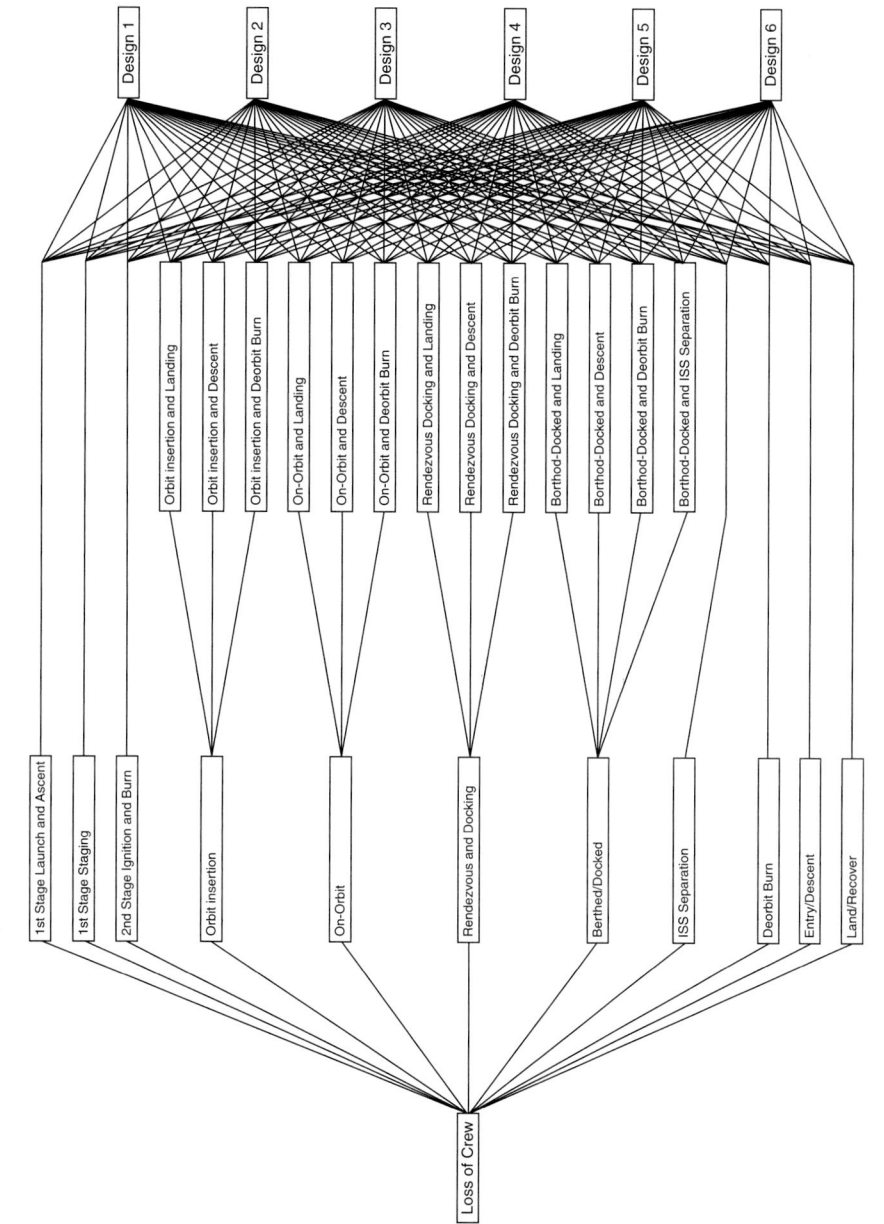

FIGURE 29.10 Hierarchy of functions for the relative comparisons.

Goal Level	Weights	Level 1	Breakdown of Level 1	Weights	Level 2
Loss of Crew	0.0532	1st Stage Launch and Ascent	1st Stage Launch and Ascent	1.0000	1st Stage Launch and Ascent
	0.0031	1st Stage Staging	1st Stage Staging	1.0000	1st Stage Staging
	0.2624	2nd Stage Ignition and Burn	2nd Stage Ignition and Burn	1.0000	2nd Stage Ignition and Burn
	0.0003	Orbit Insertion	Orbit Insertion	0.2069	Orbit Insertion and Landing
	0.0008	On-Orbit		0.1485	Orbit Insertion and Descent
	0.0015	Rendezvous and Docking	On-Orbit	0.6446	Orbit Insertion and Deorbit Burn
	0.0007	Berthed/Docked		0.2069	On-Orbit and Landing
	0.1230	ISS Separation		0.1485	On-Orbit and Descent
	0.3577	Deorbit Burn	Rendezvous and Docking	0.6446	On-Orbit and Deorbit Burn
	0.0824	Entry/Descent		0.2069	Rendezvous-Docking and Landing
	0.1148	Land/Recover		0.1485	Rendezvous-Docking and Descent
			Berthed/Docked	0.6446	Rendezvous-Docking and Deorbit Burn
				0.1693	Berthed-Docked and ISS Landing
				0.1216	Berthed-Docked and Descent
				0.5277	Berthed-Docked and Deorbit Burn
			ISS Separation	1.0000	Berthed-Docked and Separation
			Deorbit Burn	1.0000	ISS Separation
			Entry/Descent	1.0000	Deorbit Burn
			Land/Recover	1.0000	Entry/Descent
					Land/Recover

FIGURE 29.11 Importance of weights for the hierarchy.

29.4 Evaluations of the Relative Risks of Alternative Designs

For each of the level-2 criterion subfunctions, the ratio of the failure probability for an alternative design is compared to the reference design following Equation (29.19). Where the function could be decomposed into principal systems, the function failure probability is assessed as the sum of the system failure probabilities, and the ratio is taken of the sum of the system failure probabilities for a design alternative to the sum for the reference system failure probabilities. When the function could not be decomposed into principal systems, the ratio of the function failure probabilities is directly estimated. Many of the alternative designs had system designs or function designs similar to the reference design. For these cases, the ratio of the alternative function contribution to the reference design contribution is near unity. Figure 29.12 illustrates the relative ratios of the alternative designs for a particular evaluation. The reference design is labeled design 5. Relative difference ratios, that is $\left(C_i^{(2)} - C_i^{(1)}\right)/C_i^{(1)}$, are shown in Figure 29.11 to better highlight design differences. Because the relative difference ratio is 1 minus the ratio it has similar uncertainty characteristics.

An important aspect of relative risk assessments, or of any risk assessment, is the assessment of the associated uncertainties. As identified in the previous section describing the relative risk methodology, the uncertainties in risk ratios and relative risk ratios can be considerably smaller than the uncertainties in the associated absolute risk estimates. These smaller uncertainties are due to correlations between factors in the numerator and denominator of the ratio, especially the same factor or similar factors occurring in both. The smaller uncertainties also are due to the greater confidence of the assessor in estimating a relative value based on design differences as opposed to the absolute value.

When uncertainty distributions are calculable for the numerator and denominator, the uncertainty in the ratio is estimated using Monte Carlo simulations. The correlation

Level 2	Design 1	Design 2	Design 3	Design 4	Design 5	Design 6
1st Stage Launch and Ascent	7.1358	3.4393	0.9932	7.2593	0	6.7660
1st Stage Staging	−0.0055	−0.0013	−0.0068	−0.0045	0	−0.0055
2nd Stage Ignition and Burn	−0.0055	−0.0013	−0.4951	−0.0045	0	−0.5024
Orbit Insertion and Landing	−0.0055	−0.0013	0.1245	−0.0045	0	−0.0035
Orbit Insertion and Descent	−0.0055	−0.0013	−0.0048	−0.0045	0	−0.0035
Orbit Insertion and Deorbit Burn	−0.0055	−0.0013	−0.0048	−0.0045	0	−0.0035
On-Orbit and Landing	−0.0055	−0.0013	0.1245	−0.0045	0	0.9931
On-Orbit and Descent	−0.0055	−0.0013	−0.0048	−0.0045	0	0.9931
On-Orbit and Deorbit Burn	−0.0055	−0.0013	−0.0048	−0.0045	0	0.9931
Rendezvous-Docking and Landing	−0.0055	−0.0013	0.1245	−0.0045	0	−0.0050
Rendezvous-Docking and Descent	−0.0055	−0.0013	−0.0048	−0.0045	0	−0.0050
Rendezvous-Docking and Deorbit Bur	−0.0055	−0.0013	−0.0048	−0.0045	0	−0.0050
Berthed-Docked and ISS Landing	−0.0055	−0.0013	0.1245	−0.0045	0	−0.0050
Berthed-Docked and Descent	−0.0055	−0.0013	−0.0048	−0.0045	0	−0.0050
Berthed-Docked and Deorbit Burn	−0.0055	−0.0013	−0.0048	−0.0045	0	−0.0050
Berthed-Docked and Separation	−0.0055	−0.0013	−0.0048	−0.0045	0	−0.0050
ISS Separation	−0.0055	−0.0013	−0.0048	−0.0045	0	−0.0050
Deorbit Burn	−0.0055	−0.0013	−0.0048	−0.0045	0	−0.0050
Entry/Descent	−0.0055	−0.0013	−0.0048	−0.0045	0	−0.0050
Land/Recover	−0.0055	−0.0013	0.1245	−0.0045	0	−0.0050

FIGURE 29.12 Relative difference ratios for the alternative designs.

between the numerator and denominator are accounted for by using the same random number for the same term occurring in the numerator and denominator (termed *coupling in probabilistic risk assessment Monte Carlo simulations*). In a number of cases, uncertainty in the ratio is estimated directly by estimating the 25 and 75% bounds, based on the design differences. These direct estimates of the ratio uncertainty are made either when there are no detailed models for the alternative design function or additional factors are considered that were not incorporated into the model. The uncertainties on the ratios are incorporated into the software package DecisionPlus®, which is a priority setting tool using the analytic hierarchy principle. DecisionPlus is also a decision analysis tool, using various techniques such as the SMART and direct trade-offs. The analytic hierarchy process priority setting approach is used to roll up the ratio estimates following Equation (29.19). DecisionPlus is used because it can propagate uncertainties on the ratios to obtain the uncertainties on the overall result. Figure 29.13 shows the results of an evaluation for the ratio uncertainty distributions. For comparison, Figure 29.14 shows an evaluation of the uncertainty bounds for the absolute values for the probability of loss of crew for the alternative designs and reference design.

The design identifiers in Figure 29.14, which are along the *x*-axis, are ordered the way they are because of specific characteristics of the designs. The reference design is again design 5. The specific characteristics are not important here nor are the specific absolute probability values. What is important to note is that the 5 and 95% bounds for the absolute

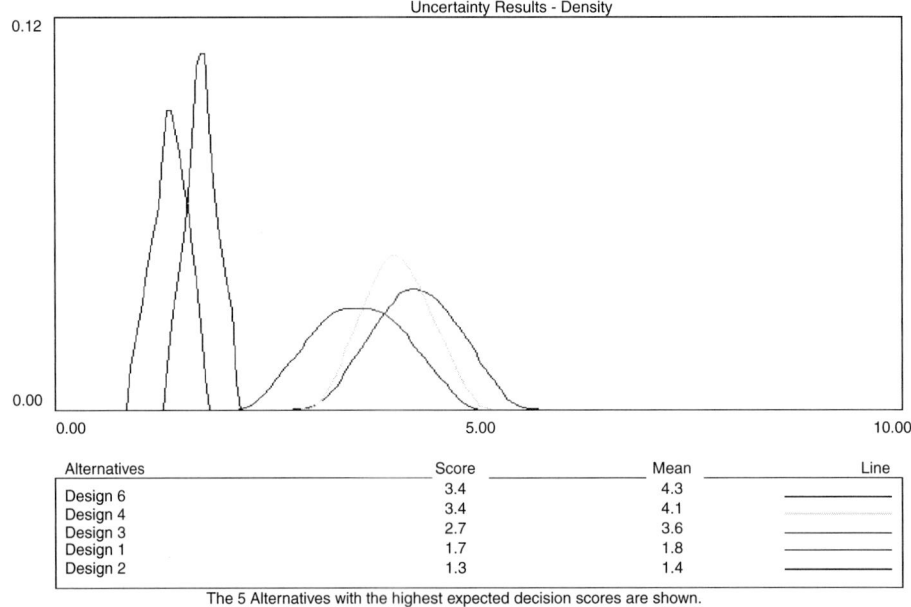

FIGURE 29.13 Uncertainty distributions of the ratios of loss of crew probabilities.

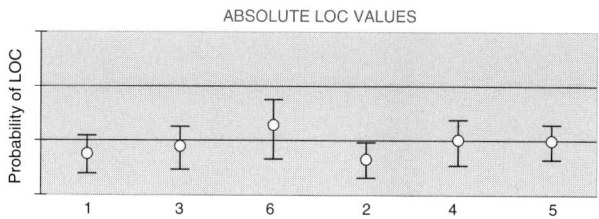

FIGURE 29.14 Absolute loss of crew probabilities for the design.

probabilities are roughly a factor of 10 apart and they generally overlap for all the designs. Therefore, the designs are not able to be clearly differentiated or clearly grouped based on their absolute probabilities. In contrast, the relative risk ratios shown in Figure 29.13 indicate a clear grouping, even accounting for uncertainties. The uncertainty curves in Figure 29.13 are shown between the 5 and 95% bounds to be consistent with the uncertainty levels for the absolute values. The score at the bottom of the figure is the median value. As seen from Figure 29.13, designs 6, 4, and 3 have clearly higher relative loss of crew probabilities compared to the reference than designs 1 and 2. These differences are significant at the uncertainty levels (confidence levels) shown.

Thus, the relative risk evaluations show clear differences in designs while the absolute risk evaluations do not. These results can be entered to assist in the decision on the design or designs to select or further pursue, considering cost, performance, and other attributes. Those designs that had the larger relative risks might be further pursued if they had other significant positive attributes. Their uncertainty decompositions could be carried out using the relative risk quantification methodology previously described. This would show what the largest uncertainty contributors were for further focus. Relative risk evaluations thus can provide an added dimension to probabilistic risk assessments, particularly in the design phase, when comparative assessments are most relevant and most understandable and when information and data are limited.

REFERENCES

Stamatelatos, M. G. (2007) Risk-informed assessment and management to support NASA's Space Exploration Program. The EPRI Risk Management Forum. Juno Beach, FL: National Aeronautics and Space Administration, Kennedy Space Center.

Stamatelatos, M. G., et al. (2001) *PRA Procedures Guide for Managers and Practitioners*. Washington, DC: National Aeronautics and Space Administration, Headquarters.

Vesely, W. E. (2006) Relative risk evaluations of commercial designs. The ESA/JAXA/NASA Trilateral Safety and Mission Assurance Meeting. Washington, DC: National Aeronautics and Space Administration, Headquarters.

Index

A

A-weighted decibel, 44-48
A weighting, 45
Abort once around, 148, 239, 853
Above ground level, 795
Absorption, 9, 17, 53, 61, 72, 85, 93, 166, 217, 343, 345, 348, 353, 372, 405, 563, 565, 581, 611, 621, 737, 794, 872, 873
Accelerated rate calorimetry, 685, 686
Acceleration characterization and analysis project, 36
Acceleration measurement system, 35-37
Acceptance test, 279, 339, 551, 556, 567, 570, 572, 590, 594, 599, 600, 639, 640, 644, 646, 647, 660, 682, 810
Acoustic emission, 47, 49, 333, 342
Active contamination control, 203, 204
Acute gastrointestinal illness, 626
Adiabatic, 383, 395, 685
Advanced communications technology satellite, 2, 460
Advanced crew escape suit, 240, 241, 245
Advanced resistance exercise device, 132
Advisory, 215, 727, 728
Air quality, 192, 198, 199, 612, 613, 618
Albedo, 72, 73, 76-79, 86
Ammonium di-nitramide, 663
Ammonium nitrate, 663
Ammonium perchlorate, 662, 663, 670
Amplitude modulation (AM), 68
Analytical hierarchical process, 903
Androgynous peripheral docking system, 275-277, 279
Annoyance, 43
Aphelion, 75
Application executive, 487
Assembly, test, and launch operations, 486
As low as reasonably achievable, 62, 824
Assured crew return vehicle, 249, 250
Atmosphere, 8-41, 53, 69, 71, 76, 77, 86, 164, 165, 185, 187-192, 195-205, 207-209, 212, 240, 251, 255, 264, 281, 293-296, 332, 333, 335, 360, 361, 363, 364, 376, 377, 431, 512, 513, 518, 534, 568, 610, 611, 616, 618-620, 656, 657, 664, 679, 727, 728, 742, 760, 764, 831, 847, 848, 853, 854, 866-871, 873
Atomic oxygen, 8, 12-15, 25, 87, 89-91, 93, 563, 721
Attitude control systems, 274
Autoignition temperature, 383, 688, 689, 696
Automated electronic defibrillator, 253, 786
Automated transfer vehicle, 212, 450, 451, 812, 813
  control center, 450
Auxiliary power unit, 363
Avionics air assembly, 351
Avionics verification model, 448

B

Background noise, 45, 47, 50
Baseline data collection, 123, 767, 771, 772, 786, 795, 796, 797, 802, 803
  equipment, 798, 800, 810

Beats per minute, 118
Beta angle, 80-85
Binary decision diagrams, 894
Biohazardous materials, 608, 621-630
Biolab, 757
Biomedical engineer, 139
Blackbody, 73-75
Block for micropurification, 205, 207
Blood forming organ, 56, 66
Board engineer, 749
Body restraint tether, 709
Boiler plate, 296, 297
Bone mineral density, 116
Built-in test, 310
Burst before leak, 578, 579

C

Canadian Space Agency, 301
Capability maturity model integration (CMMI), 486
Carbon dioxide removal assembly, 197
Carbon fiber reinforced plastic, 329, 331
Cardiopulmonary resuscitation, 761, 765
Cargo launch vehicle, 896
Cathode potential fall, 409
Caution and warning system, 169, 333, 725-732, 756, 762
Center of gravity, 144, 711, 796, 827
Center of lift, 796
Centers for Disease Control and Prevention, 625, 626, 628
Central nervous system, 63, 611
Certificate of compliance, 774-776, 803, 804
Certificate of conformance, 803, 805
Chemical oxygen generator, 38, 39, 398
Chlorotrifluoroethylene, 851
Coefficient of thermal expansion, 359, 598
Coehn's law, 420
Colony forming unit, 623
Columbia Accident Investigation Board, 148, 150
Combustion module, 39
Command and data group (CDG), 750
Command and data handling, 403, 549, 750
  computer, 485
Commander, 142, 259, 680, 748, 749, 782
Command module, 150
Commercial equipment, 2, 40, 150, 237, 244, 281, 291, 296, 298, 299, 376, 616
Commercial off the shelf, 190, 362, 400, 443, 448, 469, 471, 495, 509, 511, 774, 775, 802, 824, 848, 850, 851, 854
Committee for the Protection of Human Subjects, 629
Committee on hearing, bioacoustics, and biomechanics, 95
Common pressure vessel, 534
Comparative soot diagnostics, 39, 40, 859
Component damage, 333, 336
Composite overwrapped pressure vessel, 568, 569, 572, 575-581, 825
Computational fluid dynamics, 286
Computer aided design, 344

Computer based control system, 413, 414
Computer based training, 311
Computer software, 682
　configuration item, 307
Conducted emissions, 448, 450
Conducted susceptibility, 444
Contact electrification, 417, 418, 423
Containment, 166, 180, 203, 208, 209, 284, 287, 288,
　　307, 397, 413, 497, 500, 574, 607-649, 676-678,
　　682, 714, 716, 758, 764, 793, 848, 850, 867
　abort, 240
　exercise surface, 129
　planning, 819-820, 824
　shuttle crew support, 255
　water container, 211
Continuous noise, 44-47, 49, 342, 345, 346, 351
Continuous wave, 447, 459, 737, 741
Coronal mass ejection, 53
Corrosion, 25, 92, 221, 333, 359, 364, 367-370, 381, 400,
　　404, 405, 409, 431, 434, 513, 526, 529, 577, 578,
　　580, 597, 684, 714, 792
Countermeasure, 40, 105, 113, 116, 118, 119, 121, 123,
　　127, 128, 133-139, 141, 142, 146, 152, 448, 627,
　　628, 630, 648
Coupled loads dynamic analysis, 559, 561, 562
Crew and equipment translation aid, 720
Crew compartment, 3, 44-47, 49, 50, 146, 149, 333, 336,
　　337, 341-346, 348, 355, 356, 581, 783
Crew compartment trainer, 783
Crew egress and escape, 147, 149
Crew escape module, 149
Crew escape system, 148, 149, 163, 230, 237
Crew exploration vehicle, 674, 895-897
Crew health care system, 141, 750, 761
Crew launch vehicle, 896
Crew medical officer, 125, 141, 750, 751
Crew module, 148, 151, 236, 333, 335, 346, 631, 896
Crew qualification responsibility matrix, 749, 750
Crew resource management, 768-770
Crew return vehicle, 141, 246, 247, 249-251, 253-255,
　　655, 894
Crew safety officer, 800, 810, 811
Crew surgeon, 124, 805
Crew survival, 106, 143-152, 163, 168, 181-182,
　　229-231, 244, 254, 727, 764
　system, 144, 149, 151, 169
　working group, 148
Critical design review, 172, 177, 448, 770
Critical items list, 682
Criticality analysis, 664
Current interrupt device, 540, 541, 543, 544
Cycle ergometer with vibration isolation and stability
　　system, 129, 130

D

Damage repair, 335-336
Data management system, 750
Data parameter monitoring, 830, 861-864
Decibel, 263
Decompression sickness, 113, 114, 115, 122, 138, 139,
　　144, 146, 779, 795

Deflagration to detonation transition, 667, 703, 795, 797
Degrees of freedom, 268, 278, 292, 302, 304, 305, 313,
　　316, 317, 602
Deoxyribonucleic acid, 111, 113
Department of Defense, 487
Department of Transportation, 512, 545
Derived off the shelf, 774
Design for minimum risk, 168-169, 180, 181, 468,
　　469, 569, 639, 640, 654, 656, 659-660, 680,
　　681, 820
Development model, 773, 774, 802
Dielectric breakdown, 404
Dielectric constant, 417-420
Dielectric strength, 435
Direct current, 736
Dissipative, 348, 349, 418, 419, 424, 425
Distributed single point ground, 446
Donning, 758, 777

E

Earth departure stage, 896
Earth landing system, 296-298
Earth radiation budget experiment, 79
East coast abort landing, 240
Ecliptic true solar longitude, 80
Electric field, 70, 406, 419, 434, 665
Electrical bond, 408-410, 423, 424
Electrical bonding, 404, 405, 410, 423, 425
Electrical ground, 405-408, 410, 451, 824
Electrical power distribution system, 428
Electrical power system, 897, 905
Electrification, 417, 418, 423
Electrocardiogram, 786
Electroencephalogram, 786
Electroexplosive device, 290, 446, 447, 457-460, 673,
　　696-704
Electromagnetic compatibility, 71, 168, 404, 410, 426,
　　442-449, 466, 467, 469
Electromagnetic interference, 404, 432, 440, 442, 443,
　　445, 447-450, 456, 457, 459, 469, 665
Electromagnetic radiation, 74, 437, 697, 700, 702
Electron mobility, 420
Electrode, 424, 435, 507, 523, 524, 527-529, 531, 533,
　　534, 537
Electron spin resonance, 523
Electronic, electrical, and electromechanical, 422
Electrostatic control plan, 421, 447
Electrostatic discharge, 359, 417, 420-423, 425, 426, 437,
　　442, 447, 457, 472, 665, 672, 698-700, 702, 783,
　　796, 854
Electrostatic discharge association, 426
Emergency, caution, and warning, 169, 725-732
Emergency repair, 332, 336
End of life, 190, 588, 589, 590, 612, 677, 717, 721
Engineering model, 448, 773, 784
Engineering or electrical qualification model, 448
Entry, descent, and landing, 267, 281, 282, 475
Environmental control and life support systems, 857, 864,
　　872, 905
Epstein-barr virus, 113
Ethylene propylenediene monomer, 641

European astronaut center, 140, 757, 780
European corporation on space standardization, 293
European crew personal dosimeter, 139
European drawer rack, 863
European physiology module, 757
European retrievable carrier, 14, 88, 89
European Space Agency (ESA), 34, 35
European Space Research and Technology Center, 363
European robotic arm, 305, 306, 308, 310, 311, 316, 317
European technology exposure facility, 758
Event sequence diagram, 891-893
Event tree, 891-893, 897, 904-906
Exploding bridgewire, 703
Exploding foil, 703
Exogenous charging, 417
External tank, 87, 227, 240, 498
Extravehicular activity (EVA), 35, 41, 54, 63, 105, 110, 115, 122, 123, 130, 138, 141, 144, 171, 186, 189, 243, 244, 305-308, 311, 316, 318, 404, 423, 437, 438, 440, 441, 618, 619, 636, 637, 705-712, 714-722, 747-749, 751, 753-755, 758-761, 773, 777, 812
  mobility unit, 114, 139, 150, 376, 705, 706, 708, 711-716, 718, 719, 759-761, 779
  training assembly, 779
  training facility, 753, 760, 779
Extreme ultraviolet, 9, 11, 25

F

Facility safety manager, 800, 809
Fail safe, 165, 167, 309, 411-414, 497, 500, 558, 644, 657, 659, 680, 787, 854
Failure detection, isolation, and recovery, 727, 730, 731
Failure modes and effects analysis, 152, 179, 226, 291, 293, 401, 432, 471, 472, 479, 492, 577, 578, 592-594, 601, 604, 729, 763
Failure tolerance, 168, 179, 180-182, 279, 445, 468, 591-594, 601-604, 653, 656-660, 820, 886
Fault containment region, 413
Fault tolerance, 4, 167, 180, 254, 308, 414, 509, 569, 593, 603, 639, 640, 656, 678, 681, 820, 821, 827
Fault tree, 492, 893, 894, 897
Fault tree analysis, 492
Federal Aviation Administration, 178
Federal aviation regulations, 788
Federal aviation requirements, 775
Finite element method or model, 559, 560, 571, 645
Fire point, 687, 690
Flame velocity, 687, 689
Flammability limits, 255, 394, 609, 687, 688, 836, 837, 840, 841, 845, 846
Flash point, 512, 690
Flight acceptance review, 448
Flight engineer, 142, 398, 749, 782
Flight model, 448, 449, 562, 563, 644, 645-648, 771-774, 783, 784, 796, 797, 799-802, 811-814
Flight planning load limits, 308
Flight safety data package, 773, 802
Flight safety review, 770
Flight support equipment, 598, 773
Floater, 669

Fluid science laboratory, 757
Fluorinated ethylene propylene, 851
Flux, 14, 15, 26-30, 52, 54-56, 58, 59, 62, 74-80, 82-86, 91, 93, 324, 332, 571, 666, 834, 835, 842-844
Flywheel exercise device, 133
Force moment sensor, 314
Foreign object damage, 789
Foreign object debris, 198, 204, 360, 596
Form factor, 73, 77, 78, 85
Full face mask, 264, 265
Full fuselage trainer, 783
Functional cargo block, 192, 194, 348, 350, 355

G

Gagarin cosmonaut training center, 775, 776, 778, 782, 788, 789, 792
Galactic cosmic radiation, 52, 53, 58, 59, 60, 61, 62, 63, 64, 65, 67, 111, 112, 120
Gas density, 16, 333
Geomagnetically induced storm, 17
Geostationary operational environmental satellite, 54, 56
Geostationary orbit, 18, 19, 21, 23, 26, 319, 322, 325, 426
Geostationary transfer orbit, 23
Getaway special, 36
Global positioning system, 23, 71, 250, 251, 252, 290, 293
Glow discharge, 434
Government furnished equipment, 43, 44, 51, 348
Granular activated carbon, 204
Ground environment, 818, 821, 823, 824, 825
Ground fault current interrupters, 467
Ground handling, 289, 432, 555, 771, 827
Ground safety data package, 802
Ground Safety Review Panel, 171
Ground support equipment, 428, 429, 451, 551, 571, 772, 773, 795, 796, 799, 802, 814, 818, 820
Guidance, navigation, and control, 670, 675, 679, 682

H

H-II Transfer Vehicle, 212
Handheld laser, 741
Hardware configuration item, 307
Hazard report, 171-173, 313, 604, 633, 643, 644, 726, 727, 810
Hazardous materials, 143, 180, 607-649
Hazardous operations, 41, 183, 404, 738, 739, 821
Health stabilization program, 627
Hearing, 43, 49, 50, 110, 263, 343, 344, 355, 730, 766
Hearing loss, 43
Hearing protection, 50, 263, 343, 344, 355
High altitude ejection test, 146
High charge particles, 59
High efficiency particulate air, 201
High elliptical earth orbit, 75
High energy particle, 55, 56, 89
High explosive, 674, 683, 703
High frequency, 17, 37, 48, 69
High resolution accelerometer package, 36, 37
High vacuum, 93, 436
High voltage, 166, 404, 434, 436, 467, 511, 514, 537, 539, 541, 697, 719, 736, 737, 740, 742, 785, 786
Highly eccentric orbit, 19, 21

Hubble Space Telescope, 2, 89, 90, 91, 172, 247, 441, 533, 535, 812
Human behavior and performance, 762, 768-770, 780
Human factors, 144, 183, 401, 488, 490, 500, 501, 674, 768, 811, 822
Human-machine interaction (HMI), 500, 501
Human powered centrifuge, 136
Human Rating Board, 178
Human Rating Independent Review Team, 178
Human rating plan, 177, 178, 182, 184
Human research facility, 355
Hydrazine, 676, 683-691, 822, 893
Hydrazinium nitroformate, 663
Hydrogen chloride, 204, 614, 617, 620
Hydroxyl terminated polybutadiene, 662, 663, 668
Hypergol propellants, 683-685
Hypergolic, 678, 686, 687, 825
Hypervelocity impact, 325, 326, 331
Hypoxia, 164, 331, 333, 776

I

Immediate hazard, 414
Immediately dangerous to life or health, 264, 265
Impulse, 12, 17, 34, 47, 249, 263, 307, 675, 676, 696, 708
Incendivity, 423
Inclination, 25, 62, 79, 81, 82, 245, 247, 320, 323, 325, 456
Independent technical authority, 177, 181
Inertial measurement unit, 348, 349
Inertial upper stage, 23
Independent inhibit, 658
Individual pressure vessel, 534
Infrared emittance, 76, 85
Infrasound, 48, 49
Inhibits, 166, 168, 179, 180, 438, 440, 442, 565, 658, 679, 680, 700, 701, 703, 718, 719, 820, 827
Initiating event, 889-891
Institute for medicine and physiology in space, 745
Institute of biomedical problems, 117
Insulator, 418-420, 426, 429
Integrated circuit, 420, 537
Integrated medical group, 750
Integrated simulation, 754, 763, 814
Intercontinental ballistic missile, 679
Interface control document, 810, 885, 886
Interim resistance exercise device, 131-133, 142
Intermittent noise, 47
Intermodule ventilation, 195, 196, 348, 351, 352
Internal volume, 335, 570, 573
International partner, 44, 142, 212, 220, 303
Intravehicular activity, 636, 707, 753
Intravenous, 138, 253, 761, 765
Ionic charging, 417
Ionospheric storm, 17

J

Japanese experiment module remote manipulator system, 304
Joint aviation regulations, 788

K

Keep-out zone, 248, 309, 637, 711, 760, 794

L

$-80°$ laboratory freezer, 352, 354, 355
Laminar soot process, 39, 858, 859
Launch abort system, 182, 230, 896, 897
Launch and entry suit, 748, 756
Launch and mission related objects, 19
Launch escape system, 145, 147, 169, 231, 236, 237, 674
Launch on need, 151
Leak before burst, 166, 518, 567, 573, 574, 578-581
Leak detection, 332, 756, 757, 893
Leak location, 332-333, 337
Level of containment, 609, 619, 640, 641, 716
Light emitting diode, 438, 638, 768, 861
Liquid crystal display, 438
Liquid locking compound, 601
Local area network, 69
Local safety officer, 802, 809
Long duration exposure facility (LDEF), 25, 89, 90, 92
Loss of crew, 270, 716, 721, 727, 764, 889, 897, 900, 904, 905, 910, 911
Loss of mission, 486, 889, 897, 899
Loss of signal, 317
Low Earth orbit, 8, 13, 18, 19, 21, 23, 25-28, 41, 62-64, 66, 87, 93, 111, 120, 225, 244, 247, 319, 320, 324, 333, 417, 426, 533, 535, 628, 654, 896
Lower body negative pressure, 137, 139
Lower explosive limit, 526, 527
Lower flammability limit, 687
Lunar excursion module, 150
Lunar orbit insertion, 896
Lunar surface access module, 896, 897

M

Main arm, 304, 305
Major constituent analyzer, 189, 190
Manufacture, assembly, integration, and test, 86
Mars climate orbiter, 475
Mars polar lander, 475
Martian radiation environment experiment, 59
Master logic diagram, 890
Material compatibility, 683, 826
Material degradation, 8, 88, 622, 684-685
Material exposure and degradation experiment, 92
Materials and processes technical information system, 361, 362, 847
Materials international space station experiment, 92
Maximum contaminant level, 215, 216
Maximum design pressure, 571, 572, 609, 641, 642, 645-647, 660
Maximum expected operating pressure, 571, 575, 580, 581, 676
Maximum permissible exposure, 739
Mean time between failures, 230, 479
Mechanical systems working group, 636
Mechanically initiated explosive devices, 702
Medical board, 767, 805, 810
Medical operations, 123, 140

Index   917

Medical support system, 750
Medium earth orbit, 19, 20, 24, 319
Mercury cadmium telluride, 37
Mesophase carbon microbeads, 536
Meteoroid and space debris terrestrial environment reference, 19
Microgravity acceleration measurement system, 36, 37
Microgravity research experiment, 38
Microgravity science glovebox, 833
Micrometeoroid and orbital debris, 91, 93, 144, 228, 333, 334, 722, 753
Military standard, 71, 410, 426, 437, 443, 646, 775
Miniature electrostatic accelerometer, 37
Minimum ignition energy, 394, 423, 424, 688
Mission control center, 277, 279, 316
Mission data system, 497
Mission loss, 295, 333, 345, 346, 485
Mission related object, 19-21, 24
Mission specialist, 748, 749
Mock-up, 756, 757, 760, 773, 777, 780, 783, 786, 787, 792, 793, 801, 802
Modified off-the-shelf, 400
Module isolation, 335, 337
Monomethyl hydrazine, 691
Motor drive amplifiers, 309, 310
Multilayer insulation, 87, 89, 90, 331, 717, 753, 758
Multipurpose experiment support structure, 35
Multipurpose logistics module, 35, 301

N

Narrowband noise, 348
NASA standard initiator, 697-700
National Center for Atmospheric Research (NCAR), 17
National Council on Radiation Protection and Measurements, 12, 18, 37, 41, 47, 54, 56, 58, 61, 63, 64, 66, 67, 79, 89, 118, 120, 190, 201, 333, 344, 451-454, 468, 508, 570, 669, 681, 786, 835, 859
National Geophysical Data Center, 54
National Oceanographic and Atmospheric Agency (NOAA), 54, 55, 65
National Research Council (NRC), 62, 215, 316, 612
National Space Biomedical Research Institute, 38
Negative structural margin, 278
Neutral buoyancy facility, 753, 760, 761, 777, 780, 794
Neutral buoyancy laboratory, 756, 759, 760, 777, 794
Neutral buoyancy training facility, 777, 779
Nitrogen tetroxide, 363, 372, 691
Nitric oxide, 371, 372
Noise, 43-51, 71, 143, 145, 264, 333, 337, 341-356, 405, 425, 442, 445, 447, 456, 467, 674, 730, 766, 786, 787
    criteria, 44-47, 49, 345
    reduction rating, 263
Nominal hazard zone, 739, 740
Noncompliance report, 448
Nonconformance report, 173, 680
Nondestructive evaluation, 339, 567, 580
Nondestructive inspection, 570, 642, 644, 657
Nondestructive testing, 572-574
Nonmethane volatile organic compounds, 199

O

Obliquity of the ecliptic, 80
Occupational Safety and Health Administration, 257, 260, 261, 264, 361
Octave band, 44, 45, 48, 49, 349, 350, 354
Off-gassing, 343, 359-362, 469
Office of primary responsibility, 178
Off-nominal, 4, 401, 469-470, 706, 711, 750, 754, 755, 758, 761, 763, 764, 814, 842, 863
Onboard data file, 752, 811
Onboard training, 747, 751, 754
One-third octave band, 49
On-orbit safe haven, 150
Open circuit voltage, 439, 441, 513, 515, 516, 520, 521, 525-527, 529
Operational hazard control, 761, 763, 809, 812
Operations manual, 795, 796, 800, 810
Optical explosive devices, 673
Optical solar reflector, 89
Orbit normal vector, 81, 82
Orbital acceleration research experiment, 36
Orbital maneuvering system, 239, 905
Orbital replaceable unit, 304, 305, 786
Orbiter boom sensor system, 741
Ordnance, 695, 820, 825
Orthophthalaldehyde, 618
Orthostatic tolerance, 109, 110, 142
Outgoing longwave radiation, 79

P

Parachute test vehicle, 296-298
Part number, 773
Partial discharge, 737
Partial pressure, 138, 145, 146, 188, 189, 195, 198, 210, 337, 616, 755, 792, 794
Particulate matter, 198-201, 204, 207, 210, 261, 424
Paschen's curves, 424, 435
Passive contamination control, 204, 210
    assist module, 23
    onboard data file, 752
    safety engineer, 635
    safety review panel, 171, 172, 629
    specialist, 749
Peak voltage, 455, 671
Perihelion, 29, 75
Permissible exposure limits, 361
Permittivity, 419, 435
Personal protective equipment, 256-257, 259-262, 265, 620, 698, 702, 740, 758, 764, 765, 821, 822, 825
Personal rescue enclosure, 150, 243
Pilot operated valve, 679
Polar cap absorption, 17, 53
Polymethylmethacrylate, 830, 831, 842, 843, 845, 870, 874, 875
Polypropylene-fiberglass, 843
Polytetrafluoroethylene, 380, 381, 851
Portable computer system, 307, 727
Portable fire extinguisher, 764, 765, 866, 869
Positive temperature coefficient, 513, 514, 517, 538, 539, 541
Power spectral density, 34

Preferred speech interference level, 45, 46
Preliminary design review, 172, 177, 448, 449, 595
Preliminary hazard analysis, 491
Pressure component, 575, 609, 641
Pressure integrity, 335, 336, 338, 339
Pressure wall, 332-339
Pressure wall access, 333, 335, 337-338
Pressurized hardware, 609, 677, 712
Pressurized mating adapter, 192, 193
Probabilistic risk assessment, 170, 182, 502, 592, 680, 683, 889-897, 900, 904, 910, 911
Probability of detection, 728, 729
Probability of false alarm, 323, 728, 729
Probability of no failure, 331
Probability of no penetration, 331, 332
Procedural hazard control, 811
Protoflight model, 448, 645-647
Proton exchange membrane, 392, 398
Pump package assembly, 351-353

Q

Qualification(s)
   model, 448, 644-648, 773, 784
   process, 339, 670, 771
Quality assurance, 295, 575, 609, 644, 648, 675, 677, 678, 800, 810
Quasisteady acceleration measurement, 36

R

Radioisotope thermoelectric generator, 445, 824
Radiated emissions, 447, 448, 715
Radiated susceptibility, 444
Radar ocean reconnaissance satellite, 25
Ram and wake effect, 426
Reaction control system, 151, 298, 679, 905
Reaction rate, 685, 688, 690
Reinforced carbon-carbon, 713
Relative humidity, 86, 191, 418, 420, 423, 630, 663, 665, 672
Remote triaxial sensor, 36
Rendezvous and docking, 256, 267, 750, 751, 754, 904, 908
Rendezvous pitch maneuver, 744
Repair, 150, 332-339, 472, 570, 593, 621, 653, 707, 713, 715, 742, 812, 824
Request for proposal, 447
Request for waiver, 51, 173, 179, 448
Requirements definition review, 447
Resistance-capacitance, 421
Resistance exercise device, 131-133, 142
Return to launch site, 148, 235
Reverberation time, 50, 343
Robotics work station, 312, 313
Rodnik, 211, 220
Root mean square, 459, 690
Russian operating segment, 750
Russian service module, 209, 450-452
Russian Space Agency (RSA), 34, 138

S

Safe and arm device, 654, 672, 673, 695, 827
Safety and mission assurance, 4, 171, 178, 182-184, 186, 226, 230
Safety data package, 171, 172, 208, 629, 633, 773, 798, 800, 802, 803, 810
Safety data sheet, 774, 776, 803
Safety documentation, 819
Safety review, 170-173, 208, 441, 468, 501, 608, 629, 633, 649, 753, 770, 803, 810, 812, 819, 822
Safety review panel, 171, 172, 629, 810, 812
Satellite Pour l'Observation de la Terre, 91
Scanning electron microscopy, 543
Seal, 261, 263, 333, 342, 380, 385, 397, 512, 518, 539, 638, 639, 641-646, 853
Seal damage or failure, 333
Second flight model, 184, 234
Secondary oxygen pack, 235, 708
Self contained oxygen generators, 398
Self contained underwater breathing apparatus, 264, 265
Self sustaining discharges, 526, 535, 544
Sensor enclosure, 36
Shatterable material, 607, 608, 631-639, 714, 753, 758, 764
Short transverse, 364, 366, 368-370
Sick building syndrome, 198, 200, 202
Simplified aid for extravehicular rescue, 711, 721
Simplified general perturbations, 320
Simulator(s), 294, 311, 587, 738, 749, 751, 755, 756, 763, 775, 777, 780, 783, 786, 787, 799
Single lifetime exposure, 215
Single point ground, 446
Single system trainer, 783
Small fine arm, 304, 305
Smoke aerosol measurement experiment, 40
Software failure modes and effects analysis, 492
Solar absorptance, 84, 89
Solar particle event, 53-58, 62-67, 112, 120
Solar power platform, 305
Solar vector, 80, 81
Solid apogee boost motors, 664
Solid fuel oxygen generator, 375, 398-401, 872
Solid rocket motor, 20, 23, 24, 164, 236, 563, 665, 667, 668, 671, 826
Sound absorption, 342, 343
Sound pressure, 45, 48, 263, 342, 343, 345, 346, 349, 351, 353, 354, 554
South atlantic anomaly, 62
Space adaptation syndrome, 40, 152
Space environments, 7, 8, 332, 341, 359, 484, 851
Space medicine information system, 142
Space motion sickness, 40, 110-112, 119
Space shuttle main engine, 148, 227, 240
Space shuttle mission simulator, 13, 35, 119, 239, 256, 428, 430, 431, 754
Space shuttle remote manipulator system, 305
Space station freedom, 249, 250, 865, 868
Space station remote manipulator, system, 304, 308, 310
Space station training facility, 765
Space suit assembly, 114

Space surveillance, 18
   network, 19, 319, 320, 323
Space telescope solar array, 89
Space transportation architecture study, 149
Spacecraft charging at high altitudes, 89
Spacecraft maximum allowable concentration, 200, 202, 361
Spaceflight resource management, 768
Spacecraft water exposure guidelines, 215-217
Spark(s), 38, 165, 376, 396, 398, 429, 434, 512, 665, 688, 853-855
Speech intelligibility, 50
Sputtering, 425
Stamp analysis, 493
Statistical energy analysis, 344
Stoichiometric mixture, 394-396, 423, 688
Stress-corrosion cracking, 91, 359, 363-373, 641, 643
Stress intensity, 371, 372, 573, 578, 582, 590
Structural integrity, 233, 333, 335-337, 339, 594, 645, 647, 663
Structural life, 333, 335, 338, 339, 552, 583, 587, 721
Structural qualification model, 784
Subject loading device, 128, 129
Subject positioning device, 128
Subsystem components, 337
Sudden ionospheric disturbances, 17
Supplied air respirator, 264, 265
Surface degradation, 25, 26, 333, 423
Surface resistance, 419
Surface supplied diving system, 760
Surface supported diving system, 760
System failure(s), 21, 144, 164, 169, 181, 296, 386, 413, 435, 679, 731, 732, 893, 900, 909
System readiness review, 447
System requirements document, 184
System requirements review, 177

T

Test model, 771-773, 781, 795-797, 799-803
Thermal micrometeoroid garment, 717, 718
Thermal protective system, 905
Threshold limit values, 361
Through bulkhead initiator, 695
Time of closest approach, 321, 322, 324
Total organic carbon, 218
Toxic materials, 169, 608, 610-621, 654, 825
Trace contaminant, 188, 198-211, 361, 611
Trace contaminant control subassembly, 205, 206
Tracking, 177, 317, 320, 322, 323, 403, 431, 570, 654, 741, 820
Tracking and data relay satellite, 70, 317
Trainer, 763, 782
Training model, 757, 762, 771-773, 783, 784, 786, 795-798
Training or test safety data package, 800, 802, 803, 810
Training or test safety engineer, 800

Training or test safety officer, 810, 812
Training or test readiness review, 811, 812
Training or test safety review, 810
Transportation safety data package, 798
Trajectory control sensor, 741
Transatlantic alternate landing site, 240, 755
Transfer orbit stage, 19, 21, 23
Trans-lunar injection, 896
Transmission electron microscope, 835, 859
Treadmill with vibration isolation and stability system, 218-219
Triboelectrification, 417
True equator and mean equinox, 320
Two-line element, 319, 320, 323

U

Ultrasonic, 48, 49, 333, 337, 572, 667, 668
Ultrasound, 48, 49, 141
Ultraviolet, 9, 11, 25, 87, 89, 90, 91, 93, 293, 394, 721, 734, 735, 740, 855
Underwriters Laboratories, 857
United Nations, 226, 663, 670
Uninterruptible power source, 526
Union of Soviet Socialist Republics, 19, 674
Upper flammability limit, 687
U.S. Environmental Protection Agency, 215, 623
U.S. operating segment, 750
Urinary tract infection, 628
Utility feed through leaks, 333

V

Vacuum ultraviolet, 90, 93
Valves, leaking or stuck, 333
Velo ergometer, 130, 131, 142
Vergeltungswaffe-2, 674
Vertical motion simulator, 783
Very high vacuum, 436
Vibration isolation and stability systems, 128-129
Voltage breakdown, 736
Volume resistance, 508

W

Warning box, 763
Weightless environment test system, 258, 777
Weld, 23, 363, 367, 369, 543, 570, 572, 574
Wet mock-up, 760, 783, 792, 793
Wireless instrumentation system, 333, 519
Work function(s), 167, 418, 420, 421
World Health Organization, 121, 198

Z

Zero gravity, 31, 218, 332, 830